ADVANCED ENGINEERING MATHEMATICS

ADVANCED ENGINEERING MATHEMATICS

7th Edition

PETER V. O'NEIL

The University of Alabama at Birmingham

Australia · Brazil · Japan · Korea · Mexico · Singapore · Spain · United Kingdom · United States

Advanced Engineering Mathematics
Seventh Edition

Peter V. O'Neil

Publisher, Global Engineering: Christopher M. Shortt

Senior Acquisitions Editor: Randall Adams

Senior Developmental Editor: Hilda Gowans

Editorial Assistant: Tanya Altieri

Team Assistant: Carly Rizzo

Marketing Manager: Lauren Betsos

Media Editor: Chris Valentine

Content Project Manager: D. Jean Buttrom

Production Service: RPK Editorial Services, Inc.

Copyeditor: Shelly Gerger-Knechtl

Proofreader: Martha McMaster

Indexer: Shelly Gerger-Knechtl

Compositor: Integra

Senior Art Director: Michelle Kunkler

Cover Designer: Patti Hudepohl

Photo Credits:

B/W Image: Digital Vision

Cover Image: Shutterstock Images/Roman Sigaev

Internal Designer: Terri Wright

Senior Rights, Acquisitions Specialist: Mardell Glinski-Schultz

Text and Image Permissions Researcher: Kristiina Paul

First Print Buyer: Arethea L. Thomas

Library of Congress Control Number: 2010932700

International Edition:

ISBN-13: 978-1-111-42742-9

ISBN-10: 1-111-42742-9

Cengage Learning International Offices

Asia
www.cengageasia.com
tel: (65) 6410 1200

Australia/New Zealand
www.cengage.com.au
tel: (61) 3 9685 4111

Brazil
www.cengage.com.br
tel: (55) 11 3665 9900

India
www.cengage.co.in
tel: (91) 11 4364 1111

Latin America
www.cengage.com.mx
tel: (52) 55 1500 6000

UK/Europe/Middle East/Africa
www.cengage.co.uk
tel: (44) 0 1264 332 424

Represented in Canada by Nelson Education, Ltd.
tel: (416) 752 9100/(800) 668 0671
www.nelson.com

Cengage Learning is a leading provider of customized learning solutions with office locations around the globe, Including Singapore, the United Kingdom, Australia, Mexico, Brazil, and Japan. Locate your local office at: **www.cengage.com/global**

For product information: **www.cengage.com/international**
Visit your local office: **www.cengage.com/global**
Visit our corporate website: **www.cengage.com**

Printed in Canada
1 2 3 4 5 6 7 13 12 11

Contents

PART 6 Complex Functions 667

Preface

This seventh edition of *Advanced Engineering Mathematics* differs from the sixth in four ways.

First, based on reviews and user comments, new material has been added, including the following.

- Orthogonal projections and least squares approximations of vectors and functions. This provides a unifying theme in recognizing partial sums of eigenfunction expansions as projections onto subspaces, as well as understanding lines of best fit to data points.
- Orthogonalization and the production of orthogonal bases.
- LU factorization of matrices.
- Linear transformations and matrix representations.
- Application of the Laplace transform to the solution of Bessel's equation and to problems involving wave motion and diffusion.
- Expanded treatment of properties and applications of Legendre polynomials and Bessel functions, including a solution of Kepler's problem and a model of alternating current flow.
- Heaviside's formula for the computation of inverse Laplace transforms.
- A complex integral formula for the inverse Laplace transform, including an application to heat diffusion in a slab.
- Vector operations in orthogonal curvilinear coordinates.
- Application of vector integral theorems to the development of Maxwell's equations.
- An application of the Laplace transform convolution to a replacement scheduling problem.

The second new feature of this edition is the interaction of the text with Maple™. An appendix (called A Maple Primer) is included on the use of Maple™ and references to the use of Maple™ are made throughout the text.

Third, there is an added emphasis on constructing and analyzing models, using ordinary and partial differential equations, integral transforms, special functions, eigenfunction expansions, and matrix and complex function methods.

Finally, the answer section in the back of the book has been expanded to provide more information to the student.

Supplements for Instructors:

- A detailed and completely revised Instructor's Solutions Manual and
- PowerPoint Slides

are available through the Instructor's Resource site at www.cengage.com/international.

Supplements for Students:

CourseMate from Cengage Learning offers students book-specific interactive learning tools at an incredible value. Each CourseMate website includes an e-book and interactive learning tools. To access additional course materials (including CourseMate), please visit www.cengagebrain.com. At the resulting page, search for the ISBN of your title (from the back cover of your book) using the search box at the top of the page. This will take you to the product page where these resources can be found.

In preparing this edition, the author is indebted to many individuals, including:

Charles S. Campbell, University of Southern California
David Y. Gao, Virginia Tech
Donald Hartig, California Polytechnic State University, San Luis Obispo
Konstantin A. Lurie, Worcester Polytechnic Institute
Allen Plotkin, San Diego State University
Mehdi Pourazady, University of Toledo
Carl Prather, Virginia Tech
Scott Short, Northern Illinois University

PETER V. O'NEIL

PART 1

Ordinary Differential Equations

CHAPTER 1 First-Order Differential Equations

TERMINOLOGY AND SEPARABLE EQUATIONS LINEAR EQUATIONS EXACT EQUATIONS HOMOGENEOUS BERNOULLI AND RICCATI EQUATIONS EXISTENCE AND UNIQUENESS

1.1 Terminology and Separable Equations

Part 1 of this book deals with *ordinary differential equations*, which are equations that contain one or more derivatives of a function of a single variable. Such equations can be used to model a rich variety of phenomena of interest in the sciences, engineering, economics, ecological studies, and other areas.

We begin in this chapter with first-order differential equations, in which only the first derivative of the unknown function appears. As an example,

$$y' + xy = 0$$

is a first-order equation for the unknown function $y(x)$. A *solution* of a differential equation is any function satisfying the equation. It is routine to check by substitution that $y = ce^{-x^2/2}$ is a solution of $y' + xy = 0$ for any constant c.

We will develop techniques for solving several kinds of first-order equations which arise in important contexts, beginning with separable equations.

A differential equation is *separable* if it can be written (perhaps after some algebraic manipulation) as

$$\frac{dy}{dx} = F(x)G(y)$$

in which the derivative equals a product of a function just of x and a function just of y. This suggests a method of solution.

Step 1. For y such that $G(y) \neq 0$, write the differential form

$$\frac{1}{G(y)}\,dy = F(x)\,dx.$$

In this equation, we say that the variables have been separated.

Step 2. Integrate

$$\int \frac{1}{G(y)}\,dy = \int F(x)\,dx.$$

Step 3. Attempt to solve the resulting equation for y in terms of x. If this is possible, we have an explicit solution (as in Examples 1.1 through 1.3). If this is not possible, the solution is implicitly defined by an equation involving x and y (as in Example 1.4).

Step 4. Following this, go back and check the differential equation for any values of y such that $G(y) = 0$. Such values of y were excluded in writing $1/G(y)$ in step (1) and may lead to additional solutions beyond those found in step (3). This happens in Example 1.1.

EXAMPLE 1.1

To solve $y' = y^2 e^{-x}$, first write

$$\frac{dy}{dx} = y^2 e^{-x}.$$

If $y \neq 0$, this has the differential form

$$\frac{1}{y^2}\,dy = e^{-x}\,dx.$$

The variables have been separated. Integrate

$$\int \frac{1}{y^2}\,dy = \int e^{-x}\,dx$$

or

$$-\frac{1}{y} = -e^{-x} + k$$

in which k is a constant of integration. Solve for y to get

$$y(x) = \frac{1}{e^{-x} - k}.$$

This is a solution of the differential equation for any number k.

Now go back and examine the assumption $y \neq 0$ that was needed to separate the variables. Observe that $y = 0$ by itself satisfies the differential equation, hence it provides another solution (called a *singular solution*).

In summary, we have the general solution

$$y(x) = \frac{1}{e^{-x} - k}$$

for any number k as well as a singular solution $y = 0$, which is not contained in the general solution for any choice of k. ♦

This expression for $y(x)$ is called the *general solution* of this differential equation because it contains an arbitrary constant. We obtain *particular solutions* by making specific choices for k. In Example 1.1,

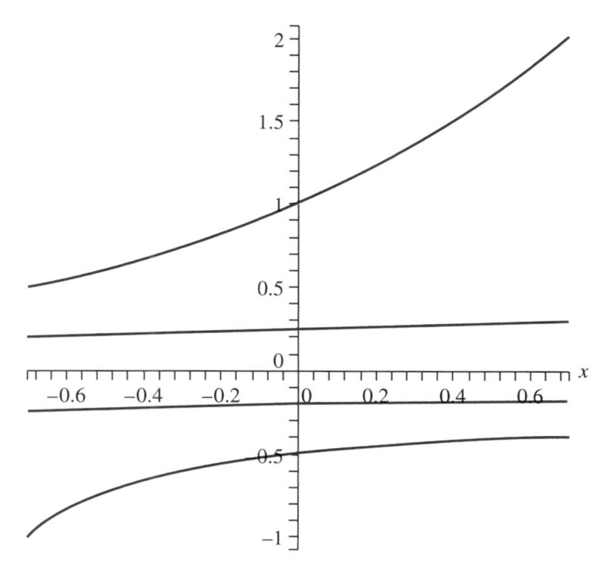

FIGURE 1.1 *Some integral curves from Example 1.1.*

$$y(x) = \frac{1}{e^{-x} - 3}, \quad y(x) = \frac{1}{e^{-x} + 3},$$

$$y(x) = \frac{1}{e^{-x} - 6}, \quad \text{and} \quad y(x) = \frac{1}{e^{-x}} = e^x$$

are particular solutions corresponding to $k = \pm 3, 6$, and 0. Particular solutions are also called *integral curves* of the differential equation. Graphs of these integral curves are shown in Figure 1.1.

EXAMPLE 1.2

$x^2 y' = 1 + y$ is separable, since we can write

$$\frac{1}{1+y}\, dy = \frac{1}{x^2}\, dx$$

if $y \neq -1$ and $x \neq 0$. Integrate to obtain

$$\ln|1+y| = -\frac{1}{x} + k$$

with k an arbitrary constant. This equation implicitly defines the solution. For a given k, we have an equation for the solution corresponding to that k, but not yet an explicit expression for this solution. In this example, we can explicitly solve for $y(x)$. First, take the exponential of both sides of the equation to get

$$|1+y| = e^k e^{-1/x} = a e^{-1/x},$$

where we have written $a = e^k$. Since k can be any number, a can be any positive number. Eliminate the absolute value symbol by writing

$$1 + y = \pm a e^{-1/x} = b e^{-1/x},$$

where the constant $b = \pm a$ can be any nonzero number. Then

$$y = -1 + b e^{-1/x}$$

with $b \neq 0$.

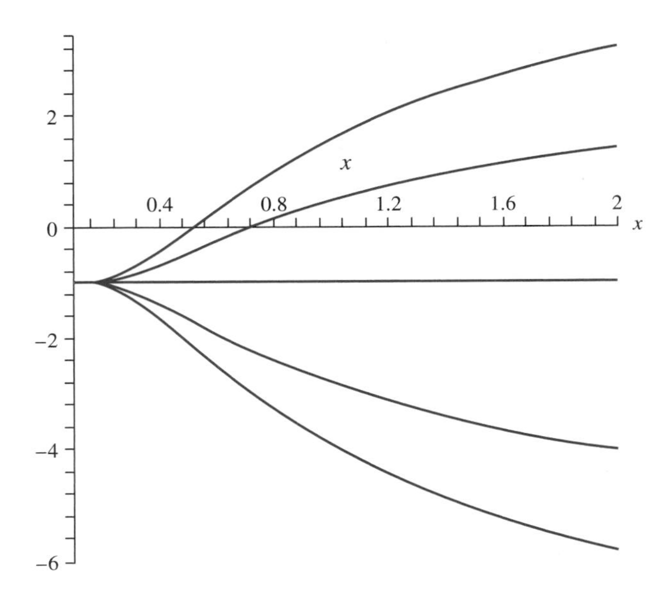

FIGURE 1.2 *Some integral curves from Example 1.2.*

Now notice that the differential equation also has the singular solution $y = -1$, which was disallowed in the separation of variables process when we divided by $y + 1$. However, unlike Example 1.1, we can include this singular solution in the solution by separation of variables by allowing $b = 0$, which gives $y = -1$. We therefore have the general solution

$$y = -1 + be^{-1/x}$$

in which b can be any real number, including zero. This expression contains all solutions. Integral curves (graphs of solutions) corresponding to $b = 0, 4, 7, -5$, and -8 are shown in Figure 1.2. ♦

Each of these examples has infinitely many solutions because of the arbitrary constant in the general solution. If we specify that the solution is to satisfy a condition $y(x_0) = y_0$ with x_0 and y_0 given numbers, then we pick out the particular integral curve passing through (x_0, y_0). The differential equation, together with a condition $y(x_0) = x_0$, is called an *initial value problem*. The condition $y(x_0) = y_0$ is called an *initial condition*.

One way to solve an initial value problem is to find the general solution and then solve for the constant to find the particular solution satisfying the initial condition.

EXAMPLE 1.3

Solve the initial value problem

$$y' = y^2 e^{-x}; \quad y(1) = 4.$$

From Example 1.1, we know that the general solution of this differential equation is

$$y(x) = \frac{1}{e^{-x} - k}.$$

Choose k so that

$$y(1) = \frac{1}{e^{-1} - k} = 4.$$

Solve this equation for k to get

$$k = e^{-1} - \frac{1}{4}.$$

The solution of the initial value problem is

$$y(x) = \frac{1}{e^{-x} + \frac{1}{4} - e^{-1}}. \quad \blacklozenge$$

It is not always possible to find an explicit solution of a differential equation, in which y is isolated on one side of an equation and some expression of x occurs on the other side. In such a case, we must be satisfied with an equation implicitly defining the general solution or the solution of an initial value problem.

EXAMPLE 1.4

We will solve the initial value problem

$$y' = y\frac{(x-1)^2}{y+3}; \qquad y(3) = -1.$$

The differential equation itself (not the algebra of separating the variables) requires that $y \neq -3$.

In differential form,

$$\frac{y+3}{y}\,dy = (x-1)^2\,dx$$

or

$$\left(1 + \frac{3}{y}\right)dy = (x-1)^2 dx.$$

Integrate to obtain

$$y + 3\ln|y| = \frac{1}{3}(x-1)^3 + k.$$

This equation implicitly defines the general solution. However, we cannot solve for y as an explicit expression of x.

This does not prevent us from solving the initial value problem. We need $y(3) = -1$, so put $x = 3$ and $y = -1$ into the implicitly defined general solution to get

$$-1 = \frac{1}{3}(2^3) + k.$$

Then $k = -11/3$, and the solution of the initial value problem is implicitly defined by

$$y + 3\ln|y| = \frac{1}{3}(x-1)^3 - \frac{11}{3}.$$

Part of this solution is graphed in Figure 1.3. $\blacklozenge$

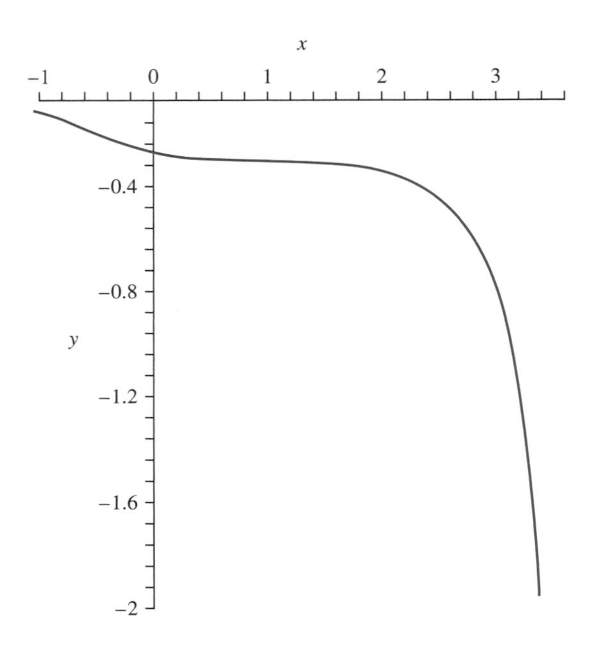

FIGURE 1.3 *Graph of the solution of Example 1.4.*

Some Applications of Separable Equations

Separable differential equations arise in many contexts. We will discuss three of these.

EXAMPLE 1.5 Estimated Time of Death

A homicide victim is discovered and a lieutenant from the forensics laboratory is summoned to estimate the time of death.

The strategy is to find an expression $T(t)$ for the body's temperature at time t, taking into account the fact that after death the body will cool by radiating heat energy into the room. $T(t)$ can be used to estimate the last time at which the victim was alive and had a "normal" body temperature. This last time was the time of death.

To find $T(t)$, some information is needed. First, the lieutenant finds that the body is located in a room that is kept at a constant 68° Fahrenheit. For some time after death, the body will lose heat into the cooler room. Assume, for want of better information, that the victim's temperature was 98.6° at the time of death.

By Newton's law of cooling, heat energy is transferred from the body into the room at a rate proportional to the temperature difference between the room and the body. If $T(t)$ is the body's temperature at time t, then Newton's law says that, for some constant of proportionality k,

$$\frac{dT}{dt} = k[T(t) - 68].$$

This is a separable differential equation, since

$$\frac{1}{T - 68}\, dT = k\, dt.$$

Integrate to obtain

$$\ln|T - 68| = kt + c.$$

To solve for T, take the exponential of both sides of this equation to get

$$|T-68| = e^{kt+c} = Ae^{kt}$$

where $A = e^c$. Then

$$T - 68 = \pm Ae^{kt} = Be^{kt},$$

so

$$T(t) = 68 + Be^{kt}.$$

Now the constants k and B must be determined. Since there are two constants, we will need two pieces of information. Suppose the lieutenant arrived at 9:40 p.m. and immediately measured the body temperature, obtaining 94.4°. It is convenient to let 9:40 p.m. be time zero in carrying out measurements. Then

$$T(0) = 94.4 = 68 + B,$$

so $B = 26.4$. So far,

$$T(t) = 68 + 26.4e^{kt}.$$

To determine k, we need another measurement. The lieutenant takes the body temperature again at 11:00 p.m. and finds it to be 89.2°. Since 11:00 p.m. is 80 minutes after 9:40 p.m., this means that

$$T(80) = 89.2 = 68 + 26.4e^{80k}.$$

Then

$$e^{80k} = \frac{21.2}{26.4},$$

so

$$80k = \ln\left(\frac{21.2}{26.4}\right).$$

Then

$$k = \frac{1}{80}\ln\left(\frac{21.2}{26.4}\right).$$

The temperature function is now completely known as

$$T(t) = 68 + 26.4e^{\ln(21.2/26.4)t/80}.$$

The time of death was the last time at which the body temperature was 98.6° (just before it began to cool). Solve for the time t at which

$$T(t) = 98.6 = 68 + 26.4e^{\ln(21.2/26.4)t/80}.$$

This gives us

$$\frac{30.6}{26.4} = e^{\ln(21.2/26.4)t/80}.$$

Take the logarithm of this equation to obtain

$$\ln\left(\frac{30.6}{26.4}\right) = \frac{t}{80}\ln\left(\frac{21.2}{26.4}\right).$$

According to this model, the time of death was

$$t = \frac{80\ln(30.6/26.4)}{\ln(21.2/26.4)},$$

which is approximately -53.8 minutes. Death occurred approximately 53.8 minutes before (because of the negative sign) the first measurement at 9:40 p.m., which was chosen as time zero in the model. This puts the murder at about 8:46 p.m.

This is an estimate, because an educated guess was made of the body's temperature before death. It is also impossible to keep the room at exactly 68°. However, the model is robust in the sense that small changes in the body's normal temperature and in the constant temperature of the room yield small changes in the estimated time of death. This can be verified by trying a slightly different normal temperature for the body, say 99.3°, to see how much this changes the estimated time of death. ♦

EXAMPLE 1.6 Radioactive Decay and Carbon Dating

In radioactive decay, mass is lost by its conversion to energy which is radiated away. It has been observed that at any time t the rate of change of the mass $m(t)$ of a radioactive element is proportional to the mass itself. This means that, for some constant of proportionality k that is unique to the element,

$$\frac{dm}{dt} = km.$$

Here k must be negative, because the mass is decreasing with time.

This differential equation for m is separable. Write it as

$$\frac{1}{m}\,dm = k\,dt.$$

A routine integration yields

$$\ln|m| = kt + c.$$

Since mass is positive, $|m| = m$ and

$$m(t) = e^{kt+c} = Ae^{kt}$$

in which A can be any positive number. Any radioactive element has its mass decrease according to a rule of this form, and this reveals an important characteristic of radioactive decay. Suppose at some time τ there are M grams. Look for h so that, at the later time $\tau + h$, exactly half of this mass has radiated away. This would mean that

$$m(\tau + h) = \frac{M}{2} = Ae^{k(\tau+h)} = Ae^{k\tau}e^{kh}.$$

But $Ae^{k\tau} = M$, so the last equation becomes

$$\frac{M}{2} = Me^{kh}.$$

Then

$$e^{kh} = \frac{1}{2}.$$

Take the logarithm of this equation to solve for h, obtaining

$$h = \frac{1}{k}\ln\left(\frac{1}{2}\right) = -\frac{1}{k}\ln(2).$$

This is positive because $k < 0$.

Notice that h, the time it takes for half of the mass to convert to energy, depends only on the number k, and not on the mass itself or the time at which we started measuring the loss. If we measure the mass of a radioactive element at any time (say in years), then h years later exactly half of this mass will have radiated away. This number h is called the *half-life* of the element. The constants h and k are both uniquely tied to the particular element and to each other by $h = -(1/k)\ln(2)$. Plutonium has one half-life, and radium has a different half-life.

Now look at the numbers A and k in the expression $m(t) = Ae^{kt}$. k is tied to the element's half-life. The meaning of A is made clear by observing that

$$m(0) = Ae^0 = A.$$

A is the mass that is present at some time designated for convenience as time zero (think of this as starting the clock when the first measurement is made). A is called the *initial mass*, usually denoted m_0. Then

$$m(t) = m_0 e^{kt}.$$

It is sometimes convenient to write this expression in terms of the half-life h. Since $h = -(1/k)\ln(2)$, then $k = -(1/h)\ln(2)$, so

$$m(t) = m_0 e^{kt} = m_0 e^{-\ln(2)t/h}. \tag{1.1}$$

This expression is the basis for an important technique used to estimate the ages of certain ancient artifacts. The Earth's upper atmosphere is bombarded by high-energy cosmic rays, producing large numbers of neutrons which collide with nitrogen, converting some of it into radioactive carbon-14, or ^{14}C. This has a half-life $h = 5,730$ years. Over the geologically short time in which life has evolved on Earth, the ratio of ^{14}C to regular carbon in the atmosphere has remained approximately constant. This means that the rate at which a plant or animal ingests ^{14}C is about the same now as in the past. When a living organism dies, it ceases its intake of ^{14}C, which then begins to decay. By measuring the ratio of ^{14}C to carbon in an artifact, we can estimate the amount of this decay and hence the time it took, giving an estimate of the last time the organism lived. This method of estimating the age of an artifact is called *carbon dating*. Since an artifact may have been contaminated by exposure to other living organisms, this is a sensitive process. However, when applied rigorously and combined with other tests and information, carbon dating has proved a valuable tool in historical and archeological studies.

If we put $h = 5730$ into equation (1.1) with $m_0 = 1$, we get

$$m(t) = e^{-\ln(2)t/5730} \approx e^{-0.000120968t}.$$

As a specific example, suppose we have a piece of fossilized wood. Measurements show that the ratio of ^{14}C to carbon is .37 of the current ratio. To calibrate our clock, say the wood died at time zero. If T is the time it would take for one gram of the radioactive carbon to decay to .37 of one gram, then T satisfies the equation

$$0.37 = e^{-0.000120968T}$$

from which we obtain

$$T = -\frac{\ln(0.37)}{0.000120968} \approx 8,219$$

years. This is approximately the age of the wood. ♦

EXAMPLE 1.7 Draining a Container

Suppose we have a container or tank that is at least partially filled with a fluid. The container is drained through an opening. How long will it take the container to empty? This is a simple enough problem for something like a soda can, but it is not so easy with a large storage tank (such as the gasoline tank at a service station).

We will derive a differential equation to model this problem. We need two principles from physics. The first is that the rate of discharge of a fluid flowing through an opening at the bottom of a container is given by

$$\frac{dV}{dt} = -kAv(t),$$

in which $V(t)$ is the volume of fluid remaining in the container at time t; $v(t)$ is the velocity of the discharge of fluid through the opening; A is the constant cross sectional area of the opening; and k is a constant determined by the viscosity of the fluid, the shape of the opening, and the fact that the cross-sectional area of fluid pouring out of the opening is in reality slightly less than the area of the opening itself. Molasses will flow at a different rate than gasoline, and the shape of the opening will obviously play some role in how the fluid empties through this opening.

The second principle we need is *Torricelli's law*, which states that $v(t)$ is equal to the velocity of a free-falling body released from a height equal to the depth of the fluid at time t. (Free-falling means influenced by gravity only.) In practice, k must be determined for the particular fluid, container, and opening and is a number between 0 and 1.

The work done by gravity in moving a body downward a distance $h(t)$ from its initial position is $mgh(t)$, and this must equal the change in the kinetic energy, which is $m(v(t)^2)/2$. Therefore,

$$v(t) = \sqrt{2gh(t)}.$$

Put the last two equations together to obtain

$$\frac{dV}{dt} = -kA\sqrt{2gh(t)}. \tag{1.2}$$

To illustrate these ideas, consider the problem of draining a hemispherical tank of radius 18 feet that is full of water and has a circular drain hole of radius 3 inches at the bottom. How long will it take for the tank to empty?

Equation (1.2) contains two unknown functions, so we must eliminate one. To do this, let $r(t)$ be the radius of the surface of the fluid at time t, and consider an interval of time from t_0 to $t_0 + \Delta t$. The volume ΔV of water draining from the tank in this time equals the volume of a disk of thickness Δh (the change in depth) and radius $r(t^*)$ for some t^* between t_0 and $t_0 + \Delta t$. Therefore,

$$\Delta V = \pi(r(t^*))^2 \Delta h.$$

Then

$$\frac{\Delta V}{\Delta t} = \pi(r(t^*))^2 \frac{\Delta h}{\Delta t}.$$

In the limit as $\Delta t \to 0$, we obtain

$$\frac{dV}{dt} = \pi r^2 \frac{dh}{dt}.$$

Substitute this into equation (1.2) to obtain

$$\pi r^2 \frac{dh}{dt} = -kA\sqrt{2gh}.$$

Now $V(t)$ has been eliminated, but at the cost of introducing $r(t)$. However, from Figure 1.4,

$$r^2 = 18^2 - (18-h)^2 = 36h - h^2.$$

Then

$$\pi(36h - h^2)\frac{dh}{dt} = -kA\sqrt{2gh}.$$

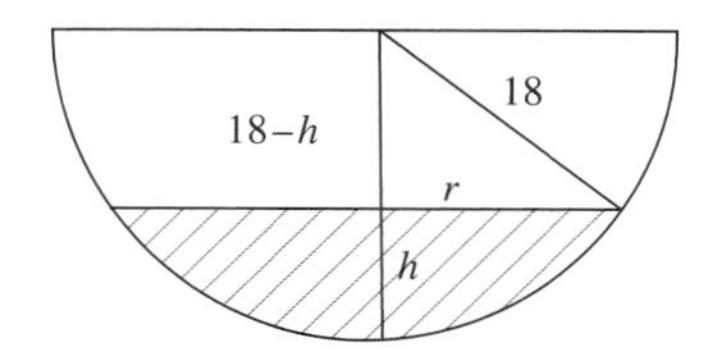

FIGURE 1.4 *Draining a hemispherical tank.*

This is a separable differential equation which we write as

$$\pi\frac{36h-h^2}{h^{1/2}}\,dh=-kA\sqrt{2g}\,dt.$$

Take $g=32$ feet per second per second. The radius of the circular opening is 3 inches (or $1/4$ feet), so its area is $A=\pi/16$. For water and an opening of this shape and size, experiment gives $k=0.8$. Therefore,

$$(36h^{1/2}-h^{3/2})\,dh=-(0.8)\frac{1}{16}\sqrt{64}\,dt,$$

or

$$(36h^{1/2}-h^{3/2})\,dh=-0.4\,dt.$$

A routine integration gives us

$$24h^{3/2}-\frac{2}{5}h^{5/2}=-\frac{2}{5}t+c$$

with c as yet an arbitrary constant. Multiply by $5/2$ to obtain

$$60h^{3/2}-h^{5/2}=-t+C$$

with C arbitrary. For the problem under consideration, the radius of the hemisphere is 18 feet, so $h(0)=18$. Therefore,

$$60(18)^{3/2}-(18)^{5/2}=C.$$

Then $C=2268\sqrt{2}$, and

$$60h^{3/2}-h^{5/2}=2268\sqrt{2}-t.$$

The tank is empty when $h=0$, and this occurs when $t=2268\sqrt{2}$ seconds or about 53 minutes, 28 seconds. This is time it takes for the tank to drain. ♦

These last three examples illustrate an important point. A differential equation or initial value problem may be used to model and describe a process of interest. However, the process usually occurs as something we observe and want to understand, not as a differential equation. This must be derived, using whatever information and fundamental principles may apply (such as laws of physics, chemistry, or economics), as well as the measurements we may take. We saw this in Examples 1.5, 1.6, and 1.7. The solution of the differential equation or initial value problem gives us a function that quantifies some part of the process and enables us to understand its behavior in the hope of being able to predict future behavior or perhaps design a process that better suits our purpose. This approach to the analysis of phenomena is called *mathematical modeling*. We see it today in studies of global warming, ecological and financial systems, and physical and biological processes.

SECTION 1.1 PROBLEMS

In each of Problems 1 through 6, determine whether $y = \varphi(x)$ is a solution of the differential equation. C is constant wherever it appears.

1. $y' = -\dfrac{2y + e^x}{2x}$ for $x > 0$; $\varphi(x) = \dfrac{C - e^x}{2x}$
2. $y' + y = 1$; $\varphi(x) = 1 + Ce^{-x}$
3. $xy' = x - y$; $\varphi(x) = \dfrac{x^2 - 3}{2x}$ for $x \neq 0$
4. $y' + y = 0$; $\varphi(x) = Ce^{-x}$
5. $2yy' = 1$; $\varphi(x) = \sqrt{x - 1}$ for $x > 1$
6. $y' = \dfrac{2xy}{2 - x^2}$ for $x \neq \pm\sqrt{2}$; $\varphi(x) = \dfrac{C}{x^2 - 2}$

In each of Problems 7 through 16, determine if the differential equation is separable. If it is, find the general solution (perhaps implicitly defined) and also any singular solutions the equation might have. If it is not separable, do not attempt a solution.

7. $x\sin(y)y' = \cos(y)$
8. $[\cos(x + y) + \sin(x - y)]y' = \cos(2x)$
9. $y + y' = e^x - \sin(y)$
10. $\dfrac{x}{y}y' = \dfrac{2y^2 + 1}{x + 1}$
11. $3y' = 4x/y^2$
12. $y' = \dfrac{(x + 1)^2 - 2y}{2^y}$
13. $\cos(y)y' = \sin(x + y)$
14. $e^{x+y}y' = 3x$
15. $xy' + y = y^2$
16. $y + xy' = 0$

In each of Problems 17 through 21, solve the initial value problem.

17. $\ln(y^x)y' = 3x^2y$; $y(2) = e^3$
18. $2yy' = e^{x-y^2}$; $y(4) = -2$
19. $yy' = 2x\sec(3y)$; $y(2/3) = \pi/3$
20. $y' = 3x^2(y + 2)$; $y(2) = 8$
21. $xy^2y' = y + 1$; $y(3e^2) = 2$
22. An object having a temperature of 90° Fahrenheit is placed in an environment kept at 60°. Ten minutes later the object has cooled to 88°. What will be the temperature of the object after it has been in this environment for 20 minutes? How long will it take for the object to cool to 65°?
23. Evaluate
$$\int_0^\infty e^{-t^2-9/t^2}\,dt.$$
Hint: Let
$$I(x) = \int_0^\infty e^{-t^2-(x/t)^2}\,dt.$$
Calculate $I'(x)$ and find a differential equation for $I(x)$. Use the standard integral $\int_0^\infty e^{-t^2}\,dt = \sqrt{\pi}/2$ to determine $I(0)$, and use this initial condition to solve for $I(x)$. Finally, evaluate $I(3)$.
24. (Draining a Hot Tub) Consider a cylindrical hot tub with a 5-foot radius and a height of 4 feet placed on one of its circular ends. Water is draining from the tub through a circular hole 5/8 inches in diameter in the base of the tub.
    (a) With $k = 0.6$, determine the rate at which the depth of the water is changing. Here it is useful to write
    $$\frac{dh}{dt} = \frac{dh}{dV}\frac{dV}{dt} = \frac{dV/dt}{dV/dh}.$$
    (b) Calculate the time T required to drain the hot tub if it is initially full. *Hint*: One way to do this is to write
    $$T = \int_H^0 \frac{dt}{dh}\,dh.$$
    (c) Determine how much longer it takes to drain the lower half than the upper half of the tub. *Hint*: Use the integral of part (b) with different limits for each half.
25. Calculate the time required to empty the hemispherical tank of Example 1.7 if the tank is inverted to lie on a flat cap across the open part of the hemisphere. The drain hole is in this cap. Take $k = 0.8$ as in the example.
26. Determine the time it takes to drain a spherical tank with a radius of 18 feet if it is initially full of water, which drains through a circular hole with a radius of 3 inches in the bottom of the tank. Use $k = 0.8$.
27. The half-life of Uranium-238 is approximately $4.5(10^9)$ years. How much of a 10 kilogram block of $U - 238$ will be present one billion years from now?
28. Given that 12 grams of a radioactive element decays to 9.1 grams in 4 minutes, what is the half-life of this element?
29. A thermometer is carried outside a house whose ambient temperature is 70° Fahrenheit. After five minutes, the thermometer reads 60°, and fifteen minutes after

this, it reads 50.4°. What is the outside temperature (which is assumed to be constant)?

30. A radioactive element has a half-life of ln(2) weeks. If e^3 tons are present at a given time, how much will be left three weeks later?

31. A tank shaped like a right circular cone, vertex down, is 9 feet high and has a diameter of 8 feet. It is initially full of water.
    (a) Determine the time required to drain the tank through a circular hole with a diameter of 2 inches at the vertex. Take $k = 0.6$.
    (b) Determine the time it takes to drain the tank if it is inverted and the drain hole is of the same size and shape as in (a), but now located in the new (flat) base.

32. Determine the rate of change of the depth of water in the tank of Problem 31 (vertex at the bottom) if the drain hole is located in the side of the cone 2 feet above the bottom of the tank. What is the rate of change in the depth of the water when the drain hole is located in the bottom of the tank? Is it possible to determine the location of the drain hole if we are told the rate of change of the depth and the depth of the water in the tank? Can this be done without knowing the size of the drain opening?

33. (Logistic Model of Population Growth) In 1837, the Dutch biologist Verhulst developed a differential equation to model changes in a population (he was studying fish populations in the Adriatic Sea). Verhulst reasoned that the rate of change of a population $P(t)$ with respect to time should be influenced by growth factors (for example, current population) and also factors tending to retard the population (such as limitations on food and space). He formed a model by assuming that growth factors can be incorporated into a term $aP(t)$ and retarding factors into a term $-bP(t)^2$ with a and b as positive constants whose values depend on the particular population. This led to his *logistic equation*

$$P'(t) = aP(t) - bP(t)^2.$$

Note that, when $b = 0$, this is the exponential model.

Solve the logistic model, subject to the initial condition $P(0) = p_0$, to obtain

$$P(t) = \frac{ap_0}{a - bp_0 + bp_0 e^{at}} e^{at}.$$

This is the *logistic model of population growth*. Show that, unlike exponential growth, the logistic model produces a population function $P(t)$ that is bounded above and increases asymptotically toward a/b as $t \to \infty$. Thus, a logistic model produces a population function that never grows beyond a certain value.

34. Continuing Problem 33, a 1920 study by Pearl and Reed (appearing in the *Proceedings of the National Academy of Sciences*) suggested the values

$$a = 0.03134, \quad b = (1.5887)10^{-10}$$

for the population of the United States. Table 1.1 gives the census data for the United States in ten year

TABLE 1.1 ***Census data for Problems 33 and 34, Section 1.1.***

Year	Population	$P(t)$	Percent error	$Q(t)$	Percent error
1790	3,929,214				
1800	5,308,483				
1810	7,239,881				
1820	9,638,453				
1830	12,886,020				
1840	17,069,453				
1850	23,191,876				
1860	31,443,321				
1870	38,558,371				
1880	50,189,209				
1890	62,979,766				
1900	76,212,168				
1910	92,228,496				
1920	106,021,537				
1930	123,202,624				
1940	132,164,569				
1950	151,325,798				
1960	179,323,175				
1970	203,302,031				
1980	226,547,042				

increments from 1790 through 1980. Taking 1790 as year zero to determine p_0, show that the logistic model for the United States population is

$$P(t) = \frac{123,141.5668}{0.03072 + 000062e^{0.03134t}} e^{0.03134t}.$$

Calculate $P(t)$ in ten year increments from 1790 to fill in the $P(t)$ column in the table. Remember that (with 1790 as the base year) 1800 is year $t = 10$ in the model, 1810 is $t = 20$, and so on. Also, calculate the percentage error in the model and fill in this column. Plot the census figures and the numbers predicted by the logistic model on the same set of axes. You should observe that the model is fairly accurate for a long period of time, then diverges from the actual census numbers. Show that the limit of the population in this model is about 197,300,000, which the United States actually exceeded in 1970.

Sometimes an exponential model $Q'(t) = kQ(t)$ is used for population growth. Use the census data (again with 1790 as year zero) to solve for $Q(t)$. Compute $Q(t)$ for the years of the census data and the percentage error in this exponential prediction of population. Plot the census data and the exponential model predicted data on the same set of axes. It should be clear that $Q(t)$ diverges rapidly from the actual census figures. Exponential models are useful for very simple populations (such as bacteria in a dish) but are not sophisticated enough for human or (in general) animal populations, despite occasional claims by experts that the population of the world is increasing exponentially.

1.2 Linear Equations

A first-order differential equation is *linear* if it has the form

$$y' + p(x)y = q(x)$$

for some functions p and q.

There is a general approach to solving a linear equation. Let

$$g(x) = e^{\int p(x)dx}$$

and notice that

$$g'(x) = p(x)e^{\int p(x)dx} = p(x)g(x). \tag{1.3}$$

Now multiply $y' + p(x)y = q(x)$ by $g(x)$ to obtain

$$g(x)y' + p(x)g(x)y = q(x)g(x).$$

In view of equation (1.3), this is

$$g(x)y' + g'(x)y = q(x)g(x).$$

Now we see the point to multiplying the differential equation by $g(x)$. The left side of the new equation is the derivative of $g(x)y$. The differential equation has become

$$\frac{d}{dx}(g(x)y) = q(x)g(x),$$

which we can integrate to obtain

$$g(x)y = \int q(x)g(x)dx + c.$$

If $g(x) \neq 0$, we can solve this equation for y:

$$y(x) = \frac{1}{g(x)} \int q(x)g(x)dx + \frac{c}{g(x)}.$$

This is the general solution with the arbitrary constant c.

We do not recommend memorizing this formula for $y(x)$. Instead, carry out the following procedure.

Step 1. If the differential equation is linear, $y' + p(x)y = q(x)$. First compute

$$e^{\int p(x)\,dx}.$$

This is called an *integrating factor* for the linear equation.

Step 2. Multiply the differential equation by the integrating factor.

Step 3. Write the left side of the resulting equation as the derivative of the product of y and the integrating factor. The integrating factor is designed to make this possible. The right side is a function of just x.

Step 4. Integrate both sides of this equation and solve the resulting equation for y, obtaining the general solution. The resulting general solution may involve integrals (such as $\int \cos(x^2)\,dx$) which cannot be evaluated in elementary form.

EXAMPLE 1.8

The equation $y' + y = x$ is linear with $p(x) = 1$ and $q(x) = x$. An integrating factor is

$$e^{\int p(x)dx} = e^{\int dx} = e^x.$$

Multiply the differential equation by e^x to get

$$e^x y' + e^x y = xe^x.$$

This is

$$(ye^x)' = xe^x$$

with the left side as a derivative. Integrate this equation to obtain

$$ye^x = \int xe^x dx = xe^x - e^x + c.$$

Finally, solve for y by multiplying this equation by e^{-x}:

$$y = x - 1 + ce^{-x}.$$

This is the general solution, containing one arbitrary constant. ♦

EXAMPLE 1.9

Solve the initial value problem

$$y' = 3x^2 - \frac{y}{x};\ y(1) = 5.$$

This differential equation is not linear. Write it as

$$y' + \frac{1}{x}y = 3x^2,$$

which is linear. An integrating factor is

$$e^{\int (1/x)dx} = e^{\ln(x)} = x$$

for $x > 0$. Multiply the differential equation by x to obtain

$$xy' + y = 3x^3$$

or

$$(xy)' = 3x^3.$$

Integrate to obtain

$$xy = \frac{3}{4}x^4 + c.$$

Solve for y to write the general solution

$$y = \frac{3}{4}x^3 + \frac{c}{x}$$

for $x > 0$. For the initial condition, we need

$$y(1) = \frac{3}{4} + c = 5.$$

Then $c = 17/4$, and the solution of the initial value problem is

$$y = \frac{3}{4}x^3 + \frac{17}{4x}. \blacklozenge$$

As suggested previously, solving a linear differential equation may lead to integrals we cannot evaluate in elementary form. As an example, consider $y' + xy = 2$. Here $p(x) = x$, and an integrating factor is

$$e^{\int x\,dx} = e^{x^2/2}.$$

Multiply the differential equation by the integrating factor:

$$y'e^{x^2/2} + xe^{x^2/2}y = 2e^{x^2/2}.$$

Write the left side as the derivative of a product:

$$\frac{d}{dx}\left(e^{x^2/2}y\right) = 2e^{x^2/2}.$$

Integrate

$$ye^{x^2/2} = 2\int e^{x^2/2}\,dx + c.$$

The general solution is

$$y = 2e^{-x^2/2}\left(\int e^{x^2/2}\,dx\right) + ce^{-x^2/2}.$$

We cannot evaluate $\int e^{x^2/2}\,dx$ in elementary terms (as a finite algebraic combination of elementary functions). We could do some additional computation. For example, if we write $e^{x^2/2}$ as a power series about 0, we could integrate this series term by term. This would yield an infinite series expression for the solution.

Here is an application of linear equations to a mixing problem.

EXAMPLE 1.10 A Mixing Problem

We want to determine how much of a given substance is present in a container in which various substances are being added, mixed, and drained out. This is a *mixing problem*, and it is encountered in the chemical industry, manufacturing processes, swimming pools and (on a more sophisticated level) in ocean currents and atmospheric activity.

As a specific example, suppose a tank contains 200 gallons of brine (salt mixed with water) in which 100 pounds of salt are dissolved. A mixture consisting of $1/8$ pound of salt per gallon

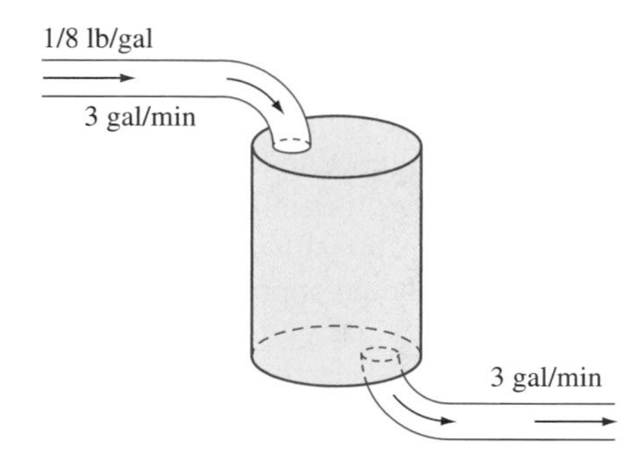

FIGURE 1.5 *Storage tank in Example 1.10.*

is pumped into the tank at a rate of 3 gallons per minute, and the mixture is continuously stirred. Brine also is allowed to empty out of the tank at the same rate of 3 gallons per minute (see Figure 1.5). How much salt is in the tank at any time?

Let $Q(t)$ be the amount of salt in the tank at time t. The rate of change of $Q(t)$ with respect to time must equal the rate at which salt is pumped in minus the rate at which it is pumped out:

$$\begin{aligned}\frac{dQ}{dt} &= (\text{rate in}) - (\text{rate out})\\ &= \left(\frac{1}{8}\frac{\text{pounds}}{\text{gallon}}\right)\left(3\frac{\text{gallons}}{\text{minute}}\right) - \left(\frac{Q(t)}{200}\frac{\text{pounds}}{\text{gallon}}\right)\left(3\frac{\text{gallons}}{\text{minute}}\right)\\ &= \frac{3}{8} - \frac{3}{200}Q(t).\end{aligned}$$

This is the linear equation

$$Q'(t) + \frac{3}{200}Q = \frac{3}{8}.$$

An integrating factor is $e^{\int(3/200)dt} = e^{3t/200}$. Multiply the differential equation by this to get

$$(Qe^{3t/200})' = \frac{3}{8}e^{3t/200}.$$

Integrate to obtain

$$Qe^{3t/200} = \frac{3}{8}\frac{200}{3}e^{3t/200} + c.$$

Then

$$Q(t) = 25 + ce^{-3t/200}.$$

Now use the initial condition

$$Q(0) = 100 = 25 + c,$$

so $c = 75$ and

$$Q(t) = 25 + 75e^{-3t/200}.$$

Notice that $Q(t) \to 25$ as $t \to \infty$. This is the *steady-state* value of $Q(t)$. The term $75e^{-3t/200}$ is called the *transient* part of the solution, and it decays to zero as t increases. $Q(t)$ is the sum of a steady-state part and a transient part. This type of decomposition of a solution is found in many settings. For example, the current in a circuit is often written as a sum of a steady-state term and a transient term.

The initial ratio of salt to brine in the tank is 100 pounds per 200 gallons or $1/2$ pound per gallon. Since the mixture pumped in has a constant ratio of $1/8$ pound per gallon, we expect the brine mixture to dilute toward the incoming ratio with a terminal amount of salt in the tank of $1/8$ pound per gallon times 200 gallons. This leads to the expectation (in the long term) that the amount of salt in the tank should approach 25, as the model verifies. ♦

SECTION 1.2 PROBLEMS

In each of Problems 1 through 5, find the general solution.

1. $y' - 2y = -8x^2$
2. $y' + \sec(x)y = \cos(x)$
3. $y' - \frac{3}{x}y = 2x^2$
4. $y' + y = \frac{1}{2}(e^x - e^{-x})$
5. $y' + 2y = x$

In each of Problems 6 through 10, solve the initial value problem.

6. $y' + \frac{5y}{9x} = 3x^3 + x;\ y(-1) = 4$
7. $y' + \frac{2}{x+1}y = 3;\ y(0) = 5$
8. $y' - y = 2e^{4x};\ y(0) = -3$
9. $y' + \frac{1}{x-2}y = 3x;\ y(3) = 4$
10. $y' + 3y = 5e^{2x} - 6;\ y(0) = 2$
11. Two tanks are connected as in Figure 1.6. Tank 1 initially contains 20 pounds of salt dissolved in 100 gallons of brine. Tank 2 initially contains 150 gallons of brine in which 90 pounds of salt are dissolved. At time zero, a brine solution containing $1/2$ pound of salt per gallon is added to tank 1 at the rate of 5 gallons per minute. Tank 1 has an output that discharges brine into tank 2 at the rate of 5 gallons per minute, and tank 2 also has an output of 5 gallons per minute. Determine the amount of salt in each tank at any time. Also, determine when the concentration of salt in tank 2 is a minimum and how much salt is in the tank at that time. *Hint*: Solve for the amount of salt in tank 1 at time t and use this solution to help determine the amount in tank 2.
12. A 500 gallon tank initially contains 50 gallons of brine solution in which 28 pounds of salt have been dissolved. Beginning at time zero, brine containing 2 pounds of salt per gallon is added at the rate of 3 gallons per minute, and the mixture is poured out of the tank at the rate of 2 gallons per minute. How much salt is in the tank when it contains 100 gallons of brine? *Hint*: The amount of brine in the tank at time t is $50 + t$.
13. Find all functions with the property that the y intercept of the tangent to the graph at (x, y) is $2x^2$.

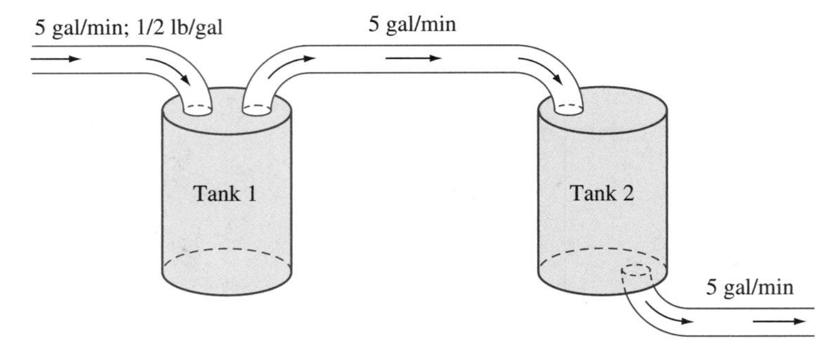

FIGURE 1.6 *Storage tank in Problem 11, Section 1.2.*

1.3 Exact Equations

A differential equation $M(x, y) + N(x, y)y' = 0$ can be written in differential form as

$$M(x, y)dx + N(x, y)dy = 0. \quad (1.4)$$

Sometimes this differential form is the key to writing a general solution. Recall that the differential of a function $\varphi(x, y)$ of two variables is

$$d\varphi = \frac{\partial\varphi}{\partial x}dx + \frac{\partial\varphi}{\partial y}dy. \quad (1.5)$$

If we can find a function $\varphi(x, y)$ such that

$$\frac{\partial\varphi}{\partial x} = M(x, y) \text{ and } \frac{\partial\varphi}{\partial y} = N(x, y), \quad (1.6)$$

then the differential equation $Mdx + Ndy = 0$ is just

$$M(x, y)dx + N(x, y)dy = d\varphi = 0.$$

But if $d\varphi = 0$, then $\varphi(x, y) =$ constant. The equation

$$\varphi(x, y) = c,$$

with c an arbitrary constant, implicitly defines the general solution of $Mdx + Ndy = 0$.

EXAMPLE 1.11

We will use these ideas to solve

$$\frac{dy}{dx} = \frac{2x - e^x \sin(y)}{e^x \cos(y) + 1}.$$

This equation is neither separable nor linear. Write it in the form of equation (1.4) as

$$M(x, y)dx + N(x, y)dy = (e^x \sin(y) - 2x)dx + (e^x \cos(y) + 1)dy = 0.$$

Now let $\varphi(x, y) = e^x \sin(y) + y - x^2$. Then

$$\frac{\partial\varphi}{\partial x} = e^x \sin(y) - 2x = M(x, y) \text{ and } \frac{\partial\varphi}{\partial y} = e^x \cos(y) + 1 = N(x, y),$$

so equations (1.6) are satisfied. The differential equation becomes just $d\varphi = 0$, with general solution defined implicitly by

$$\varphi(x, y) = e^x \sin(y) + y - x^2 = c.$$

To verify that this equation does indeed implicitly define the solution of the differential equation, differentiate it implicitly with respect to x, thinking of y as $y(x)$, to get

$$e^x \sin(y) + e^x \cos(y)y' + y' - 2x = 0$$

and solve this for y' to get

$$y' = \frac{2x - e^x \sin(y)}{e^x \cos(y) + 1},$$

which is the original differential equation. ♦

Example 1.11 suggests a method. The difficult part in applying it is finding the function $\varphi(x, y)$. One magically appeared in Example 1.11, but usually we have to do some work to find a function satisfying equations (1.6).

EXAMPLE 1.12

Consider

$$\frac{dy}{dx} = -\frac{2xy^3+2}{3x^2y^2+8e^{4y}}.$$

This is neither linear nor separable. Write

$$(2xy^3+2)dx + (3x^2y^2+8e^{4y})dy = 0$$

so

$$M(x, y) = 2xy^3 + 2 \text{ and } N(x, y) = 3x^2y^2 + 8e^{4y}.$$

From equations (1.6), we want $\varphi(x, y)$ such that

$$\frac{\partial \varphi}{\partial x} = 2xy^3 + 2 \text{ and } \frac{\partial \varphi}{\partial y} = 3x^2y^2 + 8e^{4y}.$$

Choose either of these equations and integrate it. If we choose the first equation, then integrate with respect to x:

$$\varphi(x, y) = \int \frac{\partial \varphi}{\partial x} dx$$
$$= \int (2xy^3+2)dx = x^2y^3 + 2x + g(y).$$

In this integration, we are reversing a partial derivative with respect to x, so y is treated like a constant. This means that the constant of integration may also involve y; hence it is called $g(y)$. Now we know $\varphi(x, y)$ to within this unknown function $g(y)$. To determine $g(y)$, use the fact that we know what $\partial\varphi/\partial y$ must be

$$\frac{\partial \varphi}{\partial y} = 3x^2y^2 + 8e^{4y}$$
$$= 3x^2y^2 + g'(y).$$

This means that $g'(y) = 8e^{4y}$, so $g(y) = 2e^{4y}$. This fills in the missing piece, and

$$\varphi(x, y) = x^2y^3 + 2x + 2e^{4y}.$$

The general solution of the differential equation is implicitly defined by

$$x^2y^3 + 2x + 2e^{4y} = c,$$

in which c is an arbitrary constant. In this example, we are not able to solve for y explicitly in terms of x. ♦

A function $\varphi(x, y)$ satisfying equations (1.6) is called a *potential function* for the differential equation $M + Ny' = 0$. If we can find a potential function $\varphi(x, y)$, we have at least the implicit expression $\varphi(x, y) = c$ for the solution.

The method of Example 1.12 may produce a potential function if the integrations can be carried out. However, it may also be the case that a potential function does not exist.

EXAMPLE 1.13

The equation $y' + y = 0$ is separable and linear, and the general solution is $y(x) = ce^{-x}$. However, to make a point, try to find a potential function. In differential form,

$$ydx + dy = 0.$$

Here $M(x, y) = y$ and $N(x, y) = 1$. A potential function φ would have to satisfy

$$\frac{\partial \varphi}{\partial x} = y \quad \text{and} \quad \frac{\partial \varphi}{\partial y} = 1.$$

If we integrate the first of these equations with respect to x, we get

$$\varphi(x, y) = \int y\,dx = xy + g(y).$$

Then we need

$$\frac{\partial \varphi}{\partial y} = 1 = x + g'(y).$$

But then $g'(y) = 1 - x$, which is impossible if g is a function of y only. There is no potential function for this differential equation. ♦

We call a differential equation $M + Ny' = 0$ *exact* if it has a potential function. Otherwise it is not exact.

There is a simple test to determine whether $M + Ny' = 0$ is exact for (x, y) in a rectangle R of the plane.

THEOREM 1.1 *Test for Exactness*

Suppose M, N, $\partial N/\partial x$ and $\partial M/\partial y$ are continuous for all (x, y) in some rectangle R in the $(x, y)-$ plane. Then $M + Ny' = 0$ is exact on R if and only if

$$\frac{\partial N}{\partial x} = \frac{\partial M}{\partial y}$$

for (x,y) in R. ♦

Proof If $M + Ny' = 0$ is exact, then there is a potential function φ and

$$\frac{\partial \varphi}{\partial x} = M(x, y) \text{ and } \frac{\partial \varphi}{\partial y} = N(x, y).$$

Then, for (x, y) in R,

$$\frac{\partial M}{\partial y} = \frac{\partial}{\partial y}\left(\frac{\partial \varphi}{\partial x}\right) = \frac{\partial^2 \varphi}{\partial y \partial x} = \frac{\partial^2 \varphi}{\partial x \partial y} = \frac{\partial}{\partial x}\left(\frac{\partial \varphi}{\partial y}\right) = \frac{\partial N}{\partial x}.$$

Conversely, suppose $\partial M/\partial y$ and $\partial N/\partial x$ are continuous on R, and that

$$\frac{\partial N}{\partial x} = \frac{\partial M}{\partial y}.$$

Choose any (x_0, y_0) in R, and define for (x, y) in R

$$\varphi(x, y) = \int_{x_0}^{x} M(\xi, y_0)\,d\xi + \int_{y_0}^{y} N(x, \eta)\,d\eta.$$

In these integrals, x and y are thought of as fixed, and the integration variables are ξ and η respectively. Now, y appears on the right side only in the second integral, so the fundamental theorem of calculus gives us immediately

$$\frac{\partial \varphi}{\partial y} = N(x, y).$$

Computing $\partial\varphi/\partial x$ is less straightforward, since x occurs in both integrals defining $\varphi(x, y)$. For $\partial\varphi/\partial x$, use the condition that $\partial M/\partial y = \partial N/\partial x$ to write

$$\begin{aligned}\frac{\partial \varphi}{\partial x} &= \frac{\partial}{\partial x}\int_{x_0}^{x} M(\xi, y_0)\,d\xi + \frac{\partial}{\partial x}\int_{y_0}^{y} N(x, \eta)\,d\eta \\ &= M(x, y_0) + \int_{y_0}^{y} \frac{\partial N}{\partial x}(x, \eta)\,d\eta \\ &= M(x, y_0) + \int_{y_0}^{y} \frac{\partial M}{\partial y}(x, \eta)\,d\eta \\ &= M(x, y_0) + M(x, y) - M(x, y_0) = M(x, y).\end{aligned}$$

This completes the proof. ♦

In the case of $y\,dx + dy = 0$, $M(x, y) = y$ and $N(x, y) = 1$, so

$$\frac{\partial M}{\partial y} = 1 \text{ and } \frac{\partial N}{\partial x} = 0.$$

Theorem 1.1 tells us that this differential equation is not exact on any rectangle in the plane. We saw this in Example 1.13.

EXAMPLE 1.14

We will solve the initial value problem

$$(\cos(x) - 2xy) + (e^y - x^2)y' = 0;\ y(1) = 4.$$

In differential form,

$$(\cos(x) - 2xy)\,dx + (e^y - x^2)\,dy = 0 = M\,dx + N\,dy$$

with

$$M(x, y) = \cos(x) - 2xy \text{ and } N(x, y) = e^y - x^2.$$

Compute

$$\frac{\partial M}{\partial y} = -2x = \frac{\partial N}{\partial x}$$

for all (x, y). By Theorem 1.1, the differential equation is exact over every rectangle, hence over the entire plane. A potential function $\varphi(x, y)$ must satisfy

$$\frac{\partial \varphi}{\partial x} = \cos(x) - 2xy \text{ and } \frac{\partial \varphi}{\partial y} = e^y - x^2.$$

Choose one of these to integrate. If we begin with the second, then integrate with respect to y:

$$\varphi(x, y) = \int (e^y - x^2)dy = e^y - x^2 y + h(x).$$

The "constant of integration" is $h(x)$, because x is held fixed in a partial derivative with respect to y. Now we know $\varphi(x, y)$ to within $h(x)$. Next we need

$$\frac{\partial \varphi}{\partial x} = \cos(x) - 2xy = -2xy + h'(x).$$

This requires that

$$h'(x) = \cos(x),$$

so $h(x) = \sin(x)$. A potential function is

$$\varphi(x, y) = e^y - x^2 y + \sin(x).$$

The general solution is implicitly defined by

$$e^y - x^2 y + \sin(x) = c.$$

For the initial condition, choose c so that $y(1) = 4$. We need

$$e^4 - 4 + \sin(1) = c.$$

The solution of the initial value problem is implicitly defined by

$$e^y - x^2 y + \sin(x) = e^4 - 4 + \sin(1). \quad \blacklozenge$$

SECTION 1.3 PROBLEMS

In each of Problems 1 through 5, test the differential equation for exactness. If it is exact (on some region of the plane), find a potential function and the general solution (perhaps implicitly defined). If it is not exact anywhere, do not attempt a solution.

1. $1/x + y + (3y^2 + x)y' = 0$
2. $2\cos(x + y) - 2x\sin(x + y) - 2x\sin(x + y)y' = 0$
3. $2y^2 + ye^{xy} + (4xy + xe^{xy} + 2y)y' = 0$
4. $4xy + 2x + (2x^2 + 3y^2)y' = 0$
5. $4xy + 2x^2 y + (2x^2 + 3y^2)y' = 0$

In each of Problems 6 and 7, determine α so that the equation is exact. Obtain the general solution of the exact equation.

6. $3x^2 + xy^\alpha - x^2 y^{\alpha-1} y' = 0$
7. $2xy^3 - 3y - (3x + \alpha x^2 y^2 - 2\alpha y)y' = 0$

In each of Problems 8 through 11, determine if the differential equation is exact in some rectangle containing the point where the initial condition is given. If it is exact, solve the initial value problem. If not, do not attempt a solution.

8. $1 + e^{y/x} - \frac{y}{x}e^{y/x} + e^{y/x}y' = 0$; $y(1) = -5$
9. $x\cos(2y - x) - \sin(2y - x) - 2x\cos(2y - x)y' = 0$; $y(\pi/12) = \pi/8$
10. $2y - y^2\sec^2(xy^2) + (2x - 2xy\sec^2(xy^2))y' = 0$; $y(1) = 2$
11. $3y^4 - 1 + 12xy^3 y' = 0$; $y(1) = 2$
12. $e^y + (xe^y - 1)y' = 0$; $y(5) = 0$
13. Show that

    $$x^2 y' + xy = -y^{-3/2}$$

    is not exact. Solve this equation by finding an integrating factor of the form $\mu(x, y) = x^a y^b$. *Hint*: Consider the differential equation

    $$\mu x^2 y' + \mu xy = -\mu y^{-3/2}$$

    and solve form a and b so that equations (1.6) are satisfied.
14. Try the strategy of Problem 13 on the differential equation

    $$2y^2 - 9xy + (3xy - 6x^2)y' = 0.$$

15. Let φ be a potential function for $M + Ny' = 0$. Show that $\varphi + c$ is also a potential function for any constant c. How does the general solution obtained using φ differ from that obtained using $\varphi + c$?

If $M + Ny' = 0$ is not exact, it might be possible to find a nonzero function $\mu(x, y)$ such that $\mu M + \mu N y' = 0$ is exact. The benefit to this is that $M + Ny' = 0$ and $\mu(M + Ny') = 0$ have the same solutions if $\mu(x, y) \neq 0$ for any x and

y, and the latter equation is exact (hence is solvable if we can find a potential function). Such a function $\mu(x, y)$ is called an *integrating factor* for $M + Ny' = 0$.

16. (a) Show that $y - xy' = 0$ is not exact on any rectangle in the plane.
    (b) Show that $\mu(x, y) = x^{-2}$ is an integrating factor on any rectangle over which $x \neq 0$. Use this to find the general solution of the differential equation.
    (c) Show that $\nu(x, y) = y^{-2}$ is also an integrating factor on any rectangle where $y \neq 0$, and use this to solve the differential equation.
    (d) Show that $\delta(x, y) = xy^{-3}$ is also an integrating factor on any rectangle where $x \neq 0$ and $y \neq 0$. Use this integrating factor to find the general solution.
    (e) Write the differential equation as

$$y' - \frac{1}{x}y = 0$$

and solve it as a linear differential equation.

    (f) How do the solutions found in parts (b) through (e) differ from each other?

1.4 Homogeneous, Bernoulli, and Riccati Equations

We will discuss three other types of first-order differential equations for which techniques of solution are available.

1.4.1 The Homogeneous Differential Equation

A *homogeneous* differential equation is one of the form

$$y' = f(y/x)$$

with y' isolated on one side and on the other an expression in which x and y always occur in the combination y/x. Examples are $y' = \sin(y/x) - x/y$ and $y' = x^2/y^2$.

In some instances, a differential equation can be manipulated into homogeneous form. For example, with

$$y' = \frac{y}{x + y}$$

we can divide numerator and denominator on the right by x to obtain the homogeneous equation

$$y' = \frac{y/x}{1 + y/x}.$$

This manipulation requires the assumption that $x \neq 0$.

A homogeneous differential equation can always be transformed to a separable equation by letting

$$y = ux.$$

To see this, compute $y' = u'x + u$ and write $u = y/x$ to transform

$$y' = u'x + u = f(y/x) = f(u).$$

In terms of u and x, this is

$$xu' + u = f(u)$$

or

$$x\frac{du}{dx} = f(u) - u.$$

The variables u and x separate as

$$\frac{1}{f(u) - u}\,du = \frac{1}{x}\,dx.$$

We attempt to solve this separable equation and then substitute $u = y/x$ to obtain the solution of the original homogeneous equation.

EXAMPLE 1.15

We will solve

$$xy' = \frac{y^2}{x} + y.$$

Write this as

$$y' = \left(\frac{y}{x}\right)^2 + \frac{y}{x}.$$

With $y = ux$, this becomes

$$xu' + u = u^2 + u$$

or

$$xu' = u^2.$$

The variables separate as

$$\frac{1}{u^2}\,du = \frac{1}{x}\,dx.$$

Integrate to obtain

$$-\frac{1}{u} = \ln|x| + c.$$

Then

$$u = \frac{-1}{\ln|x| + c}.$$

Then

$$y = \frac{-x}{\ln|x| + c},$$

and this is the general solution of the original homogeneous equation. ♦

1.4.2 The Bernoulli Equation

A *Bernoulli equation* is one of the form

$$y' + P(x)y = R(x)y^{\alpha}$$

in which α is constant. This equation is linear if $\alpha = 0$ and separable if $\alpha = 1$.

In about 1696, Leibniz showed that, if $\alpha \neq 1$, the Bernoulli equation transforms to a linear equation with the change of variable

$$v = y^{1-\alpha}.$$

This is routine to verify in general. We will see how this works in an example.

EXAMPLE 1.16

We will solve the Bernoulli equation

$$y' + \frac{1}{x}y = 3x^2y^3.$$

Here $P(x) = 1/x$, $R(x) = 3x^2$, and $\alpha = 3$. Let

$$v = y^{1-\alpha} = y^{-2}.$$

Then $y = v^{-1/2}$, so

$$y'(x) = -\frac{1}{2}v^{-3/2}v',$$

and the differential equation becomes

$$-\frac{1}{2}v^{-3/2}v' + \frac{1}{x}v^{-1/2} = 3x^2v^{-3/2}.$$

Upon multiplying by $-2v^{3/2}$, we obtain the linear equation

$$v' - \frac{2}{x}v = -6x^2.$$

This has integrating factor

$$e^{\int -(2/x)\,dx} = e^{\ln(x^{-2})} = x^{-2}.$$

Multiply the differential equation by x^{-2}:

$$x^{-2}v' - 2x^{-3}v = -6.$$

This is

$$(x^{-2}v)' = -6,$$

and an integration yields

$$x^{-2}v = -6x + c.$$

Then

$$v = -6x^3 + cx^2.$$

In terms of y, the original Bernoulli equation has the general solution

$$y(x) = \frac{1}{\sqrt{v(x)}} = \frac{1}{\sqrt{cx^2 - 6x^3}}. \ \blacklozenge$$

1.4.3 The Riccati Equation

A differential equation of the form

$$y' = P(x)y^2 + Q(x)y + R(x)$$

is called a *Riccati equation*.

This is linear when $P(x)$ is identically zero. If we can somehow obtain one particular solution $S(x)$ of a Riccati equation, then the change of variables

$$y = S(x) + \frac{1}{z}$$

transforms the Riccati equation to a linear equation in x and z. The strategy is to find the general solution of this linear equation and use it to write the general solution of the original Riccati equation.

EXAMPLE 1.17

We will solve the Riccati equation

$$y' = \frac{1}{x}y^2 + \frac{1}{x}y - \frac{2}{x}.$$

By inspection, $y = S(x) = 1$ is one solution. Define a new variable z by

$$y = 1 + \frac{1}{z}.$$

Then

$$y' = -\frac{1}{z^2}z',$$

so the Riccati equation transforms to

$$-\frac{1}{z^2}z' = \frac{1}{x}\left(1 + \frac{1}{z}\right)^2 + \frac{1}{x}\left(1 + \frac{1}{z}\right) - \frac{2}{x}.$$

This is the linear equation

$$z' + \frac{3}{x}z = -\frac{1}{x},$$

which has integrating factor x^3. Multiplying by x^3 yields

$$x^3z' + 3x^2z = (x^3z)' = -x^2.$$

Integrate to obtain

$$x^3z = -\frac{1}{3}x^3 + c$$

or

$$z(x) = -\frac{1}{3} + \frac{c}{x^3}.$$

The general solution of the Riccati equation is

$$y(x) = 1 + \frac{1}{z(x)} = 1 + \frac{1}{-1/3 + c/x^3}.$$

This can be written as

$$y(x) = \frac{k + 2x^3}{k - x^3}$$

in which $k = 3c$ is an arbitrary constant. ♦

SECTION 1.4 PROBLEMS

In each of Problems 1 through 14, find the general solution. These problems include all three types discussed in this section.

1. $y' = \dfrac{y}{x+y}$

2. $x^2y' = x^2 + y^2$

3. $(x - 2y)y' = 2x - y$

4. $y' = \dfrac{1}{2x}y^2 - \dfrac{1}{x}y - \dfrac{4}{x}$

5. $y' + \dfrac{1}{x}y = \dfrac{1}{x^4}y^{-3/4}$

6. $y' = \dfrac{x}{y} + \dfrac{y}{x}$

7. $y' + xy = xy^2$

8. $y' + \dfrac{1}{x} = \dfrac{2}{x^3}y^{-4/3}$

9. $y' = -\dfrac{1}{x}y^2 + \dfrac{2}{x}y$

10. $x^3y' = x^2y - y^3$

11. $y' = -e^{-x}y^2 + y + e^x$

12. $xy' = x\cos(y/x) + y$

13. $y' = \dfrac{1}{x^2}y^2 - \dfrac{1}{x}y + 1$

14. $y' + \dfrac{2}{x}y = \dfrac{3}{x}y^2$

15. Consider the differential equation

$$y' = F\left(\frac{ax + by + c}{dx + py + r}\right)$$

in which a, b, c, d, p, and r are constants. Show that this equation is homogeneous if and only if $c = r = 0$. Thus, suppose at least one of c and r is not zero. Then this differential equation is called *nearly homogeneous*. Show that if $ap - bd \neq 0$ it is possible to choose constants h and k such that the transformation $x = X + h, y = Y + k$ results in a homogeneous equation.

In each of Problems 16 through 19, use the idea from Problem 15 to solve the differential equation.

16. $y' = \dfrac{x + 2y + 7}{-2x + y - 9}$

17. $y' = \dfrac{2x - 5y - 9}{-4x + y + 9}$

18. $y' = \dfrac{y - 3}{x + y - 1}$

19. $y' = \dfrac{3x - y - 9}{x + y + 1}$

1.5 Additional Applications

This section is devoted to some additional applications of first-order differential equations. We will need Newton's second law of motion, which states that the sum of the external forces acting on an object is equal to the derivative (with respect to time) of the product of the mass and the velocity. When the mass is constant, $dm/dt = 0$, and Newton's law reduces to the familiar $F = ma$ in which $a = dv/dt$ is the acceleration.

Terminal Velocity An object is falling under the influence of gravity in a medium such as water, air, or oil. We want to analyze the motion.

Let $v(t)$ be the velocity of the object at time t. Gravity pulls the object downward, while the medium retards the downward motion. Experiment has shown that this retarding force is proportional in magnitude to the square of the velocity. Let m be the mass of the object, g the usual constant acceleration due to gravity, and α the constant of proportionality in the retarding force of the medium. Choose downward as the positive direction (this is arbitrary). Let F be the magnitude of the total external force acting on the object. By Newton's law,

$$F = mg - \alpha v^2 = m\frac{dv}{dt}.$$

Suppose that at time zero the object is dropped (not thrown) downward, so $v(0) = 0$. We now have an initial value problem for the velocity:

$$v' = g - \frac{\alpha}{m}v^2;\ v(0) = 0.$$

This differential equation is separable:

$$\frac{1}{g - (\alpha/m)v^2}\,dv = dt.$$

Integrate to get

$$\sqrt{\frac{m}{\alpha g}}\tanh^{-1}\left(\sqrt{\frac{\alpha}{mg}}v\right) = t + c.$$

This equation involves the inverse of the hyperbolic tangent function $\tanh(x)$, which is given by

$$\tanh(x) = \frac{e^{2x} - 1}{e^{2x} + 1}.$$

Solving for $v(t)$, we obtain

$$v(t) = \sqrt{\frac{mg}{\alpha}}\tanh\left(\sqrt{\frac{\alpha g}{m}}(t + c)\right).$$

Now use the initial condition:

$$v(0) = \sqrt{\frac{mg}{\alpha}}\tanh\left(c\sqrt{\frac{\alpha g}{m}}\right) = 0.$$

Since $\tanh(w) = 0$ only for $w = 0$, this requires that $c = 0$ and the solution for the velocity is

$$v(t) = \sqrt{\frac{mg}{\alpha}}\tanh\left(\sqrt{\frac{\alpha g}{m}}t\right).$$

This expression yields an interesting and perhaps nonintuitive conclusion. As $t \to \infty$, $\tanh(\sqrt{\alpha g/m}t) \to 1$. This means that

$$\lim_{t\to\infty} v(t) = \sqrt{\frac{mg}{\alpha}}.$$

An object falling under the influence of gravity through a retarding medium will not increase in velocity indefinitely, even given enough space. It will instead settle eventually into a nearly constant velocity fall, approaching the velocity $\sqrt{mg/\alpha}$ as t increases. This limiting value is called the *terminal velocity* of the object. Skydivers have experienced this phenomenon.

Sliding Motion on an Inclined Plane A block weighing 96 pounds is released from rest at the top of an inclined plane of slope length 50 feet and making an angle of $\pi/6$ radians with the horizontal. Assume a coefficient of friction of $\mu = \sqrt{3}/4$. Assume also that air resistance acts to retard the block's descent down the ramp with a force of magnitude equal to $1/2$ of the block's velocity. We want to determine the velocity $v(t)$ of the block.

Figure 1.7 shows the forces acting on the block. Gravity acts downward with magnitude $mg\sin(u)$, which is $96\sin(\pi/6)$ or 48 pounds. Here $mg = 96$ is the weight (as distinguished from mass) of the block. The drag due to friction acts in the reverse direction and in pounds is given by

$$-\mu N = -\mu mg\cos(u) = -\frac{\sqrt{3}}{4}(96)\cos(\pi/6) = -36.$$

The drag force due to air resistance is $-v/2$. The total external force acting on the block has a magnitude of

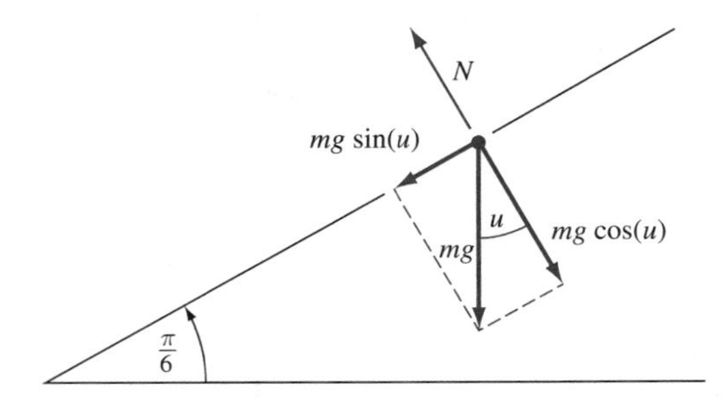

FIGURE 1.7 *Forces acting on the sliding block.*

$$F = 48 - 36 - \frac{1}{2}v = 12 - \frac{1}{2}v.$$

Since the block weighs 96 pounds, its mass is $96/32 = 3$ slugs. From Newton's second law,

$$3\frac{dv}{dt} = 12 - \frac{1}{2}v.$$

This is the linear equation

$$v' + \frac{1}{6}v = 4.$$

Compute the integrating factor $e^{\int(1/6)dt} = e^{t/6}$. Multiply the differential equation by $e^{t/6}$ to obtain

$$v'e^{t/6} + \frac{1}{6}ve^{t/6} = 4e^{t/6}$$

or

$$(ve^{t/6})' = 4e^{t/6}.$$

Integrate to get

$$ve^{t/6} = 24e^{t/6} + c$$

so

$$v(t) = 24 + ce^{-t/6}.$$

Assuming that the block starts from rest at time zero, then $v(0) = 0 = 24 + c$, so $c = -24$ and

$$v(t) = 24\left(1 - e^{-t/6}\right).$$

This gives the block's velocity at any time. We can also determine its position. Let $x(t)$ be the position of the block at time t measured from the top. Since $v(t) = x'(t)$ and $x(0) = 0$, then

$$x(t) = \int_0^t v(\tau)\,d\tau = \int_0^t 24\left(1 - e^{-\tau/6}\right)d\tau$$
$$= 24t + 144\left(e^{-t/6} - 1\right).$$

Suppose we want to know when the block reaches the bottom of the ramp. This occurs at a time T such that $x(T) = 50$. We must solve for T in

$$24T + 144\left(e^{-T/6} - 1\right) = 50.$$

This equation cannot be solved algebraically for T, but a computer approximation yields $T \approx 5.8$ seconds.

Notice that

$$\lim_{t\to\infty} v(t) = 24.$$

Of course, this limit is irrelevant in this setting, since the block reaches the bottom in about 5.8 seconds. However, if the ramp is long enough, the block will approach arbitrarily close to 24 feet per second in velocity. For practical purposes on a sufficiently long ramp, the block will appear to settle into a constant velocity slide. This is similar to the terminal velocity experienced by an object falling in a retarding medium.

Electrical Circuits An RLC circuit is one having only constant resistors, capacitors, and inductors (assumed constant here) as elements and an electromotive driving force $E(t)$. The current $i(t)$ and charge $q(t)$ are related by $i(t) = q'(t)$. The voltage drop across a resistor having resistance R is iR, the drop across a capacitor having capacitance C is q/C, and the drop across an inductor having inductance L is Li'.

We can construct differential equations for circuits by using Kirchhoff's current and voltage laws. The current law states that the algebraic sum of the currents at any junction of a circuit is zero. This means that the total current entering the junction must balance the current leaving it (conservation of energy). The voltage law states that the algebraic sum of the potential rises and drops around any closed loop in a circuit is zero.

As an example of a mathematical model of a simple circuit, consider the RL circuit of Figure 1.8 in which E is constant. Starting at an arbitrary point A, move clockwise around the circuit. First, cross the battery where there is an increase in potential of E volts. Next, there is a decrease in potential of iR volts across the resistor. Finally, there is a decrease of Li' across the inductor, after which we return to A. By Kirchhoff's voltage law,

$$E - iR - Li' = 0,$$

which is the linear equation

$$i' + \frac{E}{R}i = \frac{E}{L}$$

with the general solution

$$i(t) = \frac{E}{R} + ke^{-Rt/L}.$$

We can determine k if we have an initial condition. Even without knowing k, we have $\lim_{t\to\infty} i(t) = E/R$. This is the steady-state value of the current. The solution for the current has a form we have seen before—a steady-state term added to a transient term that decays to zero as t increases.

Often, we encounter discontinuous currents and potential functions in working with circuits. For example, switches may be turned on and off. We will solve more substantial circuit models when we have the Laplace transform at our disposal.

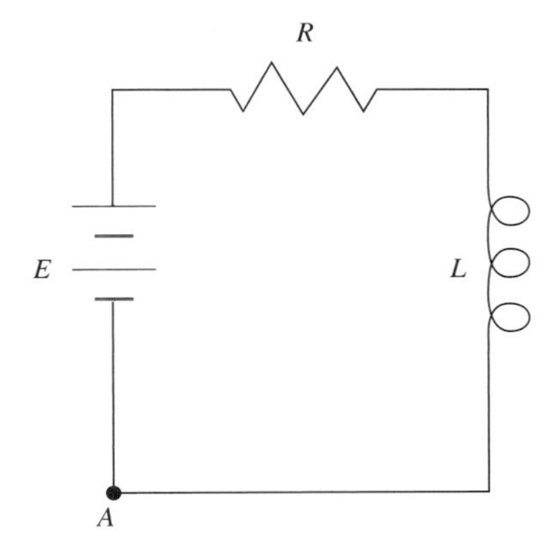

FIGURE 1.8 *A simple RL circuit.*

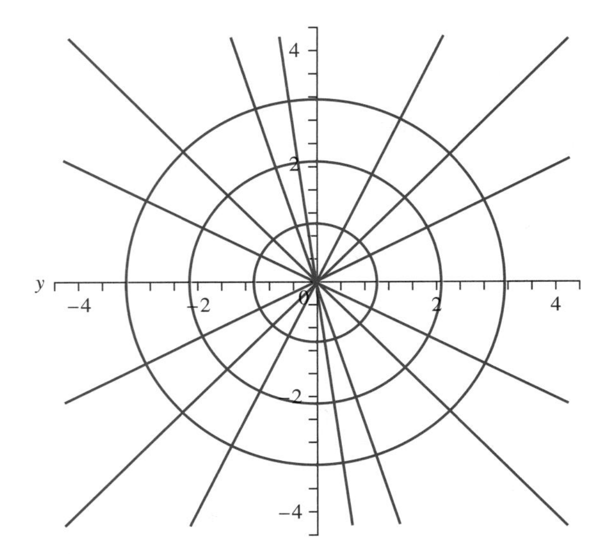

FIGURE 1.9 *Two families of orthogonal trajectories.*

Orthogonal Trajectories Two curves intersecting at a point P are *orthogonal* if their tangents are perpendicular at P. This occurs when the slopes of these tangents at P are negative reciprocals of each other.

Suppose we have two sets (or families) of curves, $\mathcal{F}$ and $\mathcal{G}$. We say that $\mathcal{F}$ is a set of *orthogonal trajectories* of $\mathcal{G}$ if, whenever a curve of $\mathcal{F}$ intersects a curve of $\mathcal{G}$, these curves are orthogonal at the point of intersection. When $\mathcal{F}$ is a family of orthogonal trajectories of $\mathcal{G}$, then $\mathcal{G}$ is also a family of orthogonal trajectories of $\mathcal{F}$.

For example, let $\mathcal{F}$ consist of all circles about the origin and $\mathcal{G}$ of all straight lines through the origin. Figure 1.9 shows some curves of these families, which are orthogonal trajectories of each other. Wherever one of the lines intersects one of the circles, the line is orthogonal to the tangent to the circle there.

Given a family $\mathcal{F}$ of curves, suppose we want to find the family $\mathcal{G}$ of orthogonal trajectories of $\mathcal{F}$. Here is a strategy to do this. The curves of $\mathcal{F}$ are assumed to be graphs of an equation $F(x, y, k) = 0$ with different choices of k giving different curves. Think of these curves as integral curves (graphs of solutions) of some differential equation $y' = f(x, y)$. The curves in the set of orthogonal trajectories are then integral curves of the differential equation $y' = -1/f(x, y)$ with the negative reciprocal ensuring that curves of one family are orthogonal to curves of the other family at points of intersection. The idea is to produce the differential equation $y' = f(x, y)$ from $\mathcal{F}$; then solve the equation $y' = -1/f(x, y)$ for the orthogonal trajectories.

EXAMPLE 1.18

Let $\mathcal{F}$ consist of curves that are graphs of

$$F(x, y, k) = y - kx^2 = 0.$$

These are parabolas through the origin. We want the family of orthogonal trajectories. First obtain the differential equation of $\mathcal{F}$. From $y - kx^2 = 0$ we can write $k = y/x^2$. Differentiate $y - kx^2 = 0$ to get $y' = 2kx$. Substitute for k in this derivative to get

$$y' - 2kx = 0 = y' - 2\left(\frac{y}{x^2}\right)x = 0.$$

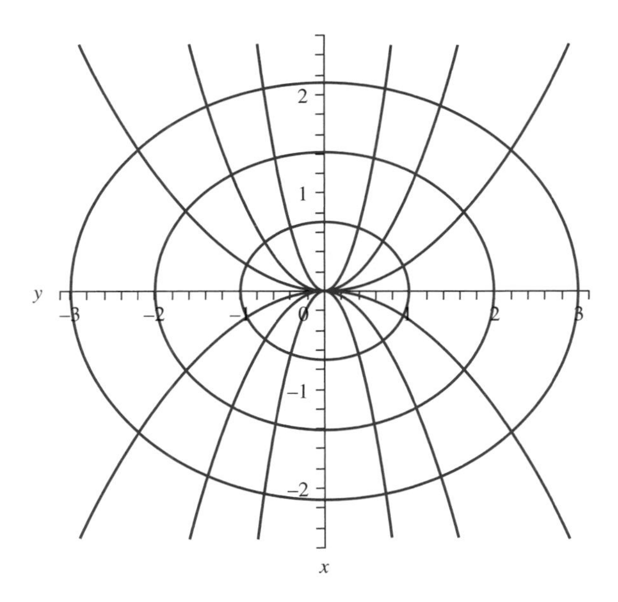

FIGURE 1.10 *Families of orthogonal trajectories in Example 1.18.*

This gives us

$$y' = 2\frac{y}{x} = f(x, y)$$

as the differential equation of $\mathcal{F}$. This means that $\mathcal{F}$ is the family of integral curves of $y' = 2y/x$. The differential equation of the family of orthogonal trajectories is therefore

$$y' = -\frac{1}{f(x, y)} = -\frac{x}{2y}.$$

This is a separable equation that can be written

$$2y\,dy = -x\,dx$$

with the general solution

$$y^2 + \frac{1}{2}x^2 = c.$$

These curves are ellipses, and they make up the family $\mathcal{G}$ of orthogonal trajectories of $\mathcal{F}$. Figure 1.10 shows some of the ellipses in $\mathcal{G}$ and the parabolas in $\mathcal{F}$. ♦

A Pursuit Problem

In a *pursuit problem*, the object is to determine a trajectory so that one object intercepts another. Examples are missiles fired at airplanes and a rendezvous of a shuttle with a space station.

We will solve the following pursuit problem. Suppose a person jumps into a canal and swims toward a fixed point directly opposite the point of entry. The person's constant swimming speed is v, and the water is moving at a constant speed of s. As the person swims, he or she always orients to face toward the target point. We want to determine the swimmer's trajectory.

Suppose the canal has a width of w. Figure 1.11 has the point of entry at $(w, 0)$, and the target point is at the origin. At time t, the swimmer is at $(x(t), y(t))$.

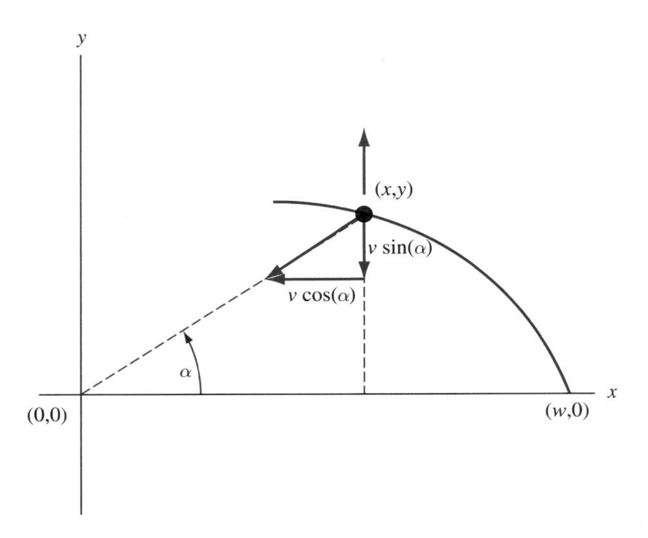

FIGURE 1.11 *The swimmer's path in the pursuit problem.*

The horizontal and vertical components of the swimmer's velocity vector are

$$x'(t) = -v\cos(\alpha) \quad \text{and} \quad y'(t) = s - v\sin(\alpha),$$

where α is the angle between the $x-$ axis and the line from the origin to $(x(t), y(t))$. From these equations,

$$\frac{dy}{dx} = \frac{y'(t)}{x'(t)} = \frac{s - v\sin(\alpha)}{-v\cos(\alpha)} = \tan(\alpha) - \frac{s}{v}\sec(\alpha).$$

From the diagram,

$$\tan(\alpha) = \frac{y}{x} \text{ and } \sec(\alpha) = \frac{1}{x}\sqrt{x^2 + y^2}.$$

Therefore,

$$\frac{dy}{dx} = \frac{y}{x} - \frac{s}{v}\frac{1}{x}\sqrt{x^2 + y^2},$$

which we write as

$$\frac{dy}{dx} = \frac{y}{x} - \frac{s}{v}\sqrt{1 + \left(\frac{y}{x}\right)^2}.$$

This is a homogeneous equation. Put $y = ux$ to obtain

$$\frac{1}{\sqrt{1+u^2}}\,du = -\frac{s}{v}\frac{1}{x}\,dx.$$

Integrate to obtain

$$\ln\left|u + \sqrt{1+u^2}\right| = -\frac{s}{v}\ln|x| + c.$$

Take the exponential of both sides of this equation to obtain

$$\left|u + \sqrt{1+u^2}\right| = e^c e^{-s(\ln|x|/v)}.$$

Write this as

$$u + \sqrt{1+u^2} = Kx^{-s/v}.$$

To solve this for u, first write

$$\sqrt{1+u^2} = Kx^{-s/v} - u$$

and square both sides to obtain

$$1+u^2 = K^2x^{-2s/v} + u^2 - 2Kux^{-s/v}.$$

Now u^2 cancels, and we can solve for u to obtain

$$u(x) = \frac{1}{2}Kx^{-s/v} - \frac{1}{2}\frac{1}{K}x^{s/v}.$$

Finally, $u = y/x$, so

$$y(x) = \frac{1}{2}Kx^{1-s/v} - \frac{1}{2}\frac{1}{K}x^{1+s/v}.$$

To determine K, notice that $y(w) = 0$, since we put the origin at the target point. Then

$$\frac{1}{2}Kw^{1-s/v} - \frac{1}{2}\frac{1}{K}w^{1+s/v} = 0,$$

and we obtain $K = w^{s/v}$. Therefore,

$$y(x) = \frac{w}{2}\left[\left(\frac{x}{w}\right)^{1-s/v} - \left(\frac{x}{w}\right)^{1+s/v}\right].$$

As might be expected, the swimmer's path depends on the width of the canal, the speed of the swimmer, and the speed of the current. Figure 1.12 shows trajectories corresponding to s/v equal to $1/5$ (lowest curve), $1/3$, $1/2$, and $3/4$ (highest curve) with $w = 1$.

Velocity of an Unwinding Chain A 40 foot chain weighing ρ pounds per foot is supported in a pile several feet above the floor. It begins to unwind when released from rest with 10 feet already played out. We want to find the velocity with which the chain leaves the support.

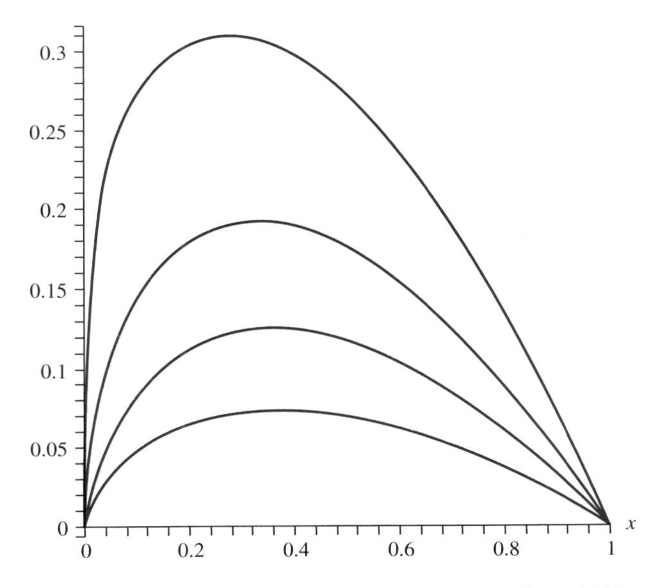

FIGURE 1.12 *Trajectories for $x/v = 1/5$, $1/3$, $1/2$, and $3/4$.*

The length of chain that is actually in motion varies with time. Let $x(t)$ be the length of that part of the chain that has left the support by time t and is currently in motion. The equation of motion is

$$m\frac{dv}{dt} + v\frac{dm}{dt} = F,$$

where F is the magnitude of the total external force acting on the chain. Now $F = \rho x = mg$, so $m = \rho x/g = \rho x/32$. Then

$$\frac{dm}{dt} = \frac{\rho}{32}\frac{dx}{dt} = \frac{\rho}{32}v.$$

Furthermore,

$$\frac{dv}{dt} = \frac{dv}{dx}\frac{dx}{dt} = v\frac{dv}{dx}.$$

Substituting this into the previous equation gives us

$$\frac{\rho x v}{32}\frac{dv}{dx} + \frac{\rho}{32}v^2 = \rho x.$$

Multiply by $32/\rho x v$ to obtain

$$\frac{dv}{dx} + \frac{1}{x}v = \frac{32}{v}. \tag{1.7}$$

This is a Bernoulli equation with $\alpha = -1$. Make the change of variable $w = v^{2-\alpha} = v^2$. Then $v = w^{1/2}$, and

$$\frac{dv}{dx} = \frac{1}{2}w^{-1/2}\frac{dw}{dx}.$$

Substitute this into equation (1.7) to obtain

$$\frac{1}{2}w^{-1/2}\frac{dw}{dx} + \frac{1}{x}w^{1/2} = 32w^{-1/2}.$$

Multiply by $2w^{1/2}$ to obtain the linear equation

$$w' + \frac{2}{x}w = 64.$$

Solve this to obtain

$$w(x) = v(x)^2 = \frac{64}{3}x + \frac{c}{x^2}.$$

Since $v = 0$ when $x = 10$,

$$\frac{64}{3}(10) + \frac{c}{100} = 0,$$

so $c = -64,000/3$. Therefore,

$$v(x)^2 = \frac{64}{3}\left[x - \frac{1000}{x^2}\right].$$

The chain leaves the support when $x = 40$, so at this time,

$$v^2 = \frac{64}{3}\left[40 - \frac{1000}{1600}\right] = 4(210).$$

The velocity at this time is $v = 2\sqrt{210}$, which is about 29 feet per second.

SECTION 1.5 PROBLEMS

1. In a constant electromotive force RL circuit, we find that the current is given by

$$i(t) = \frac{E}{R}(1 - e^{-Rt/L}) + i(0)e^{-Rt/L}.$$

Let $i(0) = 0$.

(a) Show that the current increases with time.

(b) Find a time t_0 at which the current is 63 percent of E/R. This time is called the *inductive time constant* of the circuit.

(c) Does the inductive time constant depend on $i(0)$? If so, in what way?

2. The acceleration due to gravity inside the earth is proportional to the distance from the center of the earth. An object is dropped from the surface of the earth into a hole extending straight through the planet's center. Calculate the speed the object achieves by the time it reaches the center.

3. A particle starts from rest at the highest point of a vertical circle and slides under only the influence of gravity along a chord to another point on the circle. Show that the time taken is independent of the choice of the terminal point. What is this common time?

4. Archimedes' principle of buoyancy states that an object submerged in a fluid is buoyed up by a force equal to the weight of the fluid that is displaced by the object. A rectangular box of $1 \times 2 \times 3$ feet and weighing 384 pounds is dropped into a 100-foot deep freshwater lake. The box begins to sink with a drag due to the water having a magnitude equal to $1/2$ the velocity. Calculate the terminal velocity of the box. Will the box have achieved a velocity of 10 feet per second by the time it reaches the bottom? Assume that the density of water is 62.5 pounds per cubic foot.

5. A skydiver and her equipment together weigh 192 pounds. Before the parachute is opened, there is an air drag force equal in magnitude to six times her velocity. Four seconds after stepping from the plane, the skydiver opens the parachute, producing a drag equal to three times the square of the velocity. Determine the velocity and how far the skydiver has fallen at time t. What is the terminal velocity?

6. A 48 pound box is given an initial push of 16 feet per second down an inclined plane that has a gradient of $7/24$. If there is a coefficient of friction of $1/3$ between the box and the plane and a force of air resistance equal in magnitude to $3/2$ of the velocity of the box, determine how far the box will travel down the plane before coming to rest.

7. Suppose the box in Problem 4 cracks open upon hitting the bottom of the lake, and 32 pounds of its contents spill out. Approximate the velocity with which the box surfaces.

8. Determine the currents in the circuit of Figure 1.13.

9. In the circuit of Figure 1.14, the capacitor is initially discharged. How long after the switch is closed will the capacitor voltage be 76 volts? Determine the current in the resistor at that time. The resistances are in thousands of ohms, and the capacitor is in microfarads (10^{-6} farads).

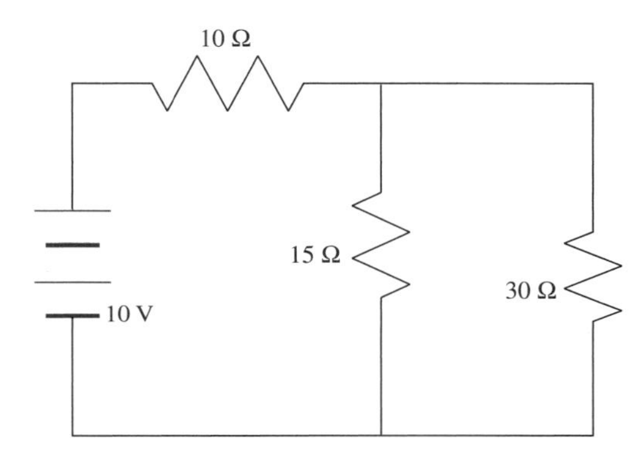

FIGURE 1.13 *Circuit of Problem 8, Section 1.5.*

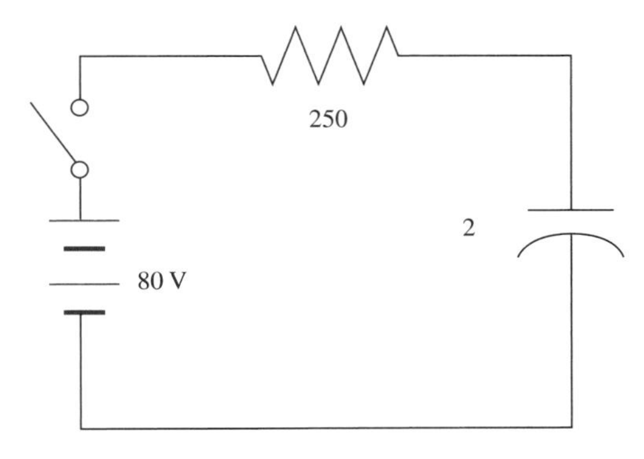

FIGURE 1.14 *Circuit of Problem 9, Section 1.5.*

10. For the circuit in Figure 1.15, find all currents immediately after the switch is closed, assuming that all of these currents and the charges on the capacitors are zero just prior to closing the switch. Resistances are in ohms, the capacitor in farads, and the inductor in henrys.

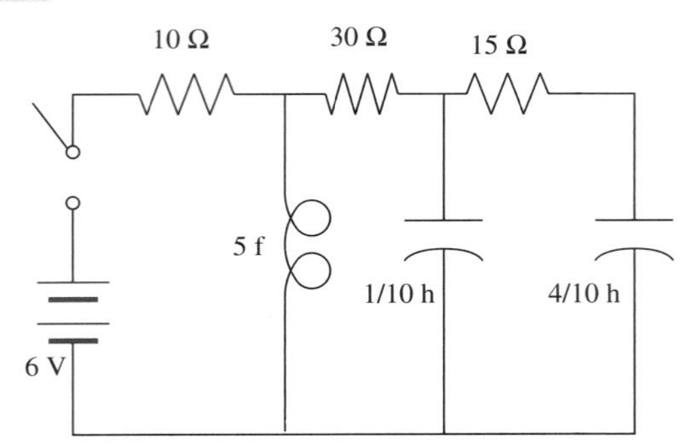

FIGURE 1.15 *Circuit of Problem 10, Section 1.5.*

11. A 10-pound ballast bag is dropped from a hot air balloon which is at an altitude of 342 feet and ascending at 4 feet per second. Assuming that air resistance is not a factor, determine the maximum height reached by the bag, how long it remains aloft, and the speed with which it eventually strikes the ground.
12. A man stands at the junction of two perpendicular roads, and his dog is watching him from one of the roads at a distance A feet away. At some time, the man starts to walk with constant speed v along the other road, and at the same time, the dog begins to run toward the man with a speed of $2v$. Determine the path the dog will take, assuming that it always moves so that it is facing the man. Also determine when the dog will eventually catch the man.

In each of Problems 13 through 17, find the family of orthogonal trajectories of the given family of curves. If software is available, graph some curves of both families.

13. $y = kx^2 + 1$
14. $x + 2y = k$
15. $y = e^{kx}$
16. $x^2 + 2y^2 = k$
17. $2x^2 - 3y = k$
18. Recall that the charge $q(t)$ in an RC circuit satisfies the linear differential equation

$$q' + \frac{1}{RC}q = \frac{1}{R}E(t).$$

(a) Solve for the charge in the case that $E(t) = E$, which is constant. Evaluate the constant of integration in this solution process by using the condition $q(0) = 0$.

(b) Determine $\lim_{t\to\infty} q(t)$, and show that this limit is independent of q_0.

(c) Determine at what time $q(t)$ is within 1 percent of its steady-state value (the limiting value requested in part (b)).

19. A 24 foot chain weighing ρ pounds per foot is stretched out on a very tall, frictionless table with 6 feet hanging off the edge. If the chain is released from rest, determine the time it takes for the end of the chain to fall off the table and also the velocity of the chain at this instant.
20. Suppose the chain in Problem 19 is placed on a table that is only 4 feet high, so that the chain accumulates on the floor as it slides off the table. Two feet of chain are already piled up on the floor at the time that the rest of the chain is released. Determine the velocity of the moving end of the chain at the instant it leaves the table top. *Hint*: Newton's law applies to the center of mass of the moving system.
21. A bug is located at each corner of a square table of side length a. At a given time, the bugs begin moving at constant speed v with each pursuing the neighbor to the right.

(a) Determine the curve of pursuit of each bug. *Hint*: Use polar coordinates with the origin at the center of the table and the polar axis containing one of the corners. When a bug is at $(f(\theta), \theta)$, its target is at $(f(\theta, \theta + \pi/2))$. Use the chain rule to write

$$\frac{dy}{dx} = \frac{dy/d\theta}{dx/d\theta}$$

where

$$y(\theta) = f(\theta)\sin(\theta) \text{ and } x(\theta) = f(\theta)\cos(\theta).$$

(b) Determine the distance traveled by each bug.

(c) Does any bug actually reach its quarry?

22. A bug steps onto the edge of a disk of radius a that is spinning at a constant angular speed of ω. The bug moves toward the center of the disk at constant speed v.

(a) Derive a differential equation for the path of the bug using polar coordinates.

(b) How many revolutions will the disk make before the bug reaches the corner? (The solution will be in terms of the angular speed and radius of the disk).

(c) Referring to part (b), what is the total distance the bug will travel, taking into account the motion of the disk?

1.6 Existence and Uniqueness Questions

There are initial value problems having no solution. One example is

$$y' = 2\sqrt{y};\ y(0) = -1.$$

The differential equation has general solution $y = (x + c)^2$, but there is no real number c such that $y(0) = -1$.

An initial value problem also may have more than one solution. In particular, the initial value problem

$$y' = 2\sqrt{y};\ y(1) = 0$$

has the zero solution $y = \varphi(x) = 0$ for all x. But it also has the solution

$$y = \psi(x) = \begin{cases} 0 & \text{for } x \leq 1 \\ (x-1)^2 & \text{for } x \geq 1. \end{cases}$$

Because existence and/or uniqueness can fail for even apparently simple initial value problems, we look for conditions that are sufficient to guarantee both existence and uniqueness of a solution. Here is one such result.

THEOREM 1.2 *Existence and Uniqueness*

Let $f(x, y)$ and $\partial f/\partial y$ be continuous for all (x, y) in a rectangle R centered at (x_0, y_0). Then there is a positive number h such that the initial value problem

$$y' = f(x, y);\ y(x_0) = y_0$$

has a unique solution defined at least for $x_0 - h < x < x_0 + h$. ♦

A proof of Theorem 1.2 is outlined in the remarks preceding Problems 6 through 9.

The theorem gives no control over h, hence it may guarantee a unique solution only on a small interval about x_0.

EXAMPLE 1.19

The problem

$$y' = e^{x^2 y} - \cos(x - y);\ y(1) = 7$$

has a unique solution on some interval $(1 - h, 1 + h)$, because $f(x, y) = e^{x^2 y} - \cos(x - y)$ and $\partial f/\partial y$ are continuous for all (x, y), hence, on any rectangle centered at $(1, 7)$. Despite this, the theorem does not give us any control over the size of h. ♦

EXAMPLE 1.20

The initial value problem

$$y' = y^2;\ y(0) = n$$

in which n is a positive integer has the solution

$$y(x) = -\frac{1}{x - \frac{1}{n}}.$$

This solution is defined only for $-1/n < x < 1/n$, hence, on smaller intervals about $x_0 = 0$ as n is chosen larger. ♦

For this reason, Theorem 1.2 is called a *local result*, giving a conclusion about a solution only on a perhaps very small interval about the given point x_0.

SECTION 1.6 PROBLEMS

In each of Problems 1 through 4, use Theorem 1.2 to show that the initial value problem has a unique solution in some interval about the value x_0 at which the initial condition is specified. Assume routine facts about continuity of standard functions of two variables.

1. $y' = x^2 - y^2 + 8x/y;\ y(3) = -1$
2. $y' = \cos(e^{xy});\ y(0) = -4$
3. $y' = \sin(xy);\ y(\pi/2) = 1$
4. $y' = \ln|x - y|;\ y(3) = \pi$
5. Consider the initial value problem

$$|y'| = 2y;\ y(x_0) = y_0,$$

in which x_0 is any number.

(a) Assuming that $y_0 > 0$, find two solutions.

(b) Explain why the conclusion of part (a) does not violate Theorem 1.2.

Theorem 1.2 can be proved using Picard iterates. Here is the idea. Consider the initial value problem

$$y' = f(x, y);\ y(x_0) = y_0.$$

For each positive integer n, define

$$y_n(x) = y_0 + \int_{x_0}^{x} f(t, y_{n-1}(t))\,dt.$$

This is a recursive definition, giving $y_1(x)$ in terms of y_0, then $y_2(x)$ in terms of $y_1(x)$, and so on. The functions $y_n(x)$ are called *Picard iterates* for the initial value problem. Under the assumptions of the theorem, the sequence of functions $y_n(x)$ converges for all x in some interval about x_0, and the limit of this sequence is the solution of the initial value problem on this interval.

In each of Problems 6 through 9:

(a) Use Theorem 1.2 to show that the problem has a solution in some interval about x_0.

(b) Find this solution.

(c) Compute Picard iterates $y_1(x)$ through $y_6(x)$, and from these, guess $y_n(x)$ in general.

(d) Find the Taylor series of the solution from part (b) about x_0.

You should find that the iterates computed in part (c) are exactly the partial sums of the series solution of part (d). Conclude that in these examples the Picard iterates converge to an infinite series representation of the solution.

6. $y' = 2x^2;\ y(1) = 3$
7. $y' = \cos(x);\ y(\pi) = 1$
8. $y' = 2 - y;\ y(0) = 1$
9. $y' = 4 + y;\ y(0) = 3$

THE LINEAR SECOND-ORDER EQUATION THE CONSTANT COEFFICIENT CASE THE NONHOMOGENEOUS EQUATION SPRING MOTION EULER'S DIFFERENTIAL EQUATION

CHAPTER 2 Linear Second-Order Equations

A second-order differential equation is one containing a second derivative but no higher derivative. The theory of second-order differential equations is vast, and we will focus on linear second-order equations, which have many important uses.

2.1 The Linear Second-Order Equation

This section lays the foundations for writing solutions of the second-order linear differential equation. Generally, this equation is

$$P(x)y'' + Q(x)y'(x) + R(x)y(x) = F(x).$$

Notice that this equation "loses" its second derivative at any point where $P(x)$ is zero, presenting technical difficulties in writing solutions. We will therefore begin by restricting the equation to intervals (perhaps the entire real line) on which $P(x) \neq 0$. On such an interval, we can divide the differential equation by $P(x)$ and confine our attention to the important case

$$y'' + p(x)y' + q(x)y = f(x). \tag{2.1}$$

We will refer to this as the *second-order linear differential equation*.

Often, we assume that p and q are continuous (at least on the interval where we seek solutions). The function f is called a *forcing function* for the differential equation, and in some applications, it can have finitely many jump discontinuities.

To get some feeling for what we are dealing with, consider a simple example

$$y'' - 12x = 0.$$

Since $y'' = 12x$, we can integrate once to obtain

$$y'(x) = \int 12x\,dx = 6x^2 + c$$

and then once again to get

$$y(x) = 2x^3 + cx + k$$

with c and k as arbitrary constants. It seems natural that the solution of a second-order differential equation, which involves two integrations, should contain two arbitrary constants. For any choices of c and k, we can graph the corresponding solution, obtaining integral curves. Figure 2.1 shows integral curves for several choices of c and k.

Unlike the first-order case, there may be many integral curves through a given point in the plane. In this example, if we specify that $y(0) = 3$, then we must choose $k = 3$, leaving c still arbitrary. These solutions through (0, 3) are

$$y(x) = 2x^3 + cx + 3.$$

Some of these curves are shown in Figure 2.2.

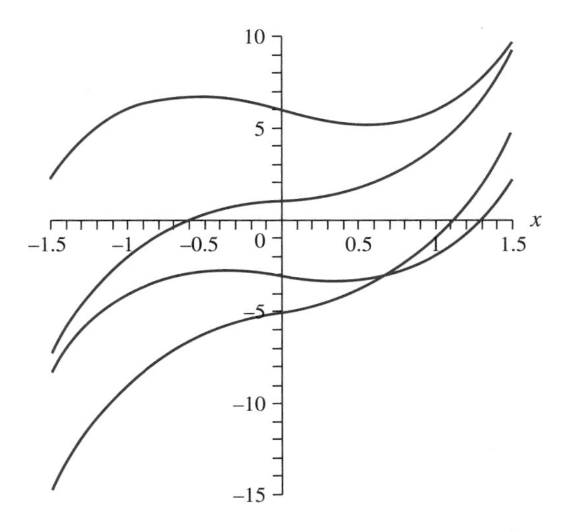

FIGURE 2.1 *Graphs of some functions* $y = 2x^3 + cx + k$.

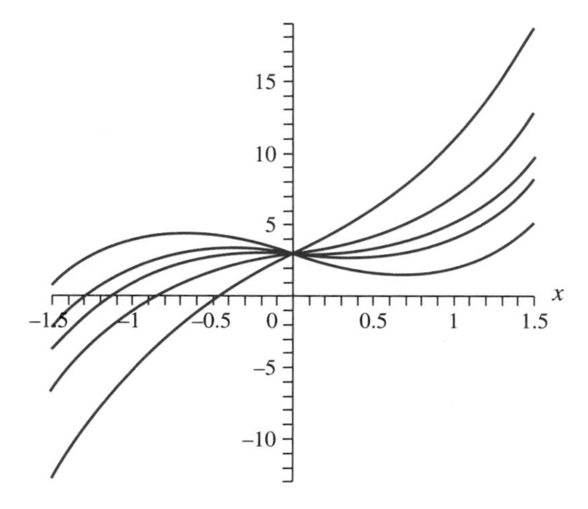

FIGURE 2.2 *Graphs of some functions* $y = 2x^3 + cx + 3$.

We single out exactly one of these curves if we specify its slope at $(0, 3)$. For example, if we specify that $y'(0) = -1$, then

$$y'(0) = c = -1,$$

so $y(x) = 2x^3 - x + 3$. This is the only solution passing through $(0, 3)$ with slope -1.

To sum up, in this example, we obtain a unique solution by specifying a point that the graph must pass through, together with the slope this solution must have at this point.

This leads us to define the *initial value problem* for the linear second-order differential equation as the problem

$$y'' + p(x)y' + q(x)y = f(x);\ y(x_0) = A,\ y'(x_0) = B$$

in which x_0, A, and B are given. We will state, without proof, an existence theorem for this initial value problem.

THEOREM 2.1 Existence of Solutions

Let p, q, and f be continuous on an open interval I. Then the initial value problem

$$y'' + p(x)y' + q(x)y = f(x);\ y(x_0) = A,\ y'(x_0) = B,$$

has a unique solution on I. ♦

We now have an idea of the kind of problem we will be solving and of some conditions under which we are guaranteed a solution. Now we want to develop a strategy to follow to solve linear equations and initial value problems. This strategy will be in two steps, beginning with the case that $f(x)$ is identically zero.

The Structure of Solutions

The *second-order linear homogeneous equation* has the form

$$y'' + p(x)y' + q(x)y = 0. \tag{2.2}$$

If y_1 and y_2 are solutions and c_1 and c_2 are numbers, we call $c_1y_1 + c_2y_2$ a *linear combination* of y_1 and y_2. It is an important property of the homogeneous linear equation (2.2) that a linear combination of solutions is again a solution.

THEOREM 2.2

Every linear combination of solutions of the homogeneous linear equation (2.2) is also a solution. ♦

Proof Let y_1 and y_2 be solutions, and let c_1 and c_2 be numbers. Substitute $c_1y_1 + c_2y_2$ into the differential equation:

$$\begin{aligned}
&(c_1y_1 + c_2y_2)'' + p(x)(c_1y_1 + c_2y_2)' + q(x)(c_1y_1 + c_2y_2) \\
&= c_1y_1'' + c_2y_2'' + c_1p(x)y_1' + c_2p(x)y_2' + c_1q(x)y_1 + c_2q(x)y_2 \\
&= c_1\left[y_1'' + p(x)y_1' + q(x)y_1\right] + c_2\left[y_2'' + p(x)y_2' + q(x)y_2\right] \\
&= c_1(0) + c_2(0) = 0
\end{aligned}$$

because of the assumption that y_1 and y_2 are both solutions. ♦

The point to taking linear combinations $c_1y_1 + c_2y_2$ is to generate new solutions from y_1 and y_2. However, if y_2 is itself a constant multiple of y_1, say $y_2 = ky_2$, then the linear combination

$$c_1y_1 + c_2y_2 = c_1y_1 + c_2ky_1 = (c_1 + c_2k)y_1$$

is just a constant multiple of y_1, so y_2 does not contribute any new information. This leads to the following definition.

Two functions are *linearly independent* on an open interval I (which can be the entire real line) if neither function is a constant multiple of the other for all x in the interval. If one function is a constant multiple of the other on the entire interval, then these functions are called *linearly dependent*.

EXAMPLE 2.1

$\cos(x)$ and $\sin(x)$ are solutions of $y'' + y = 0$ over the entire real line. These solutions are linearly independent, because there is no number k such that $\cos(x) = k\sin(x)$ or $\sin(x) = k\cos(x)$ for all x. Because these solutions are linearly independent, linear combinations $c_1\cos(x) + c_2\sin(x)$ give us new solutions, not just constant multiples of one of the known solutions. ♦

There is a simple test to determine whether two solutions of equation (2.2) are linearly independent or dependent on an open interval I. Define the *Wronskian* $W(y_1, y_2)$ of two solutions y_1 and y_2 to be the 2×2 determinant

$$W(y_1, y_2) = \begin{vmatrix} y_1 & y_2 \\ y_1' & y_2' \end{vmatrix} = y_1y_2' - y_2y_1'.$$

Often we denote this Wronskian as just $W(x)$.

THEOREM 2.3 Properties of the Wronskian

Suppose y_1 and y_2 are solutions of equation (2.2) on an open interval I.

1. Either $W(x) = 0$ for all x in I, or $W(x) \neq 0$ for all x in I.
2. y_1 and y_2 are linearly independent on I if and only if $W(x) \neq 0$ on I. ♦

Conclusion (2) is called the *Wronskian test* for linear independence. Two solutions are linearly independent on I exactly when their Wronskian is nonzero on I. In view of conclusion (1), we need only check the Wronskian at a single point of I, since the Wronskian must be either identically zero on the entire interval or nonzero on all of I. It cannot vanish for some x and be nonzero for others in I.

EXAMPLE 2.2

Check by substitution that $y_1(x) = e^{2x}$ and $y_2(x) = xe^{2x}$ are solutions of $y'' - 4y' + 4y = 0$ for all x. The Wronskian is

$$W(x) = \begin{vmatrix} e^{2x} & xe^{2x} \\ 2e^{2x} & e^{2x} + 2xe^{2x} \end{vmatrix} = e^{4x} + 2xe^{4x} - 2xe^{4x} = e^{4x},$$

and this is never zero, so y_1 and y_2 are linearly independent solutions. ♦

In many cases, it is obvious whether two functions are linearly independent or dependent. However, the Wronskian test is important, as we will see shortly (for example, in Section 2.3.1 and in the proof of Theorem 2.4).

We are now ready to determine what is needed to find all solutions of the homogeneous linear equation $y'' + p(x)y' + q(x)y = 0$. We claim that, if we can find two linearly independent solutions, then every solution must be a linear combination of these two solutions.

THEOREM 2.4

Let y_1 and y_2 be linearly independent solutions of $y'' + py' + qy = 0$ on an open interval I. Then every solution on I is a linear combination of y_1 and y_2. ♦

This provides a strategy for finding all solutions of the homogeneous linear equation on an open interval.

1. Find two linearly independent solutions y_1 and y_2.
2. The linear combination

$$y = c_1 y_1 + c_2 y_2$$

contains all possible solutions.

For this reason, we call two linearly independent solutions y_1 and y_2 a *fundamental set of solutions* on I, and we call $c_1 y_1 + c_2 y_2$ the *general solution* of the differential equation on I.

Once we have the general solution $c_1 y_1 + c_2 y_2$ of the differential equation, we can find the unique solution of an initial value problem by using the initial conditions to determine c_1 and c_2.

Proof Let φ be any solution of $y'' + py' + qy = 0$ on I. We want to show that there are numbers c_1 and c_2 such that $\varphi = c_1 y_1 + c_2 y_2$.

Choose any x_0 in I. Let $\varphi(x_0) = A$ and $\varphi'(x_0) = B$. Then φ is the unique solution of the initial value problem

$$y'' + py' + qy = 0;\ y(x_0) = A,\ y'(x_0) = B.$$

Now consider the two algebraic equations in the two unknowns c_1 and c_2:

$$y_1(x_0)c_1 + y_2(x_0)c_2 = A$$
$$y_1'(x_0)c_1 + y'(x_0)c_2 = B.$$

Because y_1 and y_2 are linearly independent, their Wronskian $W(x)$ is nonzero. These two algebraic equations therefore yield

$$c_1 = \frac{Ay_2'(x_0) - By_2(x_0)}{W(x_0)} \quad \text{and} \quad c_2 = \frac{By_1(x_0) - Ay_1'(x_0)}{W(x_0)}.$$

With these choices of c_1 and c_2, $c_1 y_1 + c_2 y_2$ is also a solution of the initial value problem. Since this problem has the unique solution φ, then $\varphi = c_1 y_1 + c_2 y_2$, as we wanted to show. ♦

EXAMPLE 2.3

e^x and e^{2x} are solutions of $y'' - 3y' + 2y = 0$. Therefore, every solution has the form

$$y(x) = c_1 e^x + c_2 e^{2x}.$$

This is the general solution of $y'' - 3y' + 2y = 0$.

If we want to satisfy the initial conditions $y(0) = -2$, $y'(0) = 3$, choose the constants c_1 and c_2 so that

$$y(0) = c_1 + c_2 = -2$$
$$y'(0) = c_1 + 2c_2 = 3.$$

Then $c_1 = -7$ and $c_2 = 5$, so the unique solution of the initial value problem is

$$y(x) = -7e^x + 5e^{2x}. \ \blacklozenge$$

The Nonhomogeneous Case

We now want to know what the general solution of equation (2.1) looks like when $f(x)$ is nonzero at least for some x. In this case, the differential equation is *nonhomogeneous*.

The main difference between the homogeneous and nonhomogeneous cases is that, for the nonhomogeneous equation, sums and constant multiples of solutions need not be solutions.

EXAMPLE 2.4

We can check by substitution that $\sin(2x) + 2x$ and $\cos(2x) + 2x$ are solutions of the nonhomogeneous equation $y'' + 4y = 8x$. However, if we substitute the sum of these solutions, $\sin(2x) + \cos(2x) + 4x$, into the differential equation, we find that this sum is not a solution. Furthermore, if we multiply one of these solutions by 2, taking, say, $2\sin(2x) + 4x$, we find that this is not a solution either. ♦

However, given any two solutions Y_1 and Y_2 of equation (2.1), we find that their *difference* $Y_1 - Y_2$ is a solution, not of the nonhomogeneous equation, but of the *associated homogeneous equation* (2.2). To see this, substitute $Y_1 - Y_2$ into equation (2.2):

$$\begin{aligned}
&(Y_1 - Y_2)'' + p(x)(Y_1 - Y_2)' + q(x)(Y_1 - Y_2) \\
&= [Y_1'' + p(x)Y_1' + q(x)Y_1] - [Y_2'' + p(x)Y_2' + q(x)Y_2] \\
&= f(x) - f(x) = 0.
\end{aligned}$$

But the general solution of the associated homogeneous equation (2.2) has the form $c_1y_1 + c_2y_2$, where y_1 and y_2 are linearly independent solutions of the homogeneous equation (2.2). Since $Y_1 - Y_2$ is a solution of this homogeneous equation, then for some numbers c_1 and c_2:

$$Y_1 - Y_2 = c_1y_1 + c_2y_2,$$

which means that

$$Y_1 = c_1y_1 + c_2y_2 + Y_2.$$

But Y_1 and Y_2 are *any* solutions of equation (2.1). This means that, given any one solution Y_2 of the nonhomogeneous equation, any other solution has the form $c_1y_1 + c_2y_2 + Y_2$ for some constants c_1 and c_2.

We will summarize this discussion as a general conclusion, in which we will use Y_p (for *particular solution*) instead of Y_2 of the discussion.

THEOREM 2.5

Let Y_p be any solution of the homogeneous equation (2.1). Let y_1 and y_2 be linearly independent solutions of equation (2.2). Then the expression

$$c_1 y_1 + c_2 y_2 + Y_p$$

contains every solution of equation (2.1). ♦

For this reason, we call $c_1 y_1 + c_2 y_2 + Y_p$ the *general solution* of equation (2.1).

Theorem 2.5 suggests a strategy for finding all solutions of the nonhomogeneous equation (2.1).

1. Find two linearly independent solutions y_1 and y_2 of the associated homogeneous equation $y'' + p(x)y' + q(x)y = 0$.
2. Find *any* particular solution Y_p of the nonhomogeneous equation $y'' + p(x)y' + q(x)y = f(x)$.
3. The general solution of $y'' + p(x)y' + q(x)y = f(x)$ is

$$y(x) = c_1 y_1(x) + c_2 y_2(x) + Y_p(x)$$

in which c_1 and c_2 can be any real numbers.

If there are initial conditions, use these to find the constants c_1 and c_2 to solve the initial value problem.

EXAMPLE 2.5

We will find the general solution of

$$y'' + 4y = 8x.$$

It is routine to verify that $\sin(2x)$ and $\cos(2x)$ are linearly independent solutions of $y'' + 4y = 0$. Observe also that $Y_p(x) = 2x$ is a particular solution of the nonhomogeneous equation. Therefore, the general solution of $y'' + 4y = 8x$ is

$$y = c_1 \sin(2x) + c_2 \cos(2x) + 2x.$$

This expression contains every solution of the given nonhomogeneous equation by choosing different values of the constants c_1 and c_2.

Suppose we want to solve the initial value problem

$$y'' + 4y = 8x;\ y(\pi) = 1,\ y'(\pi) = -6.$$

First we need

$$y(\pi) = c_2 \cos(2\pi) + 2\pi = c_2 + 2\pi = 1,$$

so $c_2 = 1 - 2\pi$. Next, we need

$$y'(\pi) = 2c_1 \cos(2\pi) - 2c_2 \sin(2\pi) + 2 = 2c_1 + 2 = -6,$$

so $c_1 = -4$. The unique solution of the initial value problem is

$$y(x) = -4\sin(2x) + (1 - 2\pi)\cos(2x) + 2x. \ ♦$$

We now have strategies for solving equations (2.1) and (2.2) and the initial value problem. We must be able to find two linearly independent solutions of the homogeneous equation and any one particular solution of the nonhomogeneous equation. We now will develop important cases in which we can carry out these steps.

SECTION 2.1 PROBLEMS

In each of Problems 1 through 5, verify that y_1 and y_2 are solutions of the homogeneous differential equation, calculate the Wronskian of these solutions, write the general solution, and solve the initial value problem.

1. $y'' - 2y' + 2y = 0$; $y(0) = 6$, $y'(0) = 1$
   $y_1(x) = e^x \cos(x)$, $y_2(x) = e^x \sin(x)$
2. $y'' - 6y' + 13y = 0$; $y(0) = -1$, $y'(0) = 1$
   $y_1(x) = e^{3x} \cos(2x)$, $y_2(x) = e^{3x} \sin(2x)$
3. $y'' + 3y' + 2y = 0$; $y(0) = -3$, $y'(0) = -1$
   $y_1(x) = e^{-2x}$, $y_2(x) = e^{-x}$
4. $y'' - 16y = 0$; $y(0) = 12$, $y'(0) = 3$
   $y_1(x) = e^{4x}$, $y_2(x) = e^{-4x}$
5. $y'' + 36y = 0$; $y(0) = -5$, $y'(0) = 2$
   $y_1(x) = \sin(6x)$, $y_2(x) = \cos(6x)$

In Problems 6 through 10, use the results of Problems 1 through 5, respectively, and the given particular solution Y_p to write the general solution of the nonhomogeneous equation.

6. $y'' - 2y' + 2y = -5x^2$; $Y_p(x) = -5x^2/2 - 5x - 4$
7. $y'' - 6y' + 13y = -e^x$; $Y_p(x) = -8e^x$
8. $y'' + 3y' + 2y = 15$; $Y_p(x) = 15/2$
9. $y'' - 16y = 4x^2$; $Y_p(x) = -x^2/4 + 1/2$
10. $y'' + 36y = x - 1$, $Y_p(x) = (x - 1)/36$
11. Let φ be a solution of $y'' + py' + qy = 0$ on an open interval I. Suppose $\varphi(x_0) = 0$ for some x_0 in this interval. Suppose $\varphi(x)$ is not identically zero. Prove that $\varphi'(x_0) \neq 0$.
12. Suppose y_1 and y_2 are solutions of equation (2.2) on (a, b) and that p and q are continuous. Suppose y_1 and y_2 both have a relative extremum at some point between a and b. Show that y_1 and y_2 are linearly dependent.
13. Show that $y_1(x) = x$ and $y_2(x) = x^2$ are linearly independent solutions of $x^2y'' - 2xy' + 2y = 0$ on $(-1, 1)$ but that $W(0) = 0$. Why does this not contradict Theorem 2.3 conclusion (1)?
14. Let y_1 and y_2 be distinct solutions of equation (2.2) on an open interval I. Let x_0 be in I, and suppose $y_1(x_0) = y_2(x_0) = 0$. Prove that y_1 and y_2 are linearly dependent on I. Thus, linearly independent solutions cannot share a common zero.
15. Here is a sketch of a proof of Theorem 2.2. Fill in the details. Denote $W(y_1, y_2) = W$ for convenience.

    For conclusion (1), use the fact that y_1 and y_2 are solutions of equation (2.2) to write

    $$y_1'' + py_1' + qy_1 = 0$$
    $$y_2'' + py_2' + qy_2 = 0.$$

    Multiply the first equation by y_2 and the second by $-y_1$ and add. Use the resulting equation to show that

    $$W' + pW = 0.$$

    Solve this linear equation to verify the conclusion of part (1).

    To prove conclusion (2), show first that, if $y_2(x) = ky_1(x)$ for all x in I, then $W(x) = 0$. A similar conclusion holds if $y_1(x) = ky_2(x)$. Thus, linear dependence implies vanishing of the Wronskian.

    Conversely, suppose $W(x) = 0$ on I. Suppose first that $y_2(x)$ does not vanish on I. Differentiate y_1/y_2 to show that

    $$y_2^2 \frac{d}{dx}\left(\frac{y_1}{y_2}\right) = -W(x) = 0$$

    on I. This means that y_1/y_2 has a zero derivative on I, hence $y_1/y_2 = c$, so $y_1 = cy_2$ and these solutions are linearly dependent. A technical argument, which we omit, covers the case that $y_2(x)$ can vanish at points of I.
16. Let $y_1(x) = x^2$ and $y_2(x) = x^3$. Show that $W(x) = x^4$. Now $W(0) = 0$, but $W(x) \neq 0$ if $x \neq 0$. Why does this not violate Theorem 2.3 conclusion (1)?

2.2 The Constant Coefficient Case

We have outlined strategies for solving second-order linear homogeneous and nonhomogeneous differential equations. In both cases, we must begin with two linearly independent solutions of a homogeneous equation. This can be a difficult problem. However, when the coefficients are constants, we can write solutions fairly easily.

Consider the constant-coefficient linear homogeneous equation

$$y'' + ay' + by = 0 \tag{2.3}$$

in which a and b are constants (numbers). A method suggests itself if we read the differential equation like a sentence. We want a function y such that the second derivative, plus a constant multiple of the first derivative, plus a constant multiple of the function itself is equal to zero for all x. This behavior suggests an exponential function $e^{\lambda x}$, because derivatives of $e^{\lambda x}$ are constant multiples of $e^{\lambda x}$. We therefore try to find λ so that $e^{\lambda x}$ is a solution.

Substitute $e^{\lambda x}$ into equation (2.3) to get

$$\lambda^2 e^{\lambda x} + a\lambda e^{\lambda x} + be^{\lambda x} = 0.$$

Since $e^{\lambda x}$ is never zero, the exponential factor cancels, and we are left with a quadratic equation for λ:

$$\lambda^2 + a\lambda + b = 0. \tag{2.4}$$

The quadratic equation (2.4) is the *characteristic equation* of the differential equation (2.3). Notice that the characteristic equation can be read directly from the coefficients of the differential equation, and we need not substitute $e^{\lambda x}$ each time. The characteristic equation has roots

$$\frac{1}{2}(-a \pm \sqrt{a^2 - 4b}),$$

leading to the following three cases.

Case 1: Real, Distinct Roots

This occurs when $a^2 - 4b > 0$. The distinct roots are

$$\lambda_1 = \frac{1}{2}(-a + \sqrt{a^2 - 4b}) \quad \text{and} \quad \lambda_2 = \frac{1}{2}(-a - \sqrt{a^2 - 4b}).$$

$e^{\lambda_1 x}$ and $e^{\lambda_2 x}$ are linearly independent solutions, and in this case, the general solution of equation (2.3) is

$$y = c_1 e^{\lambda_1 x} + c_2 e^{\lambda_2 x}.$$

EXAMPLE 2.6

From the differential equation

$$y'' - y' - 6y = 0,$$

we immediately read the characteristic equation

$$\lambda^2 - \lambda - 6 = 0$$

as having real, distinct roots 3 and -2. The general solution is

$$y = c_1 e^{3x} + c_2 e^{-2x}. \ \blacklozenge$$

Case 2: Repeated Roots

This occurs when $a^2 - 4b = 0$ and the root of the characteristic equation is $\lambda = -a/2$. One solution of the differential equation is $e^{-ax/2}$.

We need a second, linearly independent solution. We will invoke a method called *reduction of order*, which will produce a second solution if we already have one solution. Attempt a second solution $y(x) = u(x)e^{-ax/2}$. Compute

$$y' = u'e^{-ax/2} - \frac{a}{2}ue^{-ax/2}$$

and

$$y'' = u''e^{-ax/2} - au'e^{-ax/2} + \frac{a^2}{4}ue^{-ax/2}.$$

Substitute these into equation (2.3) to get

$$\begin{aligned} &u''e^{-ax/2} - au'e^{-ax/2} + \frac{a^2}{4}ue^{-ax/2} \\ &+ au'e^{-ax/2} - a\frac{a}{2}ue^{-ax/2} + bue^{-ax/2} \\ &= e^{-ax/2}\left[u'' - \left(b - \frac{a^2}{4}\right)\right] = 0. \end{aligned}$$

Since $b - a^2/4 = 0$ in this case and $e^{-ax/2}$ never vanishes, this equation reduces to

$$u'' = 0.$$

This has solutions $u(x) = cx + d$ with c and d as arbitrary constants. Therefore, any function $y = (cx + d)e^{-ax/2}$ is also a solution of equation (2.3) in this case. Since we need only one solution that is linearly independent from $e^{-ax/2}$, choose $c = 1$ and $d = 0$ to get the second solution $xe^{-ax/2}$. The general solution in this repeated roots case is

$$y = c_1e^{-ax/2} + c_2xe^{-ax/2}.$$

This is often written as $y = e^{-ax/2}(c_1 + c_2x)$.

It is not necessary to repeat this derivation every time we encounter the repeated root case. Simply write one solution $e^{-ax/2}$, and a second, linearly independent solution is $xe^{-ax/2}$.

EXAMPLE 2.7

We will solve $y'' + 8y' + 16y = 0$. The characteristic equation is

$$\lambda^2 + 8\lambda + 16 = 0$$

with repeated root $\lambda = -4$. The general solution is

$$y = c_1e^{-4x} + c_2xe^{-4x}. \ \blacklozenge$$

Case 3: Complex Roots

The characteristic equation has complex roots when $a^2 - 4b < 0$. Because the characteristic equation has real coefficients, the roots appear as complex conjugates $\alpha + i\beta$ and $\alpha - i\beta$ in which α can be zero but β is nonzero. Now the general solution is

$$y = c_1e^{(\alpha+i\beta)x} + c_2e^{(\alpha-i\beta)x}$$

or

$$y = e^{\alpha x}\left(c_1e^{i\beta x} + c_2e^{-i\beta x}\right). \tag{2.5}$$

This is correct, but it is sometimes convenient to have a solution that does not involve complex numbers. We can find such a solution using an observation made by the eighteenth century Swiss mathematician Leonhard Euler, who showed that, for any real number β,

$$e^{i\beta x} = \cos(\beta x) + i\sin(\beta x).$$

Problem 22 suggests a derivation of Euler's formula. By replacing x with $-x$, we also have

$$e^{-i\beta x} = \cos(\beta x) - i\sin(\beta x).$$

Then

$$\begin{aligned} y(x) &= e^{\alpha x}\left(c_1 e^{i\beta x} + c_2 e^{-i\beta x}\right) \\ &= c_1 e^{\alpha x}(\cos(\beta x) + i\sin(\beta x)) + c_2 e^{\alpha x}(\cos(\beta x) - i\sin(\beta x)) \\ &= (c_1 + c_2)e^{\alpha x}\cos(\beta x) + i(c_1 - c_2)e^{\alpha x}\sin(\beta x). \end{aligned}$$

Here c_1 and c_2 are arbitrary real or complex numbers. If we choose $c_1 = c_2 = 1/2$, we obtain the particular solution $e^{\alpha x}\cos(\beta x)$. And if we choose $c_1 = 1/2i = -c_2$, we obtain the particular solution $e^{\alpha x}\sin(\beta x)$. Since these solutions are linearly independent, we can write the general solution in this complex root case as

$$y(x) = c_1 e^{\alpha x}\cos(\beta x) + c_2 e^{\alpha x}\sin(\beta x) \tag{2.6}$$

in which c_1 and c_2 are arbitrary constants. We may also write this general solution as

$$y(x) = e^{\alpha x}(c_1\cos(\beta x) + c_2\sin(\beta x)). \tag{2.7}$$

Either of equations (2.6) or (2.7) is the preferred way of writing the general solution in Case 3, although equation (2.5) also is correct.

We do not repeat this derivation each time we encounter Case 3. Simply write the general solution (2.6) or (2.7), with $\alpha \pm i\beta$ the roots of the characteristic equation.

EXAMPLE 2.8

Solve $y'' + 2y' + 3y = 0$. The characteristic equation is

$$\lambda^2 + 2\lambda + 3 = 0$$

with complex conjugate roots $-1 \pm i\sqrt{2}$. With $\alpha = -1$ and $\beta = \sqrt{2}$, the general solution is

$$y = c_1 e^{-x}\cos(\sqrt{2}x) + c_2 e^{-x}\sin(\sqrt{2}x). \quad \blacklozenge$$

EXAMPLE 2.9

Solve $y'' + 36y = 0$. The characteristic equation is

$$\lambda^2 + 36 = 0$$

with complex roots $\lambda = \pm 6i$. Now $\alpha = 0$ and $\beta = 6$, so the general solution is

$$y(x) = c_1\cos(6x) + c_2\sin(6x). \quad \blacklozenge$$

We are now able to solve the constant coefficient homogeneous equation

$$y'' + ay' + by = 0$$

in all cases. Here is a summary.

Let λ_1 and λ_2 be the roots of the characteristic equation

$$\lambda^2 + a\lambda + b = 0.$$

Then:

1. If λ_1 and λ_2 are real and distinct,

$$y(x) = c_1 e^{\lambda_1 x} + c_2 e^{\lambda_2 x}.$$

2. If $\lambda_1 = \lambda_2$,

$$y(x) = c_1 e^{\lambda_1 x} + c_2 x e^{\lambda_1 x}.$$

3. If the roots are complex $\alpha \pm i\beta$,

$$y(x) = c_1 e^{\alpha x}\cos(\beta x) + c_2 e^{\alpha x}\sin(\beta x).$$

SECTION 2.2 PROBLEMS

In each of Problems 1 through 10, write the general solution.

1. $y'' + 3y' + 18y = 0$
2. $y'' + 6y' - 40y = 0$
3. $y'' + 10y' + 26y = 0$
4. $y'' + 16y' + 64y = 0$
5. $y'' - 14y' + 49y = 0$
6. $y'' - 6y' + 7y = 0$
7. $y'' + 6y' + 9y = 0$
8. $y'' - 3y' = 0$
9. $y'' - y' - 6y = 0$
10. $y'' - 2y' + 10y = 0$

In each of Problems 11 through 20, solve the initial value problem.

11. $y'' + y' - 12y = 0;\ y(2) = 2,\ y'(2) = -1$
12. $y'' - 4y' + 4y = 0;\ y(0) = 3,\ y'(0) = 5$
13. $y'' - 2y' + y = 0;\ y(1) = 12,\ y'(1) = -5$
14. $y'' - 5y' + 12y = 0;\ y(2) = 0,\ y'(2) = -4$
15. $y'' - y' + 4y = 0;\ y(-2) = 1,\ y'(-2) = 3$
16. $y'' + y' - y = 0;\ y(-4) = 7,\ y'(-4) = 1$
17. $y'' - 2y' + y = 0;\ y(1) = y'(1) = 0$
18. $y'' - 2y' - 5y = 0;\ y(0) = 0,\ y'(0) = 3$
19. $y'' + 3y' = 0;\ y(0) = 3,\ y'(0) = 6$
20. $y'' + 2y' - 3y = 0;\ y(0) = 6,\ y'(0) = -2$
21. Suppose φ is a solution of

$$y'' + ay' + by = 0;\ y(x_0) = A,\ y'(x_0) = B$$

with a, b, A, and B as given numbers and a and b positive. Show that

$$\lim_{x\to\infty} \varphi(x) = 0.$$

22. Use power series expansions to derive Euler's formula. *Hint*: Write

$$e^x = \sum_{n=0}^{\infty} \frac{1}{n!}x^n,$$

$$\sin(x) = \sum_{n=0}^{\infty} \frac{(-1)^n}{(2n+1)!}x^{2n+1},$$

and

$$\cos(x) = \sum_{n=0}^{\infty} \frac{(-1)^n}{(2n)!}x^{2n}.$$

Let $x = i\beta$ with β real, and use the fact that

$$i^{2n} = (-1)^n \text{ and } i^{2n+1} = (-1)^n i.$$

for every positive integer n.

23. This problem illustrates how small changes in the coefficients of a differential equation may cause dramatic changes in the solution.
(a) Find the general solution $\varphi(x)$ of

$$y'' - 2\alpha y' + \alpha^2 y = 0$$

with $\alpha \neq 0$.
(b) Find the general solution $\varphi_\epsilon(x)$ of

$$y'' - 2\alpha y' + (\alpha^2 - \epsilon^2)y = 0$$

with ϵ a positive constant.
(c) Show that, as $\epsilon \to 0$, the solution in part (b) does not approach the solution in part (a), even though the differential equation in part (b) would appear to more closely resemble that of part (a) as ϵ is chosen smaller.

24. (a) Find the solution ψ of the initial value problem

$$y'' - 2\alpha y' + \alpha^2 y = 0;\ y(0) = c,\ y'(0) = d$$

with $\alpha \neq 0$.
(b) Find the solution ψ_ϵ of

$$y'' - 2\alpha y' + (\alpha^2 - \epsilon^2)y = 0;\ y(0) = c,\ y'(0) = d.$$

(c) Is it true that $\psi_\epsilon(x) \to \psi(x)$ as $\epsilon \to 0$?

2.3 The Nonhomogeneous Equation

From Theorem 2.4, the keys to solving the nonhomogeneous linear equation (2.1) are to find two linearly independent solutions of the associated homogeneous equation and a particular solution Y_p for the nonhomogeneous equation. We can perform the first task at least when the coefficients are constant. We will now focus on finding Y_p, considering two methods for doing this.

2.3.1 Variation of Parameters

Suppose we know two linearly independent solutions y_1 and y_2 of the associated homogeneous equation. One strategy for finding Y_p is called the *method of variation of parameters*. Look for functions u_1 and u_2 so that

$$Y_p(x) = u_1(x)y_1(x) + u_2(x)y_2(x).$$

To see how to choose u_1 and u_2, substitute Y_p into the differential equation. We must compute two derivatives. First,

$$Y_p' = u_1 y_1' + u_2 y_2' + u_1' y_1 + u_2' y_2.$$

Simplify this derivative by imposing the condition that

$$u_1' y_1 + u_2' y_2 = 0. \tag{2.8}$$

Now

$$Y_p' = u_1 y_1' + u_2 y_2',$$

so

$$Y_p'' = u_1' y_1' + u_2' y_2' + u_1 y_1'' + u_2 y_2''.$$

Substitute Y_p into the differential equation to get

$$\begin{aligned} &u_1' y_1' + u_2' y_2' + u_1 y_1'' + u_2 y_2'' \\ &+ p(x)(u_1 y_1' + u_2 y_2') + q(x)(u_1 y_1 + u_2 y_2) = f(x). \end{aligned}$$

Rearrange terms to write

$$\begin{aligned} &u_1[y_1'' + p(x)y_1' + q(x)y_1] \\ &+ u_2[y_2'' + p(x)y_2' + q(x)y_2] \\ &+ u_1' y_1' + u_2' y_2' = f(x). \end{aligned}$$

The two terms in square brackets are zero, because y_1 and y_2 are solutions of $y'' + p(x)y' + q(x)y = 0$. The last equation therefore reduces to

$$u_1' y_1' + u_2' y_2' = f(x). \tag{2.9}$$

Equations (2.8) and (2.9) can be solved for u_1' and u_2' to get

$$u_1'(x) = -\frac{y_2(x)f(x)}{W(x)} \quad \text{and} \quad u_2'(x) = \frac{y_1(x)f(x)}{W(x)} \tag{2.10}$$

where $W(x)$ is the Wronskian of $y_1(x)$ and $y_2(x)$.

We know that $W(x) \neq 0$, because y_1 and y_2 are assumed to be linearly independent solutions of the associated homogeneous equation. Integrate equations (2.10) to obtain

$$u_1(x) = -\int \frac{y_2(x)f(x)}{W(x)}\,dx \quad \text{and} \quad u_2(x) = \int \frac{y_1(x)f(x)}{W(x)}\,dx. \tag{2.11}$$

Once we have u_1 and u_2, we have a particular solution $Y_p = u_1y_1 + u_2y_2$, and the general solution of $y'' + p(x)y' + q(x)y = f(x)$ is

$$y = c_1y_1 + c_2y_2 + Y_p.$$

EXAMPLE 2.10

Find the general solution of

$$y'' + 4y = \sec(x)$$

for $-\pi/4 < x < \pi/4$.

The characteristic equation of $y'' + 4y = 0$ is $\lambda^2 + 4 = 0$ with complex roots $\lambda = \pm 2i$. Two linearly independent solutions of the associated homogeneous equation $y'' + 4y = 0$ are

$$y_1(x) = \cos(2x) \text{ and } y_2(x) = \sin(2x).$$

Now look for a particular solution of the nonhomogeneous equation. First compute the Wronskian

$$W(x) = \begin{vmatrix} \cos(2x) & \sin(2x) \\ -2\sin(2x) & 2\cos(2x) \end{vmatrix} = 2(\cos^2(x) + \sin^2(x)) = 2.$$

Use equations (2.11) with $f(x) = \sec(x)$ to obtain

$$\begin{aligned} u_1(x) &= -\int \frac{\sin(2x)\sec(x)}{2}\,dx \\ &= -\int \frac{2\sin(x)\cos(x)\sec(x)}{2}\,dx \\ &= -\int \frac{\sin(x)\cos(x)}{\cos(x)}\,dx \\ &= -\int \sin(x)\,dx = \cos(x) \end{aligned}$$

and

$$\begin{aligned} u_2(x) &= \int \frac{\cos(2x)\sec(x)}{2}\,dx \\ &= \int \frac{(2\cos^2(x) - 1)}{2\cos(x)}\,dx \\ &= \int \left(\cos(x) - \frac{1}{2}\sec(x)\right)dx \\ &= \sin(x) - \frac{1}{2}\ln|\sec(x) + \tan(x)|. \end{aligned}$$

This gives us the particular solution

$$\begin{aligned} Y_p(x) &= u_1(x)y_1(x) + u_2(x)y_2(x) \\ &= \cos(x)\cos(2x) + \left(\sin(x) - \frac{1}{2}\ln|\sec(x) + \tan(x)|\right)\sin(2x). \end{aligned}$$

The general solution of $y'' + 4y' = \sec(x)$ is

$$\begin{aligned} y(x) &= c_1\cos(2x) + c_2\sin(2x) \\ &\quad + \cos(x)\cos(2x) + \left(\sin(x) - \frac{1}{2}\ln|\sec(x) + \tan(x)|\right)\sin(2x). \quad \blacklozenge \end{aligned}$$

2.3.2 Undetermined Coefficients

We will discuss a second method for finding a particular solution of the nonhomogeneous equation, which, however, applies only to the constant coefficient case $y'' + ay' + by = f(x)$.

The idea behind the method of undetermined coefficients is that sometimes we can guess a general form for $Y_p(x)$ from the appearance of $f(x)$.

EXAMPLE 2.11

We will find the general solution of $y'' - 4y = 8x^2 - 2x$.

It is routine to find the general solution $c_1e^{2x} + c_2e^{-2x}$ of the associated homogeneous equation. We need a particular solution $Y_p(x)$ of the nonhomogeneous equation.

Because $f(x) = 8x^2 - 2x$ is a polynomial and derivatives of polynomials are polynomials, it is reasonable to think that there might be a polynomial solution. Furthermore, no such solution can include a power of x higher than 2. If $Y_p(x)$ had an x^3 term, this term would be retained by the $-4y$ term of $y'' - 4y$, and $8x^2 - 2x$ has no such term.

This reasoning suggests that we try a particular solution $Y_p(x) = Ax^2 + Bx + C$. Compute $y'(x) = 2Ax + B$ and $y''(x) = 2A$. Substitute these into the differential equation to get

$$2A - 4(Ax^2 + Bx + C) = 8x^2 - 2x.$$

Write this as

$$(-4A - 8)x^2 + (-4B + 2)x + (2A - 4C) = 0.$$

This second-degree polynomial must be zero for all x if Y_p is to be a solution. But a second-degree polynomial has only two roots, unless all of its coefficients are zero. Therefore,

$$-4A - 8 = 0,$$

$$-4B + 2 = 0,$$

and

$$2A - 4C = 0.$$

Solve these to get $A = -2$, $B = 1/2$, and $C = -1$. This gives us the particular solution

$$Y_p(x) = -2x^2 + \frac{1}{2}x - 1.$$

The general solution is

$$y(x) = c_1e^{2x} + c_2e^{-2x} - 2x^2 + \frac{1}{2}x - 1. \ \blacklozenge$$

EXAMPLE 2.12

Find the general solution of $y'' + 2y' - 3y = 4e^{2x}$.

The general solution of $y'' + 2y' - 3y = 0$ is $c_1e^{-3x} + c_2e^x$.

Now look for a particular solution. Because derivatives of e^{2x} are constant multiples of e^{2x}, we suspect that a constant multiple of e^{2x} might serve. Try $Y_p(x) = Ae^{2x}$. Substitute this into the differential equation to get

$$4Ae^{2x} + 4Ae^{2x} - 3Ae^{2x} = 5Ae^{2x} = 4e^{2x}.$$

This works if $5A = 4$, so $A = 4/5$. A particular solution is $Y_p(x) = 4e^{2x}/5$.

The general solution is

$$y(x) = c_1 e^{-3x} + c_2 e^x + \frac{4}{5} e^{2x}. \quad \blacklozenge$$

EXAMPLE 2.13

Find the general solution of $y'' - 5y' + 6y = -3\sin(2x)$.

The general solution of $y'' - 5y' + 6y = 0$ is $c_1 e^{3x} + c_2 e^{2x}$.

We need a particular solution Y_p of the nonhomogeneous equation. Derivatives of $\sin(2x)$ are constant multiples of $\sin(2x)$ or $\cos(2x)$. Derivatives of $\cos(2x)$ are also constant multiples of $\sin(2x)$ or $\cos(2x)$. This suggests that we try a particular solution $Y_p(x) = A\cos(2x) + B\sin(2x)$. Notice that we include both $\sin(2x)$ and $\cos(2x)$ in this first attempt, even though $f(x)$ just has a $\sin(2x)$ term. Compute

$$Y_p'(x) = -2A\sin(2x) + 2B\cos(2x) \text{ and } Y_p''(x) = -4A\cos(2x) - 4B\sin(2x).$$

Substitute these into the differential equation to get

$$-4A\cos(2x) - 4B\sin(2x) - 5[-2A\sin(2x) + 2B\cos(2x)]$$
$$+ 6[A\cos(2x) + B\sin(2x)] = -3\sin(2x).$$

Rearrange terms to write

$$[2B + 10A + 3]\sin(2x) = [-2A + 10B]\cos(2x).$$

But $\sin(2x)$ and $\cos(2x)$ are not constant multiples of each other unless these constants are zero. Therefore,

$$2B + 10A + 3 = 0 = -2A + 10B.$$

Solve these to get $A = -15/52$ and $B = -3/52$. A particular solution is

$$Y_p(x) = -\frac{15}{52}\cos(2x) - \frac{3}{52}\sin(2x).$$

The general solution is

$$y(x) = c_1 e^{3x} + c_2 e^{2x} - \frac{15}{52}\cos(2x) - \frac{3}{52}\sin(2x). \quad \blacklozenge$$

The method of undetermined coefficients has a trap built into it. Consider the following.

EXAMPLE 2.14

Find a particular solution of $y'' + 2y' - 3y = 8e^x$.

Reasoning as before, try $Y_p(x) = Ae^x$. Substitute this into the differential equation to obtain

$$Ae^x + 2Ae^x - 3Ae^x = 8e^x.$$

But then $8e^x = 0$, which is a contradiction. $\blacklozenge$

The problem here is that e^x is also a solution of the associated homogeneous equation, so the left side will vanish when Ae^x is substituted into $y'' + 2y' - 3y = 8e^x$.

Whenever a term of a proposed $Y_p(x)$ is a solution of the associated homogeneous equation, multiply this proposed solution by x. If this results in another solution of the associated homogeneous equation, multiply it by x again.

EXAMPLE 2.15

Revisit Example 2.14. Since $f(x) = 8e^x$, our first impulse was to try $Y_p(x) = Ae^x$. But this is a solution of the associated homogeneous equation, so multiply by x and try $Y_p(x) = Axe^x$. Now

$$Y_p' = Ae^x + Axe^x \text{ and } Y_p'' = 2Ae^x + Axe^x.$$

Substitute these into the differential equation to get

$$2Ae^x + Axe^x + 2(Ae^x + Axe^x) - 3Axe^x = 8e^x.$$

This reduces to $4Ae^x = 8e^x$, so $A = 2$, yielding the particular solution $Y_p(x) = 2xe^x$. The general solution is

$$y(x) = c_1e^{-3x} + c_2e^x + 2xe^x. \blacklozenge$$

EXAMPLE 2.16

Solve $y'' - 6y' + 9y = 5e^{3x}$.

The associated homogeneous equation has the characteristic equation $(\lambda - 3)^2 = 0$ with repeated roots $\lambda = 3$. The general solution of this associated homogeneous equation is $c_1e^{3x} + c_2xe^{3x}$.

For a particular solution, we might first try $Y_p(x) = Ae^{3x}$, but this is a solution of the homogeneous equation. Multiply by x and try $Y_p(x) = Axe^{3x}$. This is also a solution of the homogeneous equation, so multiply by x again and try $Y_p(x) = Ax^2e^{3x}$. If this is substituted into the differential equation, we obtain $A = 5/2$, so a particular solution is $Y_p(x) = 5x^2e^{3x}/2$. The general solution is

$$y = c_1e^{3x} + c_2xe^{3x} + \frac{5}{2}x^2e^{3x}. \blacklozenge$$

The method of undetermined coefficients is limited by our ability to "guess" a particular solution from the form of $f(x)$, and unlike variation of parameters, requires that the coefficients of y' and y be constant.

Here is a summary of the method. Suppose we want to find the general solution of

$$y'' + ay' + by = f(x).$$

Step 1. Write the general solution

$$y_h(x) = c_1y_1(x) + c_2y_2(x)$$

of the associated homogeneous equation

$$y'' + ay' + by = 0$$

with y_1 and y_2 linearly independent. We can always do this in the constant coefficient case.

Step 2. We need a particular solution Y_p of the nonhomogeneous equation. This may require several steps. Make an initial attempt of a general form of a particular solution using $f(x)$ and perhaps Table 2.1 as a guide. If this is not possible, this method cannot be used. If we can solve for the constants so that this first guess works, then we have Y_p.

Step 3. If *any term* of the first attempt is a solution of the associated homogeneous equation, multiply by x. If any term of this revised attempt is a solution of the homogeneous equation, multiply by x again. Substitute this final general form of a particular solution into the differential equation and solve for the constants to obtain Y_p.

Step 4. The general solution is

$$y = y_1 + y_2 + Y_p.$$

TABLE 2.1 ***Functions to Try for $Y_p(x)$ in the Method of Undetermined Coefficients***

$f(x)$	$Y_p(x)$
$P(x)$	$Q(x)$
Ae^{cx}	Re^{cx}
$A\cos(\beta x)$	$C\cos(\beta x) + D\sin(\beta x)$
$A\sin(\beta x)$	$C\cos(\beta x) + D\sin(\beta x)$
$P(x)e^{cx}$	$Q(x)e^{cx}$
$P(x)\cos(\beta x)$	$Q(x)\cos(\beta x) + R(x)\sin(\beta x)$
$P(x)\sin(\beta x)$	$Q(x)\cos(\beta x) + R(x)\sin(\beta x)$
$P(x)e^{cx}\cos(\beta x)$	$Q(x)e^{cx}\cos(\beta x) + R(x)e^{cx}\sin(\beta x)$
$P(x)e^{cx}\sin(\beta x)$	$Q(x)e^{cx}\cos(\beta x) + R(x)e^{cx}\sin(\beta x)$

Table 2.1 provides a list of functions for a first try at $Y_p(x)$ for various functions $f(x)$ that might appear in the differential equation. In this list, $P(x)$ is a given polynomial of degree n, $Q(x)$ and $R(x)$ are polynomials of degree n with undetermined coefficients for which we must solve, and c and β are constants.

2.3.3 The Principle of Superposition

Suppose we want to find a particular solution of

$$y'' + p(x)y' + q(x)y = f_1(x) + f_2(x) + \cdots + f_N(x).$$

It is routine to check that, if Y_j is a solution of

$$y'' + p(x)y' + q(x)y = f_j(x),$$

then $Y_1 + Y_2 + \cdots + Y_N$ is a particular solution of the original differential equation.

EXAMPLE 2.17

Find a particular solution of

$$y'' + 4y = x + 2e^{-2x}.$$

To find a particular solution, consider two problems:

Problem 1: $y'' + 4y = x$

Problem 2: $y'' + 4y = 2e^{-2x}$

Using undetermined coefficients, we find a particular solution $Y_{p_1}(x) = x/4$ of Problem 1 and a particular solution $Y_{p_2}(x) = e^{-2x}/4$ of Problem 2. A particular solution of the given differential equation is

$$Y_p(x) = \frac{1}{4}x + \frac{1}{4}e^{-2x}.$$

Using this, the general solution is

$$y(x) = c_1\cos(2x) + c_2\sin(2x) + \frac{1}{4}\left(x + e^{-2x}\right). \quad \blacklozenge$$

SECTION 2.3 PROBLEMS

In each of Problems 1 through 6, find the general solution, using the method of variation of parameters for a particular solution.

1. $y'' + 9y = 12\sec(3x)$
2. $y'' - 2y' - 3y = 2\sin^2(x)$
3. $y'' - 3y' + 2y = \cos(e^{-x})$
4. $y'' - 5y' + 6y = 8\sin^2(4x)$
5. $y'' + y = \tan(x)$
6. $y'' - 4y' + 3y = 2\cos(x + 3)$

In each of Problems 7 through 16, find the general solution, using the method of undetermined coefficients for a particular solution.

7. $y'' - 3y' + 2y = 10\sin(x)$
8. $y'' - 2y' + y = 3x + 25\sin(3x)$
9. $y'' - 4y' + 13y = 3e^{2x} - 5e^{3x}$
10. $y'' - 4y' = 8x^2 + 2e^{3x}$
11. $y'' - 2y' + 10y = 20x^2 + 2x - 8$
12. $y'' - 4y' + 5y = 21e^{2x}$
13. $y'' - 6y' + 8y = 3e^x$
14. $y'' + 6y' + 9y = 9\cos(3x)$
15. $y'' - y' - 2y = 2x^2 + 5$
16. $y'' - y' - 6y = 8e^{2x}$

In each of Problems 17 through 24, solve the initial value problem.

17. $y'' + 8y' + 12y = e^{-x} + 7;\ y(0) = 1,\ y'(0) = 0$
18. $y'' + y = \tan(x);\ y(0) = 4,\ y'(0) = 3$
19. $y'' - 2y' - 8y = 10e^{-x} + 8e^{2x};\ y(0) = 1,\ y'(0) = 4$
20. $y'' - y' + y = 1;\ y(1) = 4,\ y'(1) = -2$
21. $y'' - y = 5\sin^2(x);\ y(0) = 2,\ y'(0) = -4$
22. $y'' - 3y' = 2e^{2x}\sin(x);\ y(0) = 1,\ y'(0) = 2$
23. $y'' - 4y = -7e^{2x} + x;\ y(0) = 1,\ y'(0) = 3$
24. $y'' + 4y' = 8 + 34\cos(x);\ y(0) = 3,\ y'(0) = 2$

2.4 Spring Motion

A spring suspended vertically and allowed to come to rest has a *natural length* L. An object (bob) of mass m is attached at the lower end, pulling the spring d units past its natural length. The bob comes to rest in its *equilibrium position* and is then displaced vertically a distance y_0 units (Figure 2.3) and released from rest or with some initial velocity. We want to construct a model allowing us to analyze the motion of the bob.

Let $y(t)$ be the displacement of the bob from the equilibrium position at time t, and take this equilibrium position to be $y = 0$. Down is chosen as the positive direction. Now consider the forces acting on the bob. Gravity pulls it downward with a force of magnitude mg. By Hooke's law, the spring exerts a force ky on the object. k is the *spring constant*, which is a number quantifying the "stiffness" of the spring. At the equilibrium position, the force of the spring is $-kd$, which is negative because it acts upward. If the object is pulled downward a distance y from this position, an additional force $-ky$ is exerted on it. The total force due to the spring is therefore $-kd - ky$. The total force due to gravity and the spring is $mg - kd - ky$. Since at the equilibrium point this force is zero, then $mg = kd$. The net force acting on the object due to gravity and the spring is therefore just $-ky$.

There are forces tending to retard or damp out the motion. These include air resistance or perhaps viscosity of a medium in which the object is suspended. A standard assumption (verified by observation) is that the retarding forces have magnitude proportional to the velocity y'. Then for some constant c called the *damping constant*, the retarding forces equal cy'. The total force acting on the bob due to gravity, damping, and the spring itself is $-ky - cy'$.

Finally, there may be a driving force $f(t)$ acting on the bob. In this case, the total external force is

$$F = -ky - cy' + f(t).$$

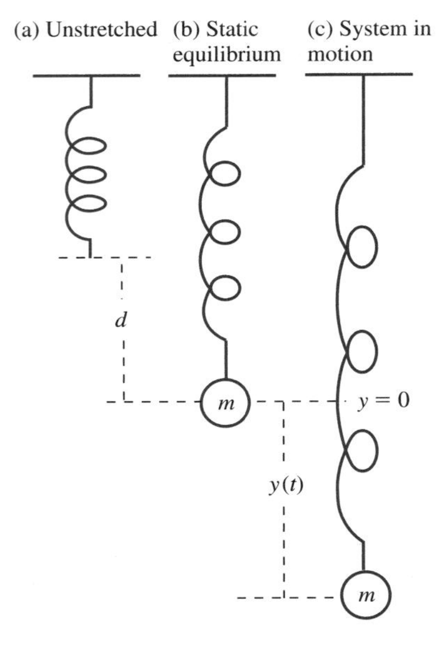

FIGURE 2.3 *Spring at natural and equilibrium lengths and in motion.*

Assuming that the mass is constant, Newton's second law of motion gives us

$$my'' = -ky - cy' + f(t)$$

or

$$y'' + \frac{c}{m}y' + \frac{k}{m}y = \frac{1}{m}f(t). \tag{2.12}$$

This is the *spring equation*. Solutions give the displacement of the bob as a function of time and enable us to analyze the motion under various conditions.

2.4.1 Unforced Motion

The motion is unforced if $f(t) = 0$. Now the spring equation is homogeneous, and the characteristic equation has roots

$$\lambda = -\frac{c}{2m} \pm \frac{1}{2m}\sqrt{c^2 - 4km}.$$

As we might expect, the solution for the displacement, and hence the motion of the bob, depends on the mass, the amount of damping, and the stiffness of the spring.

Case 1: $c^2 - 4km > 0$

Now the roots of the characteristic equation are real and distinct:

$$\lambda_1 = -\frac{c}{2m} + \frac{1}{2m}\sqrt{c^2 - 4km} \text{ and } \lambda_2 = -\frac{c}{2m} - \frac{1}{2m}\sqrt{c^2 - 4km}.$$

The general solution of the spring equation in this case is

$$y(t) = c_1 e^{\lambda_1 t} + c_2 e^{\lambda_2 t}.$$

Clearly, $\lambda_2 < 0$. Since m and k are positive, $c^2 - 4km < c^2$, so $\sqrt{c^2 - 4km} < c$ and $\lambda_1 < 0$. Therefore,

$$\lim_{t\to\infty} y(t) = 0$$

regardless of initial conditions. In the case that that $c^2 - 4km > 0$, the motion simply decays to zero as time increases. This case is called *overdamping*.

EXAMPLE 2.18 Overdamping

Let $c = 6$, $k = 5$, and $m = 1$. Now the general solution is $y(t) = c_1 e^{-t} + c_2 e^{-5t}$. Suppose the bob was initially drawn upward 4 feet from equilibrium and released downward with a speed of 2 feet per second. Then $y(0) = -4$ and $y'(0) = 2$, and we obtain

$$y(t) = \frac{1}{2} e^{-t} \left(-9 + e^{-4t}\right).$$

Figure 2.4 is a graph of this solution. Keep in mind here that down is the positive direction. Since $-9 + e^{-4t} < 0$ for $t > 0$, then $y(t) < 0$, and the bob always remains above the equilibrium point. Its velocity $y'(t) = e^{-t}(9 - 5e^{-4t})/2$ decreases to zero as t increases, so the bob moves downward toward equilibrium with decreasing velocity, approaching arbitrarily close to but never reaching this position and never coming completely to rest. ♦

Case 2: $c^2 - 4km = 0$

In this case, the general solution of the spring equation is

$$y(t) = (c_1 + c_2 t) e^{-ct/2m}.$$

This case is called *critical damping*. While $y(t) \to 0$ as $t \to \infty$, as with overdamping, now the bob can pass through the critical point, as the following example shows.

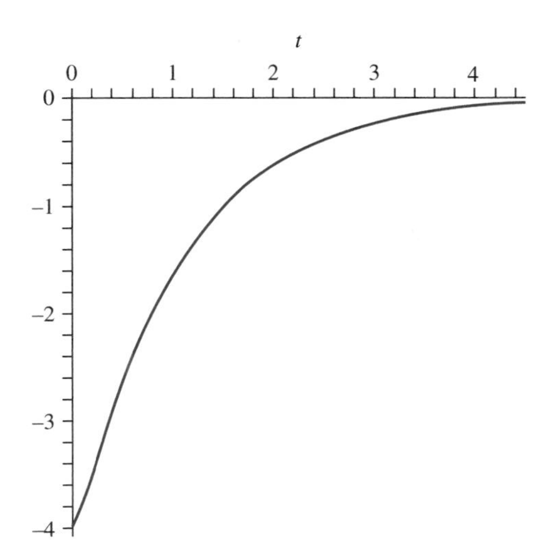

FIGURE 2.4 *Overdamped, unforced motion in Example 2.18.*

EXAMPLE 2.19 Critical Damping

Let $c = 2$ and $k = m = 1$. Then $y(t) = (c_1 + c_2 t)e^{-t}$. Suppose the bob is initially pulled up four feet above the equilibrium position and then pushed downward with a speed of 5 feet per second. Then $y(0) = -4$, and $y'(0) = 5$. So

$$y(t) = (-4 + t)e^{-t}.$$

Since $y(4) = 0$, the bob reaches the equilibrium four seconds after it was released and then passes through it. In fact, $y(t)$ reaches its maximum when $t = 5$ seconds, and this maximum value is $y(5) = e^{-5}$, which is about 0.007 units below the equilibrium point. The velocity $y'(t) = (5 - t)e^{-t}$ is negative for $t > 5$, so the bob's velocity decreases after the five second point. Since $y(t) \to 0$ as $t \to \infty$, the bob moves with decreasing velocity back toward the equilibrium point as time increases. Figure 2.5 is a graph of this displacement function for $2 \le t \le 8$. ♦

In general, when critical damping occurs, the bob either passes through the equilibrium point exactly once, as in Example 2.19, or never reaches it at all, depending on the initial conditions.

Case 3: $c^2 - 4km < 0$

Here the spring constant and mass of the bob are sufficiently large that $c^2 < 4km$ and the damping is less dominant. This is called *underdamping*. The general underdamped solution has the form

$$y(t) = e^{-ct/2m}[c_1 \cos(\beta t) + c_2 \sin(\beta t)]$$

in which

$$\beta = \frac{1}{2m}\sqrt{4km - c^2}.$$

Since c and m are positive, $y(t) \to 0$ as $t \to \infty$, as in the other two cases. This is not surprising in the absence of an external driving force. However, with underdamping, the motion is oscillatory because of the sine and cosine terms in the displacement function. The motion is not periodic however because of the exponential factor $e^{-ct/2m}$, which causes the amplitudes of the oscillations to decay as time increases.

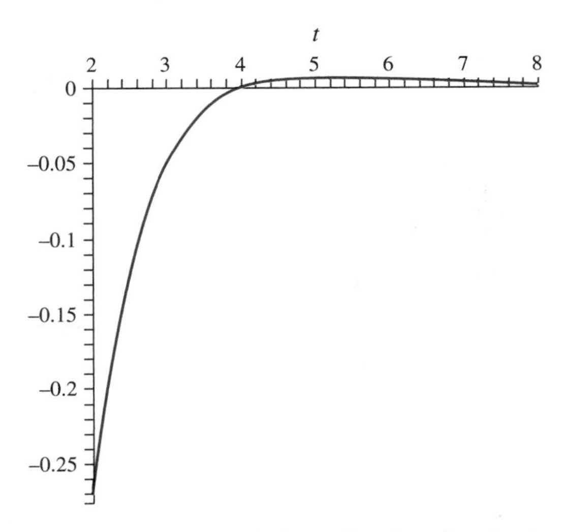

FIGURE 2.5 *Critically damped, unforced motion in Example 2.19.*

EXAMPLE 2.20 Underdamping

Let $c = k = 2$ and $m = 1$. The general solution is

$$y(t) = e^{-t}[c_1 \cos(t) + c_2 \sin(t)].$$

Suppose the bob is driven downward from a point three feet above equilibrium with an initial speed of two feet per second. Then $y(0) = -3$, and $y'(0) = 2$. The solution is

$$y(t) = -e^{-t}(3\cos(t) + \sin(t)).$$

The behavior of this solution is visualized more easily if we write it in *phase angle form*. Choose C and δ so that

$$3\cos(t) + \sin(t) = C\cos(t + \delta).$$

For this, we need

$$3\cos(t) + \sin(t) = C\cos(t)\cos(\delta) - C\sin(t)\sin(\delta).$$

Then

$$C\cos(\delta) = 3 \text{ and } C\sin(\delta) = -1,$$

so

$$\frac{C\sin(\delta)}{C\cos(\delta)} = \tan(\delta) = -\frac{1}{3}.$$

Now

$$\delta = \arctan\left(-\frac{1}{3}\right) = -\arctan\left(\frac{1}{3}\right).$$

To solve for C, write

$$C^2\cos^2(\delta) + C^2\sin^2(\delta) = C^2 = 3^2 + 1 = 10.$$

Then $C = \sqrt{10}$, and the solution can be written in phase angle form as

$$y(t) = \sqrt{10}e^{-t}\cos(t - \arctan(1/3)).$$

The graph is a cosine curve with decaying amplitude squeezed between graphs of $y = \sqrt{10}e^{-t}$ and $y = -\sqrt{10}e^{-t}$. Figure 2.6 shows $y(t)$ and these two exponential functions as reference curves.

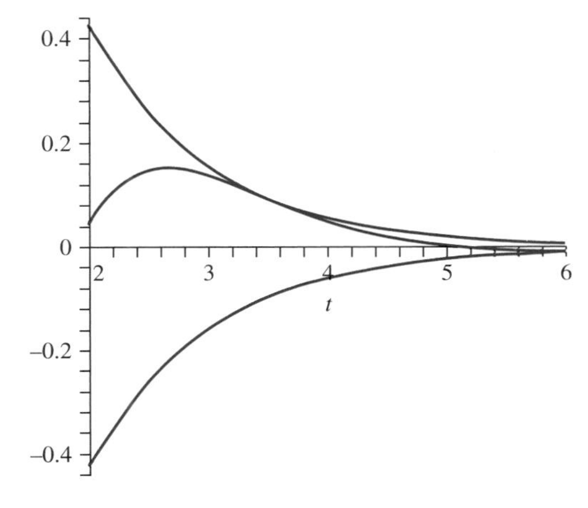

FIGURE 2.6 *Underdamped, unforced motion in Example 2.20.*

The bob passes back and forth through the equilibrium point as t increases. Specifically, it passes through the equilibrium point exactly when $y(t)=0$, which occurs at times

$$t=\arctan\left(\frac{1}{3}\right)+\frac{2n+1}{2}\pi$$

for $n=0,1,2,\cdots$. ♦

Next we will pursue the effect of a driving force on the motion of the bob.

2.4.2 Forced Motion

Different driving forces will result in different motion. We will analyze the case of a periodic driving force $f(t)=A\cos(\omega t)$. Now the spring equation (2.12) is

$$y''+\frac{c}{m}y'+\frac{k}{m}y=\frac{A}{m}\cos(\omega t). \tag{2.13}$$

We have solved the associated homogeneous equation in all cases on c, k, and m. For the general solution of equation (2.13), we need only a particular solution. Application of the method of undetermined coefficients yields the particular solution

$$\begin{aligned}Y_p(t)=&\frac{mA(k-m\omega^2)}{(k-m\omega^2)^2+\omega^2c^2}\cos(\omega t)\\&+\frac{A\omega c}{(k-m\omega^2)^2+\omega^2c^2}\sin(\omega t).\end{aligned}$$

It is customary to denote $\omega_0=\sqrt{k/m}$ to write

$$\begin{aligned}Y_p(t)=&\frac{mA(\omega_0^2-\omega^2)}{m^2(\omega_0^2-\omega^2)^2+\omega^2c^2}\cos(\omega t)\\&+\frac{A\omega c}{m^2(\omega_0^2-\omega^2)^2+\omega^2c^2}\sin(\omega t).\end{aligned}$$

We will analyze some specific cases to get some insight into the motion with this forcing function.

Case 1: Overdamped Forced Motion

Overdamping occurs when $c^2-4km>0$. Suppose $c=6$, $k=5$, $m=1$ $A=6\sqrt{5}$ and $\omega=\sqrt{5}$. If the bob is released from rest from the equilibrium position, then $y(t)$ satisfies the initial value problem

$$y''+6y'+5y=6\sqrt{5}\cos(\sqrt{5}t);\quad y(0)=y'(0)=0$$

The solution is

$$y(t)=\frac{\sqrt{5}}{4}(-e^{-t}+e^{-5t})+\sin(\sqrt{5}t).$$

A graph of this solution is shown in Figure 2.7. As time increases, the exponential terms decay to zero, and the displacement behaves increasingly like $\sin(\sqrt{5}t)$, oscillating up and down through the equilibrium point with approximate period $2\pi/\sqrt{5}$. Contrast this with the overdamped motion without the forcing function in which the bob began above the equilibrium point and moved with decreasing speed down toward it but never reached it.

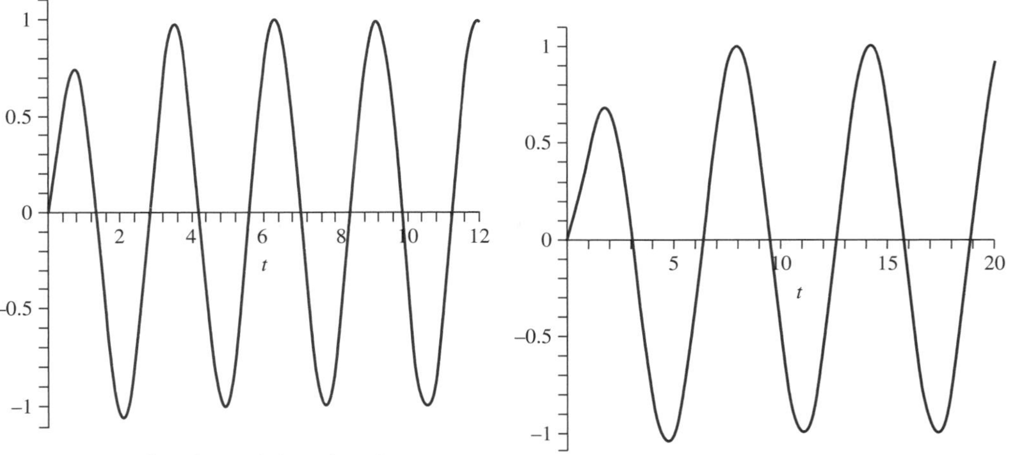

FIGURE 2.7 *Overdamped, forced motion.*

FIGURE 2.8 *Critically damped, forced motion*

Case 2: Critically Damped Forced Motion

Let $c = 2$, $m = k = 1$, $\omega = 1$, and $A = 2$. Assume that the bob is released from rest from the equilibrium point. Now the initial value problem is

$$y'' + 2y' + y = 2\cos(t);\; y(0) = y'(0) = 0$$

with the solution

$$y(t) = -te^{-t} + \sin(t).$$

Figure 2.8 is a graph of this solution, which is a case of *critically damped forced motion*. As t increases, the term with the exponential factor decays (although not as fast as in the overdamping case where there is no factor of t). Nevertheless, after sufficient time, the motion settles into nearly (but not exactly because $-te^{-t}$ is never zero for $t > 0$) a sinusoidal motion back and forth through the equilibrium point.

Case 3: Underdamped Forced Motion

Let $c = k = 2$, $m = 1$, $\omega = \sqrt{2}$, and $A = 2\sqrt{2}$, so $c^2 - 4km < 0$. Suppose the bob is released from rest at the equilibrium position. The initial value problem is

$$y'' + 2y' + 2y = 2\sqrt{2}\cos(\sqrt{2}t);\; y(0) = y'(0) = 0$$

with the solution

$$y(t) = -\sqrt{2}e^{-t}\sin(t) + \sin(\sqrt{2}t).$$

This is *underdamped forced motion*. Unlike overdamping and critical damping, the exponential term e^{-t} has a trigonometric factor $\sin(t)$. Figure 2.9 is a graph of this solution. As time increases, the term $-\sqrt{2}e^{-t}\sin(t)$ becomes less influential and the motion settles nearly into an oscillation back and forth through the equilibrium point with a period of nearly $2\pi/\sqrt{2}$.

2.4.3 Resonance

In the absence of damping, an important phenomenon called *resonance* can occur. Suppose $c = 0$, but there is still a driving force $A\cos(\omega t)$. Now the spring equation (2.12) is

$$y'' + \frac{k}{m}y = \frac{A}{m}\cos(\omega t).$$

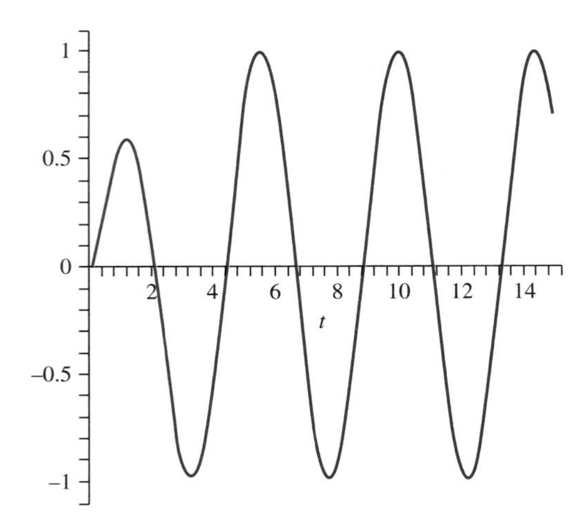

FIGURE 2.9 *Underdamped, forced motion.*

From the particular solution Y_p found in Section 2.4.2, with $c = 0$, we find that this spring equation has general solution

$$y(t) = c_1 \cos(\omega t) + c_2 \sin(\omega t) + \frac{A}{m(\omega_0^2 - \omega^2)} \cos(\omega t)$$

in which $\omega_0 = \sqrt{k/m}$. This number is called the *natural frequency* of the spring system, and it is a function of the stiffness of the spring and the mass of the bob. ω is the *input frequency* and is contained in the driving force. This general solution assumes that the natural and input frequencies are different. Of course, the closer we choose the natural and input frequencies, the larger the amplitude of the $\cos(\omega t)$ term in the solution.

Resonance occurs when the natural and input frequencies are the same. Now the differential equation is

$$y'' + \frac{k}{m} y = \frac{A}{m} \cos(\omega_0 t). \tag{2.14}$$

The solution derived for the case when $\omega \neq \omega_0$ does not apply to equation (2.14). To find the general solution in the present case, first find the general solution of the associated homogeneous equation

$$y'' + \frac{k}{m} y = 0.$$

This has the general solution

$$y_h(t) = c_1 \cos(\omega_0 t) + c_2 \sin(\omega_0 t).$$

Now we need a particular solution of equation (2.14). To use the method of undetermined coefficients, we might try a function of the form $a\cos(\omega_0 t) + b\sin(\omega_0 t)$. However, these are solutions of the associated homogeneous equation, so instead we try

$$Y_p(t) = at\cos(\omega_0 t) + bt\sin(\omega_0 t).$$

Substitute $Y_p(t)$ into equation (2.14) to get

$$-2a\omega_0 \sin(\omega_0 t) + 2b\cos(\omega_0 t) = \frac{A}{m}\cos(\omega_0 t).$$

Thus, choose

$$a = 0 \text{ and } 2b\omega_0 = \frac{A}{m}.$$

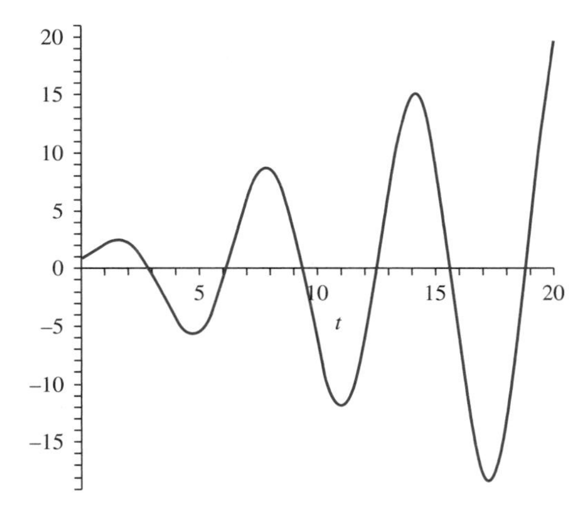

FIGURE 2.10 *Resonance.*

This gives us the particular solution

$$Y_p(t) = \frac{A}{2m\omega_0} t \sin(\omega_0 t).$$

The general solution is

$$y(t) = c_1 \cos(\omega_0 t) + c_2 \sin(\omega_0 t) + \frac{A}{2m\omega_0} t \sin(\omega_0 t).$$

This solution differs from the case $\omega \neq \omega_0$ in the factor of t in the particular solution. Because of this, solutions increase in amplitude as t increases. This phenomenon is called *resonance*.

As an example, suppose $c_1 = c_2 = \omega_0 = 1$ and $A/2m = 1$. Now the solution is

$$y(t) = \cos(t) + \sin(t) + t \sin(t).$$

Figure 2.10 displays the increasing amplitude of the oscillations with time.

While there is always some damping in the real world, if the damping constant is close to zero when compared to other factors and if the natural and input frequencies are (nearly) equal, then oscillations can build up to a sufficiently large amplitude to cause resonance-like behavior. This caused the collapse of the Broughton Bridge near Manchester, England, in 1831 when a column of soldiers marching across maintained a cadence (input frequency) that happened to closely match the natural frequency of the material of the bridge. More recently the Tacoma Narrows Bridge in the state of Washington experienced increasing oscillations driven by high winds, causing the concrete roadbed to oscillate in sensational fashion until it collapsed into Puget Sound. This occurred on November 7, 1940. At one point, one side of the roadbed was about twenty-eight feet above the other as it thrashed about. Unlike the Broughton Bridge, local news crews were on hand to film this, and motion pictures of the collapse are available in many engineering and science schools.

2.4.4 Beats

In the absence of damping, an oscillatory driving force can also cause a phenomenon called *beats*. Suppose $\omega \neq \omega_0$, and consider

$$y'' + \omega_0^2 y = \frac{A}{m} \cos(\omega t).$$

Assume that the object is released from rest from the equilibrium position, so $y(0) = y'(0) = 0$. The solution is

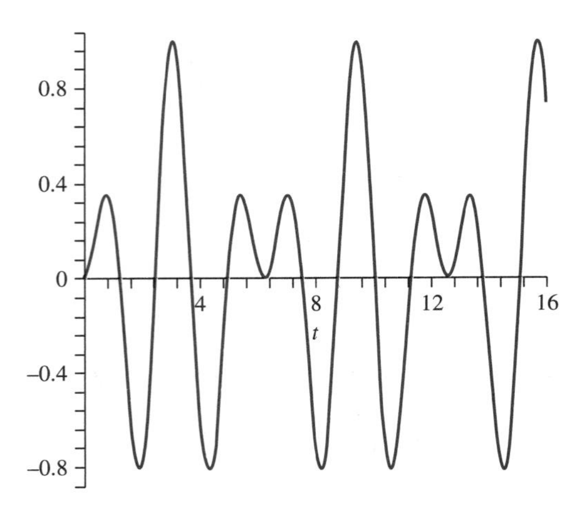

FIGURE 2.11 *Beats.*

$$y(t) = \frac{A}{m(\omega_0^2 - \omega^2)}[\cos(\omega t) - \cos(\omega_0 t)].$$

The behavior of this solution reveals itself more clearly if we write it as

$$y(t) = \frac{2A}{m(\omega_0^2 - \omega^2)} \sin\left(\frac{1}{2}(\omega_0 + \omega)t\right) \sin\left(\frac{1}{2}(\omega_0 - \omega)t\right).$$

This formulation exhibits a periodic variation of amplitude in the solution, depending on the relative sizes of $\omega_0 + \omega$ and $\omega_0 - \omega$. This periodic variation is called a *beat*. As an example, suppose $\omega_0 + \omega = 5$ and $\omega_0 - \omega = 1$, and the constants are chosen so that $2A/m(\omega_0^2 - \omega^2) = 1$. Now the displacement function is

$$y(t) = \sin\left(\frac{5t}{2}\right) \sin\left(\frac{t}{2}\right).$$

Figure 2.11 is a graph of this solution.

2.4.5 Analogy with an Electrical Circuit

In an *RLC* circuit with electromotive force $E(t)$, the differential equation for the current is

$$Li'(t) + Ri(t) + \frac{1}{C}q(t) = E(t).$$

Since $i = q'$, this is a second-order differential equation for the charge:

$$q'' + \frac{R}{L}q' + \frac{1}{LC}q = \frac{1}{L}E(t).$$

Assuming that the resistance, inductance, and capacitance are constant, this equation is exactly analogous to the spring equation with a driving force, which has the form

$$y'' + \frac{c}{m}y' + \frac{k}{m}y = \frac{1}{m}f(t).$$

This means that solutions of the spring equation immediately translate into solutions of the circuit equation with the following identifications:

Displacement function $y(t) \Longleftrightarrow$ charge $q(t)$
Velocity $y'(t) \Longleftrightarrow$ current $i(t)$
Driving force $f(t) \Longleftrightarrow$ electromotive force $E(t)$
Mass $m \Longleftrightarrow$ inductance L
Damping constant $c \Longleftrightarrow$ resistance R
Spring modulus $k \Longleftrightarrow$ reciprocal $1/C$ of the capacitance.

SECTION 2.4 PROBLEMS

1. This problem gauges the relative effects of initial position and velocity on the motion in the unforced, overdamped case. Solve the initial value problems
$$y'' + 4y' + 2y = 0;\ y(0) = 5,\ y'(0) = 0$$
and
$$y'' + 4y' + 2y = 0;\ y(0) = 0,\ y'(0) = 5.$$
Graph the solutions on the same set of axes.

2. Repeat the experiment of Problem 1, except now use the critically damped, unforced equation
$$y'' + 4y' + 4y = 0.$$

3. Repeat the experiment of Problem 1 for the underdamped, unforced equation
$$y'' + 2y' + 5y = 0.$$

Problems 4 through 9 explore the effects of changing the initial position or initial velocity on the motion of the object. In each, use the same set of axes to graph the solution of the initial value problem for the given values of A and observe the effect that these changes cause in the solution.

4. $y'' + 4y' + 4y = 0$; $y(0) = A$, $y'(0) = 0$; A has values $1, 3, 6, 10, -4$ and -7.

5. $y'' + 4y' + 4y = 0$; $y(0) = 0$, $y'(0) = A$; A has values $1, 3, 6, 10, -4$ and -7.

6. $y'' + 2y' + 5y = 0$; $y(0) = A$, $y'(0) = 0$; A has values $1, 3, 6, 10, -4$ and -7.

7. $y'' + 2y' + 5y = 0$; $y(0) = 0$, $y'(0) = A$; A has values $1, 3, 6, 10, -4$ and -7.

8. $y'' + 4y' + 2y = 0$; $y(0) = A$, $y'(0) = 0$; A has values $1, 3, 6, 10, -4$ and -7.

9. $y'' + 4y' + 2y = 0$; $y(0) = 0$, $y'(0) = A$; A has values $1, 3, 6, 10, -4$ and -7.

10. How many times can the mass pass through equilibrium in critical damping? What condition can be placed on $y(0)$ to ensure that the mass never passes through the equilibrium point? How does the initial velocity influence whether the mass passes through the equilibrium point?

11. Consider overdamped forced motion governed by
$$y'' + 6y' + 2y = 4\cos(3t).$$
(a) Find the solution satisfying $y(0) = 6$, $y' = 0$.
(b) Find the solution satisfying $y(0) = 0$, $y'(0) = 6$.

Graph these solutions on the same set of axes to compare the effects of initial displacement and velocity on the motion.

12. Suppose $y(0) = y'(0) \neq 0$. Determine the maximum displacement of the mass in critically damped, unforced motion. Show that the time at which this maximum occurs is independent of the initial displacement.

13. How many times can the mass pass through the equilibrium point in overdamped motion? What condition can be placed on the initial displacement to ensure that it never passes through equilibrium?

14. An object having a mass of 1 gram is attached to the lower end of a spring having a modulus of 29 dynes per centimeter. The mass in turn is adhered to a dashpot that imposes a damping force of $10v$ dynes, where $v(t)$ is the velocity of the mass at time t in centimeters per second. Determine the motion of the mass if it is pulled down 3 centimeters from equilibrium and then struck upward with a blow sufficient to impart a velocity of 1 centimeter per second. Graph the solution. Solve the problem when the initial velocity is (in turn) $2, 4, 7$, and 12 centimeters per second. Graph these solutions on the same axes to visualize the influence of the initial velocity on the motion.

15. In underdamped, unforced motion, what effect does the damping constant have on the frequency of the oscillations?

16. Carry out the program of Problem 11 for the critically damped, forced system governed by

$$y'' + 4y' + 4y = 4\cos(3t).$$

17. Carry out the program of Problem 11 for the underdamped, forced system governed by

$$y'' + y' + 3y = 4\cos(3t).$$

2.5 Euler's Differential Equation

If A and B are constants, the second-order differential equation

$$x^2 y'' + Axy' + By = 0 \tag{2.15}$$

is called *Euler's equation*. Euler's equation is defined on the half-lines $x > 0$ and $x < 0$. We will find solutions on $x > 0$, and a simple adjustment will yield solutions on $x < 0$.

A change of variables will convert Euler's equation to a constant coefficient linear second-order homogeneous equation, which we can always solve. Let

$$x = e^t$$

or, equivalently, $t = \ln(x)$. If we substitute $x = e^t$ into $y(x)$, we obtain a function of t as

$$Y(t) = y(e^t).$$

To convert Euler's equation to an equation in t, we need to convert derivatives of $y(x)$ to derivatives of $Y(t)$. First, by the chain rule, we have

$$\begin{aligned} y'(x) &= \frac{d}{dx}(y(x)) = \frac{d}{dx}(Y(t)) \\ &= \frac{dY}{dt}\frac{dt}{dx} = \frac{1}{x}Y'(t). \end{aligned}$$

Next,

$$\begin{aligned} y''(x) &= \frac{d}{dx}(y'(x)) \\ &= \frac{d}{dx}\left(\frac{1}{x}Y'(t)\right) \\ &= -\frac{1}{x^2}Y'(t) + \frac{1}{x}\frac{d}{dx}(Y'(t)) \\ &= -\frac{1}{x^2}Y'(t) + \frac{1}{x}\frac{dY'(t)}{dt}\frac{dt}{dx} \\ &= -\frac{1}{x^2}Y'(t) + \frac{1}{x}\frac{1}{x}Y''(t) \\ &= \frac{1}{x^2}(Y''(t) - Y'(t)). \end{aligned}$$

Therefore,

$$x^2 y''(x) = Y''(t) - Y'(t).$$

Substitute these into Euler's equation to obtain the transformed differential equation

$$Y''(t) - Y'(t) + AY'(t) + BY(t) = 0$$

or

$$Y''(t) + (A - 1)Y'(t) + BY(t) = 0. \tag{2.16}$$

This is a constant coefficient equation which we know how to solve.

We need not go through this derivation whenever we encounter an Euler equation. The coefficients of equation (2.16) can be read directly from those of the Euler equation. Solve this transformed equation for $Y(t)$, then replace $t = \ln(x)$ to obtain the solution $y(x)$ of the Euler equation. In doing this, it is useful to recall that, for any number r and for $x > 0$,

$$x^r = e^{r\ln(x)}.$$

Furthermore,

$$e^{\ln(k)} = k$$

for any positive quantity k. Thus, for example,

$$e^{3\ln(x)} = e^{\ln(x^3)} = x^3.$$

EXAMPLE 2.21

We will find the general solution of

$$x^2y'' + 2xy' - 6y = 0.$$

With $A = 2$ and $B = -6$, this Euler equation transforms to

$$Y''(t) + Y'(t) - 6Y(t) = 0.$$

This constant coefficient linear homogeneous equation has general solution

$$Y(t) = c_1e^{-3t} + c_2e^{2t}.$$

Replace $t = \ln(x)$ to obtain the general solution of the Euler equation:

$$y(x) = c_1e^{-3\ln(x)} + c_2e^{2\ln(x)} = c_1x^{-3} + c_2x^2$$

for $x > 0$. ♦

EXAMPLE 2.22

Consider the Euler equation $x^2y'' - 5xy' + 9y = 0$. The transformed equation is

$$y'' - 6y' + 9y = 0,$$

with the general solution $Y(t) = c_1e^{3t} + c_2te^{3t}$. The Euler equation has the general solution

$$y(x) = c_1e^{3\ln(x)} + c_2\ln(x)e^{3\ln(x)} = c_1x^3 + c_2x^3\ln(x)$$

for $x > 0$. ♦

EXAMPLE 2.23

Solve $x^2y'' + 3xy' + 10y = 0$.

The transformed equation is $Y'' + 2Y' + 10Y = 0$ with the general solution

$$Y(t) = c_1e^{-t}\cos(3t) + c_2e^{-t}\sin(3t).$$

Then

$$y(x) = c_1 e^{-\ln(x)} \cos(3\ln(x)) + c_2 e^{-\ln(x)} \sin(3\ln(x))$$
$$= \frac{1}{x}\left(c_1 \cos(3\ln(x)) + c_2 \sin(3\ln(x))\right). \quad \blacklozenge$$

As usual, we solve an initial value problem by finding the general solution of the differential equation and then using the initial conditions to determine the constants.

EXAMPLE 2.24

Solve

$$x^2 y'' - 5xy' + 10y = 0;\ y(1) = 4,\ y'(1) = -6.$$

The Euler equation transforms to $Y'' - 6y' + 10Y = 0$ with the general solution

$$Y(t) = c_1 e^{3t} \cos(t) + c_2 e^{3t} \sin(t)$$

for $x > 0$. Then

$$y(x) = x^3 \left(c_1 \cos(\ln(x)) + c_2 \sin(\ln(x))\right).$$

Then

$$y(1) = 4 = c_1.$$

Thus far,

$$y(x) = 4x^3 \cos(\ln(x)) + c_2 x^3 \sin(\ln(x)).$$

Compute

$$y'(x) = 12x^2 \cos(\ln(x)) - 4x^2 \sin(\ln(x))$$
$$+ 3c_2 x^2 \sin(\ln(x)) + c_2 x^2 \cos(\ln(x)).$$

Then

$$y'(1) = 12 + c_2 = -6,$$

so $c_2 = -18$. The solution of the initial value problem is

$$y(x) = 4x^3 \cos(\ln(x)) - 18x^3 \sin(\ln(x)). \quad \blacklozenge$$

SECTION 2.5 PROBLEMS

In each of Problems 1 through 10, find the general solution.

1. $x^2 y'' + 6xy' + 6y = 0$
2. $x^2 y'' - 11xy' + 35y = 0$
3. $x^2 y'' + 25xy' + 144y = 0$
4. $x^2 y'' + 3xy' + 10y = 0$
5. $x^2 y'' + xy' - 16y = 0$
6. $x^2 y'' - 5xy' + 58y = 0$
7. $x^2 y'' + xy' + 4y = 0$
8. $x^2 y'' + xy' - 4y = 0$
9. $x^2 y'' + 2xy' - 6y = 0$
10. $x^2 y'' + 3xy' + y = 0$

In each of Problems 11 through 16, solve the initial value problem.

11. $x^2 y'' - 3xy' + 4y = 0;\ y(1) = 4,\ y'(1) = 5$
12. $x^2 y'' + 25xy' + 144y = 0;\ y(1) = -4,\ y'(1) = 0$
13. $x^2 y'' - 9xy' + 24y = 0;\ y(1) = 1,\ y'(1) = 10$

14. $x^2y'' + xy' - 4y = 0$; $y(1) = 7$, $y'(1) = -3$

15. $x^2y'' + 5xy' - 21y = 0$; $y(2) = 1$, $y'(2) = 0$

16. $x^2y'' - xy' = 0$; $y(2) = 5$, $y'(2) = 8$

17. Here is another approach to solving an Euler equation. For $x > 0$, substitute $y = x^r$ into the differential equation to obtain a quadratic equation for r. Roots of this quadratic equation yield solutions $y = x^r$. Use this approach to solve the Euler equations of Examples 2.22, 2.22, and 2.23.

18. Outline a procedure for solving the Euler equation for $x < 0$. *Hint*: Let $t = \ln|x|$ in this case.

DEFINITION AND NOTATION SOLUTION OF INITIAL VALUE PROBLEMS SHIFTING AND THE HEAVISIDE FUNCTION CONVOLUTION IMPULSES AND THE DELTA FUNCTION

CHAPTER 3

The Laplace Transform

3.1 Definition and Notation

The Laplace transform is an important tool for solving certain kinds of initial value problems, particularly those involving discontinuous forcing functions, as occur frequently in areas such as electrical engineering. It is also used to solve boundary value problems involving partial differential equations to analyze wave and diffusion phenomena.

We will see that the Laplace transform converts some initial value problems to algebra problems, leading us to attempt the following procedure:

$$\begin{aligned}\text{Initial value problem} &\Longrightarrow \text{algebra problem}\\ &\Longrightarrow \text{solution of the algebra problem}\\ &\Longrightarrow \text{solution of the initial value problem.}\end{aligned}$$

This is often an effective strategy, because some algebra problems are easier to solve than initial value problems. We begin in this section with the definition and elementary properties of the transform.

The *Laplace transform* of a function f is a function $\mathcal{L}[f]$ defined by

$$\mathcal{L}[f](s) = \int_0^\infty e^{-st} f(t)dt.$$

The integration is with respect to t and defines a function of the new variable s for all s such that this integral converges.

Because the symbol $\mathcal{L}[f](s)$ may be awkward to write in computations, we will make the following convention. We will use lowercase letters for a function we put into the transform and the corresponding uppercase letters for the transformed function. In this way,

$$\mathcal{L}[f] = F, \ \mathcal{L}[g] = G, \ \text{and } \mathcal{L}[h] = H$$

and so on. If we include the variable, these would be written

$$\mathcal{L}[f](s) = F(s),\ \mathcal{L}[g](s) = G(s),\ \text{and } \mathcal{L}[h](s) = H(s)$$

It is also customary to use t (for time) as the variable of the input function and s for the variable of the transformed function.

EXAMPLE 3.1

Let a be any real number, and $f(t) = e^{at}$. The Laplace transform of f is the function defined by

$$\begin{aligned}\mathcal{L}[f](s) &= \int_0^\infty e^{-st} e^{at}\, dt \\ &= \int_0^\infty e^{(a-s)t}\, dt = \lim_{k\to\infty} \int_0^k e^{(a-s)t}\, dt \\ &= \lim_{k\to\infty} \left[\frac{1}{a-s} e^{(a-s)t} \right]_0^k \\ &= -\frac{1}{a-s} = \frac{1}{s-a}\end{aligned}$$

provided that $s > a$. The Laplace transform of $f(t) = e^{at}$ can be denoted $F(s) = 1/(s-a)$ for $s > a$. ♦

We rarely determine a Laplace transform by integration. Table 3.1 is a short table of Laplace transforms of familiar functions, and much longer tables are available. In this table, n denotes a nonnegative integer, and a and b are constants. Reading from the table (left to right), if $f(t) = \sin(3t)$ then by entry (6), we have

$$F(s) = \frac{3}{s^2+9},$$

and if $k(t) = e^{2t}\cos(5t)$ then by entry (11), we have

$$K(s) = \frac{s-2}{(s-2)^2+25}.$$

There are also software routines for transforming functions. In MAPLE, first enter

```
with(inttrans);
```

TABLE 3.1 ***Laplace Transforms of Selected Functions***

$f(t)$	$F(s)$	$f(t)$	$F(s)$
(1) 1	$\frac{1}{s}$	(8) $t\sin(at)$	$\frac{2as}{(s^2+a^2)^2}$
(2) t^n	$\frac{n!}{s^{n+1}}$	(9) $t\cos(at)$	$\frac{s^2-a^2}{(s^2+a^2)^2}$
(3) e^{at}	$\frac{1}{s-a}$	(10) $e^{at}\sin(bt)$	$\frac{b}{(s-a)^2+b^2}$
(4) $t^n e^{at}$	$\frac{n!}{(s-a)^{n+1}}$	(11) $e^{at}\cos(bt)$	$\frac{s-a}{(s-a)^2+b^2}$
(5) $e^{at}-e^{bt}$	$\frac{a-b}{(s-a)(s-b)}$	(12) $\sinh(at)$	$\frac{a}{s^2-a^2}$
(6) $\sin(at)$	$\frac{a}{s^2+a^2}$	(13) $\cosh(at)$	$\frac{s}{s^2-a^2}$
(7) $\cos(at)$	$\frac{s}{s^2+a^2}$	(14) $\delta(t-a)$	e^{-as}

to open the integral transforms package of subroutines. For the Laplace transform of $f(t)$, enter

```
laplace(f(t), t, s);
```

to obtain $F(s)$.

The Laplace transform is *linear*, which means that the transform of a sum is the sum of the transforms and that constants factor through the transform:

$$\mathcal{L}[f+g]=F+G \text{ and } \mathcal{L}[cf]=cF$$

for all s such that $F(s)$ and $G(s)$ are both defined and for any number c.

Given $F(s)$, we sometimes need to find $f(t)$ such that $\mathcal{L}[f]=F$. This is the reverse process of computing the transform of f, and we refer to it as taking an *inverse Laplace transform.* This is denoted $\mathcal{L}^{-1}$, and

$$\mathcal{L}^{-1}[F]=f \text{ exactly when } \mathcal{L}[f]=F.$$

For example, the inverse Laplace transform of $1/(s-a)$ is e^{at}.

If we use Table 3.1 to find an inverse transform, read from the right column to the left column. For example, using the table and the linearity of the Laplace transform, we can read that

$$\mathcal{L}^{-1}\left[\frac{3}{s^2+16}-7\frac{1}{(s-5)(s-12)}\right]=\frac{3}{4}\sin(4t)+e^{5t}-e^{12t}.$$

The inverse Laplace transform $\mathcal{L}^{-1}$ is linear because $\mathcal{L}$ is. This means that

$$\mathcal{L}^{-1}[F+G]=\mathcal{L}^{-1}[F]+\mathcal{L}^{-1}[G]=f+g,$$

and for any number c,

$$\mathcal{L}^{-1}[cF]=c\mathcal{L}^{-1}[F]=cf.$$

To use MAPLE to compute the inverse Laplace transform of $F(s)$, enter

```
invlaplace(F(s),s,t);
```

to obtain $f(t)$. This assumes that the integral transforms package has been opened.

SECTION 3.1 PROBLEMS

In each of Problems 1 through 5, use Table 3.1 to determine the Laplace transform of the function.

1. $k(t)=-5t^2e^{-4t}+\sin(3t)$
2. $w(t)=\cos(3t)-\cos(7t)$
3. $f(t)=3t\cos(2t)$
4. $g(t)=e^{-4t}\sin(8t)$
5. $h(t)=14t-\sin(7t)$

In each of Problems 6 through 10, use Table 3.1 to determine the inverse Laplace transform of the function.

6. $G(s)=\frac{5}{s^2+12}-\frac{4s}{s^2+8}$
7. $P(s)=\frac{1}{s+42}-\frac{1}{(s+3)^4}$
8. $F(s)=\frac{-5s}{(s^2+1)^2}$
9. $Q(s)=\frac{s}{s^2+64}$
10. $R(s)=\frac{7}{s^2-9}$

For Problems 11 through 14, suppose that $f(t)$ is defined for all $t\geq 0$ and has a period T. This means that $f(t+T)=f(t)$ for all $t\geq 0$.

11. Show that

$$\mathcal{L}[f](s)=\sum_{n=0}^{\infty}\int_{nT}^{(n+1)T}e^{-st}f(t)\,dt.$$

12. Show that

$$\int_{nT}^{(n+1)T} e^{-st} f(t)\,dt = e^{-nsT} \int_0^T e^{-st} f(t)\,dt.$$

13. From Problems 11 and 12, show that

$$\mathcal{L}[f](s) = \left[\sum_{n=0}^{\infty} e^{-nsT}\right] \int_0^T e^{-st} f(t)\,dt.$$

14. Recall the geometric series

$$\sum_{n=0}^{\infty} r^n = \frac{1}{1-r}$$

for $|r| < 1$. With this and the result of Problem 13, show that

$$\mathcal{L}[f](s) = \frac{1}{1-e^{-sT}} \int_0^T e^{-st} f(t)\,dt.$$

In each of Problems 15 through 22, a periodic function is given (sometimes by a graph). Use the result of Problem 14 to compute its Laplace transform.

15. f has period of 6, and

$$f(t) = \begin{cases} 5 & \text{for } 0 < t \le 3, \\ 0 & \text{for } 3 < t \le 6 \end{cases}$$

16. $f(t) = |E \sin(\omega t)|$ with E and ω positive numbers.

17. f has the graph of Figure 3.1.

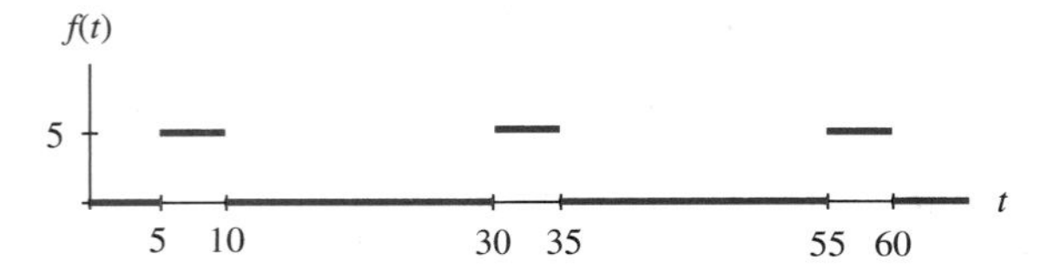

FIGURE 3.1 *Function for Problem 17, Section 3.1.*

18. f has the graph of Figure 3.2.

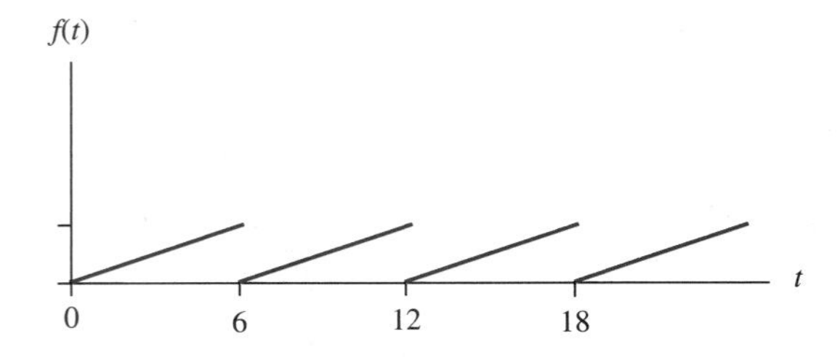

FIGURE 3.2 *Function for Problem 18, Section 3.1.*

19. f has the graph of Figure 3.3.

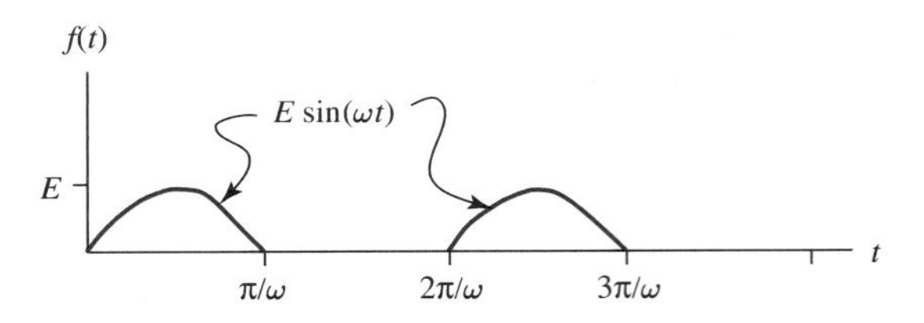

FIGURE 3.3 *Function for Problem 19, Section 3.1.*

20. f has the graph of Figure 3.4.

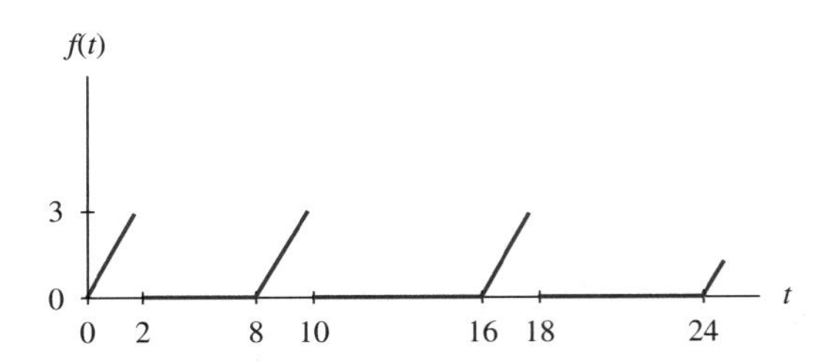

FIGURE 3.4 *Function for Problem 20, Section 3.1.*

21. f has the graph of Figure 3.5.

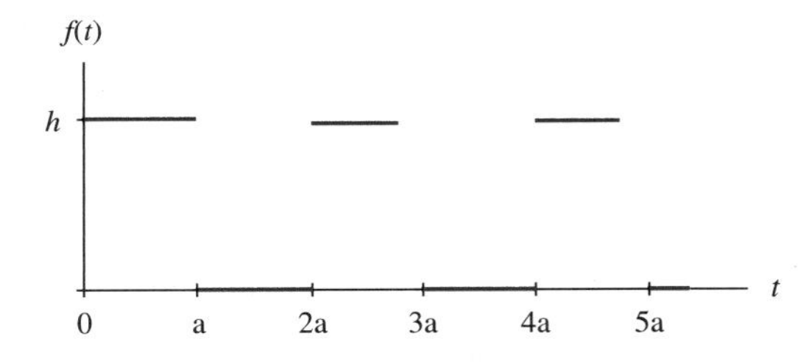

FIGURE 3.5 *Function for Problem 21, Section 3.1.*

22. f has the graph of Figure 3.6.

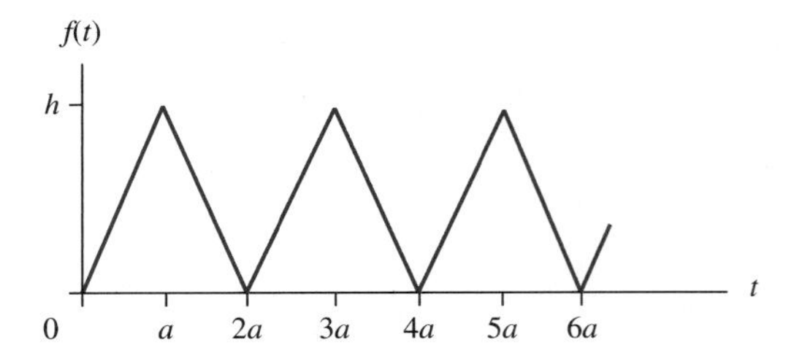

FIGURE 3.6 *Function for Problem 22, Section 3.1.*

3.2 Solution of Initial Value Problems

To apply the Laplace transform to the solution of an initial value problem, we must be able to transform a derivative. This involves the concept of a piecewise continuous function.

Suppose $f(t)$ is defined at least on $[a, b]$. Then f is *piecewise continuous* on $[a, b]$ if:

1. f is continuous at all but perhaps finitely many points of $[a, b]$.
2. If f is not continuous at t_0 in (a, b), then $f(t)$ has finite limits from both sides at t_0.
3. $f(t)$ has finite limits as t approaches a and as t approaches b from within the interval.

This means that f can have at most finitely many discontinuities on the interval, and these are all jump discontinuities. The function graphed in Figure 3.7 has jump discontinuities at t_0 and $t = t_1$. The magnitude of a jump discontinuity is the width of the gap in the graph there. In Figure 3.7, the magnitude of the jump at t_1 is

$$|\lim_{t \to t_1+} f(t) - \lim_{t \to t_1-} f(t)|.$$

By contrast, let

$$g(t) = \begin{cases} 1/t & \text{for } 0 < t \leq 1 \\ 0 & \text{for } t = 0. \end{cases}$$

Then g is continuous on $(0, 1]$, but is not piecewise continuous on $[0, 1]$, because $\lim_{t \to 0+} g(t) = \infty$.

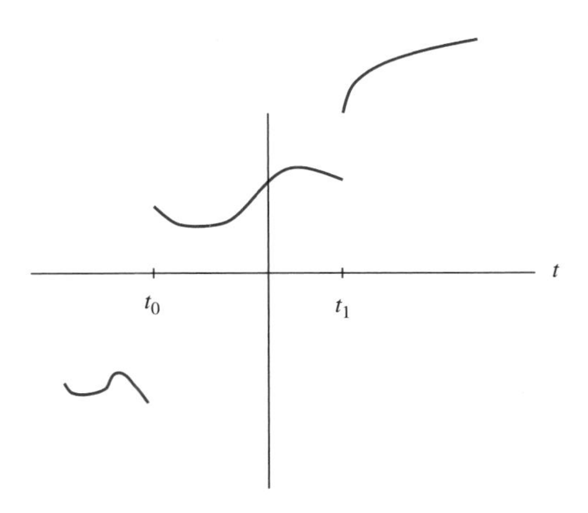

FIGURE 3.7 *Typical jump discontinuities.*

THEOREM 3.1 Transform of a Derivative

Let f be continuous for $t \geq 0$, and suppose f' is piecewise continuous on $[0, k]$ for every $k > 0$. Suppose also that $\lim_{k\to\infty} e^{-sk} f(k) = 0$ if $s > 0$. Then

$$\mathcal{L}[f'](s) = sF(s) - f(0). \quad \blacklozenge \tag{3.1}$$

This states that the transform of $f'(t)$ is s times the transform of $f(t)$, minus $f(0)$, which is the original function evaluated at $t = 0$. This can be proved by integration by parts (see Problem 11).

If f has a jump discontinuity at 0, as occurs if f is an electromotive force that is switched on at time zero, then the conclusion of the theorem must be amended to read

$$\mathcal{L}[f'](s) = sF(s) - f(0+)$$

where $f(0+) = \lim_{t\to 0+} f(t)$.

There is an extension of Theorem 3.1 to higher derivatives. If n is a positive integer, let $f^{(n)}$ denote the nth derivative of f.

THEOREM 3.2 Transform of a Higher Derivative

Let $f, f', f^{(n-1)}$ be continuous for $t > 0$, and suppose $f^{(n)}$ is piecewise continuous on $[0, k]$ for every $k > 0$. Suppose also that

$$\lim_{k\to\infty} e^{-sk} f^{(j)}(k) = 0$$

for $s > 0$ and $j = 1, 2, \cdots, n-1$. Then

$$\mathcal{L}[f^{(n)}](s) = s^n F(s) - s^{n-1} f(0) - s^{n-2} f'(0) - \cdots - s f^{(n-2)}(0) - f^{(n-1)}(0). \tag{3.2}$$

$\blacklozenge$

The second derivative case $n = 2$ occurs sufficiently often that we will record the formula separately for this case:

$$\mathcal{L}[f''](s) = s^2 F(s) - sf(0) - f'(0). \tag{3.3}$$

We are now prepared to use the Laplace transform to solve some initial value problems.

EXAMPLE 3.2

We will solve $y' - 4y = 1$; $y(0) = 1$.

We already know how to solve this problem, but we will apply the Laplace transform to illustrate the idea. Take the transform of the differential equation using the linearity of $\mathcal{L}$ and equation (3.1) to write

$$\begin{aligned}\mathcal{L}[y' - 4y](s) &= \mathcal{L}[y'](s) - 4\mathcal{L}[y](s)\\ &= (sY(s) - y(0)) - 4Y(s) = \mathcal{L}[1](s).\end{aligned}$$

Insert the initial data $y(0) = 1$, and use the table to find that $\mathcal{L}[1](s) = 1/s$. Then

$$(s-4)Y(s) - 1 = \frac{1}{s}.$$

There is no derivative in this equation! $\mathcal{L}$ has converted the differential equation into an algebraic equation for the transform $Y(s)$ of the unknown function $y(t)$. Solve for $Y(s)$ to obtain

$$Y(s) = \frac{1}{s-4} + \frac{1}{s(s-4)}.$$

This is the transform of the solution of the initial value problem. The solution is $y(t)$, which we obtain by applying the inverse transform:

$$\begin{aligned} y &= \mathcal{L}^{-1}[Y] \\ &= \mathcal{L}^{-1}\left[\frac{1}{s-4}\right] + \mathcal{L}^{-1}\left[\frac{1}{s(s-4)}\right]. \end{aligned}$$

From entry (3) of the table with $a = 4$,

$$\mathcal{L}^{-1}\left[\frac{1}{s-4}\right] = e^{4t}$$

and from entry (5) with $a = 0$ and $b = 4$,

$$\begin{aligned} \mathcal{L}^{-1}\left[\frac{1}{s(s-4)}\right] &= \frac{1}{-4}(e^{0t} - e^{4t}) \\ &= \frac{1}{4}(e^{4t} - 1). \end{aligned}$$

The solution is

$$y(t) = e^{4t} + \frac{1}{4}(e^{4t} - 1) = \frac{5}{4}e^{4t} - \frac{1}{4}. \quad \blacklozenge$$

EXAMPLE 3.3

Solve

$$y'' + 4y' + 3y = e^t;\ y(0) = 0,\ y'(0) = 2.$$

Using the linearity of $\mathcal{L}$ and equations (3.1) and (3.3), we obtain

$$\begin{aligned} &\mathcal{L}[y''] + 4\mathcal{L}[y'] + 3\mathcal{L}[y] \\ &= [s^2Y - sy(0) - y'(0)] + 4[sY - y(0)] + 3Y \\ &= [s^2Y - 2] - 4sY + 3Y \\ &= \mathcal{L}[e^t] = \frac{1}{s-1}. \end{aligned}$$

Solve for Y to get

$$\begin{aligned} Y(s) &= \frac{2s-1}{(s-1)(s^2+4s+3)} \\ &= \frac{2s-1}{(s-1)(s+1)(s+3)}. \end{aligned}$$

To read the inverse transform from the table, use a partial fractions decomposition to write the quotient on the right as a sum of simpler quotients. We will carry out the algebra of this decomposition. First write

$$\frac{2s-1}{(s-1)(s+1)(s+3)} = \frac{A}{s-1} + \frac{B}{s+1} + \frac{C}{s+3}.$$

To solve for the constants, observe that if we added the fractions on the right the numerator would have to equal the numerator $2s-1$ of the fraction on the left. Therefore,

$$A(s+1)(s+3)+B(s-1)(s+3)+C(s-1)(s+1)=2s-1.$$

We can solve for A, B, and C by inserting values of s into this equation. Put $s=1$ to get $8A=1$, so $A=1/8$. Put $s=-1$ to get $-4B=-3$, so $B=3/4$. Put $s=-3$ to get $8C=-7$, so $C=-7/8$. Then

$$Y(s)=\frac{1}{8}\frac{1}{s-1}+\frac{3}{4}\frac{1}{s+1}-\frac{7}{8}\frac{1}{s+3}.$$

Invert this to obtain the solution

$$y(t)=\frac{1}{8}e^{t}+\frac{3}{4}e^{-t}-\frac{7}{8}e^{-3t}. \quad \blacklozenge$$

Partial fractions decompositions are frequently used with the Laplace transform. The appendix at the end of this chapter reviews the algebra of this technique.

Notice that the transform method does not first produce the general solution and then solve for the constants to satisfy the initial conditions. Equations (3.1), (3.2), and (3.3) insert the initial conditions directly into an algebraic equation for the transform of the unknown function. Still, we could have solved the problem of Example 3.3 by methods from Chapter 2. The object here was to illustrate a technique. This technique extends to problems beyond the reach of methods from Chapter 2, and this is the subject of the next section.

SECTION 3.2 PROBLEMS

In each of Problems 1 through 10, use the Laplace transform to solve the initial value problem.

1. $y'+4y=\cos(t);\ y(0)=0$
2. $y''+y=1;\ y(0)=6,\ y'(0)=0$
3. $y''-4y'+4y=\cos(t);\ y(0)=1,\ y'(0)=-1$
4. $y''+9y=t^2;\ y(0)=y'(0)=0$
5. $y''+16y=1+t;\ y(0)=-2,\ y'(0)=1$
6. $y''-5y'+6y=e^{-t};\ y(0)=0,\ y'(0)=2$
7. $y'-2y=1-t;\ y(0)=4$
8. $y'-9y=t;\ y(0)=5$
9. $y'+4y=1;\ y(0)=-3$
10. $y'+2y=e^{-t};\ y(0)=1$
11. Prove Theorem 3.1. *Hint*: Write

$$\mathcal{L}[f'](s)=\int_0^\infty e^{-st}f'(t)\,dt$$

and integrate by parts.

12. Derive equation (3.3). *Hint*: Integrate by parts twice.

3.3 Shifting and the Heaviside Function

The shifting theorems of this section will enable us to solve problems involving pulses and other discontinuous forcing functions.

3.3.1 The First Shifting Theorem

We will show that the Laplace transform of $e^{at}f(t)$ is the transform of $f(t)$, shifted a units to the right. This shift is achieved by replacing s by $s-a$ in $F(s)$ to obtain $F(s-a)$.

THEOREM 3.3 First Shifting Theorem

For any number a,

$$\mathcal{L}[e^{at} f(t)](s) = F(s-a). \tag{3.4}$$

♦

This conclusion is also called *shifting in the s variable*. The proof is a straightforward appeal to the definition:

$$\begin{aligned}\mathcal{L}[e^{at} f(t)](s) &= \int_0^\infty e^{-st} e^{at} f(t)dt \\ &= \int_0^\infty e^{-(s-a)} f(t)dt = F(s-a).\end{aligned}$$

EXAMPLE 3.4

We know from the table that $\mathcal{L}[\cos(bt)] = s/(s^2+b^2) = F(s)$. For the transform of $e^{at}\cos(bt)$, replace s with $s-a$ to get

$$\mathcal{L}[e^{at}\cos(bt)](s) = \frac{s-a}{(s-a)^2+b^2}.$$ ♦

EXAMPLE 3.5

Since $\mathcal{L}[t^3] = 6/s^4$, then

$$\mathcal{L}[t^3 e^{7t}](s) = \frac{6}{(s-7)^4}.$$ ♦

Every formula for the Laplace transform of a function is also a formula for the inverse Laplace transform of a function. The inverse version of the first shifting theorem is

$$\mathcal{L}^{-1}[F(s-a)] = e^{at} f(t). \tag{3.5}$$

EXAMPLE 3.6

Compute

$$\mathcal{L}^{-1}\left[\frac{4}{s^2+4s+20}\right].$$

The idea is to manipulate the given function of s to the form $F(s-a)$ for some F and a. Then we can apply the inverse form of the shifting theorem, which is equation (3.5). Complete the square in the denominator to write

$$\frac{4}{s^2+4s+20} = \frac{4}{(s+2)^2+16} = F(s+2)$$

if

$$F(s) = \frac{4}{s^2+16}.$$

From the table, $F(s)$ has inverse $f(t) = \sin(4t)$. By equation (3.5),

$$\begin{aligned}&\mathcal{L}^{-1}\left[\frac{4}{s^2+4s+20}\right] \\ &= \mathcal{L}^{-1}[F(s+2)] \\ &= e^{-2t} f(t) = e^{-2t}\sin(4t).\end{aligned}$$ ♦

EXAMPLE 3.7

Compute

$$\mathcal{L}^{-1}\left[\frac{3s-1}{s^2-6s+2}\right].$$

Follow the strategy of Example 3.6. Manipulate $F(s)$ to a function of $s-a$ for some a:

$$\begin{aligned}\frac{3s-1}{s^2-6s+2} &= \frac{3s-1}{(s-3)^2-7}\\ &= \frac{3(s-3)+8}{(s-3)^2-7}\\ &= \frac{3(s-3)}{(s-3)^2-7}+\frac{8}{(s-3)^2-7}\\ &= G(s-3)+K(s-3)\end{aligned}$$

where

$$G(s)=\frac{3s}{s^2-7} \text{ and } K(s)=\frac{8}{s^2-7}.$$

By equation (3.5),

$$\begin{aligned}\mathcal{L}^{-1}\left[\frac{3s-1}{s^2-6s+2}\right] &= \mathcal{L}^{-1}[G(s-3)]+\mathcal{L}^{-1}[K(s-3)]\\ &= e^{3t}\mathcal{L}^{-1}[G(s)]+e^{3t}\mathcal{L}^{-1}[K(s)]\\ &= e^{3t}\mathcal{L}^{-1}\left[\frac{3s}{s^2-7}\right]+e^{3t}\mathcal{L}^{-1}\left[\frac{8}{s^2-7}\right]\\ &= 3e^{3t}\cosh(\sqrt{7}t)+\frac{8}{\sqrt{7}}e^{3t}\sinh(\sqrt{7}t).\end{aligned}$$ ♦

3.3.2 The Heaviside Function and Pulses

Functions having jump discontinuities are efficiently treated by using the *unit step function*, or *Heaviside function* H, defined by

$$H(t)=\begin{cases}0 & \text{for } t<0\\ 1 & \text{for } t\geq 0.\end{cases}$$

H is graphed in Figure 3.8. We will also use the *shifted Heaviside function* $H(t-a)$ of Figure 3.9. This is the Heaviside function shifted a units to the right:

$$H(t-a)=\begin{cases}0 & \text{for } t<a\\ 1 & \text{for } t\geq a.\end{cases}$$

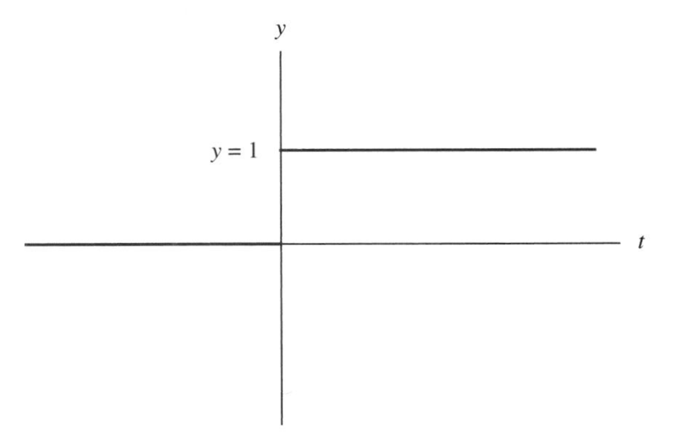

FIGURE 3.8 *The Heaviside function.*

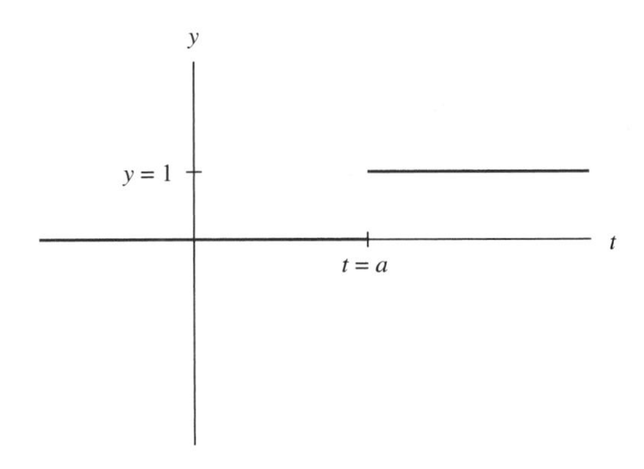

FIGURE 3.9 *Shifted Heaviside function.*

$H(t-a)$ can be used to turn a signal (function) off until time $t=a$ and then to turn it on. In particular,

$$H(t-a)g(t) = \begin{cases} 0 & \text{for } t < a \\ g(t) & \text{for } t \geq a. \end{cases}$$

To illustrate, Figure 3.10 shows $H(t-\pi)\cos(t)$. This is the familiar cosine function for $t \geq \pi$, but is turned off (equals 0) for $t < \pi$. Multiplying a function $f(t)$ by $H(t-a)$ leaves the graph of $f(t)$ unchanged for $t \geq a$, but replaces it by 0 for $t < a$.

We can also use the Heaviside function to define a *pulse*. If $a < b$, then

$$H(t-a) - H(t-b) = \begin{cases} 0 & \text{for } t < a \\ 1 & \text{for } a \leq t < b \\ 0 & \text{for } t \geq b. \end{cases}$$

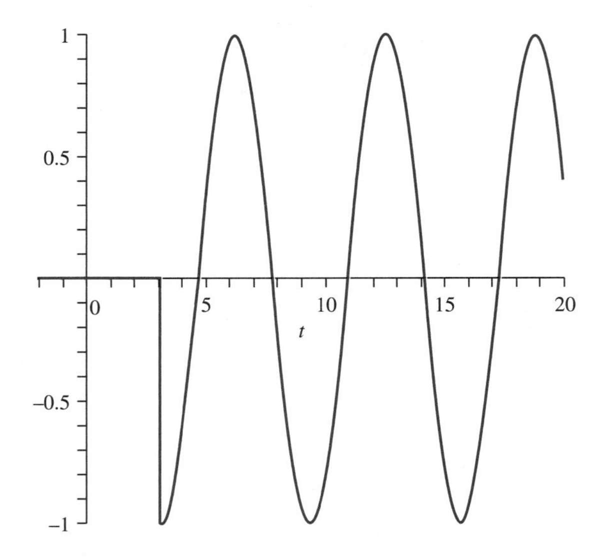

FIGURE 3.10 $H(t-\pi)\cos(t)$.

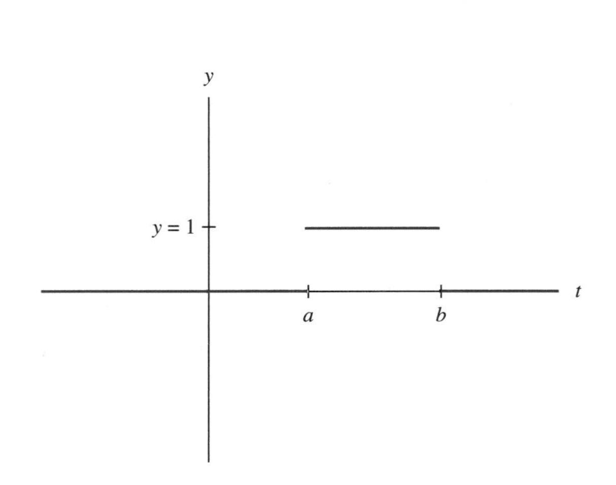

FIGURE 3.11 *A pulse* $H(t-a)-H(t-b)$.

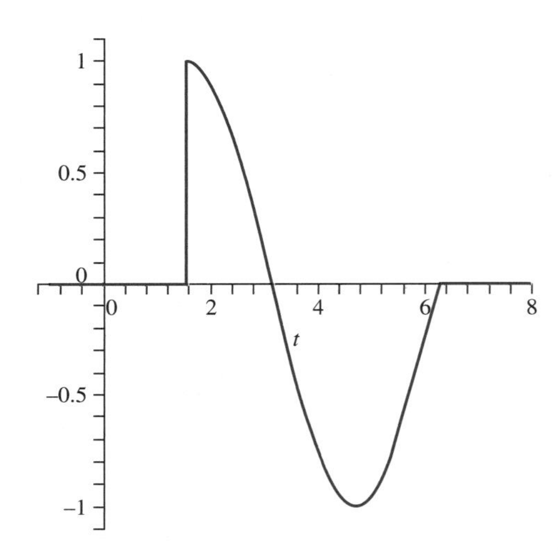

FIGURE 3.12 $(H(t-\pi/2) - H(t - 2\pi))\sin(t)$.

Figure 3.11 shows the pulse $H(t-a)-H(t-b)$ with $a < b$. Pulses are used to turn a signal off until time $t=a$ and then to turn it on until time $t=b$, after which it is switched off again. Figure 3.12 shows this effect for $[H(t-\pi/2)-H(t-2\pi)]\sin(t)$, which is zero before time $\pi/2$ and after time 2π and equals $\sin(t)$ between these times.

It is important to understand the difference between $g(t)$, $H(t - a)g(t)$ and $H(t-a)g(t-a)$. Figures 3.13, 3.14 and 3.15, show, respectively, graphs of $t\sin(t)$, $H(t-3/2)t\sin(t)$, and $H(t-4)(t-4)\sin(t-4)$. $H(t-3/2)t\sin(t)$ is zero until time $3/2$ and then equals $t\sin(t)$, while $H(t-4)(t-4)\sin(t-4)$ is zero until time 4, then is the graph of $t\sin(t)$ shifted 4 units to the right.

Using the Heaviside function, we can state the second shifting theorem, which is also called shifting in the t variable.

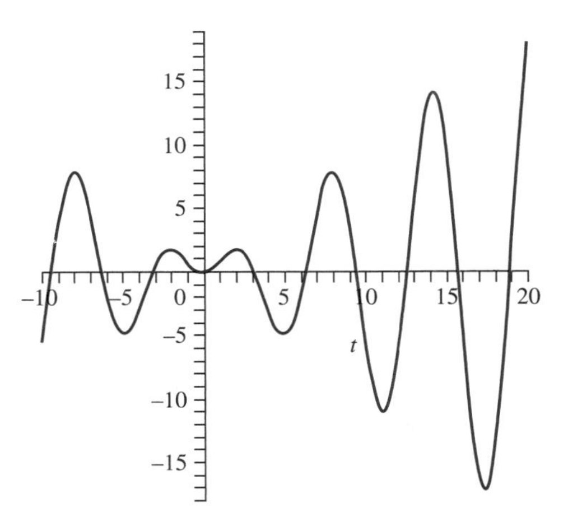

FIGURE 3.13 $t\sin(t)$.

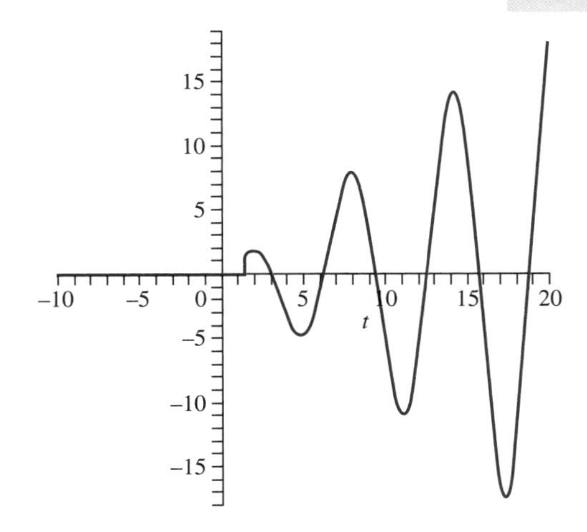

FIGURE 3.14 $H(t-3/2)t\sin(t)$.

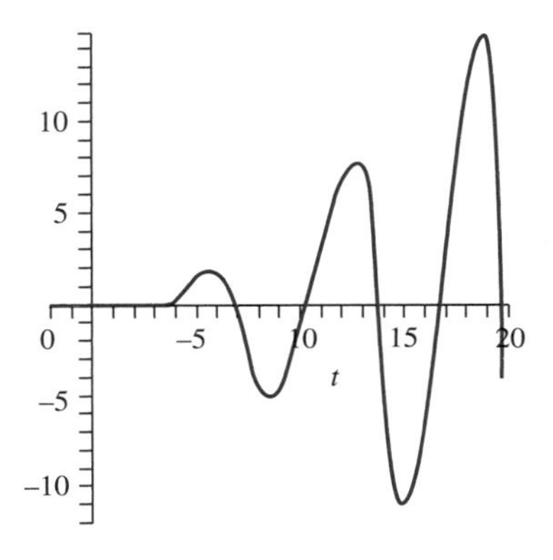

FIGURE 3.15 $H(t-4)(t-4)\sin(t-4)$.

THEOREM 3.4 Second Shifting Theorem

$$\mathcal{L}[H(t-a)f(t-a)](s)=e^{-as}F(s). \blacklozenge \tag{3.6}$$

This result follows directly from the definition of the transform and of the Heaviside function.

EXAMPLE 3.8

Suppose we want $\mathcal{L}[H(t-a)]$. Write

$$H(t-a)=H(t-a)f(t-a)$$

with $f(t)=1$ for all t. Since $F(s)=\mathcal{L}[1](s)=1/s$, then by the second shifting theorem,

$$\mathcal{L}[H(t-a)](s)=e^{-as}F(s)=\frac{1}{s}e^{-as}. \blacklozenge$$

EXAMPLE 3.9

Compute $\mathcal{L}[g]$ where

$$g(t) = \begin{cases} 0 & \text{for } t < 2 \\ t^2 + 1 & \text{for } t \geq 2. \end{cases}$$

To apply the second shifting theorem, we must write $g(t)$ as a function, or perhaps sum of functions, of the form $f(t-2)H(t-2)$. To do this, first write t^2+1 as a function of $t-2$:

$$t^2 + 1 = (t - 2 + 2)^2 + 1 = (t-2)^2 + 4(t-2) + 5.$$

Then

$$\begin{aligned} g(t) &= H(t-2)(t^2+1) \\ &= (t-2)^2 H(t-2) + 4(t-2)H(t-2) + 5H(t-2). \end{aligned}$$

Now apply the second shifting theorem to each term on the right:

$$\begin{aligned} \mathcal{L}[g] &= \mathcal{L}[(t-2)^2 H(t-2)] + 4\mathcal{L}[(t-2)H(t-2)] + 5\mathcal{L}[H(t-2)] \\ &= e^{-2s}\mathcal{L}[t^2] + 4e^{-2s}\mathcal{L}[t] + 5e^{-2s}\mathcal{L}[1] \\ &= e^{-2s}\left[\frac{2}{s^3} + \frac{4}{s^2} + \frac{5}{s}\right]. \ ♦ \end{aligned}$$

As usual, any formula for $\mathcal{L}$ can be read as a formula for $\mathcal{L}^{-1}$. The inverse version of the second shifting theorem is

$$\mathcal{L}^{-1}[e^{-as}F(s)](t) = H(t-a)f(t-a). \tag{3.7}$$

This enables us to compute the inverse transform of a known transformed function that is multiplied by an exponential e^{-as}.

EXAMPLE 3.10

Compute

$$\mathcal{L}^{-1}\left[\frac{se^{-3s}}{s^2+4}\right].$$

The presence of e^{-3s} suggests the use of equation (3.7). From the table, we read that

$$\mathcal{L}^{-1}\left[\frac{s}{s^2+4}\right] = \cos(2t).$$

Then

$$\mathcal{L}^{-1}\left[\frac{se^{-3s}}{s^2+4}\right](t) = H(t-3)\cos(2(t-3)). \ ♦$$

EXAMPLE 3.11

Solve the initial value problem

$$y'' + 4y = f(t);\ y(0) = y'(0) = 0$$

where

$$f(t) = \begin{cases} 0 & \text{for } t < 3 \\ t & \text{for } t \geq 3. \end{cases}$$

First apply $\mathcal{L}$ to the differential equation, using equations (3.1) and (3.3):

$$\mathcal{L}[y''] + 4\mathcal{L}[y] = [s^2 - sy(0) - y'(0)]Y(s) + 4Y(s)$$
$$= s^2Y(s) + 4Y(s) = (s^2+4)Y(s) = \mathcal{L}[f].$$

To compute $\mathcal{L}[f]$, use the second shifting theorem. Since $f(t) = H(t-3)t$, we can write

$$\begin{aligned}\mathcal{L}[f] &= \mathcal{L}[H(t-3)t] \\ &= \mathcal{L}[H(t-3)(t-3+3)] \\ &= \mathcal{L}[H(t-3)(t-3)] + 3\mathcal{L}[H(t-3)] \\ &= \frac{e^{-3s}}{s^2} + \frac{3e^{-3s}}{s}.\end{aligned}$$

In summary, we have

$$(s^2+4)Y(s) = \frac{1}{s^2}e^{-3s} + \frac{3}{s}e^{-3s} = \frac{3s+1}{s^2}e^{-3s}.$$

The transform of the solution is therefore

$$Y(s) = \frac{3s+1}{s^2(s^2+4)}e^{-3s}.$$

The solution is the inverse transform of $Y(s)$. To take this inverse, use a partial fractions decomposition, writing

$$\frac{3s+1}{s^2(s^2+4)} = \frac{A}{s} + \frac{B}{s^2} + \frac{Cs+D}{s^2+4}.$$

After solving for A, B, C, and D, we obtain

$$\begin{aligned}Y(s) &= \frac{3s+1}{s^2(s^2+4)}e^{-3s} \\ &= \frac{3}{4}\frac{1}{s}e^{-3s} - \frac{3}{4}\frac{s}{s^2+4}e^{-3s} + \frac{1}{4}\frac{1}{s^2}e^{-3s} - \frac{1}{4}\frac{1}{s^2+4}e^{-3s}.\end{aligned}$$

Now apply the second shifting theorem to write the solution

$$\begin{aligned}y(t) &= \frac{3}{4}H(t-3) - \frac{3}{4}H(t-3)\cos(2(t-3)) \\ &\quad + \frac{1}{4}H(t-3)(t-3) - \frac{1}{8}H(t-3)\sin(2(t-3)).\end{aligned}$$

This solution is 0 until time $t = 3$. Since $H(t-3) = 1$ for $t \geq 3$, then for these times,

$$\begin{aligned}y(t) &= \frac{3}{4} - \frac{3}{4}\cos(2(t-3)) + \frac{1}{4}(t-3) \\ &\quad - \frac{1}{8}\sin(2(t-3)).\end{aligned}$$

Upon combining terms, the solution is

$$y(t) = \begin{cases} 0 & \text{for } t < 3 \\ \frac{1}{8}[2t - 6\cos(2(t-3)) - \sin(2(t-3))] & \text{for } t \geq 3. \end{cases}$$

Figure 3.16 shows part of the graph of this solution. ♦

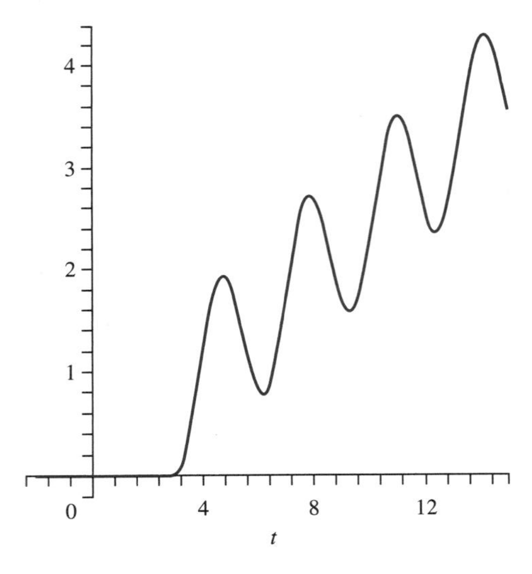

FIGURE 3.16 *Graph of the solution in Example 3.11.*

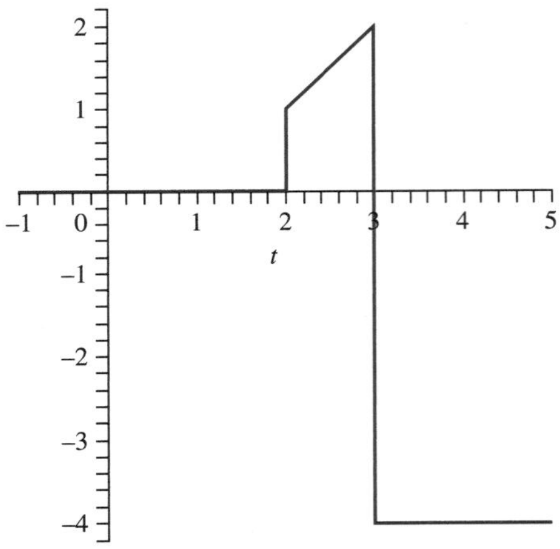

FIGURE 3.17 *$f(t)$ in Example 3.12.*

EXAMPLE 3.12

Sometimes we need to deal with a function having several jump discontinuities. Here is an example of writing such a function in terms of step functions. Let

$$f(t) = \begin{cases} 0 & \text{for } t < 2 \\ t-1 & \text{for } 2 \le t < 3 \\ -4 & \text{for } t \ge 3. \end{cases}$$

Figure 3.17 shows a graph of f. There are jump discontinuities of magnitude 1 at $t = 2$ and magnitude 6 at $t = 3$.

Think of $f(t)$ as consisting of two nonzero parts: the part that is $t - 1$ for $2 \le t < 3$ and the part that is -4 for $t \ge 3$. We want to turn on $t - 1$ at time 2 and turn it off at time 3, then turn -4 on at time 3 and leave it on.

The first effect is achieved by multiplying $t - 1$ by the pulse $H(t-2) - H(t-3)$. The second is achieved by multiplying -4 by $H(t-3)$. Thus, write

$$f(t) = [H(t-2) - H(t-3)](t-1) - 4H(t-3). \quad \blacklozenge$$

EXAMPLE 3.13

Suppose the capacitor in the circuit of Figure 3.18 initially has a charge of zero and there is no initial current. At time $t = 2$ seconds, the switch is thrown from position B to A, held there for 1 second, and then switched back to B. We want the output voltage E_{out} on the capacitor.

From the circuit, write

$$E(t) = 10[H(t-2) - H(t-3)].$$

By Kirchhoff's voltage law,

$$Ri(t) + \frac{1}{C}q(t) = E(t)$$

or

$$250,000q'(t) + 10^6 q(t) = E(t).$$

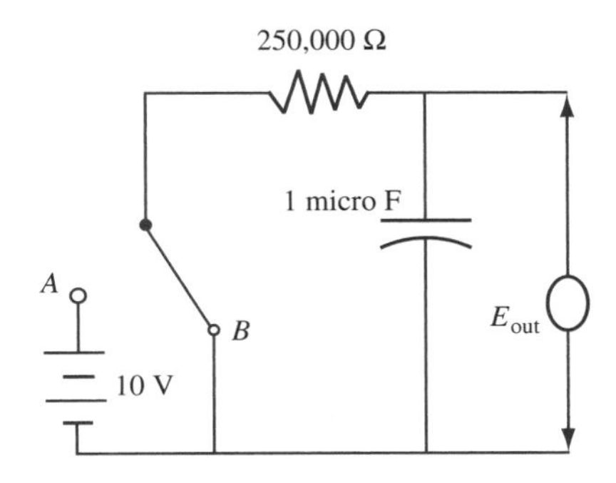

FIGURE 3.18 *The circuit of Example 3.13.*

We want to solve for $q(t)$ subject to the condition $q(0) = 0$. Take the Laplace transform of the differential equation to get

$$250,000[sQ(s) - q(0)] + 10^6 Q(s) = \mathcal{L}[E(t)].$$

Now

$$\begin{aligned}\mathcal{L}[E(t)](s) &= 10\mathcal{L}[H(t-2)](s) - 10\mathcal{L}[(t-3)](s) \\ &= \frac{10}{s}e^{-2s} - \frac{10}{s}e^{-3s}.\end{aligned}$$

Now we have an equation for Q:

$$2.5(10^5)sQ(s) + 10^6 Q(s) = \frac{10}{s}e^{-2s} - \frac{10}{s}e^{-3s}.$$

Then

$$Q(s) = 4(10^{-5})\frac{1}{s(s+4)}e^{-2s} - 4(10^{-5})\frac{1}{s(s+4)}e^{-3s}.$$

Use a partial fractions decomposition to write

$$Q(s) = 10^{-5}\left[\frac{1}{s}e^{-2s} - \frac{1}{s+4}e^{-2s}\right] - 10^{-5}\left[\frac{1}{s}e^{-3s} - \frac{1}{s+4}e^{-3s}\right].$$

Applying the second shifting theorem, we get

$$q(t) = 10^{-5}H(t-2)[1 - e^{-4(t-2)}] - 10^{-5}H(t-3)[1 - e^{-4(t-3)}].$$

Finally, the output voltage is $E_{out}(t) = 10^6 q(t)$. Figure 3.19 shows a graph of $E_{out}(t)$. ♦

3.3.3 Heaviside's Formula

There is a formula due to Heaviside that can be used to take the inverse transform of a quotient of polynomials.

Suppose $F(s) = p(s)/q(s)$ with p and q polynomials and q of higher degree than p. We assume that q can be factored into linear factors and has the form

$$q(s) = c(s - a_1)(s - a_2)\cdots(s - a_n),$$

with c a nonzero constant and the a_j's n distinct numbers (which may be real or complex). None of the a_j's are roots of $p(s)$.

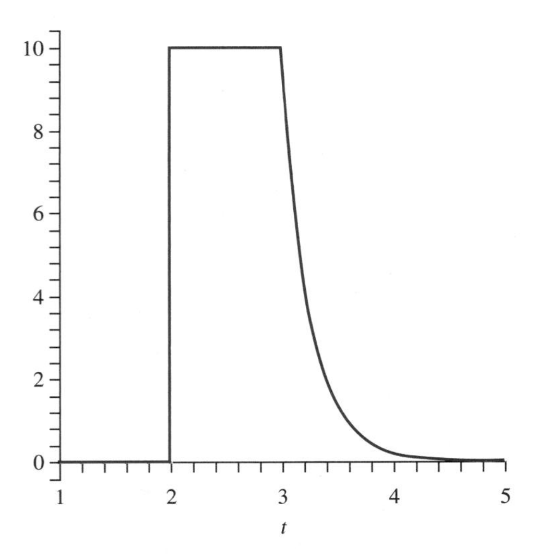

FIGURE 3.19 $E_{\text{out}}(t)$ *in Example 3.13.*

Let $q_j(s)$ be the polynomial of degree $n-1$ formed by omitting the factor $s-a_j$ from $q(s)$, for $j=1,2,\cdots,n$. For example,

$$q_1(s)=c(s-a_2)\cdots(s-a_n).$$

Then

$$\mathcal{L}^{-1}[F(s)](t)=\sum_{j=1}^{n}\frac{p(a_j)}{q_j(a_j)}e^{a_j t}.$$

This is called *Heaviside's formula*. In applying the formula, start with a_1, evaluate $p(a_1)$, then substitute a_1 into the denominator with the term $(s-a_1)$ removed. This gives the coefficient of $e^{a_1 t}$. Continue this with the other a_j's and sum to obtain $\mathcal{L}^{-1}[F]$.

Before showing why Heaviside's formula is true, here is a simple example with

$$F(s)=\frac{s}{(s^2+4)(s-1)}=\frac{s}{(s-2i)(s+2i)(s-1)}.$$

Here $p(s)=s$, and $q(s)=(s-2i)(s+2i)(s-1)$. Write $a_1=2i$, $a_2=-2i$, and $a_3=1$. Then

$$\begin{aligned}\mathcal{L}[F(s)](t)&=\frac{2i}{4i(2i-1)}e^{2it}+\frac{-2i}{-4i(-2i-1)}e^{-2it}+\frac{1}{(1-2i)(1+2i)}e^{t}\\&=\frac{-1-2i}{10}e^{2it}+\frac{-1+2i}{10}e^{-2it}+\frac{1}{5}e^{t}\\&=-\frac{1}{10}(e^{2it}+e^{-2it})-\frac{2i}{10}(e^{2it}-e^{-2it})+\frac{1}{5}e^{t}\\&=-\frac{1}{5}\cos(2t)+\frac{2}{5}\sin(2t)+\frac{1}{5}e^{t}.\end{aligned}$$

We have used the fact that

$$\cos(\theta)=\frac{1}{2}(e^{i\theta}+e^{-i\theta})\quad\text{and}\quad\sin(\theta)=\frac{1}{2i}(e^{i\theta}-e^{-i\theta}).$$

These can be obtained by solving for $\cos(\theta)$ and $\sin(\theta)$ in Euler's formulas

$$e^{i\theta}=\cos(\theta)+i\sin(\theta)\quad\text{and}\quad e^{-i\theta}=\cos(\theta)-i\sin(\theta).$$

Here is a rationale for Heaviside's formula. The partial fractions expansion of $p(s)/q(s)$ has the form

$$\frac{p(s)}{q(s)} = \frac{A_1}{s-a_1} + \frac{A_2}{s-a_2} + \cdots + \frac{A_n}{s-a_n}.$$

All we need are the numbers $A_1, \cdots, A_n$ to write

$$\mathcal{L}^{-1}[F](t) = A_1 e^{a_1 t} + \cdots + A_n e^{a_n t}.$$

We will find A_1. The other A_j's are found similarly. Notice that

$$(s-a_1)\frac{p(s)}{q(s)} = A_1 + A_2\frac{s-a_1}{s-a_1} + \cdots + A_n\frac{s-a_1}{s-a_n}.$$

Because the a_j's are assumed to be distinct, then

$$\lim_{s\to a_1}(s-a_1)\frac{p(s)}{q(s)} = A_1$$

with all the other terms on the right having zero limit as $s \to a_1$. But in this limit, $(s-a_1)p(s)/q(s)$ is exactly the quotient of $p(s)$ with the polynomial obtained by deleting $s-a_1$ from $q(s)$. This yields Heaviside's formula.

For those familiar with complex analysis, in Section 22.4, we will present a general formula for $\mathcal{L}^{-1}[F]$ as a sum of residues of $e^{tz}F(z)$ at singularities of $F(z)$. In that context, Heaviside's formula is the special case that $F(z)$ is a quotient of polynomials with simple poles at $a_1, \cdots, a_n$.

SECTION 3.3 PROBLEMS

In each of Problems 1 through 15, find the Laplace transform of the function.

1. $e^{-5t}(t^4 + 2t^2 + t)$
2. $e^{-4t}(t - \cos(t))$
3. $te^{-t}\cos(3t)$
4. $f(t) = \begin{cases} 2t - \sin(t) & \text{for } 0 \le t < \pi \\ 0 & \text{for } t \ge \pi \end{cases}$
5. $e^{-t}(1 - t^2 + \sin(t))$
6. $e^{-3t}(t - 2)$
7. $f(t) = \begin{cases} t & \text{for } 0 \le t < 3 \\ 1 - 3t & \text{for } t \ge 3 \end{cases}$
8. $f(t) = \begin{cases} -4 & \text{for } 0 \le t < 1 \\ 0 & \text{for } 1 \le t < 3 \\ e^{-t} & \text{for } t \ge 3 \end{cases}$
9. $f(t) = \begin{cases} t - 2 & \text{for } 0 \le t < 16 \\ -1 & \text{for } t \ge 16 \end{cases}$
10. $f(t) = \begin{cases} t^2 & \text{for } 0 \le t < 2 \\ 1 - t - 3t^2 & \text{for } t \ge 2 \end{cases}$
11. $f(t) = \begin{cases} 1 & \text{for } 0 \le t < 7 \\ \cos(t) & \text{for } t \ge 7 \end{cases}$
12. $f(t) = \begin{cases} 1 - \cos(2t) & \text{for } 0 \le t < 3\pi \\ 0 & \text{for } t \ge 3\pi \end{cases}$
13. $f(t) = \begin{cases} \cos(t) & \text{for } 0 \le t < 2\pi \\ 2 - \sin(t) & \text{for } t \ge 2\pi \end{cases}$
14. $e^t(1 - \cosh(t))$
15. $(t^3 - 3t + 2)e^{-2t}$

In each of Problems 16 through 25, find the inverse Laplace transform.

16. $\dfrac{1}{s-5}e^{-s}$
17. $\dfrac{1}{s(s^2+16)}e^{-21s}$
18. $\dfrac{s-4}{s^2-8s+10}$
19. $\dfrac{1}{s^2+6s+7}$
20. $\dfrac{3}{s+2}e^{-4s}$
21. $\dfrac{e^{-2s}}{s^2+9}$
22. e^{-5s}/s^3
23. $\dfrac{s+2}{s^2+6s+1}$

24. $\dfrac{1}{s^2+4s+12}$

25. $\dfrac{1}{s^2-4s+5}$

26. Determine $\mathcal{L}[e^{-2t}\int_0^t e^{2w}\cos(3w)dw]$. *Hint*: Use the first shifting theorem.

In each of Problems 27 through 32, solve the initial value problem.

27. $y''' - 8y = g(t)$; $y(0) = y'(0) = y''(0) = 0$, with

$$g(t) = \begin{cases} 0 & \text{for } 0 \le t < 6 \\ 2 & \text{for } t \ge 6 \end{cases}$$

28. $y'' - 4y' + 4y = f(t)$; $y(0) = -2$, $y'(0) = 1$, with

$$f(t) = \begin{cases} t & \text{for } 0 \le t < 3 \\ t+2 & \text{for } t \ge 3 \end{cases}$$

29. $y''' - y'' + 4y' - 4y = 0$; $y(0) = y'(0) = 0$, $y''(0) = 1$, with

$$f(t) = \begin{cases} 1 & \text{for } 0 \le t < 5 \\ 2 & \text{for } t \ge 5 \end{cases}$$

30. $y'' - 2y' - 3y = f(t)$; $y(0) = 1$, $y'(0) = 0$, with

$$f(t) = \begin{cases} 0 & \text{for } 0 \le t < 4 \\ 12 & \text{for } t \ge 4 \end{cases}$$

31. $y'' + 4y = f(t)$; $y(0) = 1$, $y'(0) = 0$, with

$$f(t) = \begin{cases} 0 & \text{for } 0 \le t < 4 \\ 3 & \text{for } t \ge 4 \end{cases}$$

32. $y'' + 5y' + 6y = f(t)$; $y(0) = y'(0) = 0$, with

$$f(t) = \begin{cases} -2 & \text{for } 0 \le t < 3 \\ 0 & \text{for } t \ge 3 \end{cases}$$

33. Determine the output voltage in the circuit of Figure 3.18, assuming that at time zero the capacitor is charged to a potential of 5 volts and the switch is opened at time zero and closed 5 seconds later. Graph this output.

34. Determine the output voltage in the RL circuit of Figure 3.20 if the current is initially zero and

$$E(t) = \begin{cases} 0 & \text{for } 0 \le t < 5 \\ 2 & \text{for } t \ge 5. \end{cases}$$

Graph this output function.

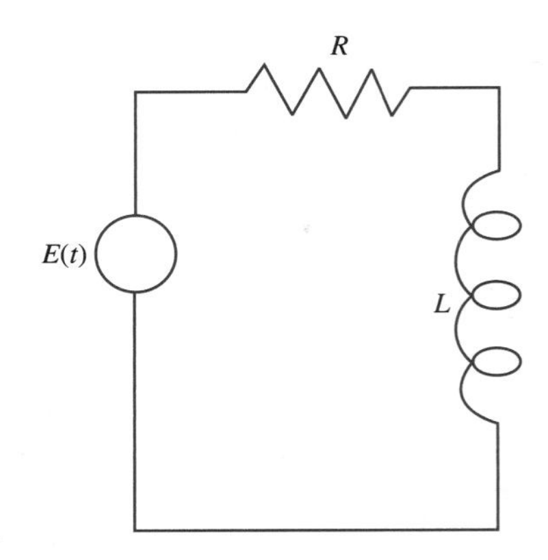

FIGURE 3.20 *The RL circuit of Problem 34, Section 3.3.*

35. Solve for the current in the RL circuit of Problem 34 if the current is initially zero and

$$E(t) = \begin{cases} k & \text{for } 0 \le t < 5 \\ 0 & \text{for } t \ge 5. \end{cases}$$

36. Show that Heaviside's formula can be written

$$\mathcal{L}^{-1}[F](t) = \sum_{j=1}^{n} \frac{p(a_j)}{q'(a_j)} e^{a_j t}.$$

Hint: Write

$$(s - a_j)\frac{p(s)}{q(s)} = \frac{p(s)}{(q(s) - q(a_j))/(s - a_j)}.$$

3.4 Convolution

If $f(t)$ and $g(t)$ are defined for $t \ge 0$, then the *convolution* $f * g$ of f with g is the function defined by

$$(f * g)(t) = \int_0^t f(t - \tau) g(\tau) d\tau$$

for $t \ge 0$ such that this integral converges.

In general the transform of a product of functions does not equal the product of their transforms. However, the transform of a convolution is the product of the transforms of the individual functions. This fact is called the *convolution theorem*, and is the rationale for the definition.

THEOREM 3.5 The Convolution Theorem

$$\mathcal{L}[f*g]=\mathcal{L}[f]\mathcal{L}[g]. \blacklozenge$$

Equivalently,

$$\mathcal{L}[f*g](s)=F(s)G(s).$$

A proof is outlined in Problem 24.

The inverse transform version of the convolution theorem is

$$\mathcal{L}^{-1}[FG]=f*g. \tag{3.8}$$

This states that the inverse transform of a product of two functions $F(s)$ and $G(s)$ is the convolution $f*g$ of the inverse transforms of the functions. This fact is sometimes useful in computing an inverse transform.

EXAMPLE 3.14

Compute

$$\mathcal{L}^{-1}\left[\frac{1}{s(s-4)^2}\right].$$

Certainly, we can do this by a partial fractions decomposition. To illustrate the use of the convolution, however, write

$$F(s)=\frac{1}{s} \text{ and } G(s)=\frac{1}{(s-4)^2}$$

so we are computing the inverse transform of a product. By the convolution theorem,

$$\mathcal{L}^{-1}\left[\frac{1}{s(s-4)^2}\right]=f*g,$$

where

$$f(t)=\mathcal{L}^{-1}\left[\frac{1}{s}\right]=1$$

and

$$g(t)=\mathcal{L}^{-1}\left[\frac{1}{(s-4)^2}\right]=te^{4t}.$$

Then

$$\begin{aligned}\mathcal{L}^{-1}\left[\frac{1}{s(s-4)^2}\right]&=f(t)*g(t)\\&=1*te^{4t}=\int_0^t \tau e^{4\tau}\,d\tau\\&=\frac{1}{4}te^{4t}-\frac{1}{16}e^{4t}+\frac{1}{16}. \blacklozenge\end{aligned}$$

Convolution is commutative:

$$f * g = g * f.$$

This can be proved by a straightforward change of variables in the integral defining the convolution.

In addition to its use in computing the inverse transform of products, convolution allows us to solve certain general initial value problems.

EXAMPLE 3.15

Solve the initial value problem

$$y'' - 2y' - 8y = f(t);\ y(0) = 1,\ y'(0) = 0.$$

We want a formula for the solution that will hold for any "reasonable" forcing function f. Apply the Laplace transform to the differential equation in the usual way, obtaining

$$s^2Y(s) - s - 2(sY(s) - 1) - 8Y(s) = F(s).$$

Then

$$(s^2 - 2s - 8)Y(s) = s - 2 + F(s).$$

Then

$$Y(s) = \frac{s-2}{s^2 - 2s - 8} + \frac{1}{s^2 - 2s - 8}F(s).$$

Factor $s^2 - 2s - 8 = (s-4)(s+2)$, and use a partial fractions decomposition to write

$$Y(s) = \frac{1}{3}\frac{1}{s-4} + \frac{2}{3}\frac{1}{s+2} + \frac{1}{6}\frac{1}{s-4}F(s) - \frac{1}{6}\frac{1}{s+2}F(s).$$

Now apply the inverse transform to obtain the solution

$$y(t) = \frac{1}{3}e^{4t} + \frac{2}{3}e^{-2t} + \frac{1}{6}e^{4t} * f(t) - \frac{1}{6}e^{-2t} * f(t),$$

which is valid for any function f for which these convolutions are defined. ♦

Convolution also enables us to solve some kinds of integral equations, which are equations in which the unknown function appears in an integral.

EXAMPLE 3.16

Solve for $f(t)$ in the integral equation

$$f(t) = 2t^2 + \int_0^t f(t-\tau)e^{-\tau}\,d\tau.$$

Recognize the integral on the right as the convolution of $f(t)$ with e^{-t}. Therefore, the integral equation has the form

$$f(t) = 2t^2 + f(t) * e^{-t}.$$

Apply the Laplace transform and the convolution theorem to this equation to get

$$F(s) = \frac{4}{s^3} + \frac{1}{s+1}F(s).$$

Then

$$F(s) = \frac{4}{s^3} + \frac{4}{s^4},$$

which we invert to obtain

$$f(t) = 2t^2 + \frac{2}{3}t^3. \blacklozenge$$

A Replacement Scheduling Problem

We will develop an integral equation that arises in the context of planning replacements for items (such as pieces of equipment that wear out or stored drugs that lose their effectiveness over time).

Suppose a company or organization uses large numbers of a certain item. An example might be portable computers for use by the military, copying machines in a business, or vaccine doses in a hospital. The organization's plan of operation includes an estimate of how many of these items it wants to have on hand at any time. We will imagine that this number is large enough that it can be approximated by a piecewise continuous *availability function* $f(t)$ that gives the number of items available for use at time t. Experience and familiarity with the items enables the organization and the supplier to produce a function $m(t)$, called a *mortality function*, that is a measure of the number of items still working satisfactorily (surviving) up to time t. We will be more explicit about $m(t)$ shortly.

Given $f(t)$ and $m(t)$ (items needed and how long items remain good), planners want to develop a *replacement function* $r(t)$ that measures the total number of replacements that must be made up to time t.

To begin the analysis, assign the time $t = 0$ to that time when these items of equipment were introduced into use, so at this initial time all the items are new. We also set $r(0) = 0$.

In a time interval from τ to $\tau + \Delta\tau$, there have been

$$r(\tau + \Delta\tau) - r(\tau) \approx r'(\tau)\Delta\tau$$

replacements. Here is where the mortality function comes in. We assume that, at any later time t, the number of surviving items, out of these replacements in this time interval, is

$$r'(\tau)(\Delta\tau)m(\Delta\tau),$$

which we write as

$$r'(\tau)m(t - \tau)\Delta\tau.$$

The total number $f(t)$ of items available for use at time t is the sum of the number of items surviving from the new items introduced at time 0 plus the number of items surviving from replacements made over every interval of length $\Delta\tau$ from $\tau = 0$ to $\tau = t$. This means that

$$f(t) = f(0)m(t) + \int_0^t r'(\tau)m(t - \tau)d\tau.$$

This is an integral equation for the derivative of the replacement function $r(t)$. Given $f(t)$ and $m(t)$, we attempt to solve this integral equation to obtain $r(t)$.

The reason this strategy works in some instances is that this integral is a convolution, suggesting the use of the Laplace transform. Application of $\mathcal{L}$ to the integral equation yields

$$\begin{aligned} F(s) &= f(0)M(s) + \mathcal{L}[r'(t)](s)\mathcal{L}[m(t)](s) \\ &= f(0)M(s) + (sR(s) - r(0))M(s) \\ &= f(0)M(s) + sR(s)M(s). \end{aligned}$$

Then

$$R(s) = \frac{F(s) - f(0)M(s)}{sM(s)}.$$

If we can invert $R(s)$, we have $r(t)$.

We will see how this model works in a specific example. Suppose we want to have $f(t) = A + Bt$ doses of a drug on hand at time t with A and B as positive constants. Thus, $f(0) = A$, and the need increases in time at the rate $f'(t) = B$. Suppose the mortality function is

$$m(t) = 1 - H(t - k)$$

in which H is the Heaviside function and k is a positive constant determined by how long doses remain effective.

Now

$$F(s) = \frac{A}{s} + \frac{B}{s^2} \text{ and } M(s) = \frac{1}{s} - \frac{1}{s}e^{-ks}.$$

The transform of the replacement function is

$$\begin{aligned} R(s) &= \frac{F(s) - F(0)M(s)}{sM(s)} \\ &= \frac{\frac{A}{s} + \frac{B}{s^2} - A\left(\frac{1}{s} - \frac{1}{s}e^{-ks}\right)}{s\left(\frac{1}{s} - \frac{1}{s}e^{-ks}\right)} \\ &= \frac{A}{s}\frac{1}{1 - e^{-ks}} + \frac{B}{s^2}\frac{1}{1 - e^{-ks}} - \frac{A}{s} \end{aligned}$$

in which we have omitted some routine algebra in going from the second line to the third. Now $0 < e^{-ks} < 1$ for $ks > 0$, so we can use the geometric series to write

$$\frac{1}{1 - e^{-ks}} = \sum_{n=0}^{\infty}(e^{-ks})^n = 1 + \sum_{n=1}^{\infty} e^{-kns}.$$

Therefore,

$$\begin{aligned} R(s) = A &\left(\frac{1}{s} + \sum_{n=1}^{\infty} \frac{1}{s}e^{-kns}\right) \\ &+ B\left(\frac{1}{s^2} + \sum_{n=1}^{\infty} \frac{1}{s^2}e^{-kns}\right) - \frac{A}{s}. \end{aligned}$$

Invert this term by term to obtain

$$\begin{aligned} r(t) &= A + A\sum_{n=1}^{\infty} H(t - nk) + Bt + B\sum_{n=1}^{\infty}(t - nk)H(t - nk) - A \\ &= Bt + \sum_{n=1}^{\infty}(A + B(t - nk))H(t - nk). \end{aligned}$$

Notice that $t - nk < 0$ (hence $H(t - nk) = 0$) if $t/n < k$. Since k is given and n increases from 1 through the positive integers, this always occurs after some time, so "most" of the terms of this series vanish for a given time.

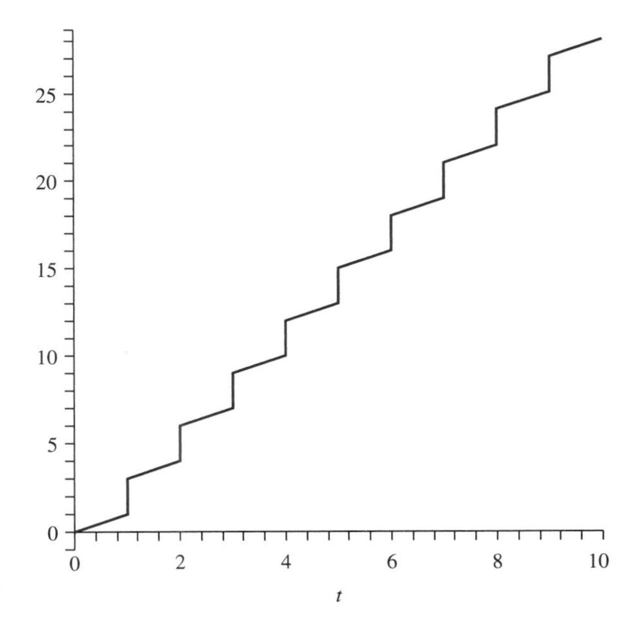

FIGURE 3.21 *Replacement function.*

Figure 3.21 is a graph of this replacement function for $A = 2$, $B = 0.001$, and $k = 1$. As expected, $r(t)$ is a strictly increasing function, because it measures total replacements up to a given time. The graph gives an indication of how the drug needs to be replenished to maintain $f(t)$ doses at time t.

SECTION 3.4 PROBLEMS

In each of Problems 1 through 8, use the convolution theorem to help compute the inverse Laplace transform of the function. Wherever they occur, a and b are positive constants.

1. $\dfrac{1}{s(s+2)}e^{-4s}$
2. $\dfrac{1}{s^4(s-5)}$
3. $\dfrac{1}{s(s^2+a^2)^2}$
4. $\dfrac{2}{s^2(s^2+5)}$
5. $\dfrac{s}{(s^2+a^2)(s^2+b^2)}$
6. $\dfrac{s^2}{(s-3)(s^2+5)}$
7. $\dfrac{1}{(s^2+4)(s^2-4)}$
8. $\dfrac{1}{s^2+16}e^{-2s}$

In each of Problems 9 through 16, use the convolution theorem to write a formula for the solution in terms of f.

9. $y'' + 9y = f(t);\ y(0) = -1,\ y'(0) = 1$
10. $y'' - k^2 y = f(t);\ y(0) = 2,\ y'(0) = -4$
11. $y^{(3)} - y'' - 4y' + 4y = f(t);\ y(0) = y'(0) = 1,\ y''(0) = 0$
12. $y^{(4)} - 11y'' + 18y = f(t);\ y(0) = y'(0) = y''(0) = y^{(3)}(0) = 0$
13. $y'' - 8y' + 12y = f(t);\ y(0) = -3,\ y'(0) = 2$
14. $y'' - 4y' - 5y = f(t);\ y(0) = 2,\ y'(0) = 1$
15. $y'' - 5y' + 6y = f(t);\ y(0) = y'(0) = 0$
16. $y'' + 10y' + 24y = f(t);\ y(0) = 1,\ y'(0) = 0$

In each of Problems 17 through 22, solve the integral equation.

17. $f(t) = e^{-t} + \int_0^t f(t-\tau)d\tau$
18. $f(t) = -1 + t - 2\int_0^t f(t-\tau)\sin(\tau)d\tau$

19. $f(t) = 3 + \int_0^t f(\tau)\cos(2(t-\tau))d\tau$
20. $f(t) = -t + \int_0^t f(t-\tau)\sin(\tau)d\tau$
21. $f(t) = -1 + \int_0^t f(t-\tau)e^{-3\tau}d\tau$
22. $f(t) = \cos(t) + e^{-2t}\int_0^t f(\tau)e^{2\tau}d\tau$
23. Solve for the replacement function $r(t)$ if $f(t) = A + Bt + Ct^2$ and $m(t) = e^{-kt}$. Graph $r(t)$.
24. Prove the convolution theorem. *Hint*: First write

$$F(s)G(s) = \int_0^\infty F(s)e^{-s\tau}g(\tau)\,d\tau.$$

Show that

$$F(s)G(s) = \int_0^\infty \mathcal{L}[H(t-\tau)f(t-\tau)](s)g(\tau)\,d\tau.$$

Use the definitions of the Heaviside function and of the transform to obtain

$$F(s)G(s) = \int_0^\infty \int_\tau^\infty e^{-st}g(\tau)f(t-\tau)\,d\tau.$$

Reverse the order of integration to obtain

$$F(s)G(s) = \int_0^\infty \int_0^t e^{-st}g(\tau)f(t-\tau)\,d\tau\,dt$$
$$= \int_0^\infty e^{-st}(f*g)(t)\,dt.$$

From this, show that $\mathcal{L}[f*g](s) = F(s)G(s)$.

25. Solve for the replacement function $r(t)$ if $f(t) = A$, constant, and $m(t) = e^{-kt}$ with k a positive constant. Graph $r(t)$.
26. Solve for the replacement function $r(t)$ if $f(t) = A + Bt$ and $m(t) = e^{-kt}$. Graph $r(t)$.

3.5 Impulses and the Delta Function

Informally, an *impulse* is a force of extremely large magnitude applied over an extremely short period of time (imagine hitting your thumb with a hammer). We can model this idea as follows. First, for any positive number ϵ consider the pulse δ_ϵ defined by

$$\delta_\epsilon(t) = \frac{1}{\epsilon}[H(t) - H(t-\epsilon)].$$

This pulse, which is graphed in Figure 3.22, has magnitude (height) of $1/\epsilon$ and duration of ϵ. The *Dirac delta function* is thought of as a pulse of infinite magnitude over an infinitely short duration and is defined to be

$$\delta(t) = \lim_{\epsilon\to 0+} \delta_\epsilon(t).$$

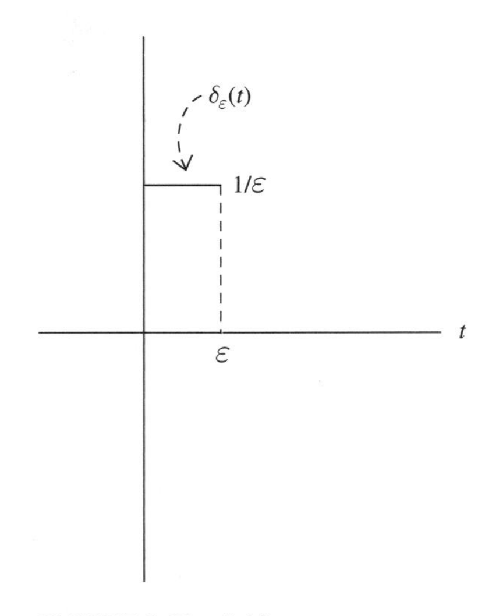

FIGURE 3.22 $\delta_\epsilon(t)$

This is not a function in the conventional sense but is a more general object called a *distribution*. For historical reasons, it continues to be known as the Dirac function after the Nobel laureate physicist P.A.M. Dirac. The *shifted delta function* $\delta(t-a)$ is zero except for $t=a$, where it has an infinite spike.

To take the Laplace transform of the delta function, begin with

$$\delta_\epsilon(t-a)=\frac{1}{\epsilon}[H(t-a)-H(t-a-\epsilon)].$$

This has transform

$$\mathcal{L}[\delta_\epsilon(t-a)]=\frac{1}{\epsilon}\left[\frac{1}{s}e^{-as}-\frac{1}{s}e^{-(a+\epsilon)s}\right]$$
$$=\frac{e^{-as}(1-e^{-\epsilon s})}{\epsilon s},$$

suggesting that we define

$$\mathcal{L}[\delta(t-a)]=\lim_{\epsilon\to 0+}\frac{e^{-as}(1-e^{-\epsilon s})}{\epsilon s}=e^{-as}.$$

In particular, we can choose $a=0$ to get

$$\mathcal{L}[\delta(t)]=1.$$

The following result is called the *filtering property* of the delta function. Suppose at time $t=a$ a signal is impacted with an impulse by mutliplying the signal by $\delta(t-a)$, and the resulting signal is then summed over all positive time by integrating it from zero to infinity. We claim that this yields exactly the value $f(a)$ of the signal at time a.

THEOREM 3.6 Filtering Property of the Delta Function

Let $a>0$ and let $\int_0^\infty f(t)dt$ converge. Suppose also that f is continuous at a. Then

$$\int_0^\infty f(t)\delta(t-a)dt=f(a). \blacklozenge$$

A proof is outlined in Problem 7.

If we apply the filtering property to $f(t)=e^{-st}$, we get

$$\int_0^\infty e^{-st}\delta(t-a)dt=e^{-as},$$

which is consistent with the definition of the Laplace transform of the delta function. Now change notation in the filtering property, and write it as

$$\int_0^\infty f(\tau)\delta(t-\tau)d\tau=f(t).$$

We recognize the convolution of f with δ. The last equation becomes

$$f*\delta=f.$$

The delta function therefore acts as an identity for the "product" defined by convolution.

In using the Laplace transform to solve an initial value problem involving the delta function, proceed as we have been doing, except that now we must use the transform of the delta function.

EXAMPLE 3.17

We will solve

$$y'' + 2y' + 2y = \delta(t-3);\ y(0) = y'(0) = 0.$$

Apply the transform to the differential equation to get

$$s^2Y(s) + 2sY(s) + 2Y(s) = e^{-3s},$$

so

$$Y(s) = \frac{1}{s^2+2s+2}e^{-3s}.$$

The solution is the inverse transform of $Y(s)$. To compute this, first write

$$Y(s) = \frac{1}{(s+1)^2+1}e^{-3s}.$$

Because $\mathcal{L}^{-1}[1/(s^2+1)] = \sin(t)$, a shift in the $s-$ variable gives us

$$\mathcal{L}^{-1}\left[\frac{1}{(s+1)^2+1}\right] = e^{-t}\sin(t).$$

Now shift in the $t-$ variable to obtain

$$y(t) = H(t-3)e^{(t-3)}\sin(t-3).$$

Figure 3.23 is a graph of this solution. ♦

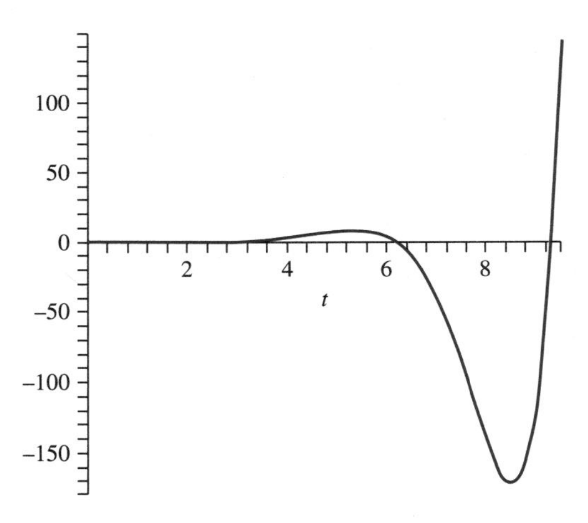

FIGURE 3.23 *Graph of the solution in Example 3.17.*

Transients can be generated in a circuit during switching and can be harmful because they contain a broad spectrum of frequencies. If one of these is near the natural frequency of the system, introducing the transient can cause resonance to occur, resulting in oscillations large enough to cause damage. For this reason, engineers sometimes use a delta function to model a transient and study its effect on a circuit being designed.

EXAMPLE 3.18

Suppose the current and charge on the capacitor in the circuit of Figure 3.24 are zero at time zero. We want to describe the output voltage response to a transient modeled by $\delta(t)$. The output voltage is $q(t)/C$, so we will determine $q(t)$. By Kirchhoff's voltage law,

$$Li' + Ri + \frac{1}{C}q = i' + 10i + 100q = \delta(t).$$

Since $i' = q$, then

$$q'' + 10q' + 100q = \delta(t).$$

Assume the initial conditions $q(0) = q'(0) = 0$. Apply the transform to the initial value problems to get

$$s^2Q(s) + 10sQ(s) + 100Q(s) = 1.$$

Then

$$Q(s) = \frac{1}{s^2 + 10s + 100} = \frac{1}{(s+5)^2 + 75}.$$

The last expression is preparation for shifting in the $s-$ variable. Since

$$\mathcal{L}^{-1}\left[\frac{1}{s^2 + 75}\right] = \frac{1}{5\sqrt{3}}\sin(5\sqrt{3}t),$$

then

$$q(t) = \mathcal{L}^{-1}\left[\frac{1}{(s+5)^2 + 75}\right] = \frac{1}{5\sqrt{3}}e^{-5t}\sin(5\sqrt{3}t).$$

The output voltage is

$$\frac{1}{C}q(t) = 100q(t) = \frac{20}{\sqrt{3}}e^{-5t}\sin(5\sqrt{3}t).$$

A graph of this output voltage is given in Figure 3.25. The circuit output displays damped oscillations at its natural frequency even though it was not explicitly forced by oscillation of this frequency. ♦

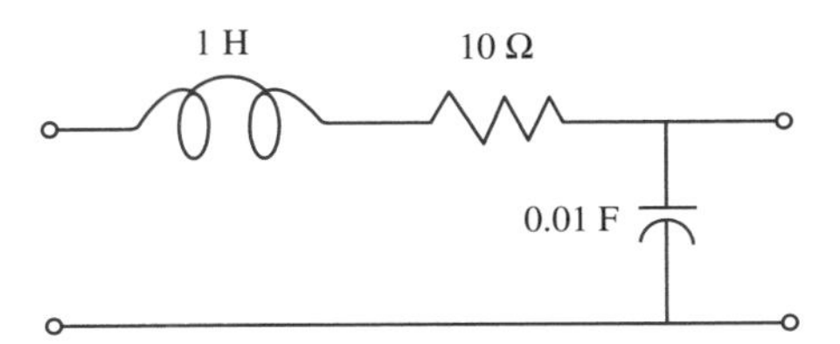

FIGURE 3.24 *Circuit of Example 3.18 with* $E_{in}(t) = \delta(t)$.

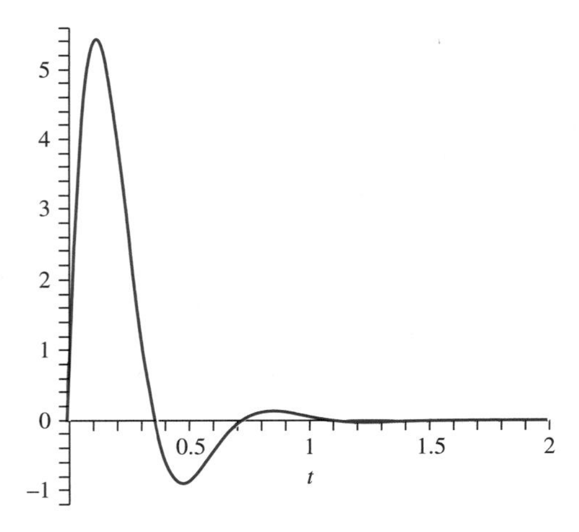

FIGURE 3.25 *Graph of the output voltage in Example 3.18.*

SECTION 3.5 PROBLEMS

In each of Problems 1 through 5, solve the initial value problem and graph the solution.

1. $y''' + 4y'' + 5y' + 2y = 6\delta(t)$; $y(0) = y'(0) = y''(0) = 0$

2. $y'' + 16y' = 12\delta(t - 5\pi/8)$; $y(0) = 3$, $y'(0) = 0$

3. $y'' + 5y' + 6y = B\delta(t)$; $y(0) = 3$, $y'(0) = 0$

4. $y'' - 4y' + 13y = 4\delta(t - 3)$; $y(0) = y'(0) = 0$

5. $y'' + 5y' + 6y = 3\delta(t - 2) - 4\delta(t - 5)$; $y(0) = y'(0) = 0$

6. An object of mass m is attached to the lower end of a spring of modulus k. Assume that there is no damping. Derive and solve an equation of motion for the object, assuming that at time zero it is pushed down from the equilibrium position with an initial velocity v_0. With what momentum does the object leave the equilibrium position?

7. Prove the filtering property of the delta function (Theorem 3.6). *Hint*: Replace $\delta(t - a)$ with
$$\lim_{\epsilon \to 0} \frac{1}{\epsilon}(H(t - a - \epsilon) - H(t - a))$$
in the integral and interchange the limit and the integral.

8. A 2 pound weight is attached to the lower end of a spring, stretching it 8/3 inches. The weight is allowed to come to rest in the equilibrium position. At some later time, which we call time 0, the weight is struck a downward blow of magnitude 1/4 pound (an impulse). Assume no damping in the system. Determine the velocity with which the weight leaves the equilibrium position as well as the frequency and magnitude of the oscillations.

9. Suppose, in the setting of Problem 6, the object is struck a downward blow of magnitude mv_0 at time 0. How does the position of this object compare with that of the object in Problem 6 at any positive time t?

3.6 Solution of Systems

Physical systems, such as circuits with multiple loops, may be modeled by systems of linear differential equations. These are often solved using the Laplace transform (and later by matrix methods).

We will illustrate the idea with a system having no particular significance, then look at a problem in mechanics and one involving a circuit.

EXAMPLE 3.19

We will solve the system (with initial conditions):

$$x'' - 2x' + 3y' + 2y = 4,$$
$$2y' - x' + 3y = 0,$$
$$x(0) = x'(0) = y(0) = 0.$$

Apply the transform to each equation of the system, making use of the initial conditions, to obtain

$$s^2X - 2sX + 3sY + 3Y = \frac{4}{s},$$
$$2sY - sX + 3Y = 0.$$

Solve these for $X(s$ and $Y(s)$:

$$X(s) = \frac{4s+6}{s^2(s+2)(s-1)} \quad \text{and} \quad Y(s) = \frac{2}{s(s+2)(s-1)}.$$

Use partial fractions to write

$$X(s) = -\frac{7}{2}\frac{1}{s} - 3\frac{1}{s^2} + \frac{1}{6}\frac{1}{s+2} + \frac{10}{3}\frac{1}{s-1}$$

and

$$Y(s) = -\frac{1}{s} + \frac{1}{3}\frac{1}{s+2} + \frac{2}{3}\frac{1}{s-1}.$$

Then

$$x(t) = -\frac{7}{2} - 3t + \frac{1}{6}e^{-2t} + \frac{10}{3}e^t$$

and

$$y(t) = -1 + \frac{1}{3}e^{-2t} + \frac{2}{3}e^t. \; \blacklozenge$$

EXAMPLE 3.20

Figure 3.26 shows a mass/spring system. Let $x_1 = x_2 = 0$ at the equilibrium position, where the weights are at rest. Choose the direction to the right as positive, and suppose the weights are at $x_1(t)$ and $x_2(t)$ at time t.

By two applications of Hooke's law, the restoring force on m_1 is

$$-k_1x_1 + k_2(x_2 - x_1)$$

and that on m_2 is

$$-k_2(x_2 - x_1) - k_3x_2.$$

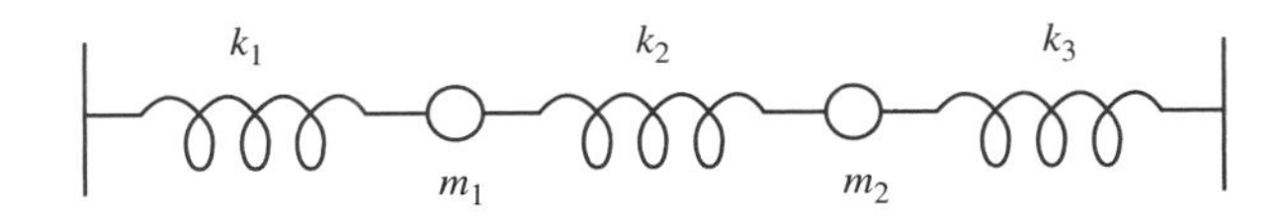

FIGURE 3.26 *Mass-spring system of Example 3.20.*

By Newton's second law,

$$m_1 x_1'' = -(k_1 + k_2)x_1 + k_2 x_2 + f_1(t)$$

and

$$m_2 x_2''(t) = k_2 x_1 - (k_2 + k_3)x_2 + f_2(t),$$

where $f_1(t)$ and $f_2(t)$ are forcing functions. We have assumed here that damping is negligible.

As a specific example, let

$$m_1 = m_2 = 1, k_1 = k_3 = 4, k_2 = 5/2.$$

Also suppose $f_2(t) = 0$ and $f_1(t) = 2(1 - H(t-3))$. This acts on the first mass with a force of constant magnitude 2 for the first three seconds, then turns off. Now the system is

$$x_1'' = -\frac{13}{2}x_1 + \frac{5}{2}x_2 + 2[1 - H(t-3)]$$

$$x_2'' = \frac{5}{2}x_1 - \frac{13}{2}x_2.$$

Suppose the masses are initially at rest in the equilibrium position:

$$x_1(0) = x_2(0) = x_1'(0) = x_2'(0) = 0.$$

Apply the transform to the system to obtain

$$s^2 X_1 = -\frac{13}{2}X_1 + \frac{5}{2}X_2 + \frac{2(1 - e^{-3s})}{s}$$

$$s^2 X_2 = \frac{5}{2}X_1 - \frac{13}{2}X_2.$$

Solve these to obtain

$$X_1(s) = \frac{2}{(s^2+9)(s^2+4)}\left(s^2 + \frac{13}{2}\right)\frac{1}{s}(1 - e^{-3s})$$

$$= \frac{13}{36}\frac{1}{s} - \frac{1}{4}\frac{s}{s^2+4}$$

$$-\frac{1}{9}\frac{s}{s^2+9} - \frac{13}{36}\frac{1}{s}e^{-3s}$$

$$+\frac{1}{4}\frac{s}{s^2+4}e^{-3s} + \frac{1}{9}\frac{s}{s^2+9}e^{-3s}$$

and

$$X_2(s) = \frac{5}{36}\frac{1}{s} - \frac{1}{4}\frac{s}{s^2+4}$$

$$+\frac{1}{9}\frac{s}{s^2+9} - \frac{5}{36}\frac{1}{s}e^{-3s}$$

$$+\frac{1}{4}\frac{s}{s^2+4}e^{-3s} - \frac{1}{9}\frac{s}{s^2+9}e^{-3s}.$$

Apply the inverse transform to obtain

$$x_1(t) = \frac{13}{36} - \frac{1}{4}\cos(2t) - \frac{1}{9}\cos(3t)$$

$$+\left[-\frac{13}{36} + \frac{1}{4}\cos(2(t-3)) - \frac{1}{9}\cos(3(t-3))\right]H(t-3)$$

and

$$x_2(t) = \frac{5}{36} - \frac{1}{4}\cos(2t) + \frac{1}{9}\cos(3t)$$
$$+ \left[-\frac{5}{36} + \frac{1}{4}\cos(2(t-3)) - \frac{1}{9}\cos(3(t-3)) \right] H(t-3). \blacklozenge$$

The next example involves an electrical circuit and will require that we know how to take the transform of a function defined by an integral. To see how to do this, suppose

$$f(t) = \int_0^t g(\tau)\,d\tau.$$

Then $f(0) = 0$, and assuming that g is continuous, $f'(t) = g(t)$, so

$$\mathcal{L}[f'] = \mathcal{L}[g] = s\mathcal{L}\left[\int_0^t g(\tau)\,d\tau\right].$$

But this means that

$$\mathcal{L}\left[\int_0^t g(\tau)\,d\tau\right] = \frac{1}{s}\mathcal{L}[g].$$

EXAMPLE 3.21

We will use this result to analyze the circuit of Figure 3.27.

Suppose the switch is closed at time zero. We want to solve for the current in each loop. Assume that both loop currents and the charges on the capacitors are initially zero, and apply Kirchhoff's laws to each loop to obtain

$$40i_1 + 120(q_1 - q_2) = 10$$
$$60i_2 + 120q_2 = 120(q_1 - q_2).$$

Since $i = q'$, we can write

$$q(t) = \int_0^t i(\tau)\,d\tau + q(0).$$

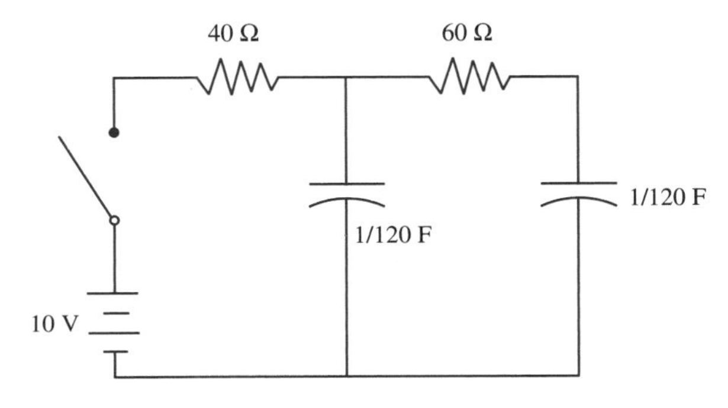

FIGURE 3.27 *Circuit in Example 3.21.*

Put these into the circuit equations to get

$$40i_1 + 120\int_0^t (i_1(\tau) - i_2(\tau))\,d\tau$$
$$+ 120(q_1(0) - q_2(0)) = 10,$$
$$60i_2 + 120\int_0^t i_2(\tau)\,d\tau + 120q_2(0)$$
$$= 120\int_0^t (i_1(\tau) - i_2(\tau))\,d\tau + 120(q_1(0) - q_2(0)).$$

Setting $q_1(0) = q_2(0) = 0$ in this system, we obtain

$$40i_1 + 120\int_0^t (i_1(\tau) - i_2(\tau))\,d\tau = 10$$
$$60i_2 + 120\int_0^t i_2(\tau)\,d\tau = 120\int_0^t (i_1(\tau) - i_2(\tau))\,d\tau.$$

Apply the transform to obtain

$$40I_1 + \frac{120}{s}I_1 - \frac{120}{s}I_2 = \frac{10}{s}$$
$$60I_2 + \frac{120}{s}I_2 = \frac{120}{s}I_1 - \frac{120}{s}I_2.$$

These can be written

$$(s+3)I_1 - 3I_2 = \frac{1}{4}$$
$$2I_1 - (s+4)I_2 = 0.$$

Solve these to obtain

$$I_1(s) = \frac{s+4}{4(s+1)(s+6)} = \frac{3}{20}\frac{1}{s+1} + \frac{1}{10}\frac{1}{s+6}$$

and

$$I_2(s) = \frac{1}{2(s+1)(s+6)} = \frac{1}{10}\frac{1}{s+1} - \frac{1}{10}\frac{1}{s+6}.$$

Then

$$i_1(t) = \frac{3}{20}e^{-t} + \frac{1}{10}e^{-6t}$$

and

$$i_2(t) = \frac{1}{10}e^{-t} - \frac{1}{10}e^{-6t}. \;\blacklozenge$$

SECTION 3.6 PROBLEMS

In each of Problems 1 through 11, use the Laplace transform to solve the initial value problem.

1. $$y_1' - 2y_2' + 3y_1 = 0$$
   $$y_1 - 4y_2' + 3y_3' = t$$
   $$y_1 - 2y_2' + 3y_3' = -1$$
   $$y_1(0) = y_2(0) = y_3(0) = 0$$
2. $x' + 2y' - x = 0, 4x' + 3y' + y = -6; x(0) = y(0) = 0$
3. $x' + y' + x - y = 0, x' + 2y' + x = 1; x(0) = y(0) = 0$
4. $x' + 4y' - y = 0, x' + 2y = e^{-t}; x(0) = y(0) = 0$
5. $x' + 2x - y' = 0, x' + y + x = t^2; x(0) = y(0) = 0$
6. $x' + 4x - y = 0, x' + y' = t; x(0) = y(0) = 0$
7. $3x' - y = 2t, x' + y' - y = 0; x(0) = y(0) = 0$
8. $x' + y' - x = \cos(t), x' + 2y' = 0; x(0) = y(0) = 0$

9. $x' + 2y' - y = 1, 2x' + y = 0; x(0) = y(0) = 0$
10. $2x' - 3y + y' = 0, x' + y' = t; x(0) = y(0) = 0$
11. $x' - 2y' = 1, x' + y - x = 0; x(0) = y(0) = 0$
12. Solve for the currents in the circuit of Figure 3.28 assuming that the currents and charges are initially zero and that $E(t) = 2H(t-4) - H(t-5)$.

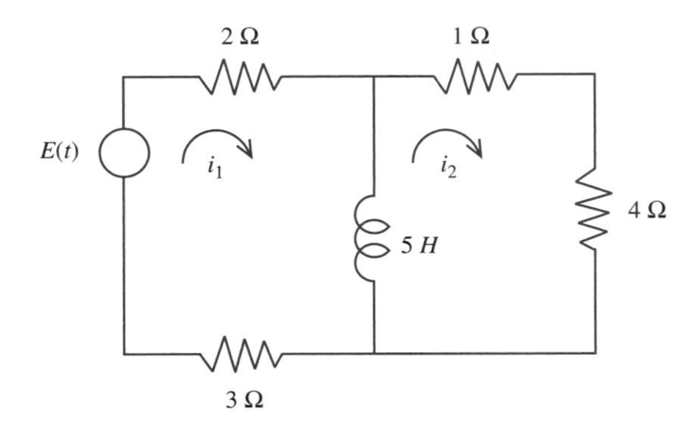

FIGURE 3.28 *Circuit in Problems 12 and 13, Section 3.6.*

13. Solve for the currents in the circuit of Figure 3.28 if the currents and charges are initially zero and $E(t) = 1 - H(t-4)\sin(2(t-4))$.
14. Solve for the displacement functions of the masses in the system of Figure 3.29. Neglect damping and assume zero initial displacements and velocities and external forces $f_1(t) = f_2(t) = 0$.

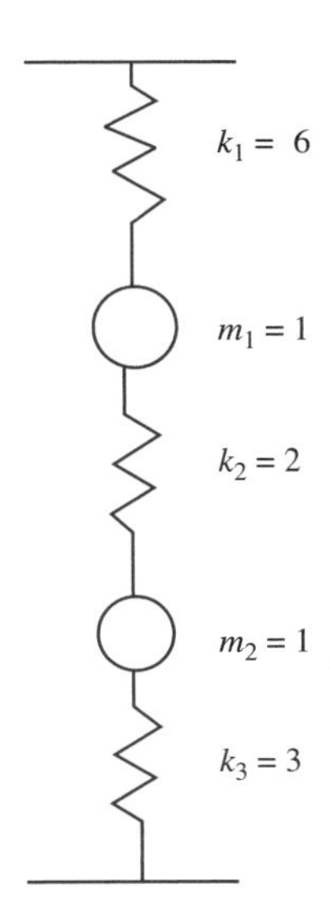

FIGURE 3.29 *Mass/spring system in Problems 14 and 15, Section 3.6.*

15. Solve for the displacement functions in the system of Figure 3.29 if

$$f_1(t) = 1 - H(t-2),\ f_2(t) = 0$$

and the initial displacements and velocities are zero.

16. Consider the system of Figure 3.30. Let M be subjected to a periodic driving force $f(t) = A\sin(\omega t)$. The masses are initially at rest in the equilibrium position.

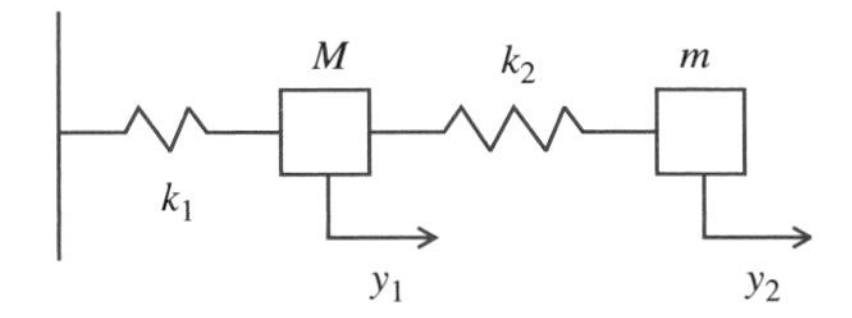

FIGURE 3.30 *Mass/spring system in Problem 16, Section 3.6.*

(a) Derive and solve the initial value problem for the displacement functions for the masses.

(b) Show that, if m and k_2 are chosen so that $\omega = \sqrt{k_2/m}$, then the mass m cancels the forced vibrations of M. In this case, we call m a *vibration absorber*.

17. Two objects of masses m_1 and m_2 are attached to opposite ends of a spring having spring constant k (Figure 3.31). The entire apparatus is placed on a highly varnished table. Show that, if the spring is stretched and released from rest, the masses oscillate with respect to each other with period

$$2\pi\sqrt{\frac{m_1 m_2}{k(m_1 + m_2)}}.$$

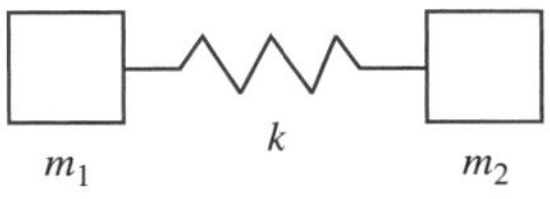

FIGURE 3.31 *Mass/spring system in Problem 17, Section 3.6.*

18. Solve for the currents in the circuit of Figure 3.32 if $E(t) = 5H(t-2)$ and the initial currents are zero.

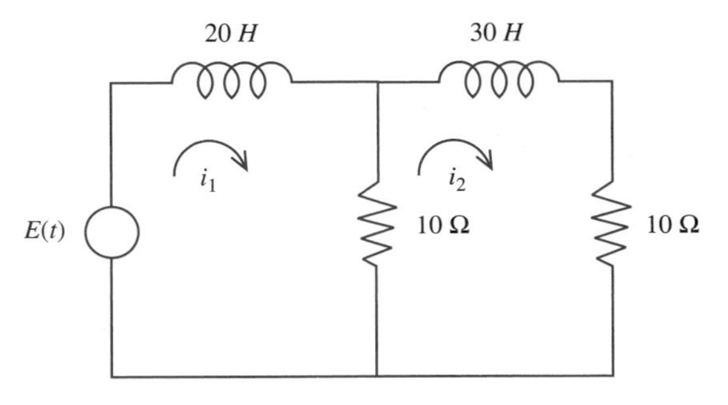

FIGURE 3.32 *Circuit of Problem 18, Section 3.6.*

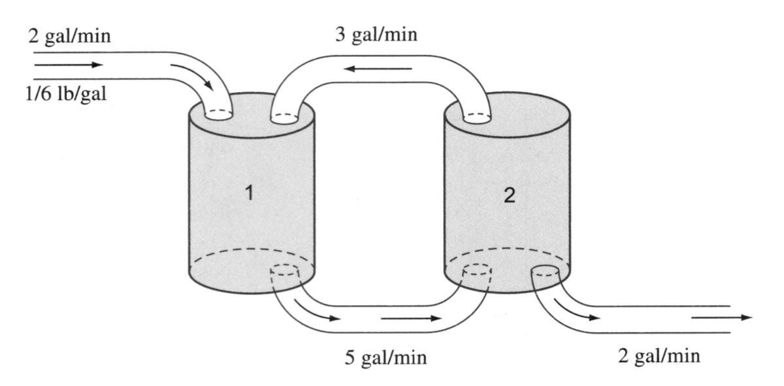

FIGURE 3.33 *System of tanks in Problem 20, Section 3.6.*

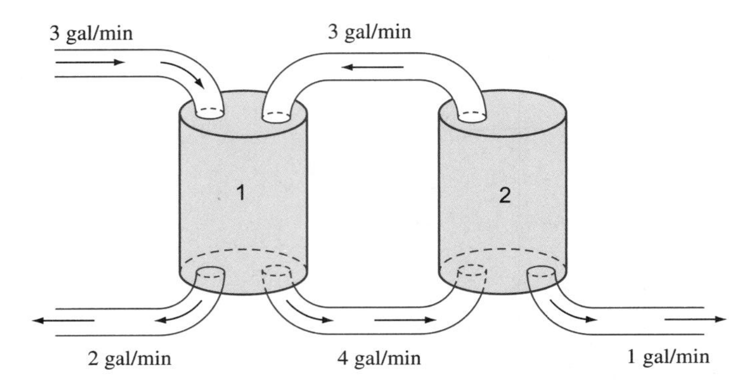

FIGURE 3.34 *System of tanks in Problem 21, Section 3.6.*

19. Solve for the currents in the circuit of Figure 3.32 if $E(t) = 5\delta(t-1)$.

20. Two tanks are connected by a series of pipes as shown in Figure 3.33. Tank 1 initially contains 60 gallons of brine in which 11 pounds of salt are dissolved. Tank 2 initially contains 7 pounds of salt dissolved in 18 gallons of brine. Beginning at time zero, a mixture containing $1/6$ pound of salt for each gallon of water is pumped into tank 1 at the rate of 2 gallons per minute, while salt water solutions are interchanged between the two tanks and also flow out of tank 2 at the rates shown in the diagram. Four minutes after time zero, salt is poured into tank 2 at the rate of 11 pounds per minute for a period of 2 minutes. Determine the amount of salt in each tank for any time $t \geq 0$.

21. Two tanks are connected by a series of pipes as shown in Figure 3.34. Tank 1 initially contains 200 gallons of brine in which 10 pounds of salt are dissolved. Tank 2 initially contains 5 pounds of salt dissolved in 100 gallons of water. Beginning at time zero, pure water is pumped into tank 1 at the rate of 3 gallons per minute, while brine solutions are interchanged between the tanks at the rates shown in the diagram. Three minutes after time zero, 5 pounds of salt are dumped into tank 2. Determine the amount of salt in each tank for any time $t \geq 0$.

3.7 Polynomial Coefficients

3.7.1 Differential Equations with Polynomial Coefficients

If a differential equation has polynomial coefficients, we can use the Laplace transform if we know how to take the transform of a function of the form $t^n f(t)$. Begin with the case $n = 1$.

THEOREM 3.7

Let $f(t)$ have Laplace transform $F(s)$ for $s > b$, and assume that $F(s)$ is differentiable. Then

$$\mathcal{L}[tf(t)](s) = -F'(s)$$

for $s > b$. ♦

Thus the transform of $tf(t)$ is the negative of the derivative of the transform of $f(t)$.

Proof Differentiate under the integral sign:

$$\begin{aligned} F'(s) &= \frac{d}{ds}\int_0^\infty e^{-st} f(t)\,dt \\ &= \int_0^\infty \frac{d}{ds}(e^{-st} f(t))\,dt \\ &= \int_0^\infty -te^{-st} f(t)\,dt \\ &= \int_0^\infty e^{-st}(-tf(t))\,dt = \mathcal{L}[-tf(t)](s). \end{aligned}$$ ♦

An induction argument yields the general result

$$\mathcal{L}[t^n f(t)](s) = (-1)^n \frac{d^n}{ds^n} F(s)$$

if $F(s)$ can be differentiated n times. ♦

We will also have use of the fact that, under certain conditions, the transform of $f(t)$ has limit 0 as $s \to \infty$.

THEOREM 3.8

Let f be piecewise continuous on $[0, k]$ for every positive number k. Suppose there are numbers M and b such that $|f(t)| \le Me^{bt}$ for $t \ge 0$. Then,

$$\lim_{s\to\infty} F(s) = 0.$$ ♦

Proof Write

$$\begin{aligned} |F(s)| &= \left|\int_0^\infty e^{-st} f(t)\,dt\right| \\ &\le \int_0^\infty e^{-st} Me^{bt}\,dt \\ &= \frac{M}{b-s} e^{-(s-b)t}\Big]_0^\infty \\ &= \frac{M}{s-b} \to 0 \end{aligned}$$

as $s \to \infty$. ♦

EXAMPLE 3.22

We will solve

$$y'' + 2ty' - 4y = 1;\ y(0) = y'(0) = 0.$$

Apply the Laplace transform to the differential equation to get

$$s^2Y(s) - sy(0) - y'(0) + 2\mathcal{L}[ty'](s) - 4Y(s) = \frac{1}{s}.$$

Now $y(0) = y'(0) = 0$, and

$$\begin{aligned}\mathcal{L}[ty'](s) &= -\frac{d}{ds}\left[\mathcal{L}[y'](s)\right] \\ &= -\frac{d}{ds}(sY(s) - y(0)) = -Y(s) - sY'(s).\end{aligned}$$

The transformed differential equation is therefore

$$s^2Y(s) - 2Y(s) - 2sY'(s) - 4Y(s) = \frac{1}{s}$$

or

$$Y' + \left(\frac{3}{s} - \frac{s}{2}\right)Y = -\frac{1}{2s^2}.$$

This is a linear first order differential equation for Y. To find an integrating factor, first compute

$$\int \left(\frac{3}{s} - \frac{s}{2}\right) ds = 3\ln(s) - \frac{1}{4}s^2.$$

The exponential of this function is a integrating factor. This is

$$e^{3\ln(s) - s^2/4}$$

or $s^3e^{-s^24}$. Multiply the differential equation for Y by this to obtain

$$(s^3e^{-s^2/4}Y)' = -\frac{1}{2}se^{-s^2/4}.$$

Then

$$s^3e^{-s^2/4}Y = e^{-s^2/4} + c.$$

Then

$$Y(s) = \frac{1}{s^3} + \frac{c}{s^3}e^{s^2/4}.$$

In order to have $\lim_{s\to 0} Y(s) = 0$, choose $c = 0$, obtaining $Y(s) = 1/s^3$. The solution is

$$y(t) = \frac{1}{2}t^2. \ \blacklozenge$$

3.7.2 Bessel Functions

If n is a nonnegative integer, the differential equation

$$t^2y'' + ty' + (t^2 - n^2)y = 0$$

is called *Bessel's equation of order* n.

This is usually considered for $t \geq 0$. Bessel's equation is second order, and the phrase *order* n refers to the parameter n in the coefficient of y. Solutions of Bessel's equation are called Bessel functions of order n, and they occur in many settings, including diffusion processes, flow of alternating current, and astronomy. Bessel functions and some applications are developed in detail in Section 15.3.

We will use the Laplace transform to derive solutions of Bessel's equation. Consider first the case $n = 0$. Bessel's equation of order zero is

$$ty'' + y' + ty = 0.$$

Apply $\mathcal{L}$ to obtain

$$\mathcal{L}[ty''] + \mathcal{L}[y'] + \mathcal{L}[ty] = 0.$$

Then

$$-\frac{d}{ds}\left(s^2Y(s) - sy(0) - y'(0)\right) + sY(s) - y(0) + \frac{d}{ds}(sY(s) - y(0)) = 0.$$

This is

$$-2sY - s^2Y' + sY - Y' = 0$$

or

$$-sY - (1 + s^2)Y' = 0.$$

This is a separable differential equation for Y. Write

$$\frac{Y'}{Y} = -\frac{s}{1+s^2}.$$

Integrate to obtain

$$\ln|Y| = -\frac{1}{2}\ln(1+s^2) + c = \ln((1+s^2)^{-1/2}) + c.$$

Take the exponential of both sides of this equation to write

$$Y(s) = e^c(1+s^2)^{-1/2} = \frac{C}{\sqrt{1+s^2}}$$

in which $C = e^c$ is constant. We have to invert this. First rewrite

$$Y(s) = \frac{C}{s}\left(1 + \frac{1}{s^2}\right)^{-1/2}.$$

The reason for doing this is to invoke the binomial series, which in general has the form

$$\begin{aligned}(1+x)^k &= 1 + kx + \frac{k(k-1)}{2!}x^2 \\ &\quad + \frac{k(k-1)(k-2)}{3!}x^2 + \cdots \\ &= \sum_{m=0}^{\infty}\binom{k}{m}x^m \text{ for } |x| < 1.\end{aligned}$$

Here

$$\binom{k}{m} = \begin{cases} 1 & \text{for } m = 0, \\ \frac{k(k-1)\cdots(k-m+1)}{m!} & \text{for } m = 1, 2, \cdots. \end{cases}$$

This is an infinite series if k is not a positive integer. Now set $k = -1/2$ and $x = 1/s^2$ in the binomial series. For $s > 1$, this gives us

$$
\begin{aligned}
Y(s) &= \frac{C}{s}\left(1 + \frac{1}{s^2}\right)^{-1/2} \\
&= \frac{C}{s}\left(1 - \frac{1}{2}\frac{1}{s^2} + \frac{(1)(3)}{2^2 2!}\frac{1}{s^4} + \cdots\right) \\
&= C\sum_{m=0}^{\infty} \frac{(-1)^m (2m)!}{(2^m m!)^2}\frac{1}{s^{2m+1}}.
\end{aligned}
$$

Then

$$
\begin{aligned}
y(t) &= C\sum_{m=0}^{\infty} \frac{(-1)^m (2m)!}{(2^m m!)^2}\mathcal{L}^{-1}\left(\frac{1}{s^{2m+1}}\right) \\
&= C\sum_{m=0}^{\infty} \frac{(-1)^m (2m)!}{(2^m m!)^2}\frac{t^{2m}}{(2m)!} \\
&= C\sum_{m=0}^{\infty} \frac{(-1)^m}{(2^m m!)^2} t^{2m}.
\end{aligned}
$$

If we impose the condition $y(0) = 1$, then $C = 1$, and we have the solution called the *Bessel function of the first kind of order zero*:

$$
J_0(t) = \sum_{m=0}^{\infty} \frac{(-1)^m}{(2^m m!)^2} t^{2m}.
$$

We will now solve Bessel's equation of any positive integer order n. Bessel's equation is

$$
t^2 y'' + t y' + (t^2 - n^2) y = 0.
$$

Change variables by setting

$$
y(t) = t^{-n} w(t).
$$

Compute y' and y'', substitute into Bessel's equation, and carry out some routine algebra to obtain

$$
t w'' + (1 - 2n) w' + t w = 0.
$$

Now apply the Laplace transform to obtain

$$
-\frac{d}{ds}\left(s^2 W - s w(0) - w'(0)\right) + (1 - 2n)(sW - w(0)) - \frac{d}{ds} W = 0.
$$

After carrying out these differentiations, we obtain

$$
(-1 - s^2) W' + (-2s + (1 - 2n)s) W + w(0) - (1 - 2n) w(0) = 0.
$$

We will seek a solution satisfying $w(0) = 0$, so this equation becomes

$$
(1 + s^2) W' + (1 + 2n) s W = 0.
$$

This is separable. Write

$$
\frac{W'}{W} = -\frac{(2n+1)s}{1 + s^2}.
$$

Integrate to obtain

$$\ln|W| = -\frac{2n+1}{2}\ln(1+s^2) = \ln\left((1+s^2)^{-(2n+1)/2}\right).$$

Here we have chosen the constant of integration to be zero to obtain a particular solution. Then

$$W(s) = (1+s^2)^{-(2n+1)/2}.$$

We must invert $W(s)$ to obtain $w(t)$ and finally $y(t)$. To carry out this inversion, write

$$W(s) = \frac{1}{s^{2n+1}}\left(1+\frac{1}{s^2}\right)^{-(2n+1)/2}$$

and use the binomial expansion to obtain

$$W(s) = \frac{1}{s^{2n+1}}\left(1 - \frac{2n+1}{2}\frac{1}{s^2} + \frac{1}{2!}\frac{-2n-1}{2}\frac{-2n-3}{2}\frac{1}{s^4} + \frac{1}{3!}\frac{-2n-1}{2}\frac{-2n-3}{2}\frac{-2n-5}{2}\frac{1}{s^4}\frac{1}{s^6} + \cdots\right).$$

Then

$$\begin{aligned}W(s) &= \frac{1}{s^{2n+1}} - \frac{2n+1}{2}\frac{1}{s^{2n+3}} \\ &+ \frac{(2n+1)(2n+3)}{2(2)(2!)}\frac{1}{s^{2n+5}} \\ &- \frac{(2n+1)(2n+3)(2n+5)}{2(2)(2)(3!)}\frac{1}{s^{2n+7}} + \cdots.\end{aligned}$$

Now we can invert this series term by term to obtain

$$\begin{aligned}w(t) &= \frac{1}{(2n)!}t^{2n} - \frac{2n+1}{2}\frac{t^{2(n+1)}}{(2(n+1))!} \\ &+ \frac{(2n+1)(2n+3)}{2(2)(2!)}\frac{t^{2(n+2)}}{(2(n+2))!} \\ &- \frac{(2n+1)(2n+3)(2n+5)}{2(2)(2)(3!)}\frac{t^{2(n+3)}}{(2(n+3))!} + \cdots.\end{aligned}$$

Finally recall that $y = t^{-n}w$ to obtain the solution

$$\begin{aligned}y(t) = t^{-n}w(t) &= \frac{1}{(2n)!}t^n - \frac{2n+1}{2(2(n+1))!}t^{n+2} \\ &+ \frac{(2n+1)(2n+3)}{2(2)(2!)((2(n+2))!)}t^{n+4} \\ &- \frac{(2n+1)(2n+3)(2n+5)}{2(2)(2)(3!)(2(n+3))!}t^{n+6} + \cdots \\ &= \sum_{k=0}^{\infty}\frac{(-1)^k}{2^{2k+n}k!(n+k)!}t^{n+2k} = J_n(t).\end{aligned}$$

This is the Bessel function of the first kind of order n, usually denoted $J_n(t)$ with the choice of constant made in the integration of the separated variables.

In Section 15.3, we will derive Bessel functions $J_\nu(t)$ of arbitrary order ν and also second, linearly independent solutions $Y_\nu(t)$ to write the general solution of Bessel's equation of order ν. We will also develop properties of Bessel functions that are needed for applications.

SECTION 3.7 PROBLEMS

Solve each of the following problems using the Laplace transform.

1. $y'' - 8ty' + 16y = 3;\ y(0) = y'(0) = 0$
2. $(1-t)y'' + ty' - y = 0;\ y(0) = 3,\ y'(0) = -1$
3. $y'' + 8ty' = 0;\ y(0) = 4,\ y'(0) = 0$
4. $y'' - 4ty' + 4y = 0;\ y(0) = 0,\ y'(0) = 10$
5. $t^2 y' - 2y = 2$ *Hint*: First set $u = 1/t$.
6. $y'' + 8ty' - 8y = 0;\ y(0) = 0,\ y'(0) = -4$
7. $ty'' + (t-1)y' + y = 0;\ y(0) = 0$
8. $y'' + 2ty' - 4y = 6;\ y(0) = y'(0) = 0$
9. $y'' - 16ty' + 32y = 0;\ y(0) = y'(0) = 0$
10. $y'' + 4ty' - 4y = 0;\ y(0) = 0,\ y'(0) = -7$
11. Review the derivation of the solution of Bessel's equation of order n for n a positive integer. Are any steps taken that would prevent n being an arbitrary positive number, not necessarily an integer? Could n be negative?

Appendix on Partial Fractions Decompositions

Partial fractions decomposition is an algebraic manipulation designed to write a quotient $P(x)/Q(x)$ of polynomials as a sum of simpler quotients, where simpler will be defined by the process.

Let P have degree m and let Q have degree n and assume that $n > m$. If this is not the case, divide Q into P. Assume that P and Q have no common roots, and that Q has been completely factored into linear and/or irreducible quadratic factors. A factor is irreducible quadratic if it is second degree with complex roots, hence it cannot be factored into linear factors with real coefficients. An example of an irreducible quadratic factor is $x^2 + 4$.

The partial fractions decomposition consisting of writing $P(x)/Q(x)$ as a sum $S(x)$ of simpler quotients is given in the following rules.

1. If $x - a$ is a factor of $Q(x)$ but $(x-a)^2$ is not, then include in $S(x)$ a term of the form

$$\frac{A}{x-a}.$$

2. If $(x-a)^k$ is a factor of $Q(x)$ with $k > 1$ but $(x-a)^{k+1}$ is not a factor, then include in $S(x)$ a sum of terms of the form

$$\frac{B_1}{x-a} + \frac{B_2}{(x-a)^2} + \cdots + \frac{B_k}{(x-a)^k}.$$

3. If $ax^2 + bx + c$ is an irreducible quadratic factor of $Q(x)$ but no higher power is a factor of $Q(x)$, then include in $S(x)$ a term of the form

$$\frac{Cx + D}{ax^2 + bx + c}.$$

4. If $(ax^2 + bx + c)^k$ is a product of irreducible factors of $Q(x)$ but $(ax^2 + bx + c)^{k+1}$ is not a factor of $Q(x)$, then include in $S(x)$ a sum of terms of the form

$$\frac{C_1 x + D_1}{ax^2 + bx + c} + \frac{C_2 x + D_2}{(ax^2 + bx + c)^2} + \cdots + \frac{C_k x + D_k}{(ax^2 + bx + c)^k}.$$

When each factor of $Q(x)$ has contributed one or more terms to $S(x)$ according to these rules, we have an expression of the form

$$\frac{P(x)}{Q(x)} = S(x),$$

with the coefficients to be determined. One way to do this is to add the terms in $S(x)$, set the numerator of the resulting quotient equal to $P(x)$, which is known, and solve for the coefficients of the terms in $S(x)$ by equating coefficients of like powers of x.

EXAMPLE 3.23

We will decompose

$$\frac{2x-1}{x^3+6x^2+5x-12}$$

into a sum of simpler fractions. First factor the denominator, then use the rules 1 through 4 to write the form of a partial fractions decomposition:

$$\frac{2x-1}{x^3+6x^2+5x-12} = \frac{2x-1}{(x-1)(x+3)(x+4)}$$
$$= \frac{A}{x-1} + \frac{B}{x+3} + \frac{C}{x+4}.$$

Once we have this template, the rest is routine algebra. If the fractions on the right are added, the numerator of the resulting quotient must equal $2x-1$, which is the numerator of the original quotient. Therefore,

$$A(x+3)(x+4) + B(x-1)(x+4) + C(x-1)(x+3) = 2x-1.$$

There are at least two ways we can find A, B, and C.
Method 1 Multiply the factors on the left and collect the coefficients of each power of x to write

$$A(x^2+7x+12) + B(x^2+3x-4) + C(x^2+2x-3)$$
$$= (A+B+C)x^2 + (7A+3B+2C)x + (12A-4B-3C) = 2x-1.$$

Equate the coefficient of each power of x on the left to the coefficient of that power of x on the right, obtaining a system of three linear equations in three unknowns:

$$A+B+C = 0 \text{ from the coefficients of } x^2,$$
$$7A+3B+2C = 2 \text{ from the coefficients of } x,$$

and

$$12A-4B-3C = -1 \text{ from the constant term.}$$

Solve these three equations obtaining $A = 1/20$, $B = 7/4$, and $C = -9/5$. Then

$$\frac{2x-1}{x^3+6x^2+5x-12} = \frac{1}{20}\frac{1}{x-1} + \frac{7}{4}\frac{1}{x+3} - \frac{9}{5}\frac{1}{x+4}.$$

Method 2 Begin with

$$A(x+3)(x+4) + B(x-1)(x+4) + C(x-1)(x+3) = 2x-1,$$

and assign values of x that make it easy to determine A, B, and C. Put $x = 1$ to get $20A = 1$, so $A = 1/20$. Put $x = -3$ to get $-4B = -7$, so $B = 7/4$. And put $x = -4$ to get $5C = -9$, so $C = -9/5$. This yields the same result as method 1, but in this example, method 2 is probably easier and quicker. ♦

EXAMPLE 3.24

Decompose

$$\frac{x^2+2x+3}{(x^2+x+5)(x-2)^2}$$

into partial fractions. First observe that x^2+x+5 has complex roots and so is irreducible. Thus, use the form

$$\frac{x^2+2x+3}{(x^2+x+5)(x-2)^2}=\frac{A}{x-2}+\frac{B}{(x-2)^2}+\frac{Cx+D}{x^2+x+5}.$$

If we add the fractions on the right, the numerator must equal x^2+2x+3. Therefore,

$$A(x-2)(x^2+x+5)+B(x^2+x+5)+(Cx+D)(x-2)^2=x^2+2x+3.$$

Expand the left side, and collect terms to write this equation as

$$(A+C)x^3+(-A+B-4C+D)x^2+(3A+B+4C-4D)x-10A+5B+4D$$
$$=x^2+2x+3.$$

Equate coefficients of like powers of x to get

$$A+C=0,$$
$$-A+B-4C+D=1,$$
$$3A+B+4C-4D=2,$$

and

$$-10A+5B+4D=3.$$

Solve these to obtain $A=1/11, B=1, C=-1/11$, and $D=-3/11$. The partial fractions decomposition is

$$\frac{x^2+2x+3}{(x^2+x+5)(x-2)^2}=\frac{1}{11}\frac{1}{x-2}+\frac{1}{(x-2)^2}-\frac{1}{11}\frac{x+3}{x^2+x+5}. \blacklozenge$$

CHAPTER 4

Series Solutions

Sometimes we can solve an initial value problem explicitly. For example, the problem

$$y' + 2y = 1;\ y(0) = 3$$

has the unique solution

$$y(x) = \frac{1}{2}(1 + 5e^{-2x}).$$

This solution is in *closed form*, which means that it is a finite algebraic combination of elementary functions (such as polynomials, exponentials, sines and cosines, and the like).

We may, however, encounter problems for which there is no closed form solution. For example,

$$y'' + e^x y = x^2;\ y(0) = 4$$

has the unique solution

$$y(x) = e^{-e^x} \int_0^x \xi^2 e^{e^\xi}\, d\xi + 4e^{-e^x}.$$

This solution (while explicit) has no elementary, closed form expression.

In such a case, we might try a numerical approximation. However, we may also be able to write a series solution that contains useful information. In this chapter, we will deal with two kinds of series solutions: power series (Section 4.1) and Frobenius series (Section 4.2).

4.1 Power Series Solutions

A function f is called *analytic* at x_0 if $f(x)$ has a power series representation in some interval $(x_0 - h, x_0 + h)$ about x_0. In this interval,

$$f(x) = \sum_{n=0}^{\infty} a_n (x - x_0)^n,$$

where the a_n's are the Taylor coefficients of $f(x)$ at x_0:

$$a_n = \frac{1}{n!} f^{(n)}(x_0).$$

Here $n!$ (n factorial) is the product of the integers from 1 through n if n is a positive integer, and $0! = 1$ by definition. The symbol $f^{(n)}(x_0)$ denotes the nth derivative of f evaluated at x_0. As examples of power series representations, $\sin(x)$ expanded about 0 is

$$\sin(x) = \sum_{n=0}^{\infty} \frac{1}{(2n+1)!} x^{2n+1}$$

for all x, and the geometric series is

$$\frac{1}{1-x} = \sum_{n=0}^{\infty} x^n$$

for $-1 < x < 1$.

An initial value problem having analytic coefficients has analytic solutions. We will state this for the first- and second-order cases when the differential equation is linear.

THEOREM 4.1

1. If p and q are analytic at x_0, then the problem

$$y' + p(x)y = q(x);\ y(x_0) = y_0$$

has a unique solution that is analytic at x_0.

2. If p, q, and f are analytic at x_0, then the problem

$$y'' + p(x)y' + q(x)y = f(x);\ y(x_0) = A,\ y'(x_0) = B$$

has a unique solution that is analytic at x_0. ♦

We are therefore justified in seeking power series solutions of linear equations having analytic coefficients. This strategy may be carried out by substituting $y = \sum_{n=0}^{\infty} a_n(x - x_0)^n$ into the differential equation and attempting to solve for the a_n's.

EXAMPLE 4.1

We will solve

$$y' + 2xy = \frac{1}{1-x}.$$

We can solve this using an integrating factor, obtaining

$$y(x) = e^{-x^2} \int_0^x \frac{1}{1-\xi} e^{-\xi^2}\, d\xi + ce^{-x^2}.$$

This is correct, but it involves an integral we cannot evaluate in closed form. For a series solution, let

$$y = \sum_{n=0}^{\infty} a_n x^n.$$

Then

$$y' = \sum_{n=1}^{\infty} n a_n x^{n-1}$$

with the summation starting at 1, because the derivative of the first term a_0 of the power series for y is zero. Substitute the series into the differential equation to obtain

$$\sum_{n=1}^{\infty} na_n x^{n-1} + \sum_{n=0}^{\infty} 2a_n x^{n+1} = \frac{1}{1-x}. \tag{4.1}$$

We would like to combine these series and factor out a common power of x to solve for the a_n's. To do this, write $1/(1-x)$ as a power series about 0 as

$$\frac{1}{1-x} = \sum_{n=0}^{\infty} x^n$$

for $-1 < x < 1$. Substitute this into equation (4.1) to obtain

$$\sum_{n=1}^{\infty} na_n x^{n-1} + \sum_{n=0}^{\infty} 2a_n x^{n+1} = \sum_{n=0}^{\infty} x^n. \tag{4.2}$$

Now rewrite the series so that they all contain powers x^n. This is like a change of variables in the summation index. First,

$$\sum_{n=1}^{\infty} na_n x^{n-1} = a_1 + 2a_2 x + 3a_3 x^2 + \cdots = \sum_{n=0}^{\infty} (n+1)a_{n+1} x^n,$$

and next,

$$\sum_{n=0}^{\infty} 2a_n x^{n+1} = 2a_0 x + 2a_1 x^2 + 2a_2 x^3 + \cdots$$

$$= \sum_{n=1}^{\infty} 2a_{n-1} x^n.$$

Now equation (4.2) can be written as

$$\sum_{n=0}^{\infty} (n+1)a_{n+1} x^n + \sum_{n=1}^{\infty} 2a_{n-1} x^n - \sum_{n=0}^{\infty} x^n = 0. \tag{4.3}$$

These rearrangements allow us to combine these summations for $n = 1, 2, \cdots$ and to write the $n = 0$ terms separately to obtain

$$\sum_{n=1}^{\infty} ((n+1)a_{n+1} + 2a_{n-1} - 1)x^n + a_1 - 1 = 0. \tag{4.4}$$

Because the right side of equation (4.4) is zero for all x in $(-1, 1)$, the coefficient of each power of x on the left, as well as the constant term $a_1 - 1$, must equal zero. This gives us

$$(n+1)a_{n+1} + 2a_{n-1} - 1 = 0 \text{ for } n = 1, 2, 3, \cdots$$

and

$$a_1 - 1 = 0.$$

Then $a_1 = 1$, and

$$a_{n+1} = \frac{1}{n+1}(1 - 2a_{n-1}) \text{ for } n = 1, 2, 3, \cdots.$$

This is a *recurrence relation* for the coefficients, giving a_{n+1} in terms of a preceding coefficient a_{n-1}. Now solve for some of the coefficients using this recurrence relation:

$$(n=1)\ a_2 = \frac{1}{2}(1 - 2a_0),$$

$$(n=2)\ a_3 = \frac{1}{3}(1 - 2a_1) = -\frac{1}{3},$$

$$(n=3)\ a_4 = \frac{1}{4}(1-2a_2)$$
$$= \frac{1}{4}(1-1+2a_0) = \frac{1}{2}a_0,$$
$$(n=4)\ a_5 = \frac{1}{5}(1-2a_3) = \frac{1}{5}(1+2/3) = \frac{1}{3},$$
$$(n=5)\ a_6 = \frac{1}{6}(1-2a_4) = \frac{1-a_0}{6},$$
$$(n=6)\ a_7 = \frac{1}{7}(1-2a_5) = \frac{1}{21},$$

and so on. With the coefficients computed thus far, the solution has the form

$$y(x) = a_0 + x + \frac{1}{2}(1-2a_0)x^2 - \frac{1}{3}x^3 + \frac{1}{2}a_0x^4 + \frac{1}{3}x^5 + \frac{1}{6}(1-a_0)x^6 + \frac{1}{21}x^7 + \cdots.$$

This has one arbitrary constant, a_0, as expected. By continuing to use the recurrence relation, we can compute as many terms of the series as we like. ♦

EXAMPLE 4.2

We will find a power series solution of

$$y'' + x^2y = 0$$

expanded about $x_0 = 0$.

Substitute $y = \sum_{n=0}^{\infty} a_nx^n$ into the differential equation. This will require that we compute

$$y' = \sum_{n=1}^{\infty} na_nx^{n-1} \text{and } y'' = \sum_{n=2}^{\infty}(n-1)na_nx^{n-2}.$$

Substitute these power series into the differential equation to obtain

$$\sum_{n=2}^{\infty}(n-1)na_nx^{n-2} + x^2\sum_{n=0}^{\infty}a_nx^n = 0$$

or

$$\sum_{n=2}^{\infty}n(n-1)a_nx^{n-2} + \sum_{n=0}^{\infty}a_nx^{n+2} = 0. \tag{4.5}$$

We will shift indices so that the power of x in both summations is the same, allowing us to combine terms from both summations. One way to do this is to write

$$\sum_{n=2}^{\infty}(n-1)na_nx^{n-2} = \sum_{n=0}^{\infty}(n+2)(n+1)a_{n+2}x^n$$

and

$$\sum_{n=0}^{\infty}a_nx^{n+2} = \sum_{n=2}^{\infty}a_{n-2}x^n.$$

Now equation (4.5) is

$$\sum_{n=0}^{\infty}(n+2)(n+1)a_{n+2}x^n + \sum_{n=2}^{\infty}a_{n-2}x^n = 0.$$

We can combine the terms for $n \geq 2$ in one summation. This requires that we write the $n = 0$ and $n = 1$ terms in the last equation separately, or else we lose terms:

$$2a_2x^0 + 2(3)a_3x + \sum_{n=2}^{\infty}[(n+2)(n+1)a_{n+2} + a_{n-2}]x^n = 0.$$

The left side can be zero for all x in some interval $(-h, h)$ only if the coefficient of each power of x is zero:

$$a_2 = a_3 = 0$$

and

$$(n+2)(n+1)a_{n+2} + a_{n-2} = 0 \text{ for } n \geq 2.$$

The last equation gives us

$$a_{n+2} = -\frac{1}{(n+2)(n+1)}a_{n-2} \text{ for } n = 2, 3, \cdots. \tag{4.6}$$

This is a recurrence relation for the coefficients of the series solution, giving us a_4 in terms of a_0, a_5 in terms of a_1, and so on. Recurrence relations always give a coefficient in terms of one or more previous coefficients, allowing us to generate as many terms of the series solution as we want. To illustrate, use $n = 2$ in equation (4.6) to obtain

$$a_4 = -\frac{1}{(4)(3)}a_0 = -\frac{1}{12}a_0.$$

With $n = 3$,

$$a_5 = -\frac{1}{(5)(4)}a_1 = -\frac{1}{20}a_1.$$

In turn, we obtain

$$a_6 = -\frac{1}{(6)(5)}a_2 = 0$$

because $a_2 = 0$,

$$a_7 = -\frac{(7)(6)}{a \quad 3} = 0$$

because $a_3 = 0$,

$$a_8 = -\frac{1}{(8)(7)}a_4 = \frac{1}{(56)(12)}a_0 = \frac{1}{672}a_0,$$

$$a_9 = -\frac{1}{(9)(8)}a_5 = \frac{1}{(72)(20)}a_1 = \frac{1}{1440}a_1,$$

and so on. Thus far, we have the first few terms of the series solution about 0:

$$y(x) = a_0 + a_1x + 0x^2 + 0x^3 - \frac{1}{12}a_0x^4$$
$$- \frac{1}{20}a_1x^5 + 0x^6 + 0x^7 + \frac{1}{672}x^8 + \frac{1}{1440}x^9 + \cdots$$

$$= a_0\left(1 - \frac{1}{12}x^4 + \frac{1}{672}x^8 + \cdots\right)$$
$$+ a_1\left(x - \frac{1}{20}x^5 + \frac{1}{1440}x^9 + \cdots\right).$$

This is the general solution, since a_0 and a_1 are arbitrary constants. Because $a_0 = y(0)$ and $a_1 = y'(0)$, a unique solution is determined by specifying these two constants. ♦

SECTION 4.1 PROBLEMS

In each of Problems 1 through 10, find the recurrence relation and use it to generate the first five terms of a power series solution about 0.

1. $y'' - xy' + y = 3$
2. $y'' + xy' + xy = 0$
3. $y'' + (1-x)y' + 2y = 1 - x^2$
4. $y' + xy = \cos(x)$
5. $y'' - x^2y' + 2y = x$
6. $y' - x^3y = 4$
7. $y' - xy = 1 - x$
8. $y'' + xy' = 1 - e^x$
9. $y' + (1-x^2)y = x$
10. $y'' + 2y' + xy = 0$

4.2 Frobenius Solutions

We will focus on the differential equation

$$P(x)y'' + Q(x)y' + R(x)y = F(x). \tag{4.7}$$

If $P(x) \neq 0$ on some interval, then we can divide by $P(x)$ to obtain the standard form

$$y'' + p(x)y' + q(x)y = f(x). \tag{4.8}$$

If $P(x_0) = 0$, we call x_0 a *singular point* of equation (4.7). This singular point *regular* if

$$(x - x_0)\frac{Q(x)}{P(x)} \text{ and } (x - x_0)^2\frac{R(x)}{P(x)}$$

are analytic at x_0. A singular point that is not regular is an *irregular singular point*.

EXAMPLE 4.3

$$x^3(x-2)^2y'' + 5(x+2)(x-2)y' + 3x^2y = 0$$

has singular points at 0 and 2. Now

$$(x - 0)\frac{Q(x)}{P(x)} = \frac{5x(x+2)(x-2)}{x^3(x-2)^2} = \frac{5}{x^2}\left(\frac{x+2}{x-2}\right)$$

is not analytic (or even defined) at 0, so 0 is an irregular singular point. But

$$(x - 2)\frac{Q(x)}{P(x)} = \frac{5(x+2)}{x^3}$$

and

$$(x-2)^2\frac{R(x)}{P(x)}=\frac{3}{x}$$

are both analytic at 2, so 2 is a regular singular point of this differential equation. ♦

We will not treat the case of an irregular singular point. If equation (4.7) has a regular singular point at x_0, there may be no power series solution about x_0, but there will be a *Frobenius series* solution, which has the form

$$y(x)=\sum_{n=0}^{\infty}c_n(x-x_0)^{n+r}$$

with $c_0\neq 0$. We must solve for the coefficients c_n and a number r to make this series a solution. We will look at an example to get some feeling for how this works and then examine the method more critically.

EXAMPLE 4.4

Zero is a regular singular point of

$$x^2y''+5xy'+(x+4)y=0.$$

Substitute $y=\sum_{n=0}^{\infty}c_nx^{n+r}$ to obtain

$$\sum_{n=0}^{\infty}(n+r)(n+r-1)c_nx^{n+r-2}+\sum_{n=0}^{\infty}5(n+r)c_nx^{n+r}$$
$$+\sum_{n=0}^{\infty}c_nx^{n+r+1}+\sum_{n=0}^{\infty}4c_nx^{n+r}=0.$$

Notice that the $n=0$ term in the proposed series solution is c_0x^r, which is not constant if $c_0\neq 0$, so the series for the derivatives begins with $n=0$ (unlike what we saw with power series). Shift indices in the third summation to write this equation as

$$\sum_{n=0}^{\infty}(n+r)(n+r-1)c_nx^{n+r-2}+\sum_{n=0}^{\infty}5(n+r)c_nx^{n+r}$$
$$+\sum_{n=1}^{\infty}c_{n-1}x^{n+r}+\sum_{n=0}^{\infty}4c_nx^{n+r}=0.$$

Combine terms to write

$$[r(r-1)+5r+4]c_0x^r$$
$$+\sum_{n=1}^{\infty}[(n+r)(n+r-1)c_n+5(n+r)c_n+c_{n-1}+4c_n]x^{n+r}=0.$$

Since we require that $c_0\neq 0$, the coefficient of x^r is zero only if

$$r(r-1)+5r+4=0.$$

This is called the *indicial equation* and is used to solve for r, obtaining the repeated root $r=-2$. Set the coefficient of x^{n+r} in the series equal to zero to obtain

$$(n+r)(n+r-1)c_n+5(n+r)c_n+c_{n-1}+4c_n=0$$

or, with $r = -2$,

$$(n-2)(n-3)c_n + 5(n-2)c_n + c_{n-1} + 4c_n = 0.$$

From this we obtain the recurrence relation

$$c_n = -\frac{1}{(n-2)(n-3)+5(n-2)+4}c_{n-1} \text{ for } n = 1, 2, \cdots.$$

This simplifies to

$$c_n = -\frac{1}{n^2}c_{n-1} \text{ for } n = 1, 2, \cdots.$$

Solve for some coefficients:

$$c_1 = -c_0$$

$$c_2 = -\frac{1}{4}c_1 = \frac{1}{4}c_0 = \frac{1}{2^2}c_0$$

$$c_3 = -\frac{1}{9}c_2 = -\frac{1}{(2\cdot 3)^2}c_0$$

$$c_4 = -\frac{1}{16}c_3 = \frac{1}{(2\cdot 3\cdot 4)^2}c_0$$

and so on. In general,

$$c_n = (-1)^n \frac{1}{(n!)^2}c_0$$

for $n = 1, 2, 3, \cdots$. We have found the Frobenius solution

$$y(x) = c_0\left[x^{-2} - x^{-1} + \frac{1}{4} - \frac{1}{36}x + \frac{1}{576}x^2 + \cdots\right]$$

$$= c_0 \sum_{n=0}^{\infty}(-1)^n \frac{1}{(n!)^2}x^{n-2}$$

for $x \neq 0$. This series converges for all nonzero x. ♦

Usually, we cannot expect the recurrence equation for c_n to have such a simple form.

Example 4.4 shows that an equation with a regular singular point may have only one Frobenius series solution about that point. A second, linearly independent solution is needed. The following theorem tells us how to produce two linearly independent solutions. For convenience, the statement is posed in terms of $x_0 = 0$.

THEOREM 4.2

Suppose 0 is a regular singular point of

$$P(x)y'' + Q(x)y' + R(x)y = 0.$$

Then

(1) The differential equation has a Frobenius solution

$$y(x) = \sum_{n=0}^{\infty} c_n x^{n+r}$$

with $c_0 \neq 0$. This series converges in some interval $(0, h)$ or $(-h, 0)$.

Suppose that the indicial equation has real roots r_1 and r_2 with $r_1 \geq r_2$. Then the following conclusions hold.

(2) If $r_1 - r_2$ is not a positive integer, then there are two linearly independent Frobenius solutions

$$y_1(x) = \sum_{n=0}^{\infty} c_n x^{n+r_1} \quad \text{and} \quad y_2(x) = \sum_{n=0}^{\infty} c_n^* x^{n+r_2}$$

with $c_0 \neq 0$ and $c_0^* \neq 0$. These solutions are valid at least in an interval $(0, h)$ or $(-h, 0)$.

(3) If $r_1 - r_2 = 0$, then there is a Frobenius solution

$$y_1(x) = \sum_{n=0}^{\infty} c_n x^{n+r_1}$$

with $c_0 \neq 0$, and there is a second solution

$$y_2(x) = y_1 \ln(x) + \sum_{n=1}^{\infty} c_n^* x^{n+r_1}.$$

These solutions are linearly independent on some interval $(0, h)$.

(4) If $r_1 - r_2$ is a positive integer, then there is a Frobenius solution

$$y_1(x) = \sum_{n=0}^{\infty} c_n x^{n+r_1}.$$

with $c_0 \neq 0$, and there is a second solution

$$y_2(x) = k y_1(x) \ln(x) + \sum_{n=0}^{\infty} c_n^* x^{n+r_2}$$

with $c_0^* \neq 0$. y_1 and y_2 are linearly independent solutions on some interval $(0, h)$. ♦

The *method of Frobenius* consists of using Frobenius series and Theorem 4.2 to solve equation (4.7) in some interval $(-h, h)$, $(0, h)$, or $(-h, 0)$, assuming that 0 is a regular singular point. Proceed as follows:

Step 1. Substitute $y = \sum_{n=0}^{\infty} c_n x^{n+r}$ into the differential equation, and solve for the roots r_1 and r_2 of the *indicial equation for* r. This yields a Frobenius solution (which may or may not be a power series).

Step 2. Depending on which of Cases (2), (3), or (4) of Theorem 4.2 applies, the theorem provides a template for a second solution which is linearly independent from the first. Once we know what this second solution looks like, we can substitute its general form into the differential equation and solve for the coefficients and, in Case (4), the constant k.

We will illustrate the Cases (2), (3), and (4) of the Frobenius theorem. For case (2), Example 4.5, we will provide all of the details. In Cases (3) and (4) (Examples 4.6, 4.7, and 4.8), we will omit some of the calculations and include just those that relate to the main point of that case.

EXAMPLE 4.5 Case 2 of the Frobenius Theorem

We will solve

$$x^2y'' + x\left(\frac{1}{2} + 2x\right)y' + \left(x - \frac{1}{2}\right)y = 0.$$

It is routine to check that 0 is a regular singular point. Substitute the Frobenius series $y = \sum_{n=0}^{\infty} c_n x^{n+r}$ to obtain

$$\sum_{n=0}^{\infty}(n+r)(n+r-1)c_n c^{n+r-2} + \sum_{n=0}^{\infty}\frac{1}{2}(n+r)c_n x^{n+r} + \sum_{n=0}^{\infty}2(n+r)x^{n+r+1}$$
$$+ \sum_{n=0}^{\infty} c_n x^{n+r+1} - \sum_{n=0}^{\infty}\frac{1}{2}c_n x^{n+r} = 0.$$

In order to be able to factor x^{n+r} from most terms, shift indices in the third and fourth summations to write this equation as

$$\sum_{n=1}^{\infty}\left[(n+r)(n+r-1)c_n + \frac{1}{2}(n+r)c_n + 2(n+r-1)c_{n-1} + c_{n-1} - \frac{1}{2}c_n\right]x^{n+r}$$
$$+ \left[r(r-1)c_0 + \frac{1}{2}c_0 r - \frac{1}{2}c_0\right]x^r = 0.$$

This equation will hold if the coefficient of each power of x is zero:

$$\left[r(r-1) + \frac{1}{2}r - \frac{1}{2}\right]c_0 = 0 \tag{4.9}$$

and for $n = 1, 2, 3, \cdots$,

$$(n+r)(n+r-1)c_n + \frac{1}{2}(n+r)c_n + 2(n+r-1)c_{n-1} + c_{n-1} - \frac{1}{2}c_n = 0. \tag{4.10}$$

Assuming that $c_0 \neq 0$, an essential requirement of the method, equation (4.9) implies that

$$r(r-1) + \frac{1}{2}r - \frac{1}{2} = 0. \tag{4.11}$$

This is the indicial equation for this differential equation. It has the roots $r_1 = 1$ and $r_2 = -1/2$. This puts us in case 2 of the Frobenius theorem. From equation (4.10), we obtain the recurrence relation

$$c_n = -\frac{1 + 2(n+r-1)}{(n+r)(n+r-1) + \frac{1}{2}(n+r) - \frac{1}{2}}c_{n-1}$$

for $n = 1, 2, 3, \cdots$.

First put $r_1 = 1$ into the recurrence relation to obtain

$$c_n = -\frac{2n+1}{n\left(n + \frac{3}{2}\right)}c_{n-1}$$

for $n = 1, 2, 3, \cdots$.

Some of these coefficients are

$$c_1 = -\frac{3}{5/2}c_0 = -\frac{6}{5}c_0,$$

$$c_2 = -\frac{5}{7}c_1 = -\frac{5}{7}\left(-\frac{6}{5}c_0\right) = \frac{6}{7}c_0,$$

$$c_3 = -\frac{7}{27/2}c_2 = -\frac{14}{27}\left(\frac{6}{7}c_0\right) = -\frac{4}{9}c_0,$$

and so on. One Frobenius solution is

$$y_1(x) = c_0\left(x - \frac{6}{5}x^2 + \frac{6}{7}x^3 - \frac{4}{9}x^4 + \cdots\right).$$

Because r_1 is a nonnegative integer, this first Frobenius series is actually a power series about 0.

For a second Frobenius solution, substitute $r = r_2 = -1/2$ into the recurrence relation. To avoid confusion with the first solution, we will denote the coefficients c_n^* instead of c_n. We obtain

$$c_n^* = -\frac{1 + 2\left(n - \frac{3}{2}\right)}{\left(n - \frac{1}{2}\right)\left(n - \frac{3}{2}\right) + \frac{1}{2}\left(n - \frac{1}{2}\right) - \frac{1}{2}}c_{n-1}^*$$

for $n = 1, 2, 3, \cdots$. This simplifies to

$$c_n^* = -\frac{2n - 2}{n\left(n - \frac{3}{2}\right)}c_{n-1}^*$$

for $n = 1, 2, 3, \cdots$. It happens in this example that $c_1^* = 0$, so each $c_n^* = 0$ for $n = 1, 2, 3, \cdots$, and the second Frobenius solution is

$$y_2(x) = \sum_{n=0}^{\infty} c_n^* x^{n-1/2} = c_0^* x^{-1/2}$$

for $x > 0$. ♦

EXAMPLE 4.6 Case 3 of the Frobenius Theorem

We will solve

$$x^2y'' + 5xy' + (x + 4)y = 0.$$

In Example 4.5, we found the indicial equation

$$r(r - 1) + 5r + 4 = 0$$

with repeated root $r_1 = r_2 = -2$ and the recurrence relation

$$c_n = -\frac{1}{n^2}c_{n-1}$$

for $n = 1, 2, \cdots$. This yielded the first Frobenius solution

$$y_1(x) = c_0 \sum_{n=0}^{\infty} (-1)^n \frac{1}{(n!)^2} x^{n-2}$$

$$= c_0\left[x^{-2} - x^{-1} + \frac{1}{4} - \frac{1}{36}x + \frac{1}{576}x^2 + \cdots\right].$$

Conclusion (3) of Theorem 4.2 tells us the general form of a second solution that is linearly independent from $y_1(x)$. Set

$$y_2(x) = y_1(x)\ln(x) + \sum_{n=1}^{\infty} c_n^* x^{n-2}.$$

Note that on the right, the series starts at $n = 1$, not $n = 0$. Substitute this series into the differential equation and find after some rearranging of terms that

$$4y_1 + 2xy_1' + \sum_{n=1}^{\infty}(n-2)(n-3)c_n^* x^{n-2} + \sum_{n=1}^{\infty} 5(n-2)c_n^* x^{n-2}$$

$$+ \sum_{n=1}^{\infty} c_n^* x^{n-1} + \sum_{n=1}^{\infty} 4c_n^* x^{n-2}$$

$$+ \ln(x)\left[x^2 y_1'' + 5xy_1' + (x+4)y_1\right] = 0.$$

The bracketed coefficient of $\ln(x)$ is zero because y_1 is a solution. Choose $c_0^* = 1$ (we need only one second solution), shift the indices to write $\sum_{n=1}^{\infty} c_n^* x^{n-1} = \sum_{n=2}^{\infty} c_{n-1}^* x^{n-2}$, and substitute the series for $y_1(x)$ to obtain

$$-2x^{-1} + c_1^* x^{-1} + \sum_{n=2}^{\infty}\left[\left[\frac{4(-1)^n}{(n!)^2} + \frac{2(-1)^n}{(n!)^2}\right](n-2) + (n-2)(n-3)c_n^*\right.$$

$$\left. + 5(n-2)c_n^* + c_{n-1}^* + 4c_n^*\right] x^{n-2} = 0.$$

Set the coefficient of each power of x equal to 0. From the coefficient of x^{-1}, we have $c_1^* = 2$. From the coefficient of x^{n-2}, we obtain (after some routine algebra)

$$\frac{2(-1)^n}{(n!)^2} n + n^2 c_n^* + c_{n-1}^* = 0$$

or

$$c_n^* = -\frac{1}{n^2} c_{n-1}^* - \frac{2(-1)^n}{n(n!)^2}$$

for $n = 2, 3, 4, \cdots$. With this, we can calculate as many coefficients as we want, yielding

$$y_2(x) = y_1(x)\ln(x) + \frac{2}{x} - \frac{3}{4} + \frac{11}{108}x$$

$$- \frac{25}{3456}x^2 + \frac{137}{432,000}x^3 + \cdots. \ \blacklozenge$$

The next two examples illustrate Case (4) of the theorem, first with $k = 0$ and then $k \neq 0$.

EXAMPLE 4.7 Case 4 of Theorem 4.2 with $k = 0$

We will solve

$$x^2 y'' + x^2 y' - 2y = 0.$$

There is a regular singular point at 0. Substitute $y = \sum_{n=0}^{\infty} c_n x^{n+r}$ to obtain

$$(r(r-1) - 2)c_0 x^r$$

$$+ \sum_{n=1}^{\infty} [(n+r)(n+r-1)c_n + (n+r-1)c_{n-1} - 2c_n] x^{n+r} = 0.$$

The indicial equation is $r^2 - r - 2 = 0$ with roots $r_1 = 2$, $r_2 = -1$. Now $r_1 - r_2 = 3$, putting us in Case (4) of the theorem. From the coefficient of x^{n+r}, we obtain the general recurrence relation

$$(n+r)(n+r-1)c_n + (+r-1)c_{n-1} - 2c_n = 0$$

for $n = 1, 2, 3, \cdots$.

For a first solution, use $r = 2$ to obtain the recurrence relation

$$c_n = -\frac{n+1}{n(n+3)}c_{n-1}$$

for $n = 1, 2, \cdots$. Using this, we obtain a first solution

$$y_1(x) = c_0 x^2 \left[1 - \frac{1}{2}x + \frac{3}{20}x^2 - \frac{1}{30}x^3 + \frac{1}{168}x^4 - \frac{1}{1120}x^5 + \cdots\right].$$

Now we need a second, linearly independent solution. Put $r = -1$ into the general recurrence relation to obtain

$$(n-1)(n-2)c_n^* + (n-2)c_{n-1}^* - 2c_n^* = 0$$

for $n = 1, 2, \cdots$. When $n = 3$, this gives $c_2^* = 0$, which forces $c_n^* = 0$ for $n = 2, 3, \cdots$. But then

$$y_2(x) = c_0^* \frac{1}{x} + c_1^*.$$

Substitute this into the differential equation to obtain

$$x^2(2c_0^* x^{-3}) + x^2(-c_0^* x^{-2}) - 2\left(c_1^* + c_0^* \frac{1}{x}\right) = -c_0^* - 2c_1^* = 0.$$

Then $c_1^* = -c_0^*/2$, and a second solution is

$$y_2(x) = c_0^* \left(\frac{1}{x} - \frac{1}{2}\right)$$

with c_0^* arbitrary but nonzero. The functions y_1 and y_2 form a fundamental set of solutions. In these solutions, there is no $y_1(x)\ln(x)$ term. ♦

EXAMPLE 4.8 Case 4 of Theorem 4.2 with $k \neq 0$

We will solve

$$xy'' - y = 0,$$

which has a regular singular point at 0. Substitute $y = \sum_{n=0}^{\infty} c_n x^{n+r}$ and rearrange terms to obtain

$$(r^2 - r)c_0 x^{r-1} + \sum_{n=1}^{\infty}[(n+r)(n+r-1)c_n - c_{n-1}]x^{n+r-1} = 0.$$

The indicial equation is $r^2 - r = 0$, with roots $r_1 = 1, r_2 = 0$. Here $r_1 - r_2 = 1$, a positive integer, putting us in Case (4) of the theorem. The general recurrence relation is

$$(n+r)(n+r-1)c_n - c_{n-1} = 0$$

for $n = 1, 2, \cdots$. With $r = 1$, this is

$$c_n = \frac{1}{n(n+1)}c_{n-1},$$

and some of the coefficients are

$$c_1 = \frac{1}{2}c_0,$$

$$c_2 = \frac{1}{2(3)}c_1 = \frac{1}{2(2)(3)}c_0,$$

$$c_3 = \frac{1}{3(4)}c_2 = \frac{1}{2(3)(2)(3)(4)}c_0,$$

and so on. In general,

$$c_n = \frac{1}{n!(n+1)!}c_0$$

for $n = 1, 2, 3, \cdots$, and one Frobenius solution is

$$y_1(x) = c_0 \sum_{n=0}^{\infty} \frac{1}{n!(n+1)!}x^{n+1}$$

$$= c_0 \left[x + \frac{1}{x}x^2 + \frac{1}{12}x^3 + \frac{1}{144}x^4 + \cdots \right].$$

For a second solution, put $r = 0$ into the general recurrence relation to obtain

$$n(n-1)c_n - c_{n-1} = 0$$

for $n = 1, 2, \cdots$. If we put $n = 1$ into this, we obtain $c_0 = 0$, violating one of the conditions for the method of Frobenius. Here we cannot obtain a second solution as a Frobenius series. Theorem 4.2, Case (4), tells us to look for a second solution of the form

$$y_2(x) = ky_1 \ln(x) + \sum_{n=0}^{\infty} c_n^* x^n.$$

Substitute this into the differential equation to obtain

$$x \left[ky_1'' \ln(x) + 2ky_1' \frac{1}{x} - ky_1 \frac{1}{x^2} + \sum_{n=2}^{\infty} n(n-1)c_n^* x^{n-2} \right]$$

$$- ky_1 \ln(x) - \sum_{n=0}^{\infty} c_n^* x^n = 0.$$

Now

$$k \ln(x)[xy_1'' - y_1] = 0,$$

because y_1 is a solution. For the remaining terms, let $c_0 = 1$ in $y_1(x)$ for convenience (we need only one more solution) to obtain

$$2k \sum_{n=0}^{\infty} \frac{1}{(n!)^2}x^n - k \sum_{n=0}^{\infty} \frac{1}{n!(n+1)!}x^n + \sum_{n=2}^{\infty} c_n^* n(n-1)x^{n-1} - \sum_{n=0}^{\infty} c_n^* x^n = 0.$$

Shift indices in the third summation to write

$$2k \sum_{n=0}^{\infty} \frac{1}{(n!)^2}x^n - k \sum_{n=0} \infty \frac{1}{n!(n+1)!}x^n$$

$$+ \sum_{n=1}^{\infty} c_{n+1} n(n+1)x^n - \sum_{n=0}^{\infty} c_n^* x^n = 0.$$

Then

$$(2k - k - c_0^*)x^0 + \sum_{n=1}^{\infty} \left[\frac{2k}{(n!)^2} - \frac{k}{n!(n+1)!} + n(n+1)c_{n+1}^* - c_n^* \right] x^n = 0.$$

This implies that $k - c_0^* = 0$, so

$$k = c_0^*.$$

Furthermore, the recurrence relation is

$$c_{n+1}^* = \frac{1}{n(n+1)}\left[c_n^* - \frac{(2n+1)k}{n!(n+1)!}\right]$$

for $n = 1, 2, \cdots$. Since c_0^* can be any nonzero number, we will for convenience let $c_0^* = 1$. For a particular solution, we may also choose $c_1^* = 1$. These give us

$$y_2(x) = y_1 \ln(x) + 1 - \frac{3}{4}x^2 - \frac{7}{36}x^3 - \frac{35}{1728}x^4 - \cdots. \quad \blacklozenge$$

SECTION 4.2 PROBLEMS

In each of Problems 1 through 10, find the first five terms of each of two linearly independent solutions.

1. $x^2y'' - 2xy' - (x^2 - 2)y = 0$
2. $x^2y'' + x(x^3 + 1)y' - y = 0$
3. $4xy'' + 2y' + 2y = 0$
4. $xy'' - y' + 2y = 0$
5. $x(2 - x)y'' - 2(x - 1)y' + 2y = 0$
6. $4x^2y'' + 4xy' + (4x^2 - 9)y = 0$
7. $x(x - 1)y'' + 3y' - 2y = 0$
8. $xy'' - 2xy' + 2y = 0$
9. $xy'' + (1 - x)y' + y = 0$
10. $4x^2y'' + 4xy' - y = 0$

CHAPTER 5
Approximation of Solutions

In this chapter, we will concentrate on the first-order initial value problem

$$y' = f(x, y);\ y(x_0) = y_0.$$

Depending on f, it may be impossible to write the solution in a form from which we can conveniently draw conclusions. For example, the problem

$$y' - \sin(x)y = 4;\ y(0) = 2$$

has the solution

$$y(x) = 4e^{-\cos(x)} \int_0^x e^{-\cos(\xi)} d\xi + 2e^{1-\cos(x)}.$$

It is unclear how this solution behaves or what its graph looks like.

In such cases, we may turn to computer-implemented methods to approximate solution values at specific points or to sketch an approximate graph. This chapter explores some techniques for doing this.

5.1 Direction Fields

Suppose $y' = f(x, y)$, with $f(x, y)$ given, at least for (x, y) in some specified region of the plane. The slope of the solution passing through (x, y) is therefore a known number $f(x, y)$. Form a rectangular grid of points (x_i, y_j). Through each grid point (x_i, y_j), draw a short line segment having slope $f(x_i, y_j)$. These line segments are called *lineal elements*. The lineal element through (x_i, y_j) is tangent to the solution through this point, and the collection of all the lineal elements is called a *direction field* for the differential equation $y' = f(x, y)$. If enough lineal elements are drawn, they trace out the shapes of integral curves of $y' = f(x, y)$, just as short tangent segments drawn along a curve give an impression of the shape of the curve. The direction field therefore provides a picture of how integral curves behave in the region over which the grid has been placed.

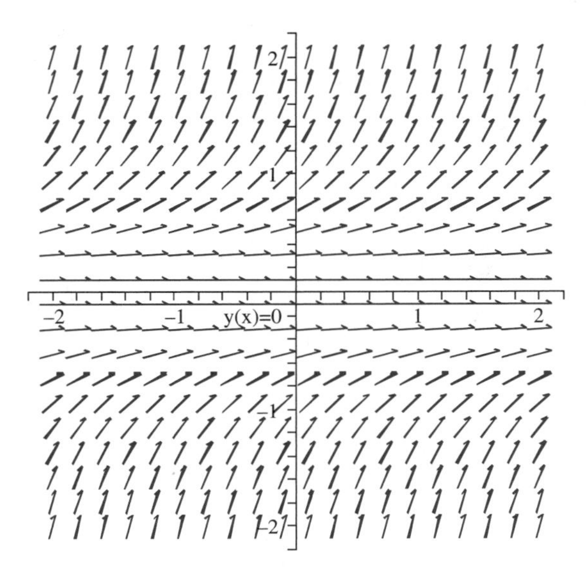

FIGURE 5.1 *Direction field for $y' = y^2$.*

If we think of the integral curves of $y' = f(x, y)$ as the trajectories of moving particles of a fluid, then the direction field is a flow pattern of this fluid.

EXAMPLE 5.1

The differential equation

$$y' = y^2.$$

has $f(x, y) = y^2$. The general solution is

$$y = -\frac{1}{x+k}$$

in which k is an arbitrary constant. Figure 5.1 shows a direction field for this differential equation for $-2 \le x \le 2$ and $-2 \le y \le 2$. Figure 5.2 shows a direction field together with four solution curves, corresponding to $y(0) = -2$, $y(0) = -1/2$, $y(0) = 1/2$ and $y(0) = 1$. These solution curves follow the flow of the tangent line segments making up the direction field. ♦

EXAMPLE 5.2

The differential equation

$$y' = \sin(xy)$$

has no nontrivial solution that can be written as a finite algebraic combination of elementary functions. Figure 5.3 shows a direction field for this equation, together with five solution curves corresponding to $y(0) = -2$, $y(0) = -1/2$, $y(0) = 1/2$, $y(0) = 1$, and $y(0) = 2$. These integral curves fit the flow of the lineal elements of the direction field. As guides in sketching integral curves, a direction field provides useful information about the behavior of solutions, which in this example we do not have explicitly in hand. ♦

It is not practical to draw direction fields by hand. Instructions for constructing direction fields using MAPLE are given in the MAPLE Primer in Appendix A.

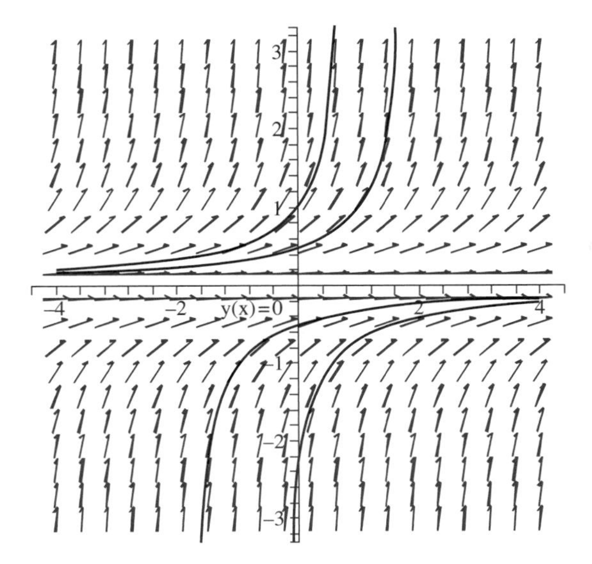

FIGURE 5.2 *Integral curves in the direction field for* $y' = y^2$.

FIGURE 5.3 *Direction field and some integral curves for* $y' = \sin(xy)$.

SECTION 5.1 PROBLEMS

In each of Problems 1 through 6, draw a direction field for the differential equation and some solution curves. Use it to sketch the integral curve of the solution of the initial value problem.

1. $y' = y\sin(x) - 3x^2$; $y(0) = 1$
2. $y' = e^x - y$; $y(-2) = 1$
3. $y' - y\cos(x) = 1 - x^2$; $y(2) = 2$
4. $y' = 2y + 3$; $y(0) = 1$
5. $y' = \sin(y)$; $y(1) = \pi/2$
6. $y' = x\cos(2x) - y$; $y(1) = 0$

5.2 Euler's Method

In this section, we present Euler's method for generating approximate numerical values of the solution of an initial value problem

$$y' = f(x, y); \quad y(x_0) = y_0$$

at selected points $x_0, x_1 = x_0 + h, x_2 = x_0 + 2h, \cdots$, and $x_n = x_0 + nh$. Here n is a positive integer (the number of iterations to be performed); and h is a (small) positive number called the *step size*. This number h is the distance between successive points at which approximate values of the solution are computed.

The idea behind Euler's method is conceptually simple. First choose n and h. We are given $y(x_0) = y_0$. Calculate $f(x_0, y_0)$ and draw the line having this slope through (x_0, y_0). This line is tangent to the solution at (x_0, y_0). Move along this tangent line to the point (x_1, y_1), where $x_1 = x_0 + h$. Use this number y_1 as the approximation to $y(x_1)$ at x_1. This is illustrated in Figure 5.4. We have some hope that this is a "good" approximation for h "small" because a tangent line at a point fits the curve closely near that point. Note that (x_1, y_1) is probably not on the integral curve through (x_0, y_0) but is on the tangent to this curve at (x_0, y_0).

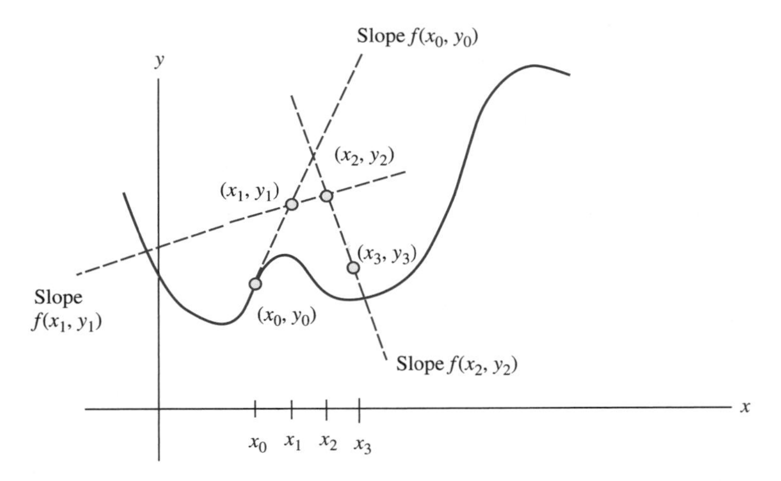

FIGURE 5.4 *The Euler approximation scheme.*

Now compute $f(x_1, y_1)$. This is the slope of the tangent to the graph of the solution passing through (x_1, y_1). Draw the line through (x_1, y_1) having this slope, and move along this line to (x_2, y_2) where $y_2 = x_1 + h = x_0 + 2h$. This determines a number y_2, which we take as an approximation to $y(x_2)$. (Figure 5.4 again).

Continue in this way. Compute $f(x_2, y_2)$, and draw the line with this slope through (x_2, y_2). Move along this line to (x_3, y_3) where $x_3 = x_2 + h = x_0 + 3h$, and use y_3 as an approximation to $y(x_3)$.

In general, once we have reached (x_k, y_k), draw the line through this point having a slope of $f(x_k, y_k)$, and move along this line to (x_{k+1}, y_{k+1}). Take y_{k+1} as an approximation to $y(x_{k+1})$.

This is the idea of the method. It is sensitive to how much $f(x, y)$ changes if x and y are varied by a small amount. The method also tends to accumulate error, since we use an approximation y_k to make the next approximation y_{k+1}. Following segments of lines is conceptually simple but is not as accurate as some other methods—two of which we will develop in the next section.

We will derive an expression for the approximate value y_k at x_k. From Figure 5.4,

$$y_1 = y_0 + f(x_0, y_0)(x_1 - x_0).$$

At the next step,

$$y_2 = y_1 + f(x_1, y_1)(x_2 - x_1).$$

After the approximation y_k has been computed, the next approximate value is

$$y_{k+1} = y_k + f(x_k, y_k)(x_{k+1} - x_k).$$

Since each $x_{k+1} - x_k = h$, the method can be summarized as follows.

Euler's method Define y_{k+1} in terms of y_k by

$$y_{k+1} = y_k + hf(x_k, y_k)$$

for $k = 0, 1, 2, \cdots, n - 1$.

TABLE 5.1 ***Euler's Method Applied to*** $y' = x\sqrt{y};\ y(2) = 4$

x	$y(x)$	Euler approximation of $y(x)$
2.0	4	4
2.05	4.205062891	4.200000000
2.1	4.42050650	4.410062491
2.15	4.646719141	4.630564053
2.2	4.88410000	4.861890566
2.25	5.133056641	5.104437213
2.3	5.394006250	5.358608481
2.35	5.667475321	5.624818168
2.4	5.953600000	6.903489382
2.45	6.253125391	6.195054550
2.5	6.566406250	6.499955415
2.55	6.893906641	6.818643042
2.6	7.236100000	7.151577819
2.65	7.593469141	7.499229462
2.7	7.966506250	7.862077016
2.75	8.355712891	8.240608856
2.8	8.761600000	8.635322690
2.85	9.184687891	9.046725564
2.9	9.625506250	9.475333860
2.95	10.08459414	9.921673298
3	10.56250000	10.38627894

EXAMPLE 5.3

Consider

$$y' = x\sqrt{y};\ y(2) = 4.$$

This problem (with separable differential equation) is easily solved exactly as

$$y(x) = \left(1 + \frac{x^2}{4}\right)^2.$$

We will apply Euler's method and use the exact solution to gauge the accuracy. Use $h = 0.05$ and $n = 20$. Then $x_0 = 2$, and $x_{20} = 2 + (20)(0.05) = 3$, so we are approximating values at points on $[2, 3]$. The approximate values are computed by

$$y_{k+1} = y_k + 0.2x_k\sqrt{y_k} \text{ for } k = 0, 1, 2, \cdots, 19.$$

Table 5.1 gives the Euler approximate values, together with values computed from the exact solution. The approximate values become less accurate as x_k moves further from x_0. ♦

It can be shown that the error in Euler's method is proportional to h. For this reason, Euler's method is called a *first-order* method. We can increase the accuracy in an Euler approximation by choosing h to be smaller (at the cost of more computing time).

SECTION 5.2 PROBLEMS

In each of Problems 1 through 6, generate approximate numerical values of the solution using $h = 0.2$ and twenty iterations ($n = 20$). In each of Problems 1 through 5, the problem can be solved exactly. Obtain this solution to compare approximate values at the x_k's with the exact solution values.

1. $y' = 3xy;\ y(0) = 5$
2. $y' = 2 - x;\ y(0) = 1$
3. $y' = y - \cos(x);\ y(1) = -2$
4. $y' = x + y;\ y(1) = -3$
5. $y' = y\sin(x);\ y(0) = 1$
6. $y' = x - y^2;\ y(0) = 4$

5.3 Taylor and Modified Euler Methods

We will develop two other numerical approximation schemes, both of which are (in general) more accurate than Euler's method.

Under certain conditions on f and h, we can use Taylor's theorem with remainder to write

$$y(x_{k+1}) = y(x_k) + hy'(x_k) + \frac{1}{2}h^2 y''(x_k) + \frac{1}{6}h^3 y^{(3)}(\xi_k)$$

for some ξ_k in $[x_k, x_{k+1}]$. If the third derivative of $y(x)$ is bounded, we can make the last term in this sum as small as we like by choosing h to be small enough, leading to the approximation

$$y_{k+1} \approx y(x_k) + hy'(x_k) + \frac{1}{2}h^2 y''(x_k). \tag{5.1}$$

Now, $y(x) = f(x, y(x))$. This suggests that in equation (5.1) we consider $f(x_k, y_k)$ as an approximation of $y'(x_k)$ if y_k is an approximation of $y(x_k)$. This leaves the term $y''(x_k)$ in equation (5.1) to approximate. To do this, differentiate the equation $y'(x) = f(x, y(x))$ with respect to x to get

$$y''(x) = \frac{\partial f}{\partial x}(x, y) + \frac{\partial f}{\partial y}(x, y)y'(x).$$

We are therefore led to approximate

$$y''(x_k) \approx \frac{\partial f}{\partial x}(x_k, y_k) + \frac{\partial f}{\partial y}(x_k, y_k)y'(x_k).$$

Insert these approximations of $y'(x_k)$ and $y''(x_k)$ into equation (5.1) to get

$$y_{k+1} \approx y_k + hf(x_k, y_k) + \frac{1}{2}h^2\left(\frac{\partial f}{\partial x}(x_k, y_k) + \frac{\partial f}{\partial y}(x_k, y_k)y'(x_k)\right).$$

The *second-order Taylor method* consists of using this expression to approximate $y(x_{k+1})$ by y_{k+1} We can simplify this expression for the approximate value of y_{k+1} by using the notation

$$f_k = f(x_k, y_k),$$

$$\frac{\partial f}{\partial x} = f_x, \quad \frac{\partial f}{\partial y} = f_y,$$

$$\frac{\partial f}{\partial x}(x_k, y_k) = f_{xk}, \quad \frac{\partial f}{\partial y}(x_k, y_k) = f_{yk}.$$

With this notation, the second-order Taylor approximation is

$$y_{k+1} \approx y_k + hf_k + \frac{1}{2}h^2(f_{xk} + f_k f_{yk}).$$

The second-order Taylor method is a *one-step method* because it approximates the solution value at x_k using the approximation made at x_{k-1}, which is just one step back. Euler's method is also one-step.

EXAMPLE 5.4

We will use the second-order Taylor method to approximate some solution values for $y' = y^2\cos(x)$; $y(0) = 1/5$. This problem can be solved exactly to obtain $y(x) = 1/(5 - \sin(x))$, so we can compare approximate values with exact values.

With $f(x, y) = y^2\cos(x)$, $f_x = -y^2\sin(x)$ and $f_y = 2y\cos(x)$. The second-order Taylor approximation formula is

$$y_{k+1} = y_k + hy_k^2\cos(x_k) + h^2y_k^2\cos^2(x_k) - \frac{1}{2}h^2y_k^2\sin(x_k).$$

Table 5.2 lists approximate values computed using $h = 0.2$ and $n = 20$. Values computed from the exact solution are included for comparison. ♦

Near the end of the nineteenth century, the German mathematician Karl Runge noticed a similarity between part of the formula for the second-order Taylor method and another Taylor polynomial approximation. Write this second-order Taylor formula as

$$y_{k+1} = y_k + h\left[f_k + \frac{1}{2}h(f_x(x_k, y_k) + f_kf_y(x_k, y_k))\right]. \tag{5.2}$$

Runge observed that the term in square brackets resembles the Taylor approximation

$$f(x_k + \alpha h, y_k + \beta h)) \approx f_k + \alpha hf_x(x_k, y_k) + \beta hf_y(x_k, y_k).$$

In fact, the term in square brackets in equation (5.2) is exactly the right side of the last equation with $\alpha = \beta = 1/2$. This suggests the following approximation scheme.

Use of the equation

$$y_{k+1} \approx y_k + hf\left(x_k + \frac{h}{2}, y_k + \frac{hf_k}{2}\right).$$

to approximate $y(x_{k+1})$ by y_{k+1} is called the *modified Euler method*. This method is in the spirit of Euler's method except that $f(x, y)$ is evaluated at $(x_k + h/2, y_k + hf_k/2)$ instead of at (x_k, y_k). Notice that $x_k + h/2$ is midway between x_k and $x_k + h$.

TABLE 5.2 ***Second-Order Taylor Method for $y' = y^2\cos(x)$; $y(0) = 1/5$***

x	Exact Value	Approximate Value	x	Exact Value	Approximate Value
0.0	0.2	0.2	2.2	0.2385778700	0.2389919589
0.2	0.2082755946	0.20832	2.4	0.2312386371	0.2315347821
0.4	0.2168923737	0.2170013470	2.6	0.2229903681	0.2231744449
0.6	0.2254609677	0.2256558280	2.8	0.2143617277	0.2144516213
0.8	0.2335006181	0.2337991830	3.0	0.2058087464	0.2058272673
1.0	0.2404696460	0.2408797598	3.2	0.197691800	0.1976613648
1.2	0.2458234042	0.2463364693	3.4	0.1902753647	0.1902141527
1.4	0.2490939041	0.2496815188	3.6	0.1837384003	0.1836603456
1.6	0.2499733530	0.2505900093	3.8	0.1781941060	0.1781084317
1.8	0.2483760942	0.2489684556	4.0	0.1737075401	0.1736197077
2.0	0.2444567851	0.2449763987			

TABLE 5.3 ***Modified Euler's Method Applied to $y' = y/x + 2x^2$; $y(1) = 4$***

x	$y(x)$	Approximate Solution	x	$y(x)$	Approximate Solution
1.0	4	4	3.0	36	35.87954731
1.2	5.328	5.320363636	3.2	42.368	42.23164616
1.4	6.944	6.927398601	3.4	49.504	49.35124526
1.6	8.896	8.869292639	3.6	57.496	57.28637379
1.8	11.232	11.19419064	3.8	66.272	66.08505841
2.0	14	13.95020013	4.0	76	75.79532194
2.2	17.248	17.18541062	4.2	86.688	86.46518560
2.4	21.024	20.94789549	4.4	98.384	98.14266841
2.6	25.376	25.25871247	4.6	111.136	110.8757877
2.8	30.352	30.24691542	4.8	124.992	124.7125592
			5.0	140	139.7009975

EXAMPLE 5.5

Consider the initial value problem

$$y' - \frac{1}{x}y = 2x^2;\ y(1) = 4.$$

Write the differential equation as

$$y' = \frac{1}{x}y + 2x^2 = f(x, y),$$

and use the Euler method with $h = 0.2$ and $n = 20$. Again, we have chosen a problem we can solve exactly, obtaining $y(x) = x^3 + 3x$. Table 5.3 lists the exact and approximate values for comparison. ♦

SECTION 5.3 PROBLEMS

In each of Problems 1 through 6, use the second-order Taylor method and the modified Euler method to approximate solution values, using $h = 0.2$ and $n = 20$. Problems 3 and 6 can be solved exactly. For these problems, list the exact solution values for comparison with the approximations.

1. $y' = \cos(y) + e^{-x}$; $y(0) = 1$
2. $y' = y^3 - 2xy$; $y(3) = 2$
3. $y' = -y + e^{-x}$; $y(0) = 4$
4. $y' = \sec(1/y) - xy^2$; $y(\pi/4) = 1$
5. $y' = \sin(x + y)$; $y(0) = 2$
6. $y' = y - x^2$; $y(1) = -4$

PART 2

Vectors, Linear Algebra, and Systems of Linear Differential Equations

VECTORS IN THE PLANE AND 3-SPACE THE DOT PRODUCT THE CROSS PRODUCT THE VECTOR SPACE R^N ORTHOGONALIZATION

CHAPTER 6

Vectors and Vector Spaces

6.1 Vectors in the Plane and 3-Space

Some quantities, such as temperature and mass, are completely specified by a number. Such quantities are called *scalars*. By contrast, a *vector* has both a magnitude and a sense of direction. If we push against an object, the effect is determined not only by the strength of the push, but its direction. Velocity and acceleration are also vectors.

We can include both both magnitude and direction in one package by representing a vector as an arrow from the origin to a point (x, y, z) in 3-space, as in Figure 6.1. The choice of the point gives the direction of the vector (when viewed from the origin), and the length is its magnitude. The greater the force, the longer the arrow representation.

To distinguish when we are thinking of a point as a vector (arrow from the origin to the point), we will denote this vector $< x, y, z >$. We call x the *first component of* $< x, y, z >$, y the *second component* and z the *third component*. These components are scalars.

Two vectors are equal exactly when their respective components are equal. That is,

$$< x_1, y_1, z_1 > = < x_2, y_2, z_2 >$$

exactly when $x_1 = x_2$, $y_1 = y_2$, and $z_1 = z_2$.

Since only direction and magnitude are important in specifying a vector, any arrow of the same length and orientation denotes the same vector. The arrows in Figure 6.2 represent the same vector.

The vector $< -x, -y, -z >$ is opposite in direction to $< x, y, z >$, as suggested in Figure 6.3.

It is convenient to denote vectors by bold-face letters (such as **F**, **G**, and **H**) and scalars (real numbers) in ordinary type.

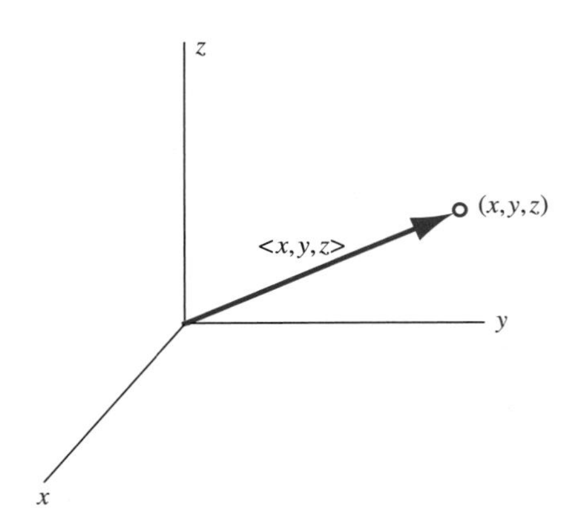

FIGURE 6.1 *Vector* $< x, y, z >$ *from the origin to the point* (x, y, z).

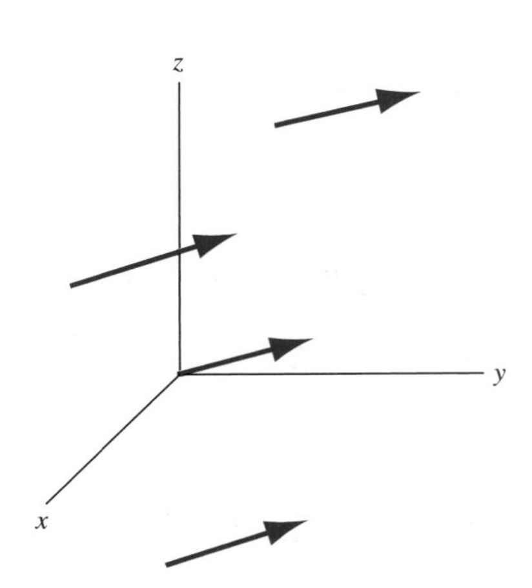

FIGURE 6.2 *Arrow representations of the same vector.*

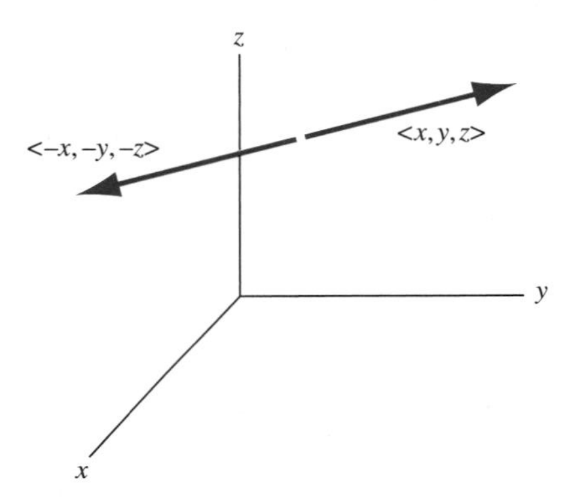

FIGURE 6.3 $< -x, -y, -z >$ *is opposite* $< x, y, z >$.

The *length* (also called the *magnitude* or *norm*) of a vector $\mathbf{F} = < x, y, z >$ is the scalar

$$\| \mathbf{F} \| = \sqrt{x^2 + y^2 + z^2}.$$

This is the distance from the origin to the point (x, y, z) and also the length of any arrow representing the vector $< x, y, z >$. For example, the norm of $\mathbf{G} = < -1, 4, 2 >$ is $\| \mathbf{G} \| = \sqrt{21}$, which is the distance from the origin to the point $(-1, 4, 2)$.

Multiply a vector $\mathbf{F} = < a, b, c >$ by a scalar α by multiplying each component of $\mathbf{F}$ by α. This produces a new vector denoted $\alpha\mathbf{F}$:

$$\alpha\mathbf{F} = < \alpha a, \alpha b, \alpha c > .$$

Then

$$\| \alpha\mathbf{F} \| = |\alpha| \, \| \mathbf{F} \|,$$

because

$$\begin{aligned}\| \alpha\mathbf{F} \| &= \sqrt{(\alpha a)^2 + (\alpha b)^2 + (\alpha c)^2} \\ &= \sqrt{(\alpha^2)(a^2+b^2+c^2)} = |\alpha|\sqrt{a^2+b^2+c^2} \\ &= |\alpha| \, \| \mathbf{F} \| .\end{aligned}$$

This means that the length of $\alpha\mathbf{F}$ is $|\alpha|$ times the length of $\mathbf{F}$. We may therefore think of multiplication of a vector by a scalar as a scaling (stretching or shrinking) operation. In particular, take the following cases:

- If $\alpha > 1$, then $\alpha\mathbf{F}$ is longer than $\mathbf{F}$ and in the same direction.
- If $0 < \alpha < 1$, then $\alpha\mathbf{F}$ is shorter than $\mathbf{F}$ and in the same direction.
- If $-1 < \alpha < 0$ then $\alpha\mathbf{F}$ is shorter than $\mathbf{F}$ and in the opposite direction.
- If $\alpha < -1$ then $\alpha\mathbf{F}$ is longer than $\mathbf{F}$ and in the opposite direction.
- If $\alpha = -1$ then $\alpha\mathbf{F}$ has the same length as $\mathbf{F}$, and exactly opposite the direction.

  For example, $\frac{1}{2}\mathbf{F}$ is a vector having the direction of $\mathbf{F}$ and half the length of $\mathbf{F}$, while $2\mathbf{F}$ has the direction of $\mathbf{F}$ and length twice that of $\mathbf{F}$, and $-\frac{1}{2}\mathbf{F}$ has direction opposite that of $\mathbf{F}$ and half the length.
- If $\alpha = 0$, then $\alpha\mathbf{F} = < 0, 0, 0 >$, which we call the *zero vector* and denote $\mathbf{O}$. This is the only vector with zero length and no direction, since it cannot be represented by an arrow.

Consistent with these interpretations of $\alpha\mathbf{F}$, we define two vectors $\mathbf{F}$ and $\mathbf{G}$ to be *parallel* if each is a nonzero scalar multiple of the other. Parallel vectors may differ in length and even be in opposite directions, but the straight lines through arrows representing them are parallel lines in 3-space.

We add two vectors by adding their respective components:
If $\mathbf{F} = < a_1, a_2, a_3 >$ and $\mathbf{G} = < b_1, b_2, b_3 >$, then

$$\mathbf{F} + \mathbf{G} = < a_1 + a_2, b_1 + b_2, c_1 + c_2 > .$$

Vector addition and multiplication by scalars have the following properties:

1. $\mathbf{F} + \mathbf{G} = \mathbf{G} + \mathbf{F}$. (commutativity)
2. $\mathbf{F} + (\mathbf{G} + \mathbf{H}) = (\mathbf{F} + \mathbf{G}) + \mathbf{H}$. (associativity)
3. $\mathbf{F} + \mathbf{O} = \mathbf{F}$.
4. $\alpha(\mathbf{F} + \mathbf{G}) = \alpha\mathbf{F} + \alpha\mathbf{G}$.
5. $(\alpha\beta)\mathbf{F} = \alpha(\beta\mathbf{F})$.
6. $(\alpha + \beta)\mathbf{F} = \alpha\mathbf{F} + \beta\mathbf{F}$.

It is sometimes useful to represent vector addition by the *parallelogram law*. If $\mathbf{F}$ and $\mathbf{G}$ are drawn as arrows from the same point, they form two sides of a parallelogram. The arrow along the diagonal of this parallelogram represents the sum $\mathbf{F} + \mathbf{G}$ (Figure 6.4). Because any arrows having the same lengths and direction represent the same vector, we can also draw the arrows in $\mathbf{F} + \mathbf{G}$ (as in Figure 6.5) with $\mathbf{G}$ drawn from the tip of $\mathbf{F}$. This still puts $\mathbf{F} + \mathbf{G}$ along the diagonal of the parallelogram.

The triangle of Figure 6.5 also suggests an important inequality involving vector sums and lengths. This triangle has sides of length $\| \mathbf{F} \|$, $\| \mathbf{G} \|$, and $\| \mathbf{F} + \mathbf{G} \|$. Because the sum of the

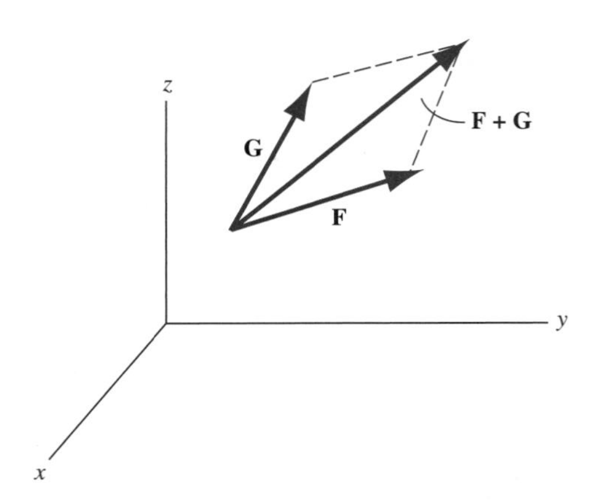

FIGURE 6.4 *Parallelogram law for vector addition.*

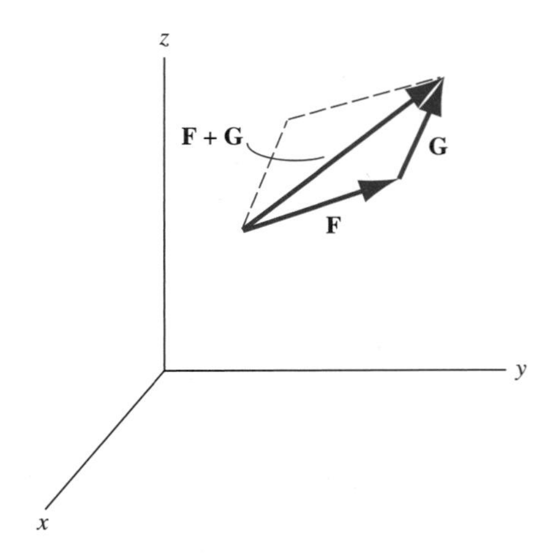

FIGURE 6.5 *Alternative view of the parallelogram law.*

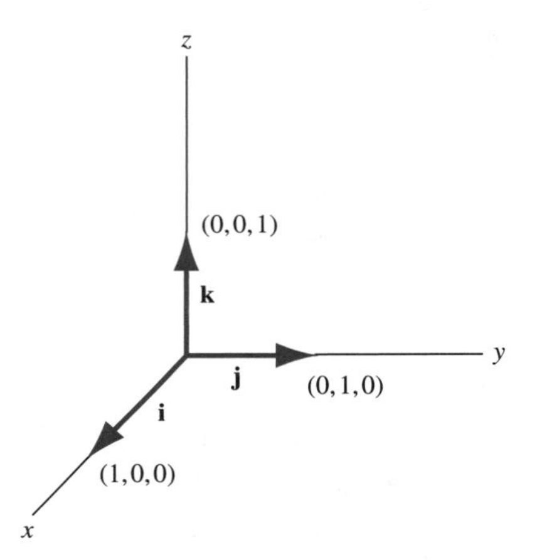

FIGURE 6.6 *Unit vectors* **i**, **j**, *and* **k**.

lengths of any two sides of a triangle must be at least as great as the length of the third side, we have the *triangle inequality*

$$\| \mathbf{F} + \mathbf{G} \| \leq \| \mathbf{F} \| + \| \mathbf{G} \| .$$

A vector of length 1 is called a *unit vector*. The unit vectors along the positive axes are shown in Figure 6.6 and are labeled

$$\mathbf{i} = < 1, 0, 0 >, \ \mathbf{j} = < 0, 1, 0 >, \ \mathbf{k} = < 0, 0, 1 > .$$

We can write any vector $\mathbf{F} = < a, b, c >$ as

$$\begin{aligned} \mathbf{F} = < a, b, c > &= a < 1, 0, 0 > + b < 0, 1, 0 > + c < 0, 0, 1 > \\ &= a\mathbf{i} + b\mathbf{j} + c\mathbf{k}. \end{aligned}$$

We call $a\mathbf{i}+b\mathbf{j}+c\mathbf{k}$ the *standard representation* of $\mathbf{F}$. When a component of a vector is zero, we usually just omit this term in the standard representation. For example, we would usually write $\mathbf{F} = < -8, 0, 3 >$ as $-8\mathbf{i}+3\mathbf{k}$ instead of $-8\mathbf{i}+0\mathbf{j}+3\mathbf{k}$.

If a vector is represented by an arrow in the x, y-plane, we often omit the third coordinate and use $\mathbf{i} = < 1, 0 >$ and $\mathbf{j} = < 0, 1 >$. For example, the vector $\mathbf{V}$ from the origin to the point $< 2, -6, 0 >$ can be represented as an arrow from the origin to the point $(2, -6)$ in the x, y-plane and can be written in standard form as

$$\mathbf{V} = 2\mathbf{i} - 6\mathbf{j}$$

where $\mathbf{i} = < 1, 0 >$ and $\mathbf{j} = < 0, 1 >$.

It is often useful use to know the components of the vector $\mathbf{V}$ represented by the arrow from one point to another, say from $P_0 = (x_0, y_0, z_0)$ to $P_1 : (x_1, y_1, z_1)$. Denote

$$\mathbf{G} = x_0\mathbf{i} + y_0\mathbf{j} + z_0\mathbf{k} \text{ and } \mathbf{F} = x_1\mathbf{i} + y_1\mathbf{j} + z_1\mathbf{k}.$$

By the parallelogram law in Figure 6.7, the vector $\mathbf{V}$ we want satisfies

$$\mathbf{G} + \mathbf{V} = \mathbf{F}.$$

Therefore,

$$\mathbf{V} = \mathbf{F} - \mathbf{G} = (x_1 - x_0)\mathbf{i} + (y_1 - y_0)\mathbf{j} + (z_1 - z_0)\mathbf{k}.$$

For example, the vector represented by the arrow from $(-1, 6, 3)$ to $(9, -1, -7)$ if $10\mathbf{i} - 7\mathbf{j} - 10\mathbf{k}$.

Using this idea, we can find a vector of any length in any given direction. For example, suppose we want a vector of length 7 in the direction from $(-1, 6, 5)$ to $(-8, 4, 9)$.

The strategy is to first find a unit vector in the given direction, then multiply it by 7 to obtain a vector of length 7 in that direction. The vector $\mathbf{V} = -7\mathbf{i} - 2\mathbf{j} + 4\mathbf{k}$ is in the direction from $(-1, 6, 5)$ to $(-8, 4, 9)$. Since $\| \mathbf{V} \| = \sqrt{69}$, a unit vector in this direction is

$$\mathbf{F} = \frac{1}{\| \mathbf{V} \|}\mathbf{V} = \frac{1}{\sqrt{69}}\mathbf{V}.$$

Then

$$7\mathbf{F} = \frac{7}{\sqrt{69}}(-7\mathbf{i} - 2\mathbf{j} + 4\mathbf{k})$$

has length 7 and is in the direction from $(-1, 6, 5)$ to $(-8, 4, 9)$.

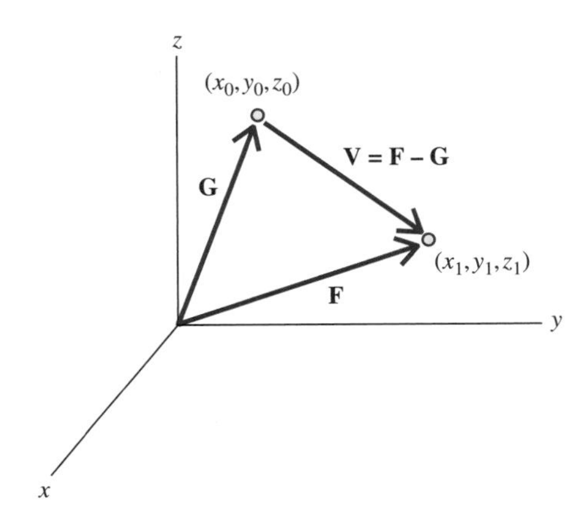

FIGURE 6.7 *Vector from* (x_0, y_0, z_0) *to* (x_1, y_1, z_1).

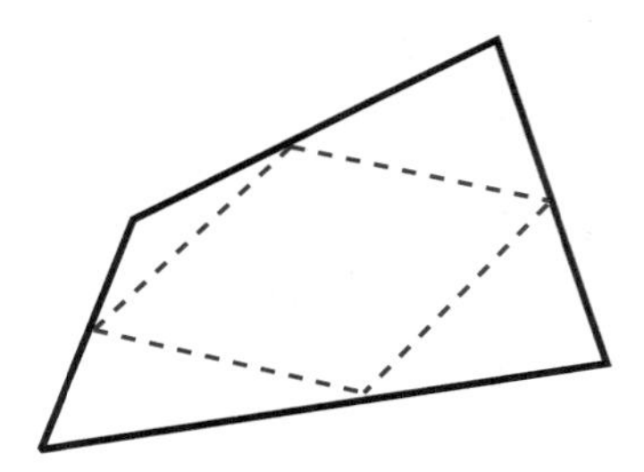

FIGURE 6.8 *Quadrilateral with lines connecting successive midpoints.*

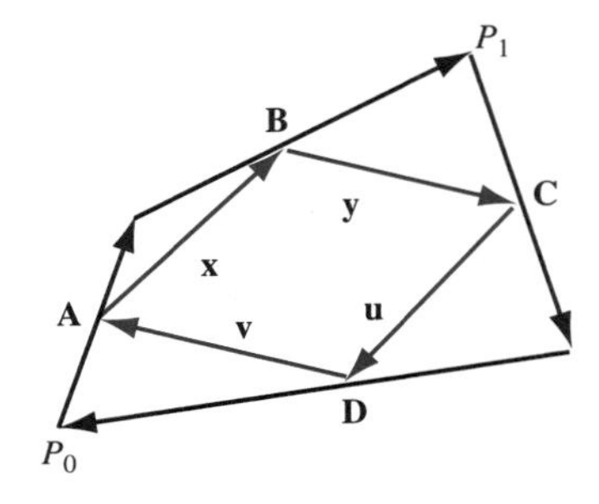

FIGURE 6.9 *Quadrilateral of Figure 6.8 with vectors as sides.*

As an example of the efficiency of vector notation, we will derive a fact about quadrilaterals: the lines formed by connecting successive midpoints of the sides of a quadrilateral form a parallelogram. Figures 6.8 and 6.9 illustrate what we want to show. Draw the quadrilateral again with vectors **A**, **B**, **C**, and **D** as the sides (Figure 6.9). The vectors **x**, **y**, **u**, and **v** connect the midpoints of successive sides. We want to show that **x** and **u** are parallel and of the same length, and the same for **y** and **v**. From the parallelogram law and the choices of these vectors,

$$\mathbf{x} = \frac{1}{2}\mathbf{A} + \frac{1}{2}\mathbf{B}$$

and

$$\mathbf{u} = \frac{1}{2}\mathbf{C} + \frac{1}{2}\mathbf{D}.$$

But also by the parallelogram law, $\mathbf{C} + \mathbf{D}$ is the vector from P_1 to P_0, while $\mathbf{A} + \mathbf{B}$ is the vector from P_0 to P_1. These vectors have the same lengths and opposite directions, so

$$\mathbf{A} + \mathbf{B} = -(\mathbf{C} + \mathbf{D}).$$

Then $\mathbf{x} = -\mathbf{u}$, so these vectors are parallel and of the same length (just opposite in direction). Similarly, **y** and **v** are parallel and of the same length.

Equation of a Line in 3-Space

We will show how to find parametric equations of a line L in 3-space containing two given points. This is more subtle than the corresponding problem in the plane, because there is no slope to exploit. To illustrate the idea, suppose the points are $(-2, -4, 7)$ and $(9, 1, -7)$. Form a vector between these two points (in either order). The arrow from the first to the second point represents the vector

$$\mathbf{V} = 11\mathbf{i} + 5\mathbf{j} - 14\mathbf{k}.$$

Because P_0 and P_1 are on L, $\mathbf{V}$ is parallel to L, hence to any other vector aligned with L. Now suppose (x, y, z) is any point on L. Then the vector $(x+2)\mathbf{i} + (y+4)\mathbf{j} + (z-7)\mathbf{k}$ from $(-2, -4, 7)$ to (x, y, z) is also parallel to L, hence to $\mathbf{V}$. This vector must therefore be a scalar multiple of $\mathbf{V}$:

$$\begin{aligned}(x+2)\mathbf{i} + (y+4)\mathbf{j} + (z-7)\mathbf{k} &= t\mathbf{V}\\ &= 11t\mathbf{i} + 5t\mathbf{j} - 14t\mathbf{k}\end{aligned}$$

for some scalar t. Since two vectors are equal only when their respective components are equal,

$$x + 2 = 11t,\ y + 4 = 5t,\ z - 7 = -14t.$$

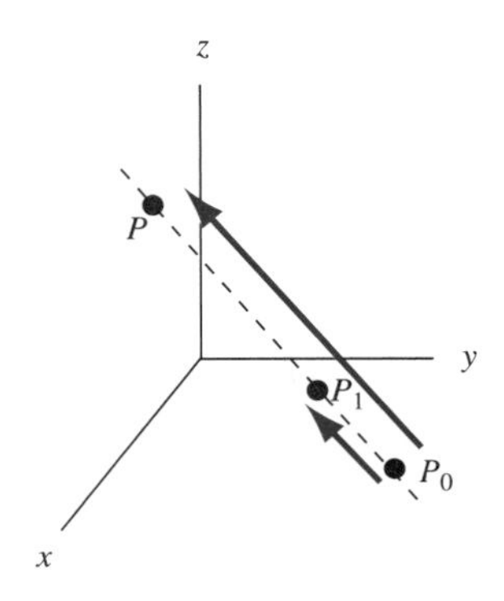

FIGURE 6.10 *Determining parametric equations of a line.*

Usually we write these equations as

$$x = -2 + 11t, y = -4 + 5t, z = 7 - 14t.$$

These are *parametric equations* of L. As t varies over the real numbers, the point $(-2 + 11t, -4 + 5t, 7 - 14t)$ varies over L. We obtain $(-2, -4, 7)$ when $t = 0$ and $(9, 1, -7)$ when $t = 1$.

The reasoning used in this example can be carried out in general. Suppose we are given points $P_0 : (x_0, y_0, z_0)$ and $P_1 : (x_1, y_1, z_1)$, and we want parametric equations of the line L through P_0 and P_1. The vector

$$(x_1 - x_0)\mathbf{i} + (y_1 - y_0)\mathbf{j} + (z_1 - z_0)\mathbf{k}$$

is along L, as is the vector

$$(x - x_0)\mathbf{i} + (y - y_0)\mathbf{j} + (z - z_0)\mathbf{k}$$

from P_0 to an arbitrary point (x, y, z) on L. These vectors (see Figure 6.10), being both along L, are parallel, hence for some real t,

$$\begin{aligned}&(x - x_0)\mathbf{i} + (y - y_0)\mathbf{j} + (z - z_0)\mathbf{k}\\&= t[(x_1 - x_0)\mathbf{i} + (y_1 - y_0)\mathbf{j} + (z_1 - z_0)\mathbf{k}].\end{aligned}$$

Then

$$x - x_0 = t(x_1 - x_0), y - y_0 = t(y_1 - y_0), z - z_0 = t(z_1 - z_0).$$

Parametric equations of the line are

$$x = x_0 + t(x_1 - x_0), y = y_0 + t(y_1 - y_0), z = z_0 + t(z_1 - z_0),$$

with t taking on all real values. We get P_0 when $t = 0$ and P_1 when $t = 1$.

EXAMPLE 6.1

Find parametric equations of the line through $(-1, -1, 7)$ and $(7, -1, 4)$.

Let one of these points be P_0 and the other P_1. To be specific, choose $P_0 = (-1, -1, 7) = (x_0, y_0, z_0)$ and $P_1 = (7, -1, 4) = (x_1, y_1, z_1)$. The line through these points has parametric equations

$$x = -1 + (7 - (-1))t, y = -1 + (-1 - (-1))t, z = 7 + (4 - 7)t$$

for t real. These parametric equations are

$$x = -1 + 8t,\ y = -1,\ z = 7 - 3t$$

for t real. We obtain P_0 when $t = 0$ and P_1 when $t = 1$. In this example, the y-coordinate of every point on the line is -1, so the line is in the plane $y = -1$.

We may also say that this line consists of all points $(-1 + 8t, -1, 7 - 3t)$ for t real. ♦

SECTION 6.1 PROBLEMS

In each of Problems 1 through 5, compute $\mathbf{F} + \mathbf{G}$, $\mathbf{F} - \mathbf{G}$, $2\mathbf{F}$, $3\mathbf{G}$, and $\| \mathbf{F} \|$.

1. $\mathbf{F} = 2\mathbf{i} - 5\mathbf{j}, \mathbf{G} = \mathbf{i} + 5\mathbf{j} - \mathbf{k}$
2. $\mathbf{F} = \sqrt{2}\mathbf{i} - \mathbf{j} - 6\mathbf{k}, \mathbf{G} = 8\mathbf{i} + 2\mathbf{k}$
3. $\mathbf{F} = \mathbf{i} + \mathbf{j} + \mathbf{k}, \mathbf{G} = 2\mathbf{i} - 2\mathbf{j} + 2\mathbf{k}$
4. $\mathbf{F} = \mathbf{i} - 3\mathbf{k}, \mathbf{G} = 4\mathbf{j}$
5. $\mathbf{F} = 2\mathbf{i} - 3\mathbf{j} + 5\mathbf{k}, \mathbf{G} = \sqrt{2}\mathbf{i} + 6\mathbf{j} - 5\mathbf{k}$

In each of Problems 6 through 9, find a vector having the given length and in the direction from the first point to the second.

6. $12, (-4, 5, 1), (6, 2, -3)$
7. $4, (0, 0, 1), (-4, 7, 5)$
8. $5, (0, 1, 4), (-5, 2, 2)$
9. $9, (1, 2, 1), (-4, -2, 3)$

In each of Problems 10 through 15, find the parametric equations of the line containing the given points.

10. $(1, 0, -4), (-2, -2, 5)$
11. $(0, 1, 3), (0, 0, 1)$
12. $(1, 0, 4), (2, 1, 1)$
13. $(2, -3, 6), (-1, 6, 4)$
14. $(2, 1, 1), (2, 1, -2)$
15. $(3, 0, 0), (-3, 1, 0)$

6.2 The Dot Product

The *dot product* $\mathbf{F} \cdot \mathbf{G}$ of $\mathbf{F}$ and $\mathbf{G}$ is the real number formed by multiplying the two first components, then the two second components, then the two third components, and adding these three numbers. If $\mathbf{F} = a_1\mathbf{i} + b_1\mathbf{j} + c_1\mathbf{k}$ and $\mathbf{G} = a_2\mathbf{i} + b_2\mathbf{j} + c_2\mathbf{k}$, then

$$\mathbf{F} \cdot \mathbf{G} = a_1a_2 + b_1b_2 + c_1c_2.$$

Again, this dot product is a number, not a vector. For example,

$$(\sqrt{3}\mathbf{i} + 4\mathbf{j} - \pi\mathbf{k}) \cdot (-2\mathbf{i} + 6\mathbf{j} + 3\mathbf{k}) = -2\sqrt{3} + 24 - 3\pi.$$

The dot product has the following properties.

1. $\mathbf{F} \cdot \mathbf{G} = \mathbf{G} \cdot \mathbf{F}$.
2. $(\mathbf{F} + \mathbf{G}) \cdot \mathbf{H} = \mathbf{F} \cdot \mathbf{H} + \mathbf{G} \cdot \mathbf{H}$.
3. $\alpha(\mathbf{F} \cdot \mathbf{G}) = (\alpha\mathbf{F}) \cdot \mathbf{G} = \mathbf{F} \cdot (\alpha\mathbf{G})$.
4. $\mathbf{F} \cdot \mathbf{F} = \| \mathbf{F} \|^2$.
5. $\mathbf{F} \cdot \mathbf{F} = 0$ if and only if $\mathbf{F} = \mathbf{O}$.
6. $\| \alpha\mathbf{F} + \beta\mathbf{G} \|^2 = \alpha^2 \| \mathbf{F} \|^2 + 2\alpha\beta\mathbf{F} \cdot \mathbf{G} + \beta^2 \| \mathbf{G} \|^2$.

Dot products of vectors can be computed using MAPLE and the `DotProduct` command, which is in the `VectorCalculus` package of subroutines. This command also applies to n-dimensional vectors, which are introduced in Section 6.4.

Conclusions (1), (2), and (3) are routine computations. Conclusion (4) is often used in computations. To verify conclusion (4), suppose

$$\mathbf{F} = a\mathbf{i} + b\mathbf{j} + c\mathbf{k}.$$

Then

$$\mathbf{F} \cdot \mathbf{F} = a^2 + b^2 + c^2 = \| \mathbf{F} \|^2.$$

Conclusion (5) follows easily from (4), since $\mathbf{O}$ is the only vector having length 0. For conclusion (6), use conclusions (1) through (4) to write

$$\begin{aligned}\| \alpha\mathbf{F} + \beta\mathbf{G} \|^2 &= (\alpha\mathbf{F} + \beta\mathbf{G}) \cdot (\alpha\mathbf{F} + \beta\mathbf{G}) \\ &= \alpha^2\mathbf{F} \cdot \mathbf{F} + \alpha\beta\mathbf{F} \cdot \mathbf{G} + \alpha\beta\mathbf{G} \cdot \mathbf{F} + \beta^2\mathbf{G} \cdot \mathbf{G} \\ &= \alpha^2 \| \mathbf{F} \|^2 + 2\alpha\beta\mathbf{F} \cdot \mathbf{G} + \beta^2 \| \mathbf{G} \|^2.\end{aligned}$$

The dot product can be used to find an angle between two vectors. Recall the law of cosines: For the upper triangle of Figure 6.11 with θ being the angle opposite the side of length c, the law of cosines states that

$$a^2 + b^2 - 2ab\cos(\theta) = c^2.$$

Apply this to the vector triangle of Figure 6.11 (lower), which has sides of length $a = \| \mathbf{G} \|$, $b = \| \mathbf{F} \|$, and $c = \| \mathbf{G} - \mathbf{F} \|$. Using property (6) of the dot product, we obtain

$$\begin{aligned}\| \mathbf{G} \|^2 + \| \mathbf{F} \|^2 - 2 \| \mathbf{F} \| \| \mathbf{G} \| \cos(\theta) &= \| \mathbf{G} - \mathbf{F} \|^2 \\ &= \| \mathbf{G} \|^2 + \| \mathbf{F} \|^2 - 2\mathbf{G} \cdot \mathbf{F}.\end{aligned}$$

Assuming that neither $\mathbf{F}$ nor $\mathbf{G}$ is the zero vector, this gives us

$$\cos(\theta) = \frac{\mathbf{F} \cdot \mathbf{G}}{\| \mathbf{F} \| \| \mathbf{G} \|}. \tag{6.1}$$

Since $|\cos(\theta)| \leq 1$ for all θ, equation (6.1) implies the *Cauchy-Schwarz inequality*:

$$|\mathbf{F} \cdot \mathbf{G}| \leq \| \mathbf{F} \| \| \mathbf{G} \|.$$

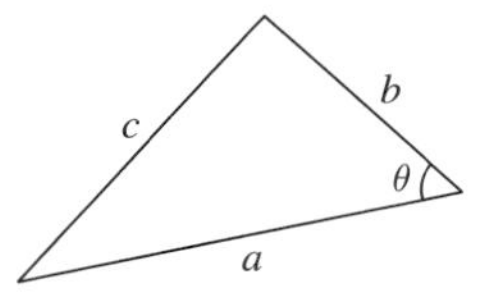

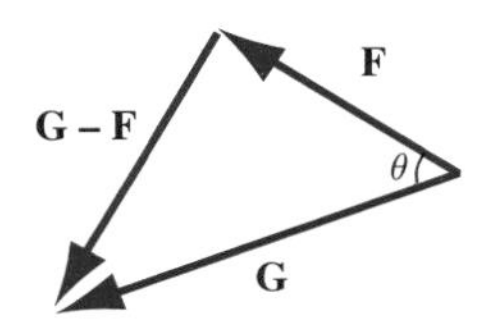

FIGURE 6.11 *The law of cosines and the angle between vectors.*

EXAMPLE 6.2

The angle θ between $\mathbf{F} = -\mathbf{i} + 3\mathbf{j} + \mathbf{k}$ and $\mathbf{G} = 2\mathbf{j} - 4\mathbf{k}$ is given by

$$\cos(\theta) = \frac{(-\mathbf{i} + 3\mathbf{j} + \mathbf{k}) \cdot (2\mathbf{j} - 4\mathbf{k})}{\| -\mathbf{i} + 3\mathbf{j} + \mathbf{k} \| \| 2\mathbf{j} - 4\mathbf{k} \|}$$
$$= \frac{(-1)(0) + (3)(2) + (1)(-4)}{\sqrt{1^2 + 3^2 + 1^2}\sqrt{2^2 + 4^2}} = \frac{2}{\sqrt{220}}.$$

Then $\theta \approx 1.436$ radians. ♦

EXAMPLE 6.3

Lines L_1 and L_2 have parametric equations

$$L_1 : x = 1 + 6t,\ y = 2 - 4t,\ z = -1 + 3t$$

and

$$L_2 : x = 4 - 3p,\ y = 2p,\ z = -5 + 4p.$$

The parameters t and p can take on any real values. We want an angle θ between these lines.

The strategy is to take a vector $\mathbf{V}_1$ along L_1 and a vector $\mathbf{V}_2$ along L_2 and find the angle between these vectors. For $\mathbf{V}_1$, find two points on L_1, say $(1, 2, -1)$ when $t = 0$ and $(7, -2, 2)$ when $t = 1$, and form

$$\mathbf{V}_1 = (7 - 1)\mathbf{i} + (-2 - 2)\mathbf{j} + (2 - (-1))\mathbf{k} = 6\mathbf{i} - 4\mathbf{j} + 3\mathbf{k}.$$

On L_2, take $(4, 0, -5)$ with $p = 0$ and $(1, 2, -1)$ with $p = 1$, forming

$$\mathbf{V}_2 = 3\mathbf{i} - 2\mathbf{j} - 4\mathbf{k}.$$

Now compute

$$\cos(\theta) = \frac{6(3) - 4(-2) + 3(-4)}{\sqrt{36 + 16 + 9}\sqrt{9 + 4 + 16}} = \frac{14}{\sqrt{1769}}.$$

An angle between L_1 and L_2 is $\arccos(14/\sqrt{1769})$, which is approximately 1.23 radians. ♦

Two nonzero vectors $\mathbf{F}$ and $\mathbf{G}$ are *orthogonal* (perpendicular) when the angle θ between them is $\pi/2$ radians. This happens exactly when

$$\cos(\theta) = 0 = \frac{\mathbf{F} \cdot \mathbf{G}}{\| \mathbf{F} \| \| \mathbf{G} \|}$$

which occurs when $\mathbf{F} \cdot \mathbf{G} = 0$. It is convenient to also agree that $\mathbf{O}$ is orthogonal to every vector. With this convention, two vectors are orthogonal if and only if their dot product is zero.

EXAMPLE 6.4

Let $\mathbf{F} = -4\mathbf{i} + \mathbf{j} + 2\mathbf{k}$, $\mathbf{G} = 2\mathbf{i} + 4\mathbf{k}$ and $\mathbf{H} = 6\mathbf{i} - \mathbf{j} - 2\mathbf{k}$. Then $\mathbf{F} \cdot \mathbf{G} = 0$, so $\mathbf{F}$ and $\mathbf{G}$ are orthogonal. But $\mathbf{F} \cdot \mathbf{H}$ and $\mathbf{G} \cdot \mathbf{H}$ are not zero, so $\mathbf{F}$ and $\mathbf{H}$ are not orthogonal and $\mathbf{G}$ and $\mathbf{H}$ are not orthogonal. ♦

Property (6) of the dot product has a particularly simple form when the vectors are orthogonal. In this case, $\mathbf{F} \cdot \mathbf{G} = 0$, and upon setting $\alpha = \beta = 1$, we have

$$\| \mathbf{F} + \mathbf{G} \|^2 = \| \mathbf{F} \|^2 + \| \mathbf{G} \|^2.$$

This is the familiar *Pythagorean theorem*, because the vectors **F** and **G** form the sides of a right triangle with hypotenuse $\mathbf{F}+\mathbf{G}$ (imagine Figure 6.5 with **F** and **G** forming a right angle).

EXAMPLE 6.5

Suppose two lines are defined parametrically by

$$L_1 : x = 2 - 4t,\ y = 6 + t,\ z = 3t$$

and

$$L_2 : x = -2 + p,\ y = 7 + 2p,\ z = 3 - 4p.$$

We want to know if these lines are orthogonal. Note that the question makes sense even if L_1 and L_2 do not intersect.

The idea is to form a vector along each line and test these vectors for orthogonality. For a vector along L_1, take two points on this line, say $(2, 6, 0)$ when $t = 0$ and $(-2, 7, 3)$ when $t = 1$. Then $\mathbf{V}_1 = -4\mathbf{i} + \mathbf{j} + 3\mathbf{k}$ is parallel to L_1. Similarly, $(-2, 7, 3)$ is on L_2 when $p = 0$, and $(-1, 9, -1)$ is on L_2 when $p = 1$, so $\mathbf{V}_2 = \mathbf{i} + 2\mathbf{j} - 4\mathbf{k}$ is parallel to L_2. Compute $\mathbf{V}_1 \cdot \mathbf{V}_2 = -14 \neq 0$. Therefore, L_1 and L_2 are not orthogonal. ♦

Orthogonality is also useful for determining the equation of a plane in 3-space. Any plane has an equation of the form

$$ax + by + cz = d.$$

As suggested by Figure 6.12, if we specify a point on the plane and a vector orthogonal to the plane, then the plane is completely determined. Example 6.6 suggests a strategy for finding the equation of this plane.

EXAMPLE 6.6

We will find the equation of the plane Π containing the point $(-6, 1, 1)$ and orthogonal to the vector $\mathbf{N} = -2\mathbf{i} + 4\mathbf{j} + \mathbf{k}$. Such a vector **N** is said to be *normal* to Π and is called a *normal vector* to Π.

Here is a strategy. Because $(-6, 1, 1)$ is on Π, a point (x, y, z) is on Π exactly when the vector between $(-6, 1, 1)$ and (x, y, z) lies in Π. But then $(x+6)\mathbf{i} + (y-1)\mathbf{j} + (z-1)\mathbf{k}$ must be orthogonal to **N**, so

$$\mathbf{N} \cdot ((x+6)\mathbf{i} + (y-1)\mathbf{j} + (z-1)\mathbf{k}) = 0.$$

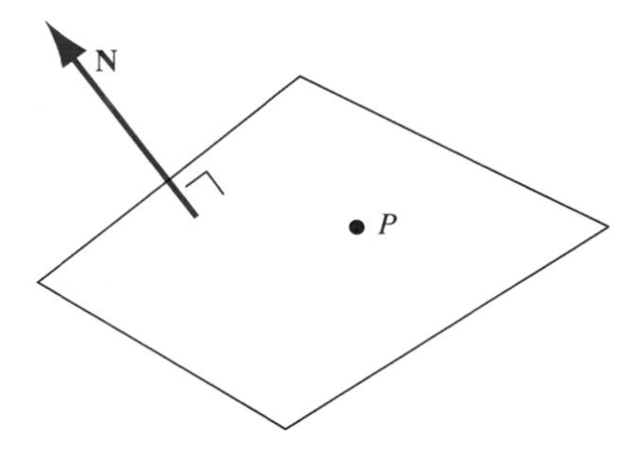

FIGURE 6.12 *A point P and a normal vector **N** determine a plane.*

Then

$$-2(x+6)+4(y-1)+(z-1)=0,$$

or

$$-2x+4y+z=17.$$

This is the equation of Π. ♦

Following this reasoning in general, the equation of a plane containing a point $P_0:(x_0, y_0, z_0)$ and having a normal vector $\mathbf{N}=a\mathbf{i}+b\mathbf{j}+c\mathbf{k}$ is

$$\mathbf{N}\cdot[(x-x_0)\mathbf{i}+(y-y_0)\mathbf{j}+(z-z_0)\mathbf{k}]=0$$

or

$$a(x-x_0)+b(y-y_0)+c(z-z_0)=0. \tag{6.2}$$

It is also sometimes convenient to notice that the vector $a\mathbf{i}+b\mathbf{j}+c\mathbf{k}$ is always a normal vector to a plane $ax+by+cz=d$, for any d. Changing the value of d moves the plane in 3-space but does not change its orientation with respect to the axes, so the normal vector remains the same and is determined by the coefficients a, b, and c only.

Another use for the dot product is in forming vector projections.

Let **u** and **v** be given, nonzero vectors, represented as arrows from a common point (for convenience). The *projection* of **v** onto **u** is a vector $\text{proj}_{\mathbf{u}}\mathbf{v}$ in the direction of **u** having magnitude equal to the length of the perpendicular projection of the arrow representing **v** onto the line along the arrow representing **u**. This projection is done by constructing a perpendicular line from the tip of **v** onto the line through **u**. The base of the right triangle having **v** as hypotenuse is the length d of $\text{proj}_{\mathbf{u}}\mathbf{v}$ (Figure 6.13).

If θ is the angle between **u** and **v**, then

$$\cos(\theta)=\frac{d}{\|\mathbf{v}\|}.$$

Then

$$d=\|\mathbf{v}\|\cos(\theta)=\|\mathbf{v}\|\frac{\mathbf{u}\cdot\mathbf{v}}{\|\mathbf{u}\|\|\mathbf{v}\|}=\frac{\mathbf{u}\cdot\mathbf{v}}{\|\mathbf{u}\|}.$$

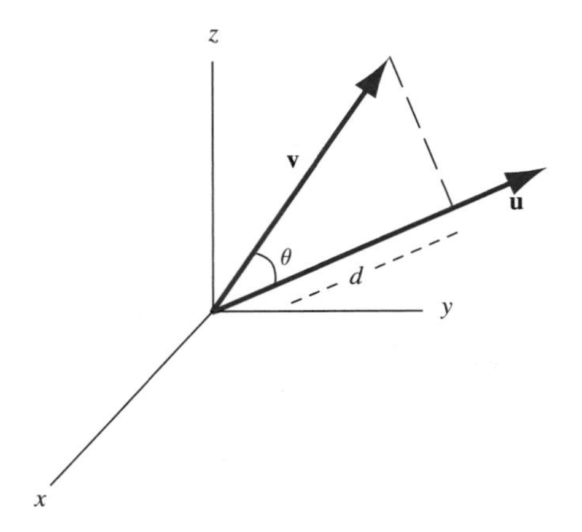

FIGURE 6.13 *Orthogonal projection of* **v** *onto* **u**.

To obtain a vector in the direction of **u** and of length d, divide **u** by its length to obtain a unit vector, then multiply this vector by d. Therefore,

$$\text{proj}_{\mathbf{u}}\mathbf{v} = d\left(\frac{\mathbf{u}}{\|\mathbf{u}\|}\right) = \frac{\mathbf{u}\cdot\mathbf{v}}{\|\mathbf{u}\|^2}\mathbf{u}.$$

As an example, suppose $\mathbf{v} = 4\mathbf{i} - \mathbf{j} + 2\mathbf{k}$ and $\mathbf{u} = \mathbf{i} - \mathbf{j} + 2\mathbf{k}$. Then

$$\mathbf{u}\cdot\mathbf{v} = 9 \text{ and } \|\mathbf{u}\|^2 = 6,$$

so

$$\text{proj}_{\mathbf{u}}\mathbf{v} = \frac{9}{6}\mathbf{u} = \frac{3}{2}(\mathbf{i} - \mathbf{j} + 2\mathbf{k}).$$

If we think of these vectors as forces, we may interpret $\text{proj}_{\mathbf{u}}\mathbf{v}$ as the effect of **v** in the direction of **u**.

SECTION 6.2 PROBLEMS

In each of Problems 1 through 6, compute the dot product of the vectors and the cosine of the angle between them. Also determine if the vectors are orthogonal.

1. $-4\mathbf{i} - 2\mathbf{i} + 3\mathbf{k}, 6\mathbf{i} - 2\mathbf{j} - \mathbf{k}$
2. $\mathbf{i} + \mathbf{j} + 2\mathbf{k}, \mathbf{i} - \mathbf{j} + 2\mathbf{k}$
3. $\mathbf{i} - 3\mathbf{k}, 2\mathbf{j} + 6\mathbf{k}$
4. $2\mathbf{i} - 6\mathbf{j} + \mathbf{k}, \mathbf{i} - \mathbf{j}$
5. $\mathbf{i}, 2\mathbf{i} - 3\mathbf{j} + \mathbf{k}$
6. $8\mathbf{i} - 3\mathbf{j} + 2\mathbf{k}, -8\mathbf{i} - 3\mathbf{j} + \mathbf{k}$

In each of Problems 7 through 12, find the equation of the plane containing the given point and orthogonal to the given vector.

7. $(2, -3, 4), 8\mathbf{i} - 6\mathbf{j} + 4\mathbf{k}$
8. $(-2, 1, -1), 4\mathbf{i} + 3\mathbf{j} + \mathbf{k}$
9. $(0, -1, 4), 7\mathbf{i} + 6\mathbf{j} - 5\mathbf{k}$
10. $(-1, 0, 0), \mathbf{i} - 2\mathbf{j}$
11. $(-1, 1, 2), 3\mathbf{i} - \mathbf{j} + 4\mathbf{k}$
12. $(-1, -1, -5), -3\mathbf{i} + 2\mathbf{j}$

In each of Problems 13, 14, and 15, find the projection of **v** onto **u**.

13. $\mathbf{v} = -\mathbf{i} + 3\mathbf{j} + 6\mathbf{k}, \mathbf{u} = 2\mathbf{i} + 7\mathbf{j} - 3\mathbf{k}$
14. $\mathbf{v} = 5\mathbf{i} + 2\mathbf{j} - 3\mathbf{k}, \mathbf{u} = \mathbf{i} - 5\mathbf{j} + 2\mathbf{k}$
15. $\mathbf{v} = \mathbf{i} - \mathbf{j} + 4\mathbf{k}, \mathbf{u} = -3\mathbf{i} + 2\mathbf{j} - \mathbf{k}$

6.3 The Cross Product

The dot product produces a scalar from two vectors. The cross product produces a vector from two vectors.

Let $\mathbf{F} = a_1\mathbf{i} + b_1\mathbf{j} + c_1\mathbf{k}$ and $\mathbf{G} = a_2\mathbf{i} + b_2\mathbf{j} + c_2\mathbf{k}$. The *cross product* of **F** with **G** is the vector $\mathbf{F} \times \mathbf{G}$ defined by

$$\mathbf{F} \times \mathbf{G} = (b_1c_2 - b_2c_1)\mathbf{i} + (a_2c_1 - a_1c_2)\mathbf{j} + (a_1b_2 - a_2b_1)\mathbf{k}.$$

Here is a simple device for remembering and computing these components. Form the determinant

$$\begin{vmatrix} \mathbf{i} & \mathbf{j} & \mathbf{k} \\ a_1 & b_1 & c_1 \\ a_2 & b_2 & c_2 \end{vmatrix}$$

having the standard unit vectors in the first row, the components of **F** in the second row, and the components of **G** in the third row. If this determinant is expanded by the first row, we get exactly $\mathbf{F} \times \mathbf{G}$:

$$\begin{vmatrix} \mathbf{i} & \mathbf{j} & \mathbf{k} \\ a_1 & b_1 & c_1 \\ a_2 & b_2 & c_2 \end{vmatrix}$$

$$= \begin{vmatrix} b_1 & c_1 \\ b_2 & c_2 \end{vmatrix} \mathbf{i} - \begin{vmatrix} a_1 & c_1 \\ a_2 & c_2 \end{vmatrix} \mathbf{j} + \begin{vmatrix} a_1 & b_1 \\ a_2 & b_2 \end{vmatrix} \mathbf{k}$$

$$= (b_1c_2 - b_2c_1)\mathbf{i} + (a_2c_1 - a_1c_2)\mathbf{j} + (a_1b_2 - a_2b_1)\mathbf{k}$$
$$= \mathbf{F} \times \mathbf{G}.$$

The cross product of two 3-vectors can be computed in MAPLE using the `CrossProduct` command, which is part of the `VectorCalculus` package.

We will summarize some properties of the cross product.

1. $\mathbf{F} \times \mathbf{G} = -\mathbf{G} \times \mathbf{F}$.
2. $\mathbf{F} \times \mathbf{G}$ is orthogonal to both **F** and **G**. This is shown in Figure 6.14.
3. $\| \mathbf{F} \times \mathbf{G} \| = \| \mathbf{F} \| \| \mathbf{G} \| \sin(\theta)$ in which θ is the angle between **F** and **G**.
4. If **F** and **G** are nonzero vectors, then $\mathbf{F} \times \mathbf{G} = \mathbf{O}$ if and only if **F** and **G** are parallel.
5. $\mathbf{F} \times (\mathbf{G} + \mathbf{H}) = \mathbf{F} \times \mathbf{G} + \mathbf{F} \times \mathbf{H}$.
6. $\alpha(\mathbf{F} \times \mathbf{G}) = (\alpha\mathbf{F}) \times \mathbf{G} = \mathbf{F} \times (\alpha\mathbf{G})$.

Property (1) of the cross product follows from the fact that interchanging two rows of a determinant changes its sign. In computing $\mathbf{F} \times \mathbf{G}$, the components of **F** are in the second row of the determinant, and those of **G** in the third row. These rows are interchanged in computing $\mathbf{G} \times \mathbf{F}$.

For property (2), compute the dot product

$$\mathbf{F} \cdot (\mathbf{F} \times \mathbf{G})$$
$$= a_1[b_1c_2 - b_2c_1] + b_1[a_2c_1 - a_1c_2] + c_1[a_1b_2 - a_2b_1] = 0.$$

Therefore, **F** is orthogonal to $\mathbf{F} \times \mathbf{G}$. Similarly, **G** is orthogonal to $\mathbf{F} \times \mathbf{G}$.

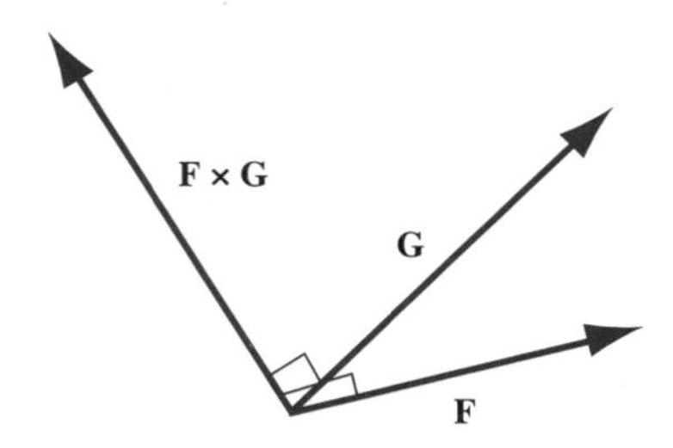

FIGURE 6.14 $\mathbf{F} \times \mathbf{G}$ *is orthogonal to* **F** *and to* **G**.

To derive property (3), suppose both vectors are nonzero and recall that $\cos(\theta) = (\mathbf{F} \cdot \mathbf{G})/ \| \mathbf{F} \| \| \mathbf{G} \|$, where θ is the angle between $\mathbf{F}$ and $\mathbf{G}$. Now write

$$\begin{aligned} &\| \mathbf{F} \|^2 \| \mathbf{G} \|^2 - (\mathbf{F} \cdot \mathbf{G})^2 \\ &= \| \mathbf{F} \|^2 \| \mathbf{G} \|^2 - \| \mathbf{F} \|^2 \| \mathbf{G} \|^2 \cos^2(\theta) \\ &= \| \mathbf{F} \|^2 \| \mathbf{G} \|^2 \sin^2(\theta). \end{aligned}$$

It is therefore enough to show that

$$\| \mathbf{F} \times \mathbf{G} \|^2 = \| \mathbf{F} \|^2 \| \mathbf{G} \|^2 - (\mathbf{F} \cdot \mathbf{G})^2,$$

and this is a tedious but routine calculation.

Property (4) follows from (3), since two nonzero vectors are parallel exactly when the angle θ between them is zero, and in this case, $\sin(\theta) = 0$. Properties (5) and (6) are routine computations.

Property (4) provides a test for three points to be *collinear*, that is, to lie on a single line. Let P, Q, and R be the points. These points will be collinear exactly when the vector $\mathbf{F}$ from P to Q is parallel to the vector $\mathbf{G}$ from P to R. By property (4), this occurs when $\mathbf{F} \times \mathbf{G} = \mathbf{O}$.

One of the primary uses of the cross product is to produce a vector orthogonal to two given vectors. This can be used to find the equation of a plane containing three given points. The strategy is to pick one of the points and write the vectors from this point to the other two. The cross product of these two vectors is normal to the plane containing the points. Now we know a normal vector and a point (in fact three points) on the plane, so we can use equation (6.2) to write the equation of the plane.

This strategy fails if the cross product is zero. But by property (4), this only occurs if the given points are collinear, hence do not determine a unique plane (there are infinitely many planes through any line in 3-space).

EXAMPLE 6.7

Find the equation of a plane containing the points $P: (-1, 4, 2)$, $Q: (6, -2, 8)$, and $R: (5, -1, -1)$.

Use the three given points to form two vectors in the plane:

$$\mathbf{PQ} = 7\mathbf{i} - 6\mathbf{j} + 6\mathbf{k} \text{ and } \mathbf{PR} = 6\mathbf{i} - 5\mathbf{j} - 3\mathbf{k}.$$

The cross product of these vectors is orthogonal to the plane of these vectors, so

$$\mathbf{N} = \mathbf{PQ} \times \mathbf{PR} = 48\mathbf{i} + 57\mathbf{j} + \mathbf{k}$$

is a normal vector. By equation (6.2), the equation of the plane is

$$48(x + 1) + 57(y - 4) + (z - 2) = 0,$$

or

$$48x + 57y + z = 182. \ \blacklozenge$$

SECTION 6.3 PROBLEMS

In each of Problems 1 through 4, compute $\mathbf{F} \times \mathbf{G}$ and $\mathbf{G} \times \mathbf{F}$ and verify the anticommutativity of the cross product.

1. $\mathbf{F} = 2\mathbf{i} - 3\mathbf{j} + 4\mathbf{k}, \mathbf{G} = -3\mathbf{i} + 2\mathbf{j}$
2. $\mathbf{F} = 8\mathbf{i} + 6\mathbf{j}, \mathbf{G} = 14\mathbf{j}$
3. $\mathbf{F} = -3\mathbf{i} + 6\mathbf{j} + \mathbf{k}, \mathbf{G} = -\mathbf{i} - 2\mathbf{j} + \mathbf{k}$
4. $\mathbf{F} = 6\mathbf{i} - \mathbf{k}, \mathbf{G} = \mathbf{j} + 2\mathbf{k}$

In each of Problems 5 through 9, determine whether the points are collinear. If they are not, determine an equation for the plane containing these points.

5. $(-4, 2, -6), (1, 1, 3), (-2, 4, 5)$
6. $(0, 0, 2), (-4, 1, 0), (2, -1, -1)$
7. $(-1, 1, 6), (2, 0, 1), (3, 0, 0)$
8. $(4, 1, 1), (-2, -2, 3), (6, 0, 1)$
9. $(1, 0, -2), (0, 0, 0), (5, 1, 1)$

In each of Problems 10, 11, and 12, find a vector normal to the given plane. There are infinitely many such vectors.

10. $x - 3y + 2z = 9$
11. $x - y + 2z = 0$
12. $8x - y + z = 12$
13. Let **F** and **G** be nonparallel vectors and let R be the parallelogram formed by representing these vectors as arrows from a common point. Show that the area of this parallelogram is $\| \mathbf{F} \times \mathbf{G} \|$.
14. Form a parallelepiped (skewed rectangular box) having as incident sides the vectors **F**, **G**, and **H** drawn as arrows from a common point. Show that the volume of this parallelepiped is

$$|\mathbf{F} \cdot (\mathbf{G} \times \mathbf{H})|.$$

This quantity is called the *scalar triple product* of **F**, **G**, and **H**.

6.4 The Vector Space R^n

For systems involving n variables we may consider n-vectors

$$< x_1, x_2, \cdots, x_n >$$

having n components. The jth component of this n-vector is x_j and this is a real number. The totality of such n-vectors is denoted R^n and is called "n-space". R^1 is the real line, consisting of all real numbers. We can think of numbers as 1-vectors, although we do not usually do this. R^2 is the familiar plane, consisting of vectors with two components. And R^3 is in 3-space. R^n has an algebraic structure which will prove useful when we consider matrices, systems of linear algebraic equations, and systems of linear differential equations.

Two n-vectors are equal exactly when their respective components are equal:

$$< x_1, x_2, \cdots, x_n > = < y_1, y_2, \cdots, y_n >$$

if and only if

$$x_1 = y_1, x_2 = y_2, \cdots, x_n = y_n.$$

Add n-vectors, and multiply them by scalars, in the natural ways:

$$< x_1, x_2, \cdots, x_n > + < y_1, y_2, \cdots, y_n > = < x_1 + y_1, x_2 + y_2, \cdots, x_n + y_n >$$

and

$$\alpha < x_1, x_2, \cdots, x_n > = < \alpha x_1, \alpha x_2, \cdots, \alpha x_n > .$$

These operations have the properties we expect of vector addition and multiplication by scalars. If **F**, **G**, and **H** are in R^n and α an β are real numbers, then

1. $\mathbf{F} + \mathbf{G} = \mathbf{G} + \mathbf{F}$.
2. $\mathbf{F} + (\mathbf{G} + \mathbf{H}) = (\mathbf{F} + \mathbf{G}) + \mathbf{H}$.
3. $\mathbf{F} + \mathbf{O} = \mathbf{F}$,

where

$$\mathbf{O} = (0, 0, \cdots, 0)$$

is the zero vector in R^n (all components zero).

4. $(\alpha + \beta)\mathbf{F} = \alpha\mathbf{F} + \beta\mathbf{F}$.
5. $(\alpha\beta)\mathbf{F} = \alpha(\beta\mathbf{F})$.
6. $\alpha(\mathbf{F} + \mathbf{G}) = \alpha\mathbf{F} + \alpha\mathbf{G}$.
7. $\alpha\mathbf{O} = \mathbf{O}$.

Because of properties (1) through (7), and the fact that $1\mathbf{F} = \mathbf{F}$ for every $\mathbf{F}$ in R^n, we refer to R^n as a *vector space*. There is a general theory of vector spaces which includes a broader class of spaces than R^n. As one example, we will touch upon the function space $C[a, b]$ in Section 6.5.

The norm (length) of an n-vector $\mathbf{F} = < x_1, x_2, \cdots, x_n >$ is

$$\| \mathbf{F} \| = \sqrt{x_1^2 + \cdots + x_n^2}.$$

This norm can be used to define a concept of distance in R^n. Given two points $P : (x_1, x_2, \cdots, x_n)$ and $Q : (y_1, y_2, \cdots, y_n)$ in R^n, think of

$$\mathbf{F} = < x_1, x_2, \cdots, x_n > \text{ and } \mathbf{G} = < y_1, y_2, \cdots, y_n >$$

as vectors from the origin to these points, respectively. The distance between the points is the norm of the difference of $\mathbf{F}$ and $\mathbf{G}$:

$$\begin{aligned}&\text{distance between } P \text{ and } Q\\ &= \| \mathbf{F} - \mathbf{G} \|\\ &= \sqrt{(x_1 - y_1)^2 + (x_2 - y_2)^2 + \cdots + (x_n - y_n)^2}.\end{aligned}$$

When $n = 3$ this is the usual distance between two points in R^3.

The dot product of two n-vectors is defined by

$$< x_1, x_2, \cdots, x_n > \cdot < y_1, y_2, \cdots, y_n > = x_1y_1 + x_2y_2 + \cdots + x_ny_n.$$

This operation is a direct generalization of the dot product in R^3. Some properties of the norm and the dot product are:

1. $\| \alpha\mathbf{F} \| = |\alpha| \, \| \mathbf{F} \|$.
2. Triangle inequality for n-vectors:

$$\| \mathbf{F} + \mathbf{G} \| \leq \| \mathbf{F} + \| \mathbf{G} \|.$$

3. $\mathbf{F} \cdot \mathbf{G} = \mathbf{G} \cdot \mathbf{F}$.
4. $(\mathbf{F} + \mathbf{G}) \cdot \mathbf{H} = \mathbf{F} \cdot \mathbf{H} + \mathbf{G}\mathbf{H}$.
5. $\alpha(\mathbf{F} \cdot \mathbf{G}) = (\alpha\mathbf{F}) \cdot \mathbf{G} = \mathbf{F} \cdot (\alpha\mathbf{G})$.
6. $\mathbf{F} \cdot \mathbf{F} = \| \mathbf{F} \|^2$.
7. $\mathbf{F} \cdot \mathbf{F} = 0$ if and only if $\mathbf{F} = \mathbf{O}$.
8. $\| \alpha\mathbf{F} + \beta\mathbf{G} \|^2 = \alpha^2 \| \mathbf{F} \|^2 + 2\alpha\beta\mathbf{F} \cdot \mathbf{G} + \beta^2 \| \mathbf{G} \|^2$.
9. Cauchy-Schwarz inequality:

$$|\mathbf{F} \cdot \mathbf{G}| \leq \| \mathbf{F} \| \| \mathbf{G} \|.$$

These conclusions are proved by straightforward manipulations. Property (8) is proved by a calculation identical to that done for vectors in R^3, thinking of $\mathbf{F}$ and $\mathbf{G}$ as vectors with n components instead of three.

To verify property (9), use (8). First observe that the conclusion is just $0 \le 0$ if either $\mathbf{F}$ or $\mathbf{G}$ is the zero vector. Thus suppose both are nonzero. In property (8), choose $\alpha = \| \mathbf{G} \|$ and $\beta = -\| \mathbf{F} \|$ to obtain

$$\begin{aligned} 0 &\le \| \alpha\mathbf{F} + \beta\mathbf{G} \|^2 \\ &= \| \mathbf{F} \|^2 \| \mathbf{G} \|^2 - 2 \| \mathbf{F} \| \| \mathbf{G} \| (\mathbf{F} \cdot \mathbf{G}) + \| \mathbf{F} \|^2 \| \mathbf{G} \|^2 \\ &= 2 \| \mathbf{F} \| \| \mathbf{G} \| (\| \mathbf{F} \| \| \mathbf{G} \| - \mathbf{F} \cdot \mathbf{G}). \end{aligned}$$

Divide this inequality by $2 \| \mathbf{F} \| \| \mathbf{G} \|$ to obtain

$$\mathbf{F} \cdot \mathbf{G} \le \| \mathbf{F} \| \| \mathbf{G} \| .$$

Now go back to conclusion (8) but this time set $\alpha = \| \mathbf{G} \|$ and $\beta = \| \mathbf{F} \|$ to obtain, by a similar computation,

$$\begin{aligned} 0 &\le \| \alpha\mathbf{F} + \beta\mathbf{G} \|^2 \\ &= 2 \| \mathbf{F} \| \| \mathbf{G} \| (\| \mathbf{F} \| \| \mathbf{G} \| + \mathbf{F} \cdot \mathbf{G}). \end{aligned}$$

Then

$$-\| \mathbf{F} \| \| \mathbf{G} \| \le \mathbf{F} \cdot \mathbf{G}.$$

Put these two inequalities together to conclude that

$$-\| \mathbf{F} \| \| \mathbf{G} \| \le \mathbf{F} \cdot \mathbf{G} \le \| \mathbf{F} \| \| \mathbf{G} \|,$$

and this is equivalent to

$$|\mathbf{F} \cdot \mathbf{G}| \le \| \mathbf{F} \| \| \mathbf{G} \| .$$

There is no cross product for n-vectors if $n > 3$.

In view of the Cauchy-Schwarz inequality, we can define the cosine of the angle between n-vectors $\mathbf{F}$ and $\mathbf{G}$ by

$$\cos(\theta) = \begin{cases} 0 & \text{if } \mathbf{F} \text{ or } \mathbf{G} \text{ is the zero vector,} \\ (\mathbf{F} \cdot \mathbf{G})/(\| \mathbf{F} \| \| \mathbf{G} \|) & \text{if both vectors are nonzero.} \end{cases}$$

This definition is motivated by the fact that this is the cosine of the angle between two vectors in R^3. We use this definition to bring some geometric intuition to vectors in R^n, which we cannot visualize if $n > 3$. For example, as in R^3, it is natural to define $\mathbf{F}$ and $\mathbf{G}$ to be *orthogonal* if their dot product is zero (so the angle between the two vectors is $\pi/2$, or one or both vectors is the zero vector).

If $\mathbf{F}$ and $\mathbf{G}$ are orthogonal, then $\mathbf{F} \cdot \mathbf{G} = 0$. Upon setting $\alpha = \beta = 1$ in (8) we obtain

$$\| \mathbf{F} + \mathbf{G} \|^2 = \| \mathbf{F} \|^2 + \| \mathbf{G} \|^2 .$$

This is the n-dimensional version of the Pythagorean theorem.

Define *standard unit vectors* along the axes in R^n by

$$\begin{aligned} \mathbf{e}_1 &= < 1, 0, 0, \cdots, 0 >, \\ \mathbf{e}_2 &= < 0, 1, 0, \cdots, 0 >, \cdots, \\ \mathbf{e}_n &= < 0, 0, \cdots, 0, 1 > . \end{aligned}$$

These vectors are *orthonormal* in the sense that each is a unit vector (length 1), and the vectors are mutually orthogonal (each is orthogonal to all of the others).

We can write any n-vector in *standard form*

$$< x_1, x_2, \cdots, x_n > = x_1\mathbf{e}_1 + x_2\mathbf{e}_2 + \cdots + x_n\mathbf{e}_n.$$

This is a direct generalization of writing a 3-vector in terms of the orthonormal 3-vectors **i**, **j** and **k**.

Suppose now that S is a set of vectors in R^n. We call S a *subspace* of R^n if the following conditions are met:

1. **O** is in S.
2. The sum of any vectors in S is in S.
3. The product of any vector in S by any real number is in S.

Conditions (2) and (3) of this definition are equivalent to asserting that $\alpha\mathbf{F} + \beta\mathbf{G}$ is in S for any numbers α and β and vectors **F** and **G** in S.

R^n is a subspace of itself, and the set $S = \{< 0, 0, \cdots, 0 >\}$ consisting of just the zero vector is a subspace of R^n. This is called the *trivial subspace*. Here are more substantial examples.

EXAMPLE 6.8

Let S consist of all vectors in R^n having norm 1. In R^2 this can be visualized as the set of points on the unit circle about the origin, and in 3-space as the set of points on the unit sphere about the origin. S is not a subspace of R^n because the zero vector is not in S, violating requirement (1) of the definition. This is enough to disqualify S from being a subspace. However, in this example, requirements (2) and (3) also fail. A sum of two vectors having length 1 does not have length 1, hence is not in S. And a scalar multiple of a vector in S is not in S unless the scalar is 1 or -1. ♦

EXAMPLE 6.9

Let K consist of all scalar multiples of $\mathbf{F} = < -1, 4, 2, 0 >$ in R^4. The zero vector is in K (this is the product of **F** with the number zero). A sum of scalar multiples of **F** is a scalar multiple of **F**, hence is in K, so requirement (2) holds. And a scalar multiple of a scalar multiple of **F** is also a scalar multiple of **F**, so requirement (3) is true. ♦

EXAMPLE 6.10

In R^6, let W consist of all vectors having second, fourth and sixth component zero. Thus S consists of all 6-vectors $< x, 0, y, 0, z, 0 >$. Then $< 0, 0, 0, 0, 0, 0 >$ is in W (choose $x = y = z = 0$). A sum of vectors in W also has second, fourth and sixth components zero, as does any scalar multiple of a vector in W. Therefore W is a subspace of R^6. ♦

EXAMPLE 6.11

Let $\mathbf{F}_1, \cdots, \mathbf{F}_k$ be any k vectors in R^n. Then the set L of all vectors of the form

$$\alpha_1\mathbf{F}_1 + \alpha_2\mathbf{F}_2 + \cdots + \alpha_k\mathbf{F}_k,$$

in which the α_j's can be any real numbers, forms a subspace of R^n. We call this subspace the *span* of $\mathbf{F}_1, \cdots, \mathbf{F}_k$ and we will say more shortly about subspaces formed in this way. ♦

In the plane and in 3-space, it is easy to visualize all of the subspaces in addition to the entire space and the trivial subspace.

First consider R^2 and look at a straight line $y = mx$ through the origin. Every point on this line has the form (x, mx). With $\mathbf{i} = <1, 0>$ and $\mathbf{j} = <0, 1>$, every vector $x\mathbf{i} + mx\mathbf{j}$, with second component m times the first, is along this line. Further, any sum of two vectors $x_1\mathbf{i} + mx_1\mathbf{j}$ and $x_2\mathbf{i} + mx_2\mathbf{j}$ has this form, as does any multiple of such a vector by a real number. Therefore the vectors $x\mathbf{i} + mx\mathbf{j}$ form a subspace of R^2.

So far we have excluded the vertical axis, which is also a line through the origin, but does not have finite slope. However, all vectors parallel to the vertical axis also form a subspace of R^2, being scalar multiples of $\mathbf{j}$.

Every line through the origin therefore determines a subspace of R^2, consisting of all vectors parallel to this line.

Are there any other subspaces of R^2 that we have missed?

Suppose S is a nontrivial subspace containing two vectors $a\mathbf{i} + b\mathbf{j}$ and $c\mathbf{i} + d\mathbf{j}$ that are not on the same line through the origin. Then $ad - bc \neq 0$, because the lines along these vectors have different slopes. We claim that this forces every 2-vector $x\mathbf{i} + y\mathbf{j}$ to be in S. To verify this, we will solve for numbers α and β such that

$$x\mathbf{i} + y\mathbf{j} = \alpha(a\mathbf{i} + b\mathbf{j}) + \beta(c\mathbf{i} + d\mathbf{j}).$$

This requires that

$$\alpha a + \beta c = x, \text{ and}$$

$$\alpha b + \beta d = y.$$

But these equations have the solutions

$$\alpha = \frac{dx - cy}{ad - bc} \text{ and } \beta = \frac{ay - bx}{ad - bc}.$$

Therefore every 2-vector $x\mathbf{i} + y\mathbf{j}$ in $R^2$is of the form

$$\alpha(a\mathbf{i} + b\mathbf{j}) + \beta(c\mathbf{i} + d\mathbf{j})$$

hence is in S. In this event $S = R^2$. We therefore know all of the subspaces of R^2. They are R^2, the trivial subspace $\{<0, 0>\}$ and, for any line L through the origin, all vectors parallel to L.

By similar reasoning, there are exactly four kinds of subspaces of R^3. These are R^3, the trivial subspace containing just the zero vector, the subspace of all vectors on any given line through the origin, and the subspace of all vectors lying on any given plane through the origin.

A *linear combination* of k vectors $\mathbf{F}_1, \cdots, \mathbf{F}_k$ in R^n is a sum of the form

$$\alpha_1\mathbf{F}_1 + \alpha_2\mathbf{F}_2 + \cdots + \alpha_k\mathbf{F}_k.$$

in which each α_j is a real number.

The *span* of vectors $\mathbf{F}_1, \mathbf{F}_2, \cdots, \mathbf{F}_k$ in R^n consists of all linear combinations of these vectors, that is, of all vectors of the form

$$\alpha_1\mathbf{F}_1 + \alpha_2\mathbf{F}_2 + \cdots + \alpha_k\mathbf{F}_k.$$

From Example 6.11, the span of any set of vectors in R^n is a subspace of R^n. We say that these vectors form a *spanning set* for this subspace.

Every nontrivial subspace has many spanning sets.

EXAMPLE 6.12

The vectors $\mathbf{i}, \mathbf{j}, \mathbf{k}$ span all of R^3. But so do

$$3\mathbf{i}, 2\mathbf{j}, -\mathbf{k}.$$

The vectors

$$\mathbf{F}_1 = \mathbf{i} + \mathbf{k}, \mathbf{F}_2 = \mathbf{i} + \mathbf{j}, \mathbf{F}_3 = \mathbf{j} + \mathbf{k}$$

also span R^3. To see this, let $\mathbf{V} = a\mathbf{i} + b\mathbf{j} + c\mathbf{k}$ be any 3− vector. Then

$$\mathbf{V} = \frac{a+b-c}{2}\mathbf{F}_1 + \frac{a-b+c}{2}\mathbf{F}_2 + \frac{3a-b-c}{2}\mathbf{F}_3. \ \blacklozenge$$

In this example these spanning sets all have three vectors in them. But a spanning set for R^3 may have more than three vectors. For example, the vectors

$$\mathbf{i}, \mathbf{j}, \mathbf{k}, -4\mathbf{i}, \sqrt{97}\mathbf{k}$$

also span R^3 because we can write any 3-vector as

$$< x, y, z >= x\mathbf{i} + y\mathbf{j} + z\mathbf{k} + 0(-4\mathbf{i}) + 0(\sqrt{97}\mathbf{k}\mathbf{j}).$$

This set of five vectors spans R^3, but does so inefficiently in the sense that two of the vectors are not needed to have a spanning set for R^3. ♦

More generally, if vectors $\mathbf{V}_1, \cdots, \mathbf{V}_k$ span a subspace S of R^n, we can adjoin any number m of other vectors of S to these k vectors, and the resulting $m + k$ vectors will still span S.

Going the other way, if $\mathbf{V}_1, \cdots, \mathbf{V}_k$ span a subspace S of R^n, it *may* be possible to remove some vectors from this set and have the smaller set of vectors still span S. This occurs when $\mathbf{V}_1, \cdots, \mathbf{V}_k$ contain redundant information and not all of them are needed to completely specify S. The efficiency of a spanning set (the idea of whether it contains unnecessary vectors) is addressed through the notions of linear dependence and independence.

A (finite) set of vectors in R^n is called *linearly dependent* if one of the vectors is a linear combination of the others. Otherwise, if no one of the vectors is a linear combination of the others, then these vectors are *linearly independent.*

EXAMPLE 6.13

The vectors

$$\mathbf{F} = < 3, -1, 0, 4 >, \mathbf{G} = < 3, -2, -1, 10 >, \mathbf{H} = < 6, -1, 1, 2 >$$

are linearly dependent in R^4 because $\mathbf{G} = 3\mathbf{F} - \mathbf{H}$. The two vectors $\mathbf{F}$ and $\mathbf{G}$ are linearly independent, because neither is a scalar multiple of the other. ♦

Think of linear independence in terms of information. Suppose $\mathbf{F}_1, \cdots, \mathbf{F}_k$ are vectors in R^n. If these vectors are linearly dependent, then at least one of them, say $\mathbf{F}_k$ for convenience, is a linear combination of $\mathbf{F}_1, \cdots, \mathbf{F}_{k-1}$. This means that any linear combination of these k vectors is really a linear combination of just the first $k - 1$ of them. Put another way, the subspace S spanned by all k of these vectors is the same as the subspace space spanned by just the first $k - 1$ of them, and $\mathbf{F}_k$ is not needed in specifying S.

EXAMPLE 6.14

Let

$$\mathbf{F}_1 = <1, 0, 1, 0>, \mathbf{F}_2 = <0, 1, 1, 0> \text{ and } \mathbf{F}_3 = <2, 3, 5, 0>.$$

These vectors are linearly dependent in R^4 because

$$\mathbf{F}_3 = 2\mathbf{F}_1 + 3\mathbf{F}_2.$$

The subspace S of R^4 spanned by $\mathbf{F}_1$ and $\mathbf{F}_2$ is the same as the subspace spanned by all three of these vectors. Indeed, any linear combination of all three vectors is a linear combination of the first two:

$$\begin{aligned} &c_1\mathbf{F}_1 + c_2\mathbf{F}_2 + c_3\mathbf{F}_3 \\ &= c_1\mathbf{F}_1 + c_2\mathbf{F}_2 + c_3(2\mathbf{F}_1 + 3\mathbf{F}_2) \\ &= (c_1 + 2c_3)\mathbf{F}_1 + (c_2 + 3c_3)\mathbf{F}_2. \end{aligned}$$

$\mathbf{F}_1$ and $\mathbf{F}_2$ contain all of the information needed to specify S. ♦

There is an important characterization of linear independence and dependence that is used frequently.

THEOREM 6.1 ***Linear Dependence and Independence***

Let $\mathbf{F}_1, \mathbf{F}_2, \cdots, \mathbf{F}_k$ be vectors in R^n. Then

1. $\mathbf{F}_1, \mathbf{F}_2, \cdots, \mathbf{F}_k$ are linearly dependent if and only if there are real numbers $\alpha_1, \alpha_2, \cdots, \alpha_k$, *not all zero*, such that

$$\alpha_1\mathbf{F}_1 + \alpha_2\mathbf{F}_2 + \cdots + \alpha_k\mathbf{F}_k = \mathbf{O}.$$

2. $\mathbf{F}_1, \mathbf{F}_2, \cdots, \mathbf{F}_k$ are linearly independent if and only if an equation

$$\alpha_1\mathbf{F}_1 + \alpha_2\mathbf{F}_2 + \cdots + \alpha_k\mathbf{F}_k = \mathbf{O},$$

can hold only if each coefficient is zero:

$$\alpha_1 = \alpha_2 = \cdots = \alpha_k = 0. \; ♦$$

Proof To prove (1), suppose first that $\mathbf{F}_1, \mathbf{F}_2, \cdots, \mathbf{F}_k$ are linearly dependent. Then at least one of these vectors is a linear combination of the others. As a convenience, suppose

$$\mathbf{F}_1 = \alpha_2\mathbf{F}_2 + \cdots + \alpha_k\mathbf{F}_k.$$

Then

$$\mathbf{F}_1 - \alpha_2\mathbf{F}_2 - \cdots - \alpha_k\mathbf{F}_k = \mathbf{O}.$$

This is a linear combination of $\mathbf{F}_1, \mathbf{F}_2, \cdots, \mathbf{F}_k$ adding up to the zero vector, and having at least one nonzero coefficient (the coefficient of $\mathbf{F}_1$ is 1).

Conversely, suppose there are real numbers $\alpha_1, \cdots, \alpha_k$, not all zero, such that

$$\alpha_1\mathbf{F}_1 + \alpha_2\mathbf{F}_2 + \cdots + \alpha_k\mathbf{F}_k = \mathbf{O}.$$

By assumption at least one of the coefficient is not zero. Suppose, for convenience, that $\alpha_k \neq 0$. Then

$$\mathbf{F}_k = -\frac{\alpha_1}{\alpha_k}\mathbf{F}_1 - \cdots - \frac{\alpha_{k-1}}{\alpha_k}\mathbf{F}_{k-1},$$

so $\mathbf{F}_k$ is a linear combination of $\mathbf{F}_1, \cdots, \mathbf{F}_{k-1}$ and $\mathbf{F}_1, \mathbf{F}_2, \cdots, \mathbf{F}_k$ are linearly dependent.

Part (2) of the theorem is proved similarly. ♦

If k and n are large, it may be difficult to tell whether a set of k vectors in R^n is linearly independent or dependent. This task is simplified if the vectors are mutually orthogonal.

THEOREM 6.2

Let $\mathbf{F}_1, \cdots \mathbf{F}_k$ be nonzero mutually orthogonal vectors in R^n. Then $\mathbf{F}_1, \cdots \mathbf{F}_k$ are linearly independent.

Proof Suppose

$$\alpha_1\mathbf{F}_1 + \alpha_2\mathbf{F}_2 + \cdots + \alpha_k\mathbf{F}_k = \mathbf{O}.$$

Take the dot product of this equation with $\mathbf{F}_1$:

$$\alpha_1\mathbf{F}_1 \cdot \mathbf{F}_1 + \alpha_2\mathbf{F}_1 \cdot \mathbf{F}_2 + \cdots + \alpha_k\mathbf{F}_1 \cdot \mathbf{F}_k = \mathbf{O} \cdot \mathbf{F}_1 = 0.$$

Because $\mathbf{F}_1 \cdot \mathbf{F}_j = 0$ for $j = 2, \cdots, k$, by the orthogonality of these vectors, this equation reduces to

$$\alpha_1\mathbf{F}_1 \cdot \mathbf{F}_1 = 0.$$

Then

$$\alpha_1 \parallel \mathbf{F}_1 \parallel^2 = 0.$$

But $\mathbf{F}_1$ is not the zero vector, so $\parallel \mathbf{F}_1 \parallel \neq 0$ and therefore $\alpha_1 = 0$. By using $\mathbf{F}_j$ in place of $\mathbf{F}_1$ in this dot product, we conclude that each $\alpha_j = 0$. By (2) of Theorem 6.1, $\mathbf{F}_1, \cdots \mathbf{F}_k$ are linearly independent. ♦

We would like to combine the notions of spanning set and linearly independence to define vector spaces and subspaces as efficiently as possible. To this end, define a *basis* for a subspace S of R^n to be a set of vectors that spans S and is linearly independent. In this definition, S may be R^n.

EXAMPLE 6.15

The vectors $\mathbf{i}, \mathbf{j}, \mathbf{k}$ in R^3 are linearly independent, and span R^3. These vectors form a basis for R^3.

In R^n, the standard unit vectors

$$\mathbf{e}_1 = < 1, 0, 0, \cdots, 0 >, \mathbf{e}_2 = < 0, 1, 0, \cdots, 0 >, \cdots, \mathbf{e}_n < 0, 0, \cdots, 0, 1 >$$

form a basis. ♦

EXAMPLE 6.16

Let S be the subspace of R^n consisting of all $n-$ vectors with first component zero. Then $\mathbf{e}_2, \cdots, \mathbf{e}_n$ form a basis for S. ♦

EXAMPLE 6.17

In R^3, let M be the subspace of all vectors parallel to the plane $x + y + z = 0$. A point is on this plane exactly when it has coordinates $(x, y, -x - y)$. Therefore every vector in M has the form $< x, y, -x - y >$. We can write this vector as

$$< x, y, -x - y > = x < 1, 0, -1 > + y < 0, 1, -1 > .$$

The vectors $< 1, 0, -1 >$ and $< 0, 1, -1 >$ span M. These vectors are also linearly independent, since neither is a scalar multiple of the other. These vectors therefore form a basis for M. Both vectors are needed to specify all vectors in S.

There is nothing unique about a basis for a subspace. For example,

$$< 2, 0, -2 > \text{ and } < 0, 2, -2 >$$

also form a basis for M, as do

$$< 1, 0, -1 > \text{ and } < 0, 4, -4 >. \blacklozenge$$

We will need some additional facts about bases. The first is that any spanning set for a subspace S of R^n contains a basis.

THEOREM 6.3

Let S be a subspace of R^n that is spanned by $\mathbf{F}_1, \cdots, \mathbf{F}_k$. Then a basis for S can be formed from some or all of the vectors $\mathbf{F}_1, \cdots, \mathbf{F}_k$. ♦

We will sketch the idea of a proof. Suppose we have a set of vectors $\mathbf{F}_1, \cdots, \mathbf{F}_k$ that span a given subspace S of R^n (perhaps all of R^n). If these vectors are also linearly independent, then they form a basis for S.

If these spanning vectors are linearly dependent, then at least one $\mathbf{F}_j$ is a linear combination of others. Remove $\mathbf{F}_j$, and the remaining set (one vector smaller) spans S. If these vectors are linearly dependent, then one is a linear combination of the others, and we can remove this one to obtain a still smaller spanning set for S. Continuing in this way, we eventually reach a spanning set for S that is linearly independent, with no one vector a linear combination of the others.

A spanning set for S is a basis if the vectors are linearly independent. If we are willing to forego linear independence, however, then we can adjoin as many vectors from S as we like to this spanning set and still have a spanning set for S. This suggests that a basis is limited in size, while a spanning set is not. The next theorem is a careful statement of this idea, and says that any spanning set for S has at least as many vectors in it as any basis for S. It is in this sense that a basis for a subspace is a "smallest possible" spanning set for this subspace.

THEOREM 6.4

Suppose $\mathbf{V}_1, \cdots, \mathbf{V}_k$ span a subspace S of R^n, and let $\mathbf{G}_1, \cdots, \mathbf{G}_t$ be a basis for S. Then $t \leq k$. ♦

Proof Since $\mathbf{V}_1, \cdots, \mathbf{V}_k$ span S and $\mathbf{G}_1$ is in S, then

$$\mathbf{G}_1 = c_1\mathbf{V}_1 + \cdots + c_k\mathbf{V}_k$$

for some numbers $c_1, \cdots, c_k$. Then

$$\mathbf{G}_1 - c_1\mathbf{V}_1 - \cdots - c_k\mathbf{V}_k = \mathbf{O}.$$

If each $c_j = 0$ then $\mathbf{G}_1 = \mathbf{O}$, impossible since $\mathbf{G}_1$ is a basis vector. Therefore some c_j is nonzero. As a notational convenience, suppose $c_1 \neq 0$. Then

$$\mathbf{V}_1 = -\frac{1}{c_1}\mathbf{G}_1 - \frac{c_2}{c_1}\mathbf{V}_2 - \cdots - \frac{c_k}{c_1}\mathbf{V}_k.$$

Further, $\mathbf{G}_1, \mathbf{V}_2, \cdots, \mathbf{V}_k$ span S. Denote this set of vectors as A_1:

$$A_1 : \mathbf{G}_1, \mathbf{V}_2, \cdots, \mathbf{V}_k.$$

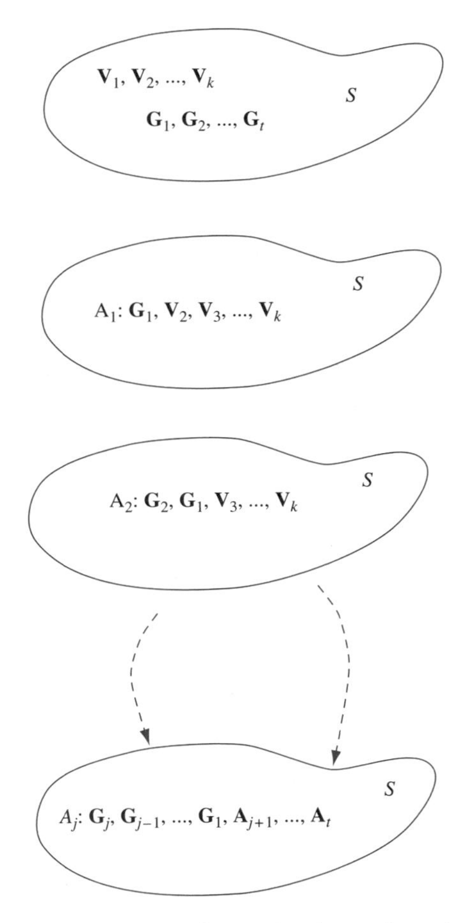

FIGURE 6.15 *The sets A_1, A_2, $\cdots$ formed in the proof of Theorem 6.4.*

Now adjoin $\mathbf{G}_2$ to this list of vectors to form

$$\mathbf{G}_2, \mathbf{G}_1, \mathbf{V}_2, \cdots, \mathbf{V}_k.$$

This set spans S and is linearly dependent because $\mathbf{G}_2$ is a linear combination of the other vectors. Arguing as we did for A_1, some $\mathbf{V}_j$ is a linear combination of the other vectors in this list. Again for notational ease, suppose this is $\mathbf{V}_2$. Deleting this vector from the list therefore yields a set of vectors that still spans S. Denote this set A_2:

$$A_2 : \mathbf{G}_2, \mathbf{G}_1, \mathbf{V}_3, \cdots, \mathbf{V}_k.$$

The vectors in A_2 span S and are linearly dependent. We can continue this process of replacing, one by one, the vectors in $\mathbf{V}_1, \cdots, \mathbf{V}_k$ with vectors in $\mathbf{G}_1, \cdots, \mathbf{G}_t$. Figure 6.15 illustrates this interchange of vectors between the basis to the spanning set that we have been carrying out.

There are two possibilities for this process to end.

First, this process may exhaust the basis vectors $\mathbf{G}_1, \cdots, \mathbf{G}_t$ with some vectors $\mathbf{V}_j$ remaining. Since we delete a $\mathbf{V}_j$ from the list exactly when we adjoin some $\mathbf{G}_i$, this would imply that $t \leq k$.

The other possibility is that at some stage we have removed all of the $\mathbf{V}_j$'s, and still have some $\mathbf{G}_i$ s left (so we would have $t > k$). At this stage, we would have the list

$$A_k : \mathbf{G}_k, \mathbf{G}_{k-1}, \cdots, \mathbf{G}_1.$$

But each time we form such a list by replacing a spanning set vector with a basis vector, we obtain a new set of vectors that spans S. This would make $\mathbf{G}_{k+1}$ a linear combination of the first k basis vectors, and then these basis vectors would be linearly dependent, a contradiction.

This proves that this possibility cannot occur, leaving the first possibility, and $t \leq k$. ♦

This theorem has a profound consequence—all bases for a given subspace of R^n have the same number of vectors in them.

COROLLARY 6.1

Let $\mathbf{G}_1, \cdots, \mathbf{G}_m$ and $\mathbf{H}_1, \cdots, \mathbf{H}_k$ be bases for a subspace S of R^n. Then $m = k$.

Proof Each basis is a spanning set, so two applications of Theorem 6.4 gives us $m \leq k$ and also $k \leq m$. ♦

The number of vectors in a basis for a subspace S of R^n is called the *dimension* of S. For example, R^n has dimension n, and the subspace of R^3 in Example 6.17 has dimension 2.

Now suppose S is a k-dimensional subspace of R^n, and $\mathbf{v}_1, \mathbf{v}_2, \cdots, \mathbf{v}_k$ form a basis for S. If $\mathbf{X}$ is in S, then there are numbers $c_1, c_2, \cdots, c_k$ such that

$$\mathbf{X} = c_1\mathbf{v}_1 + c_2\mathbf{v}_2 + \cdots + c_k\mathbf{v}_k = \sum_{j=1}^{k} c_j\mathbf{v}_k.$$

The numbers $c_1, \cdots, c_k$ are called the *coordinates* of $\mathbf{X}$ with respect to this basis. These coordinates are unique to $\mathbf{X}$ and to this basis.

For, if

$$\mathbf{X} = d_1\mathbf{v}_1 + \cdots + d_k\mathbf{v}_k$$

then

$$\mathbf{X} - \mathbf{X} = \mathbf{O} = (c_1 - d_1)\mathbf{v}_1 + \cdots + (c_k - d_k)\mathbf{v}_k = \sum_{j=1}^{k}(c_j - d_j)\mathbf{v}_j.$$

Since the vectors $\mathbf{v}_1, \cdots, \mathbf{v}_k$ are linearly independent, each $c_j - d_j = 0$, and therefore each $c_j = d_j$.

A nontrivial subspace of R^n has many bases, and each n-vector $\mathbf{X}$ has unique coordinates with respect to each basis. However, on a practical level, some bases are more convenient to work with in the sense that coordinates of vectors with respect to these bases are easier to determine. To illustrate, let S be the subspace of R^4 consisting of all vectors $< x, y, 0, 0 >$, with x and y any real numbers.

This is a two-dimensional subspace with $\mathbf{e}_1 = < 1, 0, 0, 0 >$ and $\mathbf{e}_2 = < 0, 1, 0, 0 >$ forming a basis $\mathcal{B}_1$ for S. The vectors

$$\mathbf{w}_1 = < 2, -6, 0, 0 > \text{ and } \mathbf{w}_2 = < 2, 4, 0, 0 >$$

form another basis $\mathcal{B}_2$ for S. Why does $\mathcal{B}_1$ seem more natural than $\mathcal{B}_2$? It is because, given any **X** in S, it is easy to find the coordinates of **X** with respect to $\mathcal{B}_1$. Indeed, if $\mathbf{X} = < a, b, 0, 0 >$, then immediately

$$\mathbf{X} = a\mathbf{e}_1 + b\mathbf{e}_2.$$

However, finding the coordinates of **X** with respect to $\mathcal{B}_2$ is more tedious. If these coordinates are c_1 and c_2, then we would have to have

$$\begin{aligned} < a, b, 0, 0 > &= c_1\mathbf{w}_1 + c_2\mathbf{w}_2 \\ &= c_1 < 2, -6, 0, 0 > + c_2 < 2, 4, 0, 0 > \\ &= < 2c_1 + 2c_2, -6c_1 + 4c_2, 0, 0 >, \end{aligned}$$

requiring that

$$2c_1 + 2c_2 = a \text{ and } -6c_1 + 4c_2 = b.$$

Solve for these coordinates to obtain

$$c_1 = \frac{1}{10}(2a - b),\ c_2 = \frac{1}{10}(3a + b).$$

Thus,

$$\mathbf{X} = \frac{1}{10}(2a - b)\mathbf{w}_1 + \frac{1}{10}(3a + b)\mathbf{w}_2.$$

We can tell the coordinates of any **X** in S with respect to $\mathcal{B}_1$ just by looking at **X**, while finding the coordinates of **X** with respect to $\mathcal{B}_2$ takes some work.

Another nice feature of $\mathcal{B}_1$ is that it consists of mutually orthogonal vectors. In general, a basis is an *orthogonal basis* if its vectors are mutually orthogonal. If these vectors are also unit vectors, then the basis is *orthonormal*. With any orthogonal basis for S, it is possible to write a simple formula for the coordinates of any vector **X** in S.

THEOREM 6.5 Coordinates in Orthogonal Bases

Let S be a subspace of R^n and let $\mathbf{V}_1, \cdots, \mathbf{V}_k$ be an orthogonal basis for S. If **X** is in S, then

$$\mathbf{X} = c_1\mathbf{V}_1 + c_2\mathbf{v}_2 + \cdots + c_k\mathbf{V}_k,$$

where

$$c_j = \frac{\mathbf{X} \cdot \mathbf{V}_j}{\| \mathbf{V}_j \|^2}$$

for $j = 1, 2, \cdots, k$. ♦

This gives the jth coordinate of any X with respect to these basis vectors as the dot product of **X** with $\mathbf{V}_j$, divided by the length of $\mathbf{V}_j$ squared. In terms of projections, any vector in **X** is the sum of the projections of **X** onto the orthogonal basis vectors. This is true for any orthogonal basis for S.

Proof Write

$$\mathbf{X} = c_1\mathbf{V}_1 + c_2\mathbf{V}_2 + \cdots + c_k\mathbf{V}_k.$$

We must solve for the c_j's. Take the dot product of **X** with $\mathbf{V}_j$ to obtain

$$\mathbf{X} \cdot \mathbf{V}_j = c_j\mathbf{V}_j \cdot \mathbf{V}_j = c_j \| \mathbf{V}_j \|^2,$$

since, by orthogonality, $\mathbf{V}_i \cdot \mathbf{V}_j = 0$ if $i \neq j$. This yields the expression for c_j given in the theorem. ♦

EXAMPLE 6.18

The vectors

$$\mathbf{v}_1 = < 2, 0, 1, 0, 0 >, \mathbf{v}_2 = < 0, 5, 0, 0, 0 >, \mathbf{v}_3 = < -1, 0, 2, 0, 0 >$$

form an orthogonal basis for a three-dimensional subspace S of R^5. Let $\mathbf{X} = < 12, -5, 4, 0, 0 >$, a vector in S. We will find the coordinates c_1, c_2, c_3 of $\mathbf{X}$ with respect to this basis. Compute

$$c_1 = \frac{\mathbf{X} \cdot \mathbf{v}_1}{\| \mathbf{v}_1 \|^2} = \frac{28}{5},$$

$$c_2 = \frac{\mathbf{X} \cdot \mathbf{v}_2}{\| \mathbf{v}_2 \|^2} = \frac{-25}{25} = -1,$$

$$c_3 = \frac{\mathbf{X} \cdot \mathbf{v}_3}{\| \mathbf{v}_3 \|^2} = -\frac{4}{5}. \quad \blacklozenge$$

Then

$$\mathbf{X} = c_1\mathbf{v}_1 + c_2\mathbf{v}_2 + c_3\mathbf{v}_3. \quad \blacklozenge$$

We will pursue further properties of sets of orthogonal vectors in the next section.

SECTION 6.4 PROBLEMS

In each of Problems 1 through 10, determine whether the vectors are linearly independent or dependent in the appropriate R^n.

1. $< 8, 0, 2, 0, 0, 0, 0 >, < 0, 0, 0, 0, 1, -1, 0 >$ in R^7
2. $< 3, 0, 0, 4 >, < 2, 0, 0, 8 >$ in R^4
3. $< 1, 2, -3, 1 >, < 4, 0, 0, 2 >, < 6, 4, -6, 4 >$ in R^4
4. $< 0, 1, 1, 1 >, < -3, 2, 4, 4 >, < -2, 2, 34, 2 >, < 1, 1, -6, -2 >$ in R^4
5. $< -2, 0, 0, 1, 1 >, < 1, 0, 0, 0, 0 >, < 0, 0, 0, 0, 2 >, < 1, -1, 3, 3, 1 >$ in R^5
6. $< -1, 1, 0, 0, 0 >, < 0, -1, 1, 0, 0 >, < 0, 1, 1, 1, 0 >$ in R^5
7. $3\mathbf{i} + 2\mathbf{j}, \mathbf{i} - \mathbf{j}$ in R^3
8. $< 1, 0, 0, 0 >, < 0, 1, 1, 0 >, < -4, 6, 6, 0 >$ in R^4
9. $< 1, -2 >, < 4, 1 >, < 6, 6 >$ in R^2
10. $2\mathbf{i}, 3\mathbf{j}, 5\mathbf{i} - 12\mathbf{k}, \mathbf{i} + \mathbf{j} + \mathbf{k}$ in R^3

In each of Problems 11 through 15, show that the set S is a subspace of the appropriate R^n and find a basis for this subspace and the dimension of the subspace.

11. S consists of all vectors $< 0, x, 0, 2x, 0, 3x, 0 >$ in R^7.
12. S consists of all vectors in R^6 of the form $< x, x, y, y, 0, z >$.
13. S consists of all vectors $< x, y, -y, -x >$ in R^4.
14. S consists of all vectors $< x, y, 2x, 3y >$ in R^4.
15. S consists of all vectors in R^4 with zero second component.

In each of Problems 16, 17, and 18, find the coordinates of $\mathbf{X}$ with respect to the given basis.

16. $\mathbf{X} = < -3, 1, 1, 6, 4, 5 >$, with the basis $< 4, 0, 1, 0, 0, 0 >$, $< -1, 1, 4, 0, 0, 0 >$, $< 0, 0, 0, 2, 1, 0 >$, $< 0, 0, 0, -1, 2, 5 >$, $< 0, 0, 0, 0, 0, 5 >$.
17. $\mathbf{X} = < -3, -2, 5, 1, -4 >$, with the basis $< 1, 1, 1, 1, 0 >$, $< -1, 1, 0, 0, 0 >$, $< 1, 1, -1, -1, 0 >$, $< 0, 0, 2, -2, 0 >$, $< 0, 0, 0, 0, 2 >$ of R^5.
18. $\mathbf{X} = < 4, 4, -1, 2, 0 >$ with vectors $< 2, 1, 0, 0, 0 >$, $< 1, -2, 0, 0, 0 >$, $< 0, 0, 3, -2, 0 >$, $< 0, 0, 2, -3, 0 >$ spanning a subspace S of R^5.
19. Suppose $\mathbf{V}_1, \cdots, \mathbf{V}_k$ form a basis for a subspace S of R^n. Let $\mathbf{U}$ be any other vector in S. Show that the vectors $\mathbf{V}_1, \cdots, \mathbf{V}_k, \mathbf{U}$ are linearly dependent.
20. Let $\mathbf{u}_1, \cdots, \mathbf{u}_k$ be linearly independent vectors in R^n, with $k < n$. Show that there are $n - k$ vectors $\mathbf{v}_1, \cdots, \mathbf{v}_{n-k}$ such that

$$\mathbf{u}_1, \cdots, \mathbf{u}_k, \mathbf{v}_1, \cdots, \mathbf{v}_{n-k}$$

form a basis for R^n. This states that any linearly independent set of vectors in R^n is either a basis, or can be expanded into a basis by adjoining more vectors. *Hint*: Choose $\mathbf{v}_1$ in R^n but not in the span of $\mathbf{u}, \cdots, \mathbf{u}_k$.

If $\mathbf{u}_1, \cdots, \mathbf{u}_k, \mathbf{v}_1$ span R^n, stop. Otherwise, there is some $\mathbf{v}_2$ in R^n but not in the span of $\mathbf{u}_1, \cdots, \mathbf{u}_k, \mathbf{v}_1$. If $\mathbf{u}_1, \cdots, \mathbf{u}_k, \mathbf{v}_1, \mathbf{v}_2$ span R^n, stop. Otherwise continue this process.

21. Let S be a nontrivial subspace of R^n. Show that any spanning set of S must contain a basis for S.
22. Show that any finite set of vectors that includes the zero vector is linearly dependent.
23. Let $\mathbf{X}$ and $\mathbf{Y}$ be vectors in R^n, and suppose that $\| \mathbf{X} \| = \| \mathbf{Y} \|$. Show that $\mathbf{X} - \mathbf{Y}$ and $\mathbf{X} + \mathbf{Y}$ are orthogonal. Draw a parallelogram law diagram justification for this conclusion, for the case that the vectors are in R^2.
24. Let $\mathbf{V}_1, \cdots, \mathbf{V}_k$ be mutually orthogonal vectors in R^n. Show that, for any $\mathbf{X}$ in R^n,

$$\sum_{j=1}^{k} (\mathbf{X} \cdot \mathbf{V}_j)^2 \leq \| \mathbf{X} \|^2 .$$

This is known as *Bessel's inequality* for vectors. A version for Fourier series and eigenfunction expansions will be seen in Chapter Fifteen. *Hint* Let $\mathbf{Y} = \mathbf{X} - \sum_{j=1}^{k} (\mathbf{X} \cdot \mathbf{V}_j)\mathbf{V}_j$ and compute $\| \mathbf{Y} \|^2$.

25. Suppose $\mathbf{V}_1, \cdots, \mathbf{V}_n$ are a basis for R^n, consisting of mutually orthogonal unit vectors. Show that, if $\mathbf{X}$ is any vector in R^n, then

$$\sum_{j=1}^{n} (\mathbf{X} \cdot \mathbf{V}_j)^2 = \| \mathbf{X} \|^2 .$$

This is a vector version of *Parseval's equality*.

26. Let $\mathbf{V}_1, \cdots, \mathbf{V}_k$ be mutually orthogonal vectors in R^n. Prove that

$$\| \mathbf{V}_1 + \cdots + \mathbf{V}_k \|^2 = \| \mathbf{V}_1 \|^2 + \cdots + \| \mathbf{V}_k \|^2 .$$

Hint: Write

$$\| \mathbf{V}_1 + \cdots + \mathbf{V}_k \|^2 = (\mathbf{V}_1 + \cdots + \mathbf{V}_k) \cdot (\mathbf{V}_1 + \cdots + \mathbf{V}_k).$$

6.5 Orthogonalization

Suppose $\mathbf{X}_1, \cdots, \mathbf{X}_m$ form a basis for a subspace S of R^n, with $m \geq 2$. We would like to replace this basis with an orthogonal basis $\mathbf{V}_1, \cdots, \mathbf{V}_m$ for S.

We will build an orthogonal basis one vector at a time. Begin by setting

$$\mathbf{V}_1 = \mathbf{X}_1.$$

Now look for a nonzero $\mathbf{V}_2$ that is in S and orthogonal to $\mathbf{V}_1$. One way to do this is to attempt $\mathbf{V}_2$ of the form

$$\mathbf{V}_2 = \mathbf{X}_2 - c\mathbf{V}_1.$$

Choose c so that $\mathbf{V}_2$ is orthogonal to $\mathbf{V}_1$. For this, we need

$$\mathbf{V}_2 \cdot \mathbf{V}_1 = \mathbf{X}_2 \cdot \mathbf{V}_1 - c\mathbf{V}_1 \cdot \mathbf{V}_1 = 0.$$

This will be true if

$$c = \frac{\mathbf{X}_2 \cdot \mathbf{V}_1}{\| \mathbf{V}_1 \|^2}.$$

Therefore set

$$\mathbf{V}_2 = \mathbf{X}_2 - \frac{\mathbf{X}_2 \cdot \mathbf{V}_1}{\| \mathbf{V}_1 \|^2}\mathbf{V}_1.$$

Observe that $\mathbf{V}_2$ is $\mathbf{X}_2$, minus the projection of $\mathbf{X}_2$ onto $\mathbf{V}_1$.

If $m = 2$ we are done. If $m \geq 3$, produce nonzero $\mathbf{V}_3$ in S orthogonal to $\mathbf{V}_1$ and $\mathbf{V}_2$ as follows. Try

$$\mathbf{V}_3 = \mathbf{X}_3 - d\mathbf{V}_1 - h\mathbf{V}_2.$$

We need

$$\mathbf{V}_3 \cdot \mathbf{V}_2 = \mathbf{X}_3 \cdot \mathbf{V}_2 - d\mathbf{V}_1 \cdot \mathbf{V}_2 - h\mathbf{V}_2 \cdot \mathbf{V}_2 = 0,$$

so

$$h = \frac{\mathbf{X}_3 \cdot \mathbf{V}_2}{\| \mathbf{V}_2 \|^2}.$$

And we need

$$\mathbf{V}_3 \cdot \mathbf{V}_1 = \mathbf{X}_3 \cdot \mathbf{V}_1 - d\mathbf{V}_1 \cdot \mathbf{V}_1 = 0,$$

so

$$d = \frac{\mathbf{V}_3 \cdot \mathbf{V}_1}{\mathbf{V}_1 \cdot \mathbf{V}_1} = \frac{\mathbf{V}_3 \cdot \mathbf{V}_1}{\| \mathbf{V}_1 \|^2}\mathbf{V}_1.$$

Therefore, choose

$$\mathbf{V}_3 = \mathbf{X}_3 - \frac{\mathbf{X}_3 \cdot \mathbf{V}_1}{\| \mathbf{V}_1 \|^2}\mathbf{V}_1 - \frac{\mathbf{X}_3 \cdot \mathbf{V}_2}{\| \mathbf{V}_2 \|^2}\mathbf{V}_2.$$

This is $\mathbf{X}_3$, minus the projections of $\mathbf{X}_3$ onto $\mathbf{V}_1$ and $\mathbf{V}_2$.

This pattern suggests a general procedure. Set $\mathbf{V}_1 = \mathbf{X}_1$ and, for $j = 2, \cdots, m$, $\mathbf{V}_j$ equal to $\mathbf{X}_j$ minus the projections of $\mathbf{X}_j$ onto $\mathbf{V}_1, \cdots, \mathbf{V}_{j-1}$. This gives us

$$\mathbf{V}_j = \mathbf{X}_j - \frac{\mathbf{X}_j \cdot \mathbf{V}_1}{\| \mathbf{V}_1 \|^2}\mathbf{V}_1 - \frac{\mathbf{X}_j \cdot \mathbf{V}_2}{\| \mathbf{V}_2 \|^2}\mathbf{V}_2 - \cdots - \frac{\mathbf{X}_j \cdot \mathbf{V}_{j-1}}{\| \mathbf{V}_{j-1} \|^2}\mathbf{V}_{j-1},$$

for $j = 2, \cdots, m$.

This way of forming mutually orthogonal vectors from $\mathbf{X}_1, \cdots, \mathbf{X}_m$ is called the *Gram-Schmidt orthogonalization process*. When we use it, we say that we have *orthogonalized* the given basis for S (in the sense of replacing that basis with an orthogonal basis).

The vectors $\mathbf{V}_1, \cdots, \mathbf{V}_m$ are linearly independent because they are orthogonal. Further, they span the same subspace S of R^n that $\mathbf{X}_1, \cdots, \mathbf{X}_m$ span, because each $\mathbf{V}_j$ is a linear combination of the $\mathbf{X}_j$ vectors, which span S. The vectors $\mathbf{V}_j$ therefore form an orthogonal basis for S. If we want an orthonormal basis, then divide each $\mathbf{V}_j$ by its length.

EXAMPLE 6.19

Let S be the subspace of R^7 having basis

$$\mathbf{X}_1 = <1, 2, 0, 0, 2, 0, 0>, \mathbf{X}_2 = <0, 1, 0, 0, 3, 0, 0>, \mathbf{X}_3 = <1, 0, 0, 0, -5, 0, 0>.$$

We will produce an orthogonal basis for S. First let

$$\mathbf{V}_1 = \mathbf{X}_1 = <1, 2, 0, 0, 2, 0, 0>.$$

Next let

$$\mathbf{V}_2 = \mathbf{X}_2 - \frac{\mathbf{X}_2 \cdot \mathbf{V}_1}{\| \mathbf{V}_1 \|^2}\mathbf{V}_1 = <0, 1, 0, 0, 3, 0, 0> - \frac{8}{9} <1, 2, 0, 0, 2, 0, 0> = <-8/9, -7/9, 0, 0, 11/9, 0, 0>.$$

Finally, let

$$\mathbf{V}_3 = \mathbf{X}_3 - \frac{\mathbf{X}_3 \cdot \mathbf{V}_1}{\| \mathbf{V}_1 \|^2}\mathbf{V}_1 - \frac{\mathbf{X}_3 \cdot \mathbf{V}_2}{\| \mathbf{V}_2 \|^2}\mathbf{V}_2$$

$$= < 1, 0, 0, 0, -5, 0, 0 > + < 1, 2, 0, 0, 2, 0, 0 >$$

$$+ \frac{63}{26} < -8/9, -7/9, 0, 0, 11/9, 0, 0 >$$

$$= < -2/13, 3/26, 0, 0, -1/26, 0, 0 > .$$

Then $\mathbf{V}_1, \mathbf{V}_2, \mathbf{V}_3$ form an orthogonal basis for S. ♦

SECTION 6.5 PROBLEMS

In each of Problems 1 through 8, use the Gram-Schmidt process to find an orthogonal basis spanning the same subspace of R^n as the given set of vectors.

1. $< 0, 0, 1, 1, 0, 0 >, < 0, 0, -3, 0, 0, 0 >$ in R^6
2. $< 0, -2, 0, -2, 0, -2 >, < 0, 1, 0, -1, 0, 0 >, < 0, -4, 0, 0, 0, 6 >$ in R^6
3. $< 0, 0, 2, 2, 1 >, < 0, 0, 1, -1, 5 >, < 0, 1, -2, 1, 0 >, < 0, 1, 1, 2, 0 >$ in R^5
4. $< 1, 2, 0, -1, 2, 0 >, < 3, 1, -3, -4, 0, 0 >, < 0, -1, 0, -5, 0, 0 >, < 1, -6, 4, -2, -3, 0 >$ in R^6
5. $< 0, 2, 1, -1 >, < 0, -1, 1, 6 >, < 0, 2, 2, 3 >$ in R^4
6. $< -1, 0, 3, 0, 4 >, < 4, 0, -1, 0, 3 >, < 0, 0, -1, 0, 5 >$ in R^5
7. $< 1, 4, 0 >, < 2, -5, 0 >$ in R^3.
8. $< 0, -1, 2, 0 >, < 0, 3, -4, 0 >$ in R^4

6.6 Orthogonal Complements and Projections

The Gram-Schmidt process serves as a springboard to an important concept that has practical consequences, including the rationale for least squares approximations (see Section 7.8).

Let S be a subspace of R^n. Denote by $S^\perp$ the set of all vectors in R^n that are orthogonal to every vector in S. $S^\perp$ is called the *orthogonal complement* of S in R^n.

For example, in R^3, suppose S is the two-dimensional subspace having $< 1, 0, 0 >$ and $< 0, 1, 0 >$ as basis. We think of S as the x, y - plane. Now $S^\perp$ consists of all vectors in 3-space that are perpendicular to this plane, hence all constant multiples of $\mathbf{k}$.

In this example, $S^\perp$ is a subspace of R^3. We claim that this is always true.

THEOREM 6.6

If S is a subspace of R^n, then $S^\perp$ is also a subspace of R^n. Further, the only vector in both S and $S^\perp$ is the zero vector. ♦

Proof The zero vector is certainly in $S^\perp$ because $\mathbf{O}$ is orthogonal to every vector, hence to every vector in S.

Next we will show that linear combinations of vectors in $S^\perp$ are in $S^\perp$. Suppose $\mathbf{u}$ and $\mathbf{v}$ are in $S^\perp$. Then $\mathbf{u}$ and $\mathbf{v}$ are orthogonal to every vector in S. If c and d are real numbers and $\mathbf{w}$ is in S, then

$$\mathbf{w} \cdot (c\mathbf{u} + d\mathbf{v}) = c\mathbf{w} \cdot \mathbf{u} + d\mathbf{w} \cdot \mathbf{v} = 0 + 0 = 0.$$

Therefore $\mathbf{w}$ is orthogonal to $c\mathbf{u} + d\mathbf{v}$, so $c\mathbf{u} + d\mathbf{v}$ is in $S^\perp$ and $S^\perp$ is a subspace of R^n.

Certainly $\mathbf{O}$ is in both S and $S^\perp$. If $\mathbf{u}$ is in both S and $S^\perp$, then $\mathbf{u}$ is orthogonal to itself, so

$$\mathbf{u} \cdot \mathbf{u} = \| \mathbf{u} \|^2 = 0$$

and then $\mathbf{u} = \mathbf{O}$. ♦

We will now show that, given a subspace S of R^n, containing nonzero vectors, then each vector in R^n has a unique decomposition into the sum of a vector in S and a vector in $S^\perp$. This decomposition will prove useful in developing approximation techniques in Section 7.8.

THEOREM 6.7

Let S be a nontrivial subspace of R^n and let $\mathbf{u}$ be in R^n. Then there is exactly one vector $\mathbf{u}_S$ in S and exactly one vector $\mathbf{u}^\perp$ in $S^\perp$ such that

$$\mathbf{u} = \mathbf{u}_S + \mathbf{u}^\perp. \; ♦$$

Proof We know that we can produce an orthogonal basis $\mathbf{V}_1, \cdots, \mathbf{V}_m$ for S. Define

$$\mathbf{u}_S = \frac{\mathbf{u} \cdot \mathbf{V}_1}{\| \mathbf{V}_1 \|^2}\mathbf{V}_1 + \frac{\mathbf{u} \cdot \mathbf{V}_2}{\| \mathbf{V}_2 \|^2}\mathbf{V}_2 + \cdots + \frac{\mathbf{u} \cdot \mathbf{V}_m}{\| \mathbf{V}_m \|^2}\mathbf{V}_m$$

$$= \sum_{j=1}^{m} \frac{\mathbf{u} \cdot \mathbf{V}_j}{\mathbf{V}_j \cdot \mathbf{V}_j}\mathbf{V}_j.$$

$\mathbf{u}_S$ is the sum of the projections of $\mathbf{u}$ onto each of the orthogonal basis vectors $\mathbf{V}_1, \cdots, \mathbf{V}_m$, and is in S because this is a linear combination of the basis vectors of S. Next set

$$\mathbf{u}^\perp = \mathbf{u} - \mathbf{u}_S.$$

Certainly $\mathbf{u} = \mathbf{u}_S + \mathbf{u}^\perp$. All that remains to show is that $\mathbf{u}^\perp$ is in $S^\perp$. To show this, we must show that $\mathbf{u}^\perp$ is orthogonal to every vector in S. Since every vector in S is a linear combination of $\mathbf{V}_1, \cdots, \mathbf{V}_m$, it is enough to show that $\mathbf{u}^\perp$ is orthogonal to each $\mathbf{V}_j$. Begin with $\mathbf{V}_1$. Since $\mathbf{V}_1 \cdot \mathbf{V}_j = 0$ if $j \neq 1$,

$$\mathbf{u}^\perp \cdot \mathbf{V}_1 = (\mathbf{u} - \mathbf{u}_S) \cdot \mathbf{V}_1$$

$$= \mathbf{u} \cdot \mathbf{V}_1 - \left(\sum_{j=1}^{m} \frac{\mathbf{u} \cdot \mathbf{V}_j}{\mathbf{V}_j \cdot \mathbf{V}_j}\mathbf{V}_j \right) \cdot \mathbf{V}_1$$

$$= \mathbf{u} \cdot \mathbf{V}_1 - \frac{\mathbf{u} \cdot \mathbf{V}_1}{\mathbf{V}_1 \cdot \mathbf{V}_1}(\mathbf{V}_1 \cdot \mathbf{V}_1) = 0.$$

Similarly, $\mathbf{u}^\perp \cdot \mathbf{V}_j = 0$ for $j = 2, \cdots, m$. Therefore $\mathbf{u}^\perp$ is in $S^\perp$.

Finally, we must show that $\mathbf{u}$ can be written in only one way as the sum of a vector in S and a vector in $S^\perp$. Suppose

$$\mathbf{u} = \mathbf{u}_S + \mathbf{u}^\perp = \mathbf{U} + \mathbf{U}^\perp,$$

where $\mathbf{U}$ is in S and $\mathbf{U}^\perp$ is in $S^\perp$. Then

$$\mathbf{u}_S - \mathbf{U} = \mathbf{u}^\perp - \mathbf{U}^\perp.$$

The vector on the left is in S and the vector on the right is in $S^{\perp}$. Therefore both sides equal the zero vector, so

$$\mathbf{u}_S = \mathbf{U} \text{ and } \mathbf{u}^{\perp} = \mathbf{U}^{\perp}.$$

This completes the proof. ♦

Notice in the theorem that, if $\mathbf{u}$ is actually in S, then $\mathbf{u}_S = \mathbf{u}$ and $\mathbf{u}^{\perp} = \mathbf{O}$.

The vector $\mathbf{u}_S$ produced in the proof is called the *orthogonal projection* of $\mathbf{u}$ onto S. It is the sum of the projections of $\mathbf{u}$ onto an orthogonal basis for S.

It would appear from the way $\mathbf{u}_S$ was formed that this orthogonal projection depends on the orthogonal basis specified for S. In fact it does not, and any orthogonal basis for S leads to the same orthogonal projection $\mathbf{u}_S$, justifying the term *the* orthogonal projection of $\mathbf{u}$ onto S. The reason for this is that, given $\mathbf{u}$, the orthogonal projection of $\mathbf{u}$ onto S is the unique vector in S such that $\mathbf{u}$ is the sum of this projection and a vector in $S^{\perp}$.

It is therefore true that, if we write a vector $\mathbf{u}$ as the sum of a vector in S and a vector in $S^{\perp}$, then necessarily the vector in S is $\mathbf{u}_S$ and the vector in $S^{\perp}$ is $\mathbf{u} - \mathbf{u}_S$. In particular, $\mathbf{u} - \mathbf{u}_S$ is orthogonal to every vector in S.

EXAMPLE 6.20

Let S be the subspace of R^5 consisting of all $< x, 0, y, 0, z >$ having zero second and fourth components. Let

$$\mathbf{u} = < 1, 4, 1, -1, 3 > .$$

We will determine at $\mathbf{u}_S$ and $\mathbf{u}^{\perp}$. First use the orthogonal basis

$$\mathbf{V}_1 = < 1, 0, 0, 0, 0 >, \mathbf{V}_2 = < 0, 0, 1, 0, 2 >, \mathbf{V}_3 = < 0, 0, 2, 0, -1 >$$

for S. The orthogonal projection $\mathbf{u}_S$ is

$$\begin{aligned} \mathbf{u}_S &= \frac{\mathbf{u} \cdot \mathbf{V}_1}{\mathbf{V}_1 \cdot \mathbf{V}_1}\mathbf{V}_1 + \frac{\mathbf{u} \cdot \mathbf{V}_2}{\mathbf{V}_2 \cdot \mathbf{V}_2}\mathbf{V}_2 + \frac{\mathbf{u} \cdot \mathbf{V}_3}{\mathbf{V}_3 \cdot \mathbf{V}_3}\mathbf{V}_3 \\ &= \mathbf{V}_1 + \frac{7}{5}\mathbf{V}_2 - \frac{1}{5}\mathbf{V}_3 \\ &= < 1, 0, 1, 0, 3 >, \end{aligned}$$

and

$$\mathbf{u}^{\perp} = \mathbf{u} - \mathbf{u}_S = < 0, 4, 0, -1, 0 >$$

is in the orthogonal complement of S, being orthogonal to every vector in S, and $\mathbf{u} = \mathbf{u}_S + \mathbf{u}^{\perp}$.

Suppose we used a different orthogonal basis for S, say

$$\mathbf{V}_1^* = < 1, 0, 1, 0, 0 >, \mathbf{V}_2^* = < -3, 0, 3, 0, 0 >, \mathbf{V}_3^* = < 0, 0, 0, 0, 6 > .$$

Now compute the orthogonal projection of $\mathbf{u}$ with respect to this basis:

$$\begin{aligned} &\frac{\mathbf{u} \cdot \mathbf{V}_1^*}{\mathbf{V}_1^* \cdot \mathbf{V}_1^*}\mathbf{V}_1^* + \frac{\mathbf{u} \cdot \mathbf{V}_2^*}{\mathbf{V}_2^* \cdot \mathbf{V}_2^*}\mathbf{V}_2^* + \frac{\mathbf{u} \cdot \mathbf{V}_3^*}{\mathbf{V}_3^* \cdot \mathbf{V}_3^*}\mathbf{V}_3^* \\ &= \mathbf{V}_1^* + 0\mathbf{V}_2^* + \frac{1}{2}\mathbf{V}_3^* \\ &= < 1, 0, 1, 0, 3 >, \end{aligned}$$

the same as obtained using the first orthogonal basis. This illustrates the uniqueness of $\mathbf{u}_S$, given $\mathbf{u}$ and S. ♦

We will now show that $\mathbf{u}_S$ has a remarkable property—it is the unique vector in S that is closest to $\mathbf{u}$. That is, the distance between $\mathbf{u}$ and $\mathbf{u}_S$ is less than or equal to the distance between $\mathbf{u}$ and $\mathbf{v}$ for every $\mathbf{v}$ in S:

$$\| \mathbf{u} - \mathbf{u}_S \| < \| \mathbf{u} - \mathbf{v} \| \text{ for every } \mathbf{v} \text{ in } S.$$

THEOREM 6.8

Let S be a nontrivial subspace of R^n and let $\mathbf{u}$ be in R^n. Then, for all vectors $\mathbf{v}$ in S different from $\mathbf{u}_S$,

$$\| \mathbf{u} - \mathbf{u}_S \| < \| \mathbf{u} - \mathbf{v} \| . \blacklozenge$$

Proof If $\mathbf{u}$ is in S, then $\mathbf{u} = \mathbf{u}_S$ and $\| \mathbf{u} - \mathbf{u}_S \| = 0$. Clearly $\mathbf{u}$ is the unique vector in S closest to itself.

Thus suppose that $\mathbf{u}$ is not in S. Let $\mathbf{v}$ be any vector in S different from $\mathbf{u}_S$. Write

$$\mathbf{u} - \mathbf{v} = (\mathbf{u} - \mathbf{u}_S) + (\mathbf{u}_S - \mathbf{v}).$$

Now $\mathbf{u}_S - \mathbf{v}$ is in S, being a sum of vectors in S. And we know that $\mathbf{u} - \mathbf{u}_S$ is in $S^\perp$. Therefore $\mathbf{u}_S - \mathbf{v}$ and $\mathbf{u} - \mathbf{u}_S$ are orthogonal. By the Pythagorean theorem,

$$\| \mathbf{u} - \mathbf{v} \|^2 = \| \mathbf{u} - \mathbf{u}_S \|^2 + \| \mathbf{u}_S - \mathbf{v} \|^2 .$$

But $\mathbf{u} \neq \mathbf{u}_S$, so

$$\| \mathbf{u} - \mathbf{u}_S \| > 0.$$

Therefore

$$\| \mathbf{u} - \mathbf{v} \|^2 > \| \mathbf{u}_S - \mathbf{v} \|^2$$

and this is equivalent to the conclusion of the theorem. $\blacklozenge$

EXAMPLE 6.21

Let S be the subspace of R^5 having orthogonal basis vectors

$$\mathbf{V}_1 = < 1, 0, 0, 0, 0, 0 >, \mathbf{V}_2 = < 0, 1, 0, 0, 0, 1 >, \mathbf{V}_3 = < 0, 1, 0, 0, 0, -1 > .$$

Let $\mathbf{u} = < 1, -1, 4, 1, 2, -5 >$. We will find the vector in S closest to $\mathbf{u}$. We may also think of this as the distance between $\mathbf{u}$ and S. First, the orthogonal projection of $\mathbf{u}$ onto S is

$$\begin{aligned} \mathbf{u}_S &= (\mathbf{u} \cdot \mathbf{v}_1)\mathbf{v}_1 + \frac{1}{2}(\mathbf{u} \cdot \mathbf{v}_2)\mathbf{v}_1 + \frac{1}{2}(\mathbf{u} \cdot \mathbf{v}_3)\mathbf{v}_3 \\ &= \mathbf{v}_1 - 3\mathbf{v}_2 + 2\mathbf{v}_3 \\ &= < 1, -1, 0, 0, 0, -5 > . \end{aligned}$$

Then

$$\| \mathbf{u} - \mathbf{u}_S \| = \sqrt{21}.$$

This is the distance between $\mathbf{u}$ and the vector in S closest to $\mathbf{u}$. $\blacklozenge$

Because the distance between two vectors is the square root of a sum of squares, use of Theorem 6.8 to find a vector at minimum distance from a given vector is called the *method of least squares*. We will pursue the idea of least squares approximations in the next section and in Section 7.8.

SECTION 6.6 PROBLEMS

In each of Problems 1 through 5, write $\mathbf{u}$ as a sum of a vector in S and a vector in $S^{\perp}$.

1. S has orthogonal basis $< 1, -1, 0, 1, -1 >$, $< 1, 0, 0, -1, 0 >$, $< 0, -1, 0, 0, 1 >$ in R^5, $\mathbf{u} =< 4, -1, 3, 2, -7 >$.
2. S has orthogonal basis $< 1, -1, 0, 0 >$, $< 1, 1, 6, 1 >$ in R^4, $\mathbf{u} =< 3, 9, 4, -5 >$.
3. S has orthogonal basis $< 1, 0, 1, 0, 1, 0, 0 >$, $< 0, 1, 0, 1, 0, 0, 0 >$ in R^7, $\mathbf{u} =< 8, 1, 1, 0, 0, -3, 4 >$.
4. S has orthogonal basis $< 1, 0, 0, 2, 0 >$, $< -2, 0, 0, 1, 0 >$ in R^5, $\mathbf{u} =< 0. -4, -4, 1, 3 >$.
5. S has orthogonal basis $< 1, -1, 0, 0 >$, $< 1, 1, 0, 0 >$ in R^4, $\mathbf{u} =< -2, 6, 1, 7 >$.
6. Let S be the subspace of R^6 spanned by $< 0, 1, 1, 0, 0, 1 >$, $< 0, 0, 3, 0, 0, -3 >$, and $< 0, 0, 0, 0, 0, 4 >$. Find the vector in S closest to $< 0, 1, 1, -2, -2, 6 >$.
7. Let S be the subspace of R^5 spanned by $< 1, 1, -1, 0, 0 >$, $< 0, 2, 1, 0, 0 >$ and $< 0, 1, -2, 0, 0 >$. Find the vector in S closest to $< 3, 0, 0, 1, 4 >$.
8. Let S be a subspace of R^n. Determine $(S^{\perp})^{\perp}$.
9. Suppose S is a subspace of R^n. Determine a relationship between the dimensions of S and $S^{\perp}$.
10. Let S be the subspace of R^4 spanned by $< 1, 0, 1, 0 >$ and $< 0, 0, 2, 1 >$. Find the vector in S closest to $< 1, -1, 3, -3 >$.

6.7 The Function Space $C[a, b]$

We will extend the notion of a vector space from R^n to a space of functions. This will enable us to view Theorem 6.8 as an approximation tool for functions as well as an introduction to Fourier series and eigenfunction expansions in Chapters 13 and 15.

Let $C[a, b]$ denote the set of all (real-valued) functions that are continuous on a closed interval $[a, b]$. If f and g are continuous on $[a, b]$, so is their sum $f + g$, defined by

$$(f + g)(x) = f(x) + g(x).$$

Furthermore, if c is any real number, then cf, defined by

$$(cf)(x) = cf(x)$$

is also continuous on $[a, b]$.

The zero function θ is defined by $\theta(x) = 0$ for $a \leq x \leq b$, and this is in $C[a, b]$.

These operations of addition of functions and multiplication of functions by scalars have the same properties in $C[a, b]$ as addition of vectors and multiplication of vectors by scalars in R^n. In this sense $C[a, b]$ has an algebraic structure like that of R^n, and we also refer to $C[a, b]$ as a vector space. In this space we continue to denote functions by upper and lower case letters, rather than the boldface we used for matrices and vectors in R^n.

Many of the concepts developed for vectors in R^n extend readily to this function space. We say that $f_1, f_2, \cdots, f_n$ in $C[a, b]$ are *linearly dependent* if there are numbers $c_1, \cdots, c_n$, not all zero, such that

$$c_1 f_1 + c_2 f_2 + \cdots + c_n f_n = \theta.$$

This means that

$$c_1 f_1(x) + c_2 f_2(x) + \cdots + c_n f_n(x) = 0$$

for $a \leq x \leq b$. Linear independence means that the only way a linear combination of $f_1, \cdots, f_n$ can be the zero function is for all the coefficients to be zero. This is the same as asserting that no f_j is a linear combination of the other functions. We saw this concept, without reference to the vector space context, when dealing with solutions of second order linear homogeneous differential equations in Chapter 2.

One significant difference between $C[a,b]$ and R^n is that R^n has a basis consisting of n vectors, hence has dimension n. However, $C[a,b]$ has no such finite basis. Consider, for example, the functions

$$p_0(x) = 1,\ p_1(x) = x,\ p_2(x) = x^2,\ p_3(x) = x^3, \cdots, p_n(x) = x^n,$$

with n any positive integer. These functions are all in $C[a,b]$ and are linearly independent. The reason for this is that, if

$$c_1 + c_2x + c_3x^2 + \cdots + c_nx^n = 0$$

for all x in $[a,b]$, then each $c_i = 0$ because a real polynomial of degree n can have at most n distinct roots. We can produce arbitrarily large linearly independent sets of functions in $C[a,b]$, hence $C[a,b]$ can have no finite basis.

We can introduce a dot product for functions in $C[a,b]$ as follows. Select a function p that is continuous on $[a,b]$, with $p(x) > 0$ for $a < x < b$. If f and g are in $C[a,b]$, define

$$f \cdot g = \int_a^b p(x)f(x)g(x)\,dx.$$

This operation is called a *dot product with weight function* p, and it has all of the properties we saw for dot products of vectors. In particular:

1. $f \cdot g = g \cdot f$,
2. $(f+g) \cdot h = f \cdot h + g \cdot h$,
3. $c(f \cdot g) = (cf) \cdot g = f \cdot (cg)$,
4. $f \cdot f \geq 0$, and $f \cdot f = 0$ if and only if $f(x) = 0$ for $a \leq x \leq b$.

In view of property (4), we can, as in R^n, define the *norm* or *length* of f to be

$$\| f \| = \sqrt{f \cdot f} = \sqrt{\int_a^b p(x)(f(x))^2\,dx}.$$

Once we have the norm of a function, we can define the *distance* between f and g to be the norm of $f - g$. This is

$$\| f - g \| = \sqrt{(f-g)\cdot(f-g)}$$
$$= \sqrt{\int_a^b p(x)(f(x) - g(x))^2\,dx}.$$

Continuing the analogy with R^n, define f and g to be *orthogonal* if $f \cdot g = 0$. This means that

$$\int_a^b p(x)f(x)g(x)\,dx = 0.$$

These definitions enable us to think geometrically in the function space $C[a,b]$, with concepts of distance between functions and orthogonality. The Gram-Schmidt process extends verbatim to subspaces of $C[a,b]$ using this integral dot product.

EXAMPLE 6.22

Let n and m be positive integers, and let $S_n(x) = \sin(nx)$ and $C_m(x) = \cos(mx)$. These functions are in $C[-\pi, \pi]$. Let $p(x) = 1$ to use the dot product

$$f \cdot g = \int_{-\pi}^{\pi} f(x)g(x)\,dx$$

in $C[-\pi, \pi]$. With respect to this dot product, $S_n(x)$ and $C_m(x)$ are orthogonal, because their dot product is

$$S_n \cdot C_m = \int_{-\pi}^{\pi} \sin(nx)\cos(mx)\,dx = 0,$$

by a routine integration. This type of orthogonality of functions will form the basis for Fourier series in Chapter 13, and for more general eigenfunction expansions in Chapter 15. ♦

Theorems 6.6, 6.7, and 6.8 and their proofs, while stated for vectors in R^n, depend only on the vector space structure in which they were stated, and are valid in $C[a, b]$ as well. Here is an application of Theorem 6.8.

EXAMPLE 6.23

Suppose we want to approximate $f(x) = x(\pi - x)$ on $[0, \pi]$, using a sum of the form

$$c_1 \sin(x) + c_2 \sin(2x) + c_3 \sin(3x) + c_4 \sin(4x).$$

The term "approximate" has meaning only in the context of some measure of distance, since we generally call one object a good approximation to another when the objects are close together in some sense. The necessary structure is available to us if we work in the function space $C[0, \pi]$, which contains $f(x)$ and the functions $\sin(nx)$. Using the integral dot product with $p(x) = 1$, the distance between two functions in $C[0, \pi]$ is

$$\| F - G \| = \sqrt{(F - G) \cdot (F - G)} = \sqrt{\int_0^{\pi} (F(x) - G(x))^2\,dx}.$$

To make use of Theorem 6.8, let S be the four-dimensional subspace of $C[0, \pi]$ spanned by $\sin(x)$, $\sin(2x)$, $\sin(3x)$ and $\sin(4x)$. Then S consists of exactly the linear combinations

$$c_1 \sin(x) + c_2 \sin(2x) + c_3 \sin(3x) + c_4 \sin(4x)$$

that we want to use to approximate $f(x)$. f is not in S. By Theorem 6.8, the object in S closest to f is the orthogonal projection f_S of f onto S. This is

$$f_S = \frac{f \cdot \sin(x)}{\| \sin(x) \|^2} \sin(x) + \frac{f \cdot \sin(2x)}{\| \sin(2x) \|^2} \sin(2x) + \frac{f \cdot \sin(3x)}{\| \sin(3x) \|^2} \sin(3x) + \frac{f \cdot \sin(4x)}{\| \sin(4x) \|^2} \sin(4x).$$

All that remains is to compute these coefficients. First, for $n = 1, 2, 3, 4$,

$$\| \sin(nx) \|^2 = \int_0^{\pi} \sin^2(nx)\,dx = \frac{\pi}{2}.$$

Furthermore,

$$f \cdot \sin(nx) = \int_0^{\pi} x(\pi - x)\sin(nx)\,dx = \frac{2(1 - (-1)^n)}{n^3}.$$

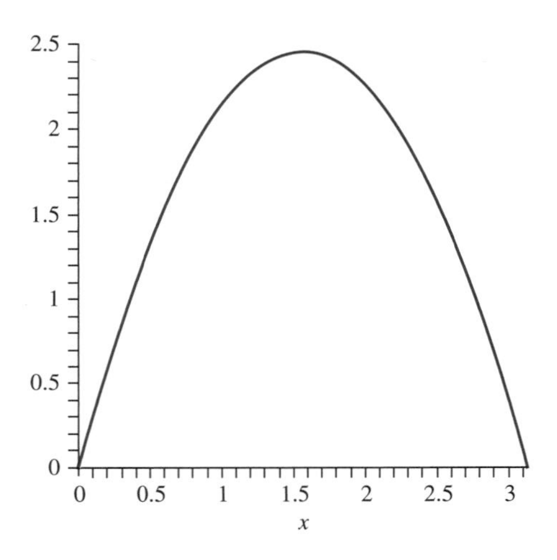

FIGURE 6.16 *$f(x)$ and $f_S(x)$ in Example 6.23.*

Therefore,

$$\frac{f(x)\cdot\sin(nx)}{\|\sin(nx)\|^2}=\frac{4(1-(-1)^n)}{\pi n^3}$$

for $n=1,2,3,4$. This number is 0 for $n=2$ and $n=4$, and equals $8/\pi$ for $n=1$ and $8/27\pi$ for $n=3$. The function in S having minimum distance (that is, the closest approximation) to $x(\pi-x)$, using this dot product metric, is

$$f_S(x)=\frac{8}{\pi}\sin(x)+\frac{8}{27\pi}\sin(3x).$$

Figure 6.16 is a graph of $f(x)$ and $f_S(x)$ on $[0,\pi]$. In the scale of the drawing, the graphs are nearly indistinguishable, so in this example the approximation appears to be quite good. More specifically, the square of the distance between $f(x)$ and $f_S(x)$ is

$$\begin{aligned}\|f-f_S\|^2&=\int_0^\pi (f(x)-f_S(x))^2\,dx\\&=\int_0^\pi \left(x(x-\pi)-\frac{8}{\pi}\sin(x)-\frac{8}{27\pi}\sin(3x)\right)^2 dx\\&\approx 0.0007674.\ \blacklozenge\end{aligned}$$

The apparent accuracy we saw in this example is not guaranteed in general, since we did no analysis to estimate errors or to determine how many terms of the form $\sin(nx)$ would have to be used to approximate $f(x)$ to within a certain tolerance. Nevertheless, Theorem 6.8 forms a starting point for some approximation schemes.

EXAMPLE 6.24

Suppose we want to approximate $f(x)=e^x$ on $[-1,1]$ by a linear combination of the first three Legendre polynomials. These polynomials are developed in Section 15.2, and the first three are

$$P_0(x)=1,\ P_1(x)=x,\ P_2(x)=\frac{1}{2}(3x^2-1).$$

These polynomials are orthogonal in $C[-1, 1]$, using the integral dot product

$$f \cdot g = \int_{-1}^{1} f(x)g(x)\,dx.$$

This means that

$$\int_{-1}^{1} P_n(x)P_m(x)\,dx = 0 \text{ if } n \neq m.$$

Let S be the subspace of $C[-1, 1]$ spanned by $P_0(x)$, $P_1(x)$, $P_2(x)$. The orthogonal projection of f onto S is

$$\begin{aligned} f_S(x) &= a_0 P_0(x) + a_1 P_1(x) + a_2 P_2(x) \\ &= a_0 + a_1 x + a_2 \frac{1}{2}(3x^2 - 1), \end{aligned}$$

where

$$\begin{aligned} a_n &= \frac{f(x) \cdot P_n(x)}{P_n(x) \cdot P_n(x)} \\ &= \frac{\int_{-1}^{1} e^x P_n(x)\,dx}{\int_{-1}^{1} P_n^2(x)\,dx} \end{aligned}$$

for $n = 0, 1, 2$. These integrals are easily done using MAPLE and we find that

$$a_0 = \frac{1}{2}(e - e^{-1}),\ a_1 = 3e^{-1},\ a_2 = -\frac{35}{2}e^{-1} + \frac{5}{2}e.$$

Using these coefficients, $f_s(x)$ is the closest approximation (in the distance defined by this dot product) to $\exp(x)$ on $[-1, 1]$. Figure 6.17 shows graphs of $f(x)$ and $f_S(x)$ on this interval.

We can improve the accuracy of this polynomial approximation by including more terms. Suppose S^* is the subspace of $C[-1, 1]$ generated by the orthogonal basis consisting of the first four Legendre polynomials. These are the three given previously, together with

$$P_3(x) = \frac{1}{2}(5x^3 - 3x).$$

S^* differs from S by the inclusion of $P_3(x)$ in the basis. Compute the orthogonal projection of $f(x)$ onto S^* to obtain

$$f_{S^*}(x) = \sum_{n=0}^{3} a_n P_n(x).$$

where a_0, a_1 and a_2 are as before, and

$$\begin{aligned} a_3 &= \frac{\int_{-1}^{1} e^x P_3(x)\,dx}{\int_{-1}^{1} P_3^2(x)\,dx} \\ &= \frac{259}{2}e^{-1} - \frac{35}{2}e. \end{aligned}$$

Figure 6.18 shows graphs of $f(x)$ and $f_{S^*}(x)$ on $[-1, 1]$. These graphs are nearly indistinguishable in the scale of the drawing.

With a little more computation we can quantify the distance between f and f_S and between f and f_{S^*}. The squares of these distances are

$$\| f - f_S \|^2 = \int_{-1}^{1} (f(x) - f_S(x))^2\,dx \approx 0.00144058$$

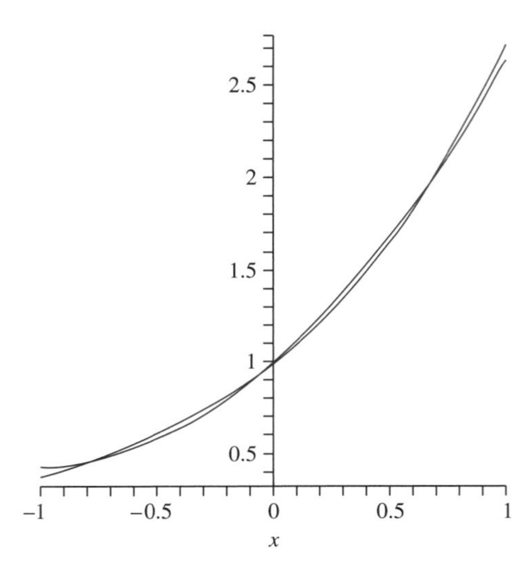

FIGURE 6.17 *f and f_S in Example 6.24.*

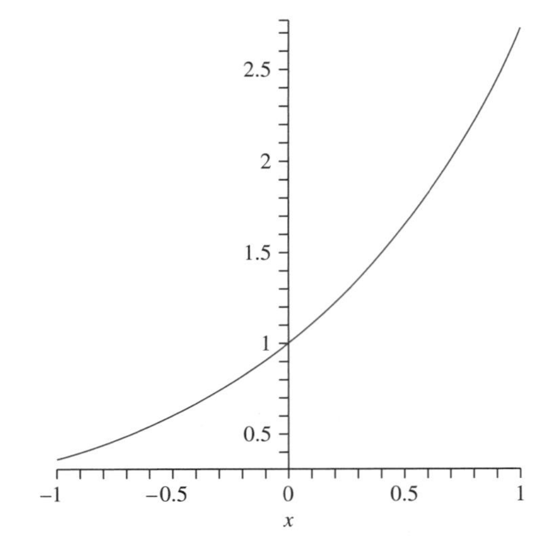

FIGURE 6.18 *f and f_{S^*} in Example 6.24.*

and

$$\| f - f_{S^*} \|^2 = \int_{-1}^{1} (f(x) - f_{S^*})^2 \, dx \approx 0.000022.$$

Then

$$\| f - f_S \| \approx 0.038 \text{ and } \| f - f_{S^*} \| \approx 0.005. \ \blacklozenge$$

SECTION 6.7 PROBLEMS

Problems 1 through 4, involve use of the Gram-Schmidt orthogonalization process in a function space $C[a, b]$.

1. In $C[0, 1]$, find an orthogonal set of functions that spans the same subspace as $1, x$ and x^2, using $p(x) = x$ in the weighted inner product.

2. In $C[0, 2]$, find an orthogonal set of functions that spans the same subspace as $1, \cos(\pi x/2)$, and $\sin(\pi x/2)$. Use $p(x) = x$ in the weighted inner product.

3. In $C[0, 1]$, find an orthogonal set of two functions that spans the same subspace as the two functions e^{-x} and e^x, using $p(x) = 1$ in the weighted inner product integral.

4. In $C[-\pi, \pi]$, find an orthogonal set of functions that spans the same subspace as $\sin(x)$, $\cos(x)$, and $\sin(2x)$. Use $p(x) = 1$ in the weighted inner product.

The following problems are in the spirit of Example 6.24.

5. Approximate $f(x) = x^2$ on $[0, \pi]$ with a linear combination of the functions $1, \cos(x), \cos(2x), \cos(3x)$, and $\cos(4x)$. Use $p(x) = 1$ in the weighted inner product on this function space. Graph $f(x)$ and the approximating linear combination on the same set of axes. *Hint*: Calculate f_S, the orthogonal projection of f onto the subspace of $C[0, \pi]$ spanned by $1, \cos(x), \cdots, \cos(4x)$.

6. Repeat Problem 5, except now use the functions $\sin(x), \cdots, \sin(5x)$.

7. Approximate $f(x) = x(2 - x)$ on $[-2, 2]$ using a linear combination of the functions $1, \cos(\pi x/2), \cos(\pi x), \cos(3\pi x/2), \sin(\pi x/2), \sin(\pi x)$, and $\sin(3\pi x/2)$. Graph f and the approximating function on the same set of axes. *Hint*: In $C[-2, 2]$, project f orthogonally onto the subspace spanned by the given functions. Use the weight function $p(x) = 1$ in the inner product for this function space.

MATRICES ELEMENTARY ROW OPERATIONS REDUCED ROW ECHELON FORM ROW AND COLUMN SPACES HOMOGENEOUS SYSTEMS NONHOMOGENEOUS SYSTEMS MATRIX

CHAPTER 7

Matrices and Linear Systems

7.1 Matrices

An n by m (or $n \times m$) *matrix* is a rectangular array of objects arranged in n rows and m columns.

We will denote matrices in boldface. For example,

$$\mathbf{A} = \begin{pmatrix} 2 & 1 & \pi \\ 1 & \sqrt{2} & -5 \end{pmatrix}$$

is a 2×3 matrix (two rows, three columns) and

$$\mathbf{B} = \begin{pmatrix} e^t & 1 & -1 & \cos(t) \\ 0 & 4t & -7 & 1-t \end{pmatrix}$$

is a 2×4 matrix.

The object located in the row i and column j place of a matrix is called its i, j *element*. Often we write $\mathbf{A} = [a_{ij}]$, meaning that the i, j element of $\mathbf{A}$ is a_{ij}. In the above matrices $\mathbf{A}$ and $\mathbf{B}$, $a_{11} = 2$, $a_{22} = \sqrt{2}$, $a_{23} = -5$, $b_{14} = \cos(t)$ and $b_{21} = 0$.

If the elements of an $n \times m$ matrix are real numbers, then each row can be thought of as a vector in R^m and each column as a vector in R^n. In the first example, $\mathbf{A}$ has two rows that are vectors in R^3 and columns forming three vectors in R^2. This vector point of view is often useful in dealing with matrices.

Two matrices $\mathbf{A} = [a_{ij}]$ and $\mathbf{B} = [b_{ij}]$ are *equal* if they have the same number of rows, the same number of columns, and for each i and j, $a_{ij} = b_{ij}$. Equal matrices have the same dimensions, and objects located in the same positions in the matrices must be equal.

There are three operations we will define for matrices: addition, multiplication by a real or complex number, and multiplication. These are defined as follows.

Addition of Matrices

If $\mathbf{A} = [a_{ij}]$ and $\mathbf{B} = [b_{ij}]$ are both $n \times m$ matrices, then their sum is defined to be the $n \times m$ matrix $\mathbf{A} + \mathbf{B} = [a_{ij} + b_{ij}]$.

We add two matrices of the same dimensions by adding objects in the same locations in the matrices. For example,

$$\begin{pmatrix} 1 & 2 & -3 \\ 4 & 0 & 2 \end{pmatrix} + \begin{pmatrix} -1 & 6 & 3 \\ 8 & 12 & 14 \end{pmatrix} = \begin{pmatrix} 0 & 8 & 0 \\ 12 & 12 & 16 \end{pmatrix}.$$

We can think of this as adding respective row vectors, or respective column vectors, of the matrix.

Multiplication by a Scalar

Multiply a matrix by a scalar quantity (say a number or function) by multiplying each matrix element by the scalar. If $\mathbf{A} = [a_{ij}]$, then $c\mathbf{A} = [ca_{ij}]$. For example,

$$\sqrt{2}\begin{pmatrix} -3 \\ 4 \\ 2t \\ \sin(2t) \end{pmatrix} = \begin{pmatrix} -3\sqrt{2} \\ 4\sqrt{2} \\ 2t\sqrt{2} \\ \sqrt{2}\sin(2t) \end{pmatrix}.$$

This is the same as multiplying each row vector, or each column vector, by c. As another example,

$$\cos(t)\begin{pmatrix} 2 & e^t \\ \sin(t) & 4 \end{pmatrix} = \begin{pmatrix} 2\cos(t) & e^t\cos(t) \\ \cos(t)\sin(t) & 4\cos(t) \end{pmatrix}.$$

Multiplication of Matrices

Let $\mathbf{A} = [a_{ij}]$ be $n \times k$ and $\mathbf{B} = [b_{ij}]$ be $k \times m$. Then the product $\mathbf{AB}$ is the $n \times m$ matrix whose i, j element is

$$a_{i1}b_{1j} + a_{i2}b_{2j} + \cdots + a_{ik}b_{kj},$$

or

$$\sum_{s=1}^{k} a_{is}b_{sj}.$$

This is the dot product of row i of $\mathbf{A}$ with column j of $\mathbf{B}$ (both are vectors in R^k):

$$\begin{aligned} &i, j \text{ element of } \mathbf{AB} = (\text{ row } i \text{ of } \mathbf{A}) \cdot (\text{ column } j \text{ of } \mathbf{B}) \\ &= (a_{i1}, a_{i2}, \cdots, a_{ik}) \cdot (b_{1j}, b_{2j}, \cdots, b_{kj}) \\ &= a_{i1}b_{1j} + a_{i2}b_{2j} + \cdots + a_{ik}b_{kj}. \end{aligned}$$

This clarifies why the number of columns of **A** must equal the number of rows of **B** for the product **AB** to be defined. We can only take the dot product of two vectors of the same dimension.

EXAMPLE 7.1

Let

$$\mathbf{A}=\begin{pmatrix}1 & 3\\ 2 & 5\end{pmatrix} \text{ and } \mathbf{B}=\begin{pmatrix}1 & 1 & 3\\ 2 & 1 & 4\end{pmatrix}.$$

Here **A** is 2×2 and **B** is 2×3, so we can compute **AB**, which is 2×3 (number of rows of **A**, number of columns of **B**). In terms of dot products of rows with columns,

$$\mathbf{AB}=\begin{pmatrix}1 & 3\\ 2 & 5\end{pmatrix}\begin{pmatrix}1 & 1 & 3\\ 2 & 1 & 4\end{pmatrix}$$

$$=\begin{pmatrix}<1,3>\cdot<1,2> & <1,3>\cdot<1,1> & <1,3>\cdot<3,4>\\ <2,5>\cdot<1,2> & <2,5>\cdot<1,1> & <2,5>\cdot<3,4>\end{pmatrix}$$

$$=\begin{pmatrix}7 & 4 & 15\\ 12 & 7 & 26\end{pmatrix}.$$

In this example, **BA** is not defined because the number of columns of **B** does not equal the number of rows of **A**. ♦

EXAMPLE 7.2

Let

$$\mathbf{A}=\begin{pmatrix}1 & 1 & 2 & 1\\ 4 & 1 & 6 & 2\end{pmatrix} \text{ and } \mathbf{B}=\begin{pmatrix}-1 & 8\\ 2 & 1\\ 1 & 1\\ 12 & 6\end{pmatrix}.$$

Because **A** is 2×4 and **B** is 4×2, then **AB** is defined and is 2×2:

$$\mathbf{AB}=\begin{pmatrix}<1,1,2,1>\cdot<-1,2,1,12> & <1,1,2,1>\cdot<8,1,1,6>\\ <4,1,6,2>\cdot<-1,2,1,12> & <4,1,6,2>\cdot<8,1,1,6>\end{pmatrix}$$

$$=\begin{pmatrix}15 & 17\\ 28 & 51\end{pmatrix}.$$

In this example, **BA** is also defined and is a 4×4 matrix:

$$\mathbf{BA}=\begin{pmatrix}-1 & 8\\ 2 & 1\\ 1 & 1\\ 12 & 6\end{pmatrix}\begin{pmatrix}1 & 1 & 2 & 1\\ 4 & 1 & 6 & 2\end{pmatrix}=\begin{pmatrix}31 & 7 & 46 & 15\\ 6 & 3 & 10 & 4\\ 5 & 2 & 8 & 3\\ 36 & 18 & 60 & 24\end{pmatrix}.$$

Even when both **AB** and **BA** are defined, these matrices may not be equal, and may not even have the same dimensions. Matrix multiplication is noncommutative. ♦

We will list some properties of these matrix operations.

THEOREM 7.1

Let **A**, **B** and **C** be matrices. Then, whenever the indicated operations are defined:

1. $\mathbf{A}+\mathbf{B}=\mathbf{B}+\mathbf{A}$ (matrix addition is commutative).
2. $\mathbf{A}(\mathbf{B}+\mathbf{C})=\mathbf{AB}+\mathbf{AC}$.
3. $(\mathbf{A}+\mathbf{B})\mathbf{C}=\mathbf{AC}+\mathbf{AC}$.
4. $(\mathbf{AB})\mathbf{C}=\mathbf{A}(\mathbf{BC})$.
5. $c\mathbf{AB}=(c\mathbf{A})\mathbf{B}=\mathbf{A}(c\mathbf{B})$ for any scalar c. ♦

Proof Proofs of these conclusions are straightforward. To illustrate, we will prove operation (3):

$$\begin{aligned} i, j \text{ element of } \mathbf{A}(\mathbf{B}+\mathbf{C}) &= (\text{row } i \text{ of } \mathbf{A}) \cdot (\text{column } j \text{ of } \mathbf{B}+\mathbf{C}) \\ &= (\text{row } i \text{ of } \mathbf{A}) \cdot (\text{column } j \text{ of } \mathbf{B}+ \text{ column } j \text{ of } \mathbf{C}) \\ &= (\text{row } i \text{ of } \mathbf{A}) \cdot (\text{column } j \text{ of } \mathbf{B}) + ((\text{row } i \text{ of } \mathbf{A}) \cdot (\text{column } j \text{ of } \mathbf{C}) \\ &= (i, j \text{ element of } \mathbf{AB}) + (i, j \text{ element of } \mathbf{AC}) \\ &= i, j \text{ element of } \mathbf{AB}+\mathbf{AC}. \end{aligned}$$ ♦

We have already noted that in some ways matrix multiplication does not behave like multiplication of real numbers. The following examples illustrate other differences.

EXAMPLE 7.3

Even when **AB** and **BA** are defined and have the same dimensions, it is possible that $\mathbf{AB} \neq \mathbf{BA}$:

$$\begin{pmatrix} 1 & 0 \\ 2 & -4 \end{pmatrix}\begin{pmatrix} -2 & 6 \\ 1 & 3 \end{pmatrix}=\begin{pmatrix} -2 & 0 \\ 8 & 0 \end{pmatrix}$$

but

$$\begin{pmatrix} -2 & 0 \\ 8 & 0 \end{pmatrix}\begin{pmatrix} 1 & 0 \\ 2 & -4 \end{pmatrix}=\begin{pmatrix} -14 & 24 \\ -5 & 12 \end{pmatrix}.$$ ♦

EXAMPLE 7.4

There is in general no cancelation in products: if $\mathbf{AB}=\mathbf{AC}$, it does not follow that $\mathbf{A}=\mathbf{C}$. To illustrate,

$$\begin{pmatrix} 1 & 1 \\ 3 & 3 \end{pmatrix}\begin{pmatrix} 4 & 2 \\ 3 & 16 \end{pmatrix}=\begin{pmatrix} 1 & 1 \\ 3 & 3 \end{pmatrix}\begin{pmatrix} 2 & 7 \\ 5 & 11 \end{pmatrix}=\begin{pmatrix} 7 & 18 \\ 21 & 54 \end{pmatrix},$$

even though

$$\begin{pmatrix} 4 & 2 \\ 3 & 16 \end{pmatrix}\neq\begin{pmatrix} 2 & 7 \\ 5 & 11 \end{pmatrix}.$$ ♦

EXAMPLE 7.5

The product of two nonzero matrices may be a zero matrix:

$$\begin{pmatrix} 1 & 2 \\ 0 & 0 \end{pmatrix}\begin{pmatrix} 6 & 4 \\ -3 & -2 \end{pmatrix}=\begin{pmatrix} 0 & 0 \\ 0 & 0 \end{pmatrix}.$$ ♦

Matrix addition and multiplication can be done in MAPLE using the `A+B` and `A.B` commands, which are in the `linalg` package of subroutines. Multiplication of **A** by a scalar c is achieved by `c*A`.

7.1.1 Matrix Multiplication from Another Perspective

Let **A** be an $n \times k$ matrix and **B** a $k \times m$ matrix. We have defined **AB** to be the $n \times m$ matrix whose i, j-element is the dot product of row i of **A** with column j of **B**.

It is sometimes useful to observe that column j of **AB** is the matrix product of **A** with column j of **B**. We can therefore compute a matrix product **AB** by multiplying an $n \times k$ matrix **A** in turn by each $k \times 1$ column of **B**.

Specifically, if the columns of **B** are $\mathbf{B}_1, \cdots, \mathbf{B}_m$, then we can think of **B** as a matrix of these columns:

$$\mathbf{B} = \begin{pmatrix} \| & \| & \cdots & \| \\ \mathbf{B}_1 & \mathbf{B}_2 & \cdots & \mathbf{B}_m \\ \| & \| & \cdots & \| \end{pmatrix}.$$

Then

$$\mathbf{AB} = \mathbf{A} \begin{pmatrix} \| & \| & \cdots & \| \\ \mathbf{B}_1 & \mathbf{B}_2 & \cdots & \mathbf{B}_m \\ \| & \| & \cdots & \| \end{pmatrix}$$

$$= \begin{pmatrix} \| & \| & \cdots & \| \\ \mathbf{AB}_1 & \mathbf{AB}_2 & \cdots & \mathbf{AB}_m \\ \| & \| & \cdots & \| \end{pmatrix}.$$

As an example, let

$$\mathbf{A} = \begin{pmatrix} 2 & -4 \\ 1 & 7 \end{pmatrix} \text{ and } \mathbf{B} = \begin{pmatrix} -3 & 6 & 7 \\ -5 & 1 & 2 \end{pmatrix}.$$

Then

$$\begin{pmatrix} 2 & -4 \\ 1 & 7 \end{pmatrix} \begin{pmatrix} -3 \\ -5 \end{pmatrix} = \begin{pmatrix} 14 \\ -38 \end{pmatrix},$$

$$\begin{pmatrix} 2 & -4 \\ 1 & 7 \end{pmatrix} \begin{pmatrix} 6 \\ 1 \end{pmatrix} = \begin{pmatrix} 8 \\ 13 \end{pmatrix},$$

and

$$\begin{pmatrix} 2 & -4 \\ 1 & 7 \end{pmatrix} \begin{pmatrix} 7 \\ 2 \end{pmatrix} = \begin{pmatrix} 8 \\ 6 \\ 21 \end{pmatrix}.$$

These are the columns of **AB**:

$$\begin{pmatrix} 2 & -4 \\ 1 & 7 \end{pmatrix} \begin{pmatrix} -3 & 6 & 7 \\ -5 & 1 & 2 \end{pmatrix} = \begin{pmatrix} 14 & 8 & 6 \\ -38 & 13 & 21 \end{pmatrix}.$$

We also will sometimes find it useful to think of a product **AX**, when **X** is a $k \times 1$ column matrix, as a linear combination of the columns $\mathbf{A}_1, \cdots, \mathbf{A}_k$ of **A**. In particular, if

$$\mathbf{X} = \begin{pmatrix} x_1 \\ x_2 \\ \vdots \\ x_k \end{pmatrix},$$

then

$$\mathbf{AX} = x_1\mathbf{A}_1 + x_2\mathbf{A}_2 + \cdots + x_k\mathbf{A}_k.$$

For example,

$$\begin{pmatrix} 6 & -3 & 4 \\ 2 & 1 & 7 \end{pmatrix} \begin{pmatrix} x_1 \\ x_2 \\ x_3 \end{pmatrix} = \begin{pmatrix} 6x_1 - 3x_2 + 4x_3 \\ 2x_1 + x_2 + 7x_3 \end{pmatrix}$$

$$= x_1 \begin{pmatrix} 6 \\ 2 \end{pmatrix} + x_2 \begin{pmatrix} -3 \\ 1 \end{pmatrix} + x_3 \begin{pmatrix} 4 \\ 7 \end{pmatrix}.$$

7.1.2 Terminology and Special Matrices

We will define some terms and special matrices that are encountered frequently.

The $n \times m$ *zero matrix* $\mathbf{O}_{nm}$ is the $n \times m$ matrix having every element equal to zero.

For example

$$\mathbf{O}_{23} = \begin{pmatrix} 0 & 0 & 0 \\ 0 & 0 & 0 \end{pmatrix}.$$

If $\mathbf{A}$ is $n \times m$ then

$$\mathbf{A} + \mathbf{O}_{nm} = \mathbf{O}_{nm} + \mathbf{A} = \mathbf{A}.$$

The negative of a matrix $\mathbf{A}$ is just the scalar product $(-1)\mathbf{A}$ formed by multiplying each matrix element by -1. We denote this matrix $-\mathbf{A}$. If $\mathbf{B}$ has the same dimensions as $\mathbf{A}$, then we denote $\mathbf{B} + (-\mathbf{A})$ as $\mathbf{B} - \mathbf{A}$, as we do with numbers.

A *square matrix* is one having the same number of rows and columns. If $\mathbf{A} = [a_{ij}]$ is $n \times n$, the *main diagonal* of $\mathbf{A}$ consists of the matrix elements $a_{11}, a_{22}, \cdots, a_{nn}$. These are the matrix elements along the diagonal from the upper left corner to the lower right corner.

The $n \times n$ *identity matrix* is the $n \times n$ matrix $\mathbf{I}_n$ having each i, j element equal to zero if $i \neq j$, and each i, i element equal to 1. For example,

$$\mathbf{I}_4 = \begin{pmatrix} 1 & 0 & 0 & 0 \\ 0 & 1 & 0 & 0 \\ 0 & 0 & 1 & 0 \\ 0 & 0 & 0 & 1 \end{pmatrix}.$$

Thus $\mathbf{I}_n$ has 1 down the main diagonal and zeros everywhere else.

THEOREM 7.2

If $\mathbf{A}$ is $n \times m$, then

$$\mathbf{I}_n\mathbf{A} = \mathbf{A}\mathbf{I}_m = \mathbf{A}. \ \blacklozenge$$

This is routine to prove. Note that the dimensions must be correct - we must multiply $\mathbf{A}$ on the left by $\mathbf{I}_n$, but on the right by $\mathbf{I}_m$, for these products to be defined.

EXAMPLE 7.6

Let

$$\mathbf{A} = \begin{pmatrix} 1 & 0 \\ 2 & 1 \\ -1 & 8 \end{pmatrix}.$$

Then

$$\mathbf{I}_3\mathbf{A} = \begin{pmatrix} 1 & 0 & 0 \\ 0 & 1 & 0 \\ 0 & 0 & 1 \end{pmatrix} \begin{pmatrix} 1 & 0 \\ 2 & 1 \\ -1 & 8 \end{pmatrix} = \begin{pmatrix} 1 & 0 \\ 2 & 1 \\ -1 & 8 \end{pmatrix} = \mathbf{A}$$

and

$$\mathbf{A}\mathbf{I}_2 = \begin{pmatrix} 1 & 0 \\ 2 & 1 \\ -1 & 8 \end{pmatrix} \begin{pmatrix} 1 & 0 \\ 0 & 1 \end{pmatrix} = \begin{pmatrix} 1 & 0 \\ 2 & 1 \\ -1 & 8 \end{pmatrix} = \mathbf{A}. \blacklozenge$$

If $\mathbf{A} = [a_{ij}]$ is an $n \times m$ matrix, the *transpose* of $\mathbf{A}$ is the $m \times n$ matrix defined by

$$\mathbf{A}^t = [a_{ji}].$$

We form the transpose by interchanging the rows and columns of $\mathbf{A}$.

EXAMPLE 7.7

Let

$$\mathbf{A} = \begin{pmatrix} -1 & 6 & 3 & -4 \\ 0 & \pi & 12 & -5 \end{pmatrix},$$

a 2×4 matrix. Then $\mathbf{A}^t$ is the 4×2 matrix

$$\mathbf{A}^t = \begin{pmatrix} -1 & 0 \\ 6 & \pi \\ 3 & 12 \\ -4 & -5 \end{pmatrix}. \blacklozenge$$

THEOREM 7.3 ***Properties of the Transpose***

1. $(\mathbf{I}_n)^t = \mathbf{I}_n$.
2. For any matrix $\mathbf{A}$,
$$(\mathbf{A}^t)^t = \mathbf{A}.$$
3. If $\mathbf{AB}$ is defined, then
$$(\mathbf{AB})^t = \mathbf{B}^t\mathbf{A}^t. \blacklozenge$$

Proof of Conclusion (2) It is obvious if we take the transpose of a transpose, then we interchange the rows and columns, then interchange them again, leaving every element in its original position.

It is less obvious that, if we take the transpose of a product, then we obtain the product of the transposes, in the reverse order, which is conclusion (3). We will prove this.

Proof of Conclusion (3) First observe that the conclusion is consistent with the definition of the matrix product. If $\mathbf{A} = [a_{ij}]$ is $n \times m$ and $\mathbf{B} = [b_{ij}]$ is $m \times k$, then $\mathbf{AB}$ is $n \times k$, so $(\mathbf{AB})^t$ is $k \times n$. However, $\mathbf{A}^t$ is $m \times n$ and $\mathbf{B}^t$ is $k \times m$, so $\mathbf{A}^t\mathbf{B}^t$ is defined only if $n = k$, while $\mathbf{B}^t\mathbf{A}^t$ is always defined and is $k \times n$.

Now, from the definition of matrix product

$$\begin{aligned} i, j \text{ element of } \mathbf{B}^t\mathbf{A}^t &= \sum_{s=1}^{k} (\mathbf{B}^t)_{is}(\mathbf{A}^t)_{sj} \\ &= \sum_{s=1}^{k} b_{si}a_{js} = \sum_{s=1}^{k} a_{js}b_{si} \\ &= j, i \text{ element of } \mathbf{AB} = i, j \text{ element of } (\mathbf{AB})^t. \end{aligned}$$

This argument can also be given conveniently in terms of dot products:

$$\begin{aligned} (\mathbf{B}^t\mathbf{A}^t)_{ij} &= (\text{ row } i \text{ of } \mathbf{B}^t) \cdot (\text{ column } j \text{ of } \mathbf{A}^t) \\ &= (\text{ column } i \text{ of } \mathbf{B}) \cdot (\text{ row } j \text{ of } \mathbf{A}) \\ &= (\text{ row } j \text{ of } \mathbf{A}) \cdot (\text{ column } i \text{ of } \mathbf{B}) \\ &= (\mathbf{AB})_{ji} = ((\mathbf{AB})^t)_{ij}. \ \blacklozenge \end{aligned}$$

In some contexts, it is useful to observe that the dot product of two n - vectors can be written as a matrix product. Write the n-vectors

$$\mathbf{X} = < x_1, x_2, \cdots, x_n > \text{ and } \mathbf{Y} = < y_1, y_2, \cdots, y_n > .$$

as $n \times 1$ column matrices

$$\mathbf{X} = \begin{pmatrix} x_1 \\ x_2 \\ \vdots \\ x_n \end{pmatrix} \text{ and } \mathbf{Y} = \begin{pmatrix} y_1 \\ y_2 \\ \vdots \\ y_n \end{pmatrix}.$$

Then $\mathbf{X}^t$ is a $1 \times n$ matrix, and $\mathbf{X}^t\mathbf{Y}$ is a 1×1 matrix, which we think of as just a scalar:

$$\begin{aligned} \mathbf{X}^t\mathbf{Y} &= \begin{pmatrix} x_1 & x_2 & \cdots & x_n \end{pmatrix} \begin{pmatrix} y_1 \\ y_2 \\ \vdots \\ y_n \end{pmatrix} \\ &= (x_1y_1 + x_2y_2 + \cdots + x_ny_n) = \mathbf{X} \cdot \mathbf{Y}. \end{aligned}$$

7.1.3 Random Walks in Crystals

We will apply matrix multiplication to the enumeration of paths through a crystal. Crystals have sites arranged in a lattice pattern. An atom may jump from a site it occupies to an adjacent, unoccupied one, and then proceed from there to other sites, making a *random walk* through the crystal.

We can represent this lattice of locations by drawing a point for each location and a line between points exactly when an atom can move directly from one to the other in the crystal. Such a diagram is called a *graph*. Figure 7.1 shows a typical graph. In this graph, an atom could move from v_1 to v_2 or v_3, to which it is connected by lines, but not directly to v_6 because there is no line between v_1 and v_6. Points connected by a line of the graph are called *adjacent*.

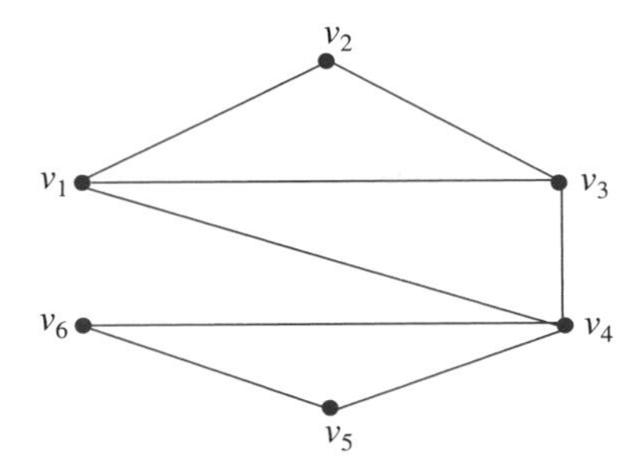

FIGURE 7.1 *A typical graph.*

A *walk* of length n in a graph is a sequence $v_1, v_2, \cdots, v_{n+1}$ of points (not necessarily different), with each v_j adjacent to v_{j+1}. Such a walk represents a possible path an atom might take over n edges (perhaps some repeated) through various sites in the crystal. A $v_i - v_j$ walk is one that begins at v_i and ends at v_j.

Physicists and materials engineers are interested in the following question: given a crystal with n sites $v_1, v_2, \cdots, v_n$, how many different walks of length k are there between two selected sites?

Define the *adjacency matrix* $\mathbf{A} = [a_{ij}]$ of the graph to be the $n \times n$ matrix having

$$a_{ij} = \begin{cases} 1 & \text{if } v_i \text{ is adjacent to } v_j \text{ in the graph} \\ 0 & \text{if there is no line between } v_i \text{ and } v_j \text{ in the graph.} \end{cases}$$

The graph of Figure 7.1 has the 6×6 adjacency matrix

$$\mathbf{A} = \begin{pmatrix} 0 & 1 & 1 & 1 & 0 & 0 \\ 1 & 0 & 1 & 0 & 0 & 0 \\ 1 & 1 & 0 & 1 & 0 & 0 \\ 1 & 0 & 1 & 0 & 1 & 1 \\ 0 & 0 & 0 & 1 & 0 & 1 \\ 0 & 0 & 0 & 1 & 1 & 0 \end{pmatrix}.$$

The main diagonal elements are zero because there is no line between any v_i and itself.

We claim that, if k be any positive integer, then the number of distinct $v_i - v_j$ walks of length k in the crystal is the i, j-element of $\mathbf{A}^k$. The elements of $\mathbf{A}^k$ therefore answer the question posed.

To see why this is true, begin with $k = 1$. If $i \neq j$, there is a walk of length 1 between v_i and v_j exactly when v_i is adjacent to v_j, and in this case $a_{ij} = 1$.

We next show that, if the result is true for walks of length k, then it must be true for walks of length $k + 1$. Consider how a $v_i - v_j$ walk of length $k + 1$ is formed. First there must be a $v_i - v_r$ walk of length 1 from v_i to some v_r adjacent to v_i, followed by a $v_r - v_j$ walk of length k (Figure 7.2). Then

$$\text{number of distinct } v_i - v_j \text{ walks of length } k+1$$

$$= \text{sum of the number of distinct } v_r - v_j \text{ walks of length } k,$$

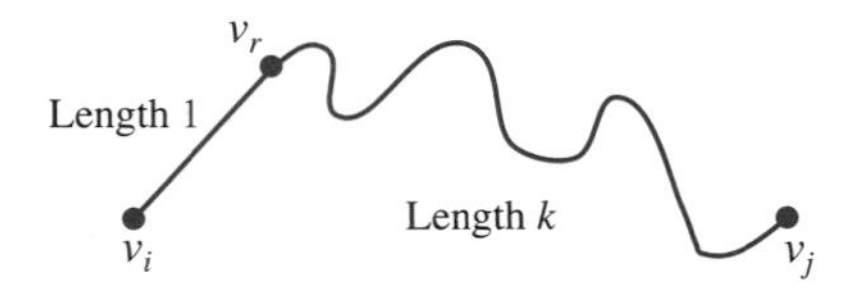

FIGURE 7.2 *Constructing* $v_i - v_j$ *walks of length* $k+1$.

with this sum taken over all points v_r adjacent to v_i. Now $a_{ir} = 1$ if v_r is adjacent to v_i, and 0 otherwise. Further, by assumption, the number of distinct $v_r - v_j$ walks of length k is the r, j-element of $\mathbf{A}^k$. Denote $\mathbf{A}^k = \mathbf{B} = [b_{ij}]$. Then, for $r = 1, \cdots, n$,

$$a_{ir}b_{rj} = 0 \text{ if } v_r \text{ is not adjacent to } v_i$$

and

$$a_{ir}b_{rj} =$$
$$\text{the number of distinct } v_i - v_j \text{ walks of length } k+1 \text{ if } v_r \text{ is adjacent to } v_i.$$

Therefore, the number of $v_i - v_j$ walks of length $k+1$ is

$$a_{i1}b_{1j} + a_{i2}b_{2j} + \cdots + a_{in}b_{nj}$$

and this is the i, j-element of $\mathbf{AB}$, which is $\mathbf{A}^{k+1}$.

EXAMPLE 7.8

The adjacency matrix of the graph of Figure 7.3 is

$$\mathbf{A} = \begin{pmatrix} 0 & 1 & 0 & 0 & 0 & 1 & 0 & 0 \\ 1 & 0 & 1 & 0 & 0 & 0 & 1 & 1 \\ 0 & 1 & 0 & 1 & 0 & 0 & 0 & 0 \\ 0 & 0 & 1 & 0 & 1 & 1 & 1 & 1 \\ 0 & 0 & 0 & 1 & 0 & 1 & 1 & 0 \\ 1 & 0 & 0 & 1 & 1 & 0 & 0 & 0 \\ 0 & 1 & 0 & 1 & 1 & 0 & 0 & 1 \\ 0 & 1 & 0 & 1 & 0 & 0 & 1 & 0 \end{pmatrix}.$$

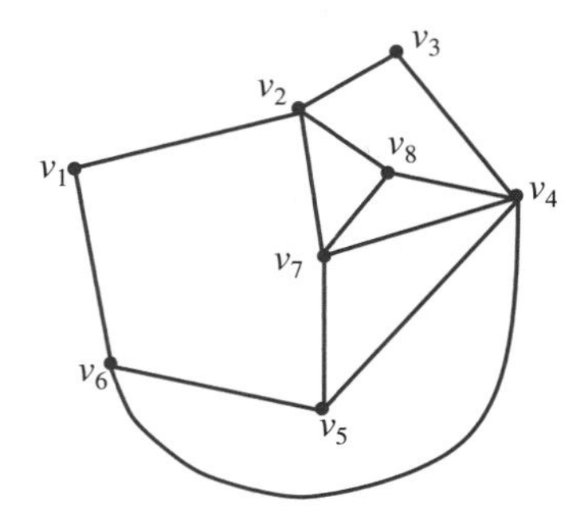

FIGURE 7.3 *Graph of Example 7.8.*

Suppose we want the number of walks of length 3 in this graph. Calculate

$$\mathbf{A}^3 = \begin{pmatrix} 0 & 5 & 1 & 4 & 2 & 4 & 3 & 2 \\ 6 & 2 & 7 & 4 & 5 & 4 & 9 & 8 \\ 1 & 7 & 0 & 8 & 3 & 2 & 3 & 2 \\ 4 & 4 & 8 & 6 & 8 & 8 & 11 & 10 \\ 2 & 5 & 3 & 8 & 4 & 6 & 8 & 4 \\ 4 & 4 & 2 & 8 & 6 & 2 & 4 & 4 \\ 3 & 9 & 3 & 11 & 8 & 4 & 6 & 7 \\ 2 & 8 & 2 & 10 & 4 & 4 & 7 & 4 \end{pmatrix}.$$

For example, the 4, 7 element of $\mathbf{A}^3$ is 11, so there are 11 walks of length 3 between v_4 and v_7. There are no walks of length 3 between v_4 and v_6.

Generally we would use a software package to compute $\mathbf{A}^k$. ♦

SECTION 7.1 PROBLEMS

In each of Problems 1 through 6, perform the requested computation.

1. $\mathbf{A} = \begin{pmatrix} 1 & -2 & 1 & 7 & -9 \\ 8 & 2 & -5 & 0 & 0 \end{pmatrix}$,

   $\mathbf{B} = \begin{pmatrix} -5 & 1 & 8 & 21 & 7 \\ 12 & -6 & -2 & -1 & 9 \end{pmatrix}$, $4\mathbf{A} + 8\mathbf{B}$

2. $\mathbf{A} = (14)$, $\mathbf{B} = (-12)$, $-3\mathbf{A} - 5\mathbf{B}$

3. $\mathbf{A} = \begin{pmatrix} 1 & -1 & 3 \\ 2 & -4 & 6 \\ -1 & 1 & 2 \end{pmatrix}$, $\mathbf{B} = \begin{pmatrix} -4 & 0 & 0 \\ -2 & -1 & 6 \\ 8 & 15 & 4 \end{pmatrix}$; $2\mathbf{A} - 3\mathbf{B}$

4. $\mathbf{A} = \begin{pmatrix} -2 & 3 \\ 1 & 1 \end{pmatrix}$, $\mathbf{B} = \begin{pmatrix} 0 & 8 \\ -5 & 1 \end{pmatrix}$, $\mathbf{A}^3 - \mathbf{B}^2$

5. $\mathbf{A} = \begin{pmatrix} x & 1-x \\ 2 & e^x \end{pmatrix}$, $\mathbf{B} = \begin{pmatrix} 1 & -6 \\ x & \cos(x) \end{pmatrix}$, $\mathbf{A}^2 + 2\mathbf{AB}$

6. $\mathbf{A} = \begin{pmatrix} -2 & 2 \\ 0 & 1 \\ 14 & 2 \\ 6 & 8 \end{pmatrix}$, $\mathbf{B} = \begin{pmatrix} 4 & 4 \\ 2 & 1 \\ 14 & 16 \\ 1 & 25 \end{pmatrix}$, $-5\mathbf{A} + 3\mathbf{B}$

In each of Problems 7 through 16, determine which of **AB** and **BA** are defined. Carry out all such products.

7. $\mathbf{A} = \begin{pmatrix} -1 & 6 & 2 & 14 & -22 \end{pmatrix}$, $\mathbf{B} = \begin{pmatrix} -3 \\ 2 \\ 6 \\ 0 \\ -4 \end{pmatrix}$

8. $\mathbf{A} = \begin{pmatrix} -3 & 1 \\ 6 & 2 \\ 18 & -22 \\ 1 & 6 \end{pmatrix}$, $\mathbf{B} = \begin{pmatrix} -16 & 0 & 0 & 28 \\ 0 & 1 & 1 & 26 \end{pmatrix}$

9. $\mathbf{A} = \begin{pmatrix} -4 & 6 & 2 \\ -2 & -2 & 3 \\ 1 & 1 & 8 \end{pmatrix}$,

   $\mathbf{B} = \begin{pmatrix} -2 & 4 & 6 & 12 & 5 \\ -3 & -3 & 1 & 1 & 4 \\ 0 & 0 & 1 & 6 & -9 \end{pmatrix}$

10. $\mathbf{A} = \begin{pmatrix} -3 & 2 \\ 0 & -2 \\ 1 & 8 \\ 3 & -3 \end{pmatrix}$, $\mathbf{B} = \begin{pmatrix} -5 & 5 & 7 & 2 \end{pmatrix}$

11. $\mathbf{A} = \begin{pmatrix} 7 & -8 \\ 1 & 6 \end{pmatrix}$, $\mathbf{B} = \begin{pmatrix} 1 & -4 & 3 \\ -4 & 7 & 0 \end{pmatrix}$

12. $\mathbf{A} = \begin{pmatrix} 3 \\ 0 \\ -1 \\ 4 \end{pmatrix}$, $\mathbf{B} = \begin{pmatrix} 3 & -2 & 4 \end{pmatrix}$

13. $\mathbf{A} = \begin{pmatrix} -21 & 4 & 8 & -3 \\ 12 & 1 & 0 & 14 \\ 1 & 16 & 0 & -8 \\ 13 & 4 & 8 & 0 \end{pmatrix}$,

    $\mathbf{B} = \begin{pmatrix} -9 & 16 & 3 & 2 \\ 5 & 9 & 14 & 0 \end{pmatrix}$

14. $\mathbf{A} = \begin{pmatrix} -2 & 4 \\ 3 & 9 \end{pmatrix}$, $\mathbf{B} = \begin{pmatrix} 1 & -3 & 7 & 2 \\ 5 & 9 & 1 & 0 \end{pmatrix}$

15. $\mathbf{A} = \begin{pmatrix} -4 & -2 & 0 \\ 0 & 5 & 3 \\ -3 & 1 & 1 \end{pmatrix}$, $\mathbf{B} = \begin{pmatrix} 1 & -3 & 4 \end{pmatrix}$

16. $\mathbf{A} = \begin{pmatrix} -2 & -4 \\ 3 & -1 \end{pmatrix}$, $\mathbf{B} = \begin{pmatrix} 6 & 8 \\ 1 & -4 \end{pmatrix}$

In each of Problems 17 through 21, determine if **AB** and/or **BA** is defined. For those products that are defined, give the dimensions of the product matrix.

17. **A** is 6×2, **B** is 4×6.
18. **A** is 1×3, **B** is 3×3.
19. **A** is 7×6, **B** is 7×7.
20. **A** is 18×4, **B** is 18×4.
21. **A** is 14×21, **B** is 21×14.
22. Let **A** be the adjacency matrix of a graph G.
    (a) Prove that the i, j-element of $\mathbf{A}^2$ equals the number of points of G that are neighbors of v_i in G. This number is called the *degree* of v_i.
    (b) Prove that the i, j-element of $\mathbf{A}^3$ equals twice the number of triangles in G containing v_i as a vertex. A triangle in G consists of three points, each a neighbor of the other.
23. For the graph G of Figure 7.4, determine the number of $v_1 - v_4$ walks of length 3, the number of $v_2 - v_3$ walks of length 3, and the number of $v_2 - v_4$ walks of length 4.
24. For the graph H of Figure 7.4, determine the number of $v_1 - v_4$ walks of length 4 and the number of $v_2 - v_3$ walks of length 2.
25. For the graph K of Figure 7.4, determine the number of $v_4 - v_5$ walks of length 2, the number of $v_2 - v_3$ walks of length 3, and the number of $v_1 - v_2$ walks and $v_4 - v_5$ walks of length 4.

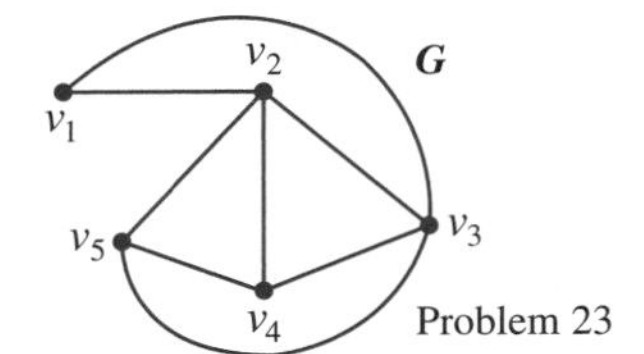

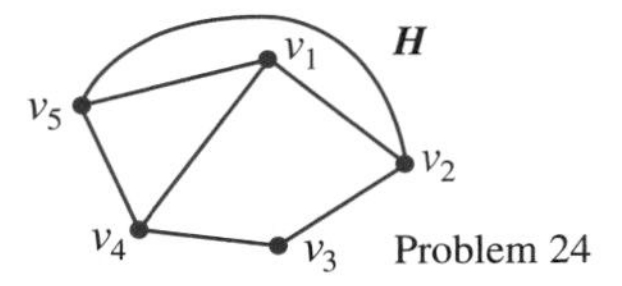

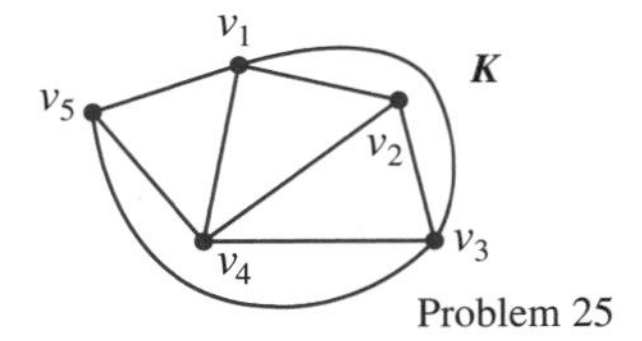

FIGURE 7.4 *Graphs of Problems 23, 24, and 25, in Section 7.1.*

26. Find nonzero 2×2 matrices **A**, **B**, and **C** such that $\mathbf{BA} = \mathbf{CA}$ but $\mathbf{B} \neq \mathbf{C}$.
27. Show that the set of all $n \times m$ matrices with real elements is a vector space, using the usual addition of matrices and multiplication of matrices by scalars as the operations. What is the dimension of this vector space?
28. Redo Problem 27 for the case that the elements in the matrices are complex numbers.

7.2 Elementary Row Operations

Some applications, as well as determining certain information about matrices, make use of *elementary row operations*. We will define three such operations. Let **A** be a matrix.

1. Type I operation: interchange two rows of **A**.
2. Type II operation: multiply a row of **A** by a nonzero number.
3. Type III operation: add a scalar multiple of one row to another row of **A**.

EXAMPLE 7.9

We will look at an example of each of these row operations. Let

$$\mathbf{A} = \begin{pmatrix} -2 & 1 & 6 & -3 \\ 1 & 1 & 2 & 5 \\ 0 & 9 & 3 & -7 \\ 2 & -3 & 4 & 11 \end{pmatrix}.$$

If we interchange rows two and three of **A**, we obtain

$$\begin{pmatrix} -2 & 1 & 6 & -3 \\ 0 & 9 & 3 & -7 \\ 1 & 1 & 2 & 5 \\ 2 & -3 & 4 & 11 \end{pmatrix}.$$

If we multiply row three of **A** by 7, we obtain

$$\begin{pmatrix} -2 & 1 & 6 & -3 \\ 1 & 1 & 2 & 5 \\ 0 & 63 & 21 & -49 \\ 2 & -3 & 4 & 11 \end{pmatrix}.$$

And if we add -6 times row one to row three of **A**, we obtain

$$\begin{pmatrix} -2 & 1 & 6 & -3 \\ 1 & 1 & 2 & 5 \\ 12 & 3 & -33 & 11 \\ 2 & -3 & 4 & 11 \end{pmatrix}. \ \blacklozenge$$

Every elementary row operation can be performed by multiplying **A** on the left by a square matrix constructed by applying that row operation to an identity matrix.

THEOREM 7.4

Let **A** be an $n \times m$ matrix. Suppose **B** is formed from **A** by an elementary row operation. Let **E** be the matrix formed by performing this row operation on $\mathbf{I}_n$. Then

$$\mathbf{B} = \mathbf{EA}. \ \blacklozenge$$

A matrix formed by performing an elementary row operation on $\mathbf{I}_n$ is called an *elementary matrix*. Theorem 7.4 says that we can perform any elementary row operation on **A** by multiplying **A** on the left by the elementary matrix formed by performing this row operation on $\mathbf{I}_n$. We leave a proof of this to Exercises 7.9, 7.10, and 7.11. However, it is instructive to see the theorem in action.

EXAMPLE 7.10

Let

$$\mathbf{A} = \begin{pmatrix} -2 & 1 & 6 & -3 \\ 1 & 1 & 2 & 5 \\ 0 & 9 & 3 & -7 \end{pmatrix}.$$

Since **A** is 3×4, we will use $\mathbf{I}_3$ to perform row operations.

First, interchange rows two and three of $\mathbf{A}$ to form

$$\mathbf{B} = \begin{pmatrix} -2 & 1 & 6 & -3 \\ 0 & 9 & 3 & -7 \\ 1 & 1 & 2 & 5 \end{pmatrix}.$$

Perform this row operation on $\mathbf{I}_3$ to obtain

$$\mathbf{E}_1 = \begin{pmatrix} 1 & 0 & 0 \\ 0 & 0 & 1 \\ 0 & 1 & 0 \end{pmatrix}.$$

Then

$$\mathbf{E}_1\mathbf{A} = \begin{pmatrix} 1 & 0 & 0 \\ 0 & 0 & 1 \\ 0 & 1 & 0 \end{pmatrix} \begin{pmatrix} -2 & 1 & 6 & -3 \\ 1 & 1 & 2 & 5 \\ 0 & 9 & 3 & -7 \end{pmatrix} = \begin{pmatrix} -2 & 1 & 6 & -3 \\ 0 & 9 & 3 & -7 \\ 1 & 1 & 2 & 5 \end{pmatrix} = \mathbf{B}.$$

Next multiply row three of $\mathbf{A}$ by 7 to form

$$\mathbf{C} = \begin{pmatrix} -2 & 1 & 6 & -3 \\ 1 & 1 & 2 & 5 \\ 0 & 63 & 21 & -49 \end{pmatrix}.$$

Perform this row operation on $\mathbf{I}_3$ to obtain

$$\mathbf{E}_2 = \begin{pmatrix} 1 & 0 & 0 \\ 0 & 1 & 0 \\ 0 & 0 & 7 \end{pmatrix}.$$

Then

$$\mathbf{E}_2\mathbf{A} = \begin{pmatrix} 1 & 0 & 0 \\ 0 & 1 & 0 \\ 0 & 0 & 7 \end{pmatrix} \begin{pmatrix} -2 & 1 & 6 & -3 \\ 1 & 1 & 2 & 5 \\ 0 & 9 & 3 & -7 \end{pmatrix} = \begin{pmatrix} -2 & 1 & 6 & -3 \\ 1 & 1 & 2 & 5 \\ 0 & 63 & 21 & -49 \end{pmatrix} = \mathbf{C}.$$

Finally, add 2 times row one to row two to form

$$\mathbf{D} = \begin{pmatrix} -2 & 1 & 6 & -3 \\ -3 & 3 & 14 & -1 \\ 0 & 9 & 3 & -7 \end{pmatrix}.$$

This operation can be achieved by the elementary matrix

$$\mathbf{E}_3 = \begin{pmatrix} 1 & 0 & 0 \\ 2 & 1 & 0 \\ 0 & 0 & 1 \end{pmatrix}.$$

As a check,

$$\mathbf{E}_3\mathbf{A} = \begin{pmatrix} 1 & 0 & 0 \\ 2 & 1 & 0 \\ 0 & 0 & 1 \end{pmatrix} \begin{pmatrix} -2 & 1 & 6 & -3 \\ 1 & 1 & 2 & 5 \\ 0 & 9 & 3 & -7 \end{pmatrix} = \begin{pmatrix} -2 & 1 & 6 & -3 \\ -3 & 3 & 14 & -1 \\ 0 & 9 & 3 & -7 \end{pmatrix} = \mathbf{D}. \ \blacklozenge$$

This result has an important consequence. Suppose we form $\mathbf{B}$ from $\mathbf{A}$ by performing a sequence of elementary row operations in succession. That is, we perform operation $\mathcal{O}_1$ on $\mathbf{A}$ to obtain $\mathbf{A}_1$, then $\mathcal{O}_2$ on $\mathbf{A}_1$ to form $\mathbf{A}_2$, and so on until we perform $\mathcal{O}_r$ on $\mathbf{A}_{r-1}$ to form $\mathbf{A}_r = \mathbf{B}$. We may envision this process

$$\mathbf{A} \xrightarrow{\mathcal{O}_1} \mathbf{A}_1 \xrightarrow{\mathcal{O}_2} \mathbf{A}_2 \xrightarrow{\mathcal{O}_3} \mathbf{A}_3 \rightarrow$$

$$\cdots \xrightarrow{\mathcal{O}_{r-1}} \mathbf{A}_{r-1} \xrightarrow{\mathcal{O}_r} \mathbf{A}_r = \mathbf{B}.$$

We can perform each elementary operation $\mathcal{O}_j$ by multiplying on the left by the elementary matrix $\mathbf{E}_j$ formed by performing that operation on $\mathbf{I}_n$. Then

$$\begin{aligned}
\mathbf{A}_1 &= \mathbf{E}_1\mathbf{A}\\
\mathbf{A}_2 &= \mathbf{E}_2\mathbf{A}_1 = \mathbf{E}_2\mathbf{E}_1\mathbf{A}\\
\mathbf{A}_3 &= \mathbf{E}_3\mathbf{A}_2 = \mathbf{E}_3\mathbf{E}_2\mathbf{E}_1\mathbf{A}\\
&\vdots\\
&\vdots\\
\mathbf{A}_{r-1} &= \mathbf{E}_{r-1}\mathbf{A}_{r-2} = \mathbf{E}_{r-1}\mathbf{E}_{r-2}\cdots\mathbf{E}_2\mathbf{E}_1\mathbf{A}\\
\mathbf{A}_r &= \mathbf{E}_r\mathbf{A}_{r-1} = \mathbf{E}_r\mathbf{E}_{r-1}\cdots\mathbf{E}_2\mathbf{E}_1\mathbf{A}.
\end{aligned}$$

If we designate

$$\Omega = \mathbf{E}_r\mathbf{E}_{r-1}\cdots\mathbf{E}_2\mathbf{E}_1$$

in this order, then

$$\mathbf{B} = \Omega\mathbf{A}.$$

Furthermore Ω is a product of elementary matrices.

We will record this as a theorem.

THEOREM 7.5

Let $\mathbf{B}$ be obtained from $\mathbf{A}$ by a sequence of elementary row operations. Then there is a matrix Ω which is a product of elementary matrices such that

$$\mathbf{B} = \Omega\mathbf{A}. \blacklozenge$$

In forming Ω as a product of elementary matrices, $\mathbf{E}_1$ performs the first row operation on $\mathbf{A}$, then $\mathbf{E}_2$ performs the second operation on $\mathbf{E}_1\mathbf{A}$, and so on. The order of the operations, hence of the factors making up Ω, is crucial.

We do not need to actually write down each $\mathbf{E}_j$ to form Ω. The same result is achieved as follows: perform the first row operation on $\mathbf{I}_n$, then the second operation on the resulting matrix, then the third operation on this matrix, and so on. After all the row operations have been performed, the end result is Ω.

EXAMPLE 7.11

Let

$$\mathbf{A} = \begin{pmatrix} 0 & -1 & 1 & 4\\ 9 & 3 & 7 & -7\\ 0 & 2 & 1 & 5 \end{pmatrix}.$$

We will form $\mathbf{B}$ by starting with $\mathbf{A}$ and performing the following operations in the order given:

$\mathcal{O}_1$: add -3 times row 2 to row 3; then
$\mathcal{O}_2$: add 2 times row 1 to row 2; then
$\mathcal{O}_3$: interchange rows 1 and 3; then
$\mathcal{O}_4$: multiply row 2 by -4.

To form Ω to perform these operations, begin

$$\mathbf{I}_3 \xrightarrow{\mathcal{O}_1} \begin{pmatrix} 1 & 0 & 0\\ 0 & 1 & 0\\ 0 & -3 & 1 \end{pmatrix} \xrightarrow{\mathcal{O}_2} \begin{pmatrix} 1 & 0 & 0\\ 2 & 1 & 0\\ 0 & -3 & 1 \end{pmatrix}$$

$$\xrightarrow{\mathcal{O}_3} \begin{pmatrix} 0 & -3 & 1 \\ 2 & 1 & 0 \\ 1 & 0 & 0 \end{pmatrix} \xrightarrow{\mathcal{O}_4} \begin{pmatrix} 0 & -3 & 1 \\ -8 & -4 & 0 \\ 1 & 0 & 0 \end{pmatrix} = \Omega.$$

Then

$$\Omega\mathbf{A} = \begin{pmatrix} 0 & -3 & 1 \\ -8 & -4 & 0 \\ 1 & 0 & 0 \end{pmatrix} \begin{pmatrix} 0 & -1 & 1 & 4 \\ 9 & 3 & 7 & -7 \\ 0 & 2 & 1 & 5 \end{pmatrix}$$

$$= \begin{pmatrix} -27 & -7 & -20 & 26 \\ -84 & -4 & -36 & -4 \\ 6 & -1 & 1 & 4 \end{pmatrix} = \mathbf{B}. \blacklozenge$$

Later it will be important to know that the effect of each elementary row operation can be reversed by an elementary row operation of the same type. To see this, look at each type in turn.

If we form **B** from **A** by interchanging rows i and j, then interchanging these rows (another type I operation) in **B** returns **A**.

If we multiply a row of **A** by a nonzero number k, then multiply that row of **B** by $1/k$ (a type II operation) to reproduce **A**.

Finally, if we add α times row i to row j of **A**, then add $-\alpha$ times row i to row j of **B** (a type III operation) to return to **A**.

Since all of these reversals are done by elementary row operations, they can also be achieved by multiplying on the left by an elementary matrix.

We say that **A** is *row equivalent* to **B** if **B** can be obtained from **A** by a sequence of elementary row operations. Row equivalence has the following properties.

THEOREM 7.6

1. Every matrix is row equivalent to itself.
2. If **A** is row equivalent to **B**, then **B** is row equivalent to **A**.
3. If **A** is row equivalent to **B**, and **B** is row equivalent to **C**, then **A** is row equivalent to **C**. ♦

Elementary row operations can be done in MAPLE using the `swaprow(A,i,j)`, `mulrow(A,2,α)`, and `addrow(A,i,j,α)` commands, within the `linalg` package of subroutines. These are discussed in the MAPLE Primer.

SECTION 7.2 PROBLEMS

In each of Problems 1 through 8, perform the elementary row operation or sequence of row operations on **A** and then produce a matrix Ω so that $\Omega\mathbf{A}$ is the end result.

1. $\mathbf{A} = \begin{pmatrix} -1 & 0 & 3 & 0 \\ 1 & 3 & 2 & 9 \\ -9 & 7 & -5 & 7 \end{pmatrix}$; multiply row 3 by 4, then add 14 times row 1 to row 2 and then interchange rows 3 and 2.

2. $\mathbf{A} = \begin{pmatrix} -4 & 6 & -3 \\ 12 & 4 & -4 \\ 1 & 3 & 0 \end{pmatrix}$; interchange rows 2 and 3, then add the negative of row 1 to row 2.

3. $\mathbf{A} = \begin{pmatrix} -2 & 1 & 4 & 2 \\ 0 & 1 & 16 & 3 \\ 1 & -2 & 4 & 8 \end{pmatrix}$; multiply row 2 by $\sqrt{3}$.

4. $\mathbf{A} = \begin{pmatrix} 0 & -9 & 14 \\ 1 & 5 & 2 \\ 9 & 15 & 0 \end{pmatrix}$; interchange rows 2 and 3, then add 3 times row 2 to row 3, then interchange rows 1 and 3 and then multiply row 3 by 4.

5. $\mathbf{A} = \begin{pmatrix} -2 & 14 & 6 \\ 8 & 1 & -3 \\ 2 & 9 & 5 \end{pmatrix}$; add $\sqrt{13}$ times row 3 to row 1, then interchange rows 2 and 1 and then multiply row 1 by 5.

6. $\mathbf{A} = \begin{pmatrix} 3 & -6 \\ 1 & 1 \\ 8 & -2 \\ 0 & 5 \end{pmatrix}$; add 6 times row 2 to row 3.

7. $\mathbf{A} = \begin{pmatrix} -3 & 15 \\ 2 & 8 \end{pmatrix}$; add $\sqrt{3}$ times row 2 to row 1, then multiply row 2 by 15, then interchange rows 1 and 2.

8. $\mathbf{A} = \begin{pmatrix} 3 & -4 & 5 & 9 \\ 2 & 1 & 3 & -6 \\ 1 & 13 & 2 & 6 \end{pmatrix}$; add row 1 to row 3, then add $\sqrt{3}$ times row 1 to row 2, then multiply row 3 by 4, then add row 2 to row 3.

In each of Problems 9, 10, and 11, $\mathbf{A}$ is an $n \times m$ matrix.

9. Let $\mathbf{B}$ be formed from $\mathbf{A}$ by adding α times row s to row t. Let $\mathbf{E}$ be formed from $\mathbf{I}_n$ by this operation. Prove that $\mathbf{B} = \mathbf{EA}$.

10. Let $\mathbf{B}$ be formed from $\mathbf{A}$ by multiplying row s by α. Let $\mathbf{E}$ be formed from $\mathbf{I}_n$ by multiplying row s by α. Prove that $\mathbf{B} = \mathbf{EA}$.

11. Let $\mathbf{B}$ be formed from $\mathbf{A}$ by interchanging rows s and t. Let $\mathbf{E}$ be formed from $\mathbf{I}_n$ by interchanging these rows. Prove that $\mathbf{B} = \mathbf{EA}$.

7.3 Reduced Row Echelon Form

Now that we know how to perform elementary row operations, we will address a reason why we should want to do this. This section establishes a special form that we will want to manipulate matrices into, and the next two sections apply this special form to the solution of systems of linear equations.

Define the *leading entry* of a row of a matrix to be its first nonzero element, reading from left to right. If all of the elements of a row are zero, then this row has no leading entry.

An $n \times m$ matrix $\mathbf{A}$ is in *reduced row echelon form* if it satisfies the following conditions.

1. The leading entry of each nonzero row is 1.
2. If any row has its leading entry in column j, then all other elements of column j are zero.
3. If row i is a nonzero row and row k is a zero row, then $i < k$.
4. If the leading entry of row r_1 is in column c_1, and the leading entry of row r_2 is in column c_2, and $r_1 < r_2$, then $c_1 < c_2$.

When a matrix satisfies these conditions, we will often shorten "reduced row echelon form" and simply say that the matrix is *reduced*, or in *reduced form*.

Condition (1) of the definition means that, if we look across any nonzero row from left to right, the first nonzero element we see is 1. Condition (2) means that, if we stand at the leading entry 1 of any nonzero row and look straight up or down that column, we see only zeros. Condition (3) means that any row of zeros in a reduced matrix must lie below *all* rows having nonzero elements. Zero rows are at the bottom of the matrix. Condition (4) means that the leading entries of a reduced matrix move downward from left to right as we look at the matrix.

EXAMPLE 7.12

The following matrices are all reduced:

$$\begin{pmatrix} 1 & -4 & 1 & 0 \\ 0 & 0 & 0 & 1 \end{pmatrix}, \begin{pmatrix} 0 & 1 & 3 & 0 \\ 0 & 0 & 0 & 1 \\ 0 & 0 & 0 & 0 \end{pmatrix}$$

$$\begin{pmatrix} 0 & 1 & 2 & 0 & 0 \\ 0 & 0 & 0 & 1 & 0 \\ 0 & 0 & 0 & 0 & 0 \\ 0 & 0 & 0 & 0 & 0 \end{pmatrix}, \begin{pmatrix} 1 & 0 & 0 & 2 & 1 \\ 0 & 1 & 0 & -2 & 4 \\ 0 & 0 & 1 & 0 & 1 \\ 0 & 0 & 0 & 0 & 0 \end{pmatrix}. \quad \blacklozenge$$

EXAMPLE 7.13

$$\mathbf{C} = \begin{pmatrix} 2 & 0 & 1 \\ 0 & -4 & 6 \\ 2 & -2 & 5 \end{pmatrix}$$

is not reduced. However, we claim that, by a sequence of elementary row operations, we can transform $\mathbf{C}$ to a reduced matrix.

First, if the matrix had one or more zero rows, we would interchange rows to place these at the bottom of the new matrix. In this example $\mathbf{C}$ has no zero rows.

In view of condition (4) of the definition, start at the upper left corner. Multiply row one by $1/2$ to obtain a matrix having a leading entry of 1 in row one:

$$\mathbf{C} \rightarrow \begin{pmatrix} 1 & 0 & 1/2 \\ 0 & -4 & 6 \\ 2 & -2 & 5 \end{pmatrix}.$$

To get zeros below the 1 in the $1,1-$ position, add -2 times row one to row three:

$$\begin{pmatrix} 1 & 0 & 1/2 \\ 0 & -4 & 6 \\ 2 & -2 & 5 \end{pmatrix} \rightarrow \begin{pmatrix} 1 & 0 & 1/2 \\ 0 & -4 & 6 \\ 0 & -2 & 4 \end{pmatrix}.$$

Now look across row two of the last matrix. The leading entry is -4, so divide this row by -4:

$$\begin{pmatrix} 1 & 0 & 1/2 \\ 0 & -4 & 6 \\ 0 & -2 & 4 \end{pmatrix} \rightarrow \begin{pmatrix} 1 & 0 & 1/2 \\ 0 & 1 & -3/2 \\ 0 & -2 & 4 \end{pmatrix}.$$

Aside from the 1 in the $2,2$ position of the last matrix, we want zeros in column two. Add 2 times row two to row three:

$$\begin{pmatrix} 1 & 0 & 1/2 \\ 0 & 1 & -3/2 \\ 0 & -2 & 4 \end{pmatrix} \rightarrow \begin{pmatrix} 1 & 0 & 1/2 \\ 0 & 1 & -3/2 \\ 0 & 0 & 1 \end{pmatrix}.$$

It happens that the leading entry of row three of the last matrix is 1. To get zeros in column three above this leading entry, add $3/2$ times row three to row one, then $-1/2$ times row three to row one:

$$\begin{pmatrix} 1 & 0 & 1/2 \\ 0 & 1 & -3/2 \\ 0 & 0 & 1 \end{pmatrix} \rightarrow \begin{pmatrix} 1 & 0 & 0 \\ 0 & 1 & 0 \\ 0 & 0 & 1 \end{pmatrix}.$$

This is a reduced matrix that is row equivalent to $\mathbf{C}$, having been obtained from it by a sequence of elementary row operations. ♦

In this example, we started with **C** and obtained a reduced matrix. If we had used a different sequence of elementary row operations, could we have reached a different reduced matrix? The answer is no.

THEOREM 7.7

Every matrix is row equivalent to a matrix in reduced form. Further, the reduced form of a matrix is completely determined by the matrix itself, not by the reduction process. That is, no matter what sequence of elementary row operations is used to produce a reduced matrix equivalent to **A**, the same reduced matrix will result. ♦

Proof Every matrix can be manipulated to reduced form by following the idea of Example 7.13. First, move any zero rows to the bottom of the matrix by row interchanges. Then start at the upper left leading entry, and by multiplying this row by a scalar, obtain a matrix having a 1 in this position. Add multiples of this row to the other rows to obtain zeros in this column below this leading entry. Then move to the second row and carry out the same procedure starting with its leading entry. After each nonzero row has been treated, a reduced matrix results. ♦

In view of the uniqueness of the reduced form of a given matrix, we will denote the reduced form of **A** as $\mathbf{A}_R$. The process of determining $\mathbf{A}_R$, by any sequence of elementary row operations, is referred to as *reducing* **A**.

EXAMPLE 7.14

Let

$$\mathbf{A} = \begin{pmatrix} 0 & 0 & 0 & 0 & 0 \\ 0 & 0 & 2 & 0 & 0 \\ 0 & 1 & 0 & 1 & 1 \\ 0 & 0 & 3 & 0 & -4 \end{pmatrix}.$$

We will reduce this matrix. First interchange rows to move the zero row to the bottom of the matrix:

$$\mathbf{A} = \begin{pmatrix} 0 & 0 & 0 & 0 & 0 \\ 0 & 0 & 2 & 0 & 0 \\ 0 & 1 & 0 & 1 & 1 \\ 0 & 0 & 3 & 0 & -4 \end{pmatrix} \rightarrow \begin{pmatrix} 0 & 0 & 2 & 0 & 0 \\ 0 & 1 & 0 & 1 & 1 \\ 0 & 0 & 3 & 0 & -4 \\ 0 & 0 & 0 & 0 & 0 \end{pmatrix}.$$

The leading entry of row one is in the $1, 3$ position, and the leading entry of row three is in the $2, 2$ position. We want the leading entries to move down the matrix from left to right, so interchange rows one and two to obtain

$$\begin{pmatrix} 0 & 0 & 2 & 0 & 0 \\ 0 & 1 & 0 & 1 & 1 \\ 0 & 0 & 3 & 0 & -4 \\ 0 & 0 & 0 & 0 & 0 \end{pmatrix} \rightarrow \begin{pmatrix} 0 & 1 & 0 & 1 & 1 \\ 0 & 0 & 2 & 0 & 0 \\ 0 & 0 & 3 & 0 & -4 \\ 0 & 0 & 0 & 0 & 0 \end{pmatrix}.$$

The leading entry of (the new) row one is 1 and already has zeros below it, so move to the second row and find its leading entry, which is 2. Multiply row two by $1/2$:

$$\begin{pmatrix} 0 & 1 & 0 & 1 & 1 \\ 0 & 0 & 2 & 0 & 0 \\ 0 & 0 & 3 & 0 & -4 \\ 0 & 0 & 0 & 0 & 0 \end{pmatrix} \rightarrow \begin{pmatrix} 0 & 1 & 0 & 1 & 1 \\ 0 & 0 & 1 & 0 & 0 \\ 0 & 0 & 3 & 0 & -4 \\ 0 & 0 & 0 & 0 & 0 \end{pmatrix}.$$

In the last matrix, add -3 times row two to row three:

$$\begin{pmatrix} 0 & 1 & 0 & 1 & 1 \\ 0 & 0 & 1 & 0 & 0 \\ 0 & 0 & 3 & 0 & -4 \\ 0 & 0 & 0 & 0 & 0 \end{pmatrix} \to \begin{pmatrix} 0 & 1 & 0 & 1 & 1 \\ 0 & 0 & 1 & 0 & 0 \\ 0 & 0 & 0 & 0 & -4 \\ 0 & 0 & 0 & 0 & 0 \end{pmatrix}.$$

The leading entry of row three is -4. Multiply row three of the last matrix by $-1/4$:

$$\begin{pmatrix} 0 & 1 & 0 & 1 & 1 \\ 0 & 0 & 1 & 0 & 0 \\ 0 & 0 & 0 & 0 & -4 \\ 0 & 0 & 0 & 0 & 0 \end{pmatrix} \to \begin{pmatrix} 0 & 1 & 0 & 1 & 1 \\ 0 & 0 & 1 & 0 & 0 \\ 0 & 0 & 0 & 0 & 1 \\ 0 & 0 & 0 & 0 & 0 \end{pmatrix}.$$

Finally, get zeros above and below the leading entry in row three of the last matrix by adding -1 times row three to row 1:

$$\begin{pmatrix} 0 & 1 & 0 & 1 & 1 \\ 0 & 0 & 1 & 0 & 0 \\ 0 & 0 & 0 & 0 & 1 \\ 0 & 0 & 0 & 0 & 0 \end{pmatrix} \to \begin{pmatrix} 0 & 1 & 0 & 1 & 0 \\ 0 & 0 & 1 & 0 & 0 \\ 0 & 0 & 0 & 0 & 1 \\ 0 & 0 & 0 & 0 & 0 \end{pmatrix} = \mathbf{A}_R.$$

If we had reduced **A** by using another sequence of elementary row operations, we would have reached the same $\mathbf{A}_R$. ♦

In all of the examples we have seen so far, observe that the nonzero rows of a reduced matrix are linearly independent m-vectors, where m is the number of columns of the matrix. This is true in general because, if row i is a nonzero row, then the leading element of row i is 1, and all rows above and below this row have 0 in this column. Thus each nonzero row vector in $\mathbf{A}_R$ has a 1 in a coordinate where all the other row vectors have zeros. Later, when we deal with rank and solve systems of equations, it will be important to know that the nonzero rows of $\mathbf{A}_R$ are linearly independent in the row space of **A**, hence they form a basis for this row space.

The elementary row operations used to reduce a matrix **A** can be achieved by multiplying **A** on the left by some elementary matrix Ω (which is a product of elementary matrices). In view of Theorems 7.5 and 7.7, we can state the following.

THEOREM 7.8

Let **A** be any matrix. Then there is a matrix Ω such that

$$\Omega\mathbf{A} = \mathbf{A}_R. \ \blacklozenge$$

Given **A**, there is a convenient notational device that allows us to find $\mathbf{A}_R$ and Ω simultaneously. Suppose **A** is $n \times m$. Then Ω will be $n \times n$. Form the $n \times (m+n)$ *augmented matrix* $[\mathbf{A} \vdots \mathbf{I}_n]$ by putting $\mathbf{I}_n$ as n additional columns to the right of **A**. The vertical dots separate the original m columns of **A** from the adjoined n columns of $\mathbf{I}_n$, and play no role in the computations. Now reduce **A**, carrying out the same operations on the adjoined rows of $\mathbf{I}_n$. When **A** (the left m columns of this augmented matrix) has been reduced to $\mathbf{A}_R$, the right n columns will be Ω, since we form Ω by starting with the identity matrix and performing the same elementary row operations used to reduce **A**.

EXAMPLE 7.15

Let

$$\mathbf{A} = \begin{pmatrix} -3 & 1 & 0 \\ 4 & -2 & 1 \end{pmatrix}.$$

We will reduce $\mathbf{A}$ and, at the same time, find a 2×2 matrix Ω such that $\Omega\mathbf{A} = \mathbf{A}_R$. Start with the augmented matrix and reduce it:

$$[\mathbf{A}\vdots\mathbf{I}_2] = \begin{pmatrix} -3 & 1 & 0 & \vdots & 1 & 0 \\ 4 & -2 & 1 & \vdots & 0 & 1 \end{pmatrix}$$

$$((-1/3) \text{ times row one }) \rightarrow \begin{pmatrix} 1 & -1/3 & 0 & \vdots & -1/3 & 0 \\ 4 & -2 & 1 & \vdots & 0 & 1 \end{pmatrix}$$

$$(\text{add } -4 \text{ times row one to row two }) \rightarrow \begin{pmatrix} 1 & -1/3 & 0 & \vdots & -1/3 & 0 \\ 0 & -2/3 & 1 & \vdots & 4/3 & 1 \end{pmatrix}$$

$$(\text{multiply row two by } -3/2) \rightarrow \begin{pmatrix} 1 & -1/3 & 0 & \vdots & -1/3 & 0 \\ 0 & 1 & -3/2 & \vdots & -2 & -3/2 \end{pmatrix}$$

$$(\text{ add } 1/3 \text{ row two to row one }) \rightarrow \begin{pmatrix} 1 & 0 & -1/2 & \vdots & -1 & -1/2 \\ 0 & 1 & -3/2 & \vdots & -2 & -3/2 \end{pmatrix}$$

$$= [\mathbf{A}\vdots\mathbf{I}_2]_R.$$

The first three columns of this reduced augmented matrix are $\mathbf{A}_R$, while the last two columns form Ω:

$$\mathbf{A}_R = \begin{pmatrix} 1 & 0 & -1/2 \\ 0 & 1 & -3/2 \end{pmatrix} \text{ and } \Omega = \begin{pmatrix} -1 & -1/2 \\ -2 & -3/2 \end{pmatrix}.$$

As a check,

$$\Omega\mathbf{A} = \begin{pmatrix} -1 & -1/2 \\ -2 & -3/2 \end{pmatrix} \begin{pmatrix} -3 & 1 & 0 \\ 4 & -2 & 1 \end{pmatrix} = \begin{pmatrix} 1 & 0 & -1/2 \\ 0 & 1 & -3/2 \end{pmatrix} = \mathbf{A}_R.$$

This is the reduced form of $[\mathbf{A}\vdots\mathbf{I}_2]$. ♦

MAPLE's `pivot` command is well suited to reducing a matrix $\mathbf{A}$ which has been entered into the program. First look for the leading entries of the nonzero rows. The location of a leading entry is called a *pivot position*. We obtain zeros above and below a leading entry by elementary row operations, adding constant multiples of this row to the other rows if necessary. This is called *pivoting* about this leading entry, and can be done in one operation which in MAPLE is called pivot. If a leading entry α occurs in the i, j position of $\mathbf{A}$, we can form a matrix $\mathbf{B}$ having zeros above and below α by entering

```
B:=pivot(A,i,j);
```

After pivoting about each leading entry, resulting in a matrix **K**, we need only make all the leading entries 1 to obtain a reduced row echelon form. If **K** has leading entry α in row i, multiply this row by $1/\alpha$ by entering

```
F:=mulrow(K, i, 1/α);
```

After this is done for all the rows containing a leading entry, the reduced row echelon form of **A** results.

SECTION 7.3 PROBLEMS

In each of Problems 1 through 12, find the reduced form of **A** and produce a matrix Ω such that $\Omega\mathbf{A} = \mathbf{A}_R$.

1. $\mathbf{A} = \begin{pmatrix} -1 & 4 & 6 \\ 2 & 3 & -5 \\ 7 & 1 & 1 \end{pmatrix}$

2. $\mathbf{A} = \begin{pmatrix} 8 & 2 & 1 & 0 \\ 0 & 1 & 1 & 3 \\ 4 & 0 & 0 & -3 \end{pmatrix}$

3. $\mathbf{A} = \begin{pmatrix} -1 & 2 & 3 & 1 \\ 1 & 0 & 0 & 0 \end{pmatrix}$

4. $\mathbf{A} = \begin{pmatrix} -3 & 4 & 4 \\ 0 & 0 & 0 \end{pmatrix}$

5. $\mathbf{A} = \begin{pmatrix} 4 & 1 & -7 \\ 2 & 2 & 0 \\ 0 & 1 & 0 \end{pmatrix}$

6. $\mathbf{A} = \begin{pmatrix} 0 \\ -3 \\ 1 \\ 1 \end{pmatrix}$

7. $\mathbf{A} = \begin{pmatrix} 0 & 1 \\ 0 & 0 \\ 1 & 3 \\ 0 & 1 \end{pmatrix}$

8. $\mathbf{A} = \begin{pmatrix} 1 & 0 & 1 & 1 & -1 \\ 0 & 1 & 0 & 0 & 2 \end{pmatrix}$

9. $\mathbf{A} = \begin{pmatrix} -1 & 4 & 1 & 1 \\ 0 & 0 & 0 & 0 \\ 0 & 0 & 0 & 0 \\ 0 & 0 & 0 & 1 \end{pmatrix}$

10. $\mathbf{A} = \begin{pmatrix} 2 & 2 \\ 1 & 1 \end{pmatrix}$

11. $\mathbf{A} = \begin{pmatrix} 1 & -1 & 3 \\ 0 & 1 & 2 \\ 0 & 0 & 0 \end{pmatrix}$

12. $\mathbf{A} = \begin{pmatrix} 3 & 1 & 1 & 4 \\ 0 & 1 & 0 & 0 \end{pmatrix}$

7.4 Row and Column Spaces

In this section, we will develop three numbers associated with matrices. These numbers play a significant role in applications such as the solution of systems of linear equations.

Let **A** be an $n \times m$ matrix of real numbers. Each of the n rows is a vector in R^m. The span of these row vectors (the set of all linear combinations of these vectors) is a subspace of R^m called the *row space* of **A**. This may or may not be all of R^m, depending on **A**. The dimension of the row space of **A** is the *row rank* of **A**.

Similarly, the m columns are vectors in R^n. The span of these column vectors is the *column space* of **A**, and is a subspace of R^n. The dimension of this column space is the *column rank* of **A**.

EXAMPLE 7.16

Let

$$\mathbf{A} = \begin{pmatrix} 5 & -1 & 5 \\ -1 & 1 & 3 \\ 1 & 1 & 7 \\ 2 & 0 & 4 \\ 1 & -3 & -6 \end{pmatrix}.$$

The row space is the subspace of R^3 spanned by the five rows of **A**. This subspace consists of all linear combinations

$$\alpha(5, 1, -5) + \beta(-1, 1, 3) + \gamma(1, 1, 7)$$
$$+ \delta(2, 0, 4) + \epsilon(1, -3, 6)$$

of the row vectors. The last three row vectors are linearly independent (none is a linear combination of the other two). The first two are linear combinations of the last three:

$$(5, -1, 5) = -(1, 1, 7) + 3(2, 0, 4)$$

and

$$(-1, 1, 3) = (1, 1, 7) - (2, 0, 4),$$

The first three row vectors therefore form a basis for the row space. This row space has dimension 3 and is all of R^3. The row rank of **A** is 3.

The column space of **A** is the subspace of R^5 consisting of all linear combinations of the column vectors, which we continue to write as columns:

$$\alpha \begin{pmatrix} 5 \\ -1 \\ 1 \\ 2 \\ 1 \end{pmatrix} + \beta \begin{pmatrix} -1 \\ 1 \\ 1 \\ 0 \\ -6 \end{pmatrix} + \gamma \begin{pmatrix} 5 \\ 3 \\ 7 \\ 4 \\ -6 \end{pmatrix}.$$

These three column vectors are linearly independent in R^5 and span a three-dimensional subspace of R^5. The column rank of **A** is 3.

In this example,

$$\text{row rank of } \mathbf{A} = \text{ column rank of } \mathbf{A} = 3. \quad \blacklozenge$$

We claim that this is not a coincidence.

THEOREM 7.9 Equality of Row and Column Rank

For any matrix, the row rank equals the column rank. ♦

Proof Although this is true in general, we will prove it when each a_{ij} is a real number, enabling us to exploit the row and column spaces of **A**. Suppose **A** is $n \times m$:

$$\mathbf{A} = \begin{pmatrix} a_{11} & a_{12} & \cdots & a_{1r} & a_{1,r+1} & \cdots & a_{1m} \\ a_{21} & a_{22} & \cdots & a_{2r} & a_{2,r+1} & \cdots & a_{2m} \\ \vdots & \vdots & \vdots & \vdots & \vdots & \vdots & \vdots \\ a_{r1} & a_{r2} & \cdots & a_{rr} & a_{r,r+1} & \cdots & a_{rm} \\ a_{r+1,1} & a_{r+1,2} & \cdots & a_{r+1,r} & a_{r+1,r+1} & \cdots & a_{r+1,m} \\ \vdots & \vdots & \vdots & \vdots & \vdots & \vdots & \vdots \\ a_{n1} & a_{n2} & \cdots & a_{nr} & a_{n,r+1} & \cdots & a_{nm} \end{pmatrix}.$$

Denote the row vectors $\mathbf{R}_1, \mathbf{R}_2, \cdots, \mathbf{R}_n$, so

$$\mathbf{R}_i = (a_{i1}, a_{i2}, \cdots, a_{im}) \text{ in } R^m.$$

Suppose the row rank of $\mathbf{A}$ is r. As a notational convenience, suppose the first r rows are linearly independent. Then each of $\mathbf{R}_{r+1}, \cdots, \mathbf{R}_n$ is a linear combination of $\mathbf{R}_1, \cdots, \mathbf{R}_r$. Write

$$\mathbf{R}_{r+1} = \beta_{r+1,1}\mathbf{R}_1 + \cdots + \beta_{r+1,r}\mathbf{R}_r$$
$$\mathbf{R}_{r+2} = \beta_{r+2,1}\mathbf{R}_1 + \cdots + \beta_{r+2,r}\mathbf{R}_r$$
$$\vdots$$
$$\mathbf{R}_n = \beta_{n,1}\mathbf{R}_1 + \cdots + \beta_{n,r}\mathbf{R}_r.$$

Now observe that column j of $\mathbf{A}$ can be written

$$\begin{pmatrix} a_{1j} \\ a_{2j} \\ \vdots \\ a_{rj} \\ a_{r+1,j} \\ \vdots \\ a_{nj} \end{pmatrix} = a_{1j}\begin{pmatrix} 1 \\ 0 \\ \vdots \\ 0 \\ \beta_{r+1,1} \\ \vdots \\ \beta_{n1} \end{pmatrix} + a_{2j}\begin{pmatrix} 0 \\ 1 \\ \vdots \\ 0 \\ \beta_{r+1,2} \\ \vdots \\ \beta_{n,2} \end{pmatrix} + \cdots + a_{rj}\begin{pmatrix} 0 \\ 0 \\ \vdots \\ 1 \\ \beta_{r+1,r} \\ \vdots \\ \beta_{n,r} \end{pmatrix}.$$

Thus, each column of $\mathbf{A}$ is a linear combination of the r n-vectors on the right side of the last equation. These r vectors therefore span the column space of $\mathbf{A}$, so the dimension of this column space is at most r (equal to r if these columns are linearly independent, less than r if they are not). This proves that

$$\text{dimension of the column space of } \mathbf{A} \leq \text{dimension of the row space.}$$

By repeating this argument, using columns instead of rows, we find that the dimension of the row space is less than or equal to the dimension of the column space. This proves the theorem. ♦

Now define the *rank* of $\mathbf{A}$ as the row rank of the matrix, which is the same as the column rank. Denote this number as rank($\mathbf{A}$). The matrix of Example 7.16 has rank 3.

Given an arbitrary real matrix $\mathbf{A}$, it may not be obvious what the rank of a $\mathbf{A}$ is. However, if $\mathbf{R}$ is a reduced matrix, then

$$\text{rank}(\mathbf{R}) = \text{ number of nonzero rows of } \mathbf{R}.$$

To see this, recall that the nonzero rows of **R** form a basis for the row space of this matrix, hence their number is the dimension of this row space.

Now suppose we perform an elementary row operation on **A** to form **B**. How does this change the row space of **A**? The answer is that it does not change it at all. This fact will be important in solving systems of linear equations

THEOREM 7.10

Let **B** be formed from an $n \times m$ matrix **A** by a sequence of elementary row operations. Then **A** and **B** have the same row space, hence also

$$\text{rank}(\mathbf{A}) = \text{rank}(\mathbf{B}). \ \blacklozenge$$

Proof It is enough to prove the theorem for the case that **B** is formed from **A** by one elementary row operation. Let the row vectors of **A** be $\mathbf{A}_1, \cdots, \mathbf{A}_n$. The row space of **A** is the subspace of R^m consisting of all linear combinations

$$\alpha_1\mathbf{A}_1 + \alpha_2\mathbf{A}_2 + \cdots + \alpha_n\mathbf{A}_n.$$

If the elementary row operation is an interchange of rows, then the rows of **A** and **B** are the same (appearing in a different order) and hence span the same subspace of R^m.

Suppose a type II elementary row operation is performed, multiplying row r of **A** by the nonzero number c. Now the row space of **B** consists of all vectors

$$\alpha_1\mathbf{A}_1 + \cdots + c\alpha_r\mathbf{A}_r + \cdots + \alpha_n\mathbf{A}_n.$$

Since the α_j's are arbitrary, this is again a linear combination of the rows of **A**, hence the row spaces of **A** and **B** are the same.

Finally, consider the case that a type III operation is performed, adding c times row i to row j to form **B**. Now the row vectors of **B** are

$$\mathbf{A}_1, \cdots, \mathbf{A}_{j-1}, c\mathbf{A}_i + \mathbf{A}_j, \mathbf{A}_{j+1}, \cdots, \mathbf{A}_n.$$

Any linear combination of these rows is again a linear combination of the rows of **A**, hence in this case the row spaces of **A** and **B** are also the same.

Finally, because the row spaces are the same, their dimension is the same and the matrices have the same rank. ♦

If we defined elementary column operations analogous to the elementary row operations, we would find that these leave the column space of a matrix unchanged.

Theorem 7.10 has several important consequences.

COROLLARY 7.1

For any real matrix **A**, **A** and $\mathbf{A}_R$ have the same row space. Thus,

$$\text{rank}(\mathbf{A}) = \text{number of nonzero rows of } \mathbf{A}_R. \ \blacklozenge$$

This follows from the fact $\mathbf{A}_R$ is formed from **A** by a sequence of elementary row operations, so

$$\begin{aligned}\text{rank}(\mathbf{A}) &= \text{rank}(\mathbf{A}_R)\\ &= \text{number of nonzero rows of } \mathbf{A}_R.\end{aligned}$$

EXAMPLE 7.17

Let

$$\mathbf{A} = \begin{pmatrix} 0 & 1 & 0 & 0 & 3 & 0 & 6 \\ 0 & 0 & 1 & 0 & -2 & 1 & 5 \\ 0 & 0 & 0 & 1 & 2 & 0 & -4 \\ 0 & 0 & 0 & 0 & 0 & 0 & 0 \end{pmatrix}.$$

Since this is a reduced matrix with three nonzero rows, $\text{rank}(\mathbf{A}) = 3$. ♦

COROLLARY 7.2

Let $\mathbf{A}$ be an $n \times n$ matrix of real numbers. Then

$$\text{rank}(\mathbf{A}) = n \text{ if and only if } \mathbf{A}_R = \mathbf{I}_n.$$

This says that the rank of a square matrix equals the number of rows exactly when the reduced form is the identity matrix.

Proof First, we know that

$$\text{rank}(\mathbf{A}) = \text{ number of nonzero rows of } \mathbf{A}_R.$$

If $\mathbf{A}_R = \mathbf{I}_n$, then $\mathbf{A}_R$ has n nonzero rows and this matrix has rank n, hence $\mathbf{A}$ also has rank n.

Conversely, if $\mathbf{A}$ has rank n, then so does $\mathbf{A}_R$, so this reduced matrix is an $n \times n$ matrix with 1 down the main diagonal and all other elements (above and below leading entries) zero. Then $\mathbf{A}_R = \mathbf{I}_n$. ♦

The MAPLE command `rank(A)` will return the rank of $\mathbf{A}$.

SECTION 7.4 PROBLEMS

In each of Problems 1 through 14, find the reduced form of the matrix and use this to determine the rank of the matrix. Also find a basis for the row space of the matrix and a basis for the column space.

1. $\begin{pmatrix} -3 & 2 & 2 \\ 1 & 0 & 5 \\ 0 & 0 & 2 \end{pmatrix}$

2. $\begin{pmatrix} -4 & -1 & 1 & 6 \\ 0 & 4 & -4 & 2 \\ 1 & 0 & 0 & 0 \end{pmatrix}$

3. $\begin{pmatrix} -4 & 1 & 3 \\ 2 & 2 & 0 \end{pmatrix}$

4. $\begin{pmatrix} -3 & 2 & 1 & 1 & 0 \\ 6 & -4 & -2 & -2 & 0 \end{pmatrix}$

5. $\begin{pmatrix} -3 & 1 \\ 2 & 2 \\ 4 & -3 \end{pmatrix}$

6. $\begin{pmatrix} 0 & -1 & 0 \\ 0 & 0 & -1 \\ 0 & 0 & 2 \end{pmatrix}$

7. $\begin{pmatrix} 8 & -4 & 3 & 2 \\ 1 & -1 & 1 & 0 \end{pmatrix}$

8. $\begin{pmatrix} 1 & 3 & 0 \\ 0 & 0 & 1 \end{pmatrix}$

9. $\begin{pmatrix} 2 & 2 & 1 \\ 1 & -1 & 3 \\ 0 & 0 & 1 \\ 4 & 0 & 7 \end{pmatrix}$

10. $\begin{pmatrix} 1 & -1 & 4 \\ 0 & 1 & 3 \\ 2 & -1 & 11 \end{pmatrix}$

11. $\begin{pmatrix} -2 & 5 & 7 \\ 0 & 1 & -3 \\ -4 & 11 & 11 \end{pmatrix}$

12. $\begin{pmatrix} 6 & 0 & 0 & 1 & 1 \\ 12 & 0 & 0 & 2 & 2 \\ 1 & -1 & 0 & 0 & 0 \end{pmatrix}$

13. $\begin{pmatrix} 0 & 4 & 3 \\ 0 & 1 & 0 \\ 2 & 2 & 2 \end{pmatrix}$

14. $\begin{pmatrix} 1 & 0 & 0 \\ 2 & 0 & 0 \\ 1 & 0 & -1 \\ 3 & 0 & 0 \end{pmatrix}$

15. Let **A** be any matrix of real numbers. Prove that

$$\text{rank}(\mathbf{A}) = \text{rank}(\mathbf{A}^t).$$

7.5 Homogeneous Systems

We want to develop a method for finding all solutions of a *linear homogeneous system* of n equations in m unknowns:

$$\begin{aligned} a_{11}x_1 + a_{12}x_2 + \cdots + a_{1m}x_m &= 0 \\ a_{21}x_1 + a_{22}x_2 + \cdots + a_{2m}x_m &= 0 \\ &\vdots \\ a_{n1}x_1 + a_{n2}x_2 + \cdots + a_{nm}x_m &= 0. \end{aligned}$$

The numbers a_{ij} are called the the *coefficients* of the system and $\mathbf{A} = [a_{ij}]$ is the *matrix of coefficients*. Row i contains the coefficients of equation i and column j contains the coefficients of x_j.

Define

$$\mathbf{X} = \begin{pmatrix} x_1 \\ x_2 \\ \vdots \\ x_m \end{pmatrix}$$

and write the $n \times 1$ zero matrix as just **O**, a column of n zeros. Then the system can be written as the matrix equation

$$\mathbf{AX} = \mathbf{O}.$$

We will develop the following strategy for solving this system.

1. We will show that $\mathbf{AX} = \mathbf{O}$ has the same solutions as the *reduced system* $\mathbf{A}_R\mathbf{X} = \mathbf{O}$.
2. We will show how to write all solutions of the reduced system directly from the reduced matrix $\mathbf{A}_R$.
3. We will also use facts about vector spaces and rank to derive additional information about solutions.

The remainder of this section consists of the details of carrying out this strategy, and examples. The first two examples give us some feeling for what to look for in solving a homogeneous system.

EXAMPLE 7.18

Consider the simple system

$$\begin{aligned} x_1 - 3x_2 + 2x_3 &= 0 \\ -2x_1 + x_2 - 3x_3 &= 0. \end{aligned}$$

Of course, we do not need matrices to solve this system, but we want to illustrate a point. The matrix of coefficients is

$$\mathbf{A}=\begin{pmatrix}1 & -3 & 2\\ -2 & 1 & -3\end{pmatrix}.$$

It is routine to find

$$\mathbf{A}_R=\begin{pmatrix}1 & 0 & 7/5\\ 0 & 1 & -1/5\end{pmatrix}.$$

The reduced system $\mathbf{A}_R\mathbf{X}=\mathbf{O}$ is

$$x_1+\frac{7}{5}x_3=0$$

$$x_2-\frac{1}{5}x_3=0.$$

This reduced system can be solved by inspection:

$$x_1=-\frac{7}{5}x_3,\ x_2=\frac{1}{5}x_3,\ x_3 \text{ is arbitrary.}$$

We can give x_3 any numerical value, and this determines x_1 and x_2 to yield a solution.

It will be useful to write this solution as a column matrix:

$$\mathbf{X}=\begin{pmatrix}x_1\\ x_2\\ x_3\end{pmatrix}=x_3\begin{pmatrix}-7/5\\ 1/5\\ 1\end{pmatrix}=\alpha\begin{pmatrix}-7/5\\ 1/5\\ 1\end{pmatrix},$$

in which we have written $x_3=\alpha$ because it looks neater. Here α can be any number.

This general solution of the reduced system is also the solution of the original system. In this example the general solution depends on one arbitrary constant, hence is, in a sense to be discussed, a one-dimensional solution. ♦

In Example 7.18, x_3 is called a *free variable*, since it can assume any value. This example had one free variable, but the general solution of a system $\mathbf{AX}=\mathbf{O}$ might have any number.

Free variables occur in columns of $\mathbf{A}_R$ *that contain no leading entry of a row.*

EXAMPLE 7.19

Consider the 3×5 system

$$x_1-3x_2+x_3-7x_4+4x_5=0$$

$$x_1+2x_2-3x_3=0$$

$$x_2-4x_3+x_5=0.$$

The matrix of coefficients is

$$\mathbf{A}=\begin{pmatrix}1 & -3 & 1 & -7 & 4\\ 1 & 2 & -3 & 0 & 0\\ 0 & 1 & -4 & 0 & 1\end{pmatrix}.$$

A routine calculation yields

$$\mathbf{A}_R=\begin{pmatrix}1 & 0 & 0 & -35/16 & 13/16\\ 0 & 1 & 0 & 28/16 & -20/16\\ 0 & 0 & 1 & 7/16 & -9/16\end{pmatrix}.$$

The reduced system $\mathbf{A}_R\mathbf{X} = \mathbf{O}$ is

$$x_1 - \tfrac{35}{16}x_4 + \tfrac{13}{16}x_5 = 0,$$
$$x_2 + \tfrac{28}{16}x_4 - \tfrac{20}{16}x_5 = 0,$$

and

$$x_3 + \tfrac{7}{16}x_4 - \tfrac{9}{16}x_5 = 0.$$

This system is easy to solve:

$$x_1 = \tfrac{35}{16}x_4 - \tfrac{13}{16}x_5,$$
$$x_2 = -\tfrac{28}{16}x_4 + \tfrac{20}{16}x_5,$$

and

$$x_3 = -\tfrac{7}{16}x_4 + \tfrac{9}{16}x_5$$

in which x_4 and x_5 (the free variables) can be given any values and these determine x_1, x_2 and x_3. Again, note that these two free variables are in the two columns of the reduced matrix that contain no leading element of any row.

We can express this solution more neatly by setting $x_4 = 16\alpha$ and $x_5 = 16\beta$ with α and β arbitrary numbers and writing

$$x_1 = 35\alpha - 13\beta,$$
$$x_2 = -28\alpha + 20\beta,$$
$$x_3 = -7\alpha + 9\beta,$$
$$x_4 = 16\alpha,$$

and

$$x_5 = 16\beta.$$

Here α and β are any numbers. This is the *general solution* of the reduced system, and it is routine to verify that it is also the solution of the original system. As a column matrix, this solution is

$$\mathbf{X} = \begin{pmatrix} 35\alpha - 13\beta \\ -28\alpha + 20\beta \\ -7\alpha + 9\beta \\ 16\alpha \\ 16\beta \end{pmatrix} = \alpha \begin{pmatrix} 35 \\ -28 \\ -7 \\ 16 \\ 0 \end{pmatrix} + \beta \begin{pmatrix} -13 \\ 20 \\ 9 \\ 0 \\ 16 \end{pmatrix}. \ \blacklozenge$$

This way of writing the general solution reveals its structure as being two dimensional, depending on two arbitrary constants. ♦

These examples illustrate the strategy outlined at the beginning of this section. This will depend on the crucial fact that the reduced system has the same solutions as the original system, as we will now verify.

THEOREM 7.11

Let $\mathbf{A}$ be $n \times m$. Then the systems $\mathbf{AX} = \mathbf{O}$ and $\mathbf{A}_R\mathbf{X} = \mathbf{O}$ have the same solutions. ♦

Proof First, we know that there is a matrix

$$\Omega = \mathbf{E}_r\mathbf{E}_{r-1}\cdots\mathbf{E}_2\mathbf{E}_1,$$

a product of elementary matrices, that reduces $\mathbf{A}$:

$$\Omega\mathbf{A} = \mathbf{A}_R.$$

Now suppose that $\mathbf{X} = \mathbf{C}$ is a solution of $\mathbf{AX} = \mathbf{O}$. Then $\mathbf{AC} = \mathbf{O}$, so

$$\Omega\mathbf{O} = \Omega(\mathbf{AC}) = (\Omega\mathbf{A})\mathbf{C} = \mathbf{A}_R\mathbf{C} = \mathbf{O},$$

so $\mathbf{C}$ is also a solution of the reduced system.

Conversely, suppose $\mathbf{K}$ is a solution of the reduced system, so $\mathbf{A}_R\mathbf{K} = \mathbf{O}$. We want to show that $\mathbf{AK} = \mathbf{O}$. Since $\Omega\mathbf{A} = \mathbf{A}_R$, we have $(\Omega\mathbf{A})\mathbf{K} = \mathbf{O}$, so

$$(\mathbf{E}_r\mathbf{E}_{r-1}\cdots\mathbf{E}_2\mathbf{E}_1)\mathbf{AK} = \mathbf{O}.$$

Now, each $\mathbf{E}_j$ is an elementary matrix, and we know that there is an elementary matrix $\mathbf{E}_j^*$ that reverses the effect of $\mathbf{E}_j$. From the last equation, we have

$$\mathbf{E}_1^*\mathbf{E}_2^*\cdots\mathbf{E}_{r-1}^*\mathbf{E}_r^*(\mathbf{E}_r\mathbf{E}_{r-1}\cdots\mathbf{E}_2\mathbf{E}_1)\mathbf{AK} = \mathbf{O}.$$

But $\mathbf{E}_r^*\mathbf{E}_r = \mathbf{I}_n$, and $\mathbf{E}_{r-1}^*\mathbf{E}_{r-1} = \mathbf{I}_n$, and so on until $\mathbf{E}_1^*\mathbf{E}_1 = \mathbf{I}_n$, so in the last product all of the elementary matrices cancel in pairs, leaving $\mathbf{AK} = \mathbf{O}$. Therefore $\mathbf{K}$ is also a solution of the original system, completing the proof. ♦

We can therefore concentrate on solving a reduced system. As we have seen in the examples, the solution of $\mathbf{A}_R\mathbf{X} = \mathbf{O}$ is easily read from this matrix, and has the added dividend that it reveals the structure of these solutions. The set of all solutions of the homogeneous system $\mathbf{AX} = \mathbf{O}$ form a vector space, which is a subspace of R^m if $\mathbf{A}$ is $n \times m$. Furthermore, the dimension of this solution space can read from $\mathbf{A}_R$, as we saw in the examples.

If $\mathbf{A}_R$ has k nonzero rows (hence rank k), then k of the x_i's are determined by the $m - k$ free variables, which can be assigned any values in writing solutions of the system. This means that $x_1, \cdots, x_k$ are determined by $x_{k+1}, \cdots, x_m$, which can be chosen arbitrarily. The general solution will have $m - k$ arbitrary constants in it.

We will summarize these observations.

THEOREM 7.12 Solution Space of a Homogeneous System

Let $\mathbf{A}$ be $n \times m$. Then

1. The set of all solutions of $\mathbf{AX} = \mathbf{O}$ forms a subspace of R^m, called the *solution space* of this system.
2. The dimension of this solution space is

$$m - \text{ number of nonzero rows of } \mathbf{A}_R,$$

which is the same as $m -$ rank $(\mathbf{A})$. ♦

Proof Let S be the set of all solutions of the system. Since

$$x_1 = x_2 = \cdots = x_m = 0$$

is a solution, the zero m-vector is in S.

Now suppose $\mathbf{X}_1$ and $\mathbf{X}_2$ are solutions, and α and β are numbers. Then

$$\mathbf{A}(\alpha\mathbf{X}_1 + \beta\mathbf{X}_2) = \alpha\mathbf{AX}_1 + \beta\mathbf{AX}_2 = \mathbf{O} + \mathbf{O} = \mathbf{O},$$

so linear combinations of solutions are solutions, and S is a subspace of R^m.

For the dimension of S, use the fact that the system has the same solution space as the reduced system. As the examples suggest, the nonzero rows of $\mathbf{A}_R$ enable us to express the general solution as a linear combination of linearly independent solutions, one for each free variable. Since the number of free variables is the number of columns of $\mathbf{A}_R$, minus the number of nonzero rows, then the dimension of S is $m -$ rank$(\mathbf{A})$. ♦

Since the number of nonzero rows of the reduced matrix is the rank of $\mathbf{A}_R$, which is also the rank of $\mathbf{A}$, then the dimension of the solution space can also be computed as

$$m - \operatorname{rank}(\mathbf{A}).$$

EXAMPLE 7.20

Solve the system

$$\begin{aligned} -x_1 + x_3 + x_4 + 2x_5 &= 0 \\ x_2 + 3x_3 + 4x_5 &= 0 \\ x_1 + 2x_2 + x_3 + x_4 + x_5 &= 0 \\ -3x_1 + x_2 + 4x_5 &= 0. \end{aligned}$$

The matrix of coefficients is

$$\mathbf{A} = \begin{pmatrix} -1 & 0 & 1 & 1 & 2 \\ 0 & 1 & 3 & 0 & 4 \\ 1 & 2 & 1 & 1 & 1 \\ -3 & 1 & 0 & 0 & 4 \end{pmatrix}.$$

Routine manipulations yield the reduced form

$$\mathbf{A}_R = \begin{pmatrix} 1 & 0 & 0 & 0 & -9/8 \\ 0 & 1 & 0 & 0 & 5/8 \\ 0 & 0 & 1 & 0 & 9/8 \\ 0 & 0 & 0 & 1 & -1/4 \end{pmatrix}.$$

In this example $\mathbf{A}$ has $m = 5$ columns, and the rank of $\mathbf{A}$ is 4 because $\mathbf{A}_R$ has four nonzero rows. The solution space will have dimension $5 - 4 = 1$.

$\mathbf{A}_R$ is the coefficient matrix of the reduced system

$$\begin{aligned} x_1 - \tfrac{9}{8}x_5 &= 0, \\ x_2 + \tfrac{5}{8}x_5 &= 0, \\ x_3 + \tfrac{9}{8}x_5 &= 0, \\ x_4 - \tfrac{1}{4}x_5 &= 0. \end{aligned}$$

Notice that x_1 through x_4 depend on the single free variable x_5, which can be chosen arbitrarily. Set $x_5 = \alpha$ to write the general solution

$$x_1 = \frac{9}{8}\alpha,\ x_2 = -\frac{5}{8}\alpha,\ x_3 = -\frac{9}{8}\alpha,\ x_4 = \frac{1}{4}\alpha,\ x_5 = \alpha.$$

If we let $\beta = \alpha/8$ (β is still any number), then

$$x_1 = 9\beta,\ x_2 = -5\beta,\ x_3 = -9\beta,\ x_4 = 2\beta,\ x_5 = 8\beta.$$

As a column matrix, this solution is

$$\mathbf{X} = \beta \begin{pmatrix} 9 \\ -5 \\ -9 \\ 2 \\ 8 \end{pmatrix}.$$

This gives the general solution as the set of all multiples of one solution, which forms a basis for the one-dimensional solution space (a subspace of R^5). ♦

EXAMPLE 7.21

We will solve the system

$$
\begin{aligned}
2x_1 - 4x_2 + x_3 + x_4 + 6x_5 + 4x_6 - 2x_7 &= 0\\
-4x_1 + x_2 + 6x_3 + 3x_4 + 10x_5 - 3x_6 + 6x_7 &= 0\\
3x_1 + x_2 - 4x_3 + 2x_4 + 5x_5 + x_6 + 3x_7 &= 0.
\end{aligned}
$$

The coefficient matrix is

$$\mathbf{A} = \begin{pmatrix} 2 & -4 & 1 & 1 & 6 & 4 & -2 \\ -4 & 1 & 6 & 3 & 10 & -3 & 6 \\ 3 & 1 & -4 & 2 & 5 & 1 & 3 \end{pmatrix}.$$

We find the reduced matrix

$$\mathbf{A}_R = \begin{pmatrix} 1 & 0 & 0 & 3 & 67/7 & 4/7 & 29/7 \\ 0 & 1 & 0 & 9/5 & 178/35 & -5/7 & 118/35 \\ 0 & 0 & 1 & 11/5 & 36/5 & 0 & 16/5 \end{pmatrix}.$$

Since $m = 7$ and $\mathbf{A}_R$ has three nonzeros, the solution space is a four-dimensional subspace of R^7. The general solution depends on the arbitrary free variables $x_4, \cdots, x_7$. Let $x_4 = \alpha$, $x_5 = \beta$, $x_6 = \gamma$ and $x_7 = \delta$ to write the general solution

$$\mathbf{X} = \alpha \begin{pmatrix} -3 \\ -9/5 \\ -11/5 \\ 1 \\ 0 \\ 0 \\ 0 \end{pmatrix} + \beta \begin{pmatrix} -67/7 \\ -178/35 \\ -36/5 \\ 0 \\ 1 \\ 0 \\ 0 \end{pmatrix} + \gamma \begin{pmatrix} -4/7 \\ 5/7 \\ 0 \\ 0 \\ 0 \\ 1 \\ 0 \end{pmatrix} + \delta \begin{pmatrix} -29/7 \\ -118/35 \\ -16/5 \\ 0 \\ 0 \\ 0 \\ 1 \end{pmatrix}. \quad ◆$$

As Example 7.21 suggests, with a little practice, the general solution can be read directly from the reduced matrix.

A homogenous system always has at least the trivial solution, and may or may not have nontrivial solutions. Here is a simple condition for a homogeneous system to have a nontrivial solution.

COROLLARY 7.3

Let $\mathbf{A}$ be $n \times m$. Then the homogeneous system $\mathbf{AX} = \mathbf{O}$ has a nontrivial solution if and only

$$m - \text{ number of nonzero rows of } (\mathbf{A}_R) > 0. \quad ◆$$

The reason for this is that the system can have a nontrivial solution only when the dimension of the solution space is positive, having something in it other than the zero vector. Since this solution space has dimension $m -$ rank$(\mathbf{A})$, there will be a nontrivial solution exactly when this number is positive.

In particular, look at the case that the system has more equations than unknowns, so $m < n$. Since the rank of $\mathbf{A}$ cannot exceed the number of rows (equations), in this case

$$\text{rank}(\mathbf{A}) \le n < m$$

so $m -$ rank$(\mathbf{A}) > 0$ and the system has a nontrivial solution.

COROLLARY 7.4

A linear homogeneous system with more unknowns than equations always has a nontrivial solution. ♦

Corollary 7.3 implies that $\mathbf{AX} = \mathbf{O}$ has only the trivial solution exactly when m minus the number of nonzero rows of the reduced matrix is zero. In particular, when $\mathbf{A}$ is square, then $m = n$ and this occurs exactly when the $n \times n$ matrix $\mathbf{A}_R$ has n nonzero rows, which in turn happens exactly when $\mathbf{A}_R = \mathbf{I}_n$.

COROLLARY 7.5

If $\mathbf{A}$ is $n \times n$, then $\mathbf{AX} = \mathbf{O}$ has only the trivial solution if and only if $\mathbf{A}_R = \mathbf{I}_n$. ♦

EXAMPLE 7.22

We will solve the system

$$\begin{aligned} -4x_1 + x_2 - 7x_3 &= 0 \\ 2x_1 + 9x_2 - 13x_3 &= 0 \\ x_1 + x_2 + 10x_3 &= 0. \end{aligned}$$

The coefficient matrix is

$$\mathbf{A} = \begin{pmatrix} -4 & 1 & -7 \\ 2 & 9 & -13 \\ 1 & 1 & 10 \end{pmatrix}.$$

We find that $\mathbf{A}_R = \mathbf{I}_3$. Therefore the system has only the trivial solution. This can also be seen from the reduced system, which is

$$\begin{aligned} x_1 &= 0 \\ x_2 &= 0 \\ x_3 &= 0. \end{aligned}$$ ♦

SECTION 7.5 PROBLEMS

In each of Problems 1 through 12, determine the dimension of the solution space and find the general solution of the system by reducing the coefficient matrix. Write the general solution in terms of one or more column matrices.

1. $$\begin{aligned} x_2 - 3x_4 + x_5 &= 0 \\ 2x_1 - x_2 + x_4 &= 0 \\ 2x_1 - 3x_2 + 4x_5 &= 0 \end{aligned}$$

2. $$\begin{aligned} 2x_1 - 4x_5 + x_7 + x_8 &= 0 \\ 2x_2 - x_6 + x_7 - x_8 &= 0 \\ x_3 - 4x_4 + x_5 &= 0 \\ x_2 - x_3 + x_4 &= 0 \\ x_2 - x_5 + x_6 - x_7 &= 0 \end{aligned}$$

3. $$\begin{aligned} x_1 - 2x_2 + x_5 - x_6 + x_7 &= 0 \\ x_3 - x_4 + x_5 - 2x_6 + 3x_7 &= 0 \\ x_1 - x_5 + 2x_6 &= 0 \\ 2x_1 - 3x_4 + x_5 &= 0 \end{aligned}$$

4. $$\begin{aligned} 8x_1 - 2x_3 + x_6 &= 0 \\ 2x_1 - x_2 + 3x_4 - x_6 &= 0 \\ x_2 + x_3 - 2x_5 - x_6 &= 0 \\ x_4 - 3x_5 + 2x_6 &= 0 \end{aligned}$$

5. $$\begin{aligned} -10x_1 - x_2 + 4x_3 - x_4 + x_5 - x_6 &= 0 \\ x_2 - x_3 + 3x_4 &= 0 \\ 2x_1 - x_2 + x_5 &= 0 \\ x_2 - x_4 + x_6 &= 0 \end{aligned}$$

6. $-3x_1 + x_2 - x_3 + x_4 + x_5 = 0$
$x_2 + x_3 + 4x_5 = 0$
$-3x_3 + 2x_4 + x_5 = 0$

7. $-2x_1 + x_2 + 2x_3 = 0$
$x_1 - x_2 = 0$
$x_1 + x_2 = 0$

8. $6x_1 - x_2 + x_3 = 0$
$x_1 - x_4 + 2x_5 = 0$
$x_1 - 2x_5 = 0$

9. $x_1 - x_2 + 3x_3 - x_4 + 4x_5 = 0$
$2x_1 - 2x_2 + x_3 + x_4 = 0$
$x_1 - 2x_3 + x_5 = 0$
$x_3 + x_4 - x_5 = 0$

10. $4x_1 + x_2 - 3x_3 + x_4 = 0$
$2x_1 - x_3 = 0$

11. $x_1 + 2x_2 - x_3 + x_4 = 0$
$x_2 - x_3 + x_4 = 0$

12. $4x_1 - 3x_2 + x_4 + x_5 - 3x_6 = 0$
$2x_2 + 4x_4 - x_5 - 6x_6 = 0$
$3x_1 - 2x_2 + 4x_5 - x_6 = 0$
$2x_1 + x_2 - 3x_3 + 4x_4 = 0$

13. Let **A** be an $n \times m$ matrix of real numbers. Let $S(\mathbf{A})$ denote the solution space of **A**. Let R be the row space and C the column space of **A**.
(a) Show that $R^{\perp} = S(\mathbf{A})$.
(b) Show that $C^{\perp} = S(\mathbf{A}^t)$.

14. Show that a system $\mathbf{AX} = \mathbf{O}$ has a nontrivial solution if and only if the columns of **A** are linearly dependent. *Hint*: This can be done using a dimension argument. Another approach is to write **AX** as a linear combination of the columns of **A**, as suggested in Section 7.1.1.

15. Can a system $\mathbf{AX} = \mathbf{O}$ having at least as many equations as unknowns, have a nontrivial solution?

7.6 Nonhomogeneous Systems

Now consider the nonhomogeneous linear system of n equations in m unknowns:

$$a_{11}x_1 + a_{12}x_2 + \cdots + a_{1m}x_m = b_1$$
$$a_{21}x_1 + a_{22}x_2 + \cdots + a_{2m}x_m = b_2$$
$$\vdots$$
$$a_{n1}x_1 + a_{n2}x_2 + \cdots + a_{nm}x_m = b_n.$$

In matrix form,

$$\mathbf{AX} = \mathbf{B} \tag{7.1}$$

where **A** is the coefficient matrix,

$$\mathbf{X} = \begin{pmatrix} x_1 \\ x_2 \\ \vdots \\ x_m \end{pmatrix} \text{ and } \mathbf{B} = \begin{pmatrix} b_1 \\ b_2 \\ \vdots \\ b_n \end{pmatrix}.$$

The system is *nonhomogeneous* if at least one $b_j \neq 0$. Nonhomogeneous systems differ from linear systems in two significant ways.

1. A nonhomogeneous system may have no solution. For example, the system

$$2x_1 - 3x_2 = 6$$
$$4x_1 - 6x_2 = 8$$

can have no solution. If $2x_1 - 3x_2 = 6$, then $4x_1 - 6x_2$ must equal 12, not 8.

We call $\mathbf{AX} = \mathbf{B}$ *consistent* if there is a solution. If there is no solution, the system is *inconsistent*.

2. A linear combination of solutions of a nonhomogeneous system $\mathbf{AX} = \mathbf{B}$ need not be a solution. Therefore the solutions do not have the vector space structure seen in the homogeneous case.

Nevertheless, solutions of $\mathbf{AX} = \mathbf{B}$ do have a property that parallels that for solutions of linear second order differential equations. We will call $\mathbf{AX} = \mathbf{O}$ the *associated homogeneous system* of the nonhomogeneous system $\mathbf{AX} = \mathbf{B}$. Although a sum of solutions of the nonhomogeneous system need not be a solution, we claim that the *difference* of any two solutions of the nonhomogeneous system is a solution, not of the system, but of the associated *homogeneous* system. The reason for this is that, if $\mathbf{AU}_1 = \mathbf{B}$ and $\mathbf{AU}_2 = \mathbf{B}$, then

$$\mathbf{A}(\mathbf{U}_1 - \mathbf{U}_2) = \mathbf{AU}_1 - \mathbf{AU}_2 = \mathbf{B} - \mathbf{B} = \mathbf{O}.$$

This is the key to the fundamental theorem for writing the general solution of $\mathbf{AX} = \mathbf{B}$.

THEOREM 7.13

Let $\mathbf{H}$ be the general solution of the associated homogeneous system. Let $\mathbf{U}_p$ be any particular solution of $\mathbf{AX} = \mathbf{B}$. Then the expression $\mathbf{H} + \mathbf{U}_p$ contains every solution of the nonhomogeneous system $\mathbf{AX} = \mathbf{B}$. ♦

Proof Suppose $\mathbf{H}_1, \cdots, \mathbf{H}_k$ form a basis for the solution space of $\mathbf{AX} = \mathbf{O}$, where $k = m -$ number of nonzero rows of $(\mathbf{A}_R)$. Then the general solution of the homogeneous system is

$$\mathbf{H} = \alpha_1 \mathbf{H}_1 + \cdots + \alpha_k \mathbf{H}_k.$$

If $\mathbf{U}$ is any solution of $\mathbf{AX} = \mathbf{B}$, then $\mathbf{U} - \mathbf{U}_p$ is a solution of the associated homogeneous system, and therefore has the form

$$\mathbf{U} - \mathbf{U}_p = c_1 \mathbf{H}_1 + \cdots + c_k \mathbf{H}_k$$

for some constants $c_1, \cdots, c_k$. But then

$$\mathbf{U} = c_1 \mathbf{H}_1 + \cdots + c_k \mathbf{H}_k + \mathbf{U}_p,$$

and this solution is contained in the general expression $\mathbf{H} + \mathbf{U}_p$. ♦

As an immediate consequence, Theorem 7.13 tells us when a nonhomogeneous system can have only one solution.

COROLLARY 7.6

A consistent nonhomogeneous system $\mathbf{AX} = \mathbf{B}$ has a unique solution if and only if the associated homogeneous system has only the trivial solution. ♦

The corollary follows from the fact that the nonhomogeneous system has a unique solution exactly when $\mathbf{H}$ is the zero vector in Theorem 7.13.

Theorem 7.13 suggests a strategy for finding all solutions of $\mathbf{AX} = \mathbf{B}$, when the system is consistent.

Step 1. Find the general solution $\mathbf{H}$ of $\mathbf{AX}=\mathbf{O}$.
Step 2. Find any one solution $\mathbf{U}_p$ of $\mathbf{AX}=\mathbf{B}$.
Step 3. The general solution $\mathbf{AX}=\mathbf{B}$ is then $\mathbf{H}+\mathbf{U}_p$.

We know how to carry out step (1). We will outline a procedure for step (2).

To find a particular solution $\mathbf{U}_p$, proceed as follows.

Step 1. Define the $n \times m+1$ *augmented matrix* $[\mathbf{A}\vdots\mathbf{B}]$ by adjoining the column matrix $\mathbf{B}$ as an additional column to $\mathbf{A}$. The augmented matrix contains the coefficients of the unknowns of the system (in the first m columns), as well as the numbers on the right side of the equations (elements of $\mathbf{B}$).

Step 2. Reduce $[\mathbf{A}\vdots\mathbf{B}]$. Since we reduce a matrix to obtain leading entries of 1 wherever possible from upper left toward the lower right, this results eventually in a reduced matrix

$$[\mathbf{A}\vdots\mathbf{B}]_R = [\mathbf{A}_R\vdots\mathbf{C}],$$

in which the first m columns are the reduced form of $\mathbf{A}$, and the last column is whatever results from $\mathbf{B}$ after the row operations used to reduce $\mathbf{A}$ have been applied to $[\mathbf{A}\vdots\mathbf{B}]$.

Solutions of the reduced system $\mathbf{A}_R\mathbf{X}=\mathbf{C}$ are the same as solutions of the original system $\mathbf{AX}=\mathbf{B}$ because the operations performed on the coefficients of the unknowns are also performed on the b_j 's.

Step 3. From $[\mathbf{A}_R\vdots\mathbf{C}]$, read a particular solution $\mathbf{U}_p$. When added to the general solution $\mathbf{H}$ of the associated homogeneous system, we have the general solution of $\mathbf{AX}=\mathbf{B}$.

We will look at some examples. Example 7.24 suggests how this augmented matrix procedure tells us when the system has no solution.

EXAMPLE 7.23

We will solve the system

$$\begin{pmatrix} -3 & 2 & 2 \\ 1 & 4 & -6 \\ 0 & -2 & 2 \end{pmatrix} \mathbf{X} = \begin{pmatrix} 8 \\ 1 \\ -2 \end{pmatrix}.$$

The first step is to reduce the augmented matrix

$$[\mathbf{A}\vdots\mathbf{B}] = \begin{pmatrix} -3 & 2 & 2 & \vdots & 8 \\ 1 & 4 & -6 & \vdots & 1 \\ 0 & -2 & 2 & \vdots & -2 \end{pmatrix}.$$

Carrying out the reduction procedure on this 3×4 augmented matrix, we obtain

$$[\mathbf{A}\vdots\mathbf{B}]_R = \begin{pmatrix} 1 & 0 & 0 & \vdots & 0 \\ 0 & 1 & 0 & \vdots & 5/2 \\ 0 & 0 & 1 & \vdots & 3/2 \end{pmatrix} = [\mathbf{A}_R\vdots\mathbf{C}] = [\mathbf{I}_3\vdots\mathbf{C}].$$

$\mathbf{C}$ is whatever results in the fourth column when we reduce $\mathbf{A}$, the first three columns of $[\mathbf{A}\vdots\mathbf{B}]$).

This reduced augmented matrix $[\mathbf{A}_R \vdots \mathbf{C}]$ represents the reduced system $\mathbf{I}_3\mathbf{X} = \mathbf{C}$, which is the system

$$x_1 = 0$$
$$x_2 = 5/2$$
$$x_3 = 3/2.$$

From this, we directly read a solution of the reduced nonhomogeneous system. In this example, this solution is unique by Corollary 7.6 because the associated homogeneous system has only the trivial solution ($\mathbf{A}_R = \mathbf{I}_3$).

Consistent with treating the system as a matrix equation, we usually write the solution in terms of column matrices. In this example,

$$\mathbf{X} = \begin{pmatrix} 0 \\ 5/2 \\ 3/2 \end{pmatrix}. \ \blacklozenge$$

EXAMPLE 7.24

We have seen that the system

$$2x_1 - 3x_2 = 6$$
$$4x_1 - 6x_2 = 8$$

has no solution. We will see how this conclusion reveals itself when we work with the augmented matrix, which is

$$[\mathbf{A} \vdots \mathbf{B}] = \begin{pmatrix} 2 & 3 & \vdots & 6 \\ 4 & -6 & \vdots & 8 \end{pmatrix}.$$

Reduce this matrix to obtain

$$[\mathbf{A} \vdots \mathbf{B}]_R = [\mathbf{A}_R \vdots \mathbf{C}] = \begin{pmatrix} 1 & -3/2 & \vdots & 2 \\ 0 & 0 & \vdots & -4 \end{pmatrix}.$$

The second equation of the reduced system is

$$0x_1 + 0x_2 = -4$$

which can have no solution. ♦

In this example, the augmented matrix has rank 2, while the matrix of the homogeneous system has rank 1. In general, whenever the rank of $\mathbf{A}$ is less than the rank of $[\mathbf{A} \vdots \mathbf{B}]$, then $\mathbf{A}_R$ will have at least one row of zeros, while the corresponding row in the reduced augmented matrix $[\mathbf{A}_R \vdots \mathbf{C}]$ has a nonzero element in this row in the $\mathbf{C}$ column. This corresponds to an equation of the form

$$0x_1 + 0x_2 + \cdots + 0x_m = c_j \neq 0$$

and this has no solution for the x_i 's. In this case the system is inconsistent.

We will record this important observation.

THEOREM 7.14

The nonhomogeneous system $\mathbf{AX} = \mathbf{B}$ is consistent if and only if $\mathbf{A}$ and $[\mathbf{A}\vdots\mathbf{B}]$ have the same rank. ♦

EXAMPLE 7.25

We will solve the system

$$\begin{aligned} x_1 - x_2 + 2x_3 &= 3 \\ -4x_1 + x_2 + 7x_3 &= -5 \\ -2x_1 - x_2 + 11x_3 &= 14. \end{aligned}$$

The augmented matrix is

$$[\mathbf{A}\vdots\mathbf{B}] = \begin{pmatrix} 1 & -1 & 2 & \vdots & 3 \\ -4 & 1 & 7 & \vdots & -5 \\ -2 & -1 & 11 & \vdots & 14 \end{pmatrix}.$$

Reduce this augmented matrix to obtain

$$[\mathbf{A}\vdots\mathbf{B}]_R = [\mathbf{A}_R\vdots\mathbf{C}] = \begin{pmatrix} 1 & 0 & -3 & \vdots & 0 \\ 0 & 1 & -5 & \vdots & 0 \\ 0 & 0 & 0 & \vdots & 1 \end{pmatrix}.$$

$\mathbf{A}$ has rank 2, because its reduced matrix has two nonzero rows. But $[\mathbf{A}\vdots\mathbf{B}]$ has rank 3 because its reduced form has three nonzero rows. Therefore, this system is inconsistent. We can also observe from the reduced system that the last equation is

$$0x_1 + 0x_2 + 0x_3 = 1$$

with no solution. ♦

EXAMPLE 7.26

Solve the system

$$\begin{aligned} x_1 - x_2 + 2x_4 + x_5 + 6x_6 &= -3 \\ x_2 + x_3 + 3x_4 + 2x_5 + 4x_6 &= 1 \\ x_1 - 4x_2 + 3x_3 + x_4 + 2x_6 &= 0. \end{aligned}$$

The augmented matrix is

$$[\mathbf{A}\vdots\mathbf{B}] = \begin{pmatrix} 1 & 0 & -1 & 2 & 1 & 6 & \vdots & -3 \\ 0 & 1 & 1 & 3 & 2 & 4 & \vdots & 1 \\ 1 & -4 & 3 & 1 & 0 & 2 & \vdots & 0 \end{pmatrix}.$$

Reduce this to obtain

$$[\mathbf{A}\vdots\mathbf{B}]_R = [\mathbf{A}_R\vdots\mathbf{C}] \begin{pmatrix} 1 & 0 & 0 & 27/8 & 15/8 & 60/8 & \vdots & -17/8 \\ 0 & 1 & 0 & 13/8 & 9/8 & 20/8 & \vdots & 1/8 \\ 0 & 0 & 1 & 11/8 & 7/8 & 12/8 & \vdots & 7/8 \end{pmatrix}.$$

The first six columns are $\mathbf{A}_R$, and we read from $[\mathbf{A} \vdots \mathbf{B}]_R$ that both $\mathbf{A}$ and $[\mathbf{A} \vdots \mathbf{B}]$ have rank 3, so the system is consistent. From the reduced augmented matrix, we see immediately that

$$x_1 + \tfrac{27}{8}x_4 + \tfrac{15}{8}x_5 + \tfrac{60}{8}x_6 = -\tfrac{17}{8}$$
$$x_2 + \tfrac{13}{8}x_4 + \tfrac{9}{8}x_5 + \tfrac{20}{8}x_6 = \tfrac{1}{8}$$
$$x_3 + \tfrac{11}{8}x_4 + \tfrac{7}{8}x_5 + \tfrac{12}{8}x_6 = \tfrac{7}{8}.$$

From these we have

$$x_1 = -\tfrac{27}{8}x_4 - \tfrac{15}{8}x_5 - \tfrac{60}{8}x_6 - \tfrac{17}{8}$$
$$x_2 = -\tfrac{13}{8}x_4 - \tfrac{9}{8}x_5 - \tfrac{20}{8}x_6 + \tfrac{1}{8}$$
$$x_3 = -\tfrac{11}{8}x_4 - \tfrac{7}{8}x_5 - \tfrac{12}{8}x_6 + \tfrac{7}{8}.$$

We could have gone directly to these equations without the intermediate step. These equations actually give the general solution, with x_1, x_2, and x_3 in terms of the arbitrary constants x_4, x_5, and x_6. The solution is

$$\mathbf{X} = \begin{pmatrix} -\tfrac{27}{8}x_4 - \tfrac{15}{8}x_5 - \tfrac{60}{8}x_6 - \tfrac{17}{8} \\ -\tfrac{13}{8}x_4 - \tfrac{9}{8}x_5 - \tfrac{20}{8}x_6 + \tfrac{1}{8} \\ -\tfrac{11}{8}x_4 - \tfrac{7}{8}x_5 - \tfrac{12}{8}x_6 + \tfrac{7}{8} \\ x_4 \\ x_5 \\ x_6 \end{pmatrix}.$$

To write this in a more revealing way, let $x_4 = 8\alpha$, $x_5 = 8\beta$, and $x_6 = 8\gamma$ to write

$$\mathbf{X} = \alpha \begin{pmatrix} -27 \\ -13 \\ -11 \\ 8 \\ 0 \\ 0 \end{pmatrix} + \beta \begin{pmatrix} -15 \\ -9 \\ -7 \\ 0 \\ 8 \\ 0 \end{pmatrix} + \gamma \begin{pmatrix} -60 \\ -20 \\ -12 \\ 0 \\ 0 \\ 8 \end{pmatrix} + \begin{pmatrix} -17/8 \\ 1/8 \\ 7/8 \\ 0 \\ 0 \\ 0 \end{pmatrix} = \mathbf{H} + \mathbf{U}_p$$

with $\mathbf{H}$ as the general solution of $\mathbf{AX} = \mathbf{O}$ and $\mathbf{U}_p$ as a particular solution of $\mathbf{AX} = \mathbf{C}$. ♦

EXAMPLE 7.27

The system

$$\begin{pmatrix} 2 & 1 & -11 \\ -5 & 1 & 9 \\ 1 & 1 & 14 \end{pmatrix} \mathbf{X} = \begin{pmatrix} -6 \\ 12 \\ -5 \end{pmatrix}.$$

has the augmented matrix

$$[\mathbf{A} \vdots \mathbf{B}] = \begin{pmatrix} 2 & 1 & -11 & \vdots & -6 \\ -5 & 1 & 9 & \vdots & 12 \\ 1 & 1 & 14 & \vdots & -5 \end{pmatrix},$$

and we reduce this to

$$[\mathbf{A} \vdots \mathbf{B}]_R = \begin{pmatrix} 1 & 0 & 0 & \vdots & -86/31 \\ 0 & 1 & 0 & \vdots & -191/155 \\ 0 & 0 & 1 & \vdots & -11/155 \end{pmatrix}.$$

The first three columns tell us that $\mathbf{A}$ has a rank of 3, so the associated homogeneous system has only the trivial solution. Since the rank of $[\mathbf{A}_R \vdots \mathbf{C}]$ is also 3, the system has a solution. This solution is unique because $\mathbf{A}_R = \mathbf{I}_3$.

From the fourth column of $[\mathbf{A} \vdots \mathbf{B}]_R$, we read the unique solution

$$\mathbf{X} = \begin{pmatrix} -86/31 \\ -191/155 \\ -11/155 \end{pmatrix}. \ \blacklozenge$$

SECTION 7.6 PROBLEMS

In each of Problems 1 through 14, find the general solution of the system or show that the system is inconsistent. Write the solution in matrix form.

1. $$\begin{aligned} 2x_1 - 3x_2 + x_4 - x_6 &= 0 \\ 3x_1 - 2x_3 + x_5 &= 1 \\ x_2 - x_4 + 6x_6 &= 3 \end{aligned}$$

2. $$\begin{aligned} -4x_1 + 5x_2 - 6x_3 &= 2 \\ 2x_1 - 6x_2 + x_3 &= -5 \\ -6x_1 + 16x_2 - 11x_3 &= 1 \end{aligned}$$

3. $$\begin{aligned} 8x_2 - 4x_3 + 10x_6 &= 1 \\ x_3 + x_5 - x_6 &= 2 \\ x_4 - 3x_5 + 2x_6 &= 0 \end{aligned}$$

4. $$\begin{aligned} -6x_1 + 2x_2 - x_3 + x_4 &= 0 \\ x_1 + 4x_2 - x_4 &= -5 \\ x_1 + x_2 + x_3 - 7x_4 &= 0 \end{aligned}$$

5. $$\begin{aligned} 4x_1 - x_2 + 4x_3 &= 1 \\ x_1 + x_2 - 5x_3 &= 0 \\ -2x_1 + x_2 + 7x_3 &= 4 \end{aligned}$$

6. $$\begin{aligned} 2x_1 - 3x_3 &= 1 \\ x_1 - x_2 + x_3 &= 1 \\ 2x_1 - 4x_2 + x_3 &= 2 \end{aligned}$$

7. $$\begin{aligned} 14x_3 - 3x_5 + x_7 &= 2 \\ x_1 + x_2 + x_3 - x_4 + x_6 &= -4 \end{aligned}$$

8. $$\begin{aligned} 3x_1 - 2x_2 &= -1 \\ 4x_1 + 3x_2 &= 4 \end{aligned}$$

9. $$\begin{aligned} 7x_1 - 3x_2 + 4x_3 &== -7 \\ 2x_1 + x_2 - x_3 + 4x_4 &= 6 \\ x_2 - 3x_4 &= -5 \end{aligned}$$

10. $$\begin{aligned} 2x_1 - 3x_2 + x_4 &= 1 \\ 3x_2 + x_3 - x_4 &= 0 \\ 2x_1 - 3x_2 + 10x_3 &= 0 \end{aligned}$$

11. $$\begin{aligned} 3x_2 - 4x_4 &= 10 \\ x_1 - 3x_2 + 4x_3 - x_6 &= 8 \\ x_2 + x_3 - 6x_4 + x_6 &= -9 \\ x_1 - x_2 + x_6 &= 0 \end{aligned}$$

12. $$\begin{aligned} 2x_1 - 3x_2 &= 1 \\ -x_1 + 3x_2 &= 0 \\ x_1 - 4x_2 &= 3 \end{aligned}$$

13. $$\begin{aligned} 3x_1 - 2x_2 + x_3 &= 6 \\ x_1 + 10x_2 - x_3 &= 2 \\ -3x_1 - 2x_2 + x_3 &= 0 \end{aligned}$$

14. $$\begin{aligned} 4x_1 - 2x_2 + 3x_3 + 10x_4 &= 1 \\ x_1 - 3x_4 &= 8 \\ 2x_1 - 3x_2 + x_4 &= 16 \end{aligned}$$

15. Show that the system $\mathbf{AX} = \mathbf{B}$ is consistent if and only if $\mathbf{B}$ is in the column space of $\mathbf{A}$.

7.7 Matrix Inverses

Let $\mathbf{A}$ be an $n \times n$ matrix. An $n \times n$ matrix $\mathbf{B}$ is an inverse of $\mathbf{A}$ if

$$\mathbf{AB} = \mathbf{BA} = \mathbf{I}_n.$$

It is easy to find matrices that have no inverse. For example, let

$$\mathbf{A} = \begin{pmatrix} 1 & 0 \\ 2 & 0 \end{pmatrix}.$$

Suppose

$$\mathbf{B} = \begin{pmatrix} a & b \\ c & d \end{pmatrix}.$$

is an inverse of **A**. Then

$$\mathbf{AB} = \begin{pmatrix} 1 & 0 \\ 2 & 0 \end{pmatrix}\begin{pmatrix} a & b \\ c & d \end{pmatrix} = \begin{pmatrix} a & b \\ 2a & 2b \end{pmatrix} = \begin{pmatrix} 1 & 0 \\ 0 & 1 \end{pmatrix},$$

implying that

$$a = 1, b = 0, 2a = 0 \text{ and } b = 1$$

and this is impossible. On the other hand, some matrices do have inverses. For example,

$$\begin{pmatrix} 2 & 1 \\ 1 & 4 \end{pmatrix}\begin{pmatrix} 4/7 & -1/7 \\ -1/7 & 2/7 \end{pmatrix} = \begin{pmatrix} 4/7 & -1/7 \\ -1/7 & 2/7 \end{pmatrix}\begin{pmatrix} 2 & 1 \\ 1 & 4 \end{pmatrix} = \begin{pmatrix} 1 & 0 \\ 0 & 1 \end{pmatrix}.$$

A matrix that has an inverse is called *nonsingular*. A matrix with no inverse is *singular*.

A matrix can have only one inverse. For suppose that **B** and **C** are inverses of **A**. Then

$$\mathbf{B} = \mathbf{BI}_n = \mathbf{B}(\mathbf{AC}) = (\mathbf{BA})\mathbf{C} = \mathbf{I}_n\mathbf{C} = \mathbf{C}.$$

In view of this, we will denote the inverse of **A** as $\mathbf{A}^{-1}$. Here are additional facts about nonsingular matrices and matrix inverses.

THEOREM 7.15

Let **A** be an $n \times n$ matrix. Then,

1. $\mathbf{I}_n$ is nonsingular and is its own inverse.
2. If **A** and **B** are nonsingular $n \times n$ matrices, then so is **AB**. Further,

$$(\mathbf{AB})^{-1} = \mathbf{B}^{-1}\mathbf{A}^{-1}.$$

The inverse of a product is the product of the inverses in the reverse order. This extends to a product of any finite number of matrices.

3. If **A** is nonsingular, so is $\mathbf{A}^{-1}$, and

$$(\mathbf{A}^{-1})^{-1} = \mathbf{A}.$$

The inverse of the inverse is the matrix itself.

4. If **A** is nonsingular, so is its transpose $\mathbf{A}^t$, and

$$(\mathbf{A}^t)^{-1} = (\mathbf{A}^{-1})^t.$$

The inverse of a transpose is the transpose of the inverse.

5. **A** is nonsingular if and only if $\mathbf{A}_R = \mathbf{I}_n$.
6. **A** is nonsingular if and only if $\text{rank}(\mathbf{A}) = n$.

7. If $\mathbf{AB}$ is nonsingular, so are $\mathbf{A}$ and $\mathbf{B}$.
8. If $\mathbf{A}$ and $\mathbf{B}$ are $n \times n$ matrices, and either one is singular, then their products $\mathbf{AB}$ and $\mathbf{BA}$ are singular.
9. Every elementary matrix is nonsingular, and its inverse is an elementary matrix of the same type.
10. An $n \times n$ matrix $\mathbf{A}$ is nonsingular if and only if $\mathbf{AX} = \mathbf{B}$ has a solution for every $n \times 1$ $\mathbf{B}$. ♦

Proof These statements use the uniqueness of the inverse of a matrix. This allows us to show that a matrix is the inverse of another matrix by showing that it behaves like the inverse (the product of the two matrices is the identity matrix).

Conclusion (2) of the theorem is true because

$$(\mathbf{B}^{-1}\mathbf{A}^{-1})(\mathbf{AB}) = \mathbf{B}^{-1}(\mathbf{A}^{-1}\mathbf{A})\mathbf{B} = \mathbf{B}^{-1}\mathbf{B} = \mathbf{I}_n.$$

Similarly

$$(\mathbf{AB})(\mathbf{B}^{-1}\mathbf{A}^{-1}) = \mathbf{I}_n.$$

This proves that $\mathbf{B}^{-1}\mathbf{A}^{-1}$ behaves like the inverse of $\mathbf{AB}$, hence this must be inverse.

For conclusion (3) observe that the equation

$$\mathbf{AA}^{-1} = \mathbf{A}^{-1}\mathbf{A} = \mathbf{I}_n$$

is symmetric in the sense that $\mathbf{A}^{-1}$ is the inverse of $\mathbf{A}$, but also $\mathbf{A}$ is the inverse of $\mathbf{A}^{-1}$. The latter phrasing means that

$$\mathbf{A} = (\mathbf{A}^{-1})^{-1}.$$

For conclusion (4), first write

$$\mathbf{I}_n = (\mathbf{I}_n)^t = (\mathbf{AA}^{-1})^t = (\mathbf{A}^{-1})^t\mathbf{A}^t.$$

Similarly,

$$\mathbf{A}^t(\mathbf{A}^{-1})^t = \mathbf{I}_n.$$

These two equations show that $(\mathbf{A}^t)^{-1} = (\mathbf{A}^{-1})^t$.

The key to conclusion (5) lies in recalling (Section 7.1.1) that the columns of $\mathbf{AB}$ are $\mathbf{A}$ times the columns of $\mathbf{B}$. Using this, we can attempt to build an inverse for $\mathbf{A}$ a column at a time. To find $\mathbf{B}$ so that $\mathbf{AB} = \mathbf{I}_n$, we must be able to choose the columns of $\mathbf{B}$ so that

$$\text{column } j \text{ of } \mathbf{AB} = \mathbf{A}\begin{pmatrix} b_{1j} \\ b_{2j} \\ \vdots \\ b_{nj} \end{pmatrix} = \text{column } j \text{ of } \mathbf{I}_n = \begin{pmatrix} 0 \\ 0 \\ \vdots \\ 1 \\ 0 \\ \vdots \\ 0 \end{pmatrix},$$

having 1 in the jth place and zeros elsewhere.

If now $\mathbf{A}_R = \mathbf{I}_n$, then the system just written for column j of $\mathbf{B}$ has a unique solution for $j = 1, \cdots, n$. These solutions form the columns of $\mathbf{B}$ such that $\mathbf{AB} = \mathbf{I}_n$, yielding $\mathbf{A}^{-1}$.

(Actually we must show that $\mathbf{BA} = \mathbf{I}_n$ also, but we will not go through these details.)

Conversely, if $\mathbf{A}$ is nonsingular, then this system has a unique solution for $j = 1, \cdots, n$ because these solutions are the columns of $\mathbf{A}^{-1}$. Then $\mathbf{A}_R = \mathbf{I}_n$. This proves conclusion (5).

Conclusion (6) follows directly from (5).

For conclusion (7), suppose $\mathbf{AB}$ is nonsingular. Then for some matrix $\mathbf{K}$, $(\mathbf{AB})\mathbf{K} = \mathbf{I}_n$. Then $\mathbf{A}(\mathbf{BK}) = \mathbf{I}_n$, so $\mathbf{A}$ is nonsingular. Similarly, $\mathbf{B}$ is nonsingular.

Conclusion (8) follows from (7).

Conclusion (9) follows immediately from the discussion preceding Theorem 7.6.

Finally, for conclusion (10), first suppose $\mathbf{AX} = \mathbf{B}$ has a solution for every $n \times 1$ matrix $\mathbf{B}$. Let $\mathbf{X}_j$ be the solution of

$$\mathbf{AX} = \begin{pmatrix} 0 \\ 0 \\ \vdots \\ 1 \\ 0 \\ \vdots \\ 0 \end{pmatrix}$$

with 1 in row j and all other elements zero. Then $\mathbf{X}_1, \cdots, \mathbf{X}_n$ form the columns of an $n \times n$ matrix $\mathbf{K}$ and it is routine to check that $\mathbf{AK} = \mathbf{I}_n$, hence $\mathbf{K} = \mathbf{A}^{-1}$ and $\mathbf{A}$ is nonsingular.

Conversely, if $\mathbf{A}$ is nonsingular, then $\mathbf{X} = \mathbf{A}^{-1}\mathbf{B}$ is the solution of $\mathbf{AX} = \mathbf{B}$ for any $n \times 1$ matrix $\mathbf{B}$. ♦

Matrix inverses relate to systems of linear equations in the following way.

THEOREM 7.16

Let $\mathbf{A}$ be $n \times n$.

1. A homogeneous system $\mathbf{AX} = \mathbf{O}$ has a nontrivial solution if and only if $\mathbf{A}$ is singular.
2. A consistent nonhomogeneous system $\mathbf{AX} = \mathbf{B}$ has a unique solution if and only if $\mathbf{A}$ is nonsingular. In this case the solution is

$$\mathbf{X} = \mathbf{A}^{-1}\mathbf{B}. \ ♦$$

Proof If $\mathbf{A}$ is singular, then $\mathbf{A}_R \neq \mathbf{I}_n$ by Theorem 7.15, conclusion (5), so the system $\mathbf{AX} = \mathbf{O}$ has a nontrivial solution by Corollary 7.3.

Conversely, suppose the system $\mathbf{AX} = \mathbf{O}$ has a nontrivial solution. Then $\text{rank}(\mathbf{A}) < n$ by Theorem 7.15, conclusion (6), so $\mathbf{A}$ is singular.

This proves conclusion (1). For conclusion (2), suppose the system is consistent. The general solution has the form $\mathbf{X} = \mathbf{H} + \mathbf{U}_p$, where $\mathbf{H}$ is the general solution of the associated homogeneous system. Therefore the given system has a unique solution exactly when the homogeneous system has only the trivial solution, which occurs if and only if $\mathbf{A}$ is nonsingular. ♦

Finding the inverse of a nonsingular matrix is most easily done using a software routine. In the `linalg` package of linear algebra routines of MAPLE, the inverse of a matrix $\mathbf{A}$ that has been entered can be found using

```
inverse(A);
```

If it happens that $\mathbf{A}$ is singular, the routine will return this conclusion.

Despite this, it is sometimes useful to understand a procedure for finding a matrix inverse. Let $\mathbf{A}$ be an $n \times n$ matrix. Form the $n \times 2n$ matrix $[\mathbf{I}_n \vdots \mathbf{A}]$ whose first n columns are $\mathbf{A}$ and whose second n columns are $\mathbf{I}_n$. For example, if

$$\mathbf{A} = \begin{pmatrix} 2 & 3 \\ -1 & 9 \end{pmatrix}$$

then

$$[\mathbf{A}\vdots\mathbf{I}_2] = \begin{pmatrix} 2 & 3 & \vdots & 1 & 0 \\ -1 & 9 & \vdots & 0 & 1 \end{pmatrix}.$$

Reduce $\mathbf{A}$, carrying out the row operations across the entire matrix $[\mathbf{A}\vdots\mathbf{I}_n]$. $\mathbf{A}$ is nonsingular exactly when $\mathbf{A}_R = \mathbf{I}_n$ turns up in the first n columns. In this event the second n columns form $\mathbf{A}^{-1}$.

EXAMPLE 7.28

Let

$$\mathbf{A} = \begin{pmatrix} 5 & -1 \\ 6 & 8 \end{pmatrix}.$$

Form

$$[\mathbf{A}\vdots\mathbf{I}_2] = \begin{pmatrix} 5 & -1 & \vdots & 1 & 0 \\ 6 & 8 & \vdots & 0 & 1 \end{pmatrix}.$$

Reduce $\mathbf{A}$, carrying out each row operation on the entire row of the augmented matrix. First multiply row one by $1/5$:

$$\begin{pmatrix} 1 & -1/5 & \vdots & 1/5 & 0 \\ 6 & 8 & \vdots & 0 & 1 \end{pmatrix}.$$

Add -6 times row one to row two:

$$\begin{pmatrix} 1 & -1/5 & \vdots & 1/5 & 0 \\ 0 & 46/5 & \vdots & -6/5 & 1 \end{pmatrix}.$$

Multiply row two by $5/46$:

$$\begin{pmatrix} 1 & -1/5 & \vdots & 1/5 & 0 \\ 6 & 1 & \vdots & -6/46 & 5/46 \end{pmatrix}.$$

Add $1/5$ times row two to row one:

$$\begin{pmatrix} 1 & 0 & \vdots & 8/46 & 1/46 \\ 0 & 1 & \vdots & -6/46 & 5/46 \end{pmatrix}.$$

This is in reduced form. The first two columns are $\mathbf{A}_R$. Since $\mathbf{A}_R = \mathbf{I}_2$, $\mathbf{A}$ is nonsingular. Further, we can read $\mathbf{A}^{-1}$ from the last two columns:

$$\mathbf{A}^{-1} = \begin{pmatrix} 8/46 & 1/46 \\ -6/46 & 5/46 \end{pmatrix}. \ \blacklozenge$$

EXAMPLE 7.29

Let

$$\mathbf{A} = \begin{pmatrix} -3 & 21 \\ 4 & -28 \end{pmatrix}.$$

Form

$$[\mathbf{A}\vdots\mathbf{I}_2] = \begin{pmatrix} -3 & 21 & \vdots & 1 & 0 \\ 4 & -28 & \vdots & 0 & 1 \end{pmatrix}.$$

Reduce this by multiplying row one by $-1/3$ and then adding -4 times row one to row two to get

$$\begin{pmatrix} 1 & -7 & \vdots & -1/3 & 0 \\ 0 & 0 & \vdots & 4/3 & 1 \end{pmatrix}.$$

The left two columns, which form $\mathbf{A}_R$, do not equal $\mathbf{I}_2$, so $\mathbf{A}$ is singular and has no inverse. ♦

We will illustrate the use of a matrix inverse to solve a nonhomogeneous system.

EXAMPLE 7.30

We will solve the system

$$2x_1 - x_2 + 3x_3 = 4$$
$$x_1 + 9x_2 - 2x_3 = -8$$
$$4x_1 - 8x_2 + 11x_3 = 15.$$

The matrix of coefficients is

$$\mathbf{A} = \begin{pmatrix} 2 & -1 & 3 \\ 1 & 9 & -2 \\ 4 & -8 & 11 \end{pmatrix}.$$

A routine reduction yields

$$[\mathbf{A}\vdots\mathbf{I}_3]_R = \begin{pmatrix} 1 & 0 & 0 & \vdots & 83/53 & -13/53 & -25/53 \\ 0 & 1 & 0 & \vdots & -19/53 & 10/53 & 7/53 \\ 0 & 0 & 1 & \vdots & -44/53 & 12/53 & 19/53 \end{pmatrix}.$$

The first three columns are $\mathbf{I}_3$, hence $\mathbf{A}$ is nonsingular and the system has a unique solution. The last three columns of the reduced augmented matrix give us

$$\mathbf{A}^{-1} = \frac{1}{53}\begin{pmatrix} 83 & -13 & -25 \\ -19 & 10 & 7 \\ -44 & 12 & 19 \end{pmatrix}.$$

The unique solution of the system is $\mathbf{A}^{-1}\mathbf{B}$:

$$\mathbf{X} = \mathbf{A}^{-1}\mathbf{B} = \frac{1}{53}\begin{pmatrix} 83 & -13 & -25 \\ -19 & 10 & 7 \\ -44 & 12 & 19 \end{pmatrix}\begin{pmatrix} 4 \\ -8 \\ 15 \end{pmatrix} = \begin{pmatrix} 61/53 \\ -51/53 \\ 13/53 \end{pmatrix}. \quad ♦$$

SECTION 7.7 PROBLEMS

In each of Problems 1 through 10, find the inverse of the matrix or show that the matrix is singular.

1. $\begin{pmatrix} -5 & 2 \\ 1 & 2 \end{pmatrix}$

2. $\begin{pmatrix} -2 & 1 & -5 \\ 1 & 1 & 4 \\ 0 & 3 & 3 \end{pmatrix}$

3. $\begin{pmatrix} -3 & 4 & 1 \\ 1 & 2 & 0 \\ 1 & 1 & 3 \end{pmatrix}$

4. $\begin{pmatrix} 12 & 1 & 14 \\ -3 & 2 & 0 \\ 0 & 9 & 14 \end{pmatrix}$

5. $\begin{pmatrix} -2 & 1 & 1 \\ 0 & 1 & 1 \\ -3 & 0 & 6 \end{pmatrix}$

6. $\begin{pmatrix} -1 & 0 \\ 4 & 4 \end{pmatrix}$

7. $\begin{pmatrix} 6 & 2 \\ 3 & 3 \end{pmatrix}$

8. $\begin{pmatrix} 1 & 1 & -3 \\ 2 & 16 & 1 \\ 0 & 0 & 4 \end{pmatrix}$

9. $\begin{pmatrix} -1 & 2 \\ 2 & 1 \end{pmatrix}$

10. $\begin{pmatrix} 12 & 3 \\ 4 & 1 \end{pmatrix}$

In each of Problems 11 through 15, use a matrix inverse to find the unique solution of the system.

11. $4x_1 + 6x_2 - 3x_3 = 0$
$2x_1 + 3x_2 - 4x_3 = 0$
$x_1 - x_2 + 3x_3 = -7$

12. $12x_1 + x_2 - 3x_3 = 4$
$x_1 - x_2 + 3x_3 = -5$
$-2x_1 + x_2 + x_3 = 0$

13. $x_1 - x_2 + 3x_3 - x_4 = 1$
$x_2 - 3x_3 + 5x_4 = 2$
$x_1 - x_3 + x_4 = 0$
$x_1 + 2x_3 - x_4 = -5$

14. $8x_1 - x_2 - x_3 = 4$
$x_1 + 2x_2 - 3x_3 = 0$
$2x_1 - x_2 + 4x_3 = 5$

15. $2x_1 - 6x_2 + 3x_3 = -4$
$-x_1 + x_2 + x_3 = 5$
$2x_1 + 6x_2 - 5x_3 = 8$

7.8 Least Squares Vectors and Data Fitting

In this section, we will develop an approach to the method of least squares as it applies to a data fitting problem.

Let **A** be an $n \times m$ matrix of numbers and **B** a vector in R^n. The system $\mathbf{AX} = \mathbf{B}$ may or may not have a solution. Define an m-vector $\mathbf{X}^*$ to be a *least squares vector* for the system $\mathbf{AX} = \mathbf{B}$ if

$$\| \mathbf{AX}^* - \mathbf{B} \| \leq \| \mathbf{AX} - \mathbf{B} \| \tag{7.2}$$

for every **X** in R^m.

Thus $\mathbf{X}^*$ is a least squares vector for $\mathbf{AX} = \mathbf{B}$ if $\mathbf{AX}^*$ is at least as close to **B** as **AX** is to **B**, for every m-vector **X**. This means that, for every **X**,

$$\| \mathbf{AX}^* - \mathbf{B} \| \leq \| \mathbf{AX} - \mathbf{B} \| .$$

We will develop a method for finding all least squares vectors for a given system $\mathbf{AX} = \mathbf{B}$. The key lies in the column space S of **A**. S is a subspace of R^n, spanned by the columns $\mathbf{C}_1, \cdots, \mathbf{C}_m$ of **A**. S consists of exactly those vectors **B** in R^n for which the system $\mathbf{AX} = \mathbf{B}$ has a solution. This is because, if

$$\mathbf{X} = \begin{pmatrix} x_1 \\ x_2 \\ \vdots \\ x_m \end{pmatrix}$$

is a matrix of numbers, then

$$\mathbf{AX} = x_1\mathbf{C}_1 + x_2\mathbf{C}_2 + \cdots + x_m\mathbf{C}_m = \mathbf{B}$$

exactly when $\mathbf{B}$ is a linear combination of the columns of $\mathbf{A}$, hence is in S.

The following lemma reveals a connection between the least squares vectors for $\mathbf{AX} = \mathbf{B}$ and orthogonal projections, as suggested by the inequality (7.2).

LEMMA 7.1

Let $\mathbf{B}$ be an n-vector. Then an m-vector $\mathbf{X}^*$ is a least squares vector for $\mathbf{AX} = \mathbf{B}$ if and only if

$$\mathbf{AX}^* = \mathbf{B}_S,$$

where $\mathbf{B}_S$ is the orthogonal projection of $\mathbf{B}$ onto S. ♦

Proof Suppose first that $\mathbf{AX}^* = \mathbf{B}_S$. Then

$$\| \mathbf{B} - \mathbf{B}_S \| = \| \mathbf{B} - \mathbf{AX}^* \|$$
$$\leq \| \mathbf{B} - \mathbf{C} \|$$

for all vectors $\mathbf{C}$ in S, because $\mathbf{B}_S$ is the vector in S closest to $\mathbf{B}$. But the vectors $\mathbf{C}$ in S are exactly the vectors $\mathbf{AX}$ for $\mathbf{X}$ in R^m, so

$$\| \mathbf{B} - \mathbf{AX}^* \| \leq \| \mathbf{B} - \mathbf{AX} \|$$

for every m-vector $\mathbf{X}$, and this proves that $\mathbf{X}^*$ is a least squares vector for $\mathbf{AX} = \mathbf{B}$.

Conversely, suppose $\mathbf{X}^*$ is a lease squares vector for $\mathbf{AX} = \mathbf{B}$. Then

$$\| \mathbf{AX}^* - \mathbf{B} \| \leq \| \mathbf{AX} - \mathbf{B} \|$$

for all $\mathbf{X}$ in S. But then $\mathbf{AX}^*$ is the vector in S closest to $\mathbf{B}$. Because $\mathbf{B}_S$ is the unique vector with this property, then $\mathbf{AX}^* = \mathbf{B}_S$. This completes the proof. ♦

We are now able to completely characterize the least squares vectors of $\mathbf{AX} = \mathbf{B}$ as the solutions of a system of linear equations obtained using $\mathbf{A}$.

THEOREM 7.17 ***Least Squares Vectors for*** $\mathbf{AX} = \mathbf{B}$

An m-vector $\mathbf{X}$ is a least squares vector of $\mathbf{AX} = \mathbf{B}$ if and only if $\mathbf{X}$ is a solution of the system

$$\mathbf{A}^t\mathbf{AX} = \mathbf{A}^t\mathbf{B}. \quad ♦$$

Proof Suppose first that $\mathbf{X}^*$ is a least squares vector of $\mathbf{AX} = \mathbf{B}$. By the lemma,

$$\mathbf{AX}^* = \mathbf{B}_S.$$

We know that $\mathbf{B} - \mathbf{B}_S$ is in $S^\perp$, so $\mathbf{B} - \mathbf{AX}^*$ is in $S^\perp$. This means that the columns of $\mathbf{A}$ are orthogonal to $\mathbf{B} - \mathbf{AX}^*$. Writing the dot product of column j of $\mathbf{A}$ with $\mathbf{B} - \mathbf{AX}^*$ as a matrix product, this orthogonality means that

$$(\mathbf{C}_j)^t(\mathbf{B} - \mathbf{AX}^*) = 0.$$

Now $(\mathbf{C}_j)^t$ is row j of $\mathbf{A}^t$, so

$$\mathbf{A}^t(\mathbf{B} - \mathbf{AX}^*) = \mathbf{O}$$

in which $\mathbf{O}$ is the $m \times 1$ zero matrix. But this equation can be written

$$\mathbf{A}^t\mathbf{AX}^* = \mathbf{A}^t(\mathbf{B})$$

and this means that $\mathbf{X}^*$ is a solution of the system $\mathbf{A}^t\mathbf{AX} = \mathbf{A}^t\mathbf{B}$.

To prove the converse, suppose $\mathbf{X}^*$ is a solution of this system. Reversing part of the argument just given shows that $\mathbf{B} - \mathbf{AX}^*$ is in $S^\perp$. But then

$$\mathbf{B} = \mathbf{AX}^* + (\mathbf{B} - \mathbf{AX}^*)$$

is a decomposition of $\mathbf{B}$ into a sum of a vector in S and a vector in $S^\perp$. Since this decomposition is unique, then $\mathbf{AX}^*$ must be the orthogonal projection of $\mathbf{B}$ onto S:

$$\mathbf{AX}^* = \mathbf{B}_S.$$

By the lemma, $\mathbf{X}^*$ is a least squares vector for $\mathbf{AX} = \mathbf{B}$. ♦

Theorem 7.17 provides a way of obtaining all least squares vectors for $\mathbf{AX} = \mathbf{B}$. These are the solutions of the linear system $\mathbf{A}^t\mathbf{AX} = \mathbf{A}^t\mathbf{B}$. Since we know how to solve linear systems, this provides a computable method for finding least squares vectors. For this reason, we will call the system

$$\mathbf{A}^t\mathbf{AX} = \mathbf{A}^t\mathbf{B}$$

the *auxiliary lsv system* of $\mathbf{AX} = \mathbf{B}$.

In addition to providing a method for finding all least squares vectors for a system, the auxiliary lsv system tells us when a system has only one least squares vector. This occurs exactly when the auxiliary system has a unique solution, which in turn occurs when $\mathbf{A}^T\mathbf{A}$ is nonsingular. In this event, the least squares vector for $\mathbf{AX} = \mathbf{B}$ is

$$\mathbf{X}^* = (\mathbf{A}^t\mathbf{A})^{-1}\mathbf{A}^t\mathbf{B}.$$

This proves the following.

COROLLARY 7.7

$\mathbf{AX} = \mathbf{B}$ has a unique least squares vector if $\mathbf{A}^t\mathbf{A}$ is nonsingular. ♦

EXAMPLE 7.31

Let

$$\mathbf{A} = \begin{pmatrix} -1 & -2 \\ 1 & 4 \\ 2 & 2 \end{pmatrix}$$

and

$$\mathbf{B} = \begin{pmatrix} 3 \\ -2 \\ 7 \end{pmatrix}.$$

We will find all of the least squares vectors for $\mathbf{AX} = \mathbf{B}$. Compute

$$\mathbf{A}^t\mathbf{A} = \begin{pmatrix} 6 & 10 \\ 10 & 24 \end{pmatrix}.$$

This is nonsingular, and we find that

$$(\mathbf{A}^t\mathbf{A})^{-1} = \begin{pmatrix} 12/22 & -5/22 \\ -5/22 & 3/22 \end{pmatrix}.$$

Finally,

$$\mathbf{A}^t\mathbf{B} = \begin{pmatrix} -1 & 1 & 2 \\ -2 & 4 & 2 \end{pmatrix} \begin{pmatrix} 3 \\ -2 \\ 7 \end{pmatrix} = \begin{pmatrix} 9 \\ 0 \end{pmatrix}.$$

The auxiliary lsv system is

$$\begin{pmatrix} 6 & 10 \\ 10 & 24 \end{pmatrix} \mathbf{X} = \begin{pmatrix} 9 \\ 0 \end{pmatrix}.$$

This has a unique solution, which is the unique least squares vector for the system:

$$\mathbf{X}^* = \begin{pmatrix} 6 & 10 \\ 10 & 24 \end{pmatrix}^{-1} \begin{pmatrix} 9 \\ 0 \end{pmatrix} = \begin{pmatrix} 12/22 & -5/22 \\ -5/22 & 3/22 \end{pmatrix} \begin{pmatrix} 9 \\ 0 \end{pmatrix} = \begin{pmatrix} 108/22 \\ -45/22 \end{pmatrix}. \quad \blacklozenge$$

We will apply least squares vectors to the problem of drawing a straight line that is, in some sense, a best fit to a set of given data points in the plane. We can see the idea by looking at an example. Suppose (perhaps by experiment or observation) we have data points

$$(0, -5.5), (1, -2.7), (2, -0.8), (3, 1.2), (5, 4.7),$$

which we will label (x_j, y_j) (from left to right) for $j = 1, 2, 3, 4, 5$. We want to draw a straight line $y = ax + b$ that is a "best fit" to these points. For each of the observed points (x_j, y_j), think of $ax_j + b$ as an approximation to y_j, so

$$\begin{aligned} ax_1 + b &\approx y_1, \\ ax_2 + b &\approx y_2, \\ &\vdots \\ ax_5 + b &\approx y_5. \end{aligned}$$

Consider the system

$$\begin{pmatrix} 1 & 0 \\ 1 & 1 \\ 1 & 2 \\ 1 & 3 \\ 1 & 5 \end{pmatrix} \begin{pmatrix} b \\ a \end{pmatrix} = \begin{pmatrix} -5.5 \\ -2.7 \\ -0.8 \\ 1.2 \\ 4.7 \end{pmatrix}.$$

This has the form $\mathbf{AX} = \mathbf{B}$ with $\mathbf{A}$ defined so that row j of the matrix product $\mathbf{AX}$ is $ax_j + b$, and this is set equal to the column matrix $\mathbf{B}$ listing the given y_j's. Of course, $ax_j + b$ is only approximately equal to y_j. We want a line that "best approximates" these points, so we obtain a and b by solving for a least squares vector $\mathbf{X}^*$ for this system.

Once we decide on this approach, the rest is arithmetic. Compute

$$\mathbf{A}^t\mathbf{A} = \begin{pmatrix} 5 & 11 \\ 11 & 39 \end{pmatrix},$$

and

$$(\mathbf{A}^t\mathbf{A})^{-1} = \begin{pmatrix} 39/74 & -11/74 \\ -11/74 & 5/74 \end{pmatrix}.$$

The unique least squares vector is

$$\mathbf{X}^* = (\mathbf{A}^t\mathbf{A})^{-1}\mathbf{A}^t\mathbf{B} = \begin{pmatrix} -5.0229729\cdots \\ 2.001351351\cdots \end{pmatrix}.$$

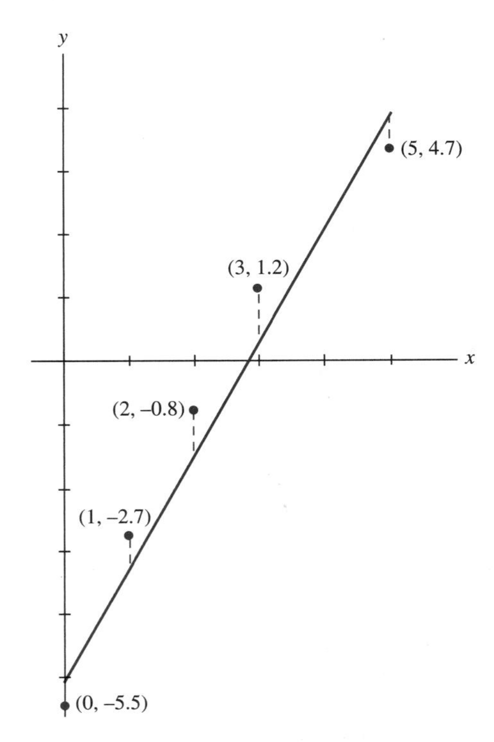

FIGURE 7.5 *Least squares fit to data.*

Choose $a = 2$ and $b = -5.02$ to obtain the line $y = 2x - 5.02$ as the line of best fit to the data. Among all lines we could draw, this minimizes the sum of the vertical distances from the line to the data points (Figure 7.5). ♦

It should not be surprising that this problem has a unique solution. The line we found is called a *least squares line* for the data. In statistics this is often referred to as the *regression line*.

SECTION 7.8 PROBLEMS

In each of Problems 1 through 6, find all least squares vectors for the given system.

1. $\begin{pmatrix} 1 & 0 \\ 6 & 2 \end{pmatrix} \mathbf{X} = \begin{pmatrix} -2 \\ -4 \end{pmatrix}$

2. $\begin{pmatrix} 1 & 1 & -2 \\ -2 & 3 & 1 \end{pmatrix} \mathbf{X} = \begin{pmatrix} 0 \\ 3 \end{pmatrix}$

3. $\begin{pmatrix} 1 & 1 & -2 & 1 \\ -2 & 3 & 0 & -4 \\ 0 & -2 & 1 & 5 \end{pmatrix} \mathbf{X} = \begin{pmatrix} 4 \\ -1 \\ 6 \end{pmatrix}$

4. $\begin{pmatrix} 1 & 1 \\ -2 & 3 \\ 0 & -1 \\ 2 & 2 \\ -3 & 7 \end{pmatrix} \mathbf{X} = \begin{pmatrix} -5 \\ 1 \\ 3 \\ 2 \\ 1 \end{pmatrix}$

5. $\begin{pmatrix} 1 & 1 \\ -2 & 3 \end{pmatrix} \mathbf{X} = \begin{pmatrix} 4 \\ -1 \end{pmatrix}$

6. $\begin{pmatrix} -5 & 2 \\ 1 & 4 \end{pmatrix} \mathbf{X} = \begin{pmatrix} 1 \\ -1 \end{pmatrix}$

In each of Problems 7 through 10, find the least squares line for the data.

7. $(-3, -23), (0, -8.2), (1, -4.6), (2, -0.5), (4, 7.3), (7, 19.2)$

8. $(-3, -7.4), (-1, -4.2), (0, -3.7), (2, -1.9), (4, 0.3), (7, 2.8), (11, 7.2)$

9. $(1, 3.8), (3, 11.7), (5, 20.6), (7, 26.5), (9, 35.2)$

10. $(-5, 21.2), (-3, 13.6), (-2, 10.7), (0, 4.2), (1, 2.4), (3, -3.7), (6, -14.2)$

7.9 LU Factorization

Let $\mathbf{A}$ be an $n \times m$ matrix of numbers. We sometimes want to factor $\mathbf{A}$ into a product of an $n \times n$ lower triangular matrix $\mathbf{L}$ and an $n \times m$ upper triangular matrix $\mathbf{U}$. We will see why this is useful shortly. First we will develop a procedure for doing this.

A matrix is *upper triangular* if its only nonzero elements lie on or above the main diagonal. Equivalently, all elements below the main diagonal are zero. A matrix is *lower triangular* if its only nonzero elements are on or below the main diagonal. In the case that the matrix is not square, main diagonal elements are the $1,1,\ 2,2,\ \cdots,\ n,n$ elements. If $m > n$, there will be columns beyond the columns containing these diagonal elements.

To see how to construct $\mathbf{L}$ and $\mathbf{U}$, consider an example. Let

$$\mathbf{A} = \begin{pmatrix} 2 & 1 & 1 & -3 & 5 \\ 2 & 3 & 6 & 1 & 4 \\ 6 & 2 & 1 & -1 & -3 \end{pmatrix}.$$

We will construct $\mathbf{U}$ using the elementary row operation of adding a scalar multiple of one row to another. We will *not* interchange rows or multiply individual rows by scalars in forming $\mathbf{U}$.

Begin with the leading entry in $\mathbf{A}$. This is 2 in the $1,1$ position. For a reason that will become clear when we construct $\mathbf{L}$, highlight column one of $\mathbf{A}$ in some way, such as boldface (or if you are writing the matrix on a piece of paper, you might circle these elements):

$$\mathbf{A} = \begin{pmatrix} \mathbf{2} & 1 & 1 & -3 & 5 \\ \mathbf{2} & 3 & 6 & 1 & 4 \\ \mathbf{6} & 2 & 1 & -1 & -3 \end{pmatrix}.$$

Now add scalar multiples of row one to the other rows to obtain zeros below the leading entry of 2. In the matrix $\mathbf{B}$, highlight the elements in column two below column one.

$$\mathbf{A} \to \mathbf{B} = \begin{pmatrix} 2 & 1 & 1 & -3 & 5 \\ 0 & \mathbf{2} & 5 & 4 & -1 \\ 0 & \mathbf{-1} & -2 & 8 & -18 \end{pmatrix}.$$

Row two has leading element 2 also. Add a scalar multiple (in this example, multiply by $1/2$) of row two to row three to obtain a zero in the $3,2$ position. After doing this, highlight the element in the $3,3$ position.

$$\mathbf{B} \to \mathbf{C} = \begin{pmatrix} 2 & 1 & 1 & -3 & 5 \\ 0 & 2 & 5 & 4 & -1 \\ 0 & 0 & \mathbf{1/2} & 10 & -37/2 \end{pmatrix}.$$

In this example $n = 3$ and $m = 5$, so the diagonal elements are the $1,1$; $2,2$; and $3,3$ elements, and there are two columns to the right of the columns containing this main diagonal. Notice that $\mathbf{C}$ is upper triangular. This is $\mathbf{U}$:

$$\mathbf{U} = \begin{pmatrix} 2 & 1 & 1 & -3 & 5 \\ 0 & 2 & 5 & 4 & -1 \\ 0 & 0 & 1/2 & 10 & -37/2 \end{pmatrix}.$$

Notice that the highlighting has played no role in producing **U**. These highlighted parts of columns will be used to form **L**, which will be 3×3 lower triangular. The highlighted elements are all in the first three columns, and all fall on or below the main diagonal. Now form the 3×3 lower triangular matrix

$$\mathbf{D} = \begin{pmatrix} 2 & 0 & 0 \\ 2 & 2 & 0 \\ 6 & -1 & 1/2 \end{pmatrix}.$$

D includes the highlighted first column from **A**, the highlighted elements of the second column in **B**, and the highlighted element of the third column in **C**, with zeros filled in above the main diagonal. This is not yet **L**. For this, we want 1 along the main diagonal. Thus, multiply column one of **D** by 1/2, the second column by 1/2, and the third column by 2. This yields **L**:

$$\mathbf{L} = \begin{pmatrix} 1 & 0 & 0 \\ 1 & 1 & 0 \\ 3 & -1/2 & 1 \end{pmatrix}.$$

It is routine to check that $\mathbf{LU} = \mathbf{A}$.

This procedure can be carried out in general. First form **U**, exploiting leading elements of columns of **A** and an elementary row operation to obtain zeros below these elements, then retaining the elements of these columns on and above the main diagonal to form the elements of **U** above its main diagonal. Fill in the rest of **U**, below the main diagonal, with zeros.

The highlighting strategy is a way of recording the elements to be used in forming columns of **L** on and below its main diagonal. After placing these elements, and filling in zeros above the main diagonal, multiply each column by a scalar to obtain 1 's along the main diagonal. The resulting matrix is **L**.

The process of factoring **A** into a product of lower and upper triangular matrices is called *LU factorization.*

What is the point to LU factorization? In real-world applications, matrices may be extremely large and the numbers will not all be small integers. A great deal of arithmetic is involved in manipulating such matrices. Upper and lower triangular matrices involve less arithmetic (hence save computer time and money), and systems of equations having triangular coefficient matrices are easier to solve.

As a specific instance of a simplification with LU factorization, suppose we want to solve a system $\mathbf{AX} = \mathbf{B}$. If we write $\mathbf{A} = \mathbf{LU}$, then the system is

$$\mathbf{AX} = (\mathbf{LU})\mathbf{X} = \mathbf{L}(\mathbf{UX}) = \mathbf{B}.$$

Let $\mathbf{UX} = \mathbf{Y}$ and solve the system

$$\mathbf{LY} = \mathbf{B}$$

for **Y**. Once we know **Y**, then the solution of $\mathbf{AX} = \mathbf{B}$ is the solution of

$$\mathbf{UX} = \mathbf{Y}.$$

Both of these systems involve triangular coefficient matrices, hence may be easier to solve than the original system.

EXAMPLE 7.32

We will solve the system $\mathbf{AX} = \mathbf{B}$, where

$$\mathbf{A} = \begin{pmatrix} 4 & 3 & 3 & -4 & 6 \\ 1 & 1 & -1 & 3 & 4 \\ 2 & 2 & -4 & 6 & 1 \\ 8 & -2 & 1 & 4 & 6 \end{pmatrix} \text{ and } \mathbf{B} = \begin{pmatrix} 4 \\ -2 \\ 6 \\ 1 \end{pmatrix}.$$

We could solve this system by finding the reduced row echelon form of **A**. To illustrate LU factorization, first factor **A**. Begin by finding **U**:

$$\mathbf{A} = \begin{pmatrix} \mathbf{4} & 3 & 3 & -4 & 6 \\ \mathbf{1} & 1 & -1 & 3 & 4 \\ \mathbf{2} & 2 & -4 & 6 & 1 \\ \mathbf{8} & -2 & 1 & 4 & 6 \end{pmatrix} \to \begin{pmatrix} 4 & 3 & 3 & -4 & 6 \\ 0 & \mathbf{1/4} & -7/4 & 4 & 5/2 \\ 0 & \mathbf{1/2} & -11/1 & 8 & -2 \\ 0 & \mathbf{-8} & -5 & 12 & -6 \end{pmatrix}$$

$$\to \begin{pmatrix} 4 & 3 & 3 & -4 & 6 \\ 0 & 1/4 & -7/4 & 4 & 5/2 \\ 0 & 0 & \mathbf{-2} & 8 & -7 \\ 0 & 0 & \mathbf{-61} & 140 & 74 \end{pmatrix} \to \begin{pmatrix} 4 & 3 & 3 & -4 & 6 \\ 0 & 1/4 & -7/4 & 4 & 5/2 \\ 0 & 0 & -2 & 0 & -7 \\ 0 & 0 & 0 & \mathbf{140} & 575/2 \end{pmatrix} = \mathbf{U}.$$

We can now form the 4×4 matrix **L** by beginning with the highlighted columns and obtaining 1's down the main diagonal:

$$\begin{pmatrix} 4 & 0 & 0 & 0 \\ 1 & 1/4 & 0 & 0 \\ 2 & 1/2 & -2 & 0 \\ 8 & -8 & -61 & 140 \end{pmatrix} \to \begin{pmatrix} 1 & 0 & 0 & 0 \\ 1/4 & 1 & 0 & 0 \\ 1/2 & 2 & 1 & 0 \\ 2 & -32 & 61/2 & 1 \end{pmatrix} = \mathbf{L}.$$

Now solve $\mathbf{LY} = \mathbf{B}$. Because **L** is lower triangular, this is the system

$$\begin{aligned} y_1 &= 4, \\ \frac{1}{4}y_1 + y_2 &= -2, \\ \frac{1}{2}y_1 + 2y_2 + y_3 &= 6 \\ 2y_1 - 32y_2 + \frac{61}{2}y_3 + y_4 &= 1 \end{aligned}$$

with solution

$$\mathbf{Y} = \begin{pmatrix} 4 \\ -3 \\ 10 \\ -408 \end{pmatrix}.$$

Now solve

$$\mathbf{UX} = \mathbf{Y}.$$

Because **U** is upper triangular, this is the system

$$4x_1 + 3x_2 + 3x_3 - 4x_4 + 6x_5 = 4,$$
$$\frac{1}{4}x_2 - \frac{7}{4}x_3 + 4x_4 + \frac{5}{2}x_5 = -3$$
$$-2x_3 - 7x_5 = 10$$
$$140x_4 + \frac{575}{2}x_5 = -408.$$

Solve this to obtain the solution of the solution of the original system:

$$\mathbf{X} = \alpha \begin{pmatrix} 523/28 \\ -183/7 \\ -7/2 \\ -115/56 \\ 1 \end{pmatrix} + \begin{pmatrix} 3971/140 \\ -1238/35 \\ -5 \\ -102/35 \\ 0 \end{pmatrix}. \blacklozenge$$

SECTION 7.9 PROBLEMS

In each of Problems 1 through 6, find an LU factorization of the matrix.

1. $\begin{pmatrix} 1 & 4 & 2 & -1 & 4 \\ 1 & -1 & 4 & -1 & 4 \\ -2 & 6 & 8 & 6 & -2 \\ 4 & 2 & 1 & 2 & -4 \end{pmatrix}$

2. $\begin{pmatrix} 1 & 7 & 2 & -1 \\ 3 & 5 & 2 & 6 \\ -3 & -7 & 10 & -4 \end{pmatrix}$

3. $\begin{pmatrix} 2 & 4 & -6 \\ 8 & 2 & 1 \\ -4 & 4 & 10 \end{pmatrix}$

4. $\begin{pmatrix} 4 & -8 & 2 \\ 2 & 24 & -2 \\ -3 & 2 & 14 \\ 0 & 1 & -5 \end{pmatrix}$

5. $\begin{pmatrix} -2 & 1 & 12 \\ 2 & -6 & 1 \\ 2 & 2 & 4 \end{pmatrix}$

6. $\begin{pmatrix} 1 & 5 & 2 \\ 3 & -4 & 2 \\ 1 & 4 & 10 \end{pmatrix}$

In each of Problems 7 through 12, solve the system $\mathbf{AX} = \mathbf{B}$ by factoring **A**. **A** is given first, then **B**

7. $\begin{pmatrix} 6 & 1 & -1 & 3 \\ 4 & 2 & 1 & 5 \\ -4 & 1 & 6 & 5 \\ 2 & -1 & -1 & 4 \end{pmatrix}, \begin{pmatrix} 4 \\ 12 \\ 2 \\ -3 \end{pmatrix}$

8. $\begin{pmatrix} 1 & 2 & 0 & 1 & 1 & 2 & -4 \\ 3 & 3 & -3 & 6 & -5 & 2 & 5 \\ 6 & 8 & 4 & 0 & -2 & 2 & 0 \end{pmatrix}, \begin{pmatrix} 0 \\ -4 \\ 2 \end{pmatrix}$

9. $\begin{pmatrix} 4 & 4 & 2 \\ 1 & -1 & 3 \\ 1 & 42 & 2 \end{pmatrix}, \begin{pmatrix} 1 \\ 0 \\ 1 \end{pmatrix}$

10. $\begin{pmatrix} 2 & 1 & 1 & 3 \\ 1 & 4 & 6 & 2 \end{pmatrix}, \begin{pmatrix} 2 \\ 4 \end{pmatrix}$

11. $\begin{pmatrix} -1 & 1 & 1 & 6 \\ 2 & 1 & 0 & 4 \\ 1 & -2 & 4 & 6 \end{pmatrix}, \begin{pmatrix} 2 \\ 1 \\ 6 \end{pmatrix}$

12. $\begin{pmatrix} 7 & 2 & -4 \\ -3 & 2 & 8 \\ 4 & 4 & 20 \end{pmatrix}, \begin{pmatrix} 7 \\ -1 \\ 3 \end{pmatrix}$

7.10 Linear Transformations

Sometimes we want to consider functions between R^n and R^m. Such a function associates with each vector in R^n a vector in R^m, according to a rule defined by the function.

A function T that maps n-vectors to m-vectors is called a *linear transformation* if the following two conditions are satisfied:

1.

$$T(\mathbf{u}+\mathbf{v}) = T(\mathbf{u}) + T(\mathbf{v})$$

for all n-vectors $\mathbf{u}$ and $\mathbf{v}$, and

2.

$$T(\alpha\mathbf{u}) = \alpha T(\mathbf{u})$$

for every real number α and all n-vectors $\mathbf{u}$

These two conditions can be rolled into the single requirement that

$$T(\alpha\mathbf{u}+\beta\mathbf{v}) = \alpha T(\mathbf{u}) + \beta T(\mathbf{v})$$

for all real numbers α and β and vectors $\mathbf{u}$ and $\mathbf{v}$ in R^n.

A linear transformation is also called a *linear mapping*.

EXAMPLE 7.33

Define T by

$$T(x, y) = < x + y, x - y, 2x > .$$

Then T maps vectors in R^2 to vectors in R^3. For example,

$$T(2, -3) = < -1, 5, 4 > \text{ and } T(1, 1) = < 2, 0, 2 > .$$

We will verify that T is a linear transformation. Let

$$\mathbf{u} = < a, b > \text{ and } \mathbf{v} = < c, d > .$$

Then

$$\mathbf{u}+\mathbf{v} = < a + c, b + d >$$

and

$$T(\mathbf{u}+\mathbf{v}) = T(a + c, b + d) = < a + c + b + d, a + c - b - d, 2a + 2c >,$$

while

$$\begin{aligned} T(\mathbf{u}) + T(\mathbf{v}) &= < a + b, a - b, 2a > + < c + d, c - d, 2c > \\ &= < a + b + c + d, a - b + c - d, 2a + 2c > \\ &= < a + c + b + d, a + c - b - d, 2a + 2c > \\ &= T(a + c, b + d) = T(\mathbf{u}+\mathbf{v}). \end{aligned}$$

This verifies condition (1) of the definition. For condition (2), let α be any number. Then

$$\begin{aligned} T(\alpha\mathbf{u}) &= T(\alpha a, \alpha b) \\ &= < \alpha a + \alpha b, \alpha a - \alpha b, 2\alpha a > \\ &= \alpha < a + b, a - b, 2a > = \alpha T(\mathbf{u}). \end{aligned}$$ ♦

It is easy to check that the function

$$P(a, b, c) = < a^2, 1, 1, \sin(a) >

from R^2 to R^4 is not linear. Generally a function is nonlinear (fails to be linear) when it involves products or powers of the coordinates, or nonlinear functions such as trigonometric functions and exponential functions, whose graphs are not straight lines.

We will use the notation

$$T : R^n \to R^m$$

to indicate that T is a linear transformation from R^n to R^m.

Every linear transformation $T : R^n \to R^m$ must map the zero vector $\mathbf{O}_n$ of R^n to the zero vector $\mathbf{O}_m$ of R^m. To see why this is true, use the linearity of T to write

$$T(\mathbf{O}_n) = T(\mathbf{O}_n + \mathbf{O}_n) = T(\mathbf{O}_n) + T(\mathbf{O}_n),$$

so

$$T(\mathbf{O}_n) = \mathbf{O}_m.$$

However, a linear transformation may take nonzero vectors to the zero vector. For example, the linear transformation

$$T(x, y) = (x - y, 0)$$

from R^2 to R^2 maps every vector $< x, x >$ to $< 0, 0 >$.

We will define two important properties that a linear transformation $T : R^n \to R^m$ may exhibit.

T is *onto* if every vector in R^m is the image of some vector in R^n under T. This means that, if $\mathbf{v}$ is in R^m, then there must be some $\mathbf{u}$ in R^n such that $T(\mathbf{u}) = \mathbf{v}$.

T is *one-to-one*, or $1 - 1$, if the only way $T(\mathbf{u}_1)$ can equal $T(\mathbf{u}_2)$ is for $\mathbf{u}_1 = \mathbf{u}_2$. This means that two vectors in R^n cannot be mapped to the same vector in R^m by T.

The notions of one-to-one and onto are independent. A linear transformation may be one-to-one and onto, one-to-one and not onto, onto and not one-to-one, or neither one-to-one or onto.

EXAMPLE 7.34

Let

$$T(x, y) = < x - y, 0, 0 > .$$

Then T is a linear transformation from R^2 to R^3. T is certainly not one-to-one, since, for example,

$$T(1, 1) = T(2, 2) = < 0, 0, 0 > .$$

In fact, $T(x, x) = < 0, 0, 0 >$ for every number x. Thus T maps many vectors to the origin in R^3.

T is also not onto R^3, since no vector in R^3 with a nonzero second or third component is the image of any vector in R^2 under T. ♦

EXAMPLE 7.35

Let $S : R^3 \to R^2$ be defined by

$$S(x, y, z) = < x, y > .$$

S is onto, since every vector in R^2 is the image of a vector in R^3 under S. For example, $< -3, \sqrt{97} >= S(-3, \sqrt{97}, 0)$. But S is not one-to-one. For example, $S(-3, \sqrt{97}, 22)$ also equals $< -3, \sqrt{97} >$. ♦

There is a convenient test to tell whether a linear transformation is one-to-one. We know that every linear transformation maps the zero vector to the zero vector. The transformation is one-to-one when this is the only vector mapping to the zero vector.

THEOREM 7.18

Let $T : R^n \to R^m$ be a linear transformation. Then T is one-to-one if and only if $T(\mathbf{u}) = \mathbf{O}_m$ occurs only if $\mathbf{u} = \mathbf{O}_n$. ♦

Proof Suppose first that T is one-to-one. If $T(\mathbf{u}) = \mathbf{O}_m$, then

$$T(\mathbf{u}) = T(\mathbf{O}_n) = \mathbf{O}_m$$

so the assumption that T is one-to-one requires that $\mathbf{u} = \mathbf{O}_n$.

Conversely, suppose $T(\mathbf{u}) = \mathbf{O}_m$ occurs only if $\mathbf{u} = \mathbf{O}_n$. To show that T is one-to-one, suppose, for some $\mathbf{u}$ and $\mathbf{v}$ in R^n,

$$T(\mathbf{u}) = T(\mathbf{v}).$$

By the linearity of T,

$$T(\mathbf{u} - \mathbf{v}) = \mathbf{O}_m.$$

By assumption, this implies that

$$\mathbf{u} - \mathbf{v} = \mathbf{O}_n.$$

But then $\mathbf{u} = \mathbf{v}$, so T is one-to-one. ♦

To illustrate, S in Example 7.35 is not one-to-one, because nonzero vectors map to the zero vector. In Example 7.34, T is not one-to-one for the same reason.

EXAMPLE 7.36

Let $T : R^4 \to R^7$ be defined by

$$T(x, y, z, w) = < x - y + 2z + 8w, y - z, x - w, y + 4w, 5x + 5y - z, 0, 0 > .$$

To see if T is one-to-one, examine whether nonzero vectors can map to the zero vector. Suppose

$$T(x, y, z, w) = \mathbf{O}_7 = < 0, 0, 0, 0, 0, 0, 0 > .$$

Then

$$< x + y + z + w, y - z, x - w, x - y + z - w, 5x + 5y - z, 0, 0 >=< 0, 0, 0, 0, 0, 0, 0 > .$$

Looking at the second and third components of both sides of this equation, we must have $y - z = 0$ and $x - w = 0$, so $y = z$ and $x = w$. From the first components,

$$x + y + z + w = 2x + 2y = 0,$$

so $y = -x$. From the fifth component,

$$5x + 5y - z = 5x - 5x - z = 0$$

yields $z = 0$. But then $y = 0$, so $x = 0$ also. Then, from the third component, $x - w = -w = 0$ implies that $w = 0$. We conclude that

$$<x, y, z, w> = <0, 0, 0, 0>.$$

The only vector T maps to the zero vector is the zero vector, so T is one-to-one. Clearly T is not onto, since T does not map any vector to a 7-vector with a nonzero sixth or seventh component. ♦

Every linear transformation $T : R^n \to R^m$ can be associated with a matrix $\mathbf{A}_T$ that carries all of the information about the transformation. Recall that the standard basis for R^n consists of the n orthogonal unit vectors

$$\mathbf{e}_1 = <1, 0, \cdots, 0>, \mathbf{e}_2 = <0, 1, 0, \cdots, 0>,$$
$$\cdots, \mathbf{e}_n = <0, 0, \cdots, 0, 1>$$

with a similar basis (with m components) for R^m. Now let $\mathbf{A}_T$ be the matrix whose columns are of the images in R^m of $T(\mathbf{e}_1), T(\mathbf{e}_2), \cdots, T(\mathbf{e}_n)$ with coordinates written in terms of the standard basis in R^m. The $\mathbf{A}_T$ is an $m \times n$ matrix that represents T in the sense that

$$T(x_1, x_2, \cdots, x_n) = \mathbf{A}_T \begin{pmatrix} x_1 \\ x_2 \\ \vdots \\ x_n \end{pmatrix}.$$

Thus we can compute $T(\mathbf{X})$ as the matrix product of $\mathbf{A}_T$ with the column matrix of the components of $\mathbf{X}$. Note that $\mathbf{A}_T$ is $m \times n$, and $\mathbf{X}$ (written as a column matrix) is $n \times 1$, so $\mathbf{A}_T\mathbf{X}$ is $m \times 1$. Hence, it is a vector in R^m.

EXAMPLE 7.37

Let $T(x, y) = <x - y, 0, 0>$, as in Example 7.34. Then

$$T(1, 0) = <1, 0, 0> \text{ and } T(0, 1) = <-1, 0, 0>$$

so

$$\mathbf{A}_T = \begin{pmatrix} 1 & -1 \\ 0 & 0 \\ 0 & 0 \end{pmatrix}.$$

Now observe that

$$\mathbf{A}_T \begin{pmatrix} x \\ y \end{pmatrix} = \begin{pmatrix} x - y \\ 0 \\ 0 \end{pmatrix},$$

giving the coordinates of $T(x, y)$ with respect to the standard basis for R^3. We can therefore read the coordinates of $T(x, y)$ as a matrix product. ♦

EXAMPLE 7.38

In Example 7.36 we had

$$T(x, y, z, w) = <x - y + 2z + 8w, y - z, x - w, y + 4w, 5x + 5y - z, 0, 0>.$$

For the matrix of T, compute

$$T(1, 0, 0, 0) = <1, 0, 1, 0, 5, 0, 0>, T(0, 1, 0, 0) = <-1, 1, 0, 1, 5, 0, 0>$$
$$T(0, 0, 1, 0) = <2, -1, 0, 0, -1, 0, 0>, T(0, 0, 0, 1) = <8, 0, -1, 4, 0, 0, 0>.$$

Then

$$\mathbf{A}_T = \begin{pmatrix} 1 & -1 & 2 & 8 \\ 0 & 1 & -1 & 0 \\ 1 & 0 & 0 & -1 \\ 0 & 1 & 0 & 4 \\ 5 & 5 & -1 & 0 \\ 0 & 0 & 0 & 0 \\ 0 & 0 & 0 & 0 \end{pmatrix}.$$

We obtain $T(x, y, z, w)$ as the matrix product

$$\mathbf{A}_T \begin{pmatrix} x \\ y \\ z \\ w \end{pmatrix}. \blacklozenge$$

$\mathbf{A}_T$ enables us to pose questions about T in terms of linear systems of equations, about which we know a good deal.

First, $T : R^n \to R^m$ is one-to-one exactly when $T(\mathbf{X}) = <0, 0, \cdots, 0>$ in R^m implies that $\mathbf{X} = <0, 0, \cdots, 0>$ in R^n. This is equivalent to asserting that the $m \times n$ system

$$\mathbf{A}_T\mathbf{X} = \mathbf{O}$$

has only the trivial solution $\mathbf{X} = \mathbf{O}$. This occurs if and only if $n - \text{rank}(\mathbf{A}_T) = 0$, which in turn occurs if and only if the n columns of $\mathbf{A}_T$ are linearly independent, since the rank of $\mathbf{A}_T$ is the dimension of its row space. This establishes the following.

THEOREM 7.19

Let $T : R^n \to R^m$ be a linear transformation. Then the following conditions are equivalent:

1. T is one-to-one.
2. $\text{rank}(\mathbf{A}_T) = n$.
3. The columns of $\mathbf{A}_T$ are linearly independent. ♦

This can be checked for T Example 7.36, with $\mathbf{A}_T$ given in Example 7.38. There $\mathbf{A}_T$ was a 7×4 matrix having rank 4, and T was one-to-one.

$\mathbf{A}_T$ will also tell us if T is onto. For T to be onto, for each $\mathbf{B}$ in R^m, there must be some $\mathbf{X}$ in R^n such that $T(\mathbf{X}) = \mathbf{B}$. This means that the $m \times n$ system $\mathbf{A}_T\mathbf{X} = \mathbf{B}$ must have a solution for each $\mathbf{B}$, and this is equivalent to the columns of $\mathbf{A}_T$ forming a spanning set for R^m. We therefore have the following.

THEOREM 7.20

Let $T : R^n \to R^m$. Then the following are equivalent.

1. T is onto.
2. The system $\mathbf{A}_T(\mathbf{X}) = \mathbf{B}$ has a solution for each $\mathbf{B}$ in R^m.
3. The columns of $\mathbf{A}_T$ span R^m.
4. $\text{rank}(\mathbf{A}_T) = \text{rank}([\mathbf{A}_T \vdots \mathbf{B}]$ for each $\mathbf{B}$ in R^m. ♦

The *null space* of a linear transformation $T : R^n \to R^m$ is the set of all vectors in R^n that T maps to the zero vector in R^m. Thus $\mathbf{X}$ in R^n is in the null space of T exactly when $T(\mathbf{X}) = <0, 0, \cdots, 0>$ in R^m.

We can determine this null space from $\mathbf{A}_T$. In terms of matrix multiplication, $T(\mathbf{X})$ is computed as $\mathbf{A}_T\mathbf{X}$, in which $\mathbf{X}$ is an $n \times 1$ column matrix. Thus $\mathbf{X}$ is in the null space of T exactly when $\mathbf{X}$ is a solution of

$$\mathbf{A}_T\mathbf{X} = \mathbf{O}.$$

The null space of T is exactly the solution space of the homogeneous linear system $\mathbf{A}_T\mathbf{X} = \mathbf{O}$. This solution space is a subspace of R^n and, because $\mathbf{A}_T$ has n columns, it has dimension $n - \text{rank}(\mathbf{A}_T)$. This proves the following.

THEOREM 7.21

Let $T : R^n \to R^m$ be a linear transformation. Then the null space of T is a subspace of R^n of dimension $n - \text{rank}(\mathbf{A}_T)$. ♦

The dimension of the null space of T is also n minus the number of nonzero rows in the reduced form of $\mathbf{A}_T$.

Algebraists often refer to the null space of a linear transformation as its *kernel*.

We have seen that every linear transformation from R^n to R^m has a matrix representation. In the other direction, every $m \times n$ matrix $\mathbf{A}$ of real numbers is the matrix of a linear transformation, defined by $T(\mathbf{X}) = \mathbf{Y}$ if $\mathbf{AX} = \mathbf{Y}$. In this sense linear transformations and matrices are equivalent bodies of information. However, matrices are better suited to computation, particularly using software packages. For example, the rank of the matrix of a linear transformation, which we can find quickly using MAPLE, tells us the dimension of the transformation's null space.

As a final note, observe that a linear transformation actually has many different matrix representations. We defined $\mathbf{A}_T$ in the most convenient way, using standard bases for R^n and R^m. If we used other bases, we could still write matrix representations, but then we would have to use coordinates of vectors with respect to these bases, and these coordinates might not be as convenient to compute.

SECTION 7.10 PROBLEMS

In each of Problems 1 through 10, determine whether or not the given function is a linear transformation. If it is, write the matrix representation of T (using the standard bases) and determine if T is onto and if T is one-to-one. Also determine the null space of T and its dimension.

1. $T(x, y, u, v, w) = <u - v - w, w + u, z, 0, 1>$
2. $T(x, y, z, u) = <x + y + 4z - 8u, y - z - x>$
3. $T(x, y) = <x - y, \sin(x - y)>$
4. $T(x, y, w) = <4y - 2x, y + 3x, 0, 0>$
5. $T(x, y, z) = <3x, x - y, 2z>$
6. $T(x, y, z, v, w) = <w, v, x - y, x - z, w - x - 3y>$
7. $T(x, y) = <x - y, x + y, 2xy, 2y, x - 2y>$
8. $T(x, y, z, v) = <3z + 8v - y, y - 4v>$
9. $T(x, y, z, u, v) = <x - u, y - z, u + v>$
10. $T(x, y, z, w) = <x - y, z - w>$

CHAPTER 8

DEFINITION OF THE DETERMINANT
EVALUATION OF DETERMINANTS I
EVALUATION OF DETERMINANTS II A
DETERMINANT FORMULA FOR A^{-1}

Determinants

8.1 Definition of the Determinant

Determinants are scalars (numbers or sometimes functions) formed from square matrices according to a rule we will develop. The Wronskian of two functions, seen in Chapter 2, is a determinant, and we will shortly see determinants in other important contexts. This chapter develops some properties of determinants that we will need to evaluate and make use of them.

Let n be an integer with $n \geq 2$. A *permutation* of the integers $1, 2, \cdots, n$ is a rearrangement of these integers. For example, if p is the permutation that rearranges

$$1, 2, 3, 4, 5, 6 \rightarrow 3, 1, 4, 5, 2, 6,$$

then $p(1) = 3$, $p(2) = 1$, $p(3) = 4$, $p(4) = 5$, $p(5) = 2$ and $p(6) = 6$.

A permutation is characterized as even or odd according to a rule we will illustrate. Consider the permutation

$$p : 1, 2, 3, 4, 5 \rightarrow 2, 5, 1, 4, 3$$

of the integers $1, 2, 3, 4, 5$. For each k in the permuted list on the right, count the number of integers to the right of k that are smaller than k. There is one number to the right of 2 smaller than 2, three numbers to the right of 5 smaller than 5, no numbers to the right of 1 smaller than 1, one number to the right of 4 smaller than 4, and no numbers to the right of 3 smaller than 3. Since $1 + 3 + 0 + 1 + 0 = 5$ is odd, p is an *odd permutation*. When this sum is even, p is an *even permutation*.

If p is a permutation on $1, 2, \cdots, n$, define

$$\sigma(p) = \begin{cases} 1 & \text{if } p \text{ is an even permutation} \\ -1 & \text{if } p \text{ is an odd permutation.} \end{cases}$$

The *determinant* of an $n \times n$ matrix $\mathbf{A}$ is defined to be

$$\det \mathbf{A} = \sum_p \sigma(p) a_{1p(1)} a_{2p(2)} \cdots a_{np(n)} \tag{8.1}$$

with this sum extending over all permutations p of $1, 2, \cdots, n$. Note that $\det \mathbf{A}$ is a sum of terms, each of which is plus or minus a product containing one element from each row and each column of $\mathbf{A}$.

We often denote $\det \mathbf{A}$ as $|\mathbf{A}|$. This is not to be confused with the absolute value, as a determinant can be negative.

EXAMPLE 8.1

We will use the definition to evaluate the general 2×2 and 3×3 determinants. For the 2×2 case, we have a matrix

$$\mathbf{A} = \begin{pmatrix} a_{11} & a_{21} \\ a_{21} & a_{22} \end{pmatrix}.$$

There are only two permutations on the numbers $1, 2$, namely

$$p_1 : 1, 2 \rightarrow 1, 2 \text{ and } p_2 : 1, 2 \rightarrow 2, 1.$$

It is easy to check that p_1 is even and p_2 is odd. Therefore

$$\begin{aligned} |\mathbf{A}| &= \sigma(p_1) a_{1p_1(1)} a_{2p_1(2)} + \sigma(p_2) a_{1p_2(1)} a_{2p_2(2)} \\ &= a_{11}a_{22} - a_{12}a_{21}. \end{aligned}$$

For the 3×3 case, suppose $\mathbf{B} = [b_{ij}]$ is a 3×3 matrix. Now we must use the six permutations of the integers $1, 2, 3$:

$$p_1 : 1, 2, 3 \rightarrow 1, 2, 3, \text{ (even)}; \; p_2 : 1, 2, 3, \rightarrow 1, 3, 2, \text{ (odd)};$$

$$p_3 : 1, 2, 3 \rightarrow 2, 3, 1, \text{ (even)}; \; p_4 : 1, 2, 3, \rightarrow 2, 1, 3, \text{ (odd)};$$

$$p_5 : 1, 2, 3, \rightarrow 3, 1, 2, \text{ (even)}; \; p_6 : 1, 2, 3, \rightarrow 3, 2, 1, \text{ (odd)}.$$

Then

$$\begin{aligned} |\mathbf{B}| &= \sum_{k=1}^{6} \sigma(p_k) b_{1p_k(1)} b_{2p_k(2)} b_{3p_k(3)} \\ &= b_{11}b_{22}b_{33} - b_{11}b_{23}b_{32} + b_{12}b_{23}b_{31} \\ &= b_{12}b_{21}b_{33} + b_{13}b_{21}b_{32} - b_{13}b_{22}b_{31}. \end{aligned}$$ ♦

There are $n! = 1 \cdot 2 \cdot 3 \cdots n$ permutations of $1, 2, \cdots, n$ (for example, 120 permutations of $1, 2, 3, 4, 5$), so the definition is not a practical method of evaluation. However, it serves as a starting point to develop the properties of determinants we will need to make use of them.

THEOREM 8.1 Some Fundamental Properties of Determinants

Let **A** be an $n \times n$ matrix. Then

1. $|\mathbf{A}^t| = |\mathbf{A}|$.
2. If **A** has a zero row or column then $|\mathbf{A}| = 0$.
3. If **B** is formed from **A** by interchanging two rows or columns (a type I operation, extended to include columns) then
$$|\mathbf{B}| = -|\mathbf{A}|.$$
4. If two rows of **A** are the same, or if two columns of **A** are the same, then $|\mathbf{A}| = 0$.
5. If **B** is formed from **A** by multiplying a row or column by a nonzero number α (a type II operation), then
$$|\mathbf{B}| = \alpha|\mathbf{A}|.$$
6. If one row (or column) of **A** is a constant multiple of another row (or column), then $|\mathbf{A}| = 0$.
7. Suppose each element of row k of **A** is written as a sum
$$a_{kj} = b_{kj} + c_{kj}.$$
Define a matrix **B** from **A** by replacing each a_{kj} of **A** by b_{kj}. Define a matrix **C** from **A** by replacing each a_{kj} by c_{kj}. Then
$$|\mathbf{A}| = |\mathbf{B}| + |\mathbf{C}|.$$
In determinant notation,
$$|\mathbf{A}| = \begin{vmatrix} a_{11} & \cdots & a_{1j} & \cdots & a_{1n} \\ \vdots & \vdots & \vdots & \vdots & \vdots \\ b_{k1}+c_{k1} & \cdots & b_{kj}+c_{kj} & \cdots & b_{kn}+c_{kn} \\ \vdots & \vdots & \vdots & \vdots & \vdots \\ a_{n1} & \cdots & a_{kj} & \cdots & a_{nn} \end{vmatrix}$$
$$= \begin{vmatrix} a_{11} & \cdots & a_{1j} & \cdots & a_{1n} \\ \vdots & \vdots & \vdots & \vdots & \vdots \\ b_{k1} & \cdots & b_{kj} & \cdots & b_{kn} \\ \vdots & \vdots & \vdots & \vdots & \vdots \\ a_{n1} & \cdots & a_{kj} & \cdots & a_{nn} \end{vmatrix} + \begin{vmatrix} a_{11} & \cdots & a_{1j} & \cdots & a_{1n} \\ \vdots & \vdots & \vdots & \vdots & \vdots \\ c_{k1} & \cdots & c_{kj} & \cdots & c_{kn} \\ \vdots & \vdots & \vdots & \vdots & \vdots \\ a_{n1} & \cdots & a_{kj} & \cdots & a_{nn} \end{vmatrix}. \tag{8.2}$$
8. If **D** is formed from **A** by adding α times one row (or column) to another row (or column) (a type III operation), then
$$|\mathbf{D}| = |\mathbf{A}|.$$
9. **A** is nonsingular if and only if $|\mathbf{A}| \neq 0$.
10. If **A** and **B** are both $n \times n$, then
$$|\mathbf{AB}| = |\mathbf{A}||\mathbf{B}|. \ \blacklozenge$$

The determinant of a product is the product of the determinants.

We will give informal arguments for these conclusions.

Proof Conclusion (1) follows from the observation that each term in the sum of equation (8.1) is a product of matrix elements, one element from each row and one from each column. We therefore obtain the same terms from both $\mathbf{A}$ and $\mathbf{A}^t$.

The reason for conclusion (2) is that a zero row or column puts a zero factor in each term of the defining sum in equation (8.1).

Conclusion (3) states that interchanging two rows, or two columns, changes the sign of the determinant. We will illustrate this for the 3×3 case. Let $\mathbf{A} = [a_{ij}]$ be 3×3 matrix and let $\mathbf{B} = [b_{ij}]$ be formed by interchanging rows one and three of $\mathbf{A}$. Then

$$b_{11} = a_{31}, b_{12} = a_{32}, b_{13} = a_{33},$$

$$b_{21} = a_{21}, b_{22} = a_{22}, b_{23} = a_{23},$$

and

$$b_{31} = a_{11}, b_{32} = a_{12}, b_{33} = a_{13}.$$

From Example 8.1,

$$\begin{aligned}
|\mathbf{B}| &= b_{11}b_{22}b_{33} - b_{11}b_{23}b_{32} + b_{12}b_{23}b_{31} \\
&= -b_{12}b_{21}b_{33} + b_{13}b_{21}b_{32} - b_{13}b_{22}b_{31} \\
&= a_{31}a_{22}a_{13} - a_{31}a_{23}a_{12} + a_{32}a_{23}a_{11} \\
&= -a_{32}a_{21}a_{13} + a_{33}a_{21}a_{12} - a_{33}a_{22}a_{11} \\
&= -|\mathbf{A}|.
\end{aligned}$$

Conclusion (4) follows immediately from (3). Form $\mathbf{B}$ from $\mathbf{A}$ by interchanging the two identical rows or columns. Since $\mathbf{A} = \mathbf{B}$, $|\mathbf{A}| = |\mathbf{B}|$. But by (3), $|\mathbf{A}| = -|\mathbf{B}| = |\mathbf{A}|$. Then $|\mathbf{A}| = 0$.

Conclusion (5) is true because multiplying a row or column of $\mathbf{A}$ by α puts a factor of α in every term of the sum (8.1) defining the determinant.

Conclusion (6) follows from (2) if $\alpha = 0$, so suppose that $\alpha \neq 0$. Now the conclusion follows from (4) and (5). Suppose that row k of $\mathbf{A}$ is α times row i. Form $\mathbf{B}$ from $\mathbf{A}$ by multiplying row k by $1/\alpha$. Then $\mathbf{B}$ has two identical rows, hence zero determinant by (4). But by (5), $|\mathbf{B}| = (1/\alpha)|\mathbf{A}| = 0$, so $|\mathbf{A}| = 0$.

Conclusion (7) follows by replacing each a_{kj} in the defining sum (8.1) with $b_{kj} + c_{kj}$. Note here that k is fixed, so only one factor in each term of (8.1) is replaced. In particular, generally the determinant of a sum is not the sum of the determinants. Conclusion (7) also holds if each element of a specified column is written as a sum of two terms.

Conclusion (8) follows from (4) and (7). To see this we will deal with rows to be specific. Suppose α times row i is added to row k of $\mathbf{A}$ to form $\mathbf{D}$. On the right side of equation (8.2), replace each b_{kj} with αa_{ij}, and each c_{kj} with a_{kj}, resulting in the following:

$$\mathbf{D} = \begin{pmatrix}
a_{11} & a_{12} & \cdots & a_{1n} \\
\cdots & \cdots & \cdots & \cdots \\
a_{i1} & a_{i2} & \cdots & a_{in} \\
\cdots & \cdots & \cdots & \cdots \\
\alpha a_{i1} + a_{k1} & \alpha a_{i2} + a_{k2} & \cdots & \alpha a_{in} + a_{kn} \\
\cdots & \cdots & \cdots & \cdots \\
a_{n1} & a_{n2} & \cdots & a_{in}
\end{pmatrix}$$

$$= \begin{pmatrix} a_{11} & a_{12} & \cdots & a_{1n} \\ \cdots & \cdots & \cdots & \cdots \\ a_{i1} & a_{i2} & \cdots & a_{in} \\ \cdots & \cdots & \cdots & \cdots \\ \alpha a_{i1} & \alpha a_{i2} & \cdots & \alpha a_{in} \\ \cdots & \cdots & \cdots & \cdots \\ a_{n1} & a_{n2} & \cdots & a_{in} \end{pmatrix} + \begin{pmatrix} a_{11} & a_{12} & \cdots & a_{1n} \\ \cdots & \cdots & \cdots & \cdots \\ a_{i1} & a_{i2} & \cdots & a_{in} \\ \cdots & \cdots & \cdots & \cdots \\ a_{k1} & a_{k2} & \cdots & a_{kn} \\ \cdots & \cdots & \cdots & \cdots \\ a_{n1} & a_{n2} & \cdots & a_{in} \end{pmatrix}.$$

Then $|\mathbf{A}|$ is the sum of the determinants of the matrices on the right. But the second determinant on the right is just $|\mathbf{A}|$ and the first is 0 by (4) because row k is a multiple of row i.

For conclusion (9), note that, by (3), (5) and (8), every time we produce **B** from **A** by an elementary row operation, $|\mathbf{B}|$ is equal to a nonzero multiple of **A**. Since we reduce a matrix by a sequence of elementary row operations, then $|\mathbf{A}|$ is always a nonzero multiple of $|\mathbf{A}_R|$. This means that $|\mathbf{A}|$ is nonzero if and only if $|\mathbf{A}_R|$ is nonzero. But this is the case exactly when **A** is nonsingular, since in this case $\mathbf{A}_R = \mathbf{I}_n$. If $\mathbf{A}_R \neq \mathbf{I}_n$, then $\mathbf{A}_R$ has at least one zero row and has determinant zero.

Vanishing or non-vanishing of the determinant is an important test for existence of an inverse, and we will use it when we discuss eigenvalues in the next chapter.

Finally, we will sketch a proof of conclusion (10). If **A** is nonsingular, then there is a product of elementary matrices that reduces **A** to $\mathbf{I}_n$:

$$\mathbf{E}_r\mathbf{E}_{r-1}\cdots\mathbf{E}_1\mathbf{A} = \mathbf{I}_n.$$

Then

$$\mathbf{A} = \mathbf{E}_1^{-1}\mathbf{E}_2^{-1}\cdots\mathbf{E}_r^{-1},$$

a product of inverses of elementary matrices, which are again elementary matrices. Since we can do this for nonsingular **B** as well, we can write **AB** as a product of elementary matrices. It is therefore sufficient to show that the determinant of a product of elementary matrices is the product of the determinants of these elementary matrices. This can be done for two elementary matrices using properties (3), (5) and (8) of determinants then extended to arbitrary products by induction.

If either **A** or **B** is singular, then so is **AB**, and in this case,

$$|\mathbf{AB}| = 0 = |\mathbf{A}||\mathbf{B}|.$$

Conclusions (3), (5), and (8) tell us the effects of elementary row operations on the determinant of a matrix. However, in the context of determinants, these operations can be applied to columns as well. When we use matrices to represent systems of equations, rows contain equations and columns contain coefficients of particular unknowns, so there is an essential difference between rows and columns. However, the determinant of a matrix does not involve these interpretations and there is no preference of rows over columns (for example, $|\mathbf{A}| = |\mathbf{A}^t|$). ♦

SECTION 8.1 PROBLEMS

1. Show that the determinant of an upper or lower triangular matrix is the product of its main diagonal elements. *Hint*: Every term but one of the sum (8.1) contains a factor a_{ij} with $i > j$ and a term a_{ij} with $i < j$, and one of these terms must be zero if the matrix is upper or lower triangular. The exceptional term corresponds to the permutation p that leaves every number $1, 2, \cdots, n$ unmoved.

2. Evaluate $|\mathbf{I}_n|$ for $n = 2, 3, \cdots$. *Hint*: In the sum of equation (8.1), the only term that does not have a zero factor corresponds to the identity permutation $p: 1, 2, \cdots, n \to 1, 2, \cdots n$.

3. Let $\mathbf{B}$ be $n \times m$. We know that we can achieve each elementary row operation by multiplying on the left by the matrix formed by performing the operation on $\mathbf{I}_n$. Show that each elementary column operation can be performed by multiplying on the right by the matrix obtained by performing the column operation on I_m.

4. Show that an upper or lower triangular matrix is nonsingular if and only if it has nonzero main diagonal elements.

5. An $n \times n$ matrix is *skew-symmetric* if $\mathbf{A} = -\mathbf{A}'$. Explain why the determinant of a skew symmetric matrix having an odd number of rows and columns must be zero.

6. Let $\mathbf{A} = [a_{ij}]$ be an $n \times n$ matrix. Let α be a nonzero number. Form the matrix $\mathbf{B} = [\alpha^{i-j} a_{ij}]$. How are $|\mathbf{A}|$ and $|\mathbf{B}|$ related? *Hint:* It is useful to look at the 2×2 and 3×3 cases to get some idea of what $\mathbf{B}$ looks like.

7. Let $\mathbf{A} = [a_{ij}]$ be an $n \times n$ matrix and let α be a number. Form $\mathbf{B} = [\alpha a_{ij}]$ by multiplying each element of $\mathbf{A}$ by α. How are $|\mathbf{A}|$ and $|\mathbf{B}|$ related?

8.2 Evaluation of Determinants I

The more zero elements a matrix has, the easier it is to evaluate its determinant. The reason for this is that every zero element causes some terms in the sum of equation (8.1) to vanish. For example, in Example 8.1, if $a_{12} = a_{13} = 0$,

$$\mathbf{A} = \begin{pmatrix} a_{11} & 0 & 0 \\ a_{21} & a_{22} & a_{23} \\ a_{31} & a_{32} & a_{33} \end{pmatrix}$$

and

$$|\mathbf{A}| = a_{11} \begin{vmatrix} a_{22} & a_{23} \\ a_{32} & a_{33} \end{vmatrix} = a_{11}(a_{22}a_{33} - a_{23}a_{32})$$

with four of the six terms of $|\mathbf{A}|$ being 0 cancelling because of the zeroes in the first row of $\mathbf{A}$.

A generalization of this observation will form the basis of a useful method for evaluating determinants.

LEMMA 8.1

Let $\mathbf{A}$ be $n \times n$, and suppose row k or column r has all zero elements, except perhaps for a_{kr}. Then

$$|\mathbf{A}| = (-1)^{k+r} a_{kr} |\mathbf{A}_{kr}|, \tag{8.3}$$

where $\mathbf{A}_{kr}$ is the $n-1 \times n-1$ matrix formed by deleting row k and column r of $\mathbf{A}$. ♦

This reduces the problem of evaluating an $n \times n$ determinant to one of evaluating a smaller, $n-1 \times n-1$, determinant. To see why the lemma is true, begin with the case that all the elements of row one, except perhaps a_{11}, are zero. Then

$$\mathbf{A} = \begin{pmatrix} a_{11} & 0 & 0 & \cdots & 0 \\ a_{12} & a_{22} & a_{23} & \cdots & a_{2n} \\ \vdots & \vdots & \vdots & \vdots & \vdots \\ a_{n1} & a_{n2} & a_{n3} & \cdots & a_{nn} \end{pmatrix}.$$

In the sum of equation (8.1), the factor $a_{1p(1)}$ is zero if $p(1) \neq 1$, because all the other elements of row one are zero. This means we need only consider the sum over permutations p of the form

$$p : 1, 2, 3, \cdots, n \to 1, p(2), p(3), \cdots, p(n).$$

But this is really just a permutation of the $n-1$ numbers $2, 3, \cdots, n$, since 1 is fixed and only $2, 3, \cdots, n$ are acted upon. In the definition of equation (8.1), we may therefore sum over only the permutations q of $2, 3, \cdots, n$, and factor a_{11} from all of the terms of the sum, to obtain

$$|\mathbf{A}| = \sum_q a_{11} \sum_q a_{2q(2)} a_{3q(3)} \cdots a_{nq(n)} = |\mathbf{A}_{11}|.$$

This is a_{11} times the determinant of the $n-1 \times n-1$ matrix formed by deleting row one and column one of **A**.

In the general case that a_{kr} is an element of a row or column whose other elements are all zero, we can interchange $k-1$ rows and then $r-1$ columns to obtain a new matrix with a_{kr} in the $1, 1$ position of a row or column having its other elements equal to zero. Since each interchange incurs a factor of -1 in the determinant, then by the preceding result,

$$|\mathbf{A}| = (-1)^{k-1+r-1} a_{kr} |\mathbf{A}_{kr}| = (-1)^{k+r} a_{kr} |\mathbf{A}_{kr}.$$

We are rarely lucky enough to encounter a matrix **A** having a row or column with all but possibly one element equal to zero. However, we can use elementary row and column operations to obtain such a matrix **B** from **A**. Furthermore from properties (3), (5), and (8) of determinants, we can track the effect of each row and column operation on the value of the determinant. This and the lemma enable us to reduce the evaluation of an $n \times n$ determinant to a constant times an $n-1 \times n-1$ determinant. We can then repeat this strategy, eventually obtaining a constant times a determinant small enough to evaluate conveniently.

EXAMPLE 8.2

Let

$$\mathbf{A} = \begin{pmatrix} 4 & 2 & -3 \\ 3 & 4 & 6 \\ 2 & -6 & 8 \end{pmatrix}.$$

We want $|\mathbf{A}|$. This is a simple example, but illustrates the point. We can get two zeros in column two by adding -2 times row one to row two, then 3 times row one to row three. Since this elementary row operation does not change the value of the determinant, then $|\mathbf{A}| = |\mathbf{B}|$, where

$$\mathbf{B} = \begin{pmatrix} 4 & 2 & -2 \\ -5 & 0 & 10 \\ 14 & 0 & 2 \end{pmatrix}.$$

Exploiting the zeros in all but the $1, 2$ place in column two, then

$$\begin{aligned} |\mathbf{A}| = |\mathbf{B}| &= (-1)^{1+2}(2)|\mathbf{B}_{12}| \\ &= -2 \begin{vmatrix} -5 & 10 \\ 14 & 2 \end{vmatrix} \\ &= -2(-10-140) = 300. \end{aligned}$$ ♦

EXAMPLE 8.3

Let

$$\mathbf{A} = \begin{pmatrix} -6 & 0 & 1 & 3 & 2 \\ -1 & 5 & 0 & 1 & 7 \\ 8 & 3 & 2 & 1 & 7 \\ 0 & 1 & 5 & -3 & 2 \\ 1 & 15 & -3 & 9 & 4 \end{pmatrix}.$$

There are many ways to evaluate $|\mathbf{A}|$. One way to begin is to exploit the 1 in the 1, 3 position to get zeros in the other locations in column 3. Add -2 times row one to row three, -5 times row one to row four, and 3 times row one to row five to get

$$\mathbf{B} = \begin{pmatrix} -6 & 0 & 1 & 3 & 2 \\ -1 & 5 & 0 & 1 & 7 \\ 20 & 3 & 0 & -5 & 3 \\ 30 & 1 & 0 & -18 & -8 \\ -17 & 15 & 0 & 18 & 10 \end{pmatrix}.$$

Adding a multiple of one row to another does not change the value of the determinant, so

$$|\mathbf{A}| = |\mathbf{B}|.$$

Furthermore, by equation (8.3),

$$|\mathbf{B}| = (-1)^{1+3}(1)|\mathbf{C}| = |\mathbf{C}|,$$

where $\mathbf{C}$ is the 4×4 matrix formed by deleting row one and column three of $\mathbf{B}$:

$$\mathbf{C} = \begin{pmatrix} -1 & 5 & 1 & 7 \\ 20 & 3 & -5 & 3 \\ 30 & 1 & -18 & -8 \\ -17 & 15 & 18 & 10 \end{pmatrix}.$$

Now work on $\mathbf{C}$. Again, there are many ways to proceed. We will use the -1 in the 1, 1 position to get zeros in row one. Add 5 times column one to column two, add column one to column three and add 7 times column one to column four of $\mathbf{C}$ to get

$$\mathbf{D} = \begin{pmatrix} -1 & 0 & 0 & 0 \\ 20 & 103 & 15 & 143 \\ 30 & 151 & 12 & 202 \\ -17 & 70 & 1 & -109 \end{pmatrix}.$$

Because we added a multiple of one column to another,

$$|\mathbf{C}| = |\mathbf{D}|.$$

And, using equation (8.3) again,

$$|\mathbf{D}| = (-1)^{1+1}(-1)|\mathbf{E}| = -|\mathbf{E}|,$$

where $\mathbf{E}$ is the 3×3 matrix formed from $\mathbf{D}$ by deleting row one and column one:

$$\mathbf{E} = \begin{pmatrix} 103 & 15 & 143 \\ 151 & 12 & 202 \\ -70 & 1 & -109 \end{pmatrix}.$$

To evaluate $\mathbf{E}$ we will use the 1 in the 3, 2 place. Add -1 times row three to row one and -12 times row three to row two to get

$$\mathbf{F} = \begin{pmatrix} 1153 & 0 & 1778 \\ 991 & 0 & 1510 \\ -70 & 1 & -109 \end{pmatrix}.$$

Then

$$|\mathbf{E}| = |\mathbf{F}|.$$

Furthermore

$$|\mathbf{F}| = (-1)^{3+2}(1)|\mathbf{G}| = -|\mathbf{G}|$$

where $\mathbf{G}$ is the 2×2 matrix obtained by deleting row three and column two of $\mathbf{F}$:

$$\mathbf{G} = \begin{pmatrix} 1153 & 1778 \\ 991 & 1510 \end{pmatrix}.$$

This is 2×2 which we evaluate easily:

$$|\mathbf{G}| = (1153)(1510) - (1778)(991) = -20,968.$$

Working back through the chain of determinants, we have

$$|\mathbf{A}| = |\mathbf{B}| = |\mathbf{C}||\mathbf{D}|$$
$$= -|\mathbf{E}| = -|\mathbf{F}| = |\mathbf{G}| = -20,968. \blacklozenge$$

SECTION 8.2 PROBLEMS

In each of Problems 1 through 10, use the method of this section to evaluate the determinant. In each problem there are many different sequences of operations that can be used to make the evaluation.

1. $\begin{vmatrix} 10 & 1 & -6 & 2 \\ 0 & 3 & 3 & 9 \\ 0 & 1 & 1 & 7 \\ -2 & 6 & 8 & 8 \end{vmatrix}$

2. $\begin{vmatrix} -7 & 16 & 2 & 4 \\ 1 & 0 & 0 & 5 \\ 0 & 3 & -4 & 4 \\ 6 & 1 & 1 & -5 \end{vmatrix}$

3. $\begin{vmatrix} 17 & -2 & 5 \\ 1 & 12 & 0 \\ 14 & 7 & -7 \end{vmatrix}$

4. $\begin{vmatrix} -3 & 3 & 9 & 6 \\ 1 & -2 & 15 & 6 \\ 7 & 1 & 1 & 5 \\ 2 & 1 & -1 & 3 \end{vmatrix}$

5. $\begin{vmatrix} -4 & 5 & 6 \\ -2 & 3 & 5 \\ 2 & -2 & 6 \end{vmatrix}$

6. $\begin{vmatrix} 2 & -5 & 8 \\ 4 & 3 & 8 \\ 13 & 0 & -4 \end{vmatrix}$

7. $\begin{vmatrix} -2 & 4 & 1 \\ 1 & 6 & 3 \\ 7 & 0 & 4 \end{vmatrix}$

8. $\begin{vmatrix} 2 & -3 & 7 \\ 14 & 1 & 1 \\ -13 & -1 & 5 \end{vmatrix}$

9. $\begin{vmatrix} 0 & 1 & 1 & -4 \\ 6 & -3 & 2 & 2 \\ 1 & -5 & 1 & -2 \\ 4 & 8 & 2 & 2 \end{vmatrix}$

10. $\begin{vmatrix} 2 & 7 & -1 & 0 \\ 3 & 1 & 1 & 8 \\ -2 & 0 & 3 & 1 \\ 4 & 8 & -1 & 0 \end{vmatrix}$

11. Fill in the details of the following argument that $|\mathbf{AB}| = |\mathbf{A}||\mathbf{B}|$.

    First, if $\mathbf{AB}$ is singular, show that at least one of $\mathbf{A}$ or $\mathbf{B}$ is singular, hence that the determinant of the product and the product of the determinants are both zero.

    Thus, suppose that $\mathbf{AB}$ is nonsingular. Show that $\mathbf{A}$ and $\mathbf{B}$ can be written as products of elementary matrices, and then show that the determinant of a product of elementary matrices equals the product of the determinants of these matrices.

8.3 Evaluation of Determinants II

In the preceding section, we evaluated determinants by using row and column operations to produce rows and/or columns with all but one entry zero. In this section we exploit this idea from a different perspective to write the determinant as a sum of numbers times smaller determinants.

This method, called *expansion by cofactors*, can be used recursively until we have determinants of small enough size to be easily evaluated.

Choose a row k of $\mathbf{A} = [a_{ij}]$. An extension of property (7) of determinants from Section 8.1 enables us to write

$$|\mathbf{A}| = |[a_{ij}]| = \begin{vmatrix} a_{11} & a_{12} & \cdots & \cdots & a_{1n} \\ \vdots & \vdots & \vdots & \vdots & \vdots \\ a_{k1} & a_{k2} & \cdots & \cdots & a_{kn} \\ \vdots & \vdots & \vdots & \vdots & \vdots \\ a_{n1} & a_{n2} & \cdots & \cdots & a_{nn} \end{vmatrix} = \begin{vmatrix} a_{11} & a_{12} & \cdots & \cdots & a_{1n} \\ \vdots & \vdots & \vdots & \vdots & \vdots \\ a_{k1} & 0 & \cdots & \cdots & 0 \\ \vdots & \vdots & \vdots & \vdots & \vdots \\ a_{n1} & a_{n2} & \cdots & \cdots & a_{nn} \end{vmatrix}$$

$$+ \begin{vmatrix} a_{11} & a_{12} & \cdots & \cdots & a_{1n} \\ \vdots & \vdots & \vdots & \vdots & \vdots \\ 0 & a_{k2} & \cdots & \cdots & 0 \\ \vdots & \vdots & \vdots & \vdots & \vdots \\ a_{n1} & a_{n2} & \cdots & \cdots & a_{nn} \end{vmatrix} + \cdots + \begin{vmatrix} a_{11} & a_{12} & \cdots & \cdots & a_{1n} \\ \vdots & \vdots & \vdots & \vdots & \vdots \\ 0 & 0 & \cdots & \cdots & a_{kn} \\ \vdots & \vdots & \vdots & \vdots & \vdots \\ a_{n1} & a_{n2} & \cdots & \cdots & a_{nn} \end{vmatrix}.$$

Each of the n determinants on the right has a row with exactly one possibly nonzero element, and can be expanded by that element, as in Section 8.2. To write this expansion, define the *minor* of a_{ij} to be the determinant of the $n-1 \times n-1$ matrix formed by deleting row i and column j of $\mathbf{A}$. This minor is denoted M_{ij}. The *cofactor* of a_{ij} is the number $(-1)^{i+j} M_{ij}$. Now this sum of determinants gives us the following theorem.

THEOREM 8.2 Cofactor Expansion by a Row

For any k with $1 \le i \le n$.

$$|\mathbf{A}| = \sum_{j=1}^{n} (-1)^{k+j} a_{kj} M_{kj}. \quad \blacklozenge \tag{8.4}$$

Equation (8.4) states that the determinant of $\mathbf{A}$ is the sum, along any row k, of the matrix elements of that row, each multiplied by its cofactor. This holds for any row of the matrix, although of course this sum is easier to evaluate if we choose a row with as many zero elements as possible. Equation (8.4) is called *expansion by cofactors along row k*. If we write out a few terms for fixed k we get

$$|\mathbf{A}| = (-1)^{k+1} a_{k1} M_{k1} + (-1)^{k+2} a_{k2} M_{k2} + \cdots + (-1)^{k+n} a_{kn} M_{kn}.$$

EXAMPLE 8.4

Let

$$\mathbf{A} = \begin{pmatrix} -6 & 3 & 7 \\ 12 & -5 & -6 \\ 2 & 4 & -6 \end{pmatrix}$$

If we expand by cofactors along row one, we get

$$
\begin{aligned}
|\mathbf{A}| &= \sum_{j=1}^{3}(-1)^{1+j}a_{1j}M_{1j} \\
&= (-1)^{1+1}(-6)\begin{vmatrix} -5 & -9 \\ 4 & -6 \end{vmatrix} + (-1)^{1+2}(3)\begin{vmatrix} 12 & -9 \\ 2 & -6 \end{vmatrix} \\
&\quad + (-1)^{1+3}(7)\begin{vmatrix} 12 & -5 \\ 2 & 4 \end{vmatrix} \\
&= (-6)(30+36) - 3(-72+18) + 7(-48+10) = 172.
\end{aligned}
$$

If we expand by row three, we get

$$
\begin{aligned}
|\mathbf{A}| &= \sum_{j=1}^{3}(-1)^{3+j}a_{3j}M_{3j} \\
&= (-1)^{3+1}(2)\begin{vmatrix} 3 & 7 \\ -5 & -9 \end{vmatrix} + (-1)^{3+2}(4)\begin{vmatrix} -6 & 7 \\ 12 & -9 \end{vmatrix} \\
&\quad + (-1)^{3+3}(-6)\begin{vmatrix} -6 & 3 \\ 12 & -5 \end{vmatrix} \\
&= (2)(-27+35) - 4(54-84) - 6(30-36) = 172. \quad \blacklozenge
\end{aligned}
$$

We can also do a cofactor expansion along a column. Now fix j and sum the elements of column j times their cofactors.

THEOREM 8.3 Cofactor Expansion by a Column

For any j with $1 \le j \le n$,

$$|\mathbf{A}| = \sum_{i=1}^{n}(-1)^{i+j}a_{ij}M_{ij}. \quad \blacklozenge \tag{8.5}$$

EXAMPLE 8.5

We will expand the determinant of the matrix of Example 8.3, using column 1:

$$
\begin{aligned}
|\mathbf{A}| &= \sum_{i=1}^{3}(-1)^{i+1}a_{i1}M_{i1} \\
&= (-1)^{1+1}(-6)\begin{vmatrix} -5 & -9 \\ 4 & -6 \end{vmatrix} + (-1)^{2+1}(12)\begin{vmatrix} 3 & 7 \\ 4 & -6 \end{vmatrix} \\
&\quad + (-1)^{3+1}(2)\begin{vmatrix} 3 & 7 \\ -5 & -9 \end{vmatrix} \\
&= (-6)(30+36) - 12(-18-28) + 2(-27+35) = 172.
\end{aligned}
$$

If we expand by column two, we get

$$
\begin{aligned}
|\mathbf{A}| &= \sum_{i=1}^{3}(-1)^{i+2}a_{i2}M_{i2} \\
&= (-1)^{1+2}(3)\begin{vmatrix} 12 & -9 \\ 2 & -6 \end{vmatrix} + (-1)^{2+2}(-5)\begin{vmatrix} -6 & 7 \\ 2 & -6 \end{vmatrix} \\
&\quad + (-1)^{3+2}(4)\begin{vmatrix} -6 & 7 \\ 12 & -9 \end{vmatrix} \\
&= (-3)(-72+18) - 5(36-14) - 4(54-84) = 172. \quad \blacklozenge
\end{aligned}
$$

Sometimes we use row and column operations to produce a row or column with some zero elements, then write a cofactor expansion by that row or column. Each zero element eliminates one term from the cofactor expansion.

SECTION 8.3 PROBLEMS

In each of Problems 1 through 10, evaluate the determinant using a cofactor expansion by a row and again by a column. Elementary row and/or column operations may be performed first to simplify the cofactor expansion.

1. $\begin{vmatrix} -5 & 4 & 1 & 7 \\ -9 & 3 & 2 & -5 \\ -2 & 0 & -1 & 1 \\ 1 & 14 & 0 & 3 \end{vmatrix}$

2. $\begin{vmatrix} -8 & 5 & 1 & 7 & 2 \\ 0 & 1 & 3 & 5 & -6 \\ 2 & 2 & 1 & 5 & 3 \\ 0 & 4 & 3 & 7 & 2 \\ 1 & 1 & -7 & -6 & 5 \end{vmatrix}$

3. $\begin{vmatrix} -5 & 0 & 1 & 6 \\ 2 & -1 & 3 & 7 \\ 4 & 4 & -5 & -8 \\ 1 & -1 & 6 & 2 \end{vmatrix}$

4. $\begin{vmatrix} 4 & 3 & -5 & 6 \\ 1 & -5 & 15 & 2 \\ 0 & -5 & 1 & 7 \\ 8 & 9 & 0 & 15 \end{vmatrix}$

5. $\begin{vmatrix} -3 & 1 & 14 \\ 0 & 1 & 16 \\ 2 & -3 & 4 \end{vmatrix}$

6. $\begin{vmatrix} 14 & 13 & -2 & 5 \\ 7 & 1 & 1 & 7 \\ 0 & 2 & 12 & 3 \\ 1 & -6 & 5 & 23 \end{vmatrix}$

7. $\begin{vmatrix} -4 & 2 & -8 \\ 1 & 1 & 0 \\ 1 & -3 & 0 \end{vmatrix}$

8. $\begin{vmatrix} 5 & -4 & 3 \\ -1 & 1 & 6 \\ -2 & -2 & 4 \end{vmatrix}$

9. $\begin{vmatrix} 7 & -3 & 1 \\ 1 & -2 & 4 \\ -3 & 1 & 0 \end{vmatrix}$

10. $\begin{vmatrix} 1 & 1 & 6 \\ 2 & -2 & 1 \\ 3 & -1 & 4 \end{vmatrix}$

11. Prove that the points (x_1, y_1), (x_2, y_2), and (x_3, y_3) in the plane are collinear (lie on a line) if and only if

$$\begin{vmatrix} 1 & x_1 & y_1 \\ 1 & x_2 & y_2 \\ 1 & x_3 & y_3 \end{vmatrix} = 0.$$

Hint: This determinant is zero exactly when one row or column is a linear combination of the others.

12. Show that

$$\begin{vmatrix} a & b & c & d \\ b & c & d & a \\ c & d & a & b \\ d & a & b & c \end{vmatrix} = (a+b+c+d)(b-a+d-c)\begin{vmatrix} 0 & 1 & -1 & 1 \\ 1 & c & d & a \\ 1 & d & a & b \\ 1 & a & b & c \end{vmatrix}.$$

13. Show that

$$\begin{vmatrix} 1 & a & a^2 \\ 1 & b & b^2 \\ 1 & c & c^2 \end{vmatrix} = (a-b)(c-a)(b-c).$$

This is called *Vandermonde's determinant*.

8.4 A Determinant Formula for $\mathbf{A}^{-1}$

When $|\mathbf{A}| \neq 0$, $\mathbf{A}$ has an inverse. Furthermore, there is a formula for the elements of this inverse in terms of determinants formed from elements of $\mathbf{A}$.

THEOREM 8.4 ***Elements of a Matrix Inverse***

Let $\mathbf{A}$ be a nonsingular $n \times n$ matrix and define an $n \times n$ matrix $\mathbf{B} = [b_{ij}]$ by

$$b_{ij} = \frac{1}{|\mathbf{A}|}(-1)^{i+j} M_{ji}.$$

Then $\mathbf{B} = \mathbf{A}^{-1}$. ♦

Note that the i, j element of $\mathbf{B}$ is defined in terms of $(-1)^{i+j} M_{ji}$, the cofactor of a_{ji} (not a_{ij}).

We can see why this construction yields $\mathbf{A}^{-1}$ by explicitly multiplying the two matrices. By the definition of matrix multiplication, the i, j element of $\mathbf{AB}$ is

$$(\mathbf{AB})_{ij} = \sum_{k=1}^{n} a_{ik} b_{kj} = \frac{1}{|\mathbf{A}|} \sum_{k=1}^{n} (-1)^{j+k} a_{ik} M_{jk}. \tag{8.6}$$

Now consider two cases. If $i = j$ the sum in equation (8.6) is exactly the cofactor expansion of $|\mathbf{A}|$ by row i. The main diagonal elements of $\mathbf{AB}$ are therefore 1.

If $i \neq j$, the sum in equation (8.6) is the cofactor expansion by row j of the determinant of the matrix formed from $\mathbf{A}$ by replacing row j by row i. But this matrix has two identical rows, so its determinant is zero and the off-diagonal elements of $\mathbf{AB}$ are all zero. This means that

$$\mathbf{AB} = \mathbf{I}_n.$$

Similarly, $\mathbf{BA} = \mathbf{I}_n$.

EXAMPLE 8.6

Let

$$\mathbf{A} = \begin{pmatrix} -2 & 4 & 1 \\ 6 & 3 & -3 \\ 2 & 9 & -5 \end{pmatrix}.$$

It is routine to compute $|\mathbf{A}| = 120$ so $\mathbf{A}$ is nonsingular. We will determine $\mathbf{A}^{-1}$ by computing the elements of the matrix $\mathbf{B}$ of Theorem 8.4:

$$b_{11} = \frac{1}{120} M_{11} = \frac{1}{120} \begin{vmatrix} 3 & -3 \\ 9 & -5 \end{vmatrix} = \frac{12}{120} = \frac{1}{10},$$

$$b_{12} = \frac{1}{120}(-1) M_{21} = -\frac{1}{120} \begin{vmatrix} 4 & 1 \\ 9 & -5 \end{vmatrix} = \frac{29}{120},$$

$$b_{13} = \frac{1}{120} M_{31} = \frac{1}{120} \begin{vmatrix} 4 & 1 \\ 3 & -3 \end{vmatrix} = -\frac{1}{8},$$

$$b_{21} = -\frac{1}{120} M_{12} = -\frac{1}{120} \begin{vmatrix} 6 & -3 \\ 2 & -5 \end{vmatrix} = \frac{1}{5},$$

$$b_{22} = \frac{1}{120}M_{22} = \frac{1}{120}\begin{vmatrix} -2 & 1 \\ 2 & -5 \end{vmatrix} = \frac{1}{15},$$

$$b_{23} = -\frac{1}{120}M_{32} = -\frac{1}{120}\begin{vmatrix} -2 & 1 \\ 6 & -3 \end{vmatrix} = 0,$$

$$b_{31} = \frac{1}{120}M_{13} = \frac{1}{120}\begin{vmatrix} 6 & 3 \\ 2 & 9 \end{vmatrix} = \frac{2}{5},$$

$$b_{32} = -\frac{1}{120}M_{23} = -\frac{1}{120}\begin{vmatrix} -2 & 4 \\ 2 & 9 \end{vmatrix} = \frac{13}{60},$$

$$b_{33} = \frac{1}{120}M_{33} = \frac{1}{120}\begin{vmatrix} -2 & 4 \\ 6 & 3 \end{vmatrix} = -\frac{1}{4}.$$

Then

$$\mathbf{B} = \mathbf{A}^{-1} = \begin{pmatrix} 1/10 & 29/120 & -1/8 \\ 1/5 & 1/15 & 0 \\ 2/5 & 13/60 & -1/4 \end{pmatrix}. \blacklozenge$$

SECTION 8.4 PROBLEMS

In each of Problems 1 through 10, test the matrix for singularity by evaluating its determinant. If the matrix is nonsingular, use Theorem 8.4 to compute the inverse.

1. $\begin{pmatrix} 6 & -1 & 3 \\ 0 & 1 & -4 \\ 2 & 2 & -3 \end{pmatrix}$

2. $\begin{pmatrix} 2 & 5 \\ -7 & -3 \end{pmatrix}$

3. $\begin{pmatrix} 0 & -4 & 3 \\ 2 & -1 & 6 \\ 1 & -1 & 7 \end{pmatrix}$

4. $\begin{pmatrix} 7 & -3 & -4 & 1 \\ 8 & 2 & 0 & 0 \\ 1 & 5 & -1 & 7 \\ 3 & -2 & -5 & 9 \end{pmatrix}$

5. $\begin{pmatrix} 3 & 1 & -2 & 1 \\ 4 & 6 & -3 & 9 \\ -2 & 1 & 7 & 4 \\ 13 & 0 & 1 & 5 \end{pmatrix}$

6. $\begin{pmatrix} 3 & 0 \\ 1 & 4 \end{pmatrix}$

7. $\begin{pmatrix} 2 & -1 \\ 1 & 6 \end{pmatrix}$

8. $\begin{pmatrix} -14 & 1 & -3 \\ 2 & -1 & 3 \\ 1 & 1 & 7 \end{pmatrix}$

9. $\begin{pmatrix} -1 & 1 \\ 1 & 4 \end{pmatrix}$

10. $\begin{pmatrix} 11 & 0 & -5 \\ 0 & 1 & 0 \\ 4 & -7 & 9 \end{pmatrix}$

8.5 Cramer's Rule

Cramer's rule is a determinant formula for the unique solution of a nonhomogeneous system $\mathbf{AX} = \mathbf{B}$ when $\mathbf{A}$ is nonsingular. Of course, this is $\mathbf{X} = \mathbf{A}^{-1}\mathbf{B}$, but the following method is sometimes convenient.

THEOREM 8.5 Cramer's Rule

Let $\mathbf{A}$ be a nonsingular $n \times n$ matrix of numbers, and $\mathbf{B}$ be an $n \times 1$ matrix of numbers. Then the unique solution of $\mathbf{AX} = \mathbf{B}$ is determined by

$$x_k = \frac{1}{|\mathbf{A}|}|\mathbf{A}(k; \mathbf{B})| \tag{8.7}$$

for $k = 1, 2, \cdots, n$, where $\mathbf{A}(k; \mathbf{B})$ is the matrix obtained from $\mathbf{A}$ by replacing column k of $\mathbf{A}$ with $\mathbf{B}$. ♦

It is easy to see why this works. Let

$$\mathbf{B} = \begin{pmatrix} b_1 \\ b_2 \\ \vdots \\ b_n \end{pmatrix}.$$

Multiply column k of $\mathbf{A}$ by x_k. This multiplies the determinant of $\mathbf{A}$ by x_k:

$$x_k|\mathbf{A}| = \begin{vmatrix} a_{11} & a_{12} & \cdots & a_{1k}x_k & \cdots & a_{1n} \\ a_{21} & a_{22} & \cdots & a_{2k}x_k & \cdots & a_{2n} \\ \vdots & \vdots & \vdots & \vdots & \vdots & \vdots \\ a_{n1} & a_{n2} & \cdots & a_{nk}x_k & \cdots & a_{nn} \end{vmatrix}.$$

For each $j \neq k$ add x_j times column j to column k in the last determinant. Since this operation does not change the value of a determinant, then

$$x_k|\mathbf{A}| = \begin{vmatrix} a_{11} & a_{12} & \cdots & a_{11}x_1 + \cdots + a_{1n}x_n & \cdots & a_{1n} \\ a_{21} & a_{22} & \cdots & a_{21}x_1 + \cdots + a_{2n}x_n & \cdots & a_{2n} \\ \vdots & \vdots & \vdots & \vdots & \vdots & \vdots \\ a_{n1} & a_{n2} & \cdots & a_{n1}x_1 + \cdots + a_{nn}x_n & \cdots & a_{nn} \end{vmatrix}$$

$$= \begin{vmatrix} a_{11} & a_{12} & \cdots & b_1 & \cdots & a_{1n} \\ a_{21} & a_{22} & \cdots & b_2 & \cdots & a_{2n} \\ \vdots & \vdots & \vdots & \vdots & \vdots & \vdots \\ a_{n1} & a_{n2} & \cdots & b_n & \cdots & a_{nn} \end{vmatrix} = |\mathbf{A}(k; \mathbf{B})|$$

and this gives us equation (8.7).

EXAMPLE 8.7

Solve the system

$$x_1 - 3x_2 - 4x_3 = 1$$
$$-x_1 + x_2 - 3x_3 = 14$$
$$x_2 - 3x_3 = 5.$$

The matrix of coefficients is

$$\mathbf{A} = \begin{pmatrix} 1 & -3 & -4 \\ -1 & 1 & -3 \\ 0 & 1 & -3 \end{pmatrix}.$$

We find that $|\mathbf{A}| = 13$, so this system has a unique solution. By Cramer's rule,

$$x_1 = \frac{1}{13}\begin{vmatrix} 1 & -3 & -4 \\ 14 & 1 & -3 \\ 5 & 1 & -3 \end{vmatrix} = -\frac{117}{13} = -9,$$

$$x_2 = \frac{1}{13}\begin{vmatrix} 1 & 1 & -4 \\ -1 & 14 & -3 \\ 0 & 5 & -3 \end{vmatrix} = -\frac{10}{13},$$

$$x_3 = \frac{1}{13}\begin{vmatrix} 1 & -3 & 1 \\ -1 & 1 & 14 \\ 0 & 1 & 5 \end{vmatrix} = -\frac{25}{13}. \ \blacklozenge$$

SECTION 8.5 PROBLEMS

In each of Problems 1 through 10, solve the system using Cramer's rule, or show that the rule does not apply because the matrix of coefficients is singular.

1. $$\begin{aligned} x_1 + x_2 - 3x_3 &= 0 \\ x_2 - 4x_3 &= 0 \\ x_1 - x_2 - x_3 &= 5 \end{aligned}$$

2. $$\begin{aligned} x_2 - 4x_4 &= 18 \\ x_1 - x_2 + 3x_3 &= -1 \\ x_1 + x_2 - 3x_3 + x_4 &= 5 \\ x_2 + 3x_4 &= 0 \end{aligned}$$

3. $$\begin{aligned} 2x_1 - 4x_2 + x_3 - x_4 &= 6 \\ x_2 - 3x_3 &= 10 \\ x_1 - 4x_3 &= 0 \\ x_2 - x_3 + 2x_4 &= 4 \end{aligned}$$

4. $$\begin{aligned} 6x_1 + 4x_2 - x_3 + 3x_4 - x_5 &= 7 \\ x_1 - 4x_2 + x_5 &= -5 \\ x_1 - 3x_2 + x_3 - 4x_5 &= 0 \\ -2x_1 + x_3 - 2x_5 &= 4 \\ x_3 - x_4 - x_5 &= 8 \end{aligned}$$

5. $$\begin{aligned} 14x_1 - 3x_3 &= 5 \\ 2x_1 - 4x_3 + x_4 &= 2 \\ x_1 - x_2 + x_3 - 3x_4 &= 1 \\ x_3 - 4x_4 &= -5 \end{aligned}$$

6. $$\begin{aligned} 2x_1 - 3x_2 + x_4 &= 2 \\ x_2 - x_3 + x_4 &= 2 \\ x_3 - 2x_4 &= 5 \\ x_1 - 3x_2 + 4x_3 &= 0 \end{aligned}$$

7. $$\begin{aligned} 8x_1 - 4x_2 + 3x_3 &= 0 \\ x_1 + 5x_2 - x_3 &= -5 \\ -2x_1 + 6x_2 + x_3 &= -4 \end{aligned}$$

8. $$\begin{aligned} 5x_1 - 6x_2 + x_3 &= 4 \\ -x_1 + 3x_2 - 4x_3 &= 5 \\ 2x_1 + 3x_2 + x_3 &= -8 \end{aligned}$$

9. $$\begin{aligned} 15x_1 - 4x_2 &= 5 \\ 8x_1 + x_2 &= -4 \end{aligned}$$

10. $$\begin{aligned} x_1 + 4x_2 &= 3 \\ x_1 + x_2 &= 0 \end{aligned}$$

8.6 The Matrix Tree Theorem

In 1847, G.R. Kirchhoff published a classic paper in which he derived many of the electrical circuit laws that bear his name, including the matrix tree theorem we will now discuss.

Figure 8.1 shows a typical electrical circuit. The underlying geometry of the circuit if shown in Figure 8.2. Such a diagram of points and interconnecting lines is called a *graph*, and was seen in the context of atoms moving through crystals in Section 7.1.3. A *labeled graph* has symbols attached to the points.

Some of Kirchhoff's results depend on geometric properties of the circuit's underlying graph. One such property is the arrangement of the closed loops. Another is the number of

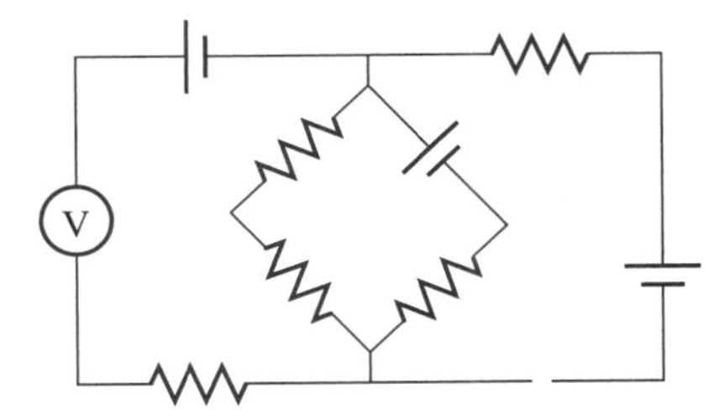

FIGURE 8.1 *Typical electrical circuit.*

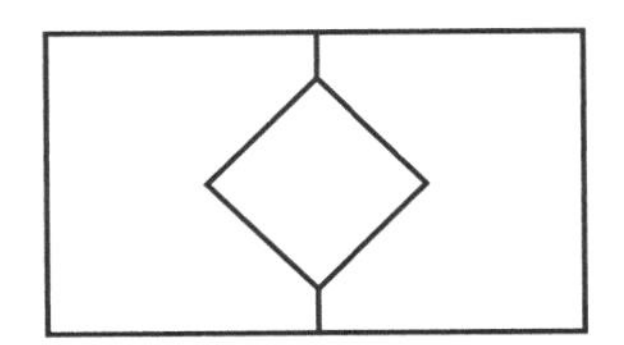

FIGURE 8.2 *Underlying graph of the circuit of Figure 8.1.*

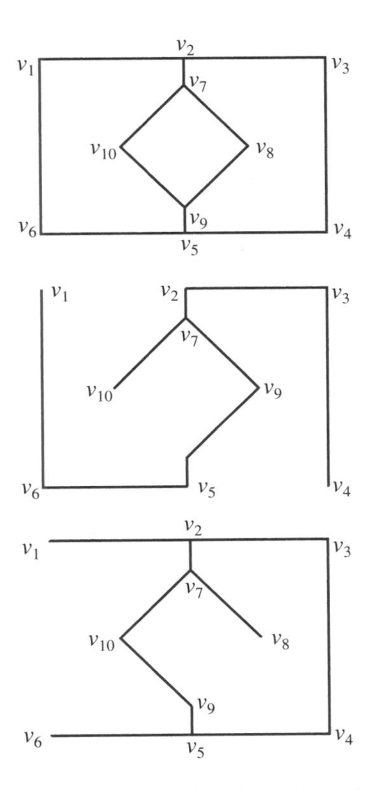

FIGURE 8.3 *Labeled graph and two spanning trees.*

spanning trees in the labeled graph. A *spanning tree* is a collection of lines in the graph forming no closed loops, but containing a path between any two points of the graph. Figure 8.3 shows a labeled graph and two spanning trees in this graph.

Kirchhoff derived a relationship between determinants and the number of spanning trees in a labeled graph.

THEOREM 8.6 The Matrix Tree Theorem

Let G be a graph with vertices labeled $v_1, v_2, \cdots, v_n$. Form an $n \times n$ matrix $\mathbf{T} = [t_{ij}]$ as follows. If $i = j$, then t_{ij} is the number of lines to v_i in the graph. If $i \neq j$, then $t_{ij} = 0$ if there is no line between v_i and v_j in G, and $t_{ij} = -1$ if there is such a line. Then all cofactors of $\mathbf{T}$ are equal and their common value is the number of spanning trees in $\mathbf{G}$. ♦

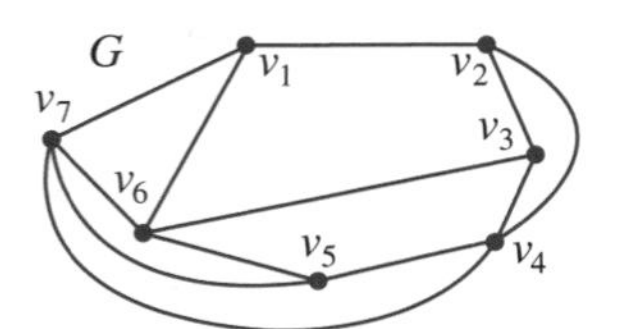

FIGURE 8.4 *Graph of Example 8.8.*

EXAMPLE 8.8

For the labeled graph of Figure 8.4, **T** is the 7×7 matrix

$$\mathbf{T} = \begin{pmatrix} 3 & -1 & 0 & 0 & 0 & -1 & -1 \\ -1 & 3 & -1 & -1 & 0 & 0 & 0 \\ 0 & -1 & 3 & -1 & 0 & -1 & 0 \\ 0 & -1 & -1 & 4 & -1 & 0 & -1 \\ 0 & 0 & 0 & -1 & 3 & -1 & -1 \\ -1 & 0 & -1 & 0 & -1 & 4 & -1 \\ -1 & 0 & 0 & -1 & -1 & -1 & 4 \end{pmatrix}.$$

Evaluate any cofactor of **T**. For example, deleting row 1 and column 1, evaluate the cofactor

$$(-1)^{1+1}M_{11} = \begin{vmatrix} 3 & -1 & -1 & 0 & 0 & 0 \\ -1 & 3 & -1 & 0 & -1 & 0 \\ -1 & -1 & 4 & -1 & 0 & -1 \\ 0 & 0 & -1 & 3 & -1 & -1 \\ 0 & -1 & 0 & -1 & 4 & -1 \\ 0 & 0 & -1 & -1 & -1 & 4 \end{vmatrix} = 386. \blacklozenge$$

Even with this small graph in Example 8.8, it would clearly be impractical to enumerate the spanning trees by listing them all.

SECTION 8.6 PROBLEMS

1. Find the number of spanning trees in the graph of Figure 8.5.

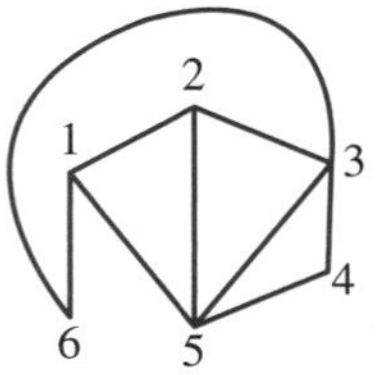

FIGURE 8.5 *Graph of Problem 1, Section 8.6.*

2. Find the number of spanning trees in the graph of Figure 8.6.

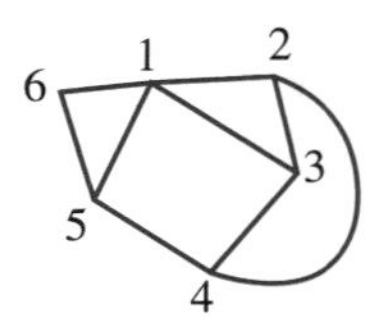

FIGURE 8.6 *Graph of Problem 2, Section 8.6.*

3. Find the number of spanning trees in the graph of Figure 8.7.

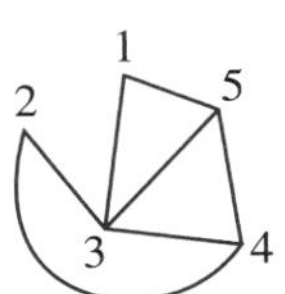

FIGURE 8.7 *Graph of Problem 3, Section 8.6.*

4. Find the number of spanning trees in the graph of Figure 8.8.

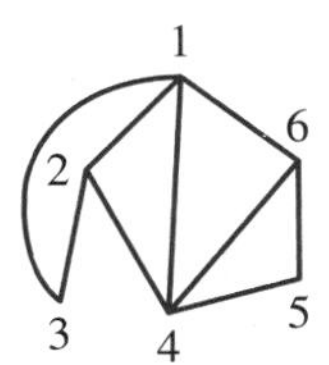

FIGURE 8.8 *Graph of Problem 4, Section 8.6.*

5. Find the number of spanning trees in the graph of Figure 8.9.

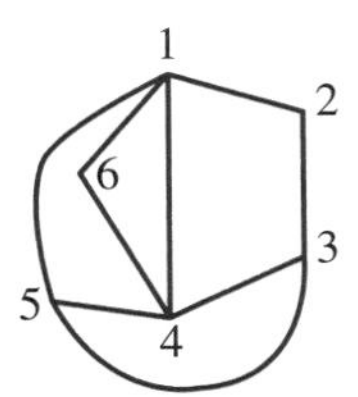

FIGURE 8.9 *Graph of Problem 5, Section 8.6.*

6. A complete graph on n points consists of n points, with a line between each pair of points. This graph is often denoted K_n. With the points labeled $1, 2, \cdots, n$, show that the number of spanning trees in K_n is n^{n-2} for $n = 3, 4, \cdots$.

EIGENVALUES AND EIGENVECTORS DIAGONALIZATION SOME SPECIAL TYPES OF MATRICES

CHAPTER 9 Eigenvalues, Diagonalization, and Special Matrices

9.1 Eigenvalues and Eigenvectors

In this chapter, the term number refers to a real or complex number. Let **A** be an $n \times n$ matrix of numbers. A number λ is an *eigenvalue* of **A** if there is a nonzero $n \times 1$ matrix **E** such that

$$\mathbf{AE} = \lambda \mathbf{E}. \tag{9.1}$$

We call **E** an *eigenvector* associated with the eigenvalue λ.

We may think of an $n \times 1$ matrix of numbers as an n-vector, with real and/or complex components. If we consider **A** as a linear transformation mapping an n-vector **X** to an n-vector **AX**, then equation (9.1) holds when **A** moves **E** to a parallel vector $\lambda\mathbf{E}$. This is the geometric significance of an eigenvector.

If c is a nonzero number and $\mathbf{AE} = \lambda\mathbf{E}$, then

$$\mathbf{A}(c\mathbf{E}) = c\mathbf{AE} = c\lambda\mathbf{E} = \lambda(c\mathbf{E}).$$

This means that nonzero constant multiples of eigenvectors are eigenvectors (with the same eigenvalue).

EXAMPLE 9.1

Let

$$\mathbf{A} = \begin{pmatrix} 1 & 0 \\ 0 & 0 \end{pmatrix}.$$

Because

$$\begin{pmatrix} 1 & 0 \\ 0 & 0 \end{pmatrix} \begin{pmatrix} 0 \\ 4 \end{pmatrix} = \begin{pmatrix} 0 \\ 0 \end{pmatrix} = 0 \begin{pmatrix} 0 \\ 4 \end{pmatrix},$$

then 0 is an eigenvalue of $\mathbf{A}$ with eigenvector

$$\mathbf{E} = \begin{pmatrix} 0 \\ 4 \end{pmatrix}.$$

For any nonzero number α,

$$\begin{pmatrix} 0 \\ 4\alpha \end{pmatrix}$$

is also an eigenvector. Zero can be an eigenvalue, but an eigenvector must be a nonzero vector (at least one nonzero component). ♦

EXAMPLE 9.2

Let

$$\mathbf{A} = \begin{pmatrix} 1 & -1 & 0 \\ 0 & 1 & 1 \\ 0 & 0 & -1 \end{pmatrix}.$$

Then

$$\mathbf{A} \begin{pmatrix} 6 \\ 0 \\ 0 \end{pmatrix} = \begin{pmatrix} 6 \\ 0 \\ 0 \end{pmatrix}.$$

Therefore 1 is an eigenvalue with eigenvector

$$\begin{pmatrix} 6 \\ 0 \\ 0 \end{pmatrix}$$

or any nonzero constant times this matrix. Similarly,

$$\mathbf{A} \begin{pmatrix} 1 \\ 2 \\ -4 \end{pmatrix} = \begin{pmatrix} -1 \\ -2 \\ 4 \end{pmatrix} = (-1) \begin{pmatrix} 1 \\ 2 \\ -4 \end{pmatrix}.$$

Therefore -1 is an eigenvalue with eigenvector

$$\begin{pmatrix} 1 \\ 2 \\ -4 \end{pmatrix},$$

or any nonzero multiple of this vector. ♦

We would like to be able to find all of the eigenvalues of a matrix. We will have $\mathbf{AE} = \lambda\mathbf{E}$, for some number λ and $n \times 1$ matrix $\mathbf{E}$, exactly when

$$\lambda\mathbf{E} - \mathbf{AE} = \mathbf{O}.$$

This is equivalent to

$$(\lambda\mathbf{I}_n - \mathbf{A})\mathbf{E} = \mathbf{O},$$

and this occurs exactly when the system

$$(\lambda\mathbf{I}_n - \mathbf{A})\mathbf{X} = \mathbf{O}$$

has a nontrivial solution $\mathbf{E}$. The condition for this is that the coefficient matrix be singular (determinant zero), hence that

$$|\lambda \mathbf{I}_n - \mathbf{A}| = 0.$$

If expanded, the determinant on the left is a polynomial of degree n in the unknown λ, and is called the *characteristic polynomial* of $\mathbf{A}$. Thus

$$p_{\mathbf{A}}(\lambda) = |\lambda \mathbf{I}_n - \mathbf{A}|.$$

This polynomial has n roots for λ (perhaps some repeated, perhaps some or all complex). These n numbers, counting multiplicities, are all of the eigenvalues of $\mathbf{A}$. Corresponding to each eigenvalue λ, a nontrivial solution of

$$(\lambda \mathbf{I}_n - \mathbf{A})\mathbf{X} = \mathbf{O}$$

is an eigenvector.

We can summarize this discussion as follows.

THEOREM 9.1 Eigenvalues and Eigenvectors of A

Let $\mathbf{A}$ be an $n \times n$ matrix of numbers. Then

1. λ is an eigenvalue of $\mathbf{A}$ if and only if λ is a root of the characteristic polynomial of $\mathbf{A}$. This occurs exactly when

$$p_{\mathbf{A}}(\lambda) = |\lambda \mathbf{I}_n - \mathbf{A}| = 0.$$

   Since $p_{\mathbf{A}}(\lambda)$ has degree n, $\mathbf{A}$ has n eigenvalues, counting each eigenvalue as many times as it appears as a root of $p_{\mathbf{A}}(\lambda)$.

2. If λ is an eigenvalue of $\mathbf{A}$, then any nontrivial solution $\mathbf{E}$ of

$$(\lambda \mathbf{I}_n - \mathbf{A})\mathbf{X} = \mathbf{O}$$

   is an eigenvector of $\mathbf{A}$ associated with λ.

3. If $\mathbf{E}$ is an eigenvector associated with the eigenvalue λ, then so is $c\mathbf{E}$ for any nonzero number c. ♦

EXAMPLE 9.3

Let

$$\mathbf{A} = \begin{pmatrix} 1 & -1 & 0 \\ 0 & 1 & 1 \\ 0 & 0 & -1 \end{pmatrix},$$

as in Example 9.2. The characteristic polynomial is

$$p_{\mathbf{A}}(\lambda) = |\lambda \mathbf{I}_3 - \mathbf{A}| = \begin{vmatrix} \lambda - 1 & 1 & 0 \\ 0 & \lambda - 1 & -1 \\ 0 & 0 & \lambda + 1 \end{vmatrix} = (\lambda - 1)^2(\lambda + 1).$$

This polynomial has roots $1, 1, -1$ and these are the eigenvalues of $\mathbf{A}$. The root 1 has multiplicity 2 and must be listed twice as an eigenvalue of $\mathbf{A}$. $\mathbf{A}$ has three eigenvalues.

To find an eigenvector associated with the eigenvalue 1, put $\lambda = 1$ in (2) of the theorem and solve the system

$$((1)\mathbf{I}_3 - \mathbf{A})\mathbf{X} = \begin{pmatrix} 0 & 1 & 0 \\ 0 & 0 & -1 \\ 0 & 0 & 2 \end{pmatrix} \begin{pmatrix} x_1 \\ x_2 \\ x_3 \end{pmatrix} = \begin{pmatrix} 0 \\ 0 \\ 0 \end{pmatrix}.$$

This system of three equations in three unknowns has the general solution

$$\begin{pmatrix} \alpha \\ 0 \\ 0 \end{pmatrix}$$

and this is an eigenvector associated with 1 for any $\alpha \neq 0$.

For eigenvectors associated with -1, put $\lambda = -1$ in (2) of the theorem and solve

$$((-1)\mathbf{I}_3 - \mathbf{A})\mathbf{X} = \begin{pmatrix} -2 & 1 & 0 \\ 0 & -2 & -1 \\ 0 & 0 & 0 \end{pmatrix} \mathbf{X} = \mathbf{O}.$$

This system has the general solution

$$\begin{pmatrix} \beta \\ 2\beta \\ -4\beta \end{pmatrix}$$

and this is an eigenvector associated with -1 for any $\beta \neq 0$. ♦

EXAMPLE 9.4

Let

$$\mathbf{A} = \begin{pmatrix} 1 & -2 \\ 2 & 0 \end{pmatrix}.$$

The characteristic polynomial is

$$p_{\mathbf{A}}(\lambda) = |\lambda \mathbf{I}_2 - \mathbf{A}| = \left| \begin{pmatrix} \lambda & 0 \\ 0 & \lambda \end{pmatrix} - \begin{pmatrix} 1 & -2 \\ 2 & 0 \end{pmatrix} \right| = \begin{vmatrix} \lambda - 1 & 2 \\ -2 & \lambda \end{vmatrix} = \lambda^2 - \lambda + 4,$$

with roots

$$\frac{1 + \sqrt{15}i}{2} \text{ and } \frac{1 - \sqrt{15}i}{2}$$

and these are the eigenvalues of $\mathbf{A}$.

For an eigenvector corresponding to $(1 + \sqrt{15}i)/2$ solve $(((1 + \sqrt{15}i)/2)\mathbf{I}_2 - \mathbf{A})\mathbf{X} = \mathbf{O}$, which is

$$\left[\frac{1 + \sqrt{15}i}{2} \begin{pmatrix} 1 & 0 \\ 0 & 1 \end{pmatrix} - \begin{pmatrix} 1 & -2 \\ 2 & 0 \end{pmatrix} \right] \mathbf{X} = \mathbf{O}.$$

This is the system

$$\begin{pmatrix} \frac{-1 + \sqrt{15}i}{2} & 2 \\ -2 & \frac{1 + \sqrt{15}i}{2} \end{pmatrix} \begin{pmatrix} x_1 \\ x_2 \end{pmatrix} = \begin{pmatrix} 0 \\ 0 \end{pmatrix}.$$

This 2×2 system has general solution

$$\alpha \begin{pmatrix} 1 \\ (1 - \sqrt{15}i)/4 \end{pmatrix}.$$

This is an eigenvector associated with $(1 + \sqrt{15}i)/2$ for any $\alpha \neq 0$.

For eigenvectors associated with $(1-\sqrt{15}i)/2$, solve the 2×2 system

$$(((1-\sqrt{15}i)/2)\mathbf{I}_2-\mathbf{A})\mathbf{X}=\mathbf{O}$$

to obtain

$$\beta\begin{pmatrix}1\\(1+\sqrt{15}i)/4\end{pmatrix}.$$

This is an eigenvector associated with $(1-\sqrt{15}i)/2$ for any $\beta\neq 0$. ♦

If $\mathbf{A}$ has real numbers as elements and $\lambda=\alpha+i\beta$ is an eigenvalue, then the conjugate $\overline{\lambda}=\alpha-i\beta$ is also an eigenvalue. This is because the characteristic polynomial of $\mathbf{A}$ has real coefficients in this case, so complex roots (eigenvalues of $\mathbf{A}$) occur in conjugate pairs. Furthermore, if $\mathbf{E}$ is an eigenvector corresponding to λ, then $\overline{\mathbf{E}}$ is an eigenvector corresponding to $\overline{\lambda}$, where we take the conjugate of a matrix by taking the conjugate of each of its elements. This can be seen by taking the conjugate of $\mathbf{AE}=\lambda\mathbf{E}$ to obtain

$$\overline{\mathbf{AE}}=\overline{\lambda}\,\overline{\mathbf{E}}.$$

Because $\mathbf{A}$ has real elements, $\overline{\mathbf{A}}=\mathbf{A}$ so

$$\mathbf{A}\overline{\mathbf{E}}=\overline{\lambda}\,\overline{\mathbf{E}}.$$

This observation can be seen in Example 9.4.

There is a general expression for the eigenvalues of a matrix that will be used soon to draw conclusions about eigenvalues of matrices having special properties.

LEMMA 9.1

Let $\mathbf{A}$ be an $n\times n$ matrix of numbers. Let λ be an eigenvalue of $\mathbf{A}$, with eigenvector $\mathbf{E}$. Then

$$\lambda=\frac{\overline{\mathbf{E}}^t\mathbf{AE}}{\overline{\mathbf{E}}^t\mathbf{E}}. \quad ♦ \tag{9.2}$$

Before giving the one line proof of this expression, examine what the right side means. Let

$$\mathbf{E}=\begin{pmatrix}e_1\\e_2\\\vdots\\e_n\end{pmatrix}.$$

Then

$$\overline{\mathbf{E}}^t\mathbf{AE}=\begin{pmatrix}\overline{e}_1 & \overline{e}_2 & \cdots & \overline{e}_n\end{pmatrix}\begin{pmatrix}a_{11} & a_{12} & \cdots & a_{1n}\\a_{21} & a_{22} & \cdots & a_{2n}\\\vdots & \vdots & \vdots & \vdots\\a_{n1} & a_{n2} & \cdots & a_{nn}\end{pmatrix}\begin{pmatrix}e_1\\e_2\\\vdots\\e_n\end{pmatrix}.$$

This is a product of a $1\times n$ matrix with an $n\times n$ matrix, then an $n\times 1$ matrix, hence is a 1×1 matrix, which we think of as a number. If we carry out this matrix product we obtain the number

$$\overline{\mathbf{E}}^t\mathbf{AE}=\sum_{i=1}^{n}\sum_{j=1}^{n}a_{ij}\overline{e}_ie_j.$$

For the denominator of equation (9.2) we have a $1\times n$ matrix multiplied by an $n\times 1$ matrix, which is also a 1×1 matrix, or number. Specifically,

$$\overline{\mathbf{E}}^t\mathbf{E} = \begin{pmatrix} \overline{e_1} & \overline{e_2} & \cdots & \overline{e_n} \end{pmatrix} \begin{pmatrix} e_1 \\ e_2 \\ \vdots \\ e_n \end{pmatrix} = \sum_{j=1}^{n} \overline{e_j} e_j = \sum_{j=1}^{n} |e_j|^2.$$

Therefore the conclusion of Lemma 9.1 can be written

$$\lambda = \frac{\sum_{i=1}^{n}\sum_{j=1}^{n} a_{ij}\overline{e_i}e_j}{\sum_{j=1}^{n}|e_j|^2}.$$

Proof of Lemma 9.1 Since $\mathbf{AE} = \lambda\mathbf{E}$, then

$$\overline{\mathbf{E}}^t\mathbf{AE} = \lambda\overline{\mathbf{E}}^t\mathbf{E},$$

yielding the conclusion of the lemma. ♦

When we discuss diagonalization, we will need to know if the eigenvectors of a matrix are linearly independent. The following theorem answers this question for the special case that the n eigenvalues of $\mathbf{A}$ are distinct (the characteristic polynomial has no repeated roots).

THEOREM 9.2

Suppose the $n \times n$ matrix $\mathbf{A}$ has n distinct eigenvalues. Then $\mathbf{A}$ has n linearly independent eigenvectors. ♦

To illustrate, in Example 9.4, $\mathbf{A}$ was 2×2 and had two distinct eigenvalues. The eigenvectors produced for each eigenvalue were linearly independent.

Proof We will show by induction that any k distinct eigenvalues have associated with them k linearly independent eigenvectors. For $k = 1$ there is nothing to show. Thus suppose $k \geq 2$ and the conclusion of the theorem is valid for any $k-1$ distinct eigenvalues. This means that any $k-1$ distinct eigenvalues have associated with them $k-1$ distinct eigenvectors. Suppose $\mathbf{A}$ has k distinct eigenvalues $\lambda_1, \cdots, \lambda_k$ with corresponding eigenvectors $\mathbf{V}_1, \cdots, \mathbf{V}_k$. We want to show that these eigenvectors are linearly independent.

If they were linearly dependent, then there would be numbers $c_1, \cdots, c_k$, not all zero, such that

$$c_1\mathbf{V}_1 + c_2\mathbf{V}_2 + \cdots + c_k\mathbf{V}_k = \mathbf{O}.$$

By relabeling if necessary, we may assume for convenience that $c_1 \neq 0$. Multiply this equation by $\lambda_1\mathbf{I}_n - \mathbf{A}$:

$$\begin{aligned}
\mathbf{O} &= (\lambda_1\mathbf{I}_n - \mathbf{A})(c_1\mathbf{V}_1 + c_2\mathbf{V}_2 + \cdots + c_k\mathbf{V}_k) \\
&= c_1(\lambda_1\mathbf{I}_n - \mathbf{A})\mathbf{V}_1 + c_2(\lambda_1\mathbf{I}_n - \mathbf{A})\mathbf{V}_2 \\
&\quad + \cdots + c_k(\lambda_1\mathbf{I}_n - \mathbf{A})\mathbf{V}_k \\
&= c_1(\lambda_1\mathbf{V}_1 - \lambda_1\mathbf{V}_1) + c_2(\lambda_1\mathbf{V}_2 - \lambda_2\mathbf{V}_2) \\
&\quad + \cdots + c_k(\lambda_1\mathbf{V}_k - \lambda_k\mathbf{V}_k) \\
&= c_2(\lambda_1 - \lambda_2)\mathbf{V}_1 + \cdots + c_k(\lambda_1 - \lambda_k)\mathbf{V}_k.
\end{aligned}$$

Now $\mathbf{V}_2, \cdots, \mathbf{V}_k$ are linearly independent by the inductive hypothesis, so these coefficients are all zero. But $\lambda_1 \neq \lambda_j$ for $j = 2, \cdots, k$ by the assumptions that the eigenvalues are distinct. Therefore

$$c_2 = \cdots = c_k = 0.$$

But then $c_1\mathbf{V}_1 = \mathbf{O}$. Since an eigenvalue cannot be the zero vector, this means that $c_1 = 0$ also. Therefore $\mathbf{V}_1, \ldots, \mathbf{V}_k$ are linearly independent. By induction, this proves the theorem. ♦

In Example 9.3, the 3×3 matrix $\mathbf{A}$ had only two distinct eigenvalues, and only two linearly independent eigenvectors. However, the matrix of the next example has three linearly independent eigenvectors even though it has only two distinct eigenvalues. When eigenvalues are repeated, a matrix may or may not have n linearly independent eigenvectors.

EXAMPLE 9.5

Let

$$\mathbf{A} = \begin{pmatrix} 5 & -4 & 4 \\ 12 & -11 & 12 \\ 4 & -4 & 5 \end{pmatrix}.$$

The eigenvalues of $\mathbf{A}$ are $-3, 1, 1$, with 1 a repeated root of the characteristic polynomial. Corresponding to -3, we find an eigenvector

$$\begin{pmatrix} 1 \\ 3 \\ 1 \end{pmatrix}.$$

Now look for an eigenvector corresponding to 1. We must solve the system

$$((1)\mathbf{I}_2 - \mathbf{A})\mathbf{X} = \begin{pmatrix} -4 & 4 & -4 \\ -12 & 12 & -12 \\ -4 & 4 & -4 \end{pmatrix} \begin{pmatrix} x_1 \\ x_2 \\ x_3 \end{pmatrix} = \begin{pmatrix} 0 \\ 0 \\ 0 \end{pmatrix}.$$

This system has the general solution

$$\alpha \begin{pmatrix} 1 \\ 0 \\ -1 \end{pmatrix} + \beta \begin{pmatrix} 0 \\ 1 \\ 1 \end{pmatrix},$$

in which α and β are any numbers. With $\alpha = 1$ and $\beta = 0$, and then with $\alpha = 0$ and $\beta = 1$, we obtain two linearly independent eigenvectors associated with eigenvalue 1:

$$\begin{pmatrix} 1 \\ 0 \\ -1 \end{pmatrix} \text{ and } \begin{pmatrix} 0 \\ 1 \\ 1 \end{pmatrix}.$$

For this matrix $\mathbf{A}$, we can produce three linearly independent eigenvectors, even though the eigenvalues are not distinct. ♦

Eigenvalues and eigenvectors of special classes of matrices may exhibit special properties. Symmetric matrices form one such class. $\mathbf{A} = [a_{ij}]$ is *symmetric* if $a_{ij} = a_{ji}$ whenever $i \neq j$. This means that $\mathbf{A} = \mathbf{A}^t$, hence that each off-diagonal element is equal to its reflection across this main diagonal. For example,

$$\begin{pmatrix} -7 & -2-i & 1 & 14 \\ -2-i & 2 & -9 & 47i \\ 1 & -9 & -4 & \pi \\ 14 & 47i & \pi & 22 \end{pmatrix}$$

is symmetric.

It is a significant property of symmetric matrices that those with real elements have all real eigenvalues.

THEOREM 9.3 Eigenvalues of Real Symmetric Matrices

The eigenvalues of a real symmetric matrix are real. ♦

Proof By Lemma 9.1 (equation (9.2)), for any eigenvalue λ of $\mathbf{A}$, with eigenvector $\mathbf{E} = (e_1, \cdots, e_n)$,

$$\lambda = \frac{\overline{\mathbf{E}}^t \mathbf{A}\mathbf{E}}{\overline{\mathbf{E}}^t \mathbf{E}}.$$

As noted previously, the denominator is

$$\overline{\mathbf{E}}^t \mathbf{E} = \sum_{j=1}^{n} |e_j|^2$$

and this is real. All we have to do is show that the numerator real, which we will do by showing that $\overline{\mathbf{E}}^t \mathbf{A}\mathbf{E}$ equals its complex conjugate. First, because elements of $\mathbf{A}$ are real, each equals its own conjugate, so $\overline{\mathbf{A}} = \mathbf{A}$. Further, because $\mathbf{A}$ is symmetric, $\mathbf{A}^t = \mathbf{A}$. Therefore

$$\overline{\overline{\mathbf{E}}^t \mathbf{A}\mathbf{E}} = \overline{\overline{\mathbf{E}}^t}\, \overline{\mathbf{A}\mathbf{E}} = \overline{\overline{\mathbf{E}}^t}\, \mathbf{A}\overline{\mathbf{E}} = \mathbf{E}^t \mathbf{A}\overline{\mathbf{E}}.$$

But the last quantity is a 1×1 matrix, which equals its own transpose. Thus, continuing the last equation,

$$\mathbf{E}^t \mathbf{A}\mathbf{E} = (\mathbf{E}^t \mathbf{A}\overline{\mathbf{E}})^t = (\overline{\mathbf{E}^t})\mathbf{A}(\mathbf{E}^t)^t = \overline{\mathbf{E}}^t \mathbf{A}\mathbf{E}.$$

The last two equations together show that $\overline{\mathbf{E}}^t \mathbf{A}\mathbf{E}$ is its own conjugate, hence is real, proving the theorem. ♦

If the eigenvalues of a real matrix are all real, then associated eigenvectors will have real elements as well. In the case that $\mathbf{A}$ is also symmetric, we claim that eigenvectors associated with distinct eigenvalues must be orthogonal.

THEOREM 9.4 Orthogonality of Eigenvectors

Let $\mathbf{A}$ be a real symmetric matrix. Then eigenvectors associated with distinct eigenvalues are orthogonal.

Proof We can derive this result by a useful interplay between matrix and vector notation. Let λ and μ be distinct eigenvalues of $\mathbf{A}$, with eigenvectors, respectively,

$$\mathbf{E} = \begin{pmatrix} e_1 \\ e_2 \\ \vdots \\ e_n \end{pmatrix} \text{ and } \mathbf{G} = \begin{pmatrix} g_1 \\ g_2 \\ \vdots \\ g_n \end{pmatrix}.$$

We have seen that

$$\mathbf{E} \cdot \mathbf{G} = e_1 g_1 + e_2 g_2 + \cdots + e_n g_n = \mathbf{E}^t \mathbf{G}.$$

Now use the facts that $\mathbf{A}\mathbf{E} = \lambda\mathbf{E}$, $\mathbf{A}\mathbf{G} = \mu\mathbf{G}$, and $\mathbf{A} = \mathbf{A}^t$ to write

$$\begin{aligned} \lambda\mathbf{E}^t\mathbf{G} &= (\mathbf{A}\mathbf{E})^t\mathbf{G} = (\mathbf{E}^t\mathbf{A}^t)\mathbf{G} \\ &= (\mathbf{E}^t\mathbf{A})\mathbf{G} = \mathbf{E}^t(\mathbf{A}\mathbf{G}) = \mathbf{E}^t(\mu\mathbf{G}) = \mu\mathbf{E}^t\mathbf{G}. \end{aligned}$$

But then

$$(\lambda - \mu)\mathbf{E}'\mathbf{G} = 0.$$

Since $\lambda \neq \mu$, then $\mathbf{E}'\mathbf{G} = \mathbf{E} \cdot \mathbf{G} = 0$. ♦

EXAMPLE 9.6

$$\mathbf{A} = \begin{pmatrix} 3 & 0 & -2 \\ 0 & 2 & 0 \\ -2 & 0 & 0 \end{pmatrix}$$

is a 3×3 symmetric matrix. The eigenvalues are 2, -1, and 4, with associated eigenvectors

$$\begin{pmatrix} 0 \\ 1 \\ 0 \end{pmatrix}, \begin{pmatrix} 1 \\ 0 \\ 2 \end{pmatrix}, \text{ and } \begin{pmatrix} 2 \\ 0 \\ -1 \end{pmatrix}.$$

These eigenvectors are mutually orthogonal. ♦

Finding eigenvalues of a matrix may be difficult because finding the roots of a polynomial can be difficult. In MAPLE, the command

```
eigenvals(A);
```

will list the eigenvalues of $\mathbf{A}$, if n is not too large. The command

```
eigenvects(A);
```

will list each eigenvalue, its multiplicity, and, for each eigenvalue, as many linearly independent eigenvectors as are associated with that eigenvalue. We can also find the characteristic polynomial of $\mathbf{A}$ by

```
charpoly(A,t);
```

in which the variable of the polynomial is called t, but could be given any designation.

There is a method due to Gershgorin that enables us to place the eigenvalues inside disks in the complex plane. This is sometimes useful to get some idea of how the eigenvalues of a matrix are distributed.

THEOREM 9.5 Gershgorin

Let $\mathbf{A}$ be an $n \times n$ matrix of numbers. For $k = 1, 2, \cdots, n$ let

$$r_k = \sum_{j=1, j\neq k}^{n} |a_{kj}|.$$

Let C_k be the circle of radius r_k centered at (α_k, β_k), where $a_{kk} = \alpha_k + \beta_k i$. Then each eigenvalue of $\mathbf{A}$, when plotted as a point in the complex plane, lies on or within one of the circles $C_1, \cdots, C_n$. ♦

C_k is the circle centered at the kth diagonal element a_{kk} of $\mathbf{A}$, having radius equal to the sum of the magnitudes of the elements across row k, excluding the diagonal element occurring in that row.

EXAMPLE 9.7

Let

$$\mathbf{A} = \begin{pmatrix} 12i & 1 & 3 \\ 2 & -6 & 2+i \\ 3 & 1 & 5 \end{pmatrix}.$$

The characteristic polynomial of $\mathbf{A}$ is

$$p_{\mathbf{A}}(\lambda) = \lambda^3 + (1 - 12i)\lambda^2 - (43 + 13i)\lambda - 68 + 381i.$$

The Gershgorin circles have centers and radii:

$$C_1 : (0, 12), r_1 = 1 + 3 = 4,$$

$$C_2 : (-6, 0), r_2 = 2 + \sqrt{5}$$

$$C_3 : (5, 0), r_3 = 3 + 1 = 4.$$

Figure 9.1 shows these Gershgorin circles. The eigenvalues are in the disks determined by these circles. ♦

Gershgorin's theorem is not a way of approximating eigenvalues, since some of the disks may have large radii. However, sometimes important information that is revealed by these disks can be useful. For example, in studies of the stability of fluid flow it is important to know whether eigenvalues occur in the right half-plane.

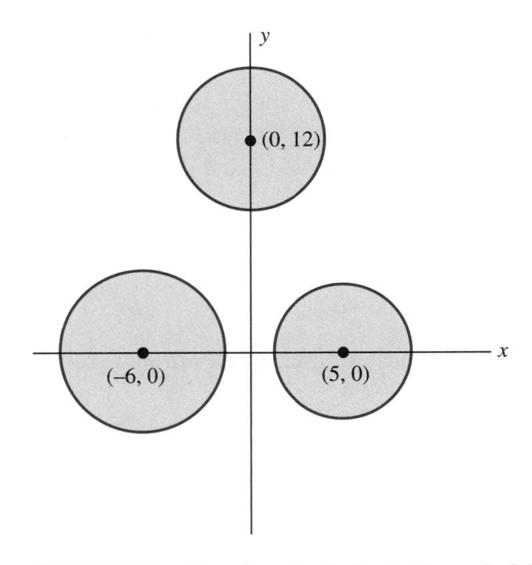

FIGURE 9.1 *Gerschgorin circles in Example 9.7.*

SECTION 9.1 PROBLEMS

In each of Problems 1 through 16, find the eigenvalues of the matrix. For each eigenvalue, find an eigenvector. Sketch the Gershgorin circles for the matrix and locate the eigenvalues as points in the plane.

1. $\begin{pmatrix} -4 & 1 & 0 & 1 \\ 0 & 1 & 0 & 0 \\ 0 & 0 & 2 & 0 \\ 1 & 0 & 0 & 3 \end{pmatrix}$

2. $\begin{pmatrix} 6 & -2 \\ -3 & 4 \end{pmatrix}$

3. $\begin{pmatrix} -14 & 1 & 0 \\ 0 & 2 & 0 \\ 1 & 0 & 2 \end{pmatrix}$

4. $\begin{pmatrix} 0 & 1 \\ 0 & 0 \end{pmatrix}$

5. $\begin{pmatrix} 2 & 0 & 0 \\ 1 & 0 & 2 \\ 0 & 0 & 3 \end{pmatrix}$

6. $\begin{pmatrix} -2 & 1 & 0 \\ 1 & 3 & 0 \\ 0 & 0 & -1 \end{pmatrix}$

7. $\begin{pmatrix} -3 & 1 & 1 \\ 0 & 0 & 0 \\ 0 & 1 & 0 \end{pmatrix}$

8. $\begin{pmatrix} -2 & 0 \\ 1 & 4 \end{pmatrix}$

9. $\begin{pmatrix} 1 & -6 \\ 2 & 2 \end{pmatrix}$

10. $\begin{pmatrix} 5 & 1 & 0 & 9 \\ 0 & 1 & 0 & 9 \\ 0 & 0 & 0 & 9 \\ 0 & 0 & 0 & 0 \end{pmatrix}$

11. $\begin{pmatrix} 1 & -2 & 0 \\ 0 & 0 & 0 \\ -5 & 0 & 7 \end{pmatrix}$

12. $\begin{pmatrix} -2 & 1 & 0 & 0 \\ 1 & 0 & 0 & 1 \\ 0 & 0 & 0 & 0 \\ 0 & 0 & 0 & 0 \end{pmatrix}$

13. $\begin{pmatrix} 1 & 3 \\ 2 & 1 \end{pmatrix}$

14. $\begin{pmatrix} 3 & 0 & 0 \\ 1 & -2 & -8 \\ 0 & -5 & 1 \end{pmatrix}$

15. $\begin{pmatrix} -5 & 0 \\ 1 & 2 \end{pmatrix}$

16. $\begin{pmatrix} 0 & 0 & -1 \\ 0 & 0 & 1 \\ 2 & 0 & 0 \end{pmatrix}$

In each of Problems 17 through 22, find the eigenvalues and associated eigenvectors of the matrix. Verify that eigenvectors associated with distinct eigenvalues are orthogonal.

17. $\begin{pmatrix} 6 & 1 \\ 1 & 4 \end{pmatrix}$

18. $\begin{pmatrix} -13 & 1 \\ 1 & 4 \end{pmatrix}$

19. $\begin{pmatrix} 0 & 1 & 0 \\ 1 & -2 & 0 \\ 0 & 0 & 3 \end{pmatrix}$

20. $\begin{pmatrix} 0 & 1 & 1 \\ 1 & 2 & 0 \\ 1 & 0 & 2 \end{pmatrix}$

21. $\begin{pmatrix} 4 & -2 \\ -2 & 1 \end{pmatrix}$

22. $\begin{pmatrix} -3 & 5 \\ 5 & 4 \end{pmatrix}$

23. Suppose λ is an eigenvalue of $\mathbf{A}$ with eigenvector $\mathbf{E}$. Let k be a positive integer. Show that λ^k is an eigenvalue of $\mathbf{A}^k$ with eigenvector $\mathbf{E}$.

24. Let $\mathbf{A}$ be an $n \times n$ matrix of numbers. Show that the constant term in the characteristic polynomial of $\mathbf{A}$ is $(-1)^n|\mathbf{A}|$. Use this to show that any singular matrix must have 0 as an eigenvalue.

9.2 Diagonalization

Recall that the elements a_{ii} of a matrix make up its *main diagonal*. All other matrix elements are called *off-diagonal elements*.

A square matrix is called a *diagonal matrix* if all the off-diagonal elements are zero. A diagonal matrix has the appearance

$$\mathbf{D} = \begin{pmatrix} d_1 & 0 & 0 & \cdots & 0 & 0 \\ 0 & d_2 & 0 & \cdots & 0 & 0 \\ 0 & 0 & d_3 & \cdots & 0 & 0 \\ \vdots & \vdots & \vdots & \vdots & \vdots & \vdots \\ 0 & 0 & 0 & \cdots & 0 & d_n \end{pmatrix}.$$

Diagonal matrices have many pleasant properties. Let **A** and **B** be $n \times n$ diagonal matrices with diagonal elements, respectively, a_{ii} and b_{ii}.

1. $\mathbf{A}+\mathbf{B}$ is diagonal with diagonal elements $a_{ii}+b_{ii}$.
2. **AB** is diagonal with diagonal elements $a_{ii}b_{ii}$.
3. $$|\mathbf{A}| = a_{11}a_{22}\cdots a_{nn},$$
   the product of the diagonal elements.
4. From (3), **A** is nonsingular exactly when each diagonal element is nonzero (so **A** has nonzero determinant). In this event, $\mathbf{A}^{-1}$ is the diagonal matrix having diagonal elements $1/a_{ii}$.
5. The eigenvalues of **A** are its diagonal elements.
6. $$\begin{pmatrix} 0 \\ 0 \\ \vdots \\ 1 \\ 0 \\ \vdots \\ 0 \end{pmatrix}$$
   with all zero elements except for 1 in the $i, 1$ place, is an eigenvector corresponding to the eigenvalue a_{ii}.

Most matrices are not diagonal. However, sometimes it is possible to transform a matrix to a diagonal one. This will enable us to transform some problems to simpler ones.

An $n \times n$ matrix **A** is *diagonalizable* if there is an $n \times n$ matrix **P** such that $\mathbf{P}^{-1}\mathbf{AP}$ is a diagonal matrix. In this case we say that **P** *diagonalizes* **A**.

We will see that not every matrix is diagonalizable. The following result not only tells us exactly when **A** is diagonalizable, but also how to choose **P** to diagonalize **A**, and what $\mathbf{P}^{-1}\mathbf{AP}$ must look like.

THEOREM 9.6 Diagonalization of a Matrix

Let $\mathbf{A}$ be $n \times n$. Then $\mathbf{A}$ is diagonalizable if and only if $\mathbf{A}$ has n linearly independent eigenvectors. Furthermore, if $\mathbf{P}$ is the $n \times n$ matrix having these eigenvectors as columns, then $\mathbf{P}^{-1}\mathbf{AP}$ is the $n \times n$ diagonal matrix having the eigenvalues of $\mathbf{A}$ down its main diagonal, in the order in which the eigenvectors were chosen as columns of $\mathbf{P}$.

In addition, if $\mathbf{Q}$ is any matrix that diagonalizes $\mathbf{A}$, then necessarily the diagonal matrix $\mathbf{Q}^{-1}\mathbf{AQ}$ has the eigenvalues of $\mathbf{A}$ along its main diagonal, and the columns of $\mathbf{Q}$ must be eigenvectors of $\mathbf{A}$, in the order in which the eigenvalues appear on the main diagonal of $\mathbf{Q}^{-1}\mathbf{AQ}$. ♦

We will prove the theorem after looking at three examples.

EXAMPLE 9.8

Let

$$\mathbf{A} = \begin{pmatrix} -1 & 4 \\ 0 & 3 \end{pmatrix}.$$

$\mathbf{A}$ has eigenvalues $-1, 3$ and corresponding linearly independent eigenvectors

$$\begin{pmatrix} 1 \\ 0 \end{pmatrix} \text{ and } \begin{pmatrix} 1 \\ 1 \end{pmatrix}.$$

Form

$$\mathbf{P} = \begin{pmatrix} 1 & 1 \\ 0 & 1 \end{pmatrix}.$$

Determine

$$\mathbf{P}^{-1} = \begin{pmatrix} 1 & -1 \\ 0 & 1 \end{pmatrix}.$$

A simple computation shows that

$$\mathbf{P}^{-1}\mathbf{AP} = \begin{pmatrix} -1 & 0 \\ 0 & 3 \end{pmatrix},$$

a diagonal matrix with the eigenvalues of $\mathbf{A}$ on the main diagonal, in the order in which the eigenvectors were used to form the columns of

If we reverse the order of these eigenvectors as columns and define

$$\mathbf{Q} = \begin{pmatrix} 1 & 1 \\ 1 & 0 \end{pmatrix},$$

then

$$\mathbf{Q}^{-1}\mathbf{AQ} = \begin{pmatrix} 3 & 0 \\ 0 & -1 \end{pmatrix}$$

with the eigenvalues along the main diagonal, but now in the order reflecting the order of the eigenvectors used in forming the columns of $\mathbf{Q}$. ♦

EXAMPLE 9.9

Let

$$\mathbf{A} = \begin{pmatrix} -1 & 1 & 3 \\ 2 & 1 & 4 \\ 1 & 0 & -2 \end{pmatrix}.$$

The eigenvalues are -1, $(-1+\sqrt{29})/2$ and $(-1-\sqrt{29})/2$, and corresponding eigenvectors are

$$\begin{pmatrix} 1 \\ -3 \\ 1 \end{pmatrix}, \begin{pmatrix} 3+\sqrt{29} \\ 10+2\sqrt{29} \\ 2 \end{pmatrix}, \begin{pmatrix} 3-\sqrt{29} \\ 10-2\sqrt{29} \\ 2 \end{pmatrix}.$$

These are linearly independent because the eigenvalues are distinct. Use these eigenvectors as columns of $\mathbf{P}$ to form

$$\mathbf{P} = \begin{pmatrix} 1 & 3+\sqrt{29} & 3-\sqrt{29} \\ -3 & 10+2\sqrt{29} & 10-2\sqrt{29} \\ 1 & 2 & 2 \end{pmatrix}.$$

We find that

$$\mathbf{P}^{-1} = \frac{\sqrt{29}}{812} \begin{pmatrix} 232/\sqrt{29} & -116/\sqrt{29} & 232/\sqrt{29} \\ 16-2\sqrt{29} & -1+\sqrt{29} & -19+5\sqrt{29} \\ -16-2\sqrt{29} & 1+\sqrt{29} & 19+5\sqrt{29} \end{pmatrix}$$

and

$$\mathbf{P}^{-1}\mathbf{A}\mathbf{P} = \begin{pmatrix} -1 & 0 & 0 \\ 0 & (-1+\sqrt{29})/2 & 0 \\ 0 & 0 & (-1-\sqrt{29})/2 \end{pmatrix},$$

with the eigenvalues down the main diagonal in the order of the eigenvalues listed for columns of $\mathbf{P}$.

In this example, $\mathbf{P}^{-1}$ is an unpleasant matrix. One of the values of Theorem 9.6 is that it tells us what $\mathbf{P}^{-1}\mathbf{A}\mathbf{P}$ looks like, without actually having to determine $\mathbf{P}^{-1}$ and carry out this product. ♦

Although n distinct eigenvalues guarantee that $\mathbf{A}$ is diagonalizable, an $n \times n$ matrix with fewer than n distinct eigenvalues *may* still be diagonalizable. This will occur if we are able to find n linearly independent eigenvectors.

EXAMPLE 9.10

Let

$$\mathbf{A} = \begin{pmatrix} 5 & -4 & 4 \\ 12 & -11 & 12 \\ 4 & -4 & 5 \end{pmatrix}$$

as in Example 9.5. We found the eigenvalues $-3, 1, 1$, with a repeated eigenvalue. Nevertheless, we were able to find three linearly independent eigenvectors. Use these as columns to form

$$\mathbf{P} = \begin{pmatrix} 1 & 1 & 0 \\ 3 & 0 & 1 \\ 1 & -1 & 1 \end{pmatrix}.$$

Then $\mathbf{P}$ diagonalizes $\mathbf{A}$:

$$\mathbf{P}^{-1}\mathbf{AP} = \begin{pmatrix} -3 & 0 & 0 \\ 0 & 1 & 0 \\ 0 & 0 & 1 \end{pmatrix}.$$

Again, we know this from Theorem 9.6, without explicitly computing the product $\mathbf{P}^{-1}\mathbf{AP}$. ♦

If $\mathbf{A}$ has fewer than n linearly independent eigenvectors, then $\mathbf{A}$ is not diagonalizable.

We will now prove Theorem 9.6.

Proof Let the eigenvalues of $\mathbf{A}$ be $\lambda_1, \lambda_2, \cdots, \lambda_n$ (not necessarily distinct). Suppose first that these eigenvalues have corresponding linearly independent eigenvectors $\mathbf{V}_1, \mathbf{V}_2, \cdots, \mathbf{V}_n$. These form the columns of $\mathbf{P}$, which we indicate by writing

$$\mathbf{P} = \begin{pmatrix} | & | & \cdots & | \\ \mathbf{V}_1 & \mathbf{V}_2 & \cdots & \mathbf{V}_n \\ | & | & \cdots & | \end{pmatrix}.$$

$\mathbf{P}$ is nonsingular because its columns are linearly independent.

Let $\mathbf{D}$ be the $n \times n$ diagonal matrix having the eigenvalues of $\mathbf{A}$, in the given order, down the main diagonal. We want to prove that

$$\mathbf{P}^{-1}\mathbf{AP} = \mathbf{D}.$$

We will prove this by showing by direct computation that

$$\mathbf{AP} = \mathbf{PD}.$$

First, recall that the product $\mathbf{AP}$ has as columns the product of $\mathbf{A}$ with the columns of $\mathbf{P}$. Thus

$$\begin{aligned} \text{column } j \text{ of } \mathbf{AP} &= \mathbf{A}(\text{column } j \text{ of } \mathbf{P}) \\ &= \mathbf{A}(\mathbf{V}_j) = \lambda_j \mathbf{V}_j. \end{aligned}$$

Now compute $\mathbf{PD}$. As a convenience in understanding the computation, write

$$\mathbf{V}_j = \begin{pmatrix} v_{1j} \\ v_{2j} \\ \cdots \\ v_{nj} \end{pmatrix}.$$

Then

$$\mathbf{PD} = \begin{pmatrix} v_{11} & v_{12} & \cdots & v_{1n} \\ v_{21} & v_{22} & \cdots & v_{2n} \\ \vdots & \vdots & \vdots & \vdots \\ v_{n1} & v_{n2} & \cdots & v_{nn} \end{pmatrix} \begin{pmatrix} \lambda_1 & 0 & \cdots & 0 \\ 0 & \lambda_2 & \cdots & 0 \\ \vdots & \vdots & \vdots & \vdots \\ 0 & 0 & \cdots & \lambda_n \end{pmatrix}$$

$$= \begin{pmatrix} \lambda_1 v_{11} & \lambda_2 v_{12} & \cdots & \lambda_n v_{1n} \\ \lambda_1 v_{21} & \lambda_2 v_{22} & \cdots & \lambda_n v_{2n} \\ \vdots & \vdots & \vdots & \vdots \\ \lambda_1 v_{n1} & \lambda_2 v_{2n} & \cdots & \lambda_n v_{nn} \end{pmatrix}$$

$$= \begin{pmatrix} | & | & \cdots & | \\ \lambda_1 \mathbf{V}_1 & \lambda_2 \mathbf{V}_2 & \cdots & \lambda_n \mathbf{V}_n \\ | & | & \cdots & | \end{pmatrix} = \mathbf{AP},$$

since column j of this matrix is $\lambda_j \mathbf{V}_j$.

Thus far we have proved that, if **A** has n linearly independent eigenvectors, then **A** is diagonalizable and $\mathbf{P}^{-1}\mathbf{AP}$ is the diagonal matrix having the eigenvalues down the main diagonal, in the order in which the eigenvectors are seen as columns of **P**.

To prove the converse, now suppose that **A** is diagonalizable. We want to show that **A** has n linearly independent eigenvectors (regardless of whether the eigenvalues are distinct). Further, we want to show that, if $\mathbf{Q}^{-1}\mathbf{AQ}$ is a diagonal matrix, then the diagonal elements of this matrix are the eigenvalues of **A**, and the columns of **Q** are corresponding eigenvectors. Thus suppose that

$$\mathbf{Q}^{-1}\mathbf{AQ} = \begin{pmatrix} d_1 & 0 & \cdots & 0 \\ 0 & d_1 & \cdots & 0 \\ \vdots & \vdots & \vdots & \vdots \\ 0 & 0 & \cdots & d_n \end{pmatrix} = \mathbf{D}.$$

Let $\mathbf{V}_j$ be column j of **Q**. These columns are linearly independent because **Q** is nonsingular. We will show that d_j is an eigenvalue of **A** with eigenvector $\mathbf{V}_j$.

From $\mathbf{Q}^{-1}\mathbf{AQ} = \mathbf{D}$, we have $\mathbf{AQ} = \mathbf{QD}$. Compute both sides of this equation separately. First, since the columns of **Q** are the $\mathbf{V}_j$'s, then

$$\mathbf{QD} = \begin{pmatrix} | & | & \cdots & | \\ \mathbf{V}_1 & \mathbf{V}_2 & \cdots & \mathbf{V}_n \\ | & | & \cdots & | \end{pmatrix} \begin{pmatrix} d_1 & 0 & \cdots & 0 \\ 0 & d_1 & \cdots & 0 \\ \vdots & \vdots & \vdots & \vdots \\ 0 & 0 & \cdots & d_n \end{pmatrix}$$

$$= \begin{pmatrix} | & | & \cdots & | \\ d_1\mathbf{V}_1 & d_2\mathbf{V}_2 & \cdots & d_n\mathbf{V}_n \\ | & | & \cdots & | \end{pmatrix},$$

which is a matrix having $d_j\mathbf{V}_j$ as column j. Now compute

$$\mathbf{AQ} = \mathbf{A}\begin{pmatrix} | & | & \cdots & | \\ \mathbf{V}_1 & \mathbf{V}_2 & \cdots & \mathbf{V}_n \\ | & | & \cdots & | \end{pmatrix} = \begin{pmatrix} | & | & \cdots & | \\ \mathbf{AV}_1 & \mathbf{AV}_2 & \cdots & \mathbf{AV}_n \\ | & | & \cdots & | \end{pmatrix},$$

which is a matrix having $\mathbf{AV}_j$ as column j. Since these matrices are equal, then

$$\mathbf{AV}_j = d_j\mathbf{V}_j$$

and this makes d_j an eigenvalue of **A** with eigenvector $\mathbf{V}_j$. ♦

Not every matrix is diagonalizable. We know from the theorem that a $n \times n$ matrix with fewer than n linearly independent eigenvectors is not diagonalizable.

EXAMPLE 9.11

Let

$$\mathbf{B} = \begin{pmatrix} 1 & -1 \\ 0 & 1 \end{pmatrix}.$$

B has eigenvalues 1, 1, and all eigenvectors are constant multiples of

$$\begin{pmatrix} 1 \\ 0 \end{pmatrix}.$$

Therefore **B** has as eigenvectors only nonzero multiples of one vector, and does not have two linearly independent eigenvectors. By the theorem, **B** is not diagonalizable.

Notice that, if **P** diagonalized **A**, then **P** would have to have eigenvectors of **B** as columns. Then **P** would have to have the form

$$\mathbf{P} = \begin{pmatrix} \alpha & \beta \\ 0 & 0 \end{pmatrix}$$

for some nonzero α and β. But this matrix is singular, with no inverse, because $|\mathbf{P}| = 0$. ♦

The key to diagonalizing **A** is the existence of n linearly independent eigenvectors. By Theorem 9.2, one circumstance in which this always happens is that **A** has n distinct eigenvalues.

COROLLARY 9.1

An $n \times n$ matrix with n distinct eigenvalues must be diagonalizable. ♦

EXAMPLE 9.12

Let

$$\mathbf{A} = \begin{pmatrix} -2 & 0 & 0 & 5 \\ 1 & 3 & 0 & 0 \\ 0 & 4 & 4 & 0 \\ 2 & 0 & 0 & -3 \end{pmatrix}.$$

A has eigenvalues $3, 4, (-5+\sqrt{41})/2$ and $(-5-\sqrt{41})/2$. Because these are distinct, **A** has 4 linearly independent eigenvectors and therefore is diagonalizable. There is a matrix **P**such that

$$\mathbf{P}^{-1}\mathbf{A}\mathbf{P} = \begin{pmatrix} 3 & 0 & 0 & 0 \\ 0 & 4 & 0 & 0 \\ 0 & 0 & (-5+\sqrt{41})/2 & 0 \\ 0 & 0 & 0 & (-5-\sqrt{41})/2 \end{pmatrix}.$$

We do not have to actually write down **P** (this would require finding eigenvectors) or compute $\mathbf{P}^{-1}$ to draw this conclusion. ♦

SECTION 9.2 *PROBLEMS*

In each of Problems 1 through 10, produce a matrix **P** that diagonalizes the given matrix, or show that the matrix is not diagonalizable. Determine $\mathbf{P}^{-1}\mathbf{A}\mathbf{P}$. *Hint*: Keep in mind that it is not necessary to compute **P** to know this product matrix.

1. $\begin{pmatrix} 5 & 0 & 0 \\ 1 & 0 & 3 \\ 0 & 0 & -2 \end{pmatrix}$

2. $\begin{pmatrix} 2 & 0 & 0 \\ 0 & 2 & 1 \\ 0 & -1 & 2 \end{pmatrix}$

3. $\begin{pmatrix} -2 & 0 & 1 \\ 1 & 1 & 0 \\ 0 & 0 & -2 \end{pmatrix}$

4. $\begin{pmatrix} 0 & 0 & 0 \\ 1 & 0 & 2 \\ 0 & 1 & 3 \end{pmatrix}$

5. $\begin{pmatrix} 1 & 0 & 0 & 0 \\ 0 & 4 & 1 & 0 \\ 0 & 0 & -3 & 1 \\ 0 & 0 & 1 & -2 \end{pmatrix}$

6. $\begin{pmatrix} -5 & 3 \\ 0 & 9 \end{pmatrix}$

7. $\begin{pmatrix} 0 & -1 \\ 4 & 3 \end{pmatrix}$

8. $\begin{pmatrix} -2 & 0 & 0 & 0 \\ -4 & -2 & 0 & 0 \\ 0 & 0 & -2 & 0 \\ 0 & 0 & 0 & -2 \end{pmatrix}$

9. $\begin{pmatrix} 1 & 0 \\ -4 & 1 \end{pmatrix}$

10. $\begin{pmatrix} 5 & 3 \\ 1 & 3 \end{pmatrix}$

11. Let $\mathbf{A}$ have eigenvalues $\lambda_1, \cdots, \lambda_n$, and suppose that $\mathbf{P}$ diagonalizes $\mathbf{A}$. Show that, for any positive integer k,

$$\mathbf{A}^k = \mathbf{P}\begin{pmatrix} \lambda_1^k & 0 & \cdots & 0 \\ 0 & \lambda_2^k & \cdots & 0 \\ \vdots & \vdots & \vdots & \vdots \\ 0 & 0 & \cdots & \lambda_n{}^k \end{pmatrix}\mathbf{P}^{-1}.$$

In each of Problems 12 through 15, use the idea of Problem 11 to compute the indicated power of the matrix.

12. $\mathbf{A} = \begin{pmatrix} -2 & 3 \\ 3 & -4 \end{pmatrix}$; $\mathbf{A}^{31}$

13. $\mathbf{A} = \begin{pmatrix} 0 & 2 \\ 1 & 0 \end{pmatrix}$; $\mathbf{A}^{43}$

14. $\mathbf{A} = \begin{pmatrix} -3 & -3 \\ -2 & 4 \end{pmatrix}$; $\mathbf{A}^{16}$

15. $\mathbf{A} = \begin{pmatrix} -1 & 0 \\ 1 & -5 \end{pmatrix}$; $\mathbf{A}^{18}$

16. Suppose $\mathbf{A}^2$ is diagonalizable. Prove that $\mathbf{A}$ is diagonalizable.

9.3 Some Special Types of Matrices

In this section, we will discuss several types of matrices having special properties.

9.3.1 Orthogonal Matrices

An $n \times n$ matrix is *orthogonal* if its transpose is its inverse:

$$\mathbf{A}^{-1} = \mathbf{A}^t.$$

In this event,

$$\mathbf{A}\mathbf{A}^t = \mathbf{A}^t\mathbf{A} = \mathbf{I}_n.$$

For example, it is routine to check that

$$\mathbf{A} = \begin{pmatrix} 0 & 1/\sqrt{5} & 2/\sqrt{5} \\ 1 & 0 & 0 \\ 0 & 2\sqrt{5} & -1\sqrt{5} \end{pmatrix}$$

is orthogonal. Just multiply this matrix by its transpose to obtain $\mathbf{I}_3$.

Because $(\mathbf{A}^t)^t = \mathbf{A}$, a matrix is orthogonal exactly when its transpose is orthogonal. It is also easy to verify that an orthogonal matrix must have determinant 1 or -1.

THEOREM 9.7

If $\mathbf{A}$ is orthogonal, then $|\mathbf{A}| = \pm 1$. ♦

Proof Because a matrix and its transpose have the same determinant,

$$|\mathbf{I}_n| = 1 = |\mathbf{A}\mathbf{A}^{-1}| = |\mathbf{A}\mathbf{A}^t| = |\mathbf{A}||\mathbf{A}^t| = |\mathbf{A}|^2. \ ♦$$

The name "orthogonal matrix" derives from the following property.

THEOREM 9.8

Let $\mathbf{A}$ be an $n \times n$ matrix of real numbers. Then

1. $\mathbf{A}$ is orthogonal if and only the row vectors are mutually orthogonal unit vectors in R^n.
2. $\mathbf{A}$ is orthogonal if and only if the column vectors are mutually orthogonal unit vectors in R^n. ♦

We say that the row vectors of an orthogonal matrix form an *orthonormal set of vectors* in R^n. The column vectors also form an orthonormal set.

Proof The i, j element of $\mathbf{AA}^t$ is the dot product of row i of $\mathbf{A}$ with column j of $\mathbf{A}^t$, and this is the dot product of row i of $\mathbf{A}$ with row j of $\mathbf{A}$.

If $i \neq j$, then this dot product is zero, because the $i, j-$ element of $\mathbf{I}_n$ is zero. And if $i = j$, then this dot product is 1 because the $i, i-$ element of $\mathbf{I}_n$ is 1. This proves that, if $\mathbf{A}$ is an orthogonal matrix, then its rows form an orthonormal set of vectors in R^n.

Conversely, suppose the rows are mutually orthogonal unit vectors in R^n. Then the i, j element of $\mathbf{AA}^t$ is 0 if $i \neq j$ and 1 if $i = j$, so $\mathbf{AA}^t = \mathbf{I}_n$.

By applying this argument to $\mathbf{A}^t$, this transpose is orthogonal if and only if its rows are orthogonal unit vectors, and these rows are the columns of $\mathbf{A}$. ♦

We now know a lot about orthogonal matrices. We will use this information to determine all 2×2 real orthogonal matrices. Suppose

$$\mathbf{Q} = \begin{pmatrix} a & b \\ c & d \end{pmatrix}$$

is orthogonal. What does this tell us about a, b, c and d? Because the row (column) vectors are mutually orthogonal unit vectors,

$$ac + bd = 0$$

$$ab + cd = 0$$

$$a^2 + b^2 = 1$$

$$c^2 + d^2 = 1.$$

Furthermore, $|\mathbf{Q}| = \pm 1$, so

$$ad - bc = 1 \text{ or } ad - bc = -1.$$

By analyzing these equations in all cases, we find that there must be some θ in $[0, 2\pi)$ such that $a = \cos(\theta)$ and $b = \sin(\theta)$, and $\mathbf{Q}$ must have one of the two forms:

$$\begin{pmatrix} \cos(\theta) & \sin(\theta) \\ -\sin(\theta) & \cos(\theta) \end{pmatrix} \text{ or } \begin{pmatrix} \cos(\theta) & \sin(\theta) \\ \sin(\theta) & -\cos(\theta) \end{pmatrix},$$

depending on whether the determinant is 1 or -1. For example, with $\theta = \pi/6$, we obtain the orthogonal 2×2 matrices

$$\begin{pmatrix} \sqrt{3}/2 & 1/2 \\ -1/2 & \sqrt{3}/2 \end{pmatrix} \text{ or } \begin{pmatrix} \sqrt{3}/2 & 1/2 \\ 1/2 & -\sqrt{3}/2 \end{pmatrix}.$$

If we put Theorems 9.4 and 9.8 together, we obtain an interesting conclusion. Suppose $\mathbf{S}$ is a real, symmetric $n \times n$ matrix with n distinct eigenvalues. Then the associated eigenvectors are orthogonal. These may not be unit vectors. However, a scalar multiple of an eigenvector is still an eigenvector. Divide each eigenvector by its length and use these unit eigenvectors as columns of an orthogonal matrix $\mathbf{Q}$ that diagonalizes $\mathbf{S}$. This proves the following.

THEOREM 9.9

An $n \times n$ real symmetric matrix with distinct eigenvalues can be diagonalized by an orthogonal matrix. ♦

EXAMPLE 9.13

Let

$$\mathbf{S} = \begin{pmatrix} 3 & 0 & -2 \\ 0 & 2 & 0 \\ -2 & 0 & 0 \end{pmatrix}.$$

This real, symmetric matrix has eigenvalues $2, -1, 4$, with corresponding eigenvectors

$$\begin{pmatrix} 0 \\ 1 \\ 0 \end{pmatrix}, \begin{pmatrix} 1 \\ 0 \\ 2 \end{pmatrix} \text{ and } \begin{pmatrix} 2 \\ 0 \\ -1 \end{pmatrix}.$$

The matrix having these eigenvectors as columns will diagonalize $\mathbf{S}$, but is not an orthogonal matrix because these eigenvectors do not all have length 1. Normalize the second and third eigenvectors by dividing them by their lengths, and then use these unit eigenvectors as columns of an orthogonal matrix $\mathbf{Q}$:

$$\mathbf{Q} = \begin{pmatrix} 0 & 1/\sqrt{5} & 2/\sqrt{5} \\ 1 & 0 & 0 \\ 0 & 2/\sqrt{5} & -1/\sqrt{5} \end{pmatrix}.$$

This orthogonal matrix also diagonalizes $\mathbf{S}$. ♦

9.3.2 Unitary Matrices

We will use the following fact. If $\mathbf{W}$ is any matrix, then the operations of taking the transpose and the complex conjugate can be performed in either order:

$$\overline{(\mathbf{W}^t)} = (\overline{\mathbf{W}})^t.$$

This is verified by a routine calculation.

It is also straightforward to verify that the operations of taking a matrix inverse, and of taking its complex conjugate, can be performed in either order.

Now let $\mathbf{U}$ be an $n \times n$ matrix with complex elements.

We say that $\mathbf{U}$ is *unitary* if the inverse is the conjugate of the transpose (which is the same as the transpose of the conjugate):

$$\mathbf{U}^{-1} = \overline{\mathbf{U}}^t.$$

This means that

$$(\overline{\mathbf{U}})^t\mathbf{U} = \mathbf{U}(\overline{\mathbf{U}})^t = \mathbf{I}_n.$$

EXAMPLE 9.14

$$\mathbf{U} = \begin{pmatrix} i/\sqrt{2} & 1/\sqrt{2} \\ -i/\sqrt{2} & 1/\sqrt{2} \end{pmatrix}.$$

It is routine to check that $\mathbf{U}$ is unitary. ♦

If $\mathbf{U}$ is a unitary matrix with real elements, then $\overline{\mathbf{U}} = \mathbf{U}$ and the condition of being unitary becomes $\mathbf{U}^{-1} = \mathbf{U}^t$. Therefore a real unitary matrix is orthogonal. In this sense unitary matrices are the extension of orthogonal matrices to allow complex matrix elements.

Since the rows (and columns) of an orthogonal matrix are mutually orthogonal unit vectors, we would expect a complex analogue of this condition for unitary matrices. If $(x_1, \cdots, x_n)$ and $(y_1, \cdots, y_n)$ are vectors in R^n, we can write

$$\mathbf{X} = \begin{pmatrix} x_1 \\ x_2 \\ \vdots \\ x_n \end{pmatrix} \text{ and } \mathbf{Y} = \begin{pmatrix} y_1 \\ y_2 \\ \vdots \\ y_n \end{pmatrix}$$

and obtain the dot product $\mathbf{X} \cdot \mathbf{Y}$ as the matrix product $\mathbf{X}^t\mathbf{Y}$, which is the 1×1 matrix (or number) $x_1y_1 + x_2y_2 + \cdots + x_ny_n$. In particular, the square of the length of $\mathbf{X}$ is

$$\mathbf{X}^t\mathbf{X} = x_1^2 + x_2^2 + \cdots + x_n^2.$$

To generalize this to the complex case, suppose we have complex $n-$ vectors $(z_1, z_2, \cdots, z_n)$ and $(w_1, w_2, \cdots, w_n)$. Let

$$\mathbf{Z} = \begin{pmatrix} z_1 \\ z_2 \\ \vdots \\ z_n \end{pmatrix} \text{ and } \mathbf{W} = \begin{pmatrix} w_1 \\ w_2 \\ \vdots \\ w_n \end{pmatrix}$$

and define the dot product $\mathbf{Z} \cdot \mathbf{W}$ by

$$\mathbf{Z} \cdot \mathbf{W} = \overline{\mathbf{Z}}^t\mathbf{W}.$$

Then

$$\mathbf{Z} \cdot \mathbf{W} = \overline{z_1}w_1 + \overline{z_2}w_2 + \cdots + \overline{z_n}w_n.$$

In this way,

$$\mathbf{Z} \cdot \mathbf{Z} = \overline{z_1}z_1 + \overline{z_2}z_2 + \cdots + \overline{z_n}z_n = \sum_{j=1}^{n} |z_j|^2,$$

a real number, consistent with the interpretation of the dot product of a vector with itself as the square of the length. With this as background, we now define the complex analogue of an orthonormal set of vectors in R^n. We will say that complex $n-$ vectors $\mathbf{F}_1, \cdots, \mathbf{F}_r$ form a *unitary system* if $\mathbf{F}_j \cdot \mathbf{F}_k = 0$ if $j \neq k$, and each $\mathbf{F}_j$ has length 1 (that is, $\mathbf{F}_j \cdot \mathbf{F}_j = 1$).

A unitary system is an orthonormal set of vectors when each of the vectors has real components. With this background, we can state the unitary version of Theorem 9.8.

THEOREM 9.10

A complex matrix $\mathbf{U}$ is unitary if and only its row (column) vectors form a unitary system. ♦

We claim that the eigenvalues of a unitary matrix must have magnitude 1.

THEOREM 9.11

Let λ be an eigenvalue of a unitary matrix $\mathbf{U}$. Then $|\lambda| = 1$. ♦

This means that the eigenvalues of **U** lie on the unit circle about the origin in the complex plane. Since a real orthogonal matrix is also unitary, this also holds for real orthogonal matrices.

Proof Let λ be an eigenvalue of **U** with eigenvector **E**. We know that $\mathbf{UE} = \lambda\mathbf{E}$. Then $\overline{\mathbf{UE}} = \overline{\lambda\mathbf{E}}$. Therefore,

$$(\overline{\mathbf{U}}\overline{\mathbf{E}})^t = \overline{\lambda}(\overline{\mathbf{E}})^t.$$

Then,

$$(\overline{\mathbf{E}})^t(\overline{\mathbf{U}})^t = \overline{\lambda}(\overline{\mathbf{E}})^t.$$

But **U** is unitary, so $\overline{\mathbf{U}}^t = \mathbf{U}^{-1}$. The last equation becomes

$$(\overline{\mathbf{E}})^t\mathbf{U}^{-1} = \overline{\lambda}(\overline{\mathbf{E}})^t.$$

Multiply both sides of this equation on the right by **UE**:

$$(\overline{\mathbf{E}})^t\mathbf{U}^{-1}\mathbf{UE} = \overline{\lambda}(\overline{\mathbf{E}})^t\mathbf{UE} = \overline{\lambda}(\overline{\mathbf{E}})^t\lambda\mathbf{E} = \overline{\lambda}\lambda\overline{\mathbf{E}}^t\mathbf{E}.$$

Now $\overline{\mathbf{E}}^t\mathbf{E}$ is the dot product of an eigenvector with itself, and so is a positive number. Dividing the last equation by $\overline{\mathbf{E}}^t\mathbf{E}$ yields the conclusion that $\overline{\lambda}\lambda = 1$. Then $|\lambda|^2 = 1$, proving the theorem. ♦

9.3.3 Hermitian and Skew-Hermitian Matrices

An $n \times n$ complex matrix **H** is *hermitian* if $\overline{\mathbf{H}} = \mathbf{H}^t$.

That is, a matrix is hermitian if its conjugate equals its transpose. If a hermitian matrix has real elements, then it must be symmetric, because then the matrix equals its conjugate, which equals its transpose.

An $n \times n$ complex matrix **S** is *skew-hermitian* if $\overline{\mathbf{S}} = -\mathbf{S}^t$.

Thus, **S** is skew-hermitian if its conjugate equals the negative of its transpose.

EXAMPLE 9.15

The matrix

$$\mathbf{H} = \begin{pmatrix} 15 & 8i & 6-2i \\ -8i & 0 & -4+i \\ 6+2i & -4-i & -3 \end{pmatrix}$$

is hermitian because

$$\overline{\mathbf{H}} = \begin{pmatrix} 15 & -8i & 6+2i \\ 8i & 0 & -4-i \\ 6-2i & -4+i & -3 \end{pmatrix} = \mathbf{H}^t.$$

The matrix

$$\mathbf{S} = \begin{pmatrix} 0 & 8i & 2i \\ 8i & 0 & 4i \\ 2i & 4i & 0 \end{pmatrix}$$

is skew-hermitian because

$$\overline{\mathbf{S}} = \begin{pmatrix} 0 & -8i & -2i \\ -8i & 0 & -4i \\ -2i & -4i & 0 \end{pmatrix} = -\mathbf{S}^t. \ \blacklozenge$$

We want to derive a result about eigenvalues of hermitian and skew-hermitian matrices. For this we need the following conclusions about the numerator of the general expression for eigenvalues in Lemma 9.1.

LEMMA 9.2

Let

$$\mathbf{Z} = \begin{pmatrix} z_1 \\ z_2 \\ \vdots \\ z_n \end{pmatrix}$$

be a complex $n \times 1$ matrix. Then

1. If $\mathbf{H}$ is $n \times n$ hermitian, then $\overline{\mathbf{Z}}^t\mathbf{HZ}$ is real.
2. If $\mathbf{S}$ is $n \times n$ skew-hermitian, then $\overline{\mathbf{Z}}^t\mathbf{HZ}$ is pure imaginary. ♦

Proof of Lemma 9.3 For condition (1), suppose $\mathbf{H}$ is hermitian, so that $\overline{\mathbf{H}}^t = \mathbf{H}$. Then

$$\overline{(\overline{\mathbf{Z}}^t\mathbf{HZ})} = (\overline{(\overline{\mathbf{Z}})^t})\overline{\mathbf{HZ}} = \mathbf{Z}^t\overline{\mathbf{HZ}}.$$

But $\overline{\mathbf{Z}}^t\mathbf{HZ}$ is a 1×1 matrix and so equals its own transpose. Continuing from the last equation, we have

$$\mathbf{Z}^t\overline{\mathbf{HZ}} = (\mathbf{Z}^t\overline{\mathbf{HZ}})^t = \overline{\mathbf{Z}}^t\overline{\mathbf{H}}^t(\mathbf{Z})^t = \overline{\mathbf{Z}}^t\mathbf{HZ}.$$

This shows that

$$\overline{(\overline{\mathbf{Z}}^t\mathbf{HZ})} = \overline{\mathbf{Z}}^t\mathbf{HZ}.$$

Since $\overline{\mathbf{Z}}^t\mathbf{HZ}$ equals its own conjugate, this quantity is real.

To prove condition (2), suppose $\mathbf{S}$ is skew-hermitian, so $\overline{\mathbf{S}}^t = -\mathbf{S}$. By an argument like that in the proof of condition (1), we find that

$$\overline{(\overline{\mathbf{Z}}^t\mathbf{SZ})} = -\overline{\mathbf{Z}}^t\mathbf{SZ}$$

If we write $\overline{\mathbf{Z}}^t\mathbf{SZ} = a + ib$, then the last equation means that

$$a - ib = -a - ib.$$

But then $a = -a$ so $a = 0$ and $\overline{\mathbf{Z}}^t\mathbf{SZ}$ is pure imaginary. This includes the possibility of a zero eigenvalue. ♦

This lemma absorbs most of the work we need for the following result, giving us information about eigenvalues.

THEOREM 9.12

1. The eigenvalues of a hermitian matrix are real.
2. The eigenvalues of a skew-hermitian are pure imaginary. ♦

Proof By Lemma 9.1, an eigenvalue λ of any $n \times n$ matrix $\mathbf{A}$, with corresponding eigenvector $\mathbf{E}$, satisfies

$$\lambda = \frac{\overline{\mathbf{E}}^t \mathbf{A}\mathbf{E}}{\overline{\mathbf{E}}^t \mathbf{E}}.$$

We know that the denominator of this quotient is a positive number. Now use Lemma 9.2. If $\mathbf{A}$ is hermitian, the numerator is real, so λ is real. If $\mathbf{A}$ is skew-hermitian then the numerator is pure imaginary, so λ is pure imaginary. ♦

Figure 9.2 shows a graphical representation of these conclusions about eigenvalues of matrices. When plotted as points in the complex plane, eigenvalues of a unitary (or orthogonal) matrix lie on the unit circle about the origin, eigenvalues of a hermitian matrix lie on the horizontal (real) axis, and eigenvalues of a skew-hermitian matrix are on the vertical (imaginary) axis.

9.3.4 Quadratic Forms

A *quadratic form* is an expression

$$\sum_{j=1}^{n} \sum_{k=1}^{n} a_{jk} \overline{z_j} z_k$$

in which the a_{jk}'s and the z_j's are complex numbers. If these quantities are all real, we say that we have a *real quadratic form.*

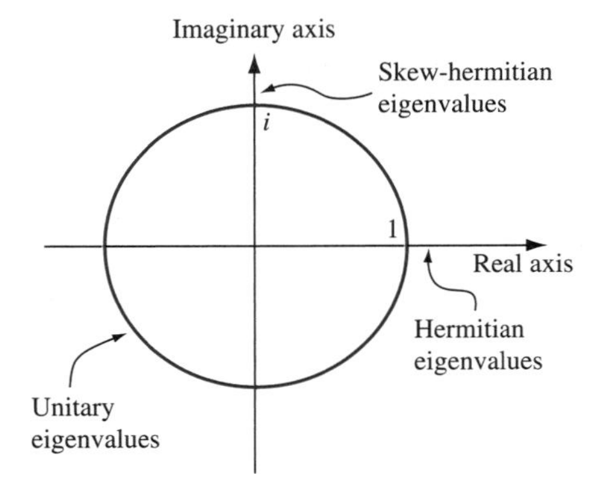

FIGURE 9.2 *Eigenvalues of unitary, hermitian, and skew-hermitian matrices.*

For $n = 2$, the quadratic form is

$$\sum_{j=1}^{2}\sum_{k=1}^{2} a_{jk}\overline{z_j}z_k = a_{11}\overline{z_1}z_1 + a_{12}\overline{z_1}z_2 + a_{21}z_1\overline{z_2} + a_{22}z_2\overline{z_2}.$$

The two middle terms are called *mixed product* terms, involving z_j and z_k with $j \neq k$.

If the quadratic form is real, then all of the numbers involved are real. In this case the conjugates play no role and this quadratic form can be written

$$\sum_{j=1}^{2}\sum_{k=1}^{2} a_{jk}x_jx_k = a_{11}x_1x_1 + a_{12}x_1x_2 + a_{21}x_1x_2 + a_{22}x_2x_2$$

$$= a_1x_1^2 + (a_{12} + a_{21})x_1x_2 + a_{22}x_2^2.$$

As we have seen previously (in the discussion immediately preceding Lemma 9.1), we can let $\mathbf{A} = [a_{jk}]$ and write the complex quadratic form as $\overline{\mathbf{Z}}^t\mathbf{AZ}$, where

$$\mathbf{Z} = \begin{pmatrix} z_1 \\ z_2 \\ \vdots \\ z_n \end{pmatrix}.$$

If all the quantities are real, we usually write this as $\mathbf{X}^t\mathbf{AX}$. In fact, any real quadratic form can be written in this way, with $\mathbf{A}$ a real symmetric matrix. We will illustrate this process.

EXAMPLE 9.16

Consider the real quadratic form

$$\begin{pmatrix} x_1 & x_2 \end{pmatrix}\begin{pmatrix} 1 & 4 \\ 3 & 2 \end{pmatrix}\begin{pmatrix} x_1 \\ x_2 \end{pmatrix} = x_1^2 + 3x_1x_2 + 4x_2x_1 + 2x_2^2$$

$$= x_1^2 + 7x_1x_2 + 2x_2^2.$$

We can write the same quadratic form as

$$\begin{pmatrix} x_1 & x_2 \end{pmatrix}\begin{pmatrix} 1 & 7/2 \\ 7/2 & 2 \end{pmatrix}\begin{pmatrix} x_1 \\ x_2 \end{pmatrix} = x_1^2 + 7x_1x_2 + 2x_2^2$$

in which $\mathbf{A}$ is a symmetric matrix. ♦

This is important in developing a standard change of variables that is used to simplify quadratic forms by eliminating cross product terms.

THEOREM 9.13 Principal Axis Theorem

Let $\mathbf{A}$ be a real symmetric matrix with distinct eigenvalues $\lambda_1, \cdots \lambda_n$. Then there is an orthogonal matrix $\mathbf{Q}$ such that the change of variables $\mathbf{X} = \mathbf{QY}$ transforms the quadratic form $\sum_{j=1}^{n}\sum_{k=1}^{n} a_{ij}x_ix_j$ to

$$\sum_{j=1}^{n} \lambda_j y_j^2.$$

Proof Let **Q** be an orthogonal matrix that diagonalizes **A**. Then

$$\sum_{j=1}^{n}\sum_{k=1}^{n} a_{ij}x_i x_j = \mathbf{X}^t\mathbf{A}\mathbf{X}$$

$$= (\mathbf{Q}\mathbf{Y})^t\mathbf{A}\mathbf{Q}\mathbf{Y} = (\mathbf{Y}^t\mathbf{Q}^t)\mathbf{A}\mathbf{Q}\mathbf{Y}$$

$$= \mathbf{Y}^t(\mathbf{Q}^t\mathbf{A}\mathbf{Q})\mathbf{Y}$$

$$= \mathbf{Y}^t(\mathbf{Q}^{-1}\mathbf{A}\mathbf{Q})\mathbf{Y}$$

$$= \begin{pmatrix} y_1 & y_2 & \cdots & y_n \end{pmatrix} \begin{pmatrix} \lambda_1 & 0 & \cdots & 0 \\ 0 & \lambda_2 & \cdots & 0 \\ \vdots & \vdots & \vdots & \vdots \\ 0 & 0 & \cdots & \lambda_n \end{pmatrix} \begin{pmatrix} y_1 \\ y_2 \\ \vdots \\ y_n \end{pmatrix}$$

$$= \lambda_1 y_1^2 + \lambda_2 y_2^2 + \cdots + \lambda_n y_n^2. \blacklozenge$$

The expression $\sum_{j=1}^{n} \lambda_j y_j^2$ is called the *standard form* of $\mathbf{X}^t\mathbf{A}\mathbf{X}$.

EXAMPLE 9.17

Consider the quadratic form

$$x_1^2 - 7x_1x_2 + x_2^2.$$

This is $\mathbf{X}^t\mathbf{A}\mathbf{X}$, where

$$\begin{pmatrix} 1 & -7/2 \\ -7/2 & 1 \end{pmatrix}.$$

In general, the real quadratic form

$$ax_1^2 + bx_1x_2 + cx_2^2$$

can always be written as $\mathbf{X}^t\mathbf{A}\mathbf{X}$, with **A** the real symmetric matrix

$$\mathbf{A} = \begin{pmatrix} a & b/2 \\ b/2 & c \end{pmatrix}.$$

In this example, the eigenvalues of **A** are $-5/2, 9/2$ with corresponding eigenvectors

$$\begin{pmatrix} 1 \\ 1 \end{pmatrix} \text{ and } \begin{pmatrix} -1 \\ 1 \end{pmatrix}.$$

Divide each eigenvector by its length to obtain columns of an orthogonal matrix **Q** that diagonalizes **A**:

$$\mathbf{Q} = \begin{pmatrix} 1/\sqrt{2} & -1/\sqrt{2} \\ 1/\sqrt{2} & 1/\sqrt{2} \end{pmatrix}.$$

The change of variables $\mathbf{X} = \mathbf{Q}\mathbf{Y}$ is equivalent to setting

$$x_1 = \frac{1}{\sqrt{2}}(y_1 - y_2)$$

$$x_2 = \frac{1}{\sqrt{2}}(y_1 + y_2).$$

This transforms the given quadratic form to its standard form

$$\lambda_1 y_1^2 + \lambda_2 y_2^2 = -\frac{5}{2}y_1^2 + \frac{9}{2}y_2^2,$$

in which there are no cross product $y_1 y_2$ terms. ♦

SECTION 9.3 PROBLEMS

In each of Problems 1 through 12, find the eigenvalues and associated eigenvectors. Check that the eigenvectors associated with distinct eigenvalues are orthogonal. Find an orthogonal matrix that diagonalizes the matrix. Note Problems 17-22, Section 9.1.

1. $\begin{pmatrix} 0 & 0 & 0 \\ 1 & 1 & -2 \\ 0 & -2 & 0 \end{pmatrix}$

2. $\begin{pmatrix} 1 & 3 & 0 \\ 3 & 0 & 1 \\ 0 & 1 & 1 \end{pmatrix}$

3. $\begin{pmatrix} 0 & 0 & 0 & 0 \\ 0 & 1 & -2 & 0 \\ 0 & -2 & 1 & 0 \\ 0 & -3 & 0 & 0 \\ 0 & 0 & 0 & 0 \end{pmatrix}$

4. $\begin{pmatrix} 5 & 0 & 0 & 0 \\ 0 & 0 & -1 & 0 \\ 0 & -1 & 0 & 0 \\ 0 & 0 & 0 & 0 \end{pmatrix}$

5. $\begin{pmatrix} 5 & 0 & 2 \\ 0 & 0 & 0 \\ 2 & 0 & 0 \end{pmatrix}$

6. $\begin{pmatrix} 2 & -4 & 0 \\ -4 & 0 & 0 \\ 0 & 0 & 0 \end{pmatrix}$

7. $\begin{pmatrix} 4 & -2 \\ -2 & 1 \end{pmatrix}$

8. $\begin{pmatrix} 0 & 1 & 1 \\ 1 & 2 & 0 \\ 0 & 0 & 3 \end{pmatrix}$

9. $\begin{pmatrix} 6 & 1 \\ 1 & 4 \end{pmatrix}$

10. $\begin{pmatrix} -3 & 5 \\ 5 & 4 \end{pmatrix}$

11. $\begin{pmatrix} 0 & 1 & 0 \\ 1 & -2 & 0 \\ 0 & 0 & 3 \end{pmatrix}$

12. $\begin{pmatrix} -13 & 1 \\ 1 & 4 \end{pmatrix}$

In each of Problems 13 through 21, determine whether the matrix is unitary, hermitian, skew-hermitian, or none of these. Find the eigenvalues and associated eigenvectors. If the matrix is diagonalizable, write a matrix that diagonalizes it. In Problems 5 and 11, eigenvalues must be approximated, so only "approximate eigenvectors" can be found. It is instructive to try to diagonalize a matrix using approximate eigenvectors.

13. $\begin{pmatrix} 0 & 1 & 0 \\ -1 & 0 & 1-i \\ 0 & -1-i & 0 \end{pmatrix}$

14. $\begin{pmatrix} 1/\sqrt{2} & i/\sqrt{2} & 0 \\ -1/\sqrt{2} & i/\sqrt{2} & 0 \\ 0 & 0 & 1 \end{pmatrix}$

15. $\begin{pmatrix} 0 & 2i \\ 2i & 4 \end{pmatrix}$

16. $\begin{pmatrix} 3 & 4i \\ 4i & -5 \end{pmatrix}$

17. $\begin{pmatrix} 8 & -1 & i \\ -1 & 0 & 0 \\ -i & 0 & 0 \end{pmatrix}$

18. $\begin{pmatrix} 3i & 0 & 0 \\ -1 & 0 & 0 \\ -i & 0 & 0 \end{pmatrix}$

19. $\begin{pmatrix} 3 & 2 & 0 \\ 2 & 0 & i \\ 0 & -i & 0 \end{pmatrix}$

20. $\begin{pmatrix} -1 & 0 & 3-i \\ 0 & 1 & 0 \\ 3+i & 0 & 0 \end{pmatrix}$

21. $\begin{pmatrix} i & 1 & 0 \\ -1 & 0 & 2i \\ 0 & 2i & 0 \end{pmatrix}$

In each of Problems 22 through 28, determine a matrix **A** so that the quadratic form is $\mathbf{X}^t\mathbf{A}\mathbf{X}$, and find the standard form of the quadratic form.

22. $-3x_1^2 + 4x_1x_2 + 7x_2^2$
23. $5x_1^2 + 4x_1x_2 + 2x_2^2$
24. $-2x_1x_2 + 2x_2^2$
25. $4x_1^2 - 12x_1x_2 + x_2^2$
26. $-5x_1^2 + 4x_1x_2 + 3x_2^2$
27. $4x_1^2 - 4x_1x_2 + x_2^2$
28. $-6x_1x_2 + 4x_2^2$
29. Prove that each main diagonal element of a skew-hermitian matrix is zero or pure imaginary.
30. Prove that the product of two unitary matrices is unitary.
31. Suppose **A** is hermitian. Show that

$$\overline{(\mathbf{A}\mathbf{A}^t)} = \overline{\mathbf{A}}\mathbf{A}.$$

32. Prove that the main diagonal elements of a hermitian matrix are real.

LINEAR SYSTEMS SOLUTION OF $X' = AX$ FOR CONSTANT A SOLUTION OF $X' = AX + G$ EXPONENTIAL MATRIX SOLUTIONS APPLICATIONS

CHAPTER 10

Systems of Linear Differential Equations

10.1 Linear Systems

We will apply matrices to the solution of a system of n linear differential equations in n unknown functions:

$$x_1'(t) = a_{11}(t)x_1(t) + a_{12}(t)x_2(t) + \cdots + a_{1n}(t)x_n(t) + g_1(t)$$
$$x_2'(t) = a_{21}(t)x_1(t) + a_{22}(t)x_2(t) + \cdots + a_{2n}(t)x_n(t) + g_2(t)$$
$$\vdots$$
$$x_n'(t) = a_{n1}(t)x_1(t) + a_{n2}(t)x_2(t) + \cdots + a_{nn}(t)x_n(t) + g_n(t).$$

The functions $a_{ij}(t)$ are continuous and $g_j(t)$ are piecewise continuous on some interval (perhaps the whole real line). Define matrices

$$\mathbf{A}(t) = [a_{ij}(t)], \mathbf{X}(t) = \begin{pmatrix} x_1(t) \\ x_2(t) \\ \vdots \\ x_n(t) \end{pmatrix} \text{ and } \mathbf{G}(t) = \begin{pmatrix} g_1(t) \\ g_2(t) \\ \vdots \\ g_n(t) \end{pmatrix}.$$

Differentiate a matrix by differentiating each element. Matrix differentiation follows the usual rules of calculus. The derivative of a sum is the sum of the derivatives, and the product rule has the same form, whenever the product is defined:

$$(\mathbf{WN})' = \mathbf{W}'\mathbf{N} + \mathbf{WN}'.$$

With this notation, the system of linear differential equations is

$$\mathbf{X}'(t) = \mathbf{A}(t)\mathbf{X}(t) + \mathbf{G}(t)$$

or

$$\mathbf{X}' = \mathbf{AX} + \mathbf{G}. \tag{10.1}$$

We will refer to this as a *linear system*. This system is *homogeneous* if $\mathbf{G}(t)$ is the $n \times 1$ zero matrix, which occurs when each $g_j(t)$ is identically zero. Otherwise the system is *nonhomogeneous*.

We have an initial value problem for this linear system if the solution is specified at some value $t = t_0$. Here is the fundamental existence/uniqueness theorem for initial value problems

THEOREM 10.1

Let I be an open interval containing t_0. Suppose $\mathbf{A}(t) = [a_{ij}(t)]$ is an $n \times n$ matrix of functions that are continuous on I, and let

$$\mathbf{G}(t) = \begin{pmatrix} g_1(t) \\ g_2(t) \\ \vdots \\ g_n(t) \end{pmatrix}$$

be an $n \times 1$ matrix of functions that are continuous on I. Let $\mathbf{X}^0$ be a given $n \times 1$ matrix of real numbers. Then the initial value problem:

$$\mathbf{X}' = \mathbf{AX} + \mathbf{G}; \mathbf{X}(t_0) = \mathbf{X}^0$$

has a unique solution that is defined for all t in I. ♦

Armed with this result, we will outline a procedure for finding all solutions of the system (10.1). This will be analogous to the theory of the second order linear differential equation $y'' + p(x)y' + q(x)y = g(x)$ in Chapter 2. We will then show how to carry out this procedure to produce solutions in the case that $\mathbf{A}$ is a constant matrix.

10.1.1 The Homogeneous System $\mathbf{X}' = \mathbf{AX}$

If Φ_1 and Φ_2 are solutions of $\mathbf{X}' = \mathbf{AX}$, then so is any linear combination

$$c_1\Phi_1 + c_2\Phi_2.$$

This is easily verified by substituting this linear combination into $\mathbf{X}' = \mathbf{AX}$. This conclusion extends to any finite sum of solutions.

A set of k solutions $\mathbf{X}_1, \cdots, \mathbf{X}_k$ is *linearly dependent* on an open interval I (which can be the entire real line) if one of these solutions is a linear combination of the others, for all t in I. This is equivalent to the assertion that there is a linear combination

$$c_1\mathbf{X}_1(t) + c_2\mathbf{X}_2(t) + \cdots + c_k\mathbf{X}_k(t) = 0$$

for all t in I, with at least one of the coefficients $c_1, \cdots, c_k$ nonzero.

We call these solutions *linearly independent* on I if they are not linearly dependent on I. This means that no one of the solutions is a linear combination of the others. Alternatively, these solutions are linearly independent if and only if the only way an equation

$$c_1\mathbf{X}_1(t) + c_2\mathbf{X}_2(t) + \cdots + c_k\mathbf{X}_k(t) = 0$$

can hold for all t in I is for each coefficient to be zero: $c_1 = c_2 = \cdots = c_k = 0$.

EXAMPLE 10.1

Consider the system

$$\mathbf{X}' = \begin{pmatrix} 1 & -4 \\ 1 & 5 \end{pmatrix} \mathbf{X}.$$

It is routine to verify by substitution that

$$\Phi_1(t) = \begin{pmatrix} -2e^{3t} \\ e^{3t} \end{pmatrix} \text{ and } \Phi_2(t) = \begin{pmatrix} (1-2t)e^{3t} \\ te^{3t} \end{pmatrix}$$

are two solutions. These are linearly independent on the entire real line, since neither is a constant multiple of the other, for all t.

A third solution is

$$\Phi_3(t) = \begin{pmatrix} (-5-6t)e^{3t} \\ (4+3t)e^{3t} \end{pmatrix}.$$

However, these three solutions are linearly dependent, since, for all real numbers t,

$$\Phi_3(t) = 4\Phi_1(t) + 3\Phi_2(t). \blacklozenge$$

There is a test for linear independence of n solutions of an $n \times n$ homogeneous system $\mathbf{X}' = \mathbf{AX}$.

THEOREM 10.2 *Test for Independence of Solutions*

Suppose that

$$\Phi_1(t) = \begin{pmatrix} \varphi_{11}(t) \\ \varphi_{21}(t) \\ \vdots \\ \varphi_{n1}(t) \end{pmatrix}, \Phi_2(t) = \begin{pmatrix} \varphi_{12}(t) \\ \varphi_{22}(t) \\ \vdots \\ \varphi_{n2}(t) \end{pmatrix}, \cdots, \Phi_n(t) = \begin{pmatrix} \varphi_{1n}(t) \\ \varphi_{2n}(t) \\ \vdots \\ \varphi_{nn}(t) \end{pmatrix}$$

are n solutions of $\mathbf{X}' = \mathbf{AX}$ on an open interval I. Let t_0 be any number in I. Then

1. $\Phi_1, \Phi_2, \cdots, \Phi_n$ are linearly independent on I if and only if $\Phi_1(t_0), \Phi_2(t_0), \cdots, \Phi_n(t_0)$ are linearly independent, when considered as vectors in R^n.
2. $\Phi_1, \Phi_2, \cdots, \Phi_n$ are linearly independent on I if and only if

$$\begin{vmatrix} \varphi_{11}(t_0) & \varphi_{12}(t_0) & \cdots & \varphi_{1n}(t_0) \\ \varphi_{21}(t_0) & \varphi_{22}(t_0) & \cdots & \varphi_{2n}(t_0) \\ \vdots & \vdots & \cdots & \vdots \\ \varphi_{n1}(t_0) & \varphi_{n2}(t_0) & \cdots & \varphi_{nn}(t_0) \end{vmatrix} \neq 0. \blacklozenge$$

Conclusion (2) is an effective test for linear independence of solutions of the homogeneous system on an interval. Evaluate each solution at any number t_0 in the interval and form the $n \times n$ determinant having $\Phi_j(t_0)$ as column j. We may choose t_0 in the interval to suit our convenience (to make this determinant as easy as possible to evaluate). If this determinant is nonzero, then the solutions are linearly independent; otherwise the solutions are linearly dependent. This is similar to the Wronskian test for second order linear differential equations.

EXAMPLE 10.2

In Example 10.1,

$$\Phi_1(t) = \begin{pmatrix} -2e^{3t} \\ e^{3t} \end{pmatrix} \text{ and } \Phi_2(t) = \begin{pmatrix} (1-2t)e^{3t} \\ te^{3t} \end{pmatrix}$$

for all t. Evaluate these at some convenient point, say $t = 0$:

$$\Phi_1(0) = \begin{pmatrix} -2 \\ 1 \end{pmatrix} \text{ and } \Phi_2(0) = \begin{pmatrix} 1 \\ 0 \end{pmatrix}.$$

Use these 2− vectors as columns of a 2×2 determinant:

$$\begin{vmatrix} -2 & 1 \\ 1 & 0 \end{vmatrix} = -1 \neq 0.$$

Therefore Φ_1 and Φ_2 are linearly independent on the real line. In this case this conclusion is obvious without the determinant test, but this is not always the case. ♦

Proof of Theorem 10.2 Conclusion (2) follows from (1) by the fact that a determinant is zero exactly when its columns are linearly dependent.

To prove conclusion (1), let t_0 be in I. Suppose first that $\Phi_1, \cdots, \Phi_n$ are linearly dependent on I. Then one of these solutions is a linear combination of the others. By relabeling if necessary, suppose Φ_1 is a linear combination of $\Phi_2, \cdots, \Phi_n$. Then there are numbers $c_2, \cdots, c_n$ such that

$$\Phi_1(t) = c_2\Phi_2(t) + \cdots + c_n\Phi_n(t)$$

for all t in I. In particular, this holds at $t = t_0$, hence the vectors

$$\Phi_1(t_0), \cdots, \Phi_n(t_0)$$

are linearly dependent.

Conversely, suppose $\Phi_1(t_0), \cdots, \Phi_n(t_0)$ are linearly dependent in R^n. Then one of these vectors is a linear combination of the others. Again, suppose for convenience that the first is a combination of the others:

$$\Phi_1(t_0) = c_2\Phi_2(t_0) + \cdots + c_n\Phi_n(t_0).$$

Define

$$\Psi(t) = \Phi_1(t) - c_2\Phi_2(t) - \cdots - c_n\Phi_n(t)$$

for t in I. Then $\Psi(t)$ is a linear combination of solutions, hence it is a solution of the system. Furthermore,

$$\Psi(t_0) = \begin{pmatrix} 0 \\ 0 \\ \vdots \\ 0 \end{pmatrix}.$$

Therefore, $\Psi(t)$ is a solution of the initial value problem

$$\mathbf{X}' = \mathbf{AX}; \mathbf{X}(0) = \mathbf{O}.$$

But the zero function $\Delta(t) = \mathbf{O}$ is also a solution of this problem. By the uniqueness of the solution of this initial value problem (Theorem 10.1),

$$\Psi(t) = \Delta(t) = \mathbf{O} = \Phi_1(t) - c_2\Phi_2(t) - \cdots - c_n\Phi_n(t)$$

for all t in I. This means that

$$\Phi_1(t) = c_2\Phi_2(t) + \cdots + c_n\Phi_n(t)$$

for all t in I, hence that $\Phi_1(t), \Phi_2(t), \cdots, \Phi_n(t)$ are linearly dependent on I. This completes the proof. ♦

Thus far, we know how to test n solutions of the homogeneous system for linear independence on an open interval. We will now show that n linearly independent solutions are all that are needed to specify all solutions.

THEOREM 10.3

Let $\mathbf{A}(t) = [a_{ij}(t)]$ be an $n \times n$ matrix of functions that are continuous on an open interval I. Then

1. The system $\mathbf{X}' = \mathbf{A}\mathbf{X}$ has n linearly independent solutions on I.
2. Given n linearly independent solutions $\Phi_1(t), , \cdots, \Phi_n(t)$ defined on I, every solution on I is a linear combination of $\Phi_1(t), , \cdots, \Phi_n(t)$.

Proof To prove that there are n linearly independent solutions, define the $n \times 1$ constant matrices

$$\mathbf{E}^{(1)} = \begin{pmatrix} 1 \\ 0 \\ 0 \\ \vdots \\ 0 \\ 0 \end{pmatrix}, \mathbf{E}^{(2)} = \begin{pmatrix} 0 \\ 1 \\ 0 \\ \vdots \\ 0 \\ 0 \end{pmatrix}, \cdots, \mathbf{E}^{(n)} = \begin{pmatrix} 0 \\ 0 \\ \vdots \\ 0 \\ 0 \\ 1 \end{pmatrix}.$$

Choose any t_0 in I. We know that the initial value problem

$$\mathbf{X}' = \mathbf{A}\mathbf{X}; \mathbf{X}(0) = \mathbf{E}^{(j)}$$

has a unique solution $\Phi_j(t)$, for $j = 1, 2, \cdots, n$. These solutions are linearly independent, because, the way the initial conditions were chosen, the $n \times n$ matrix whose columns are these solutions evaluated at t_0 is $\mathbf{I}_n$, with determinant 1. This proves part (1).

To prove conclusion (2), suppose $\Psi_1, \cdots, \Psi_n$ are n linearly independent solutions on I. Let Λ be any solution. We want to show that Λ is a linear combination of $\Psi_1, \cdots, \Psi_n$. Pick any t_0 in I. Form the $n \times n$ nonsingular matrix $\mathbf{S}$ having the linearly independent vectors $\Psi_1(t_0), \cdots, \Psi_n(t_0)$ as its columns and consider the linear system of n algebraic equations in n unknowns:

$$\mathbf{S}\begin{pmatrix} c_1 \\ c_2 \\ \vdots \\ c_n \end{pmatrix} = \Lambda(t_0).$$

Because $\mathbf{S}$ is nonsingular, this algebraic system has a unique solution for numbers $c_1, c_2, \cdots, c_n$ such that

$$\Lambda(t_0) = c_1\Psi_1(t_0) + \cdots + c_n\Psi_n(t_0).$$

Then

$$\Lambda(t) = c_1\Psi_1(t) + \cdots + c_n\Psi_n(t)$$

for all t in I, because now $\Lambda(t)$ and $c_1\Psi_1(t) + \cdots + c_n\Psi_n(t)$ are both solutions of the initial value problem

$$\mathbf{X}' = \mathbf{AX}; \mathbf{X}(t_0) = \Lambda(t_0)$$

and this solution is unique. This shows that any solution $\Lambda(t)$ of the system $\mathbf{X}' = \mathbf{AX}$ is a linear combination of $\Psi_1(t), \cdots, \Psi_n(t)$. ♦

We call

$$c_1\Psi_1(t) + \cdots + c_n\Psi_n(t)$$

the *general solution* of $\mathbf{X}' = \mathbf{AX}$ when these solutions are linearly independent. Every solution is contained in this expression by varying the choices of the constants. In the language of linear algebra, the set of all solutions of $\mathbf{X}' = \mathbf{AX}$ is a vector space of dimension n, hence any n linearly independent solutions form a basis.

EXAMPLE 10.3

We have seen that

$$\Phi_1(t) = \begin{pmatrix} -2e^{3t} \\ e^{3t} \end{pmatrix} \text{ and } \Phi_2(t) = \begin{pmatrix} (1-2t)e^{3t} \\ te^{3t} \end{pmatrix}$$

are linearly independent solutions of

$$\mathbf{X}' = \begin{pmatrix} 1 & -4 \\ 1 & 5 \end{pmatrix}\mathbf{X}.$$

The general solution is

$$\mathbf{X}(t) = c_1\Phi_1(t) + c_2\Phi_2(t). \ ♦$$

We know the general solution of $\mathbf{X}' = \mathbf{AX}$ if we have n linearly independent solutions. These solutions are $n \times 1$ matrices. We can form an $n \times n$ matrix Ω using these n solutions as columns. Such a matrix is called a *fundamental matrix* for the system. In terms of this fundamental matrix, we can write the general solution in the compact form

$$c_1\Phi_1 + c_2\Phi_2 + \cdots + c_n\Phi_n = \Omega\mathbf{C}.$$

EXAMPLE 10.4

Continuing Example 10.3, form a 2×2 matrix using the linearly independent solutions $\Phi_1(t)$ and $\Phi_2(t)$ as columns:

$$\Omega(t) = \begin{pmatrix} -2e^{3t} & (1-2t)e^{3t} \\ e^{3t} & te^{3t} \end{pmatrix}.$$

$\Omega(t)$ is a fundamental matrix for this system. The general solution $c_1\Phi_1 + c_2\Phi_2$ can be written as $\Omega\mathbf{C}$:

$$\Omega\mathbf{C} = \begin{pmatrix} -2e^{3t} + (1-2t)e^{3t} \\ e^{3t} + te^{3t} \end{pmatrix} \begin{pmatrix} c_1 \\ c_2 \end{pmatrix}$$

$$= \begin{pmatrix} c_1(-2e^{3t}) & c_2(1-2t)e^{3t} \\ c_1 e^{3t} & c_2 t e^{3t} \end{pmatrix} = c_1 \begin{pmatrix} -2e^{3t} \\ e^{3t} \end{pmatrix} + c_2 \begin{pmatrix} (1-2t)e^{3t} \\ te^{3t} \end{pmatrix}$$

$$= c_1\Phi_1(t) + c_2\Phi_2(t). \ \blacklozenge$$

In an initial value problem, $x_1(t_0), \cdots, x_n(t_0)$ are given. This information specifies the $n \times 1$ matrix $\mathbf{X}(t_0)$. We usually solve an initial value problem by finding the general solution of the system and then solving for the constants to find the particular solution satisfying the initial conditions. It is often convenient to use a fundamental matrix to carry out this plan.

EXAMPLE 10.5

Solve the initial value problem

$$\mathbf{X}' = \begin{pmatrix} 1 & -4 \\ 1 & 5 \end{pmatrix} \mathbf{X}; \mathbf{X}(0) = \begin{pmatrix} -2 \\ 3 \end{pmatrix}.$$

The general solution is $\Omega\mathbf{C}$, with Ω the fundamental matrix of Example 10.4. To solve the initial value problem we must choose $\mathbf{C}$ so that

$$\mathbf{X}(0) = \Omega(0)\mathbf{C} = \begin{pmatrix} -2 \\ 3 \end{pmatrix}.$$

This is the algebraic system

$$\begin{pmatrix} -2 & 1 \\ 1 & 0 \end{pmatrix} \mathbf{C} = \begin{pmatrix} -2 \\ 3 \end{pmatrix}.$$

The solution for $\mathbf{C}$ is

$$\mathbf{C} = \begin{pmatrix} -2 & 1 \\ 1 & 0 \end{pmatrix}^{-1} \begin{pmatrix} -2 \\ 3 \end{pmatrix} = \begin{pmatrix} 0 & 1 \\ 1 & 2 \end{pmatrix} \begin{pmatrix} -2 \\ 3 \end{pmatrix} = \begin{pmatrix} 3 \\ 4 \end{pmatrix}.$$

The unique solution of the initial value problem is

$$\mathbf{X}(t) = \Omega(t) \begin{pmatrix} 3 \\ 4 \end{pmatrix} = \begin{pmatrix} -2e^{3t} - 8te^{3t} \\ 3e^{3t} + 4te^{3t} \end{pmatrix}. \ \blacklozenge$$

10.1.2 The Nonhomogeneous System

We will develop an analog of Theorem 2.5 for the nonhomogeneous linear system $\mathbf{X}' = \mathbf{A}\mathbf{X} + \mathbf{G}$. The key observation is that, if Ψ_1 and Ψ_2 are any two solutions of this nonhomogeneous system, then their difference $\Psi_1 - \Psi_2$ is a solution of the homogeneous system $\mathbf{X}' = \mathbf{A}\mathbf{X}$. Therefore, if Ω is a fundamental matrix for this homogeneous system, then

$$\Psi_1 - \Psi_2 = \Omega\mathbf{K}$$

for some constant $n \times 1$ matrix $\mathbf{K}$, hence

$$\Psi_1 = \Psi_2 + \Omega\mathbf{K}.$$

We will state this result as a general theorem.

THEOREM 10.4

Let Ω be a fundamental matrix for the homogeneous system $\mathbf{X}' = \mathbf{AX}$. Let Ψ_p be any particular solution of the nonhomogeneous system $\mathbf{X}' = \mathbf{AX} + \mathbf{G}$. Then every solution of the nonhomogeneous system has the form

$$\mathbf{X} = \Omega\mathbf{C} + \Psi_p. \blacklozenge$$

For this reason we call

$$\Omega\mathbf{C} + \Psi_p,$$

in which $\mathbf{C}$ is an $n \times 1$ matrix of arbitrary constants, the *general solution* of $\mathbf{X}' = \mathbf{AX} + \mathbf{G}$.

We now know what to look for in solving homogeneous and nonhomogeneous $n \times n$ linear systems.

For the homogeneous system $\mathbf{X}' = \mathbf{AX}$, form a fundamental matrix Ω whose columns are n linearly independent solutions. The general solution is $\mathbf{X} = \Omega\mathbf{C}$.

For the nonhomogeneous system $\mathbf{X}' = \mathbf{AX} + \mathbf{G}$, first find the general solution $\Omega\mathbf{C}$ of the associated homogeneous system $\mathbf{X}' = \mathbf{AX}$. Then find any particular solution Ψ_p of the nonhomogeneous system. The general solution of $\mathbf{X}' = \mathbf{AX} + \mathbf{G}$ is $\mathbf{X} = \Omega\mathbf{C} + \Psi_p$.

In the next section, we will begin to carry out this strategy for the case that the coefficient matrix $\mathbf{A}$ is constant.

SECTION 10.1 PROBLEMS

In each of Problems 1 through 5, (a) verify that the given functions satisfy the system, (b) write the system in matrix form $\mathbf{X}' = \mathbf{AX}$ for an appropriate $\mathbf{A}$, (c) write n linearly independent $n \times 1$ matrix solutions $\Phi_1, \cdots, \Phi_n$, for appropriate n, (d) use the determinant test of Theorem 10.2(2) to verify that these solutions are linearly independent, (e) form a fundamental matrix for the system, and (f) use the fundamental matrix to solve the initial value problem.

1. $x_1' = 3x_1 + 8x_2, x_2' = x_1 - x_2,$
$x_1(t) = 4c_1e^{(1+2\sqrt{3})t} + 4c_2e^{(1-2\sqrt{3}t)},$
$x_2(t) = (-1+\sqrt{3})c_1e^{(1+\sqrt{3})t} + (-1-\sqrt{3})c_2e^{(1-2\sqrt{3})6t},$
$x_1(0) = 2, x_2(0) = 2$

2. $x_1' = x_1 - x_2, x_2' = 4x_1 + 2x_2,$
$x_1(t) = 2e^{3t/2}\left[c_1\cos(\sqrt{15}t/2) + c_2\sin(\sqrt{15}t/2)\right],$
$x_2(t) = c_1e^{3t/2}\left[-\cos(\sqrt{15}t/2) + \sqrt{15}\sin(\sqrt{15}t/2)\right]$
$-c_2e^{3t/2}\left[\sin(\sqrt{15}t/2) + \sqrt{15}\cos(\sqrt{15}t/2)\right],$
$x_1(0) = -2, x_2(0) = 7$

3. $x_1' = 5x_1 - 4x_2 + 4x_3, x_2' = 12x_1 - 11x_2 + 12x_3,$
$x_3'(t) = 4x_1 - 4x_2 + 5x_3$
$x_1(t) = -c_1e^t + c_3e^{-3t}, x_2(t) = c_2e^{2t} + c_3e^{-3t},$
$x_3(t) = (c_3 - c_1)e^t + c_3e^{-3t},$
$x_1(0) = 1, x_2(0) = -3, x_3(0) = 5$

4. $x_1' = 2x_1 + x_2, x_2' = -3x_1 + 6x_2,$
$x_1(t) = c_1e^{4t}\cos(t) + c_2e^{4t}\sin(t)$
$x_2(t) = 2c_1e^{4t}[\cos(t) - \sin(t)]$
$+2c_2e^{4t}[\cos(t) + \sin(t)],$
$x_1(0) = -2, x_2(0) = 1$

5. $x_1' = 5x_1 + 3x_2, x_2' = x_1 + 3x_2,$
$x_1(t) = c_1e^{2t} + 3c_2e^{6t}, x_2(t) = c_1e^{2t} + c_2e^{6t},$
$x_1(0) = 0, x_2(0) = 4$

10.2 Solution of $\mathbf{X}' = \mathbf{AX}$ for Constant $\mathbf{A}$

Now we know what to look for to solve a linear system. We must find n linearly independent solutions.

To carry out this strategy we will focus on the special case that **A** is a real, constant matrix. Taking a cue from the constant coefficient, second order differential equation, attempt solutions of the form $\mathbf{X} = \mathbf{E}e^{\lambda t}$, with **E** an $n \times 1$ matrix of numbers and λ a number. For this to be a solution, we need

$$(\mathbf{E}e^{\lambda t})' = \mathbf{E}\lambda e^{\lambda t} = \mathbf{A}\mathbf{E}e^{\lambda t}.$$

This will be true if

$$\mathbf{AE} = \lambda \mathbf{E},$$

which holds if λ is an eigenvalue of **A** with associated eigenvector **E**.

THEOREM 10.5

Let **A** be an $n \times n$ matrix of real numbers. If λ is an eigenvalue with associated eigenvector **E**, then $\mathbf{E}e^{\lambda t}$ is a solution of $\mathbf{X}' = \mathbf{AX}$. ♦

We need n linearly independent solutions to write the general solution of $\mathbf{X}' = \mathbf{AX}$. The next theorem addresses this.

THEOREM 10.6

Let **A** be an $n \times n$ matrix of real numbers. Suppose **A** has eigenvalues $\lambda_1, \cdots, \lambda_n$, and suppose there are n corresponding eigenvectors $\mathbf{E}_1, \cdots, \mathbf{E}_n$ that are linearly independent. Then $\mathbf{E}_1 e^{\lambda_1 t}, \cdots, \mathbf{E}_n e^{\lambda_n t}$ are linearly independent solutions. ♦

When the eigenvalues are distinct, we can always find n linearly independent eigenvectors. But even when the eigenvalues are not distinct, it may still be possible to find n linearly independent eigenvectors, and in this case, we have n linearly independent solutions, hence the general solution. We can also use these solutions as columns of a fundamental matrix.

EXAMPLE 10.6

We will solve the system

$$\mathbf{X}' = \begin{pmatrix} 4 & 2 \\ 3 & 3 \end{pmatrix} \mathbf{X}.$$

A has eigenvalues of $1, 6$ with corresponding eigenvectors

$$\mathbf{E}_1 = \begin{pmatrix} 1 \\ -3/2 \end{pmatrix} \text{ and } \mathbf{E}_2 = \begin{pmatrix} 1 \\ 1 \end{pmatrix}.$$

These are linearly independent (the eigenvalues are distinct), so we have two linearly independent solutions

$$\begin{pmatrix} 1 \\ -3/2 \end{pmatrix} e^{t} \text{ and } \begin{pmatrix} 1 \\ 1 \end{pmatrix} e^{6t}.$$

The general solution is

$$\mathbf{X}(t) = c_1 \begin{pmatrix} 1 \\ -3/2 \end{pmatrix} e^{t} + c_2 \begin{pmatrix} 1 \\ 1 \end{pmatrix} e^{6t}.$$

We can also write the fundamental matrix

$$\Omega(t) = \begin{pmatrix} e^t & e^{6t} \\ -3e^t/2 & e^{6t} \end{pmatrix}$$

in terms of which the general solution is $\mathbf{X}(t) = \Omega(t)\mathbf{C}$.

If we write out the components individually, the general solution is

$$x_1(t) = c_1 e^t + c_2 e^{6t}$$

$$x_2(t) = -\frac{3}{2} c_1 e^t + c_2 e^{6t}. \quad \blacklozenge$$

EXAMPLE 10.7

Consider the system

$$\mathbf{X}' = \begin{pmatrix} 5 & 14 & 4 \\ 12 & -11 & 12 \\ 4 & -4 & 5 \end{pmatrix} \mathbf{X}.$$

The eigenvalues of $\mathbf{A}$ are $-3, 1, 1$. Even though there is a repeated eigenvalue, in this example, $\mathbf{A}$ has three linearly independent eigenvectors. They are

$$\begin{pmatrix} 1 \\ 3 \\ 1 \end{pmatrix} \text{ associated with eigenvalue } -3$$

and

$$\begin{pmatrix} 1 \\ 1 \\ 0 \end{pmatrix} \text{ and } \begin{pmatrix} -1 \\ 0 \\ 1 \end{pmatrix} \text{ associated with eigenvalue } 1.$$

The general solution is

$$\mathbf{X}(t) = c_1 \begin{pmatrix} 1 \\ 3 \\ 1 \end{pmatrix} e^{-3t} + c_2 \begin{pmatrix} 1 \\ 1 \\ 0 \end{pmatrix} e^t + c_3 \begin{pmatrix} -1 \\ 0 \\ 1 \end{pmatrix} e^t.$$

We also can write the general solution $\mathbf{X}(t) = \Omega(t)\mathbf{C}$, where

$$\Omega(t) = \begin{pmatrix} e^{-3t} & e^t & -e^t \\ 3e^{-3t} & e^t & 0 \\ e^{-3t} & 0 & e^t \end{pmatrix}. \quad \blacklozenge$$

EXAMPLE 10.8 A Mixing Problem

Two tanks are connected by pipes as in Figure 10.1. Tank 1 initially contains 20 liters of water in which 150 grams of chlorine are dissolved. Tank 2 initially contains 50 grams of chlorine dissolved in 10 liters of water. Beginning at time $t = 0$, pure water is pumped into tank 1 at a rate of 3 liters per minute, while chlorine/water solutions are exchanged between the tanks and also flow out of both tanks at the rates shown. We want to determine the amount of chlorine in each tank at time t.

Let $x_j(t)$ be the number of grams of chlorine in tank j at time t. Reading from Figure 10.1,

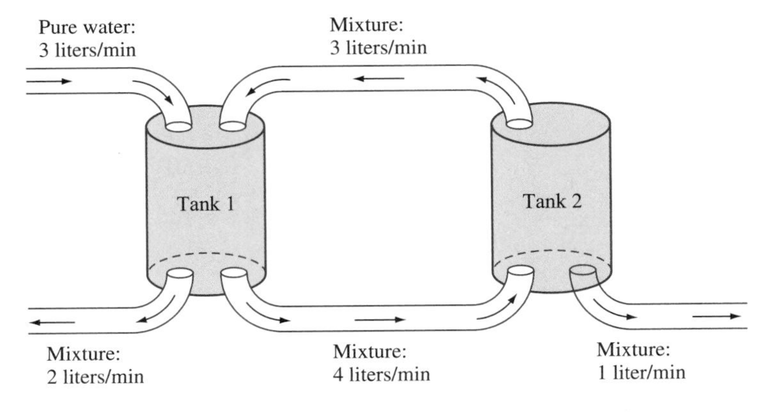

FIGURE 10.1 *Exchange of mixtures in tanks in Example 10.8.*

$$\begin{aligned}
\text{rate of change of } x_j(t) = x_j'(t) = & \text{ rate in minus rate out} \\
&= 3\left(\frac{\text{liter}}{\text{min}}\right)\cdot 0\left(\frac{\text{gram}}{\text{liter}}\right) + 3\left(\frac{\text{liter}}{\text{min}}\right)\cdot\frac{x_2}{10}\left(\frac{\text{gram}}{\text{liter}}\right) \\
&\quad - 2\left(\frac{\text{liter}}{\text{min}}\right)\cdot\frac{x_1}{20}\left(\frac{\text{gram}}{\text{liter}}\right) - 4\left(\frac{\text{liter}}{\text{min}}\right)\cdot\frac{x_1}{20}\left(\frac{\text{gram}}{\text{liter}}\right) \\
&= -\frac{6}{20}x_1 + \frac{3}{10}x_2.
\end{aligned}$$

Similarly, with the dimensions excluded,

$$x_2'(t) = 4\frac{x_1}{20} - 3\frac{x_2}{10} - \frac{x_2}{10} = \frac{4}{20}x_1 - \frac{4}{10}x_2.$$

The system we must solve is $\mathbf{X}' = \mathbf{A}\mathbf{X}$ with

$$\mathbf{A} = \begin{pmatrix} -3/10 & 3/10 \\ 1/5 & -2/5 \end{pmatrix}.$$

The initial conditions are

$$x_1(0) = 150,\ x_2(0) = 50$$

or

$$\mathbf{X}(0) = \begin{pmatrix} 150 \\ 50 \end{pmatrix}.$$

The eigenvalues of $\mathbf{A}$ are $-1/10$, $-1/5$ with corresponding eigenvalues, respectively,

$$\begin{pmatrix} 3/2 \\ 1 \end{pmatrix} \text{ and } \begin{pmatrix} -1 \\ 1 \end{pmatrix}.$$

A fundamental matrix is

$$\Omega(t) = \begin{pmatrix} (3/2)e^{-t/10} & -e^{-3t/5} \\ e^{-t/10} & e^{-3t/5} \end{pmatrix}.$$

The general solution is $\mathbf{X}(t) = \Omega(t)\mathbf{C}$. To solve the initial value problem, we need $\mathbf{C}$ so that

$$\mathbf{X}(0) = \begin{pmatrix} 150 \\ 50 \end{pmatrix} = \Omega\mathbf{C} = \begin{pmatrix} 3/2 & -1 \\ 1 & 1 \end{pmatrix}\mathbf{C}.$$

Then

$$\mathbf{C} = \begin{pmatrix} 3/2 & -1 \\ 1 & 1 \end{pmatrix}^{-1} \begin{pmatrix} 150 \\ 50 \end{pmatrix}$$

$$= \begin{pmatrix} 2/5 & 2/5 \\ -2/5 & 3/5 \end{pmatrix} \begin{pmatrix} 150 \\ 50 \end{pmatrix} = \begin{pmatrix} 80 \\ -30 \end{pmatrix}.$$

The solution of the initial value problem is

$$\mathbf{X}(t) = \begin{pmatrix} (3/2)e^{-t/10} & -e^{-3t/5} \\ e^{-t/10} & e^{-3t/5} \end{pmatrix} \begin{pmatrix} 80 \\ -30 \end{pmatrix}$$

$$= \begin{pmatrix} 120e^{-t/10} + 30e^{-3t/5} \\ 80e^{-t/10} - 30e^{-3t/5} \end{pmatrix}.$$

As $t \to \infty$, $x_1(t) \to 0$ and $x_2(t) \to 0$, as we might expect. ♦

10.2.1 Solution When A Has a Complex Eigenvalue

We used Euler's formula to write real-valued solutions of the second-order linear homogeneous constant coefficient differential equation when the characteristic equation has complex roots. We will follow a similar procedure for systems when the matrix of coefficients has (at least some) complex eigenvalues.

Since **A** is assumed to have real elements, the characteristic polynomial has real coefficients, so complex roots must occur in complex conjugate pairs. If λ is a complex eigenvalue with eigenvector ξ, then $\overline{\lambda}$ is also an eigenvalue, with complex eigenvector $\overline{\xi}$. Therefore, $\xi e^{\lambda t}$ and $\overline{\xi} e^{\overline{\lambda} t}$ are solutions. By taking linear combinations of any two such solutions, we obtain the following.

THEOREM 10.7 Solutions When Complex Eigenvalues Occur

Let **A** be an $n \times n$ matrix of real numbers. Let $\alpha + i\beta$ be a complex eigenvalue with corresponding eigenvector $\mathbf{U} + i\mathbf{V}$, in which **U** and **V** are real $n \times 1$ matrices. Then

$$e^{\alpha t}[\cos(\beta t)\mathbf{U} - \sin(\beta t)\mathbf{V}]$$

and

$$e^{\alpha t}[\sin(\beta t)\mathbf{U} + \cos(\beta t)\mathbf{V}]$$

are linearly independent solutions of $\mathbf{X}' = \mathbf{A}\mathbf{X}$. ♦

Proof We know that $\alpha - i\beta$ is also an eigenvalue with eigenvector $\mathbf{U} - i\mathbf{V}$. Write two solutions:

$$\begin{aligned}\Phi_1(t) &= e^{(\alpha+i\beta)t}(\mathbf{U} + i\mathbf{V}) \\ &= e^{\alpha t}(\cos(\beta t) + i\sin(\beta t))(\mathbf{U} + i\mathbf{V}) \\ &= e^{\alpha t}(\cos(\beta t)\mathbf{U} - \sin(\beta t)\mathbf{V}) + ie^{\alpha t}(\sin(\beta t)\mathbf{U} + \cos(\beta t)\mathbf{V})\end{aligned}$$

and

$$\begin{aligned}\Phi_2(t) &= e^{(\alpha-i\beta)t}(\mathbf{U} - i\mathbf{V}) \\ &= e^{\alpha t}(\cos(\beta t) - i\sin(\beta t))(\mathbf{U} - i\mathbf{V}) \\ &= e^{\alpha t}(\cos(\beta t)\mathbf{U} - \sin(\beta t)\mathbf{V}) + ie^{\alpha t}(-\cos(\beta t)\mathbf{V} - \sin(\beta t)\mathbf{U}).\end{aligned}$$

Linear combination of these solutions are also solutions. Thus, define two solutions

$$\frac{1}{2}(\Phi_1(t) + \Phi_2(t))$$

and

$$\frac{1}{2i}(\Phi_1(t) - \Phi_2(t)),$$

and these are the solutions given in the theorem. ♦

Theorem 10.7 enables us to replace two complex solutions

$$e^{(\alpha+i\beta)}(\mathbf{U} + i\mathbf{V}) \text{ and } e^{(\alpha-i\beta)}(\mathbf{U} - i\mathbf{V})$$

in the general solution with the two solutions given in the theorem, which involve only real quantities.

EXAMPLE 10.9

We will solve the system $\mathbf{X}' = \mathbf{AX}$ with

$$\mathbf{A} = \begin{pmatrix} 2 & 0 & 1 \\ 0 & -2 & -2 \\ 0 & 2 & 0 \end{pmatrix}.$$

The eigenvalues are $2, -1 + \sqrt{3}i, -1 - \sqrt{3}i$. Corresponding eigenvectors are, respectively,

$$\begin{pmatrix} 1 \\ 0 \\ 0 \end{pmatrix}, \begin{pmatrix} 1 \\ -2\sqrt{3}i \\ -3 + \sqrt{3}i \end{pmatrix} \text{ and } \begin{pmatrix} 1 \\ 2\sqrt{3}i \\ -3 - \sqrt{3}i \end{pmatrix}.$$

One solution is

$$\begin{pmatrix} 1 \\ 0 \\ 0 \end{pmatrix} e^{2t}.$$

Two other solutions are complex:

$$\begin{pmatrix} 1 \\ -2\sqrt{3}i \\ -3 + \sqrt{3}i \end{pmatrix} e^{(-1+\sqrt{3}i)t} \text{ and } \begin{pmatrix} 1 \\ 2\sqrt{3}i \\ -3 - \sqrt{3}i \end{pmatrix} e^{(-1-\sqrt{3}i)t}.$$

These three solutions can be used as columns of a fundamental matrix. However, if we wish, we can write a solution involving only real numbers and real-valued functions. First,

$$\begin{pmatrix} 1 \\ -2\sqrt{3}i \\ -3 + \sqrt{3}i \end{pmatrix} = \begin{pmatrix} 1 \\ 0 \\ -3 \end{pmatrix} + i \begin{pmatrix} 0 \\ -2\sqrt{3} \\ \sqrt{3} \end{pmatrix} = \mathbf{U} + i\mathbf{V}$$

with

$$\mathbf{U} = \begin{pmatrix} 1 \\ 0 \\ -3 \end{pmatrix} \text{ and } \mathbf{V} = \begin{pmatrix} 0 \\ -2\sqrt{3} \\ \sqrt{3} \end{pmatrix}.$$

By Theorem 10.7, with $\alpha = -1$ and $\beta = \sqrt{3}$, we can replace the two complex solutions with the two solutions

$$e^{-t}[\cos(\sqrt{3}t)\mathbf{U} - \sin(\sqrt{3}t)\mathbf{V}]$$

and

$$e^{-t}[\sin(\sqrt{3}t)\mathbf{U}+\cos(\sqrt{3}t)\mathbf{V}].$$

In terms of these solutions, a fundamental matrix is

$$\Omega(t)=\begin{pmatrix} e^{2t} & e^{-t}\cos(\sqrt{3}t) & e^{-t}\sin(\sqrt{3}t) \\ 0 & 2\sqrt{3}e^{-t}\sin(\sqrt{3}t) & -2\sqrt{3}e^{-t}\cos(\sqrt{3}t) \\ 0 & e^{-t}[-3\cos(\sqrt{3}t)-\sqrt{3}\sin(\sqrt{3}t)] & e^{-t}[\sqrt{3}\cos(\sqrt{3}t)-3\sin(\sqrt{3}t)] \end{pmatrix}.$$

The general solution is $\mathbf{X}(t)=\Omega(t)\mathbf{C}$. ♦

10.2.2 Solution When A Does Not Have n Linearly Independent Eigenvectors

Two examples will give us a sense of how to proceed when $\mathbf{A}$ does not have n linearly independent eigenvectors.

EXAMPLE 10.10

We will solve $\mathbf{X}'=\mathbf{AX}$ when

$$\mathbf{A}=\begin{pmatrix} 1 & 3 \\ -3 & 7 \end{pmatrix}.$$

$\mathbf{A}$ has eigenvalues 4, 4, and all eigenvectors have the form

$$\alpha\begin{pmatrix} 1 \\ 1 \end{pmatrix}$$

with $\alpha \neq 0$. $\mathbf{A}$ does not have two linearly independent eigenvectors. One solution is

$$\Phi_1(t)=\begin{pmatrix} 1 \\ 1 \end{pmatrix}e^{4t}.$$

We need a second, linearly independent solution. Let

$$\mathbf{E}_1=\begin{pmatrix} 1 \\ 1 \end{pmatrix}$$

and attempt a second solution of the form

$$\Phi_2(t)=\mathbf{E}_1te^{4t}+\mathbf{E}_2e^{4t},$$

in which $\mathbf{E}_2$ is a 2×1 constant matrix to be determined. For $\Phi_2(t)$ to be a solution, we must have $\Phi_2'(t)=\mathbf{A}\Phi_2(t)$. This is the equation

$$\mathbf{E}_1[e^{4t}+4te^{4t}]+4\mathbf{E}_2e^{4t}=\mathbf{AE}_1te^{4t}+\mathbf{AE}_2e^{4t}.$$

Divide this by e^{4t} to get

$$\mathbf{E}_1+4t\mathbf{E}_1+4\mathbf{E}_2=\mathbf{AE}_1t+\mathbf{AE}_2.$$

But $\mathbf{AE}_1=4\mathbf{E}_1$, so the terms involving t cancel, leaving

$$\mathbf{AE}_2-4\mathbf{E}_2=\mathbf{E}_1$$

or

$$(\mathbf{A}-4\mathbf{I}_2)\mathbf{E}_2=\mathbf{E}_1.$$

If

$$\mathbf{E}_2=\begin{pmatrix} a \\ b \end{pmatrix},$$

then we have the linear system of two equations in two unknowns:

$$(\mathbf{A} - 4\mathbf{I}_2)\begin{pmatrix} a \\ b \end{pmatrix} = \begin{pmatrix} 1 \\ 1 \end{pmatrix}.$$

This is the system

$$\begin{pmatrix} -3 & 3 \\ -3 & 3 \end{pmatrix}\begin{pmatrix} a \\ b \end{pmatrix} = \begin{pmatrix} 1 \\ 1 \end{pmatrix}$$

with general solution

$$\mathbf{E}_2 = \begin{pmatrix} \alpha \\ (1+3\alpha)/2 \end{pmatrix}.$$

Since we need only one $\mathbf{E}_2$, let $\alpha = 1$ to get

$$\mathbf{E}_2 = \begin{pmatrix} 1 \\ 4/3 \end{pmatrix}.$$

Therefore, a second solution is

$$\Phi_2(t) = \mathbf{E}_1 te^{4t} + \mathbf{E}_2 e^{4t} = \begin{pmatrix} 1 \\ 1 \end{pmatrix} te^{4t} + \begin{pmatrix} 1 \\ 4/3 \end{pmatrix} e^{4t} = \begin{pmatrix} 1+t \\ 4/3+t \end{pmatrix} e^{4t}.$$

Φ_1 and Φ_2 are linearly independent solutions and can be used as columns of the fundamental matrix

$$\Omega(t) = \begin{pmatrix} e^{4t} & (1+t)e^{4t} \\ e^{4t} & (4/3+t)e^{4t} \end{pmatrix}.$$

The general solution is $\mathbf{X}(t) = \Omega(t)\mathbf{C}$. ♦

EXAMPLE 10.11

We will solve $\mathbf{X}' = \mathbf{AX}$ when

$$\mathbf{A} = \begin{pmatrix} -2 & -1 & -5 \\ 25 & -7 & 0 \\ 0 & 1 & 3 \end{pmatrix}.$$

$\mathbf{A}$ has eigenvalues $-2, -2, -2$, and all eigenvectors are scalar multiples of

$$\mathbf{E}_1 = \begin{pmatrix} -1 \\ -5 \\ 1 \end{pmatrix}.$$

One solution is $\Phi_1(t) = \mathbf{E}_1 e^{-2t}$. We will try a second solution, linearly independent from the first, of the form

$$\Phi_2(t) = \mathbf{E}_1 te^{-2t} + \mathbf{E}_2 e^{-2t}$$

in which $\mathbf{E}_2$ must be determined. Substitute this proposed solution into the differential equation to get

$$\mathbf{E}_1[e^{-2t} - 2te^{-2t}] + \mathbf{E}_2[-2e^{-2t}] = \mathbf{AE}_1 te^{-2t} + \mathbf{AE}_2 e^{-2t}.$$

Divide out e^{-2t} and recall that $\mathbf{AE}_1 = -2\mathbf{E}_1$ to cancel terms in the last equation, leaving

$$\mathbf{AE}_2 + 2\mathbf{E}_2 = \mathbf{E}_1.$$

This is the system

$$(\mathbf{A} + 2\mathbf{I}_3)\mathbf{E}_2 = \mathbf{E}_1.$$

If we write

$$\mathbf{E}_2 = \begin{pmatrix} a \\ b \\ c \end{pmatrix}$$

then we have the system of algebraic equations

$$\begin{pmatrix} 0 & -1 & -5 \\ 25 & -5 & 0 \\ 0 & 1 & 5 \end{pmatrix} \begin{pmatrix} a \\ b \\ c \end{pmatrix} = \begin{pmatrix} -1 \\ -5 \\ 1 \end{pmatrix}.$$

This system has general solution

$$\begin{pmatrix} -\alpha \\ 1-5\alpha \\ \alpha \end{pmatrix}$$

for α any real number. Choose $\alpha = 1$ to get

$$\mathbf{E}_2 = \begin{pmatrix} -1 \\ -4 \\ 1 \end{pmatrix}.$$

This gives us a second solution of the differential equation

$$\Phi_2(t) = \mathbf{E}_1 te^{-2t} + \mathbf{E}_2 e^{-2t}$$
$$= \begin{pmatrix} -1 \\ -5 \\ 1 \end{pmatrix} te^{2t} + \begin{pmatrix} -1 \\ -4 \\ 1 \end{pmatrix} e^{-2t} = \begin{pmatrix} -1-t \\ -4-5t \\ 1+t \end{pmatrix} e^{-2t}.$$

We need one more solution, linearly independent from the first two. Try for a third solution of the form

$$\Phi_3(t) = \frac{1}{2}\mathbf{E}_1 t^2 e^{-2t} + \mathbf{E}_2 te^{-2t} + \mathbf{E}_3 e^{-2t}.$$

Substitute this into $\mathbf{X}' = \mathbf{AX}$ to get

$$\mathbf{E}_1[te^{-2t} - t^2e^{-2t}] + \mathbf{E}_2[e^{-2t} - 2te^{-2t}] + \mathbf{E}_3[-2e^{-2t}]$$
$$= \frac{1}{2}\mathbf{AE}_1 t^2 e^{-2t} + \mathbf{AE}_2 te^{-2t} + \mathbf{AE}_3 e^{-2t}.$$

Divide e^{-2t} and use the fact that $\mathbf{AE}_1 = -2\mathbf{E}_1$ and

$$\mathbf{AE}_2 = \begin{pmatrix} 1 \\ 3 \\ -1 \end{pmatrix}$$

to get

$$\mathbf{E}_1 t - \mathbf{E}_1 t^2 + \mathbf{E}_2 - 2\mathbf{E}_2 t - 2\mathbf{E}_3 = -\mathbf{E}_1 t^2 + \begin{pmatrix} 1 \\ 3 \\ -1 \end{pmatrix} t + \mathbf{AE}_3. \tag{10.2}$$

Now

$$\mathbf{E}_1 t - 2\mathbf{E}_2 t = (\mathbf{E}_1 - 2\mathbf{E}_2)t = \begin{pmatrix} -1-2(-1) \\ -5-2(-4) \\ 1-2(1) \end{pmatrix} t = \begin{pmatrix} 1 \\ 3 \\ -1 \end{pmatrix} t.$$

Therefore, three terms cancel in equation (10.2) and it reduces to

$$\mathbf{E}_2 - 2\mathbf{E}_1 = \mathbf{AE}_3.$$

Write this equation as

$$(\mathbf{A} + 2\mathbf{I}_3)\mathbf{E}_3 = \mathbf{E}_2.$$

This is the system

$$\begin{pmatrix} 0 & -1 & -5 \\ 25 & -5 & 0 \\ 0 & 1 & 5 \end{pmatrix} \mathbf{E}_3 = \begin{pmatrix} -11 \\ -4 \\ 1 \end{pmatrix}$$

with general solution

$$\mathbf{E}_3 = \begin{pmatrix} (1-25\alpha)/25 \\ 1-5\alpha \\ \alpha \end{pmatrix}.$$

Let $\alpha = 1$ to get

$$\mathbf{E}_3 = \begin{pmatrix} -24/25 \\ -4 \\ 1 \end{pmatrix}.$$

A third solution is

$$\begin{aligned}\Phi_3(t) &= \frac{1}{2}\begin{pmatrix} -1 \\ -5 \\ 1 \end{pmatrix} t^2 e^{-2t} + \begin{pmatrix} -1 \\ -4 \\ 1 \end{pmatrix} te^{-2t} + \begin{pmatrix} -24/25 \\ -4 \\ 1 \end{pmatrix} e^{-2t} \\ &= \begin{pmatrix} -24/25 - t - t^2/2 \\ -4 - 4t - 5t^2/2 \\ 1 + t + t^2/2 \end{pmatrix} e^{-2t}.\end{aligned}$$

We now have three linearly independent solutions and can use these as columns of the fundamental matrix

$$\Omega(t) = \begin{pmatrix} -e^{-2t} & (-1-t)e^{-2t} & (-24/25 - t - t^2/2)e^{-2t} \\ -5e^{-2t} & (-4-5t)e^{-2t} & (-4-4t-5t^2/2)e^{-2t} \\ e^{-2t} & (1+t)e^{-2t} & (1+t+t^2/2)e^{-2t} \end{pmatrix}.$$

The general solution is $\mathbf{X}(t) = \Omega(t)\mathbf{C}$. ♦

These examples suggest a procedure to follow. Suppose we know the eigenvalues of $\mathbf{A}$. If these are all distinct, the corresponding eigenvectors are linearly independent and we can write the general solution.

Thus, suppose an eigenvalue λ has multiplicity $k > 1$. If there are k linearly independent solutions associated with λ, then we can produce k linearly independent solutions corresponding to λ.

If λ only has r linearly independent associated eigenvectors and $r < k$, we need from λ a total of $r - k$ more solutions linearly independent from the others.

If $r - k = 1$, we need one more solution, which can be obtained as in Example 10.10.

If $r - k = 2$, proceed as in Example 10.11 to find another linearly independent solution.

If $r - k = 3$, follow the pattern of the previous cases, trying

$$\Phi_4(t) = \frac{1}{3!}\mathbf{E}_1 t^3 e^{\lambda t} + \frac{1}{2}\mathbf{E}_2 t^2 e^{\lambda t} + \mathbf{E}_3 t e^{\lambda t} + \mathbf{E}_4 e^{\lambda t}$$

where $\mathbf{E}_1$, $\mathbf{E}_2$, and $\mathbf{E}_3$ were found in generating preceding solutions.

If $r - k = 4$, try

$$\Phi_5(t) = \frac{1}{4!}\mathbf{E}_1 t^4 e^{\lambda t} + \frac{1}{3!}\mathbf{E}_2 t^3 e^{\lambda t} + \frac{1}{2}\mathbf{E}_3 t^2 e^{\lambda t} + \mathbf{E}_4 t e^{\lambda t} + \mathbf{E}_5 e^{\lambda t}.$$

This process must be continued until k linearly independent solutions have been obtained associated with the eigenvalue λ.

Repeat this procedure for each eigenvalue of multiplicity greater than 1. Each eigenvalue must have associated with it as many linearly independent solutions as the multiplicity of the eigenvalue. This process terminates when n linearly independent solutions have been generated.

SECTION 10.2 PROBLEMS

In each of Problems 1 through 10, find a fundamental matrix for the system and write the general solution as a matrix. If initial values are given, solve the initial value problem.

1. $x_1' = 3x_1 - x_2 + x_3, x_2' = x_1 + x_2 - x_3, x_3' = x_1 - x_2 + x_3; x_1(0) = 1, x_2(0) = 5, x_3(0) = 1$
2. $x_1' = 2x_1 + x_2 - x_3, x_2' = 3x_1 - 2x_2,$ $x_3' = 3x_1 + x_2 - 3x_3; x_1(0) = 1, x_2(0) = 7, x_3(0) = 3$
3. $x_1' = x_1 + 2x_2 + x_3, x_2' = 6x_1 - x_2, x_3' = -x_1 - 2x_2 - x_3$
4. $x_1' = 3x_1 - 4x_2, x_2' = 2x_1 - 3x_2; x_1(0) = 7, x_2(0) = 5$
5. $x_1' = x_1 + x_2, x_2' = x_1 + x_2$
6. $x_1' = 4x_1 + 2x_2, x_2' = 3x_1 + 3x_2$
7. $x_1' = 3x_1, x_2' = 5x_1 - 4x_2$
8. $x_1' = 2x_1 + x_2 - 2x_3, x_2' = 3x_1 - 2x_2,$ $x_3' = 3x_1 - x_2 - 3x_3$
9. $x_1' = x_1 - 2x_2, x_2' = -6x_1; x_1(0) = 1, x_2(0) = -19$
10. $x_1' = 2x_1 - 10x_2, x_2' = -x_1 - x_2; x_1(0) = -3, x_2(0) = 6$

In each of Problems 11 through 15, find a real-valued fundamental matrix for the system $\mathbf{X}' = \mathbf{AX}$ with the given coefficient matrix.

11. $\begin{pmatrix} 3 & -5 \\ 1 & -1 \end{pmatrix}$

12. $\begin{pmatrix} 1 & -1 & 1 \\ 1 & -1 & 0 \\ 1 & 0 & -1 \end{pmatrix}$

13. $\begin{pmatrix} -2 & 1 & 0 \\ -5 & 0 & 0 \\ 0 & 3 & -2 \end{pmatrix}$

14. $\begin{pmatrix} 0 & 5 \\ -1 & -2 \end{pmatrix}$

15. $\begin{pmatrix} 2 & -4 \\ 1 & 2 \end{pmatrix}$

In each of Problems 16 through 21, find a fundamental matrix for the system with the given coefficient matrix.

16. $\begin{pmatrix} 1 & 5 & 0 \\ 0 & 1 & 0 \\ 4 & 8 & 1 \end{pmatrix}$

17. $\begin{pmatrix} 2 & 5 & 6 \\ 0 & 8 & 9 \\ 0 & 1 & -2 \end{pmatrix}$

18. $\begin{pmatrix} 0 & 1 & 0 & 0 \\ 0 & 0 & 1 & 0 \\ 0 & 0 & 0 & 1 \\ -1 & -2 & 0 & 0 \end{pmatrix}$

19. $\begin{pmatrix} 1 & 5 & -2 & 6 \\ 0 & 3 & 0 & 4 \\ 0 & 3 & 0 & 4 \\ 0 & 0 & 0 & 1 \end{pmatrix}$

20. $\begin{pmatrix} 2 & 0 \\ 5 & 2 \end{pmatrix}$

21. $\begin{pmatrix} 3 & 2 \\ 0 & 3 \end{pmatrix}$

10.3 Solution of $\mathbf{X}' = \mathbf{AX} + \mathbf{G}$

We know that the general solution is the sum of the general solution of the homogeneous problem $\mathbf{X}' = \mathbf{AX}$ plus any particular solution of the nonhomogeneous system. We therefore need a method for finding a particular solution of the nonhomogeneous system. We will develop two methods.

10.3.1 Variation of Parameters

Variation of parameters for systems follows the same line of reasoning as variation of parameters for second order linear differential equations. If $\Omega(t)$ is a fundamental matrix for the homogeneous system $\mathbf{X}' = \mathbf{AX}$, then the general solution of the homogeneous system is $\Omega\mathbf{C}$. Using this as a template, look for a particular solution of the nonhomogeneous system of the form $\Psi_p(t) = \Omega(t)\mathbf{U}(t)$, where $\mathbf{U}(t)$ is an $n \times 1$ matrix to be determined.

Substitute this proposed particular solution into the nonhomogeneous system to obtain

$$(\Omega\mathbf{U})' = \Omega'\mathbf{U} + \Omega\mathbf{U}' = \mathbf{A}(\Omega\mathbf{U}) + \mathbf{G} = (\mathbf{A}\Omega)\mathbf{U} + \mathbf{G}. \tag{10.3}$$

Ω is a fundamental matrix for the homogeneous system, so $\Omega' = \mathbf{A}\Omega$. Therefore, $\Omega'\mathbf{U} = (\mathbf{A}\Omega)\mathbf{U}$ and equation (10.3) becomes

$$\Omega\mathbf{U}' = \mathbf{G}.$$

Since Ω is nonsingular,

$$\mathbf{U}' = \Omega^{-1}\mathbf{G}.$$

Then

$$\mathbf{U}(t) = \int \Omega^{-1}(t)\mathbf{G}(t)dt$$

in which we integrate a matrix by integrating each element of the matrix. Once we have $\mathbf{U}(t)$, we have the general solution

$$\mathbf{X}(t) = \Omega(t)\mathbf{C} + \Omega(t)\mathbf{U}(t)$$

of the nonhomogeneous system.

EXAMPLE 10.12

We will solve the system

$$\mathbf{X}' = \begin{pmatrix} 1 & -10 \\ -1 & 4 \end{pmatrix}\mathbf{X} + \begin{pmatrix} t \\ 1 \end{pmatrix}.$$

The eigenvalues of $\mathbf{A}$ are $-1, 6$ with corresponding eigenvectors

$$\begin{pmatrix} 5 \\ 1 \end{pmatrix} \text{ and } \begin{pmatrix} -2 \\ 1 \end{pmatrix}.$$

A fundamental matrix for $\mathbf{X}' = \mathbf{AX}$ is

$$\Omega(t) = \begin{pmatrix} 5e^{-t} & -2e^{6t} \\ e^{-t} & e^{6t} \end{pmatrix}.$$

Compute

$$\Omega^{-1}(t) = \frac{1}{7}\begin{pmatrix} e^{t} & 2e^{t} \\ -e^{-6t} & 5e^{-6t} \end{pmatrix}.$$

This inverse is most easily computed using MAPLE. In this 2×2 case we could also proceed as in Example 7.28 of Section 7.7.

With this inverse matrix, we have

$$\mathbf{U}'(t) = \Omega^{-1}(t)\mathbf{G}(t) = \frac{1}{7}\begin{pmatrix} e^{t} & 2e^{t} \\ -e^{-6t} & 5e^{-6t} \end{pmatrix}\begin{pmatrix} t \\ 1 \end{pmatrix}$$

$$= \frac{1}{7}\begin{pmatrix} 2e^{t} + te^{t} \\ 5e^{-6t} - te^{-6t} \end{pmatrix}.$$

Then

$$\mathbf{U}(t) = \int \Omega^{-1}(t)\mathbf{G}(t)dt$$

$$= \begin{pmatrix} (t+1)e^{t}/7 \\ (-29/252)e^{-6t} + (1/42)te^{-6t} \end{pmatrix}.$$

The general solution of the nonhomogeneous system is

$$\mathbf{X}(t) = \Omega(t)\mathbf{C} + \Omega(t)\mathbf{U}(t) = \begin{pmatrix} 5e^{-t} & -2e^{6t} \\ e^{-t} & e^{6t} \end{pmatrix}\mathbf{C}$$

$$+ \begin{pmatrix} 5e^{-t} & -2e^{6t} \\ e^{-t} & e^{6t} \end{pmatrix}\begin{pmatrix} (t+1)e^{t}/7 \\ (-29/252)e^{-6t} + (1/42)te^{-6t} \end{pmatrix}$$

$$= \begin{pmatrix} 5e^{-t} & -2e^{6t} \\ e^{-t} & e^{6t} \end{pmatrix}\mathbf{C} + \frac{1}{3}\begin{pmatrix} 17/6 + (49/7)t \\ 1/12 + t/2 \end{pmatrix}. \blacklozenge$$

Although in this example the coefficient matrix $\mathbf{A}$ was constant, this is not required to apply the method of variation of parameters.

10.3.2 Solution by Diagonalizing A

If $\mathbf{A}$ is a diagonalizable matrix of real numbers, then we can solve the system $\mathbf{X}' = \mathbf{AX} + \mathbf{G}$ by the change of variables $\mathbf{X} = \mathbf{PZ}$, where $\mathbf{P}$ diagonalizes $\mathbf{A}$.

EXAMPLE 10.13

We will solve the system

$$\mathbf{X}' = \begin{pmatrix} 3 & 3 \\ 1 & 5 \end{pmatrix}\mathbf{X} + \begin{pmatrix} 8 \\ 4e^{3t} \end{pmatrix}.$$

The eigenvalues of $\mathbf{A}$ are 2, 6, with eigenvectors, respectively,

$$\begin{pmatrix} -3 \\ 1 \end{pmatrix} \text{ and } \begin{pmatrix} 1 \\ 1 \end{pmatrix}.$$

Form $\mathbf{P}$ using these eigenvectors as columns:

$$\mathbf{P} = \begin{pmatrix} -3 & 1 \\ 1 & 1 \end{pmatrix}.$$

Then

$$\mathbf{P}^{-1}\mathbf{AP} = \mathbf{D} = \begin{pmatrix} 2 & 0 \\ 0 & 6 \end{pmatrix}$$

with the eigenvalues down the main diagonal. Compute

$$\mathbf{P}^{-1} = \begin{pmatrix} -1/4 & 1/4 \\ 1/4 & 3/4 \end{pmatrix}.$$

Now make the change of variables $\mathbf{X} = \mathbf{PZ}$ in the differential equation:

$$\mathbf{X}' = (\mathbf{PZ})' = \mathbf{PZ}' = \mathbf{A}(\mathbf{PZ}) + \mathbf{G}.$$

Then

$$\mathbf{PZ}' = (\mathbf{AP})\mathbf{Z} + \mathbf{G}.$$

Multiply this equation on the left by $\mathbf{P}^{-1}$ to get

$$\mathbf{Z}' = (\mathbf{P}^{-1}\mathbf{AP})\mathbf{Z} + \mathbf{P}^{-1}\mathbf{G}$$

or

$$\mathbf{Z}' = \mathbf{DZ} + \mathbf{P}^{-1}\mathbf{G}.$$

This is

$$\begin{pmatrix} z_1' \\ z_2' \end{pmatrix} = \begin{pmatrix} 2 & 0 \\ 0 & 6 \end{pmatrix}\begin{pmatrix} z_1 \\ z_2 \end{pmatrix} + \begin{pmatrix} -1/4 & 1/4 \\ 1/4 & 3/4 \end{pmatrix}\begin{pmatrix} 8 \\ 4e^{3t} \end{pmatrix}$$

$$= \begin{pmatrix} 2z_1 - 2 + e^{3t} \\ 6z_2 + 2 + 3e^{3t} \end{pmatrix}.$$

This is an *uncoupled system*, consisting of one differential equation for just z_1, and a second differential equation for just z_2. Solve each of these first-order linear differential equations to obtain

$$z_1(t) = c_1e^{2t} + e^{3t} + 1$$

$$z_2(t) = c_2e^{6t} - e^{3t} - \frac{1}{3}.$$

Then

$$\mathbf{X}(t) = \mathbf{PZ}(t) = \begin{pmatrix} -3 & 1 \\ 1 & 1 \end{pmatrix}\begin{pmatrix} c_1e^{2t} + e^{3t} + 1 \\ c_2e^{6t} - e^{3t} - 1/3 \end{pmatrix}$$

$$= \begin{pmatrix} -3c_1e^{2t} + c_2e^{6t} - 4e^{3t} - 10/3 \\ c_1e^{2t} + c_2e^{6t} + 2/3 \end{pmatrix} = \begin{pmatrix} -3e^{2t} & e^{6t} \\ e^{2t} & e^{6t} \end{pmatrix}\mathbf{C} + \begin{pmatrix} -4e^{3t} - 10/3 \\ 2/3 \end{pmatrix}.$$

This is the general solution in the form $\Omega(t)\mathbf{C} + \Psi_p$, which is a sum of the general solution of the associated homogeneous equation and a particular solution of the nonhomogeneous equation. ♦

SECTION 10.3 PROBLEMS

In each of Problems 1 through 9, use variation of parameters to find the general solution, with **A** and **G** given. If initial conditions are given, also satisfy the initial value problem.

1. $\begin{pmatrix} 1 & -3 & 0 \\ 3 & -5 & 0 \\ 4 & 7 & -2 \end{pmatrix}, \begin{pmatrix} te^{-2t} \\ te^{-2t} \\ t^2e^{-2t} \end{pmatrix}; \begin{pmatrix} 6 \\ 2 \\ 3 \end{pmatrix}$

2. $\begin{pmatrix} 2 & -3 & 1 \\ 0 & 2 & 4 \\ 0 & 0 & 1 \end{pmatrix}, \begin{pmatrix} 10e^{2t} \\ 6e^{2t} \\ -e^{2t} \end{pmatrix}; \begin{pmatrix} 5 \\ 11 \\ -2 \end{pmatrix}$

3. $\begin{pmatrix} 1 & 0 & 0 & 0 \\ 4 & 3 & 0 & 0 \\ 0 & 0 & 3 & 0 \\ -1 & 2 & 9 & 1 \end{pmatrix}, \begin{pmatrix} 0 \\ -2e^t \\ 0 \\ e^t \end{pmatrix}$

4. $\begin{pmatrix} 2 & 0 \\ 5 & 2 \end{pmatrix}, \begin{pmatrix} 2 \\ 10t \end{pmatrix}; \begin{pmatrix} 0 \\ 3 \end{pmatrix}$

5. $\begin{pmatrix} 5 & -4 \\ 4 & -3 \end{pmatrix}, \begin{pmatrix} 2e^t \\ 2e^t \end{pmatrix}; \begin{pmatrix} -1 \\ 3 \end{pmatrix}$

6. $\begin{pmatrix} 2 & -4 \\ 1 & -2 \end{pmatrix}, \begin{pmatrix} 1 \\ 3t \end{pmatrix}$

7. $\begin{pmatrix} 5 & 2 \\ -2 & 1 \end{pmatrix}, \begin{pmatrix} -3e^t \\ e^{3t} \end{pmatrix}$

8. $\begin{pmatrix} 2 & 0 & 0 \\ 0 & 6 & -4 \\ 0 & 4 & -2 \end{pmatrix}, \begin{pmatrix} e^{2t}\cos(3t) \\ -2 \\ -2 \end{pmatrix}$

9. $\begin{pmatrix} 7 & -1 \\ 1 & 5 \end{pmatrix}, \begin{pmatrix} 2e^{6t} \\ 6te^{6t} \end{pmatrix}$

In each of Problems 10 through 19, find a general solution of the system. If initial values are given, also solve the initial value problem.

10. $x_1' = x_1 + x_2 + 6e^{3t}, x_2' = x_1 + x_2 + 4$
11. $x_1' = 6x_1 + 5x_2 - 4\cos(3t), x_2' = x_1 + 2x_2 + 8$
12. $x_1' = -2x_1 + x_2, x_2' = -4x_1 + 3x_2 + 10\cos(t)$
13. $x_1' = 3x_1 - x_2 - x_3, x_2' = x_1 + x_2 - x_3 + t,$ $x_3' = x_1 - x_2 + x_3 + 2e^t; x_1(0) = 1,$ $x_2(0) = 2, x_3(0) = -2$
14. $x_1' = x_1 - 2x_2 + 2t, x_2' = -x_1 + 2x_2 + 5;$ $x_1(0) = 13, x_2(0) = 12$
15. $x_1' = 3x_1 + 3x_2 + 8, x_2' = x_1 + 5x_2 + 4e^{3t}$
16. $x_1' = 5x_1 - 4x_2 + 4x_3 - 3e^{-3t}, x_2' = 12x_1 - 11x_2 + 12x_3 + t, x_3' = 4x_1 - 4x_2 + 5x_3; x_1(0) = 1,$ $x_2(0) = -1, x_3(0) = 2$
17. $x_1' = x_1 + x_2 + 6e^{2t}, x_2' = x_1 + x_2 + 2e^{2t};$ $x_1(0) = 6, x_2(0) = 0$
18. $x_1' = 3x_1 - 2x_2 + 3e^{2t}, x_2' = 9x_1 - 3x_2 + e^{2t}$
19. $x_1' = 2x_1 - 5x_2 + 5\sin(t), x_2' = x_1 - 2x_2;$ $x_1(0) = 10, x_2(0) = 5$

10.4 Exponential Matrix Solutions

A differential equation $y' = ay$ with a as a constant has the general solution $y = ce^{ax}$. This leads us to ask whether there is an analogous solution for the system $\mathbf{X}' = \mathbf{AX}$ with $\mathbf{A}$ as an $n \times n$ real constant matrix.

Recall that

$$e^{ax} = 1 + ax + \frac{1}{2}(ax)^2 + \frac{1}{3!}(ax)^3 + \cdots.$$

Define the *exponential matrix* $e^{\mathbf{A}t}$ by

$$e^{\mathbf{A}t} = \mathbf{I}_n + \mathbf{A}t + \frac{1}{2}\mathbf{A}^2t^2 + \frac{1}{3!}\mathbf{A}^3t^3 + \cdots,$$

whenever the infinite series defining the i, j element on the right converges for i and j varying from 1 through n.

It is routine to verify that

$$e^{(\mathbf{A}+\mathbf{B})t} = e^{\mathbf{A}t}e^{\mathbf{B}t}$$

if $\mathbf{A}$ and $\mathbf{B}$ are $n \times n$ real matrices that commute, that is, if

$$\mathbf{AB} = \mathbf{BA}.$$

Differentiate a matrix by differentiating each element of the matrix. Using the fact that $\mathbf{A}$ is a constant matrix with derivative zero (the $n \times n$ zero matrix), we obtain from the definition that

$$(e^{\mathbf{A}t})' = \mathbf{A}e^{\mathbf{A}t},$$

which has the same form as the familiar

$$(e^{at})' = ae^{at}.$$

This derivative formula leads to the main point.

THEOREM 10.8

Let $\mathbf{A}$ be an $n \times n$ real, constant matrix and $\mathbf{K}$ be any $n \times 1$ matrix of constants. Then $e^{\mathbf{A}t}\mathbf{K}$ is a solution of $\mathbf{X}' = \mathbf{AX}$. In particular, $e^{\mathbf{A}t}$ is a fundamental matrix for this system. ♦

The proof is immediate by differentiating. Upon setting $\mathbf{X}(t) = e^{\mathbf{A}t}\mathbf{K}$, we have

$$\mathbf{X}'(t) = \frac{d}{dt}e^{\mathbf{A}t}\mathbf{K} = \mathbf{A}e^{\mathbf{A}t}\mathbf{K} = \mathbf{AX}. \quad ♦$$

We therefore have the general solution of $\mathbf{X}' = \mathbf{AX}$ if we can compute the exponential matrix $e^{\mathbf{A}t}$. Except for very simple cases this is impractical by hand and requires a computational software package. If MAPLE is used, the command

```
exponential(A,t)
```

will return $e^{\mathbf{A}t}$ if $\mathbf{A}$ has been defined and n is not "too large."

EXAMPLE 10.14

Let

$$\mathbf{A} = \begin{pmatrix} 2 & -5 \\ 1 & 4 \end{pmatrix}.$$

MAPLE returns the exponential matrix

$$e^{\mathbf{A}t} = e^{3t} \begin{pmatrix} \cos(2t) - \frac{1}{2}\sin(2t) & -\frac{5}{2}\sin(2t) \\ \frac{1}{2}\sin(2t) & \cos(2t) - \frac{1}{2}\sin(2t). \end{pmatrix}$$

This is a fundamental matrix for the system $\mathbf{X}' = \mathbf{A}\mathbf{X}$. We could also solve this system by diagonalizing $\mathbf{A}$, which has eigenvalues $3 \pm 2i$. ♦

EXAMPLE 10.15

Let

$$\mathbf{A} = \begin{pmatrix} 2 & 1 & 0 \\ 0 & 3 & -2 \\ 0 & 1 & 1 \end{pmatrix}.$$

Then

$$e^{\mathbf{A}t} = e^{2t} \begin{pmatrix} 1 & \sin(t) - \cos(t) + 1 & 2(\cos(t) - 1) \\ 0 & \sin(t) + \cos(t) & -2\sin(t) \\ 0 & \sin(t) & \cos(t) + \sin(t) \end{pmatrix}.$$

This is a fundamental matrix for $\mathbf{X}' = \mathbf{A}\mathbf{X}$. ♦

The fundamental matrix $\Omega(t) = e^{\mathbf{A}t}$ is sometimes called a *transition matrix* for the system $\mathbf{X}' = \mathbf{A}\mathbf{X}$. This is a fundamental matrix satisfying $\Omega(0) = \mathbf{I}_n$.

Variation of Parameters and the Laplace Transform

We will briefly mention a connection between the Laplace transform, the exponential matrix and the variation of parameters method for finding a particular solution $\Psi_p(t)$ of $\mathbf{X}' = \mathbf{A}\mathbf{X} + \mathbf{G}$, in which $\mathbf{A}$ is an $n \times n$ real, constant matrix.

The variation of parameters method is to write $\Psi_p(t) = \Omega(t)\mathbf{U}(t)$, where

$$\mathbf{U}(t) = \int \Omega^{-1}(t)\mathbf{G}(t)\,dt$$

and $\Omega(t)$ is a fundamental matrix for $\mathbf{X}' = \mathbf{A}\mathbf{X}$.

Write $\mathbf{U}(t)$ as a definite integral with s as the variable of integration:

$$\mathbf{U}(t) = \int_0^t \Omega^{-1}(s)\mathbf{G}(s)\,ds.$$

Then

$$\begin{aligned} \Psi_p(t) &= \Omega(t) \int_0^t \Omega^{-1}(s)\mathbf{G}(s)\,ds \\ &= \int_0^t \Omega(t)\Omega^{-1}(s)\mathbf{G}(s)\,ds. \end{aligned}$$

In this, $\Omega(t)$ can be any fundamental matrix for the system. If we choose $\Omega(t) = e^{\mathbf{A}t}$, then $\Omega^{-1}(t) = e^{-\mathbf{A}t}$ and

$$\Omega(t)\Omega^{-1}(s) = e^{\mathbf{A}t}e^{-\mathbf{A}s} = e^{\mathbf{A}(t-s)} = \Omega(t-s).$$

Now

$$\Psi_p(t) = \int_0^t \Omega(t-s)\mathbf{G}(s)\,ds = \Omega(t) * \mathbf{G}(t).$$

If we take the Laplace transform of a matrix by applying the transform to each element, then the last equation is a convolution formula for a particular solution.

To illustrate the idea, consider the system

$$\mathbf{X}' = \begin{pmatrix} 1 & -4 \\ 1 & 5 \end{pmatrix}\mathbf{X} + \begin{pmatrix} e^{2t} \\ t \end{pmatrix}.$$

Compute

$$\Omega(t) = e^{\mathbf{A}t} = \begin{pmatrix} (1-2t)e^{3t} & -4te^{3t} \\ te^{3t} & (1+2t)e^{3t} \end{pmatrix}.$$

A particular solution of the system is

$$\begin{aligned}
\Psi_p(t) &= \int_0^t \Omega(t-s)\mathbf{G}(s)\,ds \\
&= \int_0^t \begin{pmatrix} (1-2(t-s))e^{3(t-s)} & -4(t-s)e^{3(t-s)} \\ (t-s)e^{3(t-s)} & (1+2(t-s))e^{3(t-s)} \end{pmatrix}\begin{pmatrix} e^{2s} \\ s \end{pmatrix} ds \\
&= \int_0^t \begin{pmatrix} (1-2t+2s)e^{3t}e^{-s} - 4s(t-s)e^{3t}e^{-3s} \\ (t-s)e^{3t}e^{-s} + s(1+2t-2s)e^{3t}e^{-3s} \end{pmatrix} ds \\
&= \begin{pmatrix} \int_0^t [(1-2t+2s)e^{3t}e^{-s} - 4s(t-s)e^{3t}e^{-3s}]\,ds \\ \int_0^t [(t-s)e^{3t}e^{-s} + (1+2t-2s)e^{3t}e^{-3s}]\,ds \end{pmatrix} \\
&= \begin{pmatrix} -3e^{2t} + \frac{89}{27}e^{3t} - \frac{22}{9}te^{3t} - \frac{4}{9}t - \frac{8}{27} \\ e^{2t} + \frac{11}{9}te^{3t} - \frac{28}{27}e^{3t} - \frac{1}{9}t + \frac{1}{27} \end{pmatrix}.
\end{aligned}$$

The general solution is $\mathbf{X}(t) = \Omega(t)\mathbf{C} + \Psi_p(t)$, in which $\mathbf{C}$ is an $n \times 1$ matrix of constants.

SECTION 10.4 PROBLEMS

In each of the following, use a software package to compute $e^{\mathbf{A}t}$, obtaining a fundamental matrix for the system $\mathbf{X}' = \mathbf{A}\mathbf{X}$, .

1. $\mathbf{A} = \begin{pmatrix} 5 & -2 \\ 4 & 8 \end{pmatrix}$
2. $\mathbf{A} = \begin{pmatrix} 4 & -1 \\ 2 & -2 \end{pmatrix}$
3. $\mathbf{A} = \begin{pmatrix} 1 & 0 & 1 \\ -2 & 1 & 1 \\ 1 & -1 & 0 \end{pmatrix}$
4. $\mathbf{A} = \begin{pmatrix} -2 & 1 \\ 2 & -1 \end{pmatrix}$
5. $\mathbf{A} = \begin{pmatrix} -1 & 1 \\ -5 & 1 \end{pmatrix}$
6. Let $\mathbf{D}$ be an $n \times n$ diagonal matrix of numbers, with jth diagonal element d_j. Show that $e^{\mathbf{D}t}$ is the $n \times n$ diagonal matrix having $e^{d_j t}$ as its jth diagonal element.
7. Let $\mathbf{A}$ be an $n \times n$ matrix of numbers, and let $\mathbf{P}$ be an $n \times n$ nonsingular matrix of numbers. Let $\mathbf{B} = \mathbf{P}^{-1}\mathbf{A}\mathbf{P}$. Show that
$$e^{\mathbf{B}t} = \mathbf{P}^{-1}e^{\mathbf{A}t}\mathbf{P}.$$
From this, conclude that
$$e^{\mathbf{A}t} = \mathbf{P}e^{\mathbf{B}t}\mathbf{P}^{-1}.$$
8. Use the results of Problems 6 and 7 to show that, if $\mathbf{P}$ diagonalizes $\mathbf{A}$, so $\mathbf{P}^{-1}\mathbf{A}\mathbf{P} = \mathbf{D}$, which is a diagonal matrix with diagonal elements d_j. Then
$$e^{\mathbf{A}t} = \mathbf{P}e^{\mathbf{D}t}\mathbf{P}^{-1},$$
where $e^{\mathbf{D}t}$ is the diagonal matrix having $e^{d_j t}$ as main diagonal elements.
9. Use the result of Problem 8 to determine the exponential matrix in each of Problems 4 and 5.

10.5 Applications and Illustrations of Techniques

This section presents some examples involving mechanical systems and electrical circuits, whose analysis gives rise to systems of differential equations. We have previously applied the Laplace transform to solve such systems. Here we will apply matrix methods.

EXAMPLE 10.16 A Mass/Spring System

We will analyze the system of three springs and two weights shown in Figure 10.2, which displays the spring constants and the mass of each weight. At time 0, the upper weight is pulled down one unit and the lower one is raised one unit, then both are released. We want to know the position of each weight relative to its equilibrium position at any later time.

The initial value problem to be solved is

$$y_1'' = -8y_1 + 2y_2,$$
$$y_2'' = 2y_1 - 5y_2,$$
$$y_1(0) = 1,\ y_2(0) = -1,\ y_1'(0) = y_2'(0) = 0.$$

Begin by converting this system of two second-order differential equations to a system of four first-order differential equations by putting

$$x_1 = y_1,$$
$$x_2 = y_2,$$
$$x_3 = y_1',$$

and

$$x_4 = y_2'.$$

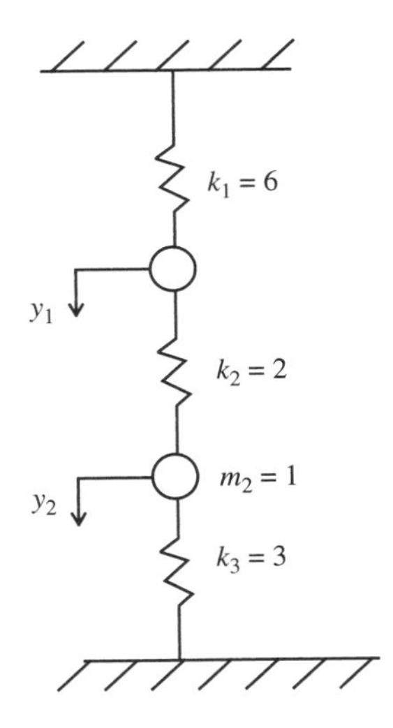

FIGURE 10.2 *Mass/spring system of Example 10.16.*

The system of two second-order equations translates to the following system in terms of $x_1, \cdots, x_4$:

$$x_1' = y_1' = x_3,$$
$$x_2' = y_2' = x_4,$$
$$x_3' = y_1'' = -8y_1 + 2y_2 = -8x_1 + 2x_2,$$
$$x_4' = y_2'' = 2y_1 - 5y_2 = 2x_1 - 5x_2,$$

and

$$x_1(0) = 1, x_2(0) = -1, x_3(0) = x_4(0) = 0.$$

This is the system $\mathbf{X}' = \mathbf{AX}$ with

$$\mathbf{A} = \begin{pmatrix} 0 & 0 & 1 & 0 \\ 0 & 0 & 0 & 1 \\ -8 & 2 & 0 & 0 \\ 2 & -5 & 0 & 0 \end{pmatrix}$$

and

$$\mathbf{X}(0) = \begin{pmatrix} 1 \\ -1 \\ 0 \\ 0 \end{pmatrix}.$$

$\mathbf{A}$ has the characteristic equation

$$(\lambda^2 + 4)(\lambda^2 + 9) = 0$$

with eigenvalues $\pm 2i$ and $\pm 3i$. Corresponding to the eigenvalues $2i$ and $3i$, we find two eigenvectors

$$\begin{pmatrix} 1 \\ 2 \\ 0 \\ 0 \end{pmatrix} + i \begin{pmatrix} 0 \\ 0 \\ 2 \\ 4 \end{pmatrix} \text{ and } \begin{pmatrix} 2 \\ -1 \\ 0 \\ 0 \end{pmatrix} + i \begin{pmatrix} 0 \\ 0 \\ 6 \\ -3 \end{pmatrix}.$$

The complex conjugates of these eigenvectors are also eigenvectors corresponding to eigenvalues $-2i$ and $-3i$. However, we will not write these other two eigenvectors, because we will use Theorem 10.8 to write the four linearly independent solutions involving only real-valued functions. From the eigenvector for $2i$, write the two solutions

$$\begin{pmatrix} 1 \\ 2 \\ 0 \\ 0 \end{pmatrix} \cos(2t) - \begin{pmatrix} 0 \\ 0 \\ 2 \\ 4 \end{pmatrix} \sin(2t) \quad \text{and} \quad \begin{pmatrix} 1 \\ 2 \\ 0 \\ 0 \end{pmatrix} \sin(2t) + \begin{pmatrix} 0 \\ 0 \\ 2 \\ 4 \end{pmatrix} \cos(2t).$$

From the eigenvector for $3i$, write the two solutions

$$\begin{pmatrix} 2 \\ -1 \\ 0 \\ 0 \end{pmatrix} \cos(3t) - \begin{pmatrix} 0 \\ 0 \\ 6 \\ -3 \end{pmatrix} \sin(3t) \quad \text{and} \quad \begin{pmatrix} 2 \\ -1 \\ 0 \\ 0 \end{pmatrix} \sin(3t) + \begin{pmatrix} 0 \\ 0 \\ 6 \\ -3 \end{pmatrix} \cos(3t).$$

Use these four linearly independent solutions as columns of the fundamental matrix

$$\Omega(t) = \begin{pmatrix} \cos(2t) & \sin(2t) & 2\cos(3t) & 2\sin(3t) \\ 2\cos(2t) & 2\sin(2t) & -\cos(3t) & -\sin(3t) \\ -2\sin(2t) & 2\cos(2t) & -6\sin(3t) & 6\cos(3t) \\ -4\sin(2t) & 4\cos(2t) & 3\sin(3t) & -3\cos(3t) \end{pmatrix}.$$

Notice that row three is the derivative of row one, and row four is the derivative of row two, consistent with the fact that $x_3 = y_1' = x_1'$ and $x_4 = y_2' = x_2'$. This serves as a partial check on the computations.

The general solution of the system is $\mathbf{X}(t) = \Omega(t)\mathbf{C}$. To solve the initial value problem, we need

$$\Omega(0)\mathbf{C} = \begin{pmatrix} 1 & 0 & 2 & 0 \\ 2 & 0 & -1 & 0 \\ 0 & 2 & 0 & 6 \\ 0 & 4 & 0 & -3 \end{pmatrix} \begin{pmatrix} c_1 \\ c_2 \\ c_3 \\ c_4 \end{pmatrix} = \begin{pmatrix} 1 \\ -1 \\ 0 \\ 0 \end{pmatrix}.$$

This has the unique solution

$$\mathbf{C} = \begin{pmatrix} -1/5 \\ 0 \\ 3/5 \\ 0 \end{pmatrix}.$$

The solution of the initial value problem is

$$\mathbf{X}(t) = \begin{pmatrix} \cos(2t) & \sin(2t) & 2\cos(3t) & 2\sin(3t) \\ 2\cos(2t) & 2\sin(2t) & -\cos(3t) & -\sin(3t) \\ -2\sin(2t) & 2\cos(2t) & -6\sin(3t) & 6\cos(3t) \\ -4\sin(2t) & 4\cos(2t) & 3\sin(3t) & -3\cos(3t) \end{pmatrix} \begin{pmatrix} -1/5 \\ 0 \\ 3/5 \\ 0 \end{pmatrix}$$

$$= \frac{1}{5}\begin{pmatrix} -\cos(2t) + 6\cos(3t) \\ -2\cos(2t) - 3\cos(3t) \\ 2\sin(2t) - 18\sin(3t) \\ 4\sin(2t) + 9\sin(3t) \end{pmatrix}.$$

Since $x_1 = y_1$ and $x_2 = y_2$, we may write, in the notation of Figure 10.2,

$$y_1(t) = -\frac{1}{5}\cos(2t) + \frac{6}{5}\cos(3t)$$

$$y_2(t) = -\frac{2}{5}\cos(2t) - \frac{3}{5}\cos(3t).$$

We could also have used the exponential matrix to produce a fundamental matrix. MAPLE may produce a different fundamental matrix than that found using Theorem 10.7, but of course, the solution of the initial value problem is the same. ♦

EXAMPLE 10.17 An Electrical Circuit

Assume that the currents and charges in the circuit of Figure 10.3 are zero until time $t = 0$, at which time the switch is closed. We want to determine the current in each loop.

Use Kirchhoff's voltage and current laws on the left and right loops to obtain

$$5i_1 + 5(i_1' - i_2') = 10 \tag{10.4}$$

and

$$5(i_1' - i_2') = 2i_2 + \frac{q_2}{5 \times 10^{-2}}. \tag{10.5}$$

Using the exterior loop (around the entire circuit), we get

$$5i_1 + 20i_2 + \frac{q_2}{5 \times 10^{-2}} = 10. \tag{10.6}$$

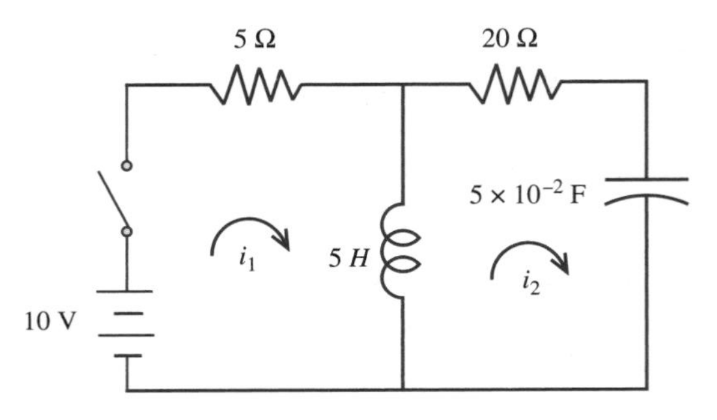

FIGURE 10.3 *Circuit of Example 10.17.*

Any two of these equations contain all of the information needed to solve the problem. We will use equations (10.4) and (10.6). The reason for this choice is that we would have to differentiate the second equation to eliminate the q_2 term in equation (10.5), producing second derivative terms in the currents. This is avoided by using the first and third equations.

Divide equations (10.4) and (10.6) by 5 and differentiate the new equation (10.6), using the fact that $q_2' = i_2$, to obtain

$$i_1' - i_2' = -i_1 + 2$$
$$i_1' + 4i_2' = -4i_2.$$

We must determine the initial conditions. We know that

$$i_1(0-) = i_2(0-) = q_2(0-) = 0.$$

Then $(i_1 - i_2)(0-) = 0$. Since the current $i_1 - i_2$ through the inductor is continuous, then $(i_1 - i_2)(0+) = 0$ also. Therefore,

$$i_1(0+) = i_2(0+) = 0.$$

Put this into equation (10.6) and use the fact that the charge on the capacitor is continuous to obtain

$$5i_1(0+) + 20i_2(0+) + 20q_2(0+) = 10$$

or

$$25i_1(0+) = 10.$$

Then

$$i_1(0+) = i_2(0+) = \frac{10}{25} = \frac{2}{5}$$

amperes. Finally, the initial value problem for the currents is

$$i_1' - i_2' = -i_1 + 2$$
$$i_1' + 4i_2' = -4i_2$$
$$i_1(0+) = i_2(0+) = \frac{2}{5}.$$

In matrix form,

$$\begin{pmatrix} 1 & -1 \\ 1 & 4 \end{pmatrix} \begin{pmatrix} i_1 \\ i_2 \end{pmatrix}' = \begin{pmatrix} -1 & 0 \\ 0 & -4 \end{pmatrix} \begin{pmatrix} i_1 \\ i_2 \end{pmatrix} + \begin{pmatrix} 2 \\ 0 \end{pmatrix}.$$

This can be written as

$$\mathbf{Bi}' = \mathbf{Ai} + \mathbf{K}.$$

This is not quite in the form we have studied. However, $|\mathbf{B}| = 5$, so $\mathbf{B}$ is nonsingular. We find that

$$\mathbf{B}^{-1} = \frac{1}{5}\begin{pmatrix} 4 & 1 \\ -1 & 1 \end{pmatrix}.$$

Multiply the system by $\mathbf{B}^{-1}$ to obtain

$$\begin{pmatrix} i_1 \\ i_2 \end{pmatrix}' = \begin{pmatrix} -4/5 & -4/5 \\ 1/5 & -4/5 \end{pmatrix}\begin{pmatrix} i_1 \\ i_2 \end{pmatrix} + \begin{pmatrix} 8/5 \\ -2/5 \end{pmatrix}.$$

This is in the standard form

$$\mathbf{i}' = \mathbf{Ai} + \mathbf{G}$$

in which

$$\mathbf{A} = \begin{pmatrix} -4/5 & -4/5 \\ 1/5 & -4/5 \end{pmatrix}.$$

We will use the exponential matrix and variation of parameters to solve for the currents, assuming availability of software to compute $e^{\mathbf{A}t}$ and carry out integrations and matrix products that are needed. First, compute the fundamental matrix

$$\Omega(t) = e^{\mathbf{A}t} = \begin{pmatrix} e^{-4t/5}\cos(2t/5) & -2e^{-4t/5}\sin(2t/5) \\ \frac{1}{2}e^{-4t/5}\sin(2t/5) & e^{-4t/5}\cos(2t/5) \end{pmatrix}.$$

Then

$$\Omega^{-1}(t) = \begin{pmatrix} e^{4t/5}\cos(2t/5) & 2e^{4t/5}\sin(2t/5) \\ -\frac{1}{2}e^{4t/5}\sin(2t/5) & e^{4t/5}\cos(2t/5) \end{pmatrix}.$$

The general solution of the associated homogeneous system $\mathbf{i}' = \mathbf{Ai}$ is $\Omega\mathbf{C}$.

For a particular solution Ψ_p of the nonhomogeneous system, first compute

$$\Omega^{-1}\mathbf{G} = \Omega\begin{pmatrix} 8/5 \\ -2/5 \end{pmatrix} = \begin{pmatrix} \frac{8}{5}e^{4t/5}\cos(2t/5) - \frac{4}{5}e^{4t/5}\sin(2t/5) \\ -\frac{4}{5}e^{4t/5}\sin(2t/5) - \frac{2}{5}e^{4t/5}\cos(2t/5) \end{pmatrix}.$$

Form

$$\mathbf{U}(t) = \int \Omega^{-1}(t)\mathbf{G}(t)\,dt = \begin{pmatrix} 2e^{4t/5}\cos(2t/5) \\ -e^{4t/5}\sin(2t/5) \end{pmatrix}.$$

A particular solution of the nonhomogeneous system is

$$\Psi(t) = \Omega(t)\mathbf{U}(t) = \begin{pmatrix} 2\cos^2(2t/5) + 2\sin^2(2t/5) \\ \cos(2t/5)\sin(2t/5) - \cos(2t/5)\sin(2t/5) \end{pmatrix} = \begin{pmatrix} 2 \\ 0 \end{pmatrix}.$$

The general solution of $\mathbf{i}' = \mathbf{Ai} + \mathbf{G}$ is

$$\mathbf{i} = \begin{pmatrix} e^{-4t/5}\cos(2t/5) & -2e^{-4t/5}\sin(2t/5) \\ \frac{1}{2}e^{-4t/5}\sin(2t/5) & e^{-4t/5}\cos(2t/5) \end{pmatrix}\begin{pmatrix} c_1 \\ c_2 \end{pmatrix} + \begin{pmatrix} 2 \\ 0 \end{pmatrix}.$$

Since $i_1(0) = i_2(0) = 2/5$, then we must choose c_1 and c_2 so that

$$\begin{pmatrix} 2/5 \\ 2/5 \end{pmatrix} = \begin{pmatrix} 1 & 0 \\ 0 & 1 \end{pmatrix}\begin{pmatrix} c_1 \\ c_2 \end{pmatrix} + \begin{pmatrix} 2 \\ 0 \end{pmatrix}.$$

Then $c_1 = -8/5$ and $c_2 = 2/5$. The solution for the currents is

$$\begin{pmatrix} i_1 \\ i_2 \end{pmatrix} = \begin{pmatrix} e^{-4t/5}\cos(2t/5) & -2e^{-4t/5}\sin(2t/5) \\ \frac{1}{2}e^{-4t/5}\sin(2t/5) & e^{-4t/5}\cos(2t/5) \end{pmatrix}\begin{pmatrix} -8/5 \\ 2/5 \end{pmatrix} + \begin{pmatrix} 2 \\ 0 \end{pmatrix}.$$

This gives us

$$i_1(t) = 2 - \frac{4}{5}e^{-4t/5}\left(2\cos(2t/5) + \sin(2t/5)\right)$$

$$i_2(t) = \frac{2}{5}e^{-4t/5}\left(\cos(2t/5) - 2\sin(2t/5)\right).$$

The currents can also be obtained by diagonalizing **A** and changing variables by setting $\mathbf{i} = \mathbf{PZ}$, where **P** diagonalizes **A**. This is a straightforward computation but is a little tedious because the eigenvalues of **A** are complex. ◆

EXAMPLE 10.18 Another Electrical Circuit

The circuit of Figure 10.4 has three connected loops (and of course, the external loop). The currents in these three loops are zero prior to $t = 0$, at which time the switch is closed. The capacitor is in a discharged state at time zero. We want to determine the current in each loop at all later times.

Apply Kirchhoff's current and voltage laws to obtain

$$4i_1 + 2i_1' - 2i_2' = 36, \tag{10.7}$$

$$2i_1' - 2i_2' = 5i_2 + 10q_2 - 10q_3, \tag{10.8}$$

$$10q_2 - 10q_3 = 5i_3, \tag{10.9}$$

$$4i_1 + 5i_2 + 10q_2 - 10q_3 = 36, \tag{10.10}$$

$$4i_1 + 5i_2 + 5i_3 = 36, \tag{10.11}$$

$$2i_1' - 2i_2' = 5i_2 + 5i_3. \tag{10.12}$$

Any three of these equations are enough to determine the currents. Because equation (10.8) involves both charge and current terms, we would have to differentiate to put everything in terms of currents (recall that $q_j' = i_j$). This would introduce second derivatives, which we want to avoid. Cross out this equation. We could use equation (10.11) to eliminate one variable and reduce the problem to a two by two system, but this would involve a lot of algebra.

Equation (10.9) is a likely candidate to retain. If we then use equations (10.7) and (10.12), we obtain a system of the form $\mathbf{Bi}' = \mathbf{Di} + \mathbf{F}$ with **B** as singular, so we would not be able to multiply by $\mathbf{B}^{-1}$ to obtain a system in standard form.

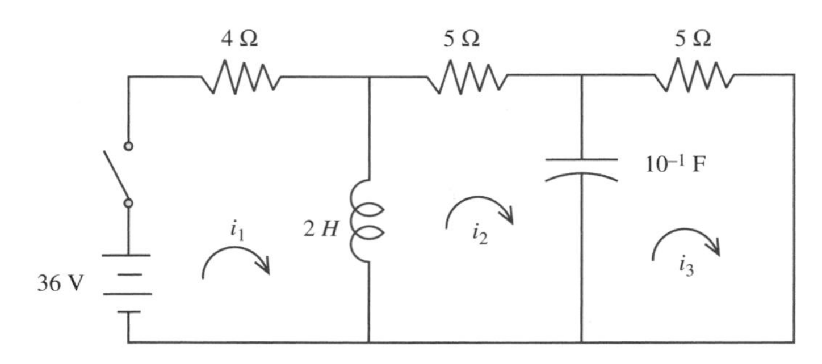

FIGURE 10.4 *Circuit of Example 10.18.*

We therefore choose to use equations (10.7), (10.9), and (10.10). In rearranged order and some manipulation, these can be rewritten as

$$i_1' - i_2' = -2i_1 + 18,$$

$$4i_1' + 5i_2' = -10i_1 + 10i_3,$$

and

$$i_3' = 2i_2 - 2i_3.$$

Henceforth, we refer to this as the system. We must determine the initial conditions. The conductor current $i_1 - i_2$ is continuous, and $i_1(0-) - i_2(0-) = 0$. Therefore,

$$i_1(0+) - i_2(0+) = 0,$$

so

$$i_1(0+) = i_2(0+).$$

The capacitor charge $q_1 - q_2$ is also continuous. Since $q_1(0-) - q_2(0-) = 0$, then

$$q_1(0+) = q_2(0+).$$

By equation (10.9),

$$i_3(0+) = 0.$$

Now use equation (10.11) to write

$$4i_1(0+) + 5i_2(0+) + 5i_3(0+) = 36.$$

Since $i_3(0+) = 0$ and $i_1(0+) = i_2(0+)$, then $9i_1(0+) = 36$. So

$$i_1(0+) = i_2(0+) = 4.$$

In summary, we now have the system

$$\begin{pmatrix} 1 & -1 & 0 \\ 4 & 5 & 0 \\ 0 & 0 & 1 \end{pmatrix} \begin{pmatrix} i_1 \\ i_2 \\ i_3 \end{pmatrix}' = \begin{pmatrix} -2 & 0 & 0 \\ 0 & -10 & 10 \\ 0 & 2 & -2 \end{pmatrix} \begin{pmatrix} i_1 \\ i_2 \\ i_3 \end{pmatrix} + \begin{pmatrix} 18 \\ 0 \\ 0 \end{pmatrix}. \tag{10.13}$$

This has the form $\mathbf{B}\mathbf{i}' = \mathbf{D}\mathbf{i} + \mathbf{F}$ with $\mathbf{B}$ nonsingular. The initial condition is

$$\mathbf{i}(0+) = \begin{pmatrix} 4 \\ 4 \\ 0 \end{pmatrix}.$$

Multiply the system (10.13) by

$$\mathbf{B}^{-1} = \frac{1}{9} \begin{pmatrix} 5 & 1 & 0 \\ -4 & 1 & 0 \\ 0 & 0 & 9 \end{pmatrix}$$

to obtain a system

$$\begin{pmatrix} i_1 \\ i_2 \\ i_3 \end{pmatrix}' = \begin{pmatrix} -10/9 & -10/9 & 10/9 \\ 8/9 & -10/9 & 10/9 \\ 0 & 2 & -2 \end{pmatrix} \begin{pmatrix} i_1 \\ i_2 \\ i_3 \end{pmatrix} + \begin{pmatrix} 10 \\ -8 \\ 0 \end{pmatrix}.$$

This is in the standard form $\mathbf{i}' = \mathbf{A}\mathbf{i} + \mathbf{G}$. $\mathbf{A}$ has eigenvalues 0, -2, and $-20/9$ with corresponding eigenvectors

$$\begin{pmatrix} 0 \\ 1 \\ 1 \end{pmatrix}, \begin{pmatrix} 5 \\ 0 \\ -4 \end{pmatrix}, \text{ and } \begin{pmatrix} 10 \\ 1 \\ -9 \end{pmatrix}.$$

The matrix

$$\mathbf{P} = \begin{pmatrix} 0 & 5 & 10 \\ 1 & 0 & 1 \\ 1 & -4 & -9 \end{pmatrix}$$

diagonalizes $\mathbf{A}$. To make the change of variables $\mathbf{i} = \mathbf{PZ}$, we will need

$$\mathbf{P}^{-1} = \frac{1}{10}\begin{pmatrix} 4 & 5 & 5 \\ 10 & -10 & 10 \\ -4 & 5 & -5 \end{pmatrix} \text{ and } \mathbf{P}^{-1}\mathbf{G} = \begin{pmatrix} 0 \\ 18 \\ -8 \end{pmatrix}.$$

Now set $\mathbf{i} = \mathbf{PZ}$ in the system to obtain

$$\mathbf{PZ}' = (\mathbf{AP})\mathbf{Z} + \mathbf{G}.$$

Multiply this system on the left by $\mathbf{P}^{-1}$ for

$$\mathbf{Z}' = (\mathbf{P}^{-1}\mathbf{AP})\mathbf{Z} + \mathbf{P}^{-1}\mathbf{G}$$

or

$$\mathbf{Z}' = \mathbf{DZ} + \mathbf{P}^{-1}\mathbf{G}$$

in which $\mathbf{D}$ is the 3×3 diagonal matrix having the eigenvalues of $\mathbf{A}$ down its main diagonal. This uncoupled system is

$$\begin{pmatrix} z_1' \\ z_2' \\ z_3' \end{pmatrix} = \begin{pmatrix} 0 & 0 & 0 \\ 0 & -2 & 0 \\ 0 & 0 & -20/9 \end{pmatrix}\begin{pmatrix} z_1 \\ z_2 \\ z_3 \end{pmatrix} + \begin{pmatrix} 0 \\ 18 \\ -8 \end{pmatrix}.$$

The uncoupled differential equations for the z_j's are

$$z_1' = 0$$
$$z_2' + 2z_2 = 18$$

and

$$z_3' + \frac{20}{9}z_3 = -8,$$

which we solve individually to obtain

$$z_1 = c_1$$
$$z_2 = c_2e^{-2t} + 9$$
$$z_3 = c_3e^{-20t/9} - \frac{18}{5}.$$

Then

$$\mathbf{i} = \begin{pmatrix} i_1 \\ i_2 \\ i_3 \end{pmatrix} = \mathbf{PZ} = \begin{pmatrix} 0 & 5 & 10 \\ 1 & 0 & 1 \\ 1 & -4 & -9 \end{pmatrix}\begin{pmatrix} z_1 \\ z_2 \\ z_3 \end{pmatrix}$$

$$= \begin{pmatrix} 0 & 5 & 10 \\ 1 & 0 & 1 \\ 1 & -4 & -9 \end{pmatrix}\begin{pmatrix} c_1 \\ c_2e^{-2t} + 9 \\ c_3e^{-20t/9} - 18/5 \end{pmatrix}$$

$$= \begin{pmatrix} 9 + 5c_2e^{-2t} + 10c_3e^{-20t/9} \\ c_1 + c_3e^{-20t/9} - 18/5 \\ c_1 - 4c_2e^{-2t} - 9c_3e^{-20t/9} - 18/5 \end{pmatrix}.$$

Now use the initial condition to solve for the constants. Solve

$$\mathbf{i}(0+) = \begin{pmatrix} 4 \\ 4 \\ 0 \end{pmatrix} = \begin{pmatrix} 5c_2 + 10c_3 + 9 \\ c_1 + c_3 - 18/5 \\ c_1 - 4c_2 - 9c_3 - 18/5 \end{pmatrix}.$$

This can be written as

$$\begin{pmatrix} 0 & 5 & 10 \\ 1 & 0 & 1 \\ 1 & -4 & -9 \end{pmatrix} \mathbf{C} = \mathbf{PC} = \begin{pmatrix} -5 \\ 38/5 \\ 18/5 \end{pmatrix}.$$

Then

$$\mathbf{C} = \mathbf{P}^{-1} \begin{pmatrix} -5 \\ 38/5 \\ 18/5 \end{pmatrix} = \frac{1}{10} \begin{pmatrix} 4 & 5 & 5 \\ 10 & -10 & 10 \\ -4 & 5 & -5 \end{pmatrix} \begin{pmatrix} -5 \\ 38/5 \\ 18/5 \end{pmatrix} = \begin{pmatrix} 18/5 \\ -9 \\ 4 \end{pmatrix}.$$

The current is

$$\mathbf{i} = \begin{pmatrix} 9 - 4te^{-2t} + 40e^{-20t/9} \\ 4e^{-20t/9} \\ 36e^{-2t} - 36e^{-20t/9} \end{pmatrix}. \blacklozenge$$

SECTION 10.5 PROBLEMS

1. Two tanks are connected as shown in Figure 10.5. Tank 1 initially contains 100 gallons of water in which 40 pounds of salt are dissolved. Tank 2 initially contains 150 gallons of pure water. Beginning at $t = 0$, a brine solution containing $1/5$ pound of salt per gallon is pumped into tank 1 at the rate of 5 gallons per minute. At this time, a solution which also contains $1/5$ pound of salt per gallon is pumped into tank 2 at the rate of 10 gallons per minute. The brine solutions are interchanged between the tanks and also flow out of both tanks at the rates shown. Determine the amount of salt in each tank for $t \geq 0$. Also calculate the time at which the brine solution in tank 1 reaches its minimum salinity (concentration of salt) and determine how much salt is in tank 1 at that time.

2. Referring to Figure 10.6, tank 1 initially contains 200 gallons of saltwater (brine), while tank 2 initially contains 300 gallons of brine. Beginning at time 0, brine is pumped into tank 1 at the rate of 4 gallons per minute, pure water is pumped into tank 2 at 6 gallons per minute, and the brine solutions are interchanged between the two tanks and also flow out of both tanks at the rates shown. The input to tank 1 contains $1/4$ pound of salt per gallon, tank 1 initially has 200 pounds of salt, and tank 2 initially has 150 pounds

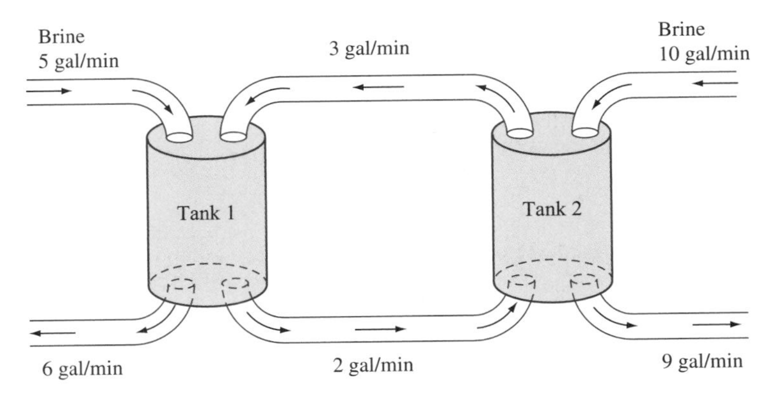

FIGURE 10.5 *Tank system for Problem 1, Section 10.5.*

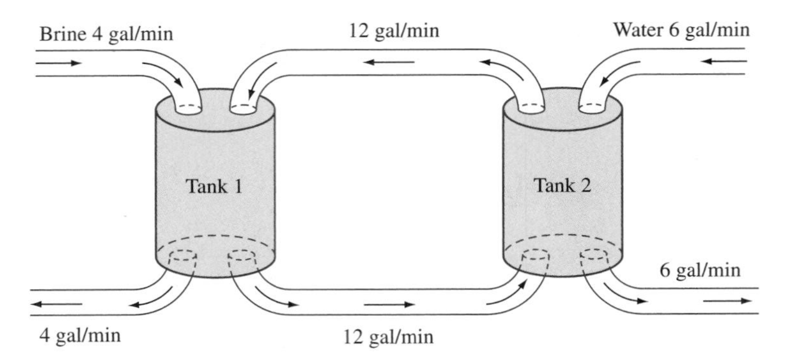

FIGURE 10.6 *Connected tank system for Problem 2, Section 10.5.*

of salt. Determine the amount of salt in each tank at time $t > 0$.

3. Referring to the circuit of Figure 10.4, determine how much time elapses between the time the switch is closed and the time the charge on the capacitor is a maximum. What is the maximum voltage on the capacitor?

4. Find the currents $i_1(t)$ and $i_2(t)$ in the circuit of Figure 10.7 for $t > 0$, assuming that the currents and charges are all zero prior to the switch being closed at $t = 0$.

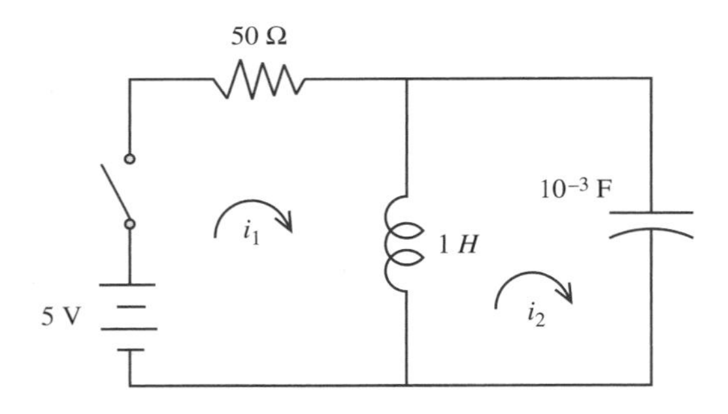

FIGURE 10.7 *Circuit for Problem 4, Section 10.5.*

Each of Problems 5 and 6 refer to the system of Figure 10.8. Derive and solve the differential equations for the motions of the masses under the assumption that there is no damping.

5. Each mass is pulled downward one unit and released from rest with no external driving forces.

6. The masses have zero initial displacement and velocity. The lower mass is subjected to an external driving force of magnitude $F(t) = 2\sin(3t)$, while the upper mass has no driving force applied to it.

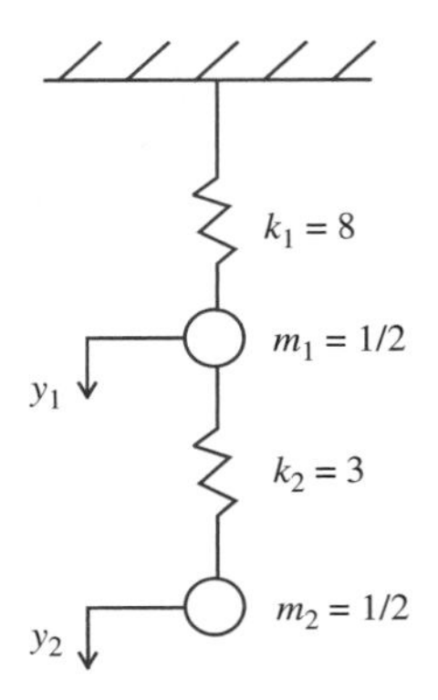

FIGURE 10.8 *Mass/spring system for Problems 5 and 6, Section 10.5.*

7. In the circuit of Figure 10.9, assume that the currents and charges are all zero prior to the switch being closed at time 0. Find the loop currents for time $t > 0$.

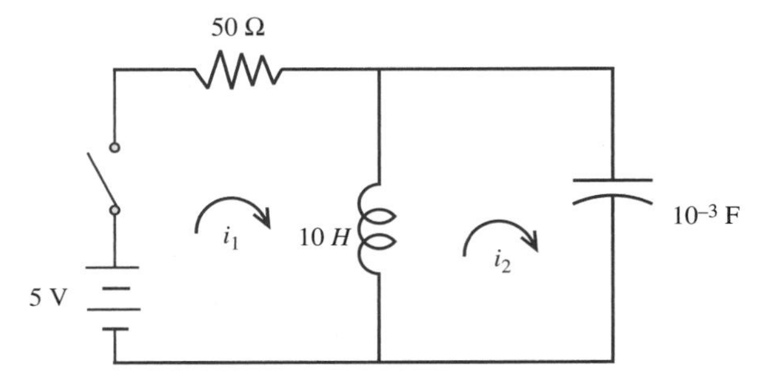

FIGURE 10.9 *Circuit for Problem 7, Section 10.5.*

8. Find the loop currents in the circuit of Figure 10.10 for $t > 0$, assuming that the currents and charge are all zero prior to the switch being closed at $t = 0$. Also

determine the maximum value of $E_{out}(t)$ and when this maximum value is reached.

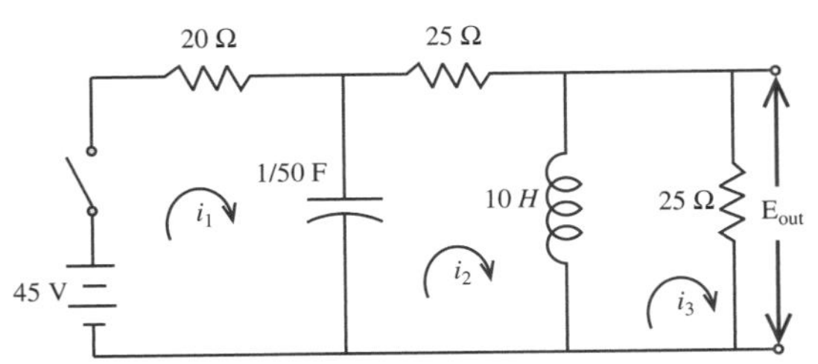

FIGURE 10.10 *Circuit for Problem 8, Section 10.5.*

9. Refer to the mechanical system of Figure 10.11. The left mass is pushed to the right one unit, and the right mass is pushed to the left one unit. Both are released from rest at time $t = 0$. Assume that there are no external driving forces. Derive and solve the differential equations with appropriate initial conditions for the displacement of the masses, assuming that there is no damping. Denote left to right as the positive direction.

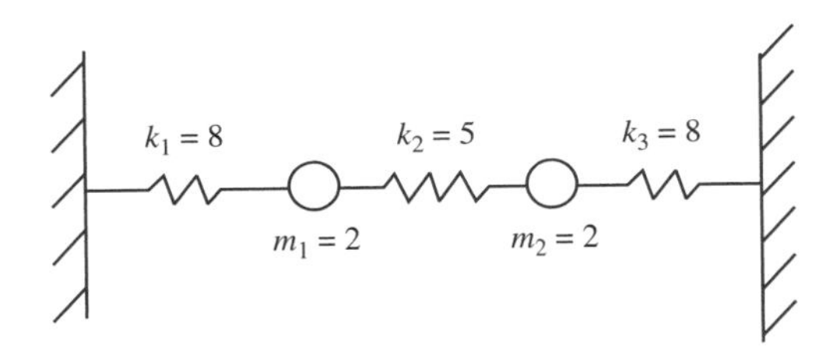

FIGURE 10.11 *Mass/spring system for Problem 9, Section 10.5.*

10. Find the currents in each loop of the circuit of Figure 10.12. Assume that the currents and charges are all zero prior to the switch being closed at $t = 0$.

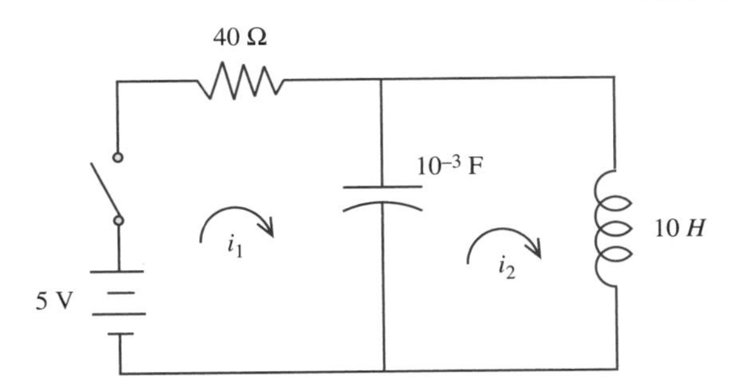

FIGURE 10.12 *Circuit for Problem 10, Section 10.5.*

11. Derive a system of differential equations for the displacement functions for the masses in Figure 10.13, in which $a = 10\sqrt{26}$. Assume that the top weight is lowered one unit and the lower one raised one unit, then both are released from rest at time 0. The upper weight is free of external driving forces, while the lower weight is subjected to an external force of magnitude $F(t) = 39\sin(t)$.

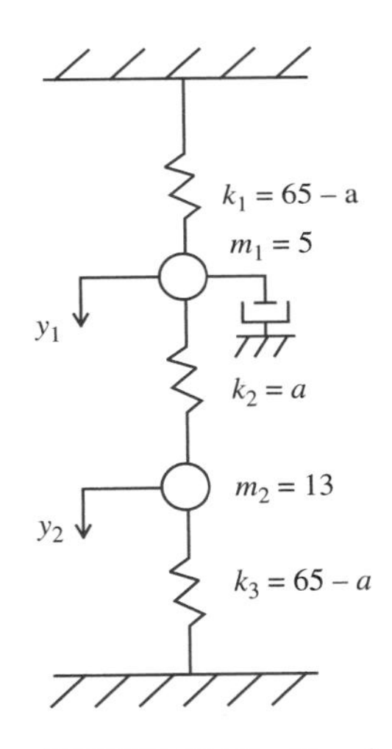

FIGURE 10.13 *Mass/spring system for Problem 11, Section 10.5.*

10.6 Phase Portraits

10.6.1 Classification by Eigenvalues

Consider the linear 2×2 system $\mathbf{X}' = \mathbf{AX}$ with $\mathbf{A}$ as a real nonsingular matrix and

$$\mathbf{X}(t) = \begin{pmatrix} x(t) \\ y(t) \end{pmatrix}.$$

We know how to solve this system. However, now we want to focus on the geometry and qualitative behavior of solutions.

Given a solution, we can think of the point $(x(t), y(t))$ as moving along a curve or *trajectory* in the plane as t, often thought of as time, increases. A copy of the plane, with trajectories drawn through various points, is called a *phase portrait* for $\mathbf{X}' = \mathbf{AX}$. Phase portraits provide visual insight into how the trajectories move and how solutions behave. Because of the uniqueness of solutions of initial value problems, there can be only one trajectory through any given point in the plane. Furthermore, two distinct trajectories cannot intersect, because at the point of intersection there would be two trajectories through the same point, and these would both be solutions of the same initial value problem.

Often phase portraits are drawn within a direction field. Recall from Chapter 5 that a direction field consists of short line segments of tangents to trajectories. These tangent segments outline the way solution curves move in the plane, and provide a flow pattern for the trajectories. Arrows drawn along these segments indicate the direction of the flow as t increases.

For the system $\mathbf{X}' = \mathbf{AX}$, the origin $(0, 0)$ plays a special role. This point is actually the graph of the constant solution

$$x(t) = 0, y(t) = 0 \text{ for all } t$$

which is the solution of the unique initial value problem

$$\mathbf{X}' = \mathbf{AX}; \mathbf{X}(0) = \begin{pmatrix} 0 \\ 0 \end{pmatrix}.$$

No other trajectory can pass through the origin, because then two distinct trajectories would intersect.

We will now examine trajectories of $\mathbf{X}' = \mathbf{AX}$, paying particular attention to their behavior near the origin. Because solutions are determined by the eigenvalues of $\mathbf{A}$, we will use these to distinguish cases.

Case 1: Real Distinct Eigenvalues λ and μ of the Same Sign

Let associated eigenvectors be $\mathbf{E}_1$ and $\mathbf{E}_2$. Because λ and μ are distinct, these eigenvectors are linearly independent and the general solution is

$$\mathbf{X}(t) = \begin{pmatrix} x(t) \\ y(t) \end{pmatrix} = c_1\mathbf{E}_1 e^{\lambda t} + c_2\mathbf{E}_2 e^{\mu t}.$$

Represent the vectors $\mathbf{E}_1$ and $\mathbf{E}_2$ as vectors from the origin, as in Figure 10.14. Draw L_1 and L_2, respectively, through the origin along these vectors. These will serve as guidelines in drawing trajectories.

Case 1(a): The Eigenvalues are Negative, say $\lambda < \mu < 0$

Now $e^{\lambda t} \to 0$ and $e^{\mu t} \to 0$ as $t \to \infty$, so $\mathbf{X}(t) \to (0, 0)$ and each trajectory approaches the origin. This can happen in three ways, depending on an initial point $P_0 : (x_0, y_0)$ we choose for a trajectory to pass through at time $t = 0$. These possibilities are as follows.

1. If P_0 is on L_1, then $c_2 = 0$ and

$$\mathbf{X}(t) = c_1 e^{\lambda t}.$$

For any t this is a scalar multiple of $\mathbf{E}_1$, so the trajectory through P_0 is part of L_1, with arrows along it pointing toward the origin because the trajectory moves toward the origin as time increases. This is the trajectory T_1 of Figure 10.15.

2. If P_0 is on L_2, then $c_1 = 0$ and now

$$\mathbf{X}(t) = c_2 e^{\mu t}.$$

This trajectory is part of the line L_2, with arrows of the direction field indicating that it also approaches the origin as t increases. This is the trajectory T_2 of Figure 10.15.

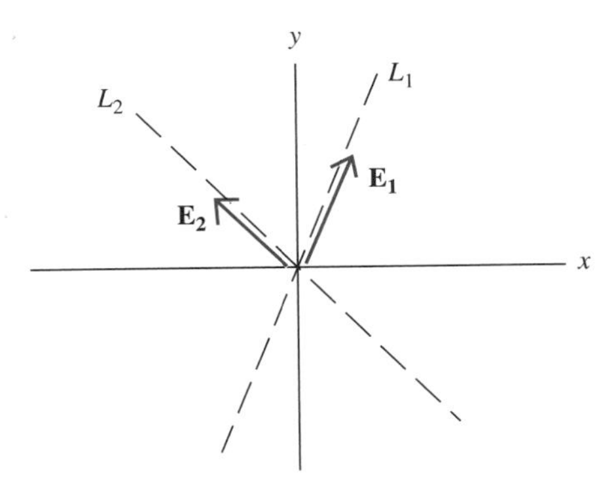

FIGURE 10.14 *Eigenvectors* $\mathbf{E}_1$, $\mathbf{E}_2$ *in Case 1.*

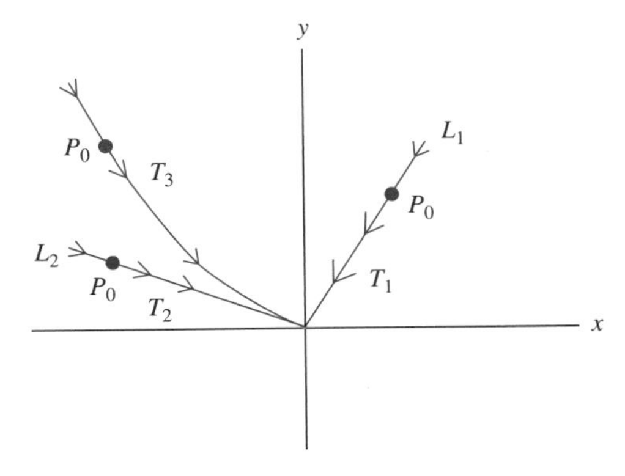

FIGURE 10.15 *Trajectories in Case 1-a.*

3. If P_0 is on neither L_1 or L_2, then the trajectory is a curve through P_0 having the parametric form

$$\mathbf{X}(t) = c_1\mathbf{E}_1 e^{\lambda t} + c_2\mathbf{E}_2 e^{\mu t}.$$

Write this as

$$\mathbf{X}(t) = e^{\mu t}[c_1\mathbf{E}_1 e^{(\lambda-\mu)t} + c_2\mathbf{E}_2].$$

Because $\lambda - \mu < 0$, $e^{(\lambda-\mu)t} \to 0$ as $t \to \infty$ and the term $c_1\mathbf{E}_1 e^{(\lambda-\mu)t}$ exerts increasingly less influence on $\mathbf{X}(t)$. The trajectory still approaches the origin, but also approaches the line L_2 asymptotically as $t \to \infty$, as with T_3 in Figure 10.15.

A phase portrait of $\mathbf{X}' = \mathbf{AX}$ in this case therefore has all trajectories approaching the origin, some along L_1, some along L_2, and all others asymptotic to L_2. In this case, the origin is called a *nodal sink* of the system. We can think of particles flowing along the trajectories toward (but never quite reaching) the origin.

EXAMPLE 10.19

Suppose

$$\mathbf{A} = \begin{pmatrix} -6 & -2 \\ 5 & 1 \end{pmatrix}.$$

$\mathbf{A}$ has eigenvalues and eigenvectors

$$-1, \begin{pmatrix} 2 \\ -5 \end{pmatrix} \text{ and } -4, \begin{pmatrix} -1 \\ 1 \end{pmatrix}.$$

Here $\lambda = -4$ and $\mu = -1$. The general solution is

$$\mathbf{X}(t) = c_1 \begin{pmatrix} -1 \\ 1 \end{pmatrix} e^{-4t} + c_2 \begin{pmatrix} 2 \\ -5 \end{pmatrix} e^{-t}.$$

L_1 is the line through the origin and $(-1, 1)$ and L_2 the line through the origin and $(2, -5)$. Figure 10.16 shows a phase portrait for this system. The origin is a nodal sink. ♦

Case 1(b): The Eigenvalues are Positive, say $0 < \mu < \lambda$

Now the trajectories are the same as in Case 1 (a), but the flow is reversed. Instead of flowing into the origin, the trajectories are directed out of and away from the origin, because now $e^{\lambda t}$ and $e^{\mu t}$ approach ∞ instead of zero as $t \to \infty$. All of the arrows on the trajectories now point away from the origin and $(0, 0)$ is called a *nodal source*.

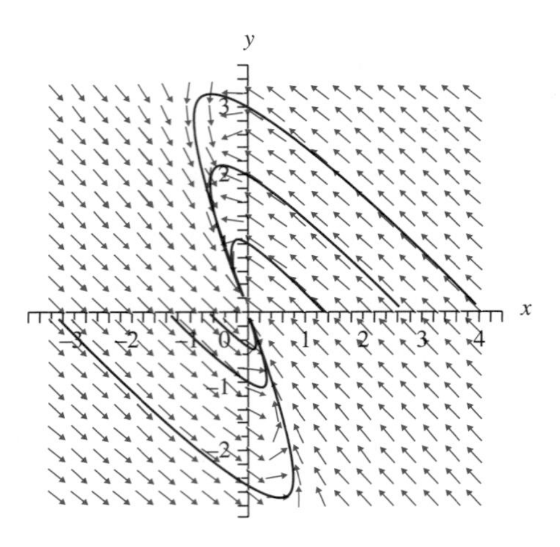

FIGURE 10.16 *Phase portrait in Example 10.19.*

EXAMPLE 10.20

The system

$$\mathbf{X}' = \begin{pmatrix} 3 & 3 \\ 1 & 5 \end{pmatrix} \mathbf{X}$$

has a nodal source at the origin because the eigenvalues are 2 and 6 and these are positive and distinct. The general solution is

$$\mathbf{X}(t) = c_1 \begin{pmatrix} -3 \\ 1 \end{pmatrix} e^{2t} + c_2 \begin{pmatrix} 1 \\ 1 \end{pmatrix} e^{6t}$$

and a phase portrait is shown in Figure 10.17. ♦

Case 2: The Eigenvalues of A are of Opposite Sign

Suppose the eigenvalues are μ and λ, with $\mu < 0 < \lambda$. The general solution will again have the form

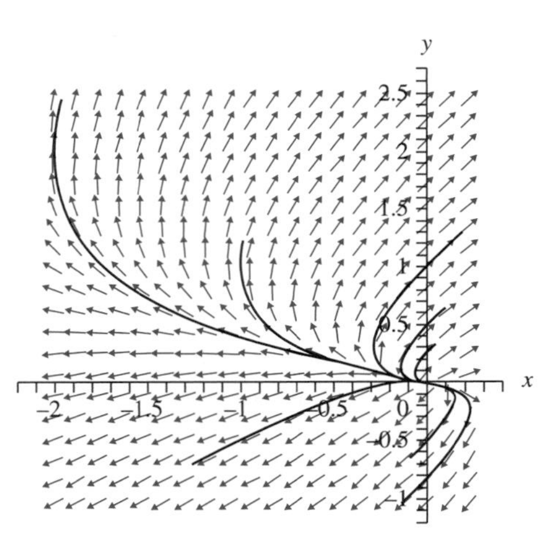

FIGURE 10.17 *Phase portrait in Example 10.20.*

$$\mathbf{X}(t) = c_1\mathbf{E}_1 e^{\lambda t} + c_2\mathbf{E}_2 e^{\mu t}.$$

Examine trajectories along an arbitrary point P_0 other than the origin.

1. If P_0 is on L_1, then $c_2 = 0$ and $\mathbf{X}(t)$ moves on part of L_1 away from the origin as t increases, because $e^{\lambda t} \to \infty$ as $t \to \infty$.
2. If P_0 is on L_2, then $c_1 = 0$ and $\mathbf{X}(t)$ moves on part of L_2 toward the origin, because $e^{\mu t} \to 0$ as $t \to \infty$. Thus, along these lines, some trajectories move toward the origin, others move away.
3. Now suppose P_0 is on neither L_1 or L_2. Then the trajectory through P_0 does not pass arbitrarily close to the origin for any times but instead moves toward the origin asymptotic to L_2 and then away from the origin asymptotic to L_1 as t increases. We may think of L_1 and L_2 as separating the plane into four regions with each trajectory confined to one region (because a trajectory starting in one of these regions cannot cross another trajectory along one of L_1 or L_2 to pass into another region). The trajectories move along L_1 away from the origin and along L_2 toward the origin or in one of the four regions these lines determine, sweeping toward and then away from the origin asymptotic to these lines. This is similar to Halley's comet entering our solar system and moving toward the Sun, then sweeping along a curve that takes it away from the Sun.

In this case, we call the origin a *saddle point*.

The behavior we have just described can be seen in the following example.

EXAMPLE 10.21

The system

$$\mathbf{X}' = \begin{pmatrix} -1 & 3 \\ 2 & -2 \end{pmatrix} \mathbf{X}$$

has general solution

$$\mathbf{X}(t) = c_1 \begin{pmatrix} -1 \\ 1 \end{pmatrix} e^{-4t} + c_2 \begin{pmatrix} 3 \\ 2 \end{pmatrix} e^{t}.$$

The eigenvalues of $\mathbf{A}$ are -4 and 1, real and of opposite sign. Figure 10.18 shows a phase portrait, with a saddle point at the origin.

Case 3: A Has Equal Eigenvalues

Suppose $\mathbf{A}$ has the eigenvalue λ of multiplicity 2. There are two possibilities.

Case 3(a): A Has Two Linearly Independent Eigenvectors $\mathbf{E}_1$ and $\mathbf{E}_2$

Now the general solution is

$$\mathbf{X} = (c_1\mathbf{E}_1 + c_2\mathbf{E}_2)e^{\lambda t}.$$

If

$$\mathbf{E}_1 = \begin{pmatrix} a \\ b \end{pmatrix} \text{ and } \mathbf{E}_2 = \begin{pmatrix} h \\ k \end{pmatrix},$$

then, in terms of components,

$$x(t) = (c_1 a + c_2 b)e^{\lambda t} \text{ and } y(t) = (c_1 h + c_2 k)e^{kt}.$$

Now

$$\frac{y(t)}{x(t)} = \text{constant}.$$

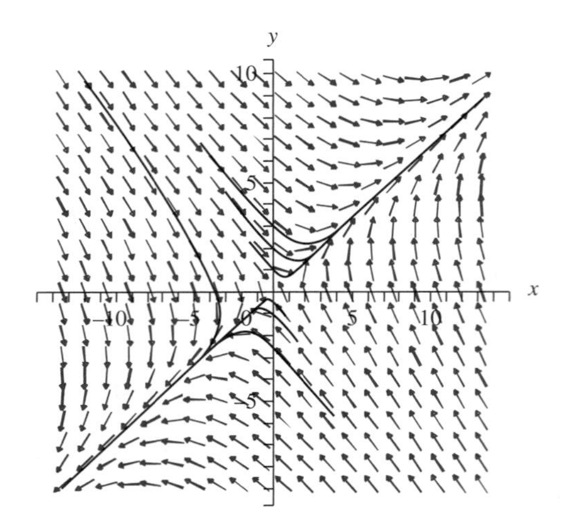

FIGURE 10.18 *Phase portrait in Example 10.21.*

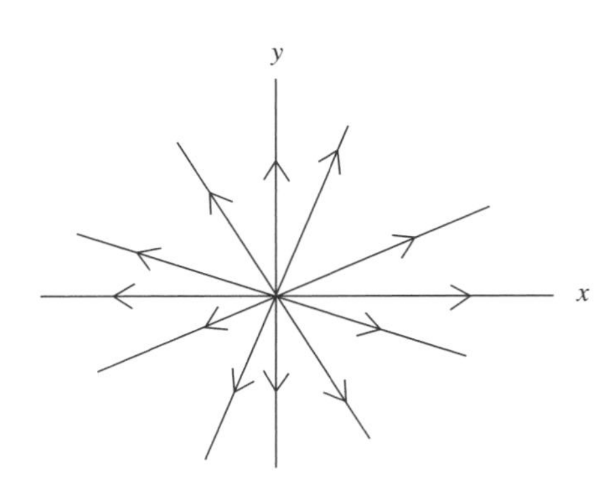

FIGURE 10.19 *Proper node in Case 3(a).*

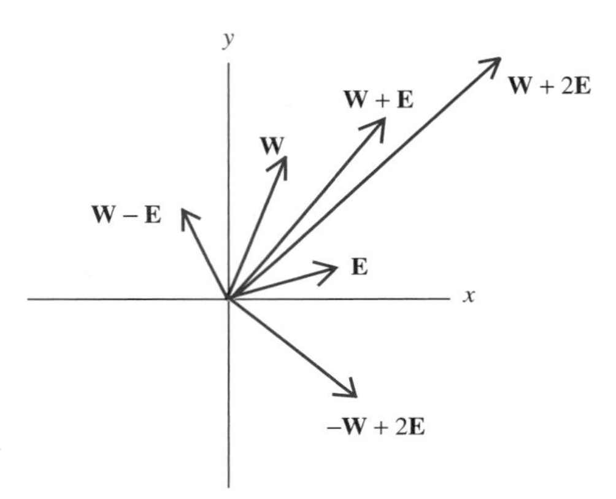

FIGURE 10.20 **W** *and* **E** *in Case 3(b).*

This means that the trajectories in this case are half-lines from the origin. If $\lambda > 0$, these move away from the origin as t increases, and if $\lambda < 0$, they move toward the origin. The origin in this case is called a *proper node*. Figure 10.19 illustrates this for trajectories moving away from the origin.

Case 3(b): A Does Not Have Two Linearly Independent Eigenvectors

In this case the general solution has the form

$$\mathbf{X}(t) = [c_1\mathbf{W} + c_2\mathbf{E}]e^{\lambda t} + c_1\mathbf{E}te^{\lambda t},$$

where **E** is an eigenvector and **W** is determined by the procedure outlined in Section 10.2.2.

To visualize the trajectories, begin with arrows from the origin representing **W** and **E**. Using these we can draw vectors $c_1\mathbf{W} + c_2\mathbf{E}$, which may have various orientations relative to **W** and **E**, depending on the signs and magnitudes of the constants. Some possibilities are shown in Figure 10.20. For given c_1 and c_2, $c_1\mathbf{W} + c_2\mathbf{E} + c_1\mathbf{E}t$ sweeps out a straight line L as t varies. For a given t, $\mathbf{X}(t)$ is $e^{\lambda t}$ times this vector (see Figure 10.21). If λ is negative, this vector shrinks to zero length as $t \to \infty$ and $\mathbf{X}(t)$ sweeps out a curve that approaches the origin tangent to **E**. If λ is positive, reverse the orientation on this trajectory.

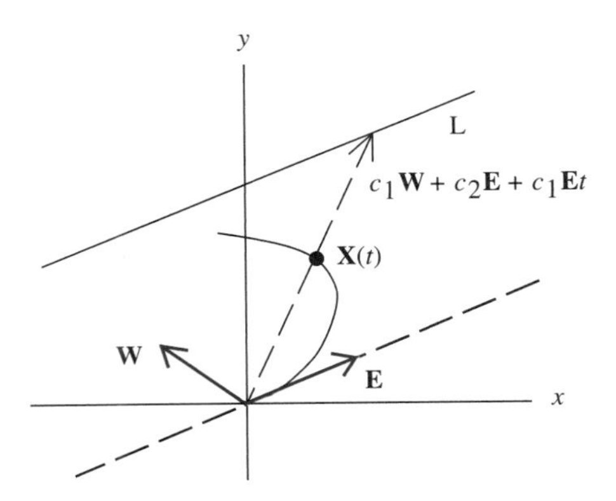

FIGURE 10.21 *Trajectory formed from* **W** *and* **E** *in Case 3(b).*

The origin in this case is called an *improper node* of $\mathbf{X}' = \mathbf{AX}$. The next example shows typical trajectories in the case of an improper node.

EXAMPLE 10.22

Let

$$\mathbf{A} = \begin{pmatrix} -10 & 6 \\ -6 & 2 \end{pmatrix}.$$

$\mathbf{A}$ has an eigenvalue of -4, and every eigenvector is a nonzero multiple of

$$\mathbf{E} = \begin{pmatrix} 1 \\ 1 \end{pmatrix}.$$

A routine calculation gives

$$\mathbf{W} = \begin{pmatrix} 1 \\ 7/6 \end{pmatrix}.$$

The general solution is

$$\mathbf{X}(t) = c_1 \begin{pmatrix} t+1 \\ t+7/6 \end{pmatrix} e^{-4t} + c_2 \begin{pmatrix} 1 \\ 1 \end{pmatrix} e^{-4t}.$$

Figure 10.22 is a phase portrait for this system. The trajectories approach the origin tangent to the line through $\mathbf{E}$, when this vector is represented as an arrow from the origin. The origin is an improper node for this system.

Case 4: A Has Complex Eigenvalues With Nonzero Real Part

Let $\lambda = \alpha + i\beta$ be an eigenvalue with $\alpha \neq 0$ and eigenvector $\mathbf{U} + i\mathbf{V}$. Then the general solution is

$$\mathbf{X}(t) = c_1 e^{\alpha t}[\mathbf{U}\cos(\beta t) - \mathbf{v}\sin(\beta t)] + c_2 e^{\alpha t}[\mathbf{U}\sin(\beta t) + \mathbf{V}\sin(\beta t)].$$

The trigonometric terms cause the solution vector $\mathbf{X}(t)$ to rotate as t increases, while if $\alpha < 0$, the length of $\mathbf{X}(t)$ decreases to zero. Thus, trajectories spiral inward toward the origin as $t \to \infty$ and the origin is called a *spiral sink.*

If $\alpha > 0$, the trajectories spiral outward from the origin as t increases, and the origin is called a *spiral source*.

In both cases, we call the origin a *spiral point*.

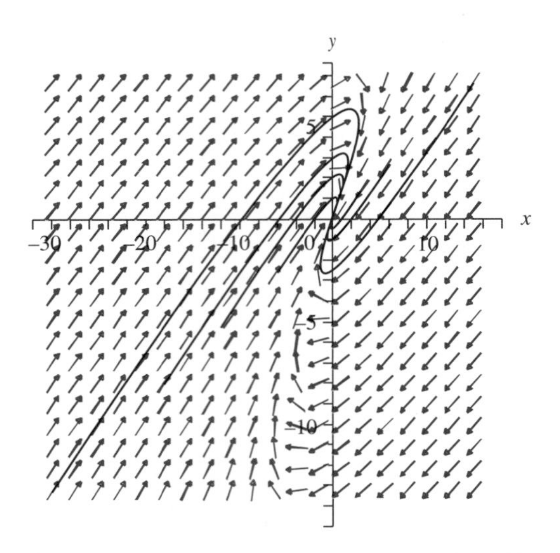

FIGURE 10.22 *Improper node in Example 10.22.*

EXAMPLE 10.23

Let

$$\mathbf{A} = \begin{pmatrix} -1 & -2 \\ 4 & 3 \end{pmatrix}.$$

Eigenvalues are $1 \pm 2i$, so $\alpha = 1$ and $\beta = 2$ in the discussion. An eigenvector for the eigenvalue $1 + 2i$ is $\mathbf{U} + i\mathbf{V}$, where

$$\mathbf{U} = \begin{pmatrix} -1 \\ 2 \end{pmatrix} \text{ and } \mathbf{V} = \begin{pmatrix} 1 \\ 0 \end{pmatrix}.$$

The general solution is

$$\begin{aligned} \mathbf{X}(t) &= c_1 e^t [\mathbf{U}\cos(2t) - \mathbf{V}\sin(2t)] \\ &+ c_2 e^t [\mathbf{u}\sin(t) + \mathbf{V}\cos(2t)]. \end{aligned}$$

Figure 10.23 is a phase portrait of this system, showing trajectories spiraling out from the origin as t increases. The origin is a spiral source. ♦

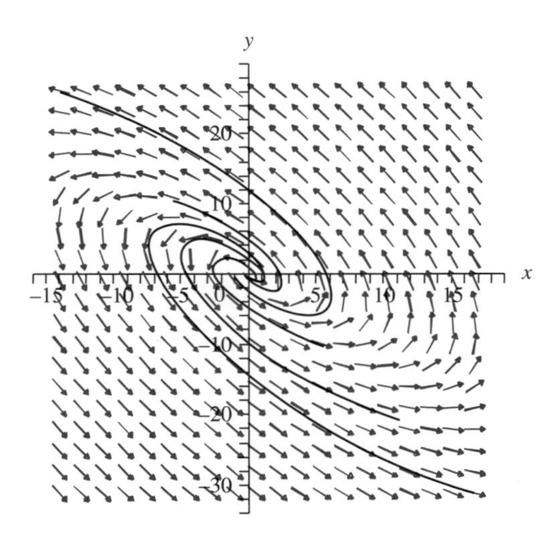

FIGURE 10.23 *Spiral source in Example 10.23.*

Case 5: A Has Pure Imaginary Eigenvalues

Now trajectories have the form

$$\mathbf{X}(t) = c_1[\mathbf{U}\cos(\beta t) - \mathbf{V}\sin(\beta t)] + c_2[\mathbf{U}\sin(\beta t) + \mathbf{V}\cos(\beta t)].$$

Without an exponential factor to increase or decrease the length of this vector, trajectories now are closed curves about the origin, representing a periodic solution. Now the origin is called a *center* of the system.

In general, closed trajectories of a system represent periodic solutions.

EXAMPLE 10.24

Let

$$\mathbf{A} = \begin{pmatrix} 3 & 18 \\ -1 & -3 \end{pmatrix}.$$

Eigenvalues of $\mathbf{A}$ are $\pm 3i$ and eigenvectors are $\mathbf{U} \pm i\mathbf{V}$, where

$$\mathbf{U} = \begin{pmatrix} -3 \\ 1 \end{pmatrix} \text{ and } \mathbf{V} = \begin{pmatrix} -3 \\ 0 \end{pmatrix}.$$

Figure 10.24 is a phase portrait, showing closed trajectories moving (in this case) clockwise about the origin, which is a center. ♦

We now have a complete description of the trajectories of the real, linear 2×2 system $\mathbf{X}' = \mathbf{AX}$. The qualitative behavior of the trajectories is determined by the eigenvalues, and we have the following classification of the origin:

- Real, distinct eigenvalues of the same sign—$(0, 0)$ is a nodal source (positive eigenvalues) or sink (negative eigenvalues);
- Real, distinct eigenvalues of the same sign—$(0, 0)$ is a saddle point;
- Equal eigenvalues, linearly independent eigenvectors—$(0, 0)$ is a proper node;
- Equal eigenvalues, all eigenvectors a multiple of a single eigenvector—$(0, 0)$ is an improper node;

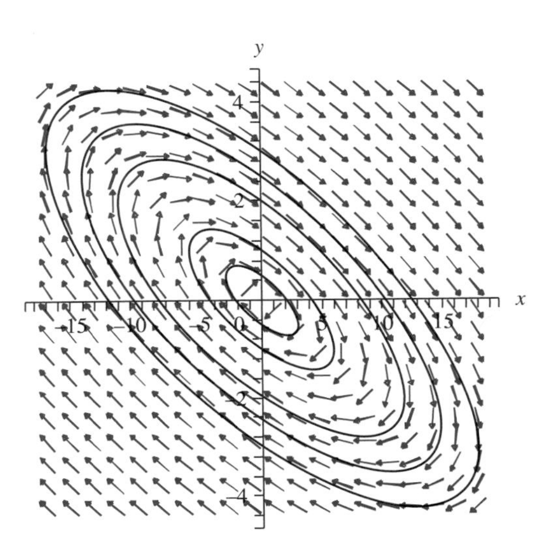

FIGURE 10.24 *Center in Example 10.24.*

- Complex eigenvalues with nonzero real part—$(0, 0)$ is a spiral source (positive real part) or spiral sink (negative real part);
- Pure imaginary eigenvalues—$(0, 0)$ is a center (periodic solutions).

10.6.2 Predator/Prey and Competing Species Models

We will show how phase portraits are used in the analysis of two types of systems that arise in important applications. These systems will be 2×2, but are nonlinear, hence they are not as easily solved as linear constant coefficient systems. Nevertheless, the phase portraits will display the qualitative behavior of solutions of the phenomena being modeled.

A Predator/Prey Model

Begin with a *predator/prey model*. Suppose an environment includes two species having populations $x(t)$ and $y(t)$ at time t. One species $y(t)$ consists of predators, whose food is in the prey $(x(t))$ population. For example, we could be looking at rabbits and foxes in a wilderness area. Or we could have birds preying on young sea turtles near an island where the turtles lay their eggs. For convenience in the discussion, we will use rabbits and foxes as a prototypical predator/prey setting.

As a simplification, assume that the rabbits have no other natural enemies in the setting, and that every encounter of a rabbit with a fox results in the fox eating the rabbit.

To model these two populations, suppose that at time t, the rabbit population increases at a rate proportional to $x(t)$, which is the number of rabbits at this time, but also decreases at a rate proportional to encounters of rabbits with foxes, which is modeled by a product $x(t)y(t)$ of the rabbit and fox populations at that time. Then, for some positive constants a and b,

$$x'(t) = ax(t) - bx(t)y(t).$$

The foxes are assumed to increase at a rate proportional to their encounters with rabbits (hence proportional to $x(t)y(t)$) but to decrease at a rate proportional to their own population (because in the absence of rabbits the foxes have no food and die). Thus, for some positive numbers c and k,

$$y'(t) = cx(t)y(t) - ky(t).$$

We now have a 2×2 system for these populations:

$$x' = ax - bxy$$
$$y' = cxy - ky.$$

This is a nonlinear system because of the xy terms.

If the initial rabbit population is $x(0) = \alpha > 0$, and there are no foxes, then $b = 0$, and the rabbit population increases exponentially with $x(t) = \alpha e^{at}$. If the initial fox population is $y(0) = \beta$ and there are no rabbits, then $c = 0$ and, with no food, the fox population dies out exponentially according to the rule $y(t) = \beta e^{-kt}$.

Phase portraits reveal an interesting characteristic of the populations in the case that α and β are both positive. Clearly all trajectories will be in the first octant of the x, y-plane, since populations must be nonnegative. Figure 10.25 is a typical trajectory of this system. The horizontal and vertical lines through $P : (k/c, a/b)$ separate the first quadrant into four regions I, II, III, and IV, and trajectories move about P through these regions. Follow a typical point $(x(t), y(t))$ around one trajectory. Suppose a population pair $(x(t_0), y(t_0))$ is in region I at some time t_0 (so the rabbit population at this time is greater than k/c and the fox population less than a/b). Now both populations may be large. This produces more encounters, hence more rabbit kills. In this region, $x'(t) < 0$ and $y'(t) > 0$, so the rabbit population is declining and the fox population increasing.

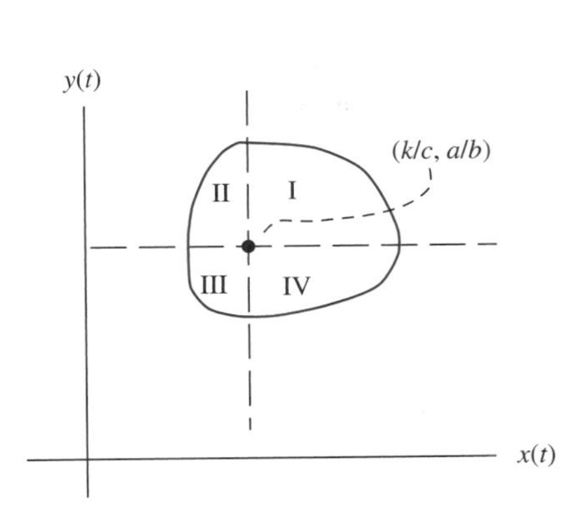

FIGURE 10.25 *Typical predator/prey trajectory.*

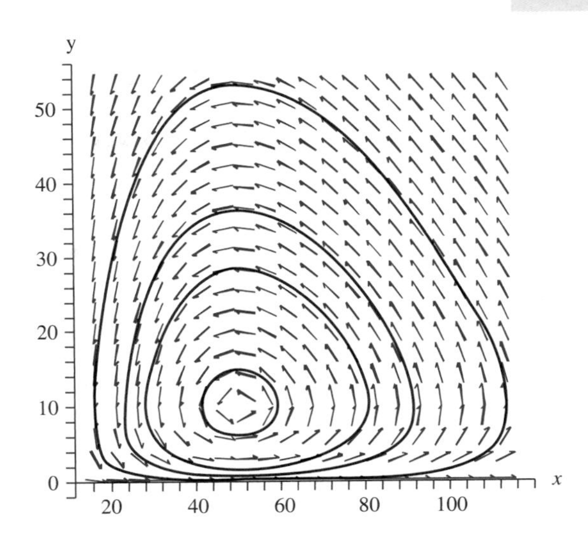

FIGURE 10.26 *Trajectories for* $x' = 0.2x - 0.02xy$, $y' = 0.02xy - 1.2y$.

Once the rabbit population reaches the value k/c, the foxes find insufficient food to sustain their population and their numbers begin to decline. Now $(x(t), y(t))$ passes into region II, where both populations are in decline.

When the fox population reaches the value a/b, their numbers are small enough that the rabbits begin to multiply much faster than they are consumed, and the point $(x(t), y(t))$ moves through region III, where the foxes decline but the rabbits increase in numbers.

When the fox population reaches its minimum value, the rabbit population is increasing at its fastest rate. Now $(x(t), y(t))$ moves into region IV, where the foxes begin to increase again in number because of the availability of more rabbits.

This process repeats cyclically, with foxes increasing any time the rabbit population can sustain them, and declining when there is a lack of food. The rabbits increase whenever the fox population falls below a certain level. Following this the foxes have more food and their population increases, so the rabbits then go into decline, and the cycle repeats.

Figure 10.26 shows several trajectories for the system

$$x' = 0.2x - 0.02xy$$
$$y' = 0.02xy - 1.2y.$$

It is possible to write an implicitly defined solution of the predator/prey model. Write

$$\frac{dy/dt}{dx/dt} = \frac{dy}{dx} = \frac{y}{x}\frac{cx - k}{a - by}$$

and separate the variables by writing

$$\frac{a - by}{y}\,dy = \frac{cx - k}{x}\,dx.$$

Integrate and rearrange terms to obtain

$$y^a e^{-by} = Kx^{-k}e^{cx},$$

in which K is a positive constant of integration.

There are predator/prey populations for which good records have been kept and against which this model can be tested. One is the lynx/snowshoe hare population in Canada. The Hudson

Bay Company has kept records of pelts traded at their stations since the middle of the nineteenth century when trappers worked these areas. Assuming that the actual populations of lynx and hare were proportional to the number of pelts obtained by trappers, the records for 1845 through 1935 exhibit cyclical variations as seen in the predator/prey model, with about a ten year cycle.

Another predator/prey setting occurred in Michigan's Isle Royale, an untamed island having a length of about 45 miles. At one time, moose abounded on the island, having no natural enemy. However, the harsh winter of 1949 caused wolves from Canada to cross over a frozen stretch of Lake Superior, searching for food. The resulting behavior of the moose/wolf populations was studied by Purdue biologist Durward Allen, who observed cyclic variations in the populations over the 1957–1993 period. In this case, the predator wolf population that came to the island had two problems which would alter the model—a very narrow genetic base coupled with the spread of a canine virus that destroyed many wolves.

More complex predator/prey models have been used in many contexts, including research into the behavior of the HIV virus. In one such model, the predators consist of the invading viruses and the prey consists of their target cells within the body. The model is complicated by the fact that the viruses mutate over time, presenting the immune system with many different predators. One study along these lines is given in a November 15, 1991 paper in the journal *Science*, entitled *Antigenic Diversity Thresholds and the Development of AIDS* and authored by Martin A. Nowak, A.R. McLean, and R.M. May of the Department of Zoology, University of Oxford; T. Wolfe and J. Goudsmit of the Human Retrovirus Laboratory, Department of Virology, Amsterdam, the Netherlands; and R.M. Anderson of the Department of Biology, Imperial College of London University.

A Competing Species Model

A *competing species model* offers a different type of population dynamic. In this model, we have some environment in which two species compete for a common resource, but neither preys on the other. In this case, it seems reasonable that an increase in either population decreases the availability of this resource for both, causing a decline in both populations. Assuming no restriction on the needed resource, one possible competing species model is given by

$$x' = ax - bxy$$

$$y' = ky - cxy,$$

in which a, b, c, and k are positive constants. Now a term proportional to the product of the populations is subtracted in both equations.

As with the predator/prey model, we can obtain an implicitly defined solution of this system. Divide the differential equations to obtain

$$\frac{dy/dt}{dx/dt} = \frac{dy}{dx} = \frac{y}{x}\frac{k - cx}{a - by}.$$

Separate variables to obtain

$$\frac{a - by}{y}\,dy = \frac{k - cx}{x}\,dx.$$

Integrate and rearrange terms to obtain

$$y^a e^{-by} = Kx^k e^{-cx}$$

in which K is the constant of integration and can be any positive number.

Typical trajectories of this model are shown in Figure 10.27. Asymptotes of these trajectories pass through $(k/c, a/b)$ and subdivide the first octant into four regions, I, II, III, and IV. If the initial population $(x(0), y(0))$ is in regions I or IV, then the x population increases with time,

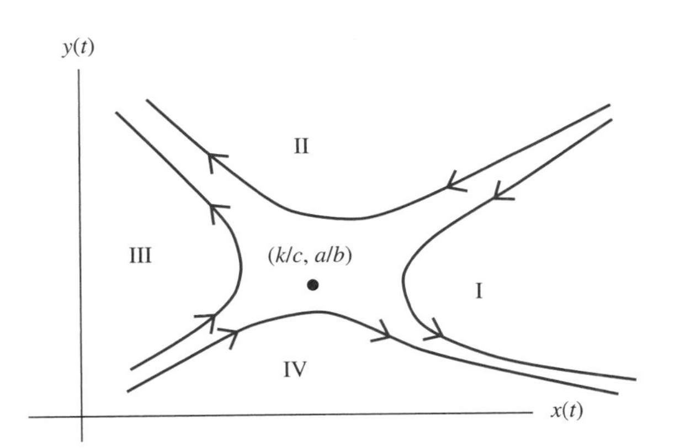

FIGURE 10.27 *Trajectories for a typical competing species model.*

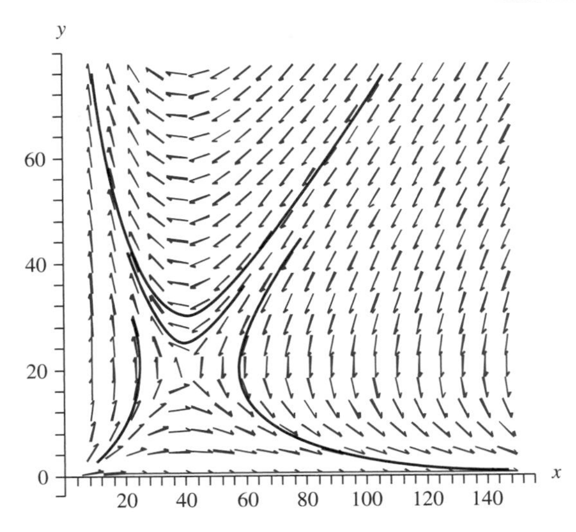

FIGURE 10.28 *Trajectories for* $x' = x - 0.01xy$, $y' = 4y - 0.1xy$.

while the y population dies out with time (decreasing to zero in the limit as $t \to \infty$). If the initial population is in II or III, then the y population wins and the x population dies out asymptotically. The coefficients a, b, c, k play a crucial role in determining the asymptotes, hence the regions, so just having a large initial population is not enough to guarantee survival.

As a specific example, Figure 10.28 shows a phase portrait for the model

$$x' = 2x - 0.1xy$$
$$y' = 4y - 0.1xy.$$

SECTION 10.6 PROBLEMS

In each of Problems 1 through 10, classify the origin of the system $\mathbf{X}' = \mathbf{A}\mathbf{X}$ for the given coefficient matrix. If software is available, produce a phase portrait.

1. $\mathbf{A} = \begin{pmatrix} -2 & -1 \\ 3 & -2 \end{pmatrix}$

2. $\mathbf{A} = \begin{pmatrix} 9 & -7 \\ 6 & -4 \end{pmatrix}$

3. $\mathbf{A} = \begin{pmatrix} 7 & -17 \\ 2 & 1 \end{pmatrix}$

4. $\mathbf{A} = \begin{pmatrix} 2 & -7 \\ 5 & -10 \end{pmatrix}$

5. $\mathbf{A} = \begin{pmatrix} 4 & -1 \\ 1 & 2 \end{pmatrix}$

6. $\mathbf{A} = \begin{pmatrix} 1 & 4 \\ 3 & 0 \end{pmatrix}$

7. $\mathbf{A} = \begin{pmatrix} 3 & -5 \\ 5 & -7 \end{pmatrix}$

8. $\mathbf{A} = \begin{pmatrix} -6 & -7 \\ 7 & -20 \end{pmatrix}$

9. $\mathbf{A} = \begin{pmatrix} 1 & -5 \\ 1 & -1 \end{pmatrix}$

10. $\mathbf{A} = \begin{pmatrix} 3 & -5 \\ 8 & -3 \end{pmatrix}$

11. Generate a phase portrait for each of the following competing species models.

    (a) $x' = 2x - xy$, $y' = y - 2xy$

    (b) $x' = 1.6y - 1.2xy$, $y' = 2y - 0.4xy$

    (c) $x' = 1.4x - 0.6xy$, $y' = 2y - 0.7xy$

    (d) $x' = 3.2x - 1.4xy$, $y' = 4.4y - 0.8xy$

12. A more sophisticated approach to a competing species model is to incorporate a logistic term, leading to the model

$$x' = ax - bx^2 - kxy,\ y' = cy - dy^2 - rxy,$$

with the coefficients positive constants. Generate phase portraits for the following systems.

(a) $x' = x(1 - x - 0.5y)$,
$y' = y(1 - 0.5y - 0.25x)$

(b) $x' = x(1 - x - 0.2y)$,
$y' = y(1 - 0.4y - 0.25x)$

(c) $x' = x(2 - x - 0.2y),\ y' = y(1 - 0.4y - x)$

(d) $x' = x(1 - 0.5x - y),\ y' = y(2 - 0.5y - 0.4x)$

13. Derive a system of differential equations modeling the predator/prey relationship in an environment with indiscriminate harvesting. Do this by assuming that there is some outside agent that removes numbers of both species from the system at a rate proportional to the populations, with the same constant of proportionality for both species.

14. Use a software package to generate a phase portrait for each of the following predator/prey models.

(a) $x' = x - 0.5xy,\ y' = 2xy - 1.2y$

(b) $x' = 3x - 1.5xy,\ y' = xy - 1.6y$

(c) $x' = 1.6x - 2.1xy,\ y' = 1.9xy - 0.4y$

(d) $x' = 1.8 - 0.2xy,\ y' = 3.1xy - 0.4y$

PART 3

Vector Analysis

$$[f](s) = \int_0^\infty e^{-st}t^{-1/2}\,dt = 2\int_0^\infty e^{-sx^2}\,dx \text{ (set } x = t^{1/2}) = \frac{2}{\sqrt{s}}\int_0^\infty e^{-z^2}\,dz \text{ (set } z = x$$

$$= \sqrt{\frac{\pi}{s}}$$

$$\mathcal{L}[f](s) = \int_0^\infty e$$

$$= 2\int_0^\infty e^{-sx}\,dx$$

$$= \sqrt{\frac{\pi}{s}}$$

$$f](s) = \int_0^\infty e^{-st}t^{-1/2}\,dt$$

$$= 2\int_0^\infty e^{-sx^2}\,dx \text{ (set } x = t^{1/2})$$

$$= \frac{2}{\sqrt{s}}\int_0^\infty e^{-z^2}\,dz \text{ (set } z = x\sqrt{s})$$

$$\mathcal{L}[f](s) = \int_0^\infty e^{-st}t^{-1/2}$$

CHAPTER 11 Vector Differential Calculus

VECTOR FUNCTIONS OF ONE VARIABLE VELOCITY AND CURVATURE VECTOR FIELDS AND STREAMLINES THE GRADIENT FIELD DIVERGENCE

11.1 Vector Functions of One Variable

A vector function of one variable is a function of the form $\mathbf{F}(t) = x(t)\mathbf{i} + y(t)\mathbf{j} + z(t)\mathbf{k}$. This vector function is continuous at t_0 if each component function is continuous at t_0.

We may think of $\mathbf{F}(t)$ as the *position vector* of a curve in 3-space. For each t for which the vector is defined, draw $\mathbf{F}(t)$ as an arrow from the origin to the point $(x(t), y(t), z(t))$. This arrow sweeps out a curve C as t varies. When thought of in this way, the coordinate functions are parametric equations of this curve.

EXAMPLE 11.1

$\mathbf{H}(t) = t^2\mathbf{i} + \sin(t)\mathbf{j} - t^2\mathbf{k}$ is the position vector for the curve given parametrically by

$$x = t^2,\ y = \sin(t),\ z = -t^2.$$

Figure 11.1 shows part of a graph of this curve. ♦

$\mathbf{F}(t) = x(t)\mathbf{i} + y(t)\mathbf{j} + z(t)\mathbf{k}$ is differentiable at t if each component function is differentiable at t, and in this case

$$\mathbf{F}'(t) = x'(t)\mathbf{i} + y'(t)\mathbf{j} + z'(t)\mathbf{k}.$$

We differentiate a vector function by differentiating each component.

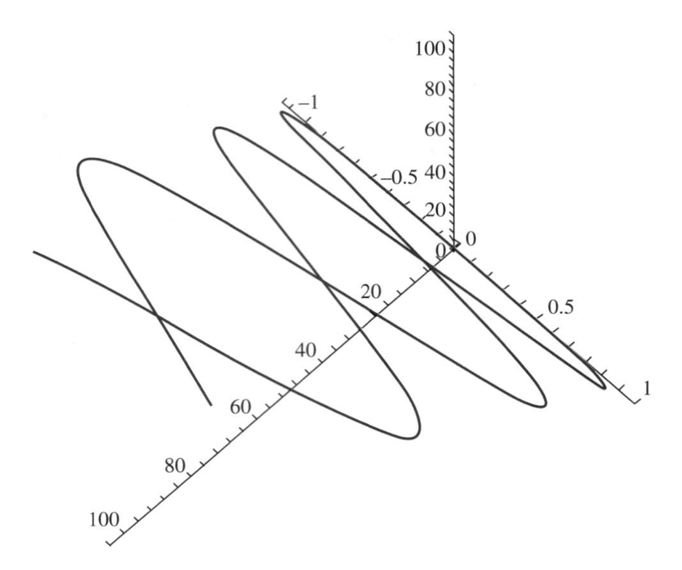

FIGURE 11.1 *Graph of the curve of Example 11.1.*

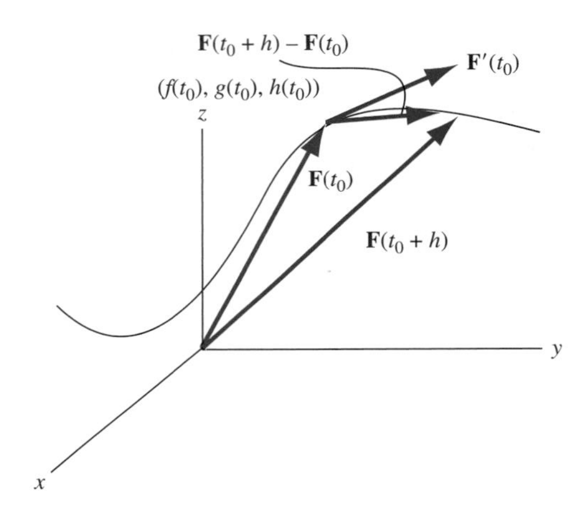

FIGURE 11.2 $\mathbf{F}'(t_0)$ *as a tangent vector.*

To give an interpretation to the vector $\mathbf{F}'(t_0)$, look at the limit of the difference quotient:

$$\begin{aligned}\mathbf{F}'(t_0) &= \lim_{h\to 0}\frac{\mathbf{F}(t_0+h)-\mathbf{F}(t_0)}{h}\\ &= \left(\lim_{h\to 0}\frac{x(t_0+h)-x(t_0)}{h}\right)\mathbf{i}+\left(\lim_{h\to 0}\frac{y(t_0+h)-y(t_0)}{h}\right)\mathbf{j}\\ &\quad+\left(\lim_{h\to 0}\frac{z(t_0+h)-z(t_0)}{h}\right)\mathbf{k}\\ &= x'(t_0)\mathbf{i}+y'(t_0)\mathbf{j}+z'(t_0)\mathbf{k}.\end{aligned}$$

Figure 11.2 shows the vectors $\mathbf{F}(t_0+h)$, $\mathbf{F}(t_0)$ and $\mathbf{F}(t_0+h)-\mathbf{F}(t_0)$, using the parallelogram law. As h is chosen smaller, the tip of the vector $\mathbf{F}(t_0+h)-\mathbf{F}(t_0)$ slides along C toward $\mathbf{F}(t_0)$, and $(1/h)[\mathbf{F}(t_0+h)-\mathbf{F}(t_0)]$ moves into the position of the *tangent vector* to C at the point $(f(t_0), g(t_0), h(t_0))$. In calculus, the derivative of a function gives the slope of the tangent to the graph at a point. In vector calculus, the derivative of the position vector of a curve gives the tangent vector to the curve at a point.

In Example 11.1,

$$\mathbf{H}'(t) = 2t\mathbf{i} + \cos(t)\mathbf{j} - 2t\mathbf{k},$$

and this vector is tangent to the curve at any point $(t^2, \sin(t), -t^2)$ on the curve. The tangent vector at $(0, 0, 0)$ is $\mathbf{H}'(0) = \mathbf{j}$, as we can visualize from Figure 11.1.

The length of a curve given parametrically by $x = x(t)$, $y = y(t)$, and $z = z(t)$ for $a \le t \le b$ is

$$\text{length} = \int_a^b \sqrt{(x'(t))^2 + (y'(t))^2 + (z'(t))^2}\, dt.$$

In vector notation, this is

$$\text{length} = \int_a^b \| \mathbf{F}'(t) \| \, dt.$$

The length of a curve is the integral (over the defining interval) of the length of the tangent vector to the curve, assuming differentiability at each t.

Now imagine starting at $(x(a), y(a), z(a))$ at time $t = a$ and moving along the curve, reaching the point $(x(t), y(t), z(t))$ at time t. Let $s(t)$ be the distance along C from the starting point to this point (Figure 11.3). Then

$$s(t) = \int_a^t \| \mathbf{F}'(\xi) \| \, d\xi.$$

This function measures length along C and is strictly increasing, hence it has an inverse. At least in theory, we can solve for $t = t(s)$, writing the parameter t in terms of arc length along C. We can substitute this function into the position function to obtain

$$\mathbf{G}(s) = \mathbf{F}(t(s)).$$

$\mathbf{G}$ is also a position vector for C, except now the variable is s and s varies from 0 to L, the length of C. Therefore, $\mathbf{G}'(s)$ is also a tangent vector to C. We claim that this tangent vector in terms of arc length is always a unit vector. To see this, observe from the fundamental theorem of calculus that

$$s'(t) = \| \mathbf{F}'(t) \| .$$

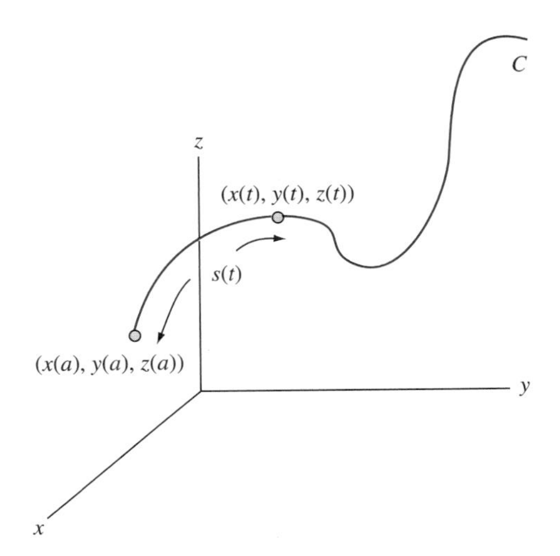

FIGURE 11.3 *Distance function along a curve.*

Then

$$\mathbf{G}'(s) = \frac{d}{ds}\mathbf{F}(t(s)) = \frac{d}{dt}\mathbf{F}(t)\frac{dt}{ds}$$
$$= \frac{1}{ds/dt}\mathbf{F}'(t) = \frac{1}{\| \mathbf{F}'(t) \|}\mathbf{F}'(t),$$

and this vector has a length of 1.

EXAMPLE 11.2

Let C be defined by

$$x = \cos(t),\ y = \sin(t),\ z = t/3$$

for $-4\pi \le t \le 4\pi$. C has the position vector

$$\mathbf{F}(t) = \cos(t)\mathbf{i} + \sin(t)\mathbf{j} + \frac{1}{3}t\mathbf{k}$$

and the tangent vector

$$\mathbf{F}'(t) = -\sin(t)\mathbf{i} + \cos(t)\mathbf{j} + \frac{1}{3}\mathbf{k}.$$

It is routine to compute $\| \mathbf{F}'(t) \| = \sqrt{10}/3$, so the distance function along C is

$$s(t) = \int_{-4\pi}^{t} \frac{1}{3}\sqrt{10}\,d\xi = \frac{1}{3}\sqrt{10}(t + 4\pi).$$

In this example, we can explicitly solve for t in terms of s:

$$t = t(s) = \frac{3}{\sqrt{10}}s - 4\pi.$$

Substitute this into $\mathbf{F}(t)$ to get

$$\mathbf{G}(s) = \mathbf{F}(t(s)) = \mathbf{F}\left(\frac{3}{\sqrt{10}}s - 4\pi\right)$$
$$= \cos\left(\frac{3}{\sqrt{10}}s - 4\pi\right)\mathbf{i} + \sin\left(\frac{3}{\sqrt{10}}s - 4\pi\right)\mathbf{j} + \frac{1}{3}\left(\frac{3}{\sqrt{10}}s - 4\pi\right)\mathbf{k}$$
$$= \cos\left(\frac{3}{\sqrt{10}}s\right)\mathbf{i} + \sin\left(\frac{3}{\sqrt{10}}s\right)\mathbf{j} + \left(\frac{1}{\sqrt{10}}s - \frac{4\pi}{3}\right)\mathbf{k}.$$

Now compute

$$\mathbf{G}'(s) = -\frac{3}{\sqrt{10}}\cos\left(\frac{3}{\sqrt{10}}s\right)\mathbf{i} + \frac{3}{\sqrt{10}}\sin\left(\frac{3}{\sqrt{10}}s\right)\mathbf{j} + \frac{1}{\sqrt{10}}\mathbf{k},$$

and this is a unit tangent vector to C. ♦

Rules for differentiating various combinations of vectors are like those for functions of one variable. If the functions and vectors are differentiable and α is a number, then

1. $[\mathbf{F}(t)+\mathbf{G}(t)]' = \mathbf{F}'(t)+\mathbf{G}'(t)$.
2. $(\alpha\mathbf{F})'(t) = \alpha\mathbf{F}'(t)$.
3. $[f(t)\mathbf{F}(t)]' = f'(t)\mathbf{F}(t) + f(t)\mathbf{F}'(t)$.
4. $[\mathbf{F}(t)\cdot\mathbf{G}(t)]' = \mathbf{F}'(t)\cdot\mathbf{G}(t)+\mathbf{F}(t)\cdot\mathbf{G}'(t)$.
5. $[\mathbf{F}(t)\times\mathbf{G}(t)]' = \mathbf{F}'(t)\times\mathbf{G}(t)+\mathbf{F}(t)\times\mathbf{G}'(t)$.
6. $[\mathbf{F}(f(t))]' = f'(t)\mathbf{F}'(f(t))$.

Rules (3), (4), and (5) are product rules, reminiscent of the rule for differentiating a product of functions of one variable. In rule (4), the order of the factors is important, since the cross product is anti-commutative. Rule (6) is a chain rule for vector differentiation.

SECTION 11.1 PROBLEMS

In each of Problems 1 through 8, compute the requested derivative in two ways, first by using rules (1) through (6) as appropriate, and second by carrying out the vector operation and then differentiating the resulting vector or scalar function.

1. $\mathbf{F}(t) = -9\mathbf{i}+t^2\mathbf{j}+t^2\mathbf{k}, \mathbf{G}(t) = e^t\mathbf{i}$; $(d/dt)[\mathbf{F}(t)\times\mathbf{G}(t)]$
2. $\mathbf{F}(t) = -4\cos(t)\mathbf{k}, \mathbf{G}(t) = -t^2\mathbf{i}+4\sin(t)\mathbf{k}$; $(d/dt)[\mathbf{F}(t)\cdot\mathbf{G}(t)]$
3. $\mathbf{F}(t) = \mathbf{i}+3t^2\mathbf{j}+2t\mathbf{k}, f(t) = 4\cos(3t)$; $(d/dt)[f(t)\mathbf{F}(t)]$
4. $\mathbf{F}(t) = t\mathbf{i} - t\mathbf{j} + t^2\mathbf{k}, \mathbf{G}(t) = \sin(t)\mathbf{i} - 4t\mathbf{j} + t^3\mathbf{k}$; $(d/dt)[\mathbf{F}(t)\cdot\mathbf{G}(t)]$
5. $\mathbf{F}(t) = t\mathbf{i}+\mathbf{j}+4\mathbf{k}, \mathbf{G}(t) = \mathbf{i}-\cos(t)\mathbf{j} + t\mathbf{k}$; $(d/dt)[\mathbf{F}(t)\times\mathbf{G}(t)]$
6. $\mathbf{F}(t) = \sinh(t)\mathbf{j} - t\mathbf{k}, \mathbf{G}(t) = t\mathbf{i}+t^2\mathbf{j} - t^2\mathbf{k}$; $(d/dt)[\mathbf{F}(t)\times\mathbf{G}(t)]$
7. $\mathbf{F}(t) = t\mathbf{i} - \cosh(t)\mathbf{j} + e^t\mathbf{k}, f(t) = 1-2t^3$; $(d/dt)[f(t)\mathbf{F}(t)]$
8. $\mathbf{F}(t) = t\mathbf{i}-3t^2\mathbf{k}, \mathbf{G}(t) = \mathbf{i}+\cos(t)\mathbf{k}$; $(d/dt)[\mathbf{F}(t)\cdot\mathbf{G}(t)]$

In each of Problems 9, 10, and 11, (a) write the position vector and tangent vector for the curve whose parametric equations are given, (b) find the length function $s(t)$ for the curve, (c) write the position vector as a function of s, and (d) verify by differentiation that this position vector in terms of s is a unit tangent to the curve.

9. $x = 2t^2, y = 3t^2, z = 4t^2; 1\le t\le 3$
10. $x = y = z = t^3; -1\le t\le 1$
11. $x = \sin(t), y = \cos(t), z = 45t; 0\le t\le 2\pi$
12. Suppose $\mathbf{F}(t) = x(t)\mathbf{i}+y(t)\mathbf{j}+z(t)\mathbf{k}$ is the position vector for a particle moving along a curve in 3-space. Suppose that $\mathbf{F}\times\mathbf{F}' = \mathbf{O}$. Show that the particle always moves in the same direction.

11.2 Velocity and Curvature

Imagine a particle or object moving along a path C having the position vector $\mathbf{F}(t) = x(t)\mathbf{i}+y(t)\mathbf{j}+z(t)\mathbf{k}$, as t varies from a to b. We want to relate $\mathbf{F}$ to the dynamics of the particle. Assume that the coordinate functions are twice differentiable.

Define the *velocity* $\mathbf{v}(t)$ of the particle at time t to be

$$\mathbf{v}(t) = \mathbf{F}'(t).$$

The *speed* $v(t)$ is the magnitude of the velocity:

$$v(t) = \|\mathbf{v}(t)\|.$$

Then

$$v(t) = \| \mathbf{F}'(t) \| = \frac{ds}{dt},$$

which is the rate of change with respect to time of the distance along the trajectory or path of motion.

The *acceleration* $\mathbf{a}(t)$ is the rate of change of the velocity with respect to time, or

$$\mathbf{a}(t) = \mathbf{v}'(t) = \mathbf{F}''(t).$$

If $\mathbf{F}'(t) \neq \mathbf{O}$, then this vector is a tangent vector to C. We obtain a unit tangent vector $\mathbf{T}(t)$ by dividing $\mathbf{F}'(t)$ by its length. This leads to various expressions for the unit tangent vector to C:

$$\mathbf{T}(t) = \frac{1}{\| \mathbf{F}'(t) \|}\mathbf{F}'(t) = \frac{1}{ds/dt}\mathbf{F}'(t)$$
$$= \frac{1}{\| \mathbf{v}(t) \|}\mathbf{v}(t) = \frac{1}{v(t)}\mathbf{v}(t).$$

Thus, the unit tangent vector is also the velocity vector divided by the speed.

The *curvature* $\kappa(s)$ of C is defined as the magnitude of the rate of change of the unit tangent with respect to arc length along C:

$$\kappa(s) = \left\| \frac{d\mathbf{T}}{ds} \right\|.$$

This definition is motivated by Figure 11.4, which suggests that the more a curve bends at a point, the faster the unit tangent vector is changing direction there. This expression for the curvature, however, is difficult to work with because we usually have the unit tangent vector as a

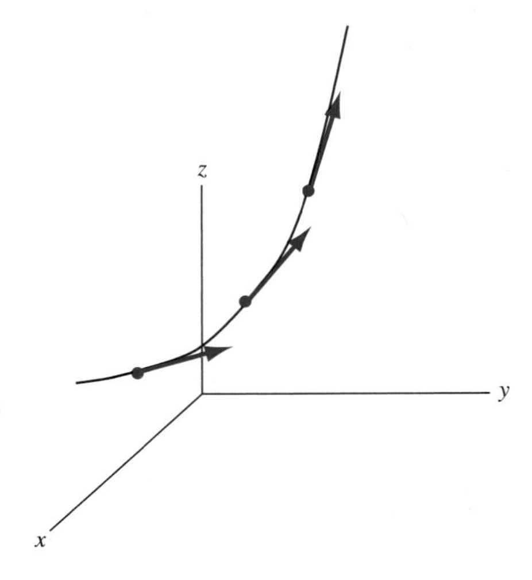

FIGURE 11.4 *Curvature as a rate of change of the tangent vector.*

function of t, not of s. We, therefore, usually compute the curvature as a function of t by using the chain rule:

$$\kappa(t) = \left\| \frac{d\mathbf{T}}{dt}\frac{dt}{ds} \right\|$$
$$= \frac{1}{\| \mathbf{F}'(t) \|} \| \mathbf{T}'(t) \| .$$

EXAMPLE 11.3

Let C have position vector

$$\mathbf{F}(t) = [\cos(t) + t\sin(t)]\mathbf{i} + [\sin(t) - t\cos(t)]\mathbf{j} + t^2\mathbf{k}.$$

for $t \geq 0$. Figure 11.5 is part of the graph of C. A tangent vector is given by

$$\mathbf{F}'(t) = t\cos(t)\mathbf{i} + t\sin(t)\mathbf{j} + 2t\mathbf{k}.$$

This tangent vector has the length

$$v(t) = \| \mathbf{F}'(t) \| = \sqrt{5}t.$$

The unit tangent vector in terms of t is

$$\mathbf{T}(t) = \frac{1}{\| \mathbf{F}'(t) \|}\mathbf{F}'(t) = \frac{1}{\sqrt{5}}[\cos(t)\mathbf{i} + \sin(t)\mathbf{j} + 2\mathbf{k}].$$

Then

$$\mathbf{T}'(t) = \frac{1}{\sqrt{5}}[-\sin(t)\mathbf{i} + \cos(t)\mathbf{j}],$$

and the curvature of C is

$$\kappa(t) = \frac{1}{\| \mathbf{F}'(t) \|} \| \mathbf{T}'(t) \|$$
$$= \frac{1}{\sqrt{5}t}\sqrt{\frac{1}{5}[\sin^2(t) + \cos^2(t)]} = \frac{1}{5t}$$

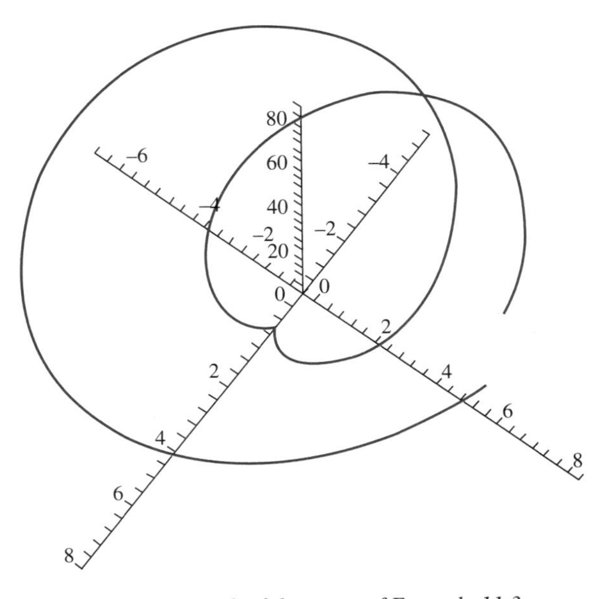

FIGURE 11.5 *Graph of the curve of Example 11.3.*

for $t > 0$. It is usually more convenient to compute such quantities as the unit tangent and the curvature in terms of the parameter t used to define C, rather than attempting to solve for t in terms of the arc length s. ♦

Given a position vector $\mathbf{F}(t)$ for a curve C, we have a unit tangent at any point where the component functions are differentiable and their derivatives are not all zero. We claim that, in terms of s, the vector

$$\mathbf{N}(s) = \frac{1}{\kappa(s)}\mathbf{T}'(s).$$

is a *unit normal vector* (orthogonal to the tangent) to C.

First, $\mathbf{N}(s)$ is a unit vector because $\kappa(s) = \| \mathbf{T}'(s) \|$, so

$$\| \mathbf{N}(s) \| = \frac{1}{\| \mathbf{T}'(s) \|} \| \mathbf{T}'(s) \| = 1.$$

We claim also that $\mathbf{N}(s)$ is orthogonal to the tangent vector $\mathbf{T}(s)$. To see this, recall that $\mathbf{T}(s)$ is a unit vector, so

$$\| \mathbf{T}(s) \|^2 = \mathbf{T}(s) \cdot \mathbf{T}(s) = 1.$$

Differentiate this equation to get

$$\mathbf{T}'(s) \cdot \mathbf{T}(s) + \mathbf{T}(s) \cdot \mathbf{T}'(s) = 2\mathbf{T}(s) \cdot \mathbf{T}'(s) = 0.$$

Therefore, $\mathbf{T}(s)$ is orthogonal to $\mathbf{T}'(s)$. But $\mathbf{N}(s)$ is a scalar multiple of $\mathbf{T}'(s)$, hence it is in the same direction as $\mathbf{T}'(s)$. Therefore, $\mathbf{T}(s)$ is orthogonal to $\mathbf{N}(s)$.

At any point where $\mathbf{F}$ is twice differentiable, we may now place a unit tangent and a unit normal vector, as in Figure 11.6. With these in hand, we claim that we can write the acceleration in terms of tangential and normal components:

$$\mathbf{a}(t) = a_T \mathbf{T}(t) + a_N \mathbf{N}(t)$$

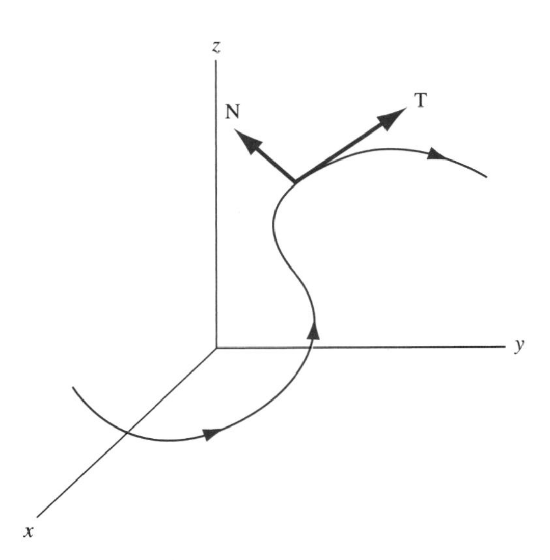

FIGURE 11.6 *Unit tangent and normal vectors to a curve.*

where

$$a_T = \text{ tangential component of the acceleration } = \frac{dv}{dt}$$

and

$$a_N = \text{ normal component of the acceleration } = v(t)^2\kappa(t).$$

To verify this decomposition of $\mathbf{a}(t)$, begin with

$$\mathbf{T}(t) = \frac{1}{\| \mathbf{F}'(t) \|}\mathbf{F}'(t) = \frac{1}{v(t)}\mathbf{v}(t).$$

Then

$$\mathbf{v} = v\mathbf{T},$$

so

$$\begin{aligned}\mathbf{a} &= \frac{d}{dt}\mathbf{v} = \frac{dv}{dt}\mathbf{T} + v\mathbf{T}' \\ &= \frac{dv}{dt}\mathbf{T} + v\frac{ds}{dt}\frac{d\mathbf{T}}{ds} \\ &= \frac{dv}{dt}\mathbf{T} + v^2\mathbf{T}'(s) \\ &= \frac{dv}{dt}\mathbf{T} + v^2\kappa\mathbf{N}.\end{aligned}$$

Because $\mathbf{T}$ and $\mathbf{N}$ are orthogonal, then

$$\begin{aligned}\| \mathbf{a} \|^2 &= \mathbf{a} \cdot \mathbf{a} = (a_T\mathbf{T} + a_N\mathbf{N}) \cdot (a_T\mathbf{T} + a_N\mathbf{N}) \\ &= a_T^2\mathbf{T} \cdot \mathbf{T} + 2a_Ta_N\mathbf{T} \cdot \mathbf{N} + a_N^2\mathbf{N} \cdot \mathbf{N} \\ &= a_T^2 + a_N^2.\end{aligned}$$

This means that, whenever two of $\| \mathbf{a} \|$, a_T, and a_N are known, we can compute the third quantity.

If a_N is known, it is sometimes convenient to compute the curvature $\kappa(t)$ as

$$\kappa(t) = \frac{a_N}{v^2}.$$

EXAMPLE 11.4

Let $\mathbf{F}(t)$ be as in Example 11.3. There we computed $\mathbf{v}(t) = \sqrt{5}t$. Therefore,

$$a_T = \frac{dv}{dt} = \sqrt{5}.$$

The acceleration is

$$\mathbf{a} = \mathbf{v}' = \mathbf{F}''(t) = [\cos(t) - t\sin(t)]\mathbf{i} + [\sin(t) + t\cos(t)]\mathbf{j} + 2\mathbf{k}.$$

Then

$$\| \mathbf{a} \| = \sqrt{5 + t^2}.$$

so

$$a_N^2 = \| \mathbf{a} \|^2 - a_T^2 = 5 + t^2 - 5 = t^2.$$

Since $t > 0$, $a_N = t$. The acceleration may be written as

$$\mathbf{a} = \sqrt{5}\mathbf{T} + t\mathbf{N}.$$

If we know a_N and v, it is easy to compute the curvature, since

$$a_N = t = \kappa v^2 = 5t^2\kappa,$$

implying that $\kappa = 1/5t$, as we found in Example 11.3. ♦

SECTION 11.2 PROBLEMS

In each of Problems 1 through 10, a position vector is given. Determine the velocity, speed, acceleration, curvature, and the tangential and normal components of the acceleration.

1. $\mathbf{F} = \alpha t^2\mathbf{i} + \beta t^2\mathbf{j} + \gamma t^2\mathbf{k}$
2. $\mathbf{F} = e^t\sin(t)\mathbf{i} - \mathbf{j} + e^t\cos(t)\mathbf{k}$
3. $\mathbf{F} = 2\sinh(t)\mathbf{j} - 2\cosh(t)\mathbf{k}$
4. $\mathbf{F} = \alpha\cos(t)\mathbf{i} + \beta t\mathbf{j} + \alpha\sin(t)\mathbf{k}$
5. $\mathbf{F} = 2t\mathbf{i} - 2t\mathbf{j} + t\mathbf{k}$
6. $\mathbf{F} = 3t\cos(t)\mathbf{j} - 3t\sin(t)\mathbf{k}$
7. $\mathbf{F} = 3e^{-t}(\mathbf{i} + \mathbf{j} - 2\mathbf{k})$
8. $\mathbf{F} = t\sin(t)\mathbf{i} + t\cos(t)\mathbf{j} + \mathbf{k}$
9. $\mathbf{F} = 3t\mathbf{i} - 2\mathbf{j} + t^2\mathbf{k}$
10. $\mathbf{F} = \ln(t)(\mathbf{i} - \mathbf{j} + 2\mathbf{k})$
11. Show that $\kappa(t) = \dfrac{||\mathbf{F}'(t) \times \mathbf{F}''(t)||}{||\mathbf{F}'(t)||^3}$.
12. Show that the curvature of a circle is constant. *Hint*: If the radius is r, show that the curvature is $1/r$.
13. Show that any straight line has curvature zero. Conversely, if a smooth curve has curvature zero, then it must be a straight line. *Hint*: For the first part, recall that any straight line has a position vector $\mathbf{F}(t) = (a + bt)\mathbf{i} + (d + ct)\mathbf{j} + (h + kt)\mathbf{k}$. For the converse, if $\kappa = 0$, then $\mathbf{T}'(t) = \mathbf{O}$.

11.3 Vector Fields and Streamlines

Vector functions $\mathbf{F}(x, y)$ in two variables, and $\mathbf{G}(x, y, z)$ in three variables, are called *vector fields*. At each point where the vector field is defined, we can draw an arrow representing the vector at that point. This suggests fields of arrows "growing" out of points in regions of the plane or 3-space.

Take partial derivatives of vector fields by differentiating each component. For example, if

$$\mathbf{F}(x, y, z) = \cos(x + y^2 + 2z)\mathbf{i} - xyz\mathbf{j} + xye^z\mathbf{k},$$

then

$$\frac{\partial \mathbf{F}}{\partial x} = -\sin(x + y^2 + 2z)\mathbf{i} - yz\mathbf{j} + ye^z\mathbf{k},$$

$$\frac{\partial \mathbf{F}}{\partial y} = -2y\sin(x + y^2 + 2z)\mathbf{i} - xz\mathbf{j} + xe^z\mathbf{k},$$

and

$$\frac{\partial \mathbf{F}}{\partial z} = -2\sin(x + y^2 + 2z)\mathbf{i} - xy\mathbf{j} + xye^z\mathbf{k}.$$

Given a vector field $\mathbf{F}$ in 3-space, a *streamline* of $\mathbf{F}$ is a curve with the property that, at each point (x, y, z) of the curve, $\mathbf{F}(x, y, z)$ is a tangent vector to the curve.

If $\mathbf{F}$ is the velocity field for a fluid flowing through some region, then the streamlines are called *flow lines* of the fluid and describe trajectories of imaginary particles moving with the fluid. If $\mathbf{F}$ is a magnetic field the streamlines are called *lines of force*. Iron filings put on a piece of cardboard held over a magnet will align themselves on the lines of force.

Given a vector field, we would like to find all of the streamlines. This is the problem of constructing a curve through each point of a region of space, given the tangent to the curve at each point. To solve this problem suppose that C is a streamline of $\mathbf{F} = f\mathbf{i} + g\mathbf{j} + h\mathbf{k}$. Let C have parametric equations $x = x(\xi), y = y(\xi), z = z(\xi)$. A position vector for C is

$$\mathbf{R}(\xi) = x(\xi)\mathbf{i} + y(\xi)\mathbf{j} + z(\xi)\mathbf{k}.$$

Now

$$\mathbf{R}'(\xi) = x'(\xi)\mathbf{i} + y'(\xi)\mathbf{j} + z'(\xi)\mathbf{k}$$

is tangent to C at $(x(\xi), y(\xi), z(\xi))$ and is therefore parallel to the tangent vector $\mathbf{F}(x(\xi), y(\xi), z(\xi))$ at this point. These vectors must therefore be scalar multiples of each other, say

$$\mathbf{R}'(\xi) = t\mathbf{F}(x(\xi), y(\xi), z(\xi)).$$

Then

$$\frac{dx}{d\xi}\mathbf{i} + \frac{dy}{d\xi}\mathbf{j} + \frac{dz}{d\xi}\mathbf{k} =$$
$$tf(x(\xi), y(\xi), z(\xi))\mathbf{i} + tg(x(\xi), y(\xi), z(\xi))\mathbf{j} + th(x(\xi), y(\xi), z(\xi))\mathbf{k}.$$

Equating respective components in this equation gives us

$$\frac{dx}{d\xi} = tf, \frac{dy}{d\xi} = tg, \frac{dz}{d\xi} = th.$$

This is a system of differential equations for the parametric equations of the streamlines. If f, g and h are nonzero this system can be written as

$$\frac{dx}{f} = \frac{dy}{g} = \frac{dz}{h}.$$

EXAMPLE 11.5

We fill find the streamlines of $\mathbf{F}(x, y, z) = x^2\mathbf{i} + 2y\mathbf{j} - \mathbf{k}$. If x and y are not zero, the streamlines satisfy

$$\frac{dx}{x^2} = \frac{dy}{2y} = \frac{dz}{-1}.$$

These differential equations can be solved in pairs. First integrate

$$\frac{dx}{x^2} = -dz$$

to get

$$-\frac{1}{x} = -z + c$$

with c an arbitrary constant. Next integrate

$$\frac{dy}{2y} = -dz$$

to get

$$\frac{1}{2}\ln|y| = -z + k.$$

It is convenient to express two of the variables in terms of the third. If we write x and y in terms of z we have

$$x = \frac{1}{z-c} \text{ and } y = ae^{-2z},$$

in which a is constant. This gives us parametric equations of the streamlines, with z as the parameter. If we want the streamline through a particular point, we must choose a and c accordingly. For example, suppose we want the streamline through $(-1, 6, 2)$. Then $z = 2$ and we need

$$-1 = \frac{1}{2-c} \text{ and } 6 = ae^{-4}.$$

Then $c = 3$ and $a = 6e^4$ so the streamline through $(-1, 6, 2)$ has parametric equations

$$x = \frac{1}{z-3}, \; y = 6e^{4-2z}, \; z = z. \; \blacklozenge$$

SECTION 11.3 PROBLEMS

1. Construct a vector field whose streamlines are circles about the origin.

In each of Problems 2 through 7, find the streamlines of the vector field and also the streamline through the given point.

2. $\mathbf{F} = \cos(y)\mathbf{i} + \sin(x)\mathbf{j}$; $(\pi/2, 0, -4)$
3. $\mathbf{F} = 2e^z\mathbf{i} - \cos(y)\mathbf{k}$; $(3, \pi/4, 0)$
4. $\mathbf{F} = 3x^2\mathbf{i} - y\mathbf{j} + z^3\mathbf{k}$; $(2, 1, 6)$
5. $\mathbf{F} = (1/x)\mathbf{i} + e^x\mathbf{j} - \mathbf{k}$; $(2, 0, 4)$
6. $\mathbf{F} = \mathbf{i} - 2\mathbf{j} + \mathbf{k}$; $(0, 1, 1)$
7. $\mathbf{F} = \mathbf{i} - y^2\mathbf{j} + z\mathbf{k}$; $(2, 1, 1)$

11.4 The Gradient Field

Let $\varphi(x, y, z)$ be a real-valued function of three variables. In the context of vector fields, φ is called a *scalar field*. The *gradient* of φ is the vector field

$$\nabla\varphi = \frac{\partial\varphi}{\partial x}\mathbf{i} + \frac{\partial\varphi}{\partial y}\mathbf{j} + \frac{\partial\varphi}{\partial z}\mathbf{k}.$$

The symbol $\nabla\varphi$ is read "del φ" and ∇ is called the *del operator*. If φ is a function of just (x, y), then $\nabla\varphi$ is a vector field in the plane.

∇ is also often called *nabla*.

For example, if $\varphi(x, y, z) = x^2 y \cos(yz)$, then

$$\nabla\varphi = 2xy\cos(yz)\mathbf{i} + [x^2\cos(yz) - x^2 z\sin(yz)]\mathbf{j} - x^2y^2\sin(yz)\mathbf{k}.$$

If P is a point, then the gradient of φ evaluated at P is denoted $\nabla\varphi(P)$.

The gradient has the obvious properties

$$\nabla(\varphi + \psi) = \nabla(\varphi) + \nabla(\psi)$$

and, for any number c,

$$\nabla(c\varphi) = c\nabla(\varphi).$$

The gradient is related to the directional derivative. Let $P_0 : (x_0, y_0, z_0)$ be a point and let $\mathbf{u} = a\mathbf{i} + b\mathbf{j} + c\mathbf{k}$ be a unit vector, represented as an arrow from P_0. We want to measure the rate of change of $\varphi(x, y, z)$ as (x, y, z) varies from P_0 in the direction of $\mathbf{u}$. To do this let $t > 0$. The point $P : (x_0 + at, y_0 + bt, z_0 + ct)$ is on the line through P_0 in the direction of $\mathbf{u}$ and P varies in this direction as t varies.

We measure the rate of change $D_{\mathbf{u}}\varphi(P_0)$ of $\varphi(x, y, z)$ in the direction of $\mathbf{u}$, at P_0, by setting

$$D_{\mathbf{u}}\varphi(P_0) = \frac{d}{dt}\left[\varphi(x_0 + at, y_0 + bt, z_0 + ct)\right]_{t=0}.$$

$D_{\mathbf{u}}\varphi(P_0)$ is the *directional derivative* of φ at P_0 in the direction of $\mathbf{u}$.

We can compute a directional derivative in terms of the gradient as follows. By the chain rule,

$$\begin{aligned} D_{\mathbf{u}}\varphi(P_0) &= \left[\frac{d}{dt}\varphi(x_0 + at, y_0 + bt, z_0 + ct)\right]_{t=0} \\ &= a\frac{\partial\varphi}{\partial x}(x_0, y_0, z_0) + b\frac{\partial\varphi}{\partial y}(x_0, y_0, z_0) + c\frac{\partial\varphi}{\partial z}(x_0, y_0, z_0) \\ &= a\frac{\partial\varphi}{\partial x}(P_0) + b\frac{\partial\varphi}{\partial y}(P_0) + c\frac{\partial\varphi}{\partial z}(P_0) \\ &= \nabla\varphi(P_0)\cdot(a\mathbf{i} + b\mathbf{j} + c\mathbf{k}) \\ &= \nabla\varphi(P_0)\cdot\mathbf{u}. \end{aligned}$$

Therefore $D_{\mathbf{u}}\varphi(P_0)$ is the dot product of the gradient of φ at the point, with the unit vector specifying the direction.

EXAMPLE 11.6

Let $\varphi(x, y, z) = x^2y - xe^z$ and $P_0 = (2, -1, \pi)$. We will compute the rate of change of $\varphi(x, y, z)$ at P_0 in the direction of $\mathbf{u} = (1/\sqrt{6})(\mathbf{i} - 2\mathbf{j} + \mathbf{k})$.

The gradient is

$$\nabla\varphi = (2xy - e^z)\mathbf{i} + x^2\mathbf{j} - xe^z\mathbf{k}.$$

Then

$$\nabla\varphi(2, -1, \pi) = (-4 - e^{\pi})\mathbf{i} + 4\mathbf{j} - 2e^{\pi}\mathbf{k}.$$

The directional derivative of φ at P_0 in the direction of $\mathbf{u}$ is

$$\begin{aligned}
&D_{\mathbf{u}}\varphi(2,-1,\pi)\\
&=((-4-e^{\pi})\mathbf{i}+4\mathbf{j}-2e^{\pi}\mathbf{k})\cdot\frac{1}{\sqrt{6}}(\mathbf{i}-2\mathbf{j}+\mathbf{k})\\
&=\frac{1}{\sqrt{6}}(-4-e^{\pi}-8-2e^{\pi})\\
&=\frac{-3}{\sqrt{6}}(4+e^{\pi}). \quad \blacklozenge
\end{aligned}$$

If a direction is specified by a vector that is not of length 1, divide it by its length before computing the directional derivative.

Now imagine standing at P_0 and observing $\varphi(x,y,z)$ as (x,y,z) moves away from P_0. In what direction will $\varphi(x,y,z)$ increase at the greatest rate? We claim that this is the direction of the gradient of φ at P_0.

THEOREM 11.1

Let φ and its first partial derivatives be continuous in some sphere about P_0, and suppose that $\nabla\varphi(P_0)\neq\mathbf{O}$. Then

1. At P_0, $\varphi(x,y,z)$ has its maximum rate of change in the direction of $\nabla\varphi(P_0)$. This maximum rate of change is $\|\nabla\varphi(P_0)\|$.
2. At P_0, $\varphi(x,y,z)$ has its minimum rate of change in the direction of $-\nabla\varphi(P_0)$. This minimum rate of change is $-\|\nabla\varphi(P_0)\|$. ♦

For condition (1), let $\mathbf{u}$ be any unit vector from P_0 and consider

$$\begin{aligned}
D_{\mathbf{u}}\varphi(P_0)&=\nabla\varphi(P_0)\cdot\mathbf{u}\\
&=\|\nabla\varphi(P_0)\|\|\mathbf{u}\|\cos(\theta)\\
&=\|\nabla\varphi(P_0)\|\cos(\theta)
\end{aligned}$$

where θ is the angle between $\mathbf{u}$ and $\nabla\varphi(P_0)$. Clearly $D_{\mathbf{u}}\varphi(P_0)$ has its maximum when $\cos(\theta)=1$, which occurs when $\theta=0$, hence when $\mathbf{u}$ is in the same direction as $\nabla\varphi(P_0)$.

For condition (2), $D_{\mathbf{u}}\varphi(P_0)$ has its minimum when $\cos(\theta)=-1$, hence when $\theta=\pi$ and $\nabla\varphi(P_0)$ is opposite $\mathbf{u}$.

EXAMPLE 11.7

Let $\varphi(x,y,z)=2xz+z^2e^y$ and $P_0:(2,1,1)$. The gradient of φ is

$$\nabla\varphi(x,y,z)=2z\mathbf{i}+z^2e^y\mathbf{j}+(2x+2ze^y)\mathbf{k}$$

so

$$\nabla\varphi(2,1,1)=2\mathbf{i}+e\mathbf{j}+(4+2e)\mathbf{k}.$$

The maximum rate of increase of $\varphi(x,y,z)$ at $(2,1,1)$ is in the direction of $2\mathbf{i}+e\mathbf{j}+(4+2e)\mathbf{k}$, and this maximum rate of change is

$$\sqrt{4+e^2+(4+2e)^2},$$

or $\sqrt{20+16e+5e^2}$. ♦

11.4.1 Level Surfaces, Tangent Planes, and Normal Lines

Depending on φ and the number k, the locus of points (x, y, z) such that $\varphi(x, y, z) = k$ may be a surface in 3-space. Any such surface is called a *level surface* of φ. For example, if $\varphi(x, y, z) = x^2 + y^2 + z^2$, then the level surface of $\varphi(x, y, z) = k$ is a sphere of radius $\sqrt{k}$ if $k > 0$, a single point $(0, 0, 0)$ if $k = 0$, and is vacuous if $k < 0$. Part of the level surface $\psi(x, y, z) = z - \sin(xy) = 0$ is shown in Figure 11.7.

Suppose $P_0 : (x_0, y_0, z_0)$ is on a level surface S given by $\varphi(x, y, z) = k$. Assume that there are smooth (having continuous tangents) curves on the surface passing through P_0, as typified by C in Figure 11.8. Each such curve has a tangent vector at P_0. The plane containing these tangent vectors is called the *tangent plane* to S at P_0. A vector orthogonal to this tangent plane at P_0 is called a *normal vector*, or *normal*, to this tangent plane at P_0. We will determine this tangent plane and normal vector. The key lies in the following fact about the gradient vector.

THEOREM 11.2 Normal to a Level Surface

Let φ and its first partial derivatives be continuous. Then $\nabla\varphi(P)$ is normal to the level surface $\varphi(x, y, z) = k$ at any point P on this surface such that $\nabla\varphi(P) \neq \mathbf{O}$. ♦

To understand this conclusion, let P_0 be on the level surface S and suppose a smooth curve C on the surface passes through P_0, as in Figure 11.8. Let C have parametric equations $x = x(t)$, $y = y(t)$, $z = z(t)$ for $a \le t \le b$. Since P_0 is on C, for some t_0,

$$x(t_0) = x_0,\ y(t_0) = y_0,\ z(t_0) = z_0.$$

Furthermore, because C lies on the level surface,

$$\varphi(x(t), y(t), z(t)) = k$$

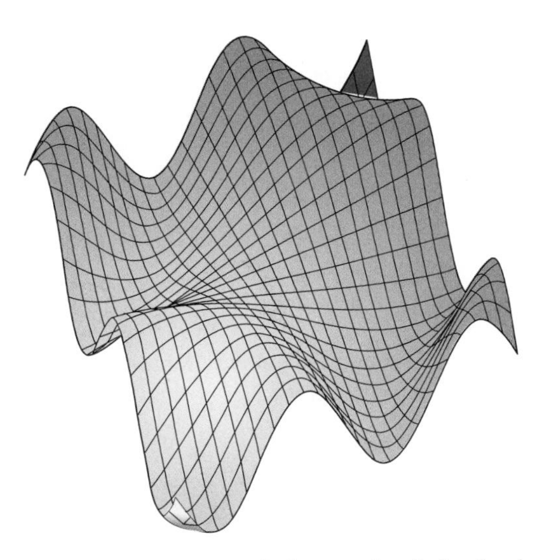

FIGURE 11.7 *Part of the graph of the level surface $z = \sin(xy)$.*

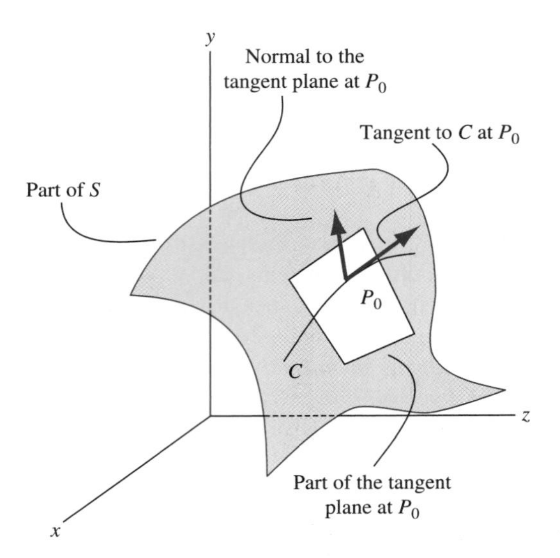

FIGURE 11.8 *Normal to a level surface.*

for $a \leq t \leq b$. Then

$$\frac{d}{dt}\varphi(x(t), y(t), z(t)) = 0 = \frac{\partial \varphi}{\partial x}x'(t) + \frac{\partial \varphi}{\partial y}y'(t) + \frac{\partial \varphi}{\partial z}z'(t)$$
$$= \nabla\varphi \cdot [x'(t)\mathbf{i} + y'(t)\mathbf{j} + z'(t)\mathbf{k}].$$

But $x'(t)\mathbf{i} + y'(t)\mathbf{j} + z'(t)\mathbf{k} = \mathbf{T}(t)$ is a tangent vector to C. Letting $t = t_0$, $\mathbf{T}(t_0)$ is tangent to C at P_0 and the last equation tells us that

$$\nabla\varphi(P_0) \cdot \mathbf{T}(t_0) = 0.$$

Therefore $\nabla\varphi(P_0)$ is normal to the tangent to C at P_0. But C is *any* smooth curve on S passing through P_0. Therefore $\nabla\varphi(P_0)$ is normal to every tangent vector at P_0 to any curve on S through P_0, and is therefore normal to the tangent plane to S at P_0.

Now we have a point P_0 on the normal plane at P_0, and a vector $\nabla\varphi(P_0)$ orthogonal to this plane. The equation of the tangent plane is

$$\nabla\varphi(P_0) \cdot [(x - x_0)\mathbf{i} + (y - y_0)\mathbf{j} + (z - z)_0\mathbf{k}] = 0,$$

or

$$\frac{\partial \varphi}{\partial x}(P_0)(x - x_0) + \frac{\partial \varphi}{\partial y}(P_0)(y - y_0) + \frac{\partial \varphi}{\partial z}(P_0)(z - z_0) = 0. \tag{11.1}$$

A straight line through P_0 and parallel to the normal vector is called the *normal line* to S at P_0. Since the gradient of φ at P_0 is a normal vector, if (x, y, z) is on this normal line, then for some scalar t,

$$(x - x_0)\mathbf{i} + (y - y_0)\mathbf{j} + (z - z_0)\mathbf{k} = t\nabla\varphi(P_0).$$

The parametric equations of the normal line to S at P_0 are

$$x = x_0 + t\frac{\partial \varphi}{\partial x}(P_0),\ y = y_0 + t\frac{\partial \varphi}{\partial y}(P_0),\ z = z_0 + t\frac{\partial \varphi}{\partial z}(P_0). \tag{11.2}$$

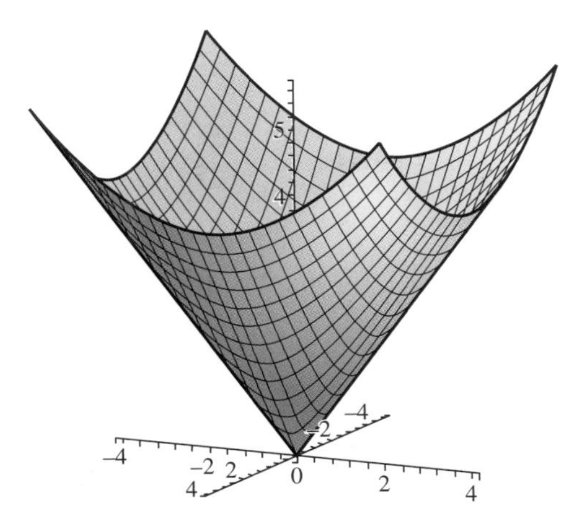

FIGURE 11.9 *Circular cone* $z=\sqrt{x^2+y^2}$.

EXAMPLE 11.8

The level surface $\varphi(x,y,z)=z-\sqrt{x^2+y^2}$ is a cone with vertex at the origin (Figure 11.9). Compute

$$\nabla\varphi(1,1,\sqrt{2})=-\frac{1}{\sqrt{2}}\mathbf{i}-\frac{1}{\sqrt{2}}\mathbf{j}+\mathbf{k}.$$

The tangent plane to the cone at $(1,1,\sqrt{2})$ has the equations

$$-\frac{1}{\sqrt{2}}(x-1)-\frac{1}{\sqrt{2}}(y-1)+z-\sqrt{2}=0$$

or

$$x+y-\sqrt{2}z=0.$$

The normal line to the cone at $(1,1,\sqrt{2})$ has parametric equations

$$x=1-\frac{1}{\sqrt{2}}t,\ y=1-\frac{1}{\sqrt{2}}t,\ z=\sqrt{2}+t.\ \blacklozenge$$

SECTION 11.4 PROBLEMS

In each of Problems 1 through 6, compute the gradient of the function and evaluate this gradient at the given point. Determine at this point the maximum and minimum rate of change of the function at this point.

1. $\varphi(x,y,z)=\cosh(2xy)-\sinh(z)$; $(0,1,1)$
2. $\varphi(x,y,z)=\cos(xyz)$; $(-1,1,\pi/2)$
3. $\varphi(x,y,z)=xyz$; $(1,1,1)$
4. $\varphi(x,y,z)=x^2y-\sin(xz)$; $(1,-1,\pi/4)$
5. $\varphi(x,y,z)=2xy+xe^z$; $(-2,1,6)$
6. $\varphi(x,y,z)=\sqrt{x^2+y^2+z^2}$; $(2,2,2)$

In each of Problems 7 through 10, compute the directional derivative of the function in the direction of the given vector.

7. $\varphi(x,y,z)=x^2yz^3$; $2\mathbf{j}+\mathbf{k}$
8. $\varphi(x,y,z)=yz+xz+xy$; $\mathbf{i}-4\mathbf{k}$
9. $\varphi(x,y,z)=8xy^2-xz$; $(1/\sqrt{3})(\mathbf{i}+\mathbf{j}+\mathbf{k})$
10. $\varphi(x,y,z)=\cos(x-y)+e^z$; $\mathbf{i}-\mathbf{j}+2\mathbf{k}$

11. Suppose that $\nabla\varphi(x, y, z) = \mathbf{i} + \mathbf{k}$. What can be said about level surfaces of φ? Show that the streamlines of $\nabla\varphi$ are orthogonal to the level surfaces of φ.

In each of Problems 12 through 17, find the equation of the tangent plane and normal line to the level surface at the point.

12. $3x^4 + 3y^4 + 6z^4 = 12$; $(1, 1, 1)$

13. $2x - \cos(xyz) = 3$; $(1, \pi, 1)$

14. $z = x^2 + y$; $(-1, 1, 2)$

15. $z^2 = x^2 - y^2$; $(1, 1, 0)$

16. $x^2 - y^2 + z^2 = 0$; $(1, 1, 0)$

17. $x^2 + y^2 + z^2 = 4$; $(1, 1, \sqrt{2})$

11.5 Divergence and Curl

The gradient operator produces a vector field from a scalar function. We will discuss two other important vector operations. One produces a scalar field from a vector field, and the other produces a vector field from a vector field. Let

$$\mathbf{F}(x, y, z) = f(x, y, z)\mathbf{i} + g(x, y, z)\mathbf{j} + h(x, y, z)\mathbf{k}.$$

The *divergence* of $\mathbf{F}$ is the scalar field

$$\text{div } \mathbf{F} = \frac{\partial f}{\partial x} + \frac{\partial g}{\partial y} + \frac{\partial h}{\partial z}.$$

The *curl* of $\mathbf{F}$ is the vector field

$$\text{curl } \mathbf{F} = \left(\frac{\partial h}{\partial y} - \frac{\partial g}{\partial z}\right)\mathbf{i} + \left(\frac{\partial f}{\partial z} - \frac{\partial h}{\partial x}\right)\mathbf{j} + \left(\frac{\partial g}{\partial x} - \frac{\partial f}{\partial y}\right)\mathbf{k}.$$

Divergence, curl and gradient can all be written as vector operations with the *del operator* ∇, which is a symbolic vector defined by

$$\nabla = \frac{\partial}{\partial x}\mathbf{i} + \frac{\partial}{\partial y}\mathbf{j} + \frac{\partial}{\partial z}\mathbf{k}.$$

The symbol ∇, which is called "del", or sometimes "nabla", is treated like a vector in carrying out calculations, and the "product" of $\partial/\partial x$, $\partial/\partial y$ and $\partial/\partial z$ with a scalar function φ is interpreted to mean, respectively, $\partial\varphi/\partial x$, $\partial\varphi/\partial y$ and $\partial\varphi/\partial z$. Now observe how gradient, divergence, and curl are obtained using this operator.

1. The product of the vector ∇ and the scalar function φ is the gradient of φ:

$$\nabla\varphi = \left(\frac{\partial}{\partial x}\mathbf{i} + \frac{\partial}{\partial y}\mathbf{j} + \frac{\partial}{\partial z}\mathbf{k}\right)\varphi$$
$$= \frac{\partial\varphi}{\partial x}\mathbf{i} + \frac{\partial\varphi}{\partial y}\mathbf{j} + \frac{\partial\varphi}{\partial z}\mathbf{k} = \text{gradient of } \varphi.$$

2. The dot product of ∇ and $\mathbf{F}$ is the divergence of $\mathbf{F}$:

$$\nabla \cdot \mathbf{F} = \left(\frac{\partial}{\partial x}\mathbf{i} + \frac{\partial}{\partial y}\mathbf{j} + \frac{\partial}{\partial z}\mathbf{k}\right) \cdot (f\mathbf{i} + g\mathbf{j} + h\mathbf{k})$$
$$= \frac{\partial f}{\partial x} + \frac{\partial g}{\partial y} + \frac{\partial h}{\partial z} = \text{divergence of } \mathbf{F}.$$

3. The cross product of ∇ with $\mathbf{F}$ is the curl of $\mathbf{F}$:

$$\nabla \times \mathbf{F} = \begin{vmatrix} \mathbf{i} & \mathbf{j} & \mathbf{k} \\ \partial/\partial x & \partial/\partial y & \partial/\partial z \\ f & g & h \end{vmatrix}$$

$$= \left(\frac{\partial h}{\partial y} - \frac{\partial g}{\partial z}\right)\mathbf{i} + \left(\frac{\partial f}{\partial z} - \frac{\partial h}{\partial x}\right)\mathbf{j} + \left(\frac{\partial g}{\partial x} - \frac{\partial f}{\partial y}\right)\mathbf{k} = \text{curl of } \mathbf{F}.$$

The del (or nabla) operator is part of the MAPLE set of routines collected under the VectorCalculus designation. Using this and the operations of scalar multiplication, dot product (DotProduct) and cross product (CrossProduct), we can carry out computations with vector fields. This package can also be used to compute divergence and curl in other coordinate systems, such as cylindrical and spherical coordinates.

There are two relationships between gradient, divergence and curl that are fundamental to vector analysis: the curl of a gradient is the zero vector, and the divergence of a curl is (the number) zero.

THEOREM 11.3

Let $\mathbf{F}$ be a continuous vector field whose components have continuous first and second partial derivatives and let φ be a continuous scalar field with continuous first and second partial derivatives. Then

1.
$$\nabla \times (\nabla \varphi) = \mathbf{O}.$$

2.
$$\nabla \cdot (\nabla \times \mathbf{F}) = 0. \quad ♦$$

These conclusions may be paraphrased:

$$\text{curl grad } = \mathbf{O}, \ \text{div curl } = 0.$$

Both of these identities can be verified by direct computation, using the equality of mixed second partial derivatives with respect to the same two variables. For example, for conclusion (1),

$$\nabla \times (\nabla \varphi) = \nabla \times \left(\frac{\partial \varphi}{\partial x}\mathbf{i} + \frac{\partial \varphi}{\partial y}\mathbf{j} + \frac{\partial \varphi}{\partial z}\mathbf{k}\right)$$

$$= \begin{vmatrix} \mathbf{i} & \mathbf{j} & \mathbf{k} \\ \partial/\partial x & \partial/\partial y & \partial/\partial z \\ \partial\varphi/\partial x & \partial\varphi/\partial y & \partial\varphi/\partial z \end{vmatrix}$$

$$= \left(\frac{\partial^2 \varphi}{\partial y \partial z} - \frac{\partial^2 \varphi}{\partial z \partial y}\right)\mathbf{i} + \left(\frac{\partial^2 \varphi}{\partial z \partial x} - \frac{\partial^2 \varphi}{\partial x \partial z}\right)\mathbf{j} + \left(\frac{\partial^2 \varphi}{\partial x \partial y} - \frac{\partial^2 \varphi}{\partial y \partial x}\right)\mathbf{k}$$

$$= \mathbf{O}$$

because the mixed partials cancel in pairs in the components of $\nabla \times (\nabla \varphi)$.

Operator notation with ∇ can simplify such calculations. In this notation, $\nabla \times (\nabla \varphi) = \mathbf{O}$ is immediate because $\nabla \times \nabla$ is the cross product of a "vector" with itself, which is always zero. Similarly, for conclusion (2), $\nabla \times \mathbf{F}$ is orthogonal to ∇, so its dot product with ∇ is zero.

11.5.1 A Physical Interpretation of Divergence

Suppose $\mathbf{F}(x, y, z, t)$ is the velocity of a fluid at point (x, y, z) and time t. Time plays no role in computing divergence, but is included here because a flow may depend on time. We will show that the divergence of $\mathbf{F}$ measures the outward flow of the fluid from any point.

Imagine a small rectangular box in the fluid, as in Figure 11.10. First look at the front and back faces II and I, respectively. The flux of the flow out of this box across II is the normal component of the velocity (dot product of $\mathbf{F}$ with $\mathbf{i}$) multiplied by the area of this face:

$$\begin{aligned}\text{flux outward across face II} &= \mathbf{F}(x+\Delta x, y, z, t)\cdot\mathbf{i}\Delta y\Delta z\\ &= f(x+\Delta x, y, z, t)\Delta y\Delta z.\end{aligned}$$

On face I the unit outer normal is $-\mathbf{i}$, so the flux outward across this face is $-f(x, y, z, t)\Delta y\Delta z$. The total outward flux across faces II and I is therefore

$$[f(x+\Delta x, y, z, t) - f(x, y, z, t)]\Delta y\Delta z.$$

A similar calculation holds for the pairs of other opposite sides. The total flux of fluid flowing out of the box across its faces is

$$\begin{aligned}\text{total flux} &= [f(x+\Delta x, y, z, t) - f(x, y, z, t)]\Delta y\Delta z\\ &\quad + [g(x, y+\Delta y, z, t) - g(x, y, z, t)]\Delta x\Delta z\\ &\quad + [h(x, y, z+\Delta z, t) - h(x, y, z, t)]\Delta x\Delta y.\end{aligned}$$

The total flux per unit volume out of the box is obtained by dividing this quantity by $\Delta x\Delta y\Delta z$, obtaining

$$\begin{aligned}\text{flux per unit volume} &= \frac{f(x+\Delta x, y, z, t) - f(x, y, z, t)}{\Delta x}\\ &\quad + \frac{g(x, y+\Delta y, z, t) - g(x, y, z, t)}{\Delta y}\\ &\quad + \frac{h(x, y, z+\Delta z, t) - h(x, y, z, t)}{\Delta z}.\end{aligned}$$

In the limit as $(\Delta x, \Delta y, \Delta t) \rightarrow (0, 0, 0)$, this sum approaches the divergence of $\mathbf{F}(x, y, z, t)$.

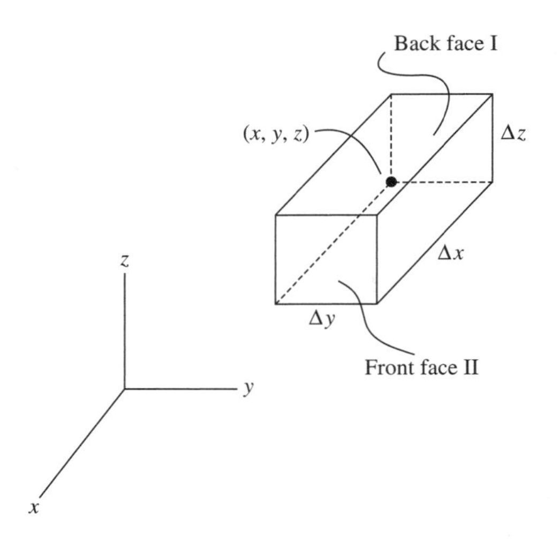

FIGURE 11.10 *Interpretation of divergence.*

11.5.2 A Physical Interpretation of Curl

The curl vector is interpreted as a measure of rotation or swirl about a point. In British literature, the curl is often called the *rot* (for rotation) of a vector field.

To understand this interpretation, suppose an object rotates with uniform angular speed ω about a line L, as in Figure 11.11. The angular velocity vector Ω has magnitude ω and is directed along L as a right-handed screw would progress if given the same sense of rotation as the object. Put L through the origin and let $\mathbf{R} = x\mathbf{i} + y\mathbf{j} + z\mathbf{k}$ for any point (x, y, z) on the rotating object. Let $\mathbf{T}(x, y, z)$ be the tangential linear velocity and $R = \| \mathbf{R} \|$. Then

$$\| \mathbf{T} \| = \omega R \sin(\theta) = \| \Omega \times \mathbf{R} \|,$$

with θ the angle between $\mathbf{R}$ and Ω. Since $\mathbf{T}$ and $\Omega \times \mathbf{R}$ have the same direction and magnitude, we conclude that $\mathbf{T} = \Omega \times \mathbf{R}$. Now write $\Omega = a\mathbf{i} + b\mathbf{j} + c\mathbf{k}$ to obtain

$$\mathbf{T} = \Omega \times \mathbf{R} = (bz - cy)\mathbf{i} + (cx - az)\mathbf{j} + (ay - bx)\mathbf{k}.$$

Then

$$\nabla \times \mathbf{T} = \begin{vmatrix} \mathbf{i} & \mathbf{j} & \mathbf{k} \\ \partial/\partial z & \partial/\partial y & \partial/\partial z \\ bz - cy & cx - az & ay - bx \end{vmatrix}$$
$$= 2a\mathbf{i} + 2b\mathbf{j} + 2c\mathbf{k} = 2\Omega.$$

Therefore,

$$\Omega = \frac{1}{2}\nabla \times \mathbf{T}.$$

The angular momentum of a uniformly rotating body is a constant times the curl of the linear velocity.

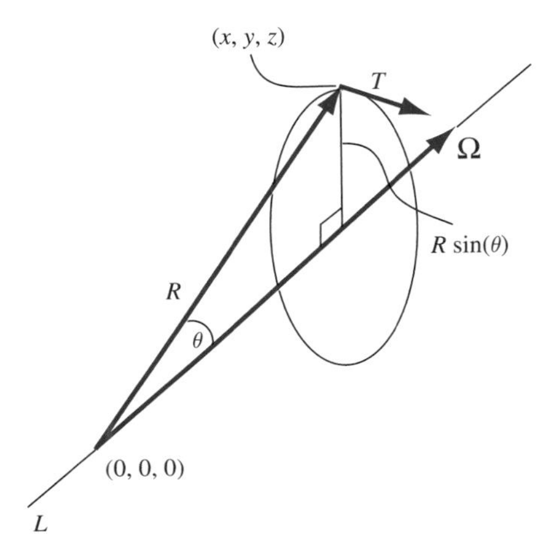

FIGURE 11.11 *Interpretation of curl.*

SECTION 11.5 PROBLEMS

In each of Problems 1 through 6, compute $\nabla \cdot \mathbf{F}$ and $\nabla \times \mathbf{F}$ and verify explicitly that $\nabla \cdot (\nabla \times \mathbf{F}) = 0$.

1. $\mathbf{F} = 2xy\mathbf{i} + xe^y\mathbf{j} + 2z\mathbf{k}$
2. $\mathbf{F} = \sinh(x - z)\mathbf{i} + 2y\mathbf{j} + (z - y^2)\mathbf{k}$
3. $\mathbf{F} = \sinh(x)\mathbf{i} + \cosh(xyz)\mathbf{j} - (x + y + z)\mathbf{k}$
4. $\mathbf{F} = \sinh(xyz)\mathbf{j}$
5. $\mathbf{F} = x\mathbf{i} + y\mathbf{j} + 2z\mathbf{k}$
6. $\mathbf{F} = x\mathbf{i} + y\mathbf{j} + 2z\mathbf{k}$

In each of Problems 7 through 12, compute $\nabla\varphi$ and verify explicitly that $\nabla \times (\nabla\varphi) = \mathbf{O}$.

7. $\varphi(x, y, z) = x\cos(x + y + z)$
8. $\varphi(x, y, z) = e^{x+y+z}$
9. $\varphi(x, y, z) = x - y + 2z^2$
10. $\varphi(x, y, z) = 18xyz + e^x$
11. $\varphi(x, y, z) = -2x^3yz^2$
12. $\varphi(x, y, z) = \sin(xz)$
13. Let φ be a scalar field and $\mathbf{F}$ a vector field. Derive expressions for $\nabla \cdot (\varphi\mathbf{F})$ and $\nabla \times (\varphi\mathbf{F})$ in terms of operations applied to $\varphi(x, y, z)$ and to $\mathbf{F}(x, y, z)$.

CHAPTER 12 Vector Integral Calculus

The primary objects of vector integral calculus are line and surface integrals and relationships between them involving the vector differential operators gradient, divergence and curl.

12.1 Line Integrals

For line integrals we need some preliminary observations about curves. Suppose a curve C has parametric equations

$$x = x(t),\ y = y(t),\ z = z(t) \text{ for } a \leq t \leq b.$$

These are the *coordinate functions* of C. It is convenient to think of t as time and C as the trajectory of an object, which at time t is at $C(t) = (x(t), y(t), z(t))$. C has an orientation, since the object starts at the *initial point* $(x(a), y(a), z(a))$ at time $t = a$ and ends at the *terminal point* $(x(b), y(b), z(b))$ at time $t = b$. We often indicate this orientation by putting arrows along the graph.

We call C:

- *continuous* if each coordinate function is continuous;
- *differentiable* if each coordinate function is differentiable;
- *closed* if the initial and terminal points coincide: $(x(a), y(a), z(a)) = (x(b), y(b), z(b))$;
- *simple* if $a < t_1 < t_2 < b$ implies that

$$(x(t_1), y)t_1), z(t_1)) \neq (x(t_2), y(t_2), z(t_2));$$

  and
- *smooth* if the coordinate functions have continuous derivatives which are never all zero for the same value of t.

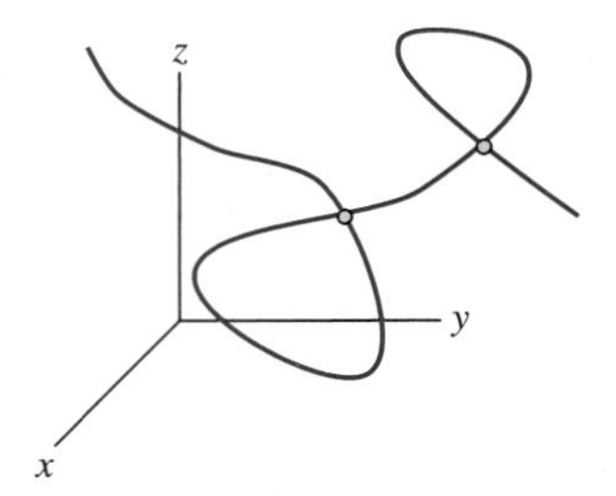

FIGURE 12.1 *A nonsimple curve.*

If C is smooth and we let $\mathbf{R}(t) = x(t)\mathbf{i} + y(t)\mathbf{i} + z(t)\mathbf{k}$ be the position function for C, then $\mathbf{R}'(t)$ is continuous tangent vector to C. Smoothness of C means that the curve has a continuous tangent vector as we move along it.

A curve is *simple* if it does not intersect itself at different times. The curve whose graph is shown in Figure 12.1 is not simple. A closed curve has the same initial terminal points, but is still called simple if it does not pass through any other point more than once.

We must be careful to distinguish between a curve and its graph, although informally we often use these terms interchangeably. The graph is a drawing, while the curve carries with it a sense of orientation from an initial to a terminal point. The graph of a curve does not carry all of this information.

EXAMPLE 12.1

Let C have coordinate functions

$$x = 4\cos(t),\ y = 4\sin(t),\ z = 9 \quad \text{for} \quad 0 \le t \le 2\pi.$$

The graph of C is a circle of radius 4 about the origin in the plane $z = 9$. C is simple, closed and smooth.

Let K be given by

$$x = 4\cos(t),\ y = 4\sin(t),\ z = 9 \quad \text{for} \quad 0 \le t \le 4\pi.$$

The graph of K is the same as the graph of K, except that a particle traversing K goes around this circle twice. K is closed and smooth but not simple. This information is not clear from the graph alone.

Let L be the curve given by

$$x(t) = 4\cos(t),\ y = 4\sin(t),\ z = 9 \quad \text{for} \quad 0 \le t \le 3\pi.$$

The graph of L is again the circle of radius 4 about the origin in the plane $z = 9$. L is smooth and not simple, but L is also not closed, since the initial point is $(4, 0, 9)$ and the terminal point is $(-4, 0, 9)$. A particle moving along L traverses the complete circle from $(4, 0, 9)$ to $(4, 0, 9)$ and then continues on to $(-4, 0, 9)$, where it stops. Again, this behavior is not clear from the graph. ♦

We are now ready to define the line integral, which is an integral over a curve.

Suppose C is a smooth curve with coordinate functions $x = x(t), y = y(t), z = z(t)$ for $a \le t \le b$. Let f, g and h be continuous at least at points on the graph of C. Then the *line integral* $\int_C f\,dx + g\,dy + h\,dz$ is defined by

$$\int_C f\,dx + g\,dy + h\,dz$$
$$= \int_a^b \left[f(x(t), y(t), z(t))\frac{dx}{dt} + g(x(t), y(t), z(t))\frac{dy}{dt} + h(x(t), y(t), z(t))\frac{dz}{dt} \right] dt.$$

$\int_C f\,dx + g\,dy + h\,dz$ is a number obtained by replacing x, y and z in $f(x, y, z)$, $g(x, y, z)$ and $h(x, y, z)$ with the coordinate functions $x(t)$, $y(t)$ and $z(t)$ of C, replacing

$$dx = x'(t)\,dt,\ dy = y'(t)\,dt,\ \text{and } dz = z'(t)\,dt,$$

and integrating the resulting function of t from a to b.

EXAMPLE 12.2

We will evaluate $\int_C x\,dx - yz\,dy + e^z\,dz$ if C is the curve with coordinate functions

$$x = t^3,\ y = -t,\ z = t^2 \text{ for } 1 \le t \le 2.$$

First,

$$dx = 3t^2\,dt,\ dy = -\,dt,\ \text{and } dz = 2t\,dt.$$

Put the coordinate functions of C into x, $-yz$ and e^z to obtain

$$\int_C x\,dx - yz\,dy + e^z\,dz$$
$$= \int_1^2 \left[t^3(3t^2) - (-t)(t^2)(-1) + e^{t^2}(2t) \right] dt$$
$$= \int_1^2 [3t^5 - t^3 + 2te^{t^2}]\,dt$$
$$= \frac{111}{4} + e^4 - e. \ ♦$$

EXAMPLE 12.3

Evaluate $\int_C xyz\,dx - \cos(yz)\,dy + xz\,dz$ along the straight line segment L from $(1, 1, 1)$ to $(-2, 1, 3)$.

Parametric equations of L are

$$x = 1 - 3t,\ y = 1,\ z = 1 + 2t \text{ for } 0 \le t \le 1.$$

Then

$$dx = -3\,dt,\ dy = 0 \text{ and } dz = 2\,dt.$$

The line integral is

$$\int_C xyz\,dx - \cos(yz)\,dy + xz\,dz$$
$$= \int_0^1 [(1 - 3t)(1 + 2t)(-3) - \cos(1 + 2t)(0) + (1 - 3t)(1 + 2t)(2)]\,dt$$
$$= \int_0^1 (-1 + t + 6t^2)\,dt = \frac{3}{2}. \ ♦$$

We have a line integral in the plane if C is in the plane and the functions involve only x and y.

EXAMPLE 12.4

Evaluate $\int_K xy\,dx - y\sin(x)\,dy$ if K has coordinate functions $x = t^2$, $y = t$ for $-1 \le t \le 2$. Here

$$dx = 2t\,dt \text{ and } dy = dt$$

so

$$\int_K xy\,dx - y\sin(x)\,dy = \int_{-1}^{2} [t^2 t(2t) - t\sin(t^2)]\,dt$$
$$= \int_{-1}^{2} [2t^4 - t\sin(t^2)]\,dt = \frac{66}{5} + \frac{1}{2}(\cos(4) - \cos(1)). \blacklozenge$$

Line integrals have properties we normally expect of integrals.

1. The line integral of a sum is the sum of the line integrals:

$$\int_C (f + f^*)\,dx + (g + g^*)\,dy + (h + h^*)\,dz$$
$$= \int_C f\,dx + g\,dy + h\,dz + \int_C f^*\,dx + g^*\,dy + h^*\,dz.$$

2. Constants factor through a line integral:

$$\int_C (cf)\,dx + (cg)\,dy + (ch)\,dz = c\int_C f\,dx + g\,dy + h\,dz.$$

For definite integrals, $\int_a^b F(x)\,dx = -\int_b^a F(x)\,dx$. The analogue of this for line integrals is that reversing the direction on C changes the sign of the line integral. Suppose C is a smooth curve from P_0 to P_1. Let C have coordinate functions

$$x = x(t),\ y = y(t),\ z = z(t) \text{ for } a \le t \le b.$$

Define K as the curve with coordinate functions

$$\tilde{x}(t) = x(a + b - t),\ \tilde{y}(t) = y(a + b - t),\ \tilde{z}(t) = z(a + b - t) \text{ for } a \le t \le b.$$

The graphs of C and K are the same, but the initial point of K is the terminal point of C, since

$$(\tilde{x}(a), \tilde{y}(a), \tilde{z}(a)) = (x(b), y(b), z(b)).$$

Similarly, the terminal point of K is the initial point of C. We denote a curve K formed from C in this way as $-C$. The effect of this reversal of orientation is to change the sign of a line integral.

3.

$$\int_C f\,dx + g\,dy + h\,dz = -\int_{-C} f\,dx + g\,dy + h\,dz.$$

This can be proved by a simple change of variables in the integrals with respect to t defining these line integrals.

The next property of line integrals reflects the fact that

$$\int_a^b F(x)\,dx = \int_a^c F(x)\,dx + \int_c^b F(x)\,dx$$

for definite integrals. A curve C is *piecewise smooth* if it has a continuous tangent at all but finitely many points. Such a curve typically has the appearance of the graph in Figure 12.2, with a finite number of "corners" at which there is no tangent.

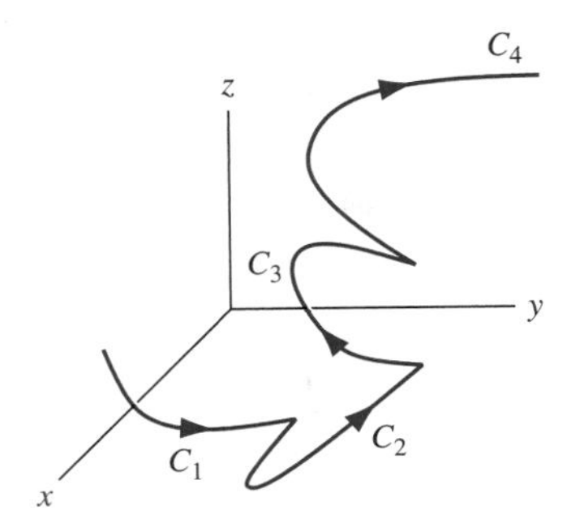

FIGURE 12.2 *A piecewise smooth curve.*

Write

$$C = C_1 \bigoplus C_2 \bigoplus \cdots \bigoplus C_n$$

if, as in Figure 12.2, C begins with a smooth piece C_1. C_2 begins where C_1 ends, C_3 begins where C_2 ends, and so on. Each C_j is smooth, but where C_j joins with C_{j+1} there may be no tangent in the resulting curve. For C formed in this way,

4.

$$\int_C f\,dx + g\,dy + h\,dz = \int_{C_1 \oplus C_2 \oplus \cdots \oplus C_n} f\,dx + g\,dy + h\,dz$$
$$= \sum_{j=1}^{n} \int_{C_j} f\,dx + g\,dy + h\,dz.$$

EXAMPLE 12.5

Let C be the curve consisting of the quarter circle $x^2 + y^2 = 1$ in the x, y - plane from $(1, 0)$ to $(0, 1)$, followed by the horizontal line segment from $(0, 1)$ to $(2, 1)$. We will compute $\int_C dx + y^2 dy$.

Write $C = C_1 \oplus C_2$, where C_1 is the quarter circle part and C_2 the line segment part. Parametrize C_1 by $x = \cos(t)$, $y = \sin(t)$ for $0 \leq t \leq \pi/2$. On C_1,

$$dx = -\sin(t)\,dt \text{ and } dy = \cos(t)\,dt,$$

so

$$\int_{C_1} dx + y^2\,dy = \int_0^{\pi/2} [-\sin(t) + \sin^2(t)\cos(t)]\,dt = -\frac{2}{3}.$$

Parametrize C_2 by $x = s$, $y = 1$ for $0 \leq s \leq 2$. On C_2,

$$dx = ds \text{ and } dy = 0$$

so

$$\int_{C_2} dx + y^2\,dy = \int_0^2 ds = 2.$$

Then

$$\int_C dx + y^2\,dy = -\frac{2}{3} + 2 = \frac{4}{3}. \blacklozenge$$

We often write a line integral in vector notation. Let $\mathbf{F} = f\mathbf{i} + g\mathbf{j} + h\mathbf{k}$ and form the position vector $\mathbf{R}(t) = x(t)\mathbf{i} + y(t)\mathbf{j} + z(t)\mathbf{k}$ for C. Then

$$d\mathbf{R} = dx\,\mathbf{i} + dy\,\mathbf{j} + dz\,\mathbf{k}$$

and

$$\mathbf{F}\cdot d\mathbf{R} = f\,dx + g\,dy + h\,dz$$

suggesting the notation

$$\int_C f\,dx + g\,dy + h\,dz = \int_C \mathbf{F}\cdot d\mathbf{R}.$$

EXAMPLE 12.6

A force $\mathbf{F}(x, y, z) = x^2\mathbf{i} - zy\mathbf{j} + x\cos(z)\mathbf{k}$ moves an object along the path C given by $x = t^2$, $y = t$, $z = \pi t$ for $0 \le t \le 3$. We want to calculate the work done by this force.

At any point on C the particle will be moving in the direction of the tangent to C at that point. We may approximate the work done along a small segment of the curve starting at (x, y, z) by $\mathbf{F}(x, y, z)\cdot d\mathbf{R}$, with the dimensions of force times distance. The work done in moving the object along the entire path is approximated by the sum of these approximations along segments of the path. In the limit as the lengths of these segments tend to zero we obtain

$$\begin{aligned}
\text{work} &= \int_C \mathbf{F}\cdot d\mathbf{R} = \int_C x^2\,dx - zy\,dy + x\cos(z)\,dz \\
&= \int_0^3 [t^4(2t) - (\pi t)(t) + t^2\cos(\pi t)(\pi)]\,dt \\
&= \int_0^3 [2t^5 - \pi t^2 + \pi t^2\cos(\pi t)]\,dt \\
&= 243 - 9\pi - \frac{6}{\pi}. \ \blacklozenge
\end{aligned}$$

12.1.1 Line Integral With Respect to Arc Length

In some contexts it is useful to have a line integral with respect to arc length along C. If $\varphi(x, y, z)$ is a scalar field and C is a smooth curve with coordinate functions $x = x(t)$, $y = y(t)$, $z = z(t)$ for $a \le t \le b$, we define

$$\int_C \varphi(x, y, z)\,ds = \int_a^b \varphi(x(t), y(t), z(t))\sqrt{x'(t)^2 + y'(t)^2 + z'(t)^2}\,dt.$$

The rationale behind this definition is that

$$ds = \sqrt{x'(t)^2 + y'(t)^2 + z'(t)^2}\,dt$$

is the differential element of arc length along C.

To see how such a line integral arises, suppose C is a thin wire having density $\delta(x, y, z)$ at (x, y, z), and we want to compute the mass. Partition $[a, b]$ into n subintervals by inserting points

$$a = t_0 < t_1 < t_2 < \cdots < t_{n-1} < t_n = b$$

of length $\Delta t = (b-a)/n$, where n is a positive integer. We can make Δt as small as we want by choosing n large, so that values of $\delta(x, y, z)$ are approximated as closely as we want on $[t_{j-1}, t_j]$ by $\delta(P_j)$, where $P_j = (x(t_j), y(t_j), z(t_j))$. The length of wire between P_{j-1} and P_j is $\Delta s = s(P_j) - s(P_{j-1}) \approx ds_j$. The density of this piece of wire is approximately $\delta(P_j)ds_j$, and $\sum_{j=1}^{n} \delta(P_j)ds_j$ approximates the mass of the wire. In the limit as $n \to \infty$, this gives

$$\text{mass of the wire } = \int_C \delta(x, y, z)\, ds.$$

A similar argument gives the coordinates $(\tilde{x}, \tilde{y}, \tilde{z})$ of the center of mass of the wire as

$$\tilde{x} = \frac{1}{m}\int_C x\delta(x, y, z)ds,\ \tilde{y} = \frac{1}{m}\int_C y\delta(x, y, z)ds,\ \tilde{z} = \frac{1}{m}\int_C z\delta(x, y, z)ds,$$

in which m is the mass.

EXAMPLE 12.7

A wire is bent into the shape of the quarter circle C given by $x = 2\cos(t)$, $y = 2\sin(t)$, $z = 3$ for $0 \le t \le \pi/2$. The density function is $\delta(x, y, z) = xy^2$. We want the mass and center of mass of the wire.

The mass is

$$\begin{aligned} m = \int_C xy^2\, ds &= \int_0^{\pi/2} 2\cos(t)[2\sin(t)]^2\sqrt{4\sin^2(t) + 4\cos^2 dt} \\ &= \int_0^{\pi/2} 16\cos(t)\sin^2(t)\, dt = \frac{16}{3}. \end{aligned}$$

Now compute the coordinates of the center of mass. First,

$$\begin{aligned} \tilde{x} &= \frac{1}{m}\int_C x\delta(x, y, z)\, ds \\ &= \frac{3}{16}\int_0^{\pi/2} [2\cos(t)]^2[2\sin(t)]^2\sqrt{4\sin^2(t) + 4\cos^2 dt} \\ &= 6\int_0^{\pi/2} \cos^2(t)\sin^2(t)\, dt = \frac{3\pi}{8}. \end{aligned}$$

Next,

$$\begin{aligned} \tilde{y} &= \frac{1}{m}\int_C y\delta(x, y, z)\, ds \\ &= \frac{3}{16}\int_0^{\pi/2} [2\cos(t)][2\sin(t)]^3\sqrt{4\sin^2(t) + 4\cos^2 dt} \\ &= 6\int_0^{\pi/2} \cos(t)\sin^3(t)\, dt = \frac{3}{2}. \end{aligned}$$

And

$$\begin{aligned} \tilde{z} &= \frac{3}{16}\int_C zxy^2\, ds \\ &= \frac{3}{16}\int_0^{16} 3[2\cos(t)][2\sin(t)]^2\sqrt{4\sin^2(t) + 4\cos^2 dt} \\ &= 9\int_0^{\pi/2} \sin^2(t)\cos(t)\, dt = 3. \end{aligned}$$

It should not be surprising that $\tilde{z} = 3$ because the wire is in the $z = 3$ plane. ♦

SECTION 12.1 PROBLEMS

In each of Problems 1 through 10, evaluate the line integral.

1. $\int_C -xyz\,dz$ with C given by $x = 1, y = \sqrt{z}$ for $4 \le z \le 9$
2. $\int_C xz\,dy$ with C given by $x = y = t, z = -4t^2$ for $1 \le t \le 3$
3. $\int_C x\,dx - dy + z\,dz$ with C given by $x = y = t, z = t^3$ for $0 \le t \le 1$
4. $\int_C 4xy\,ds$ with C given by $x = y = t, z = 2t$ for $1 \le t \le 2$
5. $\int_C (x + y)\,ds$ with C given by $x = y = t, z = t^2$ for $0 \le t \le 2$
6. $\int_C x^2 z\,ds$ with C the line segment from $(0, 1, 1)$ to $(1, 2, -1)$
7. $\int_C \mathbf{F} \cdot d\mathbf{R}$ with $\mathbf{F} = \cos(x)\mathbf{i} - y\mathbf{j} + xz\mathbf{k}$ and $\mathbf{R} = t\mathbf{i} - t^2\mathbf{j} + \mathbf{k}$ for $0 \le t \le 3$
8. $\int_C -4x\,dx + y^2\,dy - yz\,dz$ with C given by $x = -t^2, y = 0, z = -3t$ for $0 \le t \le 1$
9. $\int_C \mathbf{F} \cdot d\mathbf{R}$ with $\mathbf{F} = x\mathbf{i} + y\mathbf{j} - z\mathbf{k}$ and C the circle $x^2 + y^2 = 4, z = 0$, going around once counterclockwise.
10. $\int_C yz\,ds$ with C the parabola $z = y^2, x = 1$ for $0 \le y \le 2$
11. Show that any Riemann integral $\int_a^b f(x)dx$ is a line integral $\int_C \mathbf{F} \cdot d\mathbf{R}$ for appropriate choices of $\mathbf{F}$ and $\mathbf{R}$.
12. Find the mass and center of mass of a thin, straight wire extending from the origin to $(3, 3, 3)$ if $\delta(x, y, z) = x + y + z$ grams per centimeter.
13. Find the work done by $\mathbf{F} = x^2\mathbf{i} - 2yz\mathbf{j} + z\mathbf{k}$ in moving an object along the line segment from $(1, 1, 1)$ to $(4, 4, 4)$.

12.2 Green's Theorem

Green's theorem is a relationship between double integrals and line integrals around closed curves in the plane. It was formulated independently by the self-taught amateur British natural philosopher George Green and the Ukrainian mathematician Michel Ostrogradsky, and is used in potential theory and partial differential equations.

A closed curve C in the x, y - plane is *positively oriented* if a point on the curve moves counterclockwise as the parameter describing C increases. If the point moves clockwise, then C is *negatively oriented*. We denote orientation by placing an arrow on the graph, as in Figure 12.3.

A simple closed curve C in the plane encloses a region, called the *interior* of C. The unbounded region that remains if the interior is cut out is the *exterior* of C (Figure 12.4). If

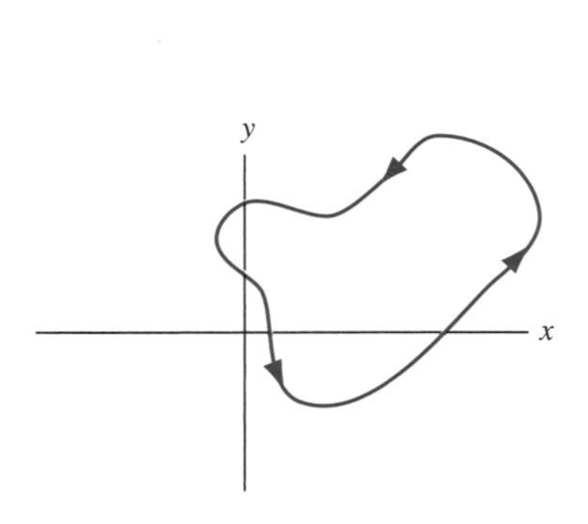

FIGURE 12.3 *Orientation on a curve.*

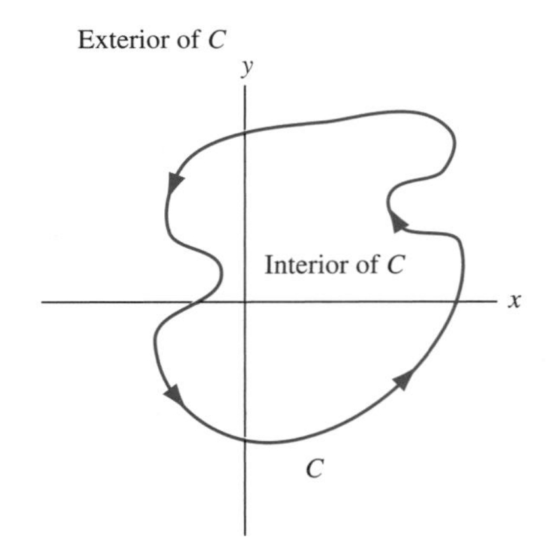

FIGURE 12.4 *Interior and exterior of a simple closed curve.*

C is positively oriented then, as we walk around C in the positive direction, the interior is over our left shoulder.

We will use the term *path* for a piecewise smooth curve. And we often denote a line integral over a closed path C as $\oint_C$, with a small oval on the integral sign. This is not obligatory and does not affect the meaning of the line integral or the way it is evaluated.

THEOREM 12.1 Green's Theorem

Let C be a simple closed positively oriented path in the plane. Let D consist of all points on C and in its interior. Let f, g, $\partial f/\partial y$ and $\partial g/\partial x$ be continuous on D. Then

$$\oint_C f(x,y)\,dx + g(x,y)\,dy = \iint_D \left(\frac{\partial g}{\partial x} - \frac{\partial f}{\partial y}\right) dA. \blacklozenge$$

A proof under special conditions on D is sketched in Problem 12.

EXAMPLE 12.8

Sometimes Green's theorem simplifies an integration. Suppose we want to compute the work done by $\mathbf{F}(x,y) = (y - x^2e^x)\mathbf{i} + (\cos(2y^2) - x)\mathbf{j}$ in moving a particle counterclockwise about the rectangular path C having vertices $(0,1)$, $(1,1)$, $(1,3)$ and $(0,3)$.

If we attempt to evaluate $\int_C \mathbf{F}\cdot d\mathbf{R}$ we encounter integrals that cannot be done in elementary form. However, by Green's theorem, with D the solid rectangle bounded by C,

$$\begin{aligned}\text{work} &= \oint_C \mathbf{F}\cdot d\mathbf{R} = \iint_D \left(\frac{\partial}{\partial x}(\cos(2y^2) - x) - \frac{\partial}{\partial y}(y - x^2e^x)\right) dA \\ &= \iint_D -2\,dA = (-2)[\text{area of} D] = -4. \blacklozenge\end{aligned}$$

EXAMPLE 12.9

Another use of Green's theorem is in deriving general results. Suppose we want to evaluate

$$\oint_C 2x\cos(2y)\,dx - 2x^2\sin(2y)\,dy$$

for every positively oriented simple closed path C in the plane.

There are infinitely many such paths. However, $f(x,y)$ and $g(x,y)$ have the special property that

$$\begin{aligned}&\frac{\partial}{\partial x}\left(-2x^2\sin(2y)\right) - \frac{\partial}{\partial y}(2x\cos(2y)) \\ &= -4x\sin(2y) + 4x\sin(2y) = 0.\end{aligned}$$

By Green's theorem, for any such closed path C in the plane,

$$\oint_C 2x\cos(2y)\,dx - 2x^2\sin(2y)\,dy = \iint_D 0\,dA = 0. \blacklozenge$$

SECTION 12.2 PROBLEMS

1. Let $u(x, y)$ be continuous with continuous first and second partial derivatives on a simple closed path C and throughout the interior D of C. Show that
$$\oint_C -\frac{\partial u}{\partial y}dx + \frac{\partial u}{\partial x}dy = \iint_D \left[\frac{\partial^2 u}{\partial x^2} + \frac{\partial^2 u}{\partial y^2}\right] dA.$$

2. Let D be the interior of a positively oriented simple closed path C.
   (a) Show that the area of D equals $\oint_C -y dx$.
   (b) Show that the area of D equals $\oint_C x dy$.
   (c) Show that the area of D equals
$$\frac{1}{2}\oint_C -y dx + x dy.$$

3. A particle moves once counterclockwise about the triangle with vertices $(0, 0)$, $(4, 0)$ and $(1, 6)$, under the influence of the force $\mathbf{F} = xy\mathbf{i} + x\mathbf{j}$. Calculate the work done by this force.

In each of Problems 4 through 11, use Green's theorem to evaluate $\oint_C \mathbf{F} \cdot d\mathbf{R}$. All curves are oriented positively.

4. $\mathbf{F} = (x + y)\mathbf{i} + (x - y)\mathbf{j}$ and C is the ellipse $x^2 + 4y^2 = 1$

5. $\mathbf{F} = e^x \cos(y)\mathbf{i} - e^x \sin(y)\mathbf{j}$ and C is any simple closed path in the plane

6. $\mathbf{F} = (x^2 - y)\mathbf{i} + (\cos(2y) - e^{3y} + 4x)\mathbf{j}$ and C is any square with sides of length 5

7. $\mathbf{F} = xy\mathbf{i} + (xy^2 - e^{\cos(y)})\mathbf{j}$ and C is the triangle with vertices $(0, 0)$, $(3, 0)$ and $(0, 5)$

8. $\mathbf{F} = x^2 y\mathbf{i} - xy^2\mathbf{j}$ and C is the boundary of the region $x^2 + y^2 \le 4, x \ge 0, y \ge 0$

9. $\mathbf{F} = 8xy^2\mathbf{j}$ and C is the circle of radius 4 about the origin

10. $\mathbf{F} = 2y\mathbf{i} - x\mathbf{j}$ and C is the circle of radius 4 about $(1, 3)$

11. $\mathbf{F} = x^2\mathbf{i} - 2xy\mathbf{j}$ and C is the triangle with vertices $(1, 1)$, $(4, 1)$, $(2, 6)$

12. Fill in the details of the following argument to prove Green's theorem under special conditions. Assume that D can be described in two ways. First, D consists of all (x, y) with $q(x) \le y \le p(x)$, for $a \le x \le b$. This means that D has an upper boundary (graph of $y = p(x)$) and a lower boundary ($y = q(x)$) for $a \le x \le b$. Also assume that D consists of all (x, y) with $\alpha(y) \le x \le \beta(y)$, with $c \le y \le d$. In this description, the graph of $x = \alpha(y)$ is a left boundary of D, and the graph of $x = \beta(y)$ is a right boundary.

    Using the first description of D, show that
$$\int_C g(x, y)\, dy = \int_c^d g(\beta(y), y)\, dy + \int_d^c g(\alpha(y), y)\, dy$$
and
$$\iint_D \frac{\partial g}{\partial x} dA = \int_c^d \int_{\alpha(y)}^{\beta(y)} \frac{\partial g}{\partial x} dA$$
$$= \int_c^c (g(\beta(y), y) - g(\alpha(y), y))\, dy.$$
Thus, conclude that
$$\int_C g(x, y)\, dy = \iint_D \frac{\partial g}{\partial x} dA.$$
Now use the other description of D to show that
$$\int_C f(x, y)\, dx = -\iint_D \frac{\partial f}{\partial y} dA.$$

13. A particle moves once counterclockwise about the rectangle with vertices $(1, 1)$, $(1, 7)$, $(3, 1)$ and $(3, 7)$, under the influence of the force $\mathbf{F} = (-\cosh(4x^4) + xy)\mathbf{i} + (e^{-y} + x)\mathbf{j}$. Calculate the work done.

14. A particle moves once counterclockwise around the circle of radius 6 about the origin, under the influence of the force $\mathbf{F} = (e^x - y + x\cosh(x))\mathbf{i} + (y^{3/2} + x)\mathbf{j}$. Calculate the work done.

12.3 An Extension of Green's Theorem

There is an extension of Green's theorem to include the case that there are finitely many points $P_1, \cdots, P_n$ enclosed by C at which f, g, $\partial f/\partial y$ and/or $\partial g/\partial x$ are not continuous, or perhaps not even defined. The idea is to excise these points by enclosing them in small disks which are thought of as cut out of D.

Enclose each P_j with a circle K_j of sufficiently small radius that no circle intersects either C or any of the other circles (Figure 12.5). Draw a channel consisting of two parallel line segments

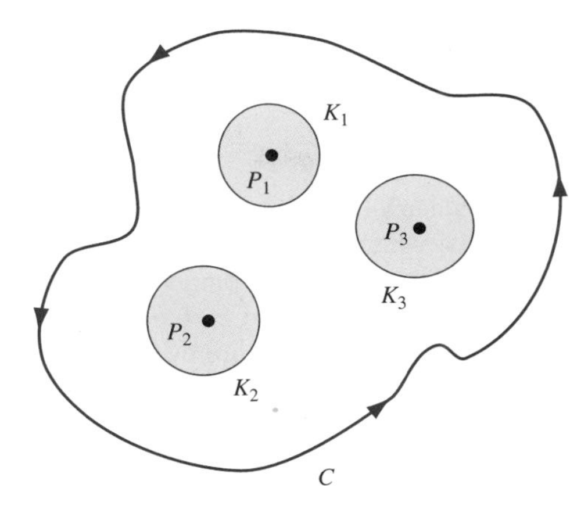

FIGURE 12.5 *Enclosing points with small circles interior to C.*

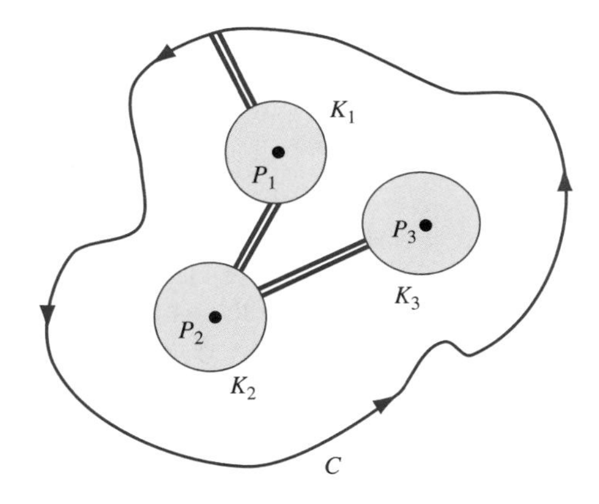

FIGURE 12.6 *Channels connecting C to K_1, K_1 to K_2, $\cdots$, K_{n-1} to K_n.*

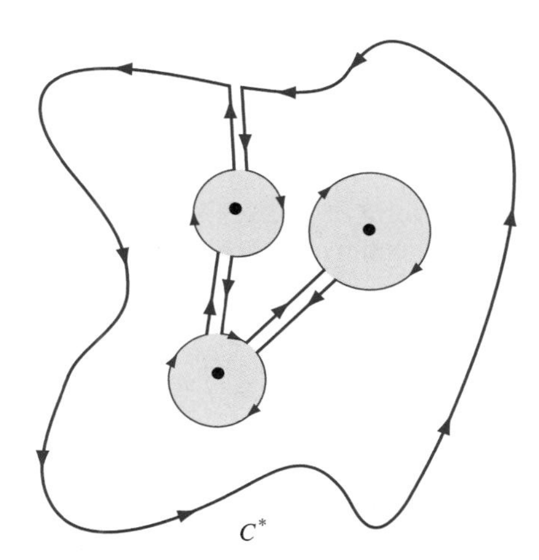

FIGURE 12.7 *The simple closed path C^*, with each P_j exterior to C^*.*

from C to K_1, then from K_1 to K_2, and so on, until the last channel is drawn from K_{n-1} to K_n. This is illustrated in Figure 12.6 for $n = 3$.

Now form the simple closed path C^* of Figure 12.7, consisting of "most of" C, "most of" each K_j, and the inserted channel lines. By "most of" C, we mean that a small arc of C and each circle between the channel cuts has been excised in forming C^*.

Each P_j is *external* to C^* and $f, g, \partial f/\partial y$ and $\partial g/\partial x$ are continuous on and in the interior of C^*. The orientation on C^* is also crucial. If we begin at a point of C just before the channel to K_1, we move counterclockwise on C until we reach the first channel cut, then go along this cut to K_1, then *clockwise* around part of K_1 until we reach a channel cut to K_2, then clockwise around K_2 until we reach a cut to K_3. After going clockwise around part of K_3, we reach the other side of the cut from this circle to K_2, move clockwise around it to the cut to K_1, then clockwise around it to the cut back to C, and then continue counterclockwise around C.

If D^* is the interior of C^*, then by Green's theorem,

$$\oint_{C^*} f\,dx + g\,dy = \iint_{D^*} \left(\frac{\partial g}{\partial x} - \frac{\partial f}{\partial y} \right) dA. \tag{12.1}$$

Now take a limit in Figure 12.7 as the channels are made narrower. The opposite sides of each channel merge to single line segments, which are integrated over in both directions in equation (12.1). The contributions to the sum in this equation from the channel cuts is therefore zero. Further, as the channels narrow, the small arcs of C and each K_j cut out in making the channels are restored, and the line integrals in equation (12.1) are over all of C and the circles K_j. Recalling that in equation (12.1) the integrations over the K_j's are clockwise, equation (12.1) can be written

$$\oint_C f\,dx + g\,dy - \sum_{j=1}^{n} \oint_{K_j} f\,dx + g\,dy = \iint_{D^*} \left(\frac{\partial g}{\partial x} - \frac{\partial f}{\partial y} \right) dA. \tag{12.2}$$

in which all integrations (over C and each K_j) are now taken in the positive, counterclockwise sense. This accounts for the minus sign on each of the integrals $\oint_{K_j} f\,dx + g\,dy$ in equation (12.2). Finally, write equation (12.2) as

$$\oint_C f\,dx + g\,dy = \sum_{j=1}^{n} \oint_{K_j} f\,dx + g\,dy + \iint_{D^*} \left(\frac{\partial g}{\partial x} - \frac{\partial f}{\partial y} \right) dA. \tag{12.3}$$

This is the extended form of Green's theorem. When D contains points at which $f, g, \partial f/\partial y$ and/or $\partial g/\partial x$ are not continuous, then $\oint_C f\,dx + g\,dy$ is the sum of the line integrals $\oint_{K_j} f\,dx + g\,dy$ about small circles centered at the P_j's, together with

$$\iint_{D^*} \left(\frac{\partial g}{\partial x} - \frac{\partial f}{\partial y} \right) dA$$

over the region D^* formed by excising from D the disks bounded by the K_j's.

EXAMPLE 12.10

We will evaluate

$$\oint_C \frac{-y}{x^2+y^2}\,dx + \frac{x}{x^2+y^2}\,dy$$

in which C is any simple closed positively oriented path in the plane, but not passing through the origin.

With

$$f(x,y) = \frac{-y}{x^2+y^2} \text{ and } g(x,y) = \frac{x}{x^2+y^2}$$

we have

$$\frac{\partial g}{\partial x} = \frac{\partial f}{\partial y} = \frac{y^2-x^2}{(x^2+y^2)^2}.$$

This suggests that we consider two cases.

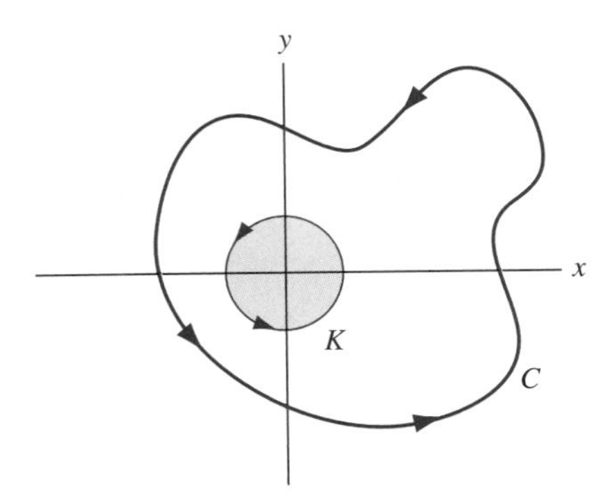

FIGURE 12.8 *Case 2 of Example 12.10.*

Case 1 If C does not enclose the origin, Green's theorem applies and

$$\oint_C \frac{-y}{x^2+y^2}\,dx + \frac{x}{x^2+y^2}\,dy = \iint_D \left(\frac{\partial g}{\partial x} - \frac{\partial f}{\partial y}\right) dA = 0.$$

Case 2 If C encloses the origin, then C encloses a point where f and g are not defined. Now use equation (12.3). Let K be a circle about the origin, with sufficiently small radius r that K does not intersect C (Figure 12.8). Then

$$\begin{aligned}
&\oint_C f\,dx + g\,dy \\
&= \oint_K f\,dx + g\,dy + \iint_{D^*} \left(\frac{\partial g}{\partial x} - \frac{\partial f}{\partial y}\right) dA \\
&= \oint_K f\,dx + g\,dy
\end{aligned}$$

where D^* is the region between D and K, including both curves. Both of these line integrals are in the counterclockwise sense. The last line integral is over a circle and can be evaluated explicitly. Parametrize K by $x = r\cos(\theta)$, $y = r\sin(\theta)$ for $0 \le \theta \le 2\pi$. Then

$$\begin{aligned}
&\oint_K f\,dx + g\,dy \\
&= \int_0^{2\pi} \left(\frac{-r\sin(\theta)}{r^2}[-r\sin(\theta)] + \frac{r\cos(\theta)}{r^2}[r\cos(\theta)]\right) d\theta \\
&= \int_0^{2\pi} d\theta = 2\pi.
\end{aligned}$$

We conclude that

$$\oint_C f\,dx + g\,dy = \begin{cases} 0 & \text{if } C \text{ does not enclose the origin} \\ 2\pi & \text{if } C \text{ encloses the origin.} \end{cases}$$ ♦

SECTION 12.3 PROBLEMS

In each of Problems 1 through 5, evaluate $\oint_C \mathbf{F} \cdot d\mathbf{R}$ over any simple closed path in the x, y-plane that does not pass through the origin. This may require cases, as in Example 12.10.

1. $\mathbf{F} = \left(\frac{-y}{x^2+y^2} + x^2\right)\mathbf{i} + \left(\frac{x}{x^2+y^2} - 2y\right)\mathbf{j}$
2. $\mathbf{F} = \left(\frac{-y}{x^2+y^2} + 3x\right)\mathbf{i} + \left(\frac{x}{x^2+y^2} - y\right)\mathbf{j}$
3. $\mathbf{F} = \left(\frac{x}{\sqrt{x^2+y^2}} + 2x\right)\mathbf{i} + \left(\frac{y}{\sqrt{x^2+y^2}} - 3y^2\right)\mathbf{j}$
4. $\mathbf{F} = \left(\frac{1}{x^2+y^2}\right)^{3/2}(x\mathbf{i} + y\mathbf{j})$
5. $\mathbf{F} = \frac{x}{x^2+y^2}\mathbf{i} + \frac{y}{x^2+y^2}\mathbf{j}$

12.4 Independence of Path and Potential Theory

A vector field $\mathbf{F}$ is *conservative* if it is derivable from a potential function. This means that for some scalar field φ,

$$\mathbf{F} = \nabla\varphi = \frac{\partial \varphi}{\partial x}\mathbf{i} + \frac{\partial \varphi}{\partial y}\mathbf{j} + \frac{\partial \varphi}{\partial z}\mathbf{k}.$$

We call φ a *potential function*, or *potential*, for $\mathbf{F}$. Of course, if φ is a potential, so is $\varphi + c$ for any constant c.

One consequence of $\mathbf{F}$ being conservative is that the value of $\int_C \mathbf{F} \cdot d\mathbf{R}$ depends only on the endpoints of C. If C has differentiable coordinate functions $x = x(t), y = y(t), z = z(t)$ for $a \le t \le b$, then

$$\begin{aligned}\int_C \mathbf{F} \cdot d\mathbf{R} &= \int_C \frac{\partial \varphi}{\partial x}dx + \frac{\partial \varphi}{\partial y}dy + \frac{\partial \varphi}{\partial z}dy \\ &= \int_a^b \left(\frac{\partial \varphi}{\partial x}\frac{dx}{dt} + \frac{\partial \varphi}{\partial y}\frac{dy}{dt} + \frac{\partial \varphi}{\partial z}\frac{dz}{dt}\right)dt \\ &= \int_a^b \frac{d}{dt}\varphi(x(t), y(t), z(t))dt \\ &= \varphi(x(b), y(b), z(b)) - \varphi(x(a), y(a), z(a))\end{aligned}$$

which requires only that we evaluate the potential function for $\mathbf{F}$ at the endpoints of C. This is the line integral version of the fundamental theorem of calculus, and applies to line integrals of conservative vector fields.

THEOREM 12.2

Let $\mathbf{F}$ be conservative in a region D (of the plane or of 3-space). Let C be a path from P_0 to P_1 in D. Then

$$\int_C \mathbf{F} \cdot d\mathbf{R} = \varphi(P_1) - \varphi(P_0). \tag{12.4}$$

In particular, if C is a closed path in D, then

$$\oint_C \mathbf{F}\cdot d\mathbf{R} = 0. \quad \blacklozenge$$

The last conclusion follows from the fact that, if the path is closed, then the initial point P_0 and the terminal point P_1 are the same and equation (12.4) yields 0 for the value of the integral.

Another consequence of $\mathbf{F}$ having a potential function is independence of path. We say that $\int_C \mathbf{F}\cdot d\mathbf{R}$ is *independent of path* in D if the value of this line integral for any path C in D depends only on the endpoints of C. Another way of putting this is that

$$\int_{C_1} \mathbf{F}\, d\mathbf{R} = \int_{C_2} \mathbf{F}\cdot d\mathbf{R}$$

for any paths C_1 and C_2 in D having the same initial point and the same terminal point in D. In this case, the route is unimportant - the only thing that matters is where we start and where we end. By equation (12.4), existence of a potential function implies independence of path.

THEOREM 12.3

If $\mathbf{F}$ is conservative in D, then $\int_C \mathbf{F}\dot{d}\mathbf{R}$ is independent of path in D. ♦

Proof Let φ be a potential function for $\mathbf{F}$ in D. If C_1 and C_2 are paths in D having initial point P_0 and terminal point P_1, then

$$\int_{C_1} \mathbf{F}\cdot d\mathbf{R} = \varphi(P_1) - \varphi(P_0) = \int_{C_2} \mathbf{F}\cdot d\mathbf{R}. \quad \blacklozenge$$

Independence of path is equivalent to the vanishing of integrals around closed paths.

THEOREM 12.4

$\int_C \mathbf{F}\cdot d\mathbf{R}$ is independent of path in D if and only if

$$\oint_C \mathbf{F}\cdot d\mathbf{R} = 0$$

for every closed path in D. ♦

Proof To go one way, suppose first that $\oint_C \mathbf{F}\cdot d\mathbf{R} = 0$ for every closed path in D and let C_1 and C_2 be paths in D from P_0 to P_1. Form a closed path C by starting at P_0, moving along C_1 to P_1, and then reversing orientation to move along $-C_2$ from P_1 to P_0. Then $C = C_1 \bigoplus (-C_2)$ and

$$\oint_C \mathbf{F}\cdot d\mathbf{R} = 0 = \int_{C_1} \mathbf{F}\cdot d\mathbf{R} - \int_{C_2} \mathbf{F}\cdot d\mathbf{R},$$

implying that

$$\int_{C_1} \mathbf{F}\cdot d\mathbf{R} = \int_{C_2} \mathbf{F}\cdot d\mathbf{R}.$$

This makes $\int_C \mathbf{F}\cdot d\mathbf{R}$ independent of path in D.

Conversely, suppose $\int_C \mathbf{F} \cdot d\mathbf{R}$ is independent of path in D and let C be any closed path in D. Choose distinct points P_0 and P_1 on C. Let C_1 be the part of C from P_0 to P_1 and C_2 the part from P_1 to P_0. Then $C = C_1 \bigoplus C_2$. Furthermore, C_1 and $-C_2$ are paths in D from P_0 to P_1, so by assumption

$$\int_{C_1} \mathbf{F} \cdot d\mathbf{R} = \int_{-C_2} \mathbf{F} \cdot d\mathbf{R} = -\int_{C_2} \mathbf{F} \cdot d\mathbf{R}.$$

Then

$$0 = \int_{C_1} \mathbf{F} \cdot d\mathbf{R} + \int_{C_2} \mathbf{F} \cdot d\mathbf{R} = \oint_C \mathbf{F} \cdot d\mathbf{R}. \quad \blacklozenge$$

Thus far, we have the following implications for $\int_C \mathbf{F} \cdot d\mathbf{R}$ over paths in some region D:

1. Conservative $\mathbf{F} \Rightarrow$ independence of path of $\int_C \mathbf{F} \cdot d\mathbf{R}$.
2. Independence of path in $D \iff$ integrals over all closed paths in D are zero.

We will improve on this table of implications shortly. First, consider the problem of finding a potential function for a conservative vector field. Sometimes this can be done by integration.

EXAMPLE 12.11

We will determine if the vector field

$$\mathbf{F}(x, y, z) = 3x^2yz^2\mathbf{i} + (x^3z^2 + e^z)\mathbf{j} + (2x^3yz + ye^z)\mathbf{k}.$$

is conservative by attempting to find a potential function.

If $\mathbf{F} = \nabla\varphi$ for some φ, then

$$\frac{\partial\varphi}{\partial x} = 3x^2yz^2, \tag{12.5}$$

$$\frac{\partial\varphi}{\partial y} = x^3z^2 + e^z, \tag{12.6}$$

$$\frac{\partial\varphi}{\partial z} = 2x^3yz + ye^z. \tag{12.7}$$

Choose one of these equations, say 12.5. To reverse $\partial\varphi/\partial x$, integrate this equation with respect to x to get

$$\varphi(x, y, z) = \int 3x^2yz^2 dx = x^3yz^2 + \alpha(y, z).$$

The "constant of integration" may involve y and z because the integration reverses a partial differentiation in which y and z were held fixed. Now we know φ to within $\alpha(y, z)$. To determine $\alpha(x, y)$, choose one of the other equations, say 12.6, to get

$$\frac{\partial\varphi}{\partial y} = x^3z^2 + e^z = \frac{\partial}{\partial y}(x^3yz^2 + \alpha(y, z))$$
$$= x^3z^2 + \frac{\partial\alpha(y, z)}{\partial y}.$$

This requires that

$$\frac{\partial\alpha(y, z)}{\partial y} = e^z.$$

Integrate this with respect to y, holding z fixed to get

$$\int \frac{\partial\alpha(y, z)}{\partial y} dy = ye^z + \beta(z),$$

with $\beta(z)$ an as yet unknown function of z. We now have

$$\varphi(x, y, z) = x^3yz^2 + \alpha(y, z) = x^3yz^2 + ye^z + \beta(z)$$

and we have only to determine $\beta(z)$. For this use the third equation, 12.7, to write

$$\frac{\partial \varphi}{\partial z} = 2x^3yz + ye^z = 2x^3yz + ye^z + \beta'(z).$$

This forces $\beta'(z) = 0$, so $\beta(z) = k$, any number. With $\varphi(x, y, z) = x^3yz^2 + ye^z + k$ for any number k (which we can choose to be 0), we have $\mathbf{F} = \nabla\varphi$ and φ is a potential function for $\mathbf{F}$.

This enables us to easily evaluate $\int_C \mathbf{F} \cdot d\mathbf{R}$. If, for example, C is a path from $(0, 0, 0)$ to $(-1, 3, -2)$, then

$$\int_C \mathbf{F} \cdot d\mathbf{R} = \varphi(-1, 3, -2) - \varphi(0, 0, 0) = -12 + 3e^{-2}.$$

And if C is a closed path, then $\oint_C \mathbf{F} \cdot d\mathbf{R} = 0$. ♦

This method for finding a potential function for a function of two variables was seen previously in solving exact differential equations (Example 1.12).

There are nonconservative vector fields.

EXAMPLE 12.12

Let $\mathbf{F} = y\mathbf{i} + e^x\mathbf{j}$, a vector field in the plane. If this is conservative, there would be a potential function $\varphi(x, y)$ such that

$$\frac{\partial \varphi}{\partial x} = y \text{ and } \frac{\partial \varphi}{\partial y} = e^x.$$

Integrate the first with respect to x, thinking of y as fixed, to get

$$\varphi(x, y) = \int y\,dx = xy + \alpha(y).$$

But then we would have to have

$$\frac{\partial \varphi}{\partial y} = e^x = \frac{\partial}{\partial y}(xy + \alpha(y)) = x + \alpha'(y).$$

This would make α depend on x. This is impossible, since $\alpha(y)$ was the "constant" of integration with respect to x. $\mathbf{F}$ has no potential and is not conservative. ♦

If a vector field is conservative, we may be able to find a potential function by integration. But in general, integration is an ineffective way to determine if a vector field is conservative, one problem being that we cannot integrate every function. The following test is simple to apply for vector fields defined over a rectangle in the plane. We will extend this test to a three dimensional version later when we have Stokes's theorem.

THEOREM 12.5 Test for a Conservative Field in the Plane

Let f and g be continuous in a region D of the plane bounded by a rectangle having its sides parallel to the axes. Then $\mathbf{F}(x, y) = f(x, y)\mathbf{i} + g(x, y)\mathbf{j}$ is conservative on D if and only if, for all (x, y) in D,

$$\frac{\partial g}{\partial x} = \frac{\partial f}{\partial y}. \quad ♦$$

Proof In one direction the proof is a simple differentiation. If $\mathbf{F}$ is conservative on D, then $\mathbf{F} = \nabla\varphi$. Then

$$f(x, y) = \frac{\partial \varphi}{\partial x} \text{ and } g(x, y) = \frac{\partial \varphi}{\partial y}$$

so

$$\frac{\partial g}{\partial x} = \frac{\partial^2 \varphi}{\partial x \partial y} = \frac{\partial^2 \varphi}{\partial y \partial x} = \frac{\partial f}{\partial y}.$$

A proof of the converse is outlined in Problem 22.

EXAMPLE 12.13

Often Theorem 12.5 is used in the following form: if

$$\frac{\partial g}{\partial x} \neq \frac{\partial f}{\partial y}$$

then $f(x, y)\mathbf{i} + g(x,)\mathbf{j}$ is not conservative. As an example of the use of this test, consider

$$\mathbf{F}(x, y) = (2xy^2 + y)\mathbf{i} + (2x^2 y + e^x y)\mathbf{j} = f(x, y)\mathbf{i} + g(x, y)\mathbf{j}.$$

This is continuous over the entire plane, hence on any rectangular region. Compute

$$\frac{\partial g}{\partial x} = 4xy + e^x y \text{ and } \frac{\partial f}{\partial y} = 4xy + 1.$$

These partial derivatives are not equal throughout any rectangular region, so **F** is not conservative.

If we attempted to find a potential function φ by integration, we would begin with

$$\frac{\partial \varphi}{\partial x} = 2xy^2 + y \text{ and } \frac{\partial \varphi}{\partial y} = 2x^2 y + e^x y.$$

Integrate the first equation with respect to x to obtain

$$\varphi(x, y) = x^2 y^2 + xy + \alpha(y),$$

in which $\alpha(y)$ is the "constant" of integration with respect to x. But then

$$\frac{\partial \varphi}{\partial y} = 2x^2 y + x + \alpha'(y) = g(x, y) = 2x^2 y + e^x y.$$

This requires that

$$\alpha'(y) = ye^x,$$

and then $\alpha'(y)$ would depend on x, not y, a contradiction. Thus **F** has no potential function, as we found with less effort using the test of Theorem 12.5. ♦

In special regions (rectangular), existence of a potential function for $f(x, y)\mathbf{i} + g(x, y)\mathbf{j}$ implies that

$$\frac{\partial g}{\partial x} = \frac{\partial f}{\partial y}.$$

We can ask whether equality of these partial derivatives implies that $f\mathbf{i} + g\mathbf{j}$ has a potential function. This is a subtle question, and the answer depends not only on the vector field, but on the set D over which this field is defined. The following example demonstrates this.

EXAMPLE 12.14

Let

$$\mathbf{F}(x, y) = \frac{-y}{x^2 + y^2}\mathbf{i} + \frac{x}{x^2 + y^2}\mathbf{j}$$

for all (x, y) except the origin. This is a vector field in the plane with the origin removed, with

$$f(x, y) = \frac{-y}{x^2 + y^2} \text{ and } g(x, y) = \frac{x}{x^2 + y^2}.$$

Routine integrations would appear to derive the potential function

$$\varphi(x, y) = -\arctan\left(\frac{x}{y}\right).$$

However, this potential is not defined for all (x, y).

If we restrict (x, y) to the right quarter plane $x > 0, y > 0$, then φ is indeed a potential function and **F** is conservative in this region. However, suppose we attempt to consider **F** over the set D consisting of the entire plane with the origin removed. Then φ is not a potential function. Further, **F** is not conservative over D because $\int_C \mathbf{F} \cdot d\mathbf{R}$ is not independent of path in D. To see this, we will evaluate this integral over two paths from $(1, 0)$ to $(-1, 0)$, shown in Figure 12.9.

First, let C_1 be the half-circle given by $x = \cos(\theta)$, $y = \sin(\theta)$ for $0 \le \theta \le \pi$. This is the upper half of the circle $x^2 + y^2 = 1$. Then

$$\begin{aligned}\int_{C_1} \mathbf{F} \cdot d\mathbf{R} &= \int_0^{\pi} [(-\sin(\theta))(-\sin(\theta)) + \cos(\theta)(\cos(\theta))]d\theta \\ &= \int_0^{\pi} d\theta = \pi.\end{aligned}$$

Next let C_2 be the half-circle from $(1, 0)$ to $(-1, 0)$ given by $x = \cos(\theta)$, $y = -\sin(\theta)$ for $0 \le \theta \le \pi$. This is the lower half of the circle $x^2 + y^2 = 1$ and

$$\begin{aligned}\int_{C_2} \mathbf{F} \cdot d\mathbf{R} &= \int_0^{\pi} [\sin(\theta)(-\sin(\theta)) + \cos(\theta)(-\cos(\theta))]d\theta \\ &= -\int_0^{\pi} d\theta = -\pi.\end{aligned}$$

In this example, $\int_C \mathbf{F} \cdot d\mathbf{R}$ depends not only on the vector field, but also on the curve, and the vector field cannot be conservative over the plane with the origin removed. ♦

In attempting a converse of the test of Theorem 12.5, Example 12.14 means that we must place some condition on the set D over which the vector field is defined. This leads us to define a set D of points in the plane to be a *domain* if it satisfies two conditions:

1. If P is a point in D, then there is a circle about P that encloses only points of D.
2. Between any two points of D there is a path lying entirely in D.

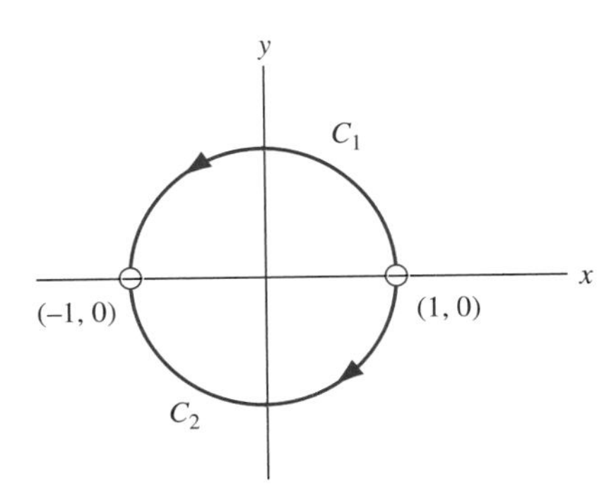

FIGURE 12.9 *Two paths of integration in Example 12.14.*

For example, the interior of a solid rectangle is a domain and the entire plane is a domain. The right quarter plane consisting of points (x, y) with $x > 0$ and $y > 0$ is also a domain. However, if we include parts of the axes, considering points (x, y) with $x \geq 0$ and $y \geq 0$, the resulting set is not a domain. For example, (0, 1) is in this set, but no circle about this point can contain only points with nonnegative coordinates, violating condition (1) for a domain.

A domain D is called *simply connected* if every simple closed path in D encloses only points of D. In Example 12.14, the plane with the origin removed is not simply connected, because a closed path about the origin encloses a point (the origin) not in the set.

Now we can improve Theorem 12.5 to obtain a necessary and sufficient condition for a vector field to be conservative.

THEOREM 12.6

Let $\mathbf{F} = f\mathbf{i} + g\mathbf{j}$ be a vector field defined over a simply connected domain D in the plane. Suppose f and g are continuous and that $\partial f/\partial y$ and $\partial g/\partial x$ are continuous. Then $\mathbf{F}$ is conservative on D if and only if

$$\frac{\partial f}{\partial y} = \frac{\partial g}{\partial x}. \quad \blacklozenge \tag{12.8}$$

Thus, under the given conditions, equality of these partials is both necessary and sufficient for the vector field to have a potential function.

In Example 12.14, the components of $\mathbf{F}$ satisfy equation (12.8), but the set (the plane with the origin removed), is not simply connected, so the theorem does not apply. In that example we saw that there is no potential function for $\mathbf{F}$ over the entire punctured plane.

In 3-space there is a similar test for a vector field to be conservative, with adjustments to accommodate the extra dimension. A set S of points in R^3 is a *domain* if it satisfies the following two conditions:

1. If P is a point in S, then there is a sphere about P that encloses only points of S.
2. Between any two points of S there is a path lying entirely in S.

For example, the interior of a cube is a domain.

Furthermore, S is *simply connected* if every simple closed path in S is the boundary of a surface in S.

With this notion of simple connectivity in 3-space, we can state a three-dimensional version of Theorem 12.6.

THEOREM 12.7

Let $\mathbf{F}$ be a vector field defined over a simply connected domain S in R^3. Then $\mathbf{F}$ is conservative on S if and only if

$$\nabla \times \mathbf{F} = \mathbf{O}. \quad \blacklozenge$$

Thus, the conservative vector fields in R^3 are those with zero curl. These are the irrotational vector fields. We will prove this theorem in Section 12.9.1 when we have Stokes's theorem.

With the right perspective, these tests in 2-space and 3-space can be combined into a single test. Given $\mathbf{F} = f\mathbf{i} + g\mathbf{j}$ in the plane, define

$$\mathbf{G}(x, y, z) = f(x, y)\mathbf{i} + g(x, y)\mathbf{j} + 0\mathbf{k}$$

to think of $\mathbf{F}$ as a vector field in 3-space. Now compute

$$\nabla \times \mathbf{G} = \begin{vmatrix} \mathbf{i} & \mathbf{j} & \mathbf{k} \\ \partial/\partial x & \partial/\partial y & \partial/\partial z \\ f(x,y) & g(x,y) & 0 \end{vmatrix} = \left(\frac{\partial g}{\partial x} - \frac{\partial f}{\partial y}\right)\mathbf{k}.$$

The 3-space condition $\nabla \times \mathbf{G} = \mathbf{O}$ therefore reduces to equation (12.8) if the vector field is in the plane.

Theorem 12.7 can be proved when Stokes's theorem is available to us.

SECTION 12.4 PROBLEMS

In each of Problems 1 through 10, determine whether $\mathbf{F}$ is conservative in the given region D. If D is not defined explicitly, it is understood to be the entire plane or 3-space. If the vector field is conservative, find a potential.

1. $\mathbf{F} = (x^2 - 2)\mathbf{i} + xyz\mathbf{j} - yz^2\mathbf{k}$
2. $\mathbf{F} = 2xy\cos(x^2)\mathbf{i} + \sin(x^2)\mathbf{j}$
3. $\mathbf{F} = \left(\frac{2x}{x^2+y^2}\right)\mathbf{i} + \left(\frac{2y}{x^2+y^2}\right)\mathbf{j}$ D is the plane with the origin removed.
4. $\mathbf{F} = e^{xyz}(1 + xyz)\mathbf{i} + x^2z\mathbf{j} + x^2y\mathbf{k}$
5. $\mathbf{F} = \mathbf{i} - 2\mathbf{j} + \mathbf{k}$
6. $\mathbf{F} = (6y + e^{xy})\mathbf{i} + (6x + xe^{xy})\mathbf{j}$
7. $\mathbf{F} = 16x\mathbf{i} + (2 - y^2)\mathbf{j}$
8. $\mathbf{F} = 2x\mathbf{i} - 2y\mathbf{j} + 2z\mathbf{k}$
9. $\mathbf{F} = y^3\mathbf{i} + (3xy^2 - 4)\mathbf{j}$
10. $\mathbf{F} = yz\cos(x)\mathbf{i} + (z\sin(x) + 1)\mathbf{j} + y\sin(x)\mathbf{k}$

In each of Problems 11 through 20, determine a potential function to evaluate $\int_C \mathbf{F}\cdot d\mathbf{R}$ for C any path from the first point to the second.

11. $\mathbf{F} = 2xy\mathbf{i} + (x^2 - 1/y)\mathbf{j}$; $(1,3)$, $(2,2)$ (The path cannot cross the x - axis).
12. $\mathbf{F} = \mathbf{i} + (6y + \sin(y))\mathbf{j}$; $(0,0)$, $(1,3)$
13. $\mathbf{F} = 6x^2e^{yz}\mathbf{i} + 2x^3ze^{yz}\mathbf{j} + 2x^3ye^{yz}\mathbf{k}$; $(0,0,0)$, $(1,2,-1)$
14. $\mathbf{F} = e^x\cos(y)\mathbf{i} - e^x\sin(y)\mathbf{j}$; $(0,0)$, $(2,\pi/4)$
15. $\mathbf{F} = \mathbf{i} - 9y^2z\mathbf{j} - 3y^3\mathbf{k}$; $(1,1,1)$, $(0,3,5)$
16. $\mathbf{F} = -8y^2\mathbf{i} - (16xy + 4z)\mathbf{j} - 4y\mathbf{k}$; $(-2,1,1)$, $(1,3,2)$
17. $\mathbf{F} = (3x^2y^2 - 6y^3)\mathbf{i} + (2x^3y - 18xy^2)\mathbf{j}$; $(0,0)$, $(1,1)$
18. $\mathbf{F} = (y - 4xz)\mathbf{i} + x\mathbf{j} + (3z^2 - 2x^2)\mathbf{k}$; $(1,1,1)$, $(3,1,4)$
19. $\mathbf{F} = 3x^2(y^2 - 4y)\mathbf{i} + (2x^3y - 4x^3)\mathbf{j}$; $(-1,1)$, $(2,3)$
20. $\mathbf{F} = (y\cos(xz) - xyz\sin(xz))\mathbf{i} + x\cos(xz)\mathbf{j} - x^2\sin(xz)\mathbf{k}$; $(1,0,\pi)$, $(1,1,7)$
21. Prove the law of conservation of energy, which states that the sum of the kinetic and potential energies of an object acted on by a conservative force is a constant. *Hint*: The kinetic energy is $(m/2)\parallel \mathbf{R}'(t)\parallel^2$, where m is the mass and $\mathbf{R}(t)$ describes the trajectory of the particle. The potential energy is $-\varphi(x,y,z)$, where $\mathbf{F} = \nabla\varphi$.
22. Complete the proof of Theorem 12.5 by filling in the details of the following argument. By differentiation, it has already been shown that, if $\mathbf{F}$ has a potential function, then
$$\frac{\partial g}{\partial x} = \frac{\partial f}{\partial y}.$$
To prove the converse, assume equality of these partial derivatives for (x,y) in D. We must produce a potential function φ for $\mathbf{F}$.

    First use Green's theorem to show that $\oint_C \mathbf{F}\cdot d\mathbf{R} = 0$ for any closed path in D. Thus conclude that $\int_C \mathbf{F}\cdot d\mathbf{R}$ is independent of path in D. Choose a point $P_0 = (a,b)$ in D. Then, for any (x,y), define
$$\varphi(x,y) = \int_{P_0}^{(x,y)} \mathbf{F}\cdot d\mathbf{R}.$$
This is a function because the integral is independent of path, hence depends only on (x,y). To show that $\partial\varphi/\partial x = f(x,y)$, first show that
$$\varphi(x+\Delta x, y) - \varphi(x,y) = \int_{(x,y)}^{(x+\Delta x,y)} f(\xi,\eta)\,d\xi + g(\xi,\eta)\,d\eta.$$
Parametrize the horizontal line segment from (x,y) to $(x+\Delta x, y)$ by $\xi = x + t\Delta x$ for $0 \le t \le 1$ to show that
$$\varphi(x+\Delta x,y) - \varphi(x,y) = \Delta x\int_0^1 f(x + t\Delta x, y)\,dt.$$
Use this to show that
$$\frac{\varphi(x+\Delta x,y) - \varphi(x,y)}{\Delta x} = f(x + t_0\Delta x, y)$$
for some t_0 in $(0,1)$. Now take the limit as $\Delta x \to 0$ to show that $\partial\varphi/\partial x = f(x,y)$. A similar argument shows that $\partial\varphi/\partial y = g(x,y)$.

12.5 Surface Integrals

Just as there are integrals of vector fields over curves, there are also integrals of vector fields over surfaces. We begin with some facts about surfaces.

A curve in R^3 is given by coordinate functions of one variable, and may be thought of as a one-dimensional object (such as a thin wire). A *surface* is defined by coordinate or parametric functions of two variables,

$$x = x(u, v), y = y(u, v), z = z(u, v)$$

for (u, v) in some specified set in the u, v-plane. We call u and v parameters for the surface.

EXAMPLE 12.15

Figure 12.10 shows part of the surface having coordinate functions

$$x = u\cos(v), y = u\sin(v), z = \frac{1}{2}u^2\sin(2v)$$

in which u and v can be any real numbers. Since $z = xy$, the surface cuts any plane $z = k$ in a hyperbola $xy = k$. However, the surface intersects a plane $y = \pm x$ in a parabola $z = \pm x^2$. For this reason the surface is called a *hyperbolic paraboloid*. ♦

Often a surface is defined as a *level surface* $f(x, y, z) = k$, with f a given function. For example

$$f(x, y, z) = (x - 1)^2 + y^2 + (z + 4)^2 = 16$$

has the sphere of radius 4 and center $(1, 0, -4)$ as its graph.

We may also express a surface as a locus of points satisfying an equation $z = f(x, y)$ or $y = h(x, z)$ or $x = w(y, z)$. Figure 12.11 shows part of the graph of $z = 6\sin(x - y)/\sqrt{1 + x^2 + y^2}$.

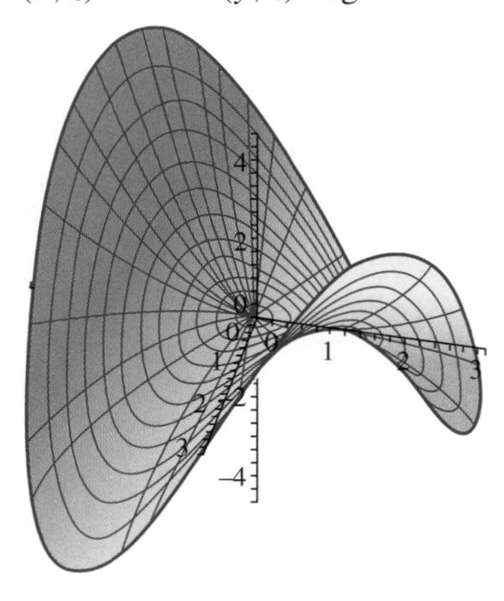

FIGURE 12.10 *The surface* $x = u\cos(v)$, $y = u\sin(v)$, $z = (u^2/2)\sin(2v)$ *in Example 12.15.*

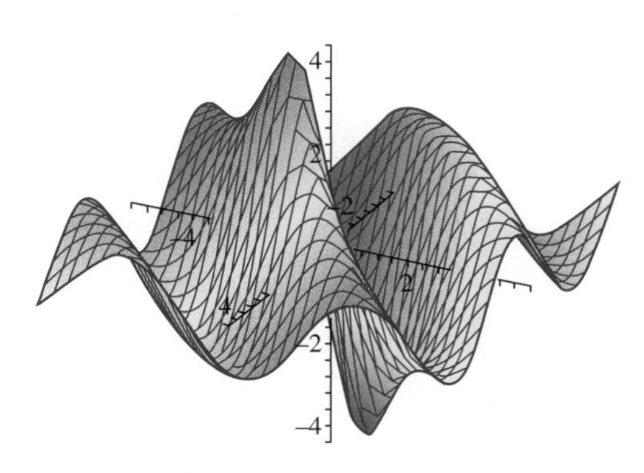

FIGURE 12.11 *The surface* $z = 6\sin(x - y)/\sqrt{1 + x^2 + y^2}$.

We often write a position vector

$$\mathbf{R}(u, v) = x(u, v)\mathbf{i} + y(u, v)\mathbf{j} + z(u, v)\mathbf{k}$$

for a surface. $\mathbf{R}(u, v)$ can be thought of as an arrow from the origin to the point $(x(u, v), y(u, v), z(u, v))$ on the surface.

Although a surface is different from its graph (the surface is a triple of coordinate functions, the graph is a geometric locus in R^3), often we will informally identify the surface with its graph, just as we sometimes identify a curve with its graph.

A surface is *simple* if it does not fold over and intersect itself. This means that $\mathbf{R}(u_1, v_1) = \mathbf{R}(u_2, v_2)$ can occur only when $u_1 = u_2$ and $v_1 = v_2$.

12.5.1 Normal Vector to a Surface

We would like to define a normal vector to a surface at a point. Previously this was done for level surfaces.

Let Σ be a surface with coordinate functions $x(u, v)$, $y(u, v)$, $z(u, v)$. Let P_0 be a point on Σ corresponding to $u = u_0$, $v = v_0$.

If we fix $v = v_0$ we can define the curve Σ_u on Σ, having coordinate functions

$$x = x(u, v_0),\ y = y(u, v_0),\ z = z(u, v_0).$$

The tangent vector to this curve at P_0 is

$$\mathbf{T}_{u0} = \frac{\partial x}{\partial u}(u_0, v_0)\mathbf{i} + \frac{\partial y}{\partial u}(u_0, v_0)\mathbf{j} + \frac{\partial z}{\partial u}(u_0, v_0)\mathbf{k}.$$

Similarly, we can fix $u = u_0$ and form the curve Σ_v on the surface. The tangent to this curve at P_0 is

$$\mathbf{T}_{v0} = \frac{\partial x}{\partial v}(u_0, v_0)\mathbf{i} + \frac{\partial y}{\partial v}(u_0, v_0)\mathbf{j} + \frac{\partial z}{\partial v}(u_0, v_0)\mathbf{k}.$$

These two curves and tangent vectors are shown in Figure 12.12.

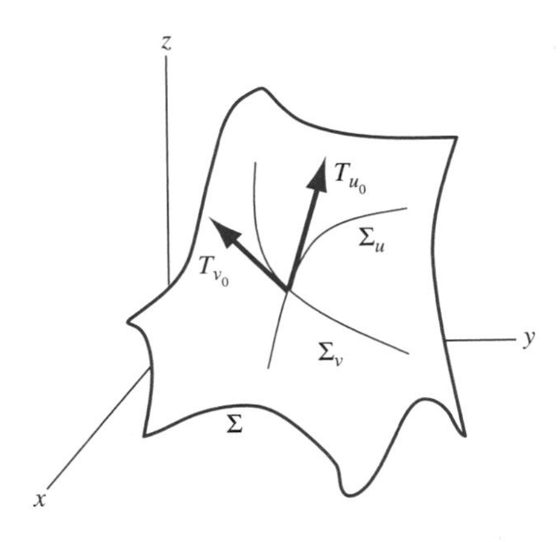

FIGURE 12.12 *Curves Σ_u and Σ_v and tangent vectors $\mathbf{T}_{u0}$ and $\mathbf{T}_{v0}$.*

Assuming that neither of these tangent vectors is the zero vector, they both lie in the tangent plane to the surface at P_0. Their cross product is therefore normal to this tangent plane. This leads us to define the *normal to the surface* at P_0 to be the vector

$$\begin{aligned}\mathbf{N}(P_0) &= \mathbf{T}_{u0} \times \mathbf{T}_{v0} \\ &= \begin{vmatrix} \mathbf{i} & \mathbf{j} & \mathbf{k} \\ \frac{\partial x}{\partial u}(u_0, v_0) & \frac{\partial y}{\partial u}(u_0, v_0) & \frac{\partial z}{\partial u}(u_0, v_0) \\ \frac{\partial x}{\partial v}(u_0, v_0) & \frac{\partial y}{\partial v}(u_0, v_0) & \frac{\partial z}{\partial v}(u_0, v_0) \end{vmatrix} \\ &= \left(\frac{\partial y}{\partial u}\frac{\partial z}{\partial v} - \frac{\partial z}{\partial u}\frac{\partial y}{\partial v}\right)\mathbf{i} + \left(\frac{\partial z}{\partial u}\frac{\partial x}{\partial v} - \frac{\partial x}{\partial u}\frac{\partial z}{\partial v}\right)\mathbf{j} + \left(\frac{\partial x}{\partial u}\frac{\partial y}{\partial v} - \frac{\partial y}{\partial u}\frac{\partial x}{\partial v}\right)\mathbf{k},\end{aligned}$$

in which all partial derivatives are evaluated at (u_0, v_0).

To make this vector easier to write, define the *Jacobian* of two functions f and g to be

$$\frac{\partial(f, g)}{\partial(u, v)} = \begin{vmatrix} \partial f/\partial u & \partial f/\partial v \\ \partial g/\partial u & \partial g/\partial v \end{vmatrix} = \frac{\partial f}{\partial u}\frac{\partial g}{\partial v} - \frac{\partial g}{\partial u}\frac{\partial f}{\partial v}.$$

Then

$$\mathbf{N}(P_0) = \frac{\partial(y, z)}{\partial(u, v)}\mathbf{i} + \frac{\partial(z, x)}{\partial(u, v)}\mathbf{j} + \frac{\partial(x, y)}{\partial(u, v)}\mathbf{k},$$

with all the partial derivatives evaluated at (u_0, v_0). This expression is easy to remember with an observation. Write x, y, z in this order. For the **i** component of **N**, delete x, leaving y, z, (in this order) in the numerator of the Jacobian. For the **j** component, delete y from x, y, z, but move to the right, getting z, x in the Jacobian. For the **k** component, delete z, leaving x, y, in this order.

EXAMPLE 12.16

The *elliptical cone* has coordinate functions

$$x = au\cos(v),\ y = au\sin(v),\ z = u$$

with a and b positive constants. Part of this surface is shown in Figure 12.13.

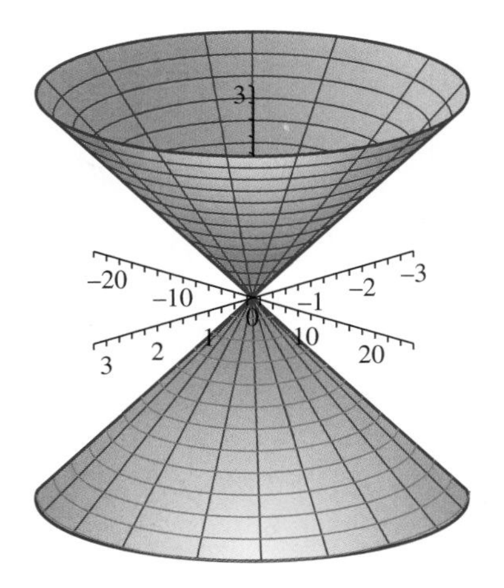

FIGURE 12.13 *Elliptical cone* $z = au\cos(v),\ y = bu\sin(v),\ z = u$.

Since

$$z^2 = \left(\frac{x}{a}\right)^2 + \left(\frac{y}{b}\right)^2.$$

this surface is a "cone" with major axis the z - axis. Planes $z = k$ parallel to the x, y - plane intersect this surface in ellipses. We will write the normal vector $\mathbf{P}_0$ at $P_0 = (a\sqrt{3}/4, b/4, 1/2)$ obtained when $u = u_0 = 1/2$ and $v = v_0 = \pi/6$. Compute the Jacobian components:

$$\begin{aligned}\frac{\partial(y, z)}{\partial(u, v)} &= \left[\frac{\partial y}{\partial u}\frac{\partial z}{\partial v} - \frac{\partial z}{\partial u}\frac{\partial y}{\partial v}\right]_{(1/2,\pi/6)} \\ &= [b\sin(v)(0) - bu\cos(v)]_{(1/2,\pi/6)} = -\sqrt{3}b/4,\end{aligned}$$

$$\begin{aligned}\frac{\partial(z, x)}{\partial(u, v)} &= \left[\frac{\partial z}{\partial u}\frac{\partial x}{\partial v} - \frac{\partial x}{\partial u}\frac{\partial z}{\partial v}\right]_{(1/2,\pi/6)} \\ &= [-au\sin(v) - a\cos(v)(0)]_{(1/2,\pi/6)} = -a/4,\end{aligned}$$

$$\begin{aligned}\frac{\partial(x, y)}{\partial(u, v)} &= \left[\frac{\partial x}{\partial u}\frac{\partial y}{\partial v} - \frac{\partial y}{\partial u}\frac{\partial x}{\partial v}\right]_{(1/2,\pi/6)} \\ &= [a\cos(v)bu\cos(v) - b\sin(v)(-au\sin(v))]_{(1/2,\pi/6)} = ab/2.\end{aligned}$$

Then

$$\mathbf{N}(P_0) = -\frac{\sqrt{3}b}{4}\mathbf{i} - \frac{a}{4}\mathbf{j} + \frac{ab}{2}\mathbf{k}. \ \blacklozenge$$

We frequently encounter the case that a surface is given by an equation $z = S(x, y)$, with $u = x$ and $v = y$ as parameters. In this case

$$\frac{\partial(y, z)}{\partial(u, v)} = \frac{\partial(y, z)}{\partial(x, y)} = \begin{vmatrix} 0 & 1 \\ \partial S/\partial x & \partial S/\partial y \end{vmatrix} = -\frac{\partial S}{\partial x},$$

$$\frac{\partial(z, x)}{\partial u, v} = \frac{\partial(z, x)}{\partial(x, y)} = \begin{vmatrix} \partial S/\partial x & \partial S/\partial y \\ 1 & 0 \end{vmatrix} = -\frac{\partial S}{\partial y},$$

and

$$\frac{\partial(x, y)}{\partial(u, v)} = \frac{\partial(x, y)}{\partial(x, y)} = \begin{vmatrix} 1 & 0 \\ 0 & 1 \end{vmatrix} = 1.$$

Now the normal vector at $P_0 : (x_0, y_0)$ is

$$\begin{aligned}\mathbf{N}(P_0) &= -\frac{\partial S}{\partial x}(x_0, y_0)\mathbf{i} - \frac{\partial S}{\partial y}(x_0, y_0)\mathbf{j} + \mathbf{k} \\ &= -\frac{\partial z}{\partial x}(x_0, y_0)\mathbf{i} - \frac{\partial z}{\partial y}(x_0, y_0)\mathbf{j} + \mathbf{k}.\end{aligned}$$

EXAMPLE 12.17

Let Σ be the hemisphere given by

$$z = \sqrt{4 - x^2 - y^2}.$$

We will find the normal vector at $P_0 : (1, \sqrt{2}, 1)$. Compute

$$\left.\frac{\partial z}{\partial x}\right|_{(1,\sqrt{2})} = -1 \text{ and } \left.\frac{\partial z}{\partial y}\right|_{(1,\sqrt{2})} = -\sqrt{2}.$$

Then

$$\mathbf{N}(P_0) = \mathbf{i} + \sqrt{2}\mathbf{j} + \mathbf{k}.$$

This result is consistent with the fact that a line from the origin through a point on this hemisphere is normal to the hemisphere at that point. ♦

12.5.2 Tangent Plane to a Surface

If a surface Σ has a normal vector $\mathbf{N}(P_0)$ at a point then it has a tangent plane at P_0. This is the plane through $P_0 : (x_0, y_0, z_0)$ having normal vector $\mathbf{N}(P_0)$. The equation of this tangent plane is

$$\mathbf{N}(P_0) \cdot [(x - x_0)\mathbf{i} + (y - y_0)\mathbf{j} + (z - z_0)\mathbf{k}] = 0,$$

or

$$\left[\frac{\partial(y, z)}{\partial(u, v)}\right]_{(u_0, v_0)} (x - x_0) + \left[\frac{\partial(z, x)}{\partial(u, v)}\right]_{(u_0, v_0)} (y - y_0) + \left[\frac{\partial(x, y)}{\partial(u, v)}\right]_{(u_0, v_0)} (z - z_0) = 0.$$

If Σ is given by $z = S(x, y)$, this tangent plane has equation

$$-\left(\frac{\partial S}{\partial x}\right)_{(x_0, y_0)} (x - x_0) - \left(\frac{\partial S}{\partial y}\right)_{(x_0, y_0)} (y - y_0) + z - z_0 = 0.$$

EXAMPLE 12.18

For the elliptical cone of Example 12.16, the tangent plane at $(\sqrt{3}a/4, b/4, 1/2)$ has equation

$$-\frac{\sqrt{3}b}{4}\left(x - \frac{\sqrt{3}a}{4}\right) - \frac{a}{4}\left(x - \frac{b}{4}\right) + \frac{ab}{2}\left(z - \frac{1}{2}\right) = 0. \ ♦$$

EXAMPLE 12.19

For the hemisphere of Example 12.17, the tangent plane at $(1, \sqrt{2}, 1)$ has equation

$$(x - 1) + \sqrt{2}(y - \sqrt{2}) + (z - 1) = 0,$$

or

$$x + \sqrt{2}y + z = 4. \ ♦$$

12.5.3 Piecewise Smooth Surfaces

A curve is *smooth* if it has a continuous tangent. A *smooth surface* is one that has a continuous normal. A *piecewise smooth surface* is one that consists of a finite number of smooth surfaces. For example, a sphere is smooth and the surface of a cube is piecewise smooth, consisting of six smooth faces. The cube does not have a normal vector (or tangent plane) along an edge.

In calculus it is shown that the area of a smooth surface Σ given by $z = S(x, y)$ is

$$\text{area of } \Sigma = \iint_D \sqrt{1 + \left(\frac{\partial S}{\partial x}\right)^2 + \left(\frac{\partial S}{\partial y}\right)^2}\, dA$$

where D is the set of points in the x, y - plane over which the surface is defined. We now recognize that this area is actually the integral of the length of the normal vector:

$$\text{area of } \Sigma = \iint_D \| \mathbf{N}(x, y) \| \, dx\, dy. \tag{12.9}$$

This is analogous to the formula for the length of a curve as the integral of the length of the tangent vector. More generally, if Σ is given by coordinate functions $x(u, v), y(u, v), z(u, v)$ for (u, v) varying over some set D in the u, v - plane, then

$$\text{area of } \Sigma = \iint_D \| \mathbf{N}(u, v) \| \, du\, dv.$$

12.5.4 Surface Integrals

The line integral of $f(x, y, z)$ over C with respect to arc length is

$$\int_C f(x, y, z)\, ds = \int_a^b f(x(t), y(t), z(t))\sqrt{x'(t)^2 + y'(t)^2 + z'(t)^2}\, dt.$$

We want to lift this idea up one dimension to integrate a function over a surface instead of over a curve. To do this, imagine that the coordinate functions are functions of two variables u and v, so $\int_a^b \cdots dt$ will be replaced by $\int\int_D \cdots du\, dv$. The differential element of arc length ds for C will be replaced by the differential element of surface area on Σ, which by equation (12.9) is $d\sigma = \| \mathbf{N}(u, v) \| \, du\, dv$.

Let Σ be a smooth surface with coordinate functions $x(u, v), y(u, v), z(u, v)$ for (u, v) in D. Let f be continuous on Σ. Then *the surface integral of f over Σ* is denoted $\int\int_\Sigma f(x, y, z) d\sigma$ and is defined by

$$\iint_\Sigma f(x, y, z)\, d\sigma = \iint_D f(x(u, v), y(u, v), z(u, v)) \, \| \mathbf{N}(u, v) \| \, du\, dv.$$

If Σ is piecewise smooth, then the line integral of f over Σ is the sum of the line integrals over the smooth pieces.

If Σ is given by $z = S(x, y)$ for (x, y) in D, then

$$\iint_\Sigma f(x, y, z)\, d\sigma = \iint_D f(x, y, S(x, y))\sqrt{1 + \left(\frac{\partial S}{\partial x}\right)^2 + \left(\frac{\partial S}{\partial y}\right)^2}\, dx\, dy.$$

EXAMPLE 12.20

We will compute the surface integral $\int\int_\Sigma xyz\, d\sigma$ over the part of the surface

$$x = u\cos(v),\ y = u\sin(v),\ z = \frac{1}{2}u^2\sin(2v)$$

corresponding to (u, v) in $D: 1 \le u \le 2, 0 \le v \le \pi$.

First we need the normal vector. The components of $\mathbf{N}(u, v)$ are:

$$\frac{\partial(y, z)}{\partial(u, v)} = \begin{vmatrix} \sin(v) & u\cos(v) \\ u\sin(2v) & u^2\cos(2v) \end{vmatrix} = u^2[\sin(v)\cos(2v) - \cos(v)\sin(2v)],$$

$$\frac{\partial(z, x)}{\partial(u, v)} = \begin{vmatrix} u\sin(2v) & u^2\cos(2v) \\ \cos(v) & -u\sin(v) \end{vmatrix} = -u^2[\sin(v)\sin(2v) + \cos(v)\cos(2v)],$$

and

$$\frac{\partial(x, y)}{\partial(u, v)} = \begin{vmatrix} \cos(v) & -u\sin(v) \\ \sin(v) & u\cos(v) \end{vmatrix} = u.$$

Then

$$\| \mathbf{N}(u, v) \|^2 = u^4[\sin(v)\cos(2v) - \cos(v)\sin(2v)]^2 + u^4[\sin(v)\sin(2v) + \cos(v)\cos(2v)]^2 + u^2 = u^2(1+u^2),$$

so

$$\| \mathbf{N}(u, v) \| = u\sqrt{1+u^2}.$$

The surface integral is

$$\iint_\Sigma xyz\,d\sigma = \iint_D [u\cos(v)][u\sin(v)]\left[\frac{1}{2}u^2\sin(2v)\right]u\sqrt{1+u^2}\,dA$$
$$= \int_0^\pi \cos(v)\sin(v)\sin(2v)\,dv \int_1^2 u^5\sqrt{1+u^2}\,du$$
$$= \frac{\pi}{4}\left(\frac{100}{21}\sqrt{21} - \frac{11}{105}\sqrt{2}\right). \blacklozenge$$

EXAMPLE 12.21

We will evaluate $\int\int_\Sigma z\,d\sigma$ over the part of the plane $x+y+z=4$ lying above the rectangle $D: 0 \le x \le 2, 0 \le 1 \le 1$. This surface is shown in Figure 12.14.

With $z = S(x, y) = 4 - x - y$ we have

$$\iint_\Sigma z\,d\sigma = \iint_D z\sqrt{1+(-1)^2+(-1)^2}\,dy\,dx$$
$$= \sqrt{3}\int_0^2\int_0^1 (4-x-y)\,dy\,dx.$$

First compute

$$\int_0^1 (4-x-y)\,dy = (4-x)y - \frac{1}{2}y^2\Big]_0^1$$
$$= 4 - x - \frac{1}{2} = \frac{7}{2} - x.$$

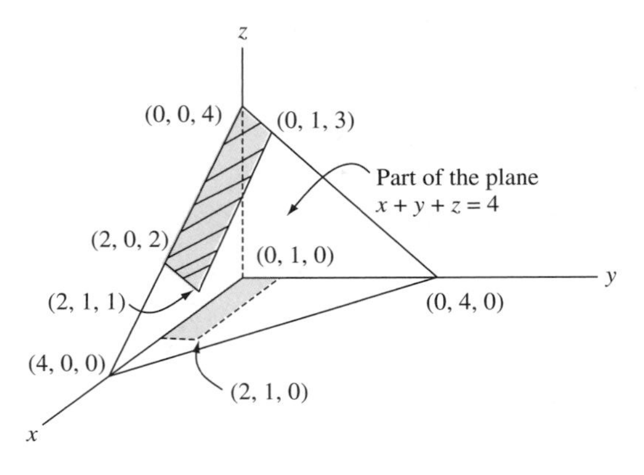

FIGURE 12.14 *Part of the plane $x+y+z=4$.*

Then

$$\iint_{\Sigma} z\,d\sigma = \sqrt{3}\int_0^2 \left(\frac{7}{2} - x\right) dx = 5\sqrt{3}. \blacklozenge$$

SECTION 12.5 PROBLEMS

In each of Problems 1 through 10, evaluate $\iint_{\Sigma} f(x, y, z) d\sigma$.

1. $f(x, y, z) = y$, Σ is the part of the cylinder $z = x^2$ for $0 \le x \le 2, 0 \le y \le 3$.
2. $f(x, y, z) = xyz$, Σ is the part of the plane $z = x + y$ with (x, y) in the square with vertices $(0, 0), (1, 0), (0, 1)$ and $(1, 1)$.
3. $f(x, y, z) = z$, Σ is the part of the cone $z = \sqrt{x^2 + y^2}$ in the first octant and between the planes $z = 2$ and $z = 4$.
4. $f(x, y, z) = xyz$, Σ is the part of the cylinder $z = 1 + y^2$ for $0 \le x \le 1, 0 \le y \le 1$.
5. $f(x, y, z) = z$, Σ is the part of the plane $z = x - y$ for $0 \le x \le 1$ and $0 \le y \le 5$.
6. $f(x, y, z) = x^2$, Σ is the part of the paraboloid $z = 4 - x^2 - y^2$ lying above the x, y - plane.
7. $f(x, y, z) = 1$, Σ is the part of the paraboloid $z = x^2 + y^2$ lying between the planes $z = 2$ and $z = 7$.
8. $f(x, y, z) = y^2$, Σ is the part of the plane $z = x$ for $0 \le x \le 2, 0 \le y \le 4$.
9. $f(x, y, z) = x$, Σ is the part of the plane $x + 4y + z = 10$ in the first octant.
10. $f(x, y, z) = x + y$, Σ is the part of the plane $4x + 8y + 10z = 25$ lying above the triangle in the x, y - plane having vertices $(0, 0)$, $(1, 0)$ and $(1, 1)$.

12.6 Applications of Surface Integrals

12.6.1 Surface Area

If Σ is a piecewise smooth surface, then

$$\iint_{\Sigma} d\sigma = \iint_{D} \| \mathbf{N}(u, v) \| \, du\, dv = \text{ area of } \Sigma.$$

This assumes a bounded surface having finite area. Clearly we do not need surface integrals to compute areas of surfaces. However, we mention this result because it is in the same spirit as other familiar mensuration formulas:

$$\int_C ds = \text{ length of } C,$$

$$\iint_D dA = \text{ area of } D,$$

$$\iiint_M dV = \text{ volume of } M.$$

12.6.2 Mass and Center of Mass of a Shell

Imagine a shell of negligible thickness in the shape of a piecewise smooth surface Σ. Let $\delta(x, y, z)$ be the density of the material of the shell at point (x, y, z). We want to compute the mass of the shell.

Let Σ have coordinate functions $x(u, v), y(u, v), z(u, v)$ for (u, v) in D. Form a grid of lines over D, as in Figure 12.15, by drawing vertical lines Δu units apart and horizontal lines

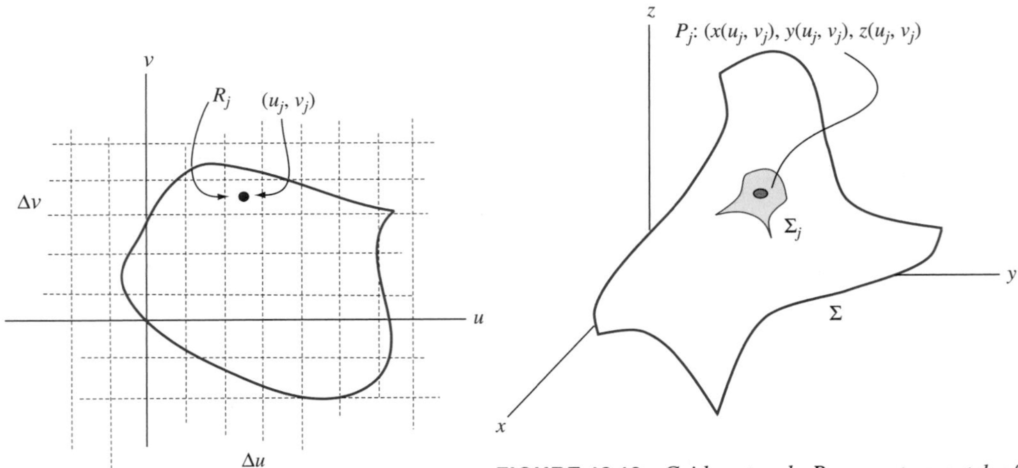

FIGURE 12.15 *Forming a grid over D.*

FIGURE 12.16 *Grid rectangle R_j maps to a patch of surface Σ_j.*

Δv apart. These lines form rectangles $R_1, \cdots, R_n$ that cover D. Each R_j corresponds to a patch of surface Σ_j, as in Figure 12.16. Let (u_j, v_j) be a point in R_j. This corresponds to a point $P_j = (x(u_j, v_j), y(u_j, v_j), z(u_j, v_j))$ on Σ_j. Approximate the mass of Σ_j by the density at P_j times the area of Σ_j. The mass of the shell is approximately the sum of the approximate masses of these patches of surface:

$$\text{mass of the shell} \approx \sum_{j=1}^{n} \delta(P_j)\left(\text{area of } \Sigma_j\right).$$

But the area of Σ_j can be approximated as the length of the normal at P_j times the area of R_j:

$$\text{area of } \Sigma_j \approx \| \mathbf{N}(P_j) \| \ \Delta u \, \Delta v.$$

Therefore,

$$\text{mass of } \Sigma \approx \sum_{j=1}^{n} \delta(P_j) \mathbf{N}(P_j) \| \ \Delta u \Delta v$$

and in the limit as $\Delta u \to 0$ and $\Delta v \to 0$ we obtain

$$\text{mass of } \Sigma = \iint_{\Sigma} \delta(x, y, z) d\sigma.$$

The center of mass of the shell is $(\overline{x}, \overline{y}, \overline{z})$, where

$$\overline{x} = \frac{1}{m} \iint_{\Sigma} x\delta(x, y, z) d\sigma, \overline{y} = \frac{1}{m} \iint_{\Sigma} y\delta(x, y, z) d\sigma,$$

and

$$\overline{z} = \frac{1}{m} \iint_{\Sigma} z\delta(x, y, z) d\sigma,$$

in which m is the mass.

If the surface is given as $z = S(x, y)$ for (x, y) in D, then the mass is

$$m = \iint_{D} \delta(x, y, z) \sqrt{1 + \left(\frac{\partial S}{\partial x}\right)^2 + \left(\frac{\partial S}{\partial y}\right)^2} \, dy \, dx.$$

EXAMPLE 12.22

We will find the mass and center of mass of the cone $z=\sqrt{x^2+y^2}$ for $x^2+y^2\leq 4$ if $\delta(x,y,z)=x^2+y^2$.

Let D be the disk of radius 2 about the origin. Compute

$$\frac{\partial z}{\partial x}=\frac{x}{z} \text{ and } \frac{\partial z}{\partial y}=\frac{y}{z}.$$

The mass is

$$\begin{aligned}
m&=\iint_{\Sigma}(x^2+y^2)d\sigma \\
&=\iint_{D}(x^2+y^2)\sqrt{1+\frac{x^2}{z^2}+\frac{y^2}{z^2}}\,dy\,dx \\
&=\int_0^{2\pi}\int_0^2 r^2\sqrt{2}r\,dr\,d\theta \\
&=2\sqrt{2}\pi\frac{1}{4}r^4\Big]_0^2=8\sqrt{2}\pi.
\end{aligned}$$

By symmetry of the surface and of the density function, we expect the center of mass to lie on the z - axis, so $\overline{x}=\overline{y}=0$. This can be verified by computation. Finally,

$$\begin{aligned}
\overline{z}&=\frac{1}{8\sqrt{2}\pi}\iint_{\Sigma}z(x^2+y^2)d\sigma \\
&=\frac{1}{8\sqrt{2}\pi}\iint_{D}\sqrt{x^2+y^2}(x^2+y^2)\sqrt{1+\frac{x^2}{z^2}+\frac{y^2}{z^2}}\,dy\,dx \\
&=\frac{1}{8\pi}\int_0^{2\pi}\int_0^2 r(r^2)r\,dr\,d\theta \\
&=\frac{1}{8\pi}(2\pi)\left[\frac{1}{5}r^5\right]_0^2=\frac{8}{5}.
\end{aligned}$$

The center of mass is $(0,0,8/5)$. ♦

12.6.3 Flux of a Fluid Across a Surface

Suppose a fluid moves in some region of 3-space with velocity $\mathbf{V}(x,y,z,t)$. In studying the flow, it is often useful to place an imaginary surface Σ in the fluid and analyze the net volume of fluid flowing across the surface per unit time. This is the *flux* of the fluid across the surface.

Let $\mathbf{n}(u,v,t)$ be the unit normal vector to the surface at time t. If we are thinking of flow out of the surface from its interior, then choose $\mathbf{n}$ to be an outer normal, oriented from a point of the surface outward away from the interior.

In a time interval Δt the volume of fluid flowing across a small piece Σ_j of Σ approximately equals the volume of the cylinder with base Σ_j and altitude $V_n\Delta t$, where V_n is the component of $\mathbf{V}$ in the direction of $\mathbf{n}$, evaluated at some point of Σ_j. This volume (Figure 12.17) is $(V_n\Delta t)A_j$, where A_j is the area of Σ_j. Because $\|\mathbf{n}\|=1$, $V_n=\mathbf{V}\cdot\mathbf{n}$. The volume of fluid across Σ_j per unit time is

$$\frac{(V_n\Delta t)A_j}{\Delta t}=V_nA_j=\mathbf{V}\cdot\mathbf{n}\Delta t.$$

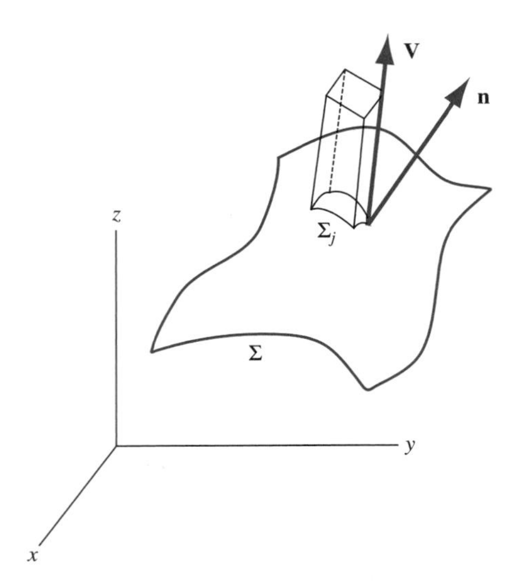

FIGURE 12.17 *Cylinder with base Σ_j and height $V_{\mathbf{n}}\Delta t$.*

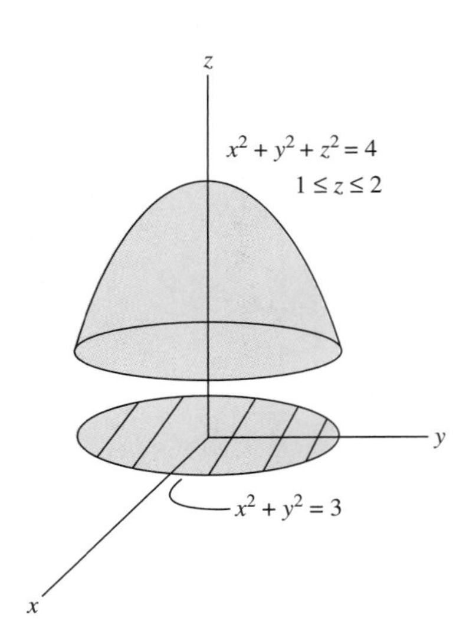

FIGURE 12.18 *Surface in Example 12.23.*

Sum these quantities over the entire surface and take a limit as the surface elements are chosen smaller, as we did for the mass of a shell. We get

$$\text{flux of } \mathbf{V} \text{ across } \Sigma \text{ in the direction of } \mathbf{n} = \iint_{\Sigma} \mathbf{V} \cdot \mathbf{n} d\sigma.$$

The flux of the fluid (or any vector field) across a surface is therefore computed as the surface integral of the normal component of the field to the surface.

EXAMPLE 12.23

We will calculate the flux of $\mathbf{F} = x\mathbf{i} + y\mathbf{j} + z\mathbf{k}$ across the part of the sphere $x^2 + y^2 + z^2 = 4$ between the planes $z = 1$ and $z = 2$.

The plane $z = 1$ intersects the sphere in the circle $x^2 + y^2 = 3, z = 1$. This circle projects onto the circle $x^2 + y^2 = 3$ in the x, y - plane. The plane $z = 2$ hits the sphere at $(0, 0, 2)$ only. Think of Σ as specified by $z = S(x, y) = \sqrt{4 - x^2 + y^2}$ for (x, y) in D, the disk $0 \le x^2 + y^2 \le 3$ (Figure 12.18).

To compute the partial derivatives $\partial z/\partial x$ and $\partial z/\partial y$ we can implicitly differentiate the equation of the sphere to get

$$2x + 2z\frac{\partial z}{\partial x} = 0$$

so

$$\frac{\partial z}{\partial x} = -\frac{x}{z}.$$

Similarly,

$$\frac{\partial z}{\partial y} = -\frac{y}{z}.$$

The vector

$$\frac{x}{z}\mathbf{i} + \frac{y}{z}\mathbf{j} + \mathbf{k}$$

is normal to the sphere and oriented outward. This vector has magnitude $2/z$, so a unit outer normal is

$$\mathbf{n} = \frac{z}{2}\left(\frac{x}{z}\mathbf{i} + \frac{y}{z}\mathbf{j} + \mathbf{k}\right) = \frac{1}{2}(x\mathbf{i} + y\mathbf{j} + z\mathbf{k}).$$

We need

$$\mathbf{F}\cdot\mathbf{n} = \frac{1}{2}(x^2 + y^2 + z^2).$$

Then

$$\begin{aligned}
\text{flux} &= \iint_{\Sigma} \frac{1}{2}(x^2+y^2+z^2)\,d\sigma \\
&= \frac{1}{2}\iint_D (x^2+y^2+z^2)\sqrt{1+\frac{x^2}{z^2}+\frac{y^2}{z^2}}\,dA \\
&= \frac{1}{2}\iint_D (x^2+y^2+z^2)\sqrt{\frac{x^2+y^2+z^2}{z^2}}\,dA \\
&= \frac{1}{2}\iint_D (x^2+y^2+z^2)^{3/2}\frac{1}{\sqrt{4-x^2-y^2}}\,dA \\
&= 4\iint_D \frac{1}{\sqrt{4-x^2-y^2}}\,dA.
\end{aligned}$$

Here we used the fact that $x^2 + y^2 + z^2 = 4$ on Σ. Converting to polar coordinates, we have

$$\begin{aligned}
\text{flux} &= 4\int_0^{2\pi}\int_0^{\sqrt{3}} \frac{1}{\sqrt{4-r^2}}\,r\,dr\,d\theta \\
&= 8\pi\left[-(4-r^2)^{1/2}\right]_0^{\sqrt{3}} = 8\pi. \ \blacklozenge
\end{aligned}$$

SECTION 12.6 PROBLEMS

In each of Problems 1 through 6, find the mass and center of mass of the shell Σ.

1. Σ is the paraboloid $z = 6 - x^2 - y^2$ for $z \geq 0$, with $\delta(x, y, z) = \sqrt{1 + 4x^2 + 4y^2}$.
2. Σ is the part of the paraboloid $z = 16 - x^2 - y^2$ in the first octant and between the cylinders $x^2 + y^2 = 1$ and $x^2 + y^2 = 9$, with $\delta(x, y, z) = xy/\sqrt{1 + 4x^2 + 4y^2}$.
3. Σ is a triangle with vertices $(1, 0, 0)$, $(0, 3, 0)$ and $(0, 0, 2)$, with $\delta(x, y, z) = xz + 1$.
4. Σ is the part of the sphere of radius 1 about the origin, lying in the first octant. The density is constant.
5. Σ is the cone $z = \sqrt{x^2 + y^2}$ for $x^2 + y^2 \leq 9$, $\delta =$ constant $= K$.
6. Σ is the part of the sphere $x^2 + y^2 + z^2 = 9$ above the plane $z = 1$, and the density function is constant.
7. Find the flux of $\mathbf{F} = x\mathbf{i} + y\mathbf{j} - z\mathbf{k}$ across the part of the plane $x + 2y + z = 8$ in the first octant.
8. Find the flux of $\mathbf{F} = xz\mathbf{i} - y\mathbf{k}$ across the part of the sphere $x^2 + y^2 + z^2 = 4$ above the plane $z = 1$.

12.7 Lifting Green's Theorem to R^3

The fundamental results of vector integral calculus are the theorems of Gauss and Stokes. In this section we will show how both can be viewed as natural generalizations of Green's theorem from two to three dimensions.

The conclusion of Green's theorem can be written

$$\oint_C f(x, y)\,dx + g(x, y)\,dy = \iint_D \left(\frac{\partial g}{\partial x} - \frac{\partial f}{\partial y}\right) dA$$

with C the simple closed path bounding the region D of the plane.

Define the vector field

$$\mathbf{F}(x, y) = g(x, y)\mathbf{i} - f(x, y)\mathbf{j}.$$

With this choice,

$$\nabla \cdot \mathbf{F} = \frac{\partial g}{\partial x} - \frac{\partial f}{\partial y}.$$

Parametrize C by arc length so the coordinate functions are $x = x(s)$, $y = y(s)$ for $0 \le s \le L$. The unit tangent vector to C is $\mathbf{T}(s) = x'(s)\mathbf{i} + y'(s)\mathbf{j}$ and the unit normal vector is $\mathbf{n}(s) = y'(s)\mathbf{i} - x'(s)\mathbf{j}$. These are shown in Figure 12.19. This normal points outward away from D, and so is called a unit outer normal. Now

$$\mathbf{F} \cdot \mathbf{n} = g(x, y)\frac{dy}{ds} + f(x, y)\frac{dx}{ds}$$

so

$$\oint_C f(x, y)\,dx + g(x, y)\,dy = \oint_C \left[f(x, y)\frac{dx}{ds} + g(x, y)\frac{dy}{ds}\right] ds$$

$$= \oint_C \mathbf{F} \cdot \mathbf{n}\,ds.$$

We may therefore write the conclusion of Green's theorem as

$$\oint_C \mathbf{F} \cdot \mathbf{n}\,ds = \iint_D \nabla \cdot \mathbf{F}\,dA.$$

In this form, Green's theorem suggests a generalization to three dimensions. Replace the closed curve C in the plane with a closed surface Σ in 3-space (closed means bounding a volume, such

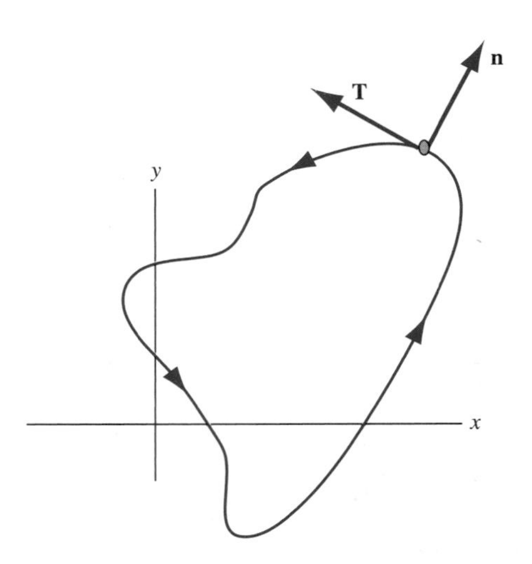

FIGURE 12.19 *Unit tangent and normal vectors to C.*

as a sphere). Replace the line integral over C with a surface integral over Σ and allow the vector field to be a function of three variables. We conjecture that Green's theorem generalizes to

$$\iint_{\Sigma} \mathbf{F} \cdot \mathbf{n}\, d\sigma = \iiint_{M} \nabla \cdot \mathbf{F}\, dV,$$

where M is the solid region bounded by Σ and $\mathbf{n}$ is a unit normal to Σ pointing out of the surface and away from M. This conclusion is Gauss's divergence theorem.

Now begin again with Green's theorem and pursue a different generalization to three dimensions. This time let

$$\mathbf{F}(x, y, z) = f(x, y)\mathbf{i} + g(x, y)\mathbf{j} + 0\mathbf{k}.$$

Including a third component allows us to take the curl:

$$\nabla \times \mathbf{F} = \begin{vmatrix} \mathbf{i} & \mathbf{j} & \mathbf{k} \\ \partial/\partial x & \partial/\partial y & \partial/\partial z \\ f & g & 0 \end{vmatrix} = \left(\frac{\partial g}{\partial x} - \frac{\partial f}{\partial y}\right)\mathbf{k}.$$

Then

$$(\nabla \times \mathbf{F}) = \frac{\partial g}{\partial x} - \frac{\partial f}{\partial y}.$$

Further, with unit tangent $\mathbf{T}(s) = x'(s)\mathbf{i} + y'(s)\mathbf{j}$ to C, we can write

$$\begin{aligned} \mathbf{F} \cdot \mathbf{T}\, ds &= [f(x, y)\mathbf{i} + g(x, y)\mathbf{j}] \cdot \left(\frac{dx}{ds}\mathbf{i} + \frac{dy}{ds}\mathbf{k}\right) \\ &= f(x, y)dx + g(x, y)dy. \end{aligned}$$

Now the conclusion of Green's theorem can be written

$$\oint_C \mathbf{F} \cdot \mathbf{T}\, ds = \iint_D (\nabla \times \mathbf{F}) \cdot \mathbf{k}\, dA.$$

Think of D as a flat surface in the x, y-plane, with unit normal $\mathbf{k}$, and bounded by the closed path C. To generalize this, allow C to be a path in 3-space, bounding a surface Σ having unit outer normal $\mathbf{N}$, as in Figure 12.20. With these changes, the last equation suggests that

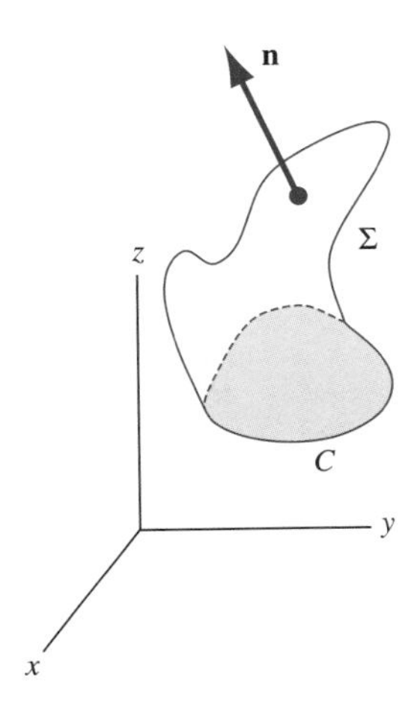

FIGURE 12.20 C *bounding a surface* Σ *having outer normal* $\mathbf{n}$.

$$\oint_C \mathbf{F} \cdot \mathbf{T}\, ds = \iint_\Sigma (\nabla \times \mathbf{F}) \cdot \mathbf{n}\, d\sigma.$$

We will see this in Section 12.9 as Stokes's theorem.

SECTION 12.7 PROBLEMS

1. Let C be a simple closed path in the x, y-plane with interior D. Let $\varphi(x, y)$ and $\psi(x, y)$ be continuous with continuous first and second partial derivatives on C and throughout D. Let

$$\nabla^2 \psi = \frac{\partial^2 \psi}{\partial x^2} + \frac{\partial^2 \psi}{\partial y^2}.$$

Prove that

$$\iint_D \varphi \nabla^2 \psi\, dA$$
$$= \oint_C -\varphi \frac{\partial \psi}{\partial y}\, dx + \varphi \frac{\partial \psi}{\partial x}\, dy - \iint_D \nabla\varphi \cdot \nabla\psi\, dA.$$

2. Under the conditions of Problem 1, show that

$$\iint_D (\varphi \nabla^2 \psi - \psi \nabla^2 \varphi)\, dA$$
$$= \oint_C \left[\psi \frac{\partial \varphi}{\partial y} - \varphi \frac{\partial \psi}{\partial y}\right] dx + \left[\varphi \frac{\partial \psi}{\partial x} - \psi \frac{\partial \varphi}{\partial x}\right] dy.$$

3. Let C be a simple closed path in the x, y-plane, with interior D. Let φ be continuous with continuous first and second partial derivatives on C and throughout D. Let $\mathbf{N}(x, y)$ be the unit outer normal to C (outer meaning pointing away from D from points on C). Prove that

$$\oint_C \varphi_{\mathbf{N}}(x, y)\, ds = \iint_D \nabla^2 \varphi(x, y)\, dA.$$

(Recall that $\varphi_{\mathbf{N}}$ is the directional derivative of φ in the direction of $\mathbf{N}$.)

12.8 The Divergence Theorem of Gauss

The discussion of the preceding section suggested a possible extension of Green's theorem to three dimensions, yielding what is known as the divergence theorem.

THEOREM 12.8 The Divergence Theorem of Gauss

Let Σ be a piecewise smooth closed surface bounding a region M of 3-space. Let Σ have unit outer normal $\mathbf{n}$. Let $\mathbf{F}$ be a vector field with continuous first and second partial derivatives on Σ and throughout M. Then

$$\iint_\Sigma \mathbf{F} \cdot \mathbf{n}\, d\sigma = \iiint_M \nabla \cdot \mathbf{F}\, dV. \quad \blacklozenge \tag{12.10}$$

The theorem is named for the great nineteenth-century German mathematician and scientist Carl Friedrich Gauss, and is actually a conservation of mass equation. Recall that the divergence of a vector field at a point is a measure of the flow of the field away from that point. Equation (12.10) states that the flux of the vector field outward from M across Σ exactly balances the flow of the field from each point in M. Whatever crosses the surface and leaves M must be accounted for by flow out of M (in the absence of sources or sinks in M).

We will look at two computational examples to get some feeling for equation (12.10), and then consider applications.

EXAMPLE 12.24

Let Σ be the piecewise smooth surface consisting of the surface Σ_1 of the cone $z=\sqrt{x^2+y^2}$ for $x^2+y^2 \leq 1$, together with the flat cap Σ_2 consisting of the disk $x^2+y^2 \leq 1$ in the plane $z=1$. Σ is shown in Figure 12.21. Let $\mathbf{F}(x,y,z)=x\mathbf{i}+y\mathbf{j}+z\mathbf{k}$. We will calculate both sides of equation (12.10).

The unit outer normal to Σ_1 is

$$\mathbf{n}_1 = \frac{1}{\sqrt{2}}\left(\frac{x}{z}\mathbf{i}+\frac{y}{z}\mathbf{j}-\mathbf{k}\right).$$

Then

$$\mathbf{F}\cdot\mathbf{n}_1 = \frac{1}{\sqrt{2}}\left(\frac{x^2}{z}+\frac{y^2}{z}-z\right)=0$$

because on Σ_1, $z^2=x^2+y^2$. Therefore

$$\iint_{\Sigma_1} \mathbf{F}\cdot\mathbf{n}_1\,d\sigma = 0.$$

The unit outer normal to Σ_2 is $\mathbf{n}_2=\mathbf{k}$, so $\mathbf{F}\cdot\mathbf{n}_2=z$. Since $z=1$ on Σ_2, then

$$\iint_{\Sigma_2} \mathbf{F}\cdot\mathbf{n}_2\,d\sigma = \iint_{\Sigma_2} z\,d\sigma = \iint_{\Sigma_2} d\sigma$$
$$= \text{area of } \Sigma_2 = \pi.$$

Therefore,

$$\iint_{\Sigma} \mathbf{F}\cdot\mathbf{n}\,d\sigma = \iint_{\Sigma_1} \mathbf{F}\cdot\mathbf{n}_1\,d\sigma + \iint_{\Sigma_2} \mathbf{F}\cdot\mathbf{n}_2\,d\sigma = \pi.$$

Now compute the triple integral. The divergence of $\mathbf{F}$ is

$$\nabla\cdot\mathbf{F} = \frac{\partial}{\partial x}x + \frac{\partial}{\partial y}y + \frac{\partial}{\partial z}z = 3,$$

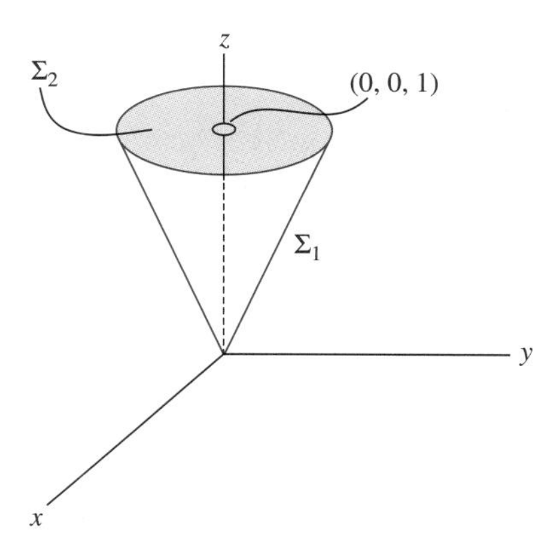

FIGURE 12.21 *Σ in Example 12.24.*

so

$$\iiint_M \nabla \cdot \mathbf{F}\, dV = \iiint_M 3\, dV$$
$$= 3[\text{volume of the cone of height 1, radius 1}]$$
$$= 3\left(\frac{1}{3}\right)\pi = \pi. \blacklozenge$$

EXAMPLE 12.25

Let Σ be the piecewise smooth surface of the cube having vertices

$$(0,0,0), (1,0,0), (0,1,0), (0,0,1)$$

$$(1,1,0), (0,1,1), (1,0,1), (1,1,1).$$

Let $\mathbf{F}(x, y, z) = x^2\mathbf{i} + y^2\mathbf{j} + z^2\mathbf{k}$. We want to compute the flux of $\mathbf{F}$ across Σ. This flux is $\int\int_\Sigma \mathbf{F} \cdot \mathbf{n}\, d\sigma$. We can certainly evaluate this integral, but this will be tedious because Σ has six smooth faces. It is easier to use the triple integral of the divergence theorem. Compute

$$\nabla \cdot \mathbf{F} = 2x + 2y + 2z.$$

Then

$$\text{flux} = \iint_\Sigma \mathbf{F} \cdot \mathbf{n}\, d\sigma$$
$$= \iiint_M \nabla \cdot \mathbf{F}\, dV = \iiint_M (2x + 2y + 2z)\, dV$$
$$= \int_0^1 \int_0^1 \int_0^1 (2x + 2y + 2z)\, dz\, dy\, dx$$
$$= \int_0^1 \int_0^1 (2x + 2y + 1)\, dy\, dx$$
$$= \int_0^1 (2x + 2)\, dx = 3. \blacklozenge$$

12.8.1 Archimedes's Principle

Archimedes's principle is that the buoyant force a fluid exerts on a solid object immersed in it, is equal to the weight of the fluid displaced. An aircraft carrier floats in the ocean if it displaces a volume of seawater whose weight at least equals that of the carrier. We will use the divergence theorem to derive this principle.

Imagine a solid object M bounded by a piecewise smooth surface Σ. Let ρ be the constant density of the fluid. Draw a coordinate system as in Figure 12.22 with M below the surface. Using the fact that pressure equals depth multiplied by density, the pressure $\rho(x, y, z)$ at a point on Σ is $p(x, y, z) = -\rho z$, the negative sign because z is negative in the downward direction and we want pressure to be positive. Now consider a piece Σ_j of Σ. The force of the pressure on Σ_j is approximately $-\rho z$ multiplied by the area A_j of Σ_j. If $\mathbf{n}$ is the unit outer normal to Σ_j, then the force caused by the pressure on Σ_j is approximately $\rho z \mathbf{n} A_j$. The vertical component of this force is the magnitude of the buoyant force acting upward on Σ_j. This vertical component is

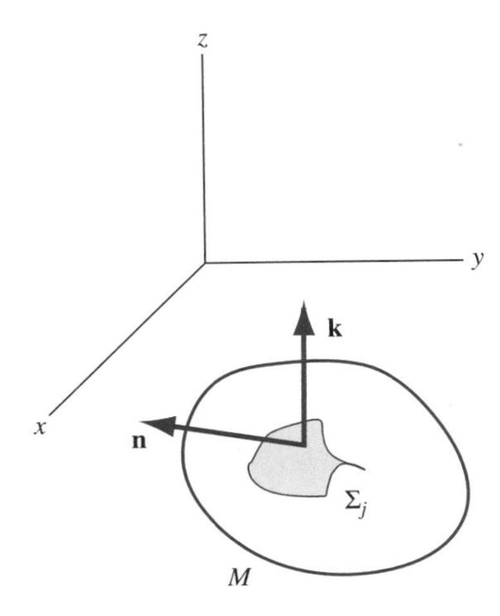

FIGURE 12.22 *Archimedes's Principle.*

$\rho z \mathbf{n} \cdot \mathbf{k} A_j$. Sum these vertical components over Σ to obtain approximately the buoyant force on the object, then take the limit as the surface elements are chosen smaller. We obtain

$$\text{net buoyant force on } \Sigma = \iint_{\Sigma} \rho z \mathbf{n} \cdot \mathbf{k} \, d\sigma.$$

Write this integral as $\int\int_{\Sigma} \rho z \mathbf{k} \cdot \mathbf{n} \, d\sigma$ and apply the divergence theorem to obtain

$$\text{net buoyant force on } \Sigma = \iiint_M \nabla \cdot (\rho z \mathbf{k}) \, dV.$$

But $\nabla \cdot (\rho z \mathbf{k}) = \rho$, so

$$\text{net buoyant force on } \Sigma = \iiint_M \rho \, dV = \rho[\text{volume of } M].$$

and ρ multiplied by the volume of M is the weight of the object.

12.8.2 The Heat Equation

We will use the divergence theorem to derive a partial differential equation that models heat conduction and diffusion processes. Suppose some medium (such as a bar of metal or water in a pool) has density $\rho(x, y, z)$, specific heat $\mu(x, y, z)$ and coefficient of thermal conductivity $K(x, y, z)$. Let $u(x, y, z, t)$ be the temperature of the medium at (x, y, z) and time t. We want an equation for u.

We will exploit an idea frequently used in constructing mathematical models. Consider an imaginary smooth closed surface Σ in the medium, bounding a solid region M. The amount of heat energy leaving M across Σ in a time interval Δt is

$$\left(\iint_{\Sigma} (K \nabla u) \cdot \mathbf{n} \, d\sigma \right) \Delta t.$$

This is the flux of $K\nabla u$ across Σ, multiplied by the length of the time interval. But, the change in temperature at (x, y, z) over Δt is approximately $(\partial u/\partial t)\Delta t$ so the resulting heat loss in M is

$$\left(\iiint_M \mu\rho\frac{\partial u}{\partial t}\,dV\right)\Delta t.$$

Assuming no heat sources or sinks in M (which occur, for example, during chemical reactions or radioactive decay), the change in heat energy in M over Δt must equal the heat exchange across Σ:

$$\left(\iint_\Sigma (K\nabla u)\cdot\mathbf{n}\,d\sigma\right)\Delta t = \left(\iiint_M \mu\rho\frac{\partial u}{\partial t}\,dV\right)\Delta t.$$

Therefore,

$$\iint_\Sigma (K\nabla u)\cdot\mathbf{n}\,d\sigma = \iiint_M \mu\rho\frac{\partial u}{\partial t}\,dV.$$

Now use the divergence theorem to convert the surface integral to a triple integral:

$$\iint_\Sigma (K\nabla u)\cdot\mathbf{n}\,d\sigma = \iiint_M \nabla\cdot(K\nabla u)\,dV.$$

Substitute this into the preceding equation to obtain

$$\iiint_M \left(\mu\rho\frac{\partial u}{\partial t} - \nabla\cdot(K\nabla u)\right)dV = 0.$$

Now Σ is an arbitrary closed surface within the medium. If the integrand in the last equation were, say, positive at some point P_0, then it would be positive throughout some (perhaps very small) sphere about P_0, and we could choose Σ as this sphere. But then the triple integral over M of a positive quantity would be positive, not zero, a contradiction. The same conclusion follows if this integrand were negative at some P_0.

Vanishing of the last integral for *every* closed surface Σ in the medium therefore forces the integrand to be identically zero:

$$\mu\rho\frac{\partial u}{\partial t} - \nabla\cdot(K\nabla u) = 0.$$

This is the partial differential equation

$$\mu\rho\frac{\partial u}{\partial t} = \nabla\cdot(K\nabla u)$$

for the temperature function. This equation is called the *heat equation*.

We can expand

$$\begin{aligned}
\nabla\cdot(K\nabla u) &= \nabla\cdot\left(K\frac{\partial u}{\partial x}\mathbf{i} + K\frac{\partial u}{\partial y}\mathbf{j} + K\frac{\partial u}{\partial z}\mathbf{k}\right)\\
&= \frac{\partial}{\partial x}\left(K\frac{\partial u}{\partial x}\right) + \frac{\partial}{\partial y}\left(K\frac{\partial u}{\partial y}\right) + \frac{\partial}{\partial z}\left(K\frac{\partial u}{\partial z}\right)\\
&= \frac{\partial K}{\partial x}\frac{\partial u}{\partial x} + \frac{\partial K}{\partial y}\frac{\partial u}{\partial y} + \frac{\partial K}{\partial z}\frac{\partial u}{\partial z} + K\left(\frac{\partial^2 u}{\partial x^2} + \frac{\partial^2 u}{\partial y^2} + \frac{\partial^2 u}{\partial z^2}\right)\\
&= \nabla K\cdot\nabla u + K\nabla^2 u.
\end{aligned}$$

Here we have introduced the symbol ∇^2, defined by

$$\nabla^2 u = \frac{\partial^2 u}{\partial x^2} + \frac{\partial^2 u}{\partial y^2} + \frac{\partial^2 u}{\partial z^2}.$$

∇^2 is called the *Laplacian*, or *Laplacian operator*, and is read "del squared."

Now the heat equation is

$$\mu\rho \frac{\partial u}{\partial t} = \nabla K \cdot \nabla u + K \nabla^2 u.$$

If K is constant then its gradient vector is zero and this equation simplifies to

$$\frac{\partial u}{\partial t} = \frac{K}{\mu\rho} \nabla^2 u.$$

In the case of one space dimension, $u = u(x, t)$ and we often write this equation as

$$\frac{\partial u}{\partial t} = k \frac{\partial^2 u}{\partial x^2}$$

in which $k = K/\mu\rho$.

The *steady-state heat equation* occurs when $\partial u/\partial t = 0$. In this case we get $\nabla^2 u = 0$, which is called *Laplace's equation.*

SECTION 12.8 PROBLEMS

In each of Problems 1 through 8, evaluate either $\int\int_\Sigma \mathbf{F} \cdot \mathbf{n}\, d\sigma$ or $\iiint_M \text{div}(\mathbf{F})\, dV$, whichever is easier.

1. $\mathbf{F} = 2yz\mathbf{i} - 4xz\mathbf{j} + xy\mathbf{k}$, Σ is the sphere of radius 5 about $(-1, 3, 1)$
2. $\mathbf{F} = x^3\mathbf{i} + y^3\mathbf{j} + z^3\mathbf{k}$, Σ is the sphere of radius 1 about the origin.
3. $\mathbf{F} = x\mathbf{i} + y\mathbf{j} - z\mathbf{k}$, Σ is the sphere of radius 4 about $(1, 1, 1)$.
4. $\mathbf{F} = 4x\mathbf{i} - 6y\mathbf{j} + z\mathbf{k}$, Σ is the surface of the solid cylinder $x^2 + y^2 \leq 4, 0 \leq z \leq 2$, including the end caps of the cylinder.
5. $\mathbf{F} = x^2\mathbf{i} + y^2\mathbf{j} + z^2\mathbf{k}$, Σ is the cone $z = \sqrt{x^2 + y^2}$ for $x^2 + y^2 \leq 2$, together with the top cap consisting of the points $(x, y, \sqrt{2})$ with $x^2 + y^2 \leq 2$.
6. $\mathbf{F} = x^2\mathbf{i} - e^z\mathbf{j} + z\mathbf{k}$, Σ is the surface bounding the cylinder $x^2 + y^2 \leq 4, 0 \leq z \leq 2$, including the top and bottom caps of the cylinder.
7. $\mathbf{F} = 4x\mathbf{i} - z\mathbf{j} + x\mathbf{k}$, Σ is the hemisphere $x^2 + y^2 + z^2 = 1, z \geq 0$, including the base consisting of points $(x, y, 0)$ with $x^2 + y^2 \leq 1$.
8. $\mathbf{F} = (x - y)\mathbf{i} + (y - 4xz)\mathbf{j} + xz\mathbf{k}$, Σ is the surface of the rectangular box bounded by the coordinate planes $x = 0, y = 0, z = 0$ and the planes $x = 4$, $y = 2, z = 3$.
9. Let Σ be a smooth closed surface and $\mathbf{F}$ a vector field that is continuous with continuous first and second partial derivatives throughout Σ and the region it bounds. Evaluate $\int\int_\Sigma (\nabla \times \mathbf{F}) \cdot \mathbf{n}\, d\sigma$.
10. Let Σ be a piecewise smooth closed surface bounding a region M. Show that

$$\text{volume of } M = \frac{1}{3} \iiint_\Sigma \mathbf{R} \cdot \mathbf{n}\, d\sigma$$

where $\mathbf{R} = x\mathbf{i} + y\mathbf{j} + z\mathbf{k}$.

12.9 Stokes's Theorem

In Section 12.7, we suggested a lifting of Green's theorem to three dimensions to arrive at Stokes's theorem. That discussion passed quickly over some subtleties which we will now address more carefully.

First we need to explore the idea of a surface and a bounding curve. Suppose Σ is a surface defined by $x = x(u, v), y = y(u, v), z = z(u, v)$ for (u, v) in some bounded region D of the u, v-plane. As (u, v) varies over D, the point $(x(u, v), y(u, v), z(u, v))$ traces out Σ. We will assume that D is bounded by a piecewise smooth curve K. As (u, v) traverses K, the corresponding point $(x(u, v), y(u, v), z(u, v))$ traverses a curve C on Σ. This is the curve we call the *boundary curve* of Σ. This is shown in Figure 12.23.

EXAMPLE 12.26

Let Σ be given by $z = x^2 + y^2$ for $0 \le x^2 + y^2 \le 4$. Here x and y are the parameters and vary over the disk of radius 2 in the x, y-plane. Figure 12.24 shows D and a graph of the surface. The boundary K of D is the circle $x^2 + y^2 = 4$ in the x, y-plane. This circle maps to the circle $x^2 + y^2 = 4, z = 4$ on Σ. This is the boundary C of Σ, and is the circle at the top of the bowl-shaped surface. ♦

We need a rule for choosing a normal to the surface at each point. We can use the standard normal vector

$$\frac{\partial(y, z)}{\partial(u, v)}\mathbf{i} + \frac{\partial(z, x)}{\partial(u, v)}\mathbf{j} + \frac{\partial(x, y)}{\partial(u, v)}\mathbf{k},$$

dividing this by its length to obtain a unit normal vector. The negative of this is also a unit normal vector. Whichever we use, call it $\mathbf{n}$ and use it throughout the surface. We cannot use $\mathbf{n}$ at some points and $-\mathbf{n}$ at others.

This choice of the normal vector $\mathbf{n}$ is used to determine an orientation on the boundary curve C of Σ. Referring to Figure 12.25, at any point on C, if you stand along $\mathbf{n}$ with your head at the tip of this normal, then the positive direction of C is the one in which you have to walk to have the surface over your left shoulder. This is admittedly informal, but a more rigorous treatment involves topological subtleties we do not wish to engage here. When this direction is chosen on C we say that C has been *oriented coherently* with $\mathbf{n}$. The choice of normal determines the

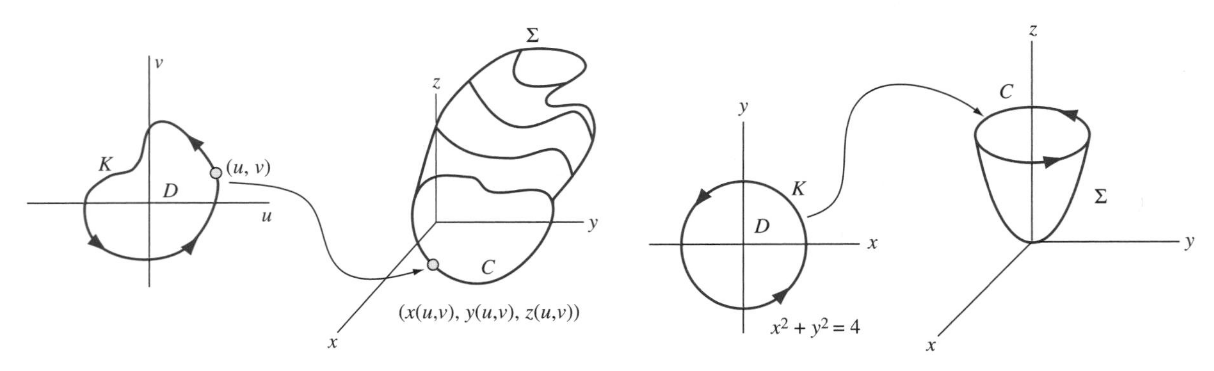

FIGURE 12.23 *Boundary curve of a surface.*

FIGURE 12.24 *The surface in Example 12.26.*

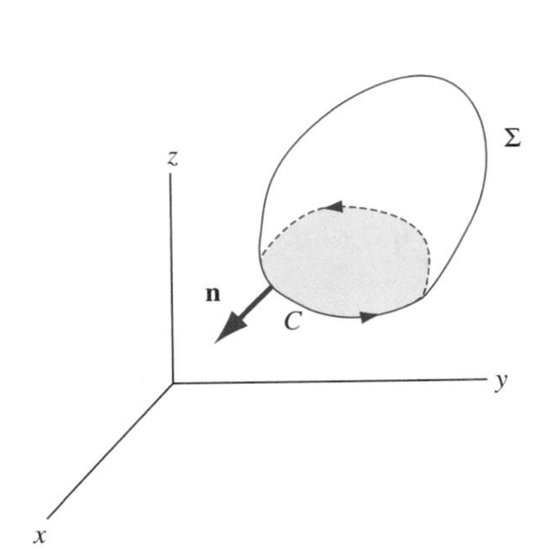

FIGURE 12.25 *Orienting the boundary curve coherently with the normal vector.*

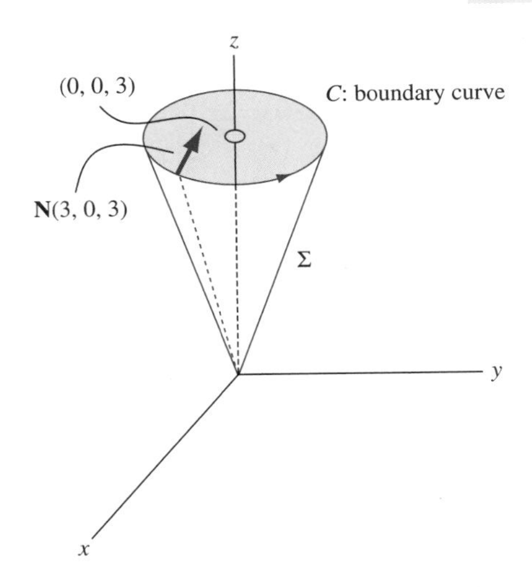

FIGURE 12.26 *The surface and boundary curve in Example 12.27.*

orientation on the boundary curve. There is no intrinsic positive or negative orientation on this curve in 3-space, simply orientation coherent with the chosen normal.

With these conventions we can state the theorem.

THEOREM 12.9 Stokes's Theorem

Let Σ be a piecewise smooth surface bounded by a piecewise smooth curve C. Suppose a unit normal $\mathbf{n}$ has been chosen on Σ and that C is oriented coherently with this normal. Let $\mathbf{F}(x, y, z)$ be a vector field that is continuous with continuous first and second partial derivatives on Σ. Then,

$$\oint_C \mathbf{F} \cdot d\mathbf{R} = \iint_\Sigma (\nabla \times \mathbf{F}) \cdot \mathbf{n}\, d\sigma. \quad \blacklozenge$$

We will illustrate the theorem with a computational example.

EXAMPLE 12.27

Let $\mathbf{F}(x, y, z) = -y\mathbf{i} + xy\mathbf{j} - xyz\mathbf{k}$. Let Σ consist of the part of the cone $z = \sqrt{x^2 + y^2}$ for $0 \leq x^2 + y^2 \leq 9$. We will compute both sides of the equation in Stokes's theorem.

The bounding curve C of Σ is the circle $x^2 + y^2 = 3, z = 3$ at the top of the cone. This is similar to Example 12.26. A routine computation yields the normal vector

$$\mathbf{N} = -\frac{x}{z}\mathbf{i} - \frac{y}{z}\mathbf{j} + \mathbf{k}.$$

There is no normal at the origin, where the cone has a sharp point.

For Stokes's theorem we need a unit normal vector, so divide $\mathbf{N}$ by $\| \mathbf{N} \|$ to get

$$\mathbf{n} = \frac{1}{\sqrt{2}z}(-x\mathbf{i} - y\mathbf{j} + \mathbf{k}).$$

Notice that this vector, if represented as an arrow from a point on the cone, points into the region bounded by the cone. Figure 12.26 shows the orientation on C coherent with $\mathbf{n}$.

We are now ready to evaluate $\oint_C \mathbf{F} \cdot d\mathbf{R}$. Parametrize C by $x = 3\cos(t)$, $y = 3\sin(t)$, $z = 3$ for $0 \leq t \leq 2\pi$. The point $(3\cos(t), 3\sin(t), 3)$ traverses C in the positive direction as t increases from 0 to 2π. Therefore

$$\begin{aligned}\oint_C \mathbf{F} \cdot d\mathbf{R} &= \oint_C -y\,dx + x\,dy - xyz\,dz \\ &= \int_0^{2\pi} [-3\sin(t)(-3\cos(t)) + 3\cos(t)(3\cos(t))]dt \\ &= \int_0^{2\pi} 9\,dt = 18\pi.\end{aligned}$$

For the surface integral, compute

$$\nabla \times \mathbf{F} = -xz\mathbf{i} + yz\mathbf{j} + 2\mathbf{k}$$

and

$$(\nabla \times \mathbf{F}) \cdot \mathbf{n} = \frac{1}{\sqrt{2}}(x^2 - y^2 + 2).$$

Then

$$\begin{aligned}\iint_\Sigma (\nabla \times \mathbf{F}) \cdot \mathbf{n}\,d\sigma &= \iint_D (\nabla \times \mathbf{F}) \cdot \mathbf{n} \,\|\mathbf{N}\|\, dx\,dy \\ &= \iint_D \frac{1}{\sqrt{2}}(x^2 - y^2 + 2)\sqrt{2}dx\,dy \\ &= \iint_D (x^2 - y^2 + 2)\,dx\,dy \\ &= \int_0^{2\pi}\int_0^3 [r^2\cos^2(\theta) - r^2\sin^2(\theta)]r\,dr\,d\theta \\ &= \int_0^{2\pi}\int_0^3 r^3\cos(2\theta)\,dr\,d\theta + \int_0^{2\pi}\int_0^3 2r\,dr\,d\theta \\ &= \left[\frac{1}{2}\sin(2\theta)\right]_0^{2\pi}\left[\frac{1}{4}r^4\right]_0^3 + 2\pi\left[r^2\right]_0^3 = 18\pi. \;\blacklozenge\end{aligned}$$

12.9.1 Potential Theory in 3-Space

In Section 12.4, we discussed conservative vector fields and used Green's theorem to derive a necessary and sufficient condition for a vector field in the plane to have a potential function. Using Stokes's theorem we can prove the three-dimensional version of this test, which was suggested in Section 12.4.

THEOREM 12.10 ***Test for a Conservative Vector Field***

Let D be a simply connected region in 3-space. Let $\mathbf{F}$ and $\nabla \times \mathbf{F}$ be continuous on D. Then $\mathbf{F}$ is conservative on D if and only if $\nabla \times \mathbf{F} = \mathbf{O}$ in D. ♦

This means that, in simply connected regions, irrotational vector fields (those with curl (rotation) zero) are exactly the fields with potential functions.

Proof In one direction the proof is simple. If $\mathbf{F} = \nabla\varphi$, then

$$\nabla \times \mathbf{F} = \nabla \times (\nabla\varphi) = \mathbf{O}.$$

For the converse, it is enough to show that, if $\mathbf{F}$ has curl zero, then $\int_C \mathbf{F} \cdot d\,\mathbf{R}$ is independent of path, since then we can define a potential function by choosing P_0 and setting

$$\varphi(x, y, z) = \int_{P_0}^{(x,y,z)} \mathbf{F} \cdot d\mathbf{R}.$$

To show this independence of path, let C and K be paths in D from P_0 to P_1. Form a closed path $L = C \bigoplus (-K)$. Since D is simply connected, there is a piecewise smooth surface Σ in D having C as boundary. By Stokes's theorem,

$$\oint_L \mathbf{F} \cdot d\mathbf{R} = \int_C \mathbf{F} \cdot d\mathbf{R} - \int_K \mathbf{F} \cdot d\mathbf{R}$$
$$= \iint_\Sigma (\nabla \times \mathbf{F}) \cdot \mathbf{n}\, d\sigma = 0. \blacklozenge$$

12.9.2 Maxwell's Equations

The theorems of Gauss and Stokes are used in the analysis of vector fields. We will illustrate this with electric and magnetic fields and Maxwell's equations. To begin, we will use the following standard notation and relationships:

$\mathbf{E}$ = electric intensity
$\mathbf{J}$ = current density
μ = permeability
Q = charge density
$\mathbf{H}$ = magnetic intensity
ϵ = permitivity of the medium
σ = conductivity
$\mathbf{D} = \epsilon \mathbf{E}$ = electric flux density
$\mathbf{B} = \mu \mathbf{H}$ = magnetic flux density

$$q = \iiint_V Q\, dV = \text{ total charge in a region } V$$

$$\varphi = \iint_\Sigma \mathbf{B} \cdot \mathbf{n}\, d\sigma = \text{ magnetic flux across } \Sigma$$

$$i = \iint_\Sigma \mathbf{J} \cdot \mathbf{n}\, d\sigma = \text{ flux of current across } \Sigma.$$

In these, flux is computed using an *outer* unit normal to the closed surface Σ.

We also have the following relationships, which have been observed and verified experimentally.

$$\text{Faraday's law } \oint_C \mathbf{E} \cdot d\mathbf{R} = -\frac{\partial \varphi}{\partial t}.$$

Here C is any piecewise smooth closed curve in the medium. We may think of this as saying that the rate of change of the magnetic flux across Σ is the negative of the measure of the tangential component of the electric intensity around any closed curve bounding Σ.

$$\text{Ampère's law } \oint \mathbf{H} \cdot d\mathbf{R} = i.$$

This says that the measure of the tangential component of the magnetic intensity about C is the current flowing through any surface bounded by C.

$$\text{Gauss's laws } \iint_{\Sigma} \mathbf{D} \cdot \mathbf{n}\, d\sigma = q \text{ and } \iint_{\Sigma} \mathbf{B} \cdot \mathbf{n}\, d\sigma = 0.$$

These say that the measure of the normal component of the electric flux density across Σ equals the total charge in the bounded region, and that the measure of the normal component of the magnetic flux density across Σ is zero.

We will now carry out arguments similar to that used to derive the heat equation using the divergence theorem. Begin by applying Stokes's theorem to Faraday's law to obtain

$$\oint_C \mathbf{E} \cdot d\mathbf{R} = \iint_{\Sigma} \nabla \times \mathbf{E} \cdot \mathbf{n}\, d\sigma = -\frac{\partial \varphi}{\partial t}$$
$$= -\frac{\partial}{\partial t} \iint_{\Sigma} \mathbf{B} \cdot \mathbf{n}\, d\sigma = \iint_{\Sigma} -\frac{\partial \mathbf{B}}{\partial t} \cdot \mathbf{n}\, d\sigma.$$

Then

$$\iint_{\Sigma} \left(\nabla \times \mathbf{E} + \frac{\partial \mathbf{B}}{\partial t} \right) \cdot \mathbf{n}\, d\sigma = 0.$$

Since this holds for any piecewise smooth closed surface Σ within the medium, then the integrand must be zero, leading to

$$\nabla \times \mathbf{E} + \frac{\partial \mathbf{B}}{\partial t} = 0.$$

A similar analysis, using Ampère's law, yields

$$\nabla \times \mathbf{H} = \mathbf{J}.$$

Maxwell had observed that

$$\mathbf{J} = \sigma \mathbf{E} + \epsilon \frac{\partial \mathbf{E}}{\partial t}.$$

Then

$$\nabla \times \mathbf{H} = \sigma \mathbf{E} + \epsilon \frac{\partial \mathbf{E}}{\partial t}.$$

Now start on a new tack. Apply Gauss's theorem to Gauss's law $q = \int\int_{\Sigma} \mathbf{D} \cdot \mathbf{n}\, d\sigma$ to obtain

$$\iint_{\Sigma} \mathbf{D} \cdot \mathbf{n}\, d\sigma = \iiint_V (\nabla \cdot \mathbf{D})\, dV = q = \iiint_V Q\, dV.$$

Again falling back on the arbitrary nature of Σ, we conclude from this that

$$\nabla \cdot \mathbf{D} = Q.$$

Now go back to $\nabla \times \mathbf{E} = -\partial \mathbf{B}/\partial t$ and take the curl of both sides:

$$\nabla \times (\nabla \times \mathbf{E}) = \nabla \times \left(-\frac{\partial \mathbf{B}}{\partial t} \right) = -\frac{\partial}{\partial t} (\nabla \times \mathbf{B}).$$

We were able to interchange ∇ and $\partial/\partial t$ here because the curl involves only the space variables. Since $\mathbf{B} = \mu \mathbf{H}$, then

$$\nabla \times (\nabla \times \mathbf{E}) = -\frac{\partial}{\partial t} (\nabla \times \mu \mathbf{H}) = -\mu \frac{\partial}{\partial t} (\nabla \times \mathbf{H}).$$

It is a routine calculation to verify that this is the same as

$$\nabla (\nabla \cdot \mathbf{E}) - (\nabla \cdot \nabla) \mathbf{E} = -\mu \frac{\partial}{\partial t} (\nabla \times \mathbf{H}).$$

In this,

$$\nabla \cdot \nabla = \frac{\partial^2}{\partial x^2} + \frac{\partial^2}{\partial y^2} + \frac{\partial^2}{\partial z^2}.$$

Since

$$\nabla \times \mathbf{H} = \sigma \mathbf{E} + \epsilon \frac{\partial \mathbf{E}}{\partial t},$$

we have finally

$$\nabla(\nabla \cdot \mathbf{E}) - (\nabla \cdot \nabla)\mathbf{E} = -\mu \frac{\partial}{\partial t}\left(\sigma \mathbf{E} + \epsilon \frac{\partial \mathbf{E}}{\partial t}\right).$$

In practice, it is often the case that $Q = 0$. In this event,

$$Q = \nabla \cdot \mathbf{D} = \nabla \cdot (c\mathbf{E}) = \epsilon \nabla \cdot \mathbf{E} = 0,$$

hence,

$$\nabla \cdot \mathbf{E} = 0.$$

We can then further conclude that

$$(\nabla \cdot \nabla)\mathbf{E} = \mu\sigma \frac{\partial \mathbf{E}}{\partial t} + \mu\epsilon \frac{\partial^2 \mathbf{E}}{\partial t^2}.$$

This is Maxwell's equation for the electric intensity field. By a similar analysis we obtain Maxwell's equation for the magnetic intensity field:

$$(\nabla \cdot \nabla)\mathbf{H} = \mu\sigma \frac{\partial \mathbf{H}}{\partial t} + \mu\epsilon \frac{\partial^2 \mathbf{H}}{\partial t^2}.$$

In the case of a perfect dielectric, $\sigma = 0$, and Maxwell's equations become

$$(\nabla \cdot \nabla)\mathbf{E} = \mu\epsilon \frac{\partial^2 (E)}{\partial t^2} \text{ and } (\nabla \cdot \nabla)\mathbf{H} = \mu\epsilon \frac{\partial^2 (H)}{\partial t^2}.$$

If, instead of $\sigma = 0$, we have $\epsilon = 0$, then we have

$$(\nabla \cdot \nabla)\mathbf{E} = \mu\sigma \frac{\partial \mathbf{E}}{\partial t} \text{ and } (\nabla \cdot \nabla)\mathbf{H} = \mu\sigma \frac{\partial \mathbf{H}}{\partial t}.$$

These are vector forms of the three-dimensional heat equation.

SECTION 12.9 PROBLEMS

In each of Problems 1 through 5, use Stokes's theorem to evaluate $\oint_C \mathbf{F} \cdot d\mathbf{R}$ or $\int\int_\Sigma (\nabla \times \mathbf{F}) \cdot \mathbf{n}\, d\sigma$, whichever appears easier.

1. $\mathbf{F} = z\mathbf{i} + x\mathbf{j} + y\mathbf{k}$ with Σ the cone $z = \sqrt{x^2 + y^2}$ for $0 \le z \le 4$.
2. $\mathbf{F} = z^2\mathbf{i} + x^2\mathbf{j} + y^2\mathbf{k}$ with Σ the part of the paraboloid $z = 6 - x^2 - y^2$ above the x, y - plane.
3. $\mathbf{F} = xy\mathbf{i} + yz\mathbf{j} + xy\mathbf{k}$ with Σ the part of the plane $2x + 4y + z = 8$ in the first octant.
4. $\mathbf{F} = xy\mathbf{i} + yz\mathbf{j} + xz\mathbf{k}$ with Σ the paraboloid $z = x^2 + y^2$ for $x^2 + y^2 \le 9$.
5. $\mathbf{F} = yx^2\mathbf{i} - xy^2\mathbf{j} + z^2\mathbf{k}$ with Σ the hemisphere $x^2 + y^2 + z^2 = 4, z \ge 0$.
6. Calculate the circulation of $\mathbf{F} = (x - y)\mathbf{i} + x^2y\mathbf{j} + axz\mathbf{k}$ counterclockwise about the circle $x^2 + y^2 = 1$. Here a is any positive number. *Hint*: Use Stokes's theorem with Σ any smooth surface having the circle as boundary.
7. Use Stokes's theorem to evaluate $\int_C \mathbf{F} \cdot \mathbf{T}\, ds$, where C is the boundary of the part of the plane $x + 4y + z = 12$ lying in the first octant, and

$$\mathbf{F} = (x - z)\mathbf{i} + (y - x)\mathbf{j} + (z - y)\mathbf{k}.$$

12.10 Curvilinear Coordinates

Thus far, we have done vector algebra and calculus in rectangular coordinates. For some settings, other coordinate systems may be more convenient. Spherical coordinates are natural when dealing with spherical surfaces, cylindrical coordinates for cylinders, and sometimes we invent systems to deal with other settings we may encounter.

Begin with the usual rectangular coordinate system with axes labeled x, y and z. Suppose we have some other coordinate system with coordinates labeled q_1, q_2 and q_3. We assume that the two systems are related by equations

$$x = x(q_1, q_2, q_3), y = y(q_1, q_2, q_3), z = z(q_1, q_2, q_3). \tag{12.11}$$

We also assume that these equations are invertible and can be solved for

$$q_1 = q_2(x, y, z), q_2 = q_2(x, y, z), q_3 = q_3(x, y, z).$$

In this way we can convert the coordinates of points back and forth from one system to the other. Finally, we assume that each point in 3-space has exactly one set of coordinates (q_1, q_2, q_3), as it does in rectangular coordinates. We call (q_1, q_2, q_3) a *system of curvilinear coordinates.*

EXAMPLE 12.28 Cylindrical Coordinates

As shown in Figure 12.27, a point P having rectangular coordinates (x, y, z) can be specified uniquely by a triple (r, θ, z), where (r, θ) are polar coordinates of the point (x, y) in the plane, and z is the same in both rectangular and cylindrical coordinates (the distance from the x, y-plane to the point).

These coordinate systems are related by

$$x = r\cos(\theta), y = r\sin(\theta), z = z$$

with $0 \le \theta < 2\pi$, $r \ge 0$ and z any real number. With some care in using the inverse function tangent function, these equations can be inverted to write

$$r = \sqrt{x^2 + y^2}, \theta = \arctan\left(\frac{y}{x}\right), z = z. \blacklozenge$$

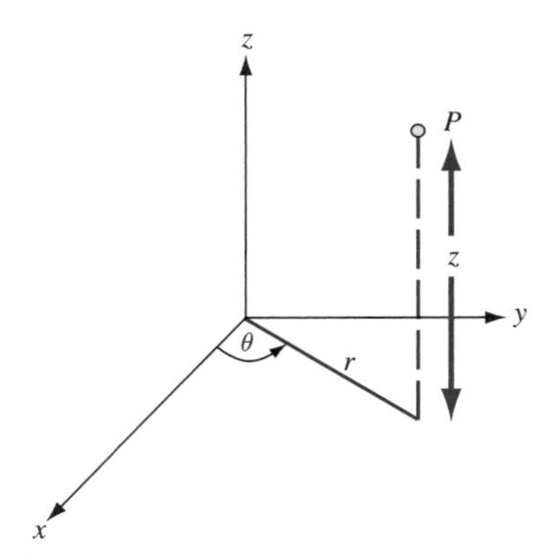

FIGURE 12.27 *Cylindrical coordinates of Example 12.28.*

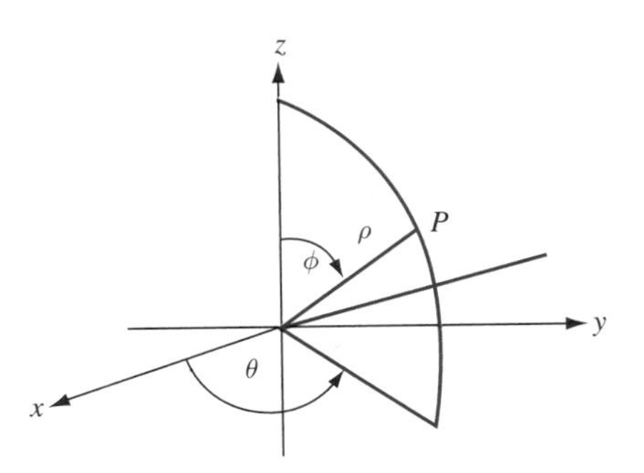

FIGURE 12.28 *Spherical coordinates of Example 12.29.*

EXAMPLE 12.29 Spherical Coordinates

Any point P having rectangular coordinates (x, y, z) also has unique spherical coordinates (ρ, θ, φ). Here ρ is the distance from the origin to P, θ is the angle of rotation from the origin to the line from $(0, 0)$ to (x, y) in the x,y-plane, and φ is the angle of declination from the positive z-axis to the line from the origin to P. These are indicated in Figure 12.28. Thus, $\rho \geq 0$, $0 \leq \theta < 2\pi$ and $0 \leq \varphi \leq \pi$.

Rectangular and spherical coordinates are related by

$$x = \rho\cos(\theta)\sin(\varphi),\ y = \rho\sin(\theta)\sin(\varphi),\ z = \rho\cos(\varphi). \quad \blacklozenge$$

Again with care in using the inverse trigonometric functions, these equations can be inverted to read

$$\rho = \sqrt{x^2 + y^2 + z^2}, \theta = \arcsin\left(\frac{y}{\sqrt{x^2 + y^2 + z^2}}\right)$$

$$\varphi = \arccos\left(\frac{y}{\sqrt{x^2 + y^2 + z^2}}\right). \quad \blacklozenge$$

The coordinate systems of these examples may appear quite dissimilar, but they share a common feature if we adopt a particular point of view. Let $P_0 : (x_0, y_0, z_0)$ be a point in rectangular coordinates. Observe that P_0 is the point of intersection of the planes $x = x_0$, $y = y_0$, and $z = z_0$, which are called *coordinate surfaces* for rectangular coordinates.

Now suppose P_0 has cylindrical coordinates (r_0, θ_0, z_0). Look at the corresponding coordinate surfaces for these coordinates. In 3-space, the surface $r = r_0$ is a cylinder of radius r_0 about the origin. The surface $\theta = \theta_0$ is a half-plane with edge on the z-axis and making an angle θ_0 with the positive x-axis. And the surface $z = z_0$ is the same as in rectangular coordinates, a plane in 3-space parallel to the x, y-plane. The point $P_0 : (r_0, \theta_0, z_0)$ is the intersection of these three cylindrical coordinate surfaces.

Spherical coordinates can be viewed in the same way. Suppose P_0 has spherical coordinates $(\rho_0, \theta_0, \varphi_0)$. The coordinate surface $\rho = \rho_0$ is a sphere of radius ρ_0 about the origin. The surface $\theta = \theta_0$ is a half-plane with one edge along the z-axis, as in cylindrical coordinates. And the surface $\varphi = \varphi_0$ is an infinite cone with vertex at the origin and making an angle φ with the z-axis. These surfaces intersect at P_0 (Figure 12.29).

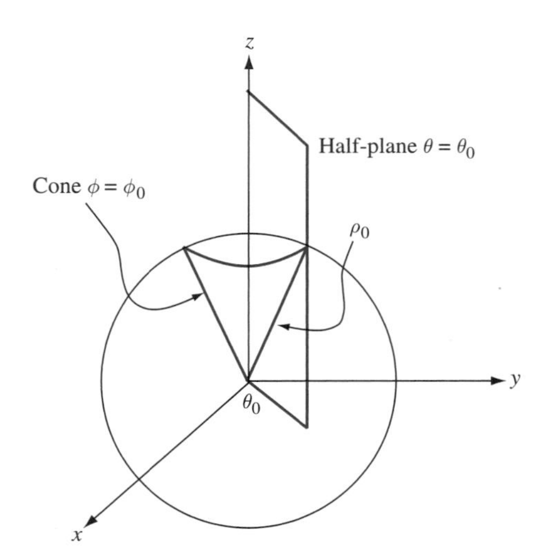

FIGURE 12.29 *Intersection of coordinate surfaces in spherical coordinates.*

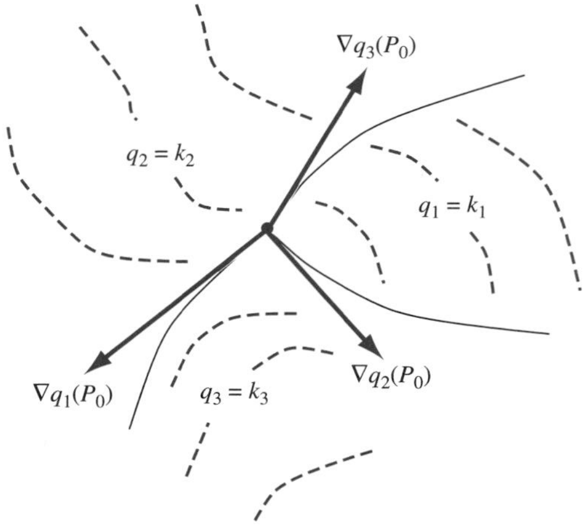

FIGURE 12.30 *Coordinate surfaces in curvilinear coordinates.*

In general curvilinear coordinates, which need not be any of these three systems, we similarly specify a point $((q_1)_0, (q_2)_0, (q_3)_0)$ as the intersection of the three coordinate surfaces $q_1 = (q_1)_0$, $q_2 = (q_2)_0$ and $q_3 = (q_3)_0$ (Figure 12.30).

In rectangular coordinates, the coordinate surfaces are planes $x = x_0$, $y = y_0$, $z = z_0$, which are mutually orthogonal. Similarly, in cylindrical and spherical coordinates, the coordinate surfaces are mutually orthogonal, in the sense that their normal vectors are mutually orthogonal at any point of intersection. Because of this, we refer to these coordinate systems as *orthogonal curvilinear coordinates*.

EXAMPLE 12.30

We will verify that cylindrical coordinates are orthogonal curvilinear coordinates. In terms of rectangular coordinates, cylindrical coordinates are given by

$$r = \sqrt{x^2 + y^2},$$

$$\theta = \arctan\left(\frac{y}{x}\right)$$

$$z = z,$$

except at the origin, which is called a *singular point* of these coordinates. Suppose P_0 is the point of intersection of the cylinder $r = r_0$, the half-plane $\theta = \theta_0$ and the half-plane $z = z_0$. To verify that these surfaces are mutually orthogonal, we will show that their normal vectors are mutually orthogonal. Compute these normal vectors using the gradient in rectangular coordinates:

$$\nabla r = \frac{1}{\sqrt{x^2 + y^2}}(x\mathbf{i} + y\mathbf{j}),$$

$$\nabla\theta = \frac{1}{x^2 + y^2}(-y\mathbf{i} + x\mathbf{j}), \nabla z = \mathbf{k}.$$

Now it is routine to verify that

$$\nabla r \cdot \nabla\theta = \nabla r \cdot \nabla z = \nabla\theta \cdot \nabla z = 0. \ \blacklozenge$$

A similar, but more complicated, calculation shows that spherical coordinates are orthogonal curvilinear coordinates.

In rectangular coordinates, the differential element ds of arc length is given by

$$(ds)^2 = (dx)^2 + (dy)^2 + (dz)^2. \tag{12.12}$$

We assume that this differential element of arc length is given in terms of the orthogonal curvilinear coordinates q_1, q_2, q_3 by the quadratic form

$$(ds)^2 = \sum_{i=1}^{3}\sum_{j=1}^{3} h_{ij}^2 dq_i dq_j.$$

The numbers h_{ij} are called *scale factors* for the curvilinear coordinate system. We want to determine these scale factors so that we can compute such quantities as arc length, area, volume, gradient, divergence and curl in curvilinear coordinates. Begin by differentiating equations (12.11):

$$dx = \frac{\partial x}{\partial q_1}dq_1 + \frac{\partial x}{\partial q_2}dq_2 + \frac{\partial x}{\partial q_3}dq_3,$$

$$dy = \frac{\partial y}{\partial q_1}dq_1 + \frac{\partial y}{\partial q_2}dq_2 + \frac{\partial y}{\partial q_3}dq_3,$$

$$dz = \frac{\partial z}{\partial q_1}dq_1 + \frac{\partial z}{\partial q_2}dq_2 + \frac{\partial z}{\partial q_3}dq_3.$$

Substitute these into equation (12.12). This is a long calculation, but after collecting the coefficients of $(dq_1)^2$, $(dq_2)^2$ and $(dq_3)^2$ (the terms in the double sum with $i = j$), and leaving the cross product terms involving $dq_i dq_j$ with $i \neq j$ within the summation notation, we obtain

$$\begin{aligned}
(ds)^2 &= \left[\left(\frac{\partial x}{\partial q_1}\right)^2 + \left(\frac{\partial y}{\partial q_1}\right)^2 + \left(\frac{\partial z}{\partial q_1}\right)^2\right](dq_1)^2 \\
&\quad + \left[\left(\frac{\partial x}{\partial q_2}\right)^2 + \left(\frac{\partial y}{\partial q_2}\right)^2 + \left(\frac{\partial z}{\partial q_2}\right)^2\right](dq_2)^2 \\
&\quad + \left[\left(\frac{\partial x}{\partial q_3}\right)^2 + \left(\frac{\partial y}{\partial q_3}\right)^2 + \left(\frac{\partial z}{\partial q_3}\right)^2\right](dq_3)^2 \\
&\quad + \sum_{i=1}^{3}\sum_{j=1, j\neq i}^{3} h_{ij} dq_i dq_j \\
&= (h_{11})^2(dq_1)^2 + (h_{22})^2(dq_2)^2 + (h_{33})^2(dq_3)^2 + \sum_{i=1}^{3}\sum_{j=1, j\neq i}^{3} h_{ij} dq_i dq_j \\
&= \sum_{i=1}^{3}\sum_{j=1}^{3} h_{ij} dq_i dq_j.
\end{aligned}$$

In this equation, equate coefficients of $(dq_i)^2$ for $i = 1, 2, 3$ to obtain

$$h_{11}^2 = \left(\frac{\partial x}{\partial q_1}\right)^2 + \left(\frac{\partial y}{\partial q_1}\right)^2 + \left(\frac{\partial z}{\partial q_1}\right)^2,$$

$$h_{22}^2 = \left(\frac{\partial x}{\partial q_2}\right)^2 + \left(\frac{\partial y}{\partial q_2}\right)^2 + \left(\frac{\partial z}{\partial q_2}\right)^2,$$

$$h_{33}^2 = \left(\frac{\partial x}{\partial q_3}\right)^2 + \left(\frac{\partial y}{\partial q_3}\right)^2 + \left(\frac{\partial z}{\partial q_3}\right)^2.$$

We left the cross product terms within the summation because all such terms are zero for orthogonal coordinates. For example,

$$h_{12}^2 = 2\left(\frac{\partial x}{\partial q_1}\frac{\partial x}{\partial q_2} + \frac{\partial y}{\partial q_1}\frac{\partial y}{\partial q_2} + \frac{\partial z}{\partial q_1}\frac{\partial z}{\partial q_2}\right)$$
$$= \nabla x(q_1, q_2, q_3) \cdot \nabla y(q_1, q_2, q_3) = 0$$

by virtue of the orthogonality of the curvilinear coordinates. Similarly, each $h_{ij} = 0$ for $i \neq j$. To simplify the notation, write $h_{ii} = h_i$ for $i = 1, 2, 3$. Finally we can write

$$(ds)^2 = (h_1 dq_1)^2 + (h_2 dq_2)^2 + (h_3 dq_3)^2. \tag{12.13}$$

with h_1, h_2, h_3 given in terms of partial derivatives of x, y, and z in terms of q_1, q_2 and q_3.

EXAMPLE 12.31

We will put these ideas into the context of cylindrical coordinates. Now $q_1 = r$, $q_2 = \theta$, and $q_3 = z$. Compute

$$h_r = \sqrt{\left(\frac{\partial x}{\partial r}\right)^2 + \left(\frac{\partial y}{\partial r}\right)^2 + \left(\frac{\partial z}{\partial r}\right)^2} = 1,$$

$$h_\theta = \sqrt{\left(\frac{\partial x}{\partial \theta}\right)^2 + \left(\frac{\partial y}{\partial \theta}\right)^2 + \left(\frac{\partial z}{\partial \theta}\right)^2} = r,$$

$$h_z = \sqrt{\left(\frac{\partial x}{\partial z}\right)^2 + \left(\frac{\partial y}{\partial z}\right)^2 + \left(\frac{\partial z}{\partial z}\right)^2} = 1.$$

In the plane, cylindrical coordinates are polar coordinates and the differential element $dx\,dy$ of area in rectangular coordinates corresponds to

$$dx\,dy = ds_1\,ds_2 = h_r h_\theta\,dr\,d\theta = r\,dr\,d\theta.$$

This accounts for the change of variables formula for transforming a double integral from rectangular to polar coordinates:

$$\iint_D f(x, y)\,dx\,dy = \iint_D f(r\cos(\theta), r\sin(\theta))r\,dr\,d\theta.$$

We can also recognize r as the Jacobian

$$\frac{\partial(x, y)}{\partial(r, \theta)} = \begin{vmatrix} \cos(\theta) & -r\sin(\theta) \\ \sin(\theta) & r\cos(\theta) \end{vmatrix} = r.$$

In 3-space,

$$dx\,dy\,dz = h_r h_\theta h_z\,dr\,d\theta\,dz = r\,dr\,d\theta\,dz.$$

This is the reason for the formula for converting a triple integral from rectangular to cylindrical coordinates:

$$\iiint_M f(x, y, z)\, dx\, dy\, dz = \iiint_M f(r\cos(\theta), r\sin(\theta), z) r\, dr\, d\theta\, dz. \ \blacklozenge$$

Again, in 3-space we can recognize the factor of r as the Jacobian

$$\frac{\partial(x, y, z)}{\partial(r, \theta, z)} = \begin{vmatrix} \cos(\theta) & -r\sin(\theta) & 0 \\ \sin(\theta) & r\cos(\theta) & 0 \\ 0 & 0 & 1 \end{vmatrix} = r. \ \blacklozenge$$

EXAMPLE 12.32

In spherical coordinates, $q_1 = \rho$, $q_2 = \theta$, and $q_3 = \varphi$. From Example 12.29 we know x, y and z in terms of ρ, θ and φ, so compute the partial derivatives to obtain

$$h_\rho = \sqrt{\left(\frac{\partial x}{\partial \rho}\right)^2 + \left(\frac{\partial y}{\partial \rho}\right)^2 + \left(\frac{\partial z}{\partial \rho}\right)^2} = 1,$$

$$h_\theta = \sqrt{\left(\frac{\partial x}{\partial \theta}\right)^2 + \left(\frac{\partial y}{\partial \theta}\right)^2 + \left(\frac{\partial z}{\partial \theta}\right)^2} = \rho\sin(\theta),$$

$$h_\varphi = \sqrt{\left(\frac{\partial x}{\partial \varphi}\right)^2 + \left(\frac{\partial y}{\partial \varphi}\right)^2 + \left(\frac{\partial z}{\partial \varphi}\right)^2} = \rho.$$

Therefore, in spherical coordinates, the differential element of arc length, squared, is

$$(ds)^2 = (d\rho)^2 + \rho^2\sin^2(\varphi)(d\theta)^2 + \rho^2(d\varphi)^2. \ \blacklozenge$$

In general, if ds_i is the differential element of arc length along the q_i axis (in the q_i direction), then

$$ds_i = h_i\, dq_i.$$

Therefore the differential elements of area are

$$ds_i ds_j = h_i h_j\, dq_i\, dq_j.$$

The differential element of volume is

$$ds_1 ds_2 ds_3 = h_1 h_2 h_3\, dq_1\, dq_2\, dq_3.$$

The formula for a differential volume element in spherical coordinates is

$$ds_\rho ds_\theta ds_\varphi = \rho^2\sin(\varphi)\, d\rho\, d\theta\, d\varphi.$$

This should look familiar. In calculus, we are told that when we convert a triple integral from rectangular to spherical coordinates we obtain

$$\iiint_M f(x, y, z)\, dx\, dy\, dz = \iiint_{M_{\rho,\theta,\varphi}} F(\rho, \theta, \varphi)\rho^2\sin(\varphi)\, d\rho\, d\theta\, d\varphi,$$

in which $F(\rho, \theta, \varphi)$ is obtained by substituting for x, y, z in terms of spherical coordinates in $f(x, y, z)$, and $M_{\rho,\theta,\varphi}$ is the region M defined in spherical coordinates. Notice that the $\rho^2\sin(\varphi)$ has shown up in the differential element of volume. That is, in terms of differentials,

$$dx\, dy\, dz = \rho^2\sin(\varphi)\, d\rho\, d\theta\, d\varphi.$$

We can also recognize $\rho^2 \sin(\varphi)$ as the Jacobian

$$\frac{\partial(x, y, z)}{\partial(\rho, \theta, \varphi)}$$

seen in the general expression for transformation of triple integrals.

Now let $\mathbf{u}_i$ be a unit vector in the direction of increasing q_i at the point $(x(q_1, q_2, q_3), y(q_1, q_2, q_3), z(q_1, q_2, q_3))$. In cylindrical coordinates, these unit vectors can be written in terms of the standard $\mathbf{i}$, $\mathbf{j}$, and $\mathbf{k}$ as

$$\mathbf{u}_r = \cos(\theta)\mathbf{i} + \sin(\theta)\mathbf{j},$$

$$\mathbf{u}_\theta = -\sin(\theta)\mathbf{i} + \cos(\theta)\mathbf{j},$$

$$\mathbf{u}_z = \mathbf{k}.$$

In spherical coordinates,

$$\mathbf{u}_\rho = \cos(\theta)\sin(\varphi)\mathbf{i} + \sin(\theta)\sin(\varphi)\mathbf{j} + \cos(\varphi)\mathbf{k},$$

$$\mathbf{u}_\theta = -\sin(\theta)\mathbf{i} + \cos(\theta)\mathbf{j},$$

$$\mathbf{u}_\varphi = \cos(\theta)\cos(\varphi)\mathbf{i} + \sin(\theta)\cos(\varphi)\mathbf{j} - \sin(\varphi)\mathbf{k}.$$

Unlike rectangular coordinates, where the standard unit vectors are constant, with orthogonal curvilinear coordinates, the vectors $\mathbf{u}_1$, $\mathbf{u}_2$, $\mathbf{u}_3$ are generally functions of the point.

A vector field in curvilinear coordinates has the form

$$\mathbf{F}(q_1, q_2, q_3) = F_1(q_1, q_2, q_3)\mathbf{u}_1 + F_2(q_1, q_2, q_3)\mathbf{u}_2 + F_3(q_1, q_2, q_3)\mathbf{u}_3.$$

We want to write expressions for the gradient, Laplacian, divergence, and curl operations in curvilinear coordinates.

Gradient

Let $\psi(q_1, q_2, q_3)$ be a scalar-valued function. At any point, we want $\nabla\psi$ to be normal to the level surface $\psi =$ constant passing through that point, and we want this gradient to have magnitude equal to the greatest rate of change of ψ from that point. Thus, the component of $\nabla\psi$ normal to $q_1 =$ constant must be $\partial\psi/\partial s_1$, or

$$\frac{1}{h_1}\frac{\partial\psi}{\partial q_1}.$$

Arguing similarly for the other components, we have

$$\nabla\psi(q_1, q_2, q_3) = \frac{1}{h_1}\frac{\partial\psi}{\partial q_1}\mathbf{u}_1 + \frac{1}{h_2}\frac{\partial\psi}{\partial q_2}\mathbf{u}_2 + \frac{1}{h_3}\frac{\partial\psi}{\partial q_3}\mathbf{u}_3.$$

Divergence

We will use the flux interpretation of divergence to obtain an expression for the divergence of a vector field in curvilinear coordinates. First write

$$\mathbf{F} = F_1\mathbf{u}_1 + F_2\mathbf{u}_2 + F_3\mathbf{u}_3.$$

Referring to Figure 12.31, the flux across the face $abcd$ is approximately

$$\mathbf{F}(q_1 + ds_1, q_2, q_3) \cdot h_2(q_1 + ds_1, q_2, q_3)h_3(q_1 + ds_1, q_2, q_3)\, dq_2 dq_3.$$

Across face $efgk$ the flux is

$$\mathbf{F}(q_1, q_2, q_3) \cdot \mathbf{u}_1 h_2(q_1, q_2, q_3)h_3(q_1, q_2, q_3)\, dq_2\, dq_q.$$

Across both of these faces the flux is approximately

$$\frac{\partial}{\partial q_1}(F_1 h_2 h_3)\, dq_1\, dq_2\, dq_3.$$

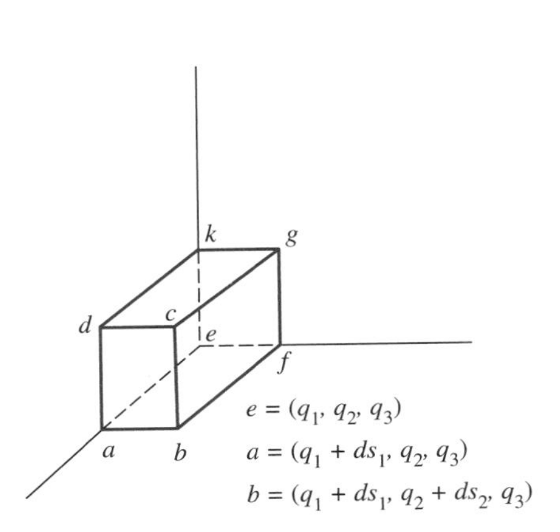

FIGURE 12.31 *Calculating the divergence in curvilinear coordinates.*

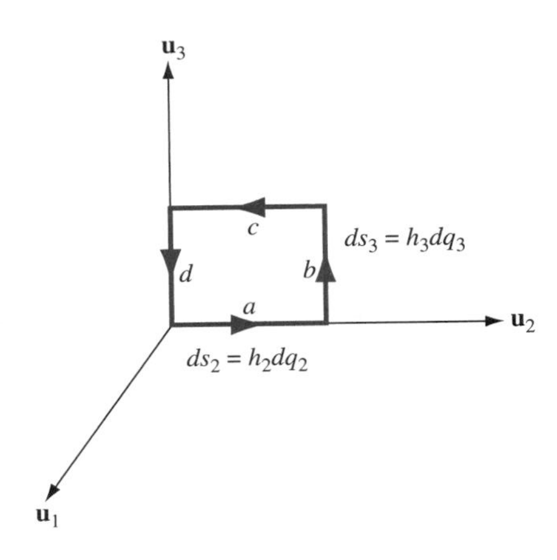

FIGURE 12.32 *Calculating the curl in curvilinear coordinates.*

Similarly, the fluxes across the other two pairs of opposite faces are

$$\frac{\partial}{\partial q_2}(F_2 h_1 h_3)\,dq_1\,dq_2\,dq_3 \text{ and } \frac{\partial}{\partial q_3}(F_3 h_1 h_2)\,dq_1\,dq_2\,dq_3.$$

We obtain the divergence, or flux per unit volume, at a point by adding these three expressions for the flux across pairs of opposite sides, and dividing by the volume $h_1 h_2 h_3\,dq_1\,dq_2\,dq_3$ to obtain

$$\begin{aligned}&\nabla\cdot\mathbf{F}(q_1,q_2,q_3)\\&=\frac{1}{h_1h_2h_3}\left(\frac{\partial}{\partial q_1}(F_1h_2h_3)\right)+\left(\frac{\partial}{\partial q_2}(F_2h_1h_3)\right)+\left(\frac{\partial}{\partial q_3}(F_3h_1h_2)\right).\end{aligned}$$

Laplacian

Knowing the divergence, we immediately have the Laplacian, since

$$\nabla^2 f = \nabla\cdot\nabla f$$

for a scalar field f. Then

$$\begin{aligned}&\nabla^2 f(q_1,q_2,q_3)=\nabla\cdot\nabla f(q_1,q_2,q_3)\\&=\frac{1}{h_1h_2h_3}\left[\frac{\partial}{\partial q_1}\left(\frac{h_2h_3}{h_1}\frac{\partial f}{\partial q_1}\right)+\frac{\partial}{\partial q_2}\left(\frac{h_1h_3}{h_2}\frac{\partial f}{\partial q_2}\right)+\frac{\partial}{\partial q_3}\left(\frac{h_1h_2}{h_3}\frac{\partial f}{\partial q_3}\right)\right].\end{aligned}$$

Curl

For the curl in curvilinear coordinates, we will use the interpretation of $(\nabla\times\mathbf{F})\cdot\mathbf{n}$ as the rotation or swirl of a fluid with velocity field $\mathbf{F}$ about a point in a plane having unit normal $\mathbf{n}$.

At P, the component of $\nabla\times\mathbf{F}$ in the direction $\mathbf{u}_1$ is

$$\lim_{A\to 0}\frac{1}{A}\int_C \mathbf{F},$$

where C may be taken as a rectangle about P in the $\mathbf{u}_2 - \mathbf{u}_3$ plane at P (Figure 12.32). Compute the integral over each side of this rectangle. On side a,

$$\int_a \mathbf{F}\approx F_2(q_1,q_2,q_3)h_2(q_1,q_2,q_3)\,dq_2,$$

since $\mathbf{u}_2$ is tangent to a. On side c, $\int_c \mathbf{F}$ is approximately

$$-F_2(q_1, q_2, q_3)h_2(q_1, q_2, q_3 + dq_3)\,dq_3.$$

The net contribution from sides a and c is approximately

$$-\frac{\partial}{\partial q_3}(F_2 h_2)\,dq_2\,dq_3.$$

Similarly, from sides b and d, the net contribution is approximately

$$\frac{\partial}{\partial q_2}(F_3 h_3)\,dq_2\,dq_3.$$

Then

$$(\nabla \times \mathbf{F}) \cdot \mathbf{u}_1 = \frac{1}{h_2 h_3\,dq_2\,dq_3}\left(\frac{\partial}{\partial q_2}(F_3 h_3) - \frac{\partial}{\partial q_3}(F_2 h_2)\right) dq_2\,dq_3.$$

Obtain the other components of $\nabla \times \mathbf{F}$ in the same way. We obtain

$$\begin{aligned}\nabla \times \mathbf{F} = &\frac{1}{h_2 h_3}\left(\frac{\partial}{\partial q_2}(F_3 h_3) - \frac{\partial}{\partial q_3}(F_2 h_2)\right)\mathbf{u}_1 \\ &+ \frac{1}{h_1 h_3}\left(\frac{\partial}{\partial q_3}(F_1 h_1) - \frac{\partial}{\partial q_1}(F_3 h_3)\right)\mathbf{u}_2 \\ &+ \frac{1}{h_1 h_2}\left(\frac{\partial}{\partial q_1}(F_2 h_2) - \frac{\partial}{\partial q_2}(F_1 h_1)\right)\mathbf{u}_3.\end{aligned}$$

This can be written in a convenient determinant form:

$$\nabla \times \mathbf{F} = \frac{1}{h_1 h_2 h_3}\begin{vmatrix} h_1\mathbf{u}_1 & h_2\mathbf{u}_2 & h_3\mathbf{u}_3 \\ \partial/\partial q_1 & \partial/\partial q_2 & \partial/\partial q_3 \\ F_1 h_1 & F_2 h_2 & F_3 h_3 \end{vmatrix}.$$

We will apply these to spherical coordinates, recalling that

$$h_\rho = 1,\ h_\theta = \rho\sin(\varphi),\ h_\varphi = \rho.$$

If

$$\mathbf{F} = F_\rho \mathbf{u}_\rho + F_\theta \mathbf{u}_\theta + F_\varphi \mathbf{u}_\varphi,$$

then divergence is given by

$$\nabla \cdot \mathbf{F} = \frac{1}{\rho^2}\frac{\partial}{\partial \rho}(\rho^2 F_\rho) + \frac{1}{\rho\sin(\varphi)}\frac{\partial}{\partial \theta}(F_\theta) + \frac{1}{\rho\sin(\varphi)}\frac{\partial}{\partial \varphi}(F_\varphi \sin(\varphi)).$$

The curl is obtained as

$$\nabla \times \mathbf{F} = \begin{vmatrix} \mathbf{u}_\rho & \rho\sin(\varphi)\mathbf{u}_\theta & \rho\mathbf{u}_\varphi \\ \partial/\partial\rho & \partial/\partial\theta & \partial/\partial\varphi \\ F_\rho & \rho\sin(\varphi)F_\theta & \rho F_\varphi \end{vmatrix}.$$

The gradient of a scalar function $f(\rho, \theta, \varphi)$ is

$$\nabla f = \frac{\partial f}{\partial \rho}\mathbf{u}_\rho + \frac{1}{\rho\sin(\varphi)}\frac{\partial f}{\partial \theta}\mathbf{u}_\theta + \frac{1}{\rho}\frac{\partial f}{\partial \varphi}\mathbf{u}_\varphi.$$

From this, we have the Laplacian

$$\nabla^2 f = \frac{1}{\rho^2}\frac{\partial}{\partial \rho}\left(\rho^2\frac{\partial f}{\partial \rho}\right) + \frac{1}{\rho^2\sin^2(\varphi)}\frac{\partial^2 f}{\partial \theta^2} + \frac{1}{\rho^2\sin(\varphi)}\frac{\partial}{\partial \varphi}\left(\sin(\varphi)\frac{\partial f}{\partial \varphi}\right).$$

The Laplacian in various coordinate systems is often encountered in connection with diffusion problems, wave motion and potential theory.

SECTION 12.10 PROBLEMS

1. Compute the scale factors for cylindrical coordinates. Use them to compute $\nabla \cdot \mathbf{F}$ and $\nabla \times \mathbf{F}$ if $\mathbf{F}(r, \theta, z)$ is a vector field in cylindrical coordinates. If $g(r, \theta, z)$ is a scalar field, compute ∇g and $\nabla^2 g$.

2. *Parabolic cylindrical coordinates* are defined by

$$x = uv,\ y = \frac{1}{2}(u^2 - v^2),\ z = z,$$

with $u \geq 0$ and v and z any real numbers.

(a) Sketch the coordinate surfaces $u =$ constant, $v =$ constant, and $z =$ constant.

(b) Determine the scale factors h_u, h_v, h_z.

(c) Determine $\nabla f(u, v, z)$ in this system.

(d) Determine $\nabla \cdot \mathbf{F}(u, v, z)$ and $\nabla \times \mathbf{F}(u, v, z)$ in this system.

(e) Determine $\nabla^2 f(u, v, z)$.

3. *Bipolar coordinates* are defined by

$$x = \frac{a\sinh(v)}{\cosh(v) - \cos(u)},\ y = \frac{a\sin(u)}{\cosh(v) - \cos(u)},\ z = z,$$

with u and z any real numbers and $0 \leq v < 2\pi$.

(a) Sketch the coordinate surfaces $u =$ constant, $v =$ constant, and $z =$ constant. Are these coordinates orthogonal?

(b) Determine the scale factors h_u, h_v, h_z.

(c) Determine $\nabla f(u, v, z)$ in this system.

(d) Determine $\nabla \cdot \mathbf{F}(u, v, z)$ and $\nabla \times \mathbf{F}(u, v, z)$ in this system.

(e) Determine $\nabla^2 f(u, v, z)$.

4. *Elliptic cylindrical coordinates* are defined by

$$x = a\cosh(u)\cos(v),\ y = a\sinh(u)\sin(v),\ z = z,$$

where $u \geq 0$, $0 \leq v < 2\pi$ and z can be any real number.

(a) Sketch the coordinate surfaces $u =$ constant, $v =$ constant, and $z =$ constant.

(b) Determine the scale factors h_u, h_v, h_z.

(c) Determine $\nabla f(u, v, z)$ in this system.

(d) Determine $\nabla \cdot \mathbf{F}(u, v, z)$ and $\nabla \times \mathbf{F}(u, v, z)$ in this system.

(e) Determine $\nabla^2 f(u, v, z)$.

PART 4

Fourier Analysis, Special Functions, and Eigen-function Expansions

CHAPTER 13 Fourier Series

WHY FOURIER SERIES? THE FOURIER SERIES OF A FUNCTION SINE AND COSINE SERIES INTEGRATION AND DIFFERENTIATION

In 1807, Joseph Fourier submitted a paper to the French Academy of Sciences in competition for a prize offered for the best mathematical treatment of heat conduction. In the course of this work Fourier shocked his contemporaries by asserting that "arbitrary" functions (such as might specify initial temperatures) could be expanded in series of sines and cosines. Consequences of Fourier's work have had an enormous impact on such diverse areas as engineering, music, medicine, and the analysis of data.

13.1 Why Fourier Series?

A Fourier series is a representation of a function as a series of constant multiples of sine and/or cosine functions of different frequencies. To see how such a series might arise, we will look at a problem of the type that concerned Fourier.

Consider a thin homogeneous bar of metal of length π, constant density and uniform cross section. Let $u(x, t)$ be the temperature in the bar at time t in the cross section at x. Then (see Section 12.8.2) u satisfies the heat equation

$$\frac{\partial u}{\partial t} = k\frac{\partial^2 u}{\partial x^2}$$

for $0 < x < \pi$ and $t > 0$. Here k is a constant depending on the material of the bar. If the left and right ends are kept at temperature zero, then

$$u(0, t) = u(\pi, t) = 0 \text{ for } t > 0.$$

These are the boundary conditions. Further, assume that the initial temperature has been specified, say

$$u(x, 0) = f(x) = x(\pi - x).$$

This is the initial condition.

Fourier found that the functions

$$u_n(x,t) = b_n \sin(nx)e^{-kn^2t}$$

satisfy the heat equation and the boundary conditions, for every positive integer n and any number b_n. However, there is no choice of n and b_n for which this function satisfies the initial condition, which would require that

$$u_n(x,0) = b_n \sin(nx) = x(\pi - x).$$

for $0 \le x \le \pi$.

We could try a finite sum of these functions, attempting a solution

$$u(x,t) = \sum_{n=1}^{N} b_n \sin(nx)e^{-kn^2t}.$$

But this would require that N and numbers $b_1, \cdots, b_N$ be found so that

$$u(x,0) = x(\pi - x) = \sum_{n=1}^{N} b_n \sin(nx)$$

for $0 \le x \le \pi$. Again, this is impossible. A finite sum of multiples of sine functions is not a polynomial.

Fourier's brilliant insight was to attempt an infinite superposition,

$$u(x,t) = \sum_{n=1}^{\infty} b_n \sin(nx)e^{-kn^2t}.$$

This function will still satisfy the heat equation and the boundary conditions $u(x,0) = u(\pi,0) = 0$. To satisfy the initial condition, the problem is to choose the numbers b_n so that

$$u(x,0) = x(\pi - x) = \sum_{n=1}^{\infty} b_n \sin(nx)$$

for $0 \le x \le \pi$. Fourier claimed not only that this could this be done, but that the right choice is

$$b_n = \frac{1}{\pi}\int_0^{\pi} x(\pi - x)\sin(nx)\,dx = \frac{4}{\pi}\frac{1-(-1)^n}{n^3}.$$

With these coefficients, Fourier wrote the solution for the temperature function:

$$u(x,t) = \frac{4}{\pi}\sum_{n=1}^{\infty}\frac{1-(-1)^n}{n^3}\sin(nx)e^{-kn^2t}.$$

The astonishing claim that

$$x(\pi - x) = \frac{4}{\pi}\sum_{n=1}^{\infty}\frac{1-(-1)^n}{n^3}\sin(nx)$$

for $0 \le x \le \pi$ was too much for Fourier's contemporaries to accept, and the absence of rigorous proofs in his paper led the Academy to reject its publication (although they awarded him the prize). However, the implications of Fourier's work were not lost on natural philosophers of his time. If Fourier was right, then many functions would have expansions as infinite series of trigonometric functions.

Although Fourier did not have the means to supply the rigor his colleagues demanded, this was provided throughout the ensuing century and Fourier's ideas are now seen in many important applications. We will use them to solve partial differential equations, beginning in Chapter 16. This and the next two chapters develop the requisite ideas from Fourier analysis.

SECTION 13.1 PROBLEMS

1. Let $p(x)$ be a polynomial. Prove that there is no finite sum $\sum_{n=1}^{N} b_n \sin(nx)$ that is equal to $p(x)$ for $0 \le x \le \pi$, for any choice of the numbers $b_1, \cdots, b_N$.

2. Let $p(x)$ be a polynomial. Prove that there is no number k such that $p(x) = k\sin(nx)$ on $[0, \pi]$ for any positive integer n.

3. Let

$$S_N(x) = \frac{4}{\pi}\sum_{n=1}^{N} \frac{1-(-1)^n}{n^3}\sin(nx).$$

Construct graphs of $S_N(x)$ and $x(\pi - x)$, for $0 \le x \le \pi$, for $N = 2$ and then $N = 10$. This will give some sense of the correctness of Fourier's claim that this polynomial could be exactly represented by the infinite series

$$\frac{4}{\pi}\sum_{n=1}^{\infty} \frac{1-(-1)^n}{n^3}\sin(nx)$$

on $[0, \pi]$.

13.2 The Fourier Series of a Function

Let $f(x)$ be defined on $[-L, L]$. We want to choose numbers $a_0, a_1, a_2 \cdots$ and $b_1, b_2, \cdots$ so that

$$f(x) = \frac{1}{2}a_0 + \sum_{k=1}^{\infty}[a_k\cos(k\pi x/L) + b_k\sin(k\pi x/L)]. \tag{13.1}$$

This is a decomposition of the function into a sum of terms, each representing the influence of a different fundamental frequency on the behavior of the function.

To determine a_0, integrate equation (13.1) term by term to get

$$\begin{aligned}\int_{-L}^{L} f(x)dx =& \frac{1}{2}\int_{-L}^{L} a_0 dx \\ &+ \sum_{k=1}^{\infty}\left(a_k\int_{-L}^{L}\cos(k\pi x/L)\,dx + b_k\int_{-L}^{L}\sin(k\pi x/L)\,dx\right) \\ =& \frac{1}{2}a_0(2L) = \pi a_0.\end{aligned}$$

because all of the integrals in the summation are zero. Then

$$a_0 = \frac{1}{L}\int_{-L}^{L} f(x)\,dx. \tag{13.2}$$

To solve for the other coefficients in the proposed equation (13.1), we will use the following three facts, which follow by routine integrations. Let m and n be integers. Then

$$\int_{-L}^{L}\cos(n\pi x/L)\sin(m\pi x/L)\,dx = 0. \tag{13.3}$$

Furthermore, if $n \ne m$, then

$$\int_{-L}^{L}\cos(n\pi x/L)\cos(m\pi x/L)\,dx = \int_{-L}^{L}\sin(n\pi x/L)\sin(m\pi x/L)\,dx = 0. \tag{13.4}$$

And, if $n \neq 0$, then

$$\int_{-L}^{L} \cos^2(n\pi x/L)\,dx = \int_{-L}^{L} \sin^2(n\pi x/L)\,dx = L. \tag{13.5}$$

Now let n be any positive integer. To solve for a_n, multiply equation (13.1) by $\cos(n\pi x/L)$ and integrate the resulting equation to get

$$\int_{-L}^{L} f(x)\cos(n\pi x/L)\,dx = \frac{1}{2}a_0 \int_{-L}^{L} \cos(n\pi x/L)\,dx$$
$$+\sum_{k=1}^{\infty}\left[a_k \int_{-L}^{L} \cos(k\pi x/L)\cos(n\pi x/L)\,dx + b_k \int_{-L}^{L} \sin(k\pi x/L)\cos(n\pi x/L)\,dx\right].$$

Because of equations (13.3) and (13.4), all of the terms on the right are zero except the coefficient of a_n, which occurs in the summation when $k = n$. The last equation reduces to

$$\int_{-L}^{L} f(x)\cos(n\pi x/L)\,dx = a_n \int_{-L}^{L} \cos^2(n\pi x/L)\,dx = a_n L$$

by equation (13.5). Therefore

$$a_n = \frac{1}{L}\int_{L}^{L} f(x)\cos(n\pi x/L)\,dx. \tag{13.6}$$

This expression contains a_0 if we let $n = 0$.

Similarly, if we multiply equation (13.1) by $\sin(n\pi x/L)$ instead of $\cos(n\pi x/L)$ and integrate, we obtain

$$b_n = \frac{1}{L}\int_{-L}^{L} f(x)\sin(n\pi x/L)\,dx. \tag{13.7}$$

The numbers

$$a_n = \frac{1}{L}\int_{-L}^{L} f(x)\cos(n\pi x/L)\,dx \text{ for } n = 0, 1, 2, \cdots \tag{13.8}$$

$$b_n = \frac{1}{L}\int_{-L}^{L} f(x)\sin(n\pi x/L)\,dx \text{ for } n = 1, 2, \cdots \tag{13.9}$$

are called the *Fourier coefficients of f on* $[L, L]$. When these numbers are used, the series (13.1) is called the *Fourier series of f on* $[L, L]$.

EXAMPLE 13.1

Let $f(x) = x - x^2$ for $-\pi \le x \le \pi$. Here $L = \pi$. Compute

$$a_0 = \frac{1}{\pi}\int_{-\pi}^{\pi} (x - x^2)\,dx = -\frac{2}{3}\pi^2,$$

$$a_n = \frac{1}{\pi}\int_{-\pi}^{\pi} (x - x^2)\cos(nx)dx$$
$$= \frac{4\sin(n\pi) - 4n\pi\cos(n\pi) - 2n^2\pi^2\sin(n\pi)}{\pi n^3}$$

$$= -\frac{4}{n^2}\cos(n\pi) = -\frac{4}{n^2}(-1)^n$$

$$= \frac{4(-1)^{n+1}}{n^2},$$

and

$$b_n = \frac{1}{\pi}\int_{-\pi}^{\pi}(x - x^2)\sin(nx)\,dx$$

$$= \frac{2\sin(n\pi) - 2n\pi\cos(n\pi)}{\pi n^2}$$

$$= -\frac{2}{n}\cos(n\pi) = -\frac{2}{n}(-1)^n$$

$$= \frac{2(-1)^{n+1}}{n}.$$

We have used the facts that $\sin(n\pi) = 0$ and $\cos(n\pi) = (-1)^n$ if n is an integer.

The Fourier series of $f(x) = x - x^2$ on $[-\pi, \pi]$ is

$$-\frac{1}{3}\pi^2 + \sum_{n=1}^{\infty}\left[\frac{4(-1)^{n+1}}{n^2}\cos(nx) + \frac{2(-1)^{n+1}}{n}\sin(nx)\right]. \ \blacklozenge$$

This example illustrates a fundamental issue. We do not know what this Fourier series converges to. We need something that establishes a relationship between the function and its Fourier series on an interval. This will require some assumptions about the function.

Recall that f is *piecewise continuous* on $[a, b]$ if f is continuous at all but perhaps finitely many points of this interval, and, at a point where the function is not continuous, f has finite limits at the point from within the interval. Such a function has at worst jump discontinuities, or finite gaps in the graph, at finitely many points. Figure 13.1 shows a typical piecewise continuous function.

If $a < x_0 < b$, denote the left limit of $f(x)$ at x_0 as $f(x_0-)$, and the right limit of $f(x)$ at x_0 as $f(x_0+)$:

$$f(x_0-) = \lim_{h\to 0+} f(x_0 - h) \text{ and } f(x_0+) = \lim_{h\to 0+} f(x_0 + h).$$

If f is continuous at x_0, then these left and right limits both equal $f(x_0)$.

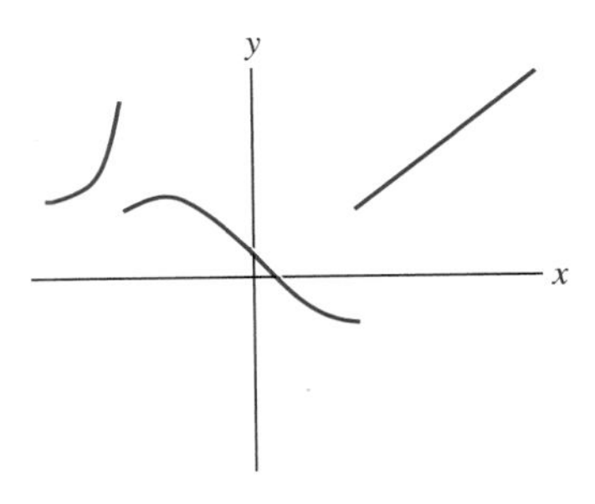

FIGURE 13.1 *A piecewise continuous function.*

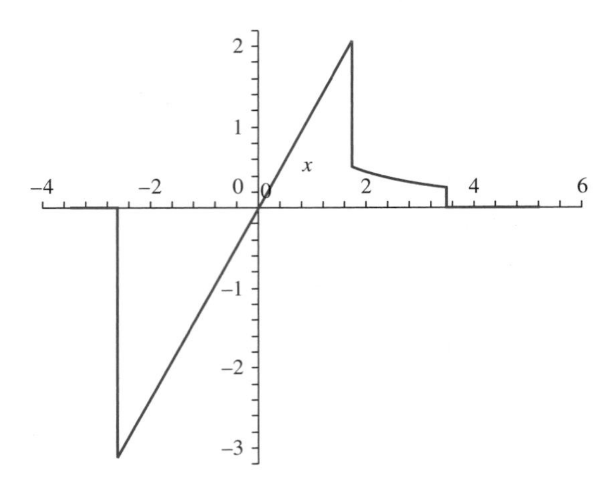

FIGURE 13.2 f in Example 13.2.

EXAMPLE 13.2

Let

$$f(x) = \begin{cases} x & \text{for } -3 \leq x < 2, \\ 1/x & \text{for } 2 \leq x \leq 4. \end{cases}$$

f is piecewise continuous on $[-3, 4]$, having a single discontinuity at $x = 2$. Furthermore, $f(2-) = 2$ and $f(2+) = 1/2$. A graph of f is shown in Figure 13.2. ♦

f is *piecewise smooth* on $[a, b]$ if f is piecewise continuous and f' exists and is continuous at all but perhaps finitely many points of (a, b).

EXAMPLE 13.3

The function f of Example 13.2 is differentiable on $(-3, 4)$ except at $x = 2$:

$$f'(x) = \begin{cases} 1 & \text{for } -3 < x < 2 \\ -1/x^2 & \text{for } 2 < x < 4. \end{cases}$$

This derivative is itself piecewise continuous. Therefore f is piecewise smooth on $[-3, 4]$. ♦

We can now state a convergence theorem.

THEOREM 13.1 Convergence of Fourier Series

Let f be piecewise smooth on $[-L, L]$. Then, for each x in $(-L, L)$, the Fourier series of f on $[-L, L]$ converges to

$$\frac{1}{2}(f(x+) + f(x-)).$$

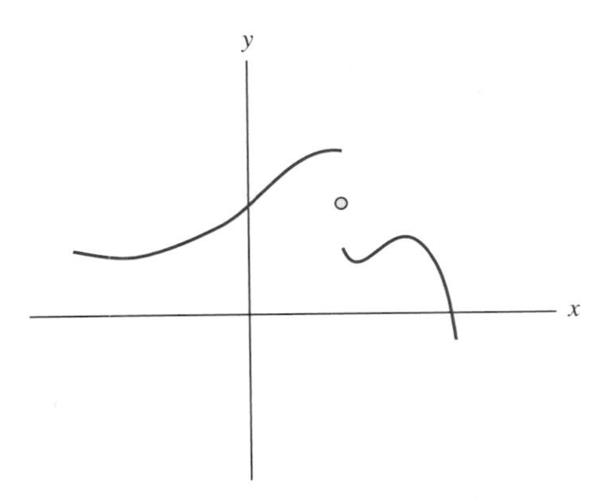

FIGURE 13.3 *Convergence of a Fourier series at a jump discontinuity.*

At both $-L$ and L this Fourier series converges to

$$\frac{1}{2}(f(L-)+f(-L+)). \blacklozenge$$

At any point in $(-L, L)$ at which $f(x)$ is continuous, the Fourier series converges to $f(x)$, because then the right and left limits at x are both equal to $f(x)$. At a point interior to the interval where f has a jump discontinuity, the Fourier series converges to the average of the left and right limits there. This is the point midway in the gap of the graph at the jump discontinuity (Figure 13.3). The Fourier series has the same sum at both ends of the interval.

EXAMPLE 13.4

Let $f(x)=x-x^2$ for $-\pi \leq x \leq \pi$. In Example 13.1 we found the Fourier series of f on $[-\pi, \pi]$. Now we can examine the relationship between this series and $f(x)$.

$f'(x)=1-2x$ is continuous for all x, hence f is piecewise smooth on $[-\pi, \pi]$. For $-\pi < x < \pi$, the Fourier series converges to $x-x^2$. At both π and $-\pi$, the Fourier series converges to

$$\frac{1}{2}(f(\pi-)+f(-\pi+)) = \frac{1}{2}((\pi-\pi^2)+(-\pi-(-\pi)^2))$$
$$= \frac{1}{2}(-2\pi^2) = -\pi^2.$$

Figures 13.4, 13.5, and 13.6 show the fifth, tenth and twentieth partial sums of this Fourier series, together with a graph of f for comparison. The partial sums are seen to approach the function as more terms are included. ♦

EXAMPLE 13.5

Let $f(x)=e^x$. The Fourier coefficients of f on $[-1, 1]$ are

$$a_0 = \int_{-1}^{1} e^x \, dx = e - e^{-1},$$

$$a_n = \int_{-1}^{1} e^x \cos(n\pi x) \, dx = \frac{(e-e^{-1})(-1)^n}{1+n^2\pi^2},$$

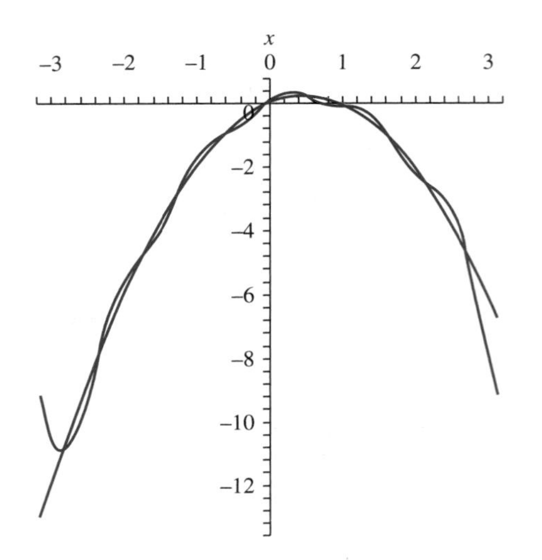

FIGURE 13.4 *Fifth partial sum of the Fourier series in Example 13.4.*

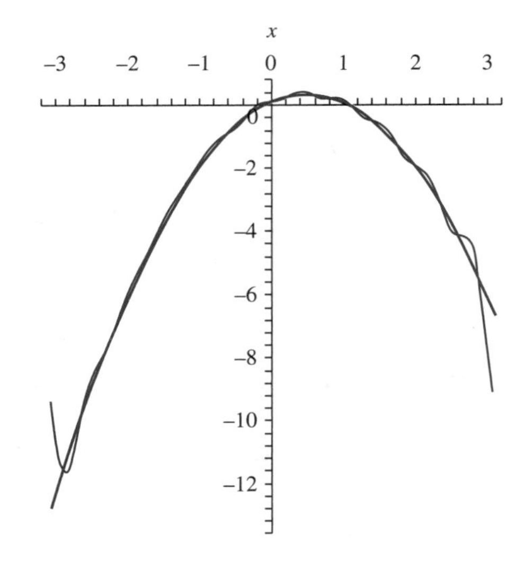

FIGURE 13.5 *Tenth partial sum in Example 13.4.*

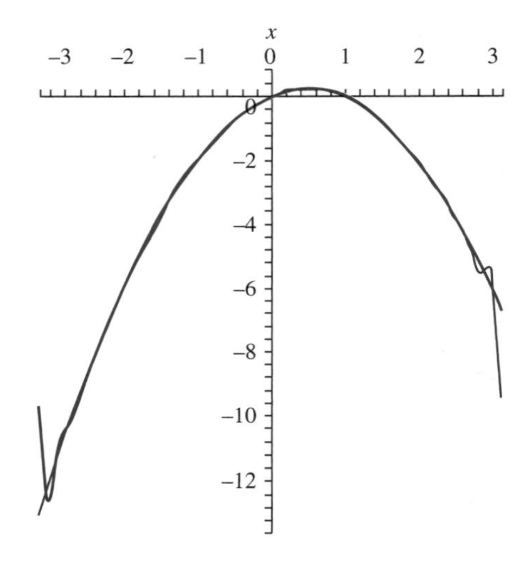

FIGURE 13.6 *Twentieth partial sum in Example 13.4.*

and

$$b_n = \int_{-1}^{1} e^x \sin(n\pi x)\,dx = -\frac{(e-e^{-1})(-1)^n(n\pi)}{1+n^2\pi^2}.$$

The Fourier series of e^x on $[-1, 1]$ is

$$\frac{1}{2}(e-e^{-1}) + (e-e^{-1})\sum_{n=1}^{\infty}\left(\frac{(-1)^n}{1+n^2\pi^2}\right)(\cos(n\pi x) - n\pi\sin(n\pi x)).$$

Because e^x is continuous with a continuous derivative for all x, this series converges to

$$\begin{cases} e^x & \text{for } -1 < x < 1 \\ \frac{1}{2}(e+e^{-1}) & \text{for } x = 1 \text{ and for } x = -1. \end{cases}$$

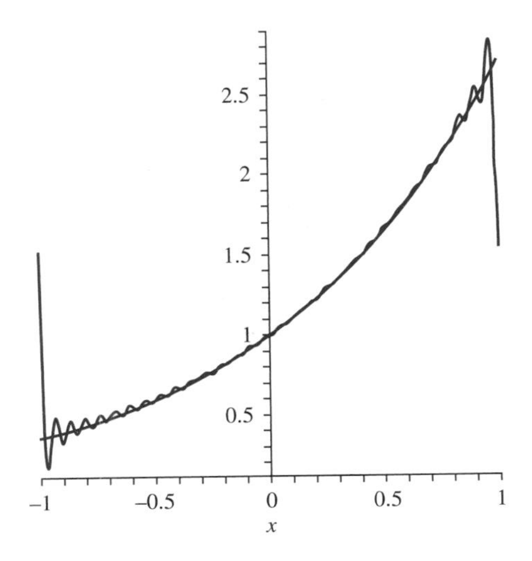

FIGURE 13.7 *Thirtieth partial sum of the Fourier series in Example 13.5.*

Figure 13.7 shows the thirtieth partial sum of this series, suggesting its convergence to the function except at the endpoints -1 and 1. ♦

EXAMPLE 13.6

Let

$$f(x) = \begin{cases} 5\sin(x) & \text{for } -2\pi \le x < -\pi/2 \\ 4 & \text{for } x = -\pi/2 \\ x^2 & \text{for } -\pi/2 < x < 2 \\ 8\cos(x) & \text{for } 2 \le x < \pi \\ 4x & \text{for } \pi \le x \le 2\pi. \end{cases}$$

f is piecewise smooth on $[-2\pi, 2\pi]$. The Fourier series of f on $[-2\pi, 2\pi]$ converges to:

$$\begin{cases} 5\sin(x) & \text{for } -2\pi < x < -\pi/2 \\ \frac{1}{2}\left(\frac{\pi^2}{4} - 5\right) & \text{for } x = -\pi/2 \\ x^2 & \text{for } -\pi/2 < x < 2 \\ \frac{1}{2}(4 + 8\cos(2)) & \text{for } x = 2 \\ 8\cos(x) & \text{for } 2 < x < \pi \\ \frac{1}{2}(4\pi - 8) & \text{for } x = \pi \\ 4x & \text{for } -\pi < x < 2\pi \\ 4\pi & \text{for } x = 2\pi \text{ and } x = -2\pi. \end{cases}$$

This conclusion does not require that we write the Fourier series. ♦

13.2.1 Even and Odd Functions

A function f is *even* on $[-L, L]$ if its graph on $[-L, 0]$ is the reflection across the vertical axis of the graph on $[0, L]$. This happens when $f(-x) = f(x)$ for $0 < x \leq L$. For example, x^{2n} and $\cos(n\pi x/L)$ are even on any $[-L, L]$ for any positive integer n.

A function f is *odd* on $[-L, L]$ if its graph on $[-L, 0)$ is the reflection through the origin of the graph on $(0, L]$. This means that f is odd when $f(-x) = -f(x)$ for $0 < x \leq L$. For example, x^{2n+1} and $\sin(n\pi x/L)$ are odd on $[-L, L]$ for any positive integer n.

Figures 13.8 and 13.9 show typical even and odd functions, respectively.

A product of two even functions is even, a product of two odd functions is even, and a product of an odd function with an even function is odd.

If f is even on $[-L, L]$ then

$$\int_{-L}^{L} f(x)\,dx = 2\int_{0}^{L} f(x)\,dx$$

and if f is odd on $[-L, L]$ then

$$\int_{-L}^{L} f(x)\,dx = 0.$$

These facts are sometimes useful in computing Fourier coefficients. If f is even, only the cosine terms and possibly the constant term will appear in the Fourier series, because $f(x)\sin(n\pi x/L)$ is odd and the integrals defining the sine coefficients will be zero. If the function is odd then $f(x)\cos(n\pi x/L)$ is odd and the Fourier series will contain only the sine terms, since the integrals defining the constant term and the coefficients of the cosine terms will be zero.

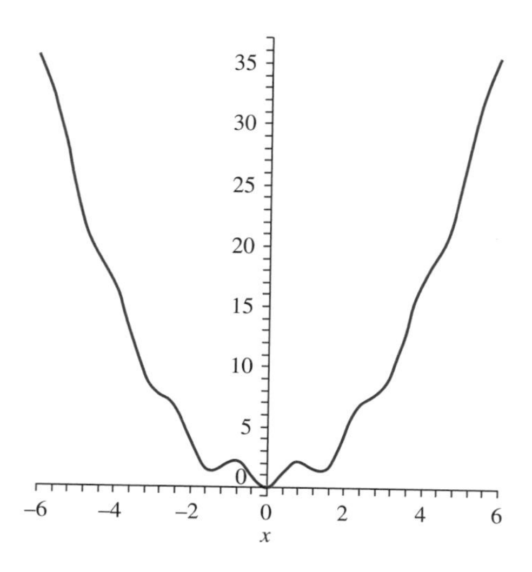

FIGURE 13.8 *A typical even function.*

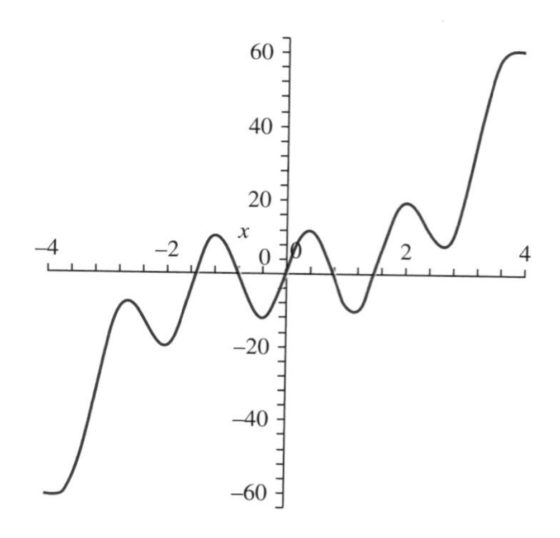

FIGURE 13.9 *A typical odd function.*

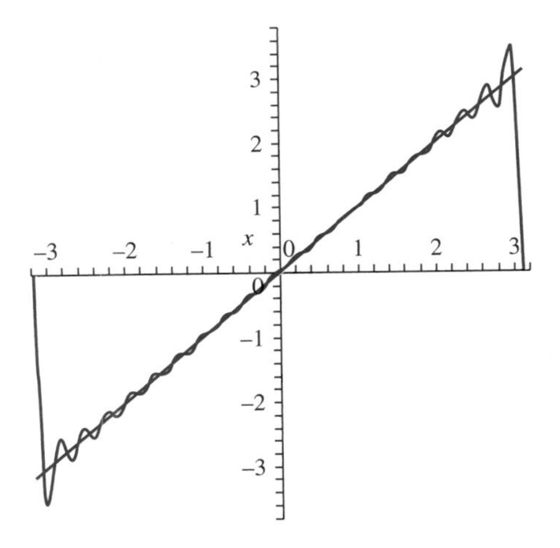

FIGURE 13.10 *Twentieth partial sum of the series in Example 13.7.*

EXAMPLE 13.7

We will compute the Fourier series of $f(x) = x$ on $[-\pi, \pi]$.

Because $x\cos(nx)$ is an odd function on $[-\pi, \pi]$ for $n = 0, 1, 2, \cdots$, each $a_n = 0$. We need only compute the b_n's:

$$\begin{aligned} b_n &= \frac{1}{\pi}\int_{-\pi}^{\pi} x\sin(nx)\,dx \\ &= \frac{2}{\pi}\int_0^{\pi} x\sin(nx)\,dx \\ &= \left[\frac{2}{n^2\pi}\sin(nx) - \frac{2x}{n\pi}\cos(nx)\right]_0^{\pi} \\ &= -\frac{2}{n}\cos(n\pi) = \frac{2}{n}(-1)^{n+1}. \end{aligned}$$

The Fourier series of x on $[-\pi, \pi]$ is

$$\sum_{n=1}^{\infty} \frac{2}{n}(-1)^{n+1}\sin(nx).$$

This converges to x for $-\pi < x < \pi$, and to 0 at $x = \pm\pi$. Figure 13.10 shows the twentieth partial sum of this Fourier series compared to the function. ♦

EXAMPLE 13.8

We will write the Fourier series of x^4 on $[-1, 1]$. Since $x^4\sin(n\pi x)$ is an odd function on $[-1, 1]$ for $n = 1, 2, \cdots$, each $b_n = 0$. Compute

$$a_0 = 2\int_0^1 x^4 dx = \frac{2}{5}$$

and

$$\begin{aligned} a_n &= 2\int_0^1 x^4\cos(n\pi x)dx \\ &= 2\left[\frac{(n\pi x)^4\sin(n\pi x) + 4(n\pi x)^3\cos(n\pi x)}{(n\pi)^5}\right]_0^1 \end{aligned}$$

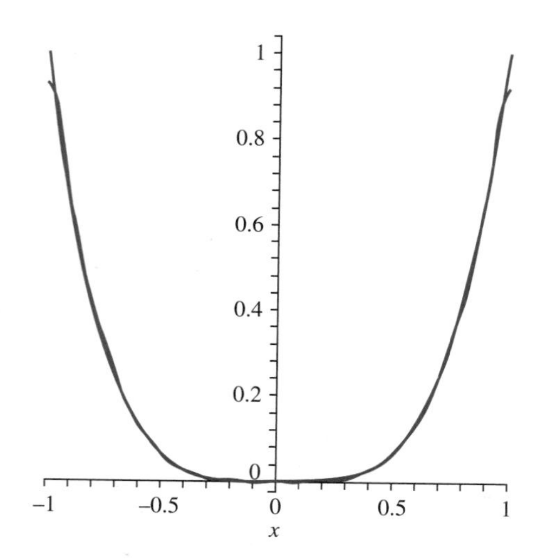

FIGURE 13.11 *Tenth partial sum of the series of Example 13.8.*

$$+2\left[\frac{-12(n\pi x)^2\sin(n\pi x)-24n\pi x\cos(n\pi x)+24\sin(n\pi x)}{(n\pi)^5}\right]_0^1$$
$$=8\frac{n^2\pi^2-6}{n^4\pi^4}(-1)^n.$$

The Fourier series is

$$\frac{1}{5}+\sum_{n=1}^{\infty}\left(\frac{8(-1)^n}{n^4\pi^4}(n^2\pi^2-6)\right)\cos(n\pi x).$$

By Theorem 13.1, this series converges to x^4 for $-1\le x\le 1$. Figure 13.11 shows the twentieth partial sum of this series, compared with the function. ◆

13.2.2 The Gibbs Phenomenon

A.A. Michelson was a Prussian-born physicist who teamed with E.W. Morley of Case-Western Reserve University to show that the postulated "ether," a fluid which supposedly permeated all of space, had no effect on the speed of light. Michelson also built a mechanical device for constructing a function from its Fourier coefficients. In one test, he used eighty coefficients for the series of $f(x)=x$ on $[-\pi,\pi]$ and noticed unexpected jumps in the graph near the endpoints. At first, he thought this was a problem with his machine. It was subsequently found that this behavior is characteristic of the Fourier series of a function at a point of discontinuity. In the early twentieth century, the Yale mathematician Josiah Willard Gibbs finally explained this behavior.

To illustrate the Gibbs phenomenon, expand f in a Fourier series on $[-\pi,\pi]$, where

$$f(x)=\begin{cases}-\pi/4 & \text{for } -\pi\le x<0\\ 0 & \text{for } x=0\\ \pi/4 & \text{for } 0<x\le\pi.\end{cases}$$

This function has a jump discontinuity at 0, but its Fourier series on $[-\pi,\pi]$ converges at 0 to

$$\frac{1}{2}(f(0+)+f(0-))=\frac{1}{2}\left(\frac{\pi}{4}-\frac{\pi}{4}\right)=0=f(0).$$

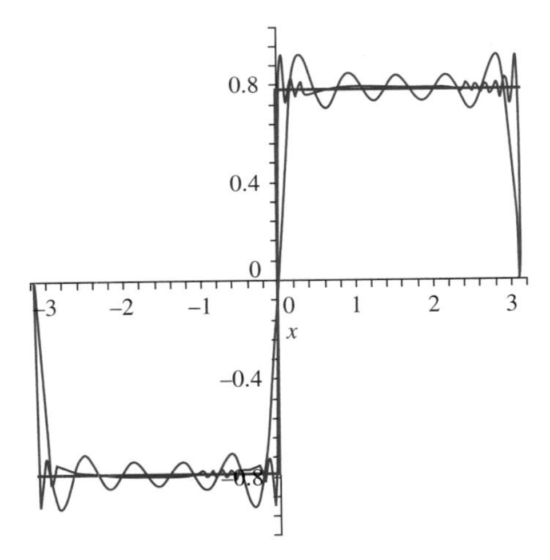

FIGURE 13.12 *The Gibbs phenomenon.*

The Fourier series therefore converges to the function on $(-\pi, \pi)$. This series is

$$\sum_{n=1}^{\infty} \frac{1}{2n-1} \sin((2n-1)x).$$

Figure 13.12 shows the fifth and twenty-fifth partial sums of this series, compared to the function. Notice that both of these partial sums show a peak near 0, the point of discontinuity of f. Since the partial sums S_N approach the function as $N \to \infty$, we might expect these peaks to flatten out, but they do not. Instead they remain roughly the same height, but move closer to the y-axis as N increases. This is the Gibbs phenomenon.

Postscript We will add two comments on the ideas of this section, the first practical and the second offering a broader perspective.

1. Writing and graphing partial sums of Fourier series are computation intensive activities. Evaluating integrals for the coefficients is most efficiently done in MAPLE using the int command, and partial sums are easily graphed using the sum command to enter the partial sum and then the plot command for the graph.
2. Partial sums of Fourier series can be viewed from the perspective of orthogonal projections onto a subspace of a vector space (Sections 6.6 and 6.7). Let $PC[-L, L]$ be the vector space of functions that are piecewise continuous on $[-L, L]$ and let S be the subspace spanned by the functions

$$C_0(x) = 1, C_n(x) = \cos(n\pi x/L) \text{ and } S_n(x) = \sin(n\pi x/L) \text{ for } n = 1, 2, \cdots, N.$$

A dot product can be defined on $PL[-L, L]$ by

$$f \cdot g = \int_{-L}^{L} f(x)g(x)\,dx.$$

Using this dot product, these functions form an orthogonal basis for S. If f if piecewise continuous on $[-L, L]$, the orthogonal projection of f onto S is

$$f_S = \frac{f \cdot C_0}{C_0 \cdot C_0} C_0 + \sum_{n=1}^{N} \left(\frac{f \cdot C_n}{C_n \cdot C_n} C_n + \frac{f \cdot S_n}{S_n \cdot S_n} S_n \right).$$

Compare these coefficients in the orthogonal projection with the Fourier coefficients of f on $[-L, L]$. First

$$\frac{f \cdot C_0}{C_0 \cdot C_0} = \frac{\int_{-L}^{L} f(x)\,dx}{\int_{-L}^{L} dx}$$
$$= \frac{1}{2L}\int_{-L}^{L} f(x)\,dx = \frac{1}{2}a_0.$$

Next,

$$\frac{f \cdot C_n}{C_n \cdot C_n} = \frac{\int_{-L}^{L} f(x)\cos(n\pi x/L)\,dx}{\int_{-L}^{L} \cos^2(n\pi x/L)\,dx}$$
$$= \frac{1}{L}\int_{-L}^{L} f(x)\cos(n\pi x/L)\,dx = a_n,$$

and similarly,

$$\frac{f \cdot S_n}{S_n \cdot S_n} = \frac{1}{L}\int_{-L}^{L} f(x)\sin(n\pi x/L)\,dx = b_n.$$

Thus, the orthogonal projection f_S of f onto S is exactly the Nth partial sum of the Fourier series of f on $[-L, L]$.

This broader perspective of Fourier series will provide a unifying theme when we consider general eigenfunction expansions in Chapter 15.

SECTION 13.2 PROBLEMS

In each of Problems 1 through 12, write the Fourier series of the function on the interval and determine the sum of the Fourier series. Graph some partial sums of the series, compared with the graph of the function.

1. $f(x) = \cos(x),\ -3 \le x \le 3$
2. $f(x) = \begin{cases} 1 - x & \text{for } -1 \le x \le 0 \\ 0 & \text{for } 0 < x \le 1 \end{cases}$
3. $f(x) = 4,\ -3 \le x \le 3$
4. $f(x) = \cos(x/2) - \sin(x),\ -\pi \le x \le \pi$
5. $f(x) = \cosh(\pi x),\ -1 \le x \le 1$
6. $f(x) = \begin{cases} -x & \text{for } -5 \le x < 0 \\ 1 + x^2 & \text{for } 0 \le x \le 5 \end{cases}$
7. $f(x) = \begin{cases} -4 & \text{for } -\pi \le x \le 0 \\ 4 & \text{for } 0 < x \le \pi \end{cases}$
8. $f(x) = \sin(2x),\ -\pi \le x \le \pi$
9. $f(x) = x^2 - x + 3,\ -2 \le x \le 2$
10. $f(x) = 1 - |x|,\ -2 \le x \le 2$
11. $f(x) = \begin{cases} 1 & \text{for } -\pi \le x < 0 \\ 2 & \text{for } 0 \le x \le \pi \end{cases}$
12. $f(x) = -x,\ -1 \le x \le 1$

In each of Problems 13 through 19, use the convergence theorem to determine the sum of the Fourier series of the function on the interval. It is not necessary to write the series to do this.

13. $f(x) = \begin{cases} -1 & \text{for } -4 \le x < 0 \\ 1 & \text{for } 0 \le x \le 4 \end{cases}$
14. $f(x) = \begin{cases} \cos(x) & \text{for } -2 \le x < 0 \\ \sin(x) & \text{for } 0 \le x \le 2 \end{cases}$
15. $f(x) = \begin{cases} -2 & \text{for } -4 \le x \le -2 \\ 1 + x^2 & \text{for } -2 < x \le 2 \\ 0 & \text{for } 2 < x \le 4 \end{cases}$

16. $f(x) = \begin{cases} 0 & \text{for } -1 \le x < 1/2 \\ 1 & \text{for } 1/2 \le x \le 3/4 \\ 2 & \text{for } 3/4 < x \le 1 \end{cases}$

17. $f(x) = \begin{cases} x^2 & \text{for } -\pi \le x \le 0 \\ 2 & \text{for } 0 < x \le \pi \end{cases}$

18. $f(x) = \begin{cases} 2x - 2 & \text{for } -\pi \le x \le 1 \\ 3 & \text{for } 1 < x \le -\pi \end{cases}$

19. $f(x) = \begin{cases} 2x & \text{for } -3 \le x < -2 \\ 0 & \text{for } -2 \le x < 1 \\ x^2 & \text{for } 1 \le x \le 3 \end{cases}$

20. Using Problem 18, write the Fourier series of the function and plot some partial sums, pointing out the occurrence of the Gibbs phenomenon at the points of discontinuity of the function.

21. Carry out the program of Problem 20 for the function of Problem 14.

13.3 Sine and Cosine Series

If f is piecewise continuous on $[-L, L]$, we can represent $f(x)$ at all but possibly finitely many points of $[-L, L]$ by its Fourier series. This series may contain just sine terms, just cosine terms, or both sine and cosine terms. We have no control over this.

If f is defined on the half interval $[0, L]$, we can write a Fourier cosine series (containing just cosine terms) and a Fourier sine series (containing just sine terms) for f on $[0, L]$.

13.3.1 Cosine Series

Suppose $f(x)$ is defined for $0 \le x \le L$. To get a pure cosine series on this interval, imagine reflecting the graph of f across the vertical axis to obtain an even function g defined on $[-L, L]$ (see Figure 13.13).

Because g is even, its Fourier series on $[-L, L]$ has only cosine terms and perhaps the constant term. But $g(x) = f(x)$ for $0 \le x \le L$, so this gives a cosine series for f on $[0, L]$. Furthermore, because g is even, the coefficients in the Fourier series of g on $[-L, L]$ are

$$\begin{aligned} a_n &= \frac{1}{L}\int_{-L}^{L} g(x)\cos(n\pi x/L)\,dx \\ &= \frac{2}{L}\int_{0}^{L} g(x)\cos(n\pi x/L)\,dx \\ &= \frac{2}{L}\int_{0}^{L} f(x)\cos(n\pi x/L)\,dx. \end{aligned}$$

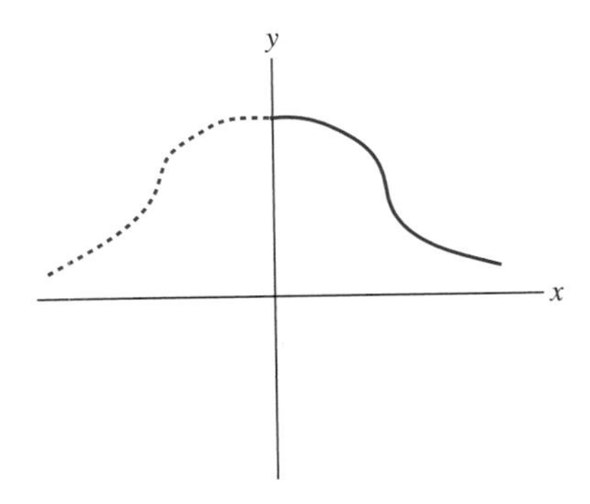

FIGURE 13.13 *Even extension of a function defined on* $[0, L]$.

for $n = 0, 1, 2, \cdots$. Notice that we can compute a_n strictly in terms of f on $[0, L]$. The construction of g showed us how to obtain this cosine series for f, but we do not need g to compute the coefficients of this series.

Based on these ideas, define the *Fourier cosine coefficients of* f *on* $[0, L]$ to be the numbers

$$a_n = \frac{2}{L}\int_0^L f(x)\cos(n\pi x/L)\,dx \tag{13.10}$$

for $n = 0, 1, 2, \cdots$. The *Fourier cosine series for* f *on* $[0, L]$ is the series

$$\frac{1}{2}a_0 + \sum_{n=1}^{\infty} a_n \cos(n\pi x/L) \tag{13.11}$$

in which the a_n's are the Fourier cosine coefficients of f on $[0, L]$.

By applying Theorem 13.1 to g, we obtain the following convergence theorem for cosine series on $[0, L]$.

THEOREM 13.2 Convergence of Fourier Cosine Series

Let f be piecewise smooth on $[0, L]$. Then

1. If $0 < x < L$, the Fourier cosine series for f on $[0, L]$ converges to

$$\frac{1}{2}(f(x+) + f(x-)).$$

2. At 0 this cosine series converges to $f(0+)$.
3. At L this cosine series converges to $f(L-)$. ♦

EXAMPLE 13.9

Let $f(x) = e^{2x}$ for $0 \le x \le 1$. We will write the cosine expansion of f on $[0, 1]$. The coefficients are

$$a_0 = 2\int_0^1 e^{2x}\,dx = e^2 - 1$$

and for $n = 1, 2, \cdots$,

$$\begin{aligned} a_n &= 2\int_0^1 e^{2x}\cos(n\pi x)\,dx \\ &= \left[\frac{4e^{2x}\cos(n\pi x) + 2n\pi e^{2x}\sin(n\pi x)}{4 + n^2\pi^2}\right]_0^1 \\ &= 4\frac{e^2(-1)^n - 1}{4 + n^2\pi^2}. \end{aligned}$$

The cosine series for e^{2x} on $[0, 1]$ is

$$\frac{1}{2}(e^2-1)+\sum_{n=1}^{\infty}4\frac{e^2(-1)^n-1}{4+n^2\pi^2}\cos(n\pi x).$$

This series converges to

$$\begin{cases} e^{2x} & \text{for } 0<x<1 \\ 1 & \text{for } x=0 \\ e^2 & \text{for } x=1. \end{cases}$$

This Fourier cosine series converges to e^{2x} for $0 \le x \le 1$. Figure 13.14 shows e^{2x} and the fifth partial sum of this cosine series. ◆

13.3.2 Sine Series

We can also write an expansion of f on $[0, L]$ that contains only sine terms. Now reflect the graph of f on $[0, L]$ through the origin to create an odd function h on $[-L, L]$, with $h(x) = f(x)$ for $0 < x \le L$ (Figure 13.15).

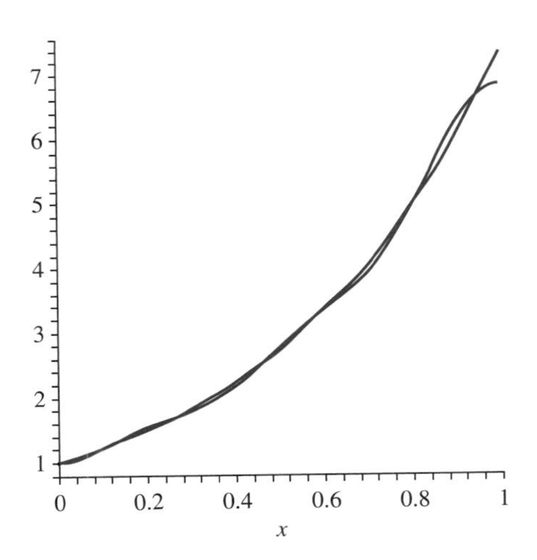

FIGURE 13.14 *Fifth partial sum of the cosine series in Example 13.9.*

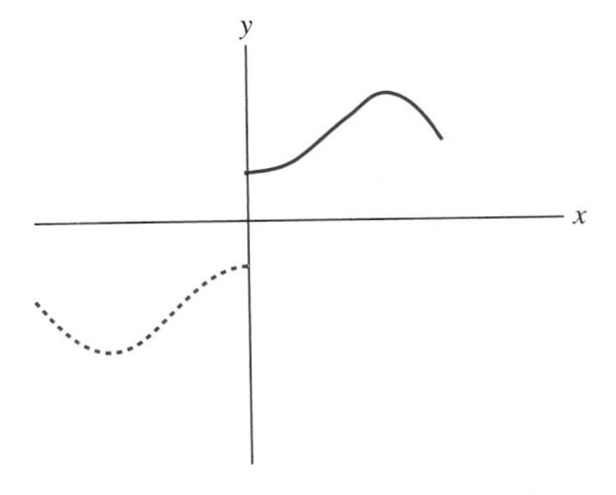

FIGURE 13.15 *Odd extension of a function defined on* $[0, L]$.

The Fourier expansion of h on $[-L, L]$ has only sine terms because h is odd on $[-L, L]$. But $h(x) = f(x)$ on $[0, L]$, so this gives a sine expansion of f on $[0, L]$. This suggests the following definitions.

The *Fourier sine coefficients of* f *on* $[0, L]$ are

$$b_n = \frac{2}{L}\int_0^L f(x)\sin(n\pi x/L)\,dx. \tag{13.12}$$

for $n = 1, 2, \cdots$. With these coefficients, the series

$$\sum_{n=1}^{\infty} b_n \sin(n\pi x/L) \tag{13.13}$$

is the *Fourier sine series for* f *on* $[0, L]$.

Again, we have a convergence theorem for sine series directly from the convergence theorem for Fourier series.

THEOREM 13.3 Convergence of Fourier Sine Series

Let f be piecewise smooth on $[0, L]$. Then

1. If $0 < x < L$, the Fourier sine series for f on $[0, L]$ converges to
$$\frac{1}{2}(f(x+) + f(x-)).$$
2. At $x = 0$ and $x = L$, this sine series converges to 0. ♦

Condition (2) is obvious because each sine term in the series vanishes at $x = 0$ and at $x = L$, regardless of the values of the function there.

EXAMPLE 13.10

Let $f(x) = e^{2x}$ for $0 \le x \le 1$. We will write the Fourier sine series of f on $[0, 1]$. The coefficients are

$$\begin{aligned} b_n &= 2\int_0^1 e^{2x}\sin(n\pi x)\,dx \\ &= \left[\frac{-2n\pi e^{2x}\cos(n\pi x) + 4e^{2x}\sin(n\pi x)}{4 + n^2\pi^2}\right]_0^1 \\ &= 2\frac{n\pi(1 - (-1)^n e^2)}{4 + n^2\pi^2}. \end{aligned}$$

The sine expansion of e^{2x} on $[0, 1]$ is

$$\sum_{n=1}^{\infty} 2\frac{n\pi(1 - (-1)^n e^2)}{4 + n^2\pi^2}\sin(n\pi x).$$

This series converges to e^{2x} for $0 < x < 1$ and to 0 at $x = 0$ and $x = 1$. Figure 13.16 shows the function and the fortieth partial sum of this sine expansion. ♦

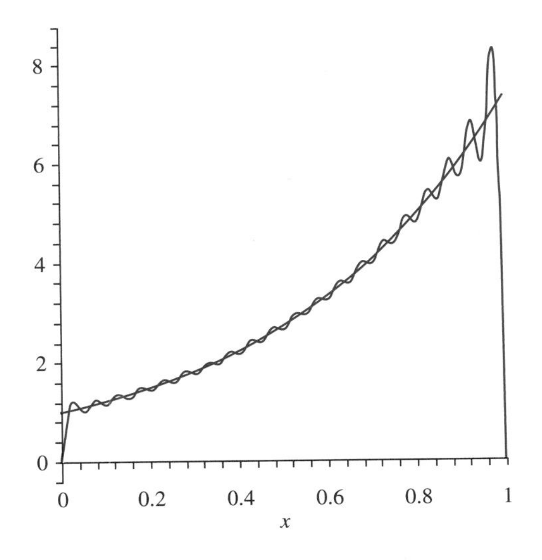

FIGURE 13.16 *Fortieth partial sum of the sine series of Example 13.10.*

SECTION 13.3 PROBLEMS

In each of Problems 1 through 10, write the Fourier cosine and sine series for f on the interval. Determine the sum of each series. Graph some of the partial sums of these series.

1. $f(x) = \begin{cases} x^2 & \text{for } 0 \le x < 1 \\ 1 & \text{for } 1 \le x \le 4 \end{cases}$
2. $f(x) = 1 - x^3, 0 \le x \le 2$
3. $f(x) = x^2, 0 \le x \le 2$
4. $f(x) = \begin{cases} 1 & \text{for } 0 \le x < 1 \\ 0 & \text{for } 1 \le x \le 3 \\ -1 & \text{for } 3 < x \le 5 \end{cases}$
5. $f(x) = \begin{cases} x & \text{for } 0 \le x \le 2 \\ 2 - x & \text{for } 2 < x \le 3 \end{cases}$
6. $f(x) = 2x, 0 \le x \le 1$
7. $f(x) = 4, 0 \le x \le 3$
8. $f(x) = \begin{cases} 1 & \text{for } 0 \le x \le 1 \\ -1 & \text{for } 1 < x \le 2 \end{cases}$
9. $f(x) = \begin{cases} 0 & \text{for } 0 \le x \le \pi \\ \cos(x) & \text{for } \pi < x \le 2\pi \end{cases}$
10. $f(x) = e^{-x}, 0 \le x \le 1$
11. Determine all functions on $[-L, L]$ that are both even and odd.
12. Let $f(x)$ be defined on $[-L, L]$. Prove that f can be written as a sum of an even function and an odd function on this interval.
13. Sum the series $\sum_{n=1}^{\infty} (-1)^n/(4n^2 - 1)$. *Hint*: Expand $\sin(x)$ in a cosine series on $[0, \pi]$ and choose an appropriate value of x.

13.4 Integration and Differentiation of Fourier Series

Term by term differentiation of a Fourier series may lead to nonsense.

EXAMPLE 13.11

Let $f(x) = x$ for $-\pi \le x \le \pi$. The Fourier series is

$$f(x) = x = \sum_{n=1}^{\infty} \frac{2}{n}(-1)^{n+1} \sin(nx)$$

for $-\pi < x < \pi$. Differentiate this series term by term to get

$$\sum_{n=1}^{\infty} 2(-1)^{n+1} \cos(nx).$$

This series does not converge on $(-\pi, \pi)$. ♦

However, under fairly mild conditions, we can integrate a Fourier series term by term.

THEOREM 13.4 ***Integration of Fourier Series***

Let f be piecewise continuous on $[-L, L]$, with Fourier series

$$\frac{1}{2}a_0 + \sum_{n=1}^{\infty}(a_n \cos(n\pi x/L) + b_n \sin(n\pi x/L)).$$

Then, for any x with $-L \leq x \leq L$,

$$\int_{-L}^{x} f(t)\,dt = \frac{1}{2}a_0(x+L)$$
$$+ \frac{L}{\pi}\sum_{n=1}^{\infty}\frac{1}{n}\left[a_n \sin(n\pi x/L) - b_n(\cos(n\pi x/L) - (-1)^n)\right]. \text{ ♦}$$

The expression on the right in this equation is exactly what we obtain by integrating the Fourier series term by term, from $-L$ to x. This means that we can integrate any piecewise continuous function f from $-L$ to x by integrating its Fourier series term by term. This is true even if the function has jump discontinuities and its Fourier series does not converge to $f(x)$ for all x in $[-L, L]$. Notice, however, that the result of this integration is not a Fourier series.

EXAMPLE 13.12

From Example 13.11,

$$f(x) = x = \sum_{n=1}^{\infty}\frac{2}{n}(-1)^{n+1}\sin(nx)$$

on $(-\pi, \pi)$. Term by term differentiation results in a series that does not converge on the interval. However, we can integrate this series term by term:

$$\int_{-\pi}^{x} t\,dt = \frac{1}{2}(x^2 - \pi^2)$$
$$= \sum_{n=1}^{\infty}\frac{2}{n}(-1)^{n+1}\int_{-\pi}^{x}\sin(nt)\,dt$$
$$= \sum_{n=1}^{\infty}\frac{2}{n}(-1)^{n+1}\left[-\frac{1}{n}\cos(nx) + \frac{1}{n}\cos(n\pi)\right]$$
$$= \sum_{n=1}^{\infty}\frac{2}{n}(-1)^{n+1}(\cos(nx) - (-1)^n). \text{ ♦}$$

Proof Define

$$F(x) = \int_{-L}^{x} f(t)\,dt - \frac{1}{2}a_0 x$$

for $-L \le x \le L$. Then F is continuous on $[-L, L]$ and

$$F(-L) = F(L) = \frac{1}{2}a_0 L.$$

Furthermore,

$$F'(x) = f(x) - \frac{1}{2}a_0$$

at every point on $(-L, L)$ at which f is continuous. Therefore $F'(x)$ is piecewise continuous on $[-L, L]$, and the Fourier series of $F(x)$ converges to $F(x)$ on this interval. This series is

$$F(x) = \frac{1}{2}A_0 + \sum_{n=1}^{\infty}\left(A_n \cos\left(\frac{n\pi x}{L}\right) + B_n \sin\left(\frac{n\pi x}{L}\right)\right).$$

Here we obtain, using integration by parts,

$$\begin{aligned}
A_n &= \frac{1}{L}\int_{-L}^{L} F(t)\cos\left(\frac{n\pi x}{L}\right) dt \\
&= \frac{1}{L}\left[F(t)\frac{L}{n\pi}\sin\left(\frac{n\pi x}{L}\right)\right]_{-L}^{L} - \frac{1}{L}\int_{-L}^{L}\frac{L}{n\pi}\sin\left(\frac{n\pi x}{L}\right)F'(t)\,dt \\
&= -\frac{1}{n\pi}\int_{-L}^{L}\left(f(t) - \frac{1}{2}a_0\right)\sin\left(\frac{n\pi x}{L}\right) dt \\
&= -\frac{1}{n\pi}\int_{-L}^{L} f(t)\sin\left(\frac{n\pi x}{L}\right) dt + \frac{1}{2n\pi}a_0\int_{-L}^{L}\sin\left(\frac{n\pi x}{L}\right) dt \\
&= -\frac{L}{n\pi}b_n.
\end{aligned}$$

Similarly,

$$B_n = \frac{1}{L}\int_{-L}^{L} F(t)\sin\left(\frac{n\pi x}{L}\right) dt = \frac{L}{n\pi}a_n,$$

where the a_n's and b_n's are the Fourier coefficients of $f(x)$ on $[-L, L]$. Therefore the Fourier series of $F(x)$ has the form

$$F(x) = \frac{1}{2}A_0 + \frac{L}{\pi}\sum_{n=1}^{\infty}\left(\frac{1}{n}\right)\left(-b_n\cos\left(\frac{n\pi x}{L}\right) + a_n\sin\left(\frac{n\pi x}{L}\right)\right)$$

for $-L \le x \le L$. To determine A_0, write

$$\begin{aligned}
F(L) &= \frac{L}{2}a_0 - \frac{L}{\pi}\sum_{n=1}^{\infty} b_n\cos(n\pi) \\
&= \frac{1}{2}A_0 - \frac{L}{\pi}\sum_{n=1}^{\infty}\left(\frac{1}{n}\right)b_n(-1)^n.
\end{aligned}$$

This gives us

$$A_0 = La_0 + \frac{2L}{\pi}\sum_{n=1}^{\infty}\left(\frac{1}{n}\right)b_n(-1)^n.$$

Upon substituting these expressions for A_0, A_n, and B_n into the Fourier series for $F(x)$, we obtain the conclusion of the theorem. ♦

Valid term by term differentiation of a Fourier series requires stronger conditions.

THEOREM 13.5 Differentiation of Fourier Series

Let f be continuous on $[-L, L]$ and suppose that $f(-L) = f(L)$. Let f' be piecewise continuous on $[-L, L]$. Then the Fourier series of f on $[-L, L]$ converges to $f(x)$ on $[-L, L]$:

$$f(x) = \frac{1}{2}a_0 + \sum_{n=1}^{\infty}(a_n \cos(n\pi x/L) + b_n \sin(n\pi x/L))$$

for $-L \leq x \leq L$. Further, at each x in $(-L, L)$ at which $f''(x)$ exists, the term by term derivative of the Fourier series converges to the derivative of the function:

$$f'(x) = \sum_{n=1}^{\infty}\frac{n\pi}{L}(-a_n \sin(n\pi x/L) + b_n \cos(n\pi x/L)).$$ ♦

The idea of a proof of this theorem is to begin with the Fourier series for $f'(x)$, noting that this series converges to $f'(x)$ at each point where f'' exists. Use integration by parts to relate the Fourier coefficients of $f'(x)$ to those for $f(x)$, similar to the strategy used in proving Theorem 13.4,

EXAMPLE 13.13

Let $f(x) = x^2$ for $-2 \leq x \leq 2$. By the Fourier convergence theorem,

$$x^2 = \frac{4}{3} + \frac{16}{\pi^2}\sum_{n=1}^{\infty}\frac{(-1)^{n+1}}{n^2}\cos(n\pi x/2)$$

for $-2 \leq x \leq 2$. Only cosine terms appear in this series because x^2 is an even function. Now, $f'(x) = 2x$ is continuous and f is twice differentiable for all x. Therefore, for $-2 < x < 2$,

$$f'(x) = 2x = \frac{8}{\pi}\sum_{n=1}^{\infty}\frac{(-1)^{n+1}}{n}\sin(n\pi x/2).$$

This can be verified by expanding $2x$ in a Fourier series on $[-2, 2]$. ♦

Fourier coefficients, and Fourier sine and cosine coefficients, satisfy an important set of inequalities called *Bessel's inequalities*.

THEOREM 13.6 Bessel's Inequalities

Suppose $\int_0^L g(x)\,dx$ exists.

1. The Fourier sine coefficients b_n of $g(x)$ on $[0, L]$ satisfy

$$\sum_{n=1}^{\infty} b_n^2 \leq \frac{2}{L}\int_0^L (g(x))^2\,dx.$$

2. The Fourier cosine coefficients a_n of $g(x)$ on $[0, L]$ satisfy

$$\frac{1}{2}a_0^2 + \sum_{n=1}^{\infty} a_n^2 \le \frac{2}{L}\int_0^L (g(x))^2\,dx.$$

3. If $\int_{-L}^{L} g(x)\,dx$ exists, then the Fourier coefficients of $f(x)$ on $[-L, L]$ satisfy

$$\frac{1}{2}a_0^2 + \sum_{n=1}^{\infty}(a_n^2 + b_n^2) \le \frac{1}{L}\int_{-L}^{L} (g(x))^2\,dx. \blacklozenge$$

In particular, the sum of the squares of the coefficients in a Fourier series (or cosine or sine series) converges.

We will prove conclusion (1). The argument is notationally simpler than that for conclusions (2) and (3), but contains the ideas involved.

Proof of (1) The Fourier sine series of $g(x)$ on $[0, L]$ is

$$\sum_{n=1}^{\infty} b_n \sin\left(\frac{n\pi x}{L}\right),$$

where

$$b_n = \frac{2}{L}\int_0^L g(x)\sin\left(\frac{n\pi x}{L}\right)dx.$$

The Nth partial sum of this sine series is

$$S_N(x) = \sum_{n=1}^{N} b_n \sin\left(\frac{n\pi x}{L}\right).$$

Then

$$\begin{aligned}
0 &\le \int_0^L (g(x) - S_N(x))^2\,dx \\
&= \int_0^L (g(x))^2\,dx - 2\int_0^L g(x)S_N(x)\,dx + \int_0^L (S_N(x))^2\,dx \\
&= \int_0^L (g(x))^2\,dx - 2\int_0^L g(x)\left(\sum_{n=1}^{N} b_n \sin\left(\frac{n\pi x}{L}\right)\right)dx \\
&\quad + \int_0^L \left(\sum_{n=1}^{N} b_n \sin\left(\frac{n\pi x}{L}\right)\right)\left(\sum_{k=1}^{N} b_k \sin\left(\frac{k\pi x}{L}\right)\right)dx \\
&= \int_0^L (g(x))^2\,dx - 2\sum_{n=1}^{N} b_n \int_0^L g(x)\sin\left(\frac{n\pi x}{L}\right)dx \\
&\quad + \sum_{n=1}^{N}\sum_{k=1}^{N} b_n b_k \int_0^L \sin\left(\frac{n\pi x}{L}\right)\sin\left(\frac{k\pi x}{L}\right)dx \\
&= \int_0^L (g(x))^2\,dx - \sum_{n=1}^{N} b_n(Lb_n) + \frac{L}{2}\sum_{n=1}^{N} b_n b_n.
\end{aligned}$$

Here we have used the fact that

$$\int_0^L \sin\left(\frac{n\pi x}{L}\right)\sin\left(\frac{k\pi x}{L}\right)dx = \begin{cases} 0 & \text{for } n \ne k, \\ L/2 & \text{for } n = k. \end{cases}$$

We therefore have

$$0 \leq \int_0^L (g(x))^2\,dx - L\sum_{n=1}^{N} b_n^2 + \frac{L}{2}\sum_{n=1}^{N} b_n^2,$$

and this gives us

$$\sum_{n=1}^{N} b_n^2 \leq \frac{2}{L}\int_0^L (f(x))^2\,dx.$$

Since this is true for all positive integers N, we can let $N \to \infty$ to obtain

$$\sum_{n=1}^{\infty} b_n^2 \leq \frac{2}{L}\int_0^L (f(x))^2\,dx. \;\blacklozenge$$

EXAMPLE 13.14

We will use Bessel's inequality to derive an upper bound for the sum of a series. Let $f(x) = x^2$ on $[-\pi, \pi]$. The Fourier series is

$$f(x) = \frac{1}{3}\pi^2 + \sum_{n=1}^{\infty} 4\frac{(-1)^n}{n^2}\cos(nx)$$

for $-\pi \leq x \leq \pi$. Here $a_0 = 2\pi^2/3$ and $a_n = 4(-1)^n/n^2$, while each $b_n = 0$ (x^2 is an even function). By Bessel's inequality,

$$\frac{1}{2}\left(\frac{2\pi}{3}\right)^2 + \sum_{n=1}^{\infty}\left(\frac{4(-1)^n}{n^2}\right)^2 \leq \frac{1}{\pi}\int_{-\pi}^{\pi} x^4\,dx = \frac{2}{5}\pi^4.$$

Then

$$16\sum_{n=1}^{\infty}\frac{1}{n^4} \leq \left(\frac{2}{5} - \frac{2}{9}\right)\pi^4 = \frac{8\pi^4}{45}.$$

Then

$$\sum_{n=1}^{\infty}\frac{1}{n^4} \leq \frac{\pi^4}{90},$$

which is approximately 1.0823. Infinite series are generally difficult to sum, so it is sometimes useful to be able to derive an upper bound. ♦

With stronger assumptions than just existence of the integral over the interval, we can derive an important equality satisfied by the Fourier coefficients of a function on $[-L, L]$ or by the Fourier sine or cosine coefficients of a function on $[0, L]$. We will state the result for $f(x)$ defined on $[-L, L]$.

THEOREM 13.7 Parseval's Theorem

Let f be continuous on $[-L, L]$ and let f' be piecewise continuous. Suppose that $f(-L) = f(L)$. Then the Fourier coefficients of f on $[-L, L]$ satisfy

$$\frac{1}{2}a_0^2 + \sum_{n=1}^{\infty}(a_n^2 + b_n^2) = \frac{1}{L}\int_{-L}^{L} f(x)^2\,dx. \;\blacklozenge$$

Proof Begin with the fact that, from the Fourier convergence theorem,

$$f(x) = \frac{1}{2}a_0 + \sum_{n=1}^{\infty}(a_n \cos(n\pi x/L) + b_n \sin(n\pi x/L)).$$

Multiply this series by $f(x)$ to get

$$f(x)^2 = \frac{1}{2}a_0 f(x) + \sum_{n=1}^{\infty}(a_n f(x)\cos(n\pi x/L) + b_n f(x)\sin(n\pi x/L)).$$

We can integrate this equation term by term (Theorem 13.4). In doing this, observe that the integrals in the series on the right are Fourier coefficients. This yields Parseval's theorem. ♦

EXAMPLE 13.15

We will apply Parseval's theorem to sum a series. The Fourier coefficients of $\cos(x/2)$ on $[-\pi, \pi]$ are

$$a_0 = \frac{1}{\pi}\int_{-\pi}^{\pi}\cos(x/2)\,dx = \frac{4}{\pi}$$

and, for $n = 1, 2, \cdots ,$,

$$a_n = \frac{1}{\pi}\int_{-\pi}^{\pi}\cos(x/2)\cos(nx)\,dx = -\frac{4}{\pi}\frac{(-1)^n}{4n^2-1}.$$

Each $b_n = 0$ because $\cos(x/2)$ is an even function. By Parseval's theorem,

$$\frac{1}{2}\left(\frac{4}{\pi}\right)^2 + \sum_{n=1}^{\infty}\left(\frac{4}{\pi}\frac{(-1)^n}{4n^2-1}\right)^2 = \frac{1}{\pi}\int_{-\pi}^{\pi}\cos^2(x/2)\,dx = 1.$$

After some routine manipulation, this yields

$$\sum_{n=1}^{\infty}\frac{1}{(4n^2-1)^2} = \frac{\pi^2-8}{16}.$$ ♦

We conclude this section with sufficient conditions for a Fourier series to converge uniformly.

THEOREM 13.8

Let f be continuous on $[-L, L]$, and let f' be piecewise continuous. Suppose $f(L) = f(-L)$. Then the Fourier series for $f(x)$ on $[-L, L]$ converges absolutely and uniformly to $f(x)$ on $[-L, L]$.

A proof is outlined in Problem 4.

SECTION 13.4 PROBLEMS

1. Let $f(x) = x\sin(x)$ for $-\pi \le x \le \pi$.
   (a) Write the Fourier series for f on $[-\pi, \pi]$
   (b) Show that this series can be differentiated term by term and use this fact to obtain the Fourier expansion of $\sin(x) + x\cos(x)$ on $[-\pi, \pi]$.
   (c) Write the Fourier series for $\sin(x) + x\cos(x)$ on $[-\pi, \pi]$ and compare this result with that of (b).
2. Let $f(x) = x^2$ for $-3 \le x \le 3$.
   (a) Write the Fourier series for f on $[-3, 3]$.
   (b) Show that this series can be differentiated term by term and use this to obtain the Fourier expansion of $2x$ on $[-3, 3]$.
   (c) Expand $2x$ in a Fourier series on $[-3, 3]$ and compare this result with that of (b).
3. Let f and f' be piecewise continuous on $[-L, L]$. Use Bessel's inequality to show that

$$\lim_{n\to\infty} \int_{-L}^{L} f(x)\cos\left(\frac{n\pi x}{L}\right) dx = \lim_{n\to\infty} \int_{-L}^{L} f(x)\sin\left(\frac{n\pi x}{L}\right) dx = 0.$$

   This result is called *Riemann's lemma.*
4. Prove Theorem 13.8 by filling in the details of the following argument. Denote the Fourier coefficients of $f(x)$ by lower case letters, and those of $f'(x)$ by upper case. Show that

$$A_0 = 0,\ A_n = \frac{n\pi}{L} b_n, \text{ and } B_n = -\frac{n\pi}{L} a_n.$$

   Show that

$$0 \le A_n^2 - \frac{2}{n}|A_n| + \frac{1}{n^2}$$

   for $n = 1, 2, \cdots$, with a similar inequality for B_n. Add these two inequalities to obtain

$$\frac{1}{n}(|A_n| + |B_n|) \le \frac{1}{2}(A_n^2 + B_n^2) + \frac{1}{n^2}.$$

   Hence show that

$$|a_n| + |b_n| \le \frac{L}{2\pi}(A_n^2 + B_n^2) + \frac{L}{\pi(n^2)}.$$

   Thus show by comparison that

$$\sum_{n=1}^{\infty}(|a_n| + |b_n|)$$

   converges. Finally, show that

$$|a_n \cos(n\pi x/L) + b_n \sin(n\pi x/L)| \le |a_n| + |b_n|$$

   and apply a theorem of Weierstrass on uniform convergence.
5. Let

$$f(x) = \begin{cases} 0 & \text{for } -\pi \le x \le 0 \\ x & \text{for } 0 < x \le \pi. \end{cases}$$

   (a) Write the Fourier series of $f(x)$ on $[-\pi, \pi]$ and show that this series converges to $f(x)$ on $(-\pi, \pi)$.
   (b) Use Theorem 12.5 to show that this series can be integrated term by term.
   (c) Use the results of (a) and (b) to obtain a trigonometric series expansion of $\int_{-\pi}^{x} f(t)dt$ on $[-\pi, \pi]$.
6. Let $f(x) = |x|$ for $-1 \le x \le 1$.
   (a) Write the Fourier series for f on $[-1, 1]$.
   (b) Show that this series can be differentiated term by term to yield the Fourier expansion of $f'(x)$ on $[-\pi, \pi]$.
   (c) Determine $f'(x)$ and expand this function in a Fourier series on $[-\pi, \pi]$. Compare this result with that of (b).

13.5 Phase Angle Form

A function f has period p if $f(x + p) = f(x)$ for all x. The smallest positive p for which this holds is the *fundamental period* of f. For example $\sin(x)$ has fundamental period 2π.

The graph of a function with fundamental period p simply repeats itself over intervals of length p. We can draw the graph for $-p/2 \le x < p/2$, then replicate this graph on $p/2 \le x < 3p/2$, $3p/2 \le x < 5p/2$, $-3p/2 \le x < -p/2$, and so on.

Now suppose f has fundamental period p. Its Fourier series on $[-p/2, p/2]$, with $L = p/2$, is

$$\frac{1}{2}a_0 + \sum_{n=1}^{\infty}(a_n \cos(2n\pi x/p) + b_n \sin(2n\pi x/p)),$$

where

$$a_n = \frac{2}{p}\int_{-p/2}^{p/2} f(x)\cos(2n\pi x/p)\,dx$$

and

$$b_n = \frac{2}{p}\int_{-p/2}^{p/2} f(x)\sin(2n\pi x/p)\,dx.$$

It is sometimes convenient to write this series in a different way. Let

$$\omega_0 = \frac{2\pi}{p}.$$

The Fourier series of $f(x)$ on $[-p/2, p/2]$ is

$$\frac{1}{2}a_0 + \sum_{n=1}^{\infty}(a_n \cos(n\omega_0 x) + b_n \sin(n\omega_0 x)),$$

where

$$a_n = \frac{2}{p}\int_{-p/2}^{p/2} f(x)\cos(n\omega_0 x)\,dx$$

and

$$b_n = \frac{2}{p}\int_{-p/2}^{p/2} f(x)\sin(n\omega_0 x)\,dx.$$

Now look for numbers c_n and δ_n so that

$$a_n \cos(n\omega_0 x) + b_n \sin(n\omega_0 x) = c_n \cos(n\omega_0 x + \delta_n).$$

To solve for these constants, use a trigonometric identity to write this equation as

$$a_n \cos(n\omega_0 x) + b_n \sin(n\omega_0 x) = c_n \cos(n\omega_0 x)\cos(\delta_n) - c_n \sin(n\omega_0 x)\sin(\delta_n).$$

One way to satisfy this equation is to put

$$c_n \cos(\delta_n) = a_n \text{ and } c_n \sin(\delta_n) = -b_n.$$

If we square both sides of these equations and add the results, we obtain

$$c_n^2 = a_n^2 + b_n^2,$$

so

$$c_n = \sqrt{a_n^2 + b_n^2}.$$

Next, divide to obtain

$$\frac{c_n \sin(\delta_n)}{c_n \cos(\delta_n)} = \tan(\delta_n) = -\frac{b_n}{a_n},$$

assuming that $a_n \neq 0$. Then

$$\delta_n = -\arctan\left(\frac{b_n}{a_n}\right).$$

When each $a_n \neq 0$, these equations enable us to write the *phase angle* form of the Fourier series of $f(x)$ on $[-p/2, p/2]$:

$$\frac{1}{2}a_0 + \sum_{n=1}^{\infty} c_n \cos(n\omega_0 x + \delta_n),$$

where

$$\omega_0 = 2\pi/p, c_n = \sqrt{a_n^2 + b_n^2}, \text{ and } \delta_n = -\arctan(b_n/a_n).$$

This phase angle form is also called the *harmonic form* of the Fourier series for $f(x)$ on $[-p/2, p/2]$. The term $\cos(n\omega_0 x + \delta_n)$ is called the nth *harmonic* of f, c_n is the nth *harmonic amplitude*, and δ_n is the nth *phase angle* of f.

If f has fundamental period p, then in the expressions for the coefficients a_n and b_n, we can compute the integrals over any interval $[\alpha, \alpha + p]$, since any interval of length p carries all of the information about a p-periodic function.

This means that the Fourier coefficients of p-periodic f can be obtained as

$$a_n = \frac{2}{p}\int_{\alpha}^{\alpha+p} f(x)\cos(n\omega_0 x)\, dx$$

and

$$b_n = \frac{2}{p}\int_{\alpha}^{\alpha+p} f(x)\sin(n\omega_0 x)\, dx$$

for any number α.

EXAMPLE 13.16

Let

$$f(x) = x^2 \text{ for } 0 \leq x < 3$$

and suppose f has fundamental period $p = 3$. A graph of f is shown in Figure 13.17.

Since f is 3-periodic, and we are given an algebraic expression for $f(x)$ only on $[0, 3)$, we will use this interval to compute the Fourier coefficients of f. That is, use $p = 3$ and $\alpha = 0$ in the preceding discussion. We also have $\omega_o = 2\pi/p = 2\pi/3$.

The Fourier coefficients are

$$a_0 = \frac{2}{3}\int_0^3 x^2\, dx = 6,$$

$$a_n = \frac{2}{3}\int_0^3 x^2 \cos\left(\frac{2n\pi x}{3}\right) dx = \frac{9}{n^2\pi^2},$$

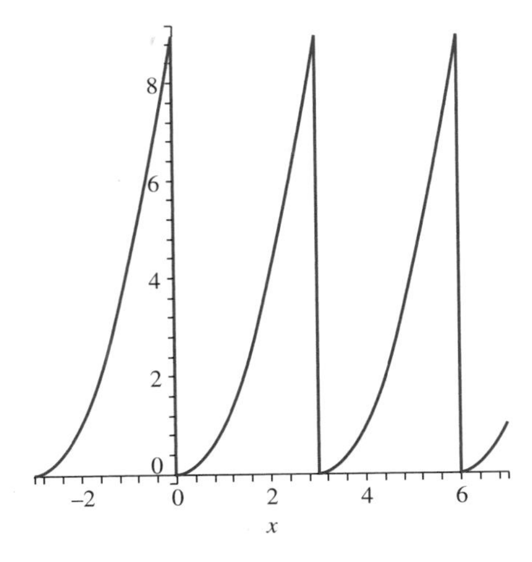

FIGURE 13.17 $f(x)$ *in Example 13.16.*

and

$$b_n = \frac{2}{3}\int_0^3 x^2 \sin\left(\frac{2n\pi x}{3}\right) dx = -\frac{9}{n\pi}.$$

The Fourier series of $f(x)$ is

$$3 + \sum_{n=1}^{\infty} \frac{9}{n\pi}\left(\frac{1}{n\pi}\cos\left(\frac{2n\pi x}{3}\right) - \sin\left(\frac{2n\pi x}{3}\right)\right).$$

We may think of this as the Fourier series of $f(x)$ on the symmetric interval $[-3/2, 3/2]$. However, keep in mind that $f(x)$ is not x^2 on this interval. We have $f(x) = x^2$ on $0 \le x < 3$, hence also on $[0, 3/2]$. But from Figure 13.17, $f(x) = (x+3)^2$ on $[-3/2, 0)$.

This Fourier series converges to

$$\begin{cases} 9/4 & \text{for } x = \pm 3/2, \\ 9/2 & \text{for } x = 0, \\ (x+3)^2 & \text{for } -3/2 < x < 0, \\ x^2 & \text{for } 0 < x < 3/2. \end{cases}$$

For the phase angle form, compute

$$c_n = \sqrt{a_n^2 + b_n^2} = \frac{9}{n^2\pi^2}\sqrt{1 + n^2\pi^2}$$

and

$$\delta_n = \arctan\left(-\frac{-9/n\pi}{9/n^2\pi^2}\right) = \arctan(n\pi) = 0.$$

The phase angle form of the Fourier series of $f(x)$ is

$$3 + \sum_{n=1}^{\infty} \frac{9}{n^2\pi^2}\sqrt{1 + n^2\pi^2}\cos\left(\frac{2n\pi x}{3}\right). \blacklozenge$$

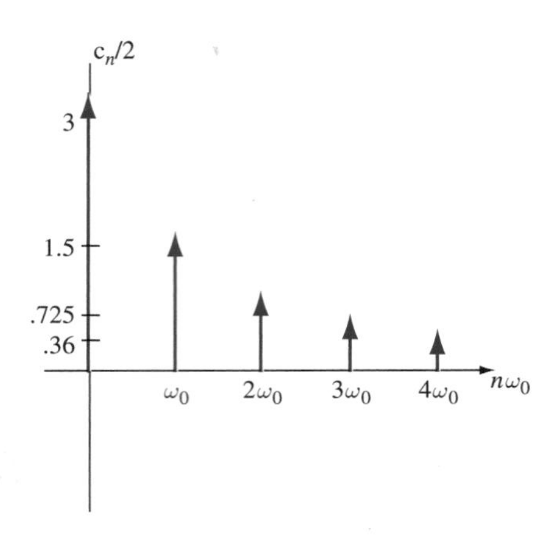

FIGURE 13.18 *Amplitude spectrum of f in Example 13.16.*

The *amplitude spectrum* of a periodic function f is a plot of points $(n\omega_0, c_n/2)$ for $n = 1, 2, \cdots$, and also the point $(0, |c_0|/2)$. For the function of Example 13.14, this is a plot of points $(0, 3)$ and, for nonzero integer n, points

$$\left(\frac{2n\pi}{3}, \frac{9}{2n^2\pi^2}\sqrt{1+n^2\pi^2}\right).$$

The amplitude spectrum for the function of Example 13.16 is shown in Figure 13.18, with the intervals on the horizontal axis of length $\omega_0 = 2\pi/3$. This graph displays the relative effects of the harmonics in the function. This is useful in signal analysis.

SECTION 13.5 PROBLEMS

In Problems 1, 2, and 3, let f be periodic of period p.

1. If f is differentiable, show that f' has period p.
2. Let α be a positive number. Show that $g(t) = f(\alpha t)$ has period p/α and $h(t) = f(t/\alpha)$ has period αp.
3. If g is also periodic of period p, show that $\alpha f + \beta g$ is periodic of period p, for any numbers α and β.

In each of Problems 4 through 12, find the phase angle form of the Fourier series of the function and plot some points of the amplitude spectrum. Some of these functions are specified by a graph.

4. Let $f(x) = 3x^2$ for $0 \le x < 4$ and let f have fundamental period 4.
5. Let
$$f(x) = \begin{cases} 1+x & \text{for } 0 \le x < 3, \\ 2 & \text{for } 2 \le x < 4, \end{cases}$$
and suppose f has fundamental period 4.
6. $f(x) = \cos(\pi x)$ for $0 \le x < 1$ and f has fundamental period 1.
7. Let
$$f(x) = \begin{cases} 1 & \text{for } 0 \le x < 1, \\ 0 & \text{for } 1 < x < 2, \end{cases}$$
and let f has fundamental period 2.
8. Let $f(x) = x$ for $0 \le x < 2$, with fundamental period 2.

9. f has the graph of Figure 13.19.

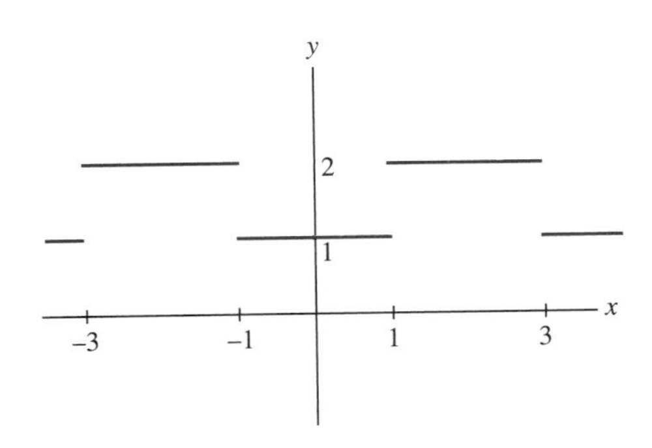

FIGURE 13.19 *$f(x)$ in Problem 9, Section 13.5.*

10. f has the graph of Figure 13.20.

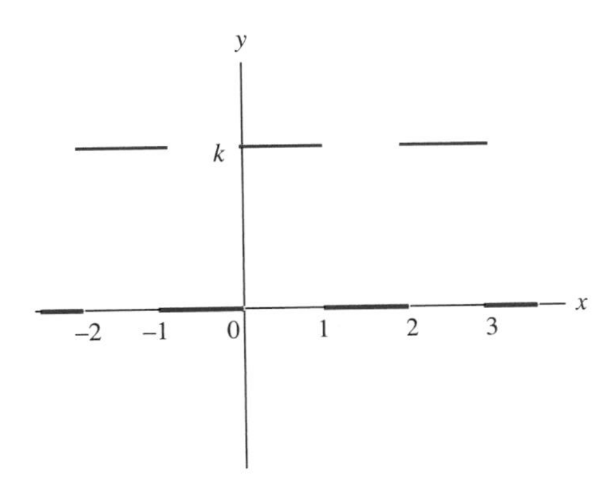

FIGURE 13.20 *$f(x)$ in Problem 10, Section 13.5.*

11. f has the graph of Figure 13.21.

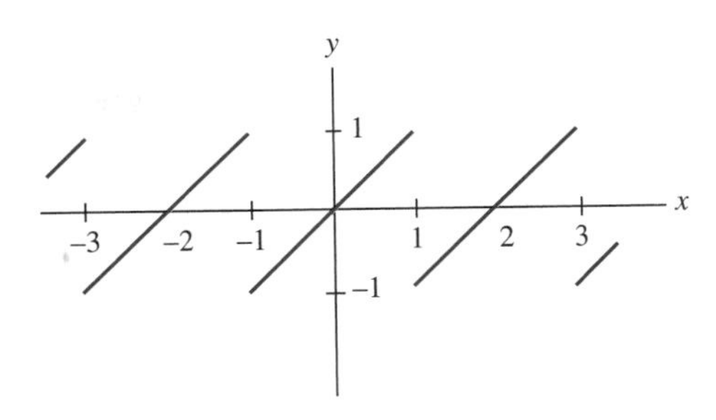

FIGURE 13.21 *$f(x)$ in Problem 11, Section 13.5.*

12. f has the graph of Figure 13.22.

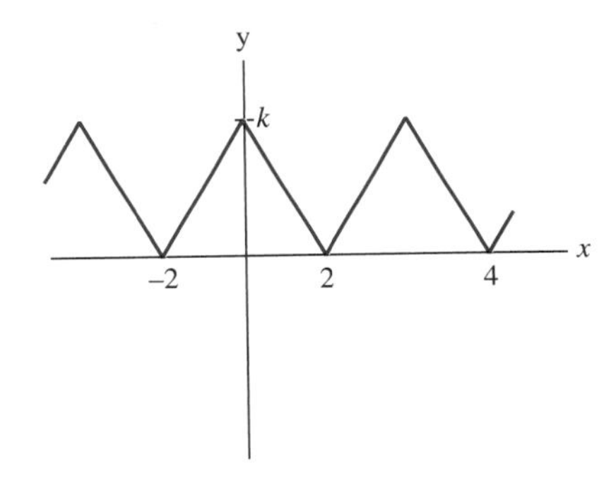

FIGURE 13.22 *$f(x)$ in Problem 12, Section 13.5.*

13.6 Complex Fourier Series

There is a complex form of Fourier series that is sometimes used. As preparation for this, recall that, in polar coordinates, a complex number (point in the plane) can be written

$$z = x + iy = r\cos(\theta) + ir\sin(\theta)$$

where

$$r = |z| = \sqrt{x^2 + y^2}$$

and θ is an *argument* of z. This is the angle (in radians) between the positive $x-$ axis and the line from the origin through (x, y), or this angle plus any integer multiple of 2π. Using Euler's formula, we obtain the *polar form* of z:

$$z = r[\cos(\theta) + i\sin(\theta)] = re^{i\theta}.$$

Now

$$e^{i\theta} = \cos(\theta) + i\sin(\theta),$$

and by replacing θ with $-\theta$, we get

$$e^{-i\theta} = \cos(\theta) - i\sin(\theta).$$

Solve these equations for $\cos(\theta)$ and $\sin(\theta)$ to obtain the complex exponential forms of the trigonometric functions:

$$\cos(\theta) = \frac{1}{2}\left(e^{i\theta} + e^{-i\theta}\right), \sin(\theta) = \frac{1}{2i}\left(e^{i\theta} - e^{-i\theta}\right).$$

We will also use the fact that, if x is a real number, then the conjugate of e^{ix} is

$$\overline{e^{ix}} = e^{-ix}.$$

This follows from Euler's formula, since

$$\overline{e^{ix}} = \overline{\cos(x) + i\sin(x)} = \cos(x) - i\sin(x) = e^{-ix}.$$

Now let f be a piecewise smooth periodic function with fundamental period $2L$. To derive a complex Fourier expansion of $f(x)$ on $[-L, L]$, begin with the Fourier series of $f(x)$. With $\omega_0 = \pi/L$, this series is

$$\frac{1}{2}a_0 + \sum_{n=1}^{\infty}(a_n\cos(n\omega_0 x) + b_n\sin(n\omega_0 x))$$

Put the complex forms of $\cos(n\omega_0 x)$ and $\sin(n\omega_0 x)$ into this expansion:

$$\begin{aligned}
&\frac{1}{2}a_0 + \sum_{n=1}^{\infty}\left[a_n\frac{1}{2}\left(e^{in\omega_0 x} + e^{-in\omega_0 x}\right) + b_n\frac{1}{2i}\left(e^{in\omega_0 x} - e^{-in\omega_0 x}\right)\right] \\
&= \frac{1}{2}a_0 + \sum_{n=1}^{\infty}\left[\frac{1}{2}(a_n - ib_n)e^{in\omega_0 x} + \frac{1}{2}(a_n + ib_n)e^{-in\omega_0 x}\right],
\end{aligned}$$

in which we used the fact that $1/i = -i$. In this series, let

$$d_0 = \frac{1}{2}a_0$$

and, for $n = 1, 2, \cdots$,

$$d_n = \frac{1}{2}(a_n - ib_n).$$

The Fourier series on $[-L, L]$ becomes

$$d_0 + \sum_{n=1}^{\infty} d_n e^{in\omega_0 x} + \sum_{n=1}^{\infty}\overline{d_n}e^{-in\omega_0 x} \tag{13.14}$$

Now

$$d_0 = \frac{1}{2}a_0 = \frac{1}{L}\int_{-L}^{L} f(x)\,dx$$

and for $n = 1, 2, \cdots$,

$$\begin{aligned}
d_n &= \frac{1}{2}(a_n - ib_n) \\
&= \frac{1}{2L}\int_{-L}^{L} f(x)\cos(n\omega_0 x)\,dx - \frac{i}{2L}\int_{-L}^{L} f(x)\sin(n\omega_0 x)\,dx \\
&= \frac{1}{2L}\int_{-L}^{L} f(x)[\cos(n\omega_0 x) - i\sin(n\omega_0 x)]\,dx \\
&= \frac{1}{2L}\int_{-L}^{L} f(x)e^{-in\omega_0 x}\,dx.
\end{aligned}$$

Then

$$\overline{d_n} = \frac{1}{2L}\int_{-L}^{L} f(x)\overline{e^{-in\omega_0 x}}\,dx = \frac{1}{2L}\int_{-L}^{L} f(x)e^{in\omega_0 x}\,dx = d_{-n}.$$

With this, the expansion of equation (13.14) becomes

$$d_0 + \sum_{n=1}^{\infty} d_n e^{in\omega_0 x} + \sum_{n=1}^{\infty} d_{-n}e^{-in\omega_0 x}$$

$$= \sum_{n=-\infty}^{\infty} d_n e^{in\omega_0 x}.$$

This leads us to define the *complex Fourier series of f on* $[-L, L]$ to be

$$\sum_{n=-\infty}^{\infty} d_n e^{in\omega_0 x},$$

with coefficients

$$d_n = \frac{1}{2L}\int_{-L}^{L} f(x)e^{-in\omega_0 x}\,dx$$

for $n = 0, \pm 1, \pm 2, \cdots$.

Because of the periodicity of f, the integral defining the coefficients can be carried out over any interval $[\alpha, \alpha + 2L]$ of length $2L$. The Fourier convergence theorem applies to this complex Fourier expansion, since it is just the Fourier series in complex form.

EXAMPLE 13.17

Let $f(x) = x$ for $-1 \le x < 1$ and suppose f has fundamental period 2, so $f(x+2) = f(x)$ for all x. Figure 13.23 is part of a graph of f. Now $\omega_0 = \pi$.

Immediately $d_0 = 0$ because f is an odd function. For $n \neq 0$,

$$d_n = \frac{1}{2}\int_{-1}^{1} xe^{-in\pi x}\,dx$$

$$= \frac{1}{2n^2\pi^2}\left[in\pi e^{in\pi} - e^{in\pi} + in\pi e^{-in\pi} + e^{-in\pi}\right]$$

$$= \frac{1}{2n^2\pi^2}\left[in\pi\left(e^{in\pi} + e^{-in\pi}\right) - \left(e^{in\pi} - e^{-in\pi}\right)\right].$$

The complex Fourier series of f is

$$\sum_{n=-\infty, n\neq 0}^{\infty} \frac{1}{2n^2\pi^2}\left[in\pi\left(e^{in\pi} + e^{-in\pi}\right) - \left(e^{in\pi} - e^{-in\pi}\right)\right]e^{in\pi x}.$$

This converges to x for $-1 < x < 1$. In this example we can simplify the series. For $n \neq 0$,

$$d_n = \frac{1}{2n^2\pi^2}[2in\pi\cos(n\pi) - 2i\sin(n\pi)]$$

$$= \frac{i}{n\pi}(-1)^n.$$

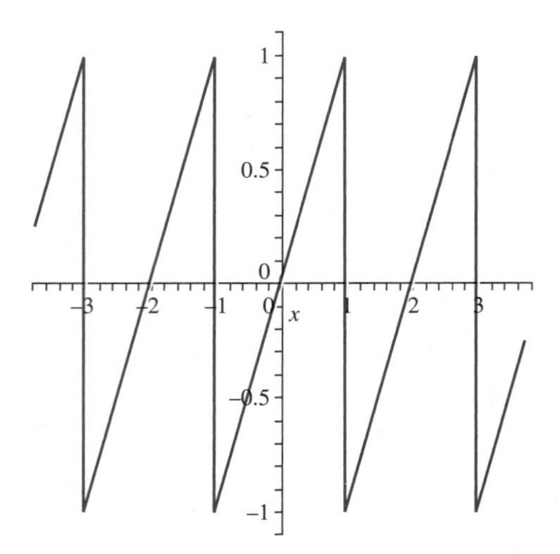

FIGURE 13.23 *Graph of f in Example 13.17.*

All the terms $\sin(n\pi)=0$. In $\sum_{n=-\infty}^{-1}$, replace n with $-n$ and sum from $n=1$ to ∞, then combine the two summations from 1 to ∞ to write

$$\sum_{n=-\infty, n\neq 0}^{\infty} \frac{i}{n\pi}(-1)^n e^{in\pi x}$$

$$= \sum_{n=1}^{\infty} \left(\frac{i}{n\pi}(-1)^n e^{in\pi x} + \frac{i}{-n\pi}(-1)^{-n} e^{-in\pi x} \right)$$

$$= \sum_{n=1}^{\infty} \frac{i}{n\pi}(-1)^n \left(e^{in\pi x} - e^{-in\pi x} \right)$$

$$= \sum_{n=1}^{\infty} \frac{2}{n\pi}(-1)^{n+1} \sin(n\pi x).$$

This is the Fourier series for $f(x)=x$ on $[-1, 1]$. ◆

The *amplitude spectrum* of a complex Fourier series of a periodic function is a graph of the points $(n\omega_0, |d_n|)$. Sometimes this graph is also referred to as a *frequency spectrum*.

SECTION 13.6 PROBLEMS

In each of Problems 1 through 7, write the complex Fourier series of f, determine the sum of the series, and plot some points of the frequency spectrum.

1. $f(x) = \begin{cases} x & \text{for } 0 \le x < 1 \\ 2-x & \text{for } 1 \le x < 2 \end{cases}$, f has period 2

2. $f(x) = 1-x$ for $0 \le x < 6$, period 6

3. $f(x) = \begin{cases} -1 & \text{for } 0 \le x < 2 \\ 2 & \text{for } 2 \le x < 4 \end{cases}$, f has period 4

4. $f(x) = e^{-x}$ for $0 \le x < 5$, period 5

5. $f(x) = \begin{cases} 0 & \text{for } 0 \le x < 1 \\ 1 & \text{for } 1 \le x < 4 \end{cases}$, f has period 4

6. $f(x) = x^2$ for $0 \le x < 2$, period 2

7. $f(x) = 2x$ for $0 \le x < 3$, period 3

13.7 Filtering of Signals

A periodic signal $f(t)$ of period $2L$ is sometimes filtered to cancel out or diminish unwanted effects, or perhaps to enhance other effects. We will briefly examine one way this is done.

Suppose f has complex Fourier series

$$\sum_{n=-\infty}^{\infty} d_n e^{n\pi it/L},$$

where

$$d_n = \frac{1}{2L}\int_{-L}^{L} f(t)e^{-n\pi it/L}\,dt.$$

The Nth partial sum of the series is

$$S_N(t) = \sum_{j=-N}^{N} d_j e^{\pi ijt/L}.$$

A *filtered partial sum* of this Fourier series is a sum of the form

$$\sum_{j=-N}^{N} Z\left(\frac{j}{N}\right) d_j e^{\pi ijt/L}. \tag{13.15}$$

Z is the *filter function* and is assumed to be a continuous even function on $[-L, L]$. In applications the object is to choose Z to achieve some specified purpose or effect.

To illustrate, we will develop a filter that damps out the Gibbs phenomenon. In the nineteenth century there was an intense effort to understand convergence properties of Fourier series. In the course of this work it was observed that the sequence of averages of partial sums of a Fourier series is in general better behaved than the sequence of partial sums of the series itself. If S_N is the Nth partial sum of the series, this average has the form

$$\sigma_N(t) = \frac{1}{N}\sum_{k=0}^{N-1} S_k(t).$$

The quantity $\sigma_N(t)$ is called the Nth *Cesàro sum* of f. It was shown that, if f is periodic of period 2π and $\int_0^{2\pi} f(t)\,dt$ exists, then $\sigma_N(t) \to f(t)$ for any t at which f is continuous, a much stronger result than holds for partial sums of Fourier series.

Inserting the summation for $S_k(t)$, we have

$$\sigma_N(t) = \frac{1}{N}\sum_{k=0}^{N-1}\left(\sum_{j=-k}^{k} d_j e^{\pi ijt/L}\right).$$

With some manipulation, this double sum can be rearranged to write

$$\sigma_N(t) = \sum_{n=-N}^{N}\left(1 - \left|\frac{n}{N}\right|\right) d_n e^{\pi int/L}.$$

This is of the form of equation (13.15) with the *Cesàro filter function*

$$Z(t) = 1 - |t| \text{ for } -1 \le t \le 1.$$

The sequence

$$\left[Z\left(\frac{n}{N}\right)\right]_{n=-N}^{N} = \left[1 - \left|\frac{n}{N}\right|\right]_{n=-N}^{N}$$

is called the *sequence of filter factors for the Cesàro filter.*

This "averaging" filter damps out the Gibbs effect in the convergence of a Fourier series. To observe this, let f have fundamental period 2π, and, on $[-\pi, \pi]$,

$$f(t) = \begin{cases} -1 & \text{for } -\pi \le t < 0 \\ 1 & \text{for } 0 \le t < \pi. \end{cases}$$

The complex Fourier series of f is

$$\sum_{n=-\infty, n\neq 0}^{\infty} \frac{i}{\pi} \frac{-1 + (-1)^n}{n} e^{int}.$$

The Nth partial sum of this series is

$$S_N(t) = \sum_{n=-N, n\neq 0}^{N} \frac{i}{\pi} \frac{-1 + (-1)^n}{n} e^{int}.$$

If we pair positive and negative values of n in this summation, we find that

$$S_N(t) = \sum_{n=1}^{N} \frac{-2}{n\pi}(-1 + (-1)^n)\sin(nt).$$

If N is even, then $-1 + (-1)^n = 0$. If N is odd, then $-1 + (-1)^n = -2$. Therefore, for odd N,

$$S_N(t) = \frac{4}{\pi}\left(\sin(t) + \frac{1}{3}\sin(3t) + \frac{1}{5}\sin(5t) + \cdots + \frac{1}{N}\sin(Nt)\right).$$

The Nth Cesàro sum is

$$\sigma_N(t) = \sum_{n=-N, n\neq 0}^{N} \left(1 - \left|\frac{n}{N}\right|\right)\left(\frac{i}{\pi}\right)\frac{-1 + (-1)^n}{n} e^{int}.$$

With some manipulation, this can be written

$$\sigma_N(t) = \sum_{n=1}^{N} \left(1 - \frac{n}{N}\right)\left(\frac{-2}{\pi}\right)\frac{-1 + (-1)^n}{n}\sin(nt).$$

Figure 13.24 shows graphs of $S_{10}(t)$ and $\sigma_{10}(t)$, and Figure 13.25 graphs of $S_{30}(t)$ and $\sigma_{30}(t)$, showing the Gibbs effect in the partial sums of the Fourier series, and this effect damped out in the smoother Cesàro sums. The Cesàro filter also damps out the higher frequency terms in the Fourier series because $1 - |n/N|$ tends to zero as n increases toward N.

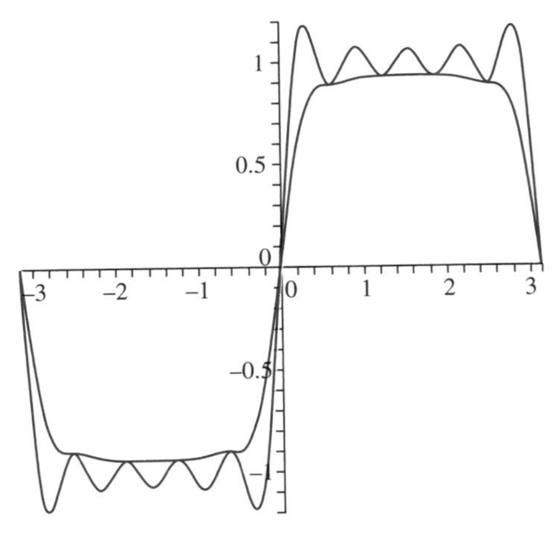

FIGURE 13.24 *Tenth partial sum and Cesàro sum of f.*

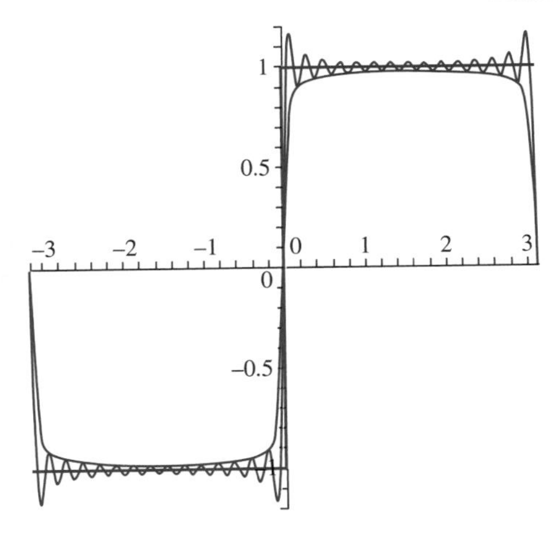

FIGURE 13.25 *Thirtieth partial sum and Cesàro sum of f.*

There are many filters used in signal analysis. Two of the more frequently used ones are the Hamming and Gauss filters.

The *Hamming filter* is named for Richard Hamming, who was a senior research scientist at Bell Labs, and is defined by

$$Z(t) = 0.54 + 0.46\cos(\pi t).$$

The *Gauss filter* is sometimes used to filter out background noise and is defined by

$$Z(t) = e^{-\alpha\pi^2 t^2},$$

with α a positive constant.

SECTION 13.7 PROBLEMS

In each of Problems 1 through 5, graph the function, the fifth partial sum of its Fourier series on the interval, and the fifth Cesàro sum, using the same set of axes. Repeat this process for the tenth and twenty-fifth partial sums. Notice in particular the graphs at points of discontinuity of the function, where the Gibbs phenomenon appears.

1. $f(t) = \begin{cases} 2+t & \text{for } -1 \le t < 0 \\ 7 & \text{for } 0 \le t < 1 \end{cases}$

2. $f(t) = \begin{cases} 0 & \text{for } -3 \le t < 0 \\ \cos(t) & \text{for } 0 \le t < 3 \end{cases}$

3. $f(t) = \begin{cases} 1 & \text{for } 0 \le t < 2 \\ -1 & \text{for } -2 \le t < 0 \end{cases}$

4. $f(t) = \begin{cases} t^2 & \text{for } -2 \le t < 1 \\ 2+t & \text{for } 1 \le t < 2 \end{cases}$

5. $f(t) = \begin{cases} -1 & \text{for } -1 \le t < -1/2 \\ 0 & \text{for } -1/2 \le t < 1/2 \\ 1 & \text{for } 1/2 \le t < 1 \end{cases}$

6. Let

$$f(t) = \begin{cases} 1 & \text{for } 0 \le t < 2 \\ -1 & \text{for } -2 \le t < 0 \end{cases}$$

Plot the fifth partial sum of the Fourier series for $f(t)$ on $[-2, 2]$, together with the fifth Cesàro sum, the fifth Hamming and Gauss filtered partial sums, using the same set of axes. Repeat this with the tenth and twenty-fifth partial sums.

7. Let

$$f(t) = \begin{cases} t & \text{for } -2 \le t < 0 \\ 2+t & \text{for } 0 \le t < 2 \end{cases}$$

Plot the fifth partial sum of the Fourier series for $f(t)$ on $[-2, 2]$, together with the fifth Cesàro sum, the fifth Hamming and Gauss filtered partial sums, using the same set of axes. Repeat this with the tenth and twenty-fifth partial sums.

CHAPTER 14

The Fourier Integral and Transforms

14.1 The Fourier Integral

If $f(x)$ is defined for $-L \leq x \leq L$, we may be able to represent $f(x)$ as a Fourier series on this interval. However, Fourier series are tied to intervals. If f is defined over the entire line and is not periodic, then the idea of a Fourier series representation is replaced with the idea of a Fourier integral representation, in which the role of $\sum_{n=0}^{\infty}$ is played by $\int_0^{\infty}$.

We will give an informal argument to suggest the form that the Fourier integral should take. Assume that f is *absolutely integrable*, which means that $\int_{-\infty}^{\infty} |f(x)|\, dx$ converges. We also assume that f is piecewise smooth on every interval $[-L, L]$.

Write the Fourier series of $f(x)$ on an arbitrary interval $[-L, L]$. With the formulas for the coefficients included, this series is

$$\frac{1}{2L}\int_{-L}^{L} f(\xi)\, d\xi + \sum_{n=1}^{\infty}\left[\left(\frac{1}{L}\int_{-L}^{L} f(\xi)\cos(n\pi\xi/L)\, d\xi\right)\cos(n\pi x/L) + \left(\frac{1}{L}\int_{-L}^{L} f(\xi)\sin(n\pi\xi/L)\, d\xi\right)\sin(n\pi x/L)\right].$$

We want to let $L \to \infty$ to obtain a representation of $f(x)$ over the entire real line. It is not clear what this quantity approaches, if anything, as $L \to \infty$, so we will rewrite some terms. First, let

$$\omega_n = \frac{n\pi}{L}$$

and

$$\omega_n - \omega_{n-1} = \frac{\pi}{L} = \Delta\omega.$$

Now the Fourier series on $[-L, L]$ can be written

$$\frac{1}{2\pi}\left(\int_{-L}^{L} f(\xi)\,d\xi\right)\Delta\omega + \frac{1}{\pi}\left[\left(\int_{-L}^{L} f(\xi)\cos(\omega_n\xi)\,d\xi\right)\cos(\omega_n x)\right.$$

$$\left. + \left(\int_{-L}^{L} f(\xi)\sin(\omega_n\xi)\,d\xi\right)\sin(\omega_n x)\right]\Delta\omega.$$

Let $L \to \infty$, so $[-L, L]$ expands to cover the entire real line. Then $\Delta\omega \to 0$. Examine what happens in the terms of the last equation. First,

$$\frac{1}{2\pi}\left(\int_{-L}^{L} f(\xi)\,d\xi\right)\Delta\omega \to 0$$

because the integral converges (hence is bounded). The other terms in this equation resemble a Riemann sum for a definite integral. As $L \to \infty$ and $\Delta\omega \to 0$, this expression approaches the limit

$$\frac{1}{\pi}\left[\left(\int_{-\infty}^{\infty} f(\xi)\cos(\omega\xi)\,d\xi\right)\cos(\omega x)\right.$$

$$\left. + \left(\int_{-\infty}^{\infty} (\xi)\sin(\omega\xi)\,d\xi\right)\sin(\omega x)\right]d\omega.$$

This is the *Fourier integral* of $f(x)$ on the real line, and it has the form

$$\int_{0}^{\infty} (A_\omega\cos(\omega x) + B_\omega\sin(\omega x))\,d\omega \tag{14.1}$$

in which the *Fourier integral coefficients of* f are

$$A_\omega = \frac{1}{\pi}\int_{-\infty}^{\infty} f(\xi)\cos(\omega\xi)\,d\xi \tag{14.2}$$

and

$$B_\omega = \frac{1}{\pi}\int_{-\infty}^{\infty} f(\xi)\sin(\omega\xi)\,d\xi. \tag{14.3}$$

The integration variable ω replaces the summation index n in this integral representation.

As with Fourier series, the relationship between the integral (14.1) and $f(x)$ must be clarified. This is done in the following theorem.

THEOREM 14.1 Convergence of the Fourier Integral

Suppose $f(x)$ is defined for all real x and that $\int_{-\infty}^{\infty}|f(x)|\,dx$ converges. Suppose f is piecewise smooth on every interval $[-L, L]$ for $L > 0$. Then at any x the Fourier integral (14.1) of f converges to

$$\frac{1}{2}(f(x+) + f(x-)).$$

In particular, if f is continuous at x, then the Fourier integral converges at x to $f(x)$. ♦

EXAMPLE 14.1

Let

$$f(x) = \begin{cases} 1 & \text{for } -1 \le x \le 1 \\ 0 & \text{for } |x| > 1. \end{cases}$$

Certainly f is absolutely integrable. The Fourier integral coefficients of f are

$$A_\omega = \frac{1}{\pi}\int_{-1}^{1}\cos(\omega\xi)\,d\xi = \frac{2\sin(\omega)}{\pi\omega}$$

and

$$B_\omega = \frac{1}{\pi}\int_{-1}^{1}\sin(\omega\xi)\,d\xi = 0.$$

The Fourier integral of f is

$$\int_0^\infty \frac{2\sin(\omega)}{\pi\omega}\cos(\omega x)\,d\omega.$$

Because f is continuous for $x \neq \pm 1$, the integral converges to $f(x)$ for $x \neq \pm 1$. At $x = 1$ the integral converges to

$$\frac{1}{2}(f(1+) + f(1-)) = \frac{1}{2}(1+0) = \frac{1}{2}.$$

Similarly, the integral converges to $1/2$ at $x = -1$. This Fourier integral is a faithful representation of the function except at 1 and -1, where it averages the ends of the jump discontinuities there.

In view of this convergence, we have

$$\frac{1}{\pi}\int_0^\infty \frac{2\sin(\omega)}{\omega}\cos(\omega x)\,d\omega = \begin{cases} 1 & \text{for } -1 < x < 1 \\ 1/2 & \text{for } x = \pm 1 \\ 0 & \text{for } |x| > 1. \end{cases}$$ ♦

There is another expression for the Fourier integral of a function that is sometimes convenient to use. Insert the coefficients into the Fourier integral:

$$\int_0^\infty [A_\omega\cos(\omega x) + B_\omega\sin(\omega x)]\,d\omega =$$

$$\int_0^\infty \left[\left(\frac{1}{\pi}\int_{-\infty}^{\infty} f(\xi)\cos(\omega\xi)\,d\xi\right)\cos(\omega x) + \left(\frac{1}{\pi}\int_{-\infty}^{\infty} f(\xi)\sin(\omega\xi)\,d\xi\right)\sin(\omega x)\right]d\omega$$

$$= \frac{1}{\pi}\int_0^\infty\int_{-\infty}^{\infty} f(\xi)[\cos(\omega\xi)\cos(\omega x) + \sin(\omega\xi)\sin(\omega x)]\,d\xi\,d\omega$$

$$= \frac{1}{\pi}\int_0^\infty\int_{-\infty}^{\infty} f(\xi)\cos(\omega(\xi - x))\,d\xi\,d\omega.$$

This gives us the equivalent Fourier integral representation

$$\frac{1}{\pi}\int_0^\infty\int_{-\infty}^{\infty} f(\xi)\cos(\omega(\xi - x))\,d\xi\,d\omega \tag{14.4}$$

of $f(x)$ on the real line.

SECTION 14.1 PROBLEMS

In each of Problems 1 through 10, write the Fourier integral representation (14.1) of the function and determine what this integral converges to.

1. $f(x) = \begin{cases} x^2 & \text{for } -100 \le x \le 100 \\ 0 & \text{for } |x| > 100 \end{cases}$

2. $f(x) = \begin{cases} |x| & \text{for } -\pi \le x \le 2\pi \\ 0 & \text{for } x < -\pi \text{ and for } x > 2\pi \end{cases}$

3. $f(x) = \begin{cases} x & \text{for } -\pi \le x \le \pi \\ 0 & \text{for } |x| > \pi \end{cases}$

4. $f(x)=\begin{cases} k & \text{for } -10\leq x\leq 10 \\ 0 & \text{for } |x|>10 \end{cases}$

5. $f(x)=e^{-|x|}$

6. $f(x)=\begin{cases} 1/2 & \text{for } -5\leq x<1 \\ 1 & \text{for } 1\leq x\leq 5 \\ 0 & \text{for } |x|>5 \end{cases}$

7. $f(x)=\begin{cases} -1 & \text{for } -\pi\leq x\leq 0 \\ 1 & \text{for } 0<x\leq\pi \\ 0 & \text{for } |x|>\pi \end{cases}$

8. $f(x)=xe^{-4|x|}$

9. $f(x)=\begin{cases} \sin(x) & \text{for } -3\pi\leq x\leq\pi \\ 0 & \text{for } x<-3\pi \text{ and for } x>\pi \end{cases}$

10. $f(x)=\begin{cases} \sin(x) & \text{for } -4\leq x\leq 0 \\ \cos(x) & \text{for } 0<x\leq 4 \\ 0 & \text{for } |x|>4 \end{cases}$

11. Show that the Fourier integral of $f(x)$ can be written

$$\lim_{\omega\to\infty}\frac{1}{\pi}\int_{-\infty}^{\infty} f(t)\frac{\sin(\omega(t-x))}{t-x}\,dt.$$

14.2 Fourier Cosine and Sine Integrals

We can define Fourier cosine and sine integral expansions for functions defined on the half-line in a manner completely analogous to Fourier cosine and sine expansions of functions defined on a half interval.

Suppose $f(x)$ is defined for $x\geq 0$. Extend f to an even function f_e on the real line. where

$$f_e(x)=\begin{cases} f(x) & \text{for } x\geq 0, \\ f(-x) & \text{for } x<0. \end{cases}$$

This reflects the graph of $f(x)$ for $x\geq 0$ back across the vertical axis to a function f_e defined on the entire line. Because f_e is an even function, its Fourier coefficients are

$$A_\omega=\frac{1}{\pi}\int_{-\infty}^{\infty} f_e(\xi)\cos(\omega\xi)\,d\xi$$
$$=\frac{2}{\pi}\int_{0}^{\infty} f(\xi)\cos(\omega\xi)\,d\xi$$

and

$$B_\omega=\frac{1}{\pi}\int_{-\infty}^{\infty} f_e(\xi)\cos(\omega\xi)\,d\xi=0.$$

The Fourier integral of $f_e(x)$ contains only cosine terms. Since $f_e(x)=f(x)$ for $x\geq 0$, this expansion may be thought of as a cosine expansion of $f(x)$, on the half-line $x\geq 0$.

This leads us to define the *Fourier cosine integral of* $f(x)$ on $x\geq 0$ to be

$$\int_0^\infty A_\omega\cos(\omega x)\,d\omega \tag{14.5}$$

in which

$$A_\omega=\frac{2}{\pi}\int_0^\infty f(\xi)\cos(\omega\xi)\,d\xi \tag{14.6}$$

is the *Fourier integral cosine coefficient*.

Similarly, we can reflect the graph of $f(x)$ through the origin to obtain an odd extension f_o defined for all real x. Now the Fourier coefficients of the Fourier expansion of f_o on the line are

$$B_\omega = \frac{1}{\pi}\int_{-\infty}^{\infty} f_o(\xi)\sin(\omega\xi)\,d\xi$$
$$= \frac{2}{\pi}\int_0^{\infty} f(\xi)\sin(\omega\xi)\,d\xi$$

and

$$A_\omega = 0.$$

This Fourier expansion of $f_o(x)$ on the whole line contains just sine terms. Furthermore $f_o(x) = f(x)$ for $x \geq 0$.

We define the *Fourier sine integral of* f on $x \geq 0$ is

$$\int_0^{\infty} B_\omega \sin(\omega x)\,d\omega \tag{14.7}$$

in which

$$B_\omega = \frac{2}{\pi}\int_0^{\infty} f(\xi)\sin(\omega\xi)\,d\xi \tag{14.8}$$

is the *Fourier integral sine coefficient*.

Theorem 14.1 immediately gives us a convergence theorem for Fourier cosine and sine integrals on the half-line.

THEOREM 14.2 *Convergence of Fourier Cosine and Sine Integrals*

Suppose $f(x)$ is defined for $x \geq 0$ and is piecewise smooth on every interval $[0, L]$ for $L > 0$. Assume that $\int_0^\infty |f(\xi)|\,d\xi$ converges. Then, at each $x > 0$, the Fourier cosine and sine integral representations converge to

$$\frac{1}{2}(f(x+) + f(x-)).$$

Further, the cosine integral converges to $f(0+)$ at $x = 0$, and the sine integral converges to 0 at $x = 0$. ♦

EXAMPLE 14.2 Laplace's Integrals

Let $f(x) = e^{-kx}$ for $x \geq 0$, with k a positive number. Then f has a continuous derivative and is absolutely integrable on $[0, \infty)$. For the Fourier cosine integral, compute the coefficients

$$A_\omega = \frac{2}{\pi}\int_0^{\infty} e^{-k\xi}\cos(\omega\xi)\,d\xi = \frac{2}{\pi}\frac{k}{k^2+\omega^2}.$$

Then, for $x \geq 0$,

$$e^{-kx} = \frac{2k}{\pi}\int_0^{\infty} \frac{1}{k^2+\omega^2}\cos(\omega x)\,d\omega.$$

Next compute the sine coefficients

$$B_\omega = \frac{2}{\pi}\int_0^\infty e^{-k\xi}\sin(\omega\xi)d\xi = \frac{2}{\pi}\frac{\omega}{k^2+\omega^2}.$$

Then, for $x > 0$, we also have

$$e^{-kx} = \frac{2}{\pi}\int_0^\infty \frac{\omega}{k^2+\omega^2}\sin(\omega x)\,d\omega.$$

However, this integral is zero for $x = 0$ and so does not represent $f(x)$ there.

These integral representations are called *Laplace's integrals* because A_ω is $2/\pi$ times the Laplace transform of $\sin(kx)$, while B_ω is $2/\pi$ times the Laplace transform of $\cos(kx)$. ♦

SECTION 14.2 PROBLEMS

In each of Problems 1 through 10, find the Fourier cosine and sine integral representations of the function. Determine what each integral representation converges to.

1. $f(x) = \begin{cases} 2x+1 & \text{for } 0 \le x \le \pi \\ 2 & \text{for } \pi < x \le 3\pi \\ 0 & \text{for } x > 3\pi. \end{cases}$

2. $f(x) = \begin{cases} x & \text{for } 0 \le x \le 1 \\ x+1 & \text{for } 1 < x \le 2 \\ 0 & \text{for } x > 2 \end{cases}$

3. $f(x) = e^{-x}\cos(x)$ for $x \ge 0$

4. $f(x) = xe^{-3x}$ for $x \ge 0$

5. $f(x) = \begin{cases} 1 & \text{for } 0 \le x \le 1 \\ 2 & \text{for } 1 < x \le 4 \\ 0 & \text{for } x > 4 \end{cases}$

6. $f(x) = \begin{cases} \cosh(x) & \text{for } 0 \le x \le 5 \\ 0 & \text{for } x > 5 \end{cases}$

7. Let k be a nonzero number and c a positive number, and

$$f(x) = \begin{cases} k & \text{for } 0 \le x \le c \\ 0 & \text{for } x > c. \end{cases}$$

8. $f(x) = e^{-2x}\cos(x)$ for $x \ge 0$.

9. $f(x) = \begin{cases} x^2 & \text{for } 0 \le x \le 10 \\ 0 & \text{for } x > 10 \end{cases}$

10. $f(x) = \begin{cases} \sin(x) & \text{for } 0 \le x \le 2\pi \\ 0 & \text{for } x > 2\pi \end{cases}$

11. Use the Laplace integrals to compute the Fourier cosine integral of $f(x) = 1/(1+x^2)$ and the Fourier sine integral of $g(x) = x/(1+x^2)$.

14.3 The Fourier Transform

We will use equation (14.4) to derive a complex form of the Fourier integral representation of a function, and then use this to define the Fourier transform.

Suppose f is absolutely integrable on the real line, and piecewise smooth on each $[-L, L]$. Then, at any x,

$$\frac{1}{2}(f(x+)+f(x-)) = \frac{1}{\pi}\int_0^\infty \int_{-\infty}^\infty f(\xi)\cos(\omega(\xi - x))\,d\xi\,d\omega.$$

Recall that

$$\cos(x) = \frac{1}{2}\left(e^{ix}+e^{-ix}\right)$$

to obtain

$$\frac{1}{2}(f(x+)+f(x-)) = \frac{1}{\pi}\int_0^\infty \int_{-\infty}^\infty f(\xi)\frac{1}{2}\left(e^{i\omega(\xi-x)}+e^{-i\omega(\xi-x)}\right)d\xi\,d\omega$$

$$= \frac{1}{2\pi}\int_0^\infty \int_{-\infty}^\infty f(\xi)e^{i\omega(\xi-x)}\,d\xi\,d\omega$$

$$+ \frac{1}{2\pi}\int_0^\infty \int_{-\infty}^\infty f(\xi)e^{-i\omega(\xi-x)}\,d\xi\,d\omega.$$

In the next-to-last integral, replace ω with $-\omega$ and compensate for this change by replacing $\int_0^\infty \cdots d\omega$ with $\int_{-\infty}^0 \cdots d\omega$. This enables us to write

$$\frac{1}{2}(f(x+)+f(x-))$$

$$= \frac{1}{2\pi}\int_{-\infty}^0 \int_{-\infty}^\infty f(\xi)e^{-i\omega(\xi-x)}\,d\xi\,d\omega + \frac{1}{2\pi}\int_0^\infty \int_{-\infty}^\infty f(\xi)e^{-i\omega(\xi-x)}\,d\xi\,d\omega.$$

Combine these integrals to obtain

$$\frac{1}{2}(f(x+)+f(x-)) = \frac{1}{2\pi}\int_{-\infty}^\infty \int_{-\infty}^\infty f(\xi)e^{-i\omega\xi}e^{i\omega x}\,d\xi\,d\omega. \tag{14.9}$$

This is the *complex Fourier integral representation of $f(x)$ on the real line*. If we let

$$C_\omega = \int_{-\infty}^\infty f(\xi)e^{-i\omega\xi}\,d\xi,$$

then this integral representation is

$$\frac{1}{2}(f(x+)+f(x-)) = \frac{1}{2\pi}\int_{-\infty}^\infty C_\omega e^{i\omega x}d\omega.$$

We call C_ω the *complex Fourier integral coefficient of f*.

We may use this complex Fourier integral as a springboard to the Fourier transform, the idea of which is contained in equation (14.9). For emphasis in how we want to think of this equation, write it as

$$\frac{1}{2}(f(x+)+f(x-)) = \frac{1}{2\pi}\int_{-\infty}^\infty \left(\int_{-\infty}^\infty f(\xi)e^{-i\omega\xi}\,d\xi\right)e^{i\omega x}\,d\omega. \tag{14.10}$$

The term in large parentheses on the right in equation (14.10) is the Fourier transform of f. We summarize this discussion as follows.

If f is absolutely integrable on the real line, then the *Fourier transform* $\mathcal{F}[f]$ of f is the function defined by

$$\mathcal{F}[f](\omega) = \int_{-\infty}^\infty f(t)e^{-i\omega t}dt.$$

Thus, the Fourier transform of f is the coefficient C_ω in the complex Fourier integral representation of f.

Because of the use of the Fourier transform in applications such as signal analysis, we usually use t (for time) as the variable in the defining integral, and ω as the variable of the transformed function $\mathcal{F}[f]$. Engineers refer to ω in the transformed function as the *frequency* of the signal f.

We also denote $\mathcal{F}[f]$ as $\hat{f}$:

$$\mathcal{F}[f](\omega) = \hat{f}(\omega).$$

EXAMPLE 14.3

We will determine the transform of $e^{-c|t|}$, with c a positive number. First, write

$$f(t) = e^{-c|t|} = \begin{cases} e^{-ct} & \text{for } t \geq 0 \\ e^{ct} & \text{for } t < 0. \end{cases}$$

Then

$$\begin{aligned}
\mathcal{F}[f](\omega) &= \int_{-\infty}^{\infty} e^{-c|t|} e^{-i\omega t}\, dt \\
&= \int_{-\infty}^{0} e^{ct} e^{-i\omega t}\, dt + \int_{0}^{\infty} e^{-ct} e^{-i\omega t}\, dt \\
&= \int_{-\infty}^{0} e^{(c-i\omega)t}\, dt + \int_{0}^{\infty} e^{-(c+i\omega)t}\, dt \\
&= \left[\frac{1}{c-i\omega} e^{(c-i\omega)t}\right]_{-\infty}^{0} + \left[\frac{-1}{c+i\omega} e^{-(c+i\omega)t}\right]_{0}^{\infty} \\
&= \left(\frac{1}{c+i\omega} + \frac{1}{c-i\omega}\right) = \frac{2c}{c^2+\omega^2}.
\end{aligned}$$

We can also write

$$\hat{f}(\omega) = \frac{2c}{c^2+\omega^2}. \quad \blacklozenge$$

EXAMPLE 14.4

Let $H(t)$ be the Heaviside function, defined by

$$H(t) = \begin{cases} 1 & \text{for } t \geq 0 \\ 0 & \text{for } t < 0. \end{cases}$$

We will compute the Fourier transform of $f(t) = H(t)e^{-5t}$. This is the function

$$f(t) = \begin{cases} e^{-5t} & \text{for } t \geq 0 \\ 0 & \text{for } t < 0. \end{cases}$$

From the definition of $\mathcal{F}$,

$$\begin{aligned}
\hat{f}(\omega) &= \int_{-\infty}^{\infty} H(t)e^{-5t}e^{-i\omega t} dt \int_{0}^{\infty} e^{-5t}e^{-i\omega t} dt = \int_{0}^{\infty} e^{-(5+i\omega)t} dt \\
&= -\frac{1}{5+i\omega}\left[e^{-(5+i\omega)t}\right]_0^{\infty} = \frac{1}{5+i\omega}. \quad \blacklozenge
\end{aligned}$$

EXAMPLE 14.5

Let a and k be positive numbers. We will determine $\hat{f}(t)$, where

$$f(t) = \begin{cases} k & \text{for } -a \leq t < a \\ 0 & \text{for } t < -a \text{ and } t \geq a. \end{cases}$$

This is the pulse

$$f(t) = k[H(t+a) - H(t-a)].$$

Then

$$\begin{aligned}\hat{f}(t) &= \int_{-\infty}^{\infty} f(t)e^{-i\omega t}\,dt \\ &= \int_{-a}^{a} ke^{-i\omega t}\,dt = \left[\frac{-k}{i\omega}e^{-i\omega t}\right]_{-a}^{a} \\ &= -\frac{k}{i\omega}[e^{-i\omega a} - e^{i\omega a}] = \frac{2k}{\omega}\sin(a\omega). \quad \blacklozenge\end{aligned}$$

These examples were done by integration. Usually the Fourier transform of a function is computed using tables or software. In MAPLE, use

```
fourier(f(t), t, ω);
```

This is in the inttrans set of subroutines, for integral transforms. The Laplace transform is also in this set.

Now suppose that f is continuous and f' is piecewise smooth on every interval $[-L, L]$. Because $\hat{f}(\omega)$ is the coefficient in the complex Fourier integral representation of f,

$$f(t) = \frac{1}{2\pi}\int_{-\infty}^{\infty} \hat{f}(\omega)e^{i\omega t}\,d\omega. \tag{14.11}$$

Equation (14.11) defines the *inverse Fourier transform*. Given f satisfying certain conditions, we can compute its Fourier transform $\hat{f}$, and, conversely, given this transform, we can recover f from equation (14.11). For this reason we call the equations

$$\hat{f}(\omega) = \int_{-\infty}^{\infty} f(t)e^{-i\omega t}\,dt \text{ and } f(t) = \frac{1}{2\pi}\int_{-\infty}^{\infty} \hat{f}(\omega)e^{i\omega t}\,d\omega$$

a *transform pair*. We also denote the inverse Fourier transform as $\mathcal{F}^{-1}$:

$$\mathcal{F}^{-1}[\hat{f}] = f \text{ exactly when } \mathcal{F}[f] = \hat{f}.$$

In MAPLE, $\mathcal{F}^{-1}[f]$ can be computed using

```
invfourier[F, ω, t];
```

EXAMPLE 14.6

Let

$$f(t) = \begin{cases} 1 - |t| & \text{for } -1 \le t \le 1 \\ 0 & \text{for } |t| > 1. \end{cases}$$

Then f is continuous and absolutely integrable, and f' is piecewise continuous. A routine integral gives us the Fourier transform of f:

$$\begin{aligned}\hat{f}(\omega) &= \int_{-\infty}^{\infty} f(t)e^{-i\omega t}\,dt \\ &= \int_{-1}^{1}(1 - |t|)e^{-i\omega t} = \frac{2(1 - \cos(\omega))}{\omega^2}.\end{aligned}$$

As an illustration, we will compute the inverse of this Fourier transform. By equation (14.11),

$$\mathcal{F}^{-1}[\hat{f}](t) = \frac{1}{2\pi}\int_{-\infty}^{\infty} \hat{f}(\omega)e^{i\omega t}\,d\omega$$
$$= \frac{1}{\pi}\int_{-\infty}^{\infty} \frac{1-\cos(\omega)}{\omega^2}e^{i\omega t}\,d\omega$$
$$= \pi(t+1)\text{sgn}(t+1) + \pi(t-1)\text{sgn}(t-1) - 2\text{sgn}(t).$$

This integration was done using MAPLE, in which

$$\text{sgn}(t) = \begin{cases} 1 & \text{for } t > 0 \\ -1 & \text{for } t < 0 \\ 0 & \text{for } t = 0. \end{cases}$$

By considering cases $t < -1$, $-1 < t < 1$ and $t > 1$, it is routine to verify that indeed $\mathcal{F}^{-1}[\hat{f}](t) = f(t)$ in this example. ♦

In the context of the Fourier transform, the *amplitude spectrum* of a signal $f(t)$ is the graph of $|\hat{f}(\omega)|$.

EXAMPLE 14.7

Let a and k be positive numbers and let

$$f(t) = \begin{cases} k & \text{for } -a \le t \le a \\ 0 & \text{for } t < -a \text{ and for } t > a. \end{cases}$$

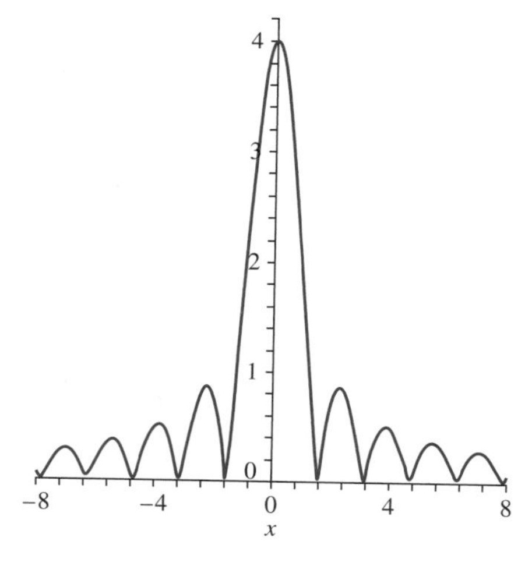

FIGURE 14.1 *Graph of $|\hat{f}(\omega)|$ in Example 14.7, for $k = 1$, $a = 2$.*

By Example 14.5,

$$\begin{aligned}\hat{f}(\omega) &= \int_{-\infty}^{\infty} f(t)e^{-i\omega t}\,dt \\ &= \int_{-a}^{a} ke^{-i\omega t}\,dt = -\frac{k}{i\omega}(e^{-i\omega t} - e^{i\omega t}) \\ &= \frac{2k}{\omega}\sin(a\omega).\end{aligned}$$

The amplitude spectrum of f is the graph of

$$|\hat{f}(\omega)| = 2k\left|\frac{\sin(a\omega)}{\omega}\right|,$$

shown in Figure 14.1 for $k = 1$ and $a = 2$. ♦

We will list some properties and computational rules for the Fourier transform.

Linearity

$$\mathcal{F}[f + g] = \mathcal{F}[f] + \mathcal{F}[g]$$

and, for any number k,

$$\mathcal{F}[kf] = k\mathcal{F}[f].$$

Time Shifting If t_0 is a real number then

$$\mathcal{F}[f(t - t_0)](\omega) = e^{-i\omega t_0}\hat{f}(\omega).$$

The Fourier transform of a shifted function $f(t - t_0)$ is the Fourier transform of f, multiplied by $e^{-i\omega t_0}$. This is similar to the second shifting theorem for the Laplace transform.

Proof From the definition of the Fourier transform,

$$\begin{aligned}\mathcal{F}[f(t - t_0)](\omega) &= \int_{-\infty}^{\infty} f(t - t_0)e^{-i\omega t}\,dt \\ &= e^{-i\omega t_0}\int_{-\infty}^{\infty} f(t - t_0)e^{-i\omega(t - t_0)}\,dt.\end{aligned}$$

Upon setting $u = t - t_0$ we have

$$\mathcal{F}[f(t - t_0)](\omega) = e^{-i\omega t_0}\int_{-\infty}^{\infty} f(u)e^{-i\omega u}\,du = e^{-i\omega t_0}\hat{f}(\omega),$$

completing the proof. ♦

The inverse version of the time shifting theorem is

$$\mathcal{F}^{-1}[e^{-i\omega t_0}\hat{f}(\omega)](t) = f(t - t_0). \tag{14.12}$$

EXAMPLE 14.8

We will compute

$$\mathcal{F}^{-1}\left[\frac{e^{2i\omega}}{5 + i\omega}\right].$$

The presence of the exponential factor $e^{2i\omega}$ suggests use of the inverse version of the time shifting theorem. Put $t_0 = -2$ and

$$\hat{f}(\omega) = \frac{1}{5 + i\omega}$$

into equation (14.12) to get

$$\mathcal{F}^{-1}[e^{2i\omega}\hat{f}(\omega)](t) = f(t - (-2)) = f(t + 2)$$

where

$$f(t) = \mathcal{F}^{-1}\left[\frac{1}{5 + i\omega}\right] = H(t)e^{-5t}$$

from Example 14.5. Then, by time shifting,

$$\mathcal{F}^{-1}\left[\frac{e^{2i\omega}}{5 + i\omega}\right] = f(t + 2) = H(t + 2)e^{-5(t+2)}. \ \blacklozenge$$

Frequency Shifting If ω_0 is any real number then

$$\mathcal{F}[e^{i\omega_0 t} f(t)] = \hat{f}(\omega - \omega_0).$$

The Fourier transform of a function multiplied by $e^{i\omega_0 t}$ is the Fourier transform of f shifted right by ω_0.

Proof To prove this result, compute

$$\mathcal{F}[e^{-\omega_0 t} f(t)](\omega) = \int_{-\infty}^{\infty} e^{i\omega_0 t} f(t)e^{-i\omega t}\, dt$$

$$= \int_{-\infty}^{\infty} e^{-i(\omega - \omega_0)t}\, dt = \hat{f}(\omega - \omega_0).$$

The inverse version of frequency shifting is that

$$\mathcal{F}^{-1}[\hat{f}(\omega - \omega_0)](t) = e^{i\omega_0 t} f(t). \ \blacklozenge$$

Time shifting and frequency shifting are reminiscent of the two shifting theorems for the Laplace transform.

Scaling If c is any nonzero real number, then

$$\mathcal{F}[f(ct)](\omega) = \frac{1}{|c|}\hat{f}(\omega/c).$$

Scaling can be verified by a change of variables $u = ct$ in the integral for the transform of $f(ct)$.

The inverse version of the scaling theorem is

$$\mathcal{F}^{-1}[\hat{f}(\omega/c)] = |c| f(ct).$$

Time Reversal

$$\mathcal{F}[f(-t)](\omega) = \hat{f}(-\omega).$$

Time reversal follows immediately from the scaling theorem upon putting $c = -1$.

Symmetry

$$\mathcal{F}[\hat{f}(t)](\omega) = 2\pi f(-\omega).$$

If we replace ω by t in the transformed function $\hat{f}$, and then take the transform of this function of t, we obtain 2π times the original function f with t replaced by $-\omega$.

Modulation If ω_0 is a real number, then

$$\mathcal{F}[f(t)\cos(\omega_0 t)](\omega) = \frac{1}{2}\left(\hat{f}(\omega+\omega_0) + \hat{f}(\omega-\omega_0)\right)$$

and

$$\mathcal{F}[f(t)\sin(\omega_0 t)](\omega) = \frac{i}{2}\left(\hat{f}(\omega+\omega_0) - \hat{f}(\omega-\omega_0)\right)$$

To prove the first expression, put

$$\cos(\omega t) = \frac{1}{2}\left(e^{i\omega t} + e^{-i\omega t}\right),$$

then use the linearity of $\mathcal{F}$ and the frequency-shifting theorem to write

$$\begin{aligned}\mathcal{F}[f(t)\cos(\omega_0 t)](\omega) &= \mathcal{F}\left[\frac{1}{2}e^{i\omega_0 t} f(t) + \frac{1}{2}e^{-i\omega_0 t} f(t)\right](\omega) \\ &= \frac{1}{2}\mathcal{F}[e^{i\omega_0 t} f(t)](\omega) + \frac{1}{2}\mathcal{F}[e^{-i\omega_0 t} f(t)](\omega) \\ &= \frac{1}{2}\hat{f}(\omega-\omega_0) + \frac{1}{2}\hat{f}(\omega+\omega_0).\end{aligned}$$

The second conclusion is proved by a similar calculation.

Operational Formula To apply the Fourier transform to a differential equation we must be able to transform a derivative. This is called an *operational rule*. Recall that the kth derivative of f is denoted $f^{(k)}$. As a convenience, we let $f^{(0)} = f$ - the zero-order derivative of a function is just the function.

Now let n be any positive integer and suppose that $f^{(n-1)}$ is continuous and $f^{(n)}$ is piecewise continuous on each interval $[-L, L]$. Suppose also that $\int_{-\infty}^{\infty} |f^{(n-1)}|\,dt$ converges and that

$$\lim_{t\to\infty} f^{(k)}(t) = \lim_{t\to-\infty} f^{(k)}(t) = 0$$

for $k = 0, 1, 2, \cdots, n-1$. Then

$$\mathcal{F}[f^{(n)}(t)](\omega) = (i\omega)^n \hat{f}(\omega).$$

That is, under the given conditions, the Fourier transform of the nth derivative of f is the nth power of $i\omega$ times the Fourier transform of f.

Proof Since

$$f^{(n)}(t) = \frac{d}{dt} f^{(n-1)}(t),$$

it is enough to derive the operational formula when $n = 1$. Integrate by parts:

$$[f'(t)](\omega) = \int_{-\infty}^{\infty} f'(t)e^{-i\omega_0 t}\,dt$$

$$= \left[f(t)e^{-i\omega t}\right]_{\infty}^{\infty} - \int_{-\infty}^{\infty} f(t)(-i\omega)e^{-i\omega t}\,dt$$

$$= i\omega \int_{-\infty}^{\infty} e^{-i\omega t} f(t)\,dt$$

$$= i\omega \hat{f}(\omega),$$

where we have used the fact that $f(t)$ has limit 0 at ∞ and at $-\infty$ to conclude that

$$\left[f(t)e^{-i\omega t}\right]_{-\infty}^{\infty} = 0. \blacklozenge$$

Now an inductive argument leads to the conclusion for the nth derivative.

EXAMPLE 14.9

We will solve the differential equation

$$y' - 4y = H(t)e^{-4t}.$$

Apply the Fourier transform to the differential equation to get

$$\mathcal{F}[y'(\omega)] - 4\hat{y}(\omega) = \mathcal{F}[H(t)e^{-4t}](\omega).$$

From the operational rule with $n = 1$,

$$\mathcal{F}[y'](\omega) = i\omega\hat{y}(\omega).$$

Further, from Example 14.4, with 4 in place of 5,

$$\mathcal{F}[H(t)e^{-4t}](\omega) = \frac{1}{4 + i\omega}.$$

Therefore

$$i\omega\hat{y} - 4\hat{y} = \frac{1}{4 + i\omega}.$$

Solve for $\hat{y}$ to get

$$\hat{y}(\omega) = \frac{-1}{16 + \omega^2}.$$

From Example 14.3,

$$y(t) = \mathcal{F}^{-1}\left[\frac{-1}{16 + \omega^2}\right] = -\frac{1}{8}e^{-4|t|}. \blacklozenge$$

The operational formula can be adjusted to accommodate a finite number of jump discontinuities of f. If these occur at $t_1, \cdots, t_M$ and if

$$\lim_{t \to -\infty} f(t) = \lim_{t \to \infty} f(t) = 0,$$

then

$$\mathcal{F}[f'(t)](\omega) = i\omega\hat{f}(\omega) - \sum_{j=1}^{M}(f(t_j+) - f(t_j-))e^{-it_j\omega}.$$

Each term $f(t_j+) - f(t_j-)$ is the magnitude of the jump discontinuity at t_j.

Frequency Differentiation The variable ω used in $\hat{f}(\omega)$ is the frequency of $f(t)$, since it occurs in the complex exponential $e^{i\omega t}$, which is $\cos(\omega t) + i\sin(\omega t)$. In this context, the process of computing $\hat{f}'(\omega)$ is called *frequency differentiation*. We will show how derivatives of $\hat{f}(\omega)$ relate to $f(t)$.

Let n be a positive integer. Let f be piecewise continuous on $[-L, L]$ for every positive L and assume that $\int_{-\infty}^{\infty} |t^n f(t)|\,dt$ converges. Then

$$\frac{d^n}{d\omega^n}\hat{f}(\omega) = i^{-n}\mathcal{F}[t^n f(t)](\omega).$$

This means that the nth derivative of the Fourier transform of f is i^{-n} times the transform of $t^n f(t)$.

We will indicate a proof for the case $n = 1$. Write

$$\begin{aligned}\frac{d}{d\omega}\hat{f}(\omega) &= \frac{d}{d\omega}\int_{-\infty}^{\infty} f(t)e^{-i\omega t}\,dt = \int_{-\infty}^{\infty}\frac{\partial}{\partial\omega}[f(t)e^{-i\omega t}]\,dt \\ &= \int_{-\infty}^{\infty} f(t)(-it)e^{-i\omega t}\,dt = -i\int_{-\infty}^{\infty}[tf(t)]e^{-i\omega t}\,dt \\ &= -i\mathcal{F}[tf(t)](\omega).\end{aligned}$$

As an example, using the result of Example 14.3, we can write

$$\mathcal{F}[t^2e^{-5|t|}](\omega) = i^2\frac{d^2}{d\omega^2}\left(\frac{10}{25+\omega^2}\right) = 20\left(\frac{25-3\omega^2}{(25+\omega^2)^2}\right).$$

The Fourier Transform of an Integral Let f be piecewise continuous on every interval $[-L, L]$. Suppose $\int_{-\infty}^{\infty} |f(t)|\,dt$ converges and that $f(0) = 0$. Then

$$\mathcal{F}\left[\int_{-\infty}^{t} f(\tau)d\tau\right](\omega) = \frac{1}{i\omega}\hat{f}(\omega).$$

To prove this, define $g(t) = \int_{-\infty}^{t} f(\tau)\,d\tau$. Then $g'(t) = f(t)$ at each point at which f is continuous. Further, $g(t) \to 0$ as $t \to -\infty$, and

$$\lim_{t\to\infty} g(t) = \int_{-\infty}^{\infty} f(\tau)\,d\tau = \hat{f}(0) = 0$$

by assumption. Therefore, applying the operational formula,

$$\begin{aligned}\hat{f}(\omega) &= \mathcal{F}[g'(t)](\omega) \\ &= i\omega[g(t)](\omega) = i\omega\mathcal{F}\left[\int_{-\infty}^{t} f(\tau)\,d\tau\right](\omega).\end{aligned}$$

Convolution

Integral transforms usually have some kind of convolution operation. We have seen a convolution for the Laplace transform. For the Fourier transform, we define the *convolution* of f with g to be the function $f * g$ given by

$$(f * g)(t) = \int_{-\infty}^{\infty} f(t-\tau)g(\tau)\,d\tau.$$

In making this definition, we assume that $\int_a^b f(t)\,dt$ and $\int_a^b g(t)\,dt$ exist for every interval $[a, b]$ and that, for every real number t, $\int_{-\infty}^{\infty} |f(t-\tau)g(t)|\,d\tau$ converges.

Convolution has the following properties.

Commutativity If $f * g$ is defined, so is $g * f$ and

$$f * g = g * f.$$

Linearity This means that, for numbers α and β and functions f, g and h,

$$(\alpha f + \beta g) * h = \alpha(f * h) + \beta(g * h)$$

provided that all these convolutions are defined.

For the next three properties of convolution, suppose that f and g are bounded and continuous on the real line and that f and g are both absolutely integrable. Then

$$\int_{-\infty}^{\infty} (f * g)(t)\, dt = \int_{-\infty}^{\infty} f(t)\, dt \int_{-\infty}^{\infty} g(t)\, dt.$$

Time Convolution

$$\mathcal{F}[f * g] = \hat{f}\hat{g}.$$

This says that the Fourier transform of the convolution of two functions is the product of the Fourier transforms of the functions. This is known as the *convolution theorem*, and a similar result holds for the Laplace transform. The ramification of convolution for the inverse Fourier transform is that

$$\mathcal{F}^{-1}[\hat{f}(\omega)\hat{g}(\omega)](t) = (f * g)(t).$$

That is, the inverse Fourier transform of a product of two transformed functions is the convolution of the functions.

Frequency Convolution

$$\mathcal{F}[fg](\omega) = \frac{1}{2\pi}(\hat{f} * \hat{g})(\omega).$$

EXAMPLE 14.10

We will compute

$$\mathcal{F}^{-1}\left[\frac{1}{(4+\omega^2)(9+\omega^2)}\right].$$

We want the inverse transform of a product, knowing the inverse of each factor:

$$\mathcal{F}^{-1}\left(\frac{1}{4+\omega^2}\right) = f(t) = \frac{1}{4}e^{-2|t|}$$

and

$$\mathcal{F}^{-1}\left(\frac{1}{9+\omega^2}\right) = g(t) = \frac{1}{6}e^{-3|t|}.$$

The inverse version of the convolution theorem tells us that

$$\mathcal{F}^{-1}\left[\frac{1}{(4+\omega^2)(9+\omega^2)}\right](t) = (f * g)(t) = \frac{1}{24}\int_{-\infty}^{\infty} e^{-2|t-\tau|}e^{-3|\tau|}\, d\tau.$$

To evaluate this integral, we must consider three cases. If $t > 0$ then

$$24(f*g)(t) = \int_{-\infty}^{0} e^{-2|t-\tau|}e^{-3|\tau|}d\tau + \int_{0}^{t} e^{-2|t-\tau|}e^{-3|\tau|}d\tau + \int_{t}^{\infty} e^{-2|t-\tau|}e^{-3|\tau|}d\tau$$

$$= \int_{-\infty}^{0} e^{-2(t-\tau)}e^{3\tau}d\tau + \int_{0}^{t} e^{-2(t-\tau)}e^{-3\tau}d\tau + \int_{t}^{\infty} e^{-2(t-\tau)}e^{-3\tau}d\tau$$

$$= \frac{6}{5}e^{-2t} - \frac{4}{5}e^{-3t}.$$

If $t < 0$, then

$$24(f*g)(t) = \int_{-\infty}^{t} e^{-2|t-\tau|}e^{-3|\tau|}d\tau + \int_{t}^{0} e^{-2|t-\tau|}e^{-3|\tau|}d\tau + \int_{0}^{\infty} e^{-2|t-\tau|}e^{-3|\tau|}d\tau$$

$$= \int_{-\infty}^{t} e^{-2(t-\tau)}e^{3\tau}d\tau + \int_{t}^{0} e^{2(t-\tau)}e^{3\tau}d\tau + \int_{0}^{\infty} e^{2(t-\tau)}e^{-3\tau}d\tau$$

$$= -\frac{4}{5}e^{3t} + \frac{6}{5}e^{2t}.$$

Finally, if $t = 0$,

$$24(f*g)(0) = \int_{-\infty}^{\infty} e^{-2|\tau|}e^{-3|\tau|}\,d\tau = \frac{2}{5}.$$

Therefore

$$\mathcal{F}^{-1}\left[\frac{1}{(4+\omega^2)(9+\omega^2)}\right](t) = \frac{1}{24}\left(\frac{6}{5}e^{-2|t|} - \frac{4}{5}e^{-3|t|}\right)$$

$$= \frac{1}{20}e^{-2|t|} - \frac{1}{30}e^{-3|t|}. \quad \blacklozenge$$

14.3.1 Filtering and the Dirac Delta Function

The *Dirac delta function* $\delta(t)$ was discussed in Section 3.5. We may think of this function as the limit of a pulse (Section 3.3.2), as the height tends to infinity and the duration to zero. In terms of the Heaviside function $H(t)$,

$$\delta(t) = \lim_{a\to 0+} \frac{1}{2a}[H(t+a) - H(t-a)].$$

In this definition, the pulse is centered at 0, extending from $t-a$ to $t+a$. Often we deal with the *shifted delta function* $H(t-t_0)$, in which the defining pulse is centered at t_0.

The filtering property of the delta function enables us to recover a function value $f(t_0)$ by "summing" function values when they are impacted with a shifted delta function.

THEOREM 14.3 *Filtering by a Delta Function*

If $f(t)$ has a Fourier transform and is continuous at t_0, then

$$\int_{-\infty}^{\infty} f(t)\delta(t-t_0)\,dt = f(t_0). \quad \blacklozenge$$

Proof To prove this, first observe that

$$H(t-t_0+a) - H(t-t_0-a) = \begin{cases} 0 & \text{for } t \le t_0 - a \text{ and for } t > t_0 + a, \\ 1 & \text{for } t_0 - a \le t < t_0 + a. \end{cases}$$

Now use the definition of $\delta(t)$ to write

$$\begin{aligned}
\int_{-\infty}^{\infty} f(t)\delta(t-t_0)\,dt &= \int_{-\infty}^{\infty} f(t)\left[\lim_{a\to 0+}\frac{1}{2a}[H(t-t_0+a)-H(t-t_0-a)]\right]dt \\
&= \lim_{a\to 0+}\frac{1}{2a}\int_{-\infty}^{\infty} f(t)[H(t-t_0+a)-H(t-t_0-a)]\,dt \\
&= \lim_{a\to 0+}\frac{1}{2a}\int_{t_0-a}^{t_0+a} f(t)\,dt.
\end{aligned}$$

By the mean value theorem for integrals, for some ξ_a,

$$\int_{t_0-a}^{t_0+a} f(t)\,dt = 2af(\xi_a)$$

where $t_0-a<\xi_a<t_0+a$. As $a\to 0+$, $\xi_a\to t_0$, so $f(\xi_a)\to f(t_0)$ and then

$$\int_{-\infty}^{\infty} f(t)\delta(t-t_0)\,dt = \lim_{a\to 0}\frac{1}{2a}(2af(\xi_a)) = f(t_0). \quad \blacklozenge$$

If f has a jump discontinuity at t_0, this argument can be modified to yield

$$\int_{-\infty}^{\infty} f(t)\delta(t-t_0)\,dt = \frac{1}{2}[f(t_0+)+f(t_0-)].$$

We will derive the Fourier transform of the delta function. Begin with

$$\begin{aligned}
\mathcal{F}[H(t+a)-H(t-a)] &= \int_{-a}^{a} e^{-i\omega t}\,dt = -\frac{1}{i\omega}e^{-i\omega t}\Big]_{-1}^{a} \\
&= \frac{1}{i\omega}\left(e^{ia\omega}-e^{-ia\omega}\right) = 2\frac{\sin(a\omega)}{\omega}.
\end{aligned}$$

By interchanging the limit and the operation of taking the Fourier transform, we have

$$\begin{aligned}
\mathcal{F}[\delta(t)](\omega) &= \mathcal{F}\left[\lim_{a\to 0+}\frac{1}{2a}[H(t+a)-H(t-a)]\right](\omega) \\
&= \lim_{a\to 0+}\frac{1}{2a}\mathcal{F}[H(t+a)-H(t-a)](\omega) \\
&= \lim_{a\to 0+}\frac{\sin(a\omega)}{a\omega} = 1.
\end{aligned}$$

This formal manipulation leads us to

$$\mathcal{F}[\delta(t)](\omega) = 1.$$

The Fourier transform of the delta function is the constant function taking on the value 1. Now use this with the convolution:

$$\mathcal{F}[\delta * f] = \mathcal{F}[\delta]\mathcal{F}[f] = \mathcal{F}[f]$$

and

$$\mathcal{F}[f * \delta] = \mathcal{F}[f]\mathcal{F}[\delta] = \mathcal{F}[f],$$

suggesting that

$$\delta * f = f * \delta = f.$$

The delta function behaves like the identity under Fourier convolution.

14.3.2 The Windowed Fourier Transform

Let $f(t)$ be a *signal* (function). We assume that $\int_{-\infty}^{\infty} |f(t)|^2\,dt$ is finite. This integral is defined to be the *energy* of the signal.

In analyzing a signal, we sometimes want to localize the frequency content with respect to time. We know that $\hat{f}(\omega)$ carries information about the frequencies ω of the signal. However, $\hat{f}(\omega)$ does not particularize this information to specific time intervals, since

$$\hat{f}(\omega) = \int_{-\infty}^{\infty} f(t)e^{-i\omega t}\,dt,$$

and this integration is over all time. From this we can compute the total amplitude spectrum $|\hat{f}(\omega)|$, but cannot look at small time intervals. If we think of $f(t)$ as a piece of music, we have to wait until the entire piece is done before computing this amplitude spectrum.

We can obtain a picture of the frequency content of $f(t)$ within a given time interval by windowing the signal before taking its transform. The idea is to use a *window function* $w(t)$ that is nonzero only on a finite interval, often $[0, T]$ or $[-T, T]$. *Window* $f(t)$ with $w(t)$ by forming the product $w(t)f(t)$, which can be nonzero only on the selected interval. The *windowed Fourier transform* of f, with respect to the particular window function w, is

$$\hat{f}_{\text{win}}(\omega) = \int_{-\infty}^{\infty} w(t)f(t)e^{-i\omega t}\,dt.$$

EXAMPLE 14.11

Let $f(t) = 6e^{-|t|}$. Then

$$\hat{f}(\omega) = \int_{-\infty}^{\infty} 6e^{-|t|}e^{-i\omega t}\,dt = \frac{12}{1+\omega^2}.$$

We will window f with the window function

$$w(t) = \begin{cases} 1 & \text{for } -2 \le t \le 2, \\ 0 & \text{for } |t| > 2. \end{cases}$$

Figures 14.2, 14.3, and 14.4 show, respectively, $f(t)$, the window function $w(t)$, and $w(t)f(t)$. The effect of windowing on this signal is to cut the signal off for times $|t| > 2$. The windowed Fourier transform is therefore an integral only over $[-2, 2]$ instead of the entire real line:

$$\begin{aligned}\hat{f}_{\text{win}}(\omega) &= \int_{-\infty}^{\infty} 6w(t)e^{-|t|}e^{-i\omega t}\,dt \\ &= \int_{-2}^{2} 6e^{-|t|}e^{-i\omega t}\,dt \\ &= \frac{12}{1+\omega^2}\left(-2e^{-2}\cos^2(\omega) + e^{-2} + e^{-2}\omega\sin(2\omega) + 1\right).\end{aligned}$$ ♦

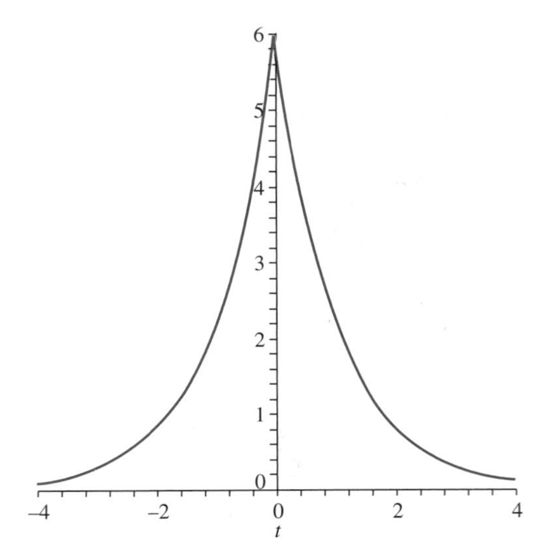

FIGURE 14.2 $f(t) = 6e^{-|t|}$

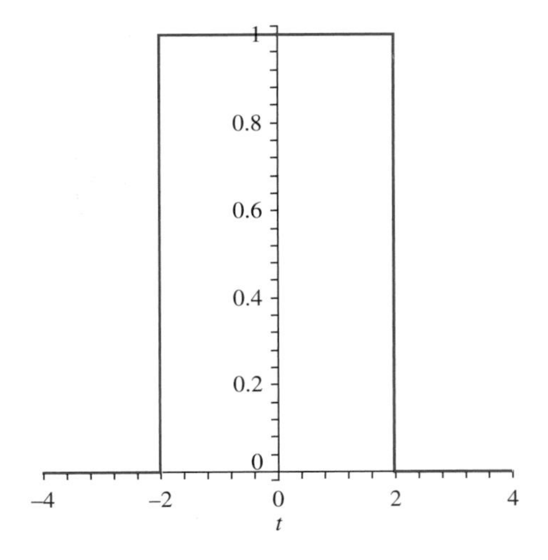

FIGURE 14.3 *Window function* $w(t)$ *in Example 14.11.*

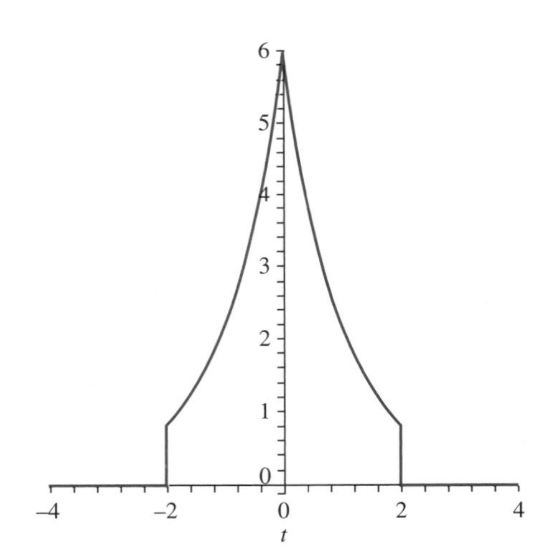

FIGURE 14.4 *Windowed function* $w(t)f(t)$ *in Example 14.11.*

Sometimes we use a *shifted window function*. If $w(t)$ is nonzero only on $[-T, T]$, then the shifted function $w(t - t_0)$ is the graph of $w(t)$ shifted t_0 units to the right and is nonzero only on $[t_0 - T, t_0 + T]$. In this case, the shifted windowed Fourier transform is the transform of $w(t - t_0)f(t)$:

$$\hat{f}_{\text{win},t_0}(\omega) = \mathcal{F}[w(t - t_0)f(t)](\omega)$$
$$= \int_{t_0 - T}^{t_0 + T} w(t - t_0)f(t)e^{-i\omega t}\,dt.$$

This gives the frequency content of the signal in the time interval $[t_0 - T, t_0 + T]$.

Engineers refer to the windowing process as *time-frequency localization*. The *center* of the window function w is defined to be

$$t_C = \frac{\int_{-\infty}^{\infty} t|w(t)|^2\,dt}{\int_{-\infty}^{\infty} |w(t)|^2\,dt}.$$

The number

$$t_R = \left(\frac{\int_{-\infty}^{\infty} (t-t_C)^2|w(t)|^2\,dt}{\int_{-\infty}^{\infty} |w(t)|^2\,dt}\right)^{1/2}$$

is the *radius* of the window function. The *width* of the window function is $2t_R$, a number referred to as the RMS *duration of the window*.

Similar terminology applies when we deal with the the Fourier transform of the window function:

$$\text{center of } \hat{w} = \omega_C = \frac{\int_{-\infty}^{\infty} \omega|\hat{w}(\omega)|^2\,d\omega}{\int_{-\infty}^{\infty} |\hat{w}(\omega)|^2\,d\omega}$$

and

$$\text{radius of } \hat{w} = \omega_R = \left(\frac{\int_{-\infty}^{\infty} (\omega-\omega_C)^2|\hat{w}(\omega)|^2\,d\omega}{\int_{-\infty}^{\infty} |\hat{w}(\omega)|^2\,d\omega}\right)^{1/2}.$$

The *width* of $\hat{w}$ is $2\omega_R$, a number referred to as the RMS *bandwidth* of the window function.

14.3.3 The Shannon Sampling Theorem

A signal $f(t)$ is *band-limited* if its Fourier transform $\hat{f}(\omega)$ has nonzero values only on some interval $[-L, L]$. If f is band-limited, the smallest positive L for which this is true is called the *bandwidth* of f. For such L we have

$$\hat{f}(\omega) = 0 \text{ if } |\omega| > L.$$

The total frequency content of such a signal lies in the band $[-L, L]$.

We will show that a band-limited signal can be reconstructed from samples taken at appropriately chosen times. Begin with the integral for the inverse Fourier transform:

$$f(t) = \frac{1}{2\pi}\int_{-\infty}^{\infty} \hat{f}(\omega)e^{i\omega t}\,d\omega.$$

Because f is assumed to have bandwidth L, we actually have

$$f(t) = \frac{1}{2\pi}\int_{-L}^{L} \hat{f}(\omega)e^{i\omega t}\,d\omega. \tag{14.13}$$

Now expand $\hat{f}(\omega)$ in a complex Fourier series on $[-L, L]$:

$$\hat{f}(\omega) = \sum_{n=-\infty}^{\infty} c_n e^{n\pi i\omega/L}, \tag{14.14}$$

where

$$c_n = \frac{1}{2L}\int_{-L}^{L} \hat{f}(\omega)e^{-n\pi i\omega/L}\,d\omega.$$

Compare c_n with $f(t)$ in equations (14.13) and (14.14) to conclude that

$$c_n = \frac{\pi}{L} f\left(\frac{-n\pi}{L}\right).$$

Substitute this into equation (14.14) to get

$$\hat{f}(\omega) = \sum_{n=-\infty}^{\infty} \frac{\pi}{L} f\left(\frac{-n\pi}{L}\right) e^{n\pi i\omega/L}.$$

Since n takes on all integer values (zero, positive and negative) in this summation, we can replace n with $-n$ without changing the sum:

$$\hat{f}(\omega) = \frac{\pi}{L}\sum_{n=-\infty}^{\infty} f\left(\frac{n\pi}{L}\right) e^{-n\pi i\omega/L}.$$

Substitute this expansion of $\hat{f}(\omega)$ into equation (14.13) to get

$$f(t) = \frac{1}{2\pi}\frac{\pi}{L}\int_{-L}^{L} f\left(\frac{n\pi}{L}\right) e^{-n\pi i\omega/L}e^{i\omega t}\,d\omega.$$

Now interchange the summation and the integral and carry out the integration to get

$$\begin{aligned}
f(t) &= \frac{1}{2L}\sum_{-\infty}^{\infty} f\left(\frac{n\pi}{L}\right)\int_{-L}^{L} e^{i\omega(t-n\pi/L)}\,d\omega \\
&= \frac{1}{2L}\sum_{n=-\infty}^{\infty} f\left(\frac{n\pi}{L}\right)\frac{1}{i(t-n\pi/L)}\left[e^{i\omega(t-n\pi/L)}\right]_{-L}^{L} \\
&= \frac{1}{2L}\sum_{n=-\infty}^{\infty} f\left(\frac{n\pi}{L}\right)\frac{1}{i(t-n\pi/L)}\left(e^{i(Lt-n\pi)} - e^{-i(Lt-n\pi)}\right) \\
&= \sum_{n=-\infty}^{\infty} f\left(\frac{n\pi}{L}\right)\frac{1}{Lt-n\pi}\frac{1}{2i}\left(e^{i(Lt-n\pi)} - e^{-(Lt-n\pi)}\right) \\
&= \sum_{n=-\infty}^{\infty} f\left(\frac{n\pi}{L}\right)\frac{\sin(Lt-n\pi)}{Lt-n\pi}.
\end{aligned}$$

This is the *Shannon sampling theorem*. It says that we know $f(t)$ at all times if we know just the function values $f(n\pi/L)$ for all integers n. An engineer would sample the signal $f(t)$ at times $0, \pm\pi/L, \pm 2\pi/L, \cdots$ and be able to reconstruct the entire signal. This is how engineers convert digital signals to analog signals, with application to technology such as that used in making compact disks.

In the case $L = \pi$ the Shannon sampling theorem is

$$f(t) = \sum_{n=-\infty}^{\infty} f(n)\frac{\sin(\pi(t-n))}{\pi(t-n)}.$$

14.3.4 Low-Pass and Bandpass Filters

If f is a signal with finite energy, then the spectrum of f is given by its Fourier transform. If ω_0 is a positive number and f is not band-limited, we can replace f with a band-limited signal f_{ω_0} having bandwidth not exceeding ω_0 by applying a low-pass filter which cuts $\hat{f}(\omega)$ off at frequencies outside the range $[-\omega_0, \omega_0]$. That is, let

$$\hat{f}_{\omega_0}(\omega) = \begin{cases} \hat{f}(\omega) & \text{for } -\omega_0 \le \omega \le \omega_0, \\ 0 & \text{for } |\omega| > \omega_0. \end{cases}$$

This defined the transform $\hat{f}_{\omega_0}$, from which we recover f_{ω_0} by the inverse Fourier transform

$$f_{\omega_0}(t) = \frac{1}{2\pi}\int_{-\infty}^{\infty} \hat{f}_{\omega_0}\omega)e^{i\omega t}\,d\omega = \frac{1}{2\pi}\int_{-\omega_0}^{\omega_0} \hat{f}_{\omega_0}(\omega)e^{i\omega t}\,d\omega.$$

Applying a low-pass filter is actually a windowing process. Define the *characteristic function* χ_I of an interval I by

$$\chi_I(t) = \begin{cases} 1 & \text{for } t \text{ in } I, \\ 0 & \text{for } t \text{ not in } I. \end{cases}$$

Then

$$\hat{f}_{\omega_0}(\omega) = \chi_{[-\omega_0,\omega_0]}(\omega)\hat{f}(\omega) \tag{14.15}$$

so we have windowed $\hat{f}(\omega)$ with the characteristic function $\chi_{[-\omega_0,\omega_0]}$. More succinctly,

$$\hat{f}_{\omega_0} = \chi_{[-\omega_0,\omega_0]}\hat{f}.$$

In this context, the window function $\chi_{[-\omega_0,\omega_0]}$ is called the *transfer function*. The inverse Fourier transform of the transfer function is

$$\mathcal{F}^{-1}[\chi_{[-\omega_0,\omega_0]}](t) = \frac{1}{2\pi}\int_{-\omega_0}^{\omega_0} e^{i\omega t}\,d\omega = \frac{\sin(\omega_0 t)}{\pi t},$$

whose graph is given in Figure 14.5 for $\omega_0 = \pi$.

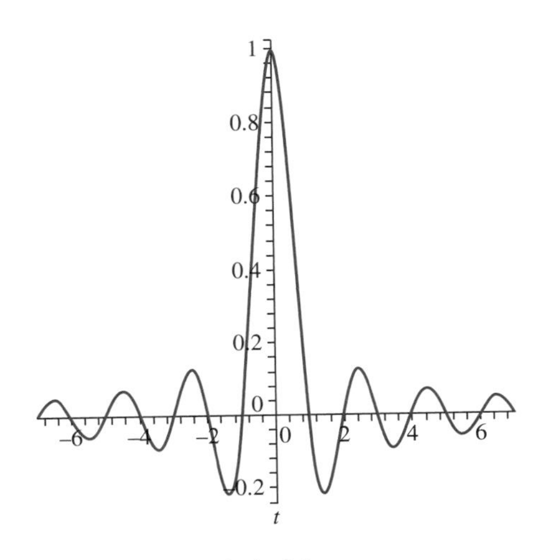

FIGURE 14.5 $sin(\pi t)/t$

Now recall time and frequency convolution for the Fourier transform. Analog filtering in the time variable is done by convolution. If $\varphi(t)$ is the filter function, then the effect of filtering f by φ is

$$f_\varphi(t) = (\varphi * f)(t) = \int_{-\infty}^{\infty} \varphi(\tau) f(t-\tau)\, d\tau.$$

Take the Fourier transform of this equation to obtain

$$\hat{f}_\varphi(\omega) = \hat{\varphi}(\omega)\hat{f}(\omega).$$

We therefore filter f in the frequency variable by taking a product of the Fourier transform of the filter function φ with the transform of f.

Using the convolution theorem, we can formulate equation (14.15) as

$$f_{\omega_0}(t) = \left(\frac{\sin(\omega_0 t)}{\pi t} * f(t)\right).$$

This gives the low-pass filtering of f as the convolution of the Shannon sampling function with f.

In low-pass filtering, we produce a band-limited signal f_{ω_0} from f. This filters out frequencies outside of $[-\omega_0, \omega_0]$. In a similar kind of filtering, called *bandpass filtering*, we want to filter out the effects of the signal outside of given bandwidths. A band-limited signal f can always be decomposed into a sum of signals, each of which carries the information content of f within a specified frequency band. To see how to do this suppose f is band-limited with bandwidth Ω. Consider a finite increasing sequence of frequencies

$$0 < \omega_1 < \omega_2 < \cdots < \omega_N = \Omega.$$

For $j = 1, 2, \cdots, N$, define a bandwidth filter function β_j by means of its transfer function:

$$\hat{\beta}_j = \chi_{[-\omega_j, -\omega_{j-1}]} + \chi_{[\omega_{j-1}, \omega_j]}.$$

This is a sum of characteristic functions of frequency intervals, and is zero outside of these intervals and 1 for $-\omega_j \le \omega \le -\omega_{j-1}$ and $\omega_{j-1} \le \omega \le \omega_j$. The bandwidth filter function $\beta_j(t)$, which filters the frequency content of $f(t)$ outside of the frequency range $[\omega_{j-1}, \omega_j]$, is obtained as the inverse Fourier transform of $\hat{\beta}_j(\omega)$. We obtain

$$\beta_j(t) = \frac{\sin(\omega_j t) - \sin(\omega_{j-1} t)}{\pi t}.$$

A graph of this function is shown in Figure 14.6. Now define functions

$$f_0(t) = \frac{\sin(\omega_0 t)}{\pi t} * f(t)$$

and, for $j = 1, 2, \cdots, N$,

$$f_j(t) = \beta_j(t) * f(t).$$

Then, for $j = 1, 2, \cdots, N$, $f_j(t)$ carries the content of $f(t)$ in the frequency range $\omega_{j-1} \le \omega \le \omega_j$, while $f_0(t)$ carries the content in $[0, \omega_0]$, the low-frequency range of $f(t)$. Furthermore,

$$f(t) = f_0(t) + f_1(t) + \cdots + f_N(t).$$

This is a decomposition of the signal into components, each carrying information about the frequency in a specific frequency interval.

We conclude this section with an observation connecting low-pass filters to a previously seen phenomenon. In signal analysis, the Gibbs phenomenon can be thought of as the step response of a low-pass filter. In this context the oscillations near the point of discontinuity are called *ringing artifacts*.

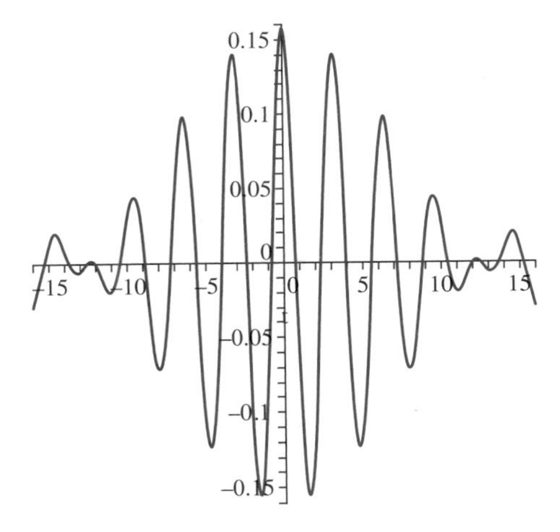

FIGURE 14.6 *Graph of* $\beta_j(t)$ *for* $\omega_j = 2.2$, $\omega_{j-1} = 1.7$.

SECTION 14.3 PROBLEMS

In each of Problems 1 through 10, find the Fourier transform of the function and graph the amplitude spectrum. Wherever k appears it is a positive constant. Use can be made of the following transforms:

$$\mathcal{F}[e^{-kt^2}](\omega) = \sqrt{\frac{\pi}{k}}e^{-\omega^2/4k}$$

and

$$\mathcal{F}\left[\frac{1}{k^2+t^2}\right](\omega) = \frac{\pi}{k}e^{-k|\omega|}$$

1. $f(t) = 5[H(t-3) - H(t-11)]$
2. $f(t) = H(t-k)t^2$
3. $f(t) = \begin{cases} 1 & \text{for } 0 \le t \le 1 \\ -1 & \text{for } -1 \le t < 0 \\ 0 & \text{for } |t| > 1 \end{cases}$
4. $f(t) = H(t-3)e^{-2t}$
5. $f(t) = 3e^{-4|t+2|}$
6. $f(t) = 5e^{-3(t-5)^2}$
7. $f(t) = H(t-k)e^{-t/4}$
8. $f(t) = \begin{cases} \sin(t) & \text{for } -k \le t \le k \\ 0 & \text{for } |t| > k \end{cases}$
9. $f(t) = 1/(1+t^2)$
10. $f(t) = 3H(t-2)e^{-3t}$

In each of Problems 11 through 15, find the inverse Fourier transform of the function.

11. $(1+i\omega)/(6-\omega^2+5i\omega)$ *Hint*: Factor the denominator and use partial fractions.
12. $10\sin(3\omega)/(\omega+\pi)$
13. $9e^{-(\omega+4)^2/32}$
14. $e^{(20-4\omega)i}/(3-(5-\omega)i)$
15. $e^{(2\omega-6)i}/(5-(3-\omega)i)$

In each of Problems 16, 17, and 18, use convolution to find the inverse Fourier transform of the function.

16. $\sin(3\omega)/\omega(2+i\omega)$
17. $1/(1+i\omega)^2$
18. $1/((1+i\omega)(2+i\omega))$
19. Prove the following version of Parseval's theorem:
$$\int_{-\infty}^{\infty} |f(t)|^2\,dt = \frac{1}{2\pi}\int_{-\infty}^{\infty} |\hat{f}(\omega)|^2\,d\omega.$$
20. Use the Fourier transform to solve
$$y'' + 6y' + 5y = \delta(t-3).$$
21. Compute the total energy of the signal $f(t) = (1/t)\sin(3t)$. *Hint*: Use Parseval's theorem, Problem 19.
22. Compute the total energy of the signal $f(t) = H(t)e^{-2t}$.

In each of Problems 23 through 28, compute the windowed Fourier transform of f for the given window function w. Also compute the center and RMS bandwidth of the window function.

23. $f(t) = (t+2)^2, \quad w(t) = \begin{cases} 1 & \text{for } -2 \le t \le 2, \\ 0 & \text{for } |t| > 2. \end{cases}$

24. $f(t)=e^t\sin(\pi t)$, $w(t)=\begin{cases}1 & \text{for } -1\le t\le 1,\\ 0 & \text{for } |t|>1.\end{cases}$

25. $f(t)=t^2$, $w(t)=\begin{cases}1 & \text{for } -5\le t\le 5,\\ 0 & \text{for } |t|>5.\end{cases}$

26. $f(t)=H(t-\pi)$, $w(t)=\begin{cases}1 & \text{for } 3\pi\le t\le 5\pi,\\ 0 & \text{for } t<3\pi \text{ or } t>5\pi.\end{cases}$

27. $f(t)=e^{-t}$, $w(t)=\begin{cases}1 & \text{for } 0\le t\le 4,\\ 0 & \text{for } t<0 \text{ or } t>4.\end{cases}$

28. $f(t)=\cos(at)$, $w(t)=\begin{cases}1 & \text{for } -4\pi\le t\le 4\pi,\\ 0 & \text{for } |t|>4\pi.\end{cases}$

14.4 Fourier Cosine and Sine Transforms

If f is piecewise smooth on each interval $[0, L]$ and $\int_0^\infty |f(t)|\,dt$ converges, then at each t where f is continuous, the Fourier cosine integral for f is

$$f(t)=\int_0^\infty a_\omega \cos(\omega t)\,d\omega,$$

where

$$a_\omega=\frac{2}{\pi}\int_0^\infty f(t)\cos(\omega t)\,dt.$$

We define the *Fourier cosine transform of* f by

$$\mathcal{F}_C[f](\omega)=\int_0^\infty f(t)\cos(\omega t)\,dt. \tag{14.16}$$

Often we denote $\mathcal{F}_C[f](\omega)=\hat{f}_C(\omega)$.

Notice that

$$\hat{f}_C(\omega)=\frac{\pi}{2}a_\omega$$

and that

$$f(t)=\frac{2}{\pi}\int_0^\infty \hat{f}_c(\omega)\cos(\omega t)d\omega. \tag{14.17}$$

Equations (14.16) and (14.17) form a transform pair for the Fourier cosine transform. Equation (14.17) is the inverse Fourier cosine transform, reproducing f from $\hat{f}_c$. This inverse is denoted $\hat{f}_C^{-1}$.

EXAMPLE 14.12

Let K be a positive number and let

$$f(t)=\begin{cases}1 & \text{for } 0\le t\le K\\ 0 & \text{for } t>K.\end{cases}$$

Then

$$\hat{f}_C(\omega)=\int_0^\infty f(t)\cos(\omega t)\,dt=\int_0^K \cos(\omega t)\,dt=\frac{\sin(K\omega)}{\omega}. \ \blacklozenge$$

Using the Fourier sine integral instead of the cosine integral, this discussion leads us to define the *Fourier sine transform* of f by

$$\mathcal{F}_S[f](\omega) = \int_0^\infty f(t)\sin(\omega t)\,dt.$$

We also denote this as $\hat{f}_S(\omega)$.

If f is continuous at $t > 0$ then the Fourier sine integral representation is

$$f(t) = \int_0^\infty b_\omega \sin(\omega t)\,d\omega,$$

where

$$b_\omega = \frac{2}{\pi}\int_0^\infty f(t)\sin(\omega t)\,dt = \frac{2}{\pi}\hat{f}_S(\omega).$$

This means that

$$f(t) = \frac{2}{\pi}\int_0^\infty \hat{f}_S(\omega)\sin(\omega t)\,d\omega$$

which provides a way of retrieving f from $\hat{f}_S$. This integral is the inverse Fourier sine transform $\hat{f}_S^{-1}$.

EXAMPLE 14.13

With f as in Example 14.12,

$$\hat{f}_S(\omega) = \int_0^\infty f(t)\sin(\omega t)dt = \int_0^K \sin(\omega t)dt = \frac{1}{\omega}[1-\cos(K\omega)]. \ ♦$$

The following operational formulas are needed when these transforms are used to solve differential equations.

Operational Formulas Let f and f' be continuous on every interval $[0, L]$ and let $\int_0^\infty |f(t)|dt$ converge. Suppose $f(t) \to 0$ and $f'(t) \to 0$ as $t \to \infty$. Suppose f'' is piecewise continuous on every $[0, L]$. Then

1. $\mathcal{F}_C[f''(t)](\omega) = -\omega^2 \hat{f}_C(\omega) - f'(0)$.
2. $\mathcal{F}_S[f''(t)](\omega) = -\omega^2 \hat{f}_S(\omega) + \omega f(0)$. ♦

SECTION 14.4 PROBLEMS

In each of Problems 1 through 6, determine the Fourier cosine transform and the Fourier sine transform of the function.

1. $f(t) = e^{-t}\cos(t)$

2. $f(t) = \begin{cases} \sinh(t) & \text{for } K \le t < 2K \\ 0 & \text{for } 0 \le t < K \text{ and for } t \ge 2K \end{cases}$

3. $f(t) = e^{-t}$

4. $f(t) = te^{-at}$ with a any positive number

5. $f(t) = \begin{cases} \cos(t) & \text{for } 0 \le t \le K \\ 0 & \text{for } t > K \end{cases}$ with K any positive number.

6. $f(t) = \begin{cases} 1 & \text{for } 0 \le t < K \\ -1 & \text{for } K \le t < 2K \\ 0 & \text{for } t \ge 2K \end{cases}$

7. Show that, under appropriate conditions on f and its derivatives,

$$\mathcal{F}_S[f^{(4)}(t)](\omega) = \omega^4 \hat{f}_S(\omega) = \omega^3 f(0) + \omega f''(0).$$

8. Show that, under appropriate conditions on f and its derivatives,

$$\mathcal{F}_C[f^{(4)}(t)](\omega) = \omega^4 \hat{f}_C(\omega) + \omega^2 f'(0) - f^{(3)}(0).$$

14.5 The Discrete Fourier Transform

If f has fundamental period p, its complex Fourier series is

$$\sum_{k=-\infty}^{\infty} d_k e^{2\pi ikt/p},$$

in which

$$d_k = \frac{1}{p}\int_0^p f(t)e^{-2\pi ikt/p}\,dt$$

for $k = 0, \pm 1, \pm 2, \cdots$.

Under certain conditions on f, this series converges at t to to $(f(t+) + f(t-))/2$.

Our objective is to define the discrete Fourier transform. To understand why this definition will take the form that it does, consider the problem of approximating the coefficients d_k in the complex Fourier series. One way is to begin by subdividing $[0, p]$ into N subintervals of equal length p/N and choosing a point in each interval of the subdivision, say

$$t_j \text{ in } \left[\frac{jp}{N}, \frac{(j+1)p}{N}\right]$$

for $j = 0, 1, 2, \cdots, N-1$. Approximate d_k by the Riemann sum

$$d_k \approx \frac{1}{p}\sum_{j=0}^{N-1} f(t_j)e^{-2\pi ikt_j/p}\frac{p}{N}.$$

This suggests the definition of the discrete Fourier transform, which acts on a sequence of N given complex numbers and produces an infinite sequence of complex numbers, one for each integer k, as follows.

Let N be a positive integer and let $u = [u_j]_{j=0}^{N-1}$ be a sequence of N complex numbers. Then the *N-point discrete Fourier transform* of u is the sequence $\mathcal{D}[u]$ defined by

$$\mathcal{D}[u](k) = \sum_{j=0}^{N-1} u_j e^{-2\pi ijk/N}$$

for $k = 0, \pm 1, \pm 2, \cdots$.

To simplify the notation, we will denote the N-point discrete Fourier transform of u by U, with lower case for the input sequence and upper case for its discrete transform. In this notation,

$$U_k = \sum_{j=0}^{N-1} u_j e^{-2\pi ijk/N}$$

for $k = 0, \pm 1, \pm 2, \cdots$. We will also abbreviate the phrase "discrete Fourier transform" to DFT.

EXAMPLE 14.14

Let $u = [c]_{j=0}^{N-1}$, a constant sequence, with c a given complex number. The N-point DFT of u is given by

$$U_k = \sum_{j=0}^{N-1} ce^{-2\pi ijk/N} = c\sum_{j=0}^{N-1} e^{-2\pi ijk/N}.$$

Observe that this is

$$U_k = c\sum_{j=0}^{N-1} \left(e^{-2\pi ik/N}\right)^j,$$

a finite geometric series. In general, for $|r| < 1$,

$$\sum_{j=0}^{N-1} r^j = \frac{1-r^N}{1-r}.$$

Then

$$U_k = \left(\frac{1-(e^{-2\pi ik/N})^N}{1-e^{-2\pi ik/N}}\right)c$$

$$= \left(\frac{1-e^{-2\pi ik}}{1-e^{-2\pi ik/N}}\right)c = 0$$

for $k = 0, \pm 1, \pm 2, \cdots$ because, for any integer k,

$$e^{-2\pi ik} = \cos(2\pi k) - i\sin(2\pi k) = 1.$$

The N-point DFT of a constant sequence is an infinite sequence of zeros. ♦

EXAMPLE 14.15

We will find the N-point DFT of $u = [\sin(ja)]_{j=0}^{N-1}$, in which N is a positive integer and a is a given complex number. To avoid trivialities, suppose a is not an integer multiple of π. We have

$$U_k = \sum_{j=0}^{N-1} \sin(ja)e^{-2\pi ijk/N}.$$

Use the fact that

$$\sin(ja) = \frac{1}{2i}\left(e^{ija} - e^{-ija}\right).$$

Then

$$U_k = \frac{1}{2i}\sum_{j=0}^{N-1} e^{ija}e^{-2\pi ijk/N} - \frac{1}{2i}\sum_{k=0}^{N-1} e^{-ija}e^{-2\pi ijk/N}$$

$$= \frac{1}{2i}\sum_{j=0}^{N-1}(e^{ia-2\pi ik/N})^j - \frac{1}{2i}\sum_{j=0}^{N-1}(e^{-ia-2\pi ik/N})^j.$$

Again recognizing geometric series in the last two terms,

$$U_k = \frac{1}{2i}\frac{1-(e^{ia-2\pi ijk/N})^N}{1-e^{ia-2\pi ijk/N}} - \frac{1}{2i}\frac{1-(e^{-ia-2\pi ijk/N})^N}{1-e^{-ia-2\pi ijk/N}}$$

$$= \frac{1}{2i}\frac{1-e^{iaN}e^{-2\pi ik}}{1-e^{ia}e^{-2\pi ik/N}} - \frac{1}{2i}\frac{1-e^{-iaN}e^{-2\pi ik}}{1-e^{-ia}e^{-2\pi ik/N}}$$

$$= \frac{1}{2i}\frac{1-e^{iaN}}{1-e^{ia-2\pi ik/N}} - \frac{1}{2i}\frac{1-e^{-iaN}}{1-e^{-ia-2\pi ik/N}},$$

in which we have used the fact that $e^{-2\pi ik}=1$.

To be specific, let $N=5$ and $a=\sqrt{2}$. Then

$$u_0=0,\ u_1=\sin(\sqrt{2}),\ u_2=\sin(2\sqrt{2}),\ u_3=\sin(3\sqrt{2}),\ u_4=\sin(4\sqrt{2}).$$

The 5-point DFT of u has kth term

$$U_k = \frac{1}{2i}\frac{1-e^{5i\sqrt{2}}}{1-e^{i\sqrt{2}-2\pi ik/5}} - \frac{1}{2i}\frac{1-e^{-5\sqrt{2}}}{1-e^{-i\sqrt{2}-2\pi ik/5}}.$$

For example,

$$U_0 = \frac{1}{2i}\frac{1-e^{5i\sqrt{2}}}{1-e^{i\sqrt{2}}} - \frac{1}{2i}\frac{1-e^{-5i\sqrt{2}}}{1-e^{-i\sqrt{2}}}$$

$$= \frac{\sin(4\sqrt{2})+\sin(\sqrt{2})-\sin(5\sqrt{2})}{2-2\cos(\sqrt{2})} \approx -0.1820207591$$

and

$$U_1 = \frac{1}{2i}\frac{1-5i\sqrt{2}}{1-e^{i\sqrt{2}-2\pi i/5}} - \frac{1}{2i}\frac{1-e^{-5i\sqrt{2}}}{1-e^{-i\sqrt{2}-2\pi i/5}}$$

$$\approx 0.4162488825 - 2.202105642i.$$ ♦

14.5.1 Linearity and Periodicity of the DFT

Linearity of the DFT is obvious because it is defined as a sum. For any numbers α and β or u and v which are N-point sequences, then

$$\mathcal{D}[\alpha u+\beta v](k) = \alpha U_k + \beta V_k.$$

We claim next that the N-point DFT is periodic of period N. This follows from the fact that $e^{-2\pi ijk}=1$, since ijk is an integer. Specifically,

$$U_{k+N} = \sum_{j=0}^{N-1} u_j e^{-2\pi ij(k+N)/N}$$

$$= \sum_{j=0}^{N-1} u_j e^{-2\pi ijk/N} e^{-2\pi ijk} = \sum_{j=0}^{N-1} u_j e^{-2\pi ijk/N} = U_k.$$

14.5.2 The Inverse N-Point DFT

Suppose we know the numbers U_k, the N-point DFT of some u. We would like to retrieve u.

We know that, whatever each u_j is,

$$U_k = \sum_{j=0}^{N-1} u_j e^{-2\pi ijk/N}.$$

We claim that

$$u_j = \frac{1}{N}\sum_{k=0}^{N-1} U_k e^{2\pi ijk/N} \tag{14.18}$$

for $j = 0, 1, \cdots, N-1$. This is the inversion formula for the N-point DFT, and it is analogous to inversion formulas for the Fourier transform and for the Fourier cosine and sine transforms, with a summation replacing an integral.

To verify equation (14.18) it is convenient to put $W = e^{-2\pi i/N}$. Then

$$W^N = 1 \text{ and } W^{-1} = e^{2\pi i/N}.$$

Write

$$\begin{aligned}
\frac{1}{N}\sum_{k=0}^{N-1} U_k e^{2\pi ijk/N} &= \frac{1}{N}\sum_{k=0}^{N-1} U_k W^{-jk} \\
&= \frac{1}{N}\sum_{k=0}^{N-1}\left(\sum_{r=0}^{N-1} u_r e^{-2\pi irk/N}\right) W^{-jk} \\
&= \frac{1}{N}\sum_{k=0}^{N-1}\sum_{r=0}^{N-1} u_r W^{rk} W^{-jk} \\
&= \frac{1}{N}\sum_{r=0}^{N-1} u_r \sum_{k=0}^{N-1} W^{rk} W^{-jk} \\
&= \frac{1}{N}\sum_{r=0}^{N-1} u_r \sum_{k=0}^{N-1} (W^{r-j})^k.
\end{aligned}$$

Now, if $r \neq j$ then (again using the finite geometric series),

$$\sum_{k=0}^{N-1} (W^{r-j})^k = \frac{1-(W^{r-j})^N}{1-W^{r-j}} = 0$$

because

$$(W^{r-j})^N = e^{-2\pi(r-j)} = 1 \text{ and } W^{r-j} = e^{2\pi i(r-j)/N} \neq 1.$$

But if $r = j$, then $(W^{r-j})^k = 1$, so

$$\sum_{k=0}^{N-1} (W^{r-j})^k = N.$$

Therefore, in the last double summation we need only retain the term when $r = j$ in the summation with respect to r, yielding

$$\begin{aligned}
\frac{1}{N}\sum_{k=0}^{N-1} U_k e^{2\pi ijk/N} &= \frac{1}{N}\sum_{r=0}^{N-1} u_r \sum_{k=0}^{N-1} (W^{r-j})^k \\
&= \frac{1}{N} u_j N = u_j.
\end{aligned}$$

14.5.3 DFT Approximation of Fourier Coefficients

We will now complete the idea suggested at the beginning of this section, approximating Fourier coefficients by a discrete Fourier transform.

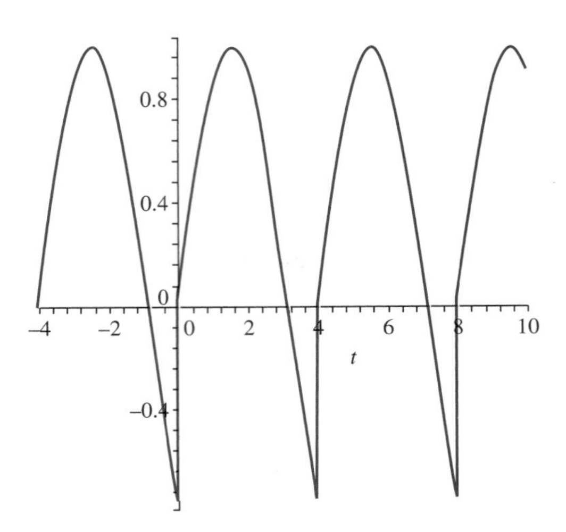

FIGURE 14.7 $f(t) = \sin(t)$ *for* $0 \le t < 4$, *period* 4.

To see the idea in action, consider the example of a function f having fundamental period 4, and $f(t) = \sin(t)$ for $0 \le t < 4$. A graph of this function is shown in Figure 14.7. With $p = 4$ the Fourier coefficients are

$$d_k = \frac{1}{4}\int_0^4 \sin(\xi)e^{-2\pi ik\xi/4}\,d\xi = \frac{1}{4}\int_0^4 \sin(\xi)e^{-\pi ik\xi/2}\,d\xi$$

$$= \frac{\cos(4) - 1}{\pi^2 k^2 - 4} + \frac{1}{2}i\frac{\pi k \sin(4)}{\pi^2 k^2 - 4}$$

for $k = 0, \pm 1, \pm 2, \cdots$.

Let N be a positive integer and subdivide $[0, 4]$ into N subintervals of equal lengths $4/N$. These subintervals are

$$\left[\frac{4j}{N}, \frac{4(j+1)}{n}\right] \text{ for } j = 0, 1, \cdots, N-1.$$

Select N numbers $4j/N$ by choosing the left endpoint of each of these subintervals. These define an N-point sequence u, where

$$u_j = \sin\left(\frac{4j}{n}\right).$$

The N-point DFT of u is

$$U_k = \sum_{j=0}^{N-1} \sin\left(\frac{4j}{N}\right) e^{-2\pi ijk/4} = \sum_{j=0}^{N-1} \sin\left(\frac{4j}{N}\right) e^{-\pi ijk/2}.$$

Now

$$\frac{!}{N}U_k = \frac{1}{N}\sum_{j=0}^{N-1} \sin\left(\frac{4j}{N}\right) e^{-\pi ijk/2}$$

is a Riemann sum for the integral defining d_k.

We now ask: To what extent does $(1/N)U_k$ approximate d_k? In this example we have an explicit expression for d_k, so we can explore this question directly. We will evaluate $(1/N)U_k$ using $a = 4/N$ in the DFT of $[\sin(ja)]_{j=0}^{N-1}$ determined in Example 14.15. This gives us

$$\frac{1}{N}U_k = \frac{1}{N}\left[\frac{1}{2i}\frac{1-e^{4i}}{1-e^{4i/N-2\pi ik/N}} - \frac{1}{2i}\frac{1-e^{-4i}}{1-e^{-4i/N-2\pi ik/N}}\right].$$

Now approximate the exponential terms in the denominator by using the first two terms of the power series of e^x about 0. This gives us

$$e^x \approx 1 + x$$

for $|x| << 1$. Then

$$\frac{1}{N}U_k = \frac{1}{N}\left[\frac{1-e^{4i}}{1-[1+(4i/N-2\pi ik/N)]} - \frac{1}{2i}\frac{1-e^{-4i}}{1-[1+(-4i/N-2\pi ik/N)]}\right]$$

$$= -\frac{1}{4}\left[\frac{1-e^{4i}}{-2+k\pi} - \frac{1-e^{-4i}}{2+k\pi}\right]$$

$$= -\frac{1}{4}\frac{1}{\pi^2k^2-4}[4-\pi i(e^{4i}-e^{-4i})-2(e^{4i}+e^{-4i})]$$

$$= -\frac{1}{4}\frac{1}{\pi^2k^2-4}[4-2\pi ki\sin(4)-4\cos(4)]$$

$$= \frac{\cos(4)-1}{\pi^2k^2-4} + \frac{1}{2}i\frac{\pi k\sin(4)}{\pi^2k^2-4}.$$

The approximation $e^x \approx 1 + x$ has led to an approximate expression for $(1/N)U_k$ that is exactly equal to d_k. This approximation cannot be valid for all k, however. First, the approximate value used for e^x assumed that x is much less than 1 in magnitude. Further, the N-point DFT is periodic of period N, so $U_{k+N} = U_k$, while there is no such periodicity in the d_k's.

In general it would be difficult to derive an estimate on relative sizes of $|k|$ and N that would result in $(1/N)U_k$ approximating d_k to within a given tolerance, and which would hold for a reasonably broad class of functions. However, for many applications encountered in practice, the empirical rule

$$|k| \le \frac{N}{8}$$

has proved to be effective.

SECTION 14.5 PROBLEMS

In each of Problems 1 through 6, compute $\mathcal{D}[u](k)$ for $k = 0, \pm 1, \cdots, \pm 4$.

1. $[j^2]_{j=0}^5$
2. $[1/(j+1)^2]_{j=0}^5$
3. $u = [\cos(j)]_{j=0}^5$
4. $[\cos(j) - \sin(j)]_{j=0}^5$
5. $[1/(j+1)]_{j=0}^5$
6. $u = [e^{ij}]_{j=0}^5$

In each of Problems 7 through 12, a sequence $[U_k]_{k=0}^N$ is given. Determine the N-point inverse discrete Fourier transform of this sequence.

7. $U_k = e^{-ik}, N = 7$
8. $U_k = \ln(k+1), N = 6$
9. $U_k = \cos(k), N = 5$
10. $U_k = i^{-k}, N = 5$
11. $U_k = (1+i)^k, N = 6$
12. $U_k = k^2, N = 5$

In each of Problems 13 through 16, compute the first seven complex Fourier coefficients $d_0, d_{\pm 1}, d_{\pm 2}$ and $d_{\pm 3}$ of $f(t)$. Then use the DFT to approximate these coefficients, with $N = 128$.

13. $f(t) = t^2$ for $0 \le t < 1$, f has period 1
14. $f(t) = te^{2t}$ for $0 \le t < 4$, f has period 4
15. $f(t) = \cos(t)$ for $0 \le t \le 2$, f has period 2
16. $f(t) = e^{-t}$ for $0 \le t < 3$, f has period 3

14.6 Sampled Fourier Series

We have just discussed the approximation of Fourier coefficients of a periodic function $f(t)$. This was done by approximating terms of an N-point DFT formed by sampling $f(t)$ at N points of $[0, p]$.

We will now discuss the use of an inverse DFT to approximate *sampled partial sums* (partial sums evaluated at selected points) of the Fourier series of a period function.

Consider the partial sum

$$S_N(t) = \sum_{k=-M}^{M} d_k e^{2\pi ikt/p}.$$

Subdivide $[0, p]$ into N subintervals of equal length p/N and choose sample points $t_j = jp/N$ for $j = 0, 1, \cdots, N-1$. Form the N-point sequence

$$u = [f(jp/N)]_{j=0}^{N-1}$$

and approximate

$$d_k \approx \frac{1}{N} U_k$$

where

$$U_k = \sum_{j=0}^{N-1} f(jp/N) e^{-2\pi ijk/N}.$$

In order to have $|k| \le N/8$, we will require that $M \le N/8$ in forming the partial sum of the Fourier series. We have

$$S_M(t) \approx \sum_{k=-M}^{M} \frac{1}{N} U_k e^{2\pi ikt/p}.$$

In particular, if we sample this partial sum at the partition point jp/N, then

$$S_M(jp/N) \approx \frac{1}{N} \sum_{k=-M}^{M} U_k e^{2\pi ijk/N}.$$

We will show that the sum on the right is actually an N-point inverse DFT for a particular N-point sequence which we will determine by exploiting the periodicity of the N-point DFT ($U_{k+N} = U_k$). Write

$$\begin{aligned}
S_M(jp/N) &\approx \frac{1}{N} \sum_{k=-M}^{-1} U_k e^{2\pi ijk/N} + \frac{1}{N} \sum_{k=0}^{M} U_k e^{2\pi ijk/N} \\
&= \frac{1}{N} \sum_{k=1}^{M} U_{-k} e^{-2\pi ijk/N} + \frac{1}{N} \sum_{k=0}^{M} U_k e^{2\pi ijk/N} \\
&= \frac{1}{N} \sum_{k=1}^{M} U_{-k+N} e^{2\pi ij(-k+N)/N} + \frac{1}{N} \sum_{k=0}^{M} U_k e^{2\pi ijk/N} \\
&= \frac{1}{N} \sum_{k=N-M}^{M-1} U_k e^{2\pi ijk/N} + \sum_{k=0}^{M} U_k e^{2\pi ijk/N}.
\end{aligned}$$

In these summations we use the $2M+1$ numbers

$$U_{N-M}, \cdots, U_{N-1}, U_0, \cdots, U_M.$$

Since $M < N/8$ we must fill in the missing values to obtain an N-point sequence. One way to do this is to fill in these places with zeros to form

$$V_k = \begin{cases} U_k & \text{for } k = 0, 1, \cdots, M \\ 0 & \text{for } k = M+1, \cdots, N-M-1 \\ U_k & \text{for } k = N-M, \cdots, N-1. \end{cases}$$

Then the Mth partial sum of the Fourier series of $f(t)$, sampled at $t = jp/N$, is approximated by

$$S_M(jp/N) \approx \frac{1}{N} \sum_{k=0}^{N-1} V_k e^{2\pi ijk/N}.$$

EXAMPLE 14.16

Let $f(t) = t$ for $0 \le t < 2$ and suppose $f(t)$ has period 2. Part of the graph of $f(t)$ is shown in Figure 14.8.

The Fourier coefficients are

$$d_k = \frac{1}{2} \int_0^2 te^{-2\pi ikt/2}\, dt = \begin{cases} i/\pi k & \text{for } k \neq 0 \\ 1 & \text{for } k = 0. \end{cases}$$

The complex Fourier series of $f(t)$ is

$$1 + \sum_{k=-\infty, k\neq 0}^{\infty} \frac{i}{\pi k} e^{\pi ikt}.$$

This converges to t on $(0, 2)$. The Mth partial sum is

$$S_M(t) = 1 + \sum_{k=-M, k\neq 0}^{M} \frac{i}{\pi k} e^{\pi ikt}$$

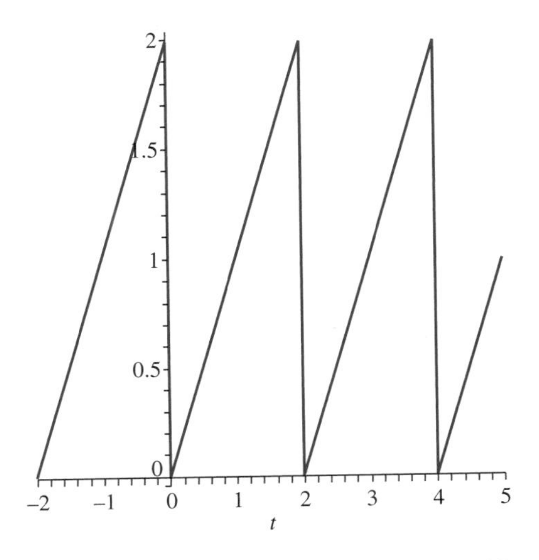

FIGURE 14.8 $f(t) = t$ for $0 \le t < 2$, period 2.

For this example, choose $N = 2^7 = 128$ and $M = 10$, so $M \leq N/8$. Sample the partial sum at points $jp/N = j/64$ for $j = 0, 1, \cdots, 127$. Thus define the finite sequence

$$u = [f(jp/N)]_{j=0}^{127} = \left[\frac{j}{64}\right]_{j=0}^{127}.$$

The 128-point DFT of u has kth term

$$U_k = \sum_{j0}^{127} \frac{j}{64} e^{-\pi ijk/64}.$$

Fill in the missing places by defining

$$V_k = \begin{cases} U_k & \text{for } k = 0, 1, \cdots, 10 \\ 0 & \text{for } k = 11, \cdots, 117 \\ U_k & \text{for } k = 117, \cdots, 127. \end{cases}$$

Then

$$S_{10}(jp/N) = S_{10}(j/64) = 1 + \sum_{k=-10, k\neq 0}^{10} \frac{i}{\pi k} e^{\pi ijk/64}$$

$$\approx \frac{1}{128} \sum_{k=0}^{127} V_k e^{\pi ijk/64}.$$

For comparison, we will compute $S_{10}(1/2)$, and then the approximate value from the last term in this equation. First,

$$S_{10}(1/2) = 1 + \sum_{k=-10, k\neq 0}^{10} \frac{i}{\pi k} e^{\pi ik/2} = 0.45847\cdots.$$

For the approximation, we first need the numbers U_k:

$$U_0 = \sum_{j=0}^{127} \frac{j}{64} = 127, U_1 = \sum_{j=0}^{127} \frac{j}{64} e^{-\pi ij/64} = -1.0 + 40.735i,$$

$$U_2 = \sum_{j=0}^{127} \frac{j}{64} e^{-\pi ij/32} = -1.0 + 20.355i, U_3 = -1.0 + 13.557i,$$

$$U_4 = -1.0 + 10.153i, U_5 = -1.0 + 8.1078i$$

$$U_6 = -1.0 + 6.7415i, U_7 = -1.0 + 5.7631i$$

$$U_8 = -1.0 + 5.0273i, U_9 = -1.0 + 4, 4532i$$

$$U_{10} = -1.0 + 3.9922i, U_{118} = -1.0 - 3.9922i$$

$$U_{119} = -1.0 - 4.4532i, U_{120} = -1.0 - 5.0273i$$

$$U_{121} = -1.0 - 5.7631i, U_{122} = -1.0 - 6.7415i$$

$$U_{123} = -1.0 - 8.1078i, U_{124} = -1.0 - 10.153i$$

$$U_{125} = -1.0 - 13.557i, U_{126} = -1.0 - 20.355i$$

$$U_{127} = -1.0 - 40.735i.$$

Using these numbers, we can compute

$$\begin{aligned}
&\sum_{k=0}^{127} V_k e^{\pi i k/2} \\
&= 127 + (-1+40.735i)e^{\pi i/2} + (-1+20.355i)e^{\pi i} \\
&\quad + (-1+13.557i)e^{3\pi i/2} + (-1+10.153i)e^{2\pi i} \\
&\quad + (-1+8.1078i)e^{5\pi i/2} + (-1+6.7415i)e^{3\pi i} \\
&\quad + (-1+5.7631i)e^{7\pi i/2} + (-1+5.0273i)e^{4\pi i} \\
&\quad + (-1+4.4532i)e^{9\pi i/2} + (-1+3.9922i)e^{5\pi i} \\
&\quad + (-1-3.9922i)e^{118\pi i/2} + (-1-4.4532i)e^{1119\pi i/2} \\
&\quad + (-1-5.0273i)e^{120\pi i/2} + (-1-5.7631i)e^{121\pi/2} \\
&\quad + (-1-6.7415i)e^{122\pi i/2} + (-1-8.1078i)e^{123\pi i/2} \\
&\quad + (-1-10.153i)e^{124\pi i/2} + (-1-13.557i)e^{125\pi i/2} \\
&\quad + (-1-20.355i)e^{126\pi i/2} + (-1-30.735i)e^{127\pi i/2} \\
&= 61.04832.
\end{aligned}$$

Then

$$\frac{1}{128}\sum_{k=0}^{127} V_k e^{\pi ijk/64} = 0.47694.$$

This gives the 128 point DFT approximation of 0.47694 to the sampled partial sum $S_{10}(1/2)$, which was computed to be 0.45847. The difference is 0.0185.

The actual sum of the Fourier series at $t = 1/2$ is $f(1/2) = 0.50000$. In practice we would achieve greater accuracy by choosing N larger, allowing larger M. ♦

SECTION 14.6 PROBLEMS

In each of Problems 1 through 6, a function is given having period p. Compute the complex Fourier series of the function and then the $10th$ partial sum of this series at the indicated t_0. Then, using $N = 128$, compute a DFT approximation to this partial sum at t_0.

1. $f(t) = \cos(t)$ for $0 \le t < 2$, $p = 2$, $t_0 = 1/8$
2. $f(t) = t\sin(t)$ for $0 \le t < 4$, $p = 4$, $t_0 = 1/8$
3. $f(t) = t^3$ for $0 \le t < 1$, $p = 1$, $t_0 = 1/4$
4. $f(t) = t^2$ for $0 \le t < 1$, $p = 1$, $t_0 = 1/2$
5. $f(t) = 1 + t$ for $0 \le t < 2$, $p = 2$, $t_0 = 1/8$
6. $f(t) = e^{-t}$ for $0 \le t < 4$, $p = 4$, $t_0 = 1/4$.

14.7 DFT Approximation of the Fourier Transform

Under certain conditions the DFT can be used to approximate the Fourier transform of a function. To begin, suppose $\hat{f}(\omega)$ can be approximated to within some acceptable tolerance by an integral over a finite interval

$$\hat{f}(\omega) = \int_{-\infty}^{\infty} f(\xi)e^{-i\omega\xi}\,d\xi \approx \int_{0}^{2\pi L} f(\xi)e^{-i\omega\xi}\,d\xi.$$

Subdivide $[0, 2\pi L]$ into N subintervals of length $2\pi L/N$ and choose partition points $\xi_j = 2\pi jL/N$ for $j = 0, 1, \cdots, N-1$. Using these, approximate

$$\begin{aligned}\hat{f}(\omega) &\approx \int_0^{2\pi L} f(\xi)e^{-i\omega\xi}\,d\xi \\ &\approx \sum_{j=0}^{N-1} f(2\pi jL/N)e^{-2\pi ijL\omega/N}\left(\frac{2\pi L}{N}\right) \\ &= \frac{2\pi L}{N}\sum_{j=0}^{N-1} f(2\pi jL/N)e^{-2\pi ijL\omega/N}.\end{aligned}$$

The sum on the right is nearly in the form of a DFT. If we put $\omega = k/L$ with k any integer, then we have

$$\hat{f}(k/L) \approx \frac{2\pi L}{N}\sum_{j=0}^{N-1} f(2\pi jL/N)e^{-2\pi ijk/N}. \tag{14.19}$$

This approximates $\hat{f}(k/L)$, the Fourier transform of f evaluated at k/L, with the N-point DFT of the sequence

$$\left[f\left(\frac{2\pi jL}{N}\right)\right]_{j=0}^{N-1}.$$

Again, we must restrict $|k| \leq N/8$ because the DFT is periodic of period N, while $\hat{f}(k/L)$ is not periodic.

EXAMPLE 14.17

We will test the approximation (14.19) for the simple case that

$$f(t) = \begin{cases} e^{-t} & \text{for } t \geq 0 \\ 0 & \text{for } t < 0. \end{cases}$$

f has Fourier transform

$$\begin{aligned}\hat{f}(\omega) &= \int_{-\infty}^{\infty} f(\xi)e^{-i\omega\xi}\,d\xi \\ &= \int_0^{\infty} e^{-\xi}e^{-i\omega\xi}\,d\xi = \frac{1-i\omega}{1+\omega^2}.\end{aligned}$$

Choose $L = 1$, $N = 2^7 = 128$ and $k = 3$, so $|k| \leq N/8$. Now $k/L = 3$ and the approximation (14.17) is

$$\begin{aligned}\hat{f}(k/L) = \hat{f}(3) &\approx \frac{2\pi}{128}\sum_{j=0}^{127} e^{-\pi j/64}e^{-6\pi ij/128} \\ &= \frac{\pi}{64}\sum_{j=0}^{127} e^{-\pi j/64}e^{-3\pi ij/64} = 0.12451 - 0.29884i.\end{aligned}$$

For comparison, the exact value is

$$\hat{f}(3) = \frac{1-3i}{10} = 0.1 - 0.3i.$$

We can improve the accuracy by choosing N larger. With $N = 2^9 = 512$ we obtain

$$\hat{f}(3) \approx \frac{2\pi}{512} \sum_{j=0}^{511} e^{-2\pi j/512} e^{-6\pi ij/512}$$
$$= 0.10595 - 0.2994i. \blacklozenge$$

EXAMPLE 14.18

We will continue the preceding example, but now carry out the approximation at more points. Use $L = 4$ and $N = 2^8 = 256$. The approximations are obtained by

$$\hat{f}(k/4) \approx \frac{\pi}{32} \sum_{j=0}^{255} e^{-\pi j/32} e^{-\pi ijk/128}.$$

To have $|k| \leq N/8 = 32$, we will only use this approximation for $\hat{f}(k/4)$ for $k = 1, 2, \cdots, 13$. Table 14.1 gives the approximate values along with the actual values of $\hat{f}(\omega)$.

The real part of $\hat{f}(\omega)$ is consistently approximated in this scheme with an error of about 0.05, while the imaginary part is approximated in many cases with an error of about 0.002. Improved accuracy is achieved by choosing N larger. ♦

This approximation was based on the assumption that $\hat{f}(\omega)$ could be approximated by an integral oer an interval $[0, 2\pi L]$. We can extend these ideas to the case that $\hat{f}(\omega)$ can be approximated by an integral $\int_{-\pi L}^{\pi L} f(\xi)e^{-i\omega\xi}\,d\xi$:

$$\hat{f}(\omega) \approx \int_{-\pi L}^{\pi L} f(\xi)e^{-i\omega\xi}\,d\xi.$$

Then

$$\hat{f}(k/L) \approx \int_{-\pi L}^{0} f(\xi)e^{-ik\xi/L}\,d\xi + \int_{0}^{L} f(\xi)e^{-ik\xi/L}\,d\xi.$$

TABLE 14.1 DFT Approximation of $\hat{f}(\omega)$ in Example 14.18

k	DFT approximation of $\hat{f}(\omega)$	$\hat{f}(\omega)$
(k = 1) $\hat{f}(1/4)$	0.99107 - 0.23509i	0.94118 - 0.23529i
(k = 2) $\hat{f}(1/2)$	0.84989 - 0.3996i	.8 - 0.4i
(k = 3) $\hat{f}(3/4)$	0.68989 - 0.4794i	0.64 - 0.48i
(k = 4) $\hat{f}(1)$	0.54989 - 0.4992i	0.5 - 0.5i
(k = 5) $\hat{f}(5/4)$	0.44013 - 0.4868i	0.39024 - 0.4878i
(k = 6) $\hat{f}(3/2)$	0.35758 - 0.46033i	0.3077 - 0.4615i
(k = 7) $\hat{f}(7/4)$	0.29605 0.42936i	0.24615 - 0.43077i
(k = 8) $\hat{f}(2)$	0.24989 - 0.39839i	0.2 - 0.4i
(k = 9) $\hat{f}(9/4)$	0.21484 - 0.36933i	0.16495 - 0.37113i
(k = 10) $\hat{f}(5/2)$	0.18782 - 0.34282i	0.13793 - 0.34483i
(k = 11) $\hat{f}(11/4)$	0.16668 - 0.31896i	0.11679 - 0.32117i
(k = 12) $\hat{f}(3)$	0.14989 - 0.29759i	0.1 - 0.3i
(k = 13) $\hat{f}(13/4)$	0.13638 - 0.27847i	0.086486 - 0.28108i

In the first integral on the right, set $\zeta = \xi + 2\pi L$ to obtain

$$\hat{f}(\omega) \approx \int_{\pi L}^{2\pi L} f(\zeta - 2\pi L)e^{-ik(\zeta-2\pi L)}\,d\zeta + \int_0^{\pi L} f(\xi)e^{-ik\xi/L}\,d\xi$$
$$= \int_{\pi L}^{2\pi L} f(\zeta - 2\pi L)e^{-ik\zeta/L}\,d\zeta + \int_0^{\pi L} f(\xi)e^{-ik\xi/L}\,d\xi.$$

Write ξ for ζ as the variable of integration to write the last approximation as

$$\hat{f}(k/L) \approx \int_{\pi L}^{2\pi L} f(\xi - 2\pi L)e^{-ik\xi/L}\,d\xi + \int_0^{\pi L} f(\xi)e^{-ik\xi/L}\,d\xi.$$

Now define

$$g(t) = \begin{cases} f(t) & \text{for } 0 \le t < \pi L \\ (f(\pi L) + f(-\pi L))/2 & \text{for } t = \pi L \\ f(t - 2\pi L) & \text{for } \pi L < t < 2\pi L. \end{cases}$$

Then

$$\hat{f}(k/L) \approx \int_0^{2\pi L} g(\xi)e^{-ik\xi/L}\,d\xi$$
$$= \int_0^{L} g(2\pi t)e^{-2\pi ikt/L}(2\pi)\,dt \text{ (let } \xi = 2\pi t)$$
$$= 2\pi \int_0^{L} g(2\pi t)e^{-2\pi ikt/L}\,dt.$$

Finally, approximate the last integral by a Riemann sum, subdividing $[0, L]$, subdividing $[0, L]$ into L/N subintervals and choosing partition points $t_j = jL/N$ for $j = 0, 1, \cdots, N-1$. Then

$$\hat{f}(k/L) \approx \frac{2\pi L}{N}\sum_{j=0}^{N-1} g\left(\frac{2\pi jL}{N}\right)e^{-2\pi ijk/N}.$$

As before, we assume that $|k| \le N/8$. This approximates $\hat{f}(k/L)$ by a constant multiple of the N-point DFT of the sequence

$$\left[g\left(\frac{2\pi jL}{N}\right)\right]_{j=0}^{N-1}.$$

SECTION 14.7 PROBLEMS

In each of Problems 1 through 4, make a DFT approximation to the Fourier transform of f at the given point, using $N = 512$ and the given L.

1. $f(t) = \begin{cases} te^{-2t} & \text{for } t \ge 0 \\ 0 & \text{for } t < 0, \end{cases}$

   $L = 3$; $\hat{f}(2)$

2. $f(t) = \begin{cases} e^{-t}\cos(t) & \text{for } t \ge 0 \\ 0 & \text{for } t < 0, \end{cases}$

   $L = 4$; $\hat{f}(4)$

3. $f(t) = \begin{cases} e^{-4t} & \text{for } t \ge 0 \\ 0 & \text{for } t < 0, \end{cases}$

   $L = 3$; $\hat{f}(4)$

4. $f(t) = \begin{cases} t\cos(t) & \text{for } 0 \le t \le 12\pi \\ 0 & \text{for } t < 0 \text{ and } t > 12\pi \end{cases}$

   $L = 6$; $\hat{f}(1)$

CHAPTER 15

Special Functions and Eigenfunction Expansions

15.1 Eigenfunction Expansions

Functions may be designated as *special functions* when they arise in important applications and contexts, often as solutions of differential equations. The most familiar special functions are $\cos(kx)$ and $\sin(kx)$, which are solutions of

$$y'' + k^2 y = 0.$$

Other special functions include Legendre polynomials and Bessel functions, which are solutions of Legendre's and Bessel's differential equations. We will develop these functions shortly.

Fourier series are used to solve many problems involving partial differential equations modeling diffusion processes and wave motion. We will see, however, that some problems require series expansions in terms of special functions. This chapter develops a framework in which to make such expansions.

Begin with the ordinary differential equation

$$y'' + R(x)y' + (Q(x) + \lambda P(x))y = 0$$

on some interval (a, b) or $[a, b]$, with λ a constant to be determined along with y. Assume that the coefficient functions are continuous on the interval. First manipulate the differential equation into a special form. Multiply it by

$$r(x) = e^{\int R(x)\,dx}$$

to obtain

$$y''e^{\int R(x)\,dx} + R(x)y'e^{\int R(x)\,dx} + (Q(x) + \lambda P(x))e^{\int R(x)\,dx}y = 0.$$

This is

$$\left(y'e^{\int R(x)dx}\right)' + \left(Q(x)e^{\int R(x)dx} + \lambda P(x)e^{\int R(x)dx}\right)y = 0,$$

which has the form

$$(ry')' + (q + \lambda p)y = 0. \tag{15.1}$$

When p, q, r, and r' are continuous on (a, b), and $r(x) > 0$ and $p(x) > 0$ on (a, b), we call equation (15.1) the *Sturm-Liouville differential equation.*

This Sturm-Liouville equation contains a quantity λ. We want to determine values of λ, called *eigenvalues*, such that there are nontrivial (not identically zero) solutions y of the differential equation which satisfy certain conditions at a and b. For a given eigenvalue λ, such a solution is an *eigenfunction* associated with λ. Conditions at a and b that solutions must satisfy are called *boundary conditions*. There are three kinds of boundary value problems for the eigenvalues and eigenfunctions, depending on the form of the boundary conditions. In each problem, we assume that $p(x) > 0$ and $r(x) > 0$ on (a, b).

The Regular Sturm-Liouville Problem

We want numbers λ for which there are nontrivial solutions of equation (15.1) satisfying *regular boundary conditions*

$$A_1 y(a) + A_2 y'(a) = 0 \text{ and } B_1 y(b) + B_2 y'(b) = 0,$$

in which A_1 and A_2 are given numbers, not both zero, and B_1 and B_2 are also given numbers, not both zero.

The Periodic Sturm-Liouville Problem

$r(a) = r(b)$ and we want numbers λ for which there are nontrivial solutions of equation (15.1) satisfying *periodic boundary conditions*

$$y(a) = y(b) \text{ and } y'(a) = y'(b).$$

The Singular Sturm-Liouville Problem

$r(a) = 0$ or $r(b) = 0$, but not both. For this problem we want numbers λ for which there are nontrivial solutions of equation (15.1), subject to the following boundary conditions at a or b.

If $r(a) = 0$ then there is the single boundary condition

$$B_1 y(b) + B_2 y'(b) = 0,$$

with B_1 and B_2 not both zero.

If $r(b) = 0$, then there is the single boundary condition

$$A_1 y(a) + A_2 y'(a) = 0,$$

with A_1 and A_2 given and not both zero.

We will derive properties of eigenvalues and eigenfunctions after looking at two examples.

EXAMPLE 15.1 A Regular Sturm-Liouville Problem

The problem

$$y'' + \lambda y = 0; \; y(0) = y(L) = 0$$

is regular on $[0, L]$. We will solve it for later use. Consider cases on λ.

Case 1: $\lambda = 0$

Then $y(x) = cx + d$ for some constants c and d. But $y(0) = 0 = d$ and then $y(L) = cL = 0$ requires that $c = 0$, so all solutions are trivial when $\lambda = 0$. Therefore λ is not an eigenvalue of this problem.

Case 2: $\lambda < 0$

Say $\lambda = -k^2$ for $k > 0$. The general solution of $y'' - k^2 y = 0$ is

$$y(x) = c_1 e^{kx} + c_2 e^{-kx}.$$

But $y(0) = c_1 + c_2 = 0$ means that $c_2 = -c_1$, so

$$y(x) = c_1 \left(e^{kx} - e^{-kx}\right).$$

Then

$$y(L) = c_1 \left(e^{kL} - e^{-kL}\right) = 0.$$

If $e^{kL} - e^{-kL} = 0$ we would have $e^{2kL} = 1$ and then $2kL = 0$, impossible if $k > 0$ and $L > 0$. Therefore $c_1 = 0$, so $c_2 = 0$ also and y is the trivial solution. This problem has no negative eigenvalue.

Case 3: $\lambda > 0$

Say $\lambda = k^2$ for $k > 0$. Now $y'' + k^2 y = 0$ has general solution

$$y(x) = c_1 \cos(kx) + c_2 \sin(kx).$$

Since $y(0) = c_1 = 0$, then $y = c_2 \sin(kx)$. Then

$$y(L) = c_2 \sin(kL) = 0.$$

To have a nontrivial solution we cannot have c_2 vanish, so we must choose k so that $\sin(kL) = 0$. Then $kL = n\pi$ for n any positive integer, so, using n as an index,

$$\lambda_n = k^2 = \left(\frac{n\pi}{L}\right)^2.$$

These are the eigenvalues of this problem, for each positive integer n. Corresponding to each such eigenvalue is the eigenfunction

$$y_n(x) = \sin\left(\frac{n\pi x}{L}\right).$$

Any nonzero constant multiple of y_n is also an eigenfunction corresponding to λ_n. ♦

EXAMPLE 15.2 A Periodic Sturm-Liouville Problem

The problem

$$y'' + \lambda y = 0;\ y(-L) = y(L),\ y'(-L) = y'(L)$$

is periodic on $[-L, L]$. If we compare $y'' + \lambda y = 0$ with the Sturm-Liouville equation we have $r(x) = 1$, so $r(-L) = r(L)$, as is required for a periodic problem on $[-L, L]$. We will also solve this problem by taking cases on λ.

Case 1: $\lambda = 0$

Then $y(x) = cx + d$. Now

$$y(-L) = -cL + d = y(L) = cL + d$$

implies that $c = 0$. The constant function $y(x) = d$ satisfies both boundary conditions. Thus 0 is an eigenvalue of this problem with constant (nonzero) functions as eigenfunctions.

Case 2: $\lambda < 0$

Say $\lambda = -k^2$. Now $y(x) = c_1 e^{kx} + c_2 e^{-kx}$. The condition $y(L) = y(-L)$ gives us

$$c_1 e^{-kL} + c_2 e^{kL} = c_1 e^{kL} + c_2 e^{-kL}.$$

Write this as

$$c_1(e^{-kL} - e^{kL}) = c_2(e^{-kL} - e^{kL}).$$

This implies that $c_1 = c_2$, so $y(x) = c_1(e^{kx} + e^{-kx})$. Now $y'(-L) = y'(L)$ gives us

$$c_1 k(e^{-kL} - e^{kL}) = c_1 k(e^{kL} - e^{-kL}).$$

This implies that $c_1 = -c_1$, hence $c_1 = 0$, so $c_2 = 0$ also. This problem has only the trivial solution, so there is no negative eigenvalue.

Case 3: $\lambda > 0$

Say $\lambda = k^2$. The general solution of $y'' + k^2 y = 0$ is

$$y(x) = c_1 \cos(kx) + c_2 \sin(kx).$$

Now

$$y(-L) = c_1 \cos(kL) - c_2 \sin(kL) = y(L) = c_1 \cos(kL) + c_2 \sin(kL).$$

This implies that $c_2 \sin(kL) = 0$. Next,

$$y'(-L) = kc_1 \sin(kL) + kc_2 \cos(kL) = y'(L) = -kc_1 \sin(kL) + kc_2 \cos(kL),$$

implying that $kc_1 \sin(kL) = 0$. If $\sin(kL) \neq 0$, then $c_1 = c_2 = 0$ and we have only the trivial solution. For a nontrivial solution, we must have $\sin(kL) = 0$, so at least one of the constants c_1 and c_2 can be chosen nonzero. But $\sin(kL) = 0$ is satisfied if $kL = n\pi$, with n a positive integer (positive because we chose $k > 0$). Since $\lambda = k^2$, the eigenvalues of this problem, indexed by n, are

$$\lambda_n = \left(\frac{n\pi}{L}\right)^2 \text{ for } n = 1, 2, \cdots$$

and corresponding eigenfunctions are

$$y_n(x) = c_1 \cos\left(\frac{n\pi x}{L}\right) + c_2 \sin\left(\frac{n\pi x}{L}\right),$$

with c_1 and c_2 any constants, not both zero.

By choosing $n = 0$ and $c_2 = 0$ but $c_1 \neq 0$, we obtain a constant eigenfunction corresponding to the eigenvalue 0. This consolidates cases 1 and 3. ♦

Bessel's equation will provide an example of a singular Sturm-Liouville problem (here we will have $r(0) = 0$).

We may also have a Sturm-Liouville problem in which $r(a) = r(b) = 0$, but there are no boundary conditions specified. In this event we impose the condition that eigenfunctions must be bounded on $[a, b]$. The Legendre differential equation will provide an example of this type of problem.

A Fourier sine series on $[0, L]$ is an expansion in the eigenfunctions of Example 15.1. A Fourier series on $[-L, L]$ is an expansion in the eigenfunctions of Example 15.2. This raises an intriguing question. If a Sturm-Liouville problem on $[a, b]$ has eigenfunctions $\varphi_1(x), \varphi_2(x), \cdots,$

we ask whether it might be possible to expand an "arbitrary" function f on $[a,b]$ in a series of these eigenfunctions,

$$f(x)=\sum_{k=1}^{\infty} c_k\varphi_k(x).$$

The key lies in the choice of the coefficients c_k, which in turn hinges on an orthogonality property of the eigenfunctions. Recall that, in seeking the coefficients a_n and b_n in the Fourier expansion

$$f(x)=\frac{1}{2}a_0+\sum_{k=1}^{\infty}(a_k\cos(k\pi x/L)+b_k\sin(k\pi x/L))$$

we multiplied the series by a particular eigenfunction $\cos(n\pi x/L)$ or $\sin(n\pi x/L)$ and integrated. Because of equations (13.3), (13.4) and (13.5), the integral of any two distinct eigenfunctions vanished, leaving simple integral expressions for a_n or b_n.

We will show that similar properties hold for the eigenfunctions of any Sturm-Liouville problem, suggesting a similar strategy for finding the coefficients in a series of eigenfunctions $\sum_{k=1}^{\infty} c_k\varphi_k(x)$.

We will also show the eigenvalues must be real numbers. Thus the absence of complex eigenvalues in Example 15.1 and 15.2 is characteristic of Sturm-Liouville problems in general.

THEOREM 15.1

Suppose we have a regular, periodic or singular Sturm-Liouville problem on an interval $[a,b]$. Then

1. There is an infinite sequence of eigenvalues λ_j which can be ordered so that

$$\lambda_1<\lambda_2<\cdots.$$

   Further, with this ordering as an increasing sequence,

$$\lim_{n\to\infty}\lambda_n=\infty.$$

2. If φ is an eigenfunction, then so is $c\varphi$ for any nonzero real number c.
3. Let λ_n and λ_m be distinct eigenvalues of the problem, with corresponding eigenfunctions φ_n and φ_m. Then

$$\int_a^b p(x)\varphi_n(x)\varphi_m(x)dx=0, \tag{15.2}$$

   where p is the coefficient of λ in the Sturm-Liouville differential equation $(ry')'+(q+\lambda p)y=0$.
4. Every eigenvalue of the Sturm-Liouville problem is real. ♦

Proof Conclusion (1) requires a lengthy analysis we will not enter into here. Note that conclusion (1) implies that the eigenvalues cannot cluster around a finite number, as, for example, the numbers $1-1/n$ fall within arbitrarily small intervals about 1 as n is chosen larger. These numbers cannot be the eigenvalues of a Sturm-Liouville problem.

For conclusion (2), suppose $\varphi(x)$ is an eigenfunction corresponding to eigenvalue λ. Then

$$(r\varphi')'+(q+\lambda p)\varphi=0.$$

Multiplication of this equation by c shows that $c\varphi$ satisfies this differential equation as well. Finally, by multiplying the appropriate boundary conditions in each case by c, it is verified that $c\varphi$ is an eigenfunction corresponding to λ. (Note: $c\varphi$ is an eigenfunction corresponding to λ, not to $c\lambda$.)

To prove equation (15.2), begin with the fact that the eigenvalues and corresponding eigenfunctions satisfy the Sturm-Liouville differential equation:

$$(r\varphi_n')' + (q + \lambda_n p)\varphi_n = 0$$

and

$$(r\varphi_m')' + (q + \lambda_m p)\varphi_m = 0.$$

Multiply the first equation by φ_m and the second by φ_n and subtract to obtain

$$(r\varphi_n')'\varphi_m - (r\varphi_m')'\varphi_n + (\lambda_n - \lambda_m)p\varphi_n\varphi_m = 0.$$

Write this equation as

$$\frac{d}{dx}\left[r\left(\varphi_m\varphi_n' - \varphi_n\varphi_m'\right)\right] = (\lambda_m - \lambda_n)p\varphi_n\varphi_m.$$

This means that

$$\begin{aligned}(\lambda_m - \lambda_n)&\int_a^b p(x)\varphi_n(x)\varphi_m(x)\,dx\\ &= \left[r\left(\varphi_m y\varphi_n' - \varphi_n\varphi_m'\right)\right]_a^b\\ &= r(b)\left[\varphi_m(b)\varphi_n'(b) - \varphi_n(b)\varphi_m'(b)\right] - r(a)\left[\varphi_m(a)\varphi_n'(a) - \varphi_n(a)\varphi_m'(a)\right].\end{aligned}$$

This gives us

$$\begin{aligned}(\lambda_m - \lambda_n)&\int_a^b p(x)\varphi_n(x)\varphi_m(x)\,dx\\ &= r(b)\left[\varphi_m(b)\varphi_n'(b) - \varphi_n(b)\varphi_m'(b)\right] - r(a)\left[\varphi_m(a)\varphi_n'(a) - \varphi_n(a)\varphi_m'(a)\right].\end{aligned} \tag{15.3}$$

Equation (15.2) is therefore verified if we can show that the right side of equation (15.3) is zero. This is done by examining the boundary conditions accompanying each type of Sturm-Liouville problem.

Suppose first that the problem is regular. Each eigenfunction must satisfy the boundary condition at a:

$$A_1\varphi_n(a) + A_2\varphi_n'(a) = 0,$$
$$A_1\varphi_m(a) + A_2\varphi_m'(a) = 0,$$

with not both A_1 and A_2 zero. Think of these boundary conditions at a as a homogeneous system of two algebraic equations in two unknowns A_1 and A_2. The fact that there is a nontrivial solution for these numbers means that the determinant of the system is zero:

$$\begin{vmatrix}\varphi_n(a) & \varphi_n'(a)\\ \varphi_m(a) & \varphi_m'(a)\end{vmatrix} = 0.$$

The same argument applies to the regular boundary condition at b:

$$\begin{vmatrix}\varphi_n(b) & \varphi_n'(b)\\ \varphi_m(b) & \varphi_m'(b)\end{vmatrix} = 0.$$

This shows that both terms in square brackets on the right side of equation (15.3) are zero. Then

$$(\lambda_m - \lambda_n)\int_a^b p(x)\varphi_n(x)\varphi_m(x)\,dx = 0.$$

Since λ_n and λ_n were assumed to be distinct eigenvalues, this means that

$$\int_a^b p(x)\varphi_n(x)\varphi_m(x)\,dx = 0,$$

proving conclusion (3) for the regular Sturm-Liouville problem.

One proceeds in similar fashion for periodic and singular boundary conditions. This proves conclusion (3) of the theorem.

For conclusion (4), let λ be any eigenvalue, with corresponding eigenfunction $\varphi(x)$. It is routine to check by taking complex conjugates that $\overline{\lambda}$ is an eigenvalue with eigenfunction $\overline{\varphi(x)}$. If $\lambda \neq \overline{\lambda}$, then by conclusion (3),

$$(\lambda - \overline{\lambda}) \int_a^b p(x)\varphi(x)\overline{\varphi(x)}\,dx = 0.$$

Then

$$(\lambda - \overline{\lambda}) \int_a^b p(x)|\varphi(x)|^2\,dx = 0.$$

But $p(x) > 0$ on (a, b), and an eigenfunction cannot be identically zero, so this integral must be positive. Therefore $\lambda = \overline{\lambda}$, implying that λ must be real, proving conclusion (4). ♦

The integral relationship (15.2) between eigenfunctions associated with distinct eigenvalues is called *orthogonality of the eigenfunctions with respect to the weight function* p. This terminology derives from the fact that $\int_a^b p(x)f(x)g(x)\,dx$ behaves like a dot product for vectors, and two vectors are called orthogonal when their dot product is zero. In Example 15.1, $p(x) = 1$ and the orthogonality relationship is the familiar

$$\int_0^L \sin(n\pi x/L)\sin(m\pi x/L)dx = 0$$

for $n \neq m$.

Using this notion of weighted orthogonality of eigenfunctions, we can solve for the coefficients in a proposed series of eigenfunctions, arguing much as we did for the coefficients of a Fourier series. Suppose the eigenvalues are λ_k and corresponding eigenfunctions are φ_k for $k = 1, 2, \cdots$, and we want to write

$$f(x) = \sum_{k=1}^{\infty} c_k \varphi_k(x).$$

Multiply this equation by $p(x)\varphi_n(x)$ and integrate to get

$$\int_a^b p(x)f(x)\varphi_n(x)\,dx = \sum_{k=1}^{\infty} \int_a^b p(x)\varphi_k(x)\varphi_n(x)\,dx.$$

By equation (15.2) all of the integrals in the summation are zero except when $k = n$, yielding

$$\int_a^b p(x)f(x)\varphi_n(x)\,dx = c_n \int_a^b p(x)\varphi_n^2(x)\,dx,$$

or

$$c_n = \frac{\int_a^b p(x)f(x)\varphi_n(x)\,dx}{\int_a^b p(x)\varphi_n^2(x)\,dx}. \tag{15.4}$$

By analogy with Fourier series, we call the numbers defined by equation (15.4) the *generalized Fourier coefficients of* f *with respect to the Sturm-Liouville problem*. With this choice of coefficients, we call

$$\sum_{k=1}^{\infty} c_k \varphi_k(x)$$

the *eigenfunction expansion of* f *with respect to the eigenfunctions of the Sturm-Liouville problem*.

As with Fourier (trigonometric) series, the question now is the relationship between the function and its eigenfunction expansion.

THEOREM 15.2 Convergence of Eigenfunction Expansions

Suppose φ_n, for $n = 1, 2, \cdots$, are the eigenfunctions of a Sturm-Liouville problem on $[a, b]$. Let f be piecewise smooth on $[a, b]$ and let c_n be given by equation (15.4). Then, for $a < x < b$, the eigenfunction expansion $\sum_{n=1}^{\infty} c_n \varphi_n(x)$ converges to

$$\frac{1}{2}(f(x+) + f(x-)).$$

In particular, if f is continuous at x, then this eigenfunction expansion converges to $f(x)$. ♦

It should not be surprising that this conclusion has the same form as that for convergent Fourier series, since Fourier series are eigenfunction expansions.

EXAMPLE 15.3

The regular Sturm-Liouville problem

$$y'' + \lambda y = 0;\ y'(0) = y'(\pi/2) = 0$$

has eigenvalues $\lambda_n = 4n^2$ and eigenfunctions $\varphi_n(x) = \cos(2nx)$ for $n = 0, 1, 2, \cdots$. Here $p(x) = 1$ and the interval is $[0, \pi/2]$.

Let $f(x) = x^2(1 - x)$ for $0 \le x \le \pi/2$. We will expand $f(x)$ in a series $\sum_{n=0}^{\infty} c_n \varphi_n(x)$ of f of the eigenfunctions of this problem, using equation (15.4) for the coefficients. First compute

$$c_0 = \frac{\int_0^{\pi/2} f(x)\varphi_0(x)dx}{\int_0^{\pi/2} \varphi_0^2 dx} = \frac{\int_0^{\pi/2} x^2(1-x)dx}{\int_0^{\pi/2} dx} = \frac{-\frac{1}{64}\pi^4 + \frac{1}{24}\pi^3}{\frac{\pi}{2}} = -\frac{1}{32}\pi^3 + \frac{1}{12}\pi^2.$$

For $n = 1, 2, \cdots$, the denominator of c_n is

$$\int_0^{\pi/2} \cos^2(2nx)dx = \frac{\pi}{4}.$$

The numerator is

$$\int_0^{\pi/2} x^2(1-x)\cos(2nx)dx = \frac{-6 + (-1)^n[6 + 4\pi n^2 - 3\pi^2 n^2]}{16n^4}.$$

Therefore

$$c_n = \frac{-6 + (-1)^n[6 + \pi n^2 - 3\pi^2 n^2]}{4\pi n^4}.$$

The eigenfunction expansion is

$$x^2(1-x) = \pi^2\left(\frac{1}{12} - \frac{1}{32}\pi\right) + \sum_{n=1}^{\infty} \frac{-6 + (-1)^n[6 + \pi n^2 - 3\pi^2 n^2]}{4\pi n^4}\cos(2nx).$$

From the convergence theorem, this eigenfunction expansion converges to $x^2(1 - x)$ for $0 < x < \pi/2$.

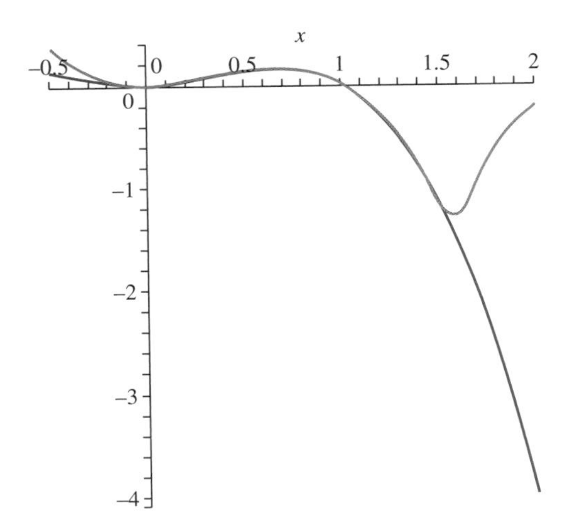

FIGURE 15.1 *Eigenfunction expansion in Example 15.3.*

Figure 15.1 shows a graph of f compared with the fifteenth partial sum of this series. The graph is drawn on a slightly larger interval than $[0, \pi/2]$. This eigenfunction expansion appears to converge rapidly to $f(x)$ on $[0, \pi/2]$, but the graphs diverge from each other outside this interval. The eigenfunction expansion is unrelated to $f(x)$ outside of the interval of the Sturm-Liouville problem. ♦

The following example illustrates an eigenfunction expansion in which the eigenfunctions are not just sines and cosines.

EXAMPLE 15.4

We will expand $f(x) = x$ in a series of the eigenfunctions of the regular Sturm-Liouville problem

$$y'' + 4y' + (3 + \lambda)y = 0;\ y(0) = y(1) = 0$$

on $[0, 1]$. First we must generate the eigenvalues and eigenfunctions. Put $y = e^{rx}$ to obtain the characteristic equation of the differential equation:

$$r^2 + 4r + (3 + \lambda) = 0,$$

with roots

$$r = -2 \pm \sqrt{1 - \lambda}.$$

Take cases on λ.

Case 1: $\lambda = 1$

Then

$$y(x) = c_1 e^{-2x} + c_2 x e^{-2x}.$$

Now

$$y(0) = c_1 = 0 \text{ and } y(1) = c_2 e^{-2} = 0$$

so $c_2 = 0$ and y is the trivial solution. Therefore, 1 is not an eigenvalue of this problem.

Case 2: $1 - \lambda > 0$

Write $1 - \lambda = \alpha^2$ with $\alpha > 0$ to obtain

$$y(x) = c_1 e^{(-2+\alpha)x} + c_2 e^{(-2-\alpha)x}.$$

Then

$$y(0) = c_1 + c_2 = 0$$

so $c_2 = -c_1$. And

$$y(1) = c_1 e^{-2}(e^{\alpha} - e^{-\alpha}) = -2c_1 e^{-2} \sinh(\alpha).$$

Since $\alpha > 0$, $\sinh(\alpha) > 0$, so $c_2 = c_1 = 0$ and again y is the trivial solution. There is no eigenvalue $\lambda < 1$.

Case 3: $1 - \lambda < 0$

Write $1 - \lambda = -\alpha^2$, with $\alpha > 0$. Now

$$y(x) = c_1 e^{-2x} \cos(\alpha x) + c_2 e^{-2x} \sin(\alpha x).$$

Then

$$y(0) = c_1 = 0.$$

This leaves us with $y(x) = c_2 e^{-2x} \sin(\alpha x)$. Now

$$y(1) = c_2 e^{-2} \sin(\alpha) = 0.$$

To have a nontrivial solution, we need $c_2 \neq 0$, so we must have $\sin(\alpha) = 0$. We can choose $\alpha = n\pi$ for n any positive integer. The eigenvalues are

$$\lambda_n = 1 + \alpha^2 = 1 + n^2\pi^2.$$

The eigenfunctions are

$$\varphi_n(x) = e^{-2x} \sin(n\pi x)$$

for $n = 1, 2, \cdots$.

To compute the coefficients in the expansion of $f(x) = x$ in a series of these eigenfunctions, we need to know the weight function p. For this, we must write the differential equation in Sturm-Liouville form. Multiply it by

$$e^{\int 4\,dx} = e^{4x}$$

to obtain

$$e^{4x}y'' + 4e^{4x}y' + (3 + \lambda)e^{4x}y = 0.$$

This is

$$\left(e^{4x}y'\right)' + (3e^{4x} + \lambda e^{4x})y = 0.$$

This is in Sturm-Liouville form with $r(x) = p(x) = e^{4x}$ and $q(x) = 3e^{4x}$. To write $f(x) = x = \sum_{n=1}^{\infty} c_n \varphi_n(x)$, choose

$$\begin{aligned} c_n &= \frac{\int_0^1 p(x) e^{-2x} \sin(n\pi x)\,dx}{\int_0^1 p(x) e^{-4x} \sin^2(n\pi x)\,dx} \\ &= \frac{\int_0^1 e^{2x} \sin(n\pi x)\,dx}{\int_0^1 \sin^2(n\pi x)\,dx} \\ &= \frac{-8n\pi - 2e^2 n^3 \pi^3 (-1)^n}{(4 + (n\pi)^2)^2}. \end{aligned}$$

Figure 15.2 compares graphs of $f(x) = x$ with the $70th$ partial sum of this expansion. This particular expansion converges fairly slowly to $f(x)$, compared to Example 15.3. The Gibbs phenomenon can be seen at $x = 1$. This behavior applies to general eigenfunction expansions, not just to Fourier series. ♦

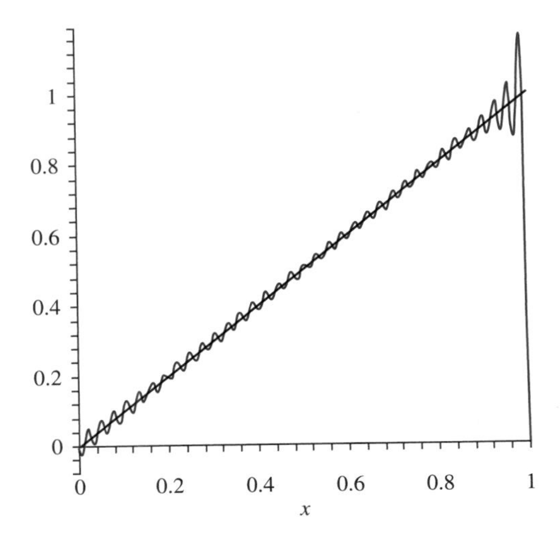

FIGURE 15.2 *Eigenfunction expansion in Example 15.4.*

15.1.1 Bessel's Inequality and Parseval's Theorem

Given a Sturm-Liouville problem on $[a, b]$, we have seen that the eigenfunctions are orthogonal with respect to the weight function p. To carry the analogy with vectors a little further, define the *weighted dot product* of functions f and g to be

$$f \cdot g = \int_a^b p(x) f(x) g(x)\, dx.$$

We may also denote

$$f \cdot f = \int_a^b p(x) f(x)^2\, dx = \| f \|^2.$$

As with vectors, define f and g to be orthogonal when $f \cdot g = 0$.

We call $\| f \|$ the *length* or *norm* of f, and we interpret

$$\begin{aligned} \| f - g \| &= \sqrt{(f-g)\cdot(f-g)} \\ &= \sqrt{\int_a^b p(x)(f(x) - g(x))^2\, dx} \end{aligned}$$

to be the distance between f and g.

Now let φ_n be eigenfunctions of a Sturm-Liouville problem, with eigenvalues λ_n. In the dot product notation, the coefficients of an expansion of f in a series of these eigenfunctions has coefficients

$$c_n = \frac{f \cdot \varphi_n}{\varphi_n \cdot \varphi_n} = \frac{f \cdot \varphi_n}{\| \varphi_n \|^2}.$$

We will use some facts from linear algebra to derive an important property of these coefficients. Suppose we write a linear combination

$$\sum_{n=1}^{N} k_n \varphi_n(x),$$

with the k_n's any numbers. We ask: how should we choose the k_n's so that $\sum_{n=1}^{N} k_n \varphi_n(x)$ best approximates $f(x)$ for $a < x < b$? This question is well posed because we have a notion of distance between functions. The problem is to choose the k_n's to minimize the distance between $f(x)$ and $\sum_{n=1}^{N} k_n \varphi_n(x)$:

$$\left\| f - \sum_{n=1}^{N} k_n \varphi_n(x) \right\|.$$

We know the answer to this question if we think in the context of orthogonal projections in a vector space. Let $PC[a, b]$ be the vector space of piecewise continuous functions defined on $[a, b]$. This is a vector space because sums and scalar multiples of piecewise continuous functions are also piecewise continuous, and the zero function is piecewise continuous. The eigenfunctions

$$\varphi_1(x), \varphi_2(x), \cdots, \varphi_N(x)$$

span a subspace S of $PC[a, b]$, and in fact form an orthogonal basis for this subspace. The linear combination of these eigenfunctions having minimum distance from f is the orthogonal projection f_S of f onto this subspace, and we know from Sections 6.6 and 6.7 that this projection is

$$f_S(x) = \sum_{n=1}^{N} \frac{f \cdot \varphi_n}{\varphi_n \cdot \varphi_n} \varphi_n(x).$$

This is exactly the Nth partial sum of the eigenfunction expansion of f.

We may therefore think of the coefficients in the eigenfunction expansion of f as those for which the Nth partial sum of the expansion is the best approximation to $f(x)$ as a linear combinations of the first N eigenfunctions.

There is also a Bessel inequality and a Parseval theorem for general eigenfunction expansion coefficients. First, let

$$\Phi_n = \frac{\varphi_n}{\| \varphi_n \|},$$

which we can also write as

$$\Phi_n = \frac{\varphi_n}{\sqrt{\varphi_n \cdot \varphi_n}}.$$

This divides each eigenfunction by its length, resulting in an eigenfunction of length 1:

$$\| \Phi_n \|^2 = \Phi_n \cdot \Phi_n = 1.$$

We say that the eigenfunctions have been *normalized*. Since nonzero constant multiples of eigenfunctions are eigenfunctions, we can expand $f(x)$ in a series of these normalized eigenfunctions

$$\sum_{n=1}^{\infty} c_n \Phi_n(x),$$

where

$$c_n = \frac{f \cdot \Phi_n}{\Phi_n \cdot \Phi_n} = f \cdot \Phi_n.$$

Now,

$$\Phi_n \cdot \Phi_m = \begin{cases} 1 & \text{if } n = m, \\ 0 & \text{if } n \neq m. \end{cases}$$

Then,

$$\begin{aligned} 0 \leq \left\| f - \sum_{n=1}^{N} c_n \Phi_n \right\|^2 &= \left(f - \sum_{n=1}^{N} c_n \Phi_n \right) \cdot \left(f - \sum_{n=1}^{N} c_m \Phi_m \right) \\ &= f \cdot f - 2 \sum_{n=1}^{N} c_n f \cdot \Phi_n + \sum_{n=1}^{N} \sum_{m=1}^{N} c_n c_m \Phi_n \cdot \Phi_m \\ &= f \cdot f - 2 \sum_{n=1}^{N} c_n^2 + \sum_{n=1}^{N} c_n^2 \\ &= f \cdot f - \sum_{n=1}^{N} c_n^2. \end{aligned}$$

Therefore,

$$\sum_{n=1}^{N} c_n^2 \leq f \cdot f.$$

This is *Bessel's inequality* for this general setting. It says that the sum of the first N squares of the generalized Fourier coefficients of f is less than or equal to the length of f. Since N can be any positive number, this implies that

$$\sum_{n=1}^{\infty} c_n^2 \leq \| f \|^2.$$

The series of squares of these coefficients converges, and the sum is bounded by the square of the length of f.

If f is continuous on $[a, b]$, then $f(x) = \sum_{n=1}^{\infty} c_n \Phi_n(x)$ on (a, b) and in this case Bessel's inequality is an equality:

$$\sum_{n=1}^{\infty} c_n^2 = \| f \|^2,$$

or, equivalently,

$$\sum_{n=1}^{\infty} c_n^2 = f \cdot f.$$

This is *Parseval's theorem*.

Both of these results were seen previously in the special case of Fourier series in Theorems 13.6 and 13.7.

We conclude this subsection with the idea of completeness, which is perhaps most easily understood in the context of R^3. The vectors **i**, **j**, and **k** are *complete* in R^3 because there is no nonzero vector that is orthogonal to all three of these vectors. In terms of axes, there is no axis perpendicular to the x-, y- and z-axes.

Now let $C'[a, b]$ be the set of functions that are continuous on $[a, b]$ with piecewise continuous derivatives. The eigenfunctions Φ_n are in $C'[a, b]$, and may be thought of as defining perpendicular axes, or directions, just as the unit vectors do in R^3. These eigenfunctions are *complete* in the sense that there is no nonzero function in $C'[a, b]$ that is orthogonal to all of the Φ_n's. It can be shown that the eigenfunctions of a Sturm-Liouville problem are complete in $C'[a, b]$.

SECTION 15.1 PROBLEMS

In each of Problems 1 through 10, classify the Sturm-Liouville problem as regular, periodic or singular, state the relevant interval, and find the eigenvalues and eigenfunctions. In some cases, the eigenvalues may be defined implicitly by a transcendental equation.

1. $y'' + \lambda y = 0;\ y(0) - 2y'(0) = 0,\ y'(1) = 0$
2. $(e^{-6x}y')' + (1+\lambda)e^{-6x}y = 0;\ y(0) = y(8) = 0$
3. $(e^{2x}y')' + \lambda e^{2x}y = 0;\ y(0) = y(\pi) = 0$
4. $y'' + \lambda y = 0;\ y(0) = 0,\ y(\pi) + 2y'(\pi) = 0$
5. $y'' + \lambda y = 0;\ y'(0) = y(4) = 0$
6. $y'' + \lambda y = 0;\ y(0) = y(\pi),\ y'(0) = y'(\pi)$
7. $y'' + \lambda y = 0;\ y(0) = y'(L) = 0$
8. $y'' + \lambda y = 0;\ y'(0) = y'(L) = 0$
9. $y'' + \lambda y = 0;\ y(-3\pi) = y(3\pi),\ y'(-3\pi) = y'(3\pi)$
10. $y'' + 2y' + (1+\lambda)y = 0;\ y(0) = y(1) = 0$

In each of Problems 11 through 16, find the eigenfunction expansion of the given function in the eigenfunctions of the Sturm-Liouville problem. In each case, determine what this expansion converges to and graph the Nth partial sum of the expansion and the function on the same set of axes. In Problem 15, do the graph for $L = 1$.

11. $f(x) = \begin{cases} -1 & \text{for } 0 \le x \le 2 \\ 1 & \text{for } 2 < x \le 4 \end{cases}$
    $y'' + \lambda y = 0;\ y'(0) = y(4) = 0;\ N = 40$
12. $f(x) = \begin{cases} 0 & \text{for } 0 \le x \le 1/2 \\ 1 & \text{for } 1/2 < x \le 1 \end{cases}$
    $y'' + 2y' + (1+\lambda)y = 0;\ y(0) = y(1) = 0;\ N = 30$
13. $f(x) = x^2$ for $-3\pi \le x \le 3\pi$
    $y'' + \lambda y = 0;\ y(-3\pi) = y(3\pi),\ y'(-3\pi) = y'(3\pi);\ N = 10$
14. $f(x) = x$ for $0 \le x \le \pi$
    $y'' + \lambda y = 0;\ y(0) = y'(\pi) = 0;\ N = 30$
15. $f(x) = 1 - x$ for $0 \le x \le L$
    $y'' + \lambda y = 0;\ y(0) = y(L) = 0;\ N = 40$
16. $f(x) = \sin(2x)$ for $0 \le x \le \pi$
    $y'' + \lambda y = 0;\ y'(0) = y'(\pi) = 0;\ N = 30$
17. Write Bessel's inequality for $f(x) = x(4-x)$ for the eigenfunctions of the Sturm-Liouville problem of Problem 5. *Hint*: Remember to normalize the eigenfunctions.
18. Write Bessel's inequality for $f(x) = e^{-x}$ for the eigenfunctions of the Sturm-Liouville problem of Problem 4.

15.2 Legendre Polynomials

We will derive a class of special functions called Legendre polynomials, which are solutions of *Legendre's differential equation*

$$((1-x^2)y')' + \lambda y = 0.$$

Adrien-Marie Legendre (1752 - 1833) was a professor at the Ecole Militaire who worked in many fields, including number theory and elliptic integrals. He encountered what would become known as Legendre polynomials in the 1780s while continuing Laplace's work on the potential equation (Chapter 18).

Legendre's differential equation is in Sturm-Liouville form, with $r(x) = 1 - x^2$, $q(x) = 0$ and $p(x) = 1$. Since $r(-1) = r(1) = 0$, Legendre's equation forms a singular Sturm-Liouville problem on $[-1, 1]$, with no boundary conditions. However, we seek solutions that are bounded on this interval.

To find such solutions, attempt a power series solution $y = \sum_{n=0}^{\infty} a_n x^n$ expanded about 0 (see Section 4.1). Substitute this series into Legendre's equation to obtain

$$\sum_{n=2}^{\infty} n(n-1)a_n x^{n-2} - \sum_{n=2}^{\infty} n(n-1)a_n x^n - \sum_{n=0}^{\infty} 2na_n x^n + \sum_{n=0}^{\infty} \lambda a_n x^n = 0.$$

Rewrite the first summation as

$$\sum_{n=2}^{\infty} n(n-1)a_n x^{n-2} = \sum_{n=0}^{\infty} (n+2)(n+1)a_{n+2} x^n$$

to obtain

$$\sum_{n=0}^{\infty} (n+2)(n+1)a_{n+2} x^n - \sum_{n=2}^{\infty} n(n-1)a_n x^n - \sum_{n=0}^{\infty} 2na_n x^n + \sum_{n=0}^{\infty} \lambda a_n x^n = 0.$$

Write this as

$$2a_2 + 6a_3 x - 2a_1 x + \lambda a_0 + \lambda a_1 x$$
$$+ \sum_{n=2}^{\infty} [(n+2)(n+1)a_{n+2} - (n^2 + n - \lambda)a_n] x^n = 0.$$

The constant term and the coefficient of each power of x on the left must be zero, so

$$2a_2 + \lambda a_0 = 0$$
$$6a_3 - 2a_1 + \lambda a_1 = 0, \text{ and}$$
$$(n+1)(n+2)a_{n+2} - [n(n+1) - \lambda]a_n = 0 \text{ for } n = 2, 3, \cdots.$$

From these equations, we obtain

$$a_2 = -\frac{\lambda}{2} a_0,$$
$$a_3 = \frac{2-\lambda}{6} a_1 = \frac{2-\lambda}{3!} a_1$$

and

$$a_{n+2} = \frac{n(n+1) - \lambda}{(n+1)(n+2)} a_n \text{ for } n = 2, 3, \cdots. \quad (15.5)$$

Equation (15.5) is a recurrence relation which gives each a_{n+2} in terms of λ and a_n. This enables us to produce the coefficients in turn from previously found coefficients. We already have expressions for a_2 and a_3 in terms of λ, a_0 and a_1. From the recurrence relation of equation (15.5) with $n = 2$,

$$a_4 = \frac{6-\lambda}{(3)(4)} a_2 = -\frac{\lambda}{2} \frac{6-\lambda}{(3)(4)} a_0 = \frac{-\lambda(6-\lambda)}{4!} a_0.$$

With $n = 4$ we obtain

$$a_6 = \frac{20-\lambda}{(5)(6)}a_4 = \frac{-\lambda(6-\lambda)(20-\lambda)}{6!}a_0,$$

and so on, with each even-indexed a_{2n} in terms of λ, n and a_0. Similarly, with $n = 3$,

$$a_5 = \frac{12-\lambda}{(4)(5)}a_3 = \frac{(2-\lambda)(12-\lambda)}{5!}a_1,$$

and with $n = 7$,

$$a_7 = \frac{30-\lambda}{(6)(7)}a_5 = \frac{(2-\lambda)(12-\lambda)(30-\lambda)}{7!}a_1$$

and so on. Each odd-indexed a_{2n+1} can be written in terms of λ, n and a_1.

Now we can write the solution

$$y(x) = \sum_{n=0}^{\infty} a_n x^n = a_0\left(1 - \frac{\lambda}{2}x^2 - \frac{\lambda(6-\lambda)}{4!}x^4 - \frac{\lambda(6-\lambda)(20-\lambda)}{6!}x^6 + \cdots\right)$$
$$+ a_1\left(x + \frac{2-\lambda}{3!}x^3 + \frac{(2-\lambda)(12-\lambda)}{5!}x^5 + \frac{(2-\lambda)(12-\lambda)(30-\lambda)}{7!}x^7 + \cdots\right).$$

These two series solutions in large parentheses are linearly independent, one containing only odd powers of x, the other only even powers. This expression therefore gives the general solution of Legendre's equation, with a_0 and a_1 arbitrary constants.

Now observe that we obtain polynomial solutions by choosing λ of the form $n(n+1)$, and either a_0 or a_1 zero. For example:

With $n = 0$ and $a_1 = 0$, we have $\lambda = 0$ and

$$y(x) = a_0 = \text{ constant };$$

with $n = 1$ and $a_0 = 0$, we have $\lambda = 2$ and

$$y(x) = a_1 x;$$

with $n = 2$ and $a_1 = 0$, we have $\lambda = 6$ and

$$y(x) = a_0(1 - 3x^2);$$

with $n = 3$ and $a_0 = 0$, we have $\lambda = 12$ and

$$y(x) = a_1\left(x - \frac{5}{3}x^3\right);$$

with $n = 4$ and $a_1 = 0$, we have $\lambda = 20$ and

$$y(x) = a_0\left(1 - 10x^2 + \frac{35}{3}x^4\right);$$

and so on. These polynomial solutions are bounded on $[-1, 1]$. If the constant is always chosen so that $y(1) = 1$, we obtain the Legendre polynomials $P_n(x)$, the first six of which are

$$P_0(x) = 1,\ P_1(x) = x,\ P_2(x) = \frac{1}{2}(3x^2 - 1),$$
$$P_3(x) = \frac{1}{2}(5x^3 - 3x),\ P_4(x) = \frac{1}{8}(35x^4 - 30x^2 + 3),\ \text{and}$$
$$P_5(x) = \frac{1}{8}(63x^5 - 70x^3 + 15x).$$

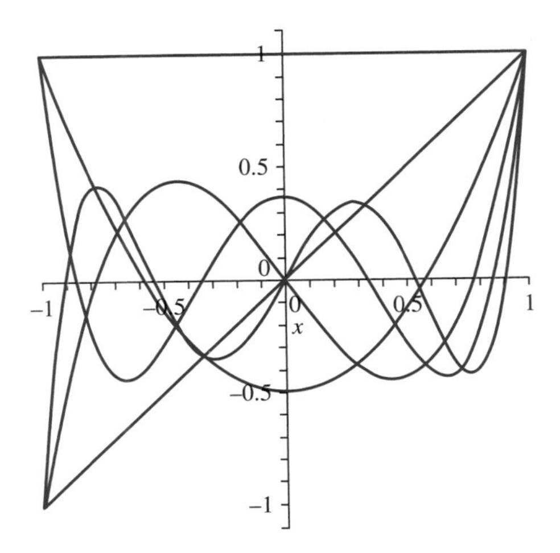

FIGURE 15.3 *The first six Legendre polynomials.*

Graphs of these first six are shown in Figure 15.3.

In summary, the numbers $\lambda = n(n+1)$ for $n = 0, 1, 2, \cdots$ are eigenvalues of Legendre's equation, for each nonnegative integer n. For each such n, $\varphi_n(x) = P_n(x)$ is an eigenfunction. This eigenfunction is an odd polynomial if n is odd and an even polynomial if n is even.

From Theorem 15.1, these Legendre polynomials (eigenfunctions) are orthogonal on $[-1, 1]$ with respect to the weight function $p(x) = 1$. This means that

$$\int_a^b P_n(x) P_m(x)\, dx = 0$$

if n and m are distinct nonnegative integers. The weight function $p(x) = 1$ can be read directly from the coefficient of y in Legendre's differential equation.

There is an extensive literature on Legendre polynomials. We will develop some frequently used facts about them.

15.2.1 A Generating Function for Legendre Polynomials

Define

$$L(x, t) = \frac{1}{\sqrt{1 - 2xt + t^2}}.$$

For a given x, we can think of $L(x, t)$ as a function of t which can be expanded in a power series about $t = 0$. We claim that, when this is done, the coefficient of t^n is $P_n(x)$. For this reason we call $L(x, t)$ a *generating function* for the Legendre polynomials.

THEOREM 15.3 *A Generating Function*

$$L(x, t) = \sum_{n=0}^{\infty} P_n(x) t^n. \quad \blacklozenge$$

This result is useful in deriving properties of Legendre polynomials. We will not give a complete proof, but we will derive some terms in the power series expansion of $L(x,t)$ about $t=0$, suggesting why the theorem is true. Begin with the Maclaurin series for $(1-w)^{-1/2}$:

$$\frac{1}{\sqrt{1-w}}=1+\frac{1}{2}w+\frac{3}{8}w^2+\frac{15}{48}w^3+\frac{105}{384}w^4+\frac{945}{3840}w^5+\cdots$$

for $-1<w<1$. Put $w=2xt-t^2$ to obtain

$$\frac{1}{\sqrt{1-2xt+t^2}}=1+\frac{1}{2}(2xt-t^2)+\frac{3}{8}(2xt-t^2)^2$$
$$+\frac{15}{48}(2xt-t^2)^3+\frac{105}{384}(2xt-t^2)^4+\frac{945}{3840}(2xt-t^2)^5+\cdots.$$

Now expand each of these powers of $2xt-t^2$ and collect the coefficient of each power of t to obtain

$$\frac{1}{\sqrt{1-2xt+t^2}}=1+xt-\frac{1}{2}t^2-\frac{3}{2}xt^3+\frac{3}{8}t^4+\frac{5}{2}x^3t^3$$
$$-\frac{15}{4}x^2t^4+\frac{15}{8}xt^5-\frac{5}{16}t^6+\frac{35}{4}x^3t^5+\frac{105}{16}x^2t^6$$
$$-\frac{35}{16}xt^7+\frac{35}{128}t^8+\frac{63}{8}x^5t^5-\frac{315}{16}x^4t^6+\frac{315}{16}x^3t^7$$
$$-\frac{315}{32}x^2t^8+\frac{315}{128}xt^9-\frac{69}{256}t^{10}+\cdots$$
$$=1+xt+\left(-\frac{1}{2}+\frac{3}{2}x^2\right)t^2+\left(-\frac{3}{2}x+\frac{5}{2}x^3\right)t^3$$
$$+\left(\frac{3}{8}-\frac{15}{4}x^2+\frac{35}{8}x^4\right)t^4+\left(\frac{15}{8}x-\frac{35}{4}x^3+\frac{63}{8}x^5\right)t^5+\cdots$$
$$=P_0(x)+P_1(x)t+P_2(x)t^2+P_3(x)t^3+P_4(x)t^4+P_5(x)t^5+\cdots.$$

We know that $P_n(1)=1$ because a_0 and a_1 in the derivation of these polynomial solutions of Legendre's equation were chosen for this purpose. Using the generating function, we can also evaluate $P_n(-1)$.

COROLLARY 15.1

For every nonnegative integer n,

$$P_n(-1)=(-1)^n.$$

Proof Evaluate

$$L(-1,t)=\frac{1}{\sqrt{1+2t+t^2}}=\frac{1}{\sqrt{(1+t)^2}}$$
$$=\frac{1}{1+t}=\sum_{n=0}^{\infty}P_n(-1)t^n$$

for $-1<t<1$. But, for these values of t, we have the geometric series

$$\frac{1}{1+t}=\sum_{n=0}^{\infty}(-1)^n t^n.$$

By comparing coefficients in these two Maclaurin series for $1/(1+t)$, we conclude that $P_n(-1)=(-1)^n$. ♦

15.2.2 A Recurrence Relation for Legendre Polynomials

There is a recurrence relation for Legendre polynomials that gives $P_{n+1}(x)$ in terms of $P_n(x)$ and $P_{n-1}(x)$.

THEOREM 15.4 ***A Recurrence Relation***

If n is a positive integer, then

$$(n+1)P_{n+1}(x)-(2n+1)xP_n(x)+nP_{n-1}(x)=0. \quad \blacklozenge \tag{15.6}$$

Proof Begin with

$$\frac{\partial L}{\partial t}=\frac{x-t}{(1-2xt+t^2)^{3/2}}.$$

Multiply this equation by $1-2xt+t^2$ to obtain

$$(1-2xt+t^2)\frac{\partial L}{\partial t}(x,t)-(x-t)L(x,t)=0.$$

Substitute $L(x,t)=\sum_{n=0}^{\infty}P_n(x)t^n$ into this equation to obtain

$$(1-2xt+t^2)\sum_{n=1}^{\infty}nP_n(x)t^{n-1}-(x-t)\sum_{n=0}^{\infty}P_n(x)t^n=0.$$

Carry out the indicated multiplications to write

$$\sum_{n=1}^{\infty}nP_nt^{n-1}-\sum_{n=1}^{\infty}2nxP_n(x)t^n+\sum_{n=1}^{\infty}nP_n(x)t^{n+1}-$$

$$\sum_{n=0}^{\infty}xP_n(x)t^n+\sum_{n=0}^{\infty}P_n(x)t^{n+1}=0.$$

Rearrange the series where necessary to have t^n as the power of t in each summation:

$$\sum_{n=0}^{\infty}(n+1)P_{n+1}(x)t^n-\sum_{n=1}^{\infty}2nxP_n(x)t^n+\sum_{n=2}^{\infty}(n-1)P_{n-1}(x)t^n$$

$$-\sum_{n=0}^{\infty}xP_n(x)t^n+\sum_{n=1}^{\infty}P_{n-1}(x)t^n=0.$$

Combine these summations from $n=2$ on, writing the terms for $n=0$ and $n=1$ separately, to obtain

$$P_1(x)+2P_2(x)t-2xP_1(x)t-xP_0(x)-xP_1(x)t+P_0(x)t$$

$$+\sum_{n=2}^{\infty}[(n+1)P_{n+1}(x)-2nxP_n(x)+(n-1)P_{n-1}(x)-xP_n(x)+P_{n-1}(x)]t^n=0.$$

For this power series in t to be zero for all t in some open interval about 0, the coefficient of each power of t must equal 0. Therefore

$$P_1(x)-xP_0(x)=0$$

$$2P_2(x)-2xP_1(x)-xP_1(x)+P_0(x)=0$$

and, for $n = 2, 3, \cdots$,

$$(n+1)P_{n+1}(x) - 2nxP_n(x) + (n-1)P_{n-1}(x) - xP_n(x) + P_{n-1}(x) = 0.$$

These give us

$$P_1(x) = xP_0(x) = x,$$

$$P_2(x) = \frac{1}{2}(3xP_1(x) - P_0(x)) = \frac{1}{2}(3x^2 - 1),$$

and, for $n = 2, 3, \cdots$,

$$(n+1)P_{n+1}(x) - (2n+1)xP_n(x) + nP_{n-1}(x) = 0.$$

Since this equation is also valid for $n = 1$, the recurrence relation is proved. ♦

When we consider eigenfunction expansions in terms of Legendre polynomials, we will need to know the coefficient of x^n in $P_n(x)$. We can obtain this using the recurrence relation.

COROLLARY 15.2

For each positive integer n, the coefficient of x^n in $P_n(x)$ is

$$\frac{1 \cdot 3 \cdots (2n-1)}{n!}. \; ♦$$

This is the product of the odd integers from 1 through $2n - 1$ inclusive, divided by the product of the integers from 1 through n.

Proof Let A_n be the coefficient of x^n in $P_n(x)$. In the recurrence relation (15.6), the highest power of x that occurs is x^{n+1}, and this power occurs only in $P_{n+1}(x)$ and in $xP_n(x)$. Therefore the coefficient of x^{n+1} on the left side of the recurrence relation is

$$(n+1)A_{n+1} - (2n+1)A_n.$$

This must equal zero, because the right side of the recurrence relation is 0, with no x^{n+1} term. Thus

$$A_{n+1} = \frac{2n+1}{n+1}A_n$$

for $n = 0, 1, 2, \cdots$. We know that $A_0 = 1$ because $P_0(x) = 1$, so we can work back from this recurrence relation for A_n to obtain:

$$\begin{aligned}
A_{n+1} &= \frac{2n+1}{n+1}A_n = \frac{2n+1}{n+1}\frac{2(n-1)+1}{(n-1)+1}A_{n-1} \\
&= \frac{2n+1}{n+1}\frac{2n-1}{n}A_{n-1} = \frac{2n+1}{n+1}\frac{2n-1}{n}\frac{2(n-2)+1}{(n-2)+1}A_{n-2} \\
&= \frac{2n+1}{n+1}\frac{2n-1}{n}\frac{2n-3}{n-1}A_{n-2} = \cdots = \frac{2n+1}{n+1}\frac{2n-1}{n}\frac{2n-3}{n-1}\cdots\frac{3}{2}A_0 \\
&= \frac{2n+1}{n+1}\frac{2n-1}{n}\frac{2n-3}{n-1}\cdots\frac{3}{2}.
\end{aligned}$$

Therefore

$$A_{n+1} = \frac{1 \cdot 3 \cdot 5 \cdots (2n-1)(2n+1)}{(n+1)!},$$

and this is the conclusion of the theorem, stated in terms of $n + 1$ instead of n. ♦

15.2.3 Fourier-Legendre Expansions

Because the Legendre polynomials are eigenfunctions of a Sturm-Liouville problem, we can write an eigenfunction expansion $\sum_{n=0}^{\infty} c_n P_n(x)$ of a function f that is piecewise smooth on $[-1, 1]$. The c_n 's are given by equation (15.4) with $p(x) = 1$ and $\varphi_n(x) = P_n(x)$:

$$c_n = \frac{\int_{-1}^{1} f(x) P_n(x)\, dx}{\int_{-1}^{1} P_n^2(x)\, dx}.$$

The resulting series $\sum_{n=0}^{\infty} c_n P_n(x)$ for $f(x)$ is called the *Fourier-Legendre expansion* of $f(x)$ on $[-1, 1]$, and the c_n 's are the *Fourier-Legendre coefficients* of $f(x)$ on this interval.

We will look at an example of a Fourier-Legendre expansion shortly. First we will observe that the Fourier-Legendre expansion of any polynomial $q(x)$ can be achieved purely by algebraic manipulation. To do this, solve for powers of x in terms of Legendre polynomials and substitute these into $q(x)$. To illustrate this process, let

$$q(x) = -4 + 2x + 9x^2.$$

We know that $x = P_1(x)$. Next solve for x^2 in $P_2(x)$. Since

$$P_2(x) = \frac{3}{2}x^2 - \frac{1}{2},$$

then

$$x^2 = \frac{2}{3}P_2(x) + \frac{1}{3} = \frac{2}{3}P_2(x) + \frac{1}{3}P_0(x).$$

Substitute these into $q(x)$ to obtain

$$\begin{aligned} q(x) &= -4 + 2x + 9x^2 \\ &= -4P_0(x) + 2P_1(x) + 9\left(\frac{2}{3}P_2(x) + \frac{1}{3}P_0(x)\right) \\ &= -P_0(x) + 2P_1(x) + 6P_2(x). \end{aligned}$$

As a simple consequence of this observation, we can show that every Legendre polynomial is orthogonal on $[-1, 1]$ to every polynomial of lower degree.

THEOREM 15.5

If $q(x)$ is a polynomial of degree m, and $n > m$, then

$$\int_{-1}^{1} q(x) P_n(x)\, dx = 0. \quad \blacklozenge$$

Proof Suppose $q(x)$ has degree m. Write the Fourier-Legendre representation

$$q(x) = \sum_{k=0}^{m} c_k P_k(x).$$

Then, if $n > m$, we have

$$\int_{-1}^{1} q(x)P_n(x)\,dx = \sum_{k=0}^{m} c_k \int_{-1}^{1} P_k(x)P_n(x)\,dx = 0$$

because each $k < n$. ♦

Using this result, we can derive a simple expression for $\int_{-1}^{1} P_n^2(x)\,dx$, the denominator in the expression for the Fourier-Legendre coefficients of any function.

THEOREM 15.6

For $n = 0, 1, 2, \cdots$,

$$\int_{-1}^{1} P_n^2(x)\,dx = \frac{2}{2n+1}. \; ♦$$

Proof Let A_n denote the coefficient of x^n in $P_n(x)$ and let

$$p_n = \int_{-1}^{1} P_n^2(x)\,dx.$$

The highest power term in $P_n(x)$ is $A_n x^n$, while the highest power term in $P_{n-1}(x)$ is $A_{n-1}x^{n-1}$. Therefore all terms involving x^n cancel in the polynomial

$$q(x) = P_n(x) - \frac{A_n}{A_{n-1}} x P_{n-1}$$

so $q(x)$ has degree at most $n-1$. Write

$$P_n(x) = q(x) + \frac{A_n}{A_{n-1}} x P_{n-1}(x).$$

Because $\int_{-1}^{1} q(x)P_n(x)\,dx = 0$ by Theorem 15.5, then

$$\begin{aligned} p_n &= \int_{-1}^{1} P_n(x)P_n(x)\,dx \\ &= \int_{-1}^{1} P_n(x)\left(q(x) + \frac{A_n}{A_{n-1}} x P_{n-1}(x)\right) dx \\ &= \frac{A_n}{A_{n-1}} \int_{-1}^{1} x P_n(x)P_{n-1}(x)\,dx. \end{aligned}$$

Now use the recurrence relation to write

$$x P_n(x) = \frac{n+1}{2n+1} P_{n+1}(x) + \frac{n}{2n+1} P_{n-1}(x).$$

Then

$$x P_n(x)P_{n-1}(x) = \frac{n+1}{2n+1} P_{n+1}(x)P_{n-1}(x) + \frac{n}{2n+1} P_{n-1}^2(x).$$

Then

$$\begin{aligned}
p_n &= \frac{A_n}{A_{n-1}} \int_{-1}^{1} x P_n(x) P_{n-1}(x)\, dx \\
&= \frac{A_n}{A_{n-1}} \left[\frac{n+1}{2n+1} \int_{-1}^{1} P_{n+1}(x) P_{n-1}(x)\, dx + \frac{n}{2n+1} \int_{-1}^{1} P_{n-1}^2(x)\, dx \right] \\
&= \frac{A_n}{A_{n-1}} \frac{n}{2n+1} p_{n-1},
\end{aligned}$$

since $\int_{-1}^{1} P_{n+1}(x) P_{n-1}(x)\, dx = 0$. But we know A_n from Corollary 15.2. Substitute this into the last expression to obtain

$$\begin{aligned}
p_n &= \frac{1 \cdot 3 \cdot 5 \cdots (2n-3)(2n-1)}{n!} \frac{(n-1)!}{1 \cdot 3 \cdot 5 \cdots (2n-3)} \frac{n}{2n+1} p_{n-1} \\
&= \frac{2n-1}{2n+1} p_{n-1}.
\end{aligned}$$

Now work forward:

$$\begin{aligned}
p_1 &= \frac{1}{3} p_0 = \frac{1}{3} \int_{-1}^{1} P_0(x)^2\, dx = \frac{1}{3} \int_{-1}^{1} dx = \frac{2}{3} \\
p_2 &= \frac{3}{5} p_1 = \frac{3}{5}\frac{2}{3} = \frac{2}{5} \\
p_3 &= \frac{5}{7} p_2 = \frac{2}{7} \\
p_4 &= \frac{7}{9} p_3 = \frac{2}{9}
\end{aligned}$$

and so on. One can complete the proof by induction and obtain

$$p_n = \frac{2}{2n+1}. \quad \blacklozenge$$

With this result, we can write the Fourier-Legendre coefficient of $f(x)$ as

$$c_n = \frac{2n+1}{2} \int_{-1}^{1} f(x) P_n(x)\, dx.$$

Now here is an example of a Fourier-Legendre expansion.

EXAMPLE 15.5

We will write the Fourier-Legendre expansion of $f(x) = \cos(\pi x/2)$. Because $\cos(\pi x/2)$ is continuous with a continuous derivative,

$$\cos(\pi x/2) = \sum_{n=0}^{\infty} c_n P_n(x)$$

for $-1 < x < 1$, where

$$c_n = \frac{2n+1}{2} \int_{-1}^{1} \cos(\pi x/2) P_n(x)\, dx.$$

There is no simple expression for c_n for arbitrary n, so we will approximate this eigenfunction expansion with the sum of the first six terms. We must therefore compute $c_0, \cdots, c_5$. Since we know $P_0(x)$ through $P_5(x)$, these integrations can be carried out explicitly.

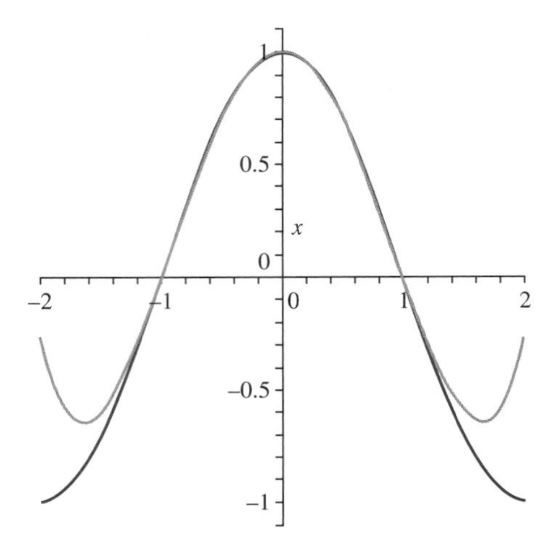

FIGURE 15.4 *Eigenfunction expansion in Example 15.5.*

First, $\cos(\pi x/2)$ is even on $[-1, 1]$, $\cos(\pi x/2)P_n(x)$ is an odd function for n odd and $\int_{-1}^{1} \cos(\pi x/2)P_n dx = 0$. This means that $c_n = 0$ if n is odd. In particular, $c_1 = c_3 = c_5 = 0$.

For c_0, c_2 and c_4, we can do the integrations exactly but it is more efficient to use a software package. If MAPLE is used, the nth Legendre polynomial is denoted *LegendreP(n,x)*. To compute c_n, use the MAPLE code:

```
((2*n + 1)/2)*int(cos(Pi*x/2)*LegendreP(n,x),x=-1..1);
```

With, respectively, $n = 0, 2, 4$, this yields

$$c_0 \approx 0.6366197722,\ c_2 \approx -0.6870852690,\ c_4 \approx 0.05177890435.$$

If we retain four decimal places, then the sixth partial sum of the Fourier-Legendre expansion of $\cos(\pi x/2)$ is

$$\cos(\pi x/2) \approx 0.6366 - 0.3435(3x^2 - 1) + 0.0065(35x^4 - 30x^2 + 3)$$

for $-1 < x < 1$.

Figure 15.4 shows a graph of $\cos(\pi x/2)$ and this partial sum on an interval slightly larger than $[-1, 1]$. Even with this small number of terms, the partial sum is nearly indistinguishable from the function on $[-1, 1]$, within the scale of the graph. Outside $[-1, 1]$, the polynomial approximation rapidly diverges from $\cos(\pi x/2)$. We would not in general expect to obtain this good an approximation with as few terms of an eigenfunction expansion. ♦

We conclude this section with some additional properties of Legendre polynomials.

15.2.4 Zeros of Legendre Polynomials

If f is a given function, a number x_0 is a *zero* of f if $f(x_0) = 0$. Not every function has real zeros. For example, $f(x) = 1 + x^2$ does not. A real-valued function of a single real variable has a zero exactly where its graph crosses the horizontal axis, and the graph of $y = 1 + x^2$ has no such crossing.

Notice, however, that for $(n = 1, 2, 3, 4, 5)$, the graph of $P_n(x)$ in Figure 15.3 crosses the x-axis exactly n times between -1 and 1, suggesting that each of these functions has n zeros between -1 and 1.

More than this can be shown. A zero x_0 of $f(x)$ is *simple* if $f(x_0)=0$ but $f'(x_0)\neq 0$. For example, $f(x)=x^2-4$ has simple zeros at $x=2$ and $x=-2$, while $g(x)=(1+x)^2$ has a zero at $x=-1$, but this zero is not simple because $g(-1)=g'(-1)=0$. We call -1 a double or repeated zero of $g(x)$.

THEOREM 15.7 Zeros of Legendre Polynomials

For each positive integer n, $P_n(x)$ has n simple zeros, all lying between -1 and 1. ♦

Proof Let n be a positive integer.
First we will show that, if $P_n(x)$ has a zero x_0 in $(-1,1)$, then this zero must be simple.

Since $P_n(x)$ satisfies Legendre's equation with $\lambda=n(n+1)$, then

$$\left[(1-x^2)P_n'\right]'+n(n+1)P_n=0.$$

If $P_n(x_0)=P_n'(x_0)=0$, then $P_n(x)$ satisfies the initial value problem

$$[(1-x^2)y']'+n(n+1)y=0;\ y(x_0)=y'(x_0)=0.$$

But $y(x)=0$, the identically zero function, is also a solution of this problem. By uniqueness of the solution to this problem, we would then have $y(x)=P_n(x)=0$ for all x, and this is false. Therefore, any zero that $P_n(x)$ has must be simple.

Next we will show that $P_n(x)$ has at least one zero in $(-1,1)$. Because $P_n(x)$ is orthogonal to $P_0(x)=1$,

$$\int_{-1}^{1}P_0(x)P_n(x)\,dx=\int_{-1}^{1}P_n(x)\,dx=0.$$

If $P_n(x)$ were strictly positive, or strictly negative, on $(-1,1)$, then this integral would be, respectively, positive or negative. Therefore for some x_0 in $(-1,1)$, $P_n(x_0)=0$.

We now know that $P_n(x)$ has at least one zero in $(-1,1)$, and any zero must be simple.

Now we will show that $P_n(x)$ has n zeros between -1 and 1. Let $x_1,\cdots,x_m$ be all the zeros of $P_n(x)$ in $(-1,1)$. Then $1\le m\le n$. Order these zeros from left to right across the interval

$$-1<x_1<x_2<\cdots<x_m<1.$$

If $m<n$, then the polynomial

$$q(x)=(x-x_1)(x-x_2)\cdots(x-x_m)$$

has degree $m<n$ and the same zeros as $P_n(x)$ in $(-1,1)$. Then $P_n(x)$ and $q(x)$ change sign at exactly the same points in $(-1,1)$. This means that $P_n(x)$ and $q(x)$ are either of the same sign on each interval $(-1,x_1),(x_1,x_2),\cdots,(x_{m-1},x_m),(x_m,1)$, or of opposite sign on each of these intervals. But then $q(x)P_n(x)$ is either strictly positive or strictly negative on $(-1,1)$, except at $x_1,\cdots,x_m$. Then

$$\int_{-1}^{1}q(x)P_n(x)\,dx$$

must be positive or negative. But this integral is zero if the degree of $q(x)$ is less than the degree of $P_n(x)$. We conclude that $m=n$, hence $P_n(x)$ has n simple zeros on $(-1,1)$, as was to be proved. ♦

The zeros of Legendre polynomials have quite fascinating properties, one of which we will now explore.

15.2.5 Distribution of Charged Particles

Suppose N charged beads are distributed on the x-axis between -1 and 1. The beads are free to move along the wire on this interval. There is a planar charge of $+1$ at the ends of the interval and each bead has a charge of $+2$. The like charges of the beads repel each other, and the ends repel the beads as well. If left alone, the beads will eventually reach an equilibrium position in which they are at rest, with the repelling forces in balance. We claim that the beads will end up at the zeros of $P_N(x)$.

To show this, we will use the fact that two particles (beads) carrying planar charge q_1 and q_2 and located r units apart exert a force of magnitude kq_1q_2/r on each other.

Begin with the simple case that $N=2$. This will help us understand the forces involved. If the beads are at x_1 and x_2, with $x_1<x_2$, then the force q_1 at x_1 is

$$F_1=2k\left[\frac{1}{x_1+1}-\frac{2}{x_2-x_1}+\frac{1}{x_1-1}\right].$$

The force on q_2 at x_2 is

$$F_2=2k\left[\frac{1}{x_2+1}+\frac{2}{x_2-x_1}+\frac{1}{x_2-1}\right].$$

To be in equilibrium it is necessary that $F_1=0$ and $F_2=0$. Add these two equations to obtain

$$\frac{2x_1}{x_1^2-1}+\frac{2x_2}{x_2^2-1}=0.$$

Simplify this to obtain

$$2(x_1x_2-1)(x_2+x_2)=0,$$

hence $x_1=-x_2$. From $F_1=0$ this yields

$$\frac{1}{x_1+1}+\frac{2}{2x_1}+\frac{1}{x_1-1}=0,$$

or $3x_1^2-1=0$. Then

$$x_1=-\frac{1}{\sqrt{3}}\text{ and }x_2=\frac{1}{\sqrt{3}}$$

and these are the zeros of $P_2(x)$.

This gives us some confidence in proceeding to the case of N beads. Suppose they are located at $x_1<x_2<\cdots<x_N$. The forces acting on q_k at x_k are $2k/(x_k+1)$ due to the charge of $+1$ at $x=-1$, $2k/(x_k-1)$ due to the charge of $+1$ at $x=+1$, $4k/(x_k-x_i)$ due to the charge of $+2$ at $x_i<x_k$, and $-4k/(x_i-x_k)$ due to the charge of $+2$ at $x_i>x_k$. The total force on q_k is

$$F_k=\frac{2k}{x_k+1}+\frac{2k}{x_k-1}+\sum_{i=1,i\neq k}^{N}\frac{4k}{x_k-x_i}.$$

The beads are in equilibrium if and only if the forces are zero, hence if and only if, after dividing out k, $x_1,\cdots,x_N$ satisfy

$$\frac{2x_k}{1-x_k^2}=\sum_{i=1,i\neq k}^{N}\frac{2}{x_k-x_i}$$

for $k=1,2,\cdots,N$. This is a necessary and sufficient condition for the beads to be in equilibrium. We will show that the N simple zeros of $P_N(x)$ (labeled in increasing order) also satisfy this condition.

First observe from Legendre's differential equation that

$$(1-x^2)P_N''(x) - 2xP_N'(x) + N(N+1)P_N(x) = 0.$$

If we put $x = x_k$, the kth zero of $P_N(x)$, then $P_N(x_k) = 0$ and $P_N'(x_k) \neq 0$, so

$$\frac{P_N''(x_k)}{P_N'(x_k)} = \frac{2x_k}{1-x_k^2}.$$

The right side of this equation is the left side of the necessary and sufficient condition for equilibrium. To complete the proof, we will show that the right side of this condition is also equal to $P_N''(x)/P_N'(x_k)$.

Since $P_N(x)$ has all simple zeros, then for some number A,

$$P_N(x) = A(x-x_1)(x-x_2)\cdots(x-x_N).$$

In product notation,

$$P_N(x) = A\Pi_{i=1}^N(x-x_i).$$

Take the derivative of this product and use the fact that $(d/dx)(x-x_i) = 1$:

$$P_N'(x) = A\sum_{i=1}^N \Pi_{m=1,m\neq i}^N(x-x_m)$$

so

$$P_N''(x) = A\sum_{j=1}^N \sum_{i=1,i\neq j}^N \Pi_{m=1,m\neq i,j}^N(x-x_m).$$

Then

$$P_N'(x_k) = A\sum_{i=1}^N \Pi_{m=1,m\neq i}^N(x_k-x_m).$$

In this expression, all terms are zero except when $m \neq k$, so

$$P_N'(x_k) = A\Pi_{m=1,m\neq k}^N(x_k-x_m).$$

Similarly compute

$$P_N''(x_k) = 2A\sum_{j=1,j\neq k}^N \Pi_{m=1,m\neq j,k}^N(x_k-x_m).$$

Since the zeros are simple, then $P_N'(x_k) \neq 0$ and

$$\frac{P_N''(x_k)}{P_N'(x_k)} = \sum_{j=1,j\neq k} \frac{2}{x_k-x_j}.$$

This is the right side of the necessary and sufficient condition for the beads to be in equilibrium. Using j for the summation index on the right, we now have

$$\frac{2x_k}{1-x_k^2} = \frac{P_N''(x_k)}{P_N'(x_k)} = \sum_{j=1,j\neq k}^N \frac{2}{x_k-x_j}.$$

Therefore the zeros satisfy the necessary and sufficient condition for the beads to be in equilibrium, completing the proof.

15.2.6 Some Additional Results

There are many other results involving Legendre polynomials. One of these is a derivative formula for $P_n(x)$, called *Rodrigues's formula*. For n any nonnegative integer,

$$P_n(x) = \frac{1}{2^n n!}\frac{d^n}{dx^n}((x^2-1)^n).$$

In this it is understood that the zero order derivative is just the function itself. This formula can be proved by induction on n.

There is also an integral formula. For any nonnegative integer n,

$$P_n(x) = \frac{1}{\pi}\int_0^{\pi}\left(x+\sqrt{x^2-1}\cos(\theta)\right)^n d\theta.$$

This integral formula can be established as follows. Let Q_n denote the integral on the right. Show that Q_n satisfies the same recurrence relation as the Legendre polynomials. This will involve an integration by parts. Finally, show directly that $Q_0 = P_0$ and $Q_1(x) = P_1(x)$. This will show that $Q_n(x) = P_n(x)$ for all nonnegative integers n.

There is a considerable literature on properties of Legendre polynomials, as well as on other weighted orthogonal polynomials arising as eigenfunctions of Sturm-Liouville problems. These include the Hermite, Laguerre and Tchebyshev polynomials, and many others.

SECTION 15.2 PROBLEMS

1. The gravitational potential at a point $P:(x, y, z)$ due to a unit mass at (x_0, y_0, z_0) is

$$\varphi(x,y,z) = \frac{1}{\sqrt{(x-x_0)^2+(y-y_0)^2+(z-z_0)^2}}.$$

In some contexts (such as astronomy), it is convenient to expand $\varphi(x, y, z)$ in powers of r or $1/r$, where

$$r = \sqrt{x^2+y^2+z^2}.$$

To do this, introduce the angle θ shown in Figure 15.5. Let

$$d = \sqrt{x_0^2+y_0^2+z_0^2}$$

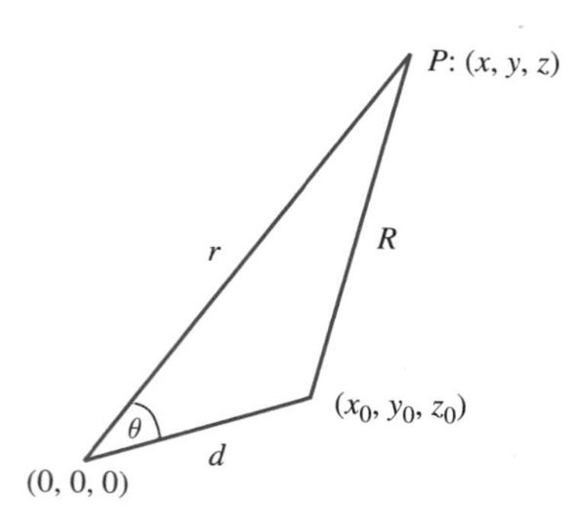

FIGURE 15.5 *Problem 1, Section 15.2.*

and

$$R = \sqrt{(x-x_0)^2+(y-y_0)^2+(z-z_0)^2}.$$

(a) Use the law of cosines to write

$$\varphi(x,y,z) = \frac{1}{d\sqrt{1-2(r/d)\cos(\theta)+(r/d)^2}}.$$

(b) If $r < d$, use the generating function for Legendre polynomials to show that

$$\varphi(r) = \sum_{n=0}^{\infty}\frac{1}{d^{n+1}}P_n(\cos(\theta))r^n.$$

(c) If $r > d$, use the generating function to show that

$$\varphi(r) = \frac{1}{r}\sum_{n=0}^{\infty} d^n P_n(\cos(\theta))r^{-n}.$$

2. Use Rodrigues's formula to derive $P_2(x)$ through $P_5(x)$.
3. Use the recurrence relation to derive $P_6(x)$, $P_7(x)$, and $P_8(x)$.
4. Expand each of the following polynomials in a series of Legendre polynomials.

   (a) $1+2x-x^2$
   (b) $2x+x^2-5x^3$
   (c) $2-x^2+4x^4$

5. It can be shown that

$$P_n(x) = \sum_{k=0}^{[n/2]} (-1)^k \frac{(2n-2k)!}{2^n k!(n-k)!(n-2k)!} x^{n-2k},$$

where, for any number t, $[t]$ denotes the largest integer not exceeding t. Use this formula to generate $P_0(x)$ through $P_5(x)$.

6. Put $\lambda = n(n+1)$ into Legendre's differential equation, and let $y(x) = u(x)P_n(x)$. Solve the resulting equation for $u(x)$ to derive the second solution

$$Q_n(x) = P_n(x) \int \frac{1}{(1-x^2)(P_n(x))^2} dx$$

of Legendre's equation. These functions $Q_n(x)$ are called *Legendre functions of the second kind*, and they are defined but unbounded on $(-1, 1)$.
Show that

$$Q_0(x) = -\frac{1}{2} \ln\left(\frac{1+x}{1-x}\right),$$

$$Q_1(x) = 1 - \frac{x}{2} \ln\left(\frac{1+x}{1-x}\right),$$

and

$$Q_2(x) = \frac{1}{4}(3x^2 - 1) \ln\left(\frac{1+x}{1-x}\right) - \frac{3}{2}x.$$

7. Let n be a nonnegative integer. Prove that

$$P_{2n+1}(0) = 0 \text{ and } P_{2n}(0) = (-1)^n \frac{(2n)!}{2^{2n}(n!)^2}.$$

8. Show that

$$\sum_{n=0}^{\infty} \left(\frac{1}{2^{n+1}}\right) P_n(1/2) = \frac{1}{\sqrt{3}}.$$

In each of Problems 9 through 14, find the first five coefficients in the Fourier-Legendre expansion of $f(x)$ on $[-1, 1]$. Graph the function and the partial sum of these first five terms on the same set of axes, for $-3 \le x \le 3$. The expansion only represents the function on $(-1, 1)$, but it is instructive to see how the partial sums and the function are unrelated outside this interval.

9. $f(x) = \sin^2(x)$
10. $f(x) = \cos(x) - \sin(x)$
11. $f(x) = \begin{cases} -1 & \text{for } -1 \le x \le 0 \\ 1 & \text{for } 0 < x \le 1. \end{cases}$
12. $f(x) = (x+1)\cos(x)$
13. $f(x) = \sin(\pi x/2)$
14. $f(x) = e^{-x}$

15.3 Bessel Functions

This section is devoted to Bessel functions, Bessel's differential equation, Fourier-Bessel eigenfunction expansions and some applications.

Friedrich Wilhelm Bessel (1784 - 1846) was a mathematician and director of the astronomical observatory in Königsberg. He obtained series which would later be known as Bessel functions in solving a problem known as Kepler's problem. The same functions had appeared in the 1730's in work by the Swiss natural philosopher Daniel Bernoulli, who was attempting to describe oscillations in a suspended heavy chain. Joseph Fourier also encountered these functions in his treatise on heat diffusion, which carried the seeds of modern day Fourier analysis. We will discuss some of these applications as we develop the functions named for Bessel.

To do this we will need the gamma function, which is used in writing Bessel functions.

15.3.1 The Gamma Function

The *gamma function* is defined by

$$\Gamma(x) = \int_0^{\infty} t^{x-1} e^{-t} dt.$$

This integral converges for $x > 0$. We will use the following property of $\Gamma(x)$.

THEOREM 15.8 Factorial Property of the Gamma Function

If $x > 0$ then

$$\Gamma(x+1) = x\Gamma(x). \quad \blacklozenge$$

Proof This is a straightforward integration by parts, with $u = t^x$ and $dv = e^{-t}dt$:

$$\begin{aligned}\Gamma(x+1) &= \int_0^\infty t^x e^{-t}dt \\ &= [t^x(-e^{-t})]_0^\infty - \int_0^\infty xt^{x-1}(-1)e^{-t}dt \\ &= x\int_0^\infty t^{x-1}e^{-t}dt = x\Gamma(x). \quad \blacklozenge\end{aligned}$$

To see why this is called the factorial property, first compute

$$\Gamma(1) = \int_0^\infty e^{-t}\,dt = 1.$$

By Theorem 15.8,

$$\begin{aligned}&\Gamma(2) = 1\Gamma(1) = 1, \Gamma(3) = 2\Gamma(2) = 2!, \\ &\Gamma(4) = 3\Gamma(3) = 3\cdot 2 = 3!, \Gamma(5) = 4\Gamma(4) = 4\cdot 3! = 4!,\end{aligned}$$

and, for any positive integer n, $\Gamma(n+1) = n!$.

The gamma function can be extended to negative noninteger values by rewriting the factorial property as

$$\Gamma(x) = \frac{1}{x}\Gamma(x+1)$$

for $x > 0$. If $-1 < x < 0$, then $x+1 > 0$, so $\Gamma(x+1)$ is defined and we can $\Gamma(x) = (1/x)\Gamma(x+1)$. Once we have defined $\Gamma(x)$ for $-1 < x < 0$, then we can use this strategy again to $\Gamma(x)$ for $-2 < x < -1$. In this way we can walk to the left over the real line, defining $\Gamma(x)$ for all negative numbers except integers.

For example,

$$\Gamma(-1/2) = \frac{1}{-1/2}\Gamma\left(-\frac{1}{2}+1\right) = -2\Gamma(1/2)$$

and

$$\Gamma(-3/2) = \frac{1}{-3/2}\Gamma\left(\frac{-3}{2}+1\right) = -\frac{2}{3}\Gamma(-1/2) = \frac{4}{3}\Gamma(1/2).$$

15.3.2 Bessel Functions of the First Kind

The differential equation

$$x^2y'' + xy' + (x^2 - \nu^2)y = 0 \tag{15.7}$$

in which $\nu \geq 0$, is called *Bessel's equation of order* ν. This differential equation is second-order, and the phrase "order ν" refers to the parameter ν appearing in it.

In Section 3.7, we used the Laplace transform to solve Bessel's equation of order n, for n a nonnegative integer. This led to solutions

$$J_n(x) = \sum_{k=0}^{\infty} \frac{(-1)^k}{2^{2k+n}k!(n+k)!} x^{2k+n}.$$

$J_n(x)$ is the Bessel function of the first kind of order n. However, we also need solutions for n not necessarily having integer values, and we need a second, linearly independent, solution for Bessel's equation. To these ends we will employ the Frobenius method of solution discussed in Section 4.2.

Bessel's equation of order ν can be written

$$y'' + \frac{1}{x}y' + \left(1 - \frac{\nu^2}{x^2}\right) y = 0,$$

from which we see that 0 is a regular singular point. We therefore attempt a Frobenius solution

$$y(x) = \sum_{n=0}^{\infty} c_n x^{n+r}.$$

Substitute the proposed Frobenius solution into the differential equation and attempt to solve for r and the coefficients c_n. Begin with the substitution of y into Bessel's equation (15.7):

$$x^2 \sum_{n=0}^{\infty} (n+r)(n+r-1)c_n x^{n+r-2} + x \sum_{n=0}^{\infty} (n+r)c_n x^{n+r-1}$$
$$+ (x^2 - \nu^2) \sum_{n=0}^{\infty} c_n x^{n+r} = 0.$$

Write this equation as

$$\sum_{n=0}^{\infty} (n+r)(n+r-1)c_n x^{n+r} + \sum_{n=0}^{\infty} (n+r)c_n x^{n+r}$$
$$+ \sum_{n=0}^{\infty} c_n x^{n+r+2} - \sum_{n=0}^{\infty} \nu^2 c_n x^{n+r} = 0.$$

Shift indices to write the third summation as

$$\sum_{n=0}^{\infty} c_n x^{n+r+2} = \sum_{n=2}^{\infty} c_{n-2} x^{n+r}$$

to write the last equation as

$$[r(r-1) + r - \nu^2]c_0 x^r + [r(r+1) + r + 1 - \nu^2]c_1 x^{r+1}$$
$$+ \sum_{n=2}^{\infty} \left[[(n+r)(n+r-1) + (n+r) - \nu^2]c_n + c_{n-2}\right] x^{n+r} = 0.$$

Set the coefficient of each power of x equal to 0. Since we require in this method that $c_0 \neq 0$, we obtain from the coefficient of x^r the indicial equation

$$r^2 - \nu^2 = 0.$$

Then $r = \pm\nu$. Set $r = \nu$ in the coefficient of x^{r+1} to obtain

$$(2\nu + 1)c_1 = 0,$$

hence $c_1 = 0$.

From the coefficient of x^{n+r}, we have

$$[(n+r)(n+r-1)+(n+r)-\nu^2]c_n + c_{n-2}$$

for $n=2,3,\cdots$. Since $r=\nu$, this reduces to

$$c_n = -\frac{1}{n(n+2\nu)}c_{n-2}$$

for $n=2,3,\cdots$. But $c_1=0$, so

$$c_3 = c_5 = c_{\text{odd}} = 0.$$

All of the odd-indexed coefficients are zero.

For the even-indexed coefficients, write

$$\begin{aligned}
c_{2n} &= -\frac{1}{2n(2n+2\nu)}c_{2n-2} = -\frac{1}{2^2 n(n+\nu)}c_{2n-2} \\
&= -\frac{1}{2^2 n(n+\nu)}\frac{-1}{2(n-1)[2(n-1)+2\nu]}c_{2n-4} \\
&= \frac{1}{2^4 n(n-1)(n+\nu)(n+\nu-1)}c_{2n-4} \\
&= \cdots = \frac{(-1)^n}{2^{2n} n(n-1)\cdots(2)(1)(n+\nu)(n-1+\nu)\cdots(1+\nu)}c_0 \\
&= \frac{(-1)^n}{2^{2n} n!(1+\nu)(2+\nu)\cdots(n+\nu)}c_0.
\end{aligned}$$

One Frobenius solution of Bessel's equation of order ν is therefore

$$y(x) = c_0 \sum_{n=0}^{\infty} \frac{(-1)^n}{2^{2n} n!(1+\nu)(2+\nu)\cdots(n+\nu)} x^{2n+\nu}.$$

For any $c_0 \neq 0$, this function is a solution of Bessel's equation of order $\nu \geq 0$. These solutions occur in many applications, including the analysis of radiation from cylindrical containers and vibrations of circular membranes.

Consider the factor

$$(1+\nu)(2+\nu)\cdots(n+\nu)$$

in the denominator of $y(x)$. Using the factorial property of the gamma function, write

$$\begin{aligned}
\Gamma(n+\nu+1) &= (n+\nu)\Gamma(n+\nu) = (n+\nu)(n+\nu-1)\Gamma(n+\nu-1) \\
&= \cdots = (n+\nu)(n+\nu-1)\cdots(n+\nu-(n-1))\Gamma(n+\nu-(n-1)) \\
&= (1+\nu)(2+\nu)\cdots(n-1+\nu)(n+\nu)\Gamma(\nu+1).
\end{aligned}$$

Then

$$(1+\nu)(2+\nu)\cdots(n+\nu) = \frac{\Gamma(n+\nu+1)}{\Gamma(\nu+1)}.$$

This enables us to write the solution of Bessel's equation as

$$y(x) = c_0 \sum_{n=0}^{\infty} \frac{(-1)^n \Gamma(\nu+1)}{2^{2n} n! \Gamma(n+\nu+1)} x^{2n+\nu}.$$

It is customary to choose

$$c_0 = \frac{1}{2^\nu \Gamma(\nu+1)}$$

to obtain the solution usually denoted $J_\nu(x)$:

$$J_\nu(x) = \sum_{n=0}^{\infty} \frac{(-1)^n}{2^{2n+\nu} n! \Gamma(n+\nu+1)} x^{2n+\nu}.$$

$J_\nu(x)$ is called a *Bessel function of the first kind of order* ν. The series defining $J_\nu(x)$ converges for all x.

Because Bessel's equation is second-order (as a differential equation), there is a second solution that is linearly independent from J_ν. The Frobenius theorem (Theorem 4.2) tells us how to proceed to find a second solution. Recall that the indicial equation of Bessel's differential equation is $r^2 - \nu^2 = 0$, with roots ν and $-\nu$. The form that a second solution will take depends on the difference 2ν of these roots. With Theorem 4.2 as a guide, we find the following second solutions by taking cases on 2ν.

Case 1 If 2ν is not an integer, then J_ν and $J_{-\nu}$ are linearly independent (neither is a constant multiple of the other). In this case, the general solution of Bessel's equation is

$$y(x) = aJ_\nu(x) + bJ_{-\nu}(x),$$

with a and b arbitrary constants.

Case 2 If 2ν is an odd positive integer, say $2\nu = 2n+1$ for some nonnegative integer n, then $\nu = n + \frac{1}{2}$ and J_ν and $J_{-\nu}$ are again linearly independent, as in Case 1. In this case, the general solution of Bessel's equation is

$$y(x) = aJ_{n+1/2}(x) + bJ_{-n-1/2}(x).$$

By manipulating the series for $J_\nu(x)$, it can be shown that in this case, $J_{n+1/2}(x)$ and $J_{-n-1/2}(x)$ can be expressed in closed form as finite sums of terms involving algebraic combinations of x, $\sin(x)$ and $\cos(x)$. For example,

$$J_{1/2}(x) = \sqrt{\frac{2}{\pi x}} \sin(x),\ J_{-1/2}(x) = \sqrt{\frac{2}{\pi x}} \cos(x),$$

$$J_{3/2}(x) = \sqrt{\frac{2}{\pi x}} \left[\frac{\sin(x)}{x} - \cos(x) \right],$$

and

$$J_{-3/2}(x) = \sqrt{\frac{2}{\pi x}} \left[-\sin(x) - \frac{\cos(x)}{x} \right].$$

Case 3 If 2ν is an integer, but not of the form $2n+1$ for any nonnegative integer n, then J_ν and $J_{-\nu}$ are solutions of Bessel's equation, but are linearly dependent:

$$J_{-\nu}(x) = (-1)^\nu J_\nu(x).$$

In this case, we cannot manufacture a second linearly independent solution from $J_\nu(x)$. This leads us to construct such a second linearly independent solution, leading us to Bessel functions of the second kind.

15.3.3 Bessel Functions of the Second Kind

We are in Case 3 of the preceding subsection. Begin with the case that $\nu = 0$. The Frobenius theorem (4.2) tells us in this case to look for a second solution

$$y_2(x) = J_\nu(x)\ln(x) + \sum_{n=1}^{\infty} c_n^* x^n.$$

Substitute this into Bessel's equation of order 0 and obtain, after a computation like that used to derive $J_\nu(x)$,

$$y_2(x) = J_0(x)\ln(x) + \sum_{n=1}^{\infty} \frac{(-1)^{n+1}}{2^{2n}(n!)^2}\phi(n)x^{2n}$$

where

$$\phi(n) = 1 + \frac{1}{2} + \cdots + \frac{1}{n}.$$

Instead of using this solution, it is customary to use a linear combination of $y_2(x)$ and $J_0(x)$, which will also be a solution. This leads to the second solution (in this case $\nu = 0$)

$$Y_0(x) = \frac{2}{\pi}[y_2(x) + (\gamma - \ln(2))J_0(x)],$$

for $x > 0$. Here γ is the *Euler constant*, defined to be

$$\gamma = \lim_{n\to\infty}(\phi(n) - \ln(n)) \approx 0.577215664901533\cdots.$$

Because of the logarithm term, Y_0 and J_0 are linearly independent solutions of Bessel's equation of order 0, which therefore has the general solution

$$y(x) = aJ_0(x) + bY_0(x)$$

for $x > 0$. Y_0 is the *Bessel function of the second kind of order* zero. With the choice of constants used to define Y_0, this function is also called *Neumann's function of order* zero.

In many applications, we can immediately choose $b = 0$ in solving Bessel's equation of order zero, because the logarithm term in $Y_0(x)$ tends to $-\infty$ as $x \to 0$. This means that a bounded solution requires $b = 0$. As we will see later, this reasoning applies when we analyze the motion of a vibrating circular membrane, since in polar coordinates the center of the membrane is $r = 0$ and the amplitudes of the vibration must be bounded.

If ν is a positive integer, say $\nu = n$, the second solution of Bessel's equation of order ν is the *Bessel function of the second kind of order* n, defined by

$$Y_n(x) = \frac{2}{\pi}\left[J_n(x)[\gamma + \ln(x/2)] + \sum_{k=1}^{\infty} \frac{(-1)^{k+1}[\phi(k) + \phi(k+1)]}{2^{2k+n+1}k!(k+n)!}x^{2k+n}\right]$$
$$-\frac{2}{\pi}\sum_{k=0}^{n-1} \frac{n-k-1!}{2^{2k-n+1}k!}x^{2k-n}.$$

This agrees with $Y_0(x)$ if $n = 0$ with the understanding that in this case the last (finite) summation is omitted.

The general solution of Bessel's equation of positive integer order $\nu = n$ is therefore

$$y(x) = aJ_n(x) + bY_n(x).$$

It is also possible to define Bessel functions of the second kind of noninteger order by setting

$$Y_\nu(x) = \frac{1}{\sin(\nu\pi)}[J_\nu(x)\cos(\nu\pi) - J_{-\nu}(x)].$$

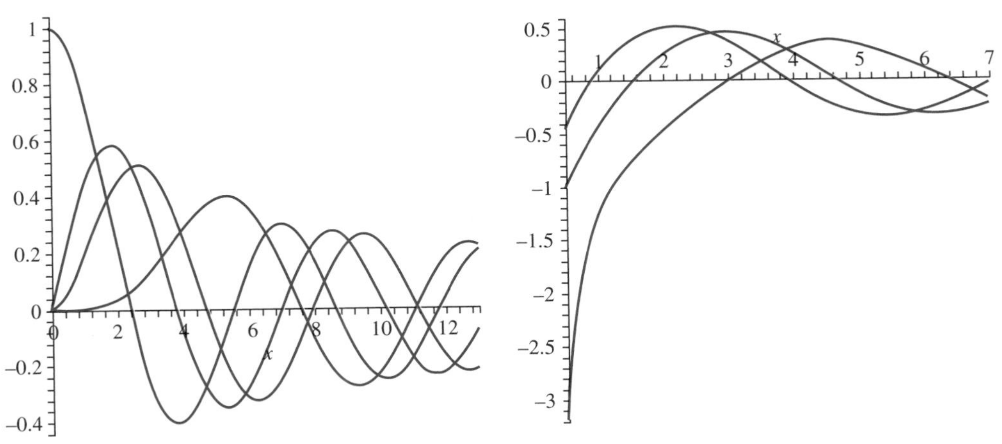

FIGURE 15.6 $J_\nu(x)$ *for* $\nu = 0, 1, 5/3, 4$.

FIGURE 15.7 $Y_\nu(x)$ *for* $\nu = 0, 1/3, 1/2$.

Figures 15.6 and 15.7 show graphs of some Bessel functions of the first and second kinds, respectively.

Sometimes we encounter disguised versions of Bessel's equation. The differential equation

$$y'' - \left(\frac{2a-1}{x}\right)y' + \left(b^2c^2x^{2c-2} + \frac{a^2-\nu^2c^2}{x^2}\right)y = 0 \qquad (15.8)$$

has the general solution

$$y(x) = c_1x^aJ_n(bx^c) + c_2x^aY_n(bx^c)$$

if $\nu = n$ is an integer, and

$$y(x) = c_1x^aJ_\nu(bx^c) + c_2x^aJ_\nu(bx^c)$$

if ν is not an integer. Verification of this is an exercise in chain rule differentiation and is requested in Problem 1.

EXAMPLE 15.6

We will solve

$$y'' - \left(\frac{2\sqrt{3}-1}{x}\right)y' + \left(784x^6 - \frac{61}{x^2}\right)y = 0.$$

Match this equation to the template (15.8). Clearly we need $a = \sqrt{3}$. Because of the x^b term, try $2c - 2 = 6$, so $c = 4$. Now we must choose b and ν so that

$$784 = b^2c^2 = 16b^2.$$

Then $b = 2$. Next,

$$a^2 - \nu^2c^2 = 3 - 16\nu^2 = -61,$$

so $\nu = 2$. The general solution of this differential equation is

$$y(x) = c_1x^{\sqrt{3}}J_2(7x^4) + c_2x^{\sqrt{3}}Y_2(7x^4).$$

Because of the Bessel function of the second kind, this solution is defined for $x > 0$. ♦

We will pause in the development of Bessel functions to look at two applications. The first is the problem studied by Daniel Bernoulli in perhaps the first appearance of a Bessel function. Notice that both solutions depend on knowledge of zeros of Bessel functions. These will also be important for eigenfunction expansions involving Bessel functions, and we will devote some time to them later.

15.3.4 Displacement of a Hanging Chain

Imagine a heavy flexible chain fixed at its upper end and free at the lower end. We want to describe oscillations caused by a small displacement of the lower end.

First, we need to model the problem. Assume that each particle of the chain oscillates in a horizontal straight line. Let m be the mass of the chain per unit length, assumed constant, L its length, and $y(x,t)$ the horizontal displacement at time t of the particle of chain at distance x from the top end of the chain. To derive an equation for y, consider an element of chain of length Δx. Let the forces acting on the ends of this segment have magnitudes T and $T+\Delta T$. The horizontal component of Newton's first law of motion (force equals mass times acceleration) requires that

$$m\Delta x\frac{\partial^2 y}{\partial t^2}=\frac{\partial}{\partial x}\left(T\frac{\partial y}{\partial x}\right)\Delta x.$$

Then

$$m\frac{\partial^2 y}{\partial t^2}=\frac{\partial}{\partial x}\left(T\frac{\partial y}{\partial x}\right).$$

We will assume at this point that

$$T=mg(L-x),$$

which has been found to be a good approximation for small disturbances. The equation for y is now

$$\frac{\partial^2 y}{\partial t^2}=-g\frac{\partial y}{\partial x}+g(L-x)\frac{\partial^2 y}{\partial x^2}.$$

This is a partial differential equation. To solve it, first change variables by putting

$$z=L-x \text{ and } u(z,t)=y(L-z,t).$$

The partial differential equation transforms to

$$\frac{\partial^2 u}{\partial t^2}=g\frac{\partial u}{\partial z}+gz\frac{\partial^2 z}{\partial z^2}.$$

We will now anticipate the method of separation of variables (Chapter 16). Look for a solution having the form of a function of z multiplied by a function of t. Not all functions of z and t have this form (for example, $\sin(zt)$ does not). However, for the equation under consideration, some thought and trial and error suggest that there might be a solution of the form

$$u(z,t)=f(z)\cos(\omega t-\delta).$$

The cosine term is suggested by the fact that we expect the motion of the free end of the chain to exhibit periodic oscillations. Substitute this expression for $u(z,t)$ into the partial differential equation to obtain

$$-\omega^2 f(z)\cos(\omega t-\delta)=gf'(z)\cos(\omega t-\delta)+gzf''(z)\cos(\omega t-\delta).$$

Dividing out the common term $\cos(\omega t-\delta)$, we obtain

$$-\omega^2 f(z)=gf'(z)+gzf''(z).$$

This can be written

$$f''(z)+\frac{1}{z}f'(z)+\frac{\omega^2}{gz}f(z)=0.$$

This is in the form of equation (15.8) if we can solve for a, b, and c so that

$$-(2a-1)=1, 2c-2=-1, b^2c^2=\frac{\omega^2}{g}, \text{ and } a^2-n^2c^2=0.$$

Choose

$$a=n=0, c=\frac{1}{2} \text{ and } b=\frac{2\omega}{\sqrt{g}}.$$

The solution for $f(z)$ is

$$f(z)=c_1J_0\left(2\omega\sqrt{\frac{z}{g}}\right)+c_2Y_0\left(2\omega\sqrt{\frac{z}{g}}\right).$$

Now $Y_0\left(2\omega\sqrt{z/g}\right)\to-\infty$ as $z\to 0$ (which occurs if $x\to L$, the lower end of the chain). This is not realistic physically, so choose $c_2=0$. Then $f(z)$ must have the form

$$f(z)=c_1J_0\left(2\omega\sqrt{\frac{z}{g}}\right).$$

Therefore,

$$u(z,t)=c_1J_0\left(2\omega\sqrt{\frac{z}{g}}\right)\cos(\omega t-\delta)$$

so $y(x,t)$ is

$$y(x,t)=c_1J_0\left(2\omega\sqrt{\frac{L-x}{g}}\right)\cos(\omega t-\delta)$$

The frequencies of the normal vibrations are determined by using the fact that the upper end of the chain does not move. This means that, for all times t, $y(0,t)=0$. Assuming that $c_1\neq 0$ (or else the solution vanishes), this requires that

$$J_0\left(2\omega\sqrt{\frac{L}{g}}\right)=0.$$

We will see shortly that the zero order Bessel function of the first kind has infinitely many positive zeros. If these zeros are labeled $\omega_1, \omega_2, \cdots$ in increasing order, then ω must satisfy

$$2\omega\sqrt{\frac{L}{g}}=\omega_j$$

for some positive integer j. This means that ω can take on the values

$$\omega=\frac{1}{2}\omega_j\sqrt{\frac{g}{L}}.$$

These are the frequencies of the normal modes of vibration of the end of the chain, one normal mode for each positive zero of J_0. The periods of the oscillation are

$$\frac{4\pi}{\omega_j}\sqrt{\frac{L}{g}}.$$

If we consult a table or use MAPLE to call up these zeros (see Example 15.7), we find that

$$\omega_1 \approx 2.40483, \omega_2 \approx 5.52008, \omega_3 \approx 8.65373,$$

and so on. Using these we can obtain approximate numerical values for the frequencies of the first three normal modes of vibration, given the length of the chain.

15.3.5 Critical Length of a Rod

We will analyze the critical bending length of a rod. Suppose we have a thin rod of constant weight w per unit length, length L and circular cross section of radius a. The rod is clamped in a vertical position. If the rod is "long enough," and the upper end is displaced and held in position until the rod is at rest, the rod will remain bent. By contrast, if the rod is "short enough", it will return to its vertical position after the end has been displaced slightly. The *critical length* L_C is the transition between these two states. If $L \geq L_C$, the rod remains bent, but if $L < L_C$, it returns to the vertical after a small displacement. We want to derive an expression for L_C.

Let E be the Young's modulus for the material of the rod. This is the ratio of stress to strain for a linear compression. Figure 15.8 shows the rod after a small displacement. The x-axis is vertical, along the original position of the rod. Downward is positive and the origin O is at the upper end. $P : (x, y)$ and $Q : (\xi, \eta)$ are points on the bent rod, as shown. The moment about P of the weight of an element $w\Delta x$ at Q is $w\Delta x[y(\xi) - y(x)]$. By integrating this we obtain the moment about P of the weight of the rod above P. From the theory of elasticity, this moment about P equals $EIy''(x)$. Assuming that the part of the rod above P is in equilibrium, then

$$EIy''(x) = \int_0^x w[y(\xi) - y(x)]\,d\xi.$$

Differentiate this equation with respect to x:

$$EIy^{(3)}(x) = w[y(x) - y(x)] - \int_0^x wy'(x)\,d\xi = -wxy'(x).$$

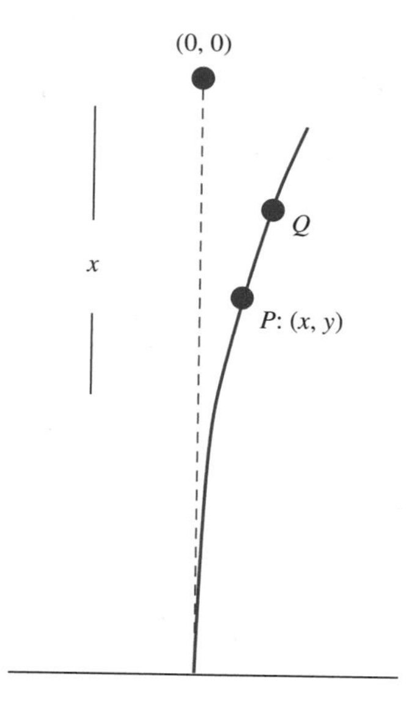

FIGURE 15.8 *Displacement of a rod.*

Then

$$y^{(3)}(x) + \frac{w}{EI}xy'(x) = 0.$$

Let $u = y'$ to write the second order differential equation

$$u'' + \frac{w}{EI}xu = 0.$$

Compare this with equation (15.8), putting

$$2a - 1 = 0, a^2 - \nu^2c^2 = 0, 2c - 3 = 1, \text{ and } b^2c^2 = \frac{w}{EI}.$$

Thus choose

$$a = \frac{1}{2}, c = \frac{3}{2}, \nu = \frac{1}{3}, b = \frac{2}{3}\sqrt{\frac{w}{EI}}.$$

The general solution for $u(x)$ is

$$u(x) = y'(x) = c_1\sqrt{x}J_{1/3}\left(\frac{2}{3}\sqrt{\frac{w}{EI}}x^{3/2}\right) + c_2\sqrt{x}J_{-1/3}\left(\frac{2}{3}\sqrt{\frac{w}{EI}}x^{3/2}\right).$$

Since there is no bending moment at the top of the rod, $y''(0) = 0$. A routine differentiation shows that this forces $c_1 = 0$. Now

$$y'(x) = c_2\sqrt{x}J_{-1/3}\left(\frac{2}{3}\sqrt{\frac{w}{EI}}x^{3/2}\right).$$

Because the lower end of the rod is clamped vertically, then $y'(L) = 0$, so

$$c_2\sqrt{L}J_{-1/3}\left(\frac{2}{3}\sqrt{\frac{w}{EI}}L^{3/2}\right) = 0.$$

We must have $c_2 \neq 0$ to have a nontrivial solution. The critical length L_C is the smallest positive value of L such that

$$J_{-1/3}\left(\frac{2}{3}\sqrt{\frac{w}{EI}}L^{3/2}\right) = 0.$$

We find from a table or from MAPLE that the smallest positive number α such that $J_{-1/3}(\alpha) = 0$ is approximately 1.8663. Therefore

$$\frac{2}{3}\sqrt{\frac{w}{EI}}L_c^{3/2} \approx 1.8663.$$

Solve for L_C:

$$L_C \approx 1.9863\left(\frac{EI}{w}\right)^{1/3}.$$

This is the critical length.

15.3.6 Modified Bessel Functions

Some applications use modified Bessel functions. We will show how these are obtained. Begin with the general solution

$$y(x) = c_1J_0(kx) + c_2Y_0(kx)$$

of the zero-order Bessel equation

$$y'' + \frac{1}{x}y' + k^2y = 0.$$

With $k = i$,

$$y(x) = c_1 J_0(ix) + c_2 Y_0(ix)$$

is the general solution of

$$y'' + \frac{1}{x}y' - y = 0,$$

for $x > 0$. This differential equation is a *modified Bessel equation of order* zero, and $J_0(ix)$ is a *modified Bessel function of the first kind of order* zero. Usually this is denoted $I_0(x)$:

$$I_0(x) = J_0(ix) = 1 + \frac{1}{2^2}x^2 + \frac{1}{2^2 4^2}x^4 + \frac{1}{2^2 4^2 6^2}x^6 + \cdots.$$

Normally $Y_0(ix)$ is not used, but instead the second solution is chosen to be

$$K_0(x) = [\ln(2) - \gamma]I_0(x) - I_0(x)\ln(x) + \frac{1}{4}x^2 + \cdots$$

for $x > 0$. Here γ is the Euler constant and K_0 is a *modified Bessel function of the second kind of order* zero. Figure 15.9 shows partial graphs of I_0 and K_0.

For $x > 0$, the general solution of

$$y'' + \frac{1}{x}y' - b^2 y = 0$$

is

$$y(x) = c_1 I_0(bx) + c_2 K_0(bx).$$

It is routine to manipulate the series for $I_0(bx)$ to obtain

$$\int x I_0(bx)\,dx = \frac{x}{b} I_0'(bx) + c, \tag{15.9}$$

for any nonzero b.

Sometimes we need to know how $I_0(x)$ behaves for large x. We can obtain an expression for $I_0(x)$, valid for large positive x, as follows. Begin with the fact that $I_0(x)$ is a solution of

$$y'' + \frac{1}{x}y' - y = 0.$$

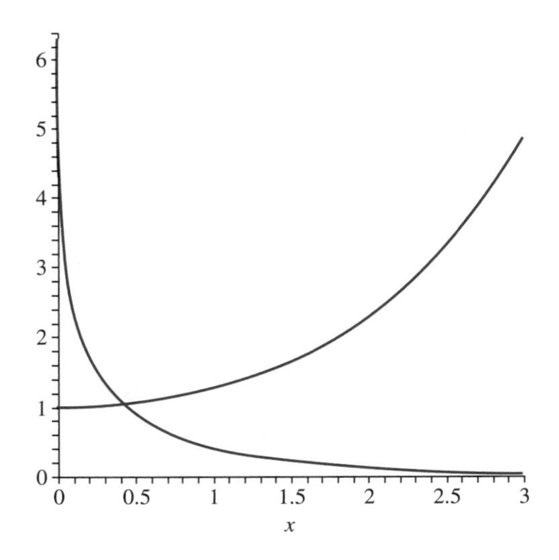

FIGURE 15.9 *$I_0(x)$ and $K_0(x)$.*

Make the change of variables $y = ux^{-1/2}$ to obtain

$$u'' = \left(1 - \frac{1}{4x^2}\right)u. \tag{15.10}$$

As x increases, $1 - 1/4x^2 \to 1$ and this differential equation for u more closely approximates $u'' - u = 0$, with solutions e^x and e^{-x}. We therefore argue intuitively that, as x increases, any solution of the differential equation for u which approaches ∞ as $x \to \infty$ will be approximated by ce^x for some c.

More generally, this suggests the further transformation $u = ve^x$. Substitute this into equation (15.10) to obtain

$$v'' + 2v' + \frac{1}{4x^2}v = 0. \tag{15.11}$$

We will attempt a solution of this equation by a series of the form

$$v(x) = 1 + c_1\frac{1}{x} + c_2\frac{1}{x^2} + c_3\frac{1}{x^3} + \cdots.$$

Substitute this into equation (15.11) and rearrange terms to obtain

$$\begin{aligned}&\left(-2c_1 + \frac{1}{4}\right)\frac{1}{x^2} + \left(2c_1 - 4c_2 + \frac{1}{4}c_1\right)\frac{1}{x^3}\\&+\left(6c_2 - 6c_3 + \frac{1}{4}c_2\right)\frac{1}{x^4}\\&+\left(12c_3 - 8c_4 + \frac{1}{4}c_3\right)\frac{1}{x^5} + \cdots = 0.\end{aligned}$$

Each coefficient of a power of $1/x$ must vanish, so

$$\begin{aligned}-2c_1 + \frac{1}{4} &= 0\\ \frac{9}{4}c_1 - 4c_2 &= 0\\ \frac{25}{4}c_2 - 6c_3 &= 0\\ \frac{49}{4}c_3 - 8c_4 &= 0\end{aligned}$$

and so on. Then

$$c_1 = \frac{1}{8},\ c_2 = \frac{9}{16}c_1 = \frac{3^2}{2\cdot 8^2},$$
$$c_3 = \frac{3^2 5^2}{3!8^3},\ c_4 = \frac{3^2 5^2 7^2}{4!8^4},$$

and so on. The pattern is clear and we write

$$v(x) = 1 + \frac{1}{8}\frac{1}{x} + \frac{3^2}{2\cdot 8^2}\frac{1}{x^2} + \frac{3^2 5^2}{3!8^3}\frac{1}{x^3} + \frac{3^2 5^2 7^2}{4!8^4}\frac{1}{x^4} + \cdots. \tag{15.12}$$

This suggests an expansion of the form

$$I_0(x) = \frac{ce^x}{\sqrt{x}}\left(1 + \frac{1}{8}\frac{1}{x} + \frac{3^2}{2\cdot 8^2}\frac{1}{x^2} + \frac{3^2 5^2}{3!8^3}\frac{1}{x^3} + \frac{3^2 5^2 7^2}{4!8^4}\frac{1}{x^4} + \cdots\right), \tag{15.13}$$

in which c is an appropriately chosen positive constant. The series on the right is actually a divergent series. However, the partial sum of the first N terms approximates $I_0(x)$ as closely as we like for x sufficiently large. Such an approximation is called an *asymptotic expansion*. By an analysis we will omit, it can be shown that $c = 1/\sqrt{2\pi}$.

15.3.7 Alternating Current and the Skin Effect

We will use a modified Bessel function to analyze the flow of alternating current in a wire of circular cross section. Begin with general principles named for Ampére and Faraday. Ampére's law states that the integral of the magnetic force around a circuit is equal to 4π times the integral of the current through the circuit. Faraday's law states that the integral of the electric force around a circuit equals the negative of the time derivative of the magnetic induction through the circuit.

We will determine the current density at radius r in a wire of radius a. Let ρ be the specific resistance of the wire, μ its permeability, $x(r,t)$ the current density, and $H(r,t)$ the magnetic intensity at radius r and time t. Apply Ampére's law to a circle of radius r having its axis along the axis of the wire:

$$2\pi r H = 4\pi \int_0^r 2\pi r x(\xi, t)\, d\xi,$$

or

$$rH = 4\pi \int_0^r r x(\xi, t)\, d\xi. \tag{15.14}$$

Then

$$\frac{\partial}{\partial r}(rH) = 4\pi x r,$$

hence

$$\frac{1}{r}\frac{\partial}{\partial r}(rH) = 4\pi x(r,t). \tag{15.15}$$

Now apply Faraday's law to the rectangular circuit having one side of length L along the axis of the wire, and the other side of length r. We get

$$\rho L x(0,t) - \rho L x(r,t) = -\frac{\partial}{\partial t}\int_0^r \mu L H(\xi, t)\, d\xi.$$

Differentiate this equation with respect to r to obtain

$$\rho \frac{\partial x}{\partial r} = \mu \frac{\partial H}{\partial t}. \tag{15.16}$$

Use equations (15.15) and (15.16) to eliminate H. First multiply (15.16) by r to obtain

$$\rho r \frac{\partial x}{\partial r} = \mu r \frac{\partial H}{\partial t}.$$

Differentiate this equation with respect to r:

$$\rho \frac{\partial}{\partial r}\left(r\frac{\partial x}{\partial r}\right) = \mu \frac{\partial}{\partial r}\left(r\frac{\partial H}{\partial t}\right) = \mu \frac{\partial}{\partial t}\left(\frac{\partial}{\partial r}(rH)\right)$$
$$= \mu \frac{\partial}{\partial t}(4\pi x r) = 4\pi \mu r \frac{\partial x}{\partial t},$$

in which we substituted from equation (15.15) at the last step. Now we have

$$\rho \frac{\partial}{\partial r}\left(r\frac{\partial x}{\partial r}\right) = 4\pi \mu r \frac{\partial x}{\partial t}. \tag{15.17}$$

The strategy is to solve this partial differential equation for $x(r,t)$, then obtain $H(r,t)$ from equation (15.14). To do this, assume that the alternating current flowing through the wire has period $2\pi/\omega$ and is given by $C\cos(\omega t)$, with C constant. It is convenient to introduce complex quantities and write

$$z(r,t) = x(r,t) + iy(r,t).$$

The quantity of interest is the real part of $z(r,t)$, denoted

$$x(r,t) = \text{Re}(z(r,t)).$$

We can also think of the current as the real part of the complex exponential function

$$Ce^{i\omega t} = C\cos(\omega t) + iC\sin(\omega t).$$

In terms of z, equation (15.17) is

$$\rho\frac{\partial}{\partial r}\left(r\frac{\partial z}{\partial r}\right) = 4\pi\mu r\frac{\partial z}{\partial t}. \tag{15.18}$$

Attempt a solution of the form $z(r,t) = f(r)e^{i\omega t}$. Substitute this into equation (15.18) to obtain

$$\rho\frac{\partial}{\partial r}(rf'(r))e^{i\omega t} = 4\pi\mu r f(r)i\omega e^{i\omega t}.$$

Upon dividing by $e^{i\omega t}$ we have

$$f''(r) + \frac{1}{r}f'(r) - b^2 f(r) = 0,$$

in which

$$b^2 = \frac{4\pi\mu\omega}{\rho}i.$$

From Section 15.3.6, this equation has general solution

$$f(r) = c_1 I_0(br) + c_2 K_0(br),$$

in which

$$b = \sqrt{\frac{4\pi\mu\omega}{\rho}}\frac{1+i}{\sqrt{2}}.$$

The logarithm term in $K_0(br)$ is unbounded as $r \to 0$, the center of the wire, so we must choose $c_2 = 0$. Then

$$f(r) = c_1 I_0(br) \text{ and } z(r,t) = c_1 I_0(br)e^{i\omega t}.$$

To determine c_1, use the fact that the current $C\cos(\omega t)$ is the real part of $Ce^{i\omega t}$. Using equation (15.9),

$$C = 2\pi c_1 \int_0^a r I_0(br)\,dr = \frac{2\pi a c_1}{b} I_0'(ba).$$

Therefore,

$$c_1 = \frac{bC}{2\pi a}\frac{1}{I_0'(ba)}.$$

Now

$$z(r,t) = \frac{bC}{2\pi a}\frac{1}{I_0'(ba)} I_0(br)e^{i\omega t}.$$

Then

$$x(r,t) = \text{Re}(z(r,t))$$

is the current density in the wire. The magnetic intensity is

$$H(r,t) = \text{Re}\left(\frac{2C}{aI_0'(ba)} I_0(br)e^{i\omega t}\right).$$

With a bit more analysis, we can observe the skin effect in this mathematical model of current flow in the wire. The entire current flowing through a cylinder of radius r within the wire, and having the same central axis as the wire, is

$$\operatorname{Re}\left[\frac{b}{2\pi a I_0'(ba)} C e^{i\omega t} \int_0^r 2\pi \xi I_0(b\xi)\, d\xi\right].$$

Again using equation (15.9), this is

$$\operatorname{Re}\left(\frac{r I_0'(br)}{a I_0'(ba)} C e^{i\omega t}\right).$$

Then

$$\frac{\text{current in the cylinder of radius } r}{\text{total current in the wire}} = \frac{r}{a}\frac{I_0'(br)}{I_0'(ba)}.$$

When the frequency ω is large, b is large in magnitude and we can use the asymptotic expansion (15.13) to write

$$\frac{r}{a}\frac{I_0'(br)}{I_0'(ba)} \approx \frac{r}{a}\frac{e^{br}}{\sqrt{br}}\frac{\sqrt{ba}}{e^{ba}} = \sqrt{\frac{b}{a}}\, e^{-b(a-r)}.$$

For any r with $0 < r < a$, the quantity on the right in this equation can be made arbitrarily small by choosing ω sufficiently large. This means that, for large frequencies of the current, "most" of the current flows in a thin layer near the outer surface of the wire. This is the *skin effect*.

15.3.8 A Generating Function for $J_n(x)$

Thus far, we have defined Bessel and modified Bessel functions as solutions to Bessel's equation, and examined some applications. Now we will carry out a program like that for Legendre polynomials. We will develop a generating function, recurrence relations, zeros of Bessel functions, and eigenfunction expansions involving Bessel functions.

THEOREM 15.9 ***Generating Function for $J_n(x)$***

$$e^{x(t-1/t)/2} = \sum_{n=-\infty}^{\infty} J_n(x) t^n. \quad ♦ \tag{15.19}$$

This means that, if we expand the exponential $e^{x(t-1/t)/2}$ in an infinite series, then the coefficient of t^n is $J_n(x)$, for any integer n. This is in the same spirit that $P_n(x)$ is the coefficient of t^n in the expansion of $L(x,t) = 1/\sqrt{1-2xt+t^2}$. One difference is that the expansion of $L(x,t)$ involves only nonnegative powers of t, while this expansion of $e^{x(t-1/t)/2}$ involves negative powers of t because of the $1/t$ term in the exponent.

To understand why equation (15.19) is true, begin with the familiar Maclaurin expansions of the exponential function:

$$\begin{aligned} e^{x(t-1/t)/2} &= e^{xt/2}e^{-x/2t} \\ &= \left(\sum_{m=0}^{\infty}\frac{1}{m!}\left(\frac{xt}{2}\right)^m\right)\left(\sum_{k=0}^{\infty}\frac{1}{k!}(-1)^k\left(\frac{x}{2t}\right)^k\right) \\ &= \left(1+\frac{xt}{2}+\frac{1}{2!}\frac{x^2t^2}{2^2}+\frac{1}{3!}\frac{x^3t^3}{2^3}+\cdots\right)\left(1-\frac{x}{2t}+\frac{1}{2!}\frac{x^2}{2^2t^2}-\frac{1}{3!}\frac{x^3}{2^3t^3}+\cdots\right). \end{aligned}$$

Now we must collect all the coefficients of t^n, for each n. To illustrate the idea, look for the coefficients of t^4 in this product. We obtain t^4 when $x^4t^4/2^4 4!$ on the left is multiplied by 1 on the right, and when $x^5t^5/2^5 5!$ on the left is multiplied by $-x/2t$ on the right, and when $x^6t^6/2^6 6!$ on the left is multiplied by $x^2/2^2 2!t^2$ on the right, and so on. In this way, we find that the coefficient of t^4 in the above product is

$$\frac{1}{2^4 4!}x^4 - \frac{1}{2^6 5!}x^5 + \frac{1}{2^8 2!6!}x^6 - \frac{1}{2^{10}3!7!}x^7 + \cdots$$
$$= \sum_{n=0}^{\infty} \frac{(-1)^n}{2^{2n+4}n!(n+4)!}x^{2n+4} = J_4(x).$$

Similar analysis shows that the coefficient of t^n in equation (15.19) is $J_n(x)$ for each nonnegative integer n. For negative integers, we can use the fact that, if n is a positive integer, then

$$J_{-n}(x) = (-1)^n J_n(x).$$

While this is not a formal proof, it is a plausibility argument in support of the generating function.

15.3.9 Recurrence Relations

We will state three recurrence relations involving Bessel functions of the first kind. In these, ν is a real number.

THEOREM 15.10

$$\frac{d}{dx}(x^{\nu}J_{\nu}(x)) = x^{\nu}J_{\nu-1}(x). \quad \blacklozenge \tag{15.20}$$

Proof Begin with the case that ν is not a negative integer. Differentiate the series for $x^{\nu}J_{\nu}(x)$ to obtain

$$\frac{d}{dx}(x^{\nu}J_{\nu}(x)) = \frac{d}{dx}\left[x^{\nu}\sum_{n=0}^{\infty}\frac{(-1)^n}{2^{2n+\nu}n!\Gamma(n+\nu+1)}x^{2n+\nu}\right]$$
$$= \frac{d}{dx}\left[\sum_{n=0}^{\infty}\frac{(-1)^n}{2^{2n+\nu}n!\Gamma(n+\nu+1)}x^{2n+2\nu}\right]$$
$$= \sum_{n=0}^{\infty}\frac{(-1)^n 2(n+\nu)}{2^{2n+\nu}n!(n+\nu)\Gamma(n+\nu)}x^{2n+2\nu-1}$$
$$= x^{\nu}\sum_{n=0}^{\infty}\frac{(-1)^n}{2^{2n+\nu-1}n!\Gamma(n+\nu)}x^{2n+\nu-1} = x^{\nu}J_{\nu-1}(x).$$

Now extend this result to negative integers by using the fact that, if $\nu = -m$, with m a positive integer, then

$$J_{\nu}(x) = J_{-m}(x) = (-1)^m J_m(x). \quad \blacklozenge$$

THEOREM 15.11

$$\frac{d}{dx}(x^{-\nu}J_{\nu}(x)) = -x^{-\nu}J_{\nu+1}(x). \quad \blacklozenge \tag{15.21}$$

The proof is like that of the preceding theorem.

THEOREM 15.12

For $x > 0$,

$$\frac{2\nu}{x} J_\nu(x) = J_{\nu+1}(x) + J_{\nu-1}(x). \quad \blacklozenge \tag{15.22}$$

Proof Carry out the indicated differentiations in equations (15.20) and (15.21) to obtain

$$x^\nu J_\nu'(x) + \nu x^{\nu-1} J_\nu(x) = x^\nu J_{\nu-1}(x)$$

and

$$x^{-\nu} J_\nu'(x) - \nu x^{-\nu-1} J_\nu(x) = -x^\nu J_{\nu+1}(x).$$

Multiply the first equation by $x^{-\nu}$ and the second by x^ν to obtain

$$J_\nu' + \frac{\nu}{x} J_\nu(x) = J_{\nu-1}(x) \tag{15.23}$$

and

$$J_\nu'(x) - \frac{\nu}{x} J_\nu(x) = -J_{\nu+1}(x). \tag{15.24}$$

Upon subtracting the second of these equations from the first, we have the conclusion of the theorem. $\blacklozenge$

As an example of how these relationships can be used, recall from Section 15.3.2 that

$$J_{1/2}(x) = \sqrt{\frac{2}{\pi x}} \sin(x), \; J_{-1/2}(x) = \sqrt{\frac{2}{\pi x}} \cos(x),$$

$$J_{3/2}(x) = \sqrt{\frac{2}{\pi x}} \left[\frac{\sin(x)}{x} - \cos(x) \right].$$

Put $\nu = 3/2$ into equation (15.22):

$$\frac{3}{x} J_{3/2}(x) = J_{5/2}(x) + J_{1/2}(x).$$

Then

$$\begin{aligned} J_{5/2}(x) &= -J_{1/2}(x) + \frac{3}{x} J_{3/2}(x) \\ &= \sqrt{\frac{2}{\pi x}} \left[-\sin(x) + \frac{3}{x^2} \sin(x) - \frac{3}{x} \cos(x) \right]. \end{aligned}$$

15.3.10 Zeros of Bessel Functions

We have seen applications in which we needed to know about zeros of a Bessel function. We will also need such information for eigenfunction expansions involving Bessel functions.

We will show that $J_\nu(x)$ has infinitely many positive zeros (positive numbers α such that $J_\nu(\alpha) = 0$). We will also derive estimates for their distribution on the half-line $x > 0$, and we will show an important relationship between zeros of $J_\nu(x)$, $J_{\nu-1}(x)$ and $J_{\nu+1}(x)$.

As a starting point, recall that $J_\nu(kx)$ is a solution of

$$x^2 y'' + xy' + (k^2 x^2 - \nu^2) y = 0.$$

Assume that $k > 1$ and substitute $u(x) = \sqrt{kx}\,J_\nu(kx)$ to obtain

$$u''(x) + \left(k^2 - \frac{\nu^2 - \frac{1}{4}}{x^2}\right)u(x) = 0. \tag{15.25}$$

Our intuition is that, as x increases, the term $(\nu^2 - 1/4)/x^2$ exerts less influence, and this differential equation for u more closely approximates $u'' + k^2u = 0$, with solutions $\cos(kx)$ and $\sin(kx)$. This suggests that, for large x, $J_\nu(kx)$ is approximated by $\cos(kx)/\sqrt{kx}$ or $\sin(kx)/\sqrt{kx}$. Since these functions have infinitely many positive zeros, we suspect that $J_\nu(kx)$ does also.

While not a proof, this informal argument suggests an approach to the question of zeros of $J_\nu(kx)$. Consider the equation

$$v''(x) + v(x) = 0, \tag{15.26}$$

which has solution $v(x) = \sin(x - \alpha)$ for any positive number α. Multiply equation (15.25) by v and equation (15.26) by u and subtract to get

$$uv'' - vu'' = \left(k^2 - \frac{\nu^2 - \frac{1}{4}}{x^2}\right)uv - uv.$$

Write this as

$$(uv' - vu')' = \left(k^2 - 1 - \frac{\nu^2 - \frac{1}{4}}{x^2}\right)uv.$$

For any positive number α, compute

$$\begin{aligned}\int_\alpha^{\alpha+\pi} (uv' - vu')\,dx &= u(\alpha+\pi)v'(\alpha+\pi) - u(\alpha)v'(\alpha) - v(\alpha+\pi)u'(\alpha+\pi) + v(\alpha)u'(\alpha)\\ &= -u(\alpha+\pi) - u(\alpha)\\ &= \int_\alpha^{\alpha+\pi}\left(k^2 - 1 - \frac{\nu^2 - \frac{1}{4}}{x^2}\right)u(x)v(x)\,dx\\ &= \int_\alpha^{\alpha+\pi}\left(k^2 - 1 - \frac{\nu^2 - \frac{1}{4}}{x^2}\right)u(x)\sin(x-\alpha)\,dx\end{aligned}$$

By the mean value theorem for integrals, there is some number τ between α and $\alpha + \pi$ such that

$$-u(\alpha+\pi) - u(\alpha) = u(\tau)\int_\alpha^{\alpha+\pi}\left(k^2 - 1 - \frac{\nu^2 - \frac{1}{4}}{x^2}\right)\sin(x-\alpha)\,dx.$$

Now $\sin(x - \alpha) > 0$ for $\alpha < x < \alpha + \pi$. Further, we can choose α large enough (depending on k and ν) so that

$$k^2 - 1 - \frac{\nu^2 - \frac{1}{4}}{x^2} > 0 \text{ for } \alpha \le x \le \alpha + \pi.$$

Therefore the integral on the right in the last equation is positive. This means that $u(\alpha+\pi)$, $u(\alpha)$ and $u(\tau)$ cannot all have the same sign. Since u is continuous, $u(x)$ must equal zero for some x between α and $\alpha + \pi$. But $u(x) = \sqrt{kx}\,J_\nu(kx)$, so wherever $u(x)$ vanishes, $J_\nu(kx)$ must be zero also. This proves that $J_\nu(kx)$ has at least one zero between α and $\alpha + \pi$.

We conclude that, if α is sufficiently large, then $J_\nu(x)$ has a zero between α and $\alpha + k\pi$.

With this as background, we will state a fundamental result on positive zeros of Bessel functions, their distribution on the half-line $x > 0$, and a relationship between zeros of $J_{\nu-1}(x)$, $J_\nu(x)$ and $J_{\nu+1}(x)$.

THEOREM 15.13 ***Zeros of $J_\nu(x)$***

Let $k > 1$ and let ν be a real number.

1. For α sufficiently large, there is a zero of $J_\nu(x)$ on each of the intervals

$$(\alpha, \alpha + k\pi), (\alpha + k\pi, \alpha + 2k\pi), (\alpha + 2k\pi, \alpha + 3k\pi), \cdots.$$

2. Every positive zero of $J_\nu(x)$ is simple.
3. The positive zeros of $J_\nu(x)$ and $J_{\nu+1}(x)$ are distinct.
4. The positive zeros of $J_\nu(x)$ and $J_{\nu-1}(x)$ are distinct.
5. If a and b are distinct positive zeros of $J_\nu(x)$, then $J_{\nu-1}(x)$ and $J_{\nu+1}(x)$ each has a zero between a and b. ♦

Proof For conclusion (1), we know from the preceding discussion that $J_\nu(x)$ has a zero between α and $\alpha + k\pi$ for α sufficiently large. But then $J_\nu(x)$ also has a zero between $\alpha + k\pi$ and $\alpha + k\pi + k\pi = \alpha + 2k\pi$, and so on.

For conclusion (2), suppose x_0 is a positive zero of $J_\nu(x)$, but is not simple. Then

$$J_\nu(x_0) = J_\nu'(x_0) = 0.$$

But then $y = J_\nu(x)$ is a solution of the initial value problem

$$y'' + \frac{1}{x}y' + \left(k^2 - \frac{\nu^2}{x^2}\right)y = 0;\ y(x_0) = y'(x_0) = 0$$

on some open interval about x_0. But $y(x) = 0$ is also a solution of this problem, so $J_\nu(x)$ would be identically zero on an open interval, and this is a contradiction.

To verify conclusion (3), suppose x_0 is a positive zero of both $J_\nu(x)$ and $J_{\nu+1}(x)$. By equation (15.24), we would also have $J_\nu'(x_0) = 0$, so x_0 would be a non-simple zero of $J_\nu(x)$, contradicting conclusion (2).

Conclusion (4) is proved by a similar argument, using equation (15.23).

To prove conclusion (5), let $f(x) = x^\nu J_\nu(x)$. Then $f(a) = f(b) = 0$. By the mean value theorem, there is some c between a and b such that $f'(c) = 0$. But by equation (15.20),

$$f'(x) = \frac{d}{dx}(x^\nu J_\nu(x)) = x^\nu J_{\nu-1}(x)$$

so $f'(c) = 0$ implies that $J_{\nu-1}(c) = 0$ also. But then c is a common positive zero of $J_\nu(x)$ and $J_{\nu-1}(x)$, contradicting conclusion (4). ♦

Conclusion (5) is called the *interlacing lemma*. It means that the graphs of $J_{\nu-1}(x)$, $J_\nu(x)$ and $J_{\nu+1}(x)$ weave about each other, as can be seen in Figure 15.10 for $J_7(x)$, $J_8(x)$ and $J_9(x)$.

We are now prepared to look at eigenfunction expansions in terms of Bessel functions.

15.3.11 Fourier-Bessel Expansions

We will write Bessel's equation in a form to fit within Sturm-Liouville theory and use equation (15.8) to determine eigenvalues and eigenfunctions. The Bessel equation

$$x^2y'' + xy' + (\lambda x^2 - \nu^2)y = 0$$

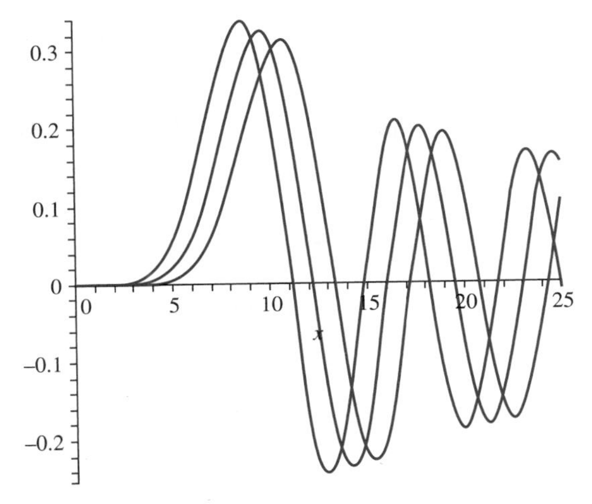

FIGURE 15.10 *Interlacing lemma for $J_7(x)$, $J_8(x)$, and $J_9(x)$.*

can be written as

$$xy'' + y' + \left(\lambda x - \frac{\nu^2}{x}\right) y = 0$$

which is the Sturm-Liouville equation

$$(xy')' + \left(\lambda x - \frac{\nu^2}{x}\right) y = 0$$

on $(0, 1)$ with $r(x) = p(x) = x$ and $q(x) = -\nu^2/x$. From equation (15.8) with $a = 0$, $c = 1$ and $\lambda = b^2$, solutions that are bounded on $(0, 1)$ are multiples of $J_\nu\left(\sqrt{\lambda}x\right)$. Further, for solutions satisfying the boundary condition $y(1) = 0$, we need $\sqrt{\lambda}$ to be a positive zero of $J_\nu(x)$. Denote these positive zeros as j_k, ordered so that

$$0 < j_1 < j_2 < j_3 < \cdots .$$

Then the numbers $\lambda_n = j_n^2$ are eigenvalues of this problem, with eigenfunctions $J_\nu(j_n x)$. Notice that ν is fixed here (occurring in the differential equation), so these eigenfunctions are all written in terms of the same Bessel function J_ν. It is j_n that changes to form the eigenfunctions $J_\nu(j_n x)$.

These eigenfunctions are orthogonal on $(0, 1)$ with weight function $p(x) = x$. This means that

$$\int_0^1 x J_\nu(j_n x) J_\nu(j_m x)\, dx = 0$$

if $n \neq m$.

If f is piecewise smooth on $(0, 1)$, then we can write the eigenfunction expansion

$$\sum_{n=1}^{\infty} c_n J_\nu(j_n x) \tag{15.27}$$

in which

$$c_n = \frac{\int_0^1 x f(x) J_\nu(j_n x)\, dx}{\int_0^1 x (J_\nu(x))^2\, dx}. \tag{15.28}$$

These are the *Fourier-Bessel coefficients* of $f(x)$ on $(0, 1)$, and the eigenfunction expansion (15.27) is called the *Fourier-Bessel expansion* or *Fourier-Bessel series* for $f(x)$ on $(0, 1)$. This expansion converges to

$$\frac{1}{2}(f(x+) + f(x-))$$

for $0 < x < 1$.

We can simplify equation (15.28) for the Fourier-Bessel coefficients of f by using the identity

$$\int_0^1 x(J_\nu(x))^2\, dx = \frac{1}{2}(J_{\nu+1}(j_n))^2. \tag{15.29}$$

To derive this, begin with the Bessel equation

$$x^2 y'' + xy' + \left(j_n^2 x^2 - \nu^2\right) y = 0$$

in which $y = J_\nu(j_n x)$. Multiply the differential equation by $2y'$ to obtain

$$2x^2 y' y'' + 2x(y')^2 + 2\left(j_n^2 x^2 - \nu^2\right) yy' = 0.$$

Write this as

$$\left[x^2(y')^2 + \left(j_n^2 x^2 - \nu^2\right) y^2\right]' - 2 j_n^2 x y^2 = 0.$$

Integrate this equation from 0 to 1, keeping in mind that $y(1) = J_\nu(j_n) = 0$. We get

$$\begin{aligned}
0 &= \left[x^2(y')^2 + \left(j_n^2 x^2 - \nu^2\right) y^2\right]_0^1 - 2 j_n^2 \int_0^1 x y^2\, dx \\
&= (y'(1))^2 - 2 j_n^2 \int_0^1 x y^2\, dx \\
&= j_n^2 \left(J_\nu'(j_n)\right) - 2 j_n^2 \int_0^1 (J_\nu(j_n x))^2\, dx.
\end{aligned}$$

Then

$$\int_0^1 x J_\nu^2(j_n x)\, dx = \frac{1}{2}\left(J_\nu'(j_n)\right)^2.$$

But, from equation (15.24),

$$J_\nu'(j_n) = -J_{\nu+1}(j_n).$$

Therefore,

$$\int_0^1 x \left(J_\nu^2(j_n x)\right)^2\, dx = \frac{1}{2} J_{\nu+1}^2(j_n),$$

as we wanted to show. Now we can write the Fourier-Bessel coefficients of f as

$$c_n = \frac{2}{J_{\nu+1}^2(j_n)} \int_0^1 x f(x) J_\nu(j_n x)\, dx. \tag{15.30}$$

These coefficients cannot be computed by hand and a software routine should be used to approximate the positive zeros j_n and then the Fourier-Bessel coefficients for a given $f(x)$. Table 15.1 gives approximate values of the first five zeros of J_0 through J_4, providing some idea of their distribution. These zeros also illustrate the interlacing property of zeros of consecutive Bessel functions (conclusion (5) of Theorem 15.13).

The eigenfunction expansion of a function f is different for each choice of ν, since the functions $J_\nu(j_n x)$ are eigenfunctions of a different Sturm-Liouville problem for each ν.

TABLE 15.1 Some Zeros of Bessel functions

J_n	j_1	j_2	j_3	j_4	j_5
J_0	2.405	5.520	8.654	11.792	14.931
J_1	3.832	7.016	10.173	13.323	16.470
J_2	5.135	8.417	11.620	14.796	17.960
J_3	6.379	9.760	13.017	16.224	19.410
J_4	7.586	11.064	14.373	17.616	20.827

EXAMPLE 15.7

We will compute some terms in the Fourier-Bessel expansion of $f(x)=x(1-x)$ on [0, 1], with $\nu=1$. In this case the eigenfunctions are $J_1(j_n x)$ and the j_n's are the positive zeros of J_1. The coefficients are

$$c_n = \frac{2\int_0^1 x^2(1-x)J_1(j_n x)\,dx}{J_2^2(j_n)}$$

and, because $x(1-x)$ is twice differentiable on [0, 1], we will have

$$x(1-x) = \sum_{n=1}^{\infty} c_n J_\nu(j_n x)\,dx$$

for $0<x<1$. We will compute the first four coefficients to illustrate the ideas involved. First we need $j_1, \cdots, j_4$. These can be obtained from tables, or from MAPLE by the command

```
evalf(BesselJZeros(ν,n));
```

with $\nu=1$ in this example and n successively chosen to be 1, 2, 3, 4. The output is

$$j_1 \approx 3.83170597,\ j_2 \approx 7.01558667,\ j_3 \approx 10.17346814,\ j_4 \approx 13.32369194.$$

Using this value for j_1, c_1 is approximately

$$c_1 \approx \frac{2}{J_2(3.83170597)^2}\int_0^1 x^2(1-x)J_1(3.83170597x)\,dx.$$

To carry out this computation, first approximate the denominator $J_2^2(j_1)$, using a software package. If MAPLE is used, the denominator is the square of the number computed as

```
evalf (BesselJ(2,3.83170597));
```

The integral in the numerator is computed as

```
evalf int((x∧2) * (1-x) * BesselJ(1,x)(3.83170597 * x),x=0..1);
```

This computation yields

$$c_1 \approx 0.45221702.$$

By repeating this calculation for $(n=2, 3, 4)$ in turn with the appropriate j_n inserted, we similarly approximate

$$c_2 \approx -0.03151859,\ c_3 \approx 0.03201789,\ c_4 \approx -0.00768864.$$

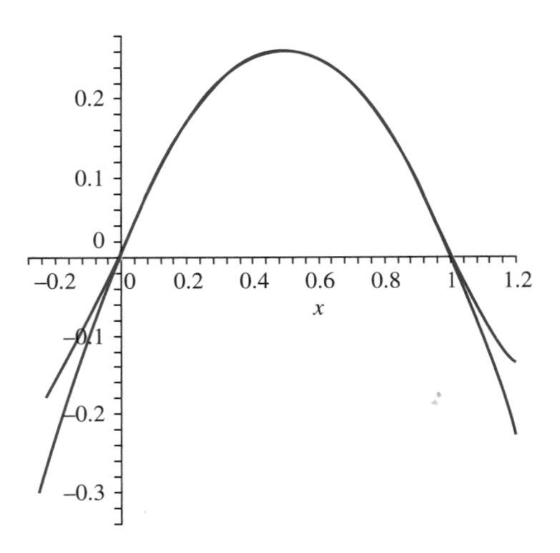

FIGURE 15.11 *Fourier-Bessel expansion in Example 15.7.*

Using the first four terms of the partial sum of the Fourier-Bessel expansion, we therefore have

$$x(1-x) \approx 0.45221702 J_1(3.83170597x) - 0.03151859 J_1(7.01558667x) + 0.03201794 J_1(10.17346814x) - 0.00768864 J_1(13.32369194x).$$

To see how good this approximation is, Figure 15.11 compares a graph of $x(1-x)$ with this sum of four terms on $[-1/4, 6/5]$. In the scale of the graph the approximation appears to be quite good on [0, 1]. Outside of [0, 1] the graphs move away from each other. Again, we cannot expect in general to obtain this good an approximation on the relevant interval with so few terms of an eigenfunction expansion. ♦

15.3.12 Bessel's Integrals and the Kepler Problem

We will use the generating function for $J_n(x)$ to derive Bessel's integrals, which are the expressions

$$J_n(x) = \frac{1}{\pi}\int_0^{\pi} \cos(n\theta - x\sin(\theta))\,d\theta$$

for $n = 0, 1, 2, \cdots$. Using these, we will solve the Kepler problem that Bessel was working on. Begin with the generating function:

$$e^{x(t-1/t)/2} = \sum_{n=-\infty}^{\infty} J_n(x)t^n.$$

If n is a positive integer, then $J_{-n}(x) = (-1)^n J_n(x)$. Therefore,

$$\begin{aligned} e^{xt/2}e^{-x/2t} &= \sum_{n=-1}^{-\infty} J_n(x)t^n + J_0(x) + \sum_{n=1}^{\infty} J_n(x)t^n \\ &= \sum_{n=1}^{\infty}(-1)^n J_n(x)t^{-n} + J_0(x) + \sum_{n=1}^{\infty} J_n(x)t^n \end{aligned}$$

$$= J_0(x) + \sum_{n=1}^{\infty} J_n(x)\left(t^n + (-1)^n \frac{1}{t^n}\right)$$

$$= J_0(x) + \sum_{n=1}^{\infty} J_{2n}(x)\left(t^{2n} + \frac{1}{t^{2n}}\right) + \sum_{n=1}^{\infty} J_{2n-1}(x)\left(t^{2n-1} - \frac{1}{t^{2n-1}}\right).$$

Now set $t = e^{i\theta}$. Then

$$t^{2n} + \frac{1}{t^{2n}} = e^{2in\theta} + e^{-2in\theta} = 2\cos(2n\theta)$$

and

$$t^{2n-1} - \frac{1}{t^{2n-1}} = e^{i(2n-1)\theta} - e^{-i(2n-1)\theta} = 2i\sin((2n-1)\theta).$$

Then

$$\begin{aligned} e^{x(t-1/t)/2} &= e^{ix\sin(\theta)} \\ &= \cos(x\sin(\theta)) + i\sin(x\sin(\theta)) \\ &= J_0(x) + 2\sum_{n=1}^{\infty} J_{2n}\cos(2n\theta) + 2i\sum_{n=1}^{\infty} J_{2n-1}(x)\sin((2n-1)\theta). \end{aligned}$$

Equating real parts on both sides and imaginary parts on both sides of this equation, we have

$$\cos(x\sin(\theta)) = J_0(x) + 2\sum_{n=1}^{\infty} J_{2n}(x)\cos(2n\theta) \tag{15.31}$$

and

$$\sin(x\sin(\theta)) = 2\sum_{n=1}^{\infty} J_{2n-1}(x)\sin((2n-1)\theta). \tag{15.32}$$

The series on the right side of each of these two equations is the Fourier series of the function on the left side, over the interval $[-\pi, \pi]$. Focusing on equation (15.31) first, we have

$$\begin{aligned} \cos(x\sin(\theta)) &= \frac{1}{2}a_0 + \sum_{k=1}^{\infty} a_k\cos(k\theta) + b_k\sin(k\theta) \\ &= J_0(x) + \sum_{n=1}^{\infty} 2J_{2n}(x)\cos(2n\theta). \end{aligned}$$

Since we know the coefficients in the Fourier expansion of a function, we conclude that

$$a_k = \frac{1}{\pi}\int_{-\pi}^{\pi}\cos(x\sin(\theta))\cos(k\theta)\,d\theta = \begin{cases} 0 & \text{if } k \text{ is odd} \\ 2J_{2n}(x) & \text{if } k = 2n \text{ is even,} \end{cases} \tag{15.33}$$

and

$$b_k = \frac{1}{\pi}\int_{-\pi}^{\pi}\cos(x\sin(\theta))\sin(k\theta)\,d\theta = 0 \tag{15.34}$$

for $k = 1, 2, 3, \cdots$.

Similarly, we know from equation (15.32) that

$$\begin{aligned} \sin(x\sin(\theta)) &= \frac{1}{2}A_0 + \sum_{k=1}^{\infty} A_k\cos(k\theta) + B_k\sin(k\theta) \\ &= \sum_{n=1}^{\infty} 2J_{2n-1}(x)\sin((2n-1)\theta). \end{aligned}$$

These Fourier coefficients are

$$A_k = \frac{1}{\pi}\int_{-\pi}^{\pi} \sin(x\sin(\theta))\cos(k(\theta))\,d\theta = 0 \tag{15.35}$$

for $k = 0, 1, 2, \cdots$, and

$$B_k = \frac{1}{\pi}\int_{-\pi}^{\pi} \sin(x\sin(\theta))\sin(k\theta)\,d\theta = \begin{cases} 0 & \text{if } k \text{ is even} \\ 2J_{2n-1}(x) & \text{if } k = 2n-1 \text{ is odd.} \end{cases} \tag{15.36}$$

Upon adding equations (15.35) and (15.36), we obtain

$$\begin{aligned} \frac{1}{\pi}\int_{-\pi}^{\pi} \cos(x\sin(\theta))\cos(k\theta)\,d\theta &+ \frac{1}{\pi}\int_{-\pi}^{\pi} \sin(x\sin(\theta))\sin(k\theta)\,d\theta \\ &= \frac{1}{\pi}\int_{-\pi}^{\pi} \cos(k\theta - x\sin(\theta))\,d\theta \\ &= \begin{cases} 2J_k(x) & \text{if } k \text{ is even} \\ 2J_k(x) & \text{if } k \text{ is odd.} \end{cases} \end{aligned}$$

In summary,

$$J_k(x) = \frac{1}{2\pi}\int_{-\pi}^{\pi} \cos(k\theta - x\sin(\theta))\,d\theta$$

for $k = 0, 1, 2, \cdots$. Finally, observe that $\cos(k\theta - x\sin(\theta))$ is an even function, so the integral over $[-\pi, \pi]$ is twice the integral over $[0, \pi]$. Then

$$J_n(x) = \frac{1}{\pi}\int_{0}^{\pi} \cos(n\theta - x\sin(\theta))\,d\theta$$

for $n = 0, 1, 2, \cdots$. These integrals are called *Bessel's integrals.*

We will apply Bessel's integrals to the solution of Kepler's problem in astronomy. A planet moves along its elliptical orbit with the sun at one focus. In Figure 15.12, the center of the orbit

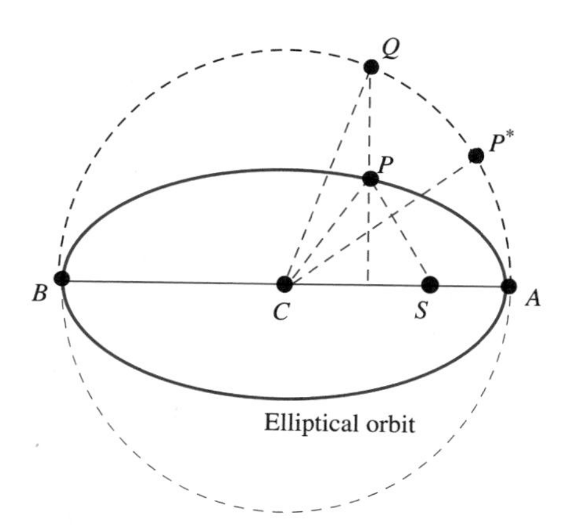

FIGURE 15.12 *Kepler's problem.*

is C and line BA is the semimajor axis, whose length we will denote $2a$. The focus is the sun at S. P is the planet at some time t, which we set at zero when P is at A. A (dashed) circle is drawn with center C and through A and B. Q is the point of intersection of this circle with the line through P perpendicular to AB. P^* is imagined to move along the circle at a constant speed so that its location coincides with that of P at both A and B.

We are interested in the two angles

$$\psi = \angle ACP^* \text{ and } \varphi = \angle ACQ.$$

ψ is the *mean anomaly* of the planet, and φ the *eccentric anomaly*. The *true anomaly* is $\theta = \angle ASP$, but we will not be concerned with θ here. The part of Kepler's problem we will discuss is to express φ in terms of ψ.

Note that ψ is proportional to the time t it takes for the planet to move from A to P. Thus, we might equivalently state the problem as expressing φ in terms of t.

As is customary, let e be the eccentricity of the ellipse. Also let T denote the period of the planet (time required for one complete orbit).

Recall that the area of an ellipse with semimajor axis of length $2a$ and semiminor axis $2b$ is πab. Now, since ψ and $\angle ASP$ are both proportional to t, then

$$\frac{t}{T} = \frac{\psi}{2\pi} = \frac{\text{area of elliptical sector } ASP}{\pi ab}$$
$$= \frac{\text{area of circular sector } ASQ}{\pi a^2}.$$

But

$$\text{area of } ASQ = \text{area of elliptical sector } ACQ$$
$$- \text{area of triangle} SCQ$$
$$= \frac{1}{2}a^2\varphi - \frac{1}{2}ae(a\sin(\varphi)).$$

Therefore

$$\frac{\psi}{2\pi} = \frac{\frac{1}{2}a^2\varphi - \frac{1}{2}a^2 e\sin(\varphi)}{\pi a^2},$$

so

$$\psi = \varphi - e\sin(\varphi).$$

We want φ in terms of ψ, so we must somehow solve this equation for φ. Observe that $\psi - \varphi$ is an odd periodic function of φ, having period 2π. We can therefore write a Fourier expansion

$$\psi - \varphi = \sum_{n=1}^{\infty} b_n \sin(n\psi)$$

in which

$$b_n = \frac{2}{\pi}\int_0^{\pi} (\varphi - \psi)\sin(n\psi)\, d\psi.$$

Integrate by parts, keeping in mind that

$$(\varphi - \psi)(0) = (\varphi - \psi)(\pi) = 0 \text{ and } \int_0^{\pi} \cos(\psi)\, d\psi = 0.$$

We obtain

$$\begin{aligned} b_n &= \frac{2}{\pi}\left[(\psi-\varphi)\frac{\cos(n\psi)}{\psi}\right]_0^{\pi} \\ &+\frac{2}{\pi}\int_0^{\pi}\frac{\cos(n\psi)}{n}d\varphi \\ &-\frac{2}{\pi}\int_0^{\pi}\frac{\cos(n\psi)}{n}d\psi \\ &=\frac{2}{n\pi}\int_0^{\pi}\cos(n\psi)d\varphi \\ &=\frac{2}{n\pi}\int_0^{\pi}\cos(n\varphi-ne\sin(\varphi))\,d\varphi \\ &=\frac{2}{n}J_n(ne). \end{aligned}$$

The last line comes from Bessel's integrals. The solution to Kepler's problem is therefore

$$\varphi=\psi+\sum_{n=1}^{\infty}\frac{2}{n}J_n(ne)\sin(n\psi).$$

SECTION 15.3 PROBLEMS

1. Show that $x^aJ_\nu(bx^c)$ is a solution of

$$y''-\left(\frac{2a-1}{x}\right)y'+\left(b^2c^2x^{2c-2}+\frac{a^2-\nu^2c^2}{x^2}\right)y=0.$$

In each of Problems 2 through 9, write the general solution of the differential equation in terms of $x^aJ_\nu(bx^x)$ and $x^aJ_{-\nu}(bx^c$ for appropriate a, b and c.

2. $y''-\dfrac{3}{x}y'+9x^4y=0$
3. $y''+\dfrac{3}{x}y'+\left(16x^2-\dfrac{5}{4x^2}\right)y=0$
4. $y''+\dfrac{1}{x}y'-\left(\dfrac{1}{16x^2}\right)y=0$
5. $y''-\dfrac{7}{x}y'+\left(36x^4+\dfrac{175}{16x^2}\right)y=0$
6. $y''-\dfrac{5}{x}y'+\left(64x^6+\dfrac{5}{x^2}\right)y=0$
7. $y''+\dfrac{5}{x}y'+\left(81x^4+\dfrac{7}{4x^2}\right)y=0$
8. $y''+\dfrac{1}{3x}y'+\left(1+\dfrac{7}{144x^2}\right)y=0$
9. $y''+\dfrac{1}{x}y'+\left(4x^2-\dfrac{4}{9x^2}\right)y=0$

10. Use the change of variables

$$bu=\frac{1}{u}\frac{du}{dx}$$

to transform the differential equation

$$\frac{dy}{dx}+by^2=cx^m$$

into the differential equation

$$\frac{d^2u}{dx^2}-bcx^mu=0.$$

Find the general solution of this differential equation in terms of Bessel functions and use this to solve the original differential equation. Assume that b is a positive constant.

In each of Problems 11 through 16, use the given change of variables to transform the differential equation into one whose general solution can be written in terms of Bessel functions. Use this to write the general solution of the original differential equation.

11. $9x^2y''+9xy'+(4x^{2/3}-16)y=0$; $z=2x^{1/3}$
12. $4x^2y''+8xy'+(4x^2-35)y=0$; $u=y\sqrt{x}$

13. $36x^2y'' - 12xy' + (36x^2 + 7)y = 0;\ u = yx^{-2/3}$
14. $4x^2y'' + 4xy' + (9x^3 - 36)y = 0;\ z = x^{3/2}$
15. $4x^2y'' + 4xy' + (x - 9)y = 0;\ z = \sqrt{x}$
16. $9x^2y'' - 27xy' + (9x^2 + 35)y = 0;\ u = yx^{-2}$
17. Show that, for any positive integer n,

$$\int x^n J_{n-1}(x)\, dx = x^n J_n(x)$$

and

$$\int \frac{J_{n+1}(x)}{x^n}\, dx = -\frac{J_n(x)}{x^n}.$$

(Here we have omitted the constants of integration). *Hint*: Use Theorem 15.10. Alternatively, one can integrate the series for $J_{n+1}(x)/x$ term by term.

18. Show that, for any positive integer n and any nonzero number α,

$$\int x^n J_{n-1}(\alpha x)\, dx = \frac{1}{\alpha} x^n J_n(\alpha x)$$

and

$$\int \frac{J_{n+1}(\alpha x)}{x^n}\, dx = -\frac{J_n(\alpha x)}{\alpha x^n}.$$

Hint: The result of Problem 17 can be used, or the series for $J_n(x)$ can be used.

19. Let α be a positive zero of J_0. Show that

$$\int_0^1 J_1(\alpha x)\, dx = \frac{1}{\alpha}.$$

20. Let $u(x) = J_0(\alpha x)$ and $v(x) = J_0(\beta x)$, with α and β positive constants.

(a) Show that $xu'' + u' + \alpha^2 xu = 0$, with a similar equation for v.

(b) Multiply the differential equation for u by v and the equation for v by u and subtract to show that

$$[x(u'v - v'u)]' = (\beta^2 - \alpha^2)xuv.$$

(c) Use the conclusion of part (b) to show that

$$(\beta^2 - \alpha^2)\int xJ_0(\alpha x)J_0(\beta x)\, dx$$
$$= x\left[\alpha J_0'(\alpha x)J_0(\beta x) - \beta J_0'(\beta x)J_0(\alpha x)\right].$$

This is one of a class of integrals known as *Lommel's integrals*.

21. For α any nonzero number, and n and k nonnegative integers, define

$$I_{n,k} = \int_0^1 (1 - x^2)^k x^{n+1} J_n(\alpha x)\, dx.$$

(a) Show that

$$I_{n,0} = \frac{1}{\alpha} J_{n+1}(\alpha).$$

Hint: Use the first integral in Problem 18.

(b) Show that

$$I_{n,k} = \int_0^1 (1 - x^2)^k \frac{d}{dx}\left(\frac{x^{n+1}}{\alpha} J_{n+1}(\alpha x)\right) dx.$$

Hint: Use the first integral from Problem 18 in the definition of $I_{n,k}$.

(c) Show that

$$I_{n,k} = \frac{2k}{\alpha} I_{n+1,k-1}.$$

This provides a recurrence relation for the quantities $I_{n,k}$. *Hint*: Integrate by parts in (b).

(d) Show that

$$I_{n,k} = \frac{2^k k!}{\alpha^k} I_{n+k,0}.$$

Hint: Apply part (c) in repetition.

(e) Show that

$$\int_0^1 (1 - x^2)^k x^{n+1} J_n(\alpha x)\, dx = \frac{2^k \Gamma(k+1)}{\alpha^{k+1}} J_{n+k+1}(\alpha).$$

Hint: Use the result of part (d).

(f) Show that

$$J_{n+k+1}(x) = \frac{x^{k+1}}{2^k \Gamma(k+1)} \int_0^1 t^{n+1}(1 - t^2)^k J_n(xt)\, dt.$$

(g) Show that, if n is a nonnegative integer and m is a positive integer with $n < m$, then

$$J_m(x) = \frac{2x^{m-n}}{2^{m-n}\Gamma(m - n)} \int_0^1 t^{n+1}(1 - t^2)^{m-n-1} J_n(xt)\, dt.$$

Hint: Let $m = n + k + 1$ in the result of part (f).

The integral expressions in parts (e), (f), and (g) are called *Sonine's integrals*.

22. Use the fact that

$$J_{-1/2}(xt) = \sqrt{\left(\frac{2}{\pi xt}\right)} \cos(xt)$$

to show that, if n is a positive integer, then

$$J_n(x) = \frac{x^n}{2^{n-1}\sqrt{\pi}\,\Gamma(n + 1/2)}$$
$$\int_0^1 (1 - t^2)^{n-1/2} \cos(xt)\, dt.$$

This is called *Hankel's integral*. *Hint*: Use Sonine's integral, Problem 21 (g).

23. Show that, if m is a positive integer, then

$$J_m(x) = \frac{x^m}{2^{m-1}\sqrt{\pi}\,\Gamma(m + 1/2)}$$
$$\int_0^{\pi/2} \cos^{2m}(\theta) \cos(x \sin(\theta))\, d\theta.$$

This expression is called *Poisson's integral*. *Hint*: Put $t = \sin(\theta)$ in Hankel's integral, Problem 22.

For each of Problems 24 through 29, find (approximately) the first five terms in the Fourier-Bessel expansion of $f(x)$

on (0, 1) in a series of the functions $J_2(j_n x)$, with j_n the nth positive zero of J_2.

24. $f(x) = x^2 e^{-x}$
25. $f(x) = xe^{-x}$
26. $f(x) = e^{-x}$
27. $f(x) = x$
28. $f(x) = x\cos(\pi x)$
29. $f(x) = \sin(\pi x)$

In each of Problems 30 through 35, find (approximately) the first five terms in the Fourier-Bessel expansion of $f(x)$ on (0, 1) in a series of the functions $J_1(j_n x)$, where j_n is the nth positive zero of J_1. Compare a graph of this partial sum with f.

30. $f(x) = e^{-x}$
31. $f(x) = x$
32. $f(x) = x^2 e^{-x}$
33. $f(x) = xe^{-x}$
34. $f(x) = x\cos(\pi x)$
35. $f(x) = \sin(\pi x)$

Problems 36 through 40 deal with the gamma and beta functions.

36. Show that, for positive x,

$$\Gamma(x) = 2\int_0^\infty e^{-t^2} t^{2x-1}\,dt.$$

Hint: Let $t = y^2$ in the definition of the gamma function.

37. Define the *beta function* by

$$B(x, y) = \int_0^1 t^{x-1}(1-t)^{y-1}\,dt.$$

It can be shown that this integral converges for x and y positive. Show that

$$B(x, y) = \int_0^\infty \frac{u^{x-1}}{(1+u)^{x+y}}\,du.$$

Hint: Let $t = u/(1+u)$ in the definition of $B(x, y)$.

38. Show that, if x and y are positive integers, then

$$B(x, y) = \frac{\Gamma(x)\Gamma(y)}{\Gamma(x+y)}.$$

Hint: Begin with $\mathcal{L}[t^x] = \frac{\Gamma(x+1)}{s^{k+1}}$. Now compute $\mathcal{L}^{-1}\left[\frac{1}{s^{x+y}}\right]$ in two ways, first by using this formula, and second, by using the convolution theorem.

39. Show that, if $r > 0$, then for any positive x,

$$\Gamma(x) = r^x \int_0^\infty e^{-rt} t^{x-1}\,dt.$$

Hint: Let $t = ry$ in the definition of $\Gamma(x)$.

40. Show that $\Gamma(1/2) = \sqrt{\pi}$, using the fact from statistics that $\int_0^\infty e^{-x^2}\,dx = \sqrt{\pi}/2$.

PART 5

Partial Differential Equations

CHAPTER 16

DERIVATION OF THE WAVE EQUATION
WAVE MOTION ON AN INTERVAL WAVE MOTION IN AN INFINITE MEDIUM

The Wave Equation

16.1 Derivation of the Wave Equation

Vibrations in a membrane or steel plate, or oscillations along a guitar string, are all modeled by the wave equation and appropriate initial and boundary conditions. We will begin with a derivation of the one-dimensional wave equation.

Suppose an elastic string has its ends fastened by two pegs. The string is displaced, released and allowed to vibrate in a plane.

Place the string along the x - axis from 0 to L and assume that it vibrates in the x, y - plane. We want a function $y(x,t)$ such that, at time t, the graph of $y(x,t)$ is the shape of the string at that time. We call $y(x,t)$ the *position function* for the string.

To take a simple case, neglect damping forces such as the weight of the string and assume that the tension $\mathbf{T}(x,t)$ acts tangent to the string, and that individual particles of the string move only vertically. Also assume that the mass ρ per unit length is constant. Consider a segment of string between x and $x+\Delta x$. By Newton's second law of motion, the net force on this segment due to the tension is equal to the acceleration of the center of mass of this segment, multiplied by its mass. This is a vector equation. Its vertical component (Figure 16.1) gives us

$$T(x+\Delta x,t)\sin(\theta+\Delta\theta)-T(x,t)\sin(\theta)=\rho\Delta x\frac{\partial^2 y}{\partial t^2}(\overline{x},t),$$

where $\overline{x}$ is the center of mass of this segment and $T(x,t)=\parallel \mathbf{T}(x,t)\parallel$. Then

$$\frac{T(x+\Delta x,t)\sin(\theta+\Delta\theta)-T(x,t)\sin(\theta)}{\Delta x}=\rho\frac{\partial^2 y}{\partial t^2}(\overline{x},t).$$

Now $v(x,t)=T(x,t)\sin(\theta)$ is the vertical component of the tension, so this equation becomes

$$\frac{v(x+\Delta x,t)-v(x,t)}{\Delta x}=\rho\frac{\partial^2 y}{\partial t^2}(\overline{x},t).$$

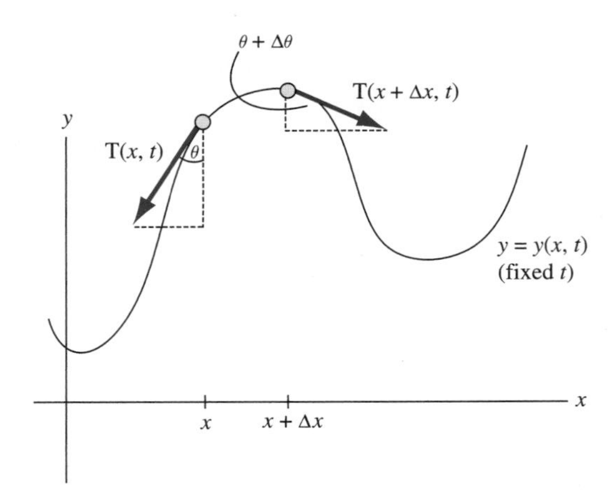

FIGURE 16.1 *Deriving the wave equation.*

As $\Delta x \to 0$, $\overline{x} \to x$ and we obtain

$$\frac{\partial v}{\partial x} = \rho \frac{\partial^2 y}{\partial t^2}.$$

The horizontal component of the tension is $h(x,t) = T(x,t)\cos(\theta)$, so

$$v(x,t) = h(x,t)\tan(\theta) = h(x,t)\frac{\partial y}{\partial x}.$$

Then

$$\frac{\partial}{\partial x}\left(h\frac{\partial y}{\partial x}\right) = \rho \frac{\partial^2 y}{\partial t^2}.$$

To compute the left side of this equation, use the fact that the horizontal component of the tension of the segment is zero, so

$$h(x + \Delta x, t) = h(x,t).$$

This means that h is independent of x, so

$$h\frac{\partial^2 y}{\partial x^2} = \rho \frac{\partial^2 y}{\partial t^2}.$$

Let $c^2 = h/\rho$ to get the *one-dimensional wave equation*

$$\frac{\partial^2 y}{\partial t^2} = c^2 \frac{\partial^2 y}{\partial x^2}. \tag{16.1}$$

The fact that the ends are held fixed is reflected in the *boundary conditions*

$$y(0,t) = y(L,t) = 0 \text{ for } t \geq 0.$$

The initial displacement of the string, and the velocity with which it is released at time 0, are the *initial conditions*, given by

$$y(x,0) = f(x) \text{ for } 0 \leq x \leq L.$$

and

$$\frac{\partial y}{\partial t}(x,0) = g(x).$$

The wave equation, together with these initial and boundary conditions, is called an *initial-boundary value problem for* $y(x,t)$.

If the string is released from rest (zero initial velocity), then $g(x)$ is identically zero. Further, if the string has its ends fixed at the same level, then the initial position function f must satisfy the compatibility condition $f(0)=f(L)=0$.

Variations on this problem can allow for moving ends with conditions such as

$$y(0,t)=\alpha(t) \text{ and } y(L,t)=\beta(t).$$

We can also have a *forcing term* in which an external force drives the motion of the string. In this case the wave equation is

$$\frac{\partial^2 y}{\partial t^2}=c^2\frac{\partial^2 y}{\partial x^2}+F(x,t).$$

The wave equation in two space dimensions is

$$\frac{\partial^2 z}{\partial t^2}=c^2\left(\frac{\partial^2 z}{\partial x^2}+\frac{\partial^2 z}{\partial y^2}\right) \tag{16.2}$$

in which $z(x,y,t)$ is the displacement function.

It is a routine exercise in chain rule differentiation to convert this two-dimensional wave equation to polar coordinates. If the displacement function is $u(r,\theta,t)$, this wave equation is

$$\frac{\partial^2 u}{\partial t^2}=c^2\left(\frac{\partial^2 u}{\partial r^2}+\frac{1}{r}\frac{\partial u}{\partial r}+\frac{1}{r^2}\frac{\partial^2 u}{\partial \theta^2}\right). \tag{16.3}$$

SECTION 16.1 PROBLEMS

1. Let f be a twice-differentiable function of one variable. Show that
$$y(x,t)=\frac{1}{2}[f(x+ct)+f(x-ct)]$$
satisfies the one-dimensional wave equation.
2. Formulate an initial-boundary value problem for the motion of an elastic string of length L fastened at both ends and released from rest with an initial position given by $f(x)$. The motion is opposed by air resistance, which has a force at each point of magnitude proportional to the square of the velocity at that point.
3. Formulate an initial-boundary value problem for vibrations of a rectangular membrane occupying $0\le x\le a, 0\le y\le b$ if the initial position is the graph of $z=f(x,y)$ and the initial velocity is $g(x,y)$. The membrane is fastened to a still frame along the rectangular boundary of the region.
4. Show that $z(x,y,t)=\sin(nx)\cos(my)\cos\left(\sqrt{n^2+m^2}t\right)$ satisfies the two-dimensional wave equation for any positive integers n and m.
5. Let $y(x,t)=\sin(n\pi x/L)\cos(n\pi ct/L)$ for $0\le x\le L$. Show that y satisfies the one-dimensional wave equation for any positive integer n.
6. Show that
$$y(x,t)=\sin(x)\cos(ct)+\frac{1}{c}\cos(x)\sin(ct)$$
satisfies the one-dimensional wave equation together with the boundary conditions
$$y(0,t)=y(2\pi,t)=\frac{1}{c}\sin(ct) \text{ for } t>0$$
and the initial conditions
$$y(x,0)=\sin(x),\ \frac{\partial y}{\partial t}(x,0)=\cos(x) \text{ for } 0<x<\pi.$$

16.2 Wave Motion on an Interval

We will solve initial-boundary value problems for the wave equation on a closed interval $[0,L]$, starting with special cases we can solve and building toward more complex wave motion.

16.2.1 Zero Initial Velocity

An elastic string of length L with fixed ends is released from rest (zero initial velocity) from an initial position given as the graph of $y=f(x)$. The initial-boundary value problem for the position function $y(x,t)$ is

$$\frac{\partial^2 y}{\partial t^2}=c^2\frac{\partial^2 y}{\partial x^2} \text{ for } 0<x<L, t>0,$$

$$y(0,t)=y(L,t)=0 \text{ for } t\geq 0,$$

$$y(x,0)=f(x) \text{ for } 0\leq x\leq L,$$

and

$$\frac{\partial y}{\partial t}(x,0)=0.$$

The *Fourier method* (or *method of separation of variables*) is to attempt a solution of the form $y(x,t)=X(x)T(t)$. Substitute this into the wave equation to get

$$XT''=c^2X''T,$$

where $X'=dX/dx$ and $T'=dT/dt$. Then

$$\frac{X''}{X}=\frac{T''}{c^2T}.$$

The left side depends only on x. We could fix x and then the right side, which depends only on t, would be constant for all t. But then the left side must equal the same constant for all x. Therefore, for some number λ, called the *separation constant*,

$$\frac{X''}{X}=\frac{T''}{c^2T}=-\lambda.$$

Calling the constant $-\lambda$ is common practice. Then

$$X''+\lambda X=0 \text{ and } T''+\lambda c^2T=0,$$

two ordinary differential equations for X and T. Next use the boundary conditions. First,

$$y(0,t)=X(0)T(t)=0 \text{ for } t\geq 0$$

implies that $X(0)=0$. Similarly, $y(x,L)=X(L)T(t)=0$ implies that $X(L)=0$. We have obtained a Sturm-Liouville problem for X:

$$X''+\lambda X=0;\ X(0)=X(L)=0.$$

In Example 15.1, we solved this problem for the values of λ (eigenvalues) and corresponding solutions for X (eigenfunctions):

$$\lambda_n=\frac{n^2\pi^2}{L^2} \text{ and } X_n(x)=\sin\left(\frac{n\pi x}{L}\right) \text{ for } n=1,2,\cdots.$$

Next focus on T. Since $\lambda_n=n^2\pi^2/L^2$, the differential equation for T is

$$T''+\frac{n^2\pi^2c^2}{L^2}T=0.$$

Because the string is released from rest,

$$\frac{\partial y}{\partial t}(x,0)=X(x)T'(0)=0,$$

so

$$T'(0)=0.$$

Solutions for $T(t)$ subject to this condition are constant multiples of $\cos(n\pi ct/L)$.

So far, for $(n = 1, 2, \cdots)$ we have a function

$$y_n(x,t) = c_n \sin(n\pi x/L)\cos(n\pi ct/L)$$

that satisfies the wave equation, both boundary conditions and the initial condition of zero initial velocity. We have yet to satisfy $y(x,0) = f(x)$. If $f(x) = K\sin(m\pi x/L)$ for some integer m, then $y(x,t) = K\sin(m\pi x/L)\cos(m\pi ct/L)$ is the solution. But $f(x)$ need not look like this. For example, if the string is picked up at its midpoint, say

$$f(x) = \begin{cases} x & \text{for } 0 \le x \le L/2 \\ L - x & \text{for } L/2 \le x \le L \end{cases} \tag{16.4}$$

then not even a finite sum $y(x,t) = \sum_{n=0}^{N} y_n(x,t)$ can satisfy $y(x,0) = f(x)$. In such a case we use an infinite superposition

$$y(x,t) = \sum_{n=1}^{\infty} c_n \sin\left(\frac{n\pi x}{L}\right)\cos\left(\frac{n\pi ct}{L}\right).$$

The condition $y(x,0) = f(x)$ is satisfied if we can choose the coefficients so that

$$y(x,0) = f(x) = \sum_{n=0}^{\infty} c_n \sin\left(\frac{n\pi x}{L}\right).$$

This is the Fourier sine expansion of $f(x)$ on $[0, L]$, hence choose

$$c_n = \frac{2}{L}\int_0^L f(\xi)\sin\left(\frac{n\pi\xi}{L}\right)d\xi.$$

The solution of the problem, with zero initial velocity and initial position given by f, is

$$y(x,t) = \frac{2}{L}\sum_{n=1}^{\infty}\left(\int_0^L f(\xi)\sin(n\pi\xi/L)\,d\xi\right)\sin\left(\frac{n\pi x}{L}\right)\cos\left(\frac{n\pi ct}{L}\right). \tag{16.5}$$

EXAMPLE 16.1

With f given by equation (16.4) and $L = \pi$, the coefficients are

$$\begin{aligned} c_n &= \frac{2}{\pi}\int_0^{\pi} f(\xi)\sin(n\xi)d\xi \\ &= \frac{2}{\pi}\int_0^{\pi/2} \xi\sin(n\xi)\,d\xi + \frac{2}{\pi}\int_{\pi/2}^{\pi}(\pi - \xi)\sin(n\xi)\,d\xi \\ &= \frac{4\sin(n\pi/2)}{\pi n^2}. \end{aligned}$$

The solution of this problem is

$$y(x,t) = \sum_{n=1}^{\infty}\frac{4\sin(n\pi/2)}{\pi n^2}\sin(nx)\cos(nct).$$

Figure 16.2 shows the string profile for $c = 2$ at times $t = 0.3, 0.6, 0.9$, and 1.2, starting at the top and moving downward in this time frame. ♦

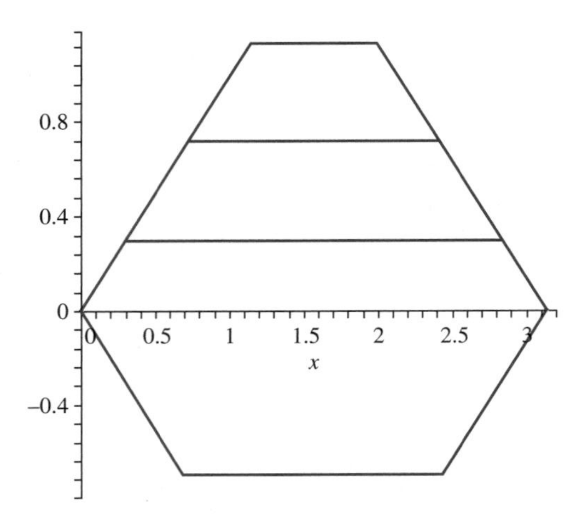

FIGURE 16.2 *Wave motion in Example 16.1.*

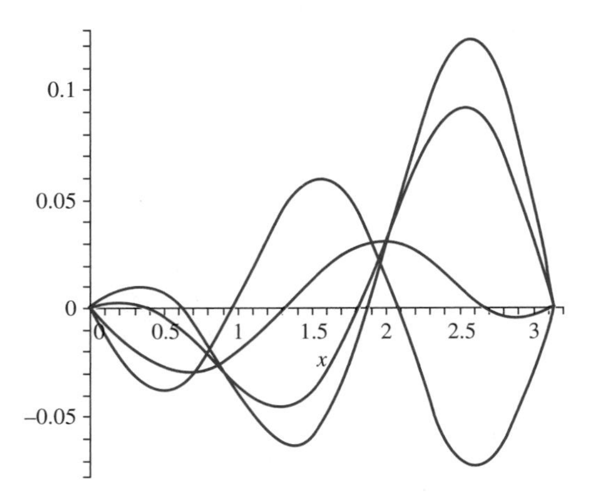

FIGURE 16.3 *Wave profiles in Example 16.2.*

EXAMPLE 16.2

We will solve for $y(x,t)$ on $[0,\pi]$ with zero initial velocity and initial position $f(x) = x\cos(5x/2)$. The coefficients are

$$c_n = \frac{2}{\pi}\int_0^{\pi} \xi\cos(5\xi/2)\sin(n\xi)\,d\xi = \frac{8}{\pi}\frac{n(-1)^{n+1}}{(5+2n)^2(5-2n)^2}.$$

The solution is

$$y(x,t) = \sum_{n=1}^{\infty} \frac{8}{\pi}\frac{n(-1)^{n+1}}{(5+2n)^2(5-2n)^2}\sin(nx)\cos(nct).$$

Figure 16.3 shows wave profiles for $c = 1$ at times $t = 0.1, 0.3, 0.6$, and 0.9, moving downward over these times. ♦

16.2.2 Zero Initial Displacement

We will solve the initial-boundary value problem for the displacement of the string if there is an initial velocity but no initial displacement. The problem is

$$\frac{\partial^2 y}{\partial t^2} = c^2 \frac{\partial^2 y}{\partial x^2} \text{ for } 0 < x < L, t > 0,$$

$$y(0, t) = y(L, t) = 0 \text{ for } t \geq 0,$$

$$y(x, 0) = 0,$$

and

$$\frac{\partial y}{\partial t}(x, 0) = g(x) \text{ for } 0 < x < L.$$

Again, let $y(x, t) = X(x)T(t)$. The problem for X is the same as before,

$$X'' + \lambda X = 0;\ X(0) = X(L) = 0$$

with eigenvalues $\lambda_n = n^2\pi^2/L^2$ and eigenfunctions $\sin(n\pi x/L)$. The problem for T, however, is different, The differential equation is still

$$T'' + \lambda c^2 T = T'' + \left(\frac{n^2\pi^2c^2}{L^2}\right)T = 0$$

but now the zero initial displacement gives us $y(x, 0) = X(x)T(0) = 0$, so $T(0) = 0$. Solutions of this problem for $T(t)$ have the form

$$T_n(t) = c_n \sin\left(\frac{n\pi ct}{L}\right).$$

Now we have functions

$$y_n(x, t) = X_n(x)T_n(t) = c_n \sin\left(\frac{n\pi x}{L}\right) \sin\left(\frac{n\pi ct}{L}\right)$$

that satisfy the wave equation, the boundary conditions, and the initial condition $y(x, 0) = 0$. To satisfy the initial velocity condition, we will generally (depending on g) need a superposition

$$y(x, t) = \sum_{n=1}^{\infty} c_n \sin\left(\frac{n\pi x}{L}\right) \sin\left(\frac{n\pi ct}{L}\right).$$

We must choose the c_n's to satisfy

$$\left.\frac{\partial y}{\partial t}\right]_{t=0} = \sum_{n=1}^{\infty} \left(\frac{n\pi c}{L}\right) c_n \sin\left(\frac{n\pi x}{L}\right) = g(x).$$

Then

$$\frac{L}{n\pi c} g(x) = \sum_{n=1}^{\infty} c_n \sin\left(\frac{n\pi x}{L}\right),$$

This is the Fourier sine expansion of $\frac{L}{n\pi c} g(x)$. Therefore choose the coefficients

$$c_n = \frac{2}{L} \frac{L}{n\pi c} \int_0^L g(\xi) \sin\left(\frac{n\pi \xi}{L}\right) d\xi,$$

or

$$c_n = \frac{2}{n\pi c} \int_0^L g(\xi) \sin\left(\frac{n\pi \xi}{L}\right) d\xi.$$

With this choice of the coefficients the solution is

$$y(x, t) = \frac{2}{\pi c} \sum_{n=1}^{\infty} \frac{1}{n} \left(\int_0^L g(\xi) \sin(n\pi\xi/L)\, d\xi \right) \sin\left(\frac{n\pi x}{L}\right) \sin\left(\frac{n\pi ct}{L}\right). \tag{16.6}$$

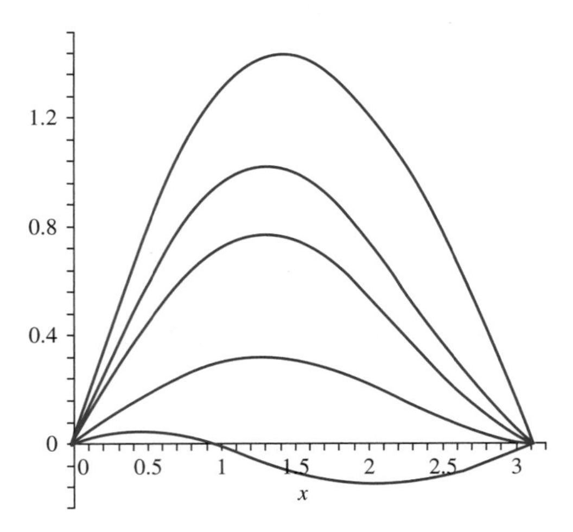

FIGURE 16.4 *Wave moving downward for increasing times in Example 16.3.*

EXAMPLE 16.3

Suppose the string is released from its horizontal position with an initial velocity given by $g(x) = x(1+\cos(\pi x/L))$. Equation (16.6) is the solution. First compute the integral

$$\int_0^L g(\xi)\sin(n\pi\xi/L)\,d\xi = \int_0^L \xi\,(1+\cos(\pi\xi/L))\sin(n\pi\xi/L)\,d\xi$$

$$= \begin{cases} \dfrac{3L^2}{4\pi} & \text{for } n = 1 \\ \dfrac{L^2(-1)^n}{n\pi(n^2-1)} & \text{for } n = 2, 3, \cdots. \end{cases}$$

The solution is

$$y(x,t) = \frac{2}{\pi c}\left(\frac{3L^2}{4\pi}\right)\sin\left(\frac{\pi x}{L}\right)\sin\left(\frac{\pi ct}{L}\right)$$
$$+ \frac{2}{\pi c}\sum_{n=2}^{\infty}\frac{L^2(-1)^n}{n^2\pi(n^2-1)}\sin\left(\frac{n\pi x}{L}\right)\sin\left(\frac{n\pi ct}{L}\right).$$

Then with $c = 1$ and $L = \pi$, this solution is

$$y(x,t) = \frac{3}{2}\sin(x)\sin(t) + \sum_{n=2}^{\infty}\frac{2(-1)^n}{n^2(n^2-1)}\sin(nx)\sin(nt).$$

Figure 16.4 shows the wave moving downward at times $t = 0.2, 0.5, 0.7, 1.3$, and 3.4. ♦

16.2.3 Nonzero Initial Displacement and Velocity

Suppose the string has an initial displacement $f(x)$ and an initial velocity $g(x)$. Write the solution $y_f(x,t)$ for the problem with initial displacement f and zero initial velocity and the solution $y_g(x,t)$ for the problem with zero initial displacement and initial velocity $g(x)$. Then the solution for the current problem is

$$y(x,t) = y_f(x,t) + y_g(x,t).$$

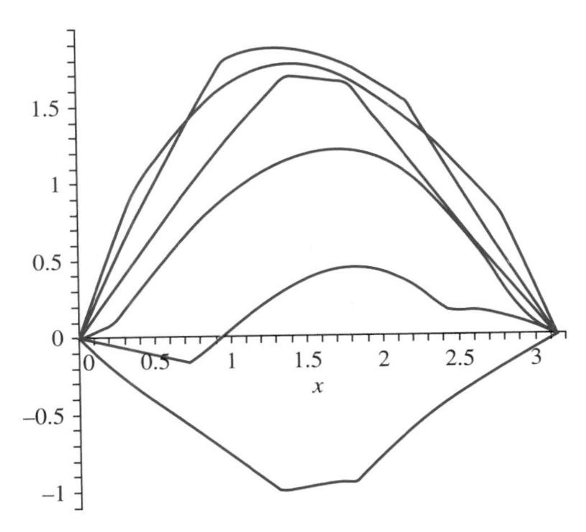

FIGURE 16.5 *Waves in Example 16.4.*

This function satisfies the wave equation and boundary conditions, because both $y_f(x,t)$ and $y_g(x,t)$ do. Furthermore,

$$y(x,0) = y_f(x,0) + y_g(x,0) = f(x) + 0 = f(x)$$

and

$$\frac{\partial y}{\partial t}(x,0) = \frac{\partial y_f}{\partial t}(x,0) + \frac{\partial y_g}{\partial t}(x,0) = 0 + g(x) = g(x).$$

EXAMPLE 16.4

We will solve the wave equation on $[0, L]$ with $y(0,t) = y(L,t) = 0$, subject to initial position

$$f(x) = \begin{cases} x & \text{for } 0 \le x \le L/2 \\ L - x & \text{for } L/2 \le x \le L. \end{cases}$$

and initial velocity $g(x) = x(1 + \cos(\pi x/L))$. The solution $y(x,t)$ is the sum of the solutions of the problems solved in Examples 16.1 and 16.3. If $c = 1$ and $L = \pi$, this solution is

$$y(x,t) = \sum_{n=1}^{\infty} \frac{4}{n^2\pi} \sin(n\pi/2)\sin(nx)\cos(nt)$$
$$+ \frac{3}{2}\sin(x)\sin(t) + \sum_{n=2}^{\infty} \frac{2(-1)^n}{n^2(n^2-1)} \sin(nx)\sin(nt).$$

In Figure 16.5, the string's position at $t = 0.2$ is the fourth graph from the top. The wave moves upward to the next graph at $t = 0.6$, then upward again at $t = 1.2$. The highest wave is at $t = 1.8$. Following this, the wave is partly below the horizontal axis at $t = 2.3$, and completely below this axis at $t = 2.9$. ♦

16.2.4 Influence of Constants and Initial Conditions

It is interesting to observe how the initial conditions and the constant c influence the wave motion.

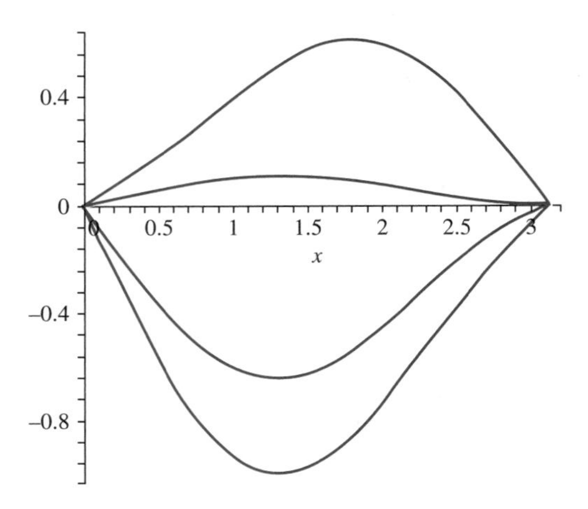

FIGURE 16.6 *Wave profiles for different values of c in Example 16.5.*

EXAMPLE 16.5

In Example 16.3, we solved the problem with zero initial displacement and initial velocity given by $g(x) = x(1 + \cos(\pi x/L))$. If $L = \pi$ this solution is

$$y(x,t) = \frac{3}{2c}\sin(x)\sin(ct) + \sum_{n=2}^{\infty} \frac{2(-1)^n}{cn^2(n^2-1)} \sin(nx)\sin(nct).$$

Figure 16.3 showed wave profiles at various times for $c = 1$. Figure 16.6 shows wave profiles at time $t = 5.3$, with $c = 1.05$, $c = 1.1$, $c = 1.2$ and $c = 1.65$. These increase in amplitude as c increases. ♦

EXAMPLE 16.6

We will examine the effects of changes in an initial condition on the wave motion with the problem

$$\frac{\partial y^2}{\partial t^2} = 1.44 \frac{\partial y^2}{\partial x^2} \text{ for } 0 < x < \pi, t > 0,$$
$$y(0,t) = y(\pi,t) = 0 \text{ for } t \geq 0,$$

and

$$y(x,0) = 0, \ \frac{\partial y}{\partial t}(x,0) = \sin(\epsilon x) \text{ for } 0 < x < \pi$$

in which ϵ is a positive number that is not an integer.

Use equation (16.6) to write the solution

$$y(x,t) = \frac{5}{3\pi} \sum_{n=1}^{\infty} \frac{\sin(\pi\epsilon)(-1)^{n+1}}{n^2 - \epsilon^2} \sin(nx)\sin(1.2t).$$

To gauge the effect of ϵ on the motion, compare graphs of this solution for different values of ϵ at given times. Figure 16.7 shows the wave profile at $t = 0.5$ for ϵ equal to 0.7 (wave above the x-axis), 1.5 (wave below the axis), and 9.3 (wave oscillating rapidly). ♦

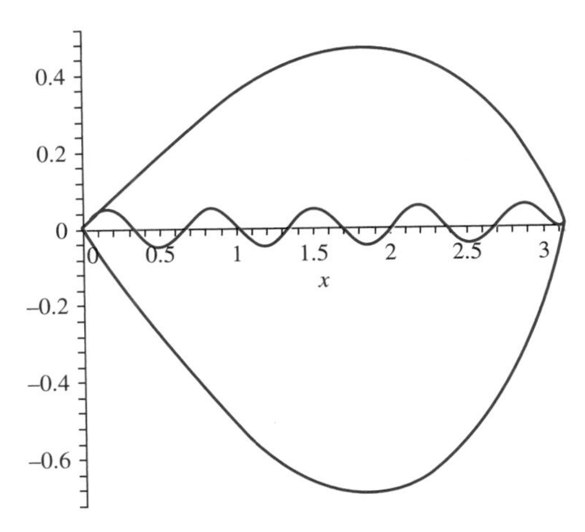

FIGURE 16.7 *Wave profiles in Example 16.6, decreasing as ϵ increases.*

16.2.5 Wave Motion with a Forcing Term

Separation of variables may fail if the partial differential equation contains terms allowing for some type of external forcing, or if the boundary conditions are nonhomogeneous. In such a case it may be possible to transform the initial-boundary value problem to one that we know how to solve.

EXAMPLE 16.7

We will solve the problem

$$\frac{\partial^2 y}{\partial t^2} = \frac{\partial y^2}{\partial x^2} + Ax \text{ for } 0 < x < L, t > 0,$$

$$y(0,t) = y(L,t) = 0 \text{ for } t \geq 0,$$

and

$$y(x,0) = 0, \ \frac{\partial y}{\partial t}(x,0) = 1 \text{ for } 0 < x < L.$$

A is a positive constant and the term Ax in the wave equation represents an external force having magnitude Ax at x. We have let $c = 1$ in this problem.

If we put $y(x,t) = X(x)T(t)$ into the partial differential equation we obtain

$$XT'' = X''T + Ax,$$

and there is no way to separate the t dependency on one side of an equation and the x dependency on the other. One strategy in such a case is to try to transform this problem into one to which separation of variables applies. Let

$$y(x,t) = Y(x,t) + \psi(x).$$

The idea is to choose ψ to obtain a problem for Y that we can solve. Substitute $y(x,t)$ into the partial differential equation to get

$$\frac{\partial^2 Y}{\partial t^2} = \frac{\partial^2 Y}{\partial x^2} + \psi''(x) + Ax.$$

This will be simplified if we choose ψ so that

$$\psi''(x) + Ax = 0.$$

Integrate twice to get

$$\psi(x) = -A\frac{x^3}{6} + Cx + D,$$

with C and D constants of integration. To find C and D look at the boundary conditions. First,

$$y(0,t) = Y(0,t) + \psi(0) = 0.$$

This will be just $y(0,t) = Y(0,t)$ if we make

$$\psi(0) = 0.$$

This requires that we choose $D = 0$. Next,

$$y(L,t) = Y(L,t) + \psi(L) = Y(L,t) - A\frac{L^3}{6} + CL = 0.$$

This reduces to $y(L,t) = Y(L,t)$ if we choose C so that

$$\psi(L) = -A\frac{L^3}{6} + CL = 0.$$

With $C = AL^2/6$, we have

$$\psi(x) = \frac{1}{6}Ax(L^2 - x^2).$$

With this ψ,

$$Y(0,t) = Y(L,t) = 0.$$

Next relate the initial conditions for y to initial conditions for Y. First,

$$Y(x,0) = y(x,0) - \psi(x) = -\psi(x) = \frac{1}{6}Ax(x^2 - L^2),$$

then

$$\frac{\partial Y}{\partial t}(x,0) = \frac{\partial y}{\partial t}(x,0) = 1.$$

Now we have an initial-boundary value problem for Y:

$$\frac{\partial^2 Y}{\partial t^2} = \frac{\partial Y^2}{\partial x^2} \text{ for } 0 < x < L, t > 0,$$

$$Y(0,t) = Y(L,t) = 0 \text{ for } t \geq 0,$$

and

$$Y(x,0) = \frac{1}{6}Ax(x^2 - L^2), \frac{\partial Y}{\partial t}(x,0) = 1 \text{ for } 0 < x < L.$$

We know the solution of this problem. By equations (16.5) and (16.6),

$$Y(x,t) = \frac{2}{L}\sum_{n=1}^{\infty}\left(\int_0^L \frac{1}{6}A\xi(\xi^2 - L^2)\sin(n\pi\xi/L)\,d\xi\right)\sin(n\pi x/L)\cos(n\pi t/L)$$

$$+ \frac{2}{\pi}\sum_{n=1}^{\infty}\frac{1}{n}\left(\int_0^L \sin(n\pi\xi/L)\,d\xi\right)\sin(n\pi x/L)\sin(n\pi t/L)$$

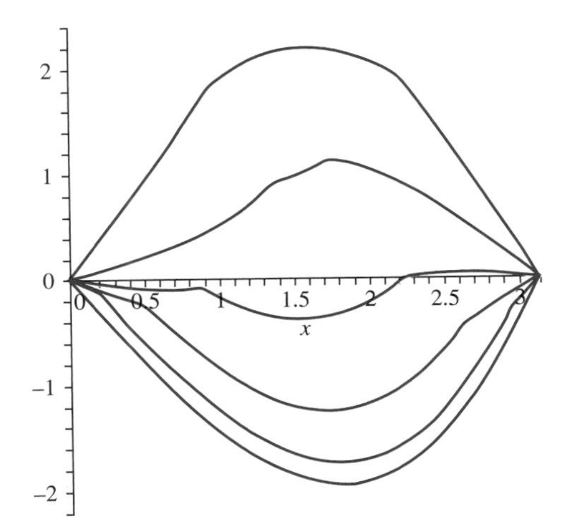

FIGURE 16.8 *Wave profiles in Example 16.7.*

$$= \frac{2AL^3}{\pi^3} \sum_{n=1}^{\infty} \frac{(-1)^n}{n^3} \sin(n\pi x/L)\cos(n\pi t/L)$$

$$+ \frac{2L}{\pi^2} \sum_{n=1}^{\infty} \frac{1-(-1)^n}{n^2} \sin(n\pi x/L)\sin(n\pi t/L).$$

The solution of the original problem is

$$y(x,t) = Y(x,t) + \frac{1}{6}Ax(L^2 - x^2).$$

Figure 16.8 shows wave profiles for $c = 1$ and $L = \pi$ at times $t = 0.03, 0.2, 0.5, 0.9, 1.4$, and 2.2. The waves move upward as t increases over these times. ♦

SECTION 16.2 PROBLEMS

In each of Problems 1 through 8, solve the initial-boundary value problem using separation of variables. Graph the fiftieth partial sum of the solution for some values of t, with $c = 1$ if c is unspecified in the problem.

1. $\frac{\partial^2 y}{\partial t^2} = 4\frac{\partial^2 y}{\partial x^2}$ for $0 < x < 3, t > 0$
   $y(0,t) = y(3,t) = 0$ for $t \geq 0$
   $y(x,0) = 0, \frac{\partial y}{\partial t}(x,0) = x(3-x)$ for $0 \leq x \leq 3$

2. $\frac{\partial^2 y}{\partial t^2} = 25\frac{\partial^2 y}{\partial x^2}$ for $0 < x < \pi, t > 0$
   $y(0,t) = y(\pi,t) = 0$ for $t \geq 0$
   $y(x,0) = \sin(2x), \frac{\partial y}{\partial t}(x,0) = \pi - x$ for $0 \leq x \leq \pi$

3. $\frac{\partial^2 y}{\partial t^2} = 9\frac{\partial^2 y}{\partial x^2}$ for $0 < x < 2, t > 0$
   $y(0,t) = y(2,t) = 0$ for $t \geq 0$
   $y(x,0) = x(x-2), \frac{\partial y}{\partial t}(x,0) = g(x)$ for $0 \leq x \leq 2$
   where $g(x) = \begin{cases} 0 & \text{for } 0 \leq x < 1/2 \text{ and } 1 < x \leq 2 \\ 3 & \text{for } 1/2 \leq x \leq 1. \end{cases}$

4. $\frac{\partial^2 y}{\partial t^2} = 9\frac{\partial^2 y}{\partial x^2}$ for $0 < x < 4, t > 0$
   $y(0,t) = y(4,t) = 0$ for $t \geq 0$
   $y(x,0) = 2\sin(\pi x), \frac{\partial y}{\partial t}(x,0) = 0$ for $0 \leq x \leq 4$

5. $\frac{\partial^2 y}{\partial t^2} = c^2\frac{\partial^2 y}{\partial x^2}$ for $0 < x < 2, t > 0$

$y(0,t) = y(2,t) = 0$ for $t \geq 0$

$y(x,0) = 0, \dfrac{\partial y}{\partial t}(x,0) = g(x)$ for $0 \leq x \leq 2$

where $g(x) = \begin{cases} 2x & \text{for } 0 \leq x \leq 1 \\ 0 & \text{for } 1 < x \leq 2. \end{cases}$

6. $\dfrac{\partial^2 y}{\partial t^2} = 9\dfrac{\partial^2 y}{\partial x^2}$ for $0 < x < \pi, t > 0$

$y(0,t) = y(\pi,t) = 0$ for $t \geq 0$

$y(x,0) = \sin(x), \dfrac{\partial y}{\partial t}(x,0) = 1$ for $0 \leq x \leq \pi$

7. $\dfrac{\partial^2 y}{\partial t^2} = 8\dfrac{\partial^2 y}{\partial x^2}$ for $0 < x < 2\pi, t > 0$

$y(0,t) = y(2\pi,t) = 0$ for $t \geq 0$

$y(x,0) = f(x), \dfrac{\partial y}{\partial t}(x,0) = 0$ for $0 \leq x \leq 2\pi$

where $f(x) = \begin{cases} 3x & \text{for } 0 \leq x \leq \pi \\ 6\pi - 3x & \text{for } \pi < x \leq 2\pi. \end{cases}$

8. $\dfrac{\partial^2 y}{\partial t^2} = 4\dfrac{\partial^2 y}{\partial x^2}$ for $0 < x < 5, t > 0$

$y(0,t) = y(5,t) = 0$ for $t \geq 0$

$y(x,0) = 0, \dfrac{\partial y}{\partial t}(x,0) = g(x)$ for $0 \leq x \leq 5$

where $g(x) = \begin{cases} 0 & \text{for } 0 \leq x < 4 \\ 5 - x & \text{for } 4 \leq x \leq 5. \end{cases}$

9. Solve

$$\frac{\partial^2 y}{\partial t^2} = \frac{\partial^2 y}{\partial x^2} - \cos(x) \text{ for } 0 < x < 2\pi, t > 0$$

$$y(0,t) = y(2\pi,t) = 0 \text{ for } t \geq 0$$

$$y(x,0) = 0, \frac{\partial y}{\partial t}(x,0) = 0 \text{ for } 0 \leq x \leq 2\pi.$$

Graph the fortieth partial sum for some values of the time.

10. Transverse vibrations in a homogeneous rod of length π are modeled by the partial differential equation

$$a^4 \frac{\partial^4 u}{\partial x^4} + \frac{\partial^2 u}{\partial t^2} = 0 \text{ for } 0 < x < \pi, t > 0.$$

Here $u(x,t)$ is the displacement at time t of the cross section through x perpendicular to the x-axis and $a^2 = EI/\rho A$, where E is *Young's modulus*, I is the moment of inertia of this cross section, ρ is the constant density and A is the constant cross-sectional area.

(a) Let $u(x,t) = X(x)T(t)$ to separate the variables.

(b) Solve for values of the separation constant and for X and T in the case of free ends, in which

$$\frac{\partial^2 u}{\partial x^2}(0,t) = \frac{\partial^2 u}{\partial x^2}(\pi,t) = \frac{\partial^3 u}{\partial x^3}(0,t) = \frac{\partial^3 y}{\partial x^3}(\pi,t) = 0$$

for $t > 0$.

(c) Solve for the separation constant and for X and T in the case of supported ends in which

$$u(0,t) = u(\pi,t) = \frac{\partial^2 u}{\partial x^2}(0,t) = \frac{\partial^2 u}{\partial x^2}(\pi,t) = 0.$$

11. Solve the *telegraph equation*

$$\frac{\partial^2 u}{\partial t^2} + A\frac{\partial u}{\partial t} + Bu = c^2\frac{\partial^2 u}{\partial x^2}$$

for $0 < x < L, t > 0$. A and B are positive constants. The boundary conditions are

$$u(0,t) = u(L,t) = 0 \text{ for } t \geq 0,$$

and the initial conditions are

$$u(x,0) = f(x), \frac{\partial u}{\partial t}(x,0) = 0 \text{ for } 0 \leq x \leq L.$$

Assume that $A^2L^2 < 4(BL^2 + c^2\pi^2)$.

12. (a) Write a series solution for

$$\frac{\partial^2 y}{\partial t^2} = 9\frac{\partial^2 y}{\partial x^2} + 5x^3 \text{ for } 0 < x < 4, t > 0$$

$$y(0,t) = y(4,t) = 0 \text{ for } t \geq 0$$

$$y(x,0) = \cos(\pi x), \frac{\partial y}{\partial t}(x,0) = 0 \text{ for } 0 \leq x \leq 4.$$

(b) Write a series solution when the forcing term $5x^3$ is removed.

(c) In order to gauge the effect of the forcing term on the motion, graph the fortieth partial sum of the solutions in parts (a) and (b) on the same set of axes when $t = 0.4$ seconds. Repeat this for $t = 0.8, 1.4, 2, 2.5, 3$, and 4.

13. (a) Write a series solution for

$$\frac{\partial^2 y}{\partial t^2} = 9\frac{\partial^2 y}{\partial x^2} - e^{-x} \text{ for } 0 < x < 4, t > 0$$

$$y(0,t) = y(4,t) = 0 \text{ for } t \geq 0$$

$$y(x,0) = \sin(\pi x), \frac{\partial y}{\partial t}(x,0) = 0 \text{ for } 0 \leq x \leq 4.$$

(b) Write a series solution when the forcing term e^{-x} is removed.

(c) In order to gauge the effect of the forcing term on the motion, graph the fortieth partial sum of the solutions in parts (a) and (b) on the same set of axes when $t = 0.4$ seconds. Repeat this for $t = 0.8, 1.4, 2, 2.5, 3$, and 4.

14. (a) Write a series solution for

$$\frac{\partial^2 y}{\partial t^2} = 9\frac{\partial^2 y}{\partial x^2} + \cos(\pi x) \text{ for } 0 < x < 4, t > 0$$

$$y(0,t) = y(4,t) = 0 \text{ for } t \geq 0$$

$$y(x,0) = x(4-x), \frac{\partial y}{\partial t}(x,0) = 0 \text{ for } 0 \leq x \leq 4.$$

(b) Write a series solution when the forcing term $\cos(\pi x)$ is removed.

(c) In order to gauge the effect of the forcing term on the motion, graph the fortieth partial sum of the solutions in (a) and (b) on the same set of axes when $t = 0.6$ seconds, then for $t = 1, 1.4, 2, 3, 5$ and 7.

15. Solve the initial-boundary value problem

$$\frac{\partial^2 y}{\partial t^2} = 3\frac{\partial^2 y}{\partial x^2} + 2x \text{ for } 0 < x < 2, t > 0$$

$$y(0, t) = y(2, t) = 0 \text{ for } t \geq 0$$

$$y(x, 0) = 0, \frac{\partial y}{\partial t}(x, 0) = 0 \text{ for } 0 \leq x \leq 2.$$

Graph the fortieth partial sum of the solution for some values of the time.

16. Solve

$$\frac{\partial^2 y}{\partial t^2} = 9\frac{\partial^2 y}{\partial x^2} + x^2 \text{ for } 0 < x < 4, t > 0$$

$$y(0, t) = y(4, t) = 0 \text{ for } t \geq 0$$

$$y(x, 0) = 0, \frac{\partial y}{\partial t}(x, 0) = 0 \text{ for } 0 \leq x \leq 4.$$

Graph the fortieth partial sum of the solution for selected values of t,

16.3 Wave Motion in an Infinite Medium

If great distances are involved (as with sound waves through the ocean or cosmic background radiation across the universe), wave motion is often modeled by the wave equation on $-\infty < x < \infty$. In this case, there is no boundary, hence no boundary condition. However, we seek bounded solutions.

The analysis is similar to that for solutions on a closed interval, except that $\int_{-\infty}^{\infty} \cdots d\omega$ replaces $\sum_{n=1}^{\infty}$. As we did on a bounded interval, we will consider separately the cases of zero initial velocity and no initial displacement.

Zero Initial Velocity The initial-boundary value problem is

$$\frac{\partial^2 y}{\partial t^2} = c^2\frac{\partial^2 y}{\partial x^2} \text{ for } -\infty < x < \infty, t > 0,$$

$$y(x, 0) = f(x), \frac{\partial y}{\partial t}(x, 0) = 0 \text{ for } -\infty < x < \infty.$$

Separate variables by putting $y(x, t) = X(x)T(t)$. Exactly as with wave motion on a bounded interval, we obtain

$$X'' + \lambda X = 0, T'' + \lambda c^2 T = 0.$$

There are three cases on λ.

Case 1: If $\lambda = 0$ then $X = ax + b$

This is bounded if $a = 0$, so 0 is an eigenvalue with constant eigenfunctions.

Case 2: If $\lambda < 0$, write $\lambda = -\omega^2$ with $\omega > 0$

Then $X(x) = c_1 e^{\omega x} + c_2 e^{-\omega x}$, and this function is unbounded on the entire real line if either constant is nonzero. This problem has no negative eigenvalue.

Case 3: If $\lambda > 0$, write $\lambda = \omega^2$ with $\omega > 0$

Now $X(x) = c_1 \cos(\omega x) + c_2 \sin(\omega x)$, a bounded function for any choices of positive ω. Every positive number $\lambda = \omega^2$ is an eigenvalue, with eigenfunctions of the form $X_\omega = c_1 \cos(\omega x) + c_2 \sin(\omega x)$.

We can consolidate cases 1 and 3 by allowing $\lambda = 0$ in case 3.

The equation for T is $T'' + c^2\omega^2 T = 0$ with solutions of the form

$$T_\omega(t) = a\cos(\omega ct) + b\sin(\omega ct).$$

But

$$\frac{\partial y}{\partial t}(x,0) = X(x)T'(0) = X(x)\omega cb = 0,$$

and this is satisfied if we choose $b = 0$. This leaves $T_\omega(t)$ as a constant multiple of $\cos(\omega ct)$.

Thus far, for every $\omega \geq 0$, we have functions

$$y_\omega(x,t) = X_\omega(x)T_\omega(t) = [a_\omega\cos(\omega x) + b_\omega\cos(\omega x)]\cos(\omega ct),$$

which satisfy the wave equation and the initial condition $(\partial y/\partial t)(x,0) = 0$ for all x. We need to satisfy the initial condition $y(x,0) = f(x)$. For the similar problem on a bounded interval, we attempted a superposition $\sum_{n=1}^{\infty} y_n(x,t)$. Now the eigenvalues fill out the entire nonnegative real line, so the superposition has the form $\int_0^\infty y_\omega(x,t)d\omega$. Thus, attempt a solution

$$y(x,t) = \int_0^\infty [a_\omega\cos(\omega x) + b_\omega\sin(\omega x)]\cos(\omega ct)\,d\omega. \qquad (16.7)$$

The initial condition requires that

$$y(x,0) = \int_0^\infty [a_\omega\cos(\omega x) + b_\omega\sin(\omega x)]\,d\omega = f(x).$$

This is a Fourier integral expansion of $f(x)$ on the real line, so choose a_ω and b_ω as the Fourier integral coefficients of f:

$$a_\omega = \frac{1}{\pi}\int_{-\infty}^{\infty} f(\xi)\cos(\omega\xi)\,d\xi$$

and

$$b_\omega = \frac{1}{\pi}\int_{-\infty}^{\infty} f(\xi)\sin(\omega\xi)\,d\xi.$$

With this choice of coefficients, equation (16.7) is the solution.

EXAMPLE 16.8

We will solve the problem

$$\frac{\partial^2 y}{\partial t^2} = c^2\frac{\partial^2 y}{\partial x^2} \text{ for } -\infty < x < \infty, t > 0,$$

$$y(x,0) = e^{-|x|},\ \frac{\partial y}{\partial t}(x,0) = 0 \text{ for } -\infty < x < \infty.$$

Compute the Fourier coefficients of the initial position function. First,

$$a_\omega = \frac{1}{\pi}\int_{-\infty}^{\infty} e^{-|\xi|}\cos(\omega\xi)\,d\xi = \frac{2}{\pi(1+\omega^2)}.$$

Because $e^{-|\xi|}\sin(\omega\xi)$ is an odd function of ξ, $b_\omega = 0$. The solution is

$$y(x,t) = \frac{2}{\pi}\int_0^\infty \frac{1}{1+\omega^2}\cos(\omega x)\cos(\omega ct)\,d\omega. \ \blacklozenge$$

The solution (16.7) is sometimes seen in a different form. If we insert the integrals for the coefficients into the solution, we have

$$y(x,t)=\int_0^\infty [a_\omega \cos(\omega x)+b_\omega \sin(\omega x)]\cos(\omega ct)\,d\omega$$
$$=\frac{1}{\pi}\int_0^\infty \left[\left(\int_{-\infty}^\infty f(\xi)\cos(\omega\xi)\,d\xi\right)\cos(\omega x)\right.$$
$$\left.+\left(\int_{-\infty}^\infty f(\xi)\sin(\omega\xi)\,d\xi\right)\sin(\omega x)\right]\cos(\omega ct)\,d\omega$$
$$=\frac{1}{\pi}\int_{-\infty}^\infty\int_0^\infty [\cos(\omega\xi)\cos(\omega x)+\sin(\omega\xi)\sin(\omega x)]f(\xi)\cos(\omega ct)\,d\omega\,d\xi.$$

Upon applying a trigonometric identity to the term in square brackets, we have

$$y(x,t)=\frac{1}{\pi}\int_{-\infty}^\infty\int_0^\infty \cos(\omega(\xi-x))f(\xi)\cos(\omega ct)\,d\omega\,d\xi. \tag{16.8}$$

We will use this form of the solution when we solve this problem using the Fourier transform.

Zero Initial Displacement We will solve

$$\frac{\partial^2 y}{\partial t^2}=c^2\frac{\partial^2 y}{\partial x^2} \text{ for } -\infty<x<\infty, t>0$$

and

$$y(x,0)=0,\ \frac{\partial y}{\partial t}(x,0)=g(x) \text{ for } -\infty<x<\infty.$$

Letting $y(x,t)=X(x)T(t)$, the analysis proceeds exactly as in the case of zero initial velocity, except now we find that $T_\omega(t)=\sin(\omega ct)$ because $T(0)=0$ instead of $T'(0)=0$. For $\omega\geq 0$, we have functions

$$y_\omega(x,t)=[a_\omega\cos(\omega x)+b_\omega\sin(\omega x)]\sin(\omega ct),$$

which satisfy the wave equation and the condition $y(x,0)=0$. To satisfy $(\partial y/\partial t)(x,0)=g(x)$, attempt a superposition

$$y(x,t)=\int_0^\infty [a_\omega\cos(\omega x)+b_\omega\sin(\omega x)]\sin(\omega ct)\,d\omega. \tag{16.9}$$

Compute

$$\frac{\partial y}{\partial t}(x,t)=\int_0^\infty [a_\omega\cos(\omega x)+b_\omega\sin(\omega x)]\omega c\cos(\omega ct)\,d\omega.$$

We must choose the coefficients so that

$$\frac{\partial y}{\partial t}(x,0)=\int_0^\infty \omega c[a_\omega\cos(\omega x)+b_\omega\sin(\omega x)]\,d\omega=g(x).$$

We can do this by choosing ωca_ω and ωcb_ω as the Fourier coefficients in the integral expansion of g. Thus, let

$$a_\omega=\frac{1}{\pi c\omega}\int_{-\infty}^\infty g(\xi)\cos(\omega\xi)\,d\xi \text{ and } b_\omega=\frac{1}{\pi c\omega}\int_{-\infty}^\infty g(\xi)\sin(\omega\xi)\,d\xi.$$

With these choices, equation (16.9) is the solution of the initial-boundary value problem.

EXAMPLE 16.9

Suppose the initial displacement is zero and the initial velocity is given by

$$g(x)=\begin{cases} e^x & \text{for } 0\leq x\leq 1\\ 0 & \text{for } x<0 \text{ and for } x>1.\end{cases}$$

Compute the coefficients

$$a_\omega = \frac{1}{\pi c\omega}\int_{-\infty}^{\infty} g(\xi)\cos(\omega\xi)\,d\xi = \frac{1}{\pi c\omega}\int_0^1 e^\xi \cos(\omega\xi)\,d\xi$$
$$= \frac{1}{\pi c\omega}\frac{e\cos(\omega)+e\omega\sin(\omega)-1}{1+\omega^2}$$

and

$$b_\omega = \frac{1}{\pi c\omega}\int_{-\infty}^{\infty} g(\xi)\sin(\omega\xi)\,d\xi = \frac{1}{\pi c\omega}\int_0^1 e^\xi \sin(\omega\xi)\,d\xi$$
$$= -\frac{1}{\pi c\omega}\frac{e\omega\cos(\omega)-e\sin(\omega)-\omega}{1+\omega^2}.$$

The solution is

$$y(x,t) = \int_0^\infty \left(\frac{1}{\pi c\omega}\frac{e\cos(\omega)+e\omega\sin(\omega)-1}{1+\omega^2}\right)\cos(\omega x)\sin(\omega ct)\,d\omega$$
$$-\int_0^\infty \left(\frac{1}{\pi c\omega}\frac{e\omega\cos(\omega)-e\sin(\omega)-\omega}{1+\omega^2}\right)\sin(\omega x)\sin(\omega ct)\,d\omega. \blacklozenge$$

As with motion on a bounded interval, the general initial-boundary value problem is solved by adding the solution with no initial velocity to the solution with no initial displacement.

Solution by Fourier Transform

We will illustrate the use of the Fourier transform to solve the initial-boundary value problem for the wave equation on the real line. Let the initial displacement be $f(x)$ and the initial velocity, $g(x)$.

Begin by taking the Fourier transform of the wave equation. This transform must be taken with respect to x, since $-\infty < x < \infty$, the appropriate range of values for this transform. Throughout this transform process, t is carried along as a symbol. Applying the transform, we have

$$\mathcal{F}\left[\frac{\partial^2 y}{\partial t^2}\right](\omega) = c^2\mathcal{F}\left[\frac{\partial^2 y}{\partial x^2}\right](\omega). \tag{16.10}$$

On the right side of equation (16.10), we must transform a second derivative with respect to x, the variable of the function being transformed. Use the operational rule for the Fourier transform to write

$$\mathcal{F}\left[\frac{\partial^2 y}{\partial x^2}\right](\omega) = (i\omega)^2\hat{y}(\omega,t) = -\omega^2\hat{y}(\omega,t).$$

For the left side of equation (16.10), the derivative with respect to t passes through the transform in the x-variable:

$$\mathcal{F}\left[\frac{\partial^2 y}{\partial t^2}\right](\omega) = \int_{-\infty}^{\infty}\frac{\partial^2 y}{\partial t^2}(x,t)e^{-i\omega x}\,dx$$
$$= \frac{\partial^2}{\partial t^2}\int_{-\infty}^{\infty} y(x,t)e^{-i\omega x}\,dx = \frac{\partial^2}{\partial t^2}\hat{f}(\omega,t).$$

The transformed wave equation is

$$\frac{\partial^2}{\partial t^2}\hat{y}(\omega,t) = -c^2\omega^2\hat{y}(x,t).$$

Write this differential equation as

$$\frac{\partial^2}{\partial t^2}\hat{y}(\omega,t)+c^2\omega^2\hat{y}(x,t)=0.$$

Think of this as an ordinary differential equation for $\hat{y}(\omega,t)$ with t as the variable and ω carried along as a parameter. The general solution is

$$\hat{y}(\omega,t)=a_\omega\cos(\omega ct)+b_\omega\sin(\omega ct).$$

To solve for the coefficients to satisfy the initial conditions, transform the initial data. First,

$$\hat{y}(\omega,0)=a_\omega=\mathcal{F}[y(x,0)](\omega)=\mathcal{F}[f(x)](\omega)=\hat{f}(\omega),$$

the transform of the initial position function. Next,

$$\begin{aligned}\frac{\partial\hat{y}}{\partial t}(\omega,0)=c\omega b_\omega&=\mathcal{F}\left[\frac{\partial y}{\partial t}(x,0)\right](\omega,0)\\&=\mathcal{F}[g(x)](\omega)=\hat{g}(\omega),\end{aligned}$$

which is the transform of the initial velocity function. Therefore,

$$b_\omega=\frac{1}{\omega c}\hat{g}(\omega).$$

Now

$$\hat{y}(\omega,t)=\hat{f}(\omega)\cos(\omega ct)+\frac{1}{\omega c}\hat{g}(\omega)\sin(\omega ct).$$

This is the Fourier transform of the solution. Invert this to find the solution

$$y(x,t)=\frac{1}{2\pi}\int_{-\infty}^{\infty}\left[\hat{f}(\omega)\cos(\omega ct)+\frac{1}{\omega c}\hat{g}(\omega)\sin(\omega ct)\right]e^{i\omega x}\,d\omega.$$

In the case that the string is released from rest, $g(x)=0$ and this solution is

$$y(x,t)=\frac{1}{2\pi}\int_{-\infty}^{\infty}\hat{f}(\omega)\cos(\omega ct)e^{i\omega x}\,d\omega. \tag{16.11}$$

We claim that this solution (16.11) obtained by using the Fourier transform agrees with the solution (16.8) obtained by separation of variables and the Fourier integral. For the moment, denote the solution (16.11) by Fourier transform as $y_{tr}(x,t)$. Manipulate $y_{tr}(x,t)$ as follows:

$$\begin{aligned}y_{tr}(x,t)&=\frac{1}{2\pi}\int_{-\infty}^{\infty}\hat{f}(\omega)\cos(\omega ct)e^{i\omega x}\,d\omega\\&=\frac{1}{2\pi}\int_{-\infty}^{\infty}\left(\int_{-\infty}^{\infty}f(\xi)e^{-i\omega\xi}\,d\xi\right)\cos(\omega ct)e^{i\omega x}\,d\omega\\&=\frac{1}{2\pi}\int_{-\infty}^{\infty}\int_{-\infty}^{\infty}e^{-i\omega(\xi-x)}\cos(\omega ct)f(\xi)\,d\omega\,d\xi\\&=\frac{1}{2\pi}\int_{-\infty}^{\infty}\int_{-\infty}^{\infty}[\cos(\omega(\xi-x))-i\sin(\omega(\xi-x))]\cos(\omega ct)f(\xi)\,d\omega\,d\xi.\end{aligned}$$

Since $y(x,t)$ must be real-valued, the solution is actually the real part of this integral. Therefore,

$$y_{tr}(x,t)=\frac{1}{2\pi}\int_{-\infty}^{\infty}\int_{-\infty}^{\infty}\cos(\omega(\xi-x))\cos(\omega ct)f(\xi)\,d\omega\,d\xi.$$

Finally, the integrand is an even function of ω, so

$$y_{tr}(x,t)=\frac{1}{\pi}\int_{-\infty}^{\infty}\int_{0}^{\infty}\cos(\omega(\xi-x))\cos(\omega ct)f(\xi)\,d\omega\,d\xi,$$

and this is the solution (16.8).

EXAMPLE 16.10

We will use the Fourier transform to solve the wave equation on the real-line subject to the initial conditions $g(x) = 0$ and

$$f(x) = \begin{cases} \cos(x) & \text{for } -\pi/2 \le x \le \pi/2 \\ 0 & \text{for } |x| > \pi/2. \end{cases}$$

With zero initial velocity, we can use the solution (16.11). We need the Fourier transform of the initial position function:

$$\hat{f}(\omega) = \int_{-\infty}^{\infty} f(\xi) e^{-i\omega\xi}\, d\xi$$

$$= \int_{-\pi/2}^{\pi/2} \cos(\xi) e^{-i\omega\xi}\, d\xi = \begin{cases} 2\cos(\pi\omega/2)/(1-\omega^2) & \text{for } \omega \ne 1 \\ \pi/2 & \text{for } \omega = 1. \end{cases}$$

$\hat{f}$ is continuous because

$$\lim_{\omega \to 1} \frac{2\cos(\pi\omega/2)}{1-\omega^2} = \frac{\pi}{2}.$$

The solution is

$$y(x,t) = \text{Re}\left(\frac{1}{\pi} \int_{-\infty}^{\infty} \frac{2\cos(\pi\omega/2)}{1-\omega^2} \cos(\omega ct) e^{i\omega x}\, d\omega \right).$$

The integral is complex because the Fourier transform is a complex quantity. The real part is the solution of the problem. In this case, we can extract the real part by writing

$$e^{i\omega x} = \cos(\omega x) + i \sin(\omega x),$$

obtaining

$$y(x,t) = \frac{1}{\pi} \int_{-\infty}^{\infty} \frac{2\cos(\pi\omega/2)}{1-\omega^2} \cos(\omega x)\cos(\omega ct)\, d\omega. \quad \blacklozenge$$

SECTION 16.3 PROBLEMS

In each of Problems 1 through 6, solve the wave equation on the real line for the given initial position f and initial velocity g, first by separation of variables and the Fourier integral, and then by using the Fourier transform. The same solution should result from both methods.

1. $c = 4$, $f(x) = 0$, and $g(x) = \begin{cases} \sin(x) & \text{for } -\pi \le x \le \pi \\ 0 & \text{for } |x| > \pi \end{cases}$

2. $c = 1$, $g(x) = 0$, and $f(x) = \begin{cases} 2 - |x| & \text{for } -2 \le x \le 2 \\ 0 & \text{for } |x| > 2 \end{cases}$

3. $c = 3$, $f(x) = 0$, and $g(x) = \begin{cases} e^{-2x} & \text{for } x \ge 1 \\ 0 & \text{for } x < 1 \end{cases}$

4. $c = 2$, $f(x) = 0$, and
$$g(x) = \begin{cases} 1 & \text{for } 0 \le x \le 2 \\ -1 & \text{for } -2 \le x < 0 \\ 0 & \text{for } x > 2 \text{ and for } x < -2 \end{cases}$$

5. $c = 12$, $f(x) = e^{-5|x|}$, $g(x) = 0$

6. $c = 8$, $g(x) = 0$,
and $f(x) = \begin{cases} 8 - x & \text{for } 0 \le x \le 8 \\ 0 & \text{for } x < 0 \text{ and for } x > 8. \end{cases}$

16.4 Wave Motion in a Semi-Infinite Medium

We will solve the wave equation on a half-line $0 \le x < \infty$. The problem is

$$\frac{\partial^2 y}{\partial t^2} = c^2 \frac{\partial^2 y}{\partial x^2} \text{ for } 0 \le x < \infty, t > 0,$$

$$y(0,t) = 0 \text{ for } t \ge 0$$

and

$$y(x,0) = f(x), \frac{\partial y}{\partial t}(x,0) = g(x) \text{ for } 0 \le x < \infty.$$

We will seek a bounded solution, considering first the case that $g(x) = 0$. Separate variables by putting $y(x,t) = X(x)T(t)$ to obtain

$$X'' + \lambda X = 0 \text{ and } T'' + \lambda c^2 T = 0.$$

Unlike the problem on the entire line, we have a boundary condition at the fixed left end. This means that $y(0,t) = X(0)T(t) = 0$, so $X(0) = 0$, and the problem for X is

$$X'' + \lambda X = 0; \ X(0) = 0.$$

Upon considering cases as we have done before, we find the eigenvalues $\lambda \ge 0$ and eigenfunctions

$$X_\omega = b_\omega \sin(\omega x).$$

Because the motion is assumed to begin from rest,

$$\frac{\partial y}{\partial t}(x,0) = X(x)T'(0) = 0,$$

so $T'(0) = 0$. The problem for T is

$$T'' + c^2\omega^2 T = 0; \ T'(0) = 0$$

with solutions that are constant multiples of $\cos(\omega ct)$. For each $\omega \ge 0$, we now have a function

$$y_\omega(x,t) = b_\omega \sin(\omega x)\cos(\omega ct)$$

that satisfies the wave equation, the boundary condition and the initial condition that the string starts from rest. To satisfy the condition $y(x,0) = f(x)$, we generally need the superposition

$$y(x,t) = \int_0^\infty b_\omega \sin(\omega x)\cos(\omega ct).$$

Then

$$y(x,0) = \int_0^\infty b_\omega \sin(\omega x) = f(x),$$

so we must choose

$$b_\omega = \frac{2}{\pi}\int_0^\infty f(\xi)\sin(\omega\xi)\,d\xi,$$

which is the coefficient in the Fourier sine integral representation of f on $[0,\infty)$. This solves the problem when $g(x) = 0$. A similar analysis allows us to write an integral solution when $f(x) = 0$ and the string has an initial velocity of $(\partial y/\partial t)(x,0) = g(x)$.

EXAMPLE 16.11

We will solve the problem

$$\frac{\partial^2 y}{\partial t^2} = 16\frac{\partial^2 y}{\partial x^2} \text{ for } 0 \le x < \infty, t > 0,$$

$$y(0,t) = 0 \text{ for } t \ge 0$$

$$\frac{\partial y}{\partial t}(x,0) = 0 \text{ for } 0 \le x < \infty$$

and

$$y(x,0) = f(x) = \begin{cases} \sin(\pi x) & \text{for } 0 \le x \le 4 \\ 0 & \text{for } x > 4. \end{cases}$$

To write the solution, we need only compute the coefficients

$$\begin{aligned} b_\omega &= \frac{2}{\pi}\int_0^\infty f(\xi)\sin(\omega\xi)\,d\xi \\ &= \frac{2}{\pi}\int_0^4 \sin(\pi\xi)\sin(\omega\xi)\,d\xi = 8\sin(\omega)\cos(\omega)\frac{2\cos^2(\omega)-1}{\omega^2-\pi^2}. \end{aligned}$$

The solution is

$$y(x,t) = \int_0^\infty 8\sin(\omega)\cos(\omega)\frac{2\cos^2(\omega)-1}{\omega^2-\pi^2}\sin(\omega x)\cos(4\omega t)\,d\omega. \ ◆$$

16.4.1 Solution by Fourier Sine or Cosine Transform

We will illustrate the use of a Fourier transform to solve the problem on a half-line. Having considered the case of zero initial velocity for the separation of variables solution, we will assume here an initial velocity $g(x)$ but zero initial position $f(x) = 0$. The problem is

$$\frac{\partial^2 y}{\partial t^2} = c^2\frac{\partial^2 y}{\partial x^2} \text{ for } 0 \le x < \infty, t > 0,$$

$$y(0,t) = 0 \text{ for } t \ge 0$$

and

$$y(x,0) = 0, \ \frac{\partial y}{\partial t}(x,0) = g(x) \text{ for } 0 \le x < \infty.$$

On the half-line, we can try a Fourier sine or a Fourier cosine transform of $y(x,t)$, thinking of x as the variable and t as a parameter. The choice depends on the operational rules for these transforms. The cosine transform requires that we know something about the derivative (with respect to x) of y at $x = 0$, while the sine transform requires information about the function itself at $x = 0$. Since we are given that $y(0,t) = 0$, this is the kind of information we have, so choose the sine transform. Apply $\mathcal{F}_S$ to the wave equation and use the fact that the derivative with respect to t goes through the transform to get

$$\frac{\partial^2}{\partial t^2}\hat{y}_S = c^2\mathcal{F}_S\left[\frac{\partial^2 y}{\partial x^2}\right] = -c^2\omega^2\hat{y}_S(\omega,t) + \omega c^2 y(0,t) = -c^2\omega^2\hat{y}_S(\omega,t).$$

This is a differential equation for $\hat{y}_S(\omega,t)$ with t as the variable:

$$\frac{\partial^2\hat{y}_S}{\partial t^2} + c^2\omega^2\hat{y}_S(\omega,t) = 0.$$

The general solution is

$$\hat{y}_S(\omega, t) = a_\omega \cos(\omega ct) + b_\omega \sin(\omega ct).$$

Since

$$a_\omega = \hat{y}_S(\omega, 0) = \mathcal{F}_S[y(x,0)](\omega) = \mathcal{F}_S[0](\omega) = 0$$

and

$$\frac{\partial \hat{y}_S}{\partial t}(\omega, 0) = \omega c b_\omega = \hat{g}_S(\omega)$$

then

$$b_\omega = \frac{1}{\omega c}\hat{g}_S(\omega).$$

This gives us

$$\hat{y}_S(\omega, t) = \frac{1}{\omega c}\hat{g}_S(\omega)\sin(\omega ct),$$

which is the sine transform of the solution. We obtain the solution by inverting:

$$y(x,t) = \frac{2}{\pi}\int_0^\infty \frac{1}{\omega c}\hat{g}_S(\omega)\sin(\omega x)\sin(\omega ct)\,d\omega.$$

If we have a wave equation on the half-line with zero initial velocity and initial position given by f, then we can proceed as we have just done, but using the Fourier cosine transform instead of the sine transform. This is because the information given now fits within the framework of the operational rule for the cosine transform. As usual, a problem with initial displacement and velocity can be solved as the sum of the solution with zero initial velocity and the solution with zero initial position.

SECTION 16.4 PROBLEMS

In each of Problems 1 through 5, solve the problem for wave equation on the half-line for the given c, initial position f and initial velocity g, first by separation of variables, then by using an appropriate Fourier transform.

1. $c = 2$, $f(x) = 0$, and

$$g(x) = \begin{cases} \cos(x) & \text{for } \pi/2 \le x \le 5\pi/2 \\ 0 & \text{for } 0 \le x < \pi/2 \text{ and for } x > 5\pi/2 \end{cases}$$

2. $c = 6$, $f(x) = -2e^{-x}$, and $g(x) = 0$

3. $c = 14$, $f(x) = 0$, and $g(x) = \begin{cases} x^2(3-x) & \text{for } 0 \le x \le 3 \\ 0 & \text{for } x > 3 \end{cases}$

4. $c = 3$, $f(x) = 0$, and $g(x) = \begin{cases} 0 & \text{for } 0 \le x \le 4 \\ 2 & \text{for } 4 < x \le 11 \\ 0 & \text{for } x > 11 \end{cases}$

5. $c = 3$, $g(x) = 0$, and $f(x) = \begin{cases} x(1-x) & \text{for } 0 \le x \le 1 \\ 0 & \text{for } x > 1 \end{cases}$

16.5 Laplace Transform Techniques

The Laplace transform is well suited to solving certain problems involving wave motion, both on closed intervals and on the half-line. We will illustrate this by solving one problem on $x > 0$ and another on a closed interval.

A Problem on a Half-Line We will solve the boundary value problem

$$\frac{\partial^2 y}{\partial t^2} = c^2 \frac{\partial^2 y}{\partial x^2} - A \text{ for } x > 0, t > 0,$$

$$y(x,0) = \frac{\partial y}{\partial t}(x,0) = 0,$$

and

$$y(0,t) = 0,$$

with A as a positive constant. Because the half-line is unbounded to the right, we also impose the condition that

$$\lim_{x\to\infty} \frac{\partial y}{\partial x}(x,t) = 0$$

for $t \geq 0$.

This problem models an infinitely long string lying on the nonnegative x-axis, with its left end ($x = 0$) fastened. The string is pulled downward by a force of constant magnitude A.

Apply the Laplace transform $\mathcal{L}$, with respect t, to the wave equation, using the boundary conditions and the operational rule for taking the transform of derivatives. In this process $\mathcal{L}$, and $\partial/\partial x$ interchange because x is independent of t. The wave equation transforms to

$$s^2 Y(x,s) - sy(x,0) - \frac{\partial y}{\partial t}(x,0) = c^2 \frac{\partial^2 Y}{\partial x^2} - \frac{A}{s}.$$

Substitute the boundary conditions into this and let $Y'(x,s)$ denote $\partial Y/\partial x$. This differential equation becomes

$$Y''(x,s) - \frac{s^2}{c^2} Y(x,s) = \frac{A}{c^2 s}.$$

Think of this as a nonhomogeneous constant coefficient second order differential equation, whose general solution is the general solution of the associated homogeneous equation plus a particular solution of the nonhomogeneous equation. By inspection, $Y_p = -A/s^3$ is a particular solution. The characteristic equation of the associated homogeneous equation is

$$\lambda^2 - \left(\frac{s}{c}\right)^2 = 0$$

with roots $\pm s/c$. The general solution of the associated homogeneous equation is

$$Y(x,s) = k_1 e^{sx/c} + k_2 e^{-sx/c}.$$

Therefore, the general solution for $Y(x,s)$ is

$$Y(x,s) = k_1 e^{sx/c} + k_2 e^{-sx/c} - \frac{A}{s^3}.$$

To solve for k_1 and k_2, we need two pieces of information. Use the boundary conditions. Take the transform of $y(0,t) = 0$ to obtain $Y(0,s) = 0$. Then

$$Y(0,s) = k_1 + k_2 - \frac{A}{s^3} = 0,$$

so $k_2 = \frac{A}{s^3} - k_1$. Thus far,

$$Y(x,s) = k_1 e^{sx/c} + \left(\frac{A}{s^3} - k_1\right) e^{-sx/c} - \frac{A}{s^3}.$$

Now use the limit boundary condition. Applying the transform, we have

$$\mathcal{L}[\lim_{x\to\infty} (\partial/\partial x) y(x,t)] = \lim_{x\to\infty} Y'(x,s) = 0.$$

Now

$$Y'(x,s) = \frac{s}{c}k_1 e^{sx/c} - \frac{s}{c}\left(\frac{A}{s^3} - k_1\right)e^{-sx/c}.$$

Since $e^{sx/c} \to \infty$ and $x \to \infty$, we must choose $k_1 = 0$. This gives us

$$Y(x,s) = \frac{A}{s^3}e^{-sx/c} - \frac{A}{s^3}.$$

The solution is $y(x,t) = \mathcal{L}^{-1}[Y(x,s)]$. Now

$$\mathcal{L}^{-1}\left[\frac{A}{s^3}\right] = \frac{A}{2}t^2$$

and, from the shifting theorems in Chapter 3,

$$\mathcal{L}^{-1}\left[\frac{A}{s^3}e^{-sx/c}\right] = \frac{A}{2}\left(t - \frac{x}{c}\right)^2 H\left(t - \frac{x}{c}\right)$$

in which H is the Heaviside function. The solution is therefore

$$y(x,t) = \frac{A}{2}\left(t - \frac{x}{c}\right)^2 H\left(t - \frac{x}{c}\right) - \frac{A}{2}t^2.$$

Since

$$\left(t - \frac{x}{c}\right)^2 H\left(t - \frac{x}{c}\right) = \begin{cases} 0 & \text{if } x > ct, \\ (t - x/c)^2 & \text{if } x \le ct, \end{cases}$$

then

$$y(x,t) = \begin{cases} -\frac{At^2}{2} & \text{for } x > ct, \\ \frac{A}{2}\frac{x^2}{c^2} - \frac{A}{c}xt & \text{for } x \le ct. \end{cases}$$

A Problem on $[0, L]$ Next consider the boundary value problem:

$$\frac{\partial^2 y}{\partial t^2} = c^2\frac{\partial^2 y}{\partial x^2} \text{ for } 0 < x < L, t > 0,$$

$$y(x,0) = \frac{\partial y}{\partial t}(x,0) = 0,$$

and

$$y(0,t) = 0, E\frac{\partial y}{\partial t}(L,t) = f(t).$$

This models vibrations in an elastic bar of constant density ρ, uniform cross section and length L. $f(t)$ is a force per unit length acting parallel to the bar at the right end $x = L$. Here $c^2 = E/\rho$, where E is Young's modulus for the material of the bar. The bar is initially at rest and lying flat along the interval $[0, L]$.

Apply the Laplace transform $\mathcal{L}$, with respect t, using the boundary conditions and the operational rule for taking the transform of derivatives. This is similar to the problem just solved and we obtain

$$Y''(x,s) - \frac{s^2}{c^2}Y(x,s) = 0,$$

in which $Y'(x,s) = \partial Y/\partial x$ and s is carried along as a parameter. This is a second order linear homogeneous constant coefficient differential equation. The characteristic equation is

$$\lambda^2 - \left(\frac{s}{c}\right)^2 = 0,$$

with roots $\pm s/c$. The general solution for $Y(s,x)$ is

$$Y(x,s) = k_1 e^{sx/c} + k_2 e^{-sx/c}.$$

We need two initial conditions to solve for k_1 and k_2. First, transform $y(0,t) = 0$ to obtain $Y(0,s) = 0$. Then

$$Y(0,s) = k_1 + k_2 = 0$$

so $k_1 = -k_2$. This means that $Y(x,s)$ has the form

$$Y(x,s) = k \sinh(sx/c),$$

in which k is an arbitrary constant. To determine k, transform the other initial condition $E(\partial y/\partial t)(L,t) = f(t)$ to obtain $EY'(L,s) = F(s)$.

Apply this to $Y(x,s) = k \sinh(sx/c)$ to obtain

$$Ek\frac{s}{c}\cosh(sL/c) = F(s),$$

or

$$k = \frac{c}{E}F(s)\frac{1}{s\cosh(sL/c)}.$$

We now have

$$Y(x,s) = \frac{c}{E}F(s)\frac{\sinh(sx/c)}{s\cosh(sL/c)}.$$

The solution to the original problem is

$$y(x,t) = \mathcal{L}^{-1}[Y(x,s)](t).$$

Because of the generality of the problem, $f(t)$ is unspecified and we cannot proceed beyond this point. However, there are special cases of interest in which we can complete the solution. We will consider two such cases.

Case 1 Suppose $f(t) = K$, constant.

Now

$$F(s) = \mathcal{L}K = \frac{K}{s}$$

so

$$Y(x,s) = \frac{cK}{E}\frac{\sinh(sx/c)}{s^2\cosh(sL/c)}.$$

We can take the inverse transform of this expression by making use of the geometric series:

$$\frac{1}{1+\xi} = \sum_{n=0}^{\infty}(-1)^n \xi^n$$

for $|\xi| < 1$. Now write

$$\begin{aligned}\frac{\sinh(sx/c)}{\cosh(sL/c)} &= \frac{e^{sx/c} - e^{-sx/c}}{e^{sL/c} + e^{-sL/c}} \\ &= \frac{e^{sx/c}e^{-sL/c} - e^{-sx/c}e^{-sL/c}}{1 + e^{-sL/c}e^{-sL/c}} \\ &= \frac{e^{-(L-x)s/c} - e^{-(L+x)s/c}}{1 + e^{-2sL/c}} \\ &= \left(e^{-(L-x)s/c} - e^{-(L+x)s/c}\right)\frac{1}{1 + e^{-2sL/c}}\end{aligned}$$

$$= \left(e^{-(L-x)s/c} - e^{-(L+x)s/c}\right) \sum_{n=0}^{\infty} (-1)^n \left(e^{-2sL/c}\right)^n$$

$$= \left(e^{-(L-x)s/c} - e^{-(L+x)s/c}\right) \sum_{n=0}^{\infty} (-1)^n e^{-2nsL/c}.$$

Here we used $\xi = e^{-2sL/c}$ in the geometric series, using the fact that this exponential is always less than 1 for $x > 0$. Finally, we can write

$$Y(x,s) = \frac{cK}{E} e^{-(L-x)s/c} \sum_{n=0}^{\infty} (-1)^n \frac{1}{s^2} e^{-2snL/c}$$

$$- \frac{cK}{E} e^{-(L+x)s/c} \sum_{n=0}^{\infty} (-1)^n \frac{1}{s^2} e^{-2nsL/c}$$

$$= \frac{cK}{E} \sum_{n=0}^{\infty} (-1)^n \frac{1}{s^2} e^{((-(2n+1)L-x)/c)s}$$

$$- \frac{cK}{E} \sum_{n=0}^{\infty} \frac{1}{s^2} e^{((-(2n+1)L+x)/c)s}.$$

Now recall that, for any nonzero number α

$$\mathcal{L}^{-1}\left(\frac{1}{s^2} e^{-\alpha s}\right) = (t-\alpha)H(t-\alpha)$$

in which H is the Heaviside function. Assuming that we can take the transform of the geometric series term by term, the solution is

$$y(x,t) = \frac{cK}{E} \sum_{n=0}^{\infty} \left(t - \frac{(2n+1)L - x}{c}\right) H\left(t - \frac{(2n+1)L - x}{c}\right)$$

$$- \frac{cK}{E} \sum_{n=0}^{\infty} \left(t - \frac{(2n+1)L + x}{c}\right) H\left(t - \frac{(2n+1)L + x}{c}\right).$$

Look at this solution for a particular choice of the constants. Suppose $c = 2$ and $K = E = L = 1$. Now the solution is

$$y(x,t) = 2\sum_{n=0}^{\infty} (-1)^n \left(t - \frac{2n+1-x}{2}\right) H\left(\left(t - \frac{2n+1-x}{2}\right)\right)$$

$$- 2\sum_{n=0}^{\infty} (-1)^n \left(t - \frac{2n+1+x}{2}\right) H\left(\left(t - \frac{2n+1+x}{2}\right)\right).$$

Graphs of this solution on $0 \le x \le 1$ are shown in Figure 16.9. The lower graph is for time $t = 1.7$. Moving up, the next graph is $t = 2.4$, then 3.4, then 1.3 and 4.7 (indistinguishable in the scale of the graphs), then 5.2, and the highest graph is 0.9.

It is interesting to look at the motion of the right end of the bar. The analysis carried out to find $Y(x,s)$ applies, except now put $x = L$ to obtain

$$Y(L,s) = \frac{cK}{E} \frac{1}{s^2} \frac{\sinh(sL/c)}{\cosh(sL/c)} = \frac{cK}{E} \frac{1}{s^2} \tanh(sL/c).$$

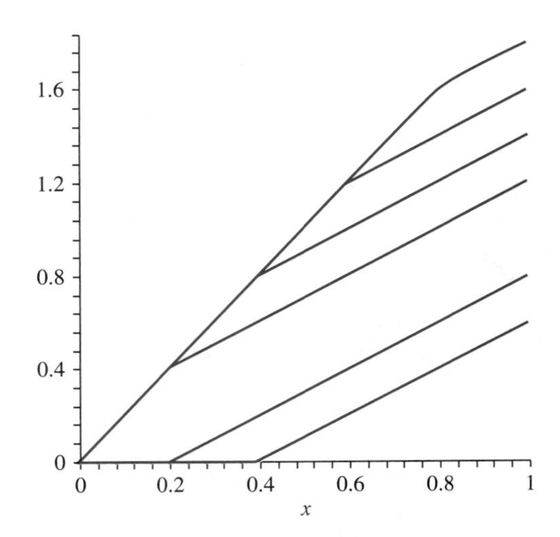

FIGURE 16.9 *Profiles of the elastic bar at different times with $f(t) = K$.*

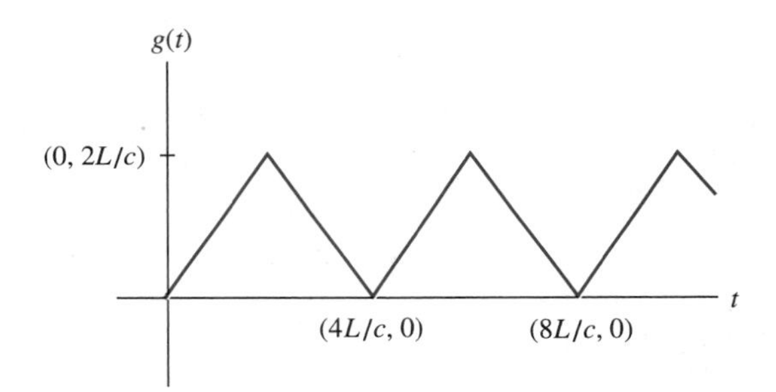

FIGURE 16.10 *Sawtooth wave.*

The interesting thing about this is that this function has a simple inverse transform. Let g be periodic of fundamental period $4L/c$, with $g(t)$ for $0 \leq t \leq 4L/c$ defined by

$$g(t) = \begin{cases} t & \text{for } 0 \leq t \leq 2L/c, \\ 4L/c & \text{for } 2L/c \leq t \leq 4L/c. \end{cases}$$

A graph of this sawtooth wave is shown in Figure 16.10.

Thus the right end of the bar moves according to the graph of g, exhibiting an up-and-down oscillation.

Case 2 Another case in which we can do a fairly complete analysis is that the end is hit with an impulse of magnitude I at time zero. Suppose $f(t) = I\delta(t)$, with δ the delta function. The analysis proceeds as in case 1, except now we obtain

$$Y(x,s) = \frac{cI}{E} \frac{\sinh(sx/c)}{s\cosh(sL/c)},$$

differing from case 1 in the power of s in the denominator. This occurs because the Laplace transform of K is K/s, while the transform of $I\delta(t)$ is just I, the delta function having transform 1. Since

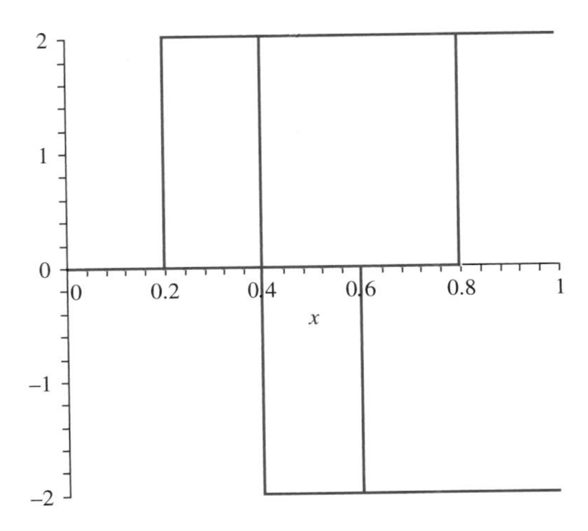

FIGURE 16.11 *Profiles of the bar at different times with* $f(t) = I\delta(t)$.

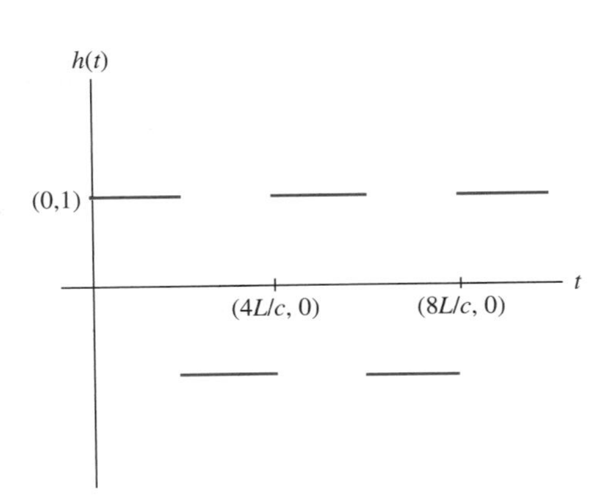

FIGURE 16.12 *Square wave of height* 1 *and period* $4L/c$.

$$\mathcal{L}\left(\frac{1}{s}e^{-\alpha s}\right) = H(t-\alpha)$$

the analysis of case 1, with this adjustment, yields

$$Y(x,s) = \frac{cI}{E}\sum_{n=0}^{\infty}(-1)^n \frac{1}{s}e^{-(((2n+1)L-x)/c)s} - \frac{cI}{E}\sum_{n=0}^{\infty}(-1)^n \frac{1}{s}e^{-(((2n+1)L+x)/c)s}.$$

Invert this term by term to obtain

$$y(x,t) = \frac{cI}{E}\sum_{n=0}^{\infty}(-1)^n H\left(t - \frac{(2n+1)L-x}{c}\right) - \frac{cI}{E}\sum_{n=0}^{\infty}(-1)^n H\left(t - \frac{(2n+1)L+x}{c}\right).$$

Figure 16.11 shows a graph of this solution for $c = 2$ and $I = E = L = 1$.

The graph shows the profile at times $t = 0.4, 0.7, 0.9$ (with maximum point 2 achieved farther to the right as t increases, and $t = 1.3, 1.8$, with minimum -2 achieved farther to the right as t increases.

As in case 1, this case also has a simple form if we focus on the right end of the bar. In this case, the Laplace transform of the solution is

$$Y(L,s) = \frac{cI}{E}\frac{1}{s}\tanh(sL/c).$$

This also has a simple inverse and we obtain $y(L,s) = h(t)$, where h is the square wave of Figure 16.12, with a height of 1 and a period of $4L/c$.

SECTION 16.5 PROBLEMS

1. Use the Laplace transform to solve:

$$\frac{\partial y}{\partial t^2} = c^2 \frac{\partial y}{\partial x^2} - Axt \text{ for } x > 0, t > 0,$$

$$y(x,0) = \frac{\partial y}{\partial t}(x,0) = 0 \text{ for } x > 0,$$

$$y(0,t) = e^{-t}, \lim_{x\to\infty} y(x,t) = 0 \text{ for } t > 0.$$

2. Use the Laplace transform to find the solution

$$y(x,t) = \frac{1}{2}(f_o(x+ct) + f_o(x-ct))$$

of the problem

$$\frac{\partial^2 y}{\partial t^2} = c^2 \frac{\partial^2 y}{\partial x^2} \text{ for } x > 0, t > 0$$

$$y(x,0) = \frac{\partial y}{\partial t}(x,0) = 0 \text{ for } x > 0,$$

$$y(0,t) = f(t), \lim_{x\to\infty} y(x,t) = 0 \text{ for } t \geq 0.$$

3. Use the Laplace transform to write the solution (in terms of $f(t)$) of the boundary value problem

$$\frac{\partial^2 y}{\partial t^2} = c^2 \frac{\partial^2 y}{\partial x^2} + K \text{ for } x > 0, t > 0$$

$$y(x,0) = \frac{\partial y}{\partial t}(x,0) = 0 \text{ for } x > 0,$$

$$y(0,t) = f(t), \lim_{x\to\infty} y(x,t) = 0 \text{ for } t \geq 0.$$

4. Use the Laplace transform to write the solution (in terms of $f(t)$) of the boundary value problem

$$9\frac{\partial^2 y}{\partial t^2} + \frac{\partial^2 y}{\partial x^2} - 6\frac{\partial^2 y}{\partial x \partial t} = 0 \text{ for } x > 0, t < 0$$

$$y(x,0) = \frac{\partial y}{\partial t}(x,0) = 0 \text{ for } x > 0, y(2,t) = f(t),$$

$$y(0,t) = 0, \lim_{x\to\infty} y(x,t) = 0 \text{ for } t \geq 0.$$

5. Use the Laplace transform to solve

$$\frac{\partial^2 y}{\partial t^2} = c^2 \frac{\partial^2 y}{\partial x^2} - At \text{ for } x > 0, t > 0,$$

$$y(x,0) = \frac{\partial y}{\partial t}(x,0) = 0,$$

$$y(0,t) = f(t), \text{ and } \lim_{x\to\infty} y(x,t) = 0,$$

in which A is a positive constant.

16.6 Characteristics and d'Alembert's Solution

In this section, we will derive d'Alembert's solution of a wave problem on the real line. We will denote partial derivatives by subscripts, $\partial u/\partial t = u_t$ and $\partial u/\partial x = u_x$. The problem we will solve is

$$u_{tt} = c^2 u_{xx} \text{ for } -\infty < x < \infty, t > 0$$

and

$$u(x,0) = f(x), u_t(x,0) = g(x) \text{ for } -\infty < x < \infty.$$

We are using $u(x,t)$ for the position function of the wave. A graph of the wave's profile at time t is the graph of $y = u(x,t)$ in the x, y-plane for that value of t.

This initial-boundary value problem is called the *Cauchy problem for the wave equation.* The lines $x - ct = k_1$ and $x + ct = k_2$ in the x, t-plane are called *characteristics* of the wave equation. These are straight lines of slope $1/c$ and $-1/c$ in the x, t-plane. Exploiting these characteristics, make the change of variables

$$\xi = x - ct, \eta = x + ct.$$

The transformation is invertible, since we can solve for x and t to get

$$x = \frac{1}{2}(\xi + \eta), t = \frac{1}{2c}(-\xi + \eta).$$

Define

$$U(\xi, \eta) = u((\xi + \eta)/2, (-\xi + \eta)/2c).$$

We must compute chain rule derivatives:

$$u_x = U_\xi \xi_x + U_\eta \eta_x = U_\xi + U_\eta,$$

and

$$\begin{aligned} u_{xx} &= U_{\xi\xi}\xi_x + U_{\xi\eta}\eta_x + U_{\eta\xi}\xi_x + U_{\eta\eta}\eta_x \\ &= U_{\xi\xi} + 2U_{\xi\eta} + U_{\eta\eta}. \end{aligned}$$

Similarly,

$$u_{tt} = c^2 U_{\xi\xi} - 2c^2 U_{\xi\eta} + c^2 U_{\eta\eta}.$$

The wave equation transforms to

$$u_{tt} - c^2 u_{xx} = 0 = [c^2 U_{\xi\xi} - 2c^2 U_{\xi\eta} + c^2 U_{\eta\eta}] - c^2[U_{\xi\xi} + 2U_{\xi\eta} + U_{\eta\eta}],$$

or

$$U_{\xi\eta} = 0.$$

Now we see the rationale for this change of variables. The transformed equation $U_{\xi\eta} = 0$ is easy to solve. First, $(U_\eta)_\xi = 0$ means that U_η is independent of ξ, so for some function h,

$$U_\eta = h(\eta).$$

Then

$$U(\xi, \eta) = \int h(\eta) d\eta + F(\xi)$$

in which this integration with respect to η may have ξ in its "constant" of integration. Now $\int h(\eta)\, d\eta$ is just another function of η, so we conclude that $U(\xi, \eta)$ must be a sum of a function just of ξ and a function just of η:

$$U(\xi, \eta) = F(\xi) + G(\eta).$$

This function satisfies the transformed wave equation for any twice differentiable functions F and G of one variable and, conversely, every solution of the transformed wave equation has this form. In terms of x and t, this means that every solution of the one-dimensional (unforced) wave equation has the form

$$u(x, t) = F(x - ct) + G(x + ct). \tag{16.12}$$

Thus far, we have dealt with just the partial differential equation. The idea now is to choose F and G to obtain a solution satisfying the initial conditions $y(x, 0) = f(x)$ and $y_t(x, 0) = g(x)$. First we need

$$u(x, 0) = F(x) + G(x) = f(x) \tag{16.13}$$

and

$$u_t(x, 0) = -cF'(x) + cG'(x) = g(x). \tag{16.14}$$

Integrate equation (16.14) and rearrange terms to get

$$-F(x) + G(x) = \frac{1}{c}\int_0^x g(w)\, dw - F(0) + G(0).$$

Add this to equation (16.13) to get

$$2G(x) = f(x) + \frac{1}{c}\int_0^x g(w)\,dw - F(0) + G(0).$$

Then

$$G(x) = \frac{1}{2}f(x) + \frac{1}{2c}\int_0^x g(w)\,dw - \frac{1}{2}F(0) + \frac{1}{2}G(0). \tag{16.15}$$

But then, from equation (16.13),

$$F(x) = f(x) - G(x) = \frac{1}{2}f(x) - \frac{1}{2c}\int_0^x g(w)\,dw + \frac{1}{2}F(0) - \frac{1}{2}G(0). \tag{16.16}$$

Finally, use equations (16.15) and (16.16) to obtain the solution

$$\begin{aligned} u(x,t) &= F(x-ct) + G(x+ct) \\ &= \frac{1}{2}f(x-ct) - \frac{1}{2c}\int_0^{x-ct} g(w)\,dw + \frac{1}{2}F(0) - \frac{1}{2}G(0) \\ &\quad + \frac{1}{2}f(x+ct) + \frac{1}{2c}\int_0^{x+ct} g(w)\,dw - \frac{1}{2}F(0) + \frac{1}{2}G(0). \end{aligned}$$

After cancellations and combining the integrals, we have the solution

$$u(x,t) = \frac{1}{2}(f(x-ct) + f(x+ct)) + \frac{1}{2c}\int_{x-ct}^{x+ct} g(w)\,dw. \tag{16.17}$$

This is *d'Alembert's solution* of the Cauchy problem for the wave equation on the real line. It gives an explicit solution in terms of the initial position and velocity functions.

EXAMPLE 16.12

We will solve the initial-boundary value problem

$$u_{tt} = 4u_{xx} \text{ for } -\infty < x < \infty, t > 0$$

and

$$u(x,0) = e^{-|x|}, u_t(x,0) = \cos(4x) \text{ for } -\infty < x < \infty.$$

Immediately,

$$\begin{aligned} u(x,t) &= \frac{1}{2}\left(e^{-|x-2t|} + e^{-|x+2t|}\right) + \frac{1}{4}\int_{x-2t}^{x+2t} \cos(4w)\,dw \\ &= \frac{1}{2}\left(e^{-|x-2t|} + e^{-|x+2t|}\right) + \frac{1}{16}(\sin(4(x+2t)) - \sin(4(x-2t))) \\ &= \frac{1}{2}\left(e^{-|x-2t|} + e^{-|x+2t|}\right) + \frac{1}{8}\sin(4x)\cos(8t). \; \blacklozenge \end{aligned}$$

16.6.1 Forward and Backward Waves

Write d'Alembert's solution as

$$\begin{aligned} u(x,t) &= \frac{1}{2}\left(f(x-ct) - \frac{1}{c}\int_0^{x-ct} g(w)\,dw\right) \\ &\quad + \frac{1}{2}\left(f(x+ct) + \frac{1}{c}\int_0^{x+ct} g(w)\,dw\right) \\ &= \varphi(x-ct) + \beta(x+ct), \end{aligned}$$

where

$$\varphi(x) = \frac{1}{2} f(x) - \frac{1}{2c} \int_0^x g(w)\, dw$$

and

$$\beta(x) = \frac{1}{2} f(x) + \frac{1}{2c} \int_0^x g(w)\, dw.$$

We call $\varphi(x - ct)$ a *forward wave*. Its graph is the graph of $\varphi(x)$ translated ct units to the right, and so may be thought of as a wave moving to the right with speed c. We call $\beta(x + ct)$ a *backward wave*. Its graph is the graph of $\beta(x)$ translated ct units to the left, and may be thought of as a wave moving to the left with speed c. This allows us to think of the wave profile $y = u(x, t)$ at any time t as a sum of a wave moving to the right and a wave moving to the left.

EXAMPLE 16.13

Suppose $g(x) = 0$, $c = 1$ and

$$f(x) = \begin{cases} 4 - x^2 & \text{for } -2 \le x \le 2 \\ 0 & \text{for } |x| > 2. \end{cases}$$

The solution is

$$u(x, t) = \varphi(x + ct) + \beta(x + ct) = \frac{1}{2}(f(x - t) + f(x + t)).$$

At any time t the wave profile consists of the initial position function translated t units to the right, superimposed on the initial position function translated t units to the left. Figures 16.13 through 16.19 show this profile at increasing times.

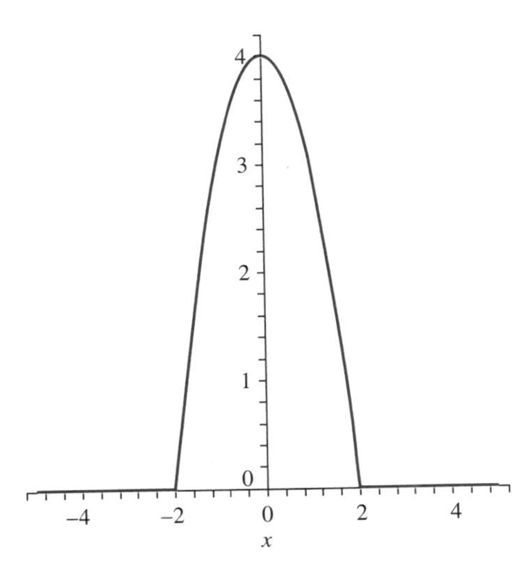

FIGURE 16.13 *Initial position in Example 16.13.*

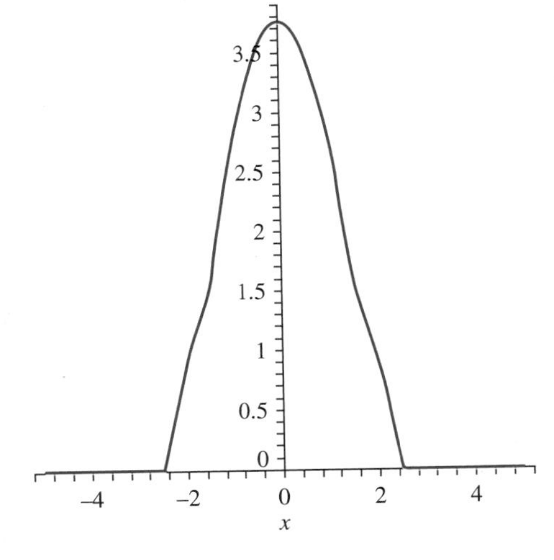

FIGURE 16.14 *Wave in Example 16.13 at time $t = 0.5$.*

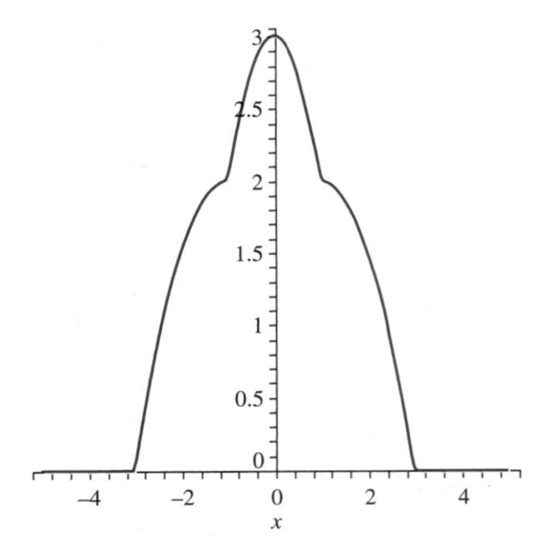

FIGURE 16.15 $t = 1$.

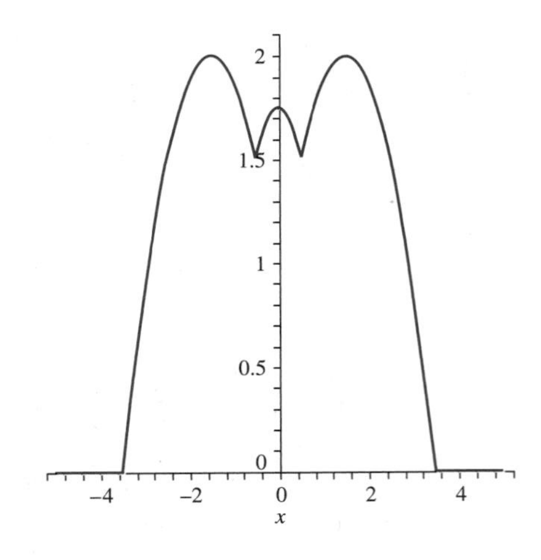

FIGURE 16.16 $t = 1.5$.

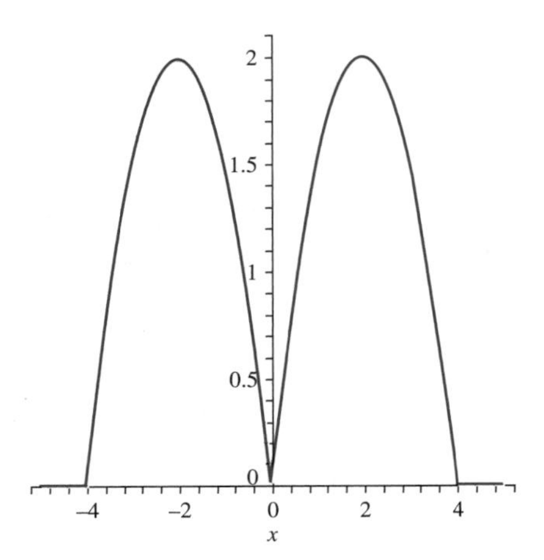

FIGURE 16.17 $t = 2$.

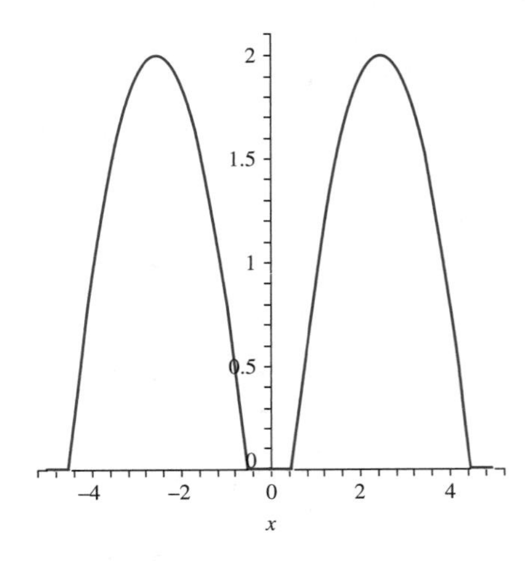

FIGURE 16.18 $t = 2.5$.

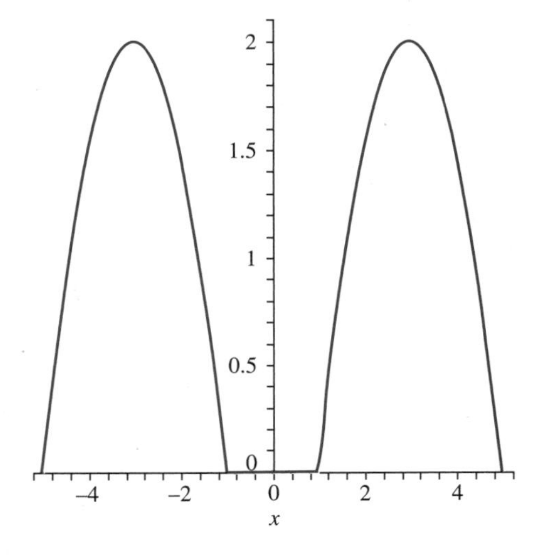

FIGURE 16.19 $t = 3$.

Notice that by time $t = 3$ the wave has split into two disjoint copies of the original ($t = 0$ wave, one continuing to move to the right, the other to the left.) This separation is because $f(x)$ in this example is nonzero on only a bounded interval. ♦

16.6.2 Forced Wave Motion

Using the characteristics we will write a solution for the Cauchy problem on the line with a forcing term:

$$u_{tt} = c^2 u_{xx} + F(x,t) \text{ for } -\infty < x < \infty, t > 0$$

and

$$u(x,0) = f(x), u_t(x,0) = g(x) \text{ for } -\infty < x < \infty.$$

Suppose we want the solution at $P_0 : (x_0, t_0)$. There are two characteristics through P_0, namely the straight lines

$$x - ct = x_0 - ct_0 \text{ and } x + ct = x_0 + ct_0.$$

Use parts of these to form the *characteristic triangle* of Figure 16.20, with vertices $(x_0 - ct_0, 0)$, $(x_0 + ct_0, 0)$ and (x_0, t_0) and sides L, M, and I. Let Δ denote this solid triangle and compute the double integral of $-F$ over Δ.

$$-\iint_\Delta F(x,t)dA = \iint_\Delta (c^2 u_{xx} - u_{tt})dA = \iint_\Delta \left(\frac{\partial}{\partial x}(c^2 u_x) - \frac{\partial}{\partial t}(u_t)\right) dA.$$

Apply Green's theorem to the last integral to obtain

$$-\iint_\Delta F(x,t)\,dA = \oint_C u_t\,dx + c^2 u_x\,dt,$$

where C is the boundary of Δ, oriented counterclockwise. C consists of the line segments L, M, and I. Compute the line integral over each side.

On I, $T = 0$, and x varies from $x_0 - ct_0$ to $x_0 + ct_0$, so

$$\int_I u_t\,dx + c^2 u_x\,dt = \int_{x_0-ct_0}^{x_0+ct_0} u_t(x,0)\,dx = \int_{x_0-ct_0}^{x_0+ct_0} g(w)\,dw.$$

On L, $x + ct = x_0 + ct_0$, so $dx = -cdt$ and

$$\int_L u_t\,dx + c^2 u_x\,dt = \int_L u_t(-c)\,dt + c^2 u_x\left(-\frac{1}{c}\right)dx = -c\int_L du$$

$$= -c[u(x_0,t_0) - u(x_0 + ct_0, 0)].$$

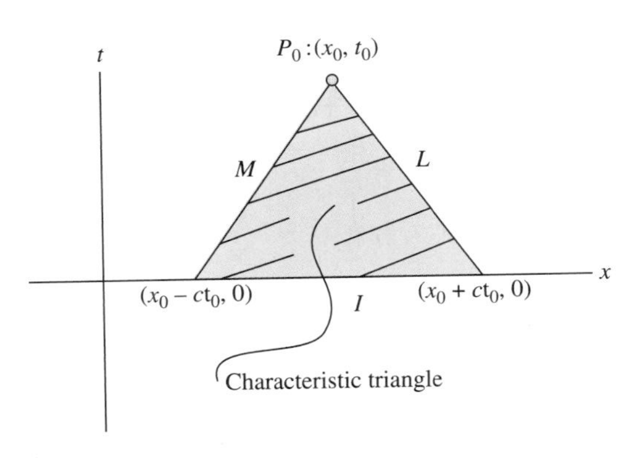

FIGURE 16.20 *The characteristic triangle.*

Finally, on M, $x - ct = x_0 - ct_0$ so $dx = c\,dt$ and

$$\int_M u_t\,dx + c^2 u_x\,dt = \int_M u_t c\,dt + c^2\left(\frac{1}{c}\right)dx = c\int_M du$$
$$= c[u(x_0 - ct_0, 0) - u(x_0, t_0)].$$

M has initial point (x_0, t_0) and terminal point $(x_0 - ct_0, 0)$ because of the counterclockwise orientation on C. Upon summing these integrals, we obtain

$$-\iint_\Delta F(x, y)\,dA = \iint_{x_0-ct_0}^{x_0+ct_0} g(w)\,dw$$
$$- c[u(x_0, t_0) - u(x_0 + ct_0, 0)] + c[u(x_0 - ct_0, 0) - u(x_0, t_0)].$$

Then

$$-\iint_\Delta F(x, y)\,dA = \int_{x_0-ct_0}^{x_0+ct_0} g(w)\,dw - 2cu(x_0, t_0) + c[f(x_0 + ct_0) + f(x_0 - ct_0)].$$

Solve this equation for $u(x_0, t_0)$ to get

$$u(x_0, t_0) = \frac{1}{2}(f(x_0 - ct_0) + f(x_0 + ct_0))$$
$$+ \frac{1}{2c}\int_{x_0-ct_0}^{x_0+ct_0} g(w)\,dw + \frac{1}{2c}\iint_\Delta F(x, t)\,dA.$$

Since x_0 is any real number and t_0 any positive number, we can drop the subscripts and write the solution $u(x, t)$ as

$$u(x, t) = \frac{1}{2}(f(x - ct) + f(x + ct)) + \frac{1}{2c}\int_{x-ct}^{x+ct} g(w)\,dw$$
$$+ \frac{1}{2c}\iint_\Delta F(X, T)\,dX\,dT$$

in which we have used X and T as the variables of integration to avoid confusion with the point (x, t) at which the solution is given.

EXAMPLE 16.14

We will solve the problem

$$u_{tt} = 25u_{xx} + x^2t^2 \text{ for } -\infty < x < \infty, t > 0$$

and

$$u(x, 0) = x\cos(x), u_t(x, 0) = e^{-x} \text{ for } -\infty < x < \infty.$$

The solution at any x and time t is

$$u(x, t) = \frac{1}{2}[(x - 5t)\cos(x - 5t) + (x + 5t)\cos(x + 5t)]$$
$$+ \frac{1}{10}\int_{x-5t}^{x+5t} e^w\,dw + \frac{1}{10}\iint_\Delta X^2T^2\,dX\,dT.$$

Compute

$$\frac{1}{10}\int_{x-5t}^{x+5t} e^{-w}\,dw = \frac{1}{10}\left(e^{-x+5t} - e^{-x-5t}\right)$$

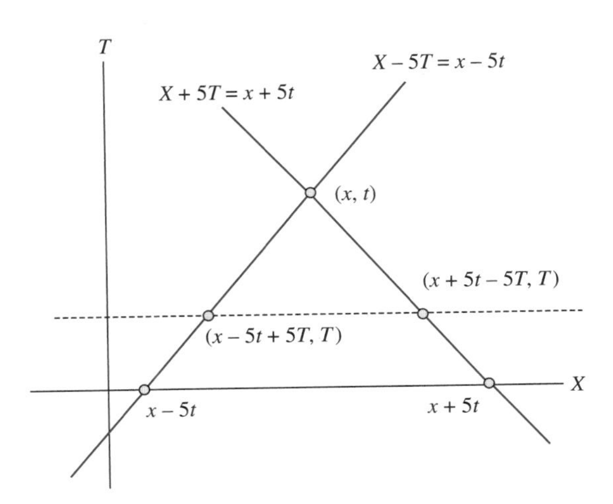

FIGURE 16.21 *Characteristic triangle in Example 16.14*

and, from Figure 16.21,

$$\frac{1}{10}\iint_{\Delta} X^2T^2 dX dT = \frac{1}{10}\int_0^T \int_{x-5t+5T}^{x+5t-5T} X^2T^2\,dX\,dT = \frac{1}{12}x^2t^4 + \frac{5}{36}t^6.$$

The solution is

$$u(x,t) = \frac{1}{2}[(x-5t)\cos(x-5t) + (x+5t)\cos(x+5t)]$$
$$+\frac{1}{10}\left(e^{-x+5t} - e^{-x-5t}\right) + \frac{1}{12}x^2t^4 + \frac{5}{36}t^6. \blacklozenge$$

SECTION 16.6 PROBLEMS

In each of Problems 1 through 6, solve the problem

$$u_{tt} = c^2 u_{xx} + F(x,t) \text{ for } -\infty < x < \infty, t > 0$$

and

$$u(x,0) = f(x), u_t(x,0) = g(x) \text{ for } -\infty < x < \infty.$$

1. $c = 3$, $f(x) = \cosh(x)$, $g(x) = 1$, $F(x,t) = 3xt^3$
2. $c = 2$, $f(x) = \sin(x)$, $g(x) = 2x$, $F(x,t) = 2xt$
3. $c = 4$, $f(x) = x$, $g(x) = e^{-x}$, $F(x,t) = x + t$
4. $c = 7$, $f(x) = 1 + x$, $g(x) = 0$, $F(x,t) = x - \cos(t)$
5. $c = 8$, $f(x) = x^2 - x$, $g(x) = \cos(2x)$, $F(x,t) = xt^2$
6. $c = 4$, $f(x) = x^2$, $g(x) = xe^{-x}$, $F(x,t) = x\sin(t)$

In each of Problems 7 through 12, write the d'Alembert solution for the problem

$$u_{tt} = c^2 u_{xx} \text{ for } -\infty < x < \infty, t > 0$$

and

$$u(x,0) = f(x), u_t(x,0) = g(x) \text{ for } -\infty < x < \infty.$$

7. $c = 1$, $f(x) = x^2$, $g(x) = -x$
8. $c = 12$, $f(x) = -5x + x^2$, $g(x) = 3$
9. $c = 14$, $f(x) = e^x$, $g(x) = x$
10. $c = 4$, $f(x) = x^2 - 2x$, $g(x) = \cos(x)$
11. $c = 7$, $f(x) = \cos(\pi x)$, $g(x) = 1 - x^2$
12. $c = 5$, $f(x) = \sin(2x)$, $g(x) = x^3$

In each of Problems 13 through 18, write the solution of the problem

$$u_{tt} = u_{xx} \text{ for } -\infty < x < \infty, t > 0$$

and

$$u(x,0) = f(x), u_t(x,0) = 0 \text{ for } -\infty < x < \infty$$

as a sum of a forward and a backward wave. Graph the initial position function and then graph the solution at selected times, showing the wave as a superposition of a wave moving to the right and a wave moving to the left.

13. $f(x) = \begin{cases} \cos(x) & \text{for } -\pi/2 \leq x \leq \pi/2 \\ 0 & \text{for } |x| > \pi/2 \end{cases}$

14. $f(x) = \begin{cases} 1 - x^2 & \text{for } |x| \leq 1 \\ 0 & \text{for } |x| > 1 \end{cases}$

15. $f(x) = \begin{cases} x^2 - x - 2 & \text{for } -1 \leq x \leq 2 \\ 0 & \text{for } x < -1 \text{ and for } x > 2 \end{cases}$

16. $f(x) = \begin{cases} x^3 - x^2 - 4x + 4 & \text{for } -2 \leq x \leq 2 \\ 0 & \text{for } |x| > 2 \end{cases}$

17. $f(x) = \begin{cases} \sin(2x) & \text{for } -\pi \leq x \leq \pi \\ 0 & \text{for } |x| > \pi \end{cases}$

18. $f(x) = \begin{cases} 1 - |x| & \text{for } -1 \leq x \leq 1 \\ 0 & \text{for } |x| > 1 \end{cases}$

16.7 Vibrations in a Circular Membrane I

Imagine an elastic membrane of radius R fastened onto a circular frame (such as a drumhead). The membrane is set in motion from a given initial position and with a given initial velocity. In polar coordinates, the membrane occupies the disk $r \leq R$. Assume that the particle of membrane at (r, θ) vibrates vertical to the x, y-plane and let the displacement of this particle at time t be $z(r, \theta, t)$.

The wave equation in polar coordinates is

$$\frac{\partial^2 z}{\partial t^2} = c^2 \left(\frac{\partial^2 z}{\partial r^2} + \frac{1}{r}\frac{\partial z}{\partial r} + \frac{1}{r^2}\frac{\partial^2 z}{\partial \theta^2} \right).$$

We will assume axial symmetry, which means that the motion is independent of θ. Then $z = z(r, t)$ and the wave equation is

$$\frac{\partial^2 z}{\partial t^2} = c^2 \left(\frac{\partial^2 z}{\partial r^2} + \frac{1}{r}\frac{\partial z}{\partial r} \right).$$

The initial position is $z(r, 0) = f(r)$ and the initial velocity is $(\partial z/\partial t)(r, 0) = g(r)$.

Attempt a solution $z(r, t) = F(r)T(t)$. A routine calculation leads to

$$F'' + \frac{1}{r}F' + \frac{\lambda}{c^2}F = 0 \text{ and } T'' + \lambda T = 0.$$

If $\lambda = \omega^2 > 0$, this equation for F is a zero-order Bessel equation with solutions (bounded on the disk $r < R$) that are multiples of

$$J_0\left(\frac{\omega}{c}r\right).$$

The equation for T is

$$T'' + \omega^2 T = 0$$

with solutions of the form

$$T(t) = a\cos(\omega t) + b\sin(\omega t).$$

For each positive number ω, we now have a function

$$z_\omega(r, t) = a_\omega J_0\left(\frac{\omega}{c}r\right)\cos(\omega t) + b_\omega J_0\left(\frac{\omega}{c}r\right)\sin(\omega t).$$

that satisfies the wave equation.

Because the membrane is fixed on a circular frame,

$$z_\omega(R,t) = a_\omega J_0\left(\frac{\omega}{c}R\right)\cos(\omega t) + b_\omega J_0\left(\frac{\omega}{c}R\right)\sin(\omega t) = 0$$

for $t > 0$. This equation will be satisfied for all $t > 0$ if $J_0(\omega R/c) = 0$. Thus, choose ω so that $\omega R/c$ is a positive zero of $J_0(x)$. Let these zeros be $j_1 < j_2 < \cdots$. For each positive integer n, choose ω_n so that $\omega_n R/c = j_n$. This gives us the eigenvalues

$$\lambda_n = \omega_n^2 = \left(\frac{j_n c}{R}\right)^2$$

and the corresponding eigenfunctions

$$J_0\left(\frac{j_n}{R}r\right).$$

The functions

$$z_n(r,t) = a_n J_0\left(\frac{j_n}{R}r\right)\cos\left(\frac{j_n ct}{R}\right) + b_n J_0\left(\frac{j_n}{R}r\right)\sin\left(\frac{j_n ct}{R}\right)$$

satisfy the wave equation and the boundary condition that $z(R,0) = 0$, for each positive integer n. To satisfy the initial condition $z(r,0) = f(r)$, use a superposition

$$z(r,t) = \sum_{n=1}^{\infty}\left[a_n J_0\left(\frac{j_n}{R}r\right)\cos\left(\frac{j_n ct}{R}\right) + b_n J_0\left(\frac{j_n}{R}r\right)\sin\left(\frac{j_n ct}{R}\right)\right]. \tag{16.18}$$

The initial condition gives us

$$z(r,0) = f(r) = \sum_{n=1}^{\infty} a_n J_0\left(\frac{j_n r}{R}\right).$$

This is a Fourier-Bessel expansion, which we developed in Chapter 15 for the interval [0, 1]. Let $s = r/R$ to convert this expansion to

$$f(Rs) = \sum_{n=1}^{\infty} a_n J_0(j_n s)$$

on $0 \le s \le 1$. Choose

$$a_n = \frac{2\int_0^1 sf(Rs)J_0(j_n s)\,ds}{J_1^2(j_n)}$$

for $n = 1, 2, \cdots$.

Next solve for the b_n's. We need

$$\frac{\partial z}{\partial t}(r,0) = g(r) = \sum_{n=1}^{\infty} b_n \frac{j_n c}{R} J_0\left(\frac{j_n r}{R}\right).$$

Again we let $s = r/R$ to normalize the interval to [0, 1] and obtain

$$b_n = \frac{2R}{j_n c}\frac{\int_0^1 sg(Rs)J_0(j_n s)\,ds}{J_1^2(j_n)}.$$

With these coefficients, equation (16.18) is the solution for $z(r,t)$.

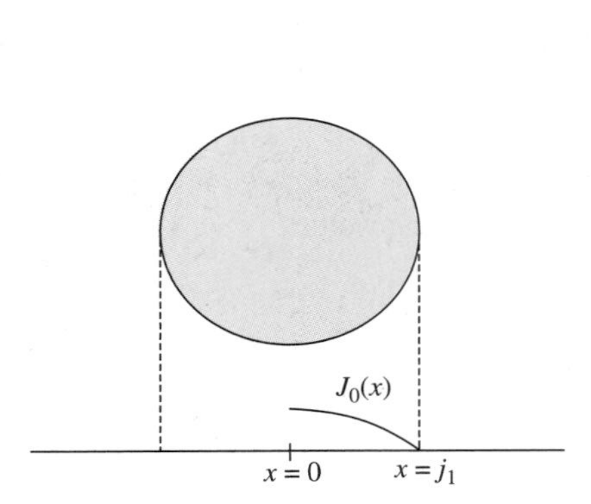

FIGURE 16.22 *First normal mode of vibration.*

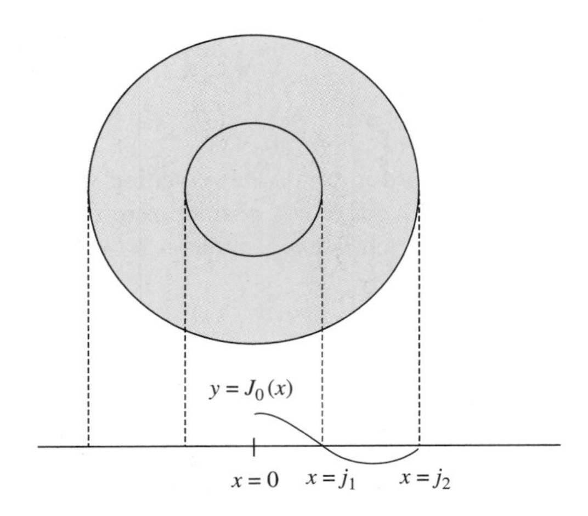

FIGURE 16.23 *Second normal mode.*

16.7.1 Normal Modes of Vibration

The numbers $\omega_n = j_n c/R$ are the *frequencies of normal modes of vibration* of the membrane with periods $2\pi/\omega_n = 2\pi R/j_n c$. The *normal modes of vibration* are the functions $z_n(r,t)$, which are often written in phase angle form as

$$z_n(r,t) = A_n J_0\left(\frac{j_n r}{R}\right)\cos(\omega_n t + \delta_n)$$

in which $\omega_n = j_n c/R$, $A_n = \sqrt{a_n^2 + b_n^2}$ and $\delta_n = \arctan(-b_n/a_n)$ if $a_n \neq 0$.

The first normal mode is

$$z_1(r,t) = A_1 J_0\left(\frac{j_1 r}{R}\right)\cos(\omega_1 t + \delta_1).$$

As r varies from 0 to R, $j_1 r/R$ varies from 0 to j_1, the first positive zero of J_0. At any time t, a radial section through the membrane takes the shape of the graph of $J_0(x)$ for $0 \le x \le j_1$ (Figure 16.22).

The second normal mode is

$$z_2(r,t) = A_2 J_0\left(\frac{j_2 r}{R}\right)\cos(\omega_2 t + \delta_2).$$

As r varies from 0 to R, $j_2 r/R$ varies from 0 to j_2, passing through j_1 along the way. Since $J_0(j_2 r/R) = 0$ when $j_2 r/R = j_1$, this mode has a nodal circle (fixed in the motion) at radius $r = j_1 R/j_2$. A section through the membrane takes the shape of the graph of $J_0(x)$ for $0 \le x \le j_2$ (Figure 16.23).

Similarly, the third normal mode is

$$z_3(r,t) = A_3 J_0\left(\frac{j_3 r}{R}\right)\cos(\omega_3 t + \delta_3)$$

and this mode has two nodes, one at $r = j_1 R/j_3$ and the second at $r = j_2 R/j_3$. Now a radial section has the shape of a graph of $J_0(x)$ for $0 \le x \le j_3$ (Figure 16.24).

In general, the nth normal mode has $N-1$ nodes (fixed circles in the motion of the membrane), occurring at $j_1 R/j_n, j_2 R/j_n, \cdots, j_{n-1} R/j_n$.

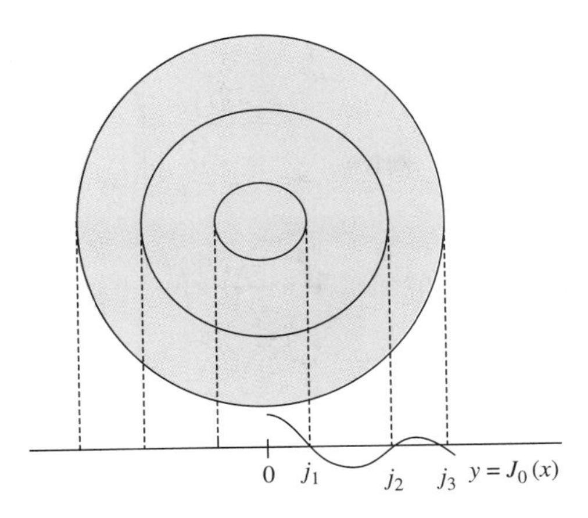

FIGURE 16.24 *Third normal mode.*

SECTION 16.7 PROBLEMS

1. Let $c = R = 1$, $f(r) = 1 - r$, and $g(r) = 0$. Approximate the coefficients a_1 through a_5 in the solution for the motion of the membrane, and graph the fifth partial sum for a selection of different times.
2. Repeat Problem 1 with $f(r) = 1 - r^2$.
3. Repeat Problem 1 with $f(r) = \sin(\pi r)$.

16.8 Vibrations in a Circular Membrane II

We will expand the discussion of vibrations in a circular elastic membrane to include a dependence of the displacement function on θ. Now the problem is

$$\frac{\partial^2 z}{\partial t^2} = c^2 \left(\frac{\partial^2 z}{\partial r^2} + \frac{1}{r}\frac{\partial z}{\partial r} + \frac{1}{r^2}\frac{\partial^2 z}{\partial \theta^2} \right)$$

$$z(r, \theta, 0) = f(r, \theta),\ \frac{\partial z}{\partial t}(r, \theta, 0) = 0 \text{ for } 0 \le r < R,\ -\pi \le \theta \le \pi,\ t > 0.$$

Thus, we assume that the membrane is released from rest with the initial displacement function $f(r, \theta)$.

In cylindrical coordinates, θ can be replaced by $\theta + 2n\pi$ for any integer n, so we also impose the *periodicity conditions*

$$z(r, -\pi, t) = z(r, \pi, t) \text{ and } \frac{\partial z}{\partial \theta}(r, -\pi, t) = \frac{\partial z}{\partial \theta}(r, \pi, t).$$

To separate the variables, set

$$z(r, \theta, t) = F(r)\Theta(\theta)T(t).$$

Upon substituting this into the wave equation we obtain

$$\frac{T''}{c^2 T} = \frac{F'' + (1/r)F'}{F} + \frac{1}{r^2}\frac{\Theta''}{\Theta} = -\lambda,$$

for some separation constant λ. The reason for this is that the left side depends only on t and the right side only on r and θ, and these variables are independent. Then

$$T'' + c^2 T = 0$$

and

$$\frac{F'' + (1/r)F'}{F} + \lambda r^2 = -\frac{\Theta''}{\Theta}.$$

Because the left side depends only on r and the right side only on θ, there is some constant μ such that

$$\frac{F'' + (1/r)F'}{F} + \lambda r^2 = -\frac{\Theta''}{\Theta} = \mu.$$

Then

$$\Theta'' + \mu\Theta = 0$$

and

$$r^2 F'' + rF' + (\lambda r^2 - \mu)F = 0.$$

These differential equations for $F(r)$, $T(t)$, and $\Theta(\theta)$ come with the conditions. First, there is the periodicity condition

$$\Theta(-\pi) = \Theta(\pi) \text{ and } \Theta'(-\pi) = \Theta'(\pi).$$

Because of the fixed frame,

$$F(R) = 0.$$

Finally, if the membrane is released from rest,

$$T'(0) = 0.$$

The problem for $\Theta(\theta)$ is a periodic Sturm-Liouville problem and was solved in Example 15.2 (here $L = \pi$). The eigenvalues are

$$\mu_n = n^2 \text{ for } n = 1, 2, \cdots$$

and the eigenfunctions are

$$\Theta_n(\theta) = a_n \cos(n\theta) + b_n \sin(n\theta).$$

Now we have $\mu_n = n^2$, so the problem for $F(r)$ is

$$r^2 F''(r) + rF'(r) + (kr^2 - n^2)F(r) = 0;\ F(R) = 0.$$

This is a Bessel equation with general solution

$$F(r) = \alpha J_n(\sqrt{\lambda} r) + \beta Y_n(\sqrt{\lambda} r),$$

with α and β as yet arbitrary constants. Because $Y_n(\sqrt{\lambda} r)$ is unbounded as $r \to 0+$ (the center of the membrane), we must choose $\beta = 0$ to have a bounded solution. We expect oscillations of a vibrating membrane to be finite in magnitude. This leaves $F(r) = \alpha J_n(\sqrt{\lambda} r)$. To find admissable values of λ, use the boundary condition to require that

$$F(R) = \alpha J_n(\sqrt{\lambda} R) = 0.$$

We must have $\alpha \neq 0$ for a nontrivial solution. Thus, choose $\sqrt{\lambda} R$ to be a positive zero of J_n. Let these positive zeros be

$$j_{n1} < j_{n2} < j_{n3} < \cdots,$$

which are double indexed because this derivation depends on the eigenvalue $\mu = n^2$. Then

$$\lambda_{nk} = \frac{j_{nk}^2}{R^2}$$

are the eigenvalues in the problem for $F(r)$. Corresponding eigenfunctions are

$$J_n\left(\frac{j_{nk}}{R}r\right) \text{ for } n=0,1,2,\cdots, k=1,2,\cdots.$$

With these eigenvalues λ, the problem for T is

$$T''+c^2\left(\frac{j_{nk}}{R}\right)^2 T=0;\ T'(0)=0.$$

Solutions are constant multiples of

$$T_{nk}(r)=\cos\left(\frac{j_{nk}ct}{R}\right).$$

We now have functions

$$z_{nk}(r,\theta,t)=[a_{nk}\cos(n\theta)+b_{nk}\sin(n\theta)]J_n\left(\frac{j_{nk}}{R}r\right)\cos\left(\frac{j_{nk}}{R}ct\right)$$

for $n=0,1,2\cdots$ and $k=1,2,3,\cdots$. Each of these functions satisfies the wave equation and the boundary conditions, together with the initial condition of zero velocity. To satisfy the condition on initial position given by f, we must in general use the superposition

$$z(r,\theta,t)=\sum_{n=0}^{\infty}\sum_{k=1}^{\infty}[a_{nk}\cos(n\theta)+b_{nk}\sin(n\theta)]J_n\left(\frac{j_{nk}}{R}r\right)\cos\left(\frac{j_{nk}}{R}ct\right).$$

We must choose the constants a_{nk} and b_{nk} to satisfy

$$z(r,\theta,0)=\sum_{n=0}^{\infty}\sum_{k=1}^{\infty}[a_{nk}\cos(n\theta)+b_{nk}\sin(n\theta)]J_n\left(\frac{j_{nk}}{R}r\right)=f(r,\theta)).$$

To see how to choose these coefficients, first write this equation in the more suggestive form

$$f(r,\theta)=\sum_{k=1}^{\infty}a_{0k}J_0\left(\frac{j_{nk}}{R}r\right)$$
$$+\sum_{n=1}^{\infty}\left(\left[\sum_{k=1}^{\infty}a_{nk}J_n\left(\frac{j_{nk}}{R}r\right)\right]\cos(n\theta)+\sum_{n=1}^{\infty}\left[\sum_{k=1}^{\infty}b_{nk}J_n\left(\frac{j_{nk}}{R}r\right)\right]\sin(n\theta)\right).$$

For a given r, $f(r,\theta)$ is a function of θ, and this is the Fourier series for this function of θ on $[-\pi,\pi]$. In this case, the coefficients are infinite series, but they are also the Fourier coefficients of $f(r,\theta)$ for a fixed r. We know these Fourier coefficients. For a given r,

$$\sum_{k=1}^{\infty}a_{0k}J_0\left(\frac{j_{nk}}{R}r\right)=\frac{1}{2\pi}\int_{-\pi}^{\pi}f(r,\theta)\,d\theta=\alpha_0(r),$$

and, for $n=1,2,\cdots$,

$$\sum_{k=1}^{\infty}a_{nk}J_n\left(\frac{j_{nk}}{R}r\right)=\frac{1}{\pi}\int_{-\pi}^{\pi}f(r,\theta)\cos(n\theta)\,d\theta=\alpha_n(r),$$

and

$$\sum_{k=1}^{\infty}b_{nk}J_n\left(\frac{j_{nk}}{R}r\right)=\frac{1}{\pi}\int_{-\pi}^{\pi}f(r,\theta)\sin(n\theta)\,d\theta=\beta_n(r),$$

Now recognize that, for $n=0,1,2,\cdots$, each of the last three equations is an expansion of a function of r in a Fourier-Bessel series, with coefficients a_{0k}, a_{nk} and b_{nk}, respectively. We know the coefficients in these expansions. Make the change of variables $r/R=\xi$ so that ξ varies from

0 to 1 as r varies from 0 to R. In this way, we can use the formula for Fourier-Bessel coefficients on $(0, 1)$. We have

$$a_{0k} = \frac{2}{[J_1(j_{0k})]^2} \int_0^1 \xi \alpha_0(R\xi) J_0(j_{0k}\xi)\, d\xi \text{ for } k = 1, 2, 3, \cdots,$$

$$a_{nk} = \frac{2}{[J_{n+1}(j_{nk})]^2} \int_0^1 \xi \alpha_n(R\xi) J_n(j_{nk}\xi)\, d\xi \text{ for } k = 1, 2, 3, \cdots \text{ and } n = 1, 2, \cdots,$$

and

$$b_{nk} = \frac{2}{[J_{n+1}(j_{nk})]^2} \int_0^1 \xi \beta_n(R\xi) J_n(j_{nk}\xi)\, d\xi.$$

For a given problem, first perform the integrations with respect to θ to obtain the functions $\alpha_0(r)$, $\alpha_n(r)$ and $\beta_n(r)$, written as Fourier-Bessel series. Then integrate to obtain the coefficients a_{nk} and b_{nk} in these expansions. These integrations require a computational software package.

SECTION 16.8 PROBLEMS

1. Approximate the vertical deflections of the center of a circular membrane of radius 2 for any time $t > 0$ by computing the first three nonzero terms in the solution for the case $c = 2$ and the initial displacement $f(r, \theta) = (4 - r^2)\sin^2(\theta)$. Assume zero initial velocity.

2. Use the general solution derived in this section to prove the plausible fact that the center of the membrane remains undeflected for all time if the initial displacement is an odd function of θ (that is, $f(r, \theta) = -f(r, -\theta)$. *Hint*: The only integer order Bessel function that is different from zero at $r = 0$ is J_0.

16.9 Vibrations in a Rectangular Membrane

Suppose an elastic membrane is attached to a rectangular frame that occupies the region $0 \le x \le L, 0 \le y \le K$. The membrane is given an initial displacement and released with a given initial velocity. We want the displacement function $z(x, y, t)$.

The initial-boundary value problem for z is

$$\frac{\partial^2 z}{\partial t^2} = c^2 \left(\frac{\partial^2 z}{\partial x^2} + \frac{\partial^2 z}{\partial y^2} \right) \text{ for } 0 < x < L, 0 < y < K, t > 0,$$

$$z(x, 0, t) = z(x, K, t) = 0 \text{ for } 0 < x < L, t > 0,$$

$$z(0, y, t) = z(L, y, t) = 0 \text{ for } 0 < y < K, t > 0,$$

$$z(x, y, 0) = f(x, y) \text{ for } 0 < x < L, 0 < y < K,$$

and

$$\frac{\partial z}{\partial t}(x, y, 0) = g(x, y) \text{ for } 0 < x < L, 0 < y < K.$$

We will solve this for the case $g(x, y) = 0$, so the membrane is displaced and released from rest. To attempt a separation of variables, substitute $z(x, y, t) = X(x)Y(y)T(t)$ into the wave equation to get

$$XYT'' = c^2(X''YT + XY''T)$$

or

$$\frac{T''}{c^2 T} - \frac{Y''}{Y} = \frac{X''}{X}.$$

The left side depends only on y and t and the right only on x, and these variables are independent, so both sides must be constant:

$$\frac{T''}{c^2T} - \frac{Y''}{Y} = \frac{X''}{X} = -\lambda.$$

Then

$$X'' + \lambda X = 0 \text{ and } \frac{T''}{c^2T} + \lambda = \frac{Y''}{Y}.$$

The equation for T and Y has one side dependent only on t and the other side only on y. Therefore, for some constant μ,

$$\frac{T''}{c^2T} + \lambda = \frac{Y''}{Y} = -\mu.$$

Then

$$Y'' + \mu Y = 0 \text{ and } T'' + c^2(\lambda + \mu)T = 0.$$

Separation of variables has introduced two separation constants. From the boundary conditions,

$$X(0) = X(L) = Y(0) = Y(K) = 0.$$

We have solved these problems for X and Y before, obtaining eigenvalues and eigenfunctions

$$\lambda_n = \frac{n^2\pi^2}{L^2},\ X_n(x) = \sin\left(\frac{n\pi x}{L}\right)$$

and

$$\mu_m = \frac{m^2\pi^2}{L^2},\ Y_m(x) = \sin\left(\frac{m\pi y}{K}\right)$$

with n and m varying independently over the positive integers. The problem for T becomes

$$T'' + c^2\left(\frac{n^2\pi^2}{L^2} + \frac{m^2\pi^2}{K^2}\right)T = 0.$$

With zero initial velocity we have $T'(0) = 0$. Therefore, $T(t)$ must be a constant multiple of $\cos(\alpha_{nm}\pi ct)$, where

$$\alpha_{nm} = \sqrt{\frac{n^2}{L^2} + \frac{m^2}{K^2}}.$$

For each positive integer n and m, we now have functions

$$z_{nm}(x, y, t) = a_{nm} \sin\left(\frac{n\pi x}{L}\right) \sin\left(\frac{m\pi y}{K}\right) \cos(\alpha_{nm}\pi ct).$$

that satisfy the wave equation and the boundary conditions, as well as the condition of zero initial velocity. To satisfy $z(x, y, 0) = f(x, y)$, attempt a superposition, which is now a double sum:

$$z(x, y, t) = \sum_{n=1}^{\infty}\sum_{m=1}^{\infty} z_{nm}(x, y, t).$$

We must choose the coefficients so that

$$z(x, y, 0) = \sum_{n=1}^{\infty}\sum_{m=1}^{\infty} a_{nm} \sin\left(\frac{n\pi x}{L}\right) \sin\left(\frac{m\pi y}{K}\right) = f(x, y).$$

If we think of y as fixed for the moment, then $f(x, y) = h_y(x)$ is a function of x. Now

$$f(x, y) = h_y(x) = \sum_{n=1}^{\infty}\left[\sum_{m=1}^{\infty} a_{nm} \sin\left(\frac{m\pi y}{K}\right)\right] \sin\left(\frac{n\pi x}{L}\right)$$

is the Fourier sine expansion in x of $f(x, y)$ on $[0, L]$. Therefore, the coefficient of $\sin(n\pi x/L)$, which is the entire sum in square brackets, is the Fourier sine coefficient of this function. For a given n,

$$\sum_{m=1}^{\infty} a_{nm} \sin\left(\frac{m\pi y}{K}\right) = \frac{2}{L}\int_0^L h_y(\xi)\sin\left(\frac{n\pi\xi}{L}\right) d\xi$$

$$= \frac{2}{L}\int_0^L f(\xi, y)\sin\left(\frac{n\pi\xi}{L}\right) d\xi$$

The integral on the right is a function of y, and on the left is a series that we can think of as its Fourier sine expansion on $[0, K]$, with the a_{nm} 's as coefficients. Therefore,

$$a_{nm} = \frac{2}{K}\int_0^K \left[\frac{2}{L}\int_0^L f(\xi, y)\sin\left(\frac{n\pi\xi}{L}\right) d\xi\right] \sin\left(\frac{m\pi\eta}{K}\right) d\eta$$

$$= \frac{4}{LK}\int_0^L \int_0^K f(\xi, \eta)\sin\left(\frac{n\pi\xi}{L}\right) \sin\left(\frac{m\pi\eta}{K}\right) d\eta\, d\xi.$$

With this choice of constants we have the solution for the displacement function $z(x, y, t)$.

EXAMPLE 16.15

Suppose the initial position function is

$$z(x, y, 0) = x(L-x)y(K-y)$$

and the initial velocity is zero. Compute

$$a_{nm} = \frac{4}{LK}\int_0^L \int_0^K \xi(L-\xi)\eta(K-\eta)\sin\left(\frac{n\pi\xi}{L}\right) \sin\left(\frac{m\pi\eta}{K}\right) d\eta\, d\xi$$

$$= \frac{16L^2K^2}{(nm\pi^2)^3}[(-1)^n - 1][(-1)^m - 1].$$

The solution is

$$z(x, y, t) =$$

$$\sum_{n=1}^{\infty}\sum_{m=1}^{\infty} \frac{16L^2K^2}{(nm\pi^2)^3}[(-1)^n - 1][(-1)^m - 1]\sin\left(\frac{n\pi x}{L}\right)\sin\left(\frac{m\pi y}{K}\right)\cos(\alpha_{nm}\pi ct).$$ ♦

SECTION 16.9 PROBLEMS

In each of Problems 1, 2, and 3, solve the problem for the rectangular membrane with the given c, L, K, $f(x, y)$ and $g(x, y)$.

1. $c=1, L=K=2\pi, f(x, y)=x^2\sin(y), g(x, y)=0$
2. $c=3, L=K=\pi, f(x, y)=0, g(x, y)=xy$
3. $c=2, L=K=2\pi, f(x, y)=0, g(x, y)=1$

INITIAL AND BOUNDARY CONDITIONS
THE HEAT EQUATION ON $[0, L]$
SOLUTIONS IN AN INFINITE MEDIUM
LAPLACE TRANSFORM TECHNIQUES HEAT

CHAPTER 17 The Heat Equation

17.1 Initial and Boundary Conditions

In Section 12.8, we used Gauss's divergence theorem to derive a partial differential equation modeling heat distribution, or diffusion. In the absence of sources or sinks within the medium, the one-dimensional heat equation is

$$\frac{\partial u}{\partial t} = k\frac{\partial^2 u}{\partial x^2} \tag{17.1}$$

in which k is a constant depending on the medium.

Equation (17.1) can be solved subject to a variety of boundary and initial conditions. For example,

$$\frac{\partial u}{\partial t} = k\frac{\partial^2 u}{\partial x^2} \text{ for } 0 < x < L, t > 0,$$

$$u(0, t) = T_1, u(L, t) = T_2 \text{ for } t \geq 0,$$

$$u(x, 0) = f(x) \text{ for } 0 \leq x \leq L$$

models the temperature distribution in a thin homogeneous bar of length L whose left end is kept at temperature T_1 and right end at temperature T_2, and having initial temperature $f(x)$ in the cross section at x.

The initial-boundary value problem

$$\frac{\partial u}{\partial t} = k\frac{\partial^2 u}{\partial x^2} \text{ for } 0 < x < L, t > 0,$$

$$\frac{\partial u}{\partial x}(0, t) = \frac{\partial u}{\partial x}(L, t) = 0 \text{ for } t \geq 0,$$

and

$$u(x, 0) = f(x) \text{ for } 0 \leq x \leq L$$

models the distribution in a bar of length L having no heat loss across its ends (*insulation conditions*) and initial temperature function f.

Other kinds of boundary conditions also can be specified. If the left end is kept at constant temperature T and the right end is insulated, then we would have

$$u(0,t)=T \text{ and } \frac{\partial u}{\partial x}(L,t)=0 \text{ for } t>0.$$

Free radiation or *convection* occurs when the bar loses energy by radiation from its ends into the surrounding medium, which is assumed to be maintained at constant temperature T. Now the boundary conditions have the form

$$\frac{\partial u}{\partial x}(0,t)=A[u(0,t)-T],\ \frac{\partial u}{\partial x}(L,t)=-A[u(L,t)-T] \text{ for } t\geq 0,$$

in which A is a positive constant. Notice that, if the bar is kept hotter than the surrounding medium, then the heat flow as measured by $\partial u/\partial x$ must be positive at one end and negative at the other.

Boundary conditions

$$u(0,t)=T_1,\ \frac{\partial u}{\partial x}(L,t)=-A[u(L,t)-T_2]$$

are used if the left end is kept at constant temperature T_1 while the right end radiates heat energy into a medium of constant temperature T_2.

As with the wave equation, we also consider the heat equation on the line or half-line, subject to various conditions.

SECTION 17.1 PROBLEMS

1. Formulate an initial-boundary value problem for the temperature distribution in a thin bar of length L if the left end is insulated and the right end is kept at temperature $\beta(t)$. The initial temperature function is f.
2. Formulate an initial-boundary value problem modeling heat conduction in a thin homogeneous bar of length L if the left end is kept at temperature $\alpha(t)$ and the right end at temperature $\beta(t)$. The initial temperature function in the cross section at x is $f(x)$.
3. Formulate an initial-boundary value problem modeling heat conduction in a thin homogeneous bar of length L if the left end is kept at temperature zero and the right end is insulated. The initial temperature function is f.

17.2 The Heat Equation on [0, L]

We will solve several initial-boundary value problems on an interval [0, L].

17.2.1 Ends Kept at Temperature Zero

If the initial temperature in the cross section at x is $f(x)$ and the ends of the bar are kept at temperature zero, the problem for the temperature distribution function is

$$\frac{\partial u}{\partial t}=k\frac{\partial^2 u}{\partial x^2} \text{ for } 0<x<L,\ t>0,$$

$$u(0,t)=u(L,t)=0 \text{ for } t\geq 0,$$

and

$$u(x,0)=f(x) \text{ for } 0\leq x\leq L.$$

Put $u(x,t) = X(x)T(t)$ into the heat equation to obtain

$$\frac{X''}{X} = \frac{T'}{kT} = -\lambda$$

in which λ is the separation constant. Now,

$$u(0,t) = X(0)T(t) = 0 = u(L,t) = X(L)T(t)$$

for all $t \geq 0$, so $X(0) = X(L) = 0$, and the problem for X is

$$X'' + \lambda X = 0;\ X(0) = X(L) = 0$$

with eigenvalues $\lambda_n = n^2\pi^2/L^2$ and eigenfunctions $\sin(n\pi x/L)$.

It is in the time dependence that the heat and wave equations differ. The equation for T is

$$T' + \frac{n^2\pi^2 k}{L^2}T = 0,$$

which is a first-order differential equation with solutions that are constant multiples of $e^{-n^2\pi^2kt/L^2}$. For $n = 1, 2, \cdots$, we have functions

$$u_n(x,t) = c_n \sin\left(\frac{n\pi x}{L}\right) e^{-n^2\pi^2kt/L^2},$$

which satisfy the heat equation and the boundary conditions $u(0,t) = u(L,t) = 0$. To satisfy the initial condition, we must (depending on f) use a superposition

$$u(x,t) = \sum_{n=1}^{\infty} c_n \sin\left(\frac{n\pi x}{L}\right) e^{-n^2\pi^2kt/L^2}$$

and choose the coefficients so that

$$u(x,0) = f(x) = \sum_{n=1}^{\infty} c_n \sin\left(\frac{n\pi x}{L}\right).$$

This is the Fourier sine expansion of f on $[0, L]$, so choose

$$c_n = \frac{2}{L}\int_0^L f(\xi)\sin\left(\frac{n\pi\xi}{L}\right) d\xi.$$

With this choice of the c_n's, the solution is

$$u(x,t) = \frac{2}{L}\sum_{n=1}^{\infty}\left(\int_0^L f(\xi)\sin(n\pi\xi/L)\,d\xi\right)\sin(n\pi x/L)e^{-n^2\pi^2kt/L^2}. \tag{17.2}$$

EXAMPLE 17.1

Suppose the ends are kept at zero temperature and the initial temperature is $f(x) = A$, which is constant. Compute

$$c_n = \frac{2}{L}\int_0^L A\sin(n\pi\xi/L)\,d\xi = \frac{2A}{n\pi}[1 - (-1)^n].$$

The solution is

$$u(x,t) = \frac{2A}{\pi}\sum_{n=1}^{\infty}\frac{1-(-1)^n}{n}\sin\left(\frac{n\pi x}{L}\right)e^{-n^2\pi^2kt/L^2}.$$

Since

$$1 - (-1)^n = \begin{cases} 2 & \text{if } n \text{ is odd} \\ 0 & \text{if } n \text{ is even,} \end{cases}$$

we can omit the even values of n in the summation to write the solution

$$u(x,t) = \frac{4A}{\pi} \sum_{n=1}^{\infty} \frac{1}{2n-1} \sin\left(\frac{(2n-1)\pi x}{L}\right) e^{-(2n-1)^2\pi^2 kt/L^2}. \quad \blacklozenge$$

17.2.2 Insulated Ends

Suppose the bar has insulated ends, hence no energy is lost across the ends. The temperature distribution is modeled by the initial-boundary value problem

$$\frac{\partial u}{\partial t} = k\frac{\partial^2 u}{\partial x^2} \text{ for } 0 < x < L, t > 0,$$

$$\frac{\partial u}{\partial x}(0,t) = \frac{\partial u}{\partial x}(L,t) = 0 \text{ for } t \geq 0,$$

and

$$u(x,0) = f(x) \text{ for } 0 \leq x \leq L.$$

As before, separation of variables yields

$$X'' + \lambda X = 0 \text{ and } T' + \lambda k T = 0.$$

The insulation conditions give us

$$\frac{\partial u}{\partial x}(0,t) = X'(0)T(t) = \frac{\partial u}{\partial x}(L,t) = X'(L)T(t) = 0,$$

so $X'(0) = X'(L) = 0$, and the problem for X is

$$X'' + \lambda X = 0; \; X'(0) = X'(L) = 0.$$

Previously, we solved this problem for the eigenvalues $\lambda_n = n^2\pi^2/L^2$ and eigenfunctions $\cos(n\pi x/L)$ for $n = 0, 1, 2, \cdots$.

The equation for T is

$$T' + \frac{n^2\pi^2 k}{L^2} T = 0.$$

For $n = 0$, we get $T_0(t) = \text{constant}$. For $n = 1, 2, \cdots$,

$$T_n(t) = e^{-n^2\pi^2 kt/L^2},$$

or constant multiples of this function. We now have a function

$$u_n(x,t) = c_n \cos\left(\frac{n\pi x}{L}\right) e^{-n^2\pi^2 kt/L^2},$$

which satisfies the heat equation and the insulation boundary conditions for $n = 0, 1, 2, \cdots$. To satisfy the initial condition, use the superposition

$$u(x,t) = \frac{1}{2}c_0 + \sum_{n=1}^{\infty} c_n \cos\left(\frac{n\pi x}{L}\right) e^{-n^2\pi^2 kt/L^2}.$$

Then

$$u(x,0) = f(x) = \frac{1}{2}c_0 + \sum_{n=1}^{\infty} c_n \cos\left(\frac{n\pi x}{L}\right),$$

so choose the c_n's to be the Fourier cosine coefficients of f on $[0, L]$:

$$c_n = \frac{2}{L}\int_0^L f(\xi)\cos\left(\frac{n\pi\xi}{L}\right) d\xi.$$

EXAMPLE 17.2

Suppose the ends of the bar are insulated and the left half of the bar is initially at constant temperature A while the right half is initially at temperature zero. Then

$$f(x) = \begin{cases} A & \text{for } 0 \le x \le L/2 \\ 0 & \text{for } L/2 < x \le L. \end{cases}$$

Compute

$$c_0 = \frac{2}{L}\int_0^{L/2} A d\xi = A$$

and, for $n = 1, 2, \cdots,$

$$c_n = \frac{2}{L}\int_0^{L/2} A\cos\left(\frac{n\pi\xi}{L}\right) d\xi = \frac{2A}{n\pi}\sin(n\pi/2).$$

The solution is

$$u(x,t) = \frac{A}{2} + \frac{2A}{\pi}\sum_{n=1}^{\infty}\frac{1}{n}\sin\left(\frac{n\pi}{2}\right)\cos\left(\frac{n\pi x}{L}\right)e^{-n^2\pi^2 kt/L^2}.$$

Since $\sin(n\pi/2) = 0$ if n is even, we can retain only odd n in this summation to write this solution as

$$u(x,t) = \frac{A}{2} + \frac{2A}{\pi}\sum_{n=1}^{\infty}\frac{1}{2n-1}\cos\left(\frac{(2n-1)\pi x}{L}\right)e^{-(2n-1)^2\pi^2 kt/L^2}. \ \blacklozenge$$

17.2.3 Radiating End

Suppose the left end of the bar is maintained at temperature zero, while the right end radiates energy into the surrounding medium, which is kept at temperature zero. If the initial temperature function is f, then the temperature distribution is modeled by the initial-boundary value problem

$$\frac{\partial u}{\partial t} = k\frac{\partial^2 u}{\partial x^2} \text{ for } 0 < x < L, t > 0,$$

$$u(0,t) = 0, \ \frac{\partial u}{\partial x}(L,t) = -Au(L,t) \text{ for } t > 0,$$

and

$$u(x,0) = f(x) \text{ for } 0 \le x \le L.$$

A is a positive constant called the *transfer coefficient*. Let $u(x,t) = X(x)T(t)$ to obtain

$$X'' + \lambda X = 0, \ T' + \lambda kT = 0.$$

Because $u(0,t) = X(0)T(t) = 0$, then $X(0) = 0$. From the radiation condition at the right end,

$$X'(L)T(t) = -AX(L)T(t) \text{ for } t > 0,$$

so

$$X'(L) + AX(L) = 0.$$

The problem for X is

$$X'' + \lambda X = 0; \ X(0) = 0, \ X'(L) + AX(L) = 0.$$

This is a regular Sturm-Liouville problem. To solve for the eigenvalues and eigenfunctions, consider cases on λ.

Case 1: $\lambda = 0$

We get $X(x) = cx + d$, so $X(0) = d = 0$. Then $X(x) = cx$, so

$$X'(L) + AX(L) = c + cL = 0$$

and this forces $c = 0$. This case yields only the trivial solution, so 0 is not an eigenvalue of this problem.

Case 2: $\lambda < 0$

Write $\lambda = -\alpha^2$ with $\alpha > 0$. Then $X'' - \alpha^2 X = 0$, with solutions

$$X(x) = ce^{\alpha x} + de^{-\alpha x}.$$

Now $X(0) = c + d = 0$ implies that $d = -c$, so

$$X(x) = c(e^{\alpha x} - e^{-\alpha x}).$$

From the radiation condition at L,

$$X'(L) = \alpha c(e^{\alpha L} + e^{-\alpha L}) = -AX(L) = -Ac(e^{\alpha L} - e^{-\alpha L}).$$

If $c \neq 0$, then

$$\alpha c(e^{\alpha L} + e^{-\alpha L}) > 0 \text{ and } -Ac\left(e^{\alpha L} - e^{-\alpha L}\right) < 0,$$

contradicting the preceding line. We conclude that $c = 0$, so this case also has only the trivial solution. This problem has no negative eigenvalue.

Case 3: $\lambda > 0$

Set $\lambda = \alpha^2$ with $\alpha > 0$. Now $X'' + \alpha^2 X = 0$, so

$$X(x) = c\cos(\alpha x) + d\sin(\alpha x).$$

Then $X(0) = c = 0$, so $X(x) = d\sin(\alpha x)$. Next,

$$X'(L) = \alpha d\cos(\alpha L) = -AX(L) = -Ad\sin(\alpha L).$$

If $d = 0$, we have only the trivial solution. If $d \neq 0$, then $\alpha\cos(\alpha L) = -A\sin(\alpha L)$, so

$$\tan(\alpha L) = -\frac{\alpha}{A}. \tag{17.3}$$

We have a nontrivial solution for X only if α is chosen to satisfy this transcendental equation, which we cannot solve algebraically for α. However, let $z = \alpha L$. Then equation (17.3) is $\tan(z) = -z/AL$. Part of the graphs of $y = \tan(z)$ and $y = -z/AL$ are shown in Figure 17.1, and they have infinitely many points of intersection for $z > 0$. The z-coordinates of these points are solutions of $\tan(z) = -z/AL$. Let these z-coordinates be $z_1, z_2, \cdots$ in increasing order. Since $\alpha = z/L$, the eigenvalues of this problem for X are

$$\lambda_n = \alpha_n^2 = \frac{z_n^2}{L^2}.$$

The eigenfunctions are functions $X_n(x) = \sin(\alpha_n x) = \sin(z_n x/L)$.

With these eigenvalues, the problem for T is

$$T' + \frac{z_n^2 k}{L^2}T = 0$$

with solutions that are constant multiples of

$$T_n(t) = e^{-z_n^2 kt/L^2}.$$

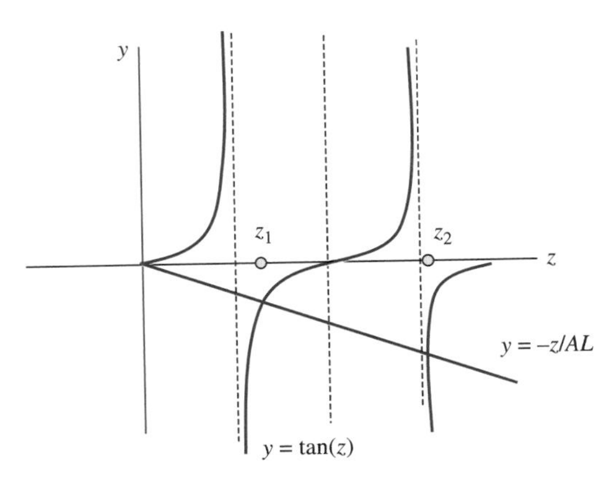

FIGURE 17.1 *Eigenvalues as the problem with radiating end.*

For each positive integer n, we now have a function

$$u_n(x,t) = X_n(x)T_n(t) = \sin\left(\frac{z_n x}{L}\right) e^{-z_n^2 kt/L^2}.$$

that satisfies the heat equation and the boundary conditions. To satisfy the initial condition, write the superposition

$$u(x,t) = \sum_{n=1}^{\infty} c_n \sin\left(\frac{z_n x}{L}\right) e^{-z_n^2 kt/L^2}.$$

We must choose the c_n's to satisfy

$$u(x,0) = f(x) = \sum_{n=1}^{\infty} c_n \sin\left(\frac{z_n x}{L}\right).$$

This is not a Fourier sine series, but it is an eigenfunction expansion in the eigenfunctions of a Sturm-Liouville problem. The coefficients are

$$c_n = \frac{\int_0^L f(\xi)\sin(z_n\xi/L)\,d\xi}{\int_0^L \sin^2(z_n\xi/L)\,d\xi}.$$

The difficulty in computing these numbers is one we frequently encounter in such problems. We do not have an explicit formula for the z_n's. However, we can compute approximate values of the z_n's for a given L. For the first four values, with $L = 1$, we obtain the approximate values

$$z_1 \approx 2.0288,\ z_2 \approx 49132,\ z_3 \approx 7.9787,\ z_4 \approx 11.0855.$$

Using these numbers, we can perform numerical integrations to approximate

$$c_1 \approx 1.9207,\ c_2 \approx 2.6593,\ c_3 \approx 4.1457,\ c_4 \approx 5.6329.$$

With just the first four terms, we have an approximation

$$u(x,t) \approx 1.9027\sin(2.0288x)e^{-4.1160kt} + 2.6593\sin(4.9132x)e^{-24.1395kt}$$
$$+ 4.1457\sin(7.9787x)e^{-63.6597kt} + 5.6329\sin(11.0855x)e^{-1,228.883kt}.$$

Depending on k, these exponentials may be decaying so fast that these first four terms would be a good enough approximation for some applications.

17.2.4 Transformation of Problems

As with the wave equation, it may be impossible to separate variables in a diffusion problem, depending on the partial differential equation and/or the boundary conditions. Sometimes we can transform such a problem into one to which separation of variables applies.

EXAMPLE 17.3

A thin, homogeneous bar extends from $x = 0$ to $x = L$. The left end is maintained at constant temperature T_1 and the right end at constant temperature T_2. The initial temperature in the cross section at x is $f(x)$.

The initial-boundary value problem modeling this setting is

$$\frac{\partial u}{\partial t} = k\frac{\partial^2 u}{\partial x^2} \text{ for } 0 < x < L, t > 0,$$

$$u(0,t) = T_1, u(L,t) = T_2 \text{ for } t > 0,$$

and

$$u(x,0) = f(x) \text{ for } 0 \le x \le L.$$

If we set $u(x,t) = X(x)T(t)$, we must have $X(0)T(t) = T_1$. If $T_1 = 0$, this is satisfied by requiring that $X(0) = 0$. If $T_1 \neq 0$, then $X(0) \neq 0$, and we must have $T(t) = T_1/X(0) =$ constant, which is not acceptable.

Perturb the temperature distribution function by setting

$$u(x,t) = U(x,t) + \psi(x).$$

The idea is to choose ψ to obtain a problem we know how to solve. Substitute U into the wave equation to get

$$\frac{\partial U}{\partial t} = k\left(\frac{\partial^2 U}{\partial x^2} + \psi''(x)\right).$$

This is the standard heat equation (17.1) if $\psi''(x) = 0$, hence, if $\psi(x) = cx + d$. Now

$$u(0,t) = T_1 = U(0,t) + \psi(0),$$

and this is the more friendly condition $U(0,t) = 0$ if $\psi(0) = T_1$. Thus, choose $d = T_1$ to have $\psi(x) = cx + T_1$. Next we need

$$u(L,t) = T_2 = U(L,t) + \psi(L),$$

and this is just $U(L,t) = 0$ if $\psi(L) = T_2$. We need $cL + T_1 = T_2$, so $c = (T_2 - T_1)/L$, and

$$\psi(x) = \frac{1}{L}(T_2 - T_1)x + T_1.$$

Finally,

$$u(x,0) = f(x) = U(x,0) + \psi(x),$$

so

$$U(x,0) = f(x) - \psi(x) = f(x) - \frac{1}{L}(T_2 - T_1)x - T_1.$$

The problem for U is:

$$\frac{\partial U}{\partial t} = k\frac{\partial^2 U}{\partial x^2} \text{ for } 0 < x < L, t > 0,$$
$$U(0,t) = U(L,t) = 0 \text{ for } t > 0,$$

and

$$U(x,0) = f(x) - \frac{1}{L}(T_2 - T_1)x - T_1 \text{ for } 0 \le x \le L.$$

We know the solution of this problem for U:

$$U(x,t) = \frac{2}{L}\sum_{n=1}^{\infty} c_n \sin\left(\frac{n\pi x}{L}\right)e^{-n^2\pi^2 kt/L^2}$$

where

$$c_n = \frac{2}{L}\int_0^L \left[f(\xi) - \frac{1}{L}(T_2 - T_1)\xi - T_1\right]\sin\left(\frac{n\pi\xi}{L}\right)d\xi.$$

The solution of the original problem is

$$u(x,t) = U(x,t) + \frac{1}{L}(T_2 - T_1)x + T_1.$$

As a specific example, suppose $T_1 = 1$, $T_2 = 2$ and $f(x) = 3/2$. Now

$$\psi(x) = \frac{1}{L}x + 1,$$

and

$$\begin{aligned} c_n &= \frac{2}{L}\int_0^L \left[\frac{3}{2} - \frac{1}{L}x - 1\right]\sin\left(\frac{n\pi\xi}{L}\right)d\xi \\ &= \frac{2}{L}\int_0^L \left(\frac{1}{2} - \frac{1}{L}x\right)\sin\left(\frac{n\pi\xi}{L}\right)d\xi \\ &= \frac{1+(-1)^n}{n\pi}. \end{aligned}$$

The solution is

$$u(x,t) = \sum_{n=1}^{\infty}\left(\frac{1+(-1)^n}{n\pi}\right)\sin\left(\frac{n\pi x}{L}\right)e^{-n^2\pi^2 kt/L^2} + \frac{1}{L}x + 1. \blacklozenge$$

Sometimes an initial-boundary value problem involving the heat equation can be transformed into a simpler problem by multiplying by an exponential function $e^{\alpha x+\beta t}$ and making appropriate choices for α and β. We will pursue this idea in the problems.

17.2.5 The Heat Equation with a Source Term

We will determine the solution of the initial-boundary value problem

$$\frac{\partial u}{\partial t} = k\frac{\partial^2 u}{\partial x^2} + F(x,t) \text{ for } 0 < x < L, t > 0,$$
$$u(0,t) = u(L,t) = 0 \text{ for } t > 0,$$

and

$$u(x,0) = f(x) \text{ for } 0 \le x \le L.$$

The term $F(x,t)$, for example, could account for a source or loss of energy within the medium.

It is routine to try separation of variables and find that it does not apply in general to this heat equation. A different approach is needed.

As a starting point, we know that in the case that $F(x,t)=0$ for all x and t the solution has the form

$$u(x,t)=\sum_{n=1}^{\infty} c_n \sin\left(\frac{n\pi x}{L}\right) e^{-n^2\pi^2 kt/L^2}.$$

Taking a cue from this case, we will attempt a solution of the current problem of the form

$$u(x,t)=\sum_{n=1}^{\infty} T_n(t)\sin\left(\frac{n\pi x}{L}\right). \tag{17.4}$$

We must find the functions $T_n(t)$. If t is fixed, the left side of equation (17.4) is a function of x and the right side is its Fourier sine expansion on [0, L], so the coefficients must be

$$T_n(t)=\frac{2}{L}\int_0^L u(\xi,t)\sin\left(\frac{n\pi\xi}{L}\right)d\xi. \tag{17.5}$$

Assume that for any $t\geq 0$, $F(x,t)$, thought of as a function of x, also can be expanded in a Fourier sine series on [0, L]:

$$F(x,t)=\sum_{n=1}^{\infty} B_n(t)\sin\left(\frac{n\pi\xi}{L}\right) \tag{17.6}$$

where

$$B_n(t)=\frac{2}{L}\int_0^L F(\xi,t)\sin\left(\frac{n\pi\xi}{L}\right)d\xi. \tag{17.7}$$

Differentiate equation (17.5) with respect to t to get

$$T_n'(t)=\frac{2}{L}\int_0^L \frac{\partial u}{\partial t}(\xi,t)\sin\left(\frac{n\pi\xi}{L}\right)d\xi. \tag{17.8}$$

Substitute for $\partial u/\partial t$ from the heat equation to get

$$T_n'(t)=\frac{2k}{L}\int_0^L \frac{\partial^2 u}{\partial x^2}(\xi,t)\sin\left(\frac{n\pi\xi}{L}\right)d\xi+\frac{2}{L}\int_0^L F(\xi,t)\sin\left(\frac{n\pi\xi}{L}\right)d\xi.$$

In view of equation (17.6), this equation becomes

$$T_n'(t)=\frac{2k}{L}\int_0^L \frac{\partial^2 u}{\partial x^2}(\xi,t)\sin\left(\frac{n\pi\xi}{L}\right)d\xi+B_n(t). \tag{17.9}$$

Integrate by parts twice, at the last step making use of the boundary conditions and equation (17.5). We get

$$\begin{aligned}
&\int_0^L \frac{\partial^2 u}{\partial x^2}(\xi,t)\sin\left(\frac{n\pi\xi}{L}\right)d\xi\\
&=\left[\frac{\partial u}{\partial x}(\xi,t)\sin\left(\frac{n\pi\xi}{L}\right)\right]_0^L-\int_0^L \frac{n\pi}{L}\frac{\partial u}{\partial x}(\xi,t)\cos\left(\frac{n\pi\xi}{L}\right)d\xi\\
&=-\frac{n\pi}{L}\left[u(\xi,t)\cos\left(\frac{n\pi\xi}{L}\right)\right]_0^L+\frac{n\pi}{L}\int_0^L -\frac{n\pi}{L}u(\xi,t)\sin\left(\frac{n\pi\xi}{L}\right)d\xi\\
&=-\frac{n^2\pi^2}{L^2}\int_0^L u(\xi,t)\sin\left(\frac{n\pi\xi}{L}\right)d\xi\\
&=-\frac{n^2\pi^2}{L^2}\frac{L}{2}T_n(t)=-\frac{n^2\pi^2}{2L}T_n(t).
\end{aligned}$$

Substitute this into equation (17.9) to get

$$T_n' = -\frac{n^2\pi^2 k}{L^2} T_n(t) + B_n(t).$$

For $n = 1, 2, \cdots$, this is a first-order differential equation for $T_n(t)$, which we write as

$$T_n'(t) + \frac{n^2\pi^2 k}{L^2} T_n(t) = B_n(t).$$

Next, use equation (17.5) to obtain the condition

$$T_n(0) = \frac{2}{L}\int_0^L u(\xi, 0)\sin\left(\frac{n\pi\xi}{L}\right) d\xi = \frac{2}{L}\int_0^L f(\xi)\sin\left(\frac{n\pi\xi}{L}\right) d\xi = b_n$$

which is the nth coefficient in the Fourier sine expansion of f on $[0, L]$. Solve the differential equation for $T_n(t)$ subject to this boundary condition to get

$$T_n(t) = \int_0^t e^{-n^2\pi^2 k(t-\tau)/L^2} B_n(\tau)\, d\tau + b_n e^{-n^2\pi^2 kt/L^2}.$$

Finally, substitute this into equation (17.4) to obtain the solution

$$u(x,t) = \sum_{n=1}^{\infty}\left(\int_0^t e^{-n^2\pi^2 k(t-\tau)/L^2} B_n(\tau)\, d\tau\right)\sin\left(\frac{n\pi x}{L}\right)$$
$$+ \frac{2}{L}\sum_{n=1}^{\infty}\left(\int_0^L f(\xi)\sin\left(\frac{n\pi\xi}{L}\right) d\xi\right)\sin\left(\frac{n\pi x}{L}\right) e^{-n^2\pi^2 kt/L^2}. \quad (17.10)$$

Notice that the last term is the solution of the problem if $F(x,t) = 0$, while the first term is the effect of this source term on the solution.

EXAMPLE 17.4

We will solve the problem

$$\frac{\partial u}{\partial t} = 4\frac{\partial^2 u}{\partial x^2} + xt \text{ for } 0 < x < \pi, t > 0,$$
$$u(0,t) = u(\pi,t) = 0 \text{ for } t \geq 0,$$

and

$$u(x,0) = f(x) = \begin{cases} 20 & \text{for } 0 \leq x \leq \pi/4 \\ 0 & \text{for } \pi/4 < x \leq \pi. \end{cases}$$

Since we have a formula (17.10) for the solution, we need only carry out the integrations. First compute

$$B_n(t) = \frac{2}{\pi}\int_0^\pi \xi t \sin(n\xi)\, d\xi = \frac{2(-1)^{n+1}}{n} t.$$

This enables us to evaluate

$$\int_0^t e^{-4n^2(t-\tau)} B_n(\tau)\, d\tau = \int_0^t \frac{2(-1)^{n+1}}{n} \tau e^{-4n^2(t-\tau)}\, d\tau$$
$$= \frac{1}{8}(-1)^{n+1}\frac{-1 + 4n^2 t + e^{-4n^2 t}}{n^5}.$$

Finally we need

$$b_n = \frac{2}{\pi}\int_0^{\pi} f(\xi)\sin(n\xi)\,d\xi = \frac{40}{\pi}\int_0^{\pi/4}\sin(n\xi)\,d\xi = \frac{40}{\pi}\frac{1-\cos(n\pi/4)}{\pi}.$$

With all the components in place, the solution is

$$u(x,t) = \sum_{n=1}^{\infty}\left(\frac{1}{8}(-1)^{n+1}\frac{-1+4n^2t+e^{-4n^2t}}{n^5}\right)\sin(nx)$$
$$+\sum_{n=1}^{\infty}\frac{40}{\pi}\frac{1-\cos(n\pi/4)}{n}\sin(nx)e^{-4n^2t}.$$

The second term on the right is the solution of the problem with the xt term omitted. Denote this "no-source" solution as

$$u_0(x,t) = \sum_{n=1}^{\infty}\frac{40}{\pi}\frac{1-\cos(n\pi/4)}{n}\sin(nx)e^{-4n^2t}.$$

Then the solution with the source term is

$$u(x,t) = u_0(x,t) + \sum_{n=1}^{\infty}\left(\frac{1}{8}(-1)^{n+1}\frac{-1+4n^2t+e^{-4n^2t}}{n^5}\right)\sin(nx).$$

Writing the solution in this way enables us to gauge the effect of the xt term on the solution. Figures 17.2 through 17.5 compare graphs of $u(x,t)$ and $u_0(x,t)$ at times $t = 0.3, 0.8, 1.2$ and 1.3. In each figure, $u_0(x,t)$ falls below $u(x,t)$. Both solutions decay to zero as t increases, but the effect of the xt term is to retard this decay. ♦

17.2.6 Effects of Boundary Conditions and Constants

We will investigate how constants and terms appearing in diffusion problems influence the behavior of solutions.

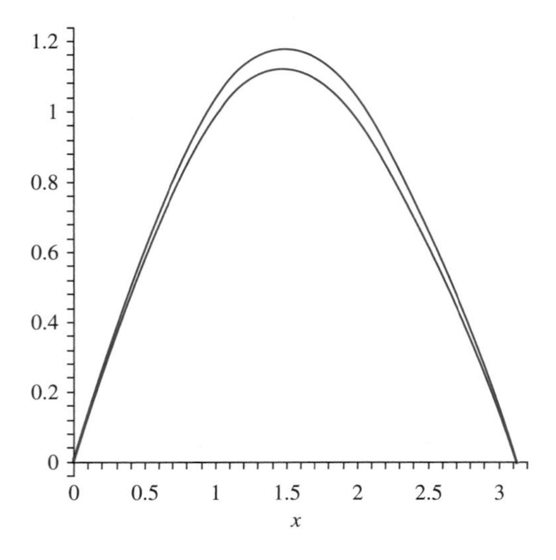

FIGURE 17.2 $u(x,t)$ and $u_0(x,t)$ at $t = 0.3$.

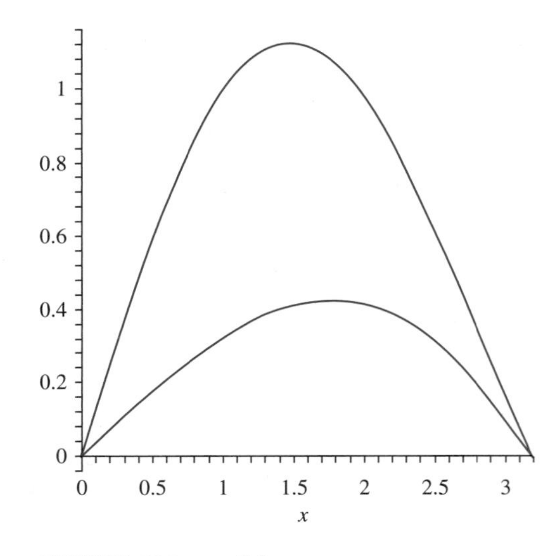

FIGURE 17.3 $t = 0.8$.

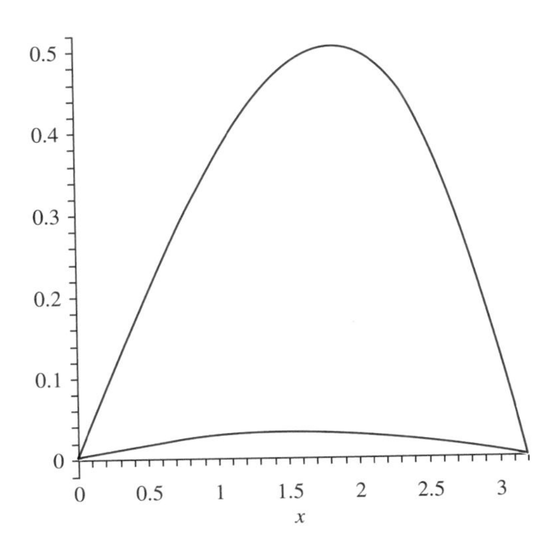

FIGURE 17.4 $t = 1.2$.

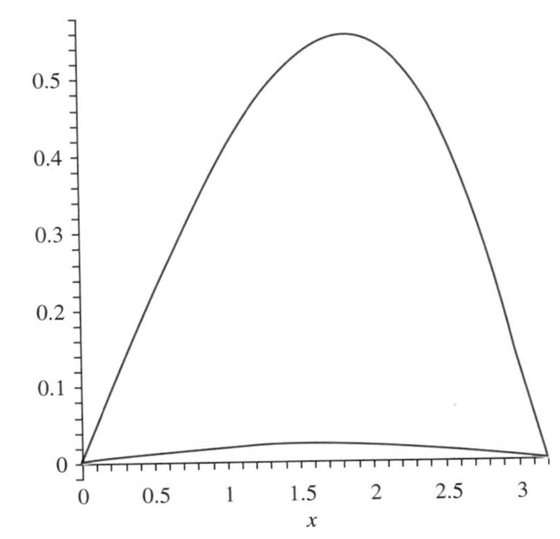

FIGURE 17.5 $t = 1.3$.

EXAMPLE 17.5

A homogeneous bar of length π has initial temperature function $f(x) = x^2\cos(x/2)$ and ends maintained at temperature zero. The temperature distribution function satisfies

$$\frac{\partial u}{\partial t} = k\frac{\partial^2 u}{\partial x^2} \text{ for } 0 < x < \pi, t > 0,$$
$$u(0,t) = u(\pi,t) = 0 \text{ for } t > 0,$$

and

$$u(x,0) = x^2\cos(x/2) \text{ for } 0 \le x \le \pi.$$

The solution is

$$\begin{aligned} u(x,t) &= \frac{2}{\pi}\sum_{n=1}^{\infty}\left(\int_0^{\pi}\xi^2\cos(\xi/2)\sin(n\xi)\,d\xi\right)\sin(nx)e^{-n^2kt} \\ &= \frac{4}{\pi}\sum_{n=1}^{\infty}\frac{16\pi n(-1)^n - 64\pi n^3(-1)^n - 48n - 64n^3}{64n^6 - 48n^4 + 12n^2 - 1}\sin(nx)e^{-n^2kt}. \end{aligned}$$

To gauge the effect of the diffusivity constant k on the solution, Figure 17.6 shows graphs of $y = u(x,t)$ for $t = 0.2$ and for $k = 0.3, 0.6, 1.1$, and 2.7. Figure 17.7 has the graphs for the same values of k, but $t = 1.2$. For each k, the temperature function decays with time, as we expect. However, for each time the temperature function has a smaller maximum as k increases. ♦

EXAMPLE 17.6

We will examine the effects on $u(x,t)$, depending on whether the ends of the bar are kept at temperature zero, or are insulated. Suppose the initial temperature function is $f(x) = x^2(\pi - x)$ and $L = \pi$ and $k = 1/4$. For the ends at temperature zero, we find that

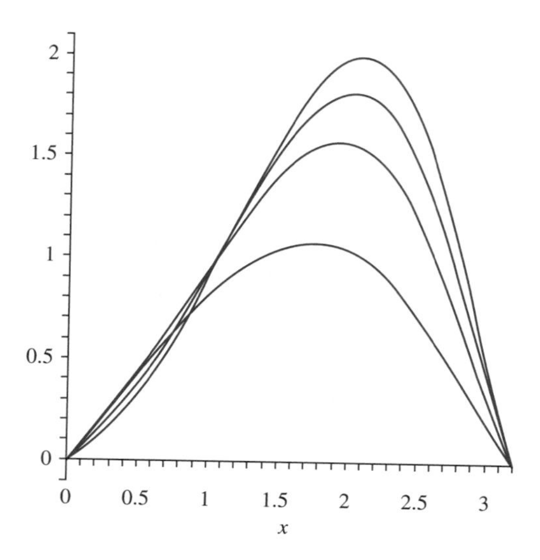

FIGURE 17.6 *The solution in Example 17.5 at* $t = 0.2$ *and* $k = 0.3, 0.6, 1.4,$ *and* 2.7.

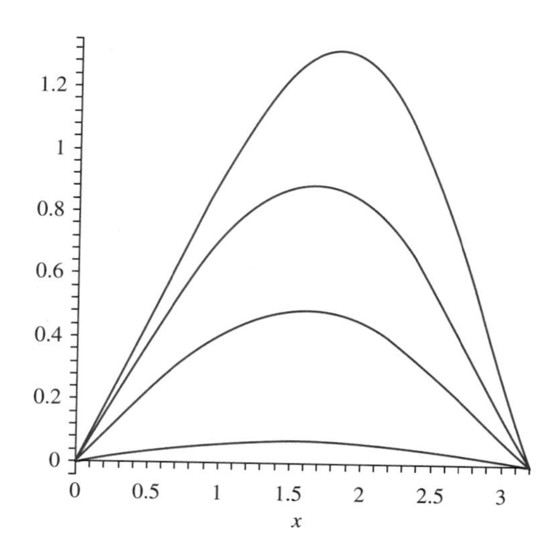

FIGURE 17.7 *The solution in Example 17.5 at* $t = 1.2$ *and* $k = 0.3, 0.6, 1.4,$ *and* 2.7.

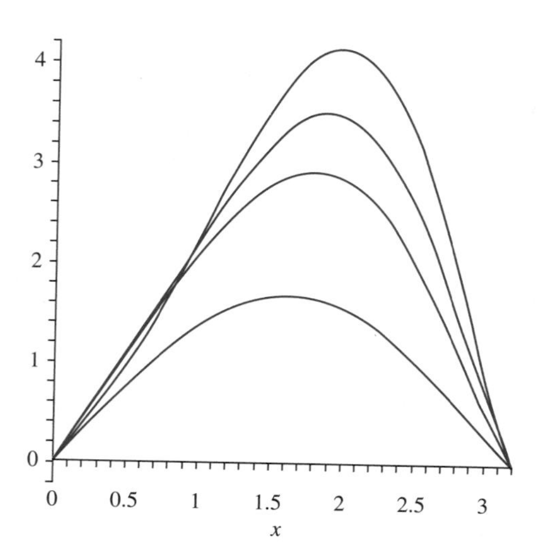

FIGURE 17.8 $u_1(x, t)$ *at times* $t = 0.4, 0.9, 1.5,$ *and* 3.6.

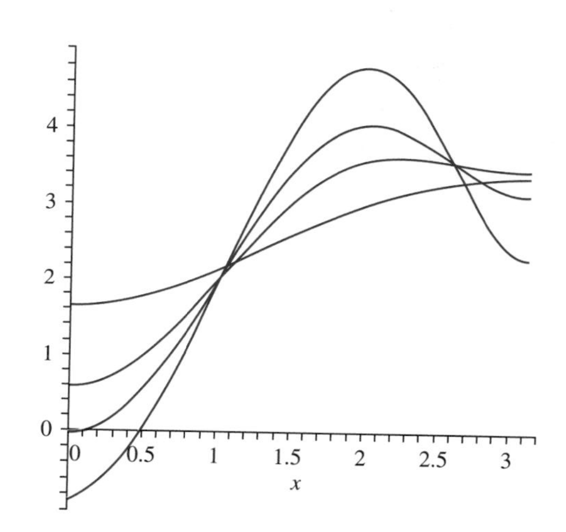

FIGURE 17.9 $u_2(x, t)$ *at times* $t = 0.4, 0.9, 1.5,$ *and* 3.6.

$$u_1(x, t) = \sum_{n=1}^{\infty} \left(\frac{8(-1)^{n+1} - 4}{n^3} \right) \sin(nx) e^{-n^2 t/4}.$$

If the ends are insulated the solution is

$$u_2(x, t) = \frac{1}{12}\pi^3 + \sum_{n=1}^{\infty} \left(\frac{n^2\pi^2(-1)^{n+1} + 6(-1)^n - 6}{n^4} \right) \cos(nx) e^{-n^2 t/4}.$$

Figure 17.8 shows $u_1(x, t)$, and Figure 17.9 shows $u_2(x, t)$, at times $t = 0.4, 0.9, 1.5$ and 3.6. Both solutions decrease as t increases. However, as $t \to \infty$, $u_2(x, t) \to \pi^3/12$, as suggested in Figure 17.9. ♦

SECTION 17.2 PROBLEMS

In each of Problems 1 through 7, write a solution of the initial-boundary value problem. Graph the twentieth partial sum of the series for $u(x,t)$ on the same set of axes for different values of the time.

1. $$\frac{\partial u}{\partial t}=3\frac{\partial^2 u}{\partial x^2} \text{ for } 0<x<L, t>0,$$
$$u(0,t)=u(L,t)=0 \text{ for } t\geq 0,$$
$$u(x,0)=L(1-\cos(2\pi x/L)) \text{ for } 0\leq x\leq L$$

2. $$\frac{\partial u}{\partial t}=4\frac{\partial^2 u}{\partial x^2} \text{ for } 0<x<3, t>0,$$
$$\frac{\partial u}{\partial x}(0,t)=\frac{\partial u}{\partial x}(3,t)=0 \text{ for } t\geq 0,$$
$$u(x,0)=x^2 \text{ for } 0\leq x\leq 3$$

3. $$\frac{\partial u}{\partial t}=2\frac{\partial^2 u}{\partial x^2} \text{ for } 0<x<6, t>0,$$
$$\frac{\partial u}{\partial x}(0,t)=\frac{\partial u}{\partial x}(6,t)=0 \text{ for } t\geq 0,$$
$$u(x,0)=e^{-x} \text{ for } 0\leq x\leq 6$$

4. $$\frac{\partial u}{\partial t}=4\frac{\partial^2 u}{\partial x^2} \text{ for } 0<x<L, t>0,$$
$$u(0,t)=u(L,t)=0 \text{ for } t\geq 0,$$
$$u(x,0)=x^2(L-x) \text{ for } 0\leq x\leq L$$

5. $$\frac{\partial u}{\partial t}=k\frac{\partial^2 u}{\partial x^2} \text{ for } 0<x<L, t>0,$$
$$u(0,t)=u(L,t)=0 \text{ for } t\geq 0,$$
$$u(x,0)=x(L-x) \text{ for } 0\leq x\leq L$$

6. $$\frac{\partial u}{\partial t}=\frac{\partial^2 u}{\partial x^2} \text{ for } 0<x<\pi, t>0,$$
$$\frac{\partial u}{\partial x}(0,t)=\frac{\partial u}{\partial x}(\pi,t)=0 \text{ for } t\geq 0,$$
$$u(x,0)=\sin(x) \text{ for } 0\leq x\leq \pi$$

7. $$\frac{\partial u}{\partial t}=4\frac{\partial^2 u}{\partial x^2} \text{ for } 0<x<2\pi, t>0,$$
$$\frac{\partial u}{\partial x}(0,t)=\frac{\partial u}{\partial x}(2\pi,t)=0 \text{ for } t\geq 0,$$
$$u(x,0)=x(2\pi-x) \text{ for } 0\leq x\leq 2\pi$$

8. A thin, homogeneous bar having thermal diffusivity of 9 and a length of 2 cm has insulated sides and its left end maintained at zero temperature, while its right end is perfectly insulated. The bar has an initial temperature $f(x)=x^2$ for $0\leq x\leq 2$. Determine the temperature distribution $u(x,t)$ and $\lim_{t\to 0}u(x,t)$.

9. Show that the partial differential equation
$$\frac{\partial u}{\partial t}=k\left(\frac{\partial^2 u}{\partial x^2}+A\frac{\partial u}{\partial x}+Bu\right)$$
can be transformed into a standard heat equation for v by letting $u(x,t)=e^{\alpha x+\beta t}v(x,t)$ and choosing α and β appropriately.

10. A thin, homogeneous bar of length L has insulated ends and initial temperature B, a positive constant. Find the temperature distribution in the bar.

11. A thin homogeneous bar of length L has initial temperature $f(x)=B$ where the right end $x=L$ is insulated, while the left end is kept at zero temperature. Find the temperature distribution in the bar.

12. Use the idea of Problem 9 to solve
$$\frac{\partial u}{\partial t}=\frac{\partial^2 u}{\partial x^2}+4\frac{\partial u}{\partial x}+2u \text{ for } 0<x<\pi, t>0,$$
$$u(0,t)=u(\pi,t)=0 \text{ for } t\geq 0,$$
$$u(x,0)=x(\pi-x) \text{ for } 0\leq x\leq \pi.$$

13. Solve
$$\frac{\partial u}{\partial t}=16\frac{\partial^2 u}{\partial x^2} \text{ for } 0<x<1, t>0,$$
$$u(0,t)=2, u(1,t)=5 \text{ for } t\geq 0,$$
$$u(x,0)=x^2 \text{ for } 0\leq x\leq 1.$$

14. Solve
$$\frac{\partial u}{\partial t}=9\frac{\partial^2 u}{\partial x^2} \text{ for } 0<x<L, t>0,$$
$$u(0,t)=T, u(L,t)=0 \text{ for } t\geq 0,$$
$$u(x,0)=0 \text{ for } 0\leq x\leq L.$$

15. Solve
$$\frac{\partial u}{\partial t}=4\frac{\partial^2 u}{\partial x^2}-Au \text{ for } 0<x<9, t>0,$$
$$u(0,t)=u(9,t)=0 \text{ for } t\geq 0,$$
$$u(x,0)=0 \text{ for } 0\leq x\leq 9.$$
Graph the twentieth partial sum of the solution for $t=0.2$ with $A=1/4$. Repeat this for $t=0.7$ and $t=1.4$

16. Solve
$$\frac{\partial u}{\partial t}=\frac{\partial^2 u}{\partial x^2}-6\frac{\partial u}{\partial x} \text{ for } 0<x<\pi, t>0,$$
$$u(0,t)=u(\pi,t)=0 \text{ for } t\geq 0,$$
$$u(x,0)=x(\pi-x) \text{ for } 0\leq x\leq \pi.$$
Graph the twentieth partial sum of the solution for selected times.

17. Solve

$$\frac{\partial u}{\partial t} = \frac{\partial^2 u}{\partial x^2} + 6\frac{\partial u}{\partial x} \text{ for } 0 < x < 4, t > 0,$$

$$u(0,t) = u(4,t) = 0 \text{ for } t \geq 0,$$

$$u(x,0) = 1 \text{ for } 0 \leq x \leq 4.$$

Graph the twentieth partial sum of the solution for selected times.

18. Solve

$$\frac{\partial u}{\partial t} = k\frac{\partial^2 u}{\partial x^2} \text{ for } 0 < x < L, t > 0,$$

$$u(0,t) = T, u(L,t) = 0 \text{ for } t \geq 0,$$

$$u(x,0) = x(L-x) \text{ for } 0 \leq x \leq L.$$

In each of Problems 19 through 23, solve the problem

$$\frac{\partial u}{\partial t} = k\frac{\partial^2 u}{\partial x^2} + F(x,t) \text{ for } 0 < x < L, t > 0,$$

$$u(0,t) = u(L,t) = 0 \text{ for } t \geq 0,$$

and

$$u(x,0) = f(x) \text{ for } 0 \leq x \leq L.$$

Choose a value of time, and using the same set of axes, graph the twentieth partial sum of the solution together with the solution of the problem with $F(x,t)$ removed. Repeat this for several times. This gives some sense of the influence of $F(x,t)$ on the temperature distribution.

19. $k = 1, F(x,t) = t\cos(x), f(x) = x^2(5-x), L = 5$
20. $k = 4, f(x) = \sin(\pi x/2), L = 2,$

$$F(x,t) = \begin{cases} K & \text{for } 0 \leq x \leq 1 \\ 0 & \text{for } 1 < x \leq 2 \end{cases}$$

21. $k = 16, F(x,t) = xt, f(x) = K, L = 3$
22. $k = 1, F(x,t) = x\sin(t), f(x) = 1, L = 4$
23. $k = 4, F(x,t) = t, f(x) = x(\pi - x), L = \pi$

17.3 Solutions in an Infinite Medium

We will consider problems involving the heat equation over the entire line or half-line.

17.3.1 Problems on the Real Line

For a setting in which one dimension is very much greater than the others, it is sometimes useful to model heat conduction or a diffusion process by imagining the space variable free to vary over the entire real line. In this case, there is no boundary condition, but we look for bounded solutions. The problem we will solve is

$$\frac{\partial u}{\partial t} = k\frac{\partial^2 u}{\partial x^2} \text{ for } -\infty < x < \infty, t > 0$$

and

$$u(x,0) = f(x) \text{ for } -\infty < x < \infty.$$

Separation of variables yields

$$X'' + \lambda X = 0, T' + \lambda k T = 0.$$

As with the wave equation on the line, the eigenvalues $\lambda = \omega^2 \geq 0$ and the eigenfunctions have the form $a_\omega \cos(\omega x) + b_\omega \sin(\omega x)$.

The problem for T is $T' + k\omega^2 T = 0$ with solutions that are constant multiples of $e^{-\omega^2 kt}$.

For $\omega \geq 0$, we now have functions

$$u_\omega(x,t) = [a_\omega \cos(\omega x) + b_\omega \sin(\omega x)]e^{-\omega^2 kt}$$

that satisfy the heat equation and are bounded for all x. To satisfy the initial condition, attempt a superposition

$$u(x,t) = \int_0^\infty [a_\omega \cos(\omega x) + b_\omega \sin(\omega x)]e^{-\omega^2 kt}\, d\omega. \tag{17.11}$$

We must choose the coefficients so that

$$u(x,0) = f(x) = \int_0^\infty [a_\omega \cos(\omega x) + b_\omega \sin(\omega x)]\, d\omega.$$

This is the Fourier integral of f on the real line, so the coefficients should be chosen as the Fourier integral coefficients of f:

$$a_\omega = \frac{1}{\pi}\int_{-\infty}^{\infty} f(\xi)\cos(\omega\xi)\, d\xi \text{ and } b_\omega = \frac{1}{\pi}\int_{-\infty}^{\infty} f(\xi)\sin(\omega\xi)\, d\xi.$$

EXAMPLE 17.7

Suppose the initial temperature function is $f(x) = e^{-|x|}$. Compute the coefficients

$$a_\omega = \frac{1}{\pi}\int_{-\infty}^{\infty} e^{-|\xi|}\cos(\omega\xi)\, d\xi = \frac{2}{\pi}\frac{1}{1+\omega^2}$$

and $b_\omega = 0$, because $e^{-|x|}\sin(\omega x)$ is an odd function. The solution is

$$u(x,t) = \frac{2}{\pi}\int_0^\infty \frac{1}{1+\omega^2}\cos(\omega x)e^{-\omega^2 kt}\, d\omega. \quad \blacklozenge$$

In general, if we put the integrals for a_ω and b_ω into the solution (17.11) and argue as we did for the solution of the wave equation on the line (equation (16.8)), we obtain an alternative form of this solution:

$$u(x,t) = \frac{1}{\pi}\int_0^\infty \int_{-\infty}^{\infty} \cos(\omega(x-\xi)) f(\xi) e^{-\omega^2 kt}\, d\omega\, d\xi. \tag{17.12}$$

It is possible to rewrite this solution in terms of a single integral. To do this, first write the solution (17.12) as

$$u(x,t) = \frac{1}{2\pi}\int_{-\infty}^{\infty} \int_{-\infty}^{\infty} \cos(\omega(x-\xi)) f(\xi) e^{-\omega^2 kt}\, d\omega\, d\xi.$$

Now we need the following integral. If α and β are real numbers with $\beta \neq 0$,

$$\int_{-\infty}^{\infty} e^{-\zeta^2}\cos\left(\frac{\alpha\zeta}{\beta}\right) d\zeta = \sqrt{\pi}e^{-\alpha^2/4\beta^2}. \tag{17.13}$$

A derivation of this integral is sketched in Problem 11. In this integral, let

$$\xi = \sqrt{kt}\,\omega, \quad \alpha = x - \xi, \quad \text{and} \quad \beta = \sqrt{kt}$$

to obtain

$$\int_{-\infty}^{\infty} e^{-\omega^2 kt}\cos(\omega(x-\xi))d\omega = \frac{\sqrt{\pi}}{\sqrt{kt}}e^{-(x-\xi)^2/4kt}.$$

Now the solution (17.12) can be written

$$u(x,t) = \frac{1}{2\sqrt{\pi kt}}\int_{-\infty}^{\infty} f(\xi)e^{-(x-\xi)^2/4kt}\, d\xi. \tag{17.14}$$

17.3.2 Solution by Fourier Transform

We will illustrate the use of the Fourier transform to solve the problem

$$\frac{\partial u}{\partial t} = k\frac{\partial^2 u}{\partial x^2} \text{ for } -\infty < x < \infty, t > 0$$

and

$$u(x,0) = f(x) \text{ for } -\infty < x < \infty.$$

Take the Fourier transform of the heat equation with respect to x to get

$$\mathcal{F}\left[\frac{\partial u}{\partial t}\right] = k\mathcal{F}\left[\frac{\partial^2 u}{\partial x^2}\right].$$

Because x and t are independent, $\mathcal{F}$ passes through $\partial/\partial t$, and

$$\mathcal{F}\left[\frac{\partial u}{\partial t}\right] = \frac{\partial}{\partial t}\hat{u}(\omega, t)$$

in which ω appears as a parameter. Next use the operational rule for the Fourier transform to write

$$\mathcal{F}\left[\frac{\partial^2 u}{\partial x^2}\right] = -\omega^2 \hat{u}(\omega, t).$$

The transformed heat equation is

$$\frac{\partial}{\partial t}\hat{u}(\omega, t) + k\omega^2 \hat{u}(\omega, t) = 0$$

with solutions

$$\hat{u}(\omega, t) = a_\omega e^{-\omega^2 kt}.$$

Since $u(x,0) = f(x)$,

$$\hat{u}(\omega, 0) = \hat{f}(\omega) = a_\omega.$$

We now have the Fourier transform of the solution:

$$\hat{u}(\omega, t) = \hat{f}(\omega) e^{-\omega^2 kt}.$$

Apply the inverse Fourier transform. Since this transform is complex valued and the solution for a temperature distribution is real, the solution is the real part of this inverse:

$$\begin{aligned}
u(x,t) &= \text{Re}\, \mathcal{F}^{-1}\left[\hat{f}(\omega) e^{-\omega^2 kt}\right](t) \\
&= \text{Re}\, \frac{1}{2\pi}\int_{-\infty}^{\infty} \hat{f}(\omega) e^{-\omega^2 kt} e^{i\omega x}\, d\omega \\
&= \text{Re}\, \frac{1}{2\pi}\int_{-\infty}^{\infty} \left(\int_{-\infty}^{\infty} f(\xi) e^{-i\omega\xi}\, d\xi\right) e^{i\omega x} e^{-\omega^2 kt}\, d\omega \\
&= \text{Re}\, \frac{1}{2\pi}\int_{-\infty}^{\infty}\int_{-\infty}^{\infty} f(\xi) e^{-i\omega(\xi - x)} e^{-\omega^2 kt}\, d\xi\, d\omega \\
&= \text{Re}\, \frac{1}{2\pi}\int_{-\infty}^{\infty}\int_{-\infty}^{\infty} f(\xi)[\cos(\omega(\xi - x)) - i\sin(\omega(\xi - x))] e^{-\omega^2 kt}\, d\xi\, d\omega \\
&= \frac{1}{2\pi}\int_{-\infty}^{\infty}\int_{-\infty}^{\infty} f(\xi)\cos(\omega(\xi - x)) e^{-\omega^2 kt}\, d\xi\, d\omega,
\end{aligned}$$

which is the same expression obtained by separation of variables.

17.3.3 Problems on the Half-Line

We will solve the problem

$$\frac{\partial u}{\partial t} = k\frac{\partial^2 u}{\partial x^2} \text{ for } 0 < x < \infty, t > 0,$$
$$u(0,t) = 0 \text{ for } t > 0,$$

and

$$u(x,0) = f(x) \text{ for } 0 < x < \infty.$$

for the half-line, taking the case that the left end at $x = 0$ is kept at zero temperature. Putting $u(x,t) = X(x)T(t)$ leads in the usual way to

$$X'' + \lambda X = 0; \ X(0) = 0$$

and

$$T' + k\lambda T = 0.$$

The eigenvalues are $\lambda = \omega^2$ with $\omega > 0$ and eigenfunctions $b_\omega \sin(\omega x)$. With these values of λ, $T(t)$ is a constant multiple of $e^{-\omega^2 kt}$. For each $\omega > 0$, the function

$$u_\omega(x,t) = b_\omega \sin(\omega x) e^{-\omega^2 kt}$$

satisfies the heat equation and the boundary condition. For the initial condition, write

$$u(x,t) = \int_0^\infty b_\omega \sin(\omega x) e^{-\omega^2 kt} \, d\omega.$$

Then

$$u(x,0) = f(x) = \int_0^\infty b_\omega \sin(\omega x) \, d\omega$$

requires that we choose the coefficients as the Fourier integral sine coefficients of f on $[0,\infty)$:

$$b_\omega = \frac{2}{\pi}\int_0^\infty f(\xi)\sin(\omega\xi)\, d\xi.$$

EXAMPLE 17.8

Suppose the initial temperature function is

$$f(x) = \begin{cases} \pi - x & \text{for } 0 \le x \le \pi \\ 0 & \text{for } x > \pi. \end{cases}$$

Compute the coefficients

$$b_\omega = \frac{2}{\pi}\int_0^\pi (\pi - \xi)\sin(\omega\xi)\, d\xi = \frac{2}{\omega}\frac{\pi\omega - \sin(\pi\omega)}{\omega^2}.$$

The solution is

$$u(x,t) = \frac{2}{\pi}\int_0^\infty \frac{\pi\omega - \sin(\pi\omega)}{\omega^2}\sin(\omega x) e^{-\omega^2 kt}\, d\omega. \ \blacklozenge$$

17.3.4 Solution by Fourier Sine Transform

We also can solve this problem on the half-line using the Fourier sine transform. Take this transform of the heat equation with respect to x to get

$$\frac{\partial u}{\partial t}\hat{u}_S(\omega,t) = -\omega^2 k\hat{u}_S(\omega,t) + \omega k u(0,t) = -\omega^2 k\hat{u}_S(\omega,t).$$

This has general solution

$$\hat{u}_S(\omega,t) = b_\omega e^{-\omega^2 kt}.$$

Furthermore,

$$\hat{u}_S(\omega,0) = \hat{f}_S(\omega) = b_\omega,$$

so

$$\hat{u}_S(\omega,t) = \hat{f}_S(\omega)e^{-\omega^2 kt}.$$

Upon taking the inverse Fourier sine transform, we obtain the solution

$$u(x,t) = \frac{2}{\pi}\int_0^\infty \hat{f}_S(\omega)e^{-\omega^2 kt}\sin(\omega x)\,d\omega.$$

It is a calculation along lines we have done previously to insert the integral for $\hat{f}_S(\omega)$ into this expression to obtain the solution by separation of variables.

SECTION 17.3 PROBLEMS

In each of Problems 1 through 4, solve the problem

$$\frac{\partial u}{\partial t} = k\frac{\partial^2 u}{\partial x^2} \text{ for } x > 0, t > 0,$$
$$u(0,t) = 0 \text{ for } t > 0,$$
$$u(x,0) = f(x) \text{ for } x > 0.$$

1. $f(x) = e^{-\alpha x}$ with α any positive constant.
2. $f(x) = \begin{cases} x & \text{for } 0 \le x \le 2 \\ 0 & \text{for } x > 2 \end{cases}$
3. $f(x) = \begin{cases} 1 & \text{for } 0 \le x \le h \\ 0 & \text{for } x > h \end{cases}$ with h any positive number.
4. $f(x) = xe^{-\alpha x}$ with α any positive constant.

In each of Problems 5 through 8, solve the problem

$$\frac{\partial u}{\partial t} = k\frac{\partial^2 u}{\partial x^2} \text{ for } -\infty < x < \infty, t > 0$$

and

$$u(x,0) = f(x) \text{ for } -\infty < x < \infty.$$

5. $f(x) = \begin{cases} x & \text{for } 0 \le x \le 4 \\ 0 & \text{for } x < 0 \text{ and for } x > 4 \end{cases}$
6. $f(x) = \begin{cases} \sin(x) & \text{for } |x| \le \pi \\ 0 & \text{for } |x| > \pi \end{cases}$
7. $f(x) = e^{-4|x|}$
8. $f(x) = \begin{cases} e^{-x} & \text{for } |x| \le 1 \\ 0 & \text{for } |x| > 1 \end{cases}$

In each of Problems 9 and 10, use a Fourier transform on the half-line to solve the problem.

9. $$\frac{\partial u}{\partial t} = \frac{\partial^2 u}{\partial x^2} - tu \text{ for } x > 0, t > 0,$$
$$u(0,t) = 0 \text{ for } t > 0,$$
$$u(x,0) = xe^{-x} \text{ for } x > 0.$$

10. $$\frac{\partial u}{\partial t} = \frac{\partial^2 u}{\partial x^2} - u \text{ for } x > 0, t > 0,$$
$$u(0,t) = 0 \text{ for } t > 0,$$
$$\frac{\partial u}{\partial x}(0,t) = f(t) \text{ for } t > 0.$$

11. Derive equation (17.13). *Hint*: This can be done using complex function theory and a contour integral. Here is another way. Let

$$F(x)=\int_0^\infty e^{-\zeta^2}\cos(x\zeta)\,d\zeta.$$

Compute $F'(x)$ by differentiating under the integral sign and show that $F'(x)=-xF(x)/2$. Solve for $F(x)$ subject to the initial condition that

$$F(0)=\int_0^\infty e^{-\zeta^2}\,d\zeta=\frac{\sqrt{\pi}}{2}.$$

This integral for $F(0)$ is familiar from statistics and is assumed to be known. Finally, let $x=\alpha/\beta$.

17.4 Laplace Transform Techniques

In this section, we make use of the Laplace transform to solve diffusion problems. As we did with the wave equation, we will look at two typical problems. First, however, we need two functions and a transform formula that occur frequently when dealing with diffusion problems.

The *error function* is defined by

$$\operatorname{erf}(t)=\frac{2}{\sqrt{\pi}}\int_0^t e^{-u^2}\,du$$

and the *complementary error function* by

$$\operatorname{erfc}(t)=\frac{2}{\sqrt{\pi}}\int_t^\infty e^{-u^2}\,du.$$

These are also used in probability and statistics. If the standard result that

$$\int_0^\infty e^{-u^2}\,du=\frac{\sqrt{\pi}}{2}$$

is used, it is routine to check that

$$\operatorname{erfc}(t)=1-\operatorname{erf}(t).$$

The transform formula that we will need is

$$\mathcal{L}\left[\operatorname{erfc}\left(\frac{k}{2\sqrt{t}}\right)\right](s)=\frac{1}{s}e^{-k\sqrt{s}}.$$

Armed with these tools, we will look at two problems.

Temperatures in a Homogeneous Slab

We will solve the following:

$$\frac{\partial u}{\partial t}=k\frac{\partial^2 u}{\partial x^2}\text{ for }0<x<L,\,t>0,$$

$$u(x,0)=T_0=\text{ constant,}$$

and

$$u(L,t)=\frac{\partial u}{\partial x}(0,t)=0\text{ for }t>0.$$

This problem models the temperature distribution in a homogeneous solid slab or bar bounded by the planes $x=0$ and $x=L$ with the left side insulated (no flow of heat energy across this face) and the right end kept at temperature zero. The initial temperature in the slab is constant.

Apply the Laplace transform with respect to t to the heat equation. As in Section 16.5, where the Laplace transform was used to analyze wave motion, the resulting differential equation will be in terms of x with the transformed t variable s carried along as a parameter. The second derivative term with respect to x passes through the transform, which is with respect to t because x and t are independent. From the partial differential equation, we obtain

$$sU(x,s) - u(x,0) = k\frac{\partial^2}{\partial x^2}U(x,s).$$

If we write

$$\frac{\partial U(x,s)}{\partial x} = U'(x,s),$$

the diffusion equation transforms to

$$sU(x,s) - T_0 = kU''(x,s).$$

This is the second-order ordinary differential equation

$$U''(x,s) - \frac{s}{k}U(x,s) = -\frac{1}{k}T_0.$$

By inspection, a particular solution of this equation is $U_p = T_0/s$. For the general solution, we need the general solution of the associated homogeneous equation

$$U''(x,s) - \frac{s}{k}U(x,s) = 0.$$

This has the characteristic equation

$$\lambda^2 - \frac{s}{k} = 0$$

with roots $\pm\sqrt{s/k}$. Therefore, the general solution for $U(x,s)$ is

$$U(x,s) = c_1 e^{\sqrt{s/k}x} + c_2 e^{-\sqrt{s/k}x} + \frac{1}{s}T_0.$$

Now use the boundary conditions. Take the transform of $u(L,t) = 0$ to obtain $U(L,s) = 0$. Similarly, $(\partial/\partial x)u(0,t) = 0$ gives us $(\partial U/\partial x)(0,s) = 0$. From the second of these conditions, we have

$$\frac{\partial U}{\partial x}(0,s) = \sqrt{\frac{s}{k}}c_1 - \sqrt{\frac{s}{k}}c_2 = 0,$$

so $c_1 = c_2$ and U has the form

$$U(x,s) = \frac{T_0}{s} + c\cosh\left(\sqrt{\frac{s}{k}}x\right)$$

in which c can be any constant. From the first boundary condition, we have

$$U(L,s) = 0 = \frac{T_0}{s} + c\cosh\left(\sqrt{\frac{s}{k}}L\right),$$

implying that

$$c = -\frac{T_0}{s\cosh\left(\sqrt{s/k}L\right)}.$$

We now have the transform of the solution:

$$U(x,s) = \frac{T_0}{s} - \frac{T_0}{s}\frac{\cosh\left(\sqrt{s/k}x\right)}{\cosh\left(\sqrt{s/k}L\right)}.$$

This can be inverted using the geometric series, as we did for the problem with the wave equation on $[0, L]$ in Section 16.5. Recall that the geometric series is

$$\frac{1}{1+u} = \sum_{n=0}^{\infty} (-1)^n u^n \text{ for } |u| < 1.$$

In the following, the third line is obtained from the second by multiplying numerator and denominator by $e^{-\sqrt{s/k}x}$:

$$\begin{aligned}
\frac{\cosh\left(\sqrt{s/k}x\right)}{\cosh\left(\sqrt{s/k}L\right)} &= \frac{e^{\sqrt{s/k}x} + e^{-\sqrt{s/k}x}}{e^{\sqrt{s/k}L} + e^{-\sqrt{s/k}L}} \\
&= \frac{e^{\sqrt{s/k}x}e^{-\sqrt{s/k}L} + e^{-\sqrt{s/k}x}e^{-\sqrt{s/k}L}}{1 + e^{-2\sqrt{s/k}L}} \\
&= \left(e^{-\sqrt{s/k}(L-x)} + e^{-\sqrt{s/k}(L+x)}\right)\left(\frac{1}{1+e^{-2\sqrt{s/k}L}}\right) \\
&= \left(e^{-\sqrt{s/k}(L-x)} + e^{-\sqrt{s/k}(L+x)}\right)\sum_{n=0}^{\infty}(-1)^n e^{-2n\sqrt{s/k}L} \\
&= \sum_{n=0}^{\infty}(-1)^n \left[e^{-\sqrt{s/k}((2n+1)L-x)} + e^{-\sqrt{s/k}((2n+1)L+x)}\right].
\end{aligned}$$

Then

$$\begin{aligned}
U(x,s) = &\frac{T_0}{s} \\
&- \frac{T_0}{s}\sum_{n=0}^{\infty}(-1)^n \left[e^{-\sqrt{s/k}((2n+1)L-x)} + e^{-\sqrt{s/k}((2n+1)L+x)}\right].
\end{aligned}$$

Taking the inverse transform term by term, we have the solution

$$\begin{aligned}
u(x,t) = &T_0 \\
&- T_0\sum_{n=0}^{\infty}(-1)^n \left[\operatorname{erfc}\left(\frac{(2n+1)L-x}{2\sqrt{kt}}\right) + \operatorname{erfc}\left(\frac{(2n+1)L+x}{2\sqrt{kt}}\right)\right].
\end{aligned}$$

Temperature Distribution in a Semi-Infinite Bar

We will solve the boundary value problem

$$\frac{\partial u}{\partial t} = k\frac{\partial^2 u}{\partial x^2} \text{ for } x > 0, t > 0$$

and

$$u(x,0) = 0, u(0,t) = f(t).$$

This models diffusion in a thin homogeneous bar lying along the x-axis with a given initial temperature function. Because there is no bound on x, we also impose the condition

$$\lim_{x\to\infty} u(x,t) = 0 \text{ for } t > 0.$$

Take the transform in the time variable to obtain

$$sU(x,s) = k\frac{\partial^2 U}{\partial x^2}$$

and

$$U(0,s)=f(s),\ \lim_{x\to\infty}U(x,t)=0.$$

This differential equation has general solution

$$U(x,s)=c_1e^{\sqrt{s/k}x}+c_2e^{-\sqrt{s/k}x}.$$

Since $u(x,t)\to 0$ as $x\to\infty$, then $U(x,s)\to 0$ as $x\to\infty$, requiring that $c_1=0$. Furthermore, $u(0,t)=f(t)$ implies that $U(0,s)=F(s)=c_2$. Note here that c_2 may depend on s, which is a parameter in the transform with respect to t. Therefore,

$$U(x,s)=F(s)e^{-\sqrt{s/k}x}.$$

The solution is

$$u(x,t)=\mathcal{L}^{-1}[u(x,s)](t)=\mathcal{L}^{-1}\left[F(s)e^{-\sqrt{s/k}x}\right].$$

We can also write this solution as a convolution

$$u(x,t)=f(t)*g(t),$$

where

$$g(t)=\mathcal{L}^{-1}\left[e^{-\sqrt{s/k}x}\right](t).$$

A Semi-Infinite Bar with Discontinuous Temperature at the Left End

We will solve the boundary value problem

$$\frac{\partial u}{\partial t}=k\frac{\partial^2 u}{\partial x^2}\text{ for }x>0,t>0,$$
$$u(x,0)=A\text{ for }x>0,$$

and

$$u(0,t)=\begin{cases}B & \text{for }0\le t\le t_0\\ 0 & \text{for }t>t_0.\end{cases}$$

Here t_0, A, and B are positive constants.

This problem models the temperature distribution in a thin, homogeneous bar extending along the nonnegative x-axis with a constant initial temperature A and a discontinuous temperature function at the left end where $x=0$. The Laplace transform is a natural approach for this problem, because this transform is well suited to treating piecewise continuous functions. Begin by writing

$$u(0,t)=B[1-H(t-t_0)]$$

in which H is the Heaviside function. Apply the Laplace transform with respect to t in the heat equation using the condition that $u(x,0)=A$ to obtain

$$\frac{\partial^2}{\partial x^2}U(x,s)-\frac{s}{k}U(x,s)=-\frac{A}{k}.$$

As usual, think of this as a differential equation in x. The general solution is

$$U(x,s)=c_1e^{\sqrt{s/k}x}+c_2e^{-\sqrt{s/k}x}+\frac{A}{s},$$

in which the "constants" c_1 and c_2 may depend on s. We will require that $U(x,s) \to 0$ as $x \to \infty$, so choose $c_2 = 0$. This leaves

$$U(x,s) = c_1 e^{\sqrt{s/k}x} + \frac{A}{s}.$$

To obtain c_1, take the Laplace transform of the boundary condition $u(0,t) = B[1 - H(t - t_0)]$ to obtain

$$U(0,s) = \mathcal{L}[B] - \mathcal{L}[B(t - t_0)] = \frac{B}{s} - B\frac{e^{-t_0 s}}{s}.$$

Then

$$U(0,s) = \frac{B}{s} - B\frac{e^{-t_0 s}}{s} = c_1 + \frac{A}{s}.$$

Solve for c_1 to obtain

$$c_1 = \frac{B - A}{s} - \frac{B}{s}e^{-t_0 s}.$$

This gives us

$$U(x,s) = \left[\frac{B - A}{s} - \frac{B}{s}e^{-t_0 s}\right] e^{-\sqrt{s/k}x} + \frac{A}{s}.$$

Invert this expression to obtain the solution in terms of the error function and the complementary error function:

$$\begin{aligned} u(x,t) = &\left(A\operatorname{erf}\left(\frac{x}{2\sqrt{kt}}\right) + B\operatorname{erfc}\left(\frac{x}{2\sqrt{kt}}\right)\right)(1 - H(t - t_0)) \\ &+ \left(A\operatorname{erf}\left(\frac{x}{2\sqrt{kt}}\right) + B\operatorname{erfc}\left(\frac{x}{2\sqrt{kt}}\right)\right) H(t - t_0) \\ &- B\operatorname{erfc}\left(\frac{x}{2\sqrt{k(t - t_0)}}\right) H(t - t_0). \end{aligned}$$

SECTION 17.4 PROBLEMS

1. $$\frac{\partial u}{\partial t} = k\frac{\partial^2 u}{\partial x^2} \text{ for } x > 0, t > 0,$$
   $$u(x,0) = e^{-x}, u(0,t) = 0, \lim_{x\to\infty} u(x,t) = 0$$

2. Apply the Laplace transform with respect to t to the problem
   $$\frac{\partial u}{\partial t} = k\frac{\partial^2 u}{\partial x^2} \text{ for } 0 < x < L, t > 0,$$
   $$u(x,0) = 1, u(0,t) = u(L,t) = 0.$$
   Then use the transform with respect to x to solve the resulting problem for $U(x,s)$.

3. Solve
   $$\frac{\partial u}{\partial t} = k\frac{\partial^2 u}{\partial x^2} \text{ for } 0 < x < L, t > 0,$$
   $$u(x,0) = 0, u(0,t) = 0, u(L,t) = T_0 = \text{ constant}.$$

4. Solve
   $$\frac{\partial u}{\partial t} = k\frac{\partial^2 u}{\partial x^2} \text{ for } x > 0, t > 0$$
   $$u(x,0) = 0, u(0,t) = t^2, \lim_{x\to\infty} u(x,t) = 0$$

17.5 Heat Conduction in an Infinite Cylinder

We will determine the temperature distribution in a solid, infinitely long, homogeneous cylinder of radius R with its axis along the z-axis in 3-space.

In cylindrical coordinates, the heat equation for the temperature distribution $U(r, \theta, z, t)$ is

$$\frac{\partial U}{\partial t} = k\left(\frac{\partial^2 U}{\partial r^2} + \frac{1}{r}\frac{\partial U}{\partial r} + \frac{1}{r^2}\frac{\partial^2 U}{\partial \theta^2} + \frac{\partial^2 U}{\partial z^2}\right).$$

This is a formidable equation to solve at this stage, and we will restrict it to the special case that the temperature at any point in the cylinder depends only on the time t and the horizontal distance r from the z-axis. This symmetry means that $\partial U/\partial\theta = \partial U/\partial z = 0$, and the heat equation becomes

$$\frac{\partial U}{\partial t} = k\left(\frac{\partial^2 U}{\partial r^2} + \frac{1}{r}\frac{\partial U}{\partial r}\right)$$

for $0 \le r < R, t > 0$. We will write $U(r, t)$, with dependence only on r and t and assume the boundary condition

$$U(R, t) = 0 \text{ for } t > 0.$$

The initial condition is

$$U(r, 0) = f(r) \text{ for } 0 \le r < R.$$

Put $U(r, t) = F(r)T(t)$ and separate the variables, obtaining

$$\frac{T'}{kT} = \frac{F'' + (1/r)F'(r)}{F(r)} = -\lambda.$$

Then

$$T' + \lambda T = 0 \text{ and } F'' + \frac{1}{r}F' + \lambda F = 0.$$

Since $U(R, t) = F(R)T(t) = 0$, then $F(R) = 0$. The problem for F is a singular Sturm-Liouville problem on $[0, R]$. To determine the eigenvalues and eigenfunctions take cases on λ.

Case 1: $\lambda = 0$

Then

$$F'' + \frac{1}{r}F' = 0$$

with general solution of the form $F(r) = c\ln(r) + d$. Since $\ln(r) \to -\infty$ as $r \to 0$ (the center of the cylinder), we must choose $c = 0$. Then $F(r) = d$. Since $T'(t) = 0$ if $\lambda = 0$, then $T(t) =$ constant also. In this case, $U(r, t) =$ constant, and this constant must be zero because $U(R, 0) = 0$. $U(r, t) = 0$ is indeed the solution if $f(r) = 0$. If $f(r)$ is not identically zero, then $\lambda = 0$ does not contribute to the solution.

Case 2: $\lambda < 0$

Write $\lambda = -\omega^2$ with $\omega > 0$. Now $T' - k\omega^2 T = 0$ has a general solution

$$T(t) = ce^{\omega^2 kt},$$

and this is unbounded as t increases. To have a bounded solution, we reject this case.

Case 3: $\lambda > 0$

Write $\lambda = \omega^2$ with $\omega > 0$. Now $T' + k\omega^2 T = 0$, with general solution

$$T(t) = ce^{-\omega^2 kt}.$$

This is a bounded function. The equation for F becomes

$$F''(r) + \frac{1}{r}F'(r) + \omega^2 F(r) = 0.$$

Write this as

$$r^2 F''(r) + rF'(r) + \omega^2 F(r) = 0.$$

This is Bessel's equation of order zero. Bounded solutions on $[0, R]$ are constant multiples of $J_0(\omega r)$.

Thus far we have

$$U_\omega(r,t) = a_\omega J_0(\omega r)e^{-\omega^2 kt}.$$

The condition $U(R, 0) = 0$ requires that

$$J_0(\omega R) = 0.$$

Let $j_1 < j_2 < \cdots$ be the positive zeros of $J_0(x)$ in ascending order. For $J_0(\omega R) = 0$, there must be some positive integer n such that $\omega R = j_n$. Denote $\omega_n = j_n/R$. This gives us the eigenvalues of this problem:

$$\lambda_n = \omega_n^2 = \frac{j_n^2}{R^2}.$$

The eigenfunctions are constant multiples of $J_0(j_n r/R)$.

Now for $n = 1, 2, \cdots$, we have functions

$$U_n(r,t) = a_n J_0\left(\frac{j_n r}{R}\right)e^{-j_n^2 kt/R^2}$$

satisfying the heat equation and the boundary condition. To satisfy the initial condition, employ a superposition

$$U(r,t) = \sum_{n=1}^{\infty} a_n J_0\left(\frac{j_n r}{R}\right)e^{-j_n^2 kt/R^2}.$$

Now we must choose the coefficients so that

$$U(r,0) = f(r) = \sum_{n=1}^{\infty} a_n J_0\left(\frac{j_n r}{R}\right).$$

Let $\xi = r/R$ to write

$$f(r\xi) = \sum_{n=1}^{\infty} a_n J_0(j_n \xi)$$

for $0 \leq \xi \leq 1$. In this framework, previous results on eigenfunction expansions apply, and we can write

$$a_n = \frac{2\int_0^1 rf(rR)J_0(j_n\xi)\,d\xi}{J_1^2(j_n)}.$$

With these coefficients, we have the solution for $U(r,t)$.

SECTION 17.5 PROBLEMS

1. Suppose $R = 1$, $k = 1$ and $f(r) = r$. Assume that $U(1, t) = 0$ for $t > 0$. Use a numerical integration to approximate the coefficients $a_1, \cdots, a_5$ and use these numbers in the fifth partial sum of the series solution to approximate $U(r, t)$. Graph this partial sum for different values of t.
2. Repeat the calculations of Problem 1 with $k = 16$, $R = 3$ and $f(r) = e^r$.
3. Repeat the calculations of Problem 1 with $k = 1/2$, $R = 3$ and $f(r) = 9 - r^2$.

17.6 Heat Conduction in a Rectangular Plate

We will solve for the temperature distribution $u(x, y, t)$ in a flat, square homogeneous plate covering the region $0 \leq x \leq 1, 0 \leq y \leq 1$ in the plane. The sides are kept at temperature zero, and the interior temperature at (x, y) at time 0 is $f(x, y)$.

To be specific, we will let $f(x, y) = x(1 - x^2)y(1 - y)$ and solve the initial-boundary value problem

$$\frac{\partial u}{\partial t} = k\left(\frac{\partial^2 u}{\partial x^2} + \frac{\partial^2 u}{\partial y^2}\right) \text{ for } 0 < x < 1, 0 < y < 1, t > 0,$$

$$u(x, 0, t) = u(x, 1, t) = 0 \text{ for } 0 < x < 1, t > 0,$$

$$u(0, y, t) = u(1, y, t) = 0 \text{ for } 0 < y < 1, t > 0,$$

and

$$u(x, y, 0) = f(x, y) = x(1 - x^2)y(1 - y).$$

Let $u(x, y, t) = X(x)Y(y)T(t)$, and separate variables to obtain

$$X'' + \lambda X = 0, Y'' + \mu Y = 0, T' + (\lambda + \mu)T = 0,$$

as in the analysis of wave motion of a membrane in Chapter 16. The boundary conditions imply that

$$X(0) = X(1) = 0, Y(0) = Y(1) = 0,$$

so the eigenvalues and eigenfunctions are

$$\lambda_n = n^2\pi^2, X_n(x) = \sin(n\pi x)$$

and

$$\mu_m = m^2\pi^2, Y_m(y) = \sin(m\pi y)$$

for $n = 1, 2, \cdots$ and $m = 1, 2, \cdots$. The equation for T becomes

$$T' + (n^2 + m^2)\pi^2 T = 0$$

with solutions that are constant multiples of $e^{-(n^2+m^2)\pi^2 kt}$. For each positive integer m and n, we now have functions

$$u_{nm}(x, y, t) = \sum_{n=1}^{\infty}\sum_{m=1}^{\infty} c_{nm} \sin(n\pi x)\sin(m\pi y)e^{-(n^2+m^2)\pi^2 kt}$$

satisfying the heat equation and the boundary conditions. Reasoning as we did with the two-dimensional wave equation, we get

$$c_{nm} = 4\int_0^1\int_0^1 \xi(1-\xi^2)\eta(1-\eta)\sin(n\pi\xi)\sin(m\pi\eta)\,d\xi\,d\eta$$
$$= 48\left(\frac{(-1)^n}{n^3\pi^3}\right)\left(\frac{(-1)^m-1}{m^3\pi^3}\right).$$

The solution is

$$u(x,y,t) = \frac{48}{\pi^6}\sum_{n=1}^{\infty}\sum_{m=1}^{\infty}\left(\frac{(-1)^n}{n^3}\right)\left(\frac{(-1)^m-1}{m^3}\right)\sin(n\pi x)\sin(m\pi y)e^{-(n^2+m^2)\pi^2 kt}.$$ ♦

SECTION 17.6 PROBLEMS

1. Write the solution for the general problem

$$\frac{\partial u}{\partial t} = k\left(\frac{\partial^2 u}{\partial x^2}+\frac{\partial^2 u}{\partial y^2}\right) \text{ for } 0<x<L,$$
$$0<y<K, t>0,$$
$$u(x,0,t)=u(x,K,t)=0 \text{ for } 0<x<L, t>0,$$
$$u(0,y,t)=u(L,y,t)=0 \text{ for } 0<y<K, t>0,$$

and

$$u(x,y,0)=f(x,y) \text{ for } 0\le x\le L, 0\le y\le K.$$

2. Solve this problem when $k=4$, $L=2$, $K=3$, and $f(x,y)=x^2(L-x)\sin(y)(K-y)$.

3. Solve this problem when $k=1$, $L=\pi$, $K=\pi$, and $f(x,y)=\sin(x)\cos(y/2)$.

LAPLACE'S EQUATION DIRICHLET PROBLEM FOR A RECTANGLE DIRICHLET PROBLEM FOR A DISK POISSON'S INTEGRAL FORMULA

CHAPTER 18
The Potential Equation

18.1 Laplace's Equation

The partial differential equation

$$\frac{\partial^2 u}{\partial x^2} + \frac{\partial^2 u}{\partial y^2} = 0$$

is called *Laplace's equation* in two dimensions. In three dimensions Laplace's equation is

$$\frac{\partial^2 u}{\partial x^2} + \frac{\partial^2 u}{\partial y^2} + \frac{\partial^2 u}{\partial z^2} = 0.$$

These equations are often written $\nabla^2 u = 0$, in which the symbol ∇ is read "del" and ∇^2 is read "del squared". We saw the del operator previously with the gradient vector field.

Laplace's equation arises in several contexts. It is the steady-state heat equation, occurring when $\partial u/\partial t = 0$. It is also called the *potential equation*. If a vector field $\mathbf{F}$ has a potential φ, then φ must satisfy Laplace's equation.

A function satisfying Laplace's equation in a region of the plane (or 3-space) is said to be *harmonic* on that region. For example, $x^2 - y^2$ and xy are both harmonic over the entire plane.

A *Dirichlet problem* for a region D consists of finding a function that is harmonic on D and assumes specified values on the boundary of D. We will be primarily concerned with Dirichlet problems in the plane, in which D is a region that is bounded by one or more piecewise smooth curves. Denote the boundary by ∂D. The Dirichlet problem for D is to solve

$$\nabla^2 u = 0 \text{ on } D$$

and

$$u(x, y) = f(x, y) \text{ for } (x, y) \text{ in } \partial D$$

with $f(x, y)$ as a given function. The function f is called *boundary data* for D.

SECTION 18.1 PROBLEMS

1. Show that if f and g are harmonic on D so are $f + g$ and, for any constant c, cf.
2. Show that the following functions are harmonic (on the entire plane, if D is not specified).

(a) $x^3 - 3xy^2$
(b) $3x^2y - y^3$
(c) $x^4 - 6x^2y^2 + y^4$
(d) $4x^3y - 4xy^3$
(e) $\sin(x)(e^y + e^{-y})$
(f) $\cos(x)(e^y - e^{-y})$
(g) $e^{-x}\cos(y)$
(h) $\ln(x^2 + y^2)$ if D is the plane with the origin removed.

18.2 Dirichlet Problem for a Rectangle

The region D exerts a great influence on our ability to explicitly solve a Dirichlet problem, or even whether a solution exists. Some regions admit solutions by Fourier methods. In this and the next section, we will treat two such cases: that D is a rectangle or disk in the plane.

Let D be the solid rectangle consisting of points (x, y) with $0 \le x \le L, 0 \le y \le K$. We will solve the Dirichlet problem for D.

This problem can be solved by separation of variables if the boundary data is nonzero on only one side of D. We will illustrate this for the case that this is the upper horizontal side of D. The problem in this case is

$$\nabla^2 u = 0 \text{ on } D,$$

$$u(x, 0) = 0 \text{ for } 0 \le x \le L,$$

$$u(0, y) = 0 \text{ for } 0 \le y \le K,$$

$$u(L, y) = 0 \text{ for } 0 \le y \le K,$$

and

$$u(x, K) = f(x) \text{ for } 0 \le x \le L.$$

Figure 18.1 shows D and the boundary data. Let $u(x, y) = X(x)Y(y)$ in Laplace's equation to obtain

$$\frac{X''}{X} = -\frac{Y''}{Y} = -\lambda$$

or

$$X'' + \lambda X = 0 \text{ and } Y'' - \lambda Y = 0.$$

From the boundary conditions, $X(0) = X(L) = Y(0) = 0$, so the problems for X and Y are

$$X'' + \lambda X = 0;\ X(0) = X(L) = 0$$

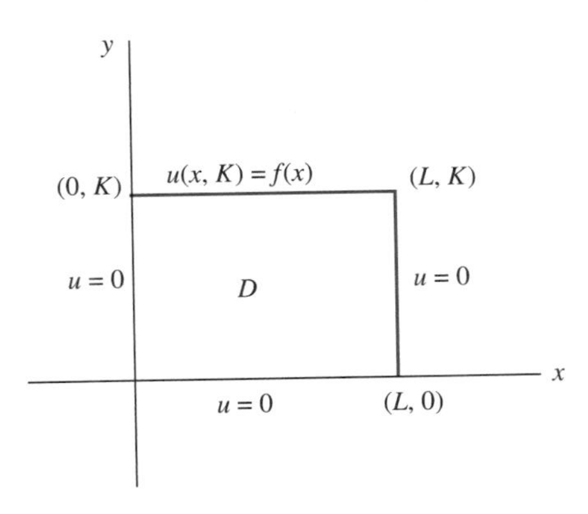

FIGURE 18.1 *D and the boundary data for this problem.*

and

$$Y'' - \lambda Y = 0;\ Y(0) = 0.$$

The problem for X has eigenvalues and eigenfunctions

$$\lambda_n = \frac{n^2\pi^2}{L^2} \text{ and } X_n(x) = \sin\left(\frac{n\pi x}{L}\right)$$

for $n = 1, 2, \cdots$.

The problem for Y becomes

$$Y'' - \frac{n^2\pi^2}{L^2} Y = 0;\ Y(0) = 0$$

with solutions that are constant multiples of

$$Y_n(y) = \sinh\left(\frac{n\pi y}{L}\right).$$

For each positive integer n, we have a function

$$u_n(x, y) = \sin\left(\frac{n\pi x}{L}\right) \sinh\left(\frac{n\pi y}{L}\right)$$

that is harmonic on D and satisfies the zero boundary conditions on the lower side and vertical sides of D. For the condition on the top side, use a superposition

$$u(x, y) = \sum_{n=1}^{\infty} b_n \sin\left(\frac{n\pi x}{L}\right) \sinh\left(\frac{n\pi y}{L}\right).$$

We need

$$u(x, K) = f(x) = \sum_{n=1}^{\infty} b_n \sin\left(\frac{n\pi x}{L}\right) \sinh\left(\frac{n\pi K}{L}\right).$$

This is a Fourier sine expansion of $f(x)$ on $[0, L]$ with coefficient $b_n \sinh(n\pi K/L)$. Thus, choose $b_n \sinh(n\pi K/L)$ to be the nth Fourier sine coefficient of $f(x)$ on $[0, L]$:

$$b_n \sinh(n\pi K/L) = \frac{2}{L} \int_0^L f(\xi) \sin\left(\frac{n\pi\xi}{L}\right) d\xi.$$

Then

$$b_n = \frac{2}{L \sinh(n\pi K/L)} \int_0^L f(\xi) \sin\left(\frac{n\pi\xi}{L}\right) d\xi.$$

FIGURE 18.2 $u(x, y) = \sum_{j=1}^{4} u_j(x, y)$.

With this choice of coefficients, the solution can be written

$$u(x, y) = \sum_{n=1}^{\infty} \frac{2}{L} \left(\int_0^{\infty} f(\xi) \sin(n\pi\xi/L)\, d\xi \right) \sin(n\pi x/L) \frac{\sinh(n\pi y/L)}{\sinh(n\pi K/L)}.$$

EXAMPLE 18.1

Suppose, in the problem just solved, $L = K = \pi$ and $f(x) = x(\pi - x)$. Compute the numbers

$$\frac{2}{\pi} \int_0^{\pi} x(\pi - \xi) \sin(n\xi)\, d\xi = \frac{4}{\pi n^3}(-1)^n.$$

The solution is

$$u(x, y) = \sum_{n=1}^{\infty} \frac{4}{\pi n^3}(-1)^n \sin(nx) \frac{\sinh(ny)}{\sinh(n\pi)}. \quad \blacklozenge$$

If nonzero boundary data is prescribed on all four sides of D, define four Dirichlet problems in each of which the data is nonzero on just one side (Figure 18.2). Each of these problems can be solved by the separation of variables. The sum of the solutions of these four problems is the solution of the original problem.

SECTION 18.2 PROBLEMS

In each of Problems 1 through 5, solve the Dirichlet problem for the given rectangle and boundary conditions.

1. $u(0, y) = u(1, y) = 0$ for $0 \le y \le 4$ and $u(x, 4) = x\cos(\pi x/2), u(x, 0) = 0$ for $0 \le x \le 1$
2. $u(0, y) = \sin(y), u(\pi, y) = 0$ for $0 \le y \le \pi$ and $u(x, 0) = x(\pi - x), u(x, \pi) = 0$ for $0 \le x \le \pi$
3. $u(0, y) = 0, u(2, y) = \sin(y)$ for $0 \le y \le \pi$ and $u(x, 0) = 0, u(x, \pi) = x\sin(\pi x)$ for $0 \le x \le 2$
4. $u(0, y) = y(2 - y), u(3, y) = 0$ for $0 \le y \le 2$ and $u(x, 0) = u(x, 2) = 0$ for $0 \le x \le 3$
5. $u(0, y) = u(1, y) = 0$ for $0 \le y \le \pi$ and $u(x, \pi) = 0, u(x, 0) = \sin(\pi x)$ for $0 \le x \le 1$.

6. Solve for the steady-state temperature distribution in a homogeneous, thin, flat plate covering the rectangle $0 \leq x \leq a$, $0 \leq y \leq b$ if the temperature on the vertical and lower sides are kept at zero and the temperature along the top side is $f(x) = x(x-a)^2$.

7. Solve for the steady-state temperature distribution in a thin, flat plate covering the rectangle $0 \leq x \leq 4$, $0 \leq y \leq 1$ if the temperature on the horizontal sides is zero while the temperature on the left side is $f(y) = \sin(\pi y)$ and on the right side, $g(y) = y(1-y)$.

8. Apply separation of variables to the problem

$$\nabla^2 u = 0 \text{ for } 0 < x < a, 0 < y < b,$$

$$u(x,0) = \frac{\partial u}{\partial y}(x,b) = 0 \text{ for } 0 \leq x \leq a,$$

$$u(0,y) = 0, u(a,y) = g(y) \text{ for } 0 \leq y \leq b.$$

9. Use separation of variables to solve

$$\nabla^2 u = 0 \text{ for } 0 < x < a, 0 < y < b,$$

$$u(x,0) = 0, u(x,b) = f(x) \text{ for } 0 \leq x \leq a,$$

$$u(0,y) = \frac{\partial u}{\partial x}(a,y) = 0 \text{ for } 0 \leq y \leq b.$$

18.3 Dirichlet Problem for a Disk

We will solve the Dirichlet problem for a disk of radius R centered at the origin. The boundary of this disk is the circle $x^2 + y^2 = R^2$. Using polar coordinates, the problem for $u(r,\theta)$ is

$$\nabla^2 u = \frac{\partial^2 u}{\partial r^2} + \frac{1}{r}\frac{\partial u}{\partial r} + \frac{1}{r^2}\frac{\partial^2 u}{\partial \theta^2} = 0 \text{ for } 0 \leq r < R, -\pi \leq \theta \leq \pi$$

and

$$u(R,\theta) = f(\theta) \text{ for } -\pi \leq \theta \leq \pi.$$

It is easy to check that the functions $1, r^n\cos(n\theta)$, and $r^n\sin(n\theta)$ are harmonic on the entire plane. Thinking ahead to the possibility of a Fourier series to satisfy the boundary condition, attempt a solution in a series of these functions:

$$u(r,\theta) = \frac{1}{2}a_0 + \sum_{n=1}^{\infty}(a_n r^n \cos(n\theta) + b_n r^n \sin(n\theta)).$$

This would require that

$$u(R,\theta) = f(\theta) = \frac{1}{2}a_0 + \sum_{n=1}^{\infty}(a_n R^n \cos(n\theta) + b_n R^n \sin(n\theta)).$$

This is a Fourier expansion of $f(\theta)$ on $[-\pi,\pi]$ if we choose the entire coefficients, $a_n R^n$ and $b_n R^n$ to be the Fourier coefficients of f on $[-\pi,\pi]$. This means that

$$a_0 = \frac{1}{\pi}\int_{-\pi}^{\pi} f(\xi)\,d\xi$$

and, for $n = 1, 2, \cdots$,

$$a_n = \frac{1}{R^n}\int_{-\pi}^{\pi} f(\xi)\cos(n\xi)\,d\xi \text{ and } b_n = \frac{1}{R^n}\int_{-\pi}^{\pi} f(\xi)\sin(n\xi)\,d\xi.$$

Another form of this solution is

$$u(r,\theta) = \frac{1}{\pi}\int_{-\pi}^{\pi} f(\xi)\,d\xi$$
$$+ \frac{1}{\pi}\sum_{n=1}^{\infty}\left(\frac{r}{R}\right)^n \left(\int_{-\pi}^{\pi} f(\xi)\cos(n\xi)\,d\xi \cos(n\theta) + \int_{-\pi}^{\pi} f(\xi)\sin(n\xi)\,d\xi \sin(n\theta)\right). \tag{18.1}$$

This can be rearranged to

$$u(r,\theta)=\frac{1}{2\pi}\int_{-\pi}^{\pi}f(\xi)\,d\xi$$
$$+\frac{1}{\pi}\sum_{n=1}^{\infty}\left(\frac{r}{R}\right)^n\int_{-\pi}^{\pi}f(\xi)\cos(n(\xi-\theta))\,d\xi. \quad (18.2)$$

If we combine terms in equation (18.2), we obtain

$$u(r,\theta)=\frac{1}{2\pi}\int_{-\pi}^{\pi}\left[1+2\sum_{n=1}^{\infty}\left(\frac{r}{R}\right)^n\cos(n(\xi-\theta))\right]f(\xi)\,d\xi, \quad (18.3)$$

which will be used to derive Poisson's integral formula in the next section.

EXAMPLE 18.2

We will solve the Dirichlet problem for the disk with a radius of 4 about the origin if $u(4,\theta)=f(\theta)=\theta^2$. Using equation (18.1), the solution is

$$u(r,\theta)=\frac{1}{2\pi}\int_{-\pi}^{\pi}\xi^2\,d\xi$$
$$+\frac{1}{\pi}\sum_{n=1}^{\infty}\left(\frac{r}{4}\right)^n\left(\int_{-\pi}^{\pi}\xi^2\cos(n\xi)\,d\xi\cos(n\theta)+\int_{-\pi}^{\pi}\xi^2\sin(n\xi)\,d\xi\sin(n\theta)\right)$$
$$=\frac{1}{3}\pi^2+\sum_{n=1}^{\infty}\frac{4(-1)^n}{n^2}\left(\frac{r}{4}\right)^n\cos(n\theta). \ \blacklozenge$$

EXAMPLE 18.3

We will solve the Dirichlet problem

$$\nabla^2 u(x,y)=0 \text{ for } x^2+y^2<9$$

and

$$u(x,y)=x^2y^2 \text{ for } x^2+y^2=9.$$

Convert this problem to polar coordinates. Let $U(r,\theta)=u(r\cos(\theta),r\sin(\theta))$. On the boundary,

$$U(3,\theta)=9\cos^2(\theta)9\sin^2(\theta)=81\sin^2(\theta)\cos^2(\theta)=f(\theta).$$

The solution in polar coordinates is

$$U(r,\theta)=\frac{1}{2\pi}\int_{-\pi}^{\pi}81\cos^2(\xi)\sin^2(\xi)\,d\xi$$
$$+\frac{1}{\pi}\sum_{n=1}^{\infty}\left(\frac{r}{3}\right)^n\left[\int_{-\pi}^{\pi}81\cos^2(\xi)\sin^2(\xi)\cos(n\xi)\,d\xi\cos(n\theta)\right.$$
$$\left.+\int_{-\pi}^{\pi}81\cos^2(\xi)\sin^2(\xi)\sin(n\xi)\,d\xi\sin(n\theta)\right].$$

There remains to evaluate the integrals:

$$\frac{1}{2\pi}\int_{-\pi}^{\pi} 81\cos^2(\xi)\sin^2(\xi)\,d\xi = \frac{81}{4}\pi,$$

$$\int_{-\pi}^{\pi} 81\cos^2(\xi)\sin^2(\xi)\cos(n\xi)\,d\xi = \begin{cases} 0 & \text{if } n \neq 4 \\ -81\pi/8 & \text{if } n = 4, \end{cases}$$

and

$$\int_{-\pi}^{\pi} 81\cos^2(\xi)\sin^2(\xi)\sin(n\xi)\,d\xi = 0.$$

The solution is

$$U(r,\theta) = \frac{1}{2\pi}\frac{81\pi}{4} - \frac{1}{\pi}\frac{81\pi}{8}\left(\frac{r}{3}\right)^4\cos(4\theta)$$
$$= \frac{81}{8} - \frac{1}{8}r^4\cos(4\theta).$$

To convert this solution to rectangular coordinates, use the fact that

$$\cos(4\theta) = 8\cos^4(\theta) - 8\cos^2(\theta) + 1$$

to obtain

$$U(r,\theta) = \frac{81}{8} - \frac{1}{8}(8r^4\cos^4(\theta) - 8r^4\cos^2(\theta) + r^4)$$
$$= \frac{81}{8} - \frac{1}{8}(8r^4\cos^4(\theta) - 8r^2r^2\cos^2(\theta) + r^4).$$

Then

$$u(x,y) = \frac{81}{8} - \frac{1}{8}(8x^4 - 8(x^2+y^2)x^2 + (x^2+y^2)^2)$$
$$= \frac{81}{8} - \frac{1}{8}(x^4 + y^4 - 6x^2y^2). \;\blacklozenge$$

SECTION 18.3 PROBLEMS

In each of Problems 1 through 4, solve the problem by converting it to polar coordinates.

1. $\nabla^2 u(x,y) = 0$ for $x^2+y^2 < 16$,
   $u(x,y) = x^2$ for $x^2+y^2 = 16$
2. $\nabla^2 u(x,y) = 0$ for $x^2+y^2 < 9$,
   $u(x,y) = x - y$ for $x^2+y^2 = 9$
3. $\nabla^2 u(x,y) = 0$ for $x^2+y^2 < 4$,
   $u(x,y) = x^2 - y^2$ for $x^2+y^2 = 4$
4. $\nabla^2 u(x,y) = 0$ for $x^2+y^2 < 25$,
   $u(x,y) = xy$ for $x^2+y^2 = 25$

In each of Problems 5 through 12, write the solution of the Dirichlet problem for the disk, with the given boundary data.

5. $R = 2$, $f(\theta) = \theta^2 - \theta$
6. $R = 1$, $f(\theta) = \sin^2(\theta)$
7. $R = 4$, $f(\theta) = e^{-\theta}$
8. $R = 3$, $f(\theta) = 8\cos(4\theta)$
9. $R = 8$, $f(\theta) = 1 - \theta^2$
10. $R = 5$, $f(\theta) = \theta\cos(\theta)$
11. $R = 3$, $f(\theta) = 1$
12. $R = 4$, $f(\theta) = \theta e^{2\theta}$

18.4 Poisson's Integral Formula

We will derive an integral formula due to Poisson for the Dirichlet problem for a disk. Suppose the disk is centered at the origin and has a radius of 1, and that $u(1, \theta) = f(\theta)$. By equation (18.3), the solution with $R = 1$ is

$$u(r, \theta) = \frac{1}{2\pi} \int_{-\pi}^{\pi} \left[1 + 2 \sum_{n=1}^{\infty} r^n \cos(n(\xi - \theta)) \right] f(\xi)\, d\xi.$$

The quantity

$$P(r, \zeta) = \frac{1}{2\pi} \left[1 + 2 \sum_{n=1}^{\infty} r^n \cos(n\zeta) \right]$$

is called the *Poisson kernel*. In terms of this kernel function, the solution is

$$u(r, \theta) = \int_{-\pi}^{\pi} P(r, \xi - \theta) f(\xi)\, d\xi.$$

We will evaluate the sum in the Poisson kernel, yielding Poisson's integral formula for the solution.

Let z be a complex number. In polar form, $z = re^{i\zeta}$ with $r < 1$ inside the unit disk and ζ an argument of z. By Euler's formula,

$$z^n = r^n e^{in\zeta} = r^n \cos(n\zeta) + ir^n \sin(n\zeta).$$

This enables us to recognize $r^n \cos(n\zeta)$, which appears in the Poisson kernel, as the real part of z^n and write

$$1 + 2 \sum_{n=1}^{\infty} r^n \cos(n\zeta) = \text{Re} \left(1 + 2 \sum_{n=1}^{\infty} z^n \right).$$

Now suppose $|z| = r < 1$. Then this sum is just a geometric series:

$$\sum_{n=1}^{\infty} z^n = \frac{z}{1 - z}.$$

Combining these observations, we have

$$1 + 2 \sum_{n=1}^{\infty} r^n \cos(n\theta) = \text{Re} \left(1 + 2 \sum_{n=1}^{\infty} z^n \right) = \text{Re} \left(1 + 2 \frac{z}{1 - z} \right)$$

$$= \text{Re} \left(\frac{1 + z}{1 - z} \right) = \text{Re} \left(\frac{1 + re^{i\zeta}}{1 - re^{i\zeta}} \right).$$

To extract this real part, compute

$$\frac{1 + re^{i\zeta}}{1 - re^{i\zeta}} = \frac{1 + re^{i\zeta}}{1 - re^{i\zeta}} \left(\frac{1 - re^{-i\zeta}}{1 - re^{-i\zeta}} \right)$$

$$= \frac{1 - r^2 + r(e^{i\zeta} - e^{-i\zeta})}{1 + r^2 - r(e^{i\zeta} + e^{-i\zeta})}$$

$$= \frac{1 - r^2 + 2ir \sin(\zeta)}{1 + r^2 - 2r \cos(\zeta)}.$$

In this form, the real part is easily identified, yielding

$$1+2\sum_{n=1}^{\infty} r^n \cos(n\zeta) = \frac{1-r^2}{1+r^2-2r\cos(\zeta)}.$$

Therefore, the solution of the Dirichlet problem for the unit disk is

$$u(r,\theta) = \frac{1}{2\pi}\int_{-\pi}^{\pi} \frac{1-r^2}{1+r^2-2r\cos(\xi-\theta)} f(\xi)\,d\xi.$$

This is *Poisson's integral formula*. For a disk of radius R, a change of variables gives us the solution

$$u(r,\theta) = \frac{1}{2\pi}\int_{-\pi}^{\pi} \frac{R^2-r^2}{R^2+r^2-2Rr\cos(\xi-\theta)} f(\xi)\,d\xi.$$

EXAMPLE 18.4

The solution of the problem of Example 18.2 also can be written

$$u(r,\theta) = \frac{1}{2\pi}\int_{-\pi}^{\pi} \frac{16-r^2}{16+r^2-8r\cos(\xi-\theta)} \xi^2\,d\xi$$

$$= \frac{16-r^2}{2\pi}\int_{-\pi}^{\pi} \frac{\xi^2}{16+r^2-8r\cos(\xi-\theta)}\,d\xi$$

for $0 \le r < 4$, $-\pi \le \theta \le \pi$. This integral solution may be more suitable than the infinite series solution if we want to approximate values at specific points. ♦

SECTION 18.4 PROBLEMS

In each of Problems 1 through 4, find an integral formula for the solution of the Dirichlet problem. Use a numerical integration routine to approximate $u(r,\theta)$ at the given points.

1. $R=15$, $f(\theta)=\theta^3-\theta$; $(4,\pi)$, $(12,\pi/6)$, $(8,\pi/4)$, $(7,\pi/3)$
2. $R=6$, $f(\theta)=e^{-\theta}$; $(5.5, 3\pi/5)$, $(4, 2\pi/7)$, $(1,\pi)$, $(4, 9\pi/4)$
3. $R=1$, $f(\theta)=\theta$; $(1/2,\pi)$, $(3/4,\pi/3)$, $(0.2,\pi/4)$
4. $R=4$, $f(\theta)=\sin(4\theta)$; $(1,\pi/6)$, $(3, 7\pi/2)$, $(1,\pi/4)$, $(2.5,\pi/12)$
5. Show that, for $0 \le r < 1$,

$$r^n \sin(n\theta) = \frac{1}{2\pi}\int_{-\pi}^{\pi} \frac{1-r^2}{1+r^2-2r\cos(\xi-\theta)} \sin(n\xi)\,d\xi.$$

Hint: Notice that $r^n \sin(n\theta)$ is harmonic (in polar coordinates) and use Poisson's formula.

18.5 Dirichlet Problem for Unbounded Regions

When D is unbounded (has points arbitrarily far from the origin), we may use a Fourier integral or transform to solve a Dirichlet problem on D.

18.5.1 The Upper Half-Plane

We will solve the problem

$$\nabla^2 u(x, y) = 0 \text{ for } -\infty < x < \infty, y > 0$$

and

$$u(x, 0) = f(x) \text{ for } -\infty < x < \infty.$$

This is a Dirichlet problem because the horizontal axis is the boundary of the upper half-plane. We seek a bounded solution.

Put $u(x, t) = X(x)T(t)$, and obtain

$$X'' + \lambda X = 0, T'' - \lambda T = 0.$$

The eigenvalues are $\lambda = \omega^2$ with $\omega \geq 0$, and the eigenfunctions are

$$X_\omega(x) = a_\omega \cos(\omega x) + b_\omega \sin(\omega x).$$

The equation for Y is $Y'' - \omega^2 Y = 0$ with constant multiples of $e^{-\omega y}$ as bounded solutions because $y \geq 0$. For each $\omega \geq 0$, we have a solution

$$u_\omega(x, y) = (a_\omega \cos(\omega x) + b_\omega \sin(\omega x))e^{-\omega y}$$

of Laplace's equation. To obtain a solution satisfying the boundary condition, use the superposition

$$u(x, y) = \int_0^\infty (a_\omega \cos(\omega x) + b_\omega \sin(\omega x))e^{-\omega y}\, d\omega.$$

We need

$$u(x, 0) = f(x) = \int_0^\infty (a_\omega \cos(\omega x) + b_\omega \sin(\omega x))\, d\omega.$$

The coefficients are the Fourier integral coefficients of f on the real line:

$$a_\omega = \frac{1}{\pi}\int_{-\infty}^\infty f(\xi)\cos(\omega\xi)\, d\xi \text{ and } b_\omega = \frac{1}{\pi}\int_{-\infty}^\infty f(\xi)\sin(\omega\xi)\, d\xi.$$

Insert these coefficients into the integral expression for $u(x, y)$:

$$\begin{aligned} u(x, y) &= \frac{1}{\pi}\int_0^\infty \int_{-\infty}^\infty [f(\xi)\cos(\omega\xi)\cos(\omega x) + f(\xi)\sin(\omega\xi)\sin(\omega x)]e^{-\omega y}\, d\xi\, d\omega \\ &= \frac{1}{\pi}\int_{-\infty}^\infty \left[\int_0^\infty \cos(\omega(\xi - x))e^{-\omega y}\, d\omega\right] f(\xi)\, d\xi. \end{aligned}$$

The inner integral can be evaluated explicitly:

$$\begin{aligned} \int_0^\infty \cos(\omega(\xi - x))e^{-\omega y}\, d\omega &= \left[\frac{e^{-\omega y}}{y^2 + (\xi - x)^2}[-y\cos(\omega(\xi - x)) + (\xi - x)\sin(\omega(\xi - x))]\right]_0^\infty \\ &= \frac{y}{y^2 + (\xi - x)^2}. \end{aligned}$$

The solution of the Dirichlet problem for the upper half-plane is therefore

$$u(x, y) = \frac{y}{\pi}\int_{-\infty}^\infty \frac{f(\xi)}{y^2 + (\xi - x)^2}\, d\xi. \tag{18.4}$$

Solution Using the Fourier Transform

We also can solve this problem for the upper half-plane using the Fourier transform. Apply the transform in the x variable to Laplace's equation to obtain

$$\mathcal{F}\left[\frac{\partial^2 u}{\partial x^2}\right]+\mathcal{F}\left[\frac{\partial^2 u}{\partial y^2}\right]=\frac{\partial^2 \hat{u}}{\partial y^2}(\omega, y)-\omega^2\hat{u}(\omega, y)=0.$$

This has the general solution

$$\hat{u}(\omega, y)=a_\omega e^{\omega y}+b_\omega e^{-\omega y}.$$

We want this to be bounded. But for $\omega > 0$, $e^{\omega y}\to\infty$ as $y\to\infty$, so $a_\omega = 0$ for positive ω. And $e^{-\omega y}\to\infty$ as $y\to\infty$ if $\omega < 0$, so for negative ω, we must have $b_\omega = 0$. Therefore,

$$\hat{u}(\omega, y)=\begin{cases} b_\omega e^{-\omega y} & \text{for } \omega > 0 \\ a_\omega e^{\omega y} & \text{for } \omega < 0. \end{cases}$$

Consolidate these cases by writing

$$\hat{u}(\omega, y)=c_\omega e^{-|\omega| y}.$$

To solve for the constants, take the transform of $u(x, 0) = f(x)$ to get

$$\hat{u}(\omega, 0)=\hat{f}(\omega)=c_\omega.$$

Then

$$\hat{u}(\omega, y)=\hat{f}(\omega)e^{-|\omega| y}.$$

Finally, apply the inverse Fourier transform to get

$$\begin{aligned} u(x, y) &= \mathcal{F}^{-1}\left[\hat{f}(\omega)e^{-|\omega| y}\right](x) \\ &= \frac{1}{2\pi}\int_{-\infty}^{\infty}\hat{f}(\omega)e^{-|\omega| y}e^{i\omega x}\,d\omega \\ &= \frac{1}{2\pi}\int_{-\infty}^{\infty}\left(\int_{-\infty}^{\infty} f(\xi)e^{-i\omega\xi}\,d\xi\right)e^{-|\omega| y}e^{i\omega x}\,d\omega \\ &= \frac{1}{2\pi}\int_{-\infty}^{\infty}\left(\int_{-\infty}^{\infty} e^{-|\omega| y}e^{-i\omega(\xi-x)}\,d\omega\right)f(\xi)\,d\xi. \end{aligned}$$

A routine integration gives us

$$\int_{-\infty}^{\infty} e^{-|\omega| y}e^{-i\omega(\xi-x)}\,d\omega=\frac{2y}{y^2+(\xi-x)^2}.$$

Then

$$\begin{aligned} u(x, y) &= \frac{1}{2\pi}\int_{-\infty}^{\infty}\left(\frac{2y}{y^2+(\xi-x)^2}\right)f(\xi)\,d\xi \\ &= \frac{y}{\pi}\int_{-\infty}^{\infty}\frac{f(\xi)}{y^2+(\xi-x)^2}\,d\xi, \end{aligned}$$

in agreement with the solution obtained by separation of variables.

18.5.2 The Right Quarter-Plane

The right quarter-plane has the nonnegative horizontal and vertical axes as boundary. The Dirichlet problem for this region is

$$\nabla^2 u(x, y) = 0 \text{ for } x > 0,\ y > 0,$$

$$u(x, 0) = f(x) \text{ for } x \geq 0,$$

and

$$u(0, y) = g(y) \text{ for } y \geq 0.$$

This problem can be treated by solving separately the cases that either $f(x)$ or $g(y)$ is identically zero. Separation of variables applies to both cases, and the solution of the given problem is the sum of the solutions of these simpler problems.

We will demonstrate a different method for the case that $g(y) = 0$. Notice that if we fold the upper half-plane across the vertical axis we obtain the right quarter-plane. This suggests that we might be able to use the solution for the upper half-plane to solve the problem for the right quarter-plane. To do this, let

$$w(x) = \begin{cases} f(x) & \text{for } x \geq 0 \\ \text{anything} & \text{for } x < 0. \end{cases}$$

where by "anything" we mean we will fill in this part shortly. We now have a Dirichlet problem for the upper half-plane with the data function $u(x, 0) = w(x)$. We know the solution u_{hp} of this problem for the upper half-plane:

$$u_{\text{hp}}(x, y) = \frac{y}{\pi} \int_{-\infty}^{\infty} \frac{w(\xi)}{y^2 + (\xi - x)^2}\, d\xi.$$

Write this as

$$u_{\text{hp}}(x, y) = \frac{y}{\pi} \left[\int_{-\infty}^{0} \frac{w(\xi)}{y^2 + (\xi - x)^2}\, d\xi + \int_{0}^{\infty} \frac{w(\xi)}{y^2 + (\xi - x)^2}\, d\xi \right].$$

Change variables in the first integral on the right by letting $\zeta = -\xi$:

$$\int_{-\infty}^{0} \frac{w(\xi)}{y^2 + (\xi - x)^2}\, d\xi = \int_{0}^{\infty} \frac{w(-\zeta)}{y^2 + (\zeta + x)^2} (-1)\, d\zeta.$$

For uniformity in notation, replace the dummy variable of integration on the right with ξ to write

$$\begin{aligned} u_{\text{hp}}(x, y) &= \frac{y}{\pi} \left[\int_{\infty}^{0} \frac{w(-\xi)}{y^2 + (\xi + x)^2} (-1)\, d\xi + \int_{0}^{\infty} \frac{w(\xi)}{y^2 + (\xi - x)^2}\, d\xi \right] \\ &= \frac{y}{\pi} \int_{0}^{\infty} \left(\frac{w(-\xi)}{y^2 + (\xi + x)^2} + \frac{f(\xi)}{y^2 + (\xi - x)^2} \right) d\xi. \end{aligned}$$

In the last integral, we used the fact that $w(\xi) = f(\xi)$ for $\xi \geq 0$. Now fill in the "anything" in the definition of w. Notice that the last integral will vanish at points $(0, y)$ on the positive y-axis if $f(\xi) + w(-\xi) = 0$ for $\xi \geq 0$. This will occur if $w(-\xi) = -f(\xi)$. Make w the odd extension of f to the entire line. In this way, we obtain the solution for the upper half-plane:

$$u_{\text{hp}}(x, y) = \frac{y}{\pi} \int_{0}^{\infty} \left(\frac{1}{y^2 + (\xi - x)^2} - \frac{1}{y^2 + (\xi + x)^2} \right) f(\xi)\, d\xi.$$

But this function is also harmonic on the right quarter-plane, vanishes when $x = 0$, and equals $f(x)$ if $x \geq 0$ and $y = 0$. Therefore, this function is also the solution of the problem for the right quarter-plane (in this case of zero data along the positive y-axis).

EXAMPLE 18.5

We have a formula for the solution of

$$\nabla^2 u(x, y) = 0 \text{ for } x > 0, y > 0,$$
$$u(0, y) = 0 \text{ for } y > 0,$$

and

$$u(x, 0) = 1 \text{ for } x > 0.$$

With $f(x) = 1$, we can write the solution

$$u(x, y) = \frac{y}{\pi}\int_0^\infty \frac{1}{y^2 + (\xi - x)^2}\, d\xi - \frac{y}{\pi}\int_0^\infty \frac{1}{y^2 + (\xi + x)^2} d\xi. \blacklozenge$$

A routine integration yields

$$\frac{y}{\pi}\int_0^\infty \frac{1}{y^2 + (\xi - x)^2}\, d\xi = \frac{1}{2} + \frac{1}{\pi}\arctan\left(\frac{x}{y}\right)$$

and

$$\frac{y}{\pi}\int_0^\infty \frac{1}{y^2 + (\xi + x)^2}\, d\xi = \frac{1}{2} - \frac{1}{\pi}\arctan\left(\frac{x}{y}\right).$$

The solution is

$$u(x, y) = \frac{2}{\pi}\arctan\left(\frac{x}{y}\right).$$

This function is harmonic on the right quarter-plane and $u(0, y) = 0$ for $y > 0$. Furthermore, if $x > 0$,

$$\lim_{y \to 0+} \frac{2}{\pi}\arctan\left(\frac{x}{y}\right) = \frac{2}{\pi}\frac{\pi}{2} = 1,$$

as required. ♦

SECTION 18.5 PROBLEMS

1. Write an integral solution for the Dirichlet problem for the upper half-plane if

$$u(x, 0) = \begin{cases} -1 & \text{for } -4 \le x < 0 \\ 1 & \text{for } 0 \le x < 4 \\ 0 & \text{for } |x| > 4. \end{cases}$$

2. Solve the Dirichlet problem for the strip $-\infty < x < \infty, 0 < y < 1$ if $u(x, 0) = 0$ for $x < 0$ and $u(x, 0) = e^{-\alpha x}$ for $x > 0$ with α a positive number.
3. Find a general formula for the solution of the Dirichlet problem for the right quarter-plane if $u(x, 0) = f(x)$ and $u(0, y) = g(y)$.
4. Write an integral solution for the Dirichlet problem for the upper half-plane if $u(x, 0) = e^{-|x|}$.
5. Find the steady-state temperature distribution in a thin, homogeneous flat plate extending over the right quarter plane if the temperature on the vertical side is e^{-y} and the temperature on the horizontal side is maintained at zero.
6. Write an integral solution for the Dirichlet problem for the right quarter-plane if $u(x, 0) = 0$ for $x > 0$ and $u(0, y) = g(y)$ for $y > 0$. Use separation of variables, and then derive a solution using an appropriate Fourier transform.
7. Write an integral solution for the Dirichlet problem for the right quarter-plane if $u(x, 0) = e^{-x}\cos(x)$ for $x > 0$ and $u(0, y) = 0$ for $y > 0$.
8. Write an integral solution for the Dirichlet problem for the lower half-plane $y < 0$ if $u(x, 0) = f(x)$.

9. Solve for the steady-state temperature distribution in a homogeneous, infinite flat plate covering the half-plane $x \geq 0$ if the temperature on the boundary $x = 0$ is $f(y)$ where

$$f(y) = \begin{cases} 1 & \text{for } |y| \leq 1 \\ 0 & \text{for } |y| > 1. \end{cases}$$

10. Write a general expression for the steady-state temperature distribution in an infinite, homogeneous flat plate covering the strip $0 \leq y \leq 1$, $x \geq 0$ if the temperature on the left boundary and on the bottom side is zero while the temperature on the top part of the boundary is $f(x)$.

11. Solve for the steady-state temperature distribution in an infinite, homogeneous flat plate covering the half-plane $y \geq 0$ if the temperature on the boundary $y = 0$ is kept at zero for $x < 4$, constant A for $4 \leq x \leq 8$, and zero for $x > 8$.

12. Solve the following problem:

$$\nabla^2 u(x, y) = 0 \text{ for } 0 < x < \pi, 0 < y < 2$$

with boundary conditions $u(0, y) = 0$ and $u(\pi, y) = 4$ for $0 < y < 2$ and

$$\frac{\partial u}{\partial y}(x, 0) = u(x, 2) = 0 \text{ for } 0 < x < \pi.$$

18.6 A Dirichlet Problem for a Cube

To illustrate a Dirichlet problem in three dimensions, we will solve:

$$\nabla^2 u(x, y, z) = 0 \text{ for } 0 < x < A, 0 < y < B, 0 < z < C,$$
$$u(x, y, 0) = u(x, y, C) = 0,$$
$$u(0, y, z) = u(A, y, z) = 0,$$

and

$$u(x, 0, z) = 0, u(x, B, z) = f(x, z).$$

Let $u(x, y, z) = X(x)Y(y)Z(z)$ to obtain

$$\frac{X''}{X} = -\frac{Y''}{Y} - \frac{Z''}{Z} = -\lambda.$$

After a second separation (of y and z), we have

$$\frac{Z''}{Z} = \lambda - \frac{Y''}{Y} = -\mu.$$

Then,

$$X'' + \lambda X = 0, \; Z'' + \mu Z = 0 \text{ and } Y'' - (\lambda + \mu)Y = 0.$$

From the boundary conditions,

$$X(0) = X(A) = 0, \; Z(0) = Z(C) = 0, \text{ and } Y(0) = 0.$$

The problems for X and Z have eigenvalues and eigenfunctions of

$$\lambda_n = \frac{n^2\pi^2}{A^2}, X_n(x) = \sin\left(\frac{n\pi x}{A}\right)$$

and

$$\mu_m = \frac{m^2\pi^2}{C^2}, Z_m(z) = \sin\left(\frac{m\pi z}{C}\right)$$

with n and m independently varying over the positive integers. The problem for Y is

$$Y'' - \left(\frac{n^2\pi^2}{A^2} + \frac{m^2\pi^2}{C^2}\right) Y = 0; Y(0) = 0$$

with solutions that are constant multiples of

$$Y_{nm}(y) = \sinh(\beta_{nm} y)$$

where

$$\beta_{nm} = \sqrt{\frac{n^2\pi^2}{A^2} + \frac{m^2\pi^2}{C^2}}.$$

Attempt a solution

$$u(x, y, z) = \sum_{n=1}^{\infty}\sum_{m=1}^{\infty} c_{nm} \sin\left(\frac{n\pi x}{A}\right) \sin\left(\frac{m\pi z}{C}\right) \sinh(\beta_{nm} y).$$

We must choose the coefficients so that

$$u(x, B, z) = f(x, z) = \sum_{n=1}^{\infty}\sum_{m=1}^{\infty} c_{nm} \sin\left(\frac{n\pi x}{A}\right) \sin\left(\frac{m\pi z}{C}\right) \sinh(\beta_{nm} B).$$

This is a double Fourier series for $f(x, z)$ on $0 \le x \le A, 0 \le z \le C$. We have seen this type of expansion before (Sections 17.7 and 18.5), leading us to choose

$$c_{nm} = \frac{2}{AC \sinh(\beta_{nm} B)} \int_0^A \int_0^C f(\xi, \zeta) \sin\left(\frac{n\pi \xi}{A}\right) \sin\left(\frac{m\pi \zeta}{C}\right) d\zeta\, d\xi.$$

As usual, if nonzero data is prescribed on more than one face, split the Dirichlet problem into a sum of problems; each of which has nonzero data on only one face.

SECTION 18.6 PROBLEMS

1. Solve

$$\nabla^2 u(x, y, z) = 0 \text{ for } 0 < x < 1, 0 < y < 2\pi, 0 < z < \pi,$$
$$u(0, y, z) = u(1, y, z) = 0,$$
$$u(x, 0, z) = u(x, y, 0) = 0,$$
$$u(x, y, \pi) = 1, u(x, 2\pi, z) = xz^2.$$

2. Solve

$$\nabla^2 u(x, y, z) = 0 \text{ for } 0 < x < 1, 0 < y < 2, 0 < z < \pi,$$
$$u(x, 0, z) = u(x, 2, z) = 0,$$
$$u(0, y, z) = u(x, y, \pi) = 0,$$
$$u(x, y, 0) = x^2(1 - x)y(2 - y), u(1, y, z) = \sin(\pi y)\sin(z).$$

3. Solve

$$\nabla^2 u(x, y, z) = 0 \text{ for } 0 < x < 1, 0 < y < 1, 0 < z < 1,$$
$$u(0, y, z) = u(1, y, z) = 0,$$
$$u(x, 0, z) = u(x, 1, z) = 0,$$
$$u(x, y, 0) = 0, u(x, y, 1) = xy.$$

4. Solve

$$\nabla^2 u(x, y, z) = 0 \text{ for } 0 < x < 2\pi, 0 < y < 2\pi, 0 < z < 1,$$
$$u(x, y, 0) = u(x, y, 1) = 0,$$
$$u(x, 0, z) = u(x, 2\pi, z) = 0,$$
$$u(0, y, z) = 0, u(2\pi, y, z) = z.$$

18.7 Steady-State Equation for a Sphere

We will solve for the steady-state temperature distribution in a solid sphere given the temperature at all times on the surface.

Let the sphere be centered at the origin and have a radius of R. Use spherical coordinates (ρ, θ, φ) in which ρ is the distance from the origin to the point, θ is the polar angle between the positive x-axis and the projection onto the x, y-plane of the line from the origin to the point, and φ is the angle of declination from the positive z-axis to this line (Figure 18.3).

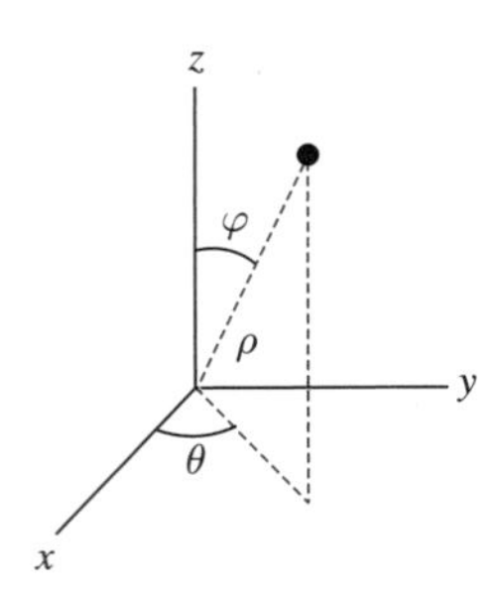

FIGURE 18.3 *Spherical coordinates.*

Assuming symmetry of the temperature function about the z-axis, the solution is independent of θ and depends only on ρ and φ. Laplace's equation in spherical coordinates (with independence from θ) is

$$\nabla^2 u(\rho,\varphi) = \frac{\partial^2 u}{\partial \rho^2} + \frac{2}{\rho}\frac{\partial u}{\partial \rho} + \frac{1}{\rho^2}\frac{\partial^2 u}{\partial \varphi^2} + \frac{\cot(\varphi)}{\rho^2}\frac{\partial u}{\partial \varphi} = 0.$$

The temperature on the surface is $u(R,\varphi) = f(\varphi)$ with f being given.

Let $u(\rho,\varphi) = X(\rho)\Phi(\varphi)$ to obtain

$$X''\Phi + \frac{2}{\rho}X'\Phi + \frac{1}{\rho^2}X\Phi'' + \frac{\cot(\varphi)}{\rho^2}X\Phi' = 0.$$

Upon dividing this equation by $X\Phi$, we can separate the variables, obtaining

$$\frac{\Phi''}{\Phi} + \cot(\varphi)\frac{\Phi'}{\Phi} = -\rho^2\frac{X''}{X} - 2\rho\frac{X'}{X} = -\lambda.$$

Then

$$\rho^2 X'' + 2\rho X' - \lambda X = 0 \text{ and } \Phi'' + \cot(\varphi)\Phi' + \lambda\Phi = 0.$$

To solve this equation for Φ, write it as

$$\frac{1}{\sin(\varphi)}[\Phi'\sin(\varphi)]' + \lambda\Phi = 0. \tag{18.5}$$

Change variables by putting $x = \cos(\varphi)$. Then $\varphi = \arccos(x)$. Let

$$G(x) = \Phi(\arccos(x)).$$

Since $0 \le \varphi \le \pi$, then $-1 \le x \le 1$. Compute

$$\begin{aligned}\Phi'(\varphi)\sin(\varphi) &= \sin(\varphi)\frac{d\Phi}{dx}\frac{dx}{d\varphi} \\ &= \sin(\varphi)G'(x)[-\sin(\varphi)] \\ &= -\sin^2(\varphi)G'(x) = -[1-\cos^2(x)]G'(x) \\ &= -(1-x^2)G'(x).\end{aligned}$$

Then

$$\frac{d}{d\varphi}[\Phi'(\varphi)\sin(\varphi) = -\frac{d}{d\varphi}[(1-x^2)G'(x)]$$
$$= -\frac{d}{dx}[(1-x^2)G'(x)]\frac{dx}{d\varphi}$$
$$= -\frac{d}{dx}[(1-x^2)G'(x)][-\sin(\varphi)].$$

We conclude that

$$\frac{1}{\sin(\varphi)}\frac{d}{d\varphi}[\Phi'(\varphi)\sin(\varphi)] = \frac{d}{dx}[(1-x^2)G'(x)].$$

The point to this calculation is that equation (18.5) transforms to

$$[(1-x^2)G'(x)]' + \lambda G(x) = 0,$$

which is Legendre's differential equation, considered on the interval $[-1, 1]$. In Section 15.2, we found the eigenvalues $\lambda_n = n(n+1)$ for $n = 0, 1, 2, \cdots$. The eigenfunctions are constant multiples of the Legendre polynomials $P_n(x)$. For nonnegative integers n, we therefore have a solution of the differential equation for Φ:

$$\Phi_n(\varphi) = G(\cos(\varphi)) = P_n(\cos(\varphi)).$$

Now that we know the eigenvalues, the differential equation for X is

$$\rho^2 X'' + 2\rho X' - n(n+1)X = 0.$$

This is an Euler equation with general solution

$$X(\rho) = a\rho^n + b\rho^{-n-1}.$$

Choose $b = 0$ to have a solution that is bounded as $\rho \to 0+$, which is the center of the sphere. For each nonnegative integer n, we now have a function

$$u_n(\rho, \varphi) = a_n \rho^n P_n(\cos(\varphi))$$

that satisfies the steady-state heat equation. To satisfy the boundary condition, write a superposition

$$u(\rho, \varphi) = \sum_{n=0}^{\infty} a_n \rho^n P_n(\cos(\varphi)).$$

We must choose the coefficients to satisfy

$$u(R, \varphi) = \sum_{n=0}^{\infty} a_n R^n P_n \cos(\varphi).$$

To put this into the context of an eigenfunction expansion in terms of Legendre polynomials, recall that $\varphi(x) = \arccos(x)$ to obtain

$$\sum_{n=0}^{\infty} a_n R^n P_n(x) = f(\arccos(x)).$$

Then

$$a_n R^n = \frac{2n+1}{2}\int_{-1}^{1} f(\arccos(x)) P_n(x)\,dx$$

so

$$a_n = \frac{2n+1}{2R^n} \int_{-1}^{1} f(\arccos(x)) P_n(x) dx.$$

The steady-state temperature function is

$$u(\rho, \varphi) = \sum_{n=0}^{\infty} \frac{2n+1}{2} \left(\int_{-1}^{1} f(\arccos(x)) P_n(x)\, dx \right) \left(\frac{\rho}{R} \right)^n P_n(\cos(\varphi)).$$

EXAMPLE 18.6

For $f(\varphi) = \varphi$, the solution is

$$u(\rho, \varphi) = \sum_{n=0}^{\infty} \frac{2n+1}{2} \left(\int_{-1}^{1} \arccos(x) P_n(x)\, dx \right) \left(\frac{\rho}{R} \right)^n P_n(\cos(\varphi)).$$

We will use numerical integrations to approximate the first six coefficients. $P_0(x), \cdots, P_5(x)$ were listed in Chapter 15. Using these and MAPLE to perform the computations, we obtain

$$\int_{-1}^{1} \arccos(x) P_0(x) dx \approx \pi,$$

$$\int_{-1}^{1} x \arccos(x)\, dx, \approx -0.7854,$$

$$\int_{-1}^{1} \frac{1}{2}(3x^2 - 1) \arccos(x)\, dx = 0,$$

$$\int_{-1}^{1} \frac{1}{2}(5x^3 - 3x) \arccos(x)\, dx \approx -.049087,$$

$$\int_{-1}^{1} \frac{1}{8}(35x^4 - 30x^2 + 3) \arccos(x)\, dx = 0,$$

$$\int_{-1}^{1} \frac{1}{8}(63x^5 - 70x^3 + 15x) \arccos(x)\, dx \approx -0.012272.$$

Then

$$u(\rho, \varphi) \approx \frac{\pi}{2} - \frac{3}{2}(0.7854)\frac{\rho}{R} \cos(\varphi) - \frac{7}{2}(0.049087)\frac{1}{2}\left(\frac{\rho}{R}\right)^3 (5\cos^3(\varphi) - 3\cos(\varphi))$$
$$- \frac{11}{2}(0.012272)\left(\frac{\rho}{R}\right)^5 \frac{1}{8}(63\cos^5(\varphi) - 70\cos^3(\varphi) + 15\cos(\varphi)). \ \blacklozenge$$

SECTION 18.7 PROBLEMS

In each of Problems 1 through 4, write a solution for the steady-state temperature distribution in the sphere if the boundary data is given by $f(\varphi)$. Use numerical integration to approximate the first six terms in the solution.

1. $f(\varphi) = \varphi^3$
2. $f(\varphi) = 2 - \varphi^2$
3. $f(\varphi) = A\varphi^2$, in which A is a positive number.
4. $f(\varphi) = \sin(\varphi)$
5. Solve for the steady-state temperature distribution in a hollowed-out sphere given in spherical coordinates by $R_1 \le \rho \le R_2$. The inner surface is kept at constant temperature T and the outer surface at temperature zero. Assume that u is a function of ρ and φ only. Approximate the solution with a sum of the first six terms, using numerical integration to approximate the coefficients.

18.8 The Neumann Problem

Recall that, if g is a function of two variables defined on a set of points D in the plane having boundary C, then the *normal derivative* $\partial g/\partial n$ of g on C is the dot product of the gradient of g with a unit normal vector **n** to C:

$$\frac{\partial g}{\partial n} = \nabla g \cdot \mathbf{n}.$$

We will assume that **n** is a unit *outer normal* to D. This means that, if drawn as an arrow from a point on C, **n** points away from D, as in Figure 12.19.

A *Neumann problem* in the plane consists of finding a function that is harmonic on a given region D, and whose normal derivative assumes given values on the boundary C of D. This problem has the form

$$\nabla^2 u(x, y) = 0 \text{ on } D,$$

$$\frac{\partial u}{\partial n} = g(x, y) \text{ for } (x, y) \text{ in } C,$$

with $g(x, y)$ a given function defined on the boundary of C.

The following lemma plays an important role in attempting to solve a Neumann problem.

LEMMA 18.1 ***Green's First Identity***

Let D be a bounded set of points in the plane, having boundary curve C. Assume that C is a simple, closed, piecewise smooth curve. Let f and g be continuous with continuous first and second partial derivatives on D and at points of C. Then

$$\oint_C g\frac{\partial f}{\partial n}\, ds = \iint_D (g\nabla^2 f + \nabla f \cdot \nabla g)\, dA. \ \blacklozenge$$

The line integral on the left is with respect to arc length along C.

Proof of Lemma 18.1 By Green's theorem,

$$\oint_C g\frac{\partial g}{\partial n}\, ds = \oint_C (g\nabla f)\cdot \mathbf{n}\, ds = \iint_D \nabla \cdot (g\nabla f)\, dA.$$

The rest of the proof consists of computing

$$\begin{aligned}
\nabla \cdot (g\nabla f) &= \nabla \cdot \left(g\frac{\partial f}{\partial x}\mathbf{i} + g\frac{\partial f}{\partial y}\mathbf{j}\right) \\
&= \frac{\partial}{\partial x}\left(g\frac{\partial f}{\partial x}\right) + \frac{\partial}{\partial y}\left(g\frac{\partial f}{\partial y}\right) \\
&= g\left(\frac{\partial^2 f}{\partial x^2} + \frac{\partial^2 f}{\partial y^2}\right) + \frac{\partial g}{\partial x}\frac{\partial f}{\partial x} + \frac{\partial g}{\partial y}\frac{\partial f}{\partial} \\
&= g\nabla^2 f + \nabla f \cdot \nabla g. \ \blacklozenge
\end{aligned}$$

Use the lemma as follows. If $g(x, y) = 1$ and $f = u$ (a harmonic function on D), then the double integral in the lemma is zero because its integrand vanishes, and the line integral is just the line integral of the normal derivative of u over the boundary C of D. For a Neumann problem, this normal derivative is a given function g, so the lemma tells us that

$$\oint_C \frac{\partial u}{\partial n}\, ds = \oint_C g\, ds = 0.$$

This means that vanishing of the integral of the given normal derivative over the boundary of the region is a necessary condition for the Neumann problem to have a solution. Put another way, if the integral of g over C is not zero, this Neumann problem has no solution.

This result can be extended to the case that D is not a bounded region and C is not a closed curve. This occurs, for example, with the right quarter plane given by $x \geq 0, y \geq 0$. Here the region is unbounded, and its boundary curve consists of the nonnegative x- and y-axes.

EXAMPLE 18.7

We will solve a Neumann problem for a square:

$$\nabla^2 u = 0 \text{ for } 0 < x <, 0 < y < 1,$$

$$\frac{\partial u}{\partial n} = \begin{cases} 0 & \text{on the left, top and lower sides,} \\ y^2 & \text{on the right side of the square.} \end{cases}$$

This means that

$$\frac{\partial u}{\partial n}(x, 0) = \frac{\partial u}{\partial n}(x, 1) = \frac{\partial u}{\partial n}(0, y) = 0$$

while

$$\frac{\partial u}{\partial n}(1, y) = y^2 \text{ for } 0 \leq x \leq 1 \text{ and } 0 \leq y \leq 1.$$

First take the line integral of $\partial u/\partial n$ about the boundary of D, which consists of four straight line segments:

$$\oint_C \frac{\partial u}{\partial n}\, ds = \int_o^2 y^2\, dy = \frac{1}{3} \neq 0.$$

Therefore this Neumann problem has no solution. ♦

Dirichlet problems may also fail to have solutions, depending on the region and the given function at values on the boundary. However, for "simple" regions such as rectangles and disks, and "reasonable" data functions on the boundary, Dirichlet problems have solutions. The last example shows that, even for a simple region (a square) and reasonably well-behaved normal derivative on the boundary, a Neumann problem may be ill posed (no solution).

We will analyze Neumann problems for rectangles and disks.

18.8.1 A Neumann Problem for a Rectangle

We will consider the Neumann problem

$$\nabla^2 u(x, y) = 0 \text{ for } 0 < x < a, 0 < y < b,$$

$$\frac{\partial u}{\partial y}(x, 0) = \frac{\partial u}{\partial y}(x, b) = 0 \text{ for } 0 \leq x \leq a,$$

$$\frac{\partial u}{\partial x}(0, y) = 0 \text{ for } 0 \leq y \leq b,$$

and

$$\frac{\partial u}{\partial x}(a, y) = g(y) \text{ for } 0 \le y \le b.$$

For the rectangle, the normal derivative is $\partial u/\partial x$ on the vertical sides and $\partial u/\partial y$ on the horizontal sides. As a necessary (but not sufficient) condition for a solution to exist, we assume that

$$\int_0^b g(y)\,dy = 0.$$

It will be instructive to see how this assumption plays a role in this problem having a solution.

Let $u(x, y) = X(x)Y(y)$, and substitute into Laplace's equation and also into the boundary conditions to obtain

$$X'' + \lambda X = 0;\ X'(0) = 0$$

and

$$Y'' - \lambda Y = 0;\ Y'(0) = Y'(b) = 0.$$

This Sturm-Liouville problem for Y has eigenvalues and eigenfunction

$$\lambda_n = -\frac{n^2\pi^2}{b^2},\ Y_n(y) = \cos\left(\frac{n\pi y}{b}\right)$$

for $n = 0, 1, 2, \cdots$. Notice that $Y(y)$ is constant for $n = 0$.

Now the problem for X is

$$X'' - \frac{n^2\pi^2}{b^2}X = 0;\ X'(0) = 0.$$

This problem for X has only a boundary condition at $x = 0$, so we must look at cases in solving for X.

If $n = 0$, the differential equation for X is just $X'' = 0$, so $X(x) = cx + d$. Then $X'(0) = d = 0$, so $X(x)$ is constant in this case.

If n is a positive integer, then the differential equation for X has the general solution

$$X(x) = ce^{n\pi x/b} + de^{-n\pi x/b}.$$

Then

$$X'(0) = \frac{n\pi}{b}c - \frac{n\pi}{b}d = 0,$$

so $c = d$. This means that $X(x)$ must be have the form

$$X(x) = c\cosh\left(\frac{n\pi}{b}x\right).$$

We now have functions

$$u_0(x, y) = \text{ constant}$$

and, for each positive integer n,

$$u_n(x, y) = X_n(x)Y_n(y) = c_n\cosh\left(\frac{n\pi}{b}x\right)\cos\left(\frac{n\pi}{b}y\right).$$

We have used the zero boundary conditions on the top, bottom, and left sides of the rectangle. To satisfy the last boundary condition (on the right side) attempt a superposition

$$u(x, y) = \sum_{n=0}^{\infty} u_n(x, y)$$

$$= c_0 + \sum_{n=1}^{\infty} c_n\cosh\left(\frac{n\pi}{b}x\right)\cos\left(\frac{n\pi}{b}y\right).$$

Now we need to choose the c_n's so that

$$\frac{\partial u}{\partial x}(a, y) = g(y) = \sum_{n=1}^{\infty} \frac{n\pi}{b} c_n \sinh\left(\frac{n\pi a}{b}\right) \cos\left(\frac{n\pi}{b} y\right).$$

This is a Fourier cosine expansion of $g(y)$ on $[0, b]$. Notice that the constant term in this expansion is zero. But this constant term is

$$\frac{1}{b}\int_0^b g(y)\,dy,$$

and we would have a contradiction if this integral were not zero. In this event, this problem would have no solution.

For the other coefficients in this cosine expansion, we have

$$\frac{n\pi}{b} c_n \sinh\left(\frac{n\pi a}{b}\right) = \frac{2}{b}\int_0^b g(\xi)\cos\left(\frac{n\pi\xi}{b}\right) d\xi,$$

so

$$c_n = \frac{2}{n\pi \sinh(n\pi a/b)}\int_0^b g(\xi)\cos\left(\frac{n\pi\xi}{b}\right) d\xi.$$

With this choice of the coefficients, the solution of the Neumann problem is

$$u(x, y) = c_0 + \sum_{n=1}^{\infty} c_n \cosh\left(\frac{n\pi x}{b}\right)\cos\left(\frac{n\pi y}{b}\right).$$

Here c_0 is an arbitrary constant. Neumann problems do not have unique solutions: If u is a solution, so is $u + c$ for any number c.

18.8.2 A Neumann Problem for a Disk

Suppose D is a disk of radius R about the origin. The boundary is the circle C of radius R about the origin. In polar coordinates, the Neumann problem for D is

$$\nabla^2 u(r, \theta) = 0 \text{ for } 0 \le r < R, -\pi \le \theta \le \pi$$

and

$$\frac{\partial u}{\partial r}(R, \theta) = f(\theta) \text{ for } -\pi \le \theta \le \pi.$$

Notice that the normal derivative to C is $\partial u/\partial r$ because the line from the origin to a point of C is perpendicular to C at that point.

A necessary condition for existence of a solution is that

$$\int_{-\pi}^{\pi} f(\theta)\,d\theta = 0,$$

which is a condition we will assume for $f(\theta)$.

Attempt a solution

$$u(r, \theta) = \frac{1}{2}a_0 + \sum_{n=1}^{\infty}[a_n r^n \cos(n\theta) + b_n r^n \sin(n\theta)].$$

We must choose the coefficients to satisfy

$$\frac{\partial u}{\partial r}(R, \theta) = f(\theta)$$

$$= \sum_{n=1}^{\infty}[n a_n R^{n-1}\cos(n\theta) + n b_n R^{n-1}\sin(n\theta)].$$

This is a Fourier expansion of $f(\theta)$ on $[-\pi, \pi]$. Notice that the constant term in this expansion is zero. But this constant term is exactly

$$\frac{1}{\pi}\int_{-\pi}^{\pi} f(\theta)\,d\theta,$$

so we would have a contradiction if this integral did not vanish, as we have assumed. For the other coefficients, we need

$$na_n R^{n-1} = \frac{1}{\pi}\int_{-\pi}^{\pi} f(\xi)\cos(n\xi)\,d\xi$$

and

$$b_n = \frac{1}{\pi}\int_{-\pi}^{\pi} f(\xi)\sin(n\xi)\,d\xi$$

for $n = 1, 2, \cdots$. Then

$$a_n = \frac{1}{n\pi R^{n-1}}\int_{-\pi}^{\pi} f(\xi)\cos(n\xi)\,d\xi$$

and

$$b_n = \frac{1}{n\pi R^{n-1}}\int_{-\pi}^{\pi} f(\xi)\sin(n\xi)\,d\xi.$$

Upon inserting these coefficients, the solution is

$$u(r,\theta) = \frac{1}{2}a_0 + \frac{R}{\pi}\sum_{n=1}^{\infty}\frac{1}{n}\left(\frac{r}{R}\right)^n \int_{0\pi}^{\pi}[\cos(n\xi)\cos(n\theta) + \sin(n\xi)\sin(n\theta)]f(\xi)\,d\xi.$$

We can also write this solution as

$$u(r,\theta) = \frac{1}{2}a_0 + \frac{R}{\pi}\sum_{n=1}^{\infty}\frac{1}{n}\left(\frac{r}{R}\right)^n \int_{-\pi}^{\pi}\cos(n(\xi-\theta))f(\xi)\,d\xi.$$

The term $a_0/2$ is an arbitrary constant, written as $a_0/2$ simply because of the context of a Fourier series.

EXAMPLE 18.8

Solve the Neumann problem for the unit disk about the origin:

$$\nabla^2 u(x,y) = 0 \text{ for } x^2 + y^2 < 1$$

and

$$\frac{\partial u}{\partial n}(x,y) = xy^2 \text{ for } x^2 + y^2 = 1.$$

Switch to polar coordinates, letting $U(r,\theta) = u(r\cos(\theta), r\sin(\theta))$. Now the problem is

$$\nabla^2 U(r,\theta) = 0 \text{ for } 0 \le r < 1, -\pi \le \theta \le \pi$$

and

$$\frac{\partial U}{\partial r}(1,\theta) = \cos(\theta)\sin^2(\theta).$$

First, compute

$$\int_{-\pi}^{\pi}\cos(\theta)\sin^2(\theta)\,d\theta = 0,$$

which is a necessary condition for this problem to have a solution. Write the solution

$$U(r,\theta) = \frac{1}{2}a_0 + \frac{1}{\pi}\sum_{n=1}^{\infty}\frac{1}{n}r^n\int_{-\pi}^{\pi}\cos(n(\xi-\theta))\cos(\xi)\sin^2(\xi)\,d\xi.$$

Evaluate this integral:

$$\int_{-\pi}^{\pi}\cos(n(\xi-\theta))\cos(\xi)\sin^2(\xi)\,d\xi =$$

$$\begin{cases} 0 & \text{for } n = 2,4,5,6,7,\cdots \\ \pi\cos(\theta)/4 & \text{for } n = 1 \\ -\pi\cos^2(\theta) + 3\pi\cos(\theta)/4 & \text{for } n = 3. \end{cases}$$

The solution is

$$\begin{aligned} U(r,\theta) &= \frac{1}{2}a_0 + \frac{1}{4}r\cos(\theta) + \frac{1}{3}r^3\left(-\cos^3(\theta) + \frac{3}{4}\cos(\theta)\right) \\ &= \frac{1}{2}a_0 + \frac{1}{4}r\cos(\theta) - \frac{1}{3}r^3\cos^3(\theta) + \frac{1}{4}r^3\cos(\theta). \end{aligned}$$

To convert this solution to rectangular coordinates, let $x = r\cos(\theta)$ and $r^2 = x^2 + y^2$ to obtain

$$u(x,y) = \frac{1}{2}a_0 + \frac{1}{4}x - \frac{1}{3}x^3 + \frac{1}{4}(x^2 + y^2)$$

with a_0 as an arbitrary constant. ◆

18.8.3 A Neumann Problem for the Upper Half-Plane

To illustrate a Neumann problem for an unbounded set, consider:

$$\nabla^2 u(x,y) = 0 \text{ for } -\infty < x < \infty,\ y > 0$$

and

$$\frac{\partial u}{\partial y}(x,0) = f(x) \text{ for } -\infty < x < \infty.$$

Again, notice that $\partial u/\partial y$ is the derivative of u normal to the horizontal axis, which is the boundary of the upper half-plane.

Assume that

$$\int_{-\infty}^{\infty} f(x)\,dx = 0$$

as a necessary condition for a solution to exist.

We can solve this problem by separation of variables. However, there is an elegant device for reducing this problem to one we have already solved. Let

$$v = \frac{\partial u}{\partial y}.$$

Then

$$\nabla^2 v = \frac{\partial^2}{\partial x^2}\left(\frac{\partial u}{\partial y}\right) + \frac{\partial^2}{\partial y^2}\left(\frac{\partial u}{\partial y}\right) = \frac{\partial}{\partial y}\left(\frac{\partial^2 u}{\partial x^2} + \frac{\partial^2 u}{\partial y^2}\right) = 0.$$

Then v is harmonic wherever u is. Furthermore,

$$v(x,0) = \frac{\partial u}{\partial y}(x,0) = f(x)$$

on the x-axis. Therefore v satisfies a Dirichlet problem for the upper half-plane. We know the solution of this problem is

$$v(x,y) = \frac{y}{\pi}\int_{-\infty}^{\infty} \frac{f(\xi)}{y^2 + (\xi - x)^2}\,d\xi.$$

Now integrate to recover u from v. To within an arbitrary constant,

$$\begin{aligned} u(x,y) &= \int \frac{\partial u}{\partial y}\,dy = \int \frac{y}{\pi}\int_{-\infty}^{\infty} \frac{f(\xi)}{y^2 + (\xi - x)^2}\,d\xi\,dy \\ &= \frac{1}{\pi}\int_{-\infty}^{\infty}\left(\int_{-\infty}^{\infty} \frac{y}{y^2 + (\xi - x)^2}\,dy\right) f(\xi)\,d\xi \\ &= \frac{1}{2\pi}\int_{-\infty}^{\infty} \ln(y^2 + (\xi - x)^2) f(\xi)\,d\xi. \end{aligned}$$

SECTION 18.8 PROBLEMS

1. $\nabla^2 u(x,y) = 0$ for $0 < x < \pi, 0 < y < \pi$,
$$\frac{\partial u}{\partial y}(x,0) = \cos(3x) \text{ for } 0 \le x \le \pi$$
$$\frac{\partial u}{\partial y}(x,\pi) = 6x - 3\pi \text{ for } 0 \le x \le \pi$$
$$\frac{\partial u}{\partial x}(0,y) = \frac{\partial u}{\partial x}(\pi,y) = 0 \text{ for } 0 \le y \le \pi$$

2. $\nabla^2 u(x,y) = 0$ for $0 < x < 1, 0 < y < \pi$,
$$\frac{\partial u}{\partial y}(x,0) = \frac{\partial u}{\partial y}(x,\pi) = 0 \text{ for } 0 \le x \le 1,$$
$$\frac{\partial u}{\partial x}(0,y) = y - \frac{\pi}{2}, \frac{\partial u}{\partial x}(\pi,y) = \cos(y) \text{ for } 0 \le y \le \pi$$

3. $\nabla^2 u(x,y) = 0$ for $0 < x < 1, 0 < y < 1$,
$$\frac{\partial u}{\partial y}(x,0) = 4\cos(\pi x), \frac{\partial u}{\partial y}(x,1) = 0 \text{ for } 0 \le x \le 1,$$
$$\frac{\partial u}{\partial x}(0,y) = \frac{\partial u}{\partial x}(1,y) = 0 \text{ for } 0 \le y \le 1$$

4. Write a series solution for
$$\nabla^2 u(r,\theta) = 0 \text{ for } 0 \le r < R, -\pi \le \theta \le \pi$$
$$\frac{\partial u}{\partial r}(R,\theta) = \sin(3\theta) \text{ for } -\pi \le \theta \le \pi$$

5. Solve the following Neumann problem for the right quarter-plane:
$$\nabla^2 u(x,y) = 0 \text{ for } x > 0, y > 0$$
$$\frac{\partial u}{\partial x}(0,y) = 0 \text{ for } y \ge 0$$
$$\frac{\partial u}{\partial y}(x,0) = f(x) \text{ for } x \ge 0$$

6. Solve the following Neumann problem for the upper half-plane:
$$\nabla^2 u(x,y) = 0 \text{ for } -\infty < x < \infty, y > 0$$
$$\frac{\partial u}{\partial y}(x,0) = xe^{-|x|} \text{ for } -\infty < x < \infty$$

7. Attempt a separation of variables to solve
$$\nabla^2 u(x,y) = 0 \text{ for } 0 < x < 1, 0 < y < 1$$
$$u(x,0) = u(x,1) = 0 \text{ for } 0 \le x \le 1$$
$$\frac{\partial u}{\partial x}(0,y) = 3y^2 - 2y, \frac{\partial u}{\partial x}(1,y) = 0 \text{ for } 0 \le y \le 1$$

8. Use separation of variables to write an expression for the solution of the mixed boundary value problem
$$\nabla^2 u(x,y) = 0 \text{ for } 0 < x < \pi, 0 < y < \pi$$
$$u(x,0) = f(x), u(x,\pi) = 0 \text{ for } 0 \le x \le \pi$$
$$\frac{\partial u}{\partial x}(0,y) = \frac{\partial u}{\partial x}(\pi,y) = 0 \text{ for } 0 \le y \le \pi$$
Does this problem have a unique solution?

9. Write a series solution for

$$\nabla^2 u(r, \theta) = 0 \text{ for } 0 \le r < R, -\pi \le \theta \le \pi$$

$$\frac{\partial u}{\partial r}(R, \theta) = \cos(2\theta) \text{ for } -\pi \le \theta \le \pi$$

10. Solve the following mixed-boundary value problem:

$$\nabla^2 u(x, y) = 0 \text{ for } x > 0, y > 0$$

$$u(0, y) = 0 \text{ for } y \ge 0$$

$$\frac{\partial u}{\partial y}(x, 0) = f(x) \text{ for } x \ge 0$$

11. Solve the following Neumann problem for the upper half-plane:

$$\nabla^2 u(x, y) = 0 \text{ for } -\infty < x < \infty, y > 0$$

$$\frac{\partial u}{\partial y}(x, 0) = e^{-|x|} \sin(x) \text{ for } -\infty < x < \infty$$

12. Write an expression for the solution of the following Neumann problem for the lower half-plane:

$$\nabla^2 u(x, y) = 0 \text{ for } -\infty < x < \infty, y < 0$$

$$\frac{\partial u}{\partial y}(x, 0) = f(x) \text{ for } -\infty < x < \infty$$

PART 6

Complex Functions

CHAPTER 19

Complex Numbers and Functions

19.1 Geometry and Arithmetic of Complex Numbers

Complex Numbers

A *complex number* is a symbol $x + iy$, or $x + yi$, where x and y are real numbers and $i^2 = -1$. Arithmetic of complex numbers is defined by:

Equality $a + ib = c + id$ exactly when $a = c$ and $b = d$.

Addition $(a + ib) + (c + id) = (a + c) + i(b + d)$.

Multiplication $(a + ib)(c + id) = (ac - bd) + i(ad + bc)$.

In multiplying two complex numbers, we proceed exactly as we would with polynomials $a + bx$ and $c + dx$ with i in place of x and $i^2 = -1$. For example,

$$(6 - 4i)(8 + 13i) = (6)(8) + (-4)(13)i^2 + i[(6)(13) + (-4)(8)] = 100 + 46i.$$

The number a is called the *real part* of $a + bi$, denoted $\mathrm{Re}(a + bi)$. We call b the *imaginary part* of $a + bi$, denoted $\mathrm{Im}(a + bi)$. For example,

$$\mathrm{Re}(-4 + 12i) = -4 \text{ and } \mathrm{Im}(-4 + 12i) = 12.$$

The real and imaginary parts of any complex numbers are themselves real.

We may think of any real number a as the complex number $a + 0i$. In this way, the complex numbers are an extension of the real numbers. This extension has profound implications in algebra and analysis. The equation $x^2 + 1 = 0$ has no solutions if we restrict x to be real. In the complex numbers, it has two solutions: i and $-i$.

Complex arithmetic obeys many of the rules we are accustomed to from working with real numbers. If z, w, and u are complex numbers, then

1. $z + w = w + z$ (addition is commutative)
2. $zw = wz$ (multiplication is commutative)
3. $z + (w + u) = (z + w) + u$ (associative law for addition)
4. $z(wu) = (zw)u$ (associative law for multiplication)
5. $z(w + u) = zw + zu$ (distributive law)
6. $z + 0 = 0 + z = z$
7. $z \cdot 1 = 1 \cdot z = z$.

The Complex Plane

Complex numbers admit two geometric interpretations.

Any complex number $z = x + iy$ can be identified with the point (x, y) in the plane (Figure 19.1(a)). In this context, the plane is called the *complex plane*, the horizontal axis is called the *real axis*, and the vertical axis is called the *imaginary axis*. Any real number x graphs as a point $(x, 0)$ on the horizontal (or real) axis, and any *pure imaginary* number yi (with y real) is a point $(0, y)$ on the imaginary axis.

A complex number $x + iy$ also can be identified with the vector $x\mathbf{i} + y\mathbf{j}$ in the plane (Figure 19.1(b)). This is consistent with addition, since we add two vectors by adding their respective components, and we add two complex numbers by adding their real and imaginary parts, respectively.

Magnitude and Conjugate

The *magnitude* of $x + iy$ is the real number

$$|x + iy| = \sqrt{x^2 + y^2}.$$

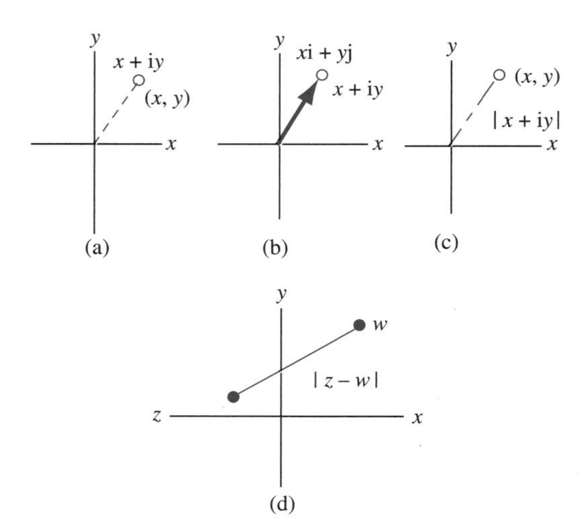

FIGURE 19.1 *$x + iy$ as a point, a vector, and its length.*

This is the distance from the origin to the point (x, y) in the complex plane or the length of the arrow representing the vector $x\mathbf{i} + y\mathbf{j}$ (Figure 19.1(c)). $|z - w|$ is the distance between the complex numbers z and w or, equivalently, between these points in the plane (Figure 19.1(d)).

The *complex conjugate*, or just *conjugate* of $x + iy$ is the complex number $x - iy$ with the sign of the imaginary part reversed. Denote the conjugate of z as $\overline{z}$.

In the complex plane, $\overline{z}$ is the reflection of z across the real axis (Figure 19.2). We have

$$\operatorname{Re}(z) = \operatorname{Re}(\overline{z}) \text{ and } \operatorname{Im}(z) = -\operatorname{Im}(\overline{z}).$$

Conjugation (the operation of taking a conjugate) and magnitude have the following properties.

1. $\overline{\overline{z}} = z$.
2. $\overline{z + w} = \overline{z} + \overline{w}$.
3. $\overline{zw} = (\overline{z})(\overline{w})$.
4. $\overline{z}/\overline{w} = \overline{z/w}$ if $w \neq 0$.
5. $|z| = |\overline{z}|$.
6. $|zw| = |z||w|$.
7. $\operatorname{Re}(z) = \frac{1}{2}(z + \overline{z})$ and $\operatorname{Im}(z) = \frac{1}{2i}(z - \overline{z})$.
8. $|z| \geq 0$, and $|z| = 0$ if and only if $z = 0$.
9. If $z = x + iy$, then $|z|^2 = z\overline{z}$.

These are established by routine calculations. For property (5), observe that $x^2 + y^2$ remains the same if y is replaced with $-y$. Equivalently, z and $\overline{z}$ are the same distance from the origin. For property (9), compute

$$|z|^2 = x^2 + y^2 = (x + iy)(x - iy) = z\overline{z}.$$

Conjugates are often used to compute a complex quotient z/w. Multiply the numerator and denominator of this quotient by the conjugate of the denominator:

$$\frac{z}{w} = \frac{z}{w}\frac{\overline{w}}{\overline{w}} = \frac{z\overline{w}}{w\overline{w}} = \frac{1}{|w|^2}(z\overline{w}).$$

This converts a division problem z/w into one of computing a product $z\overline{w}$, which is a simpler operation. For example,

$$\frac{2-7i}{8+3i} = \frac{2-7i}{8+3i}\frac{\overline{8+3i}}{\overline{8+3i}} = \frac{(2-7i)(8-3i)}{64+9} = -\frac{5}{73} - \frac{62}{73}i.$$

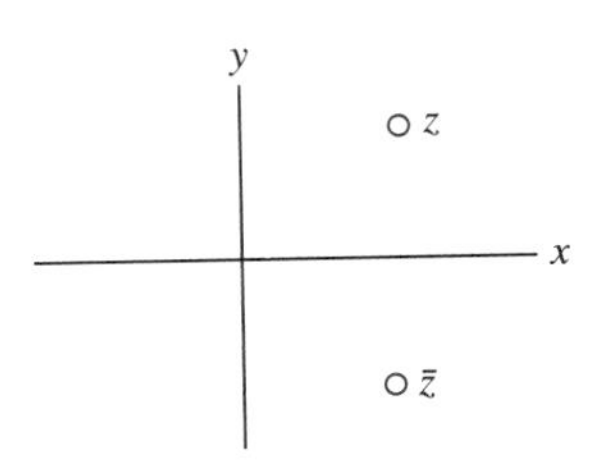

FIGURE 19.2 *Complex conjugate.*

Inequalities

There are several inequalities that are useful in dealing with complex quantities. Let z and w be complex numbers. Then

1. $|\text{Re}(z)| \leq |z|$ and $|\text{Im}(z)| \leq |z|$.
2. $|z+w| \leq |z| + |w|$.
3. $||z| - |w|| \leq |z-w|$.

Property (1) follows from the fact that $|x| \leq \sqrt{x^2+y^2}$ and $|y| \leq \sqrt{x^2+y^2}$.

Property (2) is called the *triangle inequality*. It follows immediately from the vector interpretation of complex numbers, since we already know the triangle inequality for vectors.

For property (3), use the triangle inequality to write

$$|z| = |(z+w) - w| \leq |z+w| + |w|.$$

Therefore,

$$|z| - |w| \leq |z+w|.$$

By interchanging z and w,

$$|w| - |z| \leq |z+w|.$$

Upon multiplying this inequality by -1, which reverses the inequality, we have

$$-|z+w| \leq |w| - |z|.$$

Combine inequalities to obtain

$$-|z+w| \leq |z| - |w| \leq |z+w|$$

and this implies that $||z| - |w|| \leq |z+w|$.

Argument and Polar Form

Let $z = a + ib$ be a nonzero complex number. The point (a, b) has polar coordinates (r, θ), where $r = |z|$ (Figure 19.3). We call θ an *argument* of z. Of course, given any argument θ, then $\theta + 2n\pi$ is also an argument for any integer n.

Using Euler's formula, we can write

$$z = a + ib = r\cos(\theta) + ir\sin(\theta) = r(\cos(\theta) + i\sin(\theta)) = re^{i\theta}.$$

The expression $re^{i\theta}$ is called the *polar form* of z.

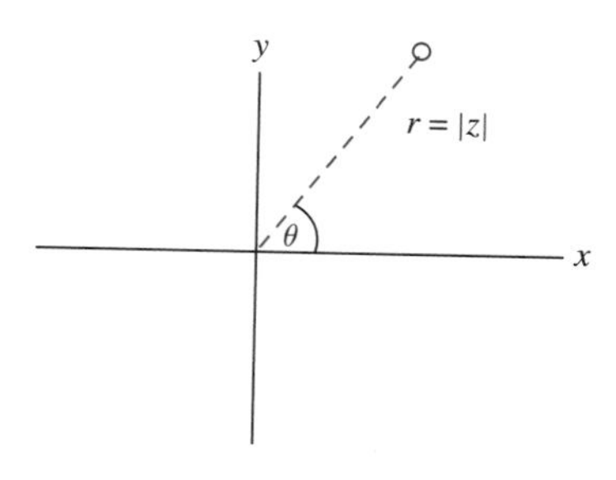

FIGURE 19.3 *Polar form of z.*

EXAMPLE 19.1

Let $z = 1 + i$. We can identify z with the point $(1, 1)$ which has polar coordinates $(\sqrt{2}, \pi/4 + 2n\pi)$ for any integer n. Any number $\pi/4 + 2n\pi$ with n an integer is an argument of $1 + i$. The polar form of z is

$$z = 1 + i = \sqrt{2}e^{i\pi/4}. \blacklozenge$$

Ordering

If a and b are distinct real numbers, then either $a < b$ (if $b - a > 0$) or $b < a$ (if $a - b > 0$). The real numbers are *ordered*. We claim that the complex numbers are not ordered.

If the complex numbers had an ordering, then i would have to be positive or negative, since i is not zero. If i were positive, then we would have $i^2 = -1$ positive in this ordering, since a product of positives is assumed to be positive. But then $(-1)i = -i$, again a product of positives, also would be positive, and we would have both i and $-i$ positive, which is a contradiction.

A similar contradiction follows if we supposed that i were negative, using the fact that a product of two negative numbers is positive.

Disks, Open Sets, and Closed Sets

A circle of radius r about a point (x_0, y_0) has the equation

$$\sqrt{(x - x_0)^2 + (y - y_0)^2} = r.$$

If we let $z = x + iy$ and $z_0 = x_0 + iy_0$, then this equation of the circle also can be written efficiently as

$$|z - z_0| = r.$$

This is the usual way of describing a circle with center z_0 and radius r.

The *open disk* of radius r about z_0 consists of all points z satisfying

$$|z - z_0| < r.$$

This open disk contains all points enclosed by the circle of radius r about z_0, but it does not include points on the boundary circle.

The *closed disk* of radius r about z_0 consists of all points z satisfying

$$|z - z_0| \leq r.$$

This closed disk consists of all points in the open disk of radius r about z_0 together with the points on the boundary circle.

In complex analysis, open disks play the role of open intervals (a, b) in calculus, and closed disks play the role of closed intervals $[a, b]$.

Now let S be any set of complex numbers, and let ζ be a complex number (which may belong to S or not).

1. ζ is an *interior point* of S if there is some open disk about ζ, all of whose points are in S. In this sense, an interior point of S is one entirely surrounded by points of S.
2. ζ is a *boundary point* of S if *every* open disk about ζ contains at least one point of S and at least one point not in S.

   A boundary point of S has points of S arbitrarily close to it and points not in S arbitrarily close to it. In this sense, a boundary point of S is "on the edge" of S and may or may not belong to S. A boundary point of S cannot be an interior point, and an interior point of S cannot be a boundary point.
3. S is *open* (an *open set*) if every point of S is an interior point.
4. S is *closed* (a *closed set*) if S contains all of its boundary points.

As the following examples show, a set may be open, closed, both open and closed, or neither open nor closed. The concepts of "open" and "closed" are not opposites—a set does not have to be either open or closed, and not being open does not make a set closed.

EXAMPLE 19.2

Let S consist of all complex numbers $z = x + iy$ with $x \geq 0$ and $y > 0$, shown in Figure 19.4. These are points (x, y) in the right quarter plane if x and y are both positive and points $(0, y)$ on the positive imaginary axis if $x = 0$. Points $(x, 0)$ on the positive real axis are not in S. These facts are indicated in the diagram by using a dashed positive real axis and a solid positive imaginary axis.

- $1 + i$ is an interior point of S. We can place a circle about $1 + i$ (say of radius $1/10$) containing only points in S.
- $2i$ is a boundary point. Every circle about $2i$ contains points of S and points not in S. Because $2i$ is in S and is not an interior point of S, S is not open.
- 2 is also a boundary point of S, because every circle drawn about 2 contains points of S and points not in S. However, 2 is not in S. S contains some of its boundary points but not all of them. S is not closed. This set is neither open nor closed. ♦

EXAMPLE 19.3

Every open disk is an open set. Every closed disk is a closed set. The open and closed disks of radius r about z_0 have the same boundary points, namely those on the circle $|z - z_0| = r$. ♦

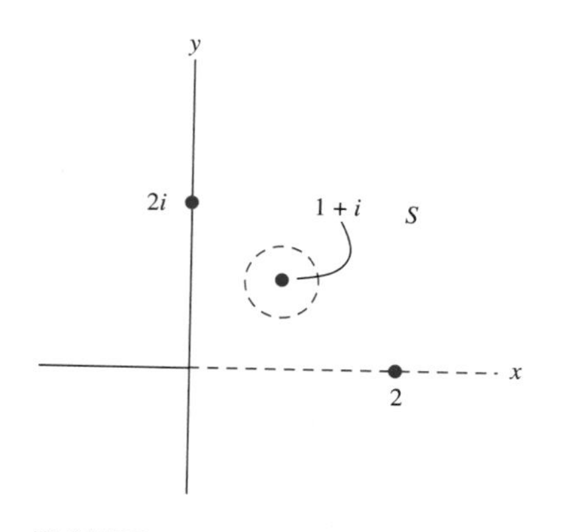

FIGURE 19.4 *The set S in Example 19.2.*

EXAMPLE 19.4

Let M consist of all complex numbers with rational real and imaginary parts. M has no interior points, because every disk about any point must contain points with irrational real and/or imaginary parts, hence it contains points not in M. This uses the fact that every real number has irrational numbers arbitrarily close to it. No points of M are interior points, and M is not open. Every complex number (whether in M or not) is a boundary point of M. M does not contain all of its boundary points and is not closed. ♦

EXAMPLE 19.5

Let K consist of all real numbers together with the number $5i$ (Figure 19.5). Because every open disk about $5i$ contains points not in K and a point in K (namely $5i$ itself), $5i$ is a boundary point in K, and K is not open. Every point of K is a boundary point, and these are its only boundary points. Therefore, K contains all of its boundary points and is closed. ♦

EXAMPLE 19.6

Any finite set of complex numbers is closed. Suppose $S = \{z_1, \cdots, z_N\}$ is a set of N numbers. Each z_j is a boundary point, and there are no other boundary points of S. Therefore, S contains all of its boundary points and is closed. ♦

THEOREM 19.1

Let S be a set of complex numbers. Then S is open if and only if S contains no boundary points. ♦

This theorem implies that a set having no boundary points must be open. It also means that containing a single boundary point is enough to disqualify a set from being open. The reason the theorem is true is that, if ζ is a boundary point of S that is in S, ζ is not an interior point, so not every point of S is an interior point and S cannot be open.

An open set can have boundary points, but these cannot be in the set. We have seen this with open disks. Every point on the bounding circle is a boundary point, but none of these are in the open disk.

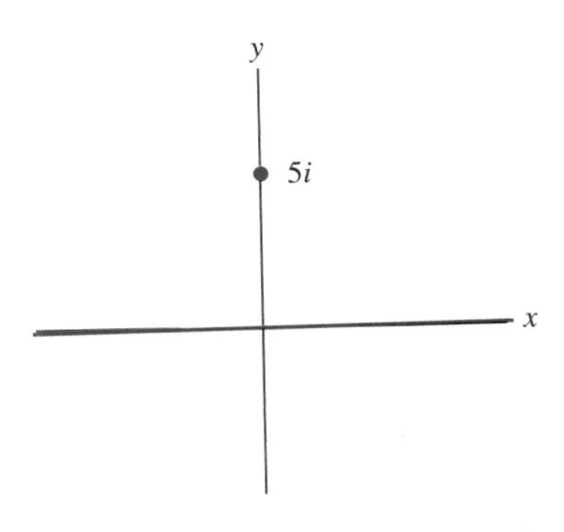

FIGURE 19.5 *K in Example 19.5.*

SECTION 19.1 PROBLEMS

In each of Problems 1 through 6, write the number in polar form.

1. $5-2i$
2. $-4-i$
3. $-2+2i$
4. $-12+3i$
5. $8+i$
6. $-7i$

In each of Problems 7 through 12, determine the magnitude and all of the arguments of z.

7. $3i$
8. $3-4i$
9. -4
10. $-2+2i$
11. $-3+2i$
12. $8+i$

In each of Problems 13 through 22, carry out the indicated calculation.

13. $(3-4i)(6+2i)$
14. $i(6-2i)+|1+i|$
15. $(2+i)/(4-7i)$
16. $((2+i)-(3-4i))/(5-i)(3+i)$
17. $(17-6i)\overline{(-4-12i)}$
18. $(3+i)^3$
19. $((-6+2i)/(1-8i))^2$
20. $|3i/(-4+8i)|$
21. i^3-4i^2+2
22. $(-3-8i)(2i)(4-i)$

In each of Problems 23 through 28, a set of complex numbers is specified. Determine whether the set is open, closed, open and closed, or not open and not closed. Specify all boundary points of the set (whether or not they belong to the set).

23. M is the set of all z satisfying $\text{Im}(z) < 7$.
24. S is the set of all z with $|z| > 2$.
25. W consists of all z with $\text{Re}(z) > (\text{Im}(z))^2$.
26. R is the set of all numbers $1/m + (1/n)i$ with m and n positive integers.
27. U is the set of all z with $1 < \text{Re}(z) \le 3$.
28. V is the set of all z with $2 < \text{Re}(z) \le 3$ and $-1 < \text{Im} z < 1$.
29. Show that, for any positive integer n,
$$i^{4n} = 1,\ i^{4n+1} = i,\ i^{4n+2} = -1,\ i^{4n+3} = -i.$$
30. Let $z = a + ib$. Determine $\text{Re}(z^2)$ and $\text{Im}(z^2)$.
31. Show that complex numbers z, w, and u form vertices of an equilateral triangle if and only if
$$z^2 + w^2 + u^2 = zw + zu + wu.$$
32. Show that $z^2 = (\overline{z})^2$ if and only if z is either real or pure imaginary.
33. Let z and w be numbers with $\overline{z}w \neq 1$. Suppose either z or w has magnitude 1. Prove that
$$\left|\frac{z-w}{1-\overline{z}w}\right| = 1.$$
Hint: Recall that $|u|^2 = u\overline{u}$ for every complex number u.
34. Show that, for any numbers z and w,
$$|z+w|^2 + |z-w|^2 = 2\left(|z|^2 + |w|^2\right).$$
Hint: Note the hint from Problem 33.

19.2 Complex Functions

A *complex function* is a function that acts on complex numbers and produces complex numbers. For example, $f(z) = z^2$ for $|z| < 1$ is a complex function acting on numbers in the open unit disk.

In this section, we will extend the calculus concepts of continuity and differentiability to complex functions and also develop complex versions of powers, exponentials, logarithms, and trigonometric functions.

19.2.1 Limits, Continuity, and Differentiability

If f is a complex function, $f(z)$ has *limit* L as z approaches z_0 if, given any positive number ϵ, there is a positive number δ such that

$$|f(z) - L| < \epsilon$$

for all z in S such that $0 < |z - z_0| < \delta$.

This means that we must be able to make the values $f(z)$ as close as we like to L by confining z to a small enough disk about z_0, excluding the center z_0. The actual value of $f(z_0)$, if this is defined, is not relevant for the limit, which has to do only with the behavior of $f(z)$ as z is taken close to z_0.

EXAMPLE 19.7

Let $f(z) = z^2$ for $z \neq i$. Even though $f(i)$ is not defined,

$$\lim_{z \to i} f(z) = i^2 = -1. \blacklozenge$$

A significant difference between limits of complex and real functions is that (on the real line) x can approach x_0 only from the left or right, while in the complex plane, z can approach z_0 along infinitely many different paths (Figure 19.6). Requiring that $f(z)$ approach the same number L along all such paths is a much stronger condition than requiring that the function approach the same value only from the left or right.

As in calculus, we rarely invoke the $\epsilon - \delta$ definition of limit to evaluate a limit. The limit of a finite sum (product, quotient) is the sum (product, quotient) of the limits whenever all the limits are defined, and in the case of a quotient, the denominator is nonzero. Furthermore, $\lim_{z \to z_0} f(z) = L$ implies that $\lim_{z \to z_0} cf(z) = cL$ for any number c.

In the special case that $\lim_{z \to z_0} f(z) = f(z_0)$, we say that f is *continuous* at z_0. This requires that $f(z_0)$ be defined and that $f(z)$ approach $f(z_0)$ as z approaches z_0 along any path. The function of Example 19.7 is not continuous at i because $f(i)$ is not defined.

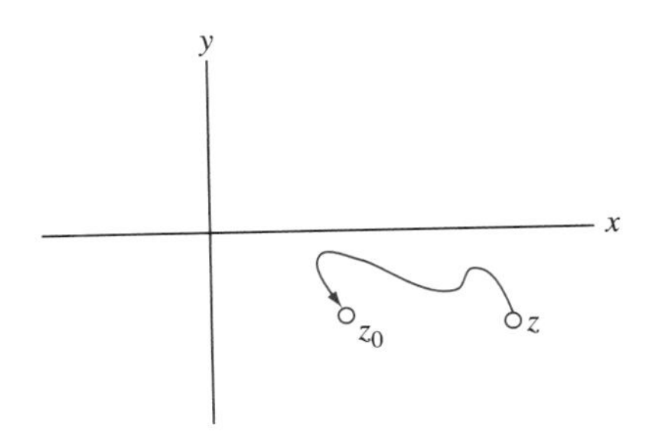

FIGURE 19.6 $z \to z_0$ *along arbitrary paths.*

If f is continuous at each point of a set S, we say that f is *continuous on* S. The function of Example 19.7 is continuous on the set S consisting of the complex plane with i removed.

A function f is *bounded* on S if for some positive number M,

$$|f(z)| \leq M \text{ for all } z \text{ in } S.$$

This means that there must be a disk about the origin containing all the numbers $f(z)$ for z in S.

A continuous function need not be bounded. For example, $f(z) = 1/z$ is continuous on the plane with the origin removed.

If, however, we place some conditions on the set S, then a function that is continuous on S will be bounded on S.

One condition we will use is that the set is bounded (not to be confused with a function being bounded). We say that a set S of complex numbers is *bounded* if, for some positive number K, $|z| \leq K$ for all z in S. This means that we can enclose all of S within a disk, if we choose the radius large enough.

THEOREM 19.2

If f is continuous on a set S that is closed and bounded, then f is a bounded function on S.

A set that is closed and bounded is called *compact*. In this terminology, the theorem says that a function that is continuous on a compact set must be a bounded function.

The complex derivative is modeled after the real derivative. We say that f is *differentiable* at z_0 if for some number L

$$\lim_{z \to z_0} \frac{f(z) - f(z_0)}{z - z_0} = L.$$

This is equivalent to requiring that

$$\lim_{h \to 0} \frac{f(z_0 + h) - f(z_0)}{h} = L$$

with the understanding that h is complex and must be allowed to approach 0 along an arbitrary path.

In this case, we call L the *derivative* of f at z_0 and denote it $f'(z_0)$, or in the Leibniz notation,

$$\left.\frac{d}{dz} f(z)\right]_{z=z_0}.$$

As with real functions, we rarely compute a complex derivative by applying the limit definition. The rules for computing derivatives of complex functions have the same form as those for real functions whenever all of the derivatives are defined.

1. $(f+g)'(z) = f'(z) + g'(z)$,
2. $(f-g)'(z) = f'(z) - g'(z)$,
3. $(cf)'(z) = cf'(z)$ for any number c,
4. $(fg)'(z) = f(z)g'(z) + f'(z)g(z)$,
5. $\left(\dfrac{f}{g}\right)'(z) = \dfrac{g(z)f'(z) - f(z)g'(z)}{(g(z))^2}$.
6. There is also a complex version of the chain rule for differentiating a composition $f \circ g$, which is defined by $(f \circ g)(z) = f(g(z))$. Assuming that the derivatives exist, then

$$(f \circ g)'(z) = f'(g(z))g'(z).$$

In the Leibniz notation,

$$\frac{d}{dz}(f(g(z))) = \frac{df}{dw}\frac{dw}{dz}$$

where $w = g(z)$.

In form, this looks just like the chain rule for real functions in calculus.

Not every function is differentiable.

EXAMPLE 19.8

Let $f(z) = \overline{z}$. Suppose f is differentiable at z. Then

$$f'(z) = \lim_{h \to 0} \frac{f(z+h) - f(z)}{h}$$

$$= \lim_{h \to 0} \frac{\overline{z} + \overline{h} - \overline{z}}{h} = \lim_{h \to 0} \frac{\overline{h}}{h}.$$

For $f'(z_0)$ to exist, we need this limit to be the same no matter how h approaches zero. If h approaches 0 along the real axis, $\overline{h} = h$ and $\overline{h}/h = 1$, so this limit would have to be 1. But if h approaches 0 along the imaginary axis, then $h = ik$ for real k, and

$$\frac{\overline{h}}{h} = \frac{\overline{ik}}{ik} = \frac{-ik}{ik} = -1.$$

Since this quotient is always -1 on the imaginary axis (no matter how close k is to zero), the limit defining $f'(z)$ would have to be -1. We conclude that the limit defining this derivative does not exist. This function has no derivative at any point. ◆

As with real functions, a complex function is continuous wherever it is differentiable.

THEOREM 19.3

If f is differentiable at z_0 then f is continuous at z_0. ◆

To see why this is true, begin with

$$f(z_0+h) - f(z_0) = h\left(\frac{f(z_0+h) - f(z_0)}{h}\right).$$

As $h \to 0$, the difference quotient in parentheses on the right approaches $f'(z_0)$, and the factor of h approaches 0, so the right side approaches 0. Therefore

$$\lim_{h \to 0}(f(z_0+h) - f(z_0)) = 0$$

and this is equivalent to $\lim_{z \to z_0} f(z) = f(z_0)$.

19.2.2 The Cauchy-Riemann Equations

We will derive a pair of partial differential equations that are intimately tied to differentiability for complex functions. These equations also play a role in potential theory and in treatments of the Dirichlet problem.

If f is a complex function and $z = x + iy$, then we can always write

$$f(z) = f(x + iy) = u(x, y) + iv(x, y)$$

where u and v are real-valued functions of two real variables:

$$u(x, y) = \text{Re}(f(z)) \text{ and } v(x, y) = \text{Im}(f(z)).$$

EXAMPLE 19.9

Let $f(z) = z^2$. Then

$$f(z) = (x + iy)^2 = x^2 - y^2 + 2ixy = u(x, y) + iv(x, y).$$

For this function, $u(x, y) = x^2 - y^2$ and $v(x, y) = 2xy$.

If $g(z) = 1/z$ for $z \neq 0$, then

$$g(z) = \frac{1}{x + iy} = \frac{x}{x^2 + y^2} - i\frac{y}{x^2 + y^2} = u(x, y) + iv(x, y).$$

For $g(z)$,

$$u(x, y) = \frac{x}{x^2 + y^2} \text{ and } v(x, y) = -\frac{y}{x^2 + y^2}. \blacklozenge$$

For f to be differentiable, partial derivatives of u and v must be related in a special way.

THEOREM 19.4 ***Cauchy-Riemann Equations***

Let $f(x + iy) = u(x, y) + iv(x, y)$ be differentiable at $z = x + iy$. Then, at (x, y),

$$\frac{\partial u}{\partial x} = \frac{\partial v}{\partial y} \text{ and } \frac{\partial v}{\partial x} = -\frac{\partial u}{\partial y}. \blacklozenge$$

These are the *Cauchy-Riemann equations* for the real and imaginary parts of f. In deriving these equations, we will also obtain expressions for $f'(z)$.

If f is differentiable at z, then

$$f'(z) = \lim_{h \to 0} \frac{f(z + h) - f(z)}{h}.$$

This difference quotient approaches $f'(z_0)$ regardless of path of approach of h to 0. Focus on two specific paths.

Path 1 Along the Real Axis.

Now h is real and moves left or right toward 0. Since h is real, $z + h = x + h + iy$ and

$$\begin{aligned} f'(z) &= \lim_{h \to 0} \frac{f(z + h) - f(z)}{h} \\ &= \lim_{h \to 0} \frac{u(x + h, y) + iv(x + h, y) - u(x, y) - iv(x, y)}{h} \end{aligned}$$

$$= \lim_{h\to 0}\left(\frac{u(x+h,y)-u(x,y)}{h} + i\frac{v(x+h,y)-v(x,y)}{h}\right)$$

$$= \frac{\partial u}{\partial x} + i\frac{\partial v}{\partial x}.$$

Path 2 Along the Imaginary Axis.

Now $h = ik$ with k real and $z + h = z + ik = x + i(y+k)$. Then

$$f'(z) = \lim_{k\to 0}\frac{u(x,y+k)+iv(x,y+k)-u(x,y)-iv(x,y)}{ik}$$

$$= \lim_{k\to 0}\left(\frac{1}{i}\frac{u(x,y+k)-u(x,y)}{k} + \frac{v(x,y+k)-v(x,y)}{k}\right)$$

$$= -i\frac{\partial u}{\partial y} + \frac{\partial v}{\partial y}.$$

Here we used the fact that $1/i = -i$. We conclude that

$$f'(z) = \frac{\partial u}{\partial x} + i\frac{\partial v}{\partial x} = -i\frac{\partial u}{\partial y} + \frac{\partial v}{\partial y}. \tag{19.1}$$

Equating real parts and then imaginary parts of opposite sides of this equation, we obtain

$$\frac{\partial u}{\partial x} = \frac{\partial v}{\partial y} \text{ and } \frac{\partial v}{\partial x} = -\frac{\partial u}{\partial y}.$$

The Cauchy-Riemann equations are a necessary condition for a complex function to be differentiable. Given f, u, and v are uniquely determined and f cannot be differentiable at any point where u and v do not satisfy the Cauchy-Riemann equations.

EXAMPLE 19.10

We already know that $f(z) = \bar{z}$ is not differentiable. To illustrate the theorem, we will use the Cauchy-Riemann equations to show this. Write

$$f(z) = \bar{z} = x - iy = u(x,y) + iv(x,y),$$

so $u(x,y) = x$ and $v(x,y) = -y$. Then

$$\frac{\partial u}{\partial x} = 1 \text{ and } \frac{\partial v}{\partial y} = -1,$$

so the Cauchy-Riemann equations fail to hold and f is not differentiable at any z. ♦

EXAMPLE 19.11

Let $f(z) = z\,\mathrm{Re}(z)$. Then

$$f(z) = f(x+iy) = (x+iy)x = x^2 + ixy,$$

so $u(x,y) = x^2$ and $v(x,y) = xy$. Compute

$$\frac{\partial u}{\partial x} = 2x,\ \frac{\partial v}{\partial y} = x$$

and

$$\frac{\partial u}{\partial y} = 0,\ \frac{\partial v}{\partial x} = 0.$$

If $z \neq 0$, the Cauchy-Riemann equations do not hold, so f cannot be differentiable at any $z \neq 0$. ♦

The theorem states that the Cauchy-Riemann equations are necessary for $f = u + iv$ to be differentiable at a point. If u, v, and their first partial derivatives are continuous, then the Cauchy-Riemann equations are also sufficient for f to be differentiable.

Using the Cauchy-Riemann equations, we can establish the following results.

THEOREM 19.5

Let f be differentiable on an open disk D. Let $f = u + iv$, and suppose that u and v are continuous with continuous first and second partial derivatives, and satisfy the Cauchy-Riemann equations on D. Then

1. If $f'(z) = 0$ on D, then f is a constant function on D.
2. If $|f(z)|$ is constant on D, so is $f(z)$. ♦

Proof Conclusion (1) is easy to prove. For each z in D,

$$f'(z) = 0 = \frac{\partial u}{\partial x} + i\frac{\partial v}{\partial x}$$

implies that $\partial u/\partial x = \partial v/\partial x = 0$ on D. By the Cauchy-Riemann equations, $\partial u/\partial y = \partial v/\partial y = 0$ on D also, so u and v are constant functions and therefore so is f.

Conclusion (2) is more involved. Suppose $|f(z)| = k$ for all z in D. Then

$$|f(z)|^2 = u(x, y)^2 + v(x, y)^2 = k^2 \tag{19.2}$$

for (x, y) in D. If $k = 0$, then $f(z) = 0$ for all z in D. If $k \neq 0$, differentiate equation (19.2) with respect to x to get

$$u\frac{\partial u}{\partial x} + v\frac{\partial v}{\partial x} = 0 \tag{19.3}$$

and with respect to y to get

$$u\frac{\partial u}{\partial y} + v\frac{\partial v}{\partial y} = 0. \tag{19.4}$$

Use the Cauchy-Riemann equations to write equations (19.3) and (19.4) as

$$u\frac{\partial u}{\partial x} - v\frac{\partial u}{\partial y} = 0 \tag{19.5}$$

and

$$u\frac{\partial u}{\partial y} + v\frac{\partial u}{\partial x} = 0. \tag{19.6}$$

Multiply equation (19.5) by u and equation (19.6) by v and add the resulting equations to get

$$(u^2 + v^2)\frac{\partial u}{\partial x} = k^2\frac{\partial u}{\partial x} = 0.$$

Therefore, $\partial u/\partial x = 0$ on D, and by the Cauchy-Riemann equations, $\partial v/\partial y = 0$ also. A similar manipulation shows that $\partial u/\partial y = \partial v/\partial x = 0$, so $u(x, y)$ and $v(x, y)$ are constant, hence $f(z)$ is constant on D. ♦

There is an intimate connection between differentiable complex functions and harmonic functions. Recall that a real-valued function $u(x, y)$ of two real variables is harmonic on a set of points in the x, y-plane if u satisfies Laplace's equation

$$\frac{\partial^2 u}{\partial x^2}+\frac{\partial^2 u}{\partial y^2}=0.$$

We claim that the real and imaginary parts of a differentiable complex function must be harmonic.

THEOREM 19.6

Let G be an open set in the complex plane, and suppose $f(z)=u(x,y)+iv(x,y)$ is differentiable on G. Then u and v are harmonic on G. ♦

Proof Begin with the fact that u and v satisfy the Cauchy-Riemann equations:

$$\frac{\partial u}{\partial x}=\frac{\partial v}{\partial y} \text{ and } \frac{\partial v}{\partial x}=-\frac{\partial u}{\partial y}.$$

Differentiate the first equation with respect to x and the second with respect to y to get

$$\frac{\partial^2 u}{\partial x^2}=\frac{\partial^2 v}{\partial y \partial x}=\frac{\partial^2 v}{\partial x \partial y}=-\frac{\partial^2 u}{\partial y^2}$$

and this implies that

$$\frac{\partial^2 u}{\partial x^2}+\frac{\partial^2 u}{\partial y^2}=0.$$

Similarly,

$$\frac{\partial^2 v}{\partial x^2}+\frac{\partial^2 v}{\partial y^2}=0.$$

Therefore, u and v are harmonic on G. ♦

Thus far, the real and imaginary parts of a differentiable complex function are harmonic. The connection also goes the other way, in the following sense. Given a function u that is harmonic on a domain D, there is a function v harmonic on D such that $f=u+iv$ is differentiable on D. We call a v a *harmonic conjugate* for u. Thus, differentiable complex functions are constructed from harmonic functions.

THEOREM 19.7

Let u be harmonic on an open disk D in the complex plane. Then, for some v defined on D, the function f defined by $f(z)=u(x,y)+iv(x,y)$ is differentiable on D. ♦

Proof Define

$$g(z)=\frac{\partial u}{\partial x}-i\frac{\partial u}{\partial y}$$

for (x,y) in D. Using the Cauchy-Riemann equations, show that g is differentiable on D. Then, for some complex function G,

$$G'(z)=g(z)$$

for z in D. Write $G(z)=U(x,y)+iV(x,y)$. Then

$$G'(z)=\frac{\partial U}{\partial x}-i\frac{\partial U}{\partial y}$$

$$= g(z) = \frac{\partial u}{\partial x} - i\frac{\partial u}{\partial y}.$$

Therefore,

$$\frac{\partial U}{\partial x} = \frac{\partial u}{\partial x} \text{ and } \frac{\partial U}{\partial y} = \frac{\partial u}{\partial y}$$

for (x, y) in D. This means that $U(x, y) - u(x, y)$ is constant on D, so for some real number K,

$$U(x, y) = u(x, y) + K.$$

Now define $f(z) = G(z) - K$. Then f is differentiable on D, and

$$f(z) = G(z) - K = U(x, y) + iV(x, y) - K = u(x, y) + iv(x, y).$$

We may therefore choose $v(x, y) = V(x, y)$, proving the theorem. ♦

Given a harmonic function defined on a domain, we are rarely interested in actually producing a harmonic conjugate. However, knowing that a harmonic conjugate exists enables us to go from harmonic u to a differentiable complex function $f = u + iv$, bringing complex function methods to bear on some problems. We will exploit this in solving Dirichlet problems by conformal mappings in Chapter 23. We will also use complex integration to derive important properties of harmonic functions in Chapter 20.

SECTION 19.2 PROBLEMS

In each of Problems 1 through 12, find u and v so that $f = u + iv$, determine all points (x, y) at which the Cauchy-Riemann equations hold, and determine all z at which f is differentiable. Familiar facts about continuity of real-valued functions of two real variables can be assumed.

1. $f(z) = -4z + 1/z$
2. $f(z) = z^3 - 8z + 2$
3. $f(z) = z - i$
4. $f(z) = (2z + 1)/z$
5. $f(z) = z/\text{Re}(z)$
6. $f(z) = (z - i)/(z + i)$
7. $f(z) = i|z|^2$
8. $f(z) = z + \text{Im}(z)$
9. $f(z) = |z|$
10. $f(z) = z^2 - iz$
11. $f(z) = (\overline{z})^2$
12. $f(z) = iz + |z|$
13. Let $z_n = a_n + ib_n$ be a sequence of complex numbers. We say that this sequence converges to $w = c + id$ if the real sequences $a_n \to c$ and $b_n \to d$. Show that, if $f(z)$ is continuous at z_0 and z_n is a sequence converging to z_0, $f(z_n)$ converges to $f(z_0)$.

19.3 The Exponential and Trigonometric Functions

The *complex exponential function* e^z is defined for all $z = x + iy$ by

$$e^z = e^x \cos(y) + ie^x \sin(y).$$

Notice that $f(iy) = e^{iy} = \cos(y) + i\sin(y)$ is just Euler's equation. The functions $u(x, y) = e^x \cos(y)$ and $v(x, y) = e^x \sin(y)$ are continuous with continuous first partial derivatives, which satisfy the Cauchy-Riemann equations. Therefore, e^z is differentiable for all z. Furthermore, using equation (19.1),

$$f'(z) = \frac{\partial u}{\partial x} + i\frac{\partial v}{\partial x} = e^x \cos(y) + ie^x \sin(y) = e^z$$

as with the real exponential function.

The following properties of e^z are straightforward consequences of the definition.

1. $e^0 = 1$.
2. $e^{z+w} = e^z e^w$.
3. $e^z \neq 0$ for all complex z.
4. $e^{-z} = 1/e^z$.
5. If t is real, then $\overline{e^{it}} = e^{-it}$.

To prove property (3), suppose

$$e^z = e^x \cos(y) + ie^x \sin(y) = 0.$$

Then

$$e^x \cos(y) = e^x \sin(y) = 0.$$

Since for real x, $e^x \neq 0$, then

$$\cos(y) = \sin(y) = 0.$$

This is impossible, because the real sine and cosine functions have no common zeros.

For property (4), use properties (1) and (2) to write

$$e^0 = 1 = e^{z-z} = e^z e^{-z}$$

implying that $1/e^z = e^{-z}$.

To verify property (5), suppose t is a real number. By Euler's formula,

$$\begin{aligned}\overline{e^{it}} &= \overline{\cos(t) + i\sin(t)} \\ &= \cos(t) - i\sin(t) = e^{-it}.\end{aligned}$$

Perhaps the first surprise we find with the complex exponential function is that e^z is periodic. This periodicity does not appear in the real exponential function because the period is pure imaginary.

THEOREM 19.8

1. $e^z = 1$ if and only if $z = 2n\pi i$ for some integer n.
2. If p is a number such that $e^{z+p} = e^z$ for all complex z, then for some integer n, $p = 2n\pi i$.
3. e^z is periodic with period $2n\pi i$ for each nonzero integer n. Furthermore, these are the only periods of the complex exponential function. ♦

Proof To prove conclusion (1) first observe that, if $z = 2n\pi i$ for some integer n, then

$$e^z = e^{2n\pi i} = \cos(2n\pi) + i\sin(2n\pi) = 1$$

because $\cos(2n\pi) = 1$ and $\sin(2n\pi) = 0$.

Conversely, if $e^z = e^{a+ib} = 1$, then

$$e^a \cos(b) + ie^a \sin(b) = 1.$$

so

$$e^a \cos(b) = 1 \text{ and } e^a \sin(b) = 0.$$

Since $e^a \neq 0$, then $\sin(b) = 0$, so $b = k\pi$ for some integer k. But then

$$e^a \cos(b) = e^a \cos(k\pi) = (-1)^k e^a = 1.$$

This requires that $(-1)^k$ be positive, so $k = 2n$ for some integer n. But then $e^a = 1$, which for real a means that $a = 0$. Therefore $z = a + bi = 2n\pi i$ for some integer n.

For conclusion (2), suppose that $e^{z+p} = e^z$ for all z. Then $e^p = 1$, so $p = 2n\pi i$ for some integer n, by (1).

For conclusion (3), let n be a nonzero integer. By conclusion (1),

$$e^{z+2n\pi i} = e^z e^{2n\pi i} = e^z,$$

for all z, so e^z has period $2n\pi i$. For the rest of conclusion (3), we need to show that any period of e^z is an integer multiple of 2π. Thus, suppose that $e^{z+p} = e^z$ for all z and for some number $p \neq 0$. Then by conclusion (2), $p = 2n\pi i$ for some nonzero integer n.

Every period of e^z is therefore pure imaginary, explaining why no period is evident for the real exponential function. ♦

Another difference between the real and complex exponential functions is that e^z can be negative. For example, it is easy to check that $e^z = -1$ exactly when $z = (2n+1)\pi$ for some integer n.

EXAMPLE 19.12

Solve the equation

$$e^z = 1 + 2i.$$

We want all z such that

$$e^z = e^x \cos(y) + i e^x \sin(y) = 1 + 2i.$$

This requires that

$$e^x \cos(y) = 1 \text{ and } e^x \sin(y) = 2.$$

If we square these equations and add, we obtain

$$e^{2x}(\cos^2(y) + \sin^2(y)) = e^{2x} = 5,$$

which implies that $x = \ln(5)/2$. Next,

$$\frac{e^x \sin(y)}{e^x \cos(y)} = \tan(y) = 2,$$

so $y = \arctan(2)$. All solutions of e^z are

$$z = \frac{1}{2}\ln(5) + i \arctan(2). \quad ♦$$

The complex sine and cosine functions are defined by

$$\cos(z) = \frac{1}{2}\left(e^{iz} + e^{-iz}\right) \text{ and } \sin(z) = \frac{1}{2i}\left(e^{-z} - e^{-iz}\right).$$

A routine calculation gives us

$$\cos(z) = \cos(x)\cosh(y) - i \sin(x)\sinh(y)$$

and

$$\sin(z) = \sin(x)\cosh(y) + i \cos(x)\sinh(y),$$

where

$$\cosh(y) = \frac{1}{2}(e^y + e^{-y}) \text{ and } \sinh(y) = \frac{1}{2}(e^y - e^{-y}).$$

To verify these, write

$$\begin{aligned}\cos(z) &= \frac{1}{2}\left(e^{iz} + e^{-iz}\right) \\ &= \frac{1}{2}\left(e^{i(x+iy)} + e^{-i(x+iy)}\right) \\ &= \frac{1}{2}\left(e^{ix}e^{-y} + e^{-ix}e^{y}\right) \\ &= \frac{1}{2}\left(e^{-y}(\cos(x) + i\sin(x)) + e^{y}(\cos(x) - i\sin(x))\right) \\ &= \frac{1}{2}\cos(x)\left(e^{y} + e^{-y}\right) + \frac{i}{2}\sin(x)\left(e^{-y} - e^{y}\right) \\ &= \cos(x)\cosh(y) - i\sin(x)\sinh(y).\end{aligned}$$

Similarly,

$$\sin(z) = \sin(x)\cosh(y) + i\cos(x)\sinh(y).$$

If $z = x$ is real, then $y = 0$, the complex sine agrees with the real sine, and the complex cosine agrees with the real cosine. In this sense, $\sin(z)$ and $\cos(z)$ are extensions of $\sin(x)$ and $\cos(x)$ to the complex plane.

A multiplication shows that, for all z,

$$\cos^2(z) + \sin^2(z) = 1,$$

as we should expect. Other identities for the real sine and cosine functions extend readily to their complex counterparts, although their derivations are very much simplified in the complex case because exponential functions are easy to compute with. For example, suppose we want to show that $\sin(2z) = 2\sin(z)\cos(z)$. Compute

$$\begin{aligned}2\sin(z)\cos(z) &= \frac{1}{2i}\left(e^{iz} - e^{-iz}\right)\left(e^{iz} + e^{-iz}\right) \\ &= \frac{1}{2i}\left(e^{2iz} - e^{-2iz}\right) = \sin(2z).\end{aligned}$$

From the Cauchy-Riemann equations, $\cos(z)$ and $\sin(z)$ are differentiable for all z. Furthermore, using equations (19.1),

$$\begin{aligned}\frac{d}{dz}\cos(z) &= \frac{\partial u}{\partial x} + i\frac{\partial v}{\partial y} \\ &= -\sin(x)\cosh(y) - i\cos(x)\sinh(y) = -\sin(z),\end{aligned}$$

and similarly,

$$\frac{d}{dz}\sin(z) = \cos(z).$$

The complex sine and cosine functions exhibit some properties that are not seen in the real case. For example, the real sine and cosine functions are bounded: $|\cos(x)| \leq 1$ and $|\sin(x)| \leq 1$ for all real x. But the complex sine and cosine are not bounded functions in the complex plane. For real y,

$$\cos(iy) = \frac{1}{2}(e^{-y} + e^{y}),$$

which can be made as large as we like by choosing y or $-y$ large. Similarly, $\sin(iy)$ can be arbitrarily large in magnitude.

In view of this new behavior, we might ask about periods and zeros of the complex sine and cosine. We claim that the extension of these functions to the complex plane does not bring any new periods or zeros.

THEOREM 19.9

1. $\sin(z) = 0$ if and only if $z = n\pi$ for some integer n.
2. $\cos(z) = 0$ if and only if $z = (2n+1)\pi/2$ for some integer n.
3. $\cos(z)$ and $\sin(z)$ are periodic with periods $2n\pi$ for every nonzero integer n. Furthermore, these are the only periods of these functions. ♦

These follow by systematic use of the real and imaginary parts of $\cos(z)$ and $\sin(z)$. For example, to find the zeros of $\sin(z)$, solve

$$\sin(z) = \sin(x)\cosh(y) + i\cos(x)\sinh(y) = 0.$$

Then

$$\sin(x)\cosh(y) = 0 \text{ and } \cos(x)\sinh(y) = 0.$$

From the first equation and the fact that $\cosh(y) \neq 0$ for real y, we have $\sin(x) = 0$, which for real x means that $x = n\pi$ for some integer n. From the second equation,

$$\cos(x)\sinh(y) = \cos(n\pi)\sinh(y) = (-1)^n \sinh(y) = 0.$$

But then $\sinh(y) = 0$ so $y = 0$. Therefore $z = n\pi$, with n an integer, proving part (1) of Theorem 19.9. Parts (2) and (3) are proved similarly.

The other trigonometric functions are defined in terms of sine and cosine in the usual way. For example, $\tan(z) = \sin(z)/\cos(z)$ for $\cos(z) \neq 0$.

SECTION 19.3 PROBLEMS

1. Determine u and v such that $e^{z^2} = u(x, y) + iv(x, y)$. Show that u and v satisfy the Cauchy-Riemann equations.
2. Find u and v such that $e^{1/z} = u(x, y) + iv(x, y)$. Show that u and v satisfy the Cauchy-Riemann equations.
3. Find u and v such that $ze^z = u(x, y) + iv(x, y)$. Show that u and v satisfy the Cauchy-Riemann equations wherever they are defined.
4. Find u and v such that $\cos^2(z) = u(x, y) + iv(x, y)$. Show that u and v satisfy the Cauchy-Riemann equations wherever they are defined.
5. Find all solutions of $e^z = 2i$.
6. Derive the following identities.

   (a) $\sin(z+w) = \sin(z)\cos(w) + \cos(z)\sin(w)$.

   (b) $\cos(z+w) = \cos(z)\cos(w) - \sin(z)\sin(w)$.
7. Find all solutions of $e^z = -2$.
8. Find all solutions of $\sin(z) = i$.

In each of Problems 9 through 18, write the function value in the form $a + bi$.

9. e^i
10. $\sin(1-4i)$
11. e^{5+2i}
12. $\cot(1-\pi i/4)$
13. $e^{\pi i/2}$
14. $\sin(e^i)$
15. $\sin^2(1+i)$
16. $\cos(2-i) - \sin(2-i)$
17. $\cos(3+2i)$
18. $\tan(3i)$

19.4 The Complex Logarithm

In real calculus, the natural logarithm is the inverse of the exponential function. For $x > 0$,

$$y = \ln(x) \text{ if and only if } x = e^y.$$

In this way, the real natural logarithm can be thought of as the solution of the equation $x = e^y$. We will use this approach in developing a complex logarithm. Given $z \neq 0$, we want to solve for w in the equation

$$e^w = z.$$

To do this, put z in polar form as $z = re^{i\theta}$, and let $w = u + iv$ to write

$$z = re^{i\theta} = e^w = e^u e^{iv}. \tag{19.7}$$

Since θ and v are real, $|e^{i\theta}| = |e^{iv}| = 1$. Taking magnitudes in equation (19.7) gives us $r = |z| = e^u$, therefore

$$u = \ln(r),$$

which is the real natural logarithm of the positive number r. But now equation (19.7) implies that $e^{i\theta} = e^{iv}$. Then $e^{iv}/e^{i\theta} = e^{i(v-\theta)} = 1$. We know all the solutions of this equation, namely

$$i(v - \theta) = 2n\pi i$$

for integer n. Then

$$v = \theta + 2n\pi.$$

In summary, given $z = re^{i\theta}$ with $r \neq 0$, there are infinitely many complex numbers w such that $e^w = z$. All of these numbers are

$$w = \ln(r) + i\theta + 2n\pi i,$$

with n any integer. This leads us to define, for $z \neq 0$,

$$\log(z) = \ln(|z|) + i\theta + 2n\pi i$$

in which θ is any argument of z and n can be any integer. The complex log is not a function in the conventional sense because each nonzero z has infinitely many different complex logarithms.

EXAMPLE 19.13

Let $z = 1 + i$. In polar form, $z = \sqrt{2}e^{i(\pi/4+2n\pi i)}$. The values of the logarithm of $1 + i$ are

$$\log(1 + iz) = \ln\sqrt{2} + i\left[\frac{\pi}{4} + 2n\pi i\right]. \blacklozenge$$

In the complex plane, we can take the logarithm of a negative number.

EXAMPLE 19.14

Let $z = -3$. In polar form, $z = 3e^{(\pi+2n\pi)i} = 3e^{(2n+1)\pi i}$. The values of the logarithm of -3 are

$$\log(-3) = \ln(3) + (2n + 1)\pi i$$

in which n can be any integer. ♦

SECTION 19.4 PROBLEMS

In each of Problems 1 through 6, determine all values of the complex logarithm of z.

1. $-9+2i$
2. $1+5i$
3. $-4i$
4. 5
5. -5
6. $2-2i$
7. Let z and w be nonzero complex numbers. Show that each value of $\log(zw)$ is equal to a value of $\log(z)$ plus a value of $\log(w)$.
8. Let z and w be nonzero complex numbers. Show that each value of $\log(z/w)$ is equal to a value of $\log(z)$ minus a value of $\log(w)$.

19.5 Powers

We want to assign a meaning to z^w when z and w are complex and $z \neq 0$. If w is a positive integer, z^n is clear. For example, $z^3 = z \cdot z \cdot z$. And $z^{-n} = 1/z^n$ if $z \neq 0$. For other powers, we will proceed in stages.

*n*th Roots

Let n be a positive integer. An *n*th *root* $z^{1/n}$ of z is a number $z^{1/n}$ whose nth power is z. We want all values of $z^{1/n}$. To find these, begin with the polar form of z,

$$z = re^{i(\theta+2k\pi)}$$

with all of the arguments $\theta + 2k\pi$ of z in the exponent. Then

$$z^{1/n} = r^{1/n} e^{i(\theta+2k\pi)/n}$$

in which $r^{1/n}$ is the real nth root of the positive number r. As k varies over the integers, the numbers on the right give all the nth roots of z.

For $k = 0, 1, \cdots, n-1$, we obtain n distinct numbers

$$r^{1/n} e^{i\theta/n},\ r^{1/n} e^{i(\theta+2\pi)/n},\ r^{1/n} e^{i(\theta+4\pi)/n} \cdots \text{ and } r^{1/n} e^{i(\theta+2(n-1)\pi)/n}. \tag{19.8}$$

These are all nth roots of z. Other choices of k reproduce numbers already in this list. For example, with $k = n$ we get

$$r^{1/n} e^{i(\theta+2n\pi)/n} = r^{1/n} e^{i\theta/n} e^{2\pi i} = r^{1/n} e^{i\theta/n}$$

because $e^{2\pi i} = 1$. Therefore $k = n$ gives us the first number in the list (19.8) corresponding to $k = 0$.

If $k = n+1$, we obtain

$$r^{1/n} e^{i(\theta+2(n+1)\pi)/n} = r^{1/n} e^{i(\theta+2\pi)/n} e^{2\pi i} = r^{1/n} e^{i(\theta+2\pi)/n},$$

which is the second number in the list (19.8), corresponding to $k = 1$.

To sum up, the nth roots of z are the n numbers

$$r^{1/n} e^{i(\theta+2k\pi)/n} \text{ for } k = 0, 1, \cdots, n-1.$$

These can be written as

$$r^{1/n} \left[\cos\left(\frac{\theta+2k\pi}{n}\right) + i \sin\left(\frac{\theta+2k\pi}{n}\right) \right] \text{ for } k = 0, 1, \cdots, n-1.$$

EXAMPLE 19.15

We will find the fourth roots of $1+i$. One argument of $1+i$ is $\pi/4$, and $|1+i| = \sqrt{2}$, so $1+i = 2^{1/2}e^{i(\pi/4+2k\pi)}$ in which k can be any integer. The fourth roots are the four numbers

$$2^{1/8}e^{\pi i/16},\ 2^{1/8}e^{i(\pi/4+2\pi)},\ 2^{1/8}e^{i(\pi/4+4\pi)/4},\ \text{and}\ 2^{1/8}e^{i(\pi/4+6\pi)/4}.$$

Other choices for k simply reproduce these numbers. The fourth roots of $1+i$ also can be written as

$$2^{1/8}\left[\cos\left(\frac{\pi}{16}\right)+i\sin\left(\frac{\pi}{16}\right)\right],$$

$$2^{1/8}\left[\cos\left(\frac{9\pi}{16}\right)+i\sin\left(\frac{9\pi}{16}\right)\right],$$

$$2^{1/8}\left[\cos\left(\frac{17\pi}{16}\right)+i\sin\left(\frac{17\pi}{16}\right)\right],$$

and

$$2^{1/8}\left[\cos\left(\frac{25\pi}{16}\right)+i\sin\left(\frac{25\pi}{16}\right)\right]. \blacklozenge$$

EXAMPLE 19.16

The nth roots of 1 are called the nth *roots of unity*. They appear in many contexts: for example, in the development of the fast Fourier transform. Since 1 has a magnitude of 1 and an argument is 0, the nth roots of unity are the n numbers

$$e^{2\pi ki/n} \text{ for } k = 0, 1, \cdots, n-1.$$

For example, the fifth roots of unity are

$$1,\ e^{2\pi i/5},\ e^{4\pi i/5},\ e^{6\pi i/5},\ \text{and}\ e^{8\pi i/5}.$$

These are the numbers

$$1,\ \cos\left(\frac{2\pi}{5}\right)+i\sin\left(\frac{2\pi}{5}\right),\ \cos\left(\frac{4\pi}{5}\right)+i\sin\left(\frac{4\pi}{5}\right),$$

$$\cos\left(\frac{6\pi}{5}\right)+i\sin\left(\frac{6\pi}{5}\right),\ \text{and}\ \cos\left(\frac{8\pi}{5}\right)+i\sin\left(\frac{8\pi}{5}\right).$$

If plotted as points in the plane, the nth roots of unity form vertices of a regular polygon with vertices on the unit circle and having one vertex at $(1, 0)$. Figure 19.7 shows the fifth roots of unity displayed in this way. ♦

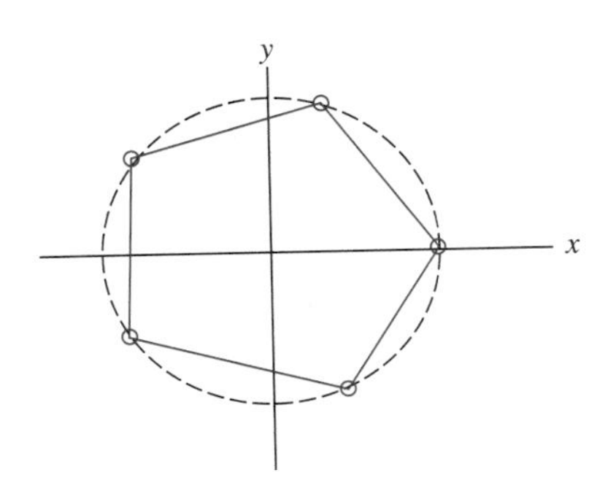

FIGURE 19.7 *Fifth roots of unity.*

Rational Powers

A rational number is a quotient of integers. If m and n are positive integers with no common factors, compute $z^{m/n}$ as $(z^m)^{1/n}$, which are the nth roots of z^m.

EXAMPLE 19.17

We will compute all values of $(2-2i)^{3/5}$. Compute $(2-2i)^3 = -16-16i$. We want the fifth roots of $-16-16i$. One argument of $-16-16i$ is $5\pi/4$ and $|-16-16i| = \sqrt{512}$, so in polar form

$$-16-16i = (512)^{1/2}e^{i(5\pi/4+2k\pi)}.$$

The fifth roots of $-16-16i$ are

$$(-16-16i)^{1/5} = (512)^{1/10}e^{i(5\pi/4+2k\pi)/5}.$$

These are the numbers

$$(512)^{1/10}e^{5\pi i/4},\ (512)^{1/10}e^{13\pi i/20},$$
$$(512)^{1/10}e^{21\pi i/20},\ (512)^{1/10}e^{29\pi i/20},\ \text{and}\ (512)^{1/10}e^{37\pi i/20}.\ \blacklozenge$$

Powers z^w

If $z \neq 0$, define for any complex number w

$$z^w = e^{w\log(z)}.$$

This definition is suggested by the fact that, if a and b are real numbers and $a \neq 0$, then $a^b = e^{b\ln(a)}$.

If w is not a rational number, then z^w has infinitely many values.

EXAMPLE 19.18

We will compute all values of $(1-i)^{1+i}$. These numbers are obtained as $e^{(1+i)\log(1-i)}$, so first determine all values of $\log(1-i)$. We have $|1-i| = \sqrt{2}$. We also need an argument of $1-i$. Any argument will do. One convenient argument is $7\pi/4$, obtained by a counterclockwise (positive) rotation from the positive real axis to the point $(1,-1)$. Another argument is $-\pi/4$, which is a clockwise (negative) rotation from the positive real axis to this point. Using the latter, we have

$$1-i = \sqrt{2}e^{i(-\pi/4+2n\pi)}.$$

Then all values of $\log(1-i)$ are given by

$$\ln\left(\sqrt{2}\right) + i\left(-\frac{\pi}{4} + 2n\pi\right)$$

in which n can be any integer. Every value of $(1-i)^{1+i}$ is contained in the expression

$$e^{(1+i)\log(1-i)} = e^{(1+i)\left[\ln\left(\sqrt{2}\right)+i(-\pi/4+2n\pi)\right]}.$$

These can be written as

$$\begin{aligned}
&e^{\ln\left(\sqrt{2}\right)+\pi/4-2n\pi}e^{i\left(\ln\left(\sqrt{2}\right)-\pi/4+2n\pi\right)}\\
&\quad = \sqrt{2}e^{\pi/4-2n\pi}\left(\cos\left(\ln\left(\sqrt{2}\right)-\pi/4+2n\pi\right)+i\sin\left(\ln\left(\sqrt{2}\right)-\pi/4+2n\pi\right)\right)\\
&\quad = \sqrt{2}e^{\pi/4-2n\pi}\left(\cos\left(\ln\left(\sqrt{2}\right)-\pi/4\right)+i\sin\left(\ln\left(\sqrt{2}\right)-\pi/4\right)\right).\ \blacklozenge
\end{aligned}$$

SECTION 19.5 PROBLEMS

In each of Problems 1 through 14, determine all values of z^w.

1. $(-4)^{2-i}$
2. $(7i)^{3i}$
3. $(-16)^{1/4}$
4. $[(1+i)/(1-i)]^{1/3}$
5. $1^{1/6}$
6. $16^{1/4}$
7. i^{1+i}
8. 6^{-2-3i}
9. i^i
10. $(1-i)^{1/3}$
11. $(-1+i)^{-3i}$
12. $(1+i)^{2-i}$
13. $i^{1/4}$
14. $(1+i)^{2i}$
15. Let $u_1, \cdots, u_n$ be the nth roots of unity with n a positive integer and $n \geq 2$. Prove that $\sum_{j=1}^{n} u_j = 0$. *Hint*: Write each nth root of unity as a power of $e^{2\pi i/n}$. A vector argument can also be made based on plotting the nth roots as vertices of a polygon.
16. Let n be a positive integer, and let $\omega = e^{2\pi i/n}$. Evaluate $\sum_{j=0}^{n-1}(-1)^j \omega^j$.

CHAPTER 20

THE INTEGRAL OF A COMPLEX FUNCTION CAUCHY'S THEOREM CONSEQUENCES OF CAUCHY'S THEOREM

Complex Integration

20.1 The Integral of a Complex Function

Real-valued functions are integrated over intervals. Complex functions are integrated over curves and have many properties in common with line integrals of vector fields. The notions of continuous, differentiable, smooth, and piecewise smooth curves were developed in Section 12.1. Here we will use complex notation and represent points (x, y) on a curve as complex numbers $x + iy$.

EXAMPLE 20.1

Let $\gamma(t) = e^{it}$ for $0 \le t \le 3\pi/2$. Then γ is a simple, smooth curve with initial point $\gamma(0) = 1$ and terminal point $\gamma(3\pi/2) = -i$. γ is therefore not closed. Since $\gamma(t) = \cos(t) + i\sin(t)$, this curve has coordinate functions $x = \cos(t)$ and $y = \sin(t)$. The graph of γ is the three-quarter circle of Figure 20.1. ♦

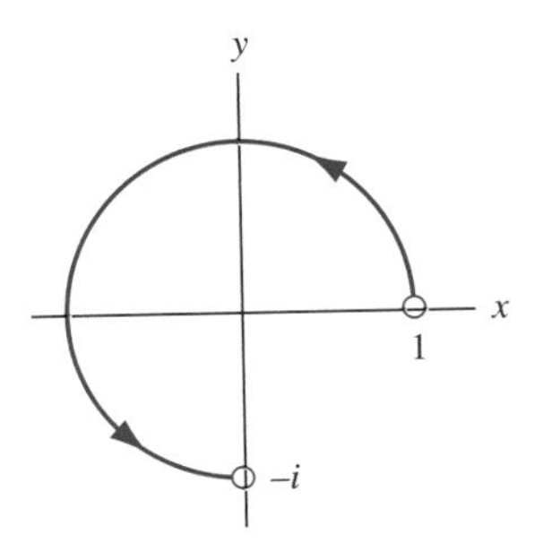

FIGURE 20.1 *γ in Example 20.1.*

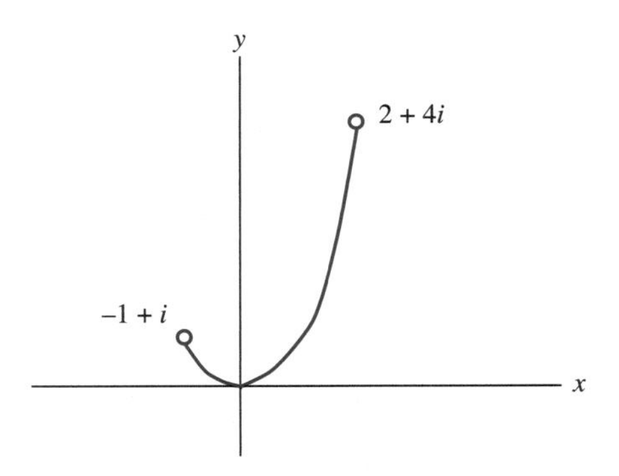

FIGURE 20.2 *δ in Example 20.2.*

EXAMPLE 20.2

Let $\delta(t) = t + it^2$ for $-1 \leq t \leq 2$. Then δ is a simple, smooth curve from $-1+i$ to $2+4i$, as shown in Figure 20.2. The coordinate functions are $x = t$ and $y = t^2$, and we can think of C as the part of the parabola $y = x^2$ from $(-1, 1)$ to $(2, 4)$. ♦

We are now ready to define the complex integral, which we will do in two stages.

First Stage—Integral Over a Closed Interval Suppose f is a complex function, and $f(x) = u(x) + iv(x)$ is defined at least for $a \leq x \leq b$. Define

$$\int_a^b f(x)\,dx = \int_a^b u(x)\,dx + i\int_a^b v(x)\,dx.$$

EXAMPLE 20.3

Let $f(x) = x - ix^2$ for $1 \leq x \leq 2$. Then

$$\int_1^2 f(x)\,dx = \int_1^2 x\,dx - i\int_1^2 x^2\,dx = \frac{3}{2} - \frac{7}{3}i. \; ♦$$

Second Stage—Integral Over a Curve Now we can define the integral of a complex function over a smooth curve in the plane. Let f be a complex function and γ a curve with $\gamma(t)$ defined for $a \leq t \leq b$. Assume that f is continuous at all points on the curve. Then we define the *integral of f over γ* by

$$\int_\gamma f(z)\,dz = \int_a^b f(\gamma(t))\gamma'(t)\,dt.$$

This integral also may be formulated as

$$\int_\gamma f(z)\,dz = \int_a^b f(z(t))z'(t)\,dt.$$

Evaluate $\int_\gamma f(z)dz$ by replacing z by $z(t) = \gamma(t)$ and integrating $f(\gamma(t))\gamma'(t)$ over $[a, b]$, according to the first stage in the definition of the integral.

EXAMPLE 20.4

We will evaluate $\int_\gamma \overline{z}\,dz$ if $\gamma(t) = e^{it}$ for $0 \leq t \leq \pi$.

The graph of γ is the upper half of the unit circle oriented counterclockwise from initial point 1 to terminal point -1. On γ, $z(t) = e^{it}$ and $z'(t) = ie^{it}$, so

$$f(z(t))z'(t) = \overline{e^{it}}ie^{it} = i.$$

Therefore,

$$\int_\gamma f(z)dz = \int_0^\pi i dt = \pi i. \ \blacklozenge$$

EXAMPLE 20.5

Evaluate $\int_\varphi z^2\, dz$ if $\varphi(t) = t + it$ for $0 \le t \le 1$.

The graph of φ is the straight line segment from 0 to $1+i$. On the curve, $z(t) = \varphi(t) = t + it = (1+i)t$. Then $z'(t) = 1+i$ and

$$f(z(t))z'(t) = ((1+i)t)^2(1+i) = (1+i)^3t^2 = (-2+2i)t^2.$$

Then

$$\int_\gamma z^2\, dz = \int_0^1 (-2+2i)t^2\, dt = \frac{2}{3}(-1+i). \ \blacklozenge$$

EXAMPLE 20.6

Evaluate $\int_\gamma z\text{Re}(z)\, dz$ if $\gamma(t) = t - it^2$ for $0 \le t \le 2$.

On this curve, $z(t) = t - it^2$, so

$$\begin{aligned} f(z(t)) &= z(t)\text{Re}(z(t)) \\ &= (t - it^2)\text{Re}(t - it^2) = (t - it^2)(t) = t^2 - it^3. \end{aligned}$$

Furthermore, $z'(t) = 1 - 2it$, so

$$f(z(t))z'(t) = (t^2 - it^3)(1 - 2it) = t^2 - 3it^3 - 2t^4.$$

Then

$$\begin{aligned} \int_\gamma f(z)\, dz &= \int_0^2 (t^2 - 3it^3 - 2t^4)\, dt \\ &= \int_0^2 (t^2 - 2t^4)\, dt - 3i\int_0^2 t^3\, dt = -\frac{152}{15} - 12i. \ \blacklozenge \end{aligned}$$

Thus far, we can only integrate over a smooth curve. Often we want to integrate over a piecewise smooth curve, which is a curve that is made up of a finite number of smooth arcs. Such a curve has a continuous tangent except perhaps at finitely many points (such as corners or "sharp points"). In this case, we think of γ as made up of curves $\gamma_1, \gamma_2, \cdots \gamma_n$, with each γ_j smooth and the terminal point of γ_j equal to the initial point of γ_{j+1}, as in Figure 20.3. We call γ the *join* of $\gamma_1, \cdots, \gamma_n$ and write

$$\gamma = \gamma_1 \bigoplus \gamma_2 \bigoplus \cdots \bigoplus \gamma_n,$$

as we did in Section 12.1.

Define

$$\int_\gamma f(z)\, dz = \sum_{j=1}^n \int_{\gamma_j} f(z)\, dz.$$

This is the analog of

$$\int_a^c + \int_c^b = \int_a^b$$

for real integrals.

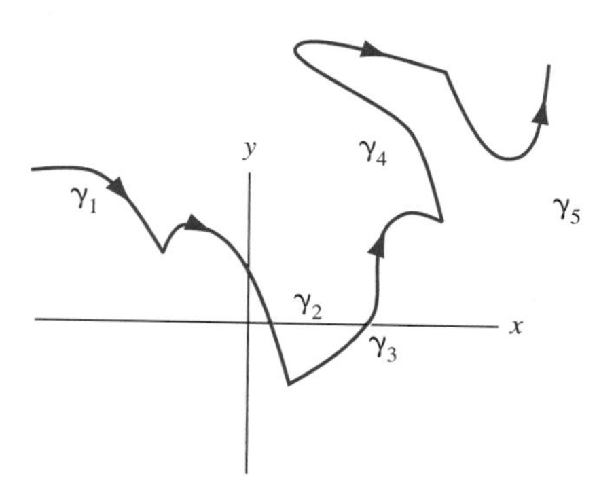

FIGURE 20.3 *The join of* $\gamma_1, \cdots, \gamma_n$.

EXAMPLE 20.7

Let $\gamma_1(t) = 3e^{it}$ for $0 \le t \le \pi/2$. Let $\gamma_2(t) = t^2 + 3(t+1)i$ for $0 \le t \le 1$. Let $\gamma = \gamma_1 \bigoplus \gamma_2$. We will compute $\int_\gamma f(z)\,dz$, where $f(z) = \operatorname{Im}(z)$.

γ is shown in Figure 20.4. γ_1 is the quarter circle part from 3 counterclockwise to $3i$, and γ_2 is part of the parabola $x = (y-3)^2/9$ from $3i$ to $1+6i$.

For the integral over γ_1, $z(t) = 3e^{it}$, so $f(z(t)) = \operatorname{Im}(3e^{it}) = 3\sin(t)$. Since

$$z'(t) = 3ie^{it} = -3\sin(t) + 3i\cos(t),$$

we have

$$\int_{\gamma_1} \operatorname{Im}(z)\,dz = \int_0^{\pi/2} 3\sin(t)[-3\sin(t) + 3i\cos(t)]\,dt$$

$$= -9\int_0^{\pi/2} \sin^2(t)\,dt + 9i\int_0^{\pi/2} \sin(t)\cos(t)\,dt$$

$$= -\frac{9}{4}\pi + \frac{9}{2}i.$$

On γ_2, $z(t) = t^2 + 3(t+1)i$, so $f(z(t)) = 3(t+1)$, and $z'(t) = 2t + 3i$. Then

$$\int_{\gamma_2} \operatorname{Im}(z)\,dz = \int_0^1 3(t+1)(2t+3i)\,dt$$

$$= \int_0^1 (6t^2 + 6t)\,dt + 9i\int_0^1 (t+1)\,dt$$

$$= 5 + \frac{27}{2}i.$$

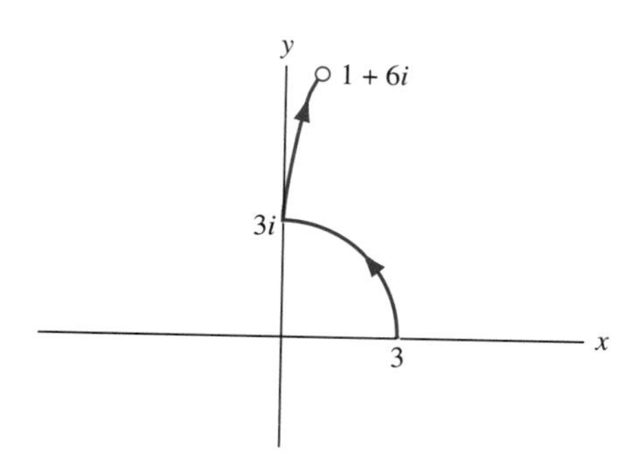

FIGURE 20.4 γ *in Example 20.7.*

Finally,

$$\int_\gamma \text{Im}(z)\,dz = -\frac{9}{4}\pi + \frac{9}{2}i + 5 + \frac{27}{2}i = 5 - \frac{9}{4}\pi + 18i. \ \blacklozenge$$

We will list some properties of the complex integral. These reflect properties of line integrals from Chapter 12.

1. $\int_\gamma (f(z) + g(z))\,dz = \int_\gamma f(z)\,dz + \int_\gamma g(z)\,dz.$
2. If c is a number, then $\int_\gamma cf(z)\,dz = c\int_\gamma f(z)\,dz.$
3. Reversing the orientation on the curve changes the sign of the integral. Specifically, given γ defined on $[a,b]$, define $\varphi(t) = \gamma(a+b-t)$ for $a \le t \le b$. Then

$$\gamma(a) = \varphi(b) \text{ and } \gamma(b) = \varphi(a).$$

   The initial point of γ is the terminal point of φ, and the terminal point of γ is the initial point of φ. We denote the curve φ formed in this way as $-\gamma$. Then

$$\int_{-\gamma} f(z)\,dz = -\int_\gamma f(z)\,dz.$$

4. There is a version of the fundamental theorem of calculus for complex integrals. Suppose f is continuous on an open set G and F is defined on G with the property that $F'(z) = f(z)$. If γ is a smooth curve in G, defined on an interval $[a,b]$, then

$$\int_\gamma f(z)\,dz = F(\gamma(b)) - F(\gamma(a)).$$

   To see why this is true, write $F(z) = U(x,y) + iV(x,y)$ to get

$$\begin{aligned}\int_\gamma f(z)\,dz &= \int_a^b f(z(t))z'(t)\,dt \\ &= \int_a^b F'(z(t))z'(t)\,dt = \int_a^b \frac{d}{dt}F(z(t))\,dt \\ &= \int_a^b \frac{d}{dt}U(x(t),y(t))\,dt + i\int_a^b \frac{d}{dt}V(x(t),y(t))\,dt \\ &= U(x(b),y(b)) + iV(x(b),y(b)) - iU(x(a),y(a)) - iV(x(a),y(a)) \\ &= F(\gamma(b)) - F(\gamma(a)).\end{aligned}$$

   One ramification of this is that, under the given conditions, the value of $\int_\gamma f(z)\,dz$ depends only on the initial and terminal points of γ and not on γ itself. This is called *independence of path*, and we saw a version of it in Chapter 12 with conservative force fields. If γ is a closed curve, the initial and terminal points are the same, and $\int_\gamma f(z)\,dz = 0$. We will consider this in more detail in the next section.

EXAMPLE 20.8

Evaluate $\int_\gamma z^2\,dz$, with γ any smooth curve from i to $1-i$. Since $F'(z) = z^2$, if $F(z) = z^3/3$,

$$\int_\gamma z^2\,dz = F(1-i) - F(i) = \frac{(1-i)^3}{3} - \frac{i^3}{3} = \frac{-2-i}{3}.$$

regardless of how γ moves about the plane between i and $1-i$. $\blacklozenge$

5. $\int_\gamma f(z)\,dz$ can be written as a sum of two real line integrals. To do this, suppose γ is defined on $[a,b]$, and write

$$f(z) = f(x+iy) = u(x,y) + iv(x,y) \text{ and } dz = (x'(t) + iy'(t))dt.$$

Then

$$\int_\gamma f(z)\,dz = \int_a^b [u(x(t),y(t)) + iv(x(t),y(t))][x'(t) + iy'(t)]\,dt$$
$$= \int_\gamma u\,dx - v\,dy + i\int_\gamma v\,dx + u\,dy. \tag{20.1}$$

6. Let γ be a smooth curve defined on $[a,b]$, and let f be continuous on γ. Suppose $|f(z)| \le M$ for all z on γ, and let L be the length of γ. Then

$$\left|\int_\gamma f(z)\,dz\right| \le ML.$$

This is the complex version of the inequality $\left|\int_a^b g(x)\,dx\right| \le M(b-a)$ if $|g(x)| \le M$ for $a \le x \le b$ for real integrals.

SECTION 20.1 PROBLEMS

In each of Problems 1 through 15, evaluate $\int_\gamma f(z)\,dz$.

1. $f(z) = |z|^2$; γ is the line segment from $-i$ to 1.
2. $f(z) = e^{iz}$; γ is any smooth curve from -2 to $-4-i$.
3. $f(z) = i\bar{z}$; γ is the line segment from 0 to $-4+3i$.
4. $f(z) = \text{Im}(z)$; γ is the circle of radius 4 about the origin (oriented positively).
5. $f(z) = (z-i)^3$; $\gamma(t) = t - it^2$ for $0 \le t \le 2$.
6. $f(z) = 1 + z^2$; γ is the part of the circle of radius 3 about the origin from $-3i$ to $3i$.
7. $f(z) = -i\cos(z)$; γ is any smooth curve from 0 to $2+i$.
8. $f(z) = |z|^2$; γ is the line segment from -4 to i.
9. $f(z) = \sin(2z)$; γ is the line segment from $-i$ to $-4i$.
10. $f(z) = iz^2$; γ is the line segment from $1+2i$ to $3+i$.
11. $f(z) = z-1$; γ is any piecewise smooth curve from $2i$ to $1-4i$.
12. $f(z) = z^2 - iz$; γ is the quarter circle about the origin from 2 to $2i$.
13. $f(z) = \text{Re}(z)$; γ is the line segment from 1 to $2+i$.
14. $f(z) = 1/z$; γ is the part of the half circle of radius 4 about the origin from $4i$ to $-4i$.
15. $f(z) = 1$; $\gamma(t) = t^2 - it$ for $1 \le t \le 3$.
16. Find a bound for $|\int_\gamma \cos(z^2)\,dz|$ if γ is the circle of radius 4 about the origin.
17. Find a bound for $|\int (1/(1+z))\,dz|$ if γ is the line segment from $2+i$ to $4+2i$.

20.2 Cauchy's Theorem

Cauchy's theorem is the cornerstone of complex integration theory. We need some terminology and preparation for its statement.

If γ is a continuous, simple, closed curve in the plane, then γ separates the plane into three parts: the graph of the curve itself, a bounded open set called the *interior* of γ, and an unbounded open set called the *exterior* of the curve (Figure 20.5). This is the *Jordan curve theorem*, which we will assume.

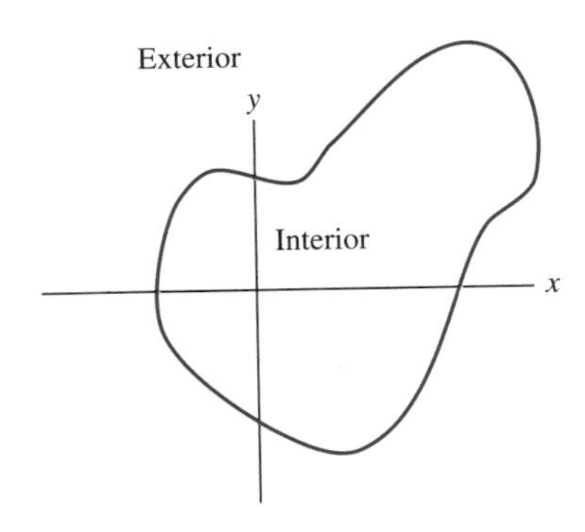

FIGURE 20.5 *Interior and exterior of a closed curve.*

We will refer to a simple, piecewise smooth curve as a *path*. A path in a set S is a path whose graph lies in S.

A set S of complex numbers is *connected* if every two points of S are endpoints of a path in S. This means that we can get from any point of S to any other point by moving along some path without leaving S. An open, connected set is called a *domain*. For example, any open disk is a domain, and the right quarter plane $x > 0$, $y > 0$ is a domain.

We encountered domains in connection with potential functions in Chapter 12.

A set S of complex numbers is *simply connected* if every closed path in S encloses only points of S.

This concept was also discussed in Chapter 12. Every open disk is simply connected. However, let S be an open disk with the center removed (a punctured disk). Then a closed path about the center in the disk encloses a point not in the set, so this set is not simply connected (although it is open and connected, hence is a domain).

We can now state the main result.

THEOREM 20.1 Cauchy

Let f be differentiable on a simply connected domain G. Then

$$\oint_{\gamma} f(z)\,dz = 0$$

for every closed path γ in G. ♦

Cauchy's theorem means that $\oint_{\gamma} f(z)\,dz = 0$ if f is differentiable on the path γ and all points enclosed by γ. Unless otherwise stated, we always understand closed curves to be oriented counterclockwise, which we take to be the positive sense.

In Theorem 20.1, the notation $\oint_\gamma$ is used. The oval on the integral is a reminder that the path is closed. This notation was also used in connection with line integrals in Chapter 12. It is not required to include this oval for integrals over closed paths.

EXAMPLE 20.9

If γ is any closed path in the plane,

$$\oint_\gamma e^{z^2}\,dz = 0,$$

because e^{z^2} is differentiable on the entire plane, which is a simply connected domain. ♦

EXAMPLE 20.10

We will evaluate

$$\oint_\gamma \frac{2z+1}{z^2+3iz}\,dz,$$

where γ is the circle $|z+3i|=2$ of radius 2 about $-3i$.

Observe that $f(z)$ is differentiable at all points except 0 and $-3i$, where the denominator vanishes. Use a partial fractions decomposition to write

$$f(z) = \frac{1}{3i}\frac{1}{z} + \frac{6+i}{3}\frac{1}{z+3i}.$$

Then

$$\oint_\gamma \frac{2z+1}{z^2+3iz}\,dz = \frac{1}{3i}\oint_\gamma \frac{1}{z}\,dz + \frac{6+i}{3}\oint_\gamma \frac{1}{z+3i}\,dz.$$

Because γ does not enclose 0, $1/z$ is differentiable on and within the simply connected domain enclosed by γ. By Cauchy's theorem,

$$\oint_\gamma \frac{1}{z}\,dz = 0.$$

However, $1/(z+3i)$ is not differentiable in the region enclosed by γ, so Cauchy's theorem does not apply to the integral of this function over γ. We will evaluate this integral directly by parametrizing $\gamma(t) = z(t) = -3i + 2e^{it}$ (polar coordinates centered at $-3i$) for $0 \le t \le 2\pi$. Now

$$\oint_\gamma \frac{1}{z+3i}\,dz = \int_0^{2\pi} \frac{1}{z(t)+3i} z'(t)\,dt$$

$$= \int_0^{2\pi} \frac{1}{2e^{it}} 2ie^{it}\,dt = \int_0^{2\pi} i\,dt = 2\pi i.$$

Therefore,

$$\oint_\gamma \frac{2z+1}{z^2+3iz}\,dz = \frac{6+i}{3}(2\pi i) = \left(-\frac{2}{3}+4i\right)\pi. \ ♦$$

A proof of Cauchy's theorem requires a delicate argument we will not engage here. However, a less general version can be established easily. Write $f = u + iv$ and assume that u and v are continuous with continuous first and second partial derivatives on G. Now we can apply Green's theorem and the Cauchy-Riemann equations. If D is the region containing the path and all points enclosed by γ, then

$$\int_\gamma f(z)\,dz = \oint_\gamma u\,dx - v\,dy + i\oint_\gamma v\,dx + u\,dy$$

$$= \iint_D \left(\frac{\partial(-v)}{\partial x} - \frac{\partial u}{\partial y}\right) dA + i\iint_D \left(\frac{\partial u}{\partial x} - \frac{\partial v}{\partial y}\right) dA = 0,$$

because by the Cauchy-Riemann equations,

$$\frac{\partial u}{\partial x} = \frac{\partial v}{\partial y} \text{ and } \frac{\partial u}{\partial y} = -\frac{\partial v}{\partial x}.$$

Cauchy's theorem has several important consequences, which are the object of the next section.

SECTION 20.2 PROBLEMS

In each of Problems 1 through 12, evaluate the integral of the function over the closed path. All curves are oriented counterclockwise. In some cases, Cauchy's theorem applies, while in others it does not, but may still be useful (as in Example 20.10).

1. $f(z) = \mathrm{Re}(z)$; γ is given by $|z| = 2$.
2. $f(z) = z^2 \sin(z)$; γ is the square having vertices $0, 1, i$ and $1 + i$.
3. $f(z) = |z|^2$; γ is the circle of radius 7 about the origin.
4. $f(z) = 1/\bar{z}$; γ is the circle of radius 5 about the origin.
5. $f(z) = ze^z$; γ is the circle $|z - 3i| = 8$.
6. $f(z) = 2z/(z - i)$; γ is the circle $|z - i| = 3$.
7. $f(z) = 1/(z - 2i)^3$; γ is given by $|z - 2i| = 2$.
8. $f(z) = z^2 + \mathrm{Im}(z)$; γ is the square with vertices $0, -2i, 2$ and $2 - 2i$.
9. $f(z) = \bar{z}$; γ is the unit circle about the origin.
10. $f(z) = z^2 - 4z + i$; γ is the rectangle with vertices $1, 8, 8 + 4i$ and $1 + 4i$.
11. $f(z) = \sin(3z)$; γ is the circle $|z| = 4$.
12. $f(z) = \sin(1/z)$; γ is the circle $|z - 1 + 2i| = 1$.

20.3 Consequences of Cauchy's Theorem

This section develops some consequences of Cauchy's theorem.

20.3.1 Independence of Path

Independence of path was mentioned briefly in connection with evaluating $\int_\gamma f(z)\,dz$ in terms of an antiderivative F of f. Independence of path can also be viewed from the perspective of Cauchy's theorem.

Suppose f is differentiable on a simply connected domain G, and z_0 and z_1 are points of G. Let γ_1 and γ_2 be paths in G from z_0 to z_1 (Figure 20.6).

If we reverse the orientation on γ_2, we form a closed path $\Gamma = \gamma_1 \bigoplus (-\gamma_2)$. By Cauchy's theorem, $\oint_\Gamma f(z)\,dz = 0$, so

$$\int_{\gamma_1} f(z)\,dz + \int_{-\gamma_2} f(z)\,dz = 0.$$

But

$$\int_{-\gamma_2} f(z)\,dz = -\int_{\gamma_2} f(z)\,dz,$$

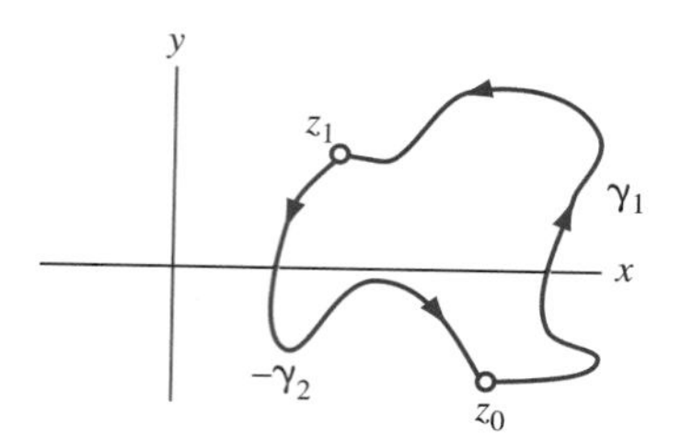

FIGURE 20.6 *Independence of path.*

so

$$\int_{\gamma_1} f(z)\,dz = \int_{\gamma_2} f(z)\,dz.$$

This means that $\int_\gamma f(z)\,dz$ is independent of path on G, because the integral over any path in G depends only on the endpoints of the path. In such a case, we sometimes write

$$\int_\gamma f(z)\,dz = \int_{z_0}^{z_1} f(z)\,dz.$$

20.3.2 The Deformation Theorem

The deformation theorem enables us, under certain conditions, to replace one closed path of integration with another, perhaps more convenient one.

THEOREM 20.2 *The Deformation Theorem*

Let Γ and γ be closed paths in the plane with γ in the interior of Γ. Suppose f is differentiable on an open set containing both paths and all points between them. Then

$$\oint_\Gamma f(z)\,dz = \oint_\gamma f(z)\,dz. \blacklozenge$$

Figure 20.7 suggests the reason for the name of the theorem. Think of γ as made of rubber, and continuously stretch and deform γ into Γ. In doing this, it is important that the intermediate stages of the deformation only cross over points at which f is differentiable, hence the assumption that f is differentiable at all points between the two curves.

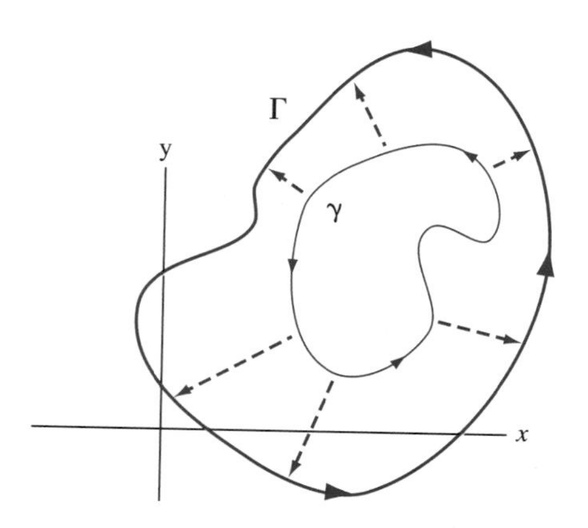

FIGURE 20.7 *The deformation theorem.*

The theorem states that the integral of f has the same value over both paths under the conditions stated. This means that we can replace the integral over one curve with the integral over the other, allowing us great flexibility in choice of paths in evaluating an integral.

EXAMPLE 20.11

Evaluate

$$\oint_\Gamma \frac{1}{z-a}\,dz$$

where Γ is any closed path enclosing the number a (Figure 20.8(a)).

We do not know Γ, so it might appear that we cannot evaluate this integral. Because $f(z) = 1/(z-a)$ is not differentiable in this region, and Cauchy's theorem does not apply. However, a is the only point at which f is not differentiable. Place a circle γ about a of sufficiently small radius r so that the two curves do not intersect (Figure 20.8(b)). Now f is differentiable on both curves and on the region between them, so

$$\oint_\Gamma f(z)\,dz = \int_\gamma f(z)\,dz.$$

The point is that we can easily evaluate the integral over γ. Using polar coordinates centered at a, write $\gamma(t) = a + re^{it}$ for $0 \le t \le 2\pi$. Then

$$\oint_\Gamma f(z)\,dz = \oint_\gamma f(z)\,dz$$
$$= \int_0^{2\pi} \frac{1}{re^{it}} ire^{it}\,dt = i\int_0^{2\pi} dt = 2\pi i. \;\blacklozenge$$

A proof of the deformation theorem is reminiscent of the argument used for the extended Green's theorem in Chapter 12. Figure 20.9(a) shows typical curves Γ and γ. Insert line segments L_1 and L_2 between these paths (Figure 20.9(b)), and use these to form two closed paths, Φ and Ψ, as in Figure 20.10. Both Φ and Ψ are oriented positively (counterclockwise), which is consistent with positive orientations on Γ and γ. Because f is differentiable on Γ and γ and all points in between, f is differentiable on Φ and Ψ and all points they enclose, so by Cauchy's theorem,

$$\oint_\Phi f(z)\,dz = \oint_\Psi f(z)\,dz = 0.$$

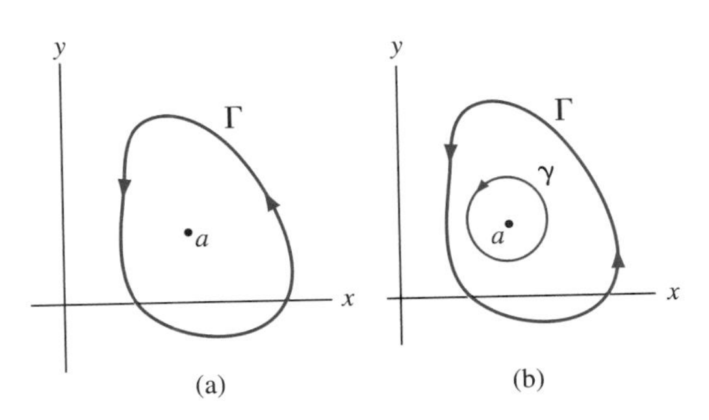

FIGURE 20.8 *Enclosing a in a circle γ interior to Γ in Example 20.11.*

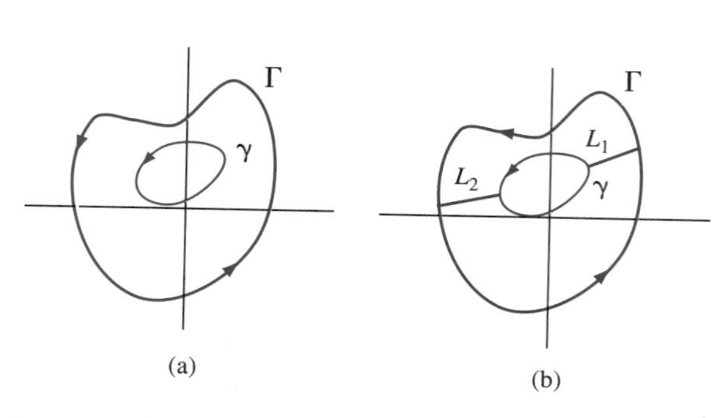

FIGURE 20.9 *Start of the proof of the deformation theorem.*

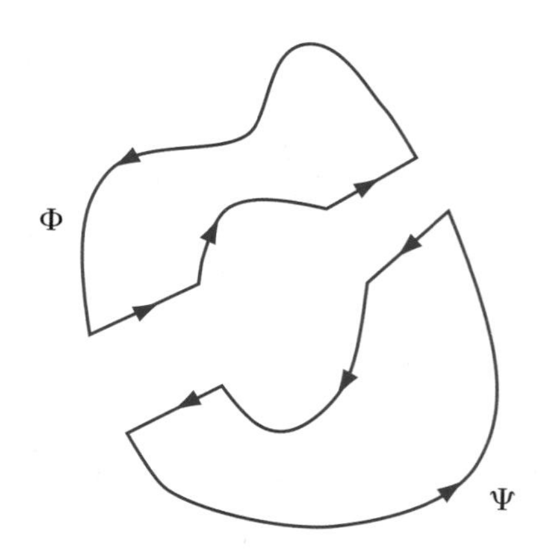

FIGURE 20.10 *Last stage in the proof of the deformation theorem.*

Then

$$\oint_{\Phi} f(z)\,dz + \oint_{\Psi} f(z)\,dz = 0. \tag{20.2}$$

In this sum of integrals, each of L_1 and L_2 is integrated over in one direction as part of Φ and in the opposite direction as part of Ψ. The contributions from these segments therefore cancel in the sum of equation (20.2). Next observe that, in adding the integrals in equation (20.2), we obtain the integral over Γ oriented counterclockwise and the integral over γ oriented clockwise (negatively). Equation (20.2) becomes

$$\oint_{\Gamma} f(z)\,dz - \oint_{\gamma} f(z)\,dz = 0$$

with both curves oriented positively. But then

$$\oint_{\Gamma} f(z)\,dz = \oint_{\gamma} f(z)\,dz.$$

The function f may have points interior to γ at which it is undefined or not differentiable, but f is differentiable on and in the region *between* Γ and γ, so $\oint_{\Gamma} f(z)\,dz = \oint_{\gamma} f(z)\,dz$.

20.3.3 Cauchy's Integral Formula

We will state a remarkable result which gives an integral formula for the values of a differentiable function.

THEOREM 20.3 *Cauchy's Integral Formula*

Let f be differentiable on an open set G. Let γ be a closed path in G enclosing only points of G. Then, for any z_0 enclosed by γ,

$$f(z_0) = \frac{1}{2\pi i}\oint_{\gamma} \frac{f(z)}{z - z_0}\,dz. \quad \blacklozenge$$

We will see many uses of this theorem. One is immediate. We can write Cauchy's formula as

$$2\pi i f(z_0) = \oint_{\gamma} \frac{f(z)}{z - z_0}\,dz,$$

thereby evaluating the integral on the right in terms of the value of $f(z)$ at a point z_0 enclosed by γ. Keep in mind, however, that we have evaluated $\oint_\gamma (f(z)/(z-z_0))\,dz$, not $\oint_\gamma f(z)\,dz$.

EXAMPLE 20.12

We will evaluate

$$\oint_\gamma \frac{e^{z^2}}{z-i}\,dz$$

for any closed path that does not pass through i. Here $f(z)=e^{z^2}$ is differentiable for all z. There are two cases.

Case 1 If γ does not enclose i, then $e^{z^2}/(z-i)$ is differentiable on and in the region enclosed by γ, so by Cauchy's theorem,

$$\oint_\gamma \frac{e^{z^2}}{z-i}\,dz=0.$$

Case 2 If γ encloses i, then by Cauchy's integral formula with $z_0=i$,

$$\oint_\gamma \frac{e^{z^2}}{z-i}\,dz=2\pi i f(i)=2\pi i e^{-1}. \blacklozenge$$

EXAMPLE 20.13

Evaluate

$$\oint_\gamma \frac{e^{2z}\sin(z^2)}{z-2}\,dz$$

over any closed path that does not pass through 2. Let $f(z)=e^{2z}\sin(z^2)$. Then f is differentiable for all z. There are two cases.

Case 1 If γ does not enclose 2, then $f(z)/(z-2)$ is differentiable on and within γ, so

$$\oint_\gamma \frac{e^{2z}\sin(z^2)}{z-2},\,dz=0.$$

Case 2 If γ encloses 2, then by Cauchy's integral formula,

$$\oint_\gamma \frac{e^{2z}\sin(z^2)}{z-2}\,dz=2\pi i f(2)=2\pi i e^4\sin(4). \blacklozenge$$

There is a version of the integral formula for derivatives.

THEOREM 20.4 ***Cauchy's Integral Formula for Higher Derivatives***

With f, G, γ, and z_0 as in Cauchy's integral formula (Theorem 20.3), then

$$f^{(n)}(z_0)=\frac{n!}{2\pi i}\oint_\gamma \frac{f(z)}{(z-z_0)^{n+1}}\,dz \tag{20.3}$$

in which n is any nonnegative integer. $\blacklozenge$

For $n = 0$, this is Cauchy's integral formula with the convention that $f^{(0)}(z) = f(z)$. In the equation, $n!$ (read n factorial) is the product of the integers 1 through n inclusive, and $f^{(n)}$ is the nth derivative of f. This integral formula for the nth derivative of $f(z)$ at z_0 is exactly what we would get if we differentiated n times with respect to z_0 under the integral sign in Cauchy's integral formula.

EXAMPLE 20.14

We will evaluate

$$\oint_\gamma \frac{e^{z^3}}{(z-i)^3}\,dz$$

with γ any closed path not passing through i. If γ does not enclose i, this integral is zero by Cauchy's theorem. Suppose that γ does enclose i. Because $z - i$ occurs to the third power in the denominator of the integral, let $n = 2$ in Cauchy's formula for higher derivatives (20.3) with $f(z) = e^{z^3}$. Compute

$$f'(z) = 3z^2 e^{z^3} \text{ and } f''(z) = (6z + 9z^4)e^{z^3}.$$

Then

$$\oint_\gamma \frac{e^{z^3}}{(z-i)^3}\,dz = \frac{2\pi i}{2!} f''(i) = (-6 + 9i)\pi e^{-i}. \ \blacklozenge$$

Proof We will outline a proof of Cauchy's integral formula. First, use the deformation theorem to replace γ with a circle C of radius r about z_0, as in Figure 20.11. Then

$$\begin{aligned}\oint_\gamma \frac{f(z)}{z-z_0}\,dz &= \oint_C \frac{f(z)}{z-z_0}\,dz = \oint_C \frac{f(z) - f(z_0) + f(z_0)}{z-z_0}\,dz \\ &= f(z_0)\oint_C \frac{1}{z-z_0}\,dz + \oint_C \frac{f(z)-f(z_0)}{z-z_0}\,dz \\ &= 2\pi i f(z_0) + \oint_C \frac{f(z)-f(z_0)}{z-z_0}\,dz\end{aligned}$$

in which we used the result of Example 20.11. We will have proved the Cauchy integral representation if we can show that the last integral is zero. On C, write $C(t) = z_0 + re^{it}$ for $0 \le t \le 2\pi$. Then

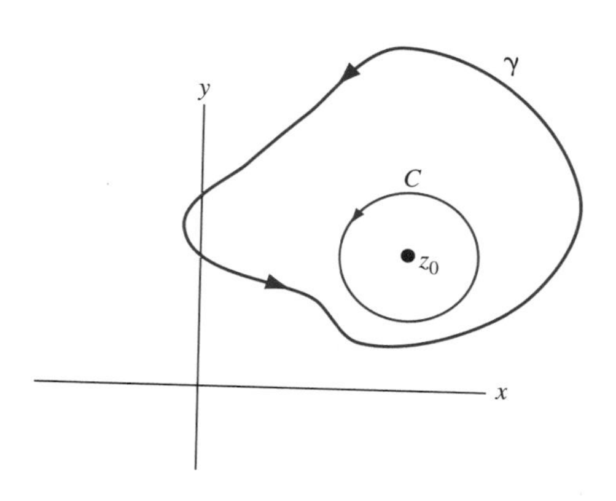

FIGURE 20.11 *A proof of the Cauchy integral formula.*

$$\left|\oint_C \frac{f(z) - f(z_0)}{z - z_0}\,dz\right| = \left|\int_0^{2\pi} \frac{f(z_0 + re^{it}) - f(z_0)}{re^{it}} ire^{it}\,dt\right|$$
$$= \left|\int_0^{2\pi} (f(z_0 + re^{it}) - f(z_0))\,dt\right|$$
$$\leq \int_0^{2\pi} | f(z_0 + re^{it}) - f(z_0) |\,dt.$$

But $| f(z_0 + re^{it}) - f(z_0) | \to 0$ as $r \to 0$ by continuity of f. We conclude that

$$\left|\oint_C \frac{f(z) - f(z_0)}{z - z_0}\,dz\right| = 0,$$

and hence, that

$$\oint_C \frac{f(z) - f(z_0)}{z - z_0}\,dz = 0,$$

establishing Cauchy's integral formula. ♦

The integral formula gives added appreciation of the power of the condition of differentiability for complex functions. The formula gives the value of $f(z)$ at all points enclosed by a closed path γ, strictly in terms of values of $f(z)$ at points on γ, because these are all that are needed to evaluate

$$\oint_\gamma \frac{f(z)}{z - z_0}\,dz.$$

By contrast, knowing the values of a differentiable real-valued function $g(x)$ at the endpoints of an interval $[a, b]$ tells us nothing about values $g(x)$ for $a < x < b$.

Another implication of the integral formula for higher derivatives is that a complex function that is differentiable on an open set has derivatives of all orders on that set. Again, real functions do not behave this well. If $g'(x)$ exists, $g''(x)$ may not.

20.3.4 Properties of Harmonic Functions

As an application of Cauchy's integral formula, we will derive two important properties of harmonic functions. This is a prime example of the use of complex functions to derive facts about real functions—a theme we will see again and which is made possible by the connection between harmonic functions and the real and imaginary parts of differentiable complex functions.

THEOREM 20.5 ***The Mean Value Property***

Let u be harmonic on a domain D. Let (x_0, y_0) be any point of D, and let C be a circle of radius r in D centered at (x_0, y_0) and enclosing only points of D. Then

$$u(x_0, y_0) = \frac{1}{2\pi}\int_0^{2\pi} u(x_0 + r\cos(\theta), y_0 + r\sin(\theta))\,d\theta. \ ♦$$

Notice that $(x_0 + r\sin(\theta), y_0 + r\sin(\theta))$ are polar coordinates of points on the circle, varying once counterclockwise over the circle as θ varies from 0 to 2π. The theorem therefore says that the value of u at the center of the circle is the average of the values of $u(x, y)$ over the circle.

Proof For some v, $f = u + iv$ is harmonic on D. Let $z_0 = x_0 + iy_0$. By Cauchy's integral formula,

$$f(z_0) = u(x_0, y_0) + iv(x_0, y_0) = \frac{1}{2\pi i}\oint_C \frac{f(z)}{z - z_0}\,dz$$
$$= \frac{1}{2\pi i}\int_0^{2\pi} \frac{f(z_0 + re^{i\theta})}{re^{i\theta}} ire^{i\theta}\,d\theta$$
$$= \frac{1}{2\pi}\int_0^{2\pi} u(x_0 + r\cos(\theta), y_0 + r\sin(\theta))\,d\theta$$
$$+ \frac{i}{2\pi}\int_0^{2\pi} v(x_0 + r\cos(\theta), y_0 + r\sin(\theta))\,d\theta.$$

By comparing the real part of the left and right sides of this equation, we have the conclusion of the theorem. ♦

If D is a bounded domain, then the set $\overline{D}$ consisting of D, together with all boundary points of D, is a closed and bounded set in the x, y-plane. If $u(x, y)$ is continuous on $\overline{D}$, then $u(x, y)$ must achieve a maximum value on $\overline{D}$. In general, this might occur at any point or number of points of $\overline{D}$. However, if u is also harmonic on D, then $u(x, y)$ must achieve its maximum value at a boundary point of D. A proof of this uses the fact that u has a harmonic conjugate v, enabling us to work with a differentiable complex function $f = u + iv$.

THEOREM 20.6 ***The Maximum Principle***

Let D be a bounded domain in the plane and suppose u is continuous on $\overline{D}$ and harmonic on D. Then $u(x, y)$ achieves its maximum value at a boundary point of D. ♦

Proof Let v be a harmonic conjugate of u in D, and let $f = u + iv$. Define

$$g(z) = e^{f(z)}.$$

Then g is differentiable on D. Now $|g(z)|$ is a continuous function of two real variables on the closed and bounded set $\overline{D}$ of points in the plane. By a theorem of calculus, $|g(z)|$ achieves a maximum value at a boundary point of **D**. But

$$|g(z)| = |e^{u(x,y)+iv(x,y)}| = |e^{u(x,y)}e^{iv(x,y)}| = e^{u(x,y)}.$$

Since $e^{u(x,y)}$ is a strictly increasing function, $e^{u(x,y)}$ and $u(x, y)$ must achieve their maximum values at the same point. Therefore, $u(x, y)$ must achieve its maximum at a boundary point. ♦

20.3.5 Bounds on Derivatives

It is possible to bound the derivatives of a complex function in terms of a bound on the function.

THEOREM 20.7 ***Bounds on Derivatives***

Suppose f is differentiable on an open set G, and z_0 is a point in G. Let the closed disk of radius r about z_0 be entirely contained in G. Suppose that $|f(z)| \le M$ for z on the circle bounding this disk. Then, for any positive integer n,

$$|f^{(n)}(z_0)| \le \frac{Mn!}{r^n}. \quad ♦$$

Theorem 20.7 can be proved by parametrizing the circle bounding the disk as $\gamma(t) = z_0 + re^{it}$ in Cauchy's integral formula for $f^{(n)}(z_0)$.

One important consequence of this bound on higher derivatives is *Liouville's theorem*, which states that a bounded function that is differentiable for all z must be constant. This means that, if f is nonconstant and differentiable for all z, then f cannot be a bounded function. We saw this with $\cos(z)$ and $\sin(z)$, which are differentiable for all z and are not bounded functions (over the entire complex plane).

To prove Liouville's theorem, suppose $|f(z)| \leq M$ for all z. By Theorem 20.7 with $n = 1$ for any number z_0,

$$|f'(z_0)| \leq \frac{M}{r}$$

in which r is the radius of a circle about z_0. Since r can be as large as we want, M/r can be made arbitrarily small, so $|f'(z_0)|$ must be zero. Then $f'(z_0) = 0$. But z_0 is any number so $f'(z) = 0$ for all z, and from this it is routine to check, using Theorem 19.5, that $f(z)$ must be constant.

Liouville's theorem provides a simple proof of the fundamental theorem of algebra, which states that if $p(z)$ is a complex polynomial of degree $n \geq 1$, then for some z_0, $p(z_0) = 0$. If this were not true, then we would have $p(z) \neq 0$ for all z. Then $1/p(z)$ would differentiable for all z. But routine estimates enable us to conclude that $1/p(z)$ is bounded on the entire plane. By Liouville's theorem, $1/p(z)$ would be constant, so $p(z)$ would be constant, which is a contradiction. Therefore, $p(z)$ must be zero for some complex number.

20.3.6 An Extended Deformation Theorem

The deformation theorem enables us, under certain conditions, to deform one closed path Γ to another γ without changing the value of $\oint_\gamma f(z)dz$. This requires that the deformation of one path into the other not pass over any points at which f is not differentiable. If γ is enclosed by Γ, this requires that f be differentiable at all points between these curves.

We will extend this result to the case that Γ encloses any finite number of disjoint closed paths. As usual, unless otherwise stated, all closed paths are oriented counterclockwise.

THEOREM 20.8 *Extended Deformation Theorem*

Let Γ be a closed path, and let $\gamma_1, \cdots, \gamma_n$ be closed paths enclosed by Γ. Assume that no two of $\Gamma, \gamma_1, \cdots \gamma_n$ intersect and no point interior to any γ_j is interior to any other γ_k. Let f be differentiable on an open set containing Γ and each γ_j and all points that are both interior to Γ and exterior to each γ_j. Then

$$\oint_\Gamma f(z)\,dz = \sum_{j=1}^{n} \oint_{\gamma_j} f(z)\,dz. \blacklozenge$$

If $n = 1$, this is the deformation theorem. Here is an example of the theorem in evaluating an integral.

EXAMPLE 20.15

We will evaluate

$$\oint_\Gamma \frac{z}{(z+2)(z-4i)}\,dz$$

where Γ is a closed path enclosing both -2 and $4i$.

As in Figure 20.12, enclose each of -2 and $4i$ by closed paths γ_1 and γ_2 small enough that they do not intersect each other or Γ. Then

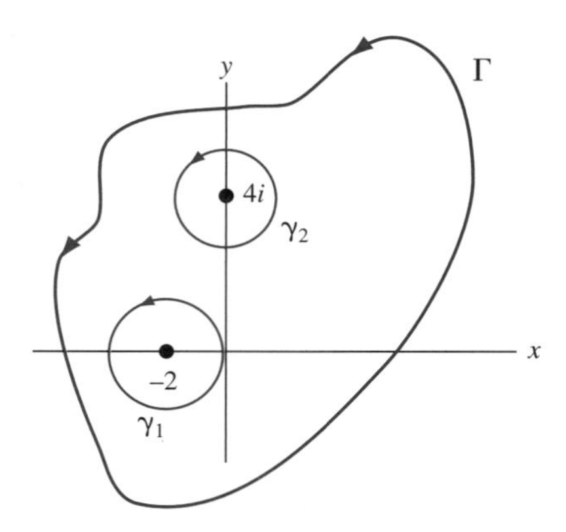

FIGURE 20.12 Γ, γ_1, and γ_2 *in Example 20.15.*

$$\oint_{\Gamma}\frac{z}{(z+2)(z-4i)}\,dz=\oint_{\gamma_1}\frac{z}{(z+2)(z-4i)}\,dz+\oint_{\gamma_2}\frac{z}{(z+2)(z-4i)}\,dz.$$

Use a partial fractions decomposition to write

$$\frac{z}{(z+2)(z-4i)}=\frac{1-2i}{5}\frac{1}{z+2}+\frac{4+2i}{5}\frac{1}{z-4i}.$$

Putting the last two equations together with Cauchy's theorem and the conclusion of Example 20.11, we have

$$\begin{aligned}\oint_{\Gamma}\frac{z}{(z+2)(z-4i)}\,dz&=\frac{1-2i}{5}\oint_{\gamma_1}\frac{1}{z+2}\,dz+\oint_{\gamma_1}\frac{4+2i}{5}\frac{1}{z-4i}\,dz\\&\quad+\frac{1-2i}{5}\oint_{\gamma_2}\frac{1}{z+2}\,dz+\frac{4+2i}{5}\oint_{\gamma_2}\frac{1}{z-4i}\,dz\\&=\frac{1-2i}{5}(2\pi i)+\frac{4+2i}{5}(2\pi i)\\&=2\pi i.\ ♦\end{aligned}$$

A proof of the extended deformation theorem can be modeled after that of the deformation theorem, except now we draw a line segment from Γ to γ_1, from γ_1 to γ_2 and so on until we come to γ_{n-1} to γ_n, and finally from γ_n to Γ, as in Figure 20.13 for $n=3$.

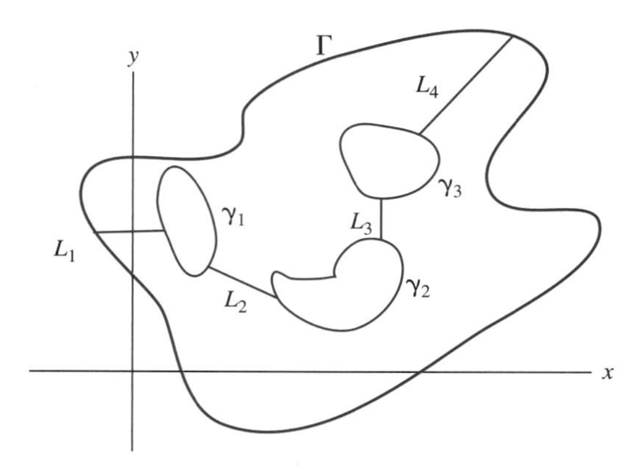

FIGURE 20.13 *Argument for the extended deformation theorem.*

20.3.7 A Variation on Cauchy's Integral Formula

We will develop another form of the Cauchy integral formula which we will use in Section 22.4 to write a complex integral formula for the inverse Laplace transform. This will be used to apply complex analysis to a diffusion problem for a solid cylinder.

THEOREM 20.9 *Cauchy Integral Formula—Second Version*

Let σ be a real number, and suppose $f(z)$ is differentiable in the half-plane $x \geq \sigma$. Suppose that there are positive numbers M and n such that

$$|z^n f(z)| \leq M$$

for $|z|$ to be sufficiently large (for example, for $|z| \geq R$ for some positive number R).

Then, for any z_0 with $\text{Re}(z_0) > \sigma$,

$$f(z_0) = -\frac{1}{2\pi} \lim_{b\to\infty} \int_{\sigma-ib}^{\sigma+ib} \frac{f(z)}{z - z_0}\, dz. \quad \blacklozenge$$

In this limit, we actually have the integral over the vertical line $x = \sigma$, oriented upward (from $-\infty$ to ∞). This integral of $f(z)/(z - z_0)$ over this line is equal to $-2\pi i f(z_0)$, hence it is in the same spirit as the Cauchy integral formula.

We will outline an argument suggesting why this formula is true. Suppose z_0 lies to the right of the line $x = \sigma$. Construct the closed rectangular path C shown in Figure 20.14 having corners $b - ib, b + ib, \sigma - ib$, and $\sigma + ib$ with b chosen to be sufficiently large, so that C encloses z_0. Let C^* be the path consisting of the upper, lower, and right sides of this rectangle, while L is the left side of the rectangle, which is the vertical line from $\sigma - ib$ to $\sigma + ib$. By the Cauchy integral formula,

$$\begin{aligned} f(z_0) &= \frac{1}{2\pi i} \oint_C \frac{f(z)}{z - z_0}\, dz \\ &= \frac{1}{2\pi i} \left[\int_{C^*} \frac{f(z)}{z - z_0}\, dz - \int_{\sigma-ib}^{\sigma+ib} \frac{f(z)}{z - z_0}\, dz \right]. \end{aligned}$$

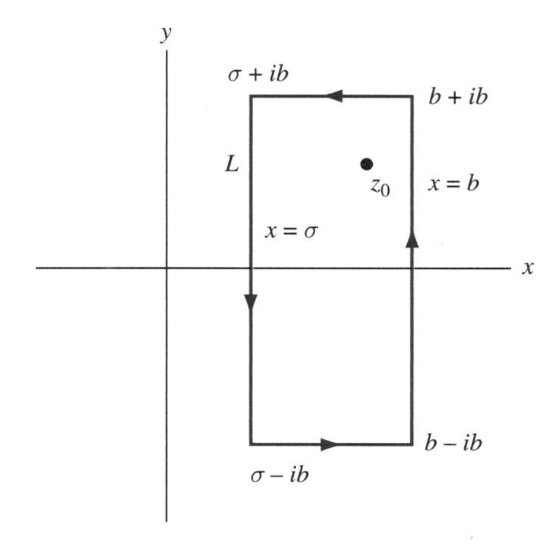

FIGURE 20.14 *C and C^* in this Cauchy integral representation.*

The negative sign on the integral from $\sigma - ib$ to $\sigma + ib$ is due to the fact that counterclockwise orientation on C requires that the integral over L be taken from $\sigma + ib$ to $\sigma - ib$. Reversing these limits of integration (as we have done) reverses the sign on the integral.

We will have the conclusion of the theorem if we can show that

$$\lim_{b\to\infty}\int_{C^*}\frac{f(z)}{z-z_0}\,dz = 0.$$

A proof of this is outlined in Problem 15.

SECTION 20.3 PROBLEMS

In each of Problems 1 through 12, evaluate $\int_\gamma f(z)\,dz$. All closed curves are positively oriented. These problems may involve Cauchy's theorem, the integral formulas, and/or the deformation theorem.

1. $f(z) = \text{Re}(z+4)$; γ is the line segment from $3+i$ to $2-5i$.
2. $f(z) = 3z^2\cosh(z)/(z+2i)^2$; γ is the circle of radius 8 about 1.
3. $f(z) = ie^z/(z-2+i)^2$; γ is the circle $|z-1| = 4$.
4. $f(z) = (z-i)^2$; γ is the semicircle of radius 1 about 0 from i to $-i$.
5. $f(z) = z\sin(3z)/(z+4)^3$; γ is the circle $|z-2i| = 9$.
6. $f(z) = 2i\bar{z}|z|$; γ is the line segment from 1 to $-i$.
7. $f(z) = (z^2-5z+i)/(z-1+2i)$; γ is the circle $|z| = 3$.
8. $f(z) = \cos(z-i)/(z+2i)^3$; γ is any closed path enclosing $-2i$.
9. $f(z) = z^4/(z-2i)$; γ is any closed path enclosing $2i$.
10. $f(z) = 2z^3/(z-2)^2$; γ is the rectangle having vertices $4 \pm i, -4 \pm i$.
11. $f(z) = -(2+i)\sin(z^4)/(z+4)^2$; γ is any closed path enclosing -4.
12. $f(z) = \sin(z^2)/(z-5)$; γ is any closed path enclosing 5.
13. Evaluate

$$\int_0^{2\pi} e^{\cos(\theta)}\cos(\sin(\theta))\,d\theta.$$

Hint: Consider $\int_\gamma (e^z/z)dz$ with γ as the unit circle about the origin. Evaluate the integral once using Cauchy's integral formula and then directly by using coordinate functions for γ.

14. Use the extended deformation theorem to evaluate

$$\oint_\gamma \frac{z-4i}{z^3+4z}\,dz$$

where γ is any closed path enclosing the origin, $-2i$ and $2i$.

15. Complete a proof of Theorem 20.9 by filling in the details of the following argument. Use a hypothesis of the theorem to show that

$$\left|\frac{z^n f(z)}{z-z_0}\right| \le M$$

for $|z|$ to be sufficiently large. Do some manipulation to show that

$$\left|\frac{f(z)}{z-z_0}\right| \le \frac{M}{|z|^{n+1}|1-(z_0/z)|}$$

for $|z|$ to be sufficiently large. Now require that b be large enough that $b > 2|z_0|$ to show that

$$\left|\frac{f(z)}{z-z_0}\right| \le \frac{M}{b^{n+1}}.$$

Taking into account the length of C^*, show that

$$\left|\int_{C^*}\frac{f(z)}{z-z_0}\,dz\right| \le \frac{2M}{b^n}\left(4-\frac{2\sigma}{b}\right).$$

Finally, show that the right side of this inequality approaches 0 as $b \to \infty$.

POWER SERIES THE LAURENT EXPANSION

CHAPTER 21

Series Representations of Functions

There are two types of series expansions that are important for working with complex functions. The first is the power series.

21.1 Power Series

We will precede a discussion of power series with some facts about series of complex numbers.

Sequences and Series of Complex Numbers

We will assume some familiarity with real sequences and series.

Suppose z_n is a complex number for each positive integer n. If we write $z_n = x_n + iy_n$, then the complex sequence z_n converges to $L = c + id$ exactly when

$$\lim_{n\to\infty} x_n = c \text{ and } \lim_{n\to\infty} y_n = d.$$

In this case, we write

$$\lim_{n\to\infty} z_n = L.$$

This reduces every complex sequence to a consideration of two real sequences.

Complex series are treated similarly in terms of their real counterparts. Suppose $\sum_{n=1}^{\infty} c_n$ is a series of complex numbers. Write $c_n = a_n + ib_n$. Then

$$\sum_{n=1}^{\infty} c_n \text{ converges to } L = A + iB$$

if and only if

$$\sum_{n=1}^{\infty} a_n = A \text{ and } \sum_{n=1}^{\infty} b_n = B.$$

This reduces questions about series of complex numbers to questions about real series to which standard tests (comparison, integral test, ratio test, and others) may apply. In particular, if $\sum_{n=1}^{\infty} c_n$

converges, then $\lim_{n\to\infty} a_n = \lim_{n\to\infty} b_n = 0$, so $\lim_{n\to\infty} c_n = 0$ also. As with real series, the general term c_n of a convergent complex series must have a limit of 0 as $n \to \infty$.

We say that $\sum_{n=1}^{\infty} c_n$ *converges absolutely* if the real series $\sum_{n=1}^{\infty} |c_n|$ converges. As with real series, absolute convergence of a complex series implies its convergence. For suppose $\sum_{n=1}^{\infty} |c_n|$ converges. Since $|a_n| \le |c_n|$, then $\sum_{n=1}^{\infty} |a_n|$ converges by comparison, so $\sum_{n=1}^{\infty} a_n$ converges. Similarly, $|b_n| \le |c_n|$, so $\sum_{n=1}^{\infty} b_n$ converges, and therefore, $\sum_{n=1}^{\infty} c_n$ converges.

Power Series and Taylor Series

A *power series* is a series of the form

$$\sum_{n=0}^{\infty} c_n(z - z_0)^n = c_0 + c_1(z - z_0) + c_2(z - z_0)^2 + \cdots.$$

The complex numbers c_n are the *coefficients* of the power series, and z_0 is its *center*. A fundamental issue about any power series is determination of those values of z for which it converges. We will show that, if a power series converges at some point z_1 different from z_0, then it must converge absolutely at all points closer to z_0 than z_1.

THEOREM 21.1 *Convergence of Power Series*

Suppose $\sum_{n=0}^{\infty} c_n(z - z_0)^n$ converges at z_1 different from z_0. Then this series converges absolutely at all z satisfying

$$|z - z_0| < |z_1 - z_0|. \ \blacklozenge$$

Proof Because $\sum_{n=0}^{\infty} c_n(z_1 - z_0)^n$ converges,

$$\lim_{n\to\infty} c_n(z_1 - z_0)^n = 0.$$

This means that we can make the terms of the series as small in magnitude as we like by choosing n to be large enough. In particular, for some N,

$$|c_n(z_1 - z_0)^n| < 1 \text{ if } n \ge N.$$

Then, for $n \ge N$,

$$\begin{aligned}
|c_n(z - z_0)^n| &= \left|\frac{(z - z_0)^n}{(z_1 - z_0)^n}\right| |c_n(z_1 - z_0)^n| \\
&\le \left|\frac{(z - z_0)^n}{(z_1 - z_0)^n}\right| \\
&= \left|\frac{z - z_0}{z_1 - z_0}\right|^n < 1
\end{aligned}$$

because $|z - z_0| < |z_1 - z_0|$. Then the geometric series

$$\sum_{n=0}^{\infty} \left|\frac{z - z_0}{z_1 - z_0}\right|^n$$

converges. By comparison,

$$\sum_{n=N}^{\infty} |c_n(z - z_0)^n|$$

converges. But then

$$\sum_{n=0}^{\infty} |c_n(z-z_0)^n|$$

converges, so $\sum_{n=0}^{\infty} c_n(z-z_0)^n$ converges absolutely. ♦

This theorem implies that there are exactly three possibilities for convergence of power series.

1. It may be that the series does not converge for any points other than z_0. In this case we say that the power series has a *radius of convergence zero*, converging only at its center.
2. The power series may converge for all z. In this case, we say that it has *infinite radius of convergence*.
3. The power series may converge for some points other than z_0 but also diverge at some points (that is, Cases (1) and (2) do not hold). Let ζ be the closest point to z_0 at which the series diverges, and let $R = |\zeta - z_0|$.
   - If $|z - z_0| < R$, then the power series must converge at z. Otherwise this open disk would contain a point at which the series diverges, and this point would be closer to z_0 than ζ.
   - If $|z - z_0| > R$, then the power series must diverge at z. For if it converged at such a z, then it would converge at all points closer to z_0 than z, hence also at ζ, a contradiction.

Therefore, in this case, there is a number R such that the power series converges within the disk $|z - z_0| < R$, and diverges outside this disk. We call the open disk $|z - z_0| < R$ the *open disk of convergence* of the power series, and R is the *radius of convergence*.

At specific points on the circle $|z - z_0| = R$, the power series may converge or diverge. This would have to be tested for each point and each power series.

These cases can be consolidated by setting $R = 0$ in Case (1) and $R = \infty$ in Case (2). In this case, the inequality

$$|z - z_0| < \infty$$

is interpreted to mean the entire complex plane, since all points z are at a finite distance from z_0.

Sometimes the radius of convergence of a power series can be computed using the ratio test.

EXAMPLE 21.1

The power series

$$\sum_{n=0}^{\infty} (-1)^n \frac{2^n}{n+1}(z-1+2i)^n$$

has center $1 - 2i$. Look at the magnitude of the ratio of successive terms:

$$\left|\frac{(-1)^{n+1}(2^{n+1}/(n+1))(z-1+2i)^{n+1}}{(-1)^n(2^n/(n+1))(z-1+2i)^n}\right| = \left|\frac{2(n+1)}{n+2}(z-1+2i)\right|$$

$$\to 2|z-1+2i| \text{ as } n \to \infty.$$

By the ratio test, this real series converges if this limit is less than 1 and diverges if this limit is greater than 1. Therefore, the power series converges if

$$|z-1+2i| < \frac{1}{2}$$

and diverges if $|z - 1 + 2i| > 1/2$. The radius of convergence is $1/2$, and the open disk of convergence is the disk $|z - 1 + 2i| < 1/2$. ♦

If we apply this method and the limit of the magnitude of successive terms is zero, then the power series has infinite radius of convergence.

A power series can be differentiated and integrated term by term within its open disk of convergence.

THEOREM 21.2 ***Differentiation and Integration of Power Series***

Let f be a function defined by

$$f(z) = \sum_{n=0}^{\infty} c_n (z - z_0)^n$$

for z in $D : |z - z_0| < R$. Then

1.

$$f'(z) = \sum_{n=1}^{\infty} n c_n (z - z_0)^{n-1} \text{ for } z \text{ in } D.$$

Furthermore, this power series for $f'(z)$ has the same radius of convergence as the power series for $f(z)$.

2. If γ is a path within D, then

$$\int_{\gamma} f(z)\, dz = \sum_{n=0}^{\infty} c_n \int_{\gamma} (z - z_0)^n\, dz. \quad \blacklozenge$$

We now want to address the possibility of representing a function as a power series about a point.

THEOREM 21.3 ***Taylor Expansion***

Suppose f is differentiable on an open disk $D : |z - z_0| < R$. Then, for z in D,

$$f(z) = \sum_{n=0}^{\infty} c_n (z - z_0)^n,$$

where for $n = 0, 1, 2, \cdots,$

$$c_n = \frac{f^{(n)}(z_0)}{n!}.$$

Furthermore, this power series converges absolutely in D. ♦

c_n is the nth *Taylor coefficient of* f *at* z_0, and this power series is called the *Taylor series or expansion of* f *about* z_0. In the case $z_0 = 0$, the Taylor series is also known as the *Maclaurin* series.

In Theorem 21.3, R can be ∞, in which case $f(z)$ is differentiable for all z, and the Taylor series representation of $f(z)$ is valid for all z.

Proof Let z be in D, and choose a number r with

$$|z - z_0| < r < R.$$

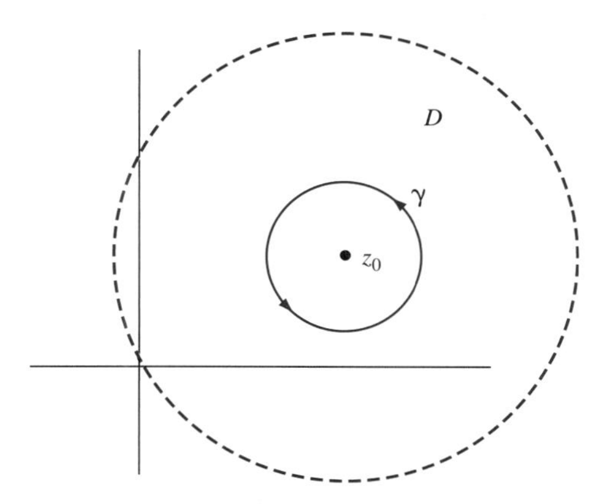

FIGURE 21.1 *γ in the proof of Theorem 21.3.*

Let γ be the circle of radius r about z_0 (Figure 21.1), so γ has a center at z_0 and encloses z. By the Cauchy integral formula,

$$f(z) = \frac{1}{2\pi i}\oint_\gamma \frac{f(w)}{w-z}\,dw.$$

An algebraic manipulation allows us to write

$$\frac{1}{w-z} = \frac{1}{w-z_0-(z-z_0)} = \frac{1}{w-z_0}\frac{1}{1-\frac{z-z_0}{w-z_0}}.$$

Now

$$\left|\frac{z-z_0}{w-z_0}\right| < 1,$$

so we can write the convergent geometric series

$$\frac{1}{w-z} = \sum_{n=0}^{\infty}\left(\frac{z-z_0}{w-z_0}\right)^n$$
$$= \sum_{n=0}^{\infty}\frac{1}{(w-z_0)^{n+1}}(z-z_0)^n.$$

Then

$$\frac{f(w)}{w-z} = \sum_{n=0}^{\infty}\frac{f(w)}{(w-z_0)^{n+1}}(z-z_0)^n.$$

It can be shown that this series can be integrated term by term, so

$$f(z) = \frac{1}{2\pi i}\oint_\gamma \frac{f(w)}{w-z}\,dw$$
$$= \frac{1}{2\pi i}\oint_C\left(\sum_{n=0}^{\infty}\frac{f(w)}{(w-z_0)^{n+1}}(z-z_0)^n\right)dw$$
$$= \sum_{n=0}^{\infty}\left(\frac{1}{2\pi i}\oint_\gamma \frac{f(w)}{(w-z_0)^{n+1}}\,dw\right)(z-z_0)^n$$
$$= \sum_{n=0}^{\infty}\frac{f^n(z_0)}{n!}(z-z_0)^n.$$

At the last line, we used the Cauchy representation formula for higher derivatives. ♦

In some cases, we have extended important real-valued functions to the complex plane. Examples are e^z; $\sin(z)$ and $\cos(z)$, which agree with e^x; and $\sin(x)$ and $\cos(x)$, respectively, when $z = x$ is real. In such cases, we can obtain the power series expansion of the complex function directly from the expansion for the real-valued function, if this is known. For example, using the familiar Maclaurin series, we can immediately write

$$e^z = \sum_{n=0}^{\infty} \frac{1}{n!} z^n,$$

$$\sin(z) = \sum_{n=0}^{\infty} \frac{(-1)^n}{(2n+1)!} z^{2n+1},$$

and

$$\cos(z) = \sum_{n=0}^{\infty} \frac{(-1)^n}{(2n)!} z^{2n}.$$

We rarely compute Taylor coefficients as $f^n(z_0)/n!$ in expanding $f(z)$ about z_0. If possible, we use known series with algebra and calculus operations. Of course, term by term differentiation and integration of power series applies to Taylor series.

EXAMPLE 21.2

$$e^{z^2} = \sum_{n=0}^{\infty} \frac{1}{n!} z^{2n}$$

by replacing z with z^2 in the Maclaurin expansion of e^z. ♦

EXAMPLE 21.3

Start with the familiar geometric series

$$\frac{1}{1-z} = \sum_{n=0}^{\infty} z^n = 1 + z + z^2 + z^3 + \cdots$$

for $|z| < 1$. Differentiate to obtain

$$\frac{1}{(1-z)^2} = \sum_{n=1}^{\infty} nz^{n=1} = 1 + 2z + 3z^2 + \cdots$$

for $|z| < 1$. Differentiate again to obtain

$$\frac{2}{(1-z)^3} = \sum_{n=2}^{\infty} n(n-1)z^{n-2} = 2 + 6z + 12z^2 + \cdots$$

for $|z| < 1$.

Replacing z with $-z$, we obtain

$$\frac{1}{1+z} = \sum_{n=0}^{\infty} (-1)^n z^n$$

for $|z| < 1$. This is also called a geometric series. Now differentiation yields

$$\frac{-1}{(1+z)^2} = \sum_{n=1}^{\infty} n(-1)^n z^{n-1}$$

for $|z| < 1$. ♦

EXAMPLE 21.4

We will use algebra and the geometric series to write the Taylor expansion of $2i/(4+iz)$ about $-3i$.

Since this expansion about $-3i$ will contain powers $(z+3i)^n$, we attempt to rearrange $2i/(4+iz)$ so that we can expand it in a geometric series involving powers of $z+3i$. Write

$$\begin{aligned}\frac{2i}{4+iz} &= \frac{2i}{4+i(z+3i)+3} \\ &= \frac{2i}{7+i(z+3i)} = \frac{2i}{7}\frac{1}{1+\frac{i(z+3i)}{7}} \\ &= \frac{2i}{7}\sum_{n=0}^{\infty}(-1)^n\left(\frac{i}{7}(z+3i)\right)^n \\ &= \sum_{n=0}^{\infty}\frac{2(-1)^{n+1}i^{n+1}}{7^{n+1}}(z+3i)^n.\end{aligned}$$

This expansion is valid for

$$\left|\frac{i}{7}(z+3i)\right| < 1$$

or

$$|z+3i| < 7.$$

This expansion has center $-3i$ and a radius of convergence of 7. ♦

We could have predicted the radius of convergence of this power series expansion without actually writing the series. The function being expanded is $f(z) = 2i/(4+iz)$, which is differentiable for all z except $z = 4i$. The radius of convergence of the expansion of $f(z)$ about $-3i$ will be the distance between the center, $-3i$, and the closest point to $-3i$ at which $f(z)$ is not differentiable, in this case $4i$. This distance is 7, which we have just seen from the expansion itself is the radius of convergence.

As a less obvious example, consider $g(z) = 1/\sin(z)$. This is differentiable for all z except integer multiples of π. We can (in theory) expand $g(z)$ in a power series about $3+i$. The radius of convergence of this series will be the distance between $3+i$ and the point nearest $3+i$ at which $g(z)$ is not differentiable. This point is π, so the radius of convergence is the distance between $3+i$ and π (or $\sqrt{(3-\pi)^2+1}$).

We will conclude this section with some consequences of power series expansions which display important properties of complex functions.

Existence of an Antiderivative

If f is differentiable on an open disk D about z_0, we claim that there must exist a differentiable function F such that $F'(z) = f(z)$ for all z in D. F is an *antiderivative* of f.

To construct $F(z)$, expand $f(z)$ in a power series about z_0 on D as

$$f(z) = \sum_{n=0}^{\infty} c_n(z-z_0)^n$$

for z in D. Now let

$$F(z) = \sum_{n=0}^{\infty} \frac{1}{n+1} c_n (z - z_0)^{n+1}.$$

It is easy to check that $F'(z) = f(z)$ for z in D.

Isolated Zeros

The Taylor expansion of a function about a point gives us important information about the zeros of the function. A number ζ is a *zero* of f if $f(\zeta) = 0$. A zero ζ is *isolated* if there is an open disk about ζ containing no other zero of f.

For example, $\sin(z)$ has isolated zeros at integer multiples of π. By contrast, let

$$g(z) = \begin{cases} \sin(1/z) & \text{for } z \neq 0 \\ 0 & \text{for } z = 0. \end{cases}$$

Then g has zeros at 0 and $1/n\pi$ for each nonzero integer n. 0 is not an isolated zero, because every disk about 0 contains zeros $1/n\pi$, which are arbitrarily close to 0 for n sufficiently large. We claim that this behavior of $g(z)$ at 0 prevents $g(z)$ from being differentiable there.

THEOREM 21.4

Let f be differentiable on a domain G, and let ζ be a zero of f in G. Then either ζ is an isolated zero or there is an open disk about ζ on which $f(z)$ is identically zero. ♦

This means that a differentiable complex function that is not identically zero on a domain can have only isolated zeros there.

Proof Write the power series expansion of f about ζ as

$$f(z) = \sum_{n=0}^{\infty} c_n (z - \zeta)^n$$

in some open disk D in G centered at ζ. There are two cases.

First, if every $c_n = 0$, then $f(z) = 0$ throughout D.

Thus, suppose some coefficients are not zero. Let m be the smallest integer such that $c_m \neq 0$. Then $c_0 = c_1 = \cdots = c_{m-1} = 0$, and for z in D,

$$f(z) = \sum_{n=m}^{\infty} c_n (z - \zeta)^n = (z - \zeta)^m \sum_{n=0}^{\infty} c_{n+m} (z - \zeta)^n.$$

Next, let $g(z) = \sum_{n=0}^{\infty} c_{n+m}(z - \zeta)^n$. Then g is differentiable on D and $g(\zeta) = c_m \neq 0$. Furthermore,

$$f(z) = (z - \zeta)^m g(z).$$

Because $g(\zeta) \neq 0$ there is some open disk K about ζ in which $g(z) \neq 0$. But then $f(z) \neq 0$ if z is in K and is different from ζ. Therefore, ζ is an isolated zero. ♦

If ζ is a zero of f, then the smallest m such that $c_m \neq 0$ in the Taylor expansion of f about ζ is called the *order* of the zero ζ. Because the Taylor coefficients preceding c_m must be zero, then $f^{(j)}(\zeta)=0$ for $j=0,1,\cdots,m-1$, while $f^{(m)}(\zeta)\neq 0$. Therefore, the order of a zero is the smallest integer such that the derivative of that order is nonzero at ζ.

In the proof of Theorem 21.4, we actually showed that, in some disk about an isolated zero of f of order m, we can write

$$f(z)=(z-z)^m g(z)$$

where $g(z)\neq 0$ on this disk. This fact is important in its own right.

EXAMPLE 21.5

Let $f(z)=z^2\sin(z)$. Then f has an isolated zero at 0. Compute

$$f'(z)=2z\sin(z)+z^2\cos(z),$$
$$f''(z)=2\sin(z)+4z\cos(z)-z^2\sin(z),$$

and

$$f'''(z)=6\cos(z)-6z\sin(z)-z^2\cos(z).$$

Now $f(0)=f'(0)=f''(0)=0$, while $f'''(0)\neq 0$. Therefore, f has a zero of order 3 at 0.

We will write $f(z)=z^3 g(z)$ where $g(z)$ is differentiable and nonzero is on some disk about 0. Use the Maclaurin expansion to write

$$\sin(z)=z+\frac{1}{3!}z^3+\frac{1}{5!}z^5+\cdots,$$

so

$$\begin{aligned} f(z)=z^2\sin(z) &= z^3+\frac{1}{3!}z^5+\frac{1}{5!}z^7+\cdots \\ &= z^3\left(1+\frac{1}{3!}z^2+\frac{1}{5!}z^4+\cdots\right) \\ &= z^3 g(z), \end{aligned}$$

where $g(z)\neq 0$ on a disk about 0. ◆

One immediate ramification of being able to write $f(z)=(z-\zeta)^m g(z)$ with $g(z)\neq 0$ in some disk about ζ is that, under certain conditions, the orders of the zeros of products add and orders of zeros of quotients subtract (reminiscent of a logarithm). To be specific, suppose h has a zero of order m at ζ, and k has a zero of order n at ζ. Then

1. $h(z)k(z)$ has a zero of order $m+n$ at ζ.
2. If $n<m$, then $h(z)/k(z)$ has a zero of order $m-n$ at ζ.

To see why statement (1) is true, write $h(z)=(z-\zeta)^m\alpha(z)$ and $k(z)=(z-\zeta)^n\beta(z)$, where $\alpha(z)$ and $\beta(z)$ are nonzero in some open disk D about ζ. Then

$$h(z)k(z)=(z-\zeta)^{m+n}\alpha(z)\beta(z)$$

and $\alpha(z)\beta(z)\neq 0$ in D, so $h(z)k(z)$ has a zero of order $m+n$ at ζ. Statement (2) is proved similarly.

These facts will be important when we consider the order of poles as singularities of functions.

EXAMPLE 21.6

$$\frac{\cos^3(z)}{(z-\pi/2)^2}$$

has a zero of order 1 at $\pi/2$, because the numerator has a zero of order of 3 there, and the denominator has a zero of order 2. ♦

SECTION 21.1 PROBLEMS

In each of Problems 1 through 6, find the radius of convergence and open disk of convergence of the power series.

1. $\sum_{n=0}^{\infty} \frac{n^n}{(n+1)^n}(z-1+3i)^n$
2. $\sum_{n=0}^{\infty} \left(\frac{2i}{5+i}\right)^n (z+3-4i)^n$
3. $\sum_{n=0}^{\infty} \frac{i^n}{2^{n+1}}(z+8i)^n$
4. $\sum_{n=0}^{\infty} \frac{(1-i)^n}{n+2}(z-3)^n$
5. $\sum_{n=0}^{\infty} \frac{n+1}{2^n}(z+3i)^n$
6. $\sum_{n=0}^{\infty}(-1)^n \frac{1}{(2n+1)^2}(z-i)^{2n}$
7. Suppose f is differentiable in an open disk about 0 and satisfies $f''(z)=2f(z)+1$. Suppose $f(0)=1$ and $f'(0)=i$. Find the first six terms of the Maclaurin expansion of $f(z)$.
8. Find the first seven terms of the Maclaurin expansion of $f(z)=\sin^2(z)$ in four ways, as follows.
   (a) First, compute the Taylor coefficients at 0.
   (b) Find the first seven terms of the product of the Maclaurin series for $\sin(z)$ with itself.
   (c) Write $\sin^2(z)$ in terms of the exponential function and use the Maclaurin expansion of this function.
   (d) Write $\sin^2(z) = (1 - \cos(2z))/2$, and use the Maclaurin expansion of $\cos(z)$.

In each of Problems 9 through 14, find the Taylor expansion of the function about the point.

9. $z^2-3z+i;\ z=2-i$
10. $e^z - i\sin(z);\ z=0$
11. $(z-9)^2;\ 1+i$
12. $\sin(z+i);\ -i$
13. $\cos(2z);\ z=0$
14. $e^{-z};\ z=-3i$
15. Is it possible for $\sum_{n=0}^{\infty} c_n(z-2i)^n$ to converge at 0 and diverge i?
16. Is it possible for $\sum_{n=0}^{\infty} c_n(z-4+2i)^n$ to converge at i and diverge at $1+i$?
17. Show that

$$\sum_{n=0}^{\infty} \frac{1}{(n!)^2} = \frac{1}{2\pi}\int_0^{2\pi} e^{2z\cos(\theta)}\,d\theta.$$

Hint: Show that

$$\left(\frac{z^n}{n!}\right)^2 = \frac{1}{2\pi i}\oint_\gamma \frac{z^n}{n!w^{n+1}}e^{zw}\,dw$$

for $n=0,1,2,\cdots$ and γ is the unit circle about the origin.

In each of Problems 18 through 24, determine the order of the zero of the function.

18. $f(z)=\frac{(z-\pi)^5}{\sin^2(z)},\ z=\pi$
19. $f(z)=\cos^3(z);\ z=3\pi/2$
20. $f(z)=z^3\cos(z);\ z=0$
21. $f(z)=\sin(z^4)/z^2,\ z=0$
22. $f(z)=(z-\pi/2)^2\cos(z);\ z=\pi/2$
23. $f(z)=z^2\sin^2(z);\ z=0$
24. $f(z)=\cos^4(z-\pi/2),\ z=0$
25. Suppose

$$f(z)=\sum_{n=0}^{\infty} a_n(z-z_0)^n = \sum_{n=0}^{\infty} b_n(z-z_0)^n$$

in some open disk D about z_0. Show that $a_n=b_n$ for $n=0,1,2,\cdots$.

21.2 The Laurent Expansion

If f is differentiable in some disk about z_0, then $f(z)$ has a Taylor series representation about z_0. If f is not differentiable at z_0, then $f(z)$ may have a different kind of series expansion about z_0, which is a Laurent expansion. This will have important applications in evaluating integrals. First we need some terminology.

The open set of points between two concentric circles is called an *annulus*. Typically, an annulus with center z_0 is described by inequalities

$$r < |z - z_0| < R,$$

in which r is the radius of the inner circle and R is the radius of the outer circle. We allow $r = 0$, in which case the annulus $0 < |z - z_0| < R$ is the open disk of radius R about z_0 with the center removed. Such an annulus is called a *punctured disk*. We also allow $R = \infty$, in which case the annulus $r < |z - z_0| < \infty$ consists of all points outside the inner circle. An annulus $0 < |z - z_0| < \infty$ is the entire plane with z_0 removed.

We can now state the fundamental result on Laurent series.

THEOREM 21.5 The Laurent Expansion

Let f be differentiable in the annulus $r < |z - z_0| < R$ where $0 \leq r < R \leq \infty$. Then, for each z in this annulus,

$$f(z) = \sum_{n=-\infty}^{\infty} c_n (z - z_0)^n,$$

where for each integer n,

$$c_n = \frac{1}{2\pi i} \oint_{\gamma} \frac{f(z)}{(z - z_0)^{n+1}} \, dz$$

with γ as any closed path in the annulus enclosing z_0. ♦

A proof is outlined in Problem 11.

The Laurent expansion about z_0 enables us to write, in some annulus about z_0,

$$f(z) = \sum_{n=-\infty}^{-1} c_n (z - z_0)^n + \sum_{n=0}^{\infty} c_n (z - z_0)^n = h(z) + g(z),$$

where

$$h(z) = \sum_{n=-\infty}^{-1} c_n (z - z_0)^n$$
$$= \cdots + \frac{c_{-3}}{(z - z_0)^3} + \frac{c_{-2}}{(z - z_0)^2} + \frac{c_{-1}}{(z - z_0)}$$

contains all of the terms in the expansion with negative powers of $z - z_0$ and

$$g(z) = \sum_{n=0}^{\infty} c_n (z - z_0)^n$$
$$= c_0 + c_1 (z - z_0) + c_2 (z - z_0)^2 + \cdots$$

contains all the nonnegative powers of $z - z_0$. The series defining $g(z)$ is a power series about z_0 and so is a differentiable function in the open disk $|z - z_0| < R$. Any "bad" behavior of $f(z)$ near z_0 is contained in $h(z)$.

It can be shown (Problem 12) that, if

$$f(z) = \sum_{n=-\infty}^{\infty} c_n (z - z_0)^n = \sum_{n=-\infty}^{\infty} d_n (z - z_0)^n$$

in some annulus about z_0, then $c_n = d_n$ for each integer n. This means that a Laurent expansion is unique to f and z_0 and will be the same no matter how it is derived. This is important, because usually, we obtain a Laurent expansion by manipulating known series and very rarely use the integral formula to compute the coefficients.

EXAMPLE 21.7

The Laurent expansion of $e^{1/z}$ about 0 is

$$\sum_{n=0}^{\infty} \frac{1}{n!}\left(\frac{1}{z}\right)^n = 1 + \frac{1}{z} + \frac{1}{2!}\frac{1}{z^2} + \cdots,$$

which is obtained by replacing z with $1/z$ in the Maclaurin expansion of e^z. This Laurent expansion is valid for $0 < |z < \infty$, hence, in the entire plane with the origin removed. ♦

EXAMPLE 21.8

$f(z) = \cos(z)/z^5$ is differentiable in the annulus $0 < |z| < \infty$, which is the entire plane with the origin removed. We know the Taylor expansion of $\cos(z)$ about 0. Therefore, we know the Laurent expansion of $f(z)$ in $0 < |z| < \infty$ is

$$\begin{aligned} f(z) &= \frac{1}{z^5}\cos(z) = \frac{1}{z^5}\sum_{n=0}^{\infty} \frac{(-1)^n}{(2n)!} z^{2n} \\ &= \sum_{n=0}^{\infty} \frac{(-1)^n}{(2n)!} z^{2n-5} \\ &= \frac{1}{z^5} - \frac{1}{2}\frac{1}{z^3} + \frac{1}{24}\frac{1}{z} - \frac{1}{720}z + \frac{1}{40,320}z^3 - \cdots \text{ for } z \neq 0. \end{aligned}$$

We can think of

$$\frac{\cos(z)}{z^5} = h(z) + g(z),$$

where

$$h(z) = \frac{1}{z^5} - \frac{1}{2!}\frac{1}{z^3} + \frac{1}{4!}\frac{1}{z},$$

and

$$g(z) = -\frac{1}{6!}z + \frac{1}{8!}z^3 - \cdots.$$

$g(z)$ is differentiable for all z and $h(z)$ is differentiable on the plane with the origin removed, so $h(z) + g(z)$ is differentiable on the plane with the origin removed. It is the behavior of $h(z)$ near the origin that determines the behavior of $\cos(z)/z^5$ there. ♦

EXAMPLE 21.9

Let

$$f(z) = \frac{1}{(z+1)(z-3i)}.$$

Then f is differentiable except at -1 and $3i$. We will find the Laurent expansion of $f(z)$ about -1. Use a partial fractions decomposition to write

$$f(z) = \frac{-1+3i}{10}\frac{1}{z+1} + \frac{1-3i}{10}\frac{1}{z-3i}.$$

On the right, the first term is itself a Laurent expansion about -1, because it is a sum (with only one term) of powers of $z+1$. Therefore, concentrate on the second term. We will manipulate it and use a geometric series, keeping in mind that we want a series of powers of $z+1$:

$$\frac{1}{z-3i} = \frac{1}{-1-3i+(z+1)} = \frac{1}{-1-3i}\frac{1}{1-\frac{z+1}{1+3i}}$$

$$= -\frac{1}{1+3i}\sum_{n=0}^{\infty}\left(\frac{z+1}{1+3i}\right)^n = \sum_{n=0}^{\infty}\frac{-1}{(1+3i)^{n+1}}(z+1)^n.$$

This expansion is valid for

$$\left|\frac{z+1}{1+3i}\right| < 1$$

or $|z+1| < \sqrt{10}$. The Laurent expansion of $f(z)$ about -1 is

$$f(z) = \frac{-1+3i}{10}\frac{1}{z+1} - \frac{1-3i}{10}\sum_{n=0}^{\infty}\frac{-1}{1+3i}^{\,n+1}(z+1)^n$$

in the annulus $0 < |z+1| < \sqrt{10}$. Behavior of $f(z)$ as z approaches -1 is determined by the $1/(z+1)$ term in this expansion. ♦

We have emphasized that we do not want to have to use the integral formula for the c_n's to compute a Laurent expansion. This is really just the tip of the iceberg. Usually we are interested in just one term of a Laurent expansion because it will enable us to evaluate integrals. This is the theme of the next chapter.

SECTION 21.2 PROBLEMS

In each of Problems 1 through 10, write the Laurent expansion of $f(z)$ in an annulus $0 < |z - z_0| < R$ about z_0, specifying R for each problem. These should all be done by manipulating known series.

1. $(z+i)/(z-i); i$
2. $(z^2+1)/(2z-1); 1/2$
3. $z^2/(1-z); 1$
4. $\sin(z)/z^2; 0$
5. $2z/(1+z^2); i$
6. $z^2\cos(i/z); 0$
7. $(1-\cos(2z))/z^2; 0$
8. $\sinh(1/z^3); 0$
9. $e^{z^2}/z^2; 0$
10. $\sin(4z)/z; 0$
11. Fill in the details of the following proof of the Laurent expansion theorem (21.5). Let z be in the annulus, and choose r_1 and r_2 such that

$$0 < r < r_1 < r_2 < R$$

and so that the circle $\gamma_1 : |z - z_0| = r_1$ does not enclose z while the circle $\gamma_2 : |z - z_0| = r_2$ encloses z. Insert

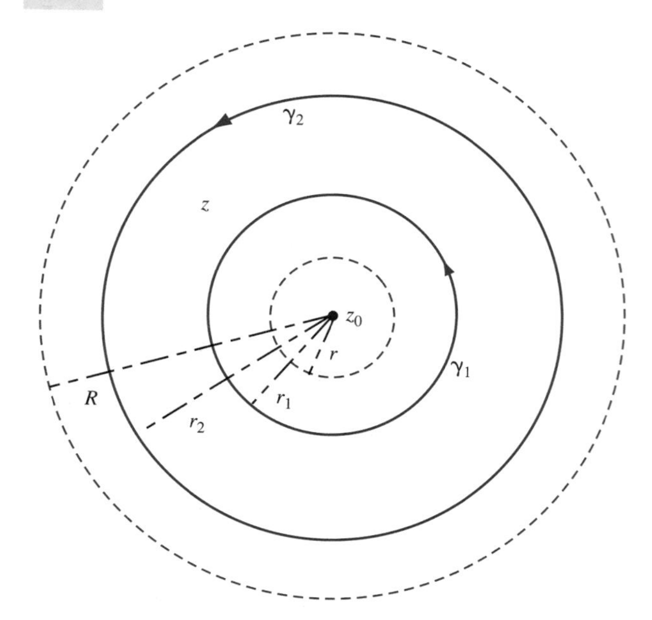

FIGURE 21.2 *Circles in the proof of Theorem 21.5 for Problem 11, Section 21.2.*

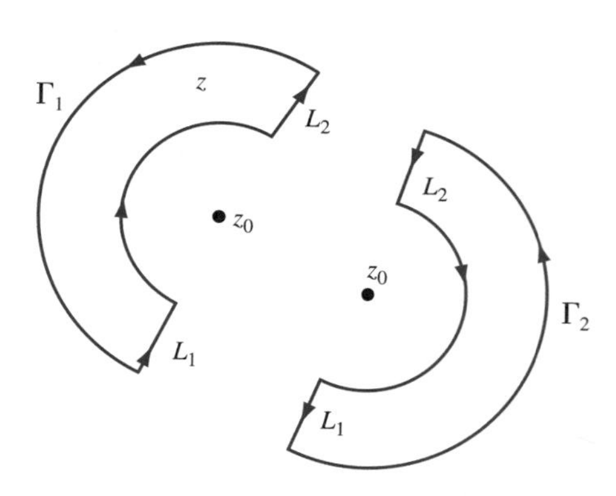

FIGURE 21.3 *Γ_1 and Γ_2 in the proof of Theorem 21.5 for Problem 11, Section 21.2.*

line segments L_1 and L_2 between these circles (Figure 21.2), forming two closed paths Γ_1 and Γ_2 in the annulus (shown separately in Figure 21.3). Show that

$$f(z) = \frac{1}{2\pi i}\oint_{\Gamma_1} \frac{f(w)}{w-z}\,dw$$

and

$$\frac{1}{2\pi i}\oint_{\Gamma_2} \frac{f(w)}{w-z}\,dw = 0.$$

Add these to obtain

$$f(w) = \frac{1}{2\pi i}\oint_{\Gamma_1} \frac{f(w)}{w-z}\,dw + \frac{1}{2\pi i}\oint_{\Gamma_2} \frac{f(w)}{w-z}\,dw$$

with counterclockwise orientation on both paths. By noting that in these integrals, the parts of the integrals over the line segments vanish (these segments are traversed in both directions), show that

$$f(z) = \frac{1}{2\pi i}\oint_{\gamma_2} \frac{f(w)}{w-z}\,dz - \frac{1}{2\pi i}\oint_{\gamma_1} \frac{f(w)}{w-z}\,dw$$

with both integrations counterclockwise over the closed path.

On γ_2, show that $|(z-z_0)/(w-z_0)| < 1$, and use the geometric series to write

$$\frac{1}{w-z} = \sum_{n=0}^{\infty} \frac{1}{(w-z_0)^{n+1}}(z-z_0)^n.$$

On γ_1, show that $|(w-z_0)/(z-z_0)| < 1$ to show that

$$\frac{1}{w-z} = -\sum_{n=0}^{\infty}(w-z_0)^n \frac{1}{(z-z_0)^{n+1}}.$$

Use these to show that

$$f(z) = \sum_{n=0}^{\infty}\left(\frac{1}{2\pi i}\oint_{\gamma_2} \frac{f(w)}{(w-z_0)^{n+1}}\,dw\right)(z-z_0)^n$$
$$+\sum_{n=0}^{\infty}\left(\frac{1}{2\pi i}\oint_{\gamma_1} f(w)(w-z_0)^n\,dw\right)\frac{1}{(z-z_0)^{n+1}}.$$

Finally, replace $n = -m-1$ in the last summation, and then use the deformation theorem to replace γ_1 and γ_2 by any closed path Γ about z_0 and in the annulus.

12. Fill in the details of the following proof of the uniqueness of Laurent expansions. Suppose, in the annulus,

$$f(z) = \sum_{n=-\infty}^{\infty} c_n(z-z_0)^n = \sum_{n=-\infty}^{\infty} b_n(z-z_0)^n,$$

where the c_n's are the Laurent coefficients. Then

$$2\pi i c_k = \oint_{\gamma} \frac{f(w)}{(w-z_0)^{k+1}}\,dw$$
$$= \oint_{\gamma}\left(\sum_{n=-\infty}^{\infty} b_n(z-z_0)^n\right)\frac{1}{(w-z_0)^{k+1}}\,dw$$
$$= \sum_{n=-\infty}^{\infty} b_n \oint_{\gamma} \frac{1}{(w-z_0)^{k-n+1}}\,dw.$$

Choose γ as a circle, and evaluate the integral in the last summation to complete the proof.

CHAPTER 22

SINGULARITIES THE RESIDUE THEOREM
EVALUATION OF REAL INTEGRALS
RESIDUES AND THE INVERSE LAPLACE TRANSFORM

Singularities and the Residue Theorem

22.1 Singularities

We will use the Laurent expansion to classify points at which complex functions are not differentiable.

We say that f has an *isolated singularity* at z_0 if f is differentiable in an annulus $0 < |z - z_0| < R$, but not at z_0 itself.

For example, $1/z$ has an isolated singularity at $z = 0$, and $\sin(z)/(z - \pi)$ has an isolated singularity at $z = \pi$.

We will identify three different kinds of isolated singularities, depending on the coefficients in the Laurent expansion.

Classification of Singularities

Suppose f has an isolated singularity at z_0. Let the Laurent expansion of $f(z)$ in an annulus $0 < |z - z_0| < R$ be

$$f(z) = \sum_{n=-\infty}^{\infty} c_n (z - z_0)^n.$$

1. z_0 is a *removable singularity* if $c_n = 0$ for every negative integer n.

2. z_0 is a *pole of order m*, with m a positive integer, if $c_{-m} \neq 0$, but

$$c_{-m-1} = c_{-m-2} = c_{-m-3} = \cdots = 0.$$

3. z_0 is an *essential singularity* if $c_{-n} \neq 0$ for infinitely many positive integers n.

Thus, z_0 is a removable singularity if the Laurent expansion about z_0 is actually a power series, a pole of order m if $1/(z - z_0)^m$ is the largest power of $1/(z - z_0)$ appearing in the Laurent expansion of $f(z)$ about z_0, and an essential singularity if the expansion of $f(z)$ about z_0 has infinitely many powers of $1/(z - z_0)$ with nonzero coefficients.

EXAMPLE 22.1

Let $f(z) = (1 - \cos(z))/z$. Then f is differentiable for all $z \neq 0$ and is not defined at 0. Using the Maclaurin series for $\cos(z)$, the Laurent expansion of $f(z)$ around zero is

$$\begin{aligned} f(z) &= \frac{1 - \cos(z)}{z} = \frac{1}{z}\left(1 - \sum_{n=0}^{\infty} \frac{(-1)^n}{(2n)!} z^{2n}\right) \\ &= \frac{1}{z}\left(\frac{1}{2!}z^2 - \frac{1}{4!}z^4 + \frac{1}{6!}z^6 - \cdots\right) \\ &= \frac{1}{2!}z - \frac{1}{4!}z^3 + \frac{1}{6!}z^5 - \cdots. \end{aligned}$$

Since this is a power series about 0, f has a removable singularity at 0. We can define $f(0) = 0$, the value of this power series at $z = 0$, and this "extended" f is differentiable at 0. ♦

EXAMPLE 22.2

Let $f(z) = 1/(z + i)^2$. This function is its own Laurent expansion about $-i$, where it is not differentiable. Because the largest power of $1/(z + i)$ appearing in this expansion is $1/(z + i)^2$, f has a pole of order 2 at $-i$. There is no way to define $f(-i)$, so the extended function is differentiable at $-i$. ♦

EXAMPLE 22.3

Let $g(z) = \sin(z)/z^5$. This function is differentiable except at $z = 0$, where it is not defined. Using the Maclaurin expansion of $\sin(z)$, the Laurent expansion of f about 0 is

$$\begin{aligned} f(z) &= \sum_{n=0}^{\infty} \frac{(-1)^n}{(2n+1)!} \frac{z^{2n+1}}{z^5} = \sum_{n=0}^{\infty} \frac{(-1)^n}{(2n+1)!} z^{2n-4} \\ &= \frac{1}{z^4} - \frac{1}{3!}\frac{1}{z^2} + \cdots \end{aligned}$$

for $z \neq 0$. The highest power of $1/z$ appearing in this expansion is 4, so f has a pole of order 4 at 0. ♦

EXAMPLE 22.4

The Laurent expansion of $e^{1/z}$ about 0 is

$$e^{1/z} = \sum_{n=0}^{\infty} \frac{1}{n!} \frac{1}{z^n}$$

for $z \neq 0$. f has an essential singularity at 0, because infinitely many powers of $1/z$ appear in this expansion. ♦

A pole of order 1 is called a *simple pole*, and a pole of order 2 is a *double pole*.

We do not want to have to resort to a Laurent expansion to classify a singularity. For the rest of this section, we will explore other ways to do this.

THEOREM 22.1 ***Condition for a Pole of Order*** m

Let f be differentiable in $0 < |z - z_0| < R$. Then f has a pole of order m at z_0 if and only if

$$\lim_{z \to z_0} (z - z_0)^m f(z)$$

exists and is finite and nonzero. ♦

Proof We can understand how this condition arises by manipulating the Laurent expansion of $f(z)$ about z_0:

$$f(z) = \sum_{n=-\infty}^{\infty} c_n (z - z_0)^n$$

for $0 < |z - z_0| < R$. If f has a pole of order m at z_0, then $c_{-m} \neq 0$ and $c_{-m-1} = c_{m-2} = \cdots = 0$, so the Laurent about the z_0 expansion is

$$f(z) = \frac{c_{-m}}{(z - z_0)^m} + \frac{c_{-m+1}}{(z - z_0)^{m-1}} + \cdots.$$

Then

$$(z - z_0)^m f(z) = c_{-m} + c_{-m+1}(z - z_0) + c_{-m+2}(z - z_0)^2 + \cdots,$$

so

$$\lim_{z \to z_0} (z - z_0)^m f(z) = c_{-m} \neq 0. \ ♦$$

We will omit the details of the proof of the converse.

To illustrate the idea, look again at Example 22.3 with $g(z) = \sin(z)/z^5$ and $z_0 = 0$. Compute

$$\lim_{z \to 0} z^4 g(z) = \lim_{z \to 0} z^4 \frac{\sin(z)}{z^5} = \lim_{z \to 0} \frac{\sin(z)}{z} = 1,$$

which is a limit that can be seen by using the Maclaurin expansion of $\sin(z)$ about 0 to write

$$\frac{\sin(z)}{z} = 1 - \frac{z^2}{3!} + \frac{z^4}{5!} - \cdots.$$

Theorem 22.1 tells us that $\sin(z)/z^5$ has a pole of order 4 at 0, as we found in Example 22.3 by examining the Laurent expansion. ♦

If $f(z)$ is a quotient of functions, it is natural to look for poles at places where the denominator vanishes, that is, where the denominator has a zero. With some care, this strategy is effective.

THEOREM 22.2 Poles of Quotients (1)

Let $f(z) = h(z)/g(z)$ where h and g are differentiable in some open disk about z_0. Suppose that $h(z_0) \neq 0$ but $g(z)$ has a zero of order m at z_0. Then f has a pole of order m at z_0. ♦

EXAMPLE 22.5

Let

$$f(z) = \frac{1 + e^{z^2} + 4z^3}{\sin^6(z)}.$$

Then f has a pole of order 6 at 0 because the numerator is differentiable and nonzero at $z = 0$, while the denominator is differentiable and has a zero of order 6 at 0. ♦

EXAMPLE 22.6

Let

$$f(z) = \frac{1}{\cos^3(z)}.$$

Then f has a pole of order 3 at each zero of $\cos(z)$, which are the numbers $z = (2n + 1)\pi/2$ for integer n. ♦

Theorem 22.2 does not apply if the numerator also vanishes at z_0. The example $f(z) = \sin(z)/z^5$ is instructive. The numerator has a zero of order 1 at 0, the denominator has a zero of order 5 at 0, but the quotient has a pole of order 4 at 0. It appears that the orders of the zeros of the numerator and denominator subtract to give the order of the pole. That is, zeros appear to cancel (recall the observations about addition and subtraction of orders of zeros in quotients at the end of Chapter 21). This is indeed the case.

THEOREM 22.3 Poles of Quotients (2)

Let $f(z) = h(z)/g(z)$, and suppose h and g are differentiable in some open disk about z_0. Let h have a zero of order k at z_0, and let g have a zero of order m at z_0 with $m > k$. Then f has a pole of order $m - k$ at z_0. ♦

If $m = k$ in Theorem 22.3, f has a removable singularity at z_0 (recall Example 22.1). If $m < k$, then f does not have a pole at z_0.

EXAMPLE 22.7

Let

$$f(z) = \frac{(z - 3\pi/2)^4}{\cos^7(z)}.$$

Here $h(z) = (z - 3\pi/2)^4$ has a zero of order 4 at $3\pi/2$, and $g(z) = \cos^7(z)$ has a zero of order 7 at $3\pi/2$. Therefore, f has a pole of order 3 at $3\pi/2$. ♦

EXAMPLE 22.8

$f(z) = \tan^3(z)/z^9$ has a pole of order 6 at 0, because the numerator has a zero of order 3 there and the denominator has a zero of order 9. ♦

EXAMPLE 22.9

Let

$$f(z) = \frac{1}{\cos^4(z)(z - \pi/2)^3}.$$

Then f has a pole of order 7 at $\pi/2$. f also has poles of order 4 at each $(2n+1)\pi/2$ with n as any nonzero integer. ♦

SECTION 22.1 PROBLEMS

In each of Problems 1 through 12, determine all singularities of the function and classify each singularity. In the case of a pole, give the order of the pole.

1. $(z - i)/(z^2 + 1)$
2. $\tan(z)$
3. $z/(z^4 - 1)$
4. $e^{1/z(z+1)}$
5. $\sec(z)$
6. $\sin(z)/\sinh(z)$
7. $e^{1/z}(z + 2i)$
8. $z/(z+1)^2$
9. $\cos(z)/z^2$
10. $\dfrac{4\sin(z+2)}{(z+i)^2(z-i)}$
11. $\dfrac{\cos(2z)}{(z-1)^2(z^2+1)}$
12. $\sin(z)/(z - \pi)$
13. Let f be differentiable at z_0 and $f(z_0) \neq 0$. Let g have a pole of order m at z_0. Show that fg has a pole of order m at z_0.

22.2 The Residue Theorem

We will use singularities and a single term of the Laurent expansion to develop a powerful method for evaluating integrals.

Suppose f is differentiable in an annulus $0 < |z - z_0| < R$ and has an isolated singularity at z_0. Let γ be a closed path in this annulus enclosing z_0. We want to evaluate $\oint_\gamma f(z)dz$. At least in theory, we can write the Laurent expansion

$$f(z) = \sum_{n=-\infty}^{\infty} c_n (z - z_0)^n.$$

Recall the formula for the c_n's:

$$c_n = \frac{1}{2\pi i}\oint_\gamma \frac{f(z)}{(z - z_0)^{n+1}}\,dz$$

for $n = \cdots, -1, -2, 0, 1, 2, \cdots$. The coefficient of $1/(z - z_0)$ in the expansion is

$$c_{-1} = \frac{1}{2\pi i}\int_\gamma f(z)\,dz.$$

This means that

$$\oint_\gamma f(z)\,dz = 2\pi i c_{-1}.$$

We will know the integral if we just know c_{-1}! This one term of the Laurent expansion wins the game!

The coefficient of $1/(z-z_0)$ in the Laurent expansion of f about z_0 is called the *residue* of f at z_0 and is denoted $\text{Res}(f, z_0)$.

What we have so far is that $\oint_\gamma f(z)\,dz = 2\pi i \text{Res}(f, z_0)$ if z_0 is the only singularity of f enclosed by γ. This is of limited value. The power of the residue theorem is that it allows for any finite number of singularities of f to be enclosed by γ.

THEOREM 22.4 The Residue Theorem

Let γ be a closed path, and suppose f is differentiable on γ and all points enclosed by γ, except for $z_1, \cdots, z_n$, which are all of the isolated singularities of f enclosed by γ. Then

$$\oint_\gamma f(z)\,dz = 2\pi i \sum_{j=1}^{n} \text{Res}(f, z_j). \blacklozenge$$

The proof is immediate. Enclose each z_j by a small circle C_j that does not intersect any of the other circles or γ. Because C_j encloses just one singularity, z_j, $\oint_{C_j} f(z)dz = 2\pi i \text{Res}(f, z_j)$. By the extended deformation theorem,

$$\oint_\gamma f(z)dz = \sum_{j=1}^{n} \oint_{C_j} f(z)dz = 2\pi i \sum_{j=1}^{n} \text{Res}(f, z_j).$$

The residue theorem is as effective as our ability to evaluate residues. We do not want to have to write a Laurent series about each singularity to pick off the coefficient of each $1/(z-z_j)$. We will now develop efficient ways to compute residues at poles.

THEOREM 22.5 Residue at a Simple Pole

If f has a simple pole at z_0, then

$$\text{Res}(f, z_0) = \lim_{z \to z_0} (z - z_0) f(z). \blacklozenge$$

To see why this is true, the Laurent expansion of f about a simple pole z_0 has the form

$$f(z) = \frac{c_{-1}}{z - z_0} + \sum_{n=0}^{\infty} c_n (z - z_0)^n$$

in some annulus about z_0. Then

$$(z - z_0) f(z) = c_{-1} + \sum_{n=0}^{\infty} c_n (z - z_0)^{n+1},$$

so $c_{-1} = \lim_{z \to z_0} (z - z_0) f(z)$.

EXAMPLE 22.10

$f(z)=\sin(z)/z^2$ has a simple pole at 0, and

$$\operatorname{Res}(f,0)=\lim_{z\to 0} zf(z)=\lim_{z\to 0}\frac{\sin(z)}{z}=1.$$

Because 0 is the only singularity of f, if γ is any closed path enclosing the origin, then

$$\oint_\gamma \frac{\sin(z)}{z^2}\,dz=2\pi i\operatorname{Res}(f,0)=2\pi i. \ \blacklozenge$$

Here is an alternative version of Theorem 22.5.

COROLLARY 22.1

Let $f(z)=h(z)/g(z)$ where h is continuous at z_0 and $h(z_0)\neq 0$. Suppose g is differentiable at z_0 and has a simple zero there. Then f has a simple pole at z_0, and

$$\operatorname{Res}(f,z_0)=\frac{h(z_0)}{g'(z_0)}. \ \blacklozenge$$

The fact that f has a simple pole at z_0 follows from Theorem 22.2. By Theorem 22.5, because $g(z_0)=0$, we can write

$$\begin{aligned}\operatorname{Res}(f,z_0)&=\lim_{z\to z_0}(z-z_0)f(z)\\&=\lim_{z\to z_0}(z-z_0)\frac{h(z)}{g(z)}\\&=\lim_{z\to z_0}\frac{h(z)}{(g(z)-g(z_0))/(z-z_0)}=\frac{h(z_0)}{g'(z_0)}.\end{aligned}$$

EXAMPLE 22.11

Let

$$f(z)=\frac{4iz-1}{\sin(z)}.$$

Then f has a simple pole at π, and by the corollary,

$$\operatorname{Res}(f,\pi)=\frac{4i\pi-1}{\cos(\pi)}=1-4\pi i.$$

In fact, f has a simple pole at $n\pi$ for each integer n, and

$$\operatorname{Res}(f,n\pi)=\frac{4in\pi-1}{\cos(n\pi)}=(-1)^n(4in\pi-1). \ \blacklozenge$$

EXAMPLE 22.12

Evaluate $\oint_\gamma f(z)dz$ where $f(z)$ is the function of Example 22.11 and γ is the closed path shown in Figure 22.1.

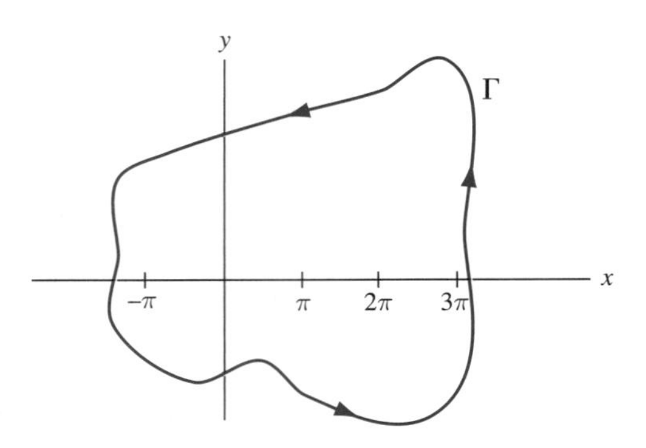

FIGURE 22.1 *γ in Example 22.12.*

The singularities of f enclosed by γ are $0, \pi, 2\pi, 3\pi$ and $-\pi$. By the residue theorem and the conclusion of Example 22.11,

$$\begin{aligned}\oint_\gamma \frac{4iz-1}{\sin(z)}\,dz &= 2\pi i \sum_{n=-1}^{3} \mathrm{Res}(f, n\pi)\\ &= 2\pi i[-(-1-4\pi i) - 1 - (-1+4\pi i) + (-1+8\pi i) - (-1+12\pi i)]\\ &= 8 + 2\pi i. \ \blacklozenge\end{aligned}$$

Next we treat poles of multiplicity greater than one.

THEOREM 22.6 ***Residue at a Pole of Order m***

Let f have a pole of order m at z_0. Then

$$\mathrm{Res}(f, z_0) = \frac{1}{(m-1)!} \lim_{z\to z_0} \frac{d^{m-1}}{dz^{m-1}}[(z-z_0)^m f(z)]. \ \blacklozenge$$

The theorem can be proved by manipulating the Laurent expansion of $f(z)$ about z_0, which in the case of a pole of order m is

$$f(z) = \frac{c_{-m}}{(z-z_0)^m} + \frac{c_{-m+1}}{(z-z_0)^{m-1}} + \cdots,$$

so that

$$(z-z_0)^m f(z) = c_{-m} + c_{-m+1}(z-z_0) + \cdots + c_{-1}(z-z_0)^{m-1} + \cdots.$$

Differentiation of this equation $m-1$ times isolates c_{-1}, yielding the residue of f at z_0.

Theorem 22.6 reduces to Corollary 22.1 if $m=1$ with the conventions that $0! = 1$ and the zero-order derivative of a function is the function itself.

By this result, to obtain the residue of f at a pole of order m,

1. Multiply $f(z)$ by $(z-z_0)^m$.
2. Differentiate $m-1$ times.
3. Take the limit as z approaches z_0.
4. Divide by $(m-1)!$.

EXAMPLE 22.13

Let

$$f(z) = \frac{\cos(z)}{(z+i)^3}.$$

f has a pole of order 3 at $-i$, and

$$\begin{aligned}\text{Res}(f, -i) &= \frac{1}{2!} \lim_{z \to -i} \frac{d^2}{dz^2} \left((z+i)^3 \frac{\cos(z)}{(z+i)^3} \right) \\ &= \frac{1}{2} \lim_{z \to -i} \frac{d^2}{dz^2} \cos(z) = -\frac{1}{2} \cos(-i) = -\frac{1}{2} \cos(i). \quad \blacklozenge\end{aligned}$$

The following example emphasizes that the value of $\oint_\gamma f(z)dz$ depends on the residues of f at singularities *enclosed by* γ. Any other singularities (outside of γ) are irrelevant for this integral.

EXAMPLE 22.14

Evaluate $\oint_\gamma f(z)\,dz$ where

$$f(z) = \frac{2iz - \cos(z)}{z^3 + z}$$

if γ is any closed path not passing through a singularity of f.

The singularities of f are simple poles at $0, i, -i$. We will need the residues:

$$\begin{aligned}\text{Res}(f, 0) &= \frac{-\cos(0)}{1} = -1, \\ \text{Res}(f, i) &= \frac{2i^2 - \cos(i)}{3i^2 + 1} = \frac{-2 - \cos(i)}{-2} = 1 + \frac{1}{2}\cos(i), \\ \text{Res}(f, -i) &= \frac{2i(-i) - \cos(-i)}{3(-i)^2 + 1} = -1 + \frac{1}{2}\cos(i).\end{aligned}$$

The following cases occur.

Case 1 If γ does not enclose any of the singularities of f, then $\oint_\gamma f(z)\,dz = 0$ by Cauchy's theorem.

Case 2 If γ encloses 0 but not $\pm i$,

$$\oint_\gamma f(z)\,dz = 2\pi i \text{Res}(f, 0) = -2\pi i.$$

Case 3 If γ encloses i but not 0 or $-i$,

$$\oint_\gamma f(z)\,dz = 2\pi i \left(1 + \frac{1}{2}\cos(i)\right).$$

Case 4 If γ encloses $-i$ but not 0 or i,

$$\oint_\gamma f(z)\,dz = 2\pi i \left(-1 + \frac{1}{2}\cos(i)\right).$$

Case 5 If γ encloses 0 and i but not $-i$,

$$\oint_\gamma f(z)\,dz = 2\pi i\left(-1+1+\frac{1}{2}\cos(i)\right) = \pi i\cos(i).$$

Case 6 If γ encloses 0 and $-i$ but not i,

$$\oint_\gamma f(z)\,dz = 2\pi i\left(-1-1+\frac{1}{2}\cos(i)\right) = 2\pi i\left(-2+\frac{1}{2}\cos(i)\right).$$

Case 7 If γ encloses i and $-i$ but not 0,

$$\oint_\gamma f(z)\,dz = 2\pi i\left(1+\frac{1}{2}\cos(i)-1+\frac{1}{2}\cos(i)\right) = 2\pi i\cos(i).$$

Case 8 If γ encloses all three singularities,

$$\int_\gamma f(z)\,dz = 2\pi i\left(-1+1+\frac{1}{2}\cos(i)-1+\frac{1}{2}\cos(i)\right) = 2\pi i(-1+\cos(i)).$$ ♦

EXAMPLE 22.15

Let

$$f(z) = \frac{\sin(z)}{z^2(z^2+4)}.$$

f has a simple poles at 0 and $\pm 2i$. Suppose γ is a closed path enclosing 0 and $2i$ but not $-2i$. Compute the residues of f at 0 and $2i$. In doing this, the corollary does not apply for the residue of f at 0 because $\sin(0)=0$. We can, if we wish, use the corollary for the residue at $2i$. Compute

$$\operatorname{Res}(f,0) = \lim_{z\to 0} zf(z) = \lim_{z\to 0}\frac{\sin(z)}{z}\frac{1}{z^2+4} = \frac{1}{4}$$

and

$$\operatorname{Res}(f,2i) = \lim_{z\to 2i}\frac{\sin(z)}{z^2(z-2i)(z+2i)} = \frac{\sin(2i)}{(-4)(4i)} = \frac{i}{16}\sin(2i).$$

Then

$$\oint_\gamma f(z)\,dz = 2\pi i\left(\frac{1}{4}+\frac{i}{16}\sin(2i)\right).$$ ♦

EXAMPLE 22.16

We will evaluate $\oint_\gamma e^{1/z}\,dz$ where γ is a closed path enclosing the origin.

The Laurent expansion of $e^{1/z}$ about 0 is

$$e^{1/z} = \sum_{n=0}^{\infty}\frac{1}{n!}\frac{1}{z^n},$$

so 0 is an essential singularity. There is no simple formula for residues at essential singularities. Because we have the Laurent expansion of $e^{1/z}$ about 0, we can read that the coefficient of $1/z$ is 1, so $\operatorname{Res}(f,0) = 1$ and $\oint_\gamma e^{1/z}\,dz = 2\pi i$. ♦

SECTION 22.2 PROBLEMS

In each of Problems 1 through 16, use the residue theorem to evaluate the integral.

1. $\oint_\gamma \frac{8z-4i+1}{z+4i}\,dz$ with γ the circle of radius 2 about $-i$.

2. $\oint_\gamma \frac{\cos(z)}{ze^z}\,dz$ with γ the circle of radius $1/2$ about $i/8$.

3. $\oint_\gamma \frac{z+i}{z^2+6}\,dz$ with γ the square of side length 8 and sides parallel to the axes centered at the origin.

4. $\oint_\gamma e^{2/z^2}\,dz$ with γ the square with sides parallel to the axes and of length 3 centered at $-i$.

5. $\oint_\gamma \frac{1+z^2}{(z-1)^2(z+2i)}\,dz$ with γ the circle of radius 7 about $-i$.

6. $\oint_\gamma \frac{z^2}{z-1+2i}\,dz$ with γ the square of side length 4 and sides parallel to the axes centered at $1-2i$.

7. $\oint_\gamma \coth(z)\,dz$ with γ the circle of radius 2 about i.

8. $\oint_\gamma \frac{(1-z)^2}{z^3-8}\,dz$ with γ the circle of radius 2 about 2.

9. $\oint_\gamma \frac{e^{2z}}{z(z-4i)}\,dz$ with γ any closed path enclosing 0 and $4i$.

10. $\oint_\gamma \left(\frac{z}{z-1}\right)^2\,dz$ with γ any closed path enclosing 1.

11. $\oint_\gamma \frac{iz}{(z^2+9)(z-i)}\,dz$ with γ the circle of radius 2 about $-3i$.

12. $\oint_\gamma \frac{2z}{(z-i)^2}\,dz$ with γ the circle of radius 3 about 1.

13. $\oint_\gamma \frac{z}{\sinh^2(z)}\,dz$ with γ the circle of radius 1 about $1/2$.

14. $\oint_\gamma \frac{\cos(z)}{4+z^2}\,dz$ with γ the square of side length 3 and sides parallel to the axes centered at $-2i$.

15. $\oint_\gamma \frac{e^z}{z}\,dz$ with γ the circle of radius 2 about $-3i$.

16. $\oint_\gamma \frac{z-i}{2z+1}\,dz$ with γ the circle of radius 1 about the origin.

17. Evaluate

$$\oint_\gamma \frac{z+1}{z^2+2z+4}\,dz$$

with γ the circle $|z|=2$, first by using the residue theorem and then by using the argument principle.

18. Let

$$p(z)=(z-z_1)(z-z_2)\cdots(z-z_n)$$

with $z_1,\cdots,z_n$ distinct complex numbers. Let γ be a positively oriented closed path enclosing each z_j. Evaluate

$$\oint_\gamma \frac{p'(z)}{p(z)}\,dz$$

first by using the residue theorem and then by using the argument principle.

19. Let h and g be differentiable at z_0 and $g(z_0)\neq 0$. Suppose h has a zero of order 2 at z_0. Show that

$$\text{Res}(g(z)/h(z),z_0)=\frac{2g'(z_0)}{h''(z_0)}-\frac{2}{3}\frac{g(z_0)h^{(3)}(z_0)}{(h''(z_0))^2}.$$

Hint: Use Theorem 22.6. Begin by writing $h(z)=(z-z_0)^2\varphi(z)$ where $\varphi(z_0)\neq 0$.

20. Suppose f is differentiable at points on a closed path γ and at all points in the region G enclosed by γ, except possibly at a finite number of poles of f in G. Let Z be the number of zeros of f in G and P be the number of poles of f in G with each zero and pole counted as many times as its multiplicity. Show that

$$\frac{1}{2\pi i}\oint_\gamma \frac{f'(z)}{f(z)}\,dz=Z-P.$$

This formula is known as the *argument principle*. *Hint*: If f has a zero of order k at z_0, show by looking at the Taylor expansion of $f(z)$ about z_0 that

$$\frac{f'(z)}{f(z)}=\frac{k}{z-z_0}+\frac{g'(z)}{g(z)}$$

where g is differentiable at z_0 and $g(z_0)\neq 0$. Use this to evaluate $\text{Res}(f'/f,z_0)$.

If f has a pole of order m at z_1, show by examining the Laurent expansion of $f(z)$ about z_1 that

$$\frac{f'(z)}{f(z)}=-\frac{m}{z-z_1}+\frac{h'(z)}{h(z)}$$

for some $h(z)$ that is differentiable and nonzero at z_1.

Use these facts and the residue theorem to derive the argument principle.

21. Evaluate

$$\oint_\gamma \frac{z}{2+z^2}\,dz$$

with γ as the circle $|z|=2$ first by using the residue theorem and then by using the argument principle.

22. Evaluate $\oint_\gamma \tan(z)\,dz$ with γ the circle $|z|=\pi$ first by using the residue theorem and then by using the argument principle.

22.3 Evaluation of Real Integrals

Complex integration can be used to evaluate some types of real integrals that are otherwise inaccessible. We will illustrate with three classes of integrals.

22.3.1 Rational Functions

We will apply complex integration to evaluate real integrals of the form

$$\int_{-\infty}^{\infty} \frac{p(x)}{q(x)}\, dx$$

in which p and q are polynomials with real coefficients. A quotient of polynomials is called a *rational function*. Assume that the degree of q exceeds that of p by at least 2, that p and q have no common factors, and that $q(x)$ has no real zeros. This ensures convergence of the improper integral.

The idea is to create a complex integral whose value is this real integral, then use the residue theorem to evaluate the complex integral. To do this, assume that we can find all the zeros of $q(z)$. Since $q(z)$ has real coefficients and no real zeros, its zeros occur in complex conjugate pairs $z_1, \overline{z_1},, z_2, \overline{z_2}, \cdots, z_m, \overline{z_m}$ with each z_j in the upper half-plane and its conjugate in the lower half-plane.

Let Γ_R be the curve of Figure 22.2 consisting of a semicircle γ_R of radius R and the segment S_R from $-R$ to R on the real axis with R large enough that Γ_R encloses $z_1, \cdots, z_m$. These are all the poles of $p(z)/q(z)$ in the upper half-plane. Then

$$\oint_{\Gamma_R} \frac{p(z)}{q(z)}\, dz = 2\pi i \sum_{j=1}^{m} \operatorname{Res}(f, z_j) = \int_{S_R} \frac{p(z)}{q(z)}\, dz + \int_{\gamma_R} \frac{p(z)}{q(z)}\, dz. \tag{22.1}$$

On S_R, $z = x$ as x varies from $-R$ to R for counterclockwise orientation on Γ_R, so

$$\int_{S_R} \frac{p(z)}{q(z)}\, dz = \int_{-R}^{R} \frac{p(x)}{q(x)}\, dx.$$

Therefore, equation (22.1) is

$$\int_{-R}^{R} \frac{p(x)}{q(x)}\, dx + \int_{\gamma_R} \frac{p(z)}{q(z)}\, dz = 2\pi i \sum_{j=1}^{m} \operatorname{Res}(f, z_j). \tag{22.2}$$

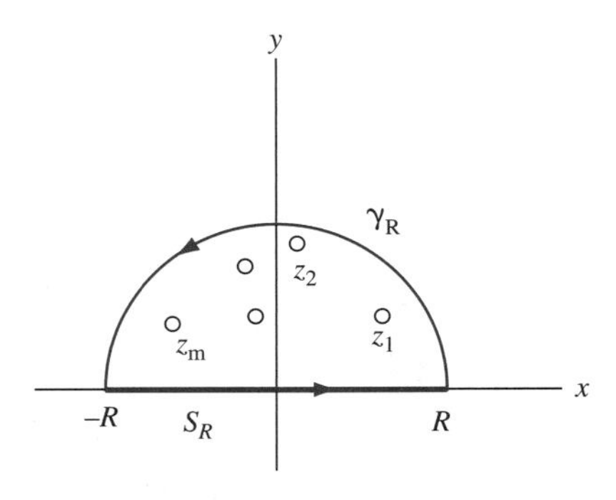

FIGURE 22.2 *Evaluation of real integrals of rational functions.*

Take the limit in this equation as $R \to \infty$. In this limit, the semicircle γ_R expands over the entire upper half-plane, and the interval $[-R, R]$ expands over the real line. Furthermore, because $q(z)$ has degrees at least 2 more than the degree of $p(z)$, the degree of $z^2 p(z)$ does not exceed that of $q(z)$. This means that, for large R, $z^2 p(z)/q(z)$ is bounded. If $|z^2 p(z)/q(z)| \leq M$ for $|z| \geq R$, then

$$\left|\frac{p(z)}{q(z)}\right| \leq \frac{M}{|z|^2} \leq \frac{M}{R^2} \text{ for } |z| \geq R.$$

But then, because γ_R has length πR,

$$\left|\int_{S_R} \frac{p(z)}{q(z)} dz\right| \leq \frac{M}{R^2}(\pi R) \to 0 \text{ as } R \to \infty.$$

Therefore, in the limit as $R \to \infty$ in equation (22.2), the integral over γ_R tends to zero and we are left with

$$\int_{-\infty}^{\infty} \frac{p(x)}{q(x)} dx = 2\pi i \sum_{j=1}^{m} \text{Res}(p(z)/q(z), z_j). \tag{22.3}$$

In sum, under the stated assumptions, $\int_{-\infty}^{\infty} p(x)/q(x)\, dx$ can be evaluated as $2\pi i$ times the sum of the residues of $p(z)/q(z)$ at the zeros of $q(z)$ occurring in the upper half-plane.

EXAMPLE 22.17

We will evaluate

$$\int_{\infty}^{\infty} \frac{1}{x^6 + 64} dx.$$

The conditions of the method are met with $p(z) = 1$ and $q(z) = z^6 + 64$. The zeros of $q(z)$ are the sixth roots of -64. To find these, put -64 in polar form as $-64 = 64e^{i(\pi+2n\pi)}$ in which n can be any integer. The sixth roots are $2e^{i(\pi+2n\pi)}$ for $n = 0, 1, 2, 3, 4, 5$. The three sixth roots in the upper half-plane correspond to $n = 0, 1, 2$ and are

$$z_1 = 2e^{\pi i/6}, \; z_2 = 2e^{\pi i/2} = 2i, \text{ and } z_3 = 2e^{5\pi i/6}.$$

We need the residue of $p(z)/q(z)$ at each of these simple poles:

$$\text{Res}(p(z)/q(z), z_1) = \frac{1}{6(2e^{\pi i/6})^5} = \frac{1}{192} e^{-5\pi i/6},$$

$$\text{Res}(p(z)/q(z), z_2) = \frac{1}{6(2i)^5} = -\frac{i}{192},$$

and

$$\text{Res}(p(z)/q(z), z_3) = \frac{1}{6(2e^{5\pi i/6})^5} = \frac{1}{192} e^{-25\pi i/6} = \frac{1}{192} e^{-\pi i/6}.$$

Then

$$\int_{-\infty}^{\infty} \frac{p(x)}{q(x)} dx = \frac{2\pi i}{192}[e^{-5\pi i/6} - i + e^{-\pi i/6}]$$
$$= \frac{\pi i}{96}\left[\cos\left(\frac{5\pi}{6}\right) - i \sin\left(\frac{5\pi}{6}\right) - i + \cos\left(\frac{\pi}{6}\right) - i \sin\left(\frac{\pi}{6}\right)\right].$$

Now

$$\cos\left(\frac{5\pi}{6}\right) + \cos\left(\frac{\pi}{6}\right) = 0 \text{ and } \sin\left(\frac{5\pi}{6}\right) = \sin\left(\frac{\pi}{6}\right) = \frac{1}{2},$$

so

$$\int_{-\infty}^{\infty} \frac{1}{x^6+64}\,dx = \frac{\pi i}{96}(-2i) = \frac{\pi}{48}. \blacklozenge$$

22.3.2 Rational Functions Times Cosine or Sine

Suppose p and q are polynomials with real coefficients, have no common zeros, and the degree of q exceeds that of p by at least 2. Suppose also that q has no real zeros, and zeros $z_1, \cdots, z_m$ are in the upper half-plane. We want to evaluate integrals of the form

$$\int_{-\infty}^{\infty} \frac{p(x)}{q(x)}\cos(cx)\,dx \quad\text{and}\quad \int_{-\infty}^{\infty} \frac{p(x)}{q(x)}\sin(cx)\,dx$$

in which c can be any positive number.

The idea is to consider

$$\oint_{\Gamma_R} \frac{p(z)}{q(z)} e^{icz}\,dz$$

with Γ_R as the curve of Section 22.3.1, consisting of the upper part of a semicircle and part of the real axis joining the ends of the semicircle. We obtain

$$\begin{aligned}
\oint_{\Gamma_R} \frac{p(z)}{q(z)} e^{icz}\,dz &= 2\pi i \sum_{j=1}^{m} \operatorname{Res}(p(z)e^{icz}/q(z), z_j) \\
&= \int_{\gamma_R} \frac{p(z)}{q(z)} e^{icz}\,dz + \int_{S_R} \frac{p(z)}{q(z)} e^{icz}\,dz \\
&= \int_{\gamma_R} \frac{p(z)}{q(z)} e^{icz}\,dz + \int_{-R}^{R} \frac{p(x)}{q(x)}\cos(cx)\,dx + i\int_{-R}^{R} \frac{p(x)}{q(x)}\sin(cx)\,dx.
\end{aligned}$$

Take the limit as $R \to \infty$. As in Section 22.3.1, the integral over γ_R tends to 0 and

$$\int_{-\infty}^{\infty} \frac{p(x)}{q(x)}\cos(cx)\,dx + i\int_{-\infty}^{\infty} \frac{p(x)}{q(x)}\sin(cx)\,dx = 2\pi i \sum_{j=1}^{m} \operatorname{Res}(p(z)e^{icz}/q(z), z_j). \qquad (22.4)$$

We actually obtain two real integrals in this calculation. After computing $2\pi i$ times the sum of the residues, the real part of this number is the integral containing $\cos(cx)$, and the imaginary part is the integral containing $\sin(cx)$.

EXAMPLE 22.18

We will evaluate

$$\int_{-\infty}^{\infty} \frac{\cos(cx)}{(x^2+\alpha^2)(x^2+\beta^2)}\,dx$$

in which c, α and β are positive numbers and $\alpha \neq \beta$. Let

$$f(z) = \frac{e^{icz}}{(z^2+\alpha^2)(z^2+\beta^2)}.$$

The poles of f in the upper half-plane are αi and βi, and

$$\operatorname{Res}(f, \alpha i) = \frac{e^{-c\alpha}}{2\alpha i(\beta^2-\alpha^2)},$$

and

$$\operatorname{Res}(f, \beta i) = \frac{e^{-c\beta}}{2\beta i(\alpha^2-\beta^2)}.$$

Then

$$\int_{-\infty}^{\infty} \frac{\cos(cx)}{(x^2+\alpha^2)(x^2+\beta^2)} dx + i \int_{-\infty}^{\infty} \frac{\sin(cx)}{(x^2+\alpha^2)(x^2+\beta^2)} dx$$
$$= 2\pi i \left[\frac{e^{-c\alpha}}{2\alpha i(\beta^2-\alpha^2)} + \frac{e^{-c\beta}}{2\beta i(\alpha^2-\beta^2)} \right]$$
$$= \frac{\pi}{\beta^2-\alpha^2} \left(\frac{e^{-c\alpha}}{\alpha} - \frac{e^{-c\beta}}{\beta} \right).$$

Separate real and imaginary parts to obtain

$$\int_{-\infty}^{\infty} \frac{\cos(cx)}{(x^2+\alpha^2)(x^2+\beta^2)} dx = \frac{\pi}{\beta^2-\alpha^2} \left(\frac{e^{-c\alpha}}{\alpha} - \frac{e^{-c\beta}}{\beta} \right)$$

and

$$\int_{-\infty}^{\infty} \frac{\sin(cx)}{(x^2+\alpha^2)(x^2+\beta^2)} dx = 0.$$

The last integral is obvious because the integrand is odd. ♦

22.3.3 Rational Functions of Cosine and Sine

Let $K(x, y)$ be a quotient of polynomials in x and y. For example,

$$K(x, y) = \frac{x^3y - 2xy^2 + x - 2y}{x^4 + xy^4 - 8}.$$

Such a function is called a *rational function of x and y*. If we replace $x = \cos(\theta)$ and $y = \sin(\theta)$, we obtain a rational function of cosine and sine. We want a way to evaluate the integral of such a function over $[0, 2\pi]$. This will be an integral of the form

$$\int_0^{2\pi} K(\cos(\theta),\ \sin(\theta))\, d\theta.$$

Again, the idea is to express this real integral as a complex integral, which is then evaluated using the residue theorem.

Let γ be the unit circle about the origin $\gamma(\theta) = e^{i\theta}$ for $0 \le \theta \le 2\pi$. On this curve, $z = e^{i\theta}$ and $\overline{z} = e^{-i\theta} = 1/z$, so

$$\cos(\theta) = \frac{1}{2}\left(z + \frac{1}{z}\right) \text{ and } \sin(\theta) = \frac{1}{2i}\left(z - \frac{1}{z}\right).$$

Furthermore, on γ,

$$dz = ie^{i\theta}\, d\theta = iz\, d\theta,$$

so

$$d\theta = \frac{1}{iz} dz.$$

Therefore,

$$\oint_\gamma K\left(\frac{1}{2}\left(z + \frac{1}{z}\right),\ \frac{1}{2i}\left(z - \frac{1}{z}\right)\right) \frac{1}{iz} dz = \int_0^{2\pi} K(\cos(\theta), \sin(\theta)) \frac{1}{ie^{i\theta}} ie^{i\theta}\, d\theta$$
$$= \int_0^{2\pi} K(\cos(\theta), \sin(\theta))\, d\theta.$$

Evaluate the integral on the left using the residue theorem, yielding the integral we want.

In summary, to evaluate $\int_0^{2\pi} K(\cos(\theta), \sin(\theta)\, d\theta$, begin by computing the function

$$f(z) = K\left(\frac{1}{2}\left(z + \frac{1}{z}\right), \frac{1}{2i}\left(z - \frac{1}{z}\right)\right)\frac{1}{iz}. \tag{22.5}$$

Then

$$\int_0^{2\pi} K(\cos(\theta), \sin(\theta))\, d\theta = 2\pi i \sum_{|z_j|<1} \text{Res}(f, z_j) \tag{22.6}$$

with this sum over all singularities z_j of $f(z)$ enclosed by the unit circle.

EXAMPLE 22.19

Evaluate

$$\int_0^{2\pi} \frac{\sin^2(\theta)}{2 + \cos(\theta)}\, d\theta.$$

Here $K(x, y) = y^2/(2 + x)$, and

$$K(\cos(\theta), \sin(\theta)) = \frac{\sin^2(\theta)}{2 + \cos(\theta)}.$$

Let $x = \cos(\theta) = (z + 1/z)/2$, and $y = \sin(\theta) = (z - 1/z)/2i$ in $K(x, y)$, and multiply by $1/iz$ to produce the complex function of equation (22.5):

$$f(z) = \left(\frac{[(z - 1/z)/2i]^2}{2 + (z + 1/z)/2}\right)\frac{1}{iz} = \frac{i}{2}\frac{z^4 - 2z^2 + 1}{z^2(z^2 + 4z + 1)}.$$

f has a double pole at 0 and simple poles at zeros of $z^2 + 4z + 1$, which are $-2 \pm \sqrt{3}$. Only the poles 0 and $2 - \sqrt{3}$ are enclosed by γ. By equation (22.6),

$$\int_0^{2\pi} \frac{\sin^2(\theta)}{2 + \cos(\theta)}\, d\theta = 2\pi i[\text{Res}(f, 0) + \text{Res}(f, -2 + \sqrt{3})].$$

Compute these residues:

$$\text{Res}(f, 0) = \lim_{z\to 0} \frac{d}{dz}(z^2 f(z)) = \lim_{z\to 0} \frac{d}{dz}\frac{i}{2}\frac{z^4 - 2z^2 + 1}{z^2 + 4z + 1}$$
$$= \frac{i}{2}\lim_{z\to 0}\left(2\frac{z^5 + 6z^4 + 2z^3 - 4z^2 - 3z - 2}{(z^2 + 4z + 1)^2}\right) = -2i$$

and

$$\text{Res}(f, -2 + \sqrt{3}) = \frac{i}{2}\left[\frac{z^4 - 2z^2 + 1}{2z(z^2 + 4z + 1) + z^2(2z + 4)}\right]_{z=-2+\sqrt{3}}$$
$$= \frac{i}{2}\frac{42 - 24\sqrt{3}}{-12 + 7\sqrt{3}}.$$

Therefore,

$$\int_0^{2\pi} \frac{\sin^2(\theta)}{2 + \cos(\theta)}\, d\theta = 2\pi i\left[-2i + \frac{i}{2}\frac{42 - 24\sqrt{3}}{-12 + 7\sqrt{3}}\right] = \left(\frac{90 - 52\sqrt{3}}{12 - 7\sqrt{3}}\right)\pi. \ \blacklozenge$$

EXAMPLE 22.20

We will evaluate

$$\int_0^{2\pi} \frac{1}{\alpha + \beta\cos(\theta)}\, d\theta$$

in which $0 < \beta < \alpha$.

Replace $\cos(\theta) = (z + 1/z)/2$, and use equation (22.5) to produce the function

$$f(z) = \frac{1}{\alpha + (\beta/2)(z + 1/z)}\frac{1}{iz} = \frac{-2i}{\beta z^2 + 2\alpha z + \beta}.$$

f has simple poles at

$$z = \frac{-\alpha \pm \sqrt{\alpha^2 - \beta^2}}{\beta}.$$

Since $\alpha > \beta$, these numbers are real. Only one of them,

$$z_1 = \frac{-\alpha + \sqrt{\alpha^2 - \beta^2}}{\beta},$$

is enclosed by γ. Therefore,

$$\int_0^{2\pi} \frac{1}{\alpha + \beta\cos(\theta)}\, d\theta = 2\pi i \operatorname{Res}(f, z_1)$$

$$= 2\pi i \frac{-2i}{2\beta z_1 + 2\alpha} = \frac{2\pi}{\sqrt{\alpha^2 - \beta^2}}. \; ◆$$

SECTION 22.3 PROBLEMS

In each of Problems 1 through 10, evaluate the integral. Wherever they appear, α and β are positive numbers.

1. $\int_{-\infty}^{\infty} \frac{\cos^2(x)}{(x^2+4)^2}\, dx$
2. $\int_{-\infty}^{\infty} \frac{\cos(\beta x)}{(x^2+\alpha^2)^2}\, dx$
3. $\int_{-\infty}^{\infty} \frac{x\sin(2x)}{x^4+16}\, dx$
4. $\int_{-\infty}^{\infty} \frac{1}{x^4+1}\, dx$
5. $\int_{-\infty}^{\infty} \frac{x^2}{(x^2+4)^2}\, dx$
6. $\int_0^{2\pi} \frac{1}{6+\sin(\theta)}\, d\theta$
7. $\int_{-\infty}^{\infty} \frac{1}{x^6+1}\, dx$
8. $\int_{-\infty}^{\infty} \frac{1}{x^2-2x+6}\, dx$
9. $\int_0^{2\pi} \frac{1}{2-\cos(\theta)}\, d\theta$
10. $\int_0^{2\pi} \frac{2\sin(\theta)}{2+\sin^2(\theta)}\, d\theta$

In Problems 11 through 18, α and β are positive numbers wherever they occur.

11. Show that $\int_0^{\infty} e^{-x^2}\cos(2\beta x)\, dx = \frac{\sqrt{\pi}}{2}e^{-\beta^2}$.

    Hint: Integrate e^{-z^2} about the rectangular path having corners at $\pm R$ and $\pm R + \beta i$. Use Cauchy's theorem to evaluate this integral, set this equal to the sum of the integrals on the sides of the rectangle, and take the limit as $R \to \infty$. Assume the standard result that

    $$\int_0^{\infty} e^{-x^2}\, dx = \frac{\sqrt{\pi}}{2}.$$

12. Show that $\int_0^{\pi/2} \frac{1}{\alpha+\sin^2(\theta)}\, d\theta = \frac{\pi}{2\sqrt{\alpha(1+\alpha)}}$.
13. Let α and β be positive numbers. Show that

    $$\int_0^{\infty} \frac{x\sin(\alpha x)}{x^4+\beta^4}\, dx = \frac{\pi}{2\beta^2}e^{-\alpha\beta/\sqrt{2}}\sin\left(\frac{\alpha\beta}{\sqrt{2}}\right).$$

14. Let $0 < \beta < \alpha$. Show that
$$\int_0^{\pi} \frac{1}{(\alpha + \beta\cos(\theta))^2}\,d\theta = \frac{\alpha\pi}{(\alpha^2 - \beta^2)^{3/2}}.$$

15. Show that $\int_{-\infty}^{\infty} \frac{\cos(\alpha x)}{x^2+1}\,dx = \pi e^{-\alpha}$.

16. Derive *Fresnel's integrals*:
$$\int_0^{\infty} \cos(x^2)\,dx = \int_0^{\infty} \sin(x^2)\,dx = \frac{1}{2}\sqrt{\frac{\pi}{2}}.$$
Hint: Integrate e^{iz^2} over the closed path bounding the sector $0 \le x \le R, 0 \le \theta \le \pi/4$, as in Figure 22.3. Use Cauchy's theorem to evaluate this integral, then evaluate it as the sum of the integrals over the boundary segments of the sector. Show that the integral over the circular arc tends to zero as $R \to \infty$, and use the integrals over the line segments to obtain Fresnel's integrals.

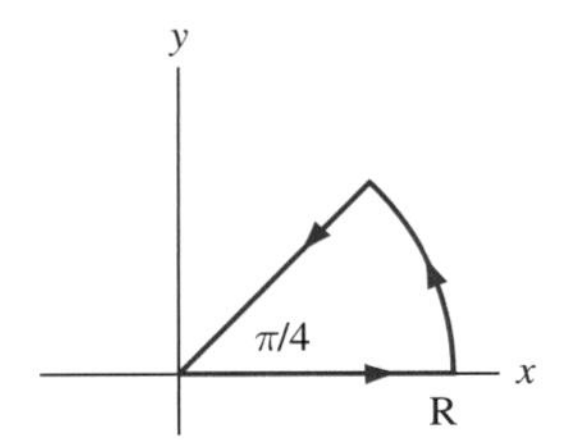

FIGURE 22.3 *Path in Problem 16, Section 22.3.*

17. Let $\alpha \ne \beta$. Show that $\int_0^{2\pi} \frac{1}{\alpha^2\cos^2(\theta) + \beta^2\sin^2(\theta)}\,d\theta = \frac{2\pi}{\alpha\beta}$.

18. Show that $\int_{-\infty}^{\infty} \frac{x^2\cos(\alpha x)}{(x^2+\beta^2)^2}\,dx = \frac{\pi}{2\beta}e^{-\alpha\beta}(1-\alpha\beta)$.

22.4 Residues and the Inverse Laplace Transform

If f is a complex function defined at least for all z on $[0, \infty)$, then the Laplace transform of f is
$$\mathcal{L}[f](z) = \int_0^{\infty} e^{-zt} f(t)\,dt$$
for all z such that this integral converges. If $\mathcal{L}[f] = F$, we write $f = \mathcal{L}^{-1}[F]$. In Chapter 3, we saw some techniques for manipulating $\mathcal{L}$ and $\mathcal{L}^{-1}$. The following theorem provides a formula for the inverse Laplace transform of $F(s)$ in terms of residues $e^{tz}F(z)$, which we can sometimes compute quite easily.

THEOREM 22.7 ***Inverse Laplace Transform***

Let F be differentiable for all z except for poles $z_1, \cdots, z_n$. Suppose for some real number σ, F is differentiable for all z with $\mathrm{Re}(z) \ge \sigma$. Suppose there are numbers M and R such that
$$|zF(z)| \le M \text{ for } |z| \ge R.$$
For $t \ge 0$, let
$$f(t) = \sum_{j=1}^{n} \mathrm{Res}\left(e^{tz}F(z), z_j\right).$$
Then
$$f = \mathcal{L}^{-1}[F]. \ \blacklozenge$$

Because F is differentiable for $\mathrm{Re}(z) \ge \sigma$, $F'(z)$ exists at least for z to the right of the vertical line $x = \sigma$. The condition that $|zF(z)| \le M$ for $|z| \ge R$ means that $zF(z)$ is bounded for all z on and outside some sufficiently large circle. This condition is satisfied by any rational function $p(z)/q(z)$ if the degree of $q(z)$ exceeds that of $p(z)$.

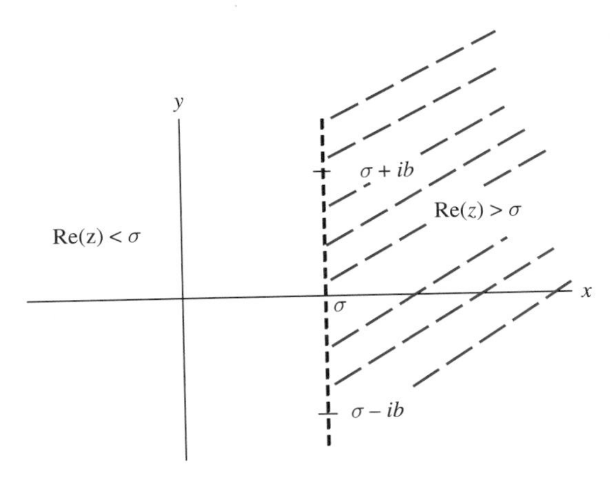

FIGURE 22.4 *σ in Theorem 22.7.*

Theorem 22.7 can be proved using the version of the Cauchy integral formula given in Section 20.3.7. Following a sketch of the argument and two examples computing inverse Laplace transforms of functions, we will use the theorem to analyze heat diffusion in a homogeneous solid cylinder.

Begin by writing, in the notation of this section,

$$F(s) = -\frac{1}{2\pi i}\lim_{b\to\infty}\int_{\sigma-ib}^{\sigma+ib}\frac{F(z)}{z-s}\,dz.$$

Referring to Figure 22.4, take $\mathcal{L}^{-1}$ through the integral (this is justified by hypotheses of the theorem) to compute

$$\begin{aligned}\mathcal{L}^{-1}[F(s)](t) &= \frac{1}{2\pi i}\lim_{b\to\infty}\int_{\sigma-ib}^{\sigma+ib}\mathcal{L}^{-1}\left[\frac{F(z)}{s-z}\right]dz \\ &= \frac{1}{2\pi i}\lim_{b\to\infty}\int_{\sigma-ib}^{\sigma+ib}e^{tz}F(z)\,dz \\ &= \sum_p \operatorname{Res}(e^{tz}F(z), p),\end{aligned}$$

with this summation extending over all of the poles of $e^{tz}F(z)$. σ is chosen so that all of these poles are to the right of σ.

EXAMPLE 22.21

Let a be a positive number. We will find the inverse Laplace transform of $F(z) = 1/(a^2 + z^2)$.

We can do this using MAPLE, a table, or a method from Chapter 3. As an illustration of the use of the theorem, note that $F(z)$ has simple poles at $\pm ai$. Compute the residues as

$$\operatorname{Res}(e^{tz}F(z), ai) = \frac{e^{ai}}{2ai} \quad \text{and} \quad \operatorname{Res}(e^{tz}F(z), -ai) = \frac{e^{-ai}}{-2ai}.$$

This can be done easily by writing

$$e^{tz}F(z) = \frac{e^{tz}}{(z-ai)(z+ai)}$$

and using Corollary 22.1. By Theorem 22.7

$$\mathcal{L}^{-1}[F](t) = \frac{1}{2ai}\left(e^{ai} - e^{-ai}\right) = \frac{1}{a}\sin(at). \quad \blacklozenge$$

EXAMPLE 22.22

Let

$$F(z) = \frac{1}{(z^2-4)(z-1)^2}.$$

Then $F(z)$ has simple poles at ± 2 and a double pole at 1. Compute

$$\text{Res}(e^{tz}F(z), 2) = \lim_{z\to 2} \frac{e^{tz}}{(z+2)(z-1)^2} = \frac{1}{4}e^{2t},$$

$$\text{Res}(e^{tz}F(z), -2) = \lim_{z\to -2} \frac{e^{tz}}{(z-2)(z-1)^2} = -\frac{1}{36}e^{-2t},$$

and

$$\begin{aligned}\text{Res}(e^{tz}F(z), 1) &= \lim_{z\to 1} \frac{d}{dz}((z^2-4)^{-1}e^{tz}) \\ &= \lim_{z\to 1}(-2z(z^2-4)^{-2}e^{tz} + te^{tz}(z^2-4)^{-1}) \\ &= -\frac{1}{3}te^t - \frac{2}{9}e^t.\end{aligned}$$

Then

$$\mathcal{L}^{-1}[F](t) = \frac{1}{4}e^{2t} - \frac{1}{36}e^{-2t} - \frac{1}{3}te^t - \frac{2}{9}e^t. \blacklozenge$$

22.4.1 Diffusion in a Cylinder

We will find the temperature distribution function for a homogeneous, solid cylinder of radius R centered along the z-axis. This problem was solved in Section 17.5 using separation of variables. We will now use the Laplace transform and Theorem 22.7 to obtain the temperature distribution function. In the course of this, we will use properties of the modified Bessel function $I_0(x)$ and the Bessel functions $J_0(x)$ and $J_1(x)$ of the first kind of orders zero and one, respectively. These functions are developed in Section 15.3.

We will assume angular independence and use the heat equation in cylindrical coordinates. The boundary value problem is

$$\frac{\partial u}{\partial t} = \frac{\partial^2 u}{\partial r^2} + \frac{1}{r}\frac{\partial u}{\partial r} \text{ for } 0 \le r \le R, t > 0$$

$$u(r, 0) = 0, u(R, t) = T_0.$$

Apply the Laplace transform with respect to t to this problem to obtain

$$\frac{\partial^2 U}{\partial r^2} + \frac{1}{r}\frac{\partial U}{\partial r} - sU(r, s) = 0.$$

This is a modified Bessel equation of order zero. A solution that is bounded at $r = 0$, which is the center of the cylinder, is given by

$$U(r, s) = cI_0(\sqrt{s}r),$$

where $I_0(z) = J_0(iz)$. Transform the condition $u(R, t) = T_0$ to obtain $U(R, s) = T_0/s$. Then

$$U(R, s) = cI_0(\sqrt{s}R) = \frac{T_0}{s}.$$

This means that

$$c = \frac{T_0}{sI_0(\sqrt{s}R)}.$$

The transform of the solution is therefore

$$U(r,s) = \frac{T_0 I_0(\sqrt{s}r)}{sI_0(\sqrt{s}R)}.$$

We must invert this to obtain $u(r,t)$. To use Theorem 22.7, we need the singularities of

$$e^{tz}U(r,z) = e^{tz}\frac{T_0 I_0(\sqrt{z}r)}{zI_0(\sqrt{z}R)}.$$

Singularities of $e^{tz}U(r,z)$ occur at zeros of the denominator. There is a simple pole at $z=0$ because $I_0(0)=1\neq 0$. Furthermore,

$$I_0(\sqrt{z}R) = J_0(i\sqrt{z}R) = 0$$

if $i\sqrt{z}R$ is a zero of J_0. These zeros are real, simple, and nonzero. If the positive zeros are labeled $j_1, j_2, \cdots$, then all the zeros are $\pm j_1, \pm j_2, \cdots$. Therefore, $I_0(\sqrt{z}R)=0$ if $\sqrt{z}R = \pm i j_n$ for some n. Then

$$z = -j_n^2/R^2.$$

Therefore, $e^{tz}U(r,z)$ has simple poles at 0 and $-j_n^2/R^2$ for $n=1,2,\cdots$. Inverting $U(r,s)$ by Theorem 22.7 yields the solution

$$u(r,t) = \text{Res}(e^{tz}U(r,z), z=0) + \sum_{n=1}^{\infty} \text{Res}\left(e^{tz}F(z), z=-j_n^2/R^2\right).$$

There remains to compute these residues. First,

$$\begin{aligned}
\text{Res}(e^{tz}U(r,z),0) &= \lim_{z\to 0} ze^{tz}\frac{I_0(\sqrt{z}r)}{zI_0(\sqrt{z}R)} \\
&= \lim_{z\to 0} e^{tz}\frac{I_0(\sqrt{z}r)}{I_0(\sqrt{z}R)} \\
&= \frac{I_0(0)}{I_0(0)} = 1.
\end{aligned}$$

For the residues at the other poles, use Corollary 22.1, since the poles are simple zeros of the denominator of a function of the form $g(z)/h(z)$ with

$$g(z) = \frac{e^{tz}I_0(\sqrt{z}r)}{z} \text{ and } h(z) = I_0(\sqrt{z}R).$$

Then

$$\begin{aligned}
&\text{Res}\left(g(z)/h(z), -j_n^2/R^2\right) \\
&= \frac{e^{-j_n^2 t/R^2} I_0(-j_n ri/R)}{-j_n^2/R^2}\left[\frac{1}{\frac{d}{dz}I_0(\sqrt{z}R)}\right]_{-j_n^2/R^2} \\
&= \frac{e^{-j_n^2 t/R^2} I_0(-j_n ri/R)}{-j_n^2/R^2}\left[\frac{2\sqrt{z}}{RI_0'(\sqrt{z}R)}\right]_{z=-j_n^2/R^2} \\
&= \frac{-2Ri}{j_n}e^{-j_n^2 t/R^2}\frac{I_0(j_n ri/R)}{I_0'(j_n i)}.
\end{aligned}$$

Now use the facts that

$$I_0'(z) = i J_0'(iz)$$

and

$$J_0'(z) = -J_1(z) = J_1(-z)$$

to obtain

$$\begin{aligned}&\operatorname{Res}(g(z)/h(z), j_n)\\&= \frac{-2R}{j_n}\frac{J_0(j_n r/R)}{J_1(j_n)}e^{-j_n^2 t/R^2}.\end{aligned}$$

The solution is therefore

$$u(r,t) = T_0\left(1 - 2\sum_{n=1}^{\infty}\frac{-2R}{j_n}\frac{J_0(j_n r/R)}{J_1(j_n)}e^{-j_n^2 t/R^2}\right).$$

SECTION 22.4 PROBLEMS

In each of Problems 1 through 10, use Theorem 22.7 to find the inverse Laplace transform of the function.

1. $\dfrac{z^2}{(z-2)^3}$
2. $\dfrac{z+3}{(z^3-1)(z+2)}$
3. $\dfrac{z}{z^2+9}$
4. $\dfrac{1}{(z+3)^2}$
5. $\dfrac{1}{(z-2)^2(z+4)}$
6. $\dfrac{1}{(z^2+9)(z-2)^2}$
7. $\dfrac{1}{(z+5)^3}$
8. $\dfrac{1}{z^3+8}$
9. $\dfrac{1}{z^4+1}$
10. $\dfrac{1}{e^z(z-1)}$

CHAPTER 23

CONFORMAL MAPPINGS CONSTRUCTION OF CONFORMAL MAPPINGS CONFORMAL MAPPING SOLUTIONS OF DIRICHLET PROBLEMS MODELS OF PLANE FLUID

Conformal Mappings and Applications

In this chapter, we will discuss conformal mappings and applications of complex functions to the solution of Dirichlet problems and the analysis of fluid flow.

23.1 Conformal Mappings

It is sometimes useful to think of a complex function as a *mapping*. If $f(z)$ is defined for all z in some set S of complex numbers and each $f(z)$ is in some set K if z is in S, we write $f : S \to K$ and say that f *maps* S *into* K. If every point in K is the image of some point in S, then f maps S *onto* K. This is the same notion of onto encountered in Chapter 7 for linear transformations.

$f(S)$ denotes the set of numbers $f(z)$ for z in S. Then $f : S \to K$ is an onto mapping exactly when $f(S) = K$.

EXAMPLE 23.1

Let $w = f(z) = z/|z|$ for $z \neq 0$. Then $f(z)$ is defined on the set S consisting of the plane with the origin removed. If $z \neq 0$, then $|f(z)| = 1$ because

$$\left|\frac{z}{|z|}\right| = 1.$$

If K is the set of all points of magnitude 1, then f maps S into K. In this case, f maps S onto K because every number in K is the image of some number under this mapping. Indeed, if z is in K, then $|z| = 1$ and $f(z) = z$. This mapping contracts the entire plane (with origin removed) onto the unit circle. ♦

In visualizing the action of a mapping f it is convenient to make two copies of the complex plane, the z-copy and the w-copy, as in Figure 23.1. Picture S in the z-plane, and K in the w-plane.

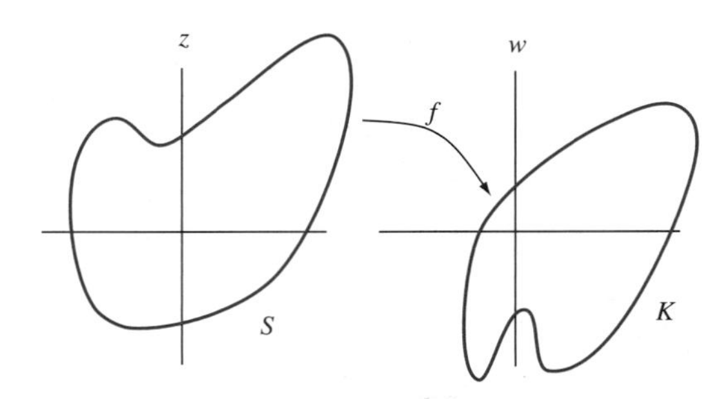

FIGURE 23.1 *$f(z)$ as a mapping.*

We will examine this mapping perspective for two familiar functions.

EXAMPLE 23.2

Let $f(z) = e^z$. Then

$$w = u + iv = e^z = e^{x+iy} = e^x \cos(y) + ie^x \sin(y),$$

so

$$u = e^x \cos(y) \text{ and } v = e^x \sin(y).$$

Consider a vertical line $x = a$ in the z-plane. The image of this line under the exponential mapping consists of points $u + iv$ with

$$u = e^a \cos(y) \text{ and } v = e^a \sin(y).$$

Now

$$u^2 + v^2 = e^{2a},$$

so this vertical line $x = a$ maps to the circle of radius e^a about the origin in the w-plane (Figure 23.2). As the point $z = a + iy$ moves along the line, the image point $e^{a+iy} = e^a e^{iy}$ moves around this circle, making one complete circuit every time y varies over an interval of length 2π. We may think of this vertical line as made up of infinitely many intervals of length 2π strung together, and the exponential function maps each of these segments once around the image circle.

The image of a point $z = x + ib$ on a horizontal line $y = b$ is the point $u + iv$ with

$$u = e^x \sin(b) \text{ and } v = e^x \cos(b).$$

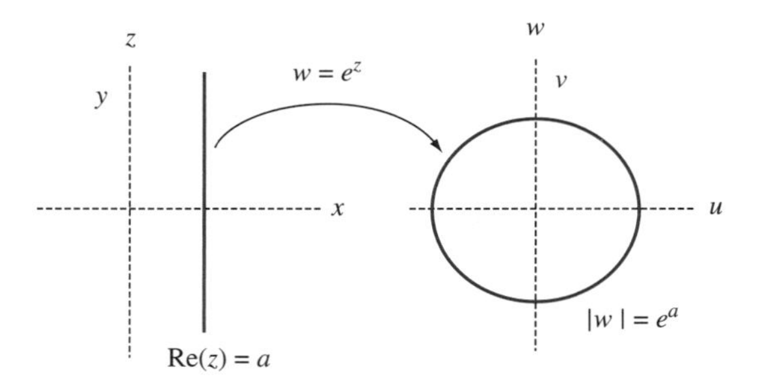

FIGURE 23.2 *Image of a vertical line under the mapping $f(z) = e^z$ in Example 23.2.*

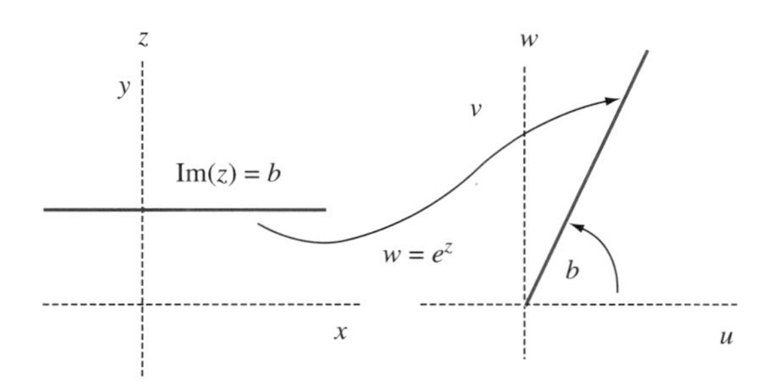

FIGURE 23.3 *Image of a horizontal line under $f(z) = e^z$ in Example 23.2.*

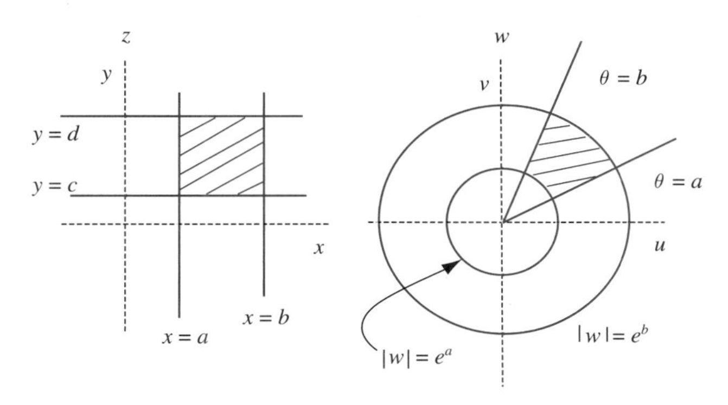

FIGURE 23.4 *Image of a rectangle under $f(z) = e^z$ for Example 23.2.*

Because $e^x > 0$ for all real x, as $x + ib$ varies over the horizontal line, the image point $e^x \sin(b) + ie^x \cos(b)$ moves along a half-line from the origin to infinity, making an angle b radians with the positive real axis (Figure 23.3). In polar coordinates, this is the half-line $\theta = b$.

We can put these results together to find the image of any rectangle in the z-plane having sides parallel to the axes. Let the rectangle have sides on the lines $x = a$, $x = b$, $y = c$, and $y = d$ (Figure 23.4). These lines map, respectively, to the circles

$$u^2 + v^2 = e^{2a} \quad \text{and} \quad u^2 + v^2 = e^{2b}$$

and the half-lines

$$\theta = c \text{ and } \theta = d.$$

The wedge in the w-plane in Figure 23.4 is the image of this rectangle under the exponential map. ♦

EXAMPLE 23.3

We will determine the image, under the mapping $w = f(z) = \sin(z)$, of the strip S consisting of all z with $-\pi/2 \leq \text{Re}(z) \leq \pi/2$ and $\text{Im}(z) \geq 0$. Write

$$w = u + iv = \sin(x)\cosh(y) + i\cos(x)\sinh(y).$$

If $z = x + iy$ is interior to S, then $y > 0$ so $\sinh(y) > 0$. Furthermore, since $-\pi/2 < x < \pi/2$ for z interior to S, $\cos(x) > 0$. Therefore, the image point has positive imaginary part and lies in the upper half-plane in the w-plane, so f maps the interior of S to the upper half-plane.

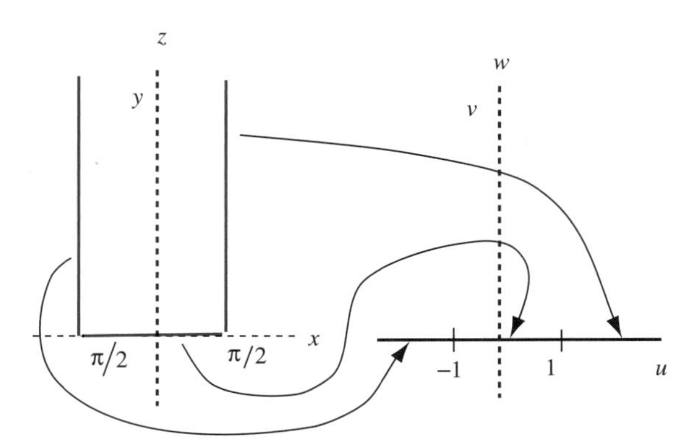

FIGURE 23.5 *Image of a strip under $f(z) = \sin(z)$ in Example 23.3.*

Now look at boundary points of S. The boundary of S consists of the segment $-\pi/2 \leq x \leq \pi/2$ on the real axis together with the half-lines $x = -\pi/2$ and $x = \pi/2$ with $y \geq 0$. Imagine a point z moving counterclockwise (positive orientation) around the boundary of S, and keep a record of how the image point w moves. Follow z and its image in Figure 23.5.

To start, put z on the left vertical boundary of S, which is the half-line $x = -\pi/2$ with $y \geq 0$. This z maps to

$$w = u + iv = -\cosh(y).$$

Furthermore, as z moves *down* this half-line toward the real axis so y is decreasing toward zero, $w = -\cosh(y)$ moves from left to right on the negative real axis and approaches -1 from the left.

Now z has reached $-\pi/2$ and turns to move from $-\pi/2$ to $\pi/2$ along the bottom boundary segment of S. On this segment, $z = x$ and $y = 0$, so the image point is $w = \sin(x)$, which moves from -1 to 1 as z moves toward $\pi/2$.

When z reaches $\pi/2$, it turns and moves up the right side of S, which is the half-line $x = \pi/2$, $y \geq 0$. The image point is $w = \cosh(y)$, which moves from 1 to the right out the real axis in the w-plane as y increases.

Therefore, f maps the boundary of S to the boundary of the upper half-plane, which is the real axis in the w-plane. If we imagine walking around the boundary of S in a positive sense with S over our left shoulder, the image point moves in a positive sense over the boundary of the image of S, which is the upper half-plane (which is over our left shoulder in that plane if we walk left to right along the real axis). ♦

We will find two properties enjoyed by some mappings to be particularly important.

1. *f preserves angles* if it satisfies the following requirement. If L_1 and L_2 are smooth curves in S intersecting at a point z_0 in S and θ is the angle between their tangents at this point, the images $f(L_1)$ and $f(L_2)$ in the w-plane have the same angle θ between their tangents at $f(z_0)$. This is indicated in Figure 23.6.
2. *f preserves orientation* if the following is true. In the scenario of property (1), if the sense of orientation from L_1 to L_2 at any point z_0 is counterclockwise, the sense of orientation from $f(L_1)$ to $f(L_2)$ in the w-plane also must be counterclockwise. Figure 23.7 shows the idea of an orientation preserving map, while Figure 23.8 suggests an orientation reversing map.

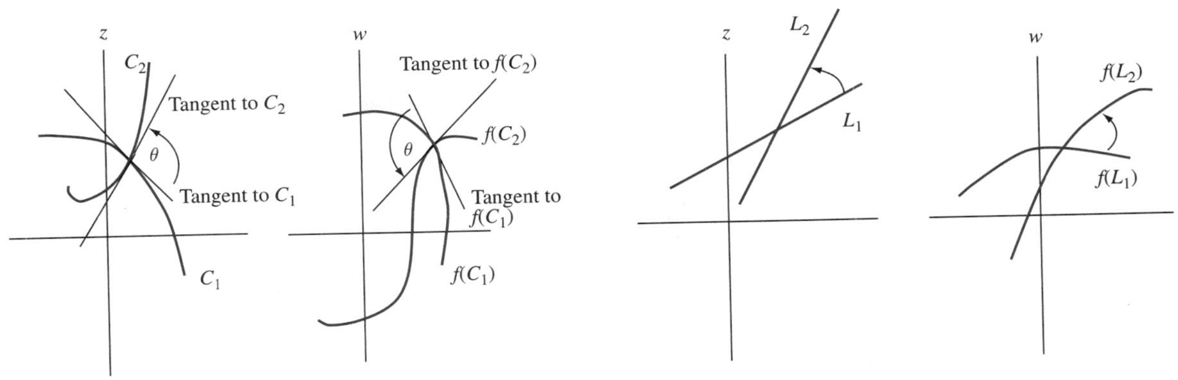

FIGURE 23.6 *f preserves angles.*

FIGURE 23.7 *f preserves orientation.*

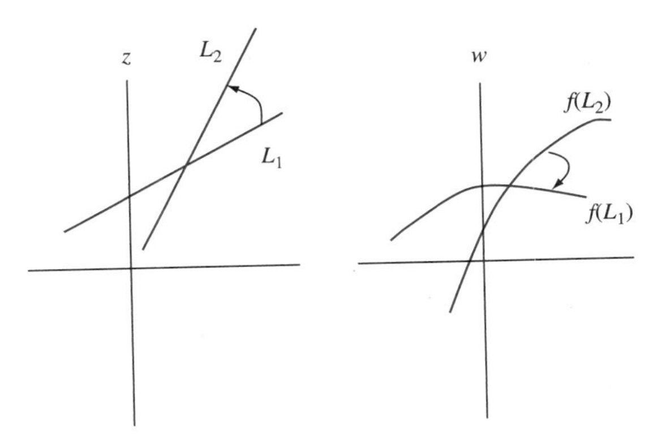

FIGURE 23.8 *f fails to preserve orientation.*

Not every function preserves angles and orientation. For example, the function $f(z) = \overline{z}$ reverses a rotation from counterclockwise to clockwise orientation. In particular, if we think of the counterclockwise rotation from $z = 0$ to $z = i$, the image points $f(0) = 0$ and $f(i) = \overline{i} = -i$ has a reversed, clockwise sense of rotation.

A function that is both orientation and angle preserving on a domain is said to be *conformal* on this domain. We refer to such a function as a *conformal mapping*.

The next theorem provides a large number of conformal mappings, namely, differentiable functions with nonvanishing derivatives.

THEOREM 23.1 Conformal Mappings

Let D and D^* be domains and let $f : D \to D^*$. Suppose f is differentiable on D and $f'(z) \neq 0$ on D. Then f is a conformal mapping. ♦

We will sketch a geometric argument to suggest why this result is true. Let z_0 be in D and let γ be a smooth curve in D through z_0. Then $f(\gamma)$, which consists of all points $f(z)$

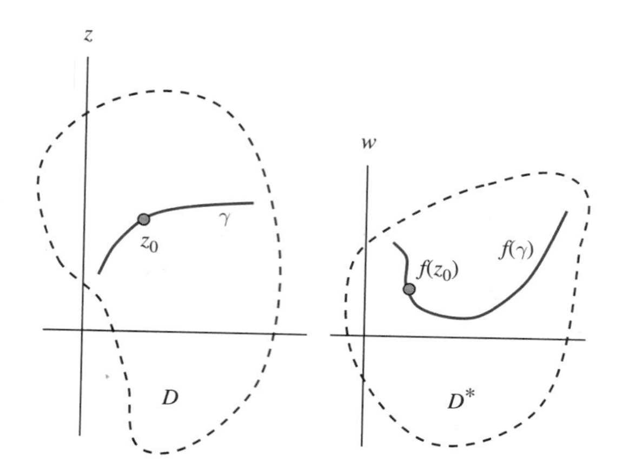

FIGURE 23.9 *A curve in a domain and its image under f.*

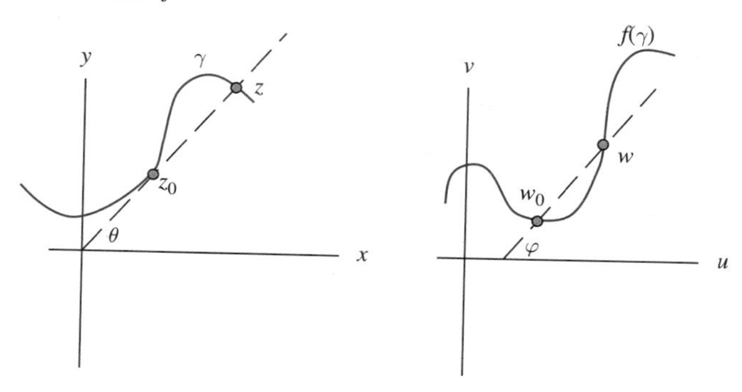

FIGURE 23.10 *Argument θ of the line through z and z_0.*

for z in γ, is a smooth curve through $f(z_0)$ in D^* (Figure 23.9). Let z be another point on γ, and write $w = f(z)$ and $w_0 = f(z_0)$. Then

$$w - w_0 = \frac{f(z) - f(z_0)}{z - z_0}(z - z_0).$$

Now, the argument of a product of two numbers is the sum of the arguments of the numbers (to within integer multiples of 2π). Therefore, to within $2n\pi$,

$$arg(w - w_0) = arg\left(\frac{f(z) - f(z_0)}{z - z_0}\right) + arg(z - z_0).$$

In Figure 23.10, θ is the angle between the positive real axis and the line through z and z_0 and is an argument of $z - z_0$. The angle φ between the positive real axis and the line through w and w_0 is an argument of $w - w_0$. In the limit, as $z \to z_0$, the last equation implies that

$$\varphi = arg(f'(z_0)) + \theta.$$

Repeat this discussion for any other smooth curve γ^* through z_0 with z^* a point on γ^*. By the same reasoning,

$$\varphi^* = arg(f'(z_0)) + \theta^*.$$

But then

$$\varphi - \varphi^* = \theta - \theta^*.$$

Since $\theta-\theta^*$ is the angle between the tangents to γ and γ^* at z_0 and $\varphi-\varphi^*$ is the angle between the tangents to $f(\gamma)$ and $f(\gamma^*)$ at $f(z_0)$, this shows that f preserves angles. The assumption that $f'(z_0)\neq 0$ is required to take the argument of $f'(z_0)$.

This equation also shows that f preserves orientation, since the sense of rotation from γ to γ^* is the same as that from $f(\gamma)$ to $f(\gamma^*)$, otherwise we would have

$$\varphi-\varphi^* = -(\theta-\theta^*).$$

This shows that f is conformal.

A mapping $f: D\to D^*$ is *one-to-one*, often written $1-1$, if two different numbers in D cannot be mapped to the same number in D^*.

This means that $f(z_1)\neq f(z_2)$ if z_1 and z_2 are different points of D. Thus, for example, $w=\sin(z)$ is not a one-to-one mapping of the complex plane to the complex plane, because $f(0)=f(2\pi)=0$.

We say that f is *onto* D^* if every number in D^* is the image of some number in D under the mapping. This means that if w is in D^* there must be some z in D such that $f(z)=w$.

If f is a one-to-one mapping of a domain D onto a domain D^*, then there is a unique pairing of each z in D with exactly one w in D^*, and conversely. This enables us to define the inverse mapping $f^{-1}: D^*\to D$ by setting

$$f^{-1}(w)=z \text{ exactly when } f(z)=w.$$

It is possible to show that f^{-1} is conformal if f is.

If $f: D\to D^*$ and $g: D^*\to D^{**}$ are both conformal, then their composition $g\circ f: D\to D^{**}$ is also conformal, since angles and orientation are preserved at each stage of the mapping (Figure 23.11).

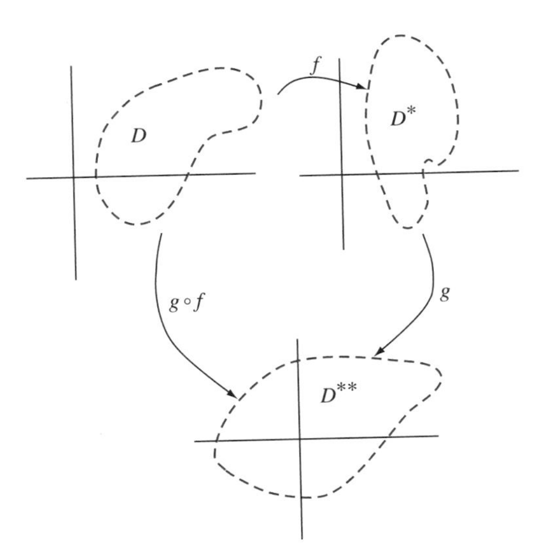

FIGURE 23.11 *A composition of conformal mappings is conformal.*

Sometimes we want to construct a conformal mapping between two sets—often between two domains D and D^*. We will see that this is one approach to solving Dirichlet problems. Depending on D and D^*, constructing a conformal mapping $f : D \to D^*$ may be very difficult. There is a special class of relatively simple mappings that can sometimes be used.

A mapping T is called a *bilinear transformation*, or *bilinear mapping* if it has the form

$$w = T(z) = \frac{az+b}{cz+d}$$

for given constants a, b, c, and d with $ad - bc \neq 0$. This condition insures that we can solve for z and invert the mapping

$$z = T^{-1}(w) = \frac{dw-b}{-cw+a},$$

which is again a bilinear transformation. Bilinear transformations are also known as *linear fractional transformations* or *Möbius transformations*, although these terms sometimes carry slight variations in the definition.

Because

$$T'(z) = \frac{ad-bc}{(cz+d)^2}$$

is not zero, a bilinear transformation and its inverse are conformal.

We will look at some special kinds of bilinear transformations, with a view toward dissecting general bilinear transformations into simple components.

EXAMPLE 23.4

Let $w = T(z) = z + b$, with b constant. This mapping is called a *translation*, because T shifts z horizontally by $\text{Re}(z)$ and vertically by $\text{Im}(z)$.

As an example, let $T(z) = 2 - i$. T takes z and moves it two units to the right and one unit down (Figure 23.12). For example, T maps

$$0 \to 2-i, \; 1 \to 3-i, \; i \to 2, \; 4+3i \to 6+2i. \; \blacklozenge$$

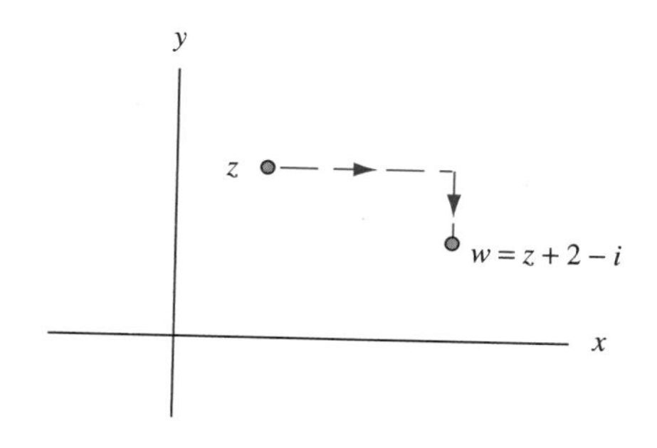

FIGURE 23.12 *A translation.*

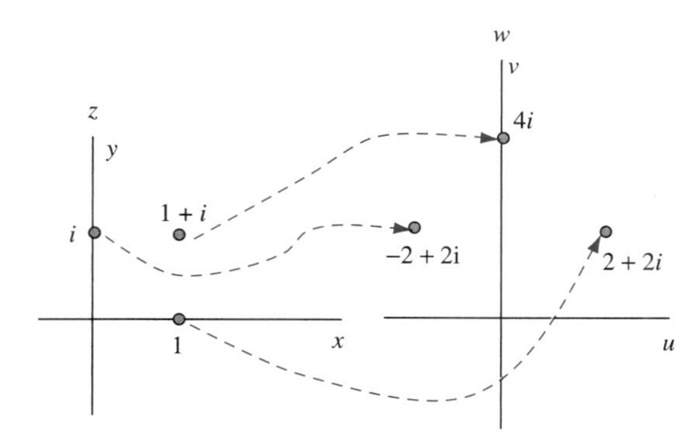

FIGURE 23.13 *Effect of the mapping $T(z) = (2+2i)z$ on specific points of Example 23.5.*

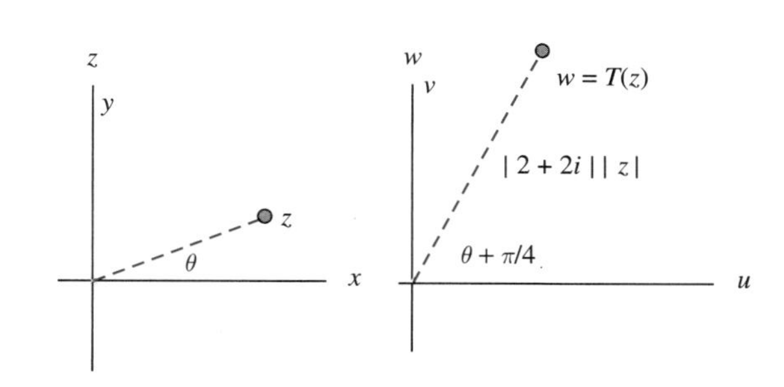

FIGURE 23.14 *Action of $T(z) = (2+2i)z$ on arbitrary z in Example 23.5.*

EXAMPLE 23.5

Let $w = T(z) = az$ with a a nonzero constant. This mapping is called a *rotation/magnification*. To see why this name applies, write

$$|w| = |a||z|.$$

If $|a| > 1$, then $T(z)$ is further from the origin than z is for $z \neq 0$. If $|a| < 1$, then T moves z closer to the origin. Thus the term *magnification*.

But T does more than this. If we write the polar forms $z = re^{i\theta}$ and $a = Ae^{i\alpha}$, then

$$T(z) = are^{i(\theta+\alpha)}$$

adding the constant angle α to the argument of z. This corresponds to a rotation by α radians—counterclockwise if $\alpha > 0$ and counterclockwise if $\alpha < 0$.

The total effect of this transformation is therefore a scaling and a rotation. We can see this effect in the mapping

$$T(z) = (2+2i)z.$$

This will map, for example,

$$i \to -2+2i, \ 1 \to 2+2i, \ 1+i \to -4i,$$

as shown in Figure 23.13. Figure 23.14 shows the action of T on an arbitrary nonzero z, multiplying the magnitude of z by $|2+2i| = \sqrt{8}$ and adding $\arg(2+2i) = \pi/4$ to the argument of z for a counterclockwise (positive) rotation through $\pi/4$ radians. ♦

If $|a| = 1$, $T(z) = az$ is called a *pure rotation*. In this case, there is no magnification effect, just a rotation through an argument of a.

EXAMPLE 23.6

Let

$$T(z) = \frac{1}{z} \text{ for } z \neq 0.$$

This mapping is called an *inversion*. If $z \neq 0$, then

$$|w| = \frac{1}{|z|}$$

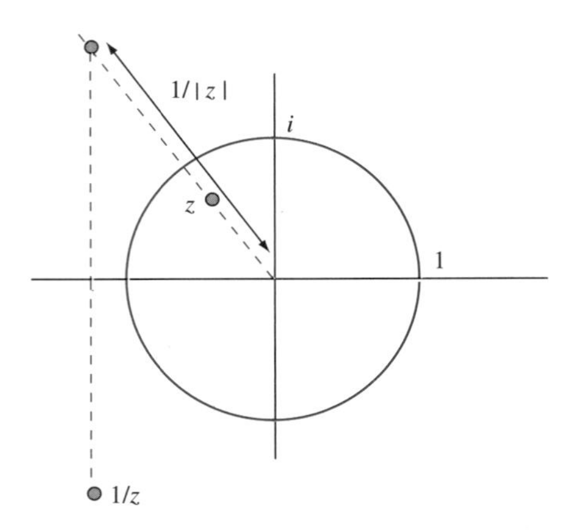

FIGURE 23.15 *Image of z under an inversion in Example 23.6.*

and

$$\arg(w) = \arg(1) - \arg(z) = -\arg(z)$$

(within integer multiples of 2π). This means that we arrive at $T(z)$ by moving $1/|z|$ units from the origin along the line from 0 to z and then reflecting this point across the real axis (Figure 23.15). Points enclosed by the unit circle map outside this circle, and points exterior to the unit circle map to the inside. Points on the unit circle remain on this circle but are moved around it, except for 1 and -1, which remain fixed under T. For example, T maps $(1+i)/\sqrt{2}$, which has argument $\pi/4$, to $(1-i)/\sqrt{2}$, which is still on the unit circle but has argument $-\pi/4$. ♦

We will now show that translations, rotation/magnifications, and inversions are the fundamental bilinear transformations in the sense that the effect of any bilinear transformation can be achieved as a sequence of mappings of these three types. To see how to do this, begin with

$$T(z) = \frac{az+b}{cz+d}.$$

If $c = 0$, then

$$T(z) = \frac{a}{d}z + \frac{b}{d},$$

which is a rotation/magnification followed by a translation:

$$z \xrightarrow{\text{rot/mag}} \frac{a}{d}z \xrightarrow{\text{trans}} +\frac{b}{d}.$$

If $c \neq 0$, then T is the result of the following sequence:

$$\begin{aligned} z &\xrightarrow{\text{rot/mag}} cz \xrightarrow{\text{trans}} cz+d \\ &\xrightarrow{\text{inv}} \frac{1}{cz+d} \xrightarrow{\text{rot/mag}} \frac{bc-ad}{c}\frac{1}{cz+d} \\ &\xrightarrow{\text{trans}} \frac{bc-ad}{c}\frac{1}{cz+d} + \frac{a}{c} \\ &= \frac{az+b}{cz+d} = T(z). \end{aligned}$$

This way of breaking a bilinear transformation into simpler components has two purposes. First, we can analyze general properties of these transformations by analyzing properties of the components. Second, and perhaps more important, we can sometimes use this sequence to build conformal mappings between given domains.

To illustrate the first point just made, we will state a general result about the action of a bilinear transformation. In this statement, the term line refers to a straight line in the plane.

THEOREM 23.2

A bilinear transformation maps any circle to a circle or line, and any line to a circle or line. ◆

Proof We need only verify the theorem for each of the three basis types: translations, rotation/magnifications, and inversions. Then the theorem will be true for any compositions of these mappings.

It is obvious geometrically that a translation maps a circle to a circle and a line to a line. Similarly, a rotation/magnification maps a circle to a circle (perhaps of different radius) and a line to a line. The issue comes down to the effect of an inversion on a circle or line. Begin with the fact that any circle or line in the plane is the graph of an equation

$$A(x^2+y^2)+Bx+Cy+R=0$$

with A, B, C, and R as real numbers. This graph is a circle if $A \neq 0$ and a line if $A=0$ and B and C are not both zero. Let $z=x+iy$. This equation becomes

$$A|z|^2+\frac{B}{2}(z+\overline{z})+\frac{C}{2i}(z-\overline{z})+R=0.$$

Now let $w=T(z)=1/z$, which is an inversion. The image in the w plane of the locus of this equation is the locus of

$$A\frac{1}{|w|^2}+\frac{B}{2}\left(\frac{1}{w}+\frac{1}{\overline{w}}\right)+\frac{C}{2i}\left(\frac{1}{w}-\frac{1}{\overline{w}}\right)+R=0.$$

Multiply this equation by $w\overline{w}$ (the same as $|w|^2$) to obtain

$$R|w|^2+\frac{B}{2}(w+\overline{w})-\frac{C}{2i}(w-\overline{w})+A=0.$$

In the w-plane, this is the equation of a circle if $R \neq 0$ and a line if $A=0$ and B and C are not both zero. This proves the theorem. ◆

As the proof shows, translations and rotation/magnifications actually map lines to lines and circles to circles, while an inversion may map a circle to a circle or line and a line to a circle or line.

EXAMPLE 23.7

We will examine the action of the inversion $w=1/z$ on the vertical line $\text{Re}(z)=a \neq 0$. This is the line $x=a$ in the x,y-plane, and it consists of all complex numbers $z=a+iy$. The image of such a point under the inversion is

$$w=\frac{1}{z}=\frac{a}{a^2+y^2}-\frac{y}{z^2+y^2}i=u+iv.$$

It is routine to check that

$$\left(u-\frac{1}{2a}\right)^2+v^2=\frac{1}{4a^2}.$$

The image of the line $x=a$ is therefore the circle of radius $1/2a$ with center $(1/2a, 0)$. ◆

In preparation for constructing mappings between given domains, we will show that it is always possible to produce a bilinear transformation mapping three given points to three given images.

THEOREM 23.3 *The Three Point Theorem*

Let z_1, z_2, and z_3 be three distinct points in the z-plane, and let w_1, w_2 and w_3 be three distinct points in the w-plane. Then there is a bilinear transformation T of the z-plane to the w-plane such that

$$T(z_1) = w_1, \; T(z_2) = w_2, \text{ and } T(z_3) = w_3. \; \blacklozenge$$

Proof We will provide a method for producing T. Let $w = T(z)$ be the solution for w in terms of z and the six given points in the equation

$$(w_1 - w)(w_3 - w_2)(z_1 - z_2)(z_3 - z) = (z_1 - z)(z_3 - z_2)(w_1 - w_2)(w_3 - w). \tag{23.1}$$

Substitution of $z = z_j$ into this equation yields $w = w_j$ for $j = 1, 2, 3$. ◆

EXAMPLE 23.8

We will produce a bilinear transformation mapping

$$3 \to i, \; 1 - i \to 4, \; 2 - i \to 6 + 2i.$$

Label

$$z_1 = 3, \; z_2 = 1 - i, \; z_3 = 2 - i$$

and

$$w_1 = i, \; w_2 = 4, \; w_3 = 6 + 2i.$$

Put these into equation (23.1):

$$(i - w)(2 + 2i)(2 + i)(2 - i - z) = (3 - z)(1)(i - 4)(6 + 2i - w).$$

Solve for w:

$$w = T(z) = \frac{(20 + 4i)z - (68 + 16i)}{(6 + 5i)z - (22 + 7i)}.$$

Then each $T(z_j) = w_j$. ◆

It is sometimes convenient to replace the complex plane with the complex sphere. This is done as follows. Let S be the sphere of radius 1 about $(0, 0, 1)$ in x, y, z-space R^3. Then S has the equation

$$x^2 + y^2 + (z - 1)^2 = 1.$$

Let N denote the point $(0, 0, 2)$. As Figure 23.16 suggests, given any point (x, y), or $x + iy$ in the plane, the straight line through N and (x, y) intersects the sphere in exactly one point $S(x, y)$. This associates with every point on the sphere, except N, exactly one point in the complex plane, and conversely, every point in the plane is associated with a unique point on the sphere. This mapping is called the *stereographic projection* of the sphere, minus the point N, onto the complex plane. For this reason, this punctured sphere is called the *complex sphere*. N does not correspond to any complex number under the stereographic projection. However, observe that, as the complex number $x + iy$ is chosen farther from the origin, the line from N to $x + iy$ intersects the sphere at a point closer to N. Thus, N plays the role of the *point at infinity*. It is not a point on the complex sphere any more than ∞ is a point in the complex plane. However, it is an identifiable

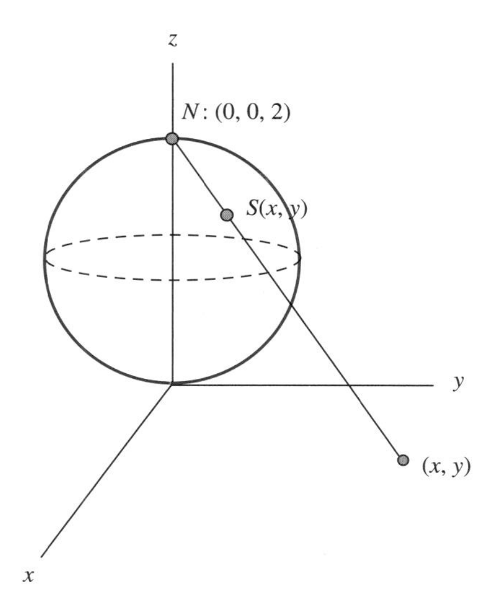

FIGURE 23.16 *Stereographic projection and the complex sphere.*

point on a simple surface, and we can envision a point on the complex sphere approaching N on the sphere more easily and concretely than we can envision $z \to \infty$ in the plane.

In defining bilinear transformations, it is sometimes convenient to map one of the three given points in Theorem 23.3 to infinity. The next theorem and its proof tell how to do this.

THEOREM 23.4

Let z_1, z_2, z_3 be three distinct numbers in the z-plane, and let w_1, w_2 be distinct numbers in the w-plane. Then there is a bilinear transformation T that maps

$$T(z_1) = w_1, \ T(z_2) = w_2, \ T(z_3) = \infty. \ \blacklozenge$$

Proof T is obtained by solving for w in the equation

$$(w_1 - w)(z_1 - z_2)(z_3 - z) = (z_1 - z)(z_1 - z_2)(w_1 - w_2). \qquad (23.2)$$

We can therefore map z_3 to infinity by deleting the factors in equation (23.1) that involve w_3.

EXAMPLE 23.9

We will find a bilinear transformation mapping

$$i \to 4i, 1 \to 3 - i, 2 + i \to \infty.$$

Solve for w in the equation

$$(4i - w)(i - 1)(2 + i - z) = (i - z)(-3 + 5i)(1 + i)$$

to obtain

$$w = T(z) = \frac{(5 - i)z - 1 + 3i}{-z + 2 + i}.$$

It is routine to check that $T(i) = 4i$, $T(1) = 3 - i$, and $T(2 + i) = \infty$. $\blacklozenge$

The next section is devoted to the construction of conformal mappings between domains.

SECTION 23.1 PROBLEMS

1. Determine the image of the sector $\pi/6 \le \theta \le \pi/3$ under the mapping $w = z^3$. Sketch the sector and its image.
2. Show that the mapping

$$w = \frac{1}{2}\left(z + \frac{1}{z}\right)$$

maps the circle $|z| = r$ onto an ellipse with foci ± 1 in the w-plane. Sketch a typical circle and its image.

3. In each of parts (a) through (e), find the image of the rectangle under the mapping $w = e^z$. Sketch the rectangle in the z-plane and its image in the w-plane.
   (a) $0 \le x \le \pi, 0 \le y \le \pi$
   (b) $-1 \le x \le 1, -\pi/2 \le y \le \pi/2$
   (c) $0 \le x \le 1, 0 \le y \le \pi/4$
   (d) $1 \le x \le 2, 0 \le y \le \pi$
   (e) $-1 \le x \le 2, -\pi/2 \le y \le \pi/2$
4. In each of parts (a) through (e), find the image of the rectangle under the mapping $w = \cos(z)$. Sketch the rectangle in the z-plane and its image in the w-plane.
   (a) $0 \le x \le 1, 1 \le y \le 2$
   (b) $\pi/2 \le x \le \pi, 1 \le y \le 3$
   (c) $0 \le x \le \pi, \pi/2 \le y \le \pi$
   (d) $\pi \le x \le 2\pi, 1 \le y \le 2$
   (e) $0 \le x \le \pi/2, 0 \le y \le 1$
5. In each of parts (a) through (e), find the image of the rectangle under the mapping $w = 4\sin(z)$. Sketch the rectangle in the z-plane and its image in the w-plane.
   (a) $0 \le x \le \pi/2, 0 \le y \le \pi/2$
   (b) $\pi/4 \le x \le \pi/2, 0 \le y \le \pi/2$
   (c) $0 \le x \le 1, 0 \le y \le \pi/6$
   (d) $\pi/2 \le x \le 3\pi/2, 0 \le y \le \pi/2$
   (e) $1 \le x \le 2, 1 \le y \le 2$
6. Determine the image of the sector $\pi/4 \le \theta \le 5\pi/4$ under the mapping $w = z^2$. Sketch the sector and its image.
7. Determine the image of D under the mapping $w = 2z^2$. Sketch this image.
8. Determine the image of the infinite strip $0 \le \text{Im} \le 2\pi$ under the mapping $w = e^z$.
9. Show that the mapping of Problem 2 maps a half-line $\theta = k$ onto a hyperbola with foci ± 1 in the w-plane, assuming that $\sin(k) \ne 0$ and $\cos(k) \ne 0$. Sketch a typical half-line and its image.
10. Let D consist of all z in the rectangle having vertices $\pm \alpha i$ and $\pi \pm \alpha i$, with α a positive number.
    (a) Determine the image of D under the mapping $w = \cos(z)$. Sketch D and its image.
    (b) Determine the image of D under the mapping $w = \sin(z)$. Sketch this image.

In each of Problems 11 through 16, find the image of the given circle or line under the bilinear transformation.

11. $w = (2z - 5)(z + i)$; $z + \overline{z} - (3/2i)(z - \overline{z}) - 5 = 0$
12. $w = (z - 1 + i)/(2z + 1)$; $|z| = 4$
13. $w = 2i/z$; $\text{Re}(z) = -4$
14. $w = ((1 + 3i)z - 2)/z$; $|z - i| = 1$
15. $w = (z - i)/iz$; $(z + \overline{z})/2 + (z - \overline{z})/2i = 4$
16. $w = 2iz - 4$; $\text{Re}(z) = 5$

In each of Problems 17 through 21, find a bilinear transformation taking the given points to the indicated images.

17. $6 + i \to 2 - i, i \to 3i, 4 \to -i$
18. $-5 + 2i \to 1, 3i \to 0, -1 \to \infty$
19. $1 \to 1, 2 \to -i, 3 \to 1 + i$
20. $i \to i, 1 \to -i, 2 \to 0$
21. $1 \to 1 + i, 2i \to 3 - i, 4 \to \infty$
22. Let T and S be bilinear mappings that agree at three points. Show that $T = S$.
23. Define the *cross ratio* of four complex numbers z_1, z_2, z_3, and z_4 to be the image of z_1 under the bilinear transformation that maps

$$z_2 \to 1,\ z_3 \to 0,\ z_4 \to \infty.$$

Denote this cross ratio $[z_1, z_2, z_3, z_4]$. Suppose T is any bilinear transformation. Show that T preserves this cross product. This means that

$$[z_1, z_2, z_3, z_4] = [T(z_1), T(z_2), T(z_3), T(z_4)].$$

24. Show that the cross ratio $[z_1, z_2, z_3, z_4]$ is the image of z_1 under the bilinear transformation

$$w = 1 - \frac{(z_3 - z_4)(z - z_2)}{(z_3 - z_2)(z - z_4)}.$$

25. Show that a cross ratio $[z_1, z_2, z_3, z_4]$ is real if and only if the z_j's are on a circle or a line.

26. Suppose T is a bilinear transformation that is not the identity mapping or a translation. Show that T must have either one or two fixed points. Why does this fail for translations?

27. Show that the mapping $w = T(z) = \overline{z}$ is not conformal.

28. Prove that the composition of two conformal mappings is conformal.

29. A point z_0 is a *fixed point* of a bilinear transformation T if $T(z_0) = z_0$. Suppose a bilinear transformation T has three fixed points. Show that T must be the identity mapping, sending each z to itself.

23.2 Construction of Conformal Mappings

One strategy for solving some kinds of problems (such as Dirichlet problems) is to find the solution for a "simple" domain (for example, a disk) and then map this domain conformally to the domain D on which we want to solve the problem. The idea is that this mapping may take the solution for the simple domain to a solution for D. This requires that we be able to construct conformal mappings between two domains.

Depending on the domains, this can be a daunting task, and it may not even be obvious that such a conformal mapping exists. The following theorem settles this issue, with one exception.

THEOREM 23.5 ***The Riemann Mapping Theorem***

Let D be the unit disk $|z| < 1$. Let D^* be a domain in the w-plane, and assume that D^* is not the entire w-plane. Then there exists a one-to-one conformal mapping $f : D \to D^*$ of D onto D^*. ♦

This powerful result implies the existence of a conformal mapping between two given domains. Suppose we want a conformal mapping from a domain D onto a domain D^* with neither domain the entire plane. Put the unit disk U in between, as in Figure 23.17. The Riemann mapping theorem ensures the existence of one-to-one, onto conformal mappings

$$f : U \to D \text{ and } g : U \to D^*.$$

Then the inverse mapping of f as

$$f^{-1} : D \to U$$

is also conformal, and the composition $F = g \circ f^{-1}$ is a one-to-one, conformal mapping of D onto D^*.

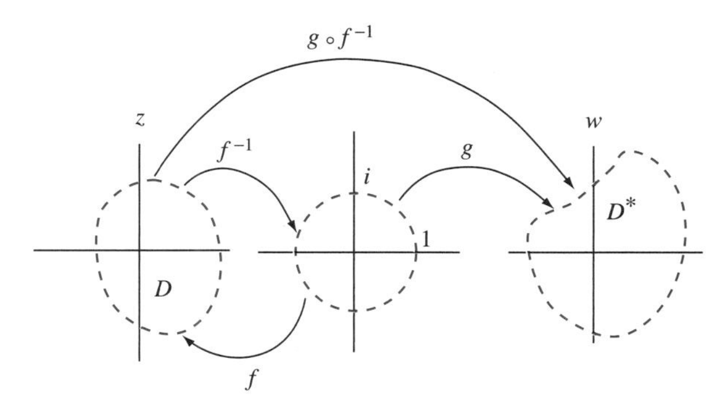

FIGURE 23.17 *Mapping D onto D^* through the unit disk.*

In theory then, two domains D and D^* (neither of which is the entire plane) can be mapped conformally onto each other in a one-to-one fashion. This does not, however, make such a conformal mapping easy to find. Bilinear transformations are conformal and will be suitable for some domains, but they will not be enough in general. However, here is an observation that is central to constructing a conformal mapping between two domains.

A conformal mapping $f : D \to D^*$ will map the boundary of D into the boundary of D^*. This can be exploited as follows. Suppose D is bounded by a path C (not necessarily closed), which separates the z-plane into two *complementary domains* D and $\mathcal{D}$. Similarly, suppose D^* is bounded by a path C^*, which separates the w-plane into complementary domains D^* and $\mathcal{D}^*$ (Figure 23.18). Try to find a conformal mapping f that sends points of C to points of C^*. This may be easier than mapping the entire domains to each other. This mapping f will then send D to either D^* or $\mathcal{D}^*$. To see which it is, choose any point z_0 in D and see whether $f(z_0)$ is in D^* or $\mathcal{D}^*$. If $f(z_0)$ is in D^*, then $f : D \to D^*$ (Figure 23.19). If $f(z_0)$ is in $\mathcal{D}^*$, then $f : D \to \mathcal{D}^*$ (Figure 23.20). In the first case, we have our conformal mapping. In the second, we do not, but in some cases, it is possible to take another step from f and construct a conformal mapping of $D \to D^*$.

We will illustrate these ideas with some examples, starting with very simple ones and building to more difficult problems.

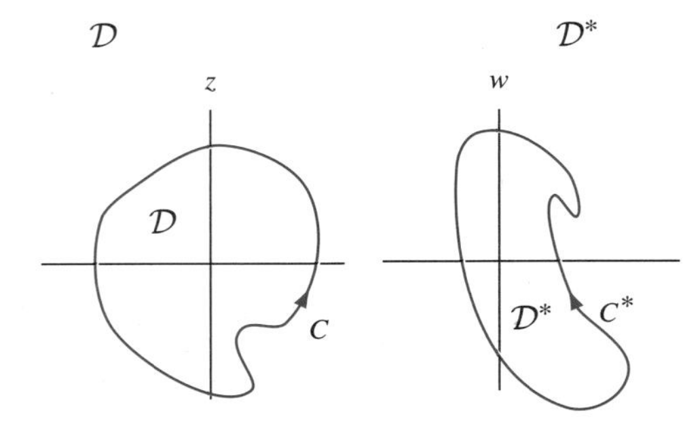

FIGURE 23.18 *Domains and complementary domains.*

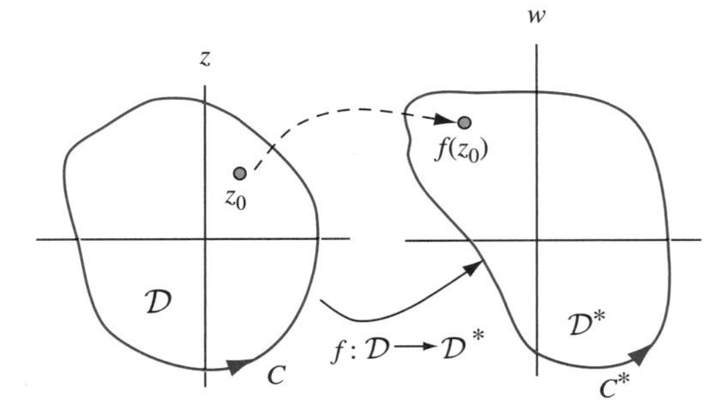

FIGURE 23.19 $f : D \to D^*$ *if* $f(x_0)$ *is in* D^*.

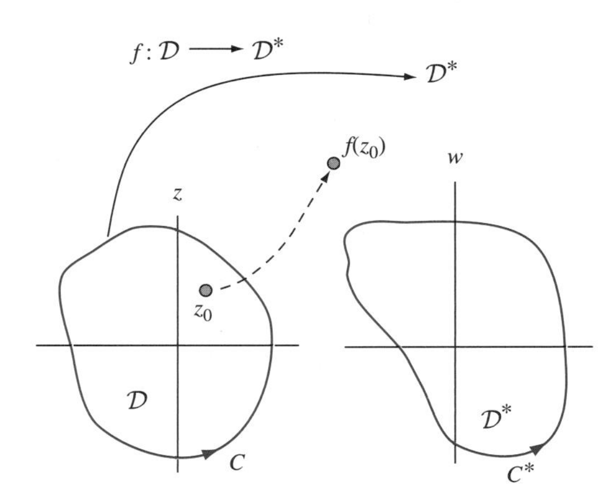

FIGURE 23.20 $f : D \to \mathcal{D}^*$ *if* $f(z_0)$ *is in* $\mathcal{D}^*$.

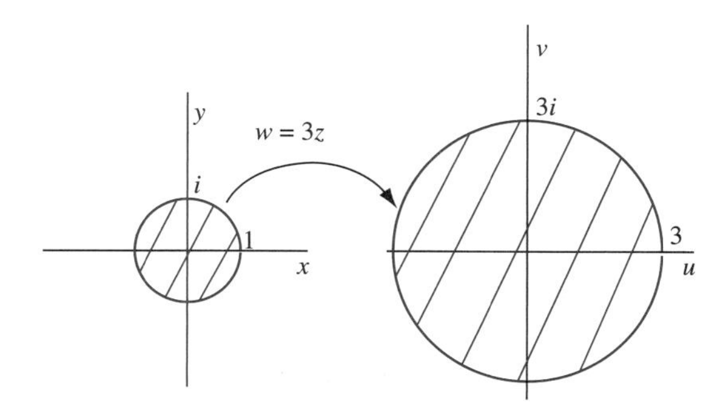

FIGURE 23.21 *Mapping $|z| < 1$ onto $|w| < 3$ in Example 23.10.*

EXAMPLE 23.10

We will map the open unit disk $D : |z| < 1$ conformally onto the open disk $D^* : |w| < 3$.

Clearly this is just a magnification, so $w = f(z) = 3z$ will do, expanding the unit disk to a disk of radius 3 while leaving the origin at the center (Figure 23.21). Observe that f carries the boundary of D onto the boundary of D^*. ♦

EXAMPLE 23.11

Map the open unit disk conformally onto the domain $D^* : |w| > 3$.

Here we are mapping the open unit disk to the complementary domain of the preceding example. We know that $f(z) = 3z$ maps D conformally onto $|w| < 3$. We also know that inversions map the domains interior to circles to domains exterior to circles. Thus, combine this map with an inversion, letting

$$g(z) = f(1/z) = \frac{3}{z}.$$

This maps $|z| < 1$ onto $|w| > 3$ (Figure 23.22). Again observe that g maps the boundary to the boundary. ♦

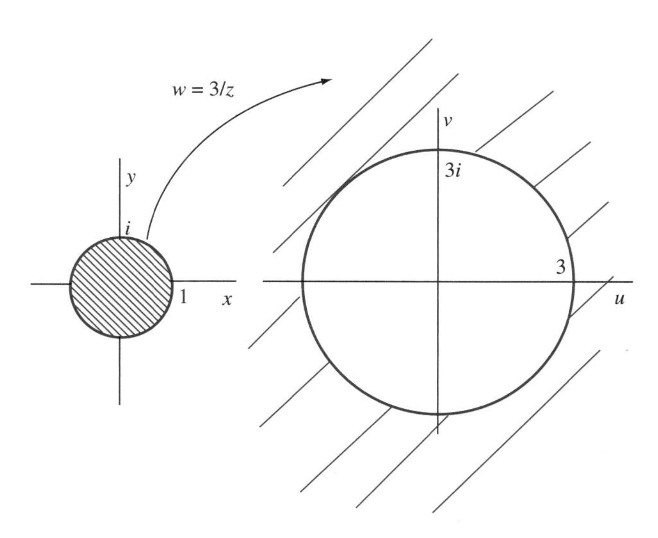

FIGURE 23.22 *Mapping $|z|$ onto $|w| > 3$ in Example 23.11.*

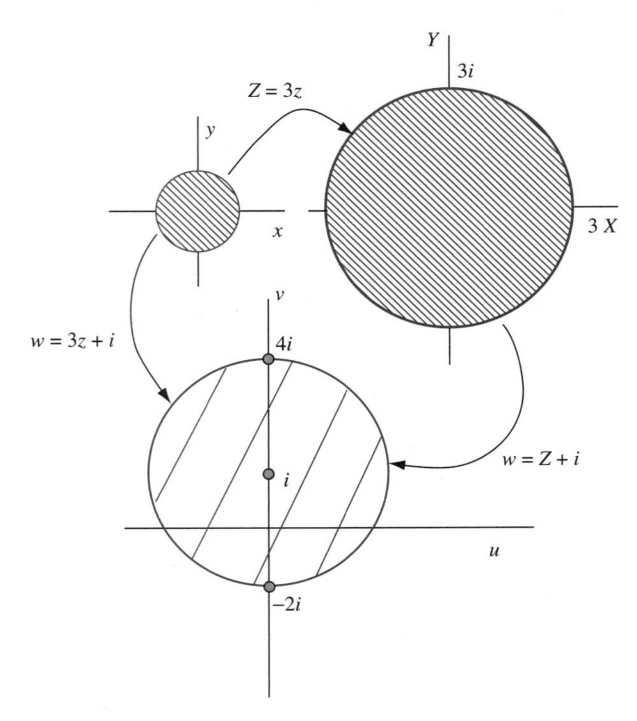

FIGURE 23.23 *Mapping $|z| < 1$ onto $|w - i| < 3$ in Example 23.12.*

EXAMPLE 23.12

Map the open unit disk onto the disk $|w - i| < 3$ or radius 3 centered at i in the w-plane.

Figure 23.23 suggests one way to construct this mapping. We want to expand the unit disk's radius by a factor of 3, then translate the resulting disk up one unit. Put an intermediary $Z = X + iY$ plane between the z-plane and the final w-plane, and map in steps:

$$z \to Z = 3z \to w = Z + i = 3z + i = w = f(z).$$

Note that the boundaries map to each other: the unit circle $|z| = 1$ maps to

$$|w - i| = 3|z| = 3,$$

which is the circle of radius 3 about i. ◆

EXAMPLE 23.13

We will find a conformal mapping of the right half-plane $\text{Re}(z) > 0$ to the unit disk $|w| < 1$.

Let S denote the right half-plane in the z-plane, and K denote the unit disk in the w-plane (Figure 23.24). The boundary of S is the imaginary axis, and the boundary of K is the unit circle. A bilinear transformation may work here, since the boundaries are a line and a circle. Pick three points on the imaginary axis (boundary of S) in order down the axis for positive orientation of this axis as the boundary of the right half-plane (walking in this direction, the right half-plane is over our left shoulder). We will use $z_1 = i$, $z_2 = 0$, and $z_3 = -i$, although other choices will do. Now choose three image points in order counterclockwise (positive orientation) on the unit circle in the w-plane, say $w_1 = 1$, $w_2 = i$, and $w_3 = -1$. This is the direction we have to walk around the unit circle to have the unit disk over our left shoulder.

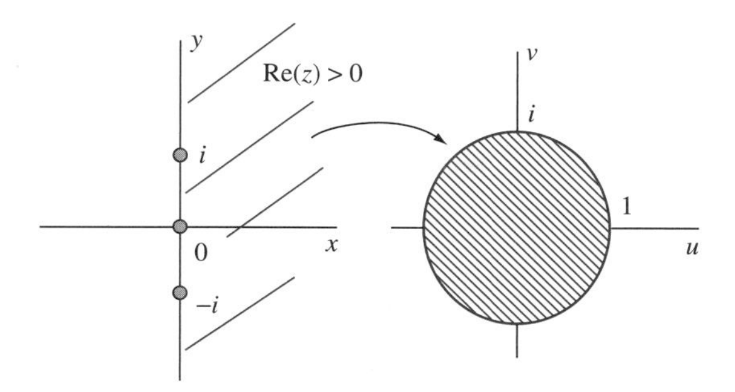

FIGURE 23.24 *Mapping the right half-plane onto the unit disk in Example 23.13.*

Put these z_j's and w_j's into equation (23.1):

$$(1 - w)(-1 - i)(i)(-i - z) = (i - z)(-i)(1 - i)(-1 - w).$$

Solve for w to get

$$w = T(z) = -i\left(\frac{z-1}{z+1}\right).$$

To ensure that T maps S onto the unit disk (not its complement), pick a point in S, say $z = 1$, and compute

$$T(1) = 0$$

as the center of the unit disk. Thus, we have a bilinear transformation (hence a conformal mapping) of the right half-plane onto the unit disk. ♦

Later (Example 23.16) we will want a mapping from the unit disk to the right half-plane. This is just the inverse of the mapping of Example 23.13, where we went the other way (inverse mapping) from the right half-plane to the unit disk. To find this inverse, solve for z in terms of w in Example 23.13 to obtain

$$z = \frac{i - w}{i + w}.$$

This is the mapping in the reverse direction from that of the last example. If we started from scratch, and wanted a mapping from the unit disk in the z-plane to the right half-plane in the w-plane, then we would interchange z and w in this mapping to write

$$w = \frac{i - z}{i + z}.$$

EXAMPLE 23.14

We will find a conformal mapping of the upper half-plane onto the exterior of the unit disk $|w| = 1$ in the w-plane. Again, since the boundaries are a line (the real axis) and a circle, we can attempt a bilinear transformation.

Choose three points on the boundary of the upper half-plane, which is the real axis oriented from left to right (so the upper half-plane is over the left shoulder as we walk along the line). Say we choose $z_1 = -1$, $z_2 = 0$, and $z_3 = 1$, as in Figure 23.25. Choose three points on the boundary of the disk of radius 1, but select them in order *clockwise* so that, as we walk along the circle in this direction, the *exterior* of the circle is over our left shoulder. Say we pick $w_1 = -1$, $w_2 = i$, and $w_3 = 1$. Now solve for w in

$$(-1 - w)(1 - i)(-1)(1 - z) = (-1 - z)(1)(-1 - i)(1 - w)$$

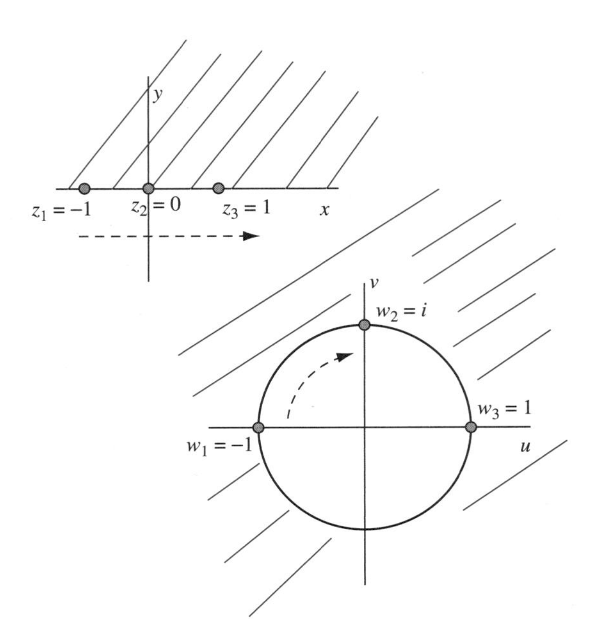

FIGURE 23.25 *Mapping the upper half-plane onto the exterior of the unit disk in Example 23.14.*

to get

$$w = T(z) = \frac{(1+i)z - 1 + i}{(-1+i)z + 1 + i}.$$

To ensure that the upper half-plane maps to the exterior (not the interior) of the unit circle, pick a point in the upper half-plane and verify that it maps to the exterior (not the interior) of the unit circle. For example, choosing $z = 2i$, we find that $T(2i) = -3i$ with a magnitude greater than 1. ♦

Thus far, we have gotten by with bilinear transformations. These are limited by the fact that they map circles and lines to circles and lines.

EXAMPLE 23.15

We will map the horizontal infinite strip $S: -\pi/2 < \text{Im}(z) < \pi/2$ onto the unit disk $|w| < 1$.

The boundary of S is not a line but consists of two lines, so a bilinear transformation is out of the question. To get our hands on a beginning, recall from Example 23.2 that the exponential function $w = e^z$ maps horizontal lines to half-rays from the origin. The boundary of S consists of two horizontal lines: $\text{Im}(z) = -\pi/2$ and $\text{Im}(z) = \pi/2$. On the lower boundary line, $z = x - i\pi/2$, so

$$w = e^z = e^x e^{-i\pi/2} = -ie^x,$$

which varies over the negative imaginary axis as x takes on all real values. On the upper boundary line of S, $z = x + i\pi/2$ and

$$w = e^z = e^x e^{i\pi/2} = ie^x,$$

which varies over the positive part of the imaginary axis as x varies over the real line. Now the imaginary axis forms the boundary of the right half-plane $\text{Re}(w) > 0$ in the w-plane and also the

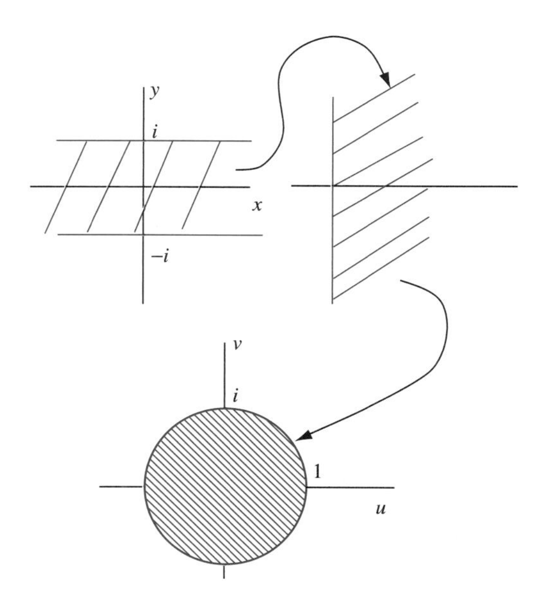

FIGURE 23.26 *Mapping a horizontal strip onto the unit disk in Example 23.15.*

boundary of the left half-plane, which is $\text{Im}(w) < 0$. The conformal mapping $w = e^z$ must send S to one of these half-planes. Since $z = 0$ is mapped to $w = 1$ in the right half-plane, then $w = e^z$ maps S to the right half-plane.

This is a start, but we want to map S onto the unit disk. However, we know of a conformal mapping of S onto the right half-plane, and now we also know of a mapping of the right half-plane onto the unit disk. This suggests that we put these together (Figure 23.26), mapping S first to the right half-plane, then the right half-plane to the open unit disk.

This will involve some change in notation, since proceeding in two steps requires that we insert an intermediary Z plane between the z- and w-planes. First map

$$Z = f(z) = e^z.$$

This takes S to the right half-plane in the Z-plane. Second, map this right half-plane $\text{Re}(Z) > 0$ onto the unit disk $|w| < 1$. We know how to do this, mapping

$$w = g(Z) = -i\left(\frac{Z-1}{Z+1}\right).$$

Compose these mappings:

$$w = F(z) = (g \circ f)(z) = g(f(z)) = g(e^z) = -i\left(\frac{e^z - 1}{e^z + 1}\right).$$

Notice if we pick a point in S, say $z = 0$, we obtain $w = F(0) = 0$ interior to $|w| < 1$, so F maps the right half-plane onto the interior (not the exterior) of the w-plane.

If we wish, we can write this mapping in terms of the hyperbolic tangent function

$$w = F(z) = -i\tanh(z/2). \quad \blacklozenge$$

EXAMPLE 23.16

We will map the disk $D : |z| < 2$ onto the domain $D^* : u + v > 0$ in the $w = u + iv$ plane. These domains are shown in Figure 23.27. D^* consists of points above the line $v = -u$ in the w-plane.

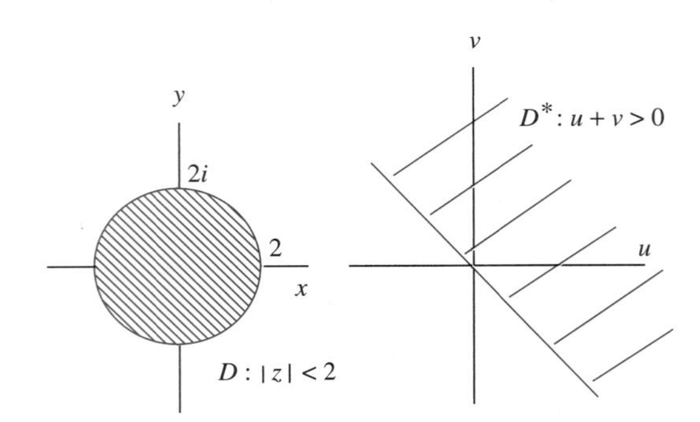

FIGURE 23.27 *The domains $|z| < 2$ and $u + v > 0$ in Example 23.16.*

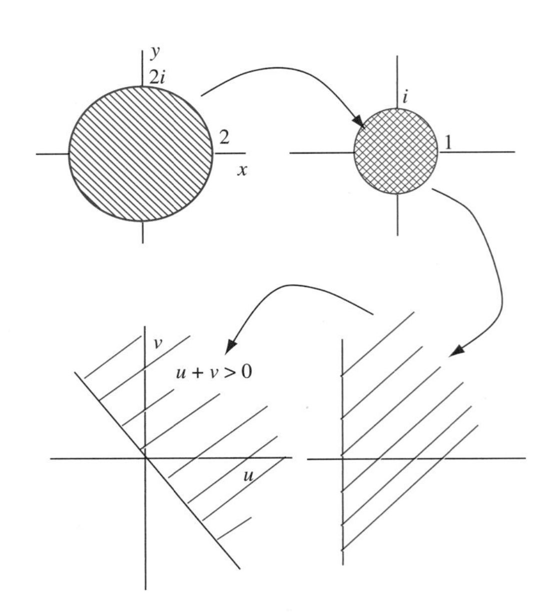

FIGURE 23.28 *Mapping $|z| < 2$ onto the rotated half-plane $u + v > 0$ in Example 23.16.*

As usual, first look for mappings we already have that relate to this problem. We can map $|z| < 2$ to the unit disk $|Z| < 1$ (multiply by $1/2$). We also know a mapping from the unit disk to the right half-plane, say from the Z-plane to the W-plane. Finally, we can obtain D^* from the right half-plane by a counterclockwise rotation through $\pi/4$ radians, which is an effect achieved by multiplying by $e^{i\pi/4}$ (a straight rotation). This suggests that we construct the mapping we want in the stages shown in Figure 23.28:

$$|z| < 2 \rightarrow |Z| < 1 \rightarrow \text{Re}(W) > 0 \rightarrow u + v > 0.$$

For the first step, let

$$Z = \frac{1}{2}z.$$

Next, following Example 23.13, we write a mapping between the unit circle and the right half-plane. In the current notation for the planes, this is

$$W = \frac{i - Z}{i + Z}.$$

Finally, we want to rotate the effect of this mapping counterclockwise by $\pi/4$ (last stage of Figure 23.28). For this, set

$$w = We^{i\pi/4}.$$

Putting everything together, a conformal mapping from D to D^* is given by

$$\begin{aligned} w = We^{i\pi/4} &= \left(\frac{i - Z}{i + Z}\right)e^{i\pi/4} \\ &= \left(\frac{i - z/2}{i + z/2}\right)e^{i\pi/4} \\ &= \left(\frac{2i - z}{2i + z}\right)e^{i\pi/4} = f(z). \end{aligned}$$

As an example, 0 is in D, and

$$f(0)=e^{i\pi/4}=\frac{\sqrt{2}}{2}(1+i)$$

is in the rotated half-plane $u+v>0$. ♦

23.2.1 The Schwarz-Christoffel Transformation

The Schwarz-Christoffel transformation is designed to produce a conformal mappings of the upper half-plane $\mathcal{U}$ in the z-plane to the region $\mathcal{P}$ bounded by a polygon P in the w-plane. This polygon could be a triangle, rectangle, pentagon, or other polyhedron. Because of the corners on the boundary polygon P of $\mathcal{P}$, such a mapping will be unlike anything we have seen up to this point.

Let P be an n-sided polygon with vertices $w_1, w_2, \cdots, w_n$ in the w-plane (Figure 23.29) with exterior angles $\alpha_1\pi, \alpha_2\pi, \cdots, \alpha_n\pi$.

The Schwarz-Christoffel conformal mapping f of the upper half-plane to the interior of P has the form

$$f(z)=a\int_{z_0}^{z}(\xi-x_1)^{-\alpha_1}(\xi-x_2)^{-\alpha_2}\cdots(\xi-x_n)^{-\alpha_n}\,d\xi+b \qquad (23.3)$$

in which $x_1, x_2, \cdots, x_n$ are real numbers labeled in increasing order, a and b are complex numbers, and z_0 is a complex number with $\text{Im}(z_0)>0$. These numbers must be chosen to suit P. The integral is taken over any path in $\mathcal{U}$ from z_0 to z in $\mathcal{U}$. The factors $(\xi-x_j)^{-\alpha_j}$ are defined using the complex logarithm obtained by restricting the argument to $[0, 2\pi)$.

To dissect this expression for $f(z)$ and see the ideas behind the various components, let

$$g(z)=a(\xi-x_1)^{-\alpha_1}(\xi-x_2)^{-\alpha_2}\cdots(\xi-x_n)^{-\alpha_n}$$

for z in $\mathcal{U}$. Then $f'(z)=g(z)$, and

$$\arg(f'(z))=\arg(a)-\alpha_1\arg(z-x_1)-\cdots-\alpha_n\arg(z-x_n).$$

As we saw in the proof of Theorem 23.1, $\arg(f'(z))$ is the number of radians by which f (as a mapping) rotates tangent lines if $f'(z)\neq 0$.

Now imagine z moving from left to right along the real axis, which is the boundary of $\mathcal{U}$. On $(-\infty, x_1)$ to the left of x_1, $f(z)$ moves along a straight line (no change in the angle). As z passes over x_1, $\arg(f'(z))$ changes by $\alpha_1\pi$. This angle remains fixed as z moves from x_1 to x_2. As z

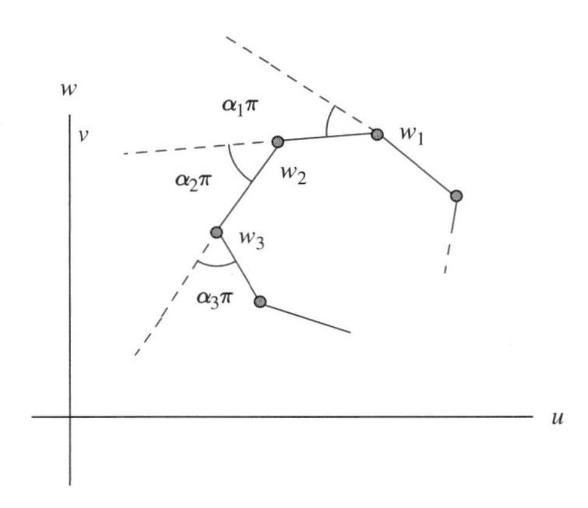

FIGURE 23.29 *A polygon and its exterior angles.*

passes over x_2, $\arg(f'(z))$ increases by $\alpha_2\pi$, then remains at this value until z passes over x_3, at which $\arg(f'(z))$ increases by $\alpha_3\pi$. In general, $\arg(f'(z))$ increases by $\alpha_j\pi$ as z passes over x_j. The net result (as z traverses the entire line) is that the real axis is mapped to a polygon P^* having exterior angles $\alpha_1\pi, \alpha_2\pi, \cdots, \alpha_n\pi$, which are the same as P. In fact $\alpha_1, \cdots, \alpha_{n-1}$ determine all n of these exterior angles, since

$$\sum_{j=1}^{n} \alpha_j\pi = 2\pi.$$

Now P^* has the same exterior angles as P but need not be the same as P because of its location and size. We may have to rotate and/or magnify P^* to obtain P. These effects are achieved by choosing $x_1, \cdots, x_n$ to make P^* similar to P, then choosing a (a rotation/magnification) and b (a translation) to superimpose P^* onto P.

If we choose $x_n = \infty$, then $x_1, x_2, \cdots, x_{n-1}, \infty$ are mapped to the vertices of P. In this case, the Schwarz-Christoffel transformation has the possibly simpler form as

$$f(z) = a\int_{z_0}^{z} (\xi - x_1)^{-\alpha_1}(\xi - x_2)^{-\alpha_2}\cdots(\xi - x_{n-1})^{-\alpha_{n-1}}\, d\xi + b \tag{23.4}$$

It can be shown that any conformal mapping of $\mathcal{U}$ onto a region bounded by a polygon must have the form of a Schwarz-Christoffel mapping. In practice, it may be impossible to carry out the integration needed to write a Schwarz-Christoffel transformation in closed form.

EXAMPLE 23.17

Suppose we want to map the upper half-plane onto a rectangle. Choose $x_1 = 0$, $x_2 = 1$, and x_3 as any number greater than 1. The corresponding Schwarz-Christoffel transformation equation (23.4) has the form

$$f(z) = a\int_{z_0}^{z} \frac{1}{\sqrt{\xi(\xi - 1)(\xi - x_3)}}\, d\xi + b$$

with a and b chosen to fit the dimensions and orientation of the original rectangle. The radical appears because the exterior angles of a rectangle are all $\pi/2$, so each $\alpha_j = 1/2$. This integral is an elliptic integral and cannot be evaluated in closed form. ♦

EXAMPLE 23.18

We will map $\mathcal{U}$ onto the strip S defined by $\text{Im}(w) > 0$, $-c < \text{Re}(z) < c$ in the w-plane. Here c is a positive constant.

$\mathcal{U}$ and S are shown in Figure 23.30. To use the Schwarz-Christoffel transformation, think of S as a polygon with vertices $-c$, c, and ∞. Choose $x_1 = -1$ to map $-c$ to -1 and $x_2 = 1$ to map to c. Map ∞ to ∞. The exterior angles of S are $\pi/2$ and $\pi/2$, so $\alpha_1 = \alpha_2 = 1/2$. The transformation has the form

$$w = f(z) = a\int_{z_0}^{z} (\xi + 1)^{-1/2}(\xi - 1)^{-1/2}\, d\xi + b.$$

Choose $z_0 = 0$ and $b = 0$, and write

$$(\xi - 1)^{-1/2} = [-(1 - \xi)]^{-1/2} = -i(1 - \xi)^{-1/2}.$$

Writing $-ai = A$, we have

$$w = f(z) = A\int_{0}^{z} \frac{1}{(1 - \xi^2)^{1/2}}\, d\xi.$$

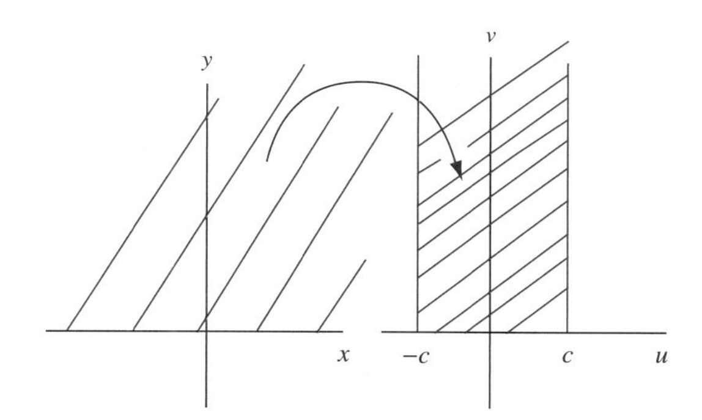

FIGURE 23.30 *Mapping the upper half-plane onto a vertical strip in Example 23.18.*

This integral reminds us of the real integral representation of the inverse sine function. Indeed, write

$$w = A\arcsin(z)$$

to mean that

$$z = \sin\left(\frac{w}{A}\right).$$

To choose A so that -1 maps to $-c$ and 1 to c, we need

$$\sin\left(\frac{c}{A}\right) = 1.$$

Thus, choose $c/A = \pi/2$, so $A = 2c/\pi$. Then

$$w = \frac{2c}{\pi}\arcsin(z).$$

If we choose $c = \pi/2$, this mapping is exactly $w = \arcsin(z)$, mapping $\mathcal{U}$ onto the strip $\text{Im}(z) > 0, -\pi/2 < \text{Re}(w) < \pi/2$. This is consistent with Example 23.3. ♦

SECTION 23.2 PROBLEMS

In each of Problems 1 through 6, find a bilinear transformation of the first domain onto the second.

1. $\text{Re}(z) < 0$ onto $|w| < 4$
2. $\text{Re}(z) > 1$ onto $\text{Im}(w) > -1$
3. $|z| < 3$ onto $|w - 1 + i| < 6$
4. $\text{Im}(z) > -4$ onto $|w - i| > 2$
5. $|z + 2i| < 1$ onto $|w - 3| > 2$
6. $|z| < 3$ onto $|w - 1 + i| > 6$
7. Show that the Schwarz-Christoffel transformation

$$f(z) = 2i\int_0^z (\xi + 1)^{-1/2}(\xi - 1)^{-1/2}\xi^{-1/2}\,d\xi$$

maps the upper half-plane onto the rectangle with vertices $0, c, c + ic$, and ic where $c = \Gamma(1/2)\Gamma(1/4)/\Gamma(3/4)$. For this problem, it is necessary to know the integral formula

$$B(x, y) = \int_0^1 u^{x-1}(1 - u)^{y-1}\,du$$

for the beta function $B(x, y)$, and also to know that, in terms of the gamma function,

$$B(x, y) = \frac{\Gamma(x)\Gamma(y)}{\Gamma(x + y)}.$$

Hint: See Section 15.3 (Problems 37 and 38).

8. Let $w = f(z) = \log(z)$ be defined by restricting the argument of z to lie in $[0, 2\pi]$. Show that f takes the upper half-plane onto the strip $0 < \text{Re}(w) < \pi$.
9. Find a conformal mapping of the upper half-plane onto the wedge $9 < \arg(w) < \pi/3$.

23.3 Conformal Mapping Solutions of Dirichlet Problems

The intimate connection between harmonic functions and differentiable complex functions allows a conformal mapping approach to certain Dirichlet problems.

The strategy behind this approach is to first solve the Dirichlet problem for a simple domain, the unit disk $\tilde{D}$. Given another domain D, we attempt to find a conformal mapping $T : \tilde{D} \to D$. T will then map the solution for $\tilde{D}$ to a solution for D.

We have Poisson's integral to solve the Dirichlet problem for the unit disk, but this is not well suited to the strategy we have outlined. We will derive another integral solution that will be more useful. Thus, suppose we want u that is harmonic on $\tilde{D}$ and $u(x, y) = g(x, y)$ on the unit circle bounding $\tilde{D}$.

If v is the harmonic conjugate of u, $f = u + iv$ is differentiable on the disk. We can assume (by adding a constant if necessary) that $v(0, 0) = 0$. Expand $f(z)$ in a Maclaurin series

$$f(z) = \sum_{n=0}^{\infty} a_n z^n.$$

Then

$$u(x, y) = \operatorname{Re}(f(x+iy)) = \frac{1}{2}\left(f(z) + \overline{f(z)}\right)$$
$$= \frac{1}{2}\sum_{n=0}^{\infty}(a_n z^n + \overline{a_n z^n}) = a_0 + \sum_{n=1}^{\infty}\frac{1}{2}(a_n z^n + \overline{a_n}\,\overline{z}^n).$$

Let ζ be on the unit circle γ. Then $|\zeta|^2 = \zeta\overline{\zeta} = 1$, so $\overline{\zeta} = 1/\zeta$ and

$$u(\zeta) = a_0 + \frac{1}{2}\sum_{n=1}^{\infty}(a_n\zeta^n + \overline{a_n}\zeta^{-n}).$$

Multiply this equation by $z^m/2\pi i$, and integrate over γ with γ as the variable of integration:

$$\frac{1}{2\pi i}\oint_\gamma u(\zeta)\zeta^m\,d\zeta = \frac{a_0}{2\pi i}\oint_\gamma \zeta^m\,d\zeta$$
$$+\frac{1}{2}\frac{1}{2\pi i}\sum_{n=1}^{\infty}\left(a_n\oint_\gamma \zeta^{n+m}\,d\zeta + \overline{a_n}\oint_\gamma z^{-n+m}\,d\zeta\right). \tag{23.5}$$

But

$$\oint_\gamma \zeta^k\,d\zeta = \begin{cases} 0 & \text{if } k \neq -1 \\ 2\pi i & \text{if } k = -1. \end{cases}$$

Therefore, if $m = -1$ in equation (23.5), we have

$$\frac{1}{2\pi i}\oint_\gamma u(\zeta)\frac{1}{\zeta}\,d\zeta = a_0.$$

If $m = -n - 1$ with $n = 1, 2, 3, \cdots$, we obtain

$$\frac{1}{2\pi i}\oint_\gamma u(\zeta)\zeta^{-n-1}\,d\zeta = \frac{1}{2}a_n.$$

Substitute these expressions for the coefficients into the Maclaurin series for $f(z)$ to get

$$f(z)=\sum_{n=0}^{\infty}a_nz^n=\frac{1}{2\pi i}\oint_\gamma u(\zeta)\frac{1}{\zeta}\,d\zeta+\sum_{n=1}^{\infty}\left(\frac{1}{\pi i}\oint_\gamma u(\zeta)\zeta^{-n-1}\right)d\zeta\, z^n$$

$$=\frac{1}{2\pi i}\oint_\gamma\left[1+2\sum_{n=1}^{\infty}\left(\frac{z}{\zeta}\right)^n\right]\frac{u(\zeta)}{\zeta}\,d\zeta.$$

Since $|z|<1$ and $|\zeta|=1$, then $|z/\zeta|<1$, and the geometric series in this equation converges:

$$\sum_{n=1}^{\infty}\left(\frac{z}{\zeta}\right)^n=\frac{z/\zeta}{1-z/\zeta}=\frac{z}{\zeta-z}.$$

Insert this into the last equation to obtain

$$f(z)=\frac{1}{2\pi i}\oint_\gamma\left[1+\frac{2z}{\zeta-z}\right]\frac{1}{\zeta}\,d\zeta=\frac{1}{2\pi i}\oint_\gamma\left(\frac{\zeta+z}{\zeta-z}\right)\frac{1}{\zeta}\,d\zeta.$$

Since ζ is on γ, $u(\zeta)=g(\zeta)$. Therefore, for $|z|<1$,

$$u(x,y)=\operatorname{Re}(f(z))=\operatorname{Re}\left(\frac{1}{2\pi i}\oint_\gamma g(\zeta)\left(\frac{\zeta+z}{\zeta-z}\right)\frac{1}{\zeta}\,d\zeta\right). \tag{23.6}$$

This is an integral solution of the Dirichlet problem for the unit disk. If $z=re^{i\theta}$ and $\zeta=e^{i\varphi}$ are inserted into this solution, the Poisson integral formula results.

Equation (23.6) is well suited to solving certain Dirichlet problems using conformal mappings. Suppose we know a conformal mapping $T:D\to\tilde{D}$ where $\tilde{D}$ is the unit disk $|w|<1$ (Figure 23.31). Assume that T maps C, which is the boundary of D, onto the unit circle $\tilde{C}$ bounding $\tilde{D}$, and that T^{-1} is also a conformal mapping.

To clarify the discussion, we will use ζ for a point of $\tilde{C}$, ξ for a point on C, and $(\tilde{x},\tilde{y})$ for a point in the w-plane.

Now consider a Dirichlet problem for D:

$$\frac{\partial^2u}{\partial x^2}+\frac{\partial^2u}{\partial^2 y}=0 \text{ for } (x,y) \text{ in } D$$

and

$$u(x,y)=g(x,y) \text{ for } (x,y) \text{ in } C=\partial D.$$

The idea is to map this problem to one for the unit disk $\tilde{D}$ for which we have the solution of equation (23.6), then use the inverse map to convert this integral into the solution for D.

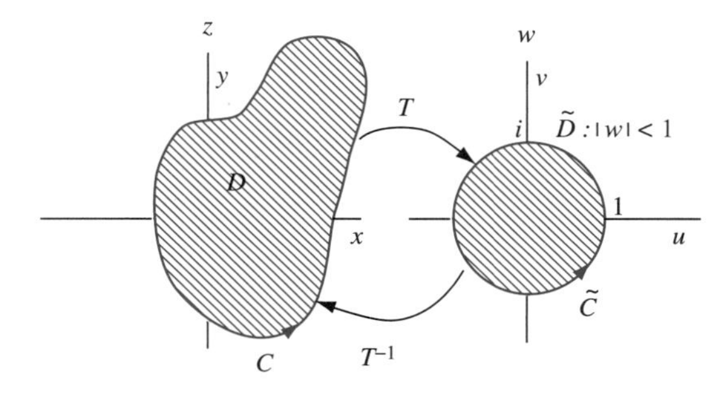

FIGURE 23.31 *Solving a Dirichlet problem by a conformal mapping.*

If $w = T(z)$, then $z = T^{-1}(w)$, and we define

$$\tilde{g}(w) = g(T^{-1}(w)) = g(z).$$

In the w-plane, we now have a Dirichlet problem for the unit disk:

$$\frac{\partial^2 \tilde{u}}{\partial x^2} + \frac{\partial^2 \tilde{u}}{\partial^2 y} = 0 \text{ for } (\tilde{x}, \tilde{y}) \text{ in } \tilde{D}$$

and

$$\tilde{u}(\tilde{x}, \tilde{y}) = \tilde{g}(\tilde{x}, \tilde{y}) \text{ for } (\tilde{x}, \tilde{y}) \text{ in } \tilde{C}.$$

By equation (23.6), the solution is

$$\tilde{u}(w) = \text{Re}(\tilde{f}(w)) = \text{Re}\left[\frac{1}{2\pi i}\oint_{\tilde{C}} \tilde{g}(\zeta)\left(\frac{\zeta + w}{\zeta - w}\right)\frac{1}{\zeta}\,d\zeta\right].$$

Because T maps C onto $\tilde{C}$ and $\zeta = T(\xi)$ for ξ on C, the solution of the Dirichlet problem for D is

$$u(x, y) = \text{Re}(f(z)) = \text{Re}\left[\frac{1}{2\pi i}\int_{\tilde{C}} \tilde{g}(T(\xi))\left(\frac{T(\xi) + T(z)}{T(\xi) - T(z)}\right)\frac{1}{T(\xi)}T'(\xi)\,d\xi\right].$$

Since $\tilde{g}(T(\xi)) = g(T^{-1}(T(\xi))) = g(\xi)$, we can write the solution

$$u(x, y) = \text{Re}(f(z)) = \text{Re}\left[\frac{1}{2\pi i}\int_{C} g(\xi)\left(\frac{T(\xi) + T(z)}{T(\xi) - T(z)}\right)\frac{T'(\xi)}{T(\xi)}\,d\xi\right]. \qquad (23.7)$$

In this integral, C need not be a closed curve. If D is (for example) the right quarter-plane $x > 0$, $y > 0$, then C will consist of two segments of the real and imaginary axes.

EXAMPLE 23.19

We will demonstrate this technique for a Dirichlet problem for the right half-plane:

$$\frac{\partial^2 u}{\partial x^2} + \frac{\partial^2 u}{\partial^2 y} = 0 \text{ for } x > 0, -\infty < y < \infty$$

and

$$u(0, y) = g(y) \text{ for } -\infty < y < \infty.$$

We need a conformal mapping from the right half-plane to the unit disk. There are many such mappings, but we found one in Example 23.13:

$$w = T(z) = -i\left(\frac{z-1}{z+1}\right).$$

Compute

$$T'(z) = \frac{-2i}{(z+1)^2}.$$

From equation (23.7), the solution is the real part of

$$f(z) =$$

$$\frac{1}{2\pi i}\int_C u(\xi)\left(\frac{-i(\xi-1)/(\xi+1) - i(z-1)/(z+1)}{-i(\xi-1)/(\xi+1) + i(z-1)/(z+1)}\right)\frac{1}{-i(\xi-1)/(\xi+1)}\frac{-2i}{(\xi+1)^2}\,d\xi$$

$$= \frac{1}{\pi i}\int_C u(\xi)\left(\frac{\xi z - 1}{\xi - z}\right)\frac{1}{\xi^2 - 1}\,d\xi.$$

C is the boundary of the right half-plane, which is the imaginary axis. Parametrize C as $\xi = (0, t) = it$ with t varying from ∞ to $-\infty$ for positive orientation (if we walk *down* this axis, the right half-plane is over our left shoulder). Now

$$f(z) = \frac{1}{\pi i}\int_{\infty}^{-\infty} u(0,t)\left(\frac{itz-1}{it-z}\right)\left(\frac{-1}{1+t^2}\right) i\,dt$$
$$= \frac{1}{\pi}\int_{-\infty}^{\infty} u(0,t)\left(\frac{itz-1}{it-z}\right)\left(\frac{1}{1+t^2}\right) dt.$$

The solution is the real part of this integral. Now t, $u(0,t)$ and $1/(1+t^2)$ are real, so we must pull out the real part of the terms containing i and $z = x+iy$ in the integral. Write

$$\frac{itz-1}{it-z} = \frac{itx-ty-1}{it-x-iy} = \left(\frac{itx-ty-1}{it-x-iy}\right)\left(\frac{-it-x+iy}{-it-x+iy}\right)$$
$$= \frac{tx(t-y) - itx^2 + ity(t-y) + txy + i(t-y) + x}{x^2+(t-y)^2}.$$

The real part of this expression is

$$\frac{x(1+t^2)}{x^2+(t-y)^2}.$$

Therefore,

$$u(x,y) = \operatorname{Re}(f(z)) = \frac{1}{\pi}\int_{-\infty}^{\infty} u(0,t)\frac{x(1+t^2)}{x^2+(t-y)^2}\frac{1}{1+t^2}\,dt$$
$$= \frac{1}{\pi}\int_{-\infty}^{\infty} g(t)\frac{x}{x^2+(t-y)^2}\,dt. \quad \blacklozenge$$

SECTION 23.3 PROBLEMS

In each of Problems 1 through 6, use equation (23.4) to solve the Dirichlet problem for the given domain with the given boundary data.

1. The unit disk if $u(x,y) = x - y$ for (x,y) on the boundary circle.
2. The unit disk with
$$u(e^{i\theta}) = \begin{cases} 1 & \text{for } 0 \le \theta \le \pi/4 \\ 0 & \text{for } \pi/4 < \theta < 2\pi. \end{cases}$$
3. Upper half-plane: $u(x,0) = f(x)$.
4. Right quarter plane: $\operatorname{Re}(z) > 0$, $u(x,0) = f(x)$, and $u(0,y) = 0$.
5. The disk $|z - z_0| < R$ if $u(x,y) = xy$ for (x,y) on the boundary.
6. Right half-plane with boundary condition:
$$u(0,y) = \begin{cases} 1 & \text{for } -1 \le x \le 1 \\ 0 & \text{for } |y| > 1. \end{cases}$$
7. Solve the Dirichlet problem for the strip $-1 < \operatorname{Im}(z) < 1$, $\operatorname{Re}(z) > 0$ with the boundary conditions
$$u(x,1) = u(x,-1) = 0 \text{ for } x > 0$$
and
$$u(0,y) = 1 - |y| \text{ for } -1 \le y \le 1.$$

23.4 Models of Plane Fluid Flow

This section is an introduction to complex function models of fluid flow. Suppose an incompressible fluid moves with velocity field $\mathbf{V}(x,y)$. By assuming that the velocity depends only on two variables, we are taking the flow to be the same in all planes parallel to the complex plane. Such a flow is called *plane-parallel*. This velocity vector is also assumed to be independent of time, in which case we say that the flow is *stationary*. Write

$$\mathbf{V}(x, y) = u(x, y)\mathbf{i} + v(x, y)\mathbf{j}.$$

Since we identify vectors in the plane with complex numbers, we will write this as

$$V(z) = V(x + iy) = u(x, y) + iv(x, y).$$

Given $V(z)$, think of the complex plane as divided into two sets. The first is the domain D on which V is defined. The *complement of* D consists of the points not in D. Think of this complement as comprising channels or barriers confining the fluid to D. This enables us to model fluid flow through a variety of configurations and around barriers of various shapes.

Suppose γ is a closed path in D. Write $\gamma(s) = x(s) + iy(s)$, using arc length s as a parameter. Then $x'(s)\mathbf{i} + y'(s)\mathbf{j}$ is the unit tangent vector to γ. Furthermore,

$$(u\mathbf{i} + v\mathbf{j}) \cdot \left(\frac{dx}{ds}\mathbf{i} + \frac{dy}{ds}\mathbf{j}\right) ds = u\,dx + v\,dy.$$

Since $u\,dx + v\,dy$ is the dot product of the velocity with the unit tangent along the trajectory γ, we interpret

$$\oint_\gamma u\,dx + v\,dy$$

as a measure of the velocity of the fluid along γ. This integral is the *circulation* of the fluid around γ.

The vector $-y'(s)\mathbf{i} + x'(s)\mathbf{j}$ is a unit normal vector to γ (Figure 23.32). Then

$$-\oint_\gamma (u\mathbf{i} + v\mathbf{j}) \cdot \left(\frac{dy}{ds}\mathbf{i} + \frac{dx}{ds}\mathbf{j}\right) ds = \oint_\gamma -v\,dx + u\,dy$$

is the negative of the integral of the normal component of the velocity along γ. When this integral is not zero, it is called the *flux* of the fluid across the path. This gives a measure of fluid flow across γ out of the region enclosed by γ. When this flux is zero for every closed path in the domain of the fluid, we call the fluid *solenoidal*.

A point $z_0 = x_0 + iy_0$ is a *vortex* of the fluid if the circulation has a nonzero value k that is the same for every closed path about z_0 in the interior of some annulus $0 < |z - z_0| < r$. We call $|k|$ the *strength of the vortex*. If $k > 0$ the vortex is a *source*, and if $k < 0$ it is a *sink*.

The following result is the key to using complex functions to analyze fluid flow.

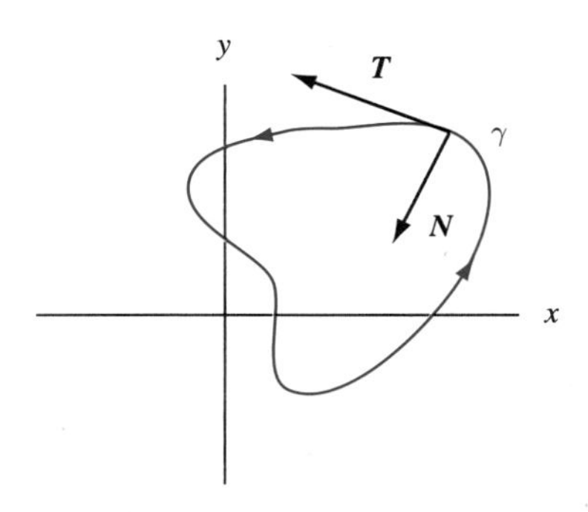

FIGURE 23.32 *Unit tangent and normal vectors to C.*

THEOREM 23.6

Let u and v be continuous with continuous first and second partial derivatives in a simply connected domain D. Suppose $u\mathbf{i}+v\mathbf{j}$ is irrotational (has curl $\mathbf{O}$) and solenoidal. Then u and $-v$ satisfy the Cauchy-Riemann equations on D, and $f(z)=u(x,y)-iv(x,y)$ is differentiable on D. Conversely, if u and $-v$ satisfy the Cauchy-Riemann equations on D, then $u\mathbf{i}+v\mathbf{j}$ defines an irrotational, solenoidal flow on D. ♦

With zero curl, the fluid experiences no swirling, although there can be translations and distortions in the motion. If the flow is solenoidal, then

$$\operatorname{div}(u\mathbf{i}+v\mathbf{j})=\frac{\partial u}{\partial x}+\frac{\partial v}{\partial y}=0.$$

A further connection between flows and complex functions is provided by the following.

THEOREM 23.7

Let f be differentiable on a domain D. Then $\overline{f'(z)}$ is an irrotational solenoidal flow on D. Conversely, if $\mathbf{V}=u\mathbf{i}+v\mathbf{j}$ is an irrotational, solenoidal vector field on a simply connected domain D, then there is a differentiable complex function f defined on D such that $\overline{f'(z)}=V(z)$. Furthermore, if $f(z)=\varphi(x,y)=i\psi(x,y)$, then

$$\frac{\partial\varphi}{\partial x}=u=\frac{\partial\psi}{\partial y}\text{ and }\frac{\partial\varphi}{\partial y}=v=-\frac{\partial\psi}{\partial x}.\ ♦$$

In view of the fact that $\overline{f'(z)}$ is the velocity of a flow, we call f a *complex potential* for the flow. Theorem 23.7 implies that any differentiable function $f(z)=\varphi(x,y)+i\psi(x,y)$ defined on a simply connected domain D determines an irrotational, solenoidal flow

$$\overline{f'(z)}=\overline{\frac{\partial\varphi}{\partial x}+i\frac{\partial\psi}{\partial x}}=\overline{u(x,y)-iv(x,y)}=u(x,y)+iv(x,y).$$

We call φ the *velocity potential* of the flow, and curves $\varphi(x,y)=k$ are *equipotential curves*. The function ψ is the *stream function* of the flow, and curves $\psi(x,y)=c$ are the *streamlines*.

At any z at which $f'(z)\neq 0$, f is a conformal mapping. A point at which $f'(z)=0$ is called a *stagnation point*. Thinking of f as a mapping of the z-plane to the w-plane, we have

$$w=f(z)=\varphi(x,y)+i\psi(x,y)=\alpha+i\beta.$$

Equipotential curves $\varphi(x,y)=k$ map to vertical lines $\alpha=k$, and streamlines $\psi(x,y)=c$ map to horizontal lines $\beta=c$. Because these vertical and horizontal lines are orthogonal in the w-plane and f is conformal, the streamlines and equipotential curves in the z-plane also form orthogonal families. Every streamline is orthogonal to each equipotential curve at any point of intersection. This condition fails at a stagnation point, where the mapping may not be conformal.

Along an equipotential curve $\varphi(x,y)=k$,

$$d\varphi=\frac{\partial\varphi}{\partial x}dx+\frac{\partial\varphi}{\partial y}dy=u\,dx+v\,dy=0.$$

Now $u\mathbf{i}+v\mathbf{j}$ is the velocity of the flow at (x,y), and $x'(s)\mathbf{i}+y'(s)\mathbf{j}$ is a unit tangent to the equipotential curve through (x,y). Since the dot product of these two vectors is zero (from the fact that $d\varphi=0$ along the equipotential curve), we conclude that the velocity is orthogonal to the equipotential curve through (x,y)—provided that (x,y) is not a stagnation point.

Along a streamline $\psi(x, y) = c$,

$$d\psi = \frac{\partial \psi}{\partial x}dx + \frac{\partial \psi}{\partial y}dy = -v\,dx + u\,dy = 0,$$

so the normal to the velocity vector is orthogonal to the streamline. This means that the velocity is tangent to the streamline and justifies the interpretation that the particle of fluid at (x, y) is moving in the direction of the streamline at this point. We therefore interpret streamlines as the trajectories of the particles in the fluid. For this reason, graphs of the streamlines form a picture of the motion of the fluid.

EXAMPLE 23.20

Let $f(z) = -Ke^{i\theta}z$ with K as a positive constant and θ is fixed with $0 \le \theta \le 2\pi$. Write

$$\begin{aligned} f(z) &= -K(\cos(\theta) + i\sin(\theta))(x + iy) \\ &= -K(x\cos(\theta) - y\sin(\theta)) - iK(y\cos(\theta) + x\sin(\theta)). \end{aligned}$$

If $f(z) = \varphi(x, y) + i\psi(x, y)$, then

$$\varphi(x, y) = -K(x\cos(\theta) - y\sin(\theta))$$

and

$$\psi(x, y) = -K(y\cos(\theta) + x\sin(\theta)).$$

Since K is constant, equipotential curves are graphs of

$$x\cos(\theta) - y\sin(\theta) = k$$

or

$$y = \cot(\theta) + b$$

in which $b = k\sec(\theta)$ is constant. These are straight lines with slope $\cot(\theta)$.

Streamlines are graphs of

$$\psi(x, y) = -\tan(\theta)x + d,$$

which are straight lines of slope $-\tan(\theta)$. These lines make an angle $\pi - \theta$ with the positive real axis, as in Figure 23.33. These are the trajectories of the flow. The streamlines and equipotential lines are orthogonal with slopes that are negative reciprocals of each other.

Now compute

$$\overline{f'(z)} = \overline{-Ke^{i\theta}} = -Ke^{-i\theta},$$

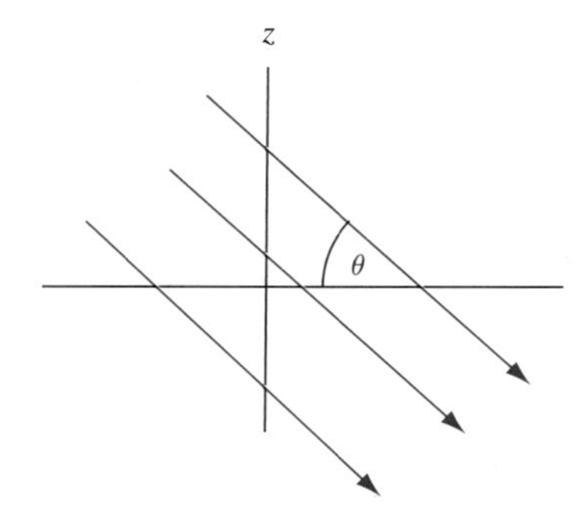

FIGURE 23.33 *Streamlines in Example 23.20.*

since θ is real. This implies that the velocity has constant magnitude K. In sum, f models a uniform flow with velocity of a constant magnitude K moving at an angle $\pi - \theta$ with the positive real axis. ♦

EXAMPLE 23.21

We will analyze the flow having complex potential $f(z) = z^2$. Since $f'(z) = 2z = 0$ exactly when $z = 0$, the origin is the only stagnation point. We will determine the trajectories of the flow.

With $z = x + iy$, $f(z) = x^2 - y^2 + 2ixy$, so

$$\varphi(x, y) = x^2 - y^2 \text{ and } \psi(x, y) = 2xy.$$

Equipotential curves are hyperbolas $x^2 - y^2 = k$ if $k \neq 0$, and straight lines $y = \pm x$ if $k = 0$. These are asymptotes of the hyperbolic equipotential curves. Streamlines are hyperbolas $xy = c$ if $c \neq 0$, and the axes are $x = 0$ and $y = 0$ if $c = 0$. Some streamlines and equipotential lines are shown in Figure 23.34.

The velocity of the flow is $\overline{f'(z)} = 2\bar{z}$. f models a nonuniform flow having velocity of magnitude $2|z|$ at z. This flow moves along the hyperbolic streamlines with the axes acting as barriers of the flow (think of sides of a container holding the fluid). ♦

EXAMPLE 23.22

Consider the flow associated with the complex potential

$$f(z) = \frac{iK}{2\pi} \text{Log}(z),$$

for $z \neq 0$, where $\text{Log}(z)$ is the logarithm of z having argument between 0 and 2π. Thus,

$$\text{Log}(z) = \frac{1}{2} \ln(x^2 + y^2) + i\theta,$$

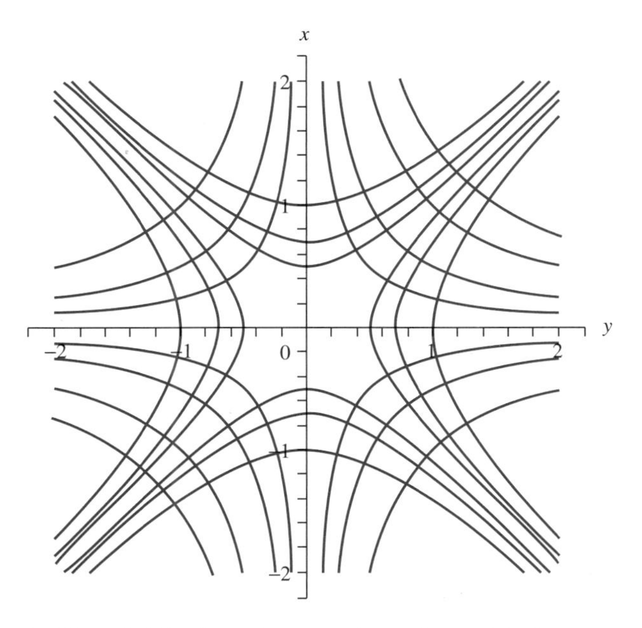

FIGURE 23.34 *Streamlines and equipotential lines in Example 23.21.*

where θ is the argument of z lying in $[0, 2\pi)$. We therefore may write

$$f(z) = \frac{iK}{2\pi}\left(\frac{1}{2}\ln(x^2+y^2) + i\theta\right) = -\frac{K\theta}{2\pi} + \frac{iK}{4\pi}\ln(x^2+y^2).$$

Then

$$\varphi(x, y) = -\frac{K}{2\pi}\theta \text{ and } \psi(x, y) = \frac{K}{2\pi}\ln(x^2+y^2).$$

Equipotential curves are graphs of $\theta =$ constant, which are half-lines from the origin making angle θ with the positive real axis. Streamlines are graphs of $\psi(x, y) = c$, and these are circles about the origin. These are trajectories of the fluid, which can be envisioned as moving in circular paths about the origin. Some streamlines and equipotential curves are shown in Figure 23.35.

Compute $f'(z) = (iK/2\pi z)$ if $z \neq 0$. On the circle $|z| = r$, the magnitude of the velocity is

$$\overline{|f'(z)|} = \frac{K}{2\pi}\frac{1}{|z|} = \frac{K}{2\pi r}.$$

This velocity increases as $r \to 0$, so the fluid swirls about the origin with increasing velocity toward the center (origin). The origin is a vortex of this flow.

To calculate the circulation of the flow about the origin, write

$$\overline{f'(z)} = -\frac{iK}{2\pi}\frac{1}{\overline{z}} = -\frac{iK}{2\pi}\frac{z}{|z|^2} = \frac{K}{2\pi}\frac{y}{x^2+y^2} - \frac{iK}{2\pi}\frac{x}{x^2+y^2} = u + iv.$$

If γ is the circle of radius r about the origin, then on γ, we have $x = r\cos(t)$ and $y = r\sin(t)$ for $0 \leq t \leq 2\pi$, so

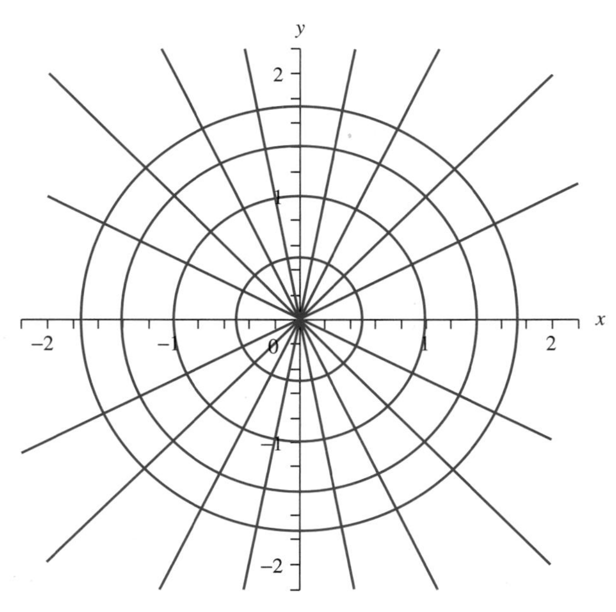

FIGURE 23.35 *Streamlines and equipotential lines in Example 23.22.*

$$\oint_\gamma u dx + v dy = \int_0^{2\pi} \left[\frac{K}{2\pi} \frac{r\sin(t)}{r^2}(-r\sin(t)) - \frac{K}{2\pi}\frac{r\cos(t)}{r^2}(r\cos(t)) \right] dt$$
$$= \frac{K}{2\pi}\int_0^{2\pi} -dt = -K.$$

This is the value of the circulation on any circle about the origin.

By a similar calculation,

$$\oint_\gamma -v\,dx + u\,dy = 0.$$

The origin is therefore neither a source nor a sink.

If we wish, we may restrict $|z| > R$ to model flow having a solid cylinder about the origin as a barrier with the flow swirling about the cylinder. ♦

EXAMPLE 23.23

Interchange the roles of streamlines and equipotential curves in Example 23.22 by setting

$$f(z) = K\,\mathrm{Log}(z)$$

with K a positive constant. Now

$$f(z) = \frac{K}{2}\ln(x^2 + y^2) + iK\theta,$$

so

$$\varphi(x, y) = \frac{K}{2}\ln(x^2 + y^2) \text{ and } \psi(x, y) = K\theta.$$

The equipotential curves are circles about the origin, and the streamlines are half-lines emanating from the origin. The velocity of the flow is

$$\overline{f'(z)} = \frac{K}{\overline{z}} = K\frac{x}{x^2 + y^2} + iK\frac{y}{x^2 + y^2} = u + iv.$$

If γ is a circle of radius r about the origin, then a straightforward calculation yields

$$\oint_\gamma u\,dx + v\,dy = 0 \text{ and } \oint_\gamma -v\,dx + u\,dy = 2\pi K.$$

The origin is a source of strength $2\pi K$. This flow streams out from the origin, moving along straight lines with velocity that decreases with distance from the source. ♦

EXAMPLE 23.24

We will model flow around an elliptical barrier using a conformal mapping. From Example 23.22, $f(z) = (iK/2\pi)\mathrm{Log}(z)$ for $|z| > R$ models flow with circulation $-K$ about a cylindrical barrier of radius R centered at the origin. If the barrier is elliptical, we can combine this complex potential with a mapping taking the circle $|z| = R$ to an ellipse. To do this, consider the mapping

$$w = z + \frac{a^2}{z}$$

in which a is a positive constant. This is a *Joukowski transformation*, and it is used in modeling fluid flow around airplane wings because of the different images of the circle that occur by making different choices for a.

Let $z = x + iy$, and $w = X + iY$. The circle $x^2 + y^2 = R^2$ is mapped to the ellipse

$$\frac{X^2}{1+(a/R)^2} + \frac{Y^2}{1-(a/R)^2} = R^2$$

if $a \neq R$. If $a = R$, the circle maps to the interval $[-2a, 2a]$ on the real axis.

Solve for z in the Joukowski transformation to get

$$z = \frac{w + \sqrt{w^2 - 4a^2}}{2}.$$

Compose this mapping with the complex potential function for the circular barrier to get

$$F(w) = f(z(w)) = \frac{iK}{2\pi}\text{Log}\left(\frac{w + \sqrt{w^2 - 4a^2}}{2}\right).$$

This is the complex potential for flow in the w-plane about an elliptical barrier if $R > a$ and about the flat plate $-2a \leq X \leq 2a$, $Y = 0$ if $R = a$. ♦

We conclude this section with an application of complex integration to fluid flow. Suppose f is a complex potential for a flow about a barrier whose boundary is a closed path γ. Let $A\mathbf{i} + B\mathbf{j}$ be the thrust of the fluid outside the barrier. *Blasius's theorem* states that

$$\overline{A + iB} = \frac{i\rho}{2}\oint_\gamma (f'(z))^2\,dz,$$

in which ρ is the constant density of the fluid. Furthermore, the moment of the thrust about the origin is

$$\text{Re}\left(-\frac{\rho}{2}\oint_\gamma z(f'(z))^2\,dz\right).$$

SECTION 23.4 PROBLEMS

In each of Problems 1 through 4, analyze the flow having the given complex potential. Sketch some equipotential curves and streamlines and determine the velocity, any stagnation points, and whether the flow has any sources or sinks.

1. $f(z) = \cos(z)$
2. $f(z) = z + iz^2$
3. $f(z) = az$ with a a nonzero complex number.
4. $f(z) = z^3$
5. Let

$$f(z) = K\left(z + \frac{1}{z}\right)$$

with K as a nonzero real number. Sketch some equipotential curves and streamlines. Show that f models flow around the upper half of the unit circle.

6. Analyze the flow having potential

$$f(z) = iKa\sqrt{3}\text{Log}\left(\frac{2z - ia\sqrt{3}}{2z + ia\sqrt{3}}\right)$$

with K and a positive constants. Show that f models an irrotational flow around a cylinder $4x^2 + 4(y - a)^2 = a^2$ with a flat boundary along the y-axis. Sketch some equipotential curves and streamlines.

7. Use Blasius's theorem to show that the force per unit width on the cylinder in Problem 6 has vertical component of $2\sqrt{3}\pi\rho a K^2$ with ρ as the constant density of the fluid.
8. Let

$$f(z) = K\text{Log}\left(\frac{z-a}{z-b}\right)$$

with K as a nonzero real numbers and a and b as distinct complex numbers. Sketch some equipotential curves and streamlines and determine any stagnation points, sources, and sinks.

9. $f(z) = K\text{Log}(z - z_0)$ with K as a nonzero real constant and z_0 as a complex number. Show that z_0 is a source if $K > 0$ and a sink if $K < 0$. Sketch some equipotential curves and streamlines. $\text{Log}(z)$ is defined here to be the branch of $\log(z)$ obtained by restricting $-\pi \leq \arg(z) < \pi$.

10. Let

$$f(z) = \frac{m - ik}{2\pi}\text{Log}\left(\frac{z-a}{z-b}\right)$$

with m and k as nonzero, real numbers and a and b as distinct, complex numbers. Show that this flow has a source or sink of strength m and a vortex of strength k at both a and b. A point combining properties of a source or sink and a vortex is called a *spiral vortex*. Sketch some equipotential curves and streamlines for this flow.

11. Analyze the flow having complex potential

$$f(z) = K\left(z + \frac{1}{z}\right) + \frac{ib}{2\pi}\text{Log}(z)$$

with k and b as nonzero, real numbers. Sketch some equipotential curves and streamlines.

$f](s) = \int_0^\infty e^{-st}t^{-1/2}\,dt = 2\int_0^\infty e^{-sx^2}\,dx \text{ (set } x = t^{1/2}) = \frac{2}{\sqrt{s}}\int_0^\infty e^{-z^2}\,dz \text{ (set } z = x$

$= \sqrt{\frac{\pi}{s}}$

$\mathcal{L}[f](s) = \int_0^\infty e$

$= 2\int_0^\infty e^{-sx^2}\,dx$

$= \sqrt{\frac{\pi}{s}}$

PART 7

Probability and Statistics

$](s) = \int_0^\infty e^{-st}t^{-1/2}\,dt$

$= 2\int_0^\infty e^{-sx^2}\,dx \text{ (set } x = t^{1/2})$

$= \frac{2}{\sqrt{s}}\int_0^\infty e^{-z^2}\,dz \text{ (set } z = x\sqrt{s})$

$\mathcal{L}[f](s) = \int_0^\infty e^{-st}t^{-1/2}$

EVENTS, SAMPLE SPACES AND PROBABILITY FOUR COUNTING PRINCIPLES COMPLEMENTARY EVENTS CONDITIONAL PROBABILITY INDEPENDENT

CHAPTER 24

Probability

Few areas of expertise have as profound an influence on us as statistics. Manufactured products are tested for reliability using statistical analysis of samples, insurance companies base premium costs on actuarial studies, and medical drug trials are designed to follow rigorous statistical protocols.

In order to develop some ideas from statistics, we will discuss some prerequisite ideas involving probability.

24.1 Events, Sample Spaces, and Probability

In the context of probability, an *experiment* is any activity that produces outcomes. This could be flipping coins, rolling dice, dealing cards, taking samples out of a warehouse, giving patients an experimental drug - anything that has outcomes we can observe and tabulate in some way.

An experiment may have infinitely many outcomes. This is true if we choose numbers between 0 and 1, or observe the point where a thrown dart hits a wall. For this chapter we will restrict our attention to experiments with finitely many outcomes.

Suppose an experiment has outcomes $o_1, o_2, \cdots, o_n$. The set of all of these outcomes is called the *sample space* of the experiment. If we call this sample space S, then

$$S = \{o_1, o_2, \cdots, o_n\}.$$

A *probability function* for this experiment is an assignment of a number $\Pr(o_j)$ to each outcome, subject to the conditions:

(1) $0 < \Pr(o_j) < 1$ for $j = 1, 2, \cdots, n$.

(2) $\sum_{j=1}^{n} \Pr(o_j) = 1$.

We call $\Pr(o_j)$ the *probability* of o_j, and require that the probability of each outcome be a number between 0 and 1. We also require that the sum of the probabilities of all the outcomes must equal 1.

EXAMPLE 24.1

Suppose the experiment is to flip two coins. There are four outcomes, namely HH, HT, TH and TT, where H denotes a head and T a tail. If the coins are honest, it is reasonable to assume that each outcome is equally likely, in which case we assign to each outcome a probability of $1/4$:

$$\Pr(HH) = \Pr(HT) = \Pr(TH) = \Pr(TT) = \frac{1}{4}.$$ ♦

EXAMPLE 24.2

The experiment is to roll two dice. Each die has six faces, numbered 1 through 6. Denote an outcome as (a, b), where a is the number that came up on the first die and b the number on the second die. We would expect there to be 36 possible outcomes of this experiment, because there are six possibilities for a, and, for each of these, six possibilities for b. In this fairly simple experiment, these 36 outcomes can be explicitly listed:

$$(1,1), (1,2), (1,3), (1,4), (1,5), (1,6),$$

$$(2,1), (2,2), (2,3), (2,4), (2,5), (2,6),$$

$$\cdots, \cdots, \cdots, \cdots, \cdots, \cdots,$$

$$(6,1), (6,2), (6,3), (6,4), (6,5), (6,6).$$

Assuming that the dice are honest, any outcome is as likely as any other and we assign the probability hence $\Pr(o) = 1/36$ to each outcome o. ♦

These two examples share a special feature that is not true of all experiments. If an experiment has N outcomes, all equally likely, then each outcome should have the same probability p. Since the sum of the probabilities of the N outcomes must equal 1, then $Np = 1$, so $p = 1/N$. In this case, we therefore assign each outcome o a probability of $\Pr(o) = 1/N$. Such an experiment is called an *equally likely outcome experiment.*

Not every experiment has this characteristic.

EXAMPLE 24.3

Suppose we have a dishonest coin, and on any flip it is twice as likely that a head will come up as a tail. If we denote the two outcomes of a single flip as H and T, then

$$\Pr(H) = 2\Pr(T).$$

The sum of their probabilities of all the outcomes must be 1, so

$$\Pr(H) + \Pr(T) = 1 = 2\Pr(T) + \Pr(T) = 3\Pr(T).$$

This means that

$$\Pr(T) = \frac{1}{3} \text{ and } \Pr(H) = \frac{2}{3}.$$

This yields a probability function for this experiment. ♦

Sometimes we want the probability not of just an individual outcome, but of a set of outcomes. An *event* is a set of outcomes, that is, a subset of the sample space. Given a probability function for this experiment, the probability of any event E is defined to be the sum of the probabilities of the outcomes in E.

EXAMPLE 24.4

A coin is flipped three times. The sample space is

$$S = \{HHH, HHT, HTH, THH, HTT, THT, TTH, TTT\}.$$

Assuming an honest coin, each outcome is equally likely, so the probability of each outcome is $1/8$.

Now consider the event that exactly two heads come up. This is the event

$$E = \{HHT, HTH, THH\}$$

because these are all the ways that exactly two heads can occur in three flips. The probability of E is

$$\Pr(E) = \Pr(HHT) + \Pr(HTH) + \Pr(THH) = \frac{3}{8}.$$

Let K be the event that the third toss comes up tails. This is the event $K = \{HHT, HTT, THT, TTT\}$, so $\Pr(K) = 4/8 = 1/2$.

Let W be the event that the first two tosses come up tails. Then $W = \{TTH, TTT\}$, so $\Pr(W) = 2/8 = 1/4$. ♦

EXAMPLE 24.5

The experiment is to roll three honest dice. What is the probability that they total 17?

Because each outcome is equally likely, we can determine a probability function for this experiment if we know how many outcomes there are.

The first roll has six possible results. With each of these, the second roll has six possible results, for 36 possibilities from the first two rolls (recall Example 24.2). With each of these, there are six possible results of the third roll, for $6(36) = 216$ possible outcomes of the experiment. For this equally likely outcome experiment, each outcome has probability $1/216$.

The event we are interested in consists of the outcomes totaling 17. There are three such outcomes, namely $(6, 6, 5)$, $(6, 5, 6)$ and $(5, 6, 6)$. The probability that the dice total 17 is $3/216$. ♦

These examples suggest that computing probabilities can involve counting things. So far this has been relatively simple. However, it is easy to imagine experiments in which computing a probability involves a difficult counting challenge. For example, what is the probability that exactly seven heads come up in twenty-five flips of a coin? This will be the number of outcomes having exactly seven heads, divided by the total number of outcomes. We will see that this probability is $480,700/33,554,432$, a non-obvious conclusion at this juncture.

In the next section we will develop some counting techniques that will enable us to deal with more complicated probability questions.

SECTION 24.1 PROBLEMS

1. Three dice are rolled. What is the probability that (a) all dice come up the same number, (b) the dice total exactly 15, (c) the first die comes up 1 and the last comes up 4, (d) the first two dice come up even?

2. Two cards are drawn from a standard 52 card deck. What is the probability that (a) the first card is a face card or ace (jack, queen, king or ace), (b) both cards are red (hearts or diamonds), (c) one card is a club and the other is a diamond, (d) both cards are aces?

3. Two coins are flipped and a die is thrown. Write out the entire sample space for this experiment. Compute the probability that (a) one head, one tail and an even number come up (b) both coins come up heads and the die comes up 1 or 4, (c) at least one tail comes up, together with a 4, 5 or 6.
4. Two dice are rolled, and a coin is flipped. Write out the entire sample space for this experiment. Compute the probability that (a) the dice total at least 9, (b) the dice total at least 9 and the coin comes up heads (c) the coin is a tail and both dice come up the same number (d) the coin comes up tails and the dice both come up even.
5. Three coins are flipped. What is the probability that (a) exactly two heads come up, and (b) at least two heads come up?
6. Two dice are rolled. What is the probability that (a) both dice come up even (b) one die is even and the other is odd (c) the dice total exactly 6, (d) the dice total at least 14?

24.2 Four Counting Principles

We will develop four counting principles that are used in counting the outcomes of experiments that are more complicated than the simple ones of Examples 1 - 5.

24.2.1 The Multiplication Principle

Suppose some process consists of in n independent stages. Independent means that the outcome of one stage is not influenced by the outcomes of the others. Suppose the first stage can be done in w_1 ways, the second in w_2 ways, and so on, until the nth stage can be done in w_n ways. Then the total number of ways the entire process can be carried out is

$$w_1 w_2 \cdots w_n$$

the product of the number of ways of carrying out each stage. This is the *multiplication principle*. We saw this in several simple instances (such as counting the outcomes of rolling two dice) in Section 1.

EXAMPLE 24.6

A game consists of flipping a coin four times, then rolling five dice. How many outcomes are there?

This game has nine stages, the first four each being a coin toss, the next five each being a roll of a die. Each coin toss has two outcomes, and each roll of a die, six outcomes. The total number of outcomes is $2^4 6^5 = 124,416$. ♦

EXAMPLE 24.7

We want to form identification codes by choosing seven integers from 1 through 9 inclusive and listing them in order in a seven digit number. How many different codes can be written in this way.?

This problem has a wrinkle to it. We need to know whether all the integers in a particular code must be different (as with 9136254, or if duplicates are allowed (such as occurs with 2219947).

Suppose first we choose the seven integers *with replacement*. This means that we can reuse any integer. Now there are nine choices for the first integer, nine for the second, and so on, for a total of $9^7 = 4,782,969$ possible ID numbers. These include ID's such as 3241169 and 8888888.

If we choose *without replacement*, then each integer can be used at most once. Now there are nine choices for the first integer, but only eight for the second, seven for the third, and so on, down to three for the seventh integer. The total number of such codes is

$$9 \cdot 8 \cdot 7 \cdot 6 \cdot 5 \cdot 4 \cdot 3 = 152,640. \blacklozenge$$

24.2.2 Counting Permutations

A *permutation* of n objects (for example, the numbers $1, 2, \cdots, n$) is a rearrangement of these objects, taking account their order. We want to know how many rearrangements there are of n objects.

Imagine n boxes next to each other, placed from left to right, say. We will put a different object in each box, forming a permutation of the objects. We can choose any one of the n objects to put into the first box, leaving $n-1$ objects from which to choose one for the second box, $n-2$ choices for the third box, all the way down to 2 choices for the next to last box, and then only one left to put in the last box. By the multiplication principle, the number of ways of filling the boxes is the product

$$n(n-1)(n-2)\cdots(2)(1).$$

This is the product of the positive integers from 1 through n. This produce is denoted $n!$, which is read as "n factorial". As a notational convenience we let $0! = 1$.

For example, there are $3! = 6$ ways of arranging the letters a, b and c in different orders. These orderings are

$$abc, acb, bac, bca, cab, cba.$$

24.2.3 Choosing r Objects from n Objects, With Order

Suppose we have n objects. We want to know how many ways we can choose r of them. The answer depends on whether or not the order in which the objects are chosen is taken into account. First consider the case that we do take the order into account.

In this case we have n objects, and we want to choose r of them, with two choices considered different if they either have different objects, or the objects are chosen in a different order.

To count the number of ways to do this, imagine r boxes laid out in order from left to right. There are n ways we can choose an object for the first box, $n-1$ ways for the second box, and so on until there are $n-r+1$ objects left to choose one for the last box r. By the multiplication principle,

number of ways of choosing$r$ objects from n objects, with order

$$= n(n-1)\cdots(n-r+1).$$

EXAMPLE 24.8

How many ways are there of choosing two of the letters a, b, c, d, taking the order of the choices into account?

With $n = 4$ and $r = 2$, the number of ways is $4(3) = 12$. In this small example, we can list these ways explicitly:

$$ab, ba, ac, ca, ad, da, bc, cd, bd, db, cd, dc. \blacklozenge$$

The number of ways of choosing r objects from n objects, with order, is denoted ${}_nP_r$ and is called the *number of permutations of r objects chosen from n objects*. (P stands for permutation in this symbol). Observe that

$$
\begin{aligned}
{}_nP_r &= n(n-1)(n-2)\cdots(n-r+1) \\
&= \frac{1\cdot 2\cdot 3\cdots(n-r-1)(n-r)(n-r+1)\cdots(n-2)(n-1)n}{1\cdot 2\cdot 3\cdots(n-r)} \\
&= \frac{n!}{(n-r)!}.
\end{aligned}
$$

In Example 24.8, with $n=4$ and $r=2$,

$$
{}_4P_2 = \frac{4!}{(4-2)!} = \frac{4!}{2!} = \frac{24}{2} = 12,
$$

as we obtained there by listing all of the possibilities.

EXAMPLE 24.9

An election is being held and there are eight nominations for three offices. A ballot consists of three lines. The name written in the first line receives a vote for president, the name placed on the second line receives a vote for vice-president, and the name on line three receives one vote for custodian. How many different ballots are possible in this election?

Think of this as choosing three objects from eight. Since the order in which candidates are listed makes a difference, we must take the order into account. The number of different ballots is

$$
{}_8P_3 = \frac{8!}{5!} = 6\cdot 7\cdot 8 = 336. \blacklozenge
$$

24.2.4 Choosing r Objects from n Objects, Disregarding Order

Now suppose we want to choose r objects from n objects, but the order in which the objects are selected is unimportant. This number is denoted ${}_nC_r$ and is called the *number of combinations of r objects chosen from n objects*. Whenever the word permutation is used, order is important. When order is discounted, the word combination is used.

We already know how to compute ${}_nC_r$. Given n and r, ${}_nP_r$ differs from ${}_nC_r$ only in taking the orderings into account. There are $r!$ ways of ordering any r of the n objects. Therefore each combination of r of the n objects counted in ${}_nC_r$ gives rise to $r!$ different orderings of these r objects, and these are all included in the number ${}_nP_r$. This means that

$$
{}_nP_r = r!\,{}_nC_r.
$$

Therefore

$$
{}_nC_r = \frac{1}{r!}({}_nP_r) = \frac{n!}{r!(n-r)!}.
$$

For example, suppose we want to choose 2 objects out of the four objects a, b, c, d. If order is taken into account, we have seen that there are ${}_4P_2 = 4!/2! = 12$ possibilities. If order is not taken into account, then, for example, ab and ba are counted as one choice, not the two distinct choices when order is disregarded. Similarly, ac and ca are the same, and so on. Without order, there are

$$
\frac{1}{2!}\,{}_4P_2 = \frac{1}{2!}\frac{4!}{2!} = 6
$$

possible choices. This makes sense, because every ordered choice of two of the objects (for example, ab and ba) corresponds to one unordered choice of these two particular objects.

EXAMPLE 24.10

A poker game begins with seven cards dealt face down. After all cards are dealt, each player picks up his or her hand and bets or folds. What is the number of possible hands?

The order in which the cards is dealt is irrelevant, since the players pick up only after the deal is complete. The number of hands is the number of ways of picking seven cards out of fifty two cards, without order. This number is

$$_{52}C_7 = \frac{52!}{7!45!} = \frac{46 \cdot 47 \cdot 48 \cdot 49 \cdot 50 \cdot 51 \cdot 52}{2 \cdot 3 \cdot 4 \cdot 5 \cdot 6 \cdot 7} = 133,784,560.$$

Now suppose the game is changed and there is a round of betting after each card is dealt. Suddenly order has become important. A player receiving two aces in the first two cards may bet differently than a player receiving the aces last. Now the number of hands is

$$_{52}P_7 = \frac{52!}{45!} = 46 \cdot 47 \cdot 48 \cdot 49 \cdot 50 \cdot 51 \cdot 52 = 674,274,182,400.$$

This is $7!_{52}C_7$. ♦

$_nC_r$ is often written $\binom{n}{r}$, which is standard notation for a binomial coefficient, because these numbers occur in the binomial expansion

$$(x+y)^n = \sum_{k=0}^{n} \binom{n}{k} x^k y^{n-k}.$$

We will illustrate the use of these counting techniques with two typical probability calculations. Recall that, if an experiment has N equally likely outcomes, then the probability of each outcome is $1/N$. Further, if and E is an event containing k outcomes, then in this case the probability of E is k/N. Stated generally, for an equally likely outcome experiment,

$$\Pr(E) = \frac{\text{number of outcomes in } E}{\text{total number of outcomes}}.$$

EXAMPLE 24.11

A player is dealt five cards, without regard to order, from a standard deck of 52 cards. What is the probability of getting four aces?

The sample space consists of all five card hands that can be drawn from fifty two cards, without regard to order. This sample space has $_{52}C_5 = 2,598,960$ outcomes in it.

Let F be the event consisting of all hands with four aces. How many outcomes are in F? If four of the cards in the five card hand are aces, the fifth card can be any of the remaining 48 cards. Therefore (without regard to order) F has 48 outcomes in it. Then

$$\Pr(F) = \frac{48}{2,598,960}.$$

This is approximately 0.0000185. The probability of drawing four aces in a five card hand is very small, as we would expect. ♦

EXAMPLE 24.12

Flip 25 honest coins. What is the probability that at least twenty two of the coins come up heads?

Since each flip has two outcomes, the experiment has 2^{25} outcomes.

Let F be the event consisting of those outcomes in which at least 22 of the flips come up heads. We want the probability of F. This probability is

$$\Pr(F) = \frac{\text{number of outcomes in } F}{2^{25}}.$$

All that remains is to count the outcomes in F. First observe that the phrase "at least 22 heads" means there can be exactly 22 heads, or 23 heads, or 24 heads, or 25 heads. We must count the number of ways each of these can occur:

$$\begin{aligned}
&\text{number of outcomes with at least 22 heads}\\
&= \text{number with exactly 22 heads}\\
&+ \text{number with exactly 23 heads}\\
&+ \text{number with exactly 24 heads}\\
&+ \text{number with exactly 25 heads}\\
&= {}_{25}C_{22} + {}_{25}C_{23} + {}_{25}C_{24} + {}_{25}C_{25}\\
&= \frac{25!}{3!22!} + \frac{25!}{2!23!} + \frac{25!}{1!24!} + \frac{25!}{0!25!}\\
&= 2,626.
\end{aligned}$$

Then

$$\Pr(F) = \frac{2,626}{2^{25}} = \frac{2,626}{33,554,432} \approx 0.000078.$$

As intuition might suggest, betting on at least 22 heads out of 25 flips is not a likely win. ♦

SECTION 24.2 PROBLEMS

The first sixteen problems are counting problems. The next ten deal with probability.

1. How many different ways can four drumsticks be chosen from a barrel containing twenty drumsticks, if the order of selection does not matter?
2. A company is selecting twelve of its forty employees to lay off. How many ways can such a selection be made, if the order of the choices is unimportant?
3. How many different ways can a ten card hand be dealt from a standard deck, if the order is unimportant?
4. How many different nine man lineups can be formed from a roster of 17 players, if the order of selection does not matter? How many can be formed if the order is significant?
5. A car dealer has a lottery. First 22 names are selected at random from a data base. Six of these names will be winners. The first name chosen can go through the lot and pick any car. The second can do the same, but perhaps the first winner got the best car. The third person can then select a car, and so on. How many different possibilities are there for lists of people to go through the lot?
6. A game consists of choosing and listing, in order, three numbers from the integers from 1 to 20, inclusive.

   (a) How many different choices are there?

   (b) What percentage of the choices begins with the number 4? Would the answer change if the first number is 17 instead of 4?

   (c) What percentage of the choices ends with 9. Would this percentage change if the last number is 11?

   (d) How many choices are there if the first number must be 3 and the last number, 15?
7. Seven members of an audience of twenty five are to be chosen to win a prize. The first name drawn will win half of the planet, the second will win an airplane, the third a new home, and so on down the list. The last name drawn will win fifty cents worth of merchandise at the nearest convenience store. How many different outcomes are there of this drawing?

8. There are five positions open on the board of a swimming club, and sixteen people are eligible for election to the board. A ballot consists of a list of names of five of the eligible members, with order being important because new board members are assigned positions of decreasing importance, depending on how far they are down the list. How many different ballots are possible in this election?

9. A lottery is run as follows. Twelve slips of paper are placed in a bowl. Each slip has a different symbol in it. A player makes an ordered list of these twelve symbols, and wins if they are drawn from the bowl in this order. How many possible different outcomes are there of this lottery?

10. The letters a through l are to be arranged in some order. How many arrangements are there that have a in the second place, d in the fifth place and k in the seventh place?

11. How many different arrangements are there of the symbols a, b, c, d, f, g, h? How many arrangements are there if we insist on using only lists that begin with a? How many arrangements are there if a must be first and g must be fifth?

12. We want to form ID numbers by using n distinct symbols and allowing any order for their arrangement. How large must n be to accommodate $20,000$ people? How many for $1,000,000$ people?

13. An ID number for each employee in a company consists of a string of nine numbers, each number chosen from the integers 1 through 9. How many different ID numbers can be formed? Hint: Allow ID numbers with repeated digits.

14. Suppose we have five symbols available, say a, b, c, d, e. One plan is to form passwords by using different orderings of these five symbols. A second plan is to use ordered strings of length five, with each symbol in the string chosen from these symbols, in which any symbol can be used one or more times. How many different passwords are possible in each plan?

15. How many ways can the first nine letters of the alphabet be arranged in different orders?

16. How many different codes can be formed from the lower case letters of the English alphabet, if a code consists of seventeen distinct letters, with different orders counted as different codes.

17. Five cards are drawn, without regard to order, from a standard deck. Find the probability that:

    (a) the hand contains exactly one jack and exactly one king,

    (b) the hand contains at least two aces.

18. Twenty integers are chosen at random, and without regard to order, from the integers $0, 1, 2, \cdots, 100$. Find the probability that:

    (a) all of the numbers chosen are larger than 79

    (b) one of the numbers chosen is 5.

19. Eight bowling balls are in a bin. Two are defective, having been manufactured as cubes instead of the traditional spherical shape. A person uses a remote gripping device to pick three of the balls out of the bin, sight unseen. The order of the choice does not matter. Find the probability that:

    (a) none of the balls chosen is defective,

    (b) exactly one defective ball is chosen,

    (c) both defective balls are chosen.

20. Seven pyramid-shaped (tetrahedron) dice are tossed. The faces on each are numbered 1 through 4. Find the probability that:

    (a) all seven dice come up 3,

    (b) five dice come up 1 and two come up 4,

    (c) the sum of the numbers that come up is 26,

    (d) the sum of the numbers that come up is at least 26.

21. Twenty balls in an urn are numbered 1 through 20. A blindfolded person draws five balls from the urn, with the results recorded in the order in which they were drawn. Find the probability that:

    (a) the balls $1, 2, 3, 4, 5$ were drawn, in that order,

    (b) the number 3 ball was drawn,

    (c) an even-numbered ball was drawn.

22. Seven drawers in a desk each contain a fifty cent piece while two other drawers each contain a thousand dollar bill. A person chooses three drawers. What is the probability that:

    (a) the person gets at least $1,000$,

    (b) the person ends up with less than one dollar,

    (c) the probability that the person ends up with 1.50.

23. Two cards are selected from a standard deck. The order of the draw is unimportant. Find the probability that:

    (a) both cards were kings,

    (b) neither card was a face card (jack, queen, king or ace).

24. Four letters are selected from the lower case English alphabet. The order of the selection is recorded. Find the probability that:

(a) the first letter is q,

(b) a and b are two of the letters,

(c) the letters chosen are a, b, d, z, in this order.

25. Five honest coins are flipped.

(a) Find the probability of getting exactly two heads.

(b) Find the probability of getting at least two heads.

26. Roll four dice. Find the probability that:

(a) exactly two $4's$ come up,

(b) exactly three $4's$ come up,

(c) at least two $4's$ come up,

(d) the dice total 22.

24.3 Complementary Events

An experiment is performed and E is an event. The *complement* of E is the event E^C consisting of all outcomes not in E. Because E and E^C have no outcomes in common, and together contain all the outcomes, then $\Pr(E) + \Pr(E^C)$ is the sum of the probabilities of all of the outcomes in the sample space, and this must equal 1:

$$\Pr(E) + \Pr(E^C) = 1.$$

This means that we know either one of these probabilities, if we know the other one. This is called the *Principle of Complementarity*. It gives us a choice of computing either $\Pr(E)$ or $\Pr(E^C)$, whichever is easier.

The principle of complementarity is often written

$$\Pr(E) = 1 - \Pr(E^C).$$

EXAMPLE 24.13

Roll three dice. What is the probability that the dice total at least 5?

Let E be this event. Since three dice can total up to 18, E contains all of the outcomes in which the total 5, or 6, or 7, $\cdots$, all the way up to 18. We can compute $\Pr(E)$ as the sum of the probability of each of these fourteen outcomes. Or we can look at the simpler event E^C, which is the event that the dice total less than 5. Since three dice must total at least 3, E^C consists of the outcomes in which the dice total either 3 or 4. They can total 3 in one way (they all come up 1), and they can total 4 in three ways (they come up 2, 1, 1 or 1, 2, 1 or 1, 1, 2). Therefore E^C has four outcomes in it. Since the total number of outcomes in a roll of three dice is 6^3, then

$$\Pr(E^C) = \frac{4}{6^3} = \frac{4}{216} = \frac{1}{54}.$$

Therefore

$$\Pr(E) = 1 - \Pr(E^C) = 1 - \frac{1}{54} = \frac{53}{54}.$$

This is approximately 0.98. Rolling a total of at least 5 with three dice is very likely. ♦

SECTION 24.3 PROBLEMS

1. Five cards are drawn from a fifty two card deck, without regard to order. What is the probability that at least one is a jack, queen, king or ace, or that the card is numbered 4 or higher?

2. Two coins are tossed and five dice are rolled. What is the probability that two heads and at least one 4 come up?

3. Four numbers are chosen, without regard to order, from among the integers 1 through 55 inclusive. What is the probability that at least one of the numbers is greater than 4?

4. Here is the famous *birthday problem*. In a room of N people, what is the probability that at least two have the same birthday? What is the smallest N for which this probability is at least $1/2$?

5. Seven dice are rolled. What is the probability that at least two of them come up 4?

6. Fourteen coins are tossed. What is the probability that at least three come up heads?

24.4 Conditional Probability

A probability can depend on how much we know. Consider the following scenario. We are watching a card game. A player is dealt a card. What is the probability that it is a diamond? Since one fourth of the cards are diamonds, this probability would appear to be $1/4$. However, suppose we can see from a mirror behind the player that the drawn card is red. This eliminates clubs and spades from the calculation and leads us to consider only the red cards. Since half of the red cards are diamonds, the probability that the card is a diamond is $1/2$ (once the clubs and spades are eliminated).

Which probability is correct? They both are. The first is based on a random draw from a fifty-two card deck. The second probability is based on the additional knowledge that the card came from only the red cards. Knowing this cuts the sample space from fifty-two cards to twenty-six, only the red cards, of which half are diamonds. This is an example of a conditional probability, in which additional information changes the sample space by eliminating some outcomes, thereby changing the calculation of the probability.

To put this on a firm footing, suppose we have an experiment with sample space S. Let U be some given event (think of this as the information in the preceding example). For any event E, let $U \cap E$ denote the set of those outcomes common to U and E. If this is empty (no outcomes in common), we write $U \cap E = \phi$. We agree to let $\Pr(\phi) = 0$, on the reasonable grounds that the probability of an event with no outcomes in it should be zero.

Now imagine that by some means (such as the mirror in the example) we know that U occurs. The *conditional probability* of E, knowing U, is denoted $\Pr(E|U)$, and is computed as

$$\Pr(E|U) = \frac{\Pr(E \cap U)}{\Pr(U)}. \tag{24.1}$$

Knowing that U occurs eliminates all outcomes not in U, hence shrinks the sample space of the experiment-with-condition down to U. Divide the probability of the outcomes in E that are also in the known U, by the probability of U.

We can formulate this as a simple counting quotient as follows. We know that

$$\Pr(E \cap U) = \frac{\text{number of outcomes of } E \text{ also in } U}{\text{number of outcomes in } S}.$$

and

$$\Pr(U) = \frac{\text{number of outcomes in } U}{\text{number of outcomes in } S}.$$

Combine these with the definition to obtain

$$\begin{aligned}\Pr(E|U) &= \frac{\Pr(E|U)}{\Pr(U)} \\ &= \frac{\text{number of outcomes common to } E \text{ and } U}{\text{number of outcomes in } U}.\end{aligned} \tag{24.2}$$

This is exactly the result we would expect if we think of U as the new sample space (for the experiment with U as given information), and count only those outcomes in E that are in U, since the information tells us that only outcomes in U can occur.

EXAMPLE 24.14

Three dice are tossed. A person sitting to the side knows that two of the dice are loaded and always come up 2. What is the probability of rolling a 5?

Let U consist of the outcomes with at least two of the dice coming up 2. These outcomes have the form

$$U = \{(x, 2, 2), (2, x, 2), (2, 2, x)\}$$

in which the die x can come up any of $1, 2, 3, 4, 5, 6$. There are 16 such outcomes (not 18, because, when $x = 2$, all three triples in U are the same).

E is the event that the dice total 5, so $E \cap U$ consists of those triples in which $x = 1$:

$$E \cap U = \{(1, 2, 2), (2, 1, 2), (2, 2, 1)\}.$$

Thus $E \cap U$ contains three outcomes. By equation (24.2),

$$\Pr(E|U) = \frac{3}{16}.$$

Alternatively, if we want to use the expression (1), we need to first compute

$$\Pr(E \cap U) = \frac{3}{6^3} \text{ and } \Pr(U) = \frac{16}{6^3}$$

Then

$$\Pr(E|U) = \frac{\Pr(E \cap U)}{\Pr(U)} = \frac{3/216}{16/216} = \frac{3}{16}.$$

This is the probability that the dice will total 5, to the person who knows about the loaded dice. ♦

SECTION 24.4 PROBLEMS

1. Find the probability that four dice will all come up odd. What is this probability if we know that one came up 1 and another came up 5? What is the probability if we know that the second die came up 6?

2. What is the probability that a toss of seven coins will produce at least five heads, if we know that four of the coins came up heads?

3. What is the probability that four rolls of the dice total exactly 19? What is this probability if we saw one die came up 1?

4. What is the probability that two rolled dice sum to at least 9, if we saw one die come up even?

5. What is the probability that a five-card poker hand (unordered) has four aces if we know that a four of spaces was dealt?

6. Six cards are dealt from a standard deck without regard to order.

   (a) What is the probability that exactly two of the cards dealt were from the jack, king, queen, ace cards?

   (b) What is the probability that exactly two of the cards were from the jack, king, queen, ace cards, if we know that a king was dealt?

7. Four coins are tossed.

   (a) What is the probability that at least three came up tails?

   (b) What is the probability that at least three came up tails if we saw two coins land tails?

8. Four coins are flipped.
   (a) What is the probability that exactly three came up heads?
   (b) What is the probability that exactly three came up heads, if we know that exactly one came up tails?

9. Two coins are flipped.
   (a) Find the probability that the first one came up heads.
   (b) What is probability that the first one came up heads, if we know that at least one came up heads?

24.5 Independent Events

Two events are thought of as independent if the knowledge that one occurs provides no information at all about the probability of the other occurring. In keeping with this idea, define events E and U to be *independent* if

$$\Pr(E|U) = \Pr(E) \text{ or } \Pr(U|E) = \Pr(U).$$

If E and U are not independent, then they are *dependent*.

EXAMPLE 24.15

Flip a coin once. The sample space is just $S = \{H, T\}$. We claim that the events $E = \{H\}$ and $U = \{T\}$ are dependent. This seems obvious, since knowing that the coin came up heads would tell us that it did not come up tails. To check the definition, compute

$$\Pr(U) = \Pr(E) = \frac{1}{2}$$

while

$$\Pr(E|U) = \Pr(U|E) = 0.$$

Thus the definition of independence does not hold, and E and U are dependent. ♦

EXAMPLE 24.16

Draw two cards from a deck by first drawing one card, then drawing the second from the remaining cards. An outcome is a pair (a, b) of cards, with order important because, for example, if we draw a king of diamonds first, then we cannot draw it second. The sample space S consists of all pairs (a, b) with a and b distinct cards. There are $(52)(51) = 2,652$ outcomes. This can also be computed as ${}_{52}P_2$.

Now consider the event E that a king is drawn first, and the event U that a jack, queen, king or ace is drawn second. Are E and U independent?

Outcomes in E are pairs of the form (king, b). There are 4 ways to draw a king, and then there are 51 ways to draw a second card, so E has $4(51) = 204$ outcomes in it.

U contains all (a, b) with two possibilities: (1) a and b are both drawn from the jacks, queens, kings and aces, and (2) only b is drawn from the jacks, kings, queens and aces. In the first category there are 16 possibilities for a and 15 for b, for $(16)(15) = 240$ outcomes. In category (2) there are 36 choices for a and 16 for b, for $(36)(16) = 576$ outcomes. Therefore U contains $240 + 576 = 816$ outcomes.

Next, $E \cap U$ consists of all outcomes (a, b) with a a king and b drawn from the jacks, queens, kings and aces. There are four choices for a and then 15 for b, so there are $4(15) = 60$ outcomes in $E \cap U$. Therefore

$$\Pr(E|U) = \frac{\text{number of outcomes in } E \text{ and } U}{\text{number of outcomes in } U} = \frac{60}{816}.$$

Since $204/2652 \neq 60/86$, then

$$\Pr(E|U) \neq \Pr(E).$$

Similarly, it is routine to check that $\Pr(U) = 816/2652$ and $\Pr(U|E) = 60/204$, so

$$\Pr(U|E) \neq \Pr(U).$$

Therefore E and U are dependent events. This makes sense. If we know U, then we know that a jack, queen, king or ace was drawn second, changing the possibilities for a king having been drawn first. ♦

If this experiment is repeated, but this time with replacement (draw a card, replace it and draw the second card), we find that $\Pr(E|U) = 64/832 = \Pr(E)$. Because of the replacement, E and U are independent. Knowing one event no longer tells us anything about the probability of the other event, because the replacement removes the effect of the first draw on the pool of cards from which the second draw is made.

Equation (24.1) can be written

$$\Pr(E \cap U) = \Pr(E|U)\Pr(U). \tag{24.3}$$

This is called the *product rule*, and it is particularly important in the case of independent events. If $\Pr(E|U) = \Pr(E)$, then the product rule states that the probability of the event of outcomes common to E and U equals the product of the probabilities of E and U.

More generally, the following can be proved.

THEOREM 24.1

Events E and U are independent if and only if

$$\Pr(E \cap U) = \Pr(E)\Pr(U). \quad ♦$$

This can be read: the probability of E and U is the probability of E multiplied by the probability of U, exactly when the events are independent.

EXAMPLE 24.17

Suppose the probability of giving birth to a boy is 0.49, and that for a girl is 0.51. A family has four children. What is the probability that exactly three are boys? Much larger problems of this type are important, for example, in actuarial statistics.

The experiment is to have four children and outcomes are strings $abcd$, where each letter is a b for boy or a g for girl. The event E that we are interested in is that exactly three children are boys:

$$E = \{bbbg, bbgb, bgbb, gbbb\}.$$

The probability of any one of these is the product of the probability of each letter, g or b. The reason for this is that any letter in a string $abcd$ can be a b or a g, independent of the other letters. Therefore

$$\Pr(bbbg) = (0.49)(0.49)(0.49)(0.51) = 0.06$$
$$= P(bbgb) = \Pr(bgbb) = \Pr(gbbb).$$

Therefore

$$\Pr(E) = 4(0.06) = 0.24. \quad ♦$$

SECTION 24.5 PROBLEMS

1. A jar contains twenty marbles, with eight red, eight blue and four magenta. The probability of drawing a blue marble is twice that of a magenta, and three times that of a red. What is the probability of drawing exactly two red marbles if three are drawn, without replacement?
2. A dishonest die comes up only 1, 4 or 6. This die is rolled three times. What is the probability that the total is 6?
3. Suppose a coin has been shaved so that the probability of tossing a head is 0.4. This coin is flipped four times. What is the probability of getting at least two heads? What is the probability of getting exactly two heads?
4. An experiment consists of picking two cards from a standard deck, with replacement. E is the event that the first card drawn was a king. U is the event that the second card drawn was an ace. Are these events independent?
5. Two cards are dealt from a standard deck, without replacement. E is the event that the first card was a jack of diamonds. U is the event that the second card was a club or spade. Are these events independent?
6. Four coins are flipped. E is the event that the first coin comes up heads. U is the event that the last coin comes up tails. Determine whether these events are independent.
7. An experiment consists of flipping two coins and then rolling two dice. E is the event that at least one coin comes up heads. U is the event that at least one die comes up 6. Determine whether these events are independent.
8. A family has six children. E is the event that at least three are girls and U is the event that at least two are girls. Determine whether these events are independent.
9. Two dice are rolled. E is the event that the dice total more than 11. U is the event that at least one die comes up an even number. Determine whether these events are independent.
10. Two cards are drawn from a standard deck without replacement. E is the event that both cards are aces. U is the event that one card is a diamond and the other a spade. Determine whether these events are independent.
11. Four coins are flipped. E is the event that exactly one coin comes up heads. U is the event that at least three coins come up tails. Determine whether E and U are independent. Hint: Sometimes equation (24.3) is handy.

24.6 Tree Diagrams

Tree diagrams, in conjunction with the product rule, provide a convenient tool for computing probabilities of outcomes of an experiment which can be broken down into a sequence of steps.

EXAMPLE 24.18

A room contains three urns. Urn 1 has three baskets, one with 1 dollar, one with 50 dollars, and one with 100 dollars. Urn 2 has two baskets, one with 1 and one with 50 dollars. Urn three has 4 baskets, one with 100 dollars and three with 1 dollar. A person chooses an urn, then, from it, a basket. We want to determine the probability of ending up with 1, 50 or 100 dollars.

Figure 24.1 shows a tree diagram summarizing the information. Start with a dot which we label P. Draw a line from P ending in a dot for each urn. The number on each line is the probability of choosing that urn (which, in this experiment, is $1/3$ for each urn).

From the dot for Urn 1, draw three lines, each ending in a dot representing a basket. The numbers on each of these lines is the probability of choosing that basket, if we had chosen Urn 1. Repeat this for the other two urns.

The dots at the end (the baskets) represent the final outcomes. Each basket is labeled with the dollar amount it contains. Further, each basket is at the end of a path of two lines, from P

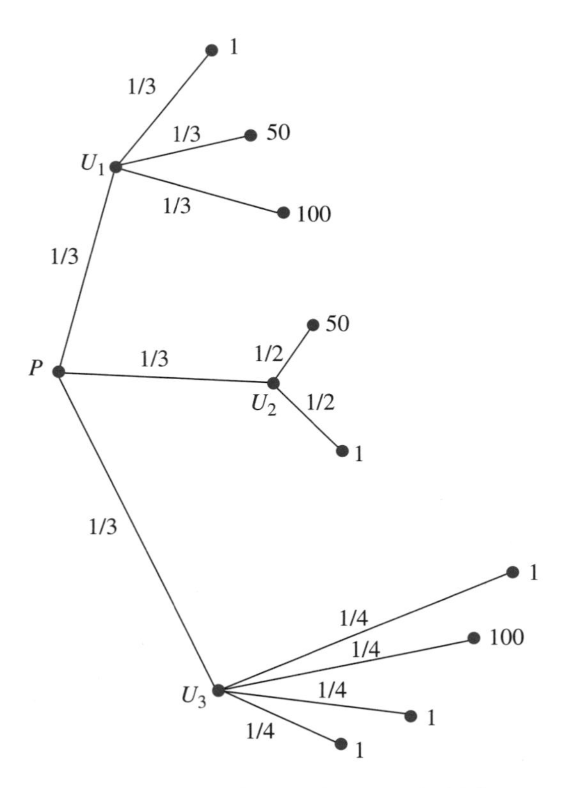

FIGURE 24.1 *Tree diagram for Example 24.6*

to an urn, then to a basket. The product of the two numbers on the lines of this path is the probability of choosing that basket. For each basket, this provides a way of computing the probability of ending up with that particular basket.

We can now determine the probability of obtaining each dollar amount. To illustrate, suppose we want the probability of getting 1 dollar. There are five paths ending with a basket containing 1 dollar. For each such path, multiply the two probabilities on the edges of that path, obtaining five numbers. The sum of these five numbers is the probability of choosing a basket with 1 dollar in it, hence of obtaining one dollar. This calculation is

$$\Pr(1) = \frac{1}{3}\frac{1}{3} + \frac{1}{3}\frac{1}{2} + \frac{1}{3}\frac{1}{4} + \frac{1}{3}\frac{1}{4} + \frac{1}{3}\frac{1}{4} = \frac{38}{72}.$$

Proceeding in the same way for the other dollar amounts, we obtain

$$\Pr(50) = \frac{1}{3}\frac{1}{3} + \frac{1}{3}\frac{1}{2} = \frac{5}{18}$$

and

$$\Pr(100) = \frac{1}{3}\frac{1}{3} + \frac{1}{3}\frac{1}{4} = \frac{7}{36}.$$

As a check, these probabilities sum to 1, as they must. ♦

More complicated experiments of this type can be treated by adding more points and lines. For example, if each basket contains a certain number of envelopes, then we would draw a line from each basket to represent the each envelope in that basket. Paths would now be three lines long (urn to basket to envelope), and the probability of choosing that envelope would be the product of the probabilities on each of the lines in the path that end with that envelope.

SECTION 24.6 PROBLEMS

1. A wealthy sultan has six automobile sheds which look identical from the outside. Each shed contains a number of identical containers, with each container closed and holding one vehicle. One shed has two identical Fords and a Chevrolet (each in its container). A second shed has a VW Beetle and a Porsche. A third shed has a Lamborghini and a very nice tricycle. A fourth shed has two Mercedes SUV's and a Honda Civic. A fifth shed has a World War II tank (partially destroyed in Africa) and a Porsche. And the sixth shed has three mountain bikes and a mint condition Stanley Steamer. A person can pick a shed and then any container in that shed. Determine the outcomes and their probabilities.
2. A traveler can choose to fly in a Piper cub that seats one passenger, a company jet seating eight passengers, or a jumbo jet seating seven hundred passengers. The traveler can pick any plane and any seat on that plane. What is the probability that the passenger picks an odd-numbered seat on the jumbo jet?
3. A person can choose any of five houses, each having four bedrooms. In one of the houses, three bedrooms are empty and one contains an antique chair worth fifty thousand dollars. In another house, two bedrooms are empty, one contains a thousand dollar bill and one contains a newly minted nickel. In the last two houses, two bedrooms contain fifteen hundred dollars each, one contains twenty dollars and one contains a hungry person-eating lion. If the person is to pick a house and then a bedroom, what are the outcomes and their probabilities?
4. A room has three urns in it. Urn 1 has two compartments, one empty and one with a key that can open any of four safes. One safe has diamonds, one has stocks, one has cash and one has a Cracker Jack whistle. The other two urns each have three compartments. One of these urns has two empty compartments and one filled with Confederate currency. The third urn has one compartment filled with expensive perfume, one filled with stock certificates and the third with the deed to a mansion. A person can pick an urn and any compartment and, if the key comes up, any one of the safes. Determine the outcomes and their probabilities.
5. A cabinet has four drawers. Two of the drawers each contain two envelopes, each containing 10 dollars, one drawer containing one envelope with 5 dollars, and one envelope with 50 dollars, and one draw contains one empty envelope and one envelope with 1200 dollars. A person must choose a drawer and then an envelope. Determine the outcomes and their probabilities.

24.7 Bayes's Theorem

Bayes's theorem is named for the Reverend Thomas Bayes (1702 - 1761). It enables us to determine the conditional probability of E knowing U, if we know the individual probabilities of E and $E \cap U$, as well as the conditional probabilities of U knowing E and U knowing E^C.

Begin with an experiment and events E and U. Since $E \cap U = U \cap E$, then

$$\Pr(E|U) = \frac{\Pr(E \cap U)}{\Pr(U)} = \frac{\Pr(U \cap E)}{\Pr(U)} = \frac{\Pr(U|E)\Pr(E)}{\Pr(U)} \tag{24.4}$$

in which we used the product rule in the numerator of the next to last equality.

Figure 24.2 shows typical E and U. The outcomes in U can be split into those in U and in E, hence are in $U \cap E$, together with those in U and not in E, hence are in $U \cap E^C$. The sets $U \cap E$ and $U \cap E^C$ not only contain all outcomes in U, but have no outcome in common. Therefore

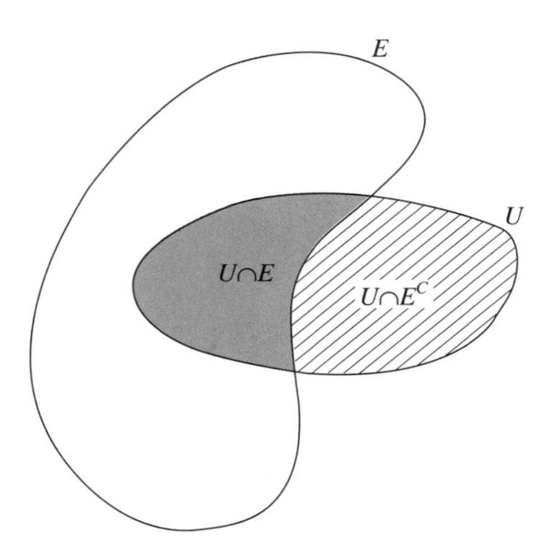

FIGURE 24.2 *E and U in Bayes's theorem (elementary version).*

$$\Pr(U) = \Pr(U \cap E) + \Pr(U \cap E^C).$$

Apply the product rule to each term on the right side of this equation to get

$$\Pr(U) = \Pr(U|E)\Pr(E) + \Pr(U|E^C)\Pr(E^C).$$

Finally, substitute this into the denominator in equation (24.4) to get

$$\Pr(E|U) = \frac{\Pr(U|E)\Pr(E)}{\Pr(U|E)\Pr(E) + \Pr(U|E^C)\Pr(E^C)}. \tag{24.5}$$

Equation (24.5) is *Bayes's theorem*, although it is a special case of a more general result we will state shortly. The tree diagram in Figure 24.3 shows points representing the events E and E^C, which together contain all outcomes. Outcome U is given. From point E (the case that E occurs), the two conditional events $U|E$ and $U^C|E$ are shown. And from vertex E^C, the case that E does not occur, the two conditional events $U|E^C$ and $U^C|E^C$ are shown. Four of the branches are labeled with probabilities. The numerator in Bayes's theorem is the product of the probabilities on branches 1 and 2. The denominator is the sum of this product, and the product of the probabilities on branches 3 and 4. The paths ending in conditional probabilities involving U^C are not relevant in computing $\Pr(E|U)$ in this way.

Bayes's theorem is all about drawing probability inferences from certain kinds of information.

EXAMPLE 24.19

A factory produces wombles. On a given day the probability of producing a defective womble is 0.04. Suppose the probability that a defective womble will result in injury to the user is 0.02, while the probability that a womble user is injured through no fault of the womble is 0.06. A lawsuit is in progress in which the plaintiff has been injured and wants a settlement. The defense attorney wants some measure of whether the womble was to blame. The defense therefore wants to know the probability that the product was defective, if it is known that an injury occurred.

Let U be the event that an injury occurs and E the event that a defective womble was produced. From the given information,

$$\Pr(E) = 0.04, \Pr(U|E) = 0.02, \text{ and } \Pr(U|E^C) = 0.06.$$

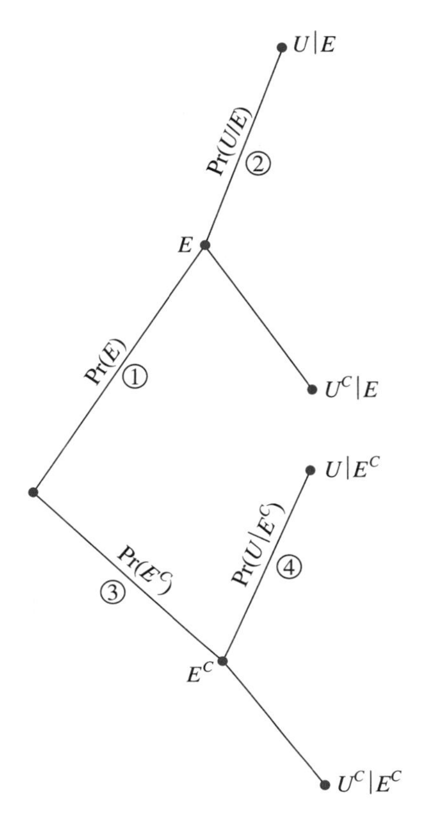

FIGURE 24.3 *Tree diagram for Bayes's theorem.*

Figure 24.4 shows these probabilities together with $\Pr(E^C) = 1 - 0.04 = 0.96$. We now have all the terms on the right side of equation (24.5) and can compute

$$\Pr(E|U) = \frac{\Pr(U|E)\Pr(E)}{\Pr(U|E)\Pr(E) + \Pr(U|E^C)\Pr(E^C)}$$
$$= \frac{(002)(0.04)}{(0.02)(0.04) + (0.06)(0.96)} \approx 0.0137. \blacklozenge$$

There is a more general form of Bayes's theorem in which the single event E is replaced by events $E_1, \cdots E_k$:

$$\Pr(E_j) = \frac{\Pr(E_j)\Pr(U|E_j)}{\sum_{j=1}^{k} \Pr(E_j)\Pr(U|E_j)}. \qquad (24.6)$$

Figure 24.5 is the generalization of Figure 24.3 to this statement.

EXAMPLE 24.20

An airplane manufacturing company receives shipments of parts from five companies. Table 24.1 gives the percent of the total parts need filled by each company and the probability for each company that a part is defective (data taken over some fixed period of time). A defective part is found. What is the probability that it came from Company 4?

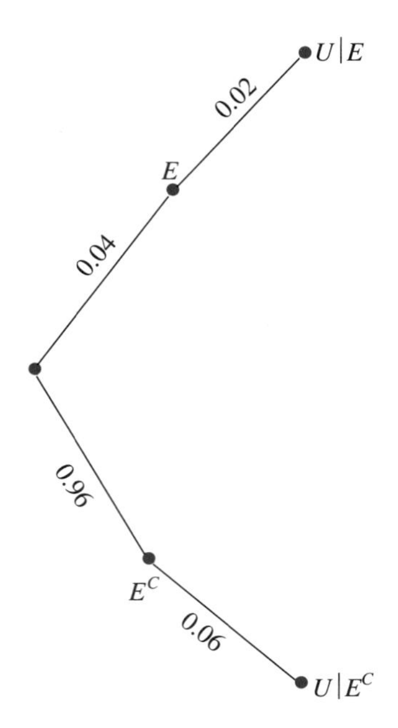

FIGURE 24.4 *Tree diagram for Example 24.19.*

Imagine a part has been chosen from inventory. Let E_j be the event that the part came from company j and let U be the event that the part is defective. We want $\Pr(E_4|U)$. From equation (24.6), this probability is

$$\begin{aligned}\Pr(E_4|U) &= \frac{\Pr(E_4)\Pr(U|E_4)}{\Pr(E_1)\Pr(U|E_1)+\Pr(E_2)\Pr(U|E_2)+\cdots\Pr(E_5)\Pr(U|E_5)}\\ &= \frac{(0.48)(0.02)}{(0.15)(0.03)+(0.02)(0.04)+(0.04)(0.04)+(0.48)(0.02)+(0.27)(0.08)}\\ &= 0.25,\end{aligned}$$

rounded to two decimal places. Notice that the percent supplied by each company must be converted to a decimal for use in Bayes's formula.

For comparison, we will compute the probability that the defective part came from the next largest supplier, Company 4. The denominator is the same and we have

$$\begin{aligned}\Pr(E_5|U) &= \frac{(0.27)(0.08)}{(0.15)(0.03)+(0.02)(0.04)+(0.04)(0.04)+(0.48)(0.02)+(0.27)(0.08)}\\ &= 0.57,\end{aligned}$$

to two decimal places. The disparity with $\Pr(E_4|U)$ is due to the fact that, while Company 5 produces a little more than half the parts Company 4 does, it has four times the failure rate. ♦

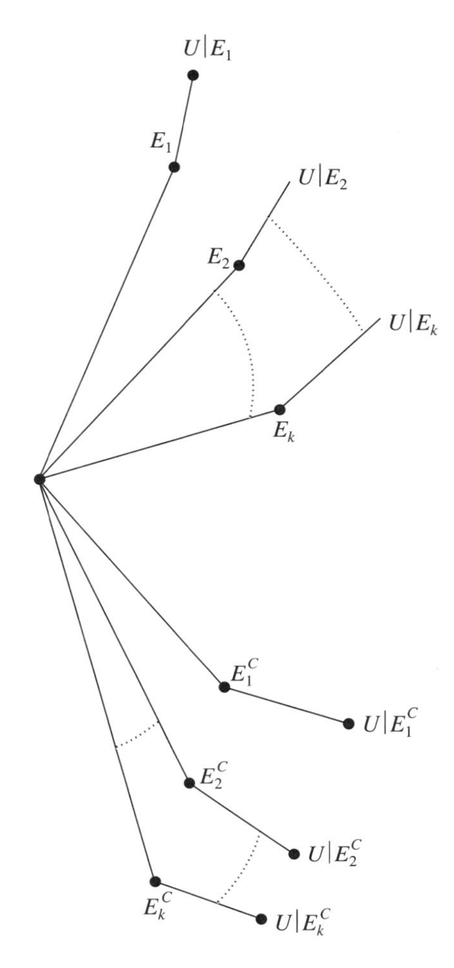

FIGURE 24.5 *Generalized Bayes theorem.*

TABLE 24.1 ***Airplane parts supplier probabilities, Example 24.7***

Company	Percent Supplied	Probability of Defect
1	15	0.03
2	2	0.04
3	4	0.01
4	48	0.02
5	27	0.08

SECTION 24.7 PROBLEMS

1. A drug trial performed with ill patients produces the data recorded in Table 24.2.

   (a) If a patient is randomly selected from this pool, what is the probability of that an adult male was chosen, if we are told that the person survived at least two years?

   (b) What is the probability of choosing a girl if we are told that the person survived less than one year?

TABLE 24.2 *Drug trial data in Problem 1, Section 24.7.*

Grouping	Adult Men	Adult Women	Boys	Girls
Survived less than 2 yrs	257	320	104	152
Survived less than 1 yr	619	471	51	38

TABLE 24.3 *Herb data in Problem 2, Section 24.7.*

Region	Percent of Total	Percent Fatal
Lower Amazon	62	14
Sub-Saharan Africa	25	6
Newark, New Jersey	13	1/2

TABLE 24.4 *Pistol data for Problem 3, Section 24.7.*

City	Percent of Total	Probability of Exploding
1	15	0.02
2	7	0.01
3	21	0.06
4	4	0.02
5	9	0.03
6	44	0.09

2. An herb is grown in three places on Earth. It is known that this herb has medicinal value, but that it is also sometimes fatal. In Table 24.3, column two gives the percent of total herb production of the world that is grown in each location, and column three gives a percent of the herb from that location that proved fatal.

   For each location, calculate the probability that a sample of the herb came from that location if it was found to be fatal.

3. Target pistols are manufactured in six cities. Table 24.4 gives the percent of the total production from each city, together with the probability that the gun explodes upon firing. Suppose a gun explodes. Calculate, for each city, the probability that the gun came from there.

4. Suppose 37 percent of a company's computer keyboards are manufactured in Los Angeles and 60 percent are made in Detroit. Suppose 2 percent of those made in Los Angeles have some problem, and 5.3 percent of those from Detroit are defective. Calculate the probability that a given defective keyboard was manufactured in Los Angeles, and also the probability that it came from Detroit.

5. A company hires four different groupings of people to install automobile carpet. The company has found that people with at least ten years experience generally produce jobs that are defective 1 percent of the time. People with five to ten years of experience produce defects 3 percent of the time, those with one to five years experience, 5.7 percent of the time, and people with less than one year experience, 11.3 percent. Of the total workforce installing carpet, those with ten years experience form 45 percent of the total, those with five to ten years, 37 percent, those with one to five years, 7 percent, and those with less than one year, 11 percent. For each of these categories, calculate the probability that a defective installation was done by a worker in that group.

24.8 Expected Value

We will define a notion of payoff, or expected value, of an experiment. An example will suggest a reasonable way to do this.

EXAMPLE 24.21

A carnival wheel has the numbers 1 through 50 equally spaced around the outer rim. It costs one dollar to bet on which number will come up when the wheel is spun and allowed to come to rest. A winner receives twenty five dollars. From the customer's perspective, is this a good game to play?

In view of the payoff, it is tempting to play. On the other hand, we should be suspicious, since the carnival is a profit-making enterprise and the owner must be making money or he or she would change the game.

We will try to define how much a player should expect to win on each spin of the wheel. Since there are 50 numbers, the probability of guessing the winning number on any spin is $1/50$. The probability of losing is $49/50$. Since winning pays twenty five dollars and losing costs one dollar, compute

$$\frac{1}{50}(25) - \frac{49}{50}(1) = -0.48.$$

This is the expected value of this game from the player's perspective. A player should expect on average to lose (because of the negative sign) forty eight cents per game. The player's loss is the carnival's gain. Although the owner will occasionally make a payout, the carnival stands on average to make nearly fifty cents per player per game.

Thought of in a different way, the player will on average win 25 dollars, for $1/50$, or 2 percent of the time, and lose 1 dollar for $49/50$, or 98 percent of the time. Although the payoff is twenty five times the cost of playing, a person will likely lose so many more times than he or she wins, on average each person will contribute 48 cents to the carnival for each game. ♦

In general, we compute the *expected value* of an experiment having payoffs by multiplying the probability of each outcome by the value of that outcome, with a plus sign for value gained and a negative sign for value lost. These numbers are then added to get the expected value. Clearly this number has a perspective, since a win for one player is a loss for another.

EXAMPLE 24.22

Based on performance over the past thirty years, a salesman estimates the following probabilities for commissions on sales over the coming year: 0.25 probability of getting 400 dollars from she sales, 0.15 probability of getting 200 dollars from trouser sales, 0.35 probability of getting 600 dollars from sports jackets, 0.20 probability of getting 1000 dollars from golf outfits, and 0.05 probability of getting 300 dollars from derby hat sales. He wants to calculate his expected commission.

The information is summarized in Table 24.5.

The expected commission is

$$400(0.25) + 200(0.15) + 600(0.35) + 1000(0.20) + 300(0.05) = 555$$

dollars. ♦

TABLE 24.5 *Commission probabilities in Example 24.22.*

Commission	400	200	600	1000	300
Probability	0.25	0.15	0.35	0.20	0.05

SECTION 24.8 PROBLEMS

1. Twenty marbles labeled 1 through 20 are in a jar. A game consists of reaching in blind and pulling out three marbles. If at least two are even, the player wins three dollars. Otherwise the player loses seven dollars. What is the player's expected value?
2. Sixteen dice are rolled and unordered results are recorded. If at least five sixes come up the player wins seventy dollars. Otherwise the player pays four dollars. What is the player's expected value?
3. Seven hats are in a closet. Four are black and three are red. A person reaches in and pulls out four hats, without regard to order. If at least two are red, the player wins ten dollars. Otherwise the player loses five dollars. What is the player's expected value?
4. Six coins are flipped, then four dice are rolled. If four heads or a total of at least 24 on the dice result, the player wins twenty five dollars. Otherwise the player loses three dollars. What is the player's expected value?
5. Six cards are dealt from a standard deck, without regard to order. If one or more aces comes up, the player wins forty five dollars. Otherwise he or she loses thirty dollars. What is the player's expected value?
6. Nine coins are flipped. If seven or more come up tails the player wins fifty dollars. Otherwise the player loses three dollars. What is the player's expected value?
7. A game consists of flipping seven coins, without regard to order. If at least three heads come up, the player wins five dollars. If not, the player loses nine dollars. What is the player's expected value?
8. Five dice are tossed. If at least three come up 5 or 6, the house pays fifty dollars. If not, the house wins four dollars. What is the player's expected value?

CHAPTER 25

MEASURES OF CENTER AND VARIATION RANDOM VARIABLES AND PROBABILITY DISTRIBUTIONS THE BINOMIAL AND POISSON DISTRIBUTIONS NORMALLY

Statistics

25.1 Measures of Center and Variation

Many processes or studies generate tables of numbers called *data*. A *statistic* is a measure that is used in the analysis of data to enable us to draw conclusions from the data. We will begin with two elementary kinds of statistics which are measures of center and measures of variation of data.

25.1.1 Measures of Center

Given a list or table x of real numbers $A_1, A_2, \cdots, A_N$, which need not all be distinct, the *mean*, or *average* of these numbers is denoted $\overline{x}$, or μ, and is defined to be

$$\mu = \overline{x} = \frac{1}{N}(A_1 + A_2 + \cdots + A_N).$$

For example, the mean of 16, −25, 27, 12, 15, 31 is

$$\overline{x} = \frac{1}{6}(16 - 25 + 27 + 12 + 15 + 31) = \frac{38}{3}.$$

This number behaves like a center of gravity of the data. If we mark these numbers on a line and place a one-pound weight at each mark, then $38/3$ is where we should place a fulcrum so that the weighted line balances.

Means are useful but must be interpreted with care.

EXAMPLE 25.1

Last year the graduates of a very prestigious engineering program received the starting salary offers of Table 25.1. The average starting salary is 824, 000, which gives a misleading impression of the success of the graduates. The list does not mention that Seagram is a baseball pitcher with 427 home runs in 428 at bats and eight consecutive no-hitters. ♦

TABLE 25.1 Engineering graduates starting salaries in Example 25.1

Graduates	Starting Salary in Dollars
Jones	80,000
Smith	85,000
Connolly	78,000
Andrews	90,000
Finch	80,000
Hatton	87,000
Douglas	92,000
Seagram	6,000,000

In this example, Seagram is an *outlier*, which is a number so far removed from the other data that it skews the mean toward that number. This can invite misinterpretations of the data.

To define the median of a set of numbers, first order (sort) the numbers. Suppose they have been labeled so that

$$A_1 \leq A_2 \leq \cdots \leq A_N.$$

The *median* of this ordered list is a number M having the property that half of the numbers in the ordered data set are less than or equal to M, and half are greater than or equal to M.

As with the mean, the median may or may not be a number in the set. We find that

$$\text{if } N \text{ is odd, then } M = A_{(N+1)/2}$$

$$\text{if } N \text{ is even, then } M = \frac{1}{2}(A_{N/2} + A_{(N/2)+1}).$$

In the case of odd N, there is a number at the exact center of the ordered set, and this is the median. If N is even, the median is the average of number $N/2$ from the left, and number $(N/2)+1$ from the left. This average of $A_{N/2}$ and $A_{1+N/2}$ may or may not be one of the numbers of the data set.

EXAMPLE 25.2

Order the data set

$$-4, 18, 1, 2, 2, 3, 7, 5, 15, -4, 2, 1, 1, 3, 9, 8$$

to obtain

$$-4, -4, 1, 1, 1, 2, 2, 2, 3, 3, 5, 7, 8, 9, 15, 18.$$

There are $N = 16$ numbers in this data set. The median is

$$M = \frac{1}{2}(A_8 + A_9) = \frac{1}{2}(2+3) = 2.5.$$

This is not a number in the set. However, there are eight numbers in the set to the left of 2.5, and eight numbers to the right. ♦

EXAMPLE 25.3

The ordered data set

$$-4, -4, 1, 1, 1, 1, 2, 2, 2, 3, 3, 5, 7, 8, 9, 15, 16.$$

has $N = 17$ numbers. Since $(N+1)/2 = 9$, the median is $A_9 = 2$, and this is the third 2 from the left, not one of the other numbers 2 in the list. ♦

TABLE 25.2 ***Frequency table in Example 25.4.***

Data Point	−3	−1	4	7	12	17
Frequency	5	2	12	6	12	21

While the mean is a center of gravity (or weighted center) of the data set, the median is the geometric center of the list. It is determined completely by the location of the "center", not by how far numbers to the left and right are from this center.

If N is large, data is often given by a frequency table, which lists, for each number in the data set, the number of times it occurs.

EXAMPLE 25.4

Data (already ordered) is given Table 25.2, which is a frequency table.

The data includes -3 five times, -1 two times, 17 twenty one times, and so on. The number of data entries is the sum of the frequencies, $N = 58$. The mean is the sum of all the data entries divided by 58:

$$\overline{x} = \frac{1}{58}((-3)(5) + (-1)(2) + (4)(12) + (7)(6) + (12)(12) + (17)(21)) = \frac{287}{29},$$

which is approximately 9.897. The median is $(A_{29} + A_{30})/2$. From the left, making use of the frequencies, the twenty-ninth data point is one of the $12's$, as is the thirtieth data point. The median is $(12 + 12)/2 = 12$. ♦

25.1.2 Measures of Variation

We will define two measures of variation, which are indicators of how spread out the data is, as opposed to an indicator of where a center of the data is. The first measure of variation is the *range*, which is simply difference between the largest and smallest numbers in the data set. This is in a sense the width of the data set. In Example 25.4, the range is $17 - (-3) = 20$.

A second measure of variation is the standard deviation. For the *standard deviation* of $A_1, \cdots, A_N$, suppose $\overline{x}$ is the mean. Two calculations of standard deviation are routinely made in statistics.

(1) If $A_1, \cdots, A_N$ is a sample of data within a much larger population, then the standard deviation of this sample is

$$s = \sqrt{\frac{1}{N-1}\sum_{j=1}^{N}(A_j - \overline{x})^2}.$$

It is an exercise in algebra to show that

$$s = \sqrt{\frac{N\sum_{j=1}^{N} A_j^2 - (\sum_{j=1}^{N} A_j)^2}{N(N-1)}}.$$

(2) If $A_1, \cdots, A_N$ are all of the data points, then the standard deviation is

$$\sigma = \sqrt{\frac{1}{N}\sum_{j=1}^{N}(A_j - \overline{x})^2}.$$

For example, a CAT scan may take data at 10, 000 points. These would constitute the entire data population and σ would be used for the standard deviation. A physician might periodically examine samples of 50 of these points and s would be used for the standard deviation of this sample.

The numbers s^2 and σ^2 are called the *variance* of the data.

The standard deviation is zero if each $A_j = \overline{x}$. In this case all the data points are the same and there is no variation. The standard deviation increases as data points differ by larger amounts from the mean, and this is seen in the variation of the data (how "spread out" it is).

EXAMPLE 25.5

Consider a sample of data consisting of the numbers

$$1.1, 1.3, 1.6, 2, 2.1, 2.2, 2.4, 2.5.$$

The average is $\overline{x} = 2.025$. For the standard deviation first compute the sum of the squares of the differences of the numbers from the mean:

$$(1.1 - 2.025)^2 + (1.3 - 2.025)^2 + (1.6 - 2.025)^2 + (2 - 2.025)^2$$
$$+ (2.1 - 2.025)^2 + (2.2 - 2.025)^2 + (2.4 - 2.025)^2 + (2.5 - 2.025)^2 = 1.9650.$$

The standard deviation is

$$s = \sqrt{\frac{1.9650}{7}} = 0.5298,$$

to four decimal places. The small standard deviation is due to the fact that the numbers in the data set are fairly close to one another (range $2.5 - 1.1 = 1.4$).

By contrast, for the sample data set $1.1, 4.6, 9.2, 15.7, 28$, a routine calculation yields $s = \sqrt{481.19/4} = 10.968$. This a much larger standard deviation than the preceding data set occurs because this data is more "spread out", with range 26.9 for only five data points. ♦

As an example of a data set with a large standard deviation, $s = 2, 091, 400$ in Example 25.1. Seagram causes a huge standard deviation in this data.

SECTION 25.1 PROBLEMS

1. Find the mean, median and standard deviation σ for the data in Table 25.3.
2. Find the mean, median and standard deviation σ for the data in Table 25.4.
3. Compute the mean, median and standard deviation σ of each of the following data sets.

(a) $-4, -6, 2.5, 3, 8, 5, -3, -4, 8, 3, 2.2$

(b) $1, 1, 1, -1, 2, 3, -1, 4, 2$

(c) $3, -4, 2, 1.5, -4, -4, 2, 1, 7$

(d) $9.3, 9.5, 9.7, 10, 8.4, 8.7, 8.8, 8.8, 4.1$

(e) $-16, -14, -10, 0, 0, 1, 1, 3, 5, 7$

TABLE 25.3 *Frequency table for Problem 1, Section 25.1.*

Data Point	−3	−1	0	1	3	4
Frequency	4	2	6	4	12	3

TABLE 25.4 *Frequency table for Problem 2, Section 25.1.*

Data Point	−12	−9.7	−8	−7.6	−5.1	4
Frequency	4	2	6	4	12	3

25.2 Random Variables and Probability Distributions

A *random variable* on an experiment is a function that assigns a real number to each outcome of the experiment, based in some specified way on the random outcomes of the experiment. If X is a random variable, then $X(o)$ is a real number for each outcome o of the experiment.

EXAMPLE 25.6

Flip three coins. Each outcome is a string of three letters, with each letter an H or T. If o is an outcome, let

$$X(o) = \text{ number of heads in } o.$$

X assigns a number to each outcome. If we list the outcomes

$$o_1 = HHH, o_2 = HHT, o_3 = HTH, o_4 = THH,$$

$$o_5 = HTT, o_6 = THT, o_7 = TTH, o_8 = TTT,$$

then this random variable has values

$$X(o_1) = 3, X(o_2) = X(o_3) = X(o_4) = 2,$$

$$X(o_5) = X(o_6) = X(o_7) = 1, \text{ and } X(o_8) = 0. \blacklozenge$$

A *probability distribution* on a random variable X is a function P which assigns a probability $P(x)$ to each value x that the random variable can assume.

If $\sum_x$ denotes a summation over all values x of the random variable, then a probability distribution must satisfy

$$\sum_x P(x) = 1 \text{ and } 0 \leq P(x) \leq 1 \text{ for each outcome } x.$$

P acts on values of the random variable, not directly on the outcomes of the experiment. $P(X(o))$ is the probability, not of the outcome o, but of $X(o)$, the value of the random variable at o. We could write $\sum_x P(x)$ as $\sum_o P(X(o))$, with this sum taken as o ranges over all outcomes of the experiment.

We will use P for a probability distribution on a random variable, and Pr for the probability of an outcome.

EXAMPLE 25.7

Flip three coins. There are 8 outcomes, each with probability $\Pr(o) = 1/8$. Let X be the random variable of Example 25.6. We will define a probability function on X. Thus P will act on the values 0, 1, 2, 3 that X takes on.

Since X takes on the value 3 once out of the eight outcomes, assign

$$P(3) = 1/8.$$

Since X assumes each of the values 2 and 1 at three outcomes, let

$$P(2) = P(1) = \frac{3}{8}.$$

Finally, X assumes the value 0 only once, at o_8, so

$$P(0) = \frac{1}{8}.$$

Then $0 \le P(x) \le 1$ for each value x that X takes on, and

$$\sum_x P(x) = \frac{1}{8} + \frac{3}{8} + \frac{3}{8} + \frac{1}{8} = 1,$$

with this summation over all values x that X can assume. ♦

Now suppose X is a random variable with probability distribution P. Define the *mean* of P to be the number

$$\mu = \sum_x x P(x).$$

The *standard deviation* σ of P is

$$\sigma = \sqrt{\sum_x (x-\mu)^2 P(x)}.$$

The expression for μ is reminiscent of the expected value of a game, being a sum of terms, each of which is the product of a payoff x with the probability $P(x)$ that the random variable achieves at x. μ also has a form reminiscent of a weighted moment in physics.

As a special case, if the numerical values achieved by X are $x_1, \cdots, x_N$, and if X is equally likely to achieve each value, then $P(x_j) = 1/N$ and

$$\mu = \sum_{j=1}^{N} x_j \frac{1}{N} = \text{mean of } x_1, \cdots, x_N.$$

Further, in this case,

$$\sigma = \sqrt{\frac{1}{N} \sum_{j=1}^{N} (x_j - \overline{x})^2},$$

the familiar formula for the standard deviation of a population. It is routine to show that

$$\sigma = \sqrt{\sum_x (x^2 P(x)) - \mu^2}.$$

EXAMPLE 25.8

Flip four coins. There are sixteen outcomes:

$$HHHH, TTTT, THHH, HTHH, HHTH, HHHT, HTHT, HHTT,$$
$$HTTH, THHT, THTH, TTHH, HTTT, THTT, TTHT, TTTH.$$

Define a random variable X by letting $X(o)$ be the number of tails in the outcome o. Then:

$X(o) = 0$ one time ; $X(o) = 4$ one time ;

$X(o) = 1$ four times ; $X(o) = 2$ six times ; and

$X(o) = 3$ four times.

Thus x can take on five numerical values. Assign probabilities to these values by putting:

$$P(0) = P(4) = \frac{1}{16},\ P(1) = P(3) = \frac{4}{16},\ P(2) = \frac{6}{16}.$$

The mean of X is

$$\mu = \sum_x xP(x)$$
$$= 0\left(\frac{1}{16}\right) + 4\left(\frac{1}{16}\right) + (1)\left(\frac{4}{16}\right) + 2\left(\frac{6}{16}\right) + 3\left(\frac{4}{16}\right) = 2.$$

On average, with four flips we expect to see two tails.

The square of the standard deviation for X is

$$\sigma^2 = \sum_x (x-\mu)^2 P(x)$$
$$= (0-2)^2\left(\frac{1}{16}\right) + (4-2)^2\left(\frac{1}{16}\right) + (1-2)^2\left(\frac{4}{16}\right)$$
$$+ (2-2)^2\left(\frac{6}{16}\right) + (3-2)^2\left(\frac{4}{16}\right) = 1,$$

so $\sigma = 1$. ♦

A random variable that assumes only finitely many values is called *discrete*. Not every random variable is discrete. For example, the experiment could be to pick a positive integer, and we could define $X(k) = k$.

SECTION 25.2 PROBLEMS

1. A wheel having the numbers 1 through 20 equally spaced about the circumference is spun and the number that is on top when the wheel comes to rest is recorded. If o is an outcome, let $X(o)$ equal the number of prime factors of o. For example, $X(7) = 1$ because 7 is prime and has only one prime factor, while $X(6) = 2$ because $6 = 3(2)$. By convention, 1 is not prime and has no prime factors. Determine the probability distribution of X and the mean and standard deviation of X.

2. Suppose two dice are rolled. If o is an outcome and the two dice show different numbers, let $X(o)$ be the quotient of the larger number divided by the smaller. If o is an outcome in which both dice come up the same, let $X(o)$ be this common value. Determine the probability distribution of X and its mean and standard deviation.

3. Two cards are picked in succession, without replacement, from a standard deck. If o is an outcome, let $X(o)$ equal the sum of the numbers on the cards if both cards are numbered; $X(o) = 11$ if exactly one card is a jack, king, queen or ace; and $X(o) = 12$ if both cards are in the jack, king, queen or ace group. Determine the probability distribution of X, as well as the mean and standard deviation.

4. An experiment is to choose an integer from 1 to 30 inclusive. If o is an outcome, let $X(o) = 1$ if o is divisible by 2, $X(o) = 2$ if X is divisible by 3 but not by 2, and $X(o) = \pi$ otherwise. Determine the probability distribution of X and the mean and standard deviation.

5. Suppose two dice are rolled. If o is an outcome, let $X(o)$ equal the sum of the numbers on the dice. Determine the probability distribution of X, as well as the mean and standard deviation of X.

6. Four coins are tossed. If o is an outcome, let

$$X(o) = \begin{cases} 1 & \text{if two or more tails come up} \\ 3 & \text{otherwise.} \end{cases}$$

Determine the probability distribution of X, as well as the mean and standard deviation.

25.3 The Binomial and Poisson Distributions

Not every distribution we can define is interesting. The binomial and Poisson distributions, however, have important applications.

25.3.1 The Binomial Distribution

The binomial distribution is used in very specific settings. Suppose some experiment or procedure involves a fixed number of trials or repetitions. For example, in flipping a coin fifty times, the experiment would be the coin flips, and each individual flip would be a trial.

Assume that the experiment satisfies the following three criteria:

(1) Each trial must have exactly two outcomes

(2) The trials are independent. The outcome of one trial does not influence the outcome of any other trial.

(3) The probabilities of the outcomes of all the trials are the same. For example, we cannot flip an honest coin and then, as another trial, flip a dishonest coin.

Now suppose N trials are performed, and the result of each trial is S (for success), with probability p, or F (for failure), with probability $q = 1 - p$. Define the random variable X by setting $X(k) =$ the number of times S occurs exactly k times in the N trials. X can take on the values $0, 1, \cdots, N$.

We will show that the probability distribution P of X is the *binomial distribution*

$$P(x) = \binom{N}{x} p^x q^{N-x} \text{ for } x = 0, 1, \cdots, N. \tag{25.1}$$

Recall that the binomial coefficient $\binom{N}{x}$, or ${}_N C_x$, is the number of ways of choosing x objects from N objects, without regard to order, and is given by

$$\binom{N}{x} = \frac{N!}{x!(N-x)!}.$$

To understand equation (25.1), consider the probability of getting S exactly x times in the N trials. Imagine choosing x objects from N objects, without order. There are $\binom{N}{x}$ ways of doing this. For each such way, the probability of that way occurring is $p^x q^{N-x}$, because S comes up x times, each time with probability p, and F comes up $N - x$ times, each time with probability q. Therefore the probability of getting S exactly x times is

$$\begin{aligned} P(x) &= \text{(number of ways this outcome occurs)(probability of each occurrence)} \\ &= \binom{N}{x} p^x q^{N-x} \end{aligned}$$

and this is exactly equation (25.1).

This expression for $P(x)$ is the term having p to the power x in the binomial expansion of $(p+q)^N$. Since $p + q = 1$, then

$$\sum_x P(x) = \sum_{x=0}^{N} \binom{N}{x} p^x q^{N-x} = (p+q)^N = 1^N = 1,$$

a requirement for the probability distribution of a random variable.

For the binomial distribution, the mean and standard deviation have particularly simple forms.

THEOREM 25.1

For a binomial distribution on N trials,

$$\mu = Np \text{ and } \sigma = \sqrt{Npq}. \blacklozenge$$

We may also write

$$\sigma = \sqrt{Np(1-p)}.$$

These can be derived by straightforward algebraic manipulations.

EXAMPLE 25.9

A district attorney is deciding a strategy for a plea bargain. The D.A. knows that jurors make a correct vote for guilt or innocence about 79 percent of the time. There are twelve jurors, who vote independently, and it takes a vote of at least ten to convict. What is the probability that at least ten jurors will reach a correct decision?

We want the probability of at least ten out of twelve votes being correct (as decided by the evidence and presentation). Let $N = 12$, $p = 0.79$ and $q = 0.21$. Since at least ten jurors means exactly ten or exactly eleven or exactly twelve, the probability that at least ten jurors reach a correct decision is

$$\begin{aligned}
&P(10) + P(11) + P(12) \\
&= \binom{12}{10}(0.79)^{10}(0.21)^2 + \binom{12}{11}(0.79)^{11}(0.21) + \binom{12}{12}(0.79)^{12} \\
&= \frac{(12)(11)}{2}(0.79)^{10}(0.21)^2 + 12(0.79)^{11}(0.21) + (0.79)^{12} \approx 0.523.
\end{aligned}$$

The probability that exactly 8 jurors reach a correct decision is

$$P(8) = \binom{12}{8}(0.79)^8(0.21)^4 \approx 0.146. \blacklozenge$$

EXAMPLE 25.10

An assembly line produces automatic veebles. It has been found that the probability that a veeble is defective (fails to veebulate) is 0.02. If 40 veebles are pulled at random from a warehouse, what is the probability that no more than three are defective?

Let $N = 40$, $p = 0.02$ and $q = 0.98$. The random variable X is defined by $X(x) =$ number of veebles that are defective, for $x = 0, 1, \cdots, 40$. The probability distribution is binomial and $P(x)$ equals the probability that x of the forty veebles are defective. If no more than three are defective, then zero, one, two or three can be defective. Compute

$$\begin{aligned}
&P(0) + P(1) + P(2) + P(4) \\
&= \binom{40}{0}p^0q^{40} + \binom{40}{1}pq^{39} + \binom{40}{2}p^2q^{38} + \binom{40}{3}p^3q^{37} \\
&= (0.98)^{40} + 40(0.02)(0.98)^{39} + (20)(39)(0.02)^2(0.98)^{38} \\
&\quad + (20)(13)(38)(0.02)^3(0.98)^{37} \\
&\approx 0.992.
\end{aligned}$$

This probability makes sense intuitively. The probability of a veeble being defective is small, so it should be likely that fewer than four (out of forty) are defective.

The mean of this distribution is $\mu = Np = 40(0.02) = 0.8$. And the standard deviation is $\sigma = \sqrt{Np(1-p)} \approx 0.885$. ♦

25.3.2 The Poisson Distribution

Suppose we have a specific interval or segment of something in mind. This could be a time interval, a length or distance, a section of a floor or wall, or something else. We are interested in the occurrence of some event falling within this interval.

Let $X(x) =$ the number of times x occurs in the given interval. Then X is a random variable, assuming that the occurrence of falling or not falling in the interval is a random event. The probability that the event falls in the interval x times can be shown to have the form

$$P(x) = \frac{\mu^x e^{-\mu}}{x!}$$

in which μ is a ratio of certain occurrences per interval, according to the setting, and varies over different experiments. P is a probability distribution on the random variable X and is called the *Poisson distribution.*

The binomial distribution is discrete, since it deals with the probability of some event occurring a certain (finite) number of times out of a given number. By contrast, the Poisson distribution is not finite, since it can take on the value of any nonnegative integer.

In the Poisson distribution it is assumed that the occurrences under observation are independent of each other and are uniformly distributed over the interval of interest. The occurrences are therefore not confined to one part of the interval.

EXAMPLE 25.11

A game consists of throwing darts onto a large flat mat that has been divided into 450 squares having 6 inch sides. In one session 370 darts were thrown. We want the probability that one square was hit exactly twice or exactly four times.

The intervals here are the squares on the mat. The experiment involves an event (dart hitting the mat) occurring over an interval (dart falls within one square). We therefore use the Poisson distribution. First, compute the value of μ for this experiment:

$$\mu = \frac{370}{450} = 0.822,$$

the ratio of darts thrown to squares on the mat. For this experiment,

$$P(x) = \frac{(0.822)^x e^{-0.822}}{x!}.$$

We want the probability of a square being hit by exactly two or exactly four darts (out of the 370 thrown). Compute

$$P(2) + P(4) = \frac{(0.822)^2 e^{-0.822}}{2!} + \frac{(0.822)^4 e^{-0.822}}{4!} = 0.157,$$

to three decimal places. Since $(450)(0.157) = 70.65$, we expect that on average, 71 squares will be hit exactly two or exactly four times.

For contrast, suppose only 80 darts are thrown. Now

$$\mu = \frac{80}{450} = 0.178,$$

so now

$$P(2)+P(4)=\frac{(0.178)^2e^{-0.178}}{2!}+\frac{(0.178)^4e^{-0.178}}{4!}=0.0133.$$

Since $(450)(0.0133)=5.985$, we would expect six of the squares to be hit exactly two or exactly four times, with 80 darts thrown. ♦

With the Poisson distribution, the mean is the number μ in the definition of $P(x)$, and $\sigma=\sqrt{\mu}$.

SECTION 25.3 PROBLEMS

Problems Involving A Binomial Distribution

1. An auto manufacturer purchases starting mechanism from an independent contractor. Every week, a quality control officer pulls one hundred starters out of the shipment and inspects each one. A starter is either defective or not, with no range in between. If two or fewer are defective, the total shipment is accepted. Otherwise it is returned. Further, it is known that overall the starter company has a 3.2 percent rate of production of defective starters. What is the probability that the next shipment will be accepted?
2. In a clinical trial, it was found that 7 percent of those in the control group, who did not receive the drug, experienced a particular side effect commonly associated with this drug. Suppose the same 7 percent rate was found in those taking the drug in the trial. Find the probability that among the 1000 subjects taking the drug exactly 92 exhibited the side effect.
3. In each of (a) through (h), use a binomial distribution to compute the probability of x successes (outcome S) for the given values.
   (a) $N=8,\ p=0.43,\ x=2$
   (b) $N=4,\ p=0.7,\ x=3$
   (c) $N=6,\ p=0.5,\ x=3$
   (d) $N=10,\ p=0.6,\ 2\le x\le 5$
   (e) $N=8,\ p=0.4,\ x\ge 7$
   (f) $N=10,\ p=0.58,\ 2\le x\le 4$
   (g) $N=10,\ p=0.35,\ x=3$ or 7
   (h) $N=7,\ p=0.24,\ x=1,3,$ or 5
4. A professor gives a test consisting of six multiple choice questions, each with four answers from which to choose. A student who knows less than nothing takes the test by the method of random choice, guessing on each answer. Calculate the probability that the student
   (a) gets exactly one answer correct,
   (b) gets exactly two correct,
   (c) gets four or five correct,
   (d) gets all six answers right.
5. In Problem 4, change the number of choices for each question to three. Now answer questions (a) through (d).
6. In Problem 4, keep the number of choices with each question to four, but suppose now there are ten questions. Answer the questions (a) through (d) and also determine the probability of getting exactly five, six or seven questions correct.

Problems Involving the Poisson Distribution

7. A golf driving range subdivides the area between 100 and 300 yards from the tee into 500 regions. In one four hour period, 476 balls were driven into this area. Assuming a Poisson distribution, compute the probability that any region was hit exactly 3 times. What is the probability that a region was hit from two to six times, inclusive?
8. A new computer chip has one million microdots etched on its surface. In the manufacturing process, 907, 850 dots are imprinted with information. Assuming a Poisson distribution, determine the probability that the process will incorrectly etch exactly three information packets on the same dot.
9. In a splatter test over a full day, a car windshield is subdivided into 320 pieces and it is found that 295 were hit by some material kicked up from the road by other cars. Assuming a Poisson distribution, determine
   (a) the probability that some piece is hit three times,
   (b) the probability that some piece is hit from two to five times, inclusive.
10. Assuming a Poisson distribution, determine $P(x)$ for each of the following.

(a) $\mu = 0.9, x = 6$
(b) $\mu = 0.85, x = 10$
(c) $\mu = 0.92, x = 4$
(d) $\mu = 0.87, x = 8$
(e) $\mu = 0.94, 1 \leq x \leq 5$
(f) $\mu = 0.64, 3 \leq x \leq 5$
(g) $\mu = 0.75, x = 3$ or 8
(h) $\mu = 0.97, x = 1$ or 3 or 10

25.4 Normally Distributed Data and Bell Curves

We will illustrate normally distributed data with a simple example. Let a trial consist of flipping a coin twelve times and recording the number of heads. For the experiment, carry out N trials (each trial is 12 flips). Record, over all N trials, the frequency of occurrence of each number $0, 1, \cdots, 12$ of heads. Thus, record the number of trials in which there were 0 heads, and the number of trials in which there was 1 head, and the number in which there were 2 heads, and so on.

Is there any pattern that becomes apparent as N is chosen larger? Since the number of heads on any trial is any number from 0 to 12, and each trial has a random outcome, we might think not. However, consider the bar graphs in Figures 25.1 through 25.6, which resulted from an actual run of N trials, for various N. Each graph plots the possible number of heads (horizontal, 0 through 12) versus the frequency of this number of heads that occurred over the N trials. For these bar graphs, N increases respectively through 21 repetitions, then 86 repetitions, 254 repetitions, 469, 920 and 1414 repetitions.

As N increases, a curve drawn through the tops of the bars, as in Figure 25.6, in these graphs approaches a bell-shaped curve like that of Figure 25.7. When this occurs as N increases, we say that the data is *normally distributed.* Of course, for any finite N the fit of the frequency graph of normally distributed data to a bell curve will only be approximate, but the difference decreases to zero as N increases, and for practical purposes, such as doing statistical studies, it is sufficient to choose N large - how large depends on the experiment.

If this experiment is repeated, different numbers of heads would probably appear in the tosses, than were recorded in this run. Nevertheless, as the number N of trials increases, the data would approach a normal distribution.

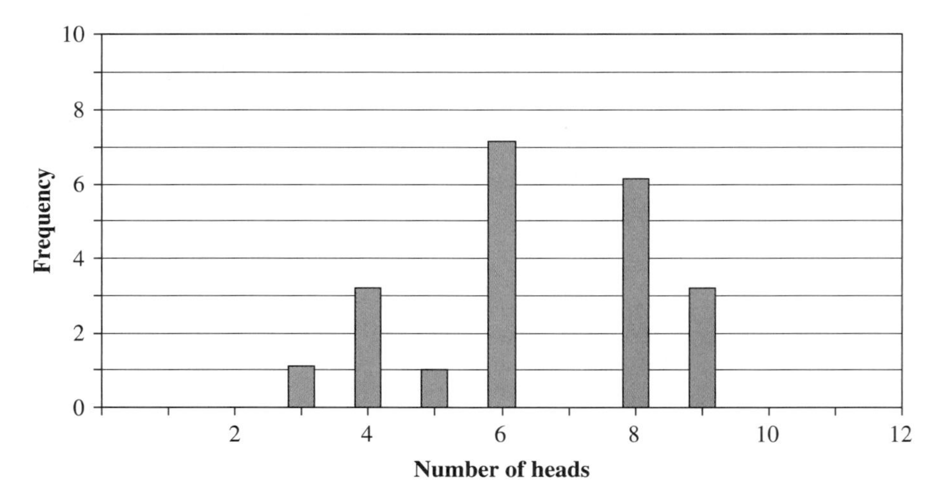

FIGURE 25.1 12 *coin flips,* 21 *repetitions.*

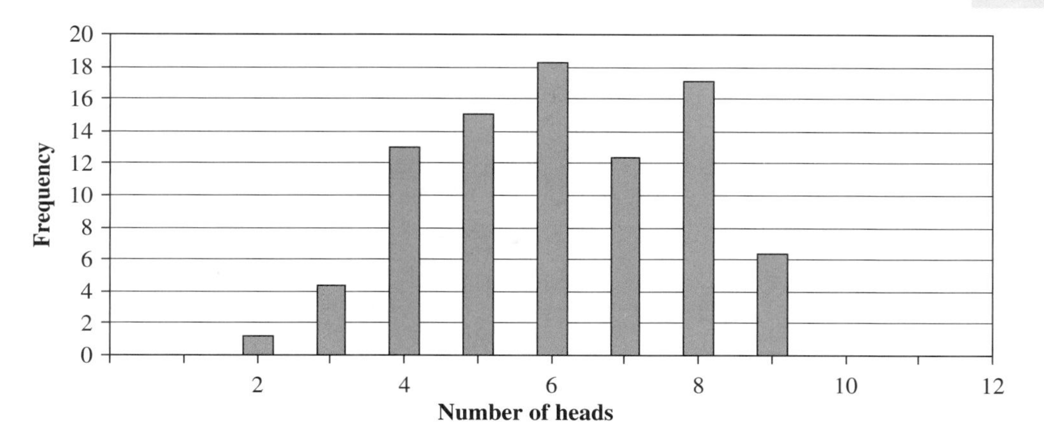

FIGURE 25.2 86 *repetitions.*

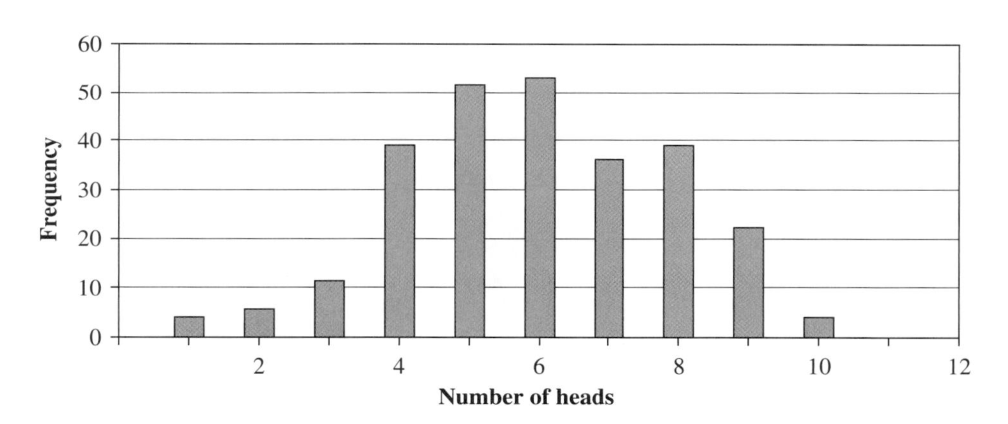

FIGURE 25.3 254 *repetitions.*

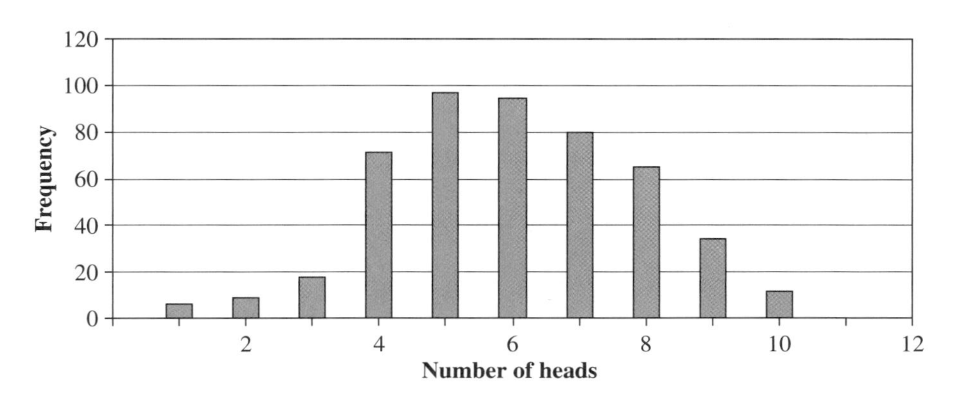

FIGURE 25.4 469 *repetitions.*

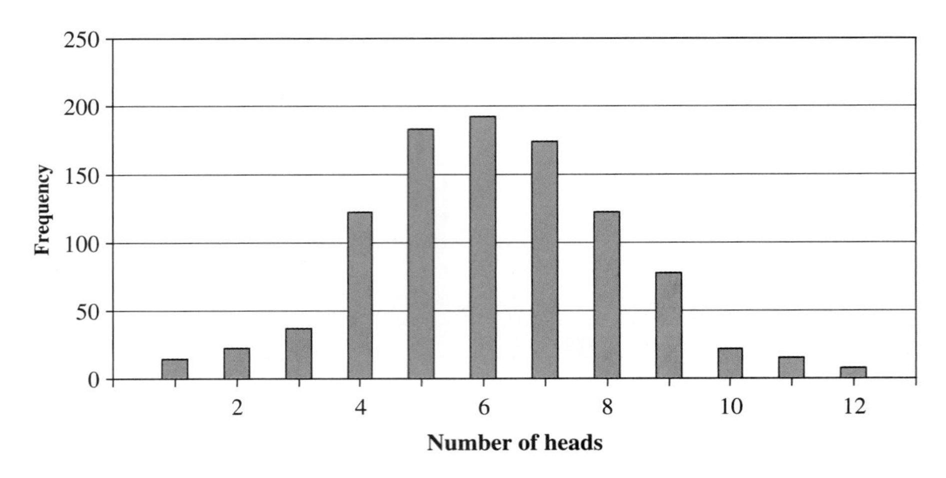

FIGURE 25.5 920 *repetitions.*

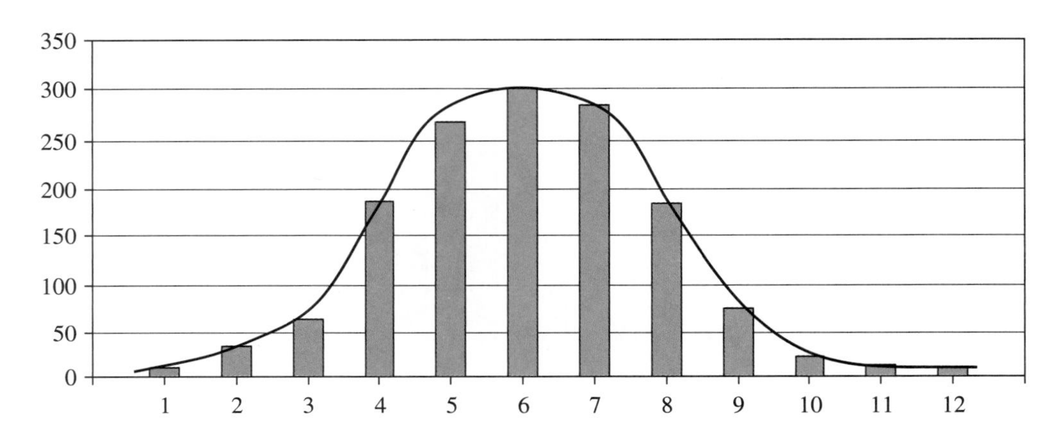

FIGURE 25.6 1414 *repetitions.*

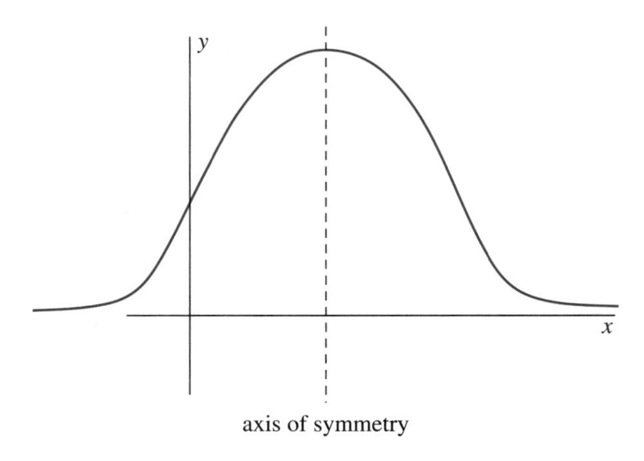

FIGURE 25.7 *A bell-curve.*

EXAMPLE 25.12

Table 25.5 gives ranges for scores on a standardized college-entrance test, together with percentages (for one year) of students taking the test whose scores fell in each range. Figure 25.8 plots this data, which is approximately bell-shaped. This test data (at least for that year) is normally distributed. ♦

Thus far the notion of a bell curve is completely intuitive. More formally, a *bell curve* is the graph of an exponential function

$$y = \frac{1}{\sqrt{2\pi}\,\sigma} e^{-(x-\mu)^2/2\sigma^2},$$

in which σ is a positive number and μ is any number. Figure 25.9 shows bell curves for $\mu = 6$ and σ equal to 3.5, 4 and 4.5. The peaks on these bell curves decrease as σ increases.

TABLE 25.5 *Standardized test scores in Example 25.12.*

Range	Percent of Students	Range	Percent of Students
200 - 240	0.9	500 - 540	17.6
250 - 290	1.8	550 - 590	14.3
300 - 340	4.6	600 - 640	10.6
350 - 390	8.9	650 - 690	6.4
400 - 440	13.5	700 - 740	2.5
450 - 490	17.6	750 - 800	1.4

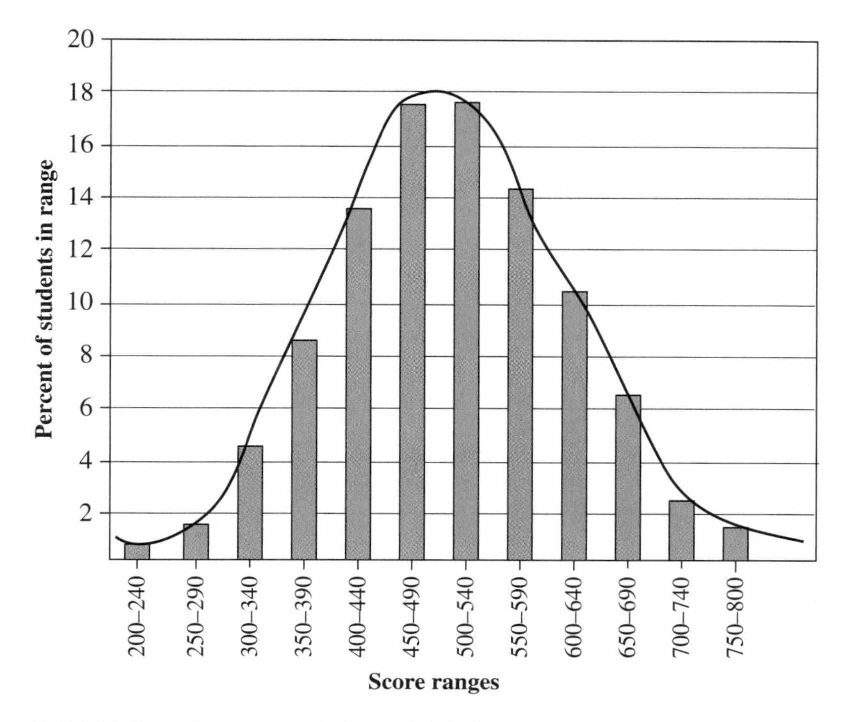

FIGURE 25.8 *Score ranges in Example 25.12.*

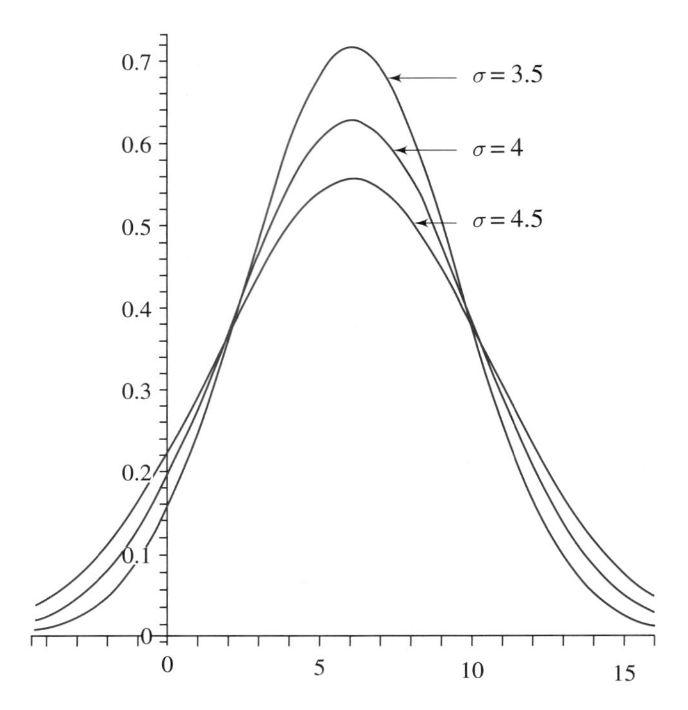

FIGURE 25.9 *Bell curves for $\mu = 6$ and $\sigma = 3.5, 4$ and $4/5$.*

Here are some of the important properties of a bell curve.

(1) The curve lies entirely above the horizontal axis and has a single maximum at $x = \mu$.

(2) The curve is symmetric about the vertical line $x = \mu$. This line is called the *axis of symmetry* of the bell curve.

(3) The curve is asymptotic to the horizontal axis as $x \to \infty$ and as $x \to -\infty$.

(4) The curve has exactly two points of inflection, occurring at $x = \mu \pm \sigma$.

(5) The area bounded by the bell curve and the x - axis is 1. This means that

$$\int_{-\infty}^{\infty} \frac{1}{\sqrt{2\pi}\,\sigma} e^{-(x-\mu)^2/2\sigma^2}\, dx = 1.$$

If we define the continuous random variable

$$X(x) = \frac{1}{\sqrt{2\pi}\,\sigma} e^{-(x-\mu)^2/2\sigma^2}$$

then property (5) of the bell curve enables us to define a probability distribution for X having mean μ and standard deviation σ. In particular, if $X(x)$ is a continuous random variable, then:

(1) If $a < b$ and $P(a \le x \le b)$ is the probability that an observed data value $X(x)$ falls within $[a, b]$, then

$$P(a \le x \le b) = \frac{1}{\sqrt{2\pi}\,\sigma} \int_a^b e^{-(x-\mu)^2/2\sigma^2}\, dx.$$

$P(a \le x \le b)$ is the hatched area in Figure 25.10.

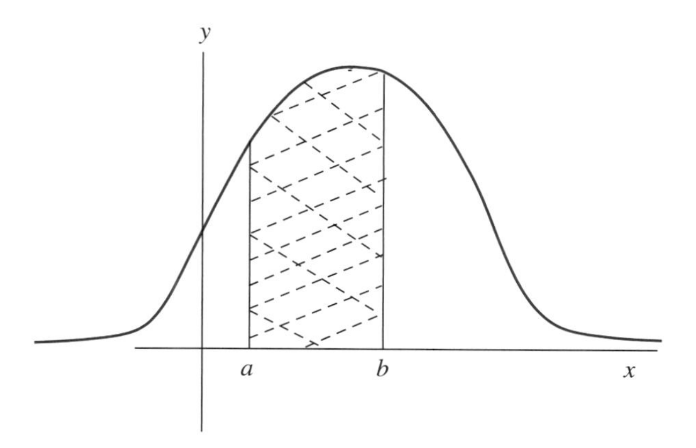

FIGURE 25.10 $P(a \leq x \leq b)$.

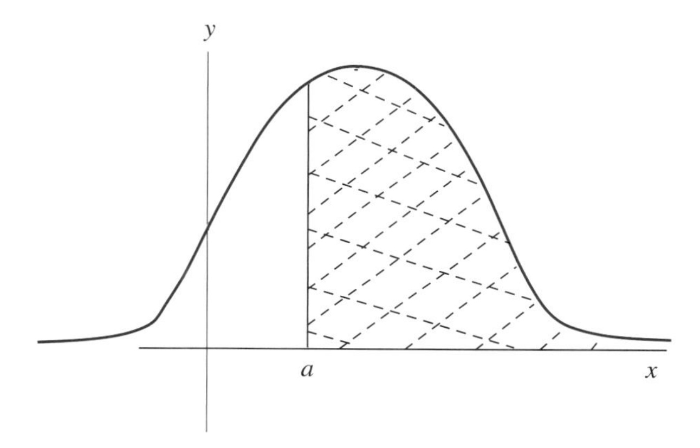

FIGURE 25.11 $P(x \geq a)$.

(2) If $P(x \geq a)$ is the probability that an observed value $X(x)$ is at least as large as a, then

$$P(x \geq a) = \frac{1}{\sqrt{2\pi}\sigma} \int_a^\infty e^{-(x-\mu)^2/2\sigma^2}\, dx.$$

This is the hatched area in Figure 25.11.

(3) If $P(x \leq b)$ is the probability that an observed data value $X(x)$ is no larger than b, then

$$P(x \leq b) = \frac{1}{\sqrt{2\pi}\sigma} \int_{-\infty}^b e^{-(x-\mu)^2/2\sigma^2}\, dx$$

This is the hatched area of Figure 25.12.

These integrals can be approximately evaluated using tables, and mathematics computation software packages can also be used.

We will show how the mean and standard deviation of normally distributed data can be used to associate this data with a particular bell curve by specifying μ and σ, and then how conclusions about the probability of outcomes related to the data can be drawn from the bell curve.

Suppose we carry out some set of trials, and each trial has exactly two possible outcomes, called A and B. Suppose also that the outcomes of separate trials are independent of each other. We therefore have a binomial random variable which we will call X, where $X(x)$ is the number of times A comes up x times in the experiment.

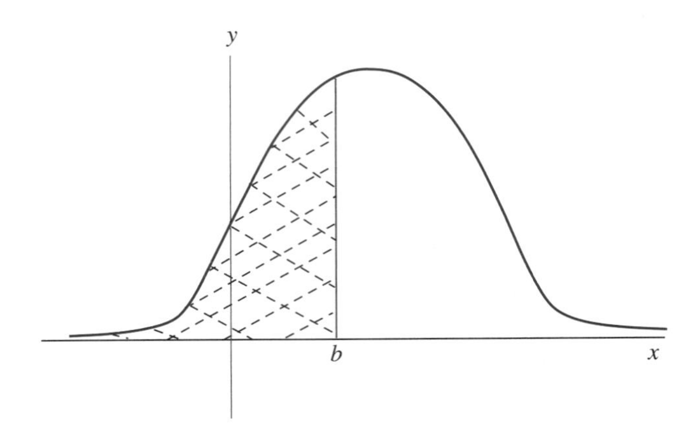

FIGURE 25.12 $P(x \leq b)$.

Let the probability that A occurs in any trial be p. Then the probability that B occurs in a trial is $q = 1 - p$. The frequency of A after N trials, as N increases, approaches a normal distribution given by a bell curve. By Theorem 1, the mean of the frequencies of any set of N trials given by this bell curve is

$$\mu = Np$$

and the standard deviation is

$$\sigma = \sqrt{Np(1-p)}.$$

The bell curve associated with this experiment is therefore the graph of

$$y = \frac{1}{\sqrt{2\pi}\,\sigma} e^{-(x-\mu)^2/2\sigma^2}. \tag{25.2}$$

Now select an interval on the x - axis. We want the probability that $X(x)$ falls within this interval (that is, that A occurs between a times and b times, inclusive). The graph of the continuous bell curve is only approximated by the graph of the discrete random variable. We must therefore make an adjustment in relating area under this curve to a probability. This is done by what is called a *continuity adjustment* in the limits of integration, achieved by adding $1/2$ to the upper limit and subtracting $1/2$ from the lower limit. This is described in the following theorem.

THEOREM 25.2

Let X be a discrete random variable having a binomial distribution.

(1) Let $a \leq x \leq b$. Let $P(a \leq x \leq b)$ be the probability that an observed data value $X(x)$ falls within $[a, b]$. Then

$$P(a \leq x \leq b) \approx \int_{a-1/2}^{b+1/2} \frac{1}{\sqrt{2\pi}\,\sigma} e^{-(x-\mu)^2/2\sigma^2}\, dx.$$

(2) Let $P(x \geq a)$ be the probability that an observed data value $X(x)$ is at least as large as a. Then

$$P(x \geq a) \approx \int_{a-1/2}^{\infty} \frac{1}{\sqrt{2\pi}\,\sigma} e^{-(x-\mu)^2/2\sigma^2}\, dx.$$

(3) Let $P(x \leq b)$ be the probability that an observed data value $X(x)$ is no larger than b. Then

$$P(x \leq b) \approx \int_{-\infty}^{b+1/2} \frac{1}{\sqrt{2\pi}\sigma} e^{-(x-\mu)^2/2\sigma^2}\, dx.$$

The reason these conclusions are stated as approximations is that the discrete random variable X is approximated by a continuous bell curve, which is chosen using μ and σ determined by X from Theorem 1. The continuity adjustment in the limits of integration is itself an approximation.

EXAMPLE 25.13

Flip a coin twelve times. Previously we performed many repetitions of this trial, charting the frequencies of the occurrences of heads. Here $N = 12$, $p = 1/2$ and $q = 1/2$, assuming an honest coin. The mean of the approximately normally distributed data is

$$\mu = Np = (12)(1/2) = 6,$$

not a surprising result. If we flip a coin twelve times, we expect on average to obtain 6 heads, although in any specific trial we may get from 0 through 12 heads.

The standard deviation is

$$\sigma = \sqrt{Np(1-p)} = \sqrt{12\left(\frac{1}{2}\right)\left(\frac{1}{2}\right)} = \sqrt{3}.$$

The bell curve approximating this experiment is

$$\begin{aligned} y &= \frac{1}{\sqrt{2\pi}\sigma} e^{-(x-\mu)^2/2\sigma^2} \\ &= \frac{1}{\sqrt{6\pi}} e^{-(x-6)^2/6}. \end{aligned}$$

This bell curve has maximum at $x = 6$ and the line $x = 6$ is the axis of symmetry. Suppose we want the probability that the number of heads that comes up falls between 4 and 8 inclusive. Taking care to use the continuity adjustment, this probability is approximated by

$$P(4 \leq x \leq 8) \approx \frac{1}{\sqrt{6\pi}} \int_{3.5}^{8.5} e^{-(x-6)^2/6}\, dx \approx 0.85109.$$

The probability is about 0.85 that a head will come up 4, 5, 6, 7 or 8 times.

In this example we can compute $P(4 \leq x \leq 8)$ exactly, using a binomial distribution. This will give us some feeling for how good the approximation is. Compute

$$\begin{aligned} &P(4) + P(5) + P(6) + P(7) + P(8) \\ &= \binom{12}{4}\left(\frac{1}{2}\right)^4\left(\frac{1}{2}\right)^8 + \binom{12}{5}\left(\frac{1}{2}\right)^5\left(\frac{1}{2}\right)^7 + \binom{12}{6}\left(\frac{1}{2}\right)^6\left(\frac{1}{2}\right)^6 \\ &\quad + \binom{12}{7}\left(\frac{1}{2}\right)^7\left(\frac{1}{2}\right)^5 + \binom{12}{8}\left(\frac{1}{2}\right)^8\left(\frac{1}{2}\right)^4 \\ &= \frac{1}{2^{12}}(495 + 792 + 924 + 792 + 495) = 0.854, \end{aligned}$$

differing from the approximate value by about 0.003.

What is the probability that the number of heads will be 10 or more? This is approximated by

$$P(x \geq 10) \approx \frac{1}{\sqrt{6\pi}} \int_{10-1/2}^{\infty} e^{-(x-6)^2/6}\, dx \approx 0.02165.$$

As we might expect, the probability of getting ten or more heads in twelve flips is very small.

Next we will compute the probability of getting from 0 to 6 heads, inclusive. This is

$$P(x \leq 6) \approx \frac{1}{\sqrt{6\pi}} \int_{-\infty}^{6+1/2} e^{-(x-6)^2/6}\, dx \approx 0.61359.$$

It is a common error to think that this probability should be $1/2$. However, there are seven ways of getting from zero to six heads, inclusive, and only six ways of getting more than six heads.

Finally, what is the probability that there will be exactly 7 heads? This probability is

$$P(x = 7) \approx \frac{1}{\sqrt{2\pi}} \int_{7-1/2}^{7+1/2} e^{-(x-6)^2/6}\, dx \approx 0.193.$$

Again, we can compute this exactly using the binomial distribution:

$$P(7) = \binom{12}{7} \left(\frac{1}{2}\right)^7 \left(\frac{1}{2}\right)^5 = 0.193. \ \blacklozenge$$

EXAMPLE 25.14

Roll fifty dice. The outcome of interest in a single trial (fifty rolls) is that 4 comes up.

Think of each roll has having two outcomes, either 4 or not 4. The probability of rolling 4 is $p = 1/6$ and the probability of not 4 is $q = 5/6$. With $N = 50$ for a typical trial, the mean and standard deviation are given by

$$\mu = (50)\frac{1}{6} = \frac{25}{3} \text{ and } \sigma = \sqrt{50\left(\frac{1}{6}\right)\left(\frac{1}{6}\right)} = \frac{5}{6}\sqrt{10}.$$

The bell curve associated with these trials is the graph of

$$y = \frac{1}{\sqrt{2\pi}\sigma} e^{-(x-\mu)^2/2\sigma^2} = \frac{6}{5\sqrt{20\pi}} e^{-9(x-25/3)^2/125}.$$

Suppose we want the probability that, on 50 rolls, the number of times 4 comes up is between 2 and 15 inclusive. This is

$$P(2 \leq x \leq 15) \approx \frac{6}{5\sqrt{20\pi}} \int_{3/2}^{31/2} e^{-9(x-25/3)^2/125}\, dx \approx 0.99198. \ \blacklozenge$$

EXAMPLE 25.15

An airline knows that its planes fly an average of 1500 miles per month per passenger, with a standard deviation of 90 miles. We want the probability that a typical passenger flies between 1400 and 1600 miles per month.

Assuming that the data is normally distributed, then $\mu = 1500$ and $\sigma = 90$. The bell curve to use is the graph of

$$y = \frac{1}{90\sqrt{2\pi}} e^{-(x-1500)^2/16200}.$$

Then

$$P(1400 \leq x \leq 1600) = \frac{1}{90\sqrt{2\pi}} \int_{1400-1/2}^{1600+1/2} e^{-(x-1500)^2/16200}\, dx \approx 0.73586. \ \blacklozenge$$

25.4.1 The Standard Bell Curve

The bell curve with $\mu = 0$ and $\sigma = 1$ is the graph of

$$y = \frac{1}{\sqrt{2\pi}} e^{-x^2/2},$$

shown in Figure 25.13. This is called the *standard* or *normal bell curve*. It is common practice to use z as the variable, and statisticians refer to values on the horizontal axis as z - *scores*.

Tables used in statistics are often compiled using the standard bell curve as a reference. For example, suppose we have in mind a particular probability and we want to know a value $z = z_0$ such that $P(z \leq z_0)$ is this number. This means that, given k, we want to solve for z_0 such that

$$\frac{1}{\sqrt{2\pi}} \int_{-\infty}^{z_0} \frac{1}{\sqrt{1\pi}} e^{-z^2}\, dz = k.$$

Using the standard bell curve makes it possible to make a table of values for k and z_0, a short version of which is given in Table 25.6.

Given any bell curve, we can by a change of variables refer it to the standard bell curve.

25.4.2 The 68, 95, 99.7 Rule

Suppose we have an experiment satisfying the requirements for a binomial distribution, with N trials, $\mu = Np$ and $\sigma = \sqrt{Np(1-p)}$. As usual, $X(x)$ is the number of occurrences of the trial result of interest in outcome x. The bell curve associated with this experiment is

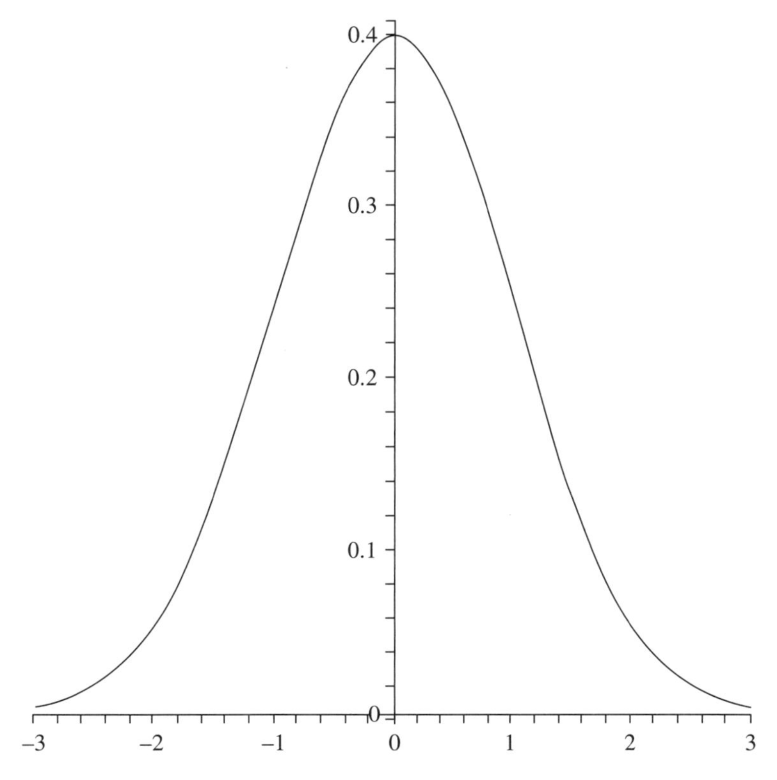

FIGURE 25.13 *Standard, or normal, bell curve.*

TABLE 25.6 ***z - scores for given probabilities.***

area left of z_0	z_0
0.99	2.33
0.95	1.645
0.90	1.29
0.85	1.04
0.80	0.84
0.75	0.67

$$y = \frac{1}{\sqrt{2\pi}\sigma} e^{-(x-\mu)^2/2\sigma^2}.$$

We know that this bell curve has two points of inflection, as $\mu \pm \sigma$, one standard deviation to the left and right of the mean. Moving one, two or three standard deviations to either side of this mean has a particular significance.

The 68, 95, 99.7 rule consists of the following three conclusions.

The area under the bell curve between $\mu - \sigma$ and $\mu + \sigma$ is approximately 68 percent of the total area.

The area under the bell curve between $\mu - 2\sigma$ and $\mu + 2\sigma$ is approximately 95 percent of the total area.

The area under the bell curve between $\mu - 3\sigma$ and $\mu + 3\sigma$ is approximately 99.7 percent of the total area.

In terms of probabilities:

The probability that the outcome falls between $\mu - \sigma$ and $\mu + \sigma$ is approximately 0.68.

The probability that the outcome falls between $\mu - 2\sigma$ and $\mu + 2\sigma$ is approximately 0.95.

The probability that the outcome falls between $\mu - 3\sigma$ and $\mu + 3\sigma$ is approximately 0.997.

Thus it is almost certain that any outcome of the experiment will fall within three standard deviations of the mean.

EXAMPLE 25.16

Suppose we have a dishonest coin and the probability of a head on any toss is $p = 0.8$. Flip the coin 1000 times. Compute $\mu = 1000(0.8) = 800$ and

$$\sigma = \sqrt{1000(0.8)(0.2)} = 12.65.$$

Now $\mu - \sigma = 787.35$ and $\mu + \sigma = 812.65$, so in the thousand flips there is a probability of 0.68 of seeing between 787 and 813 heads. There is a probability of 0.95 of seeing between 774 and 826 heads (two standard deviations to both sides of the mean). And there is a probability of 0.997 of seeing between 762 and 838 heads, three standard deviations to both sides of the median. ♦

SECTION 25.4 PROBLEMS

1. An honest die is rolled and we are interested in the outcome that an even number comes up. A total of 550 tosses are made.

   (a) Determine a and b so that there is a probability of about 0.68 of seeing between a and b even tosses.

   (b) Determine a and b so that there is a probability of about 0.95 of seeing between a and b even tosses.

   (c) Determine a and b so that there is a probability of about 0.99 of seeing between a and b even tosses.

2. A coin has probability of 0.42 of coming up heads. The coin is flipped 2000 times.

   (a) Determine a and b so that there is a probability of about 0.68 of seeing between a and b heads.

   (b) Determine a and b so that there is a probability of about 0.95 of seeing between a and b heads.

   (c) Determine a and b so that there is a probability of about 0.99 of seeing between a and b heads.

3. A bus company finds that its buses drive an average of 700 miles per month for each passenger, with a standard deviation of 65 miles. Assume that the data consisting of the number of miles driven per passenger per month is normally distributed. Determine the probability that a typical passenger rides between 250 and 600 miles per month. What is the probability that a passenger rides between 600 and 900 miles per month?

4. An auto rental firm finds that its cars travel an average of 940 miles for each rental, with a standard deviation of 76 miles. Find the probability that a customer who rents the car drives it between 750 and 1000 miles. What is the probability that the car is driven between 300 and 500 miles?

5. In a recent year in the American League hitters averaged .247 with a standard deviation of 0.021. Determine the probability that a hitter averaged between .245 and .270. What is the probability that a hitter averaged over .260? What is the probability that a batter averages .300 or more?

6. When a stratigather is produced by a company, it may be either left handed or right handed. The probability for righthanded is 0.58, and for lefthanded, 0.42. A thousand stratigathers are chosen at random from the canyon where they are stored. Compute the probability

   (a) that at least 400 are righthanded,

   (b) that between 400 and 600 are righthanded,

   (c) that no more than 450 are righthanded,

   (d) that exactly 520 are righthanded.

7. In a certain kingdom, babies are born with a probability of 0.48 for boys and 0.52 for girls. In a particular year it is expected that 350 babies will be born. We are interested in the outcome that a newborn is a girl.

   (a) Determine μ and σ for this experiment.

   (b) Determine the exponential whose graph is the bell curve.

   Determine the probability that

   (c) at least 220 will be girls,

   (d) at least 150 will be girls,

   (e) the number of girls will fall between 120 and 250, inclusive,

   (f) the number of girl babies will be exactly 180.

8. A coin is flipped 100 times.

   (a) Determine μ and σ for this experiment.

   (b) Determine the exponential whose graph is the bell curve.

   Determine the probability that a head comes up

   (c) between 20 and 40 times,

   (d) between 10 and 50 times,

   (e) between 45 and 55 times,

   (f) at least 30 times,

   (g) fewer than 60 times,

   (h) exactly 55 times.

9. Roll a die 90 times. We are interested in the outcome that a 6 comes up.

   (a) Determine μ and σ for this experiment.

   (b) Determine the exponential whose graph is the bell curve.

   Determine the probability that a 6 comes up

   (c) between 30 and 60 times,

   (d) between 2 and 80 times,

   (e) at least 35 times,

   (f) fewer than 20 times,

   (g) exactly 45 times.

25.5 The Central Limit Theorem

In a statistical study, the term *population* refers to the set of all objects under study. For example, in a study of defects in newly manufactured automobiles, the autos are the population.

Often we want to draw a conclusion about a population by examining a sample. When we do this, we want some way of estimating the confidence we can have that the conclusion drawn from the sample extends to the entire population. In particular, we would like to be able to begin with a particular level of confidence that we want to achieve, and know how large a sample to test to have this confidence level in drawing conclusions about the population.

To approach this issue, we will agree to draw samples from a population *with replacement*. We select an object and test it, then replace it, select another object (possibly the same one) and test it, replace it, and so on. Because very large numbers of objects are usually involved, sampling with or without replacement makes little practical difference. But it has the important consequence that it makes the selection of each sample independent of the selection of the other samples, avoiding some technical difficulties. We will also agree on a fixed sample size in any particular setting.

When we take samples involving numerical quantities, each sample has a mean (the mean of the numbers in the sample). The *sampling distribution of the mean* is the probability distribution of the means of the samples. We may also compute other statistics from samples, such as medians or standard distributions, and average these.

EXAMPLE 25.17

A contractor looks at the number of buildings his company has completed over the past five years. The numbers are $1, 4, 6, 8$ and 9. These numbers constitute the population under consideration. Of course, we would not take samples from such a small population, but we want to illustrate some ideas related to samples.

With replacement there are $5^2 = 25$ samples of two of the numbers. The samples are listed in Table 25.7, together with each sample mean and the sample standard deviation.

TABLE 25.7 ***Sample means and sample standard deviations in Example 25.17.***

sample x	sample mean $\overline{x}$	sample standard deviation
1,1	1	0
1,4	2.5	2.1213
1,6	3.5	3.5355
1,8	4.5	4.3012
1,9	5	5.6569
4,1	2.5	2.1213
4,4	4	0
4,6	5	1.4142
4,8	6	2.8284
4,9	6.5	3.5355
6,1	3.5	3.5355
6,4	5	1.4142
6,6	6	0
6,8	7	1.4142
6,9	6.5	2.1213
8,1	4.5	4.9497
8,4	6	2.8284
8,6	7	1.4142
8,8	8	0
8,9	8.5	0.7071
9,1	5	5.6569
9,2	6.5	3.5355
9,6	7.5	2.1213
9,8	8.5	0.7071
9,9	9	0

The average of all fifty numbers in the samples is 5.6, and this is also the average of the twenty five sample means. This will always occur. We say that this sample statistic (mean of the samples) *targets* the population parameter (the mean of the all the items appearing in samples). By contrast, the standard deviation of the fifty numbers in the samples is 2.8705, while the mean of the sample standard deviations (column three) is 2.2368. The standard deviations of the samples is not targeted by the mean of the sample standard deviations.

The *sampling distribution of the mean* is the probability distribution of the sample means, which are in column two. There are 25 samples, and the number 4.5 occurs two times in column two, so $P(4.5) = 2/25$. Reasoning this way with the other sample means, we find the probability distribution

$$P(1) = 1/25,\ P(2.5) = P(3.5) = P(4.5) = 2/25,\ P(5) = 4/25,$$
$$P(4) = 1/25,\ P(6) = 3/25,\ P(6.5) = P(7) = P(7.5) = 2/25,$$
$$P(8) = 1/25,\ P(8) = 2/25 \text{ and } P(9) = 1/25.$$

As required, these probabilities sum to 1. ♦

In making inferences about a population by testing samples, the sample distribution of a population plays a role. Suppose a sample of n objects from the population has k of one type of object. Then *the sample proportion of that sample for that item is* k/n. The *sampling distribution of the population*, with respect to that item, is the probability distribution of the sample population for that mean.

EXAMPLE 25.18

Suppose ten samples of twenty items each are taken off a production line for inspection. Table 25.8 shows the data and the sample ratios (number defective divided by sample size). The sample ratio for each sample, with respect to defective items, is the number of defective items in the sample, divided by $n = 20$. Since $1/20$ occurs four times out of ten as a sample ratio, $P(1/20) = 4/10 = 0.4$. In this way we compute the sampling distribution of this proportion:

$$P(1/20) = 0.4,\ P(0) = 0.4,\ P(1/10) = 0.2.$$ ♦

We are now prepared to state the central limit theorem, which relates a distribution of the original population itself to a distribution of the sample means.

TABLE 25.8 ***Sample ratios in Example 25.18.***

sample	number defective	sample ratio
1	1	1/20
2	0	0
3	2	1/10
4	0	0
5	0	0
6	1	1/20
7	2	1/10
8	1	1/20
9	1	1/20
10	0	0

THEOREM 25.3 Central Limit Theorem

Suppose a random variable X has mean μ and standard deviation σ. Then the distribution of sample means will approach a normal distribution as the sample size increases, regardless of whether or not the original data is normally distributed. Further,

(1) The mean of all the sample means is μ.

(2) The standard deviation of all the sample means is approximately $\sigma/\sqrt{n}$, where n is the sample size. This approximation can be improved by increasing n. ♦

Conclusion (1) means that the mean of the samples equals the mean of the population. Conclusion (2) justifies the sense in which we can approximate the standard deviation of the entire population using the standard deviation of the sample means.

The theorem justifies the approximation of a distribution by a normal distribution. Even when the origin random variable does not have a normal distribution, the sample means will approach a normal distribution. As a rule of thumb, it has been found that the sample means can be satisfactorily approximated by a normal distribution if $n > 30$.

Example 25.19 illustrates the crucial difference between the distribution of the original random variable and the distribution of the sample means.

EXAMPLE 25.19

Imagine that forty samples of six numbers each have been drawn, with replacement, from the integers $0, 1, \cdots, 9$. Table 25.9 shows the samples actually drawn, together with the sample means.

The sample column has forty entries of six numbers each for 240 numbers (the population). Each is an integer from 0 through 9, inclusive. The random variable X is defined by $X(k)$ equals

TABLE 25.9 *Samples and sample means in Example 25.19.*

sample	sample mean	sample	sample mean
289186	5.66	188635	5.16
957888	7.5	405396	4.5
231332	2.33	148235	3.83
767099	6.33	332674	4.16
686505	5	962669	6.33
859901	5.33	445186	4.66
061359	4	637269	5.50
501547	3.66	521802	3
855909	6	040848	4
161078	3.83	358298	5.83
809973	6	568735	5.60
290800	3.16	630761	3.83
952079	5.33	607495	5.16
732111	2.5	188965	6.16
795556	6.16	201037	2.16
145776	5	749018	4.83
276535	4.66	180629	4.33
605812	3.66	291554	4.33
633966	5.50	958087	6.16
747358	5.60	799365	6.5

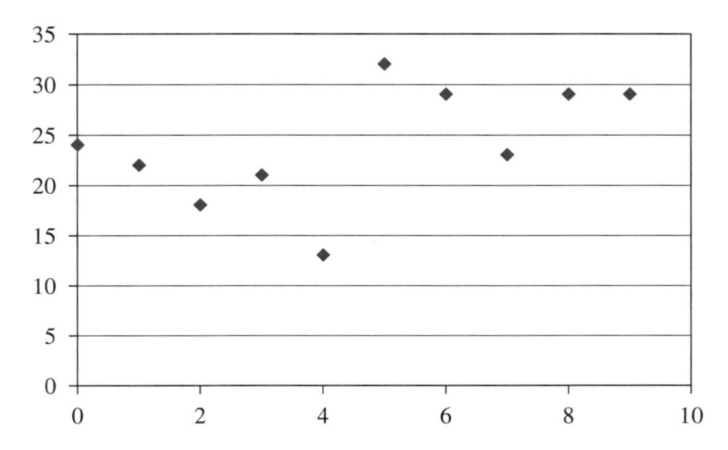

FIGURE 25.14 $X(k)$ *in Example 25.19.*

the number of occurrences of k in this set of 240 numbers. Figure 25.14 is a graph of the values of $X(k)$. Clearly this is not an approximately normal distribution, nor should we expect it to be with $n = 6$. If n were chosen larger, the sample means would more closely approximate a normal distribution.

The mean of the population is

$$\mu = \frac{387}{80}.$$

The mean of the sample means is also 387/80, as we expect.

The standard deviation of the population is

$$\sigma = \frac{\sqrt{492,159}}{240} \approx 2.9231.$$

The standard deviation of the sample means is

$$\sigma_{\text{means}} = \frac{\sqrt{86,919}}{240} \approx 1.2284.$$

Now compute $\sigma/\sqrt{n}$, which is $\sigma/\sqrt{6}$ for this example. We get 1.1933. According to the central limit theorem, this should approximate σ_{means}, and indeed this number is within 0.035 of σ_{means}. Even with small n, this approximation is good to about three hundredths in this example. ♦

EXAMPLE 25.20

The headmaster of an exclusive high school is interested in the probability that all twenty two of his seniors will score at least 670 out of 900 on a standardized test used by colleges in making admissions decisions. The national average on this test is 640, with a standard deviation of 120.

Think of taking sample sizes of size $n = 22$ from the national pool of students taking this test and use the approximately distributed sample means to draw a conclusion about the sample. For the entire population, $\mu = 640$ and $\sigma = 120$. For the sample means,

$$\mu_{\text{means}} = \mu = 640 \text{ and } \sigma_{\text{means}} = \frac{120}{\sqrt{22}}.$$

The bell curve for the sample means is the graph of

$$y = \frac{1}{\sqrt{2\pi}(120)/\sqrt{22}} e^{-(x-640)^2/(2(120)/\sqrt{22})}$$
$$= \frac{\sqrt{11}}{120\sqrt{\pi}} e^{-11(x-640)^2/14400}.$$

We want the area under this bell curve for $x \geq 670$. This is approximately

$$\frac{\sqrt{11}}{120\sqrt{\pi}} \int_{670}^{900} e^{-11(x-640)^2/14400}\, dx \approx 0.12048.$$

On average the headmaster can expect the entire class to score this high once in ten years, although it could also happen two consecutive years, or never in this headmaster's tenure.

What is the probability that all twenty two students score between 640 and 700? This is approximately

$$\frac{\sqrt{11}}{120\sqrt{\pi}} \int_{640}^{700} e^{-11(x-640)^2/14400}\, dx \approx 0.49049,$$

nearly one-half.

What is the probability that ten of the students score better than 660? Now use $n = 10$ to form the samples, so the bell curve to use is

$$y = \frac{\sqrt{5}}{120\sqrt{\pi}} e^{-5(x-640)^2/14400}$$

and the probability we want is

$$\frac{\sqrt{5}}{120\sqrt{\pi}} \int_{660}^{900} e^{-5(x-640)^2/14400}\, dx \approx 0.29908,$$

about $3/10$. ♦

SECTION 25.5 PROBLEMS

1. A recruiter for a company wants his candidates to be successful on a test administered to all prospective employees. He would like his group of thirty recruits to score at least 700 out of the possible 950. Over the time this test has been given the average score has been 670 with a standard deviation of 105.
   (a) Estimate the probability that all thirty of the recruits will score in the $700 - 750$ range.
   (b) Estimate the probability that all will score at least 670.
   (c) The recruiter gets a bonus if seven of the candidates score 800 or higher. Should the recruiter look forward to a bonus?
2. A uniform maker has a large order from a new branch of the military. On average, the pants legs are 31 inches, with a standard deviation of 3.2 inches. Suppose six new recruits are chosen at random.
   (a) What is the probability that their mean length is at least 29 inches?
   (b) Find the probability that their length is in the $29 - 33$ inch range.
   (c) Find the probability that their length does not exceed 32 inches.
3. A state uses standard braces for its highway bridges. The braces can handle a mean pressure of 5000 pounds per square inch, with a standard deviation of 800. Suppose twenty braces are tested at random throughout the state.
   (a) What is the probability that their mean pressure tolerance is at least 4800 pounds per square inch?

TABLE 25.10 *Samples in Problem 5, Section 25.5.*

sample	sample	sample	sample	sample
74974	78365	94489	84937	33933
76693	57733	58438	59786	58393
43856	86934	96649	88533	48576
85994	43865	67339	59643	58834
36496	59378	47937	75488	84938
48659	43958	48896	69537	78953

TABLE 25.11 *Samples in Problem 6, Section 25.5.*

sample	sample	sample	sample	sample
9388204	7939720	5773902	1912038	1749823
8902817	3022850	5492018	4491002	4910292
7921082	5910293	9017839	4491029	4019283
8109283	4911029	3019402	5616209	9201898
5501829	3371932	0002910	3611390	4092179
4039112	9571120	9912118	8401938	7109032

(b) What is the probability that their mean pressure tolerance falls in the 4700 − 5100 range?

(c) What is the probability that their mean pressure tolerance is not less than 4500 pounds per square inch?

4. A product coming off an assembly line has been found to weigh on average 17 pounds with a standard deviation of 1.8. Thirty samples are chosen at random from the line, with replacement.

   (a) What is the probability that these samples will weigh on average between 16.8 and 17.2?

   (b) What is the probability that they weigh on average at least 16.7 pounds?

   (c) Find the probability that they do not weigh on average less than 17.2 pounds

5. Thirty samples are taken of five digit numbers, each digit of which is chosen with replacement from $3, 4, \cdots, 9$. The samples are given in Table 25.10.

   (a) Compute the sample means, the mean of the sample means, and the mean of the population.

   (b) Compute the standard deviation of the population and the standard deviation of the sample means and verify that $\sigma_{\text{means}} \approx \sigma/\sqrt{n}$ for the appropriate n.

6. Thirty samples are taken of seven digit numbers, each digit of which is chosen with replacement from $0, 1, \cdots, 9$. The samples are given in Table 25.11.

   (a) Compute the sample means, the mean of the sample means, and the mean of the population.

   (b) Compute the standard deviation of the population and the standard deviation of the sample means and verify that $\sigma_{\text{means}} \approx \sigma/\sqrt{n}$ for the appropriate n.

25.6 Confidence Intervals and Population Proportion

Suppose we want to use samples to draw conclusions about an entire population (such as manufactured items). In this section we will focus on how to estimate certain statistical measures, as well as how to determine a confidence interval. This will measure, in a sense we will define, ranges of accuracy of our estimates.

Suppose we have some population proportion p. We are examining some issue with exactly two outcomes (success or failure, defective or nondefective, or the like). Take random samples of size n and consider the sample proportion $\tilde{p} = x/n$, where x is the number of successes in the sample. This number is between 0 and 1 and may be thought of as a probability. The complementary probability is $\tilde{q} = 1 - \tilde{p}$. For example, we might draw samples of 900 trout from a stream over a day and find that 47 percent are male. The sample size is $n = 900$ and $\tilde{p} = 0.47$.

We will assume that there is a fixed number of independent trials, and that the resulting distribution can be approximated by a normal distribution.

In making estimates of various statistics associated with the population, we will use the notion of a *confidence interval*, which is an interval of values used to estimate the statistic. The corresponding *confidence level* of the estimate is the ratio of the times that this interval actually does contain the statistic of interest, taken over a large number of repetitions or samples. This confidence level is routinely denoted $1 - \alpha$.

The most common confidence levels, expressed as percentages, are 90, 95 or 99. Below 90 percent is often unacceptable as a useful measure of the accuracy of an estimate. Associate with 90 percent the probability 0.9, so $\alpha = 0.1$ and $1 - \alpha = 0.9$. Similarly, for 95 percent use $\alpha = 0.05$ so $1 - \alpha = 0.95$, and for 99 percent, use $\alpha = 0.01$. If, for example, we are using 95 percent as a confidence level to estimate the population proportion p, and this determines (in a way we will describe), an interval $a < p < b$, then we say we are 95 percent confident that the estimated value of p falls between a and b.

When these percentages are written as decimals we think of them as probabilities, or area bounded by the standard bell curve, which is the graph of

$$y = \frac{1}{\sqrt{2\pi}} e^{-z^2/2}$$

shown in Figure 25.13. For a given α, there is a point Z such that the area under the graph to the right of Z and to the left of $-Z$ is α, which means that the area between $-Z$ and Z is $1 - \alpha$. This number Z, which depends on α, is denoted $z_{\alpha/2}$ and is called a *critical value*, separating those sample proportions (between $-z_{\alpha/2}$ and $z_{\alpha/2}$) that are likely from those that are unlikely (to the right beyond $z_{\alpha/2}$ or to the left of $-z_{\alpha/2}$).

Table 25.12 gives critical values corresponding to confidence levels of 85, 90, 95 and 99 percent, which must be written as decimals to have area interpretations relative to the standard bell curve. For example, if $\alpha = 0.15$, then $1 - \alpha = 0.85$ and we read from the table that $z_{\alpha/2} = 1.44$. A numerical integration now yields

$$\frac{1}{2\pi} \int_{-1.44}^{1.44} e^{-z^2/2}\, dz \approx 0.85013,$$

approximately 0.85, or 85 percent. This would mean that there is 0.85 probability that a sample proportion falls between -1.44 and 1.44 on the standard bell curve.

We know that

$$\mu = np \text{ and } \sigma = \sqrt{npq}.$$

TABLE 25.12 ***Critical values.***

confidence level (as a decimal)	α	$z_{\alpha/2}$
0.85	0.15	1.44
0.90	0.10	1.645
0.95	0.05	1.96
0.99	0.01	2.575

Since these results are for n trials we can write the mean of the sample proportions as

$$\mu = \frac{np}{n} = p$$

with the standard deviations of the sample proportions as

$$\sigma = \frac{\sqrt{npq}}{n} = \sqrt{\frac{pq}{n}}.$$

Generally we do not know p (or q) and we want to estimate these values from $\tilde{p}$ or $\tilde{q}$. Replace p by $\tilde{p}$ and q by $\tilde{q}$ in the expression for σ to get $\sqrt{\tilde{p}\tilde{q}/n}$. The probability is $1-\alpha$ that a sample proportion will differ from the actual population proportion (which we want to estimate) by no more than $z_{\alpha/2}\sqrt{\tilde{p}\tilde{q}/n}$. This leads us to define the *maximum error of the estimate* to be

$$\epsilon = z_{\alpha/2}\sqrt{\frac{\tilde{p}\tilde{q}}{n}}, \tag{25.3}$$

or

$$\epsilon = z_{\alpha/2}\sqrt{\frac{\tilde{p}(1-\tilde{p})}{n}}.$$

The confidence interval associated with α is

$$\tilde{p} - \epsilon < p < \tilde{p} + \epsilon.$$

EXAMPLE 25.21

Suppose samples of 900 fish are drawn from a stream each day. Suppose on a particular day it is found that 47 percent were male. Here $n = 900$ and $\tilde{p} = 0.47$. Based on the information available, 0.47 is the best estimate we can make for $\tilde{p}$, the proportion of the total fish population that is male.

Suppose we want the maximum error associated with a 95 percent confidence level. For this, $\alpha = 0.05$ and we read from the table that $z_{\alpha/2} = 1.96$. The maximum error is

$$\epsilon = (1.96)\sqrt{\frac{(0.47)(0.53)}{900}} = 0.0326078.$$

Generally we round maximum error calculations to three places, taking $\epsilon = 0.033$. The 95 percent confidence interval is

$$0.47 - 0.033 < p < 0.47 + 0.033.$$

There is a probability of $1-\alpha$ (or 0.95) that the actual population proportion falls in this interval. ♦

Sometimes we want to determine the size of the sample we need to estimate some statistic with a certain level of confidence. To do this for the population proportion begin with the expression (25.3) for the maximum error and solve for n to get

$$n = \frac{(z_{\alpha/2})^2\tilde{p}\tilde{q}}{\epsilon^2}. \tag{25.4}$$

This is used to estimate the size that n should be to estimate p with a confidence interval $\tilde{p} - \epsilon < p < \tilde{p} + \epsilon$, assuming that $\tilde{p}$ is known. Since equation (25.4) will generally not produce an integer, in practice we take the smallest integer larger than $(z_{\alpha/2})^2\tilde{p}\tilde{q}/\epsilon^2$.

In many cases that occur in practice we do now know $\tilde{p}$. In such a case, replace $\tilde{p}\tilde{q}$ with 0.25 to write

$$n = \frac{(0.25)(z_{\alpha/2})^2}{\epsilon^2}.$$

The rationale for this is that 0.25 is the maximum value that $\tilde{p}(1-\tilde{p})$ can assume for $0 \le \tilde{p} \le 1$. This maximum occurs when $\tilde{p} = \tilde{q} = 1/2$.

EXAMPLE 25.22

A member of a United Nations team wants to survey the population of a country to determine the commonality of households having at least two telephones. She wants to be 90 percent confident of her results, and to be in error by no more than 3 percent. How many households should she survey?

Take 0.9 to be the confidence level that is wanted, and $\epsilon = 0.03$. There is no information about $\tilde{p}$ so use $\tilde{p}\tilde{q} = 0.25$ to compute

$$n = \frac{(0.25)(z^2_{\alpha/2})^2}{\epsilon^2}.$$

For a 90 percent confidence level the table gives us $z_{\alpha/2} = 1.645$. Thus

$$n = \frac{(0.25)(1.645)^2}{0.03^2} = 751.67.$$

She should survey 752 households.

Suppose, from another study, it is estimated that $\tilde{p} = 0.32$. Now use equation (25.4) to compute

$$n = \frac{(1.645)^2(0.32)(0.68)}{0.03^2} = 654.26.$$

With this estimate, she can get by surveying 655 households. However, this survey assumes the reliability of the recent study claiming that $\tilde{p} = 0.32$. If this study is, say, ten years old, this may not be reliable current information and it might be safer to go with the estimate based on the maximum value of $\tilde{p}(1-\tilde{p})$, even though it requires that a larger number of households be surveyed.

Suppose, instead of 90 percent, we want 95 percent confidence in the conclusion. Again, assuming no information about $\tilde{p}$, we would use $\alpha = 0.05$ and $z_{\alpha/2} = 1.96$ to compute

$$n = \frac{(0.25)(1.96)^2}{0.03^2} = 1067.11.$$

For a 95 percent confidence level she should survey 1068 households. It is not surprising that the number of households that should be surveyed increases if we raise the confidence level. ♦

SECTION 25.6 PROBLEMS

1. A professional pollster is charged with taking a survey throughout a large state to estimate the number of qualified voters who favor the Freeload Party candidate.

   (a) Suppose the pollster wants an 85 percent confidence level, with an error of no more than 7 percent. How many people should be surveyed?

   (b) How many people should be surveyed if, in (a), an error of no more than 12 percent is acceptable?

   (c) Suppose, in the previous election, it was found that $\tilde{p} = 0.06$. Estimate how many people need to be surveyed for an 85 percent confidence level and an error of no more than 7 percent.

2. A plant produces packaged chicken for distribution to grocery stores. In a sample of 100 packages of chicken taken over one work day, it is found that seven percent contain the blue virus, which turns the person consuming the chicken permanently light blue.

(a) Determine an estimate with a 99 percent confidence interval of the proportion of one day's chicken packages that contain the virus.

(b) Determine an estimate with 90 percent confidence of the proportion of one day's chicken packages containing the virus.

3. A marketing executive wants to survey a city to determine how many people have purchased his company's product over the past twelve months.

(a) Suppose he wants a 95 percent confidence in the results of the survey with an error of no more than 2 percent. How many people should he survey?

(b) How many people should he survey if, in (a), he is willing to have an error of no more than 5 percent?

(c) Suppose the previous year's study gives $\tilde{p} = 0.37$. Make an estimate of how many people he should survey for a 95 percent confidence level and an error of no more than 2 percent, assuming this value for $\tilde{p}$.

4. A public health official is charged with completing an annual survey of a county population, checking for incidents of a viral infection over the past year.

(a) Suppose the official wants a 99 percent confidence level, with an error of no more than 3 percent. How many people must be surveyed?

(b) How many people should be surveyed if, in (a), a 95 percent confidence level is specified?

(c) Suppose, in the previous year's study, it was found that $\tilde{p} = 0.12$. Estimate how many people should be surveyed for a 99 percent confidence level and an error of no more than 2 percent.

5. A survey taken over a week has inspected 200 fish in a lake and found that 87 were man-eating guppies.

(a) Determine an estimate with 95 percent confidence of the proportion of the fish population that are man-eating guppies.

(b) Determine an estimate with 85 percent confidence of the proportion of the fish population that are man-eating guppies.

6. A drug being tested for a certain illness appears to have side effects. In a large clinical trial, 750 patients who took the drug are selected at random and it is found that 22 percent develop skin rashes.

(a) Determine an estimate with a 99 percent confidence interval of the proportion of the entire patient population that will be expected to develop a skin rash.

(b) Redo (a) for a 95 percent confidence level.

7. A parachute assembly and packing plant packages thousands of parachutes each month. One month, 1200 are inspected and it is found that 0.2 percent have a defect.

(a) Find an estimate with a 99 percent confidence of the proportion of parachutes produced in a month that are defective.

(b) Redo (a) with a 95 percent confidence level.

(c) Redo (a) with a 90 percent confidence level.

(d) Suppose only 800 parachutes are inspected instead of 1200, and it is still found that 0.2 percent are defective. Now determine an estimate with 95 percent confidence of the proportion of parachutes produced in a month that are defective. How has decreasing the size of the sample influenced the confidence interval?

25.7 Population Mean and the Student Distribution

This section continues the theme of Section 25.6, except there we wanted to estimate population proportion and now we want to estimate the population mean μ. Otherwise the setting of the analysis is the same.

Denote the sample mean of sample x by $\overline{x}$. In the absence of specific information, estimate the population mean as the sample mean. The issue now is to develop a concept of maximum error enabling us to state confidence intervals. Previously we saw that, if σ is the standard deviation of the population, then $\sigma/\sqrt{n}$ is the standard deviation of the sample means, with n the sample size. Equation (25.3) gives the maximum error in the estimate of population proportions. The adaptation of this to maximum error of the population mean is

$$\epsilon = z_{\alpha/2}\frac{\sigma}{\sqrt{n}}. \tag{25.5}$$

Here we are again using ϵ for the maximum error, taking from context whether we are referring to population proportion or mean. The corresponding confidence interval in estimating μ from $\overline{x}$ is

$$\overline{x}-\epsilon<\mu<\overline{x}+\epsilon.$$

Solve equation (25.5) to get

$$n=\left(\frac{\sigma z_{\alpha/2}}{\epsilon}\right)^2.$$

This is used to estimate the size of samples that should be taken to have a maximum error of ϵ and a confidence level determined by the critical point $z_{\alpha/2}$.

EXAMPLE 25.23

A study of blood pressure is being planned. Sample groups of size $n=210$ are to be used and a 99 percent confidence level is sought It is known from other studies that $\sigma=0.82$ and the sample mean is $\overline{x}=127$. Estimate the population mean and the confidence interval.

From Table 25.12, choose $z_{\alpha/2}=2.575$. Then

$$\epsilon=(2.575)\frac{0.82}{\sqrt{210}}=0.14571,$$

so

$$\overline{x}-0.14571<\mu<\overline{x}+0.14571,$$

or

$$126.85<\mu<127.15.$$

Go back to the planning stage and imagine that we have not yet decided on the number of samples to take. We want n for a 95 percent confidence level. Now use $z_{\alpha/2}=1.96$. We need some information about the maximum error. Suppose we want $\epsilon=0.2$. Estimate

$$n=\left(\frac{(0.82)(1.96)}{0.2}\right)^2=64.577$$

so use $n=65$ as the sample size. ♦

This example took the ideal position that knew σ, and in practical applications this may not be the case. In such a circumstance we may use the student t distribution instead of the normal distribution. Assuming that the population is approximately normal, the *student t distribution*, or just *t distribution*, is defined by

$$t(x)=\frac{\overline{x}-\mu}{s/\sqrt{n}}.$$

There is therefore a different t distribution for each n. Each is bell-shaped in general appearance, but is not a normal distribution. However, as n increases, the corresponding student t distributions approach the standard bell curve. Critical numbers $t_{\alpha/2}$ for the t distribution are defined similarly to $z_{\alpha/2}$. The area under the t distribution for $x\geq t_{\alpha/2}$ and for $x\leq -t_{\alpha/2}$ is α, so the area under the t distribution for $-t_{\alpha/2}\leq x\leq t_{\alpha/2}$ is $1-\alpha$.

In determining $t_{\alpha/2}$, n plays a role, since the t distribution is different for each n. With estimation of population means our main objective in this section, note that $\overline{x}$ is a sum of n numbers divided by n. If $\overline{x}$ is known, any $n-1$ of these numbers can be chosen as having any value, and the nth number is then determined. We therefore say that $t_{\alpha/2}$ has $n-1$ *degrees of freedom*. In a table of critical points for t distributions, there will be a row or column associated with the number of degrees of freedom, and from this $t_{\alpha/2}$ can be read for given α and n.

In practice we use t - critical points just as we used z - critical points previously. Given n and α, which in turn is determined by a stated confidence level, define the maximum error ϵ_t by

$$\epsilon_t = t_{\alpha/2}\frac{s}{\sqrt{n}}$$

in which s is the standard deviation of a sample. The corresponding confidence interval is

$$\overline{x} - \epsilon_t < \mu < \overline{x} + \epsilon_t.$$

EXAMPLE 25.24

An international gymnastic oversight committee is checking blood levels of athletes for the performance enhancing chemical hyperjump. Random samples of 35 gymnasts are taken and it is known that the sample mean is $\overline{x} = 0.4$. We want to construct the maximum error if a confidence level of 99 percent is needed, and then find the associated confidence interval about the population mean.

For a 99 percent confidence level , $\alpha = 0.01$. With $n = 35$ the number of degrees of freedom is 34 and we find from a standard table that $t_{\alpha/2} = 2.728$. Suppose from the sample we compute the standard deviation $s = 0.03$. Then the maximum error is

$$\epsilon_t = (2.728)\frac{0.03}{\sqrt{35}} = 0.013833.$$

The confidence interval is

$$0.4 - 0.013833 < \mu < 0.4 + 0.013833,$$

or

$$0.386 < \mu < 0.414,$$

to three decimal places. ♦

SECTION 25.7 PROBLEMS

1. Samples of 100 patients are being drawn randomly from a clinical trial. It is estimated that $\sigma = 2.4$ and $\overline{x} = 106$.
   (a) Estimate the population mean and confidence interval for a 95 percent confidence level.
   (b) Suppose that the sample size has not been set, but n needs to be chosen to have a 99 percent confidence level with an error of no more than 0.1. Estimate the sample size that should be used.
2. Random samples of fifty lawnmower motors are taken from a plant each week, checking for a certain amount of contaminant in the fuel line. Assume that $\overline{x} = 0.8$ and $s = 0.05$. Use the t distribution to construct a 95 percent confidence interval in estimating the population mean. Hint: A table estimates the critical value for this problem as 2.009.
3. Random samples of 75 students are drawn from the senior year population of a large city public school system in studying absences per month. With $\overline{x} = 3$ and $s = 0.7$, construct a 95 percent confidence interval in estimating the population mean. Hint: The critical value is approximately 1.992.
4. Random samples of 200 fish are taken from waters near an island. With $\overline{x} = 7$ and $s = 0.01$, construct a 95 percent confidence interval in estimating the population mean. Hint: The critical value is approximately 1.971.
5. A study of blood concentration of a certain protein is being done. Groups of 350 patients are sampled. It is known that $\sigma = 1.04$ and the sample means are $\overline{x} = 72$.
   (a) For a 99 percent confidence level, estimate the population mean and the confidence interval.

(b) Suppose the sample size has not yet been determined, but n should be chosen to have a 99 percent confidence level and an error of no more than 0.2. Estimate the sample size that should be used.

6. Samples of size 50 are being drawn from a company's production line. It has been found that $\overline{x} = 119$ and $\sigma = 0.7$.

(a) If a confidence level of 95 percent is wanted, estimate the population mean and determine the confidence interval.

(b) Suppose the sample size has not yet been determined, but n should be chosen to have a 99 percent confidence level and an error of no more than 0.2. Estimate the sample size that should be used.

25.8 Correlation and Regression

Suppose we have a collection of n data pairs (x_j, y_j), which we can plot as points in the plane. A *correlation* exists between the x - values and the y - values if there is some specific relationship between them. This could be almost anything, but we will focus here on *linear correlation*, in which the points (x_j, y_j) lie approximately (in a sense we will define) along a straight line. A linear correlation is *positive* if this line has positive slope and *negative* if the slope is negative. As examples, the points in Figures 25.15 and 25.16 appear to have a linear correlation (first positive, then negative), while the points in Figure 25.17 do not.

To put this on more solid footing, begin with the number

$$c = \frac{\sum_{j=1}^{n}(x_j - \overline{x})(y_j - \overline{y})}{(n-1)s_x s_y}. \tag{25.6}$$

We call c the *linear correlation coefficient* of the data. $\overline{x}$ is the mean of the $x_j's$, $\overline{y}$ the mean of the $y_j's$, s_x is the standard deviation of the $x_j's$ and s_y the standard deviation of the $y_j's$. It is a routine exercise in algebra to show that

$$c = \frac{n\sum_{j=1}^{n} x_j y_j - (\sum_{j=1}^{n} x_j)(\sum_{j=1}^{n} y_j)}{\sqrt{n(\sum_{j=1}^{n} x_j^2) - (\sum_{j=1}^{n} x_j)^2}\sqrt{n(\sum_{j=1}^{n} y_j^2) - (\sum_{j=1}^{n} y_j)^2}}. \tag{25.7}$$

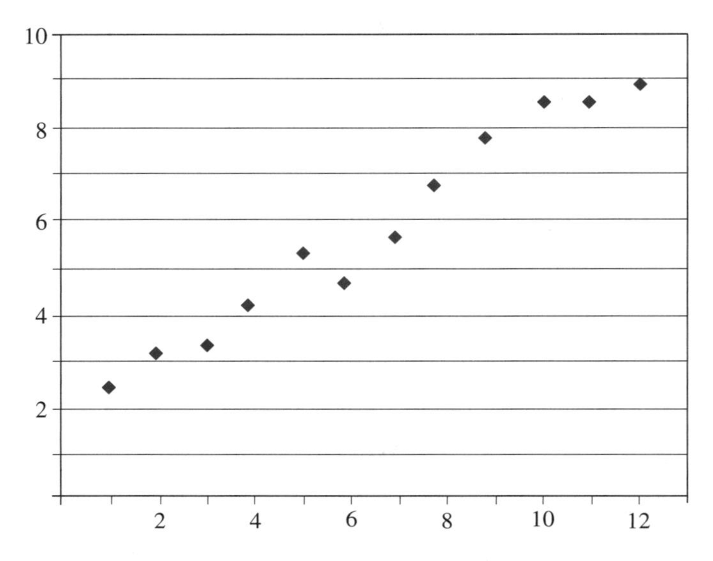

FIGURE 25.15 *Data with a positive linear correlation.*

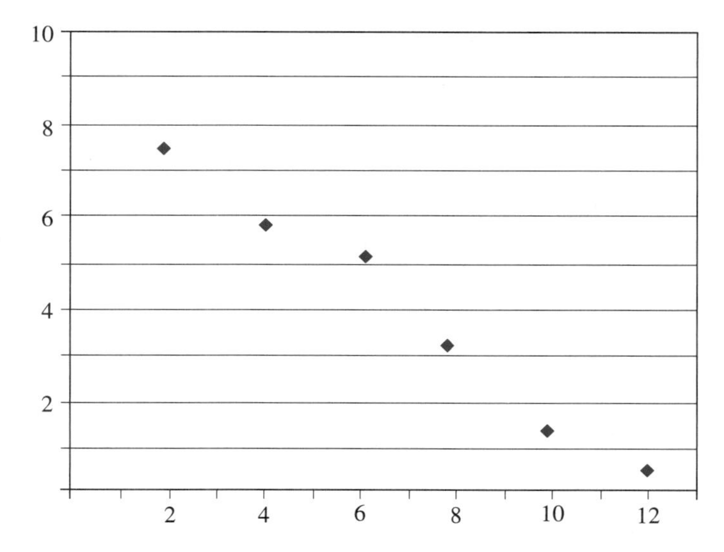

FIGURE 25.16 *Data with a negative linear correlation.*

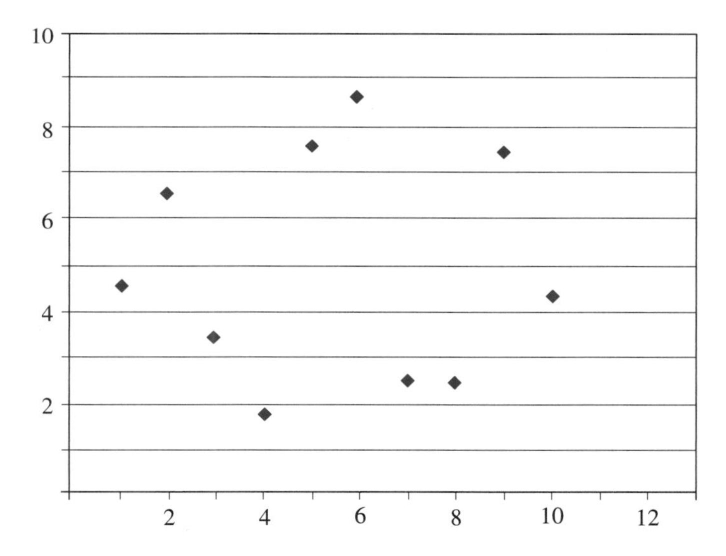

FIGURE 25.17 *Data with no significant linear correlation.*

While this expression for c may appear more complicated than the first, it is more explicit in the sense that it depends only on the $x_j's$ and $y_j's$ themselves, without requiring means and standard deviations of these data sets.

It is easy to check that $-1 \leq c \leq 1$. Further, if the points lie on a line, say $y_i = ax_i$ for $c = 1, 2, \cdots, n$, then $c = 1$. Roughly speaking, if c is near -1 or 1 we say that there is a *significant linear correlation* between the $x_i's$ and $y_i's$. If c is near 0 we say that there is *no significant linear correlation* between these pairs of numbers.

These terms have a more rigorous meaning when referred to tables such as Table 25.13, which has been constructed for a 95 percent ($\alpha = 0.05$) confidence level. Corresponding to the number n of points, there is a critical value of c for this value of α, with the following

TABLE 25.13 *Critical values of c for $\alpha = 0.05$.*

n	critical value of c for $\alpha = 0.05$
10	0.632
15	0.514
20	0.444
25	0.396
30	0.361
35	0.335
40	0.312
45	0.294
50	0.279
60	0.254
80	0.220
100	0.196

TABLE 25.14 *Olympic pole vaulting gold medal heights.*

Year	Height	Year	Height
1932	14.15	1972	18.04
1936	14.27	1976	18.04
1948	14.10	1980	18.96
1952	14.92	1984	18.85
1956	14.96	1988	19.77
1960	15.42	1992	19.02
1964	16.73	1996	19.42
1968	17.71		

meaning. If $|c|$ is greater than the critical value, then there is a $1 - 0.05 = .95$ probability (confidence) that there is a significant correlation between the $x_i's$ and $y_i's$. For example, we read from the table that, with $n = 15$ points, if $|c| > 0.514$ there is a 0.95 probability of a significant linear correlation. If $n = 50$ and $|c| > 0.270$, there is a 0.95 probability of a significant linear correlation.

If $|c|$ is less than or equal to the critical value, then there is only a 0.05 probability of a linear correlation between the $x_i's$ and $y_i's$.

As the number of sample points increases, we can infer linear correlation with smaller values of $|c|$, while, if n is small (and therefore provide less information) it takes a larger $|c|$ to draw this inference. Intuitively, it is easier to draw a straight line approximation to a small number of points, than to a large number.

EXAMPLE 25.25

Table 25.14 gives gold medal heights in the pole vault for the Olympics, for some years extending from 1932 through 1996. Heights are in feet.

The points (year, height) are plotted in Figure 25.18 and appear to be on an approximately straight line. To check this intuition, compute $\overline{x} = 1966.9333$, $\overline{y} = 16.9573$, $x_x = 19.6813$ and $x_y = 2.1138$, and finally, $c = 0.9546$. Since $n = 15$, we have $|c| = 0.9546 > 0.514$, the latter number being obtained from the table of critical values. Therefore there is indeed a significant linear correlation between years and heights. ♦

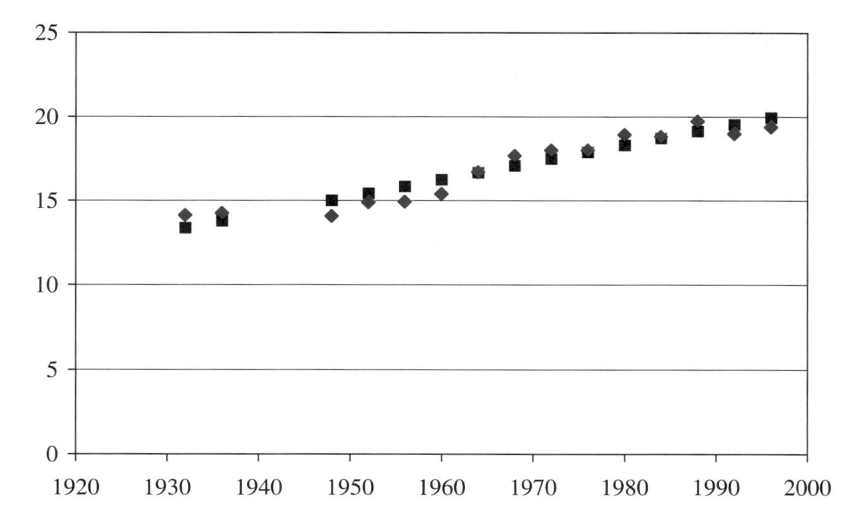

FIGURE 25.18 *Olympic pole vaulting data in Example 25.25.*

TABLE 25.15 ***Data set for Example 25.26.***

x_i	0	0.5	1.0	1.5	2.0	2.5	3.0	3.5	4.0
y_i	1	2.12	6.24	2.44	11.58	3.72	1.04	25.63	2.63

EXAMPLE 25.26

For the data of Table 25.15, a computation yields $c = 0.3884$. Table 25.13 of critical values does not have an entry for $n = 9$ data points, so we use the value corresponding to $n = 10$ as a best estimate. This critical value is 0.632. Since $|c| < 0.632$, there is no significant linear correlation between the $x_i's$ and $y_i's$. ♦

It is important not to confuse correlation with causality. Even if there is a significant linear correlation between $x_i's$ and $y_i's$, it does not necessarily follow that the $y_i's$ are caused by the $x_i's$ (or the $x_i's$ by the $y_i's$). In Example 25.25, the winning heights correlate linearly with advancing years, but are not caused by the years themselves.

We will now develop the idea of regression. Suppose the n points (x_i, y_i) have a significant linear correlation. This means that the points, when plotted, form an "approximate" straight line. It is possible to determine a line L that is a best fit to the data. This line is called the *regression line* for the data. It is defined to be the best least squares fit to the data, in the sense of minimizing the sum of the squares of the vertical distances between the $y_i's$ in the data set and the respective y - values on L corresponding to the $x_i's$.

If we write the equation of this regression line as $y = a + bx$, then the standard least squares method yields

$$a = \frac{(\sum_{i=1}^{n} y_i)(\sum_{i=1}^{n} x_i^2) - (\sum_{i=1}^{n} x_i)(\sum_{i=1}^{n} x_i y_i)}{n(\sum_{i=1}^{n} x_i^2) - (\sum_{i=1}^{n} x_i)^2} \tag{25.8}$$

and

$$b = \frac{n \sum_{i=1}^{n} x_i y_i - (\sum_{i=1}^{n} x_i)(\sum_{i=1}^{n} y_i)}{n(\sum_{i=1}^{n} x_i^2) - (\sum_{i=1}^{n} x_i)^2}. \tag{25.9}$$

We can also compute b first and then obtain a as

$$a = \overline{y} - b\overline{x}. \tag{25.10}$$

EXAMPLE 25.27

For the data of Example 25.25, we find that $a = -185.637$ and $b = 0.103$, so the regression line has equation

$$y = -185.637 + 0.103x.$$

Notice that the data ends at 1996. If we use the regression line to try to project information we are on dangerous ground. For example, moving out the line until 2008, we project a winning pole vault of 21.187 feet. This may be possible. Projecting to the 3000 Olympics, we see a winning vault of 123.36 feet. Barring an evolutionary miracle, this will not happen. ♦

As Example 25.27 suggests, we must not use a regression line to attempt to project conclusions about y - values that are "too far" outside the data set from which the line was derived.

There is a useful interpretation that can be made for the square of the linear correlation coefficient c. Suppose the sample data points are $(x_1, y_1), \cdots, (x_n, y_n)$. Suppose there is a significant linear correlation with regression line $y = a + bx$. For a given x, write $(y_r)_i = a + bx_i$ to distinguish between y_i and the y - coordinate on the regression line of the point corresponding to x_i.

Think of the regression line as defining a sense of how the data "should" be related: $(y_r)_i$ is the y - data point that "should go" with x_i, while y_i is the data point that actually does go with x_i in the linear correlation. It is reasonable to attempt to compute the ratio of differences of the data from the mean that are explained by the regression line, to the total differences of the data from the mean.

To quantify these ideas, interpret $\sum_{i=1}^{n}(y_i - \overline{y})$ as the total deviation of the $y_i's$ from their mean, and $\sum_{i=1}^{n}((y_r)_i - \overline{y})$ as the deviation of regression line y - values from the mean of the $y_i's$. This is a deviation explained by the regression line. The ratio we are interested in, which we will temporarily call k, is therefore

$$k = \frac{\sum_{i=1}^{n}((y_r)_i - \overline{y})}{\sum_{i=1}^{n}(y_i - \overline{y})}.$$

Next,

$$\sum_{i=1}^{n}(y_i - (y_r)_i)$$

is the deviation of the regression line values from the actual data values. Think of this as the deviation for which the regression line presents no rationale. Now observe that

$$\sum_{i=1}^{n}(y_i - \overline{y}) = \sum_{i=1}^{n}((y_r)_i - \overline{y}) + \sum_{i=1}^{n}(y_i - (y_r)_i).$$

Therefore

$$k = \frac{\sum_{i=1}^{n}((y_r)_i - \overline{y})}{\sum_{i=1}^{n}((y_r)_i - \overline{y}) + \sum_{i=1}^{n}(y_i - (y_r)_i)}$$

$$= \frac{\text{differences of the } y_i's \text{ from the mean explained by the regression line}}{\text{total differences of } y_i's \text{ from the mean}}.$$

Now comes what may be a surprise. If the algebra is carried out, we find that $k = c^2$. Therefore c enables us to detect a significant linear correlation and c^2 gives a measure of how well the deviation of the $y_i's$ from the norm are explained by the regression line.

EXAMPLE 25.28

A state compiles statistics on deer hunters and deaths in the deer population each year. For a period of ten years, these statistics are given in Table 25.16.

We find that $\overline{x} = 255.50$, $\overline{y} = 45.10$, $s_x = 97.961$, and $x_y = 12.28$. We obtain $c = 0.8488$. With $n = 10$ the critical value is 0.632. Because $c > 0.632$, there is a 0.95 probability that there is a significant linear correlation between the x - and y - values.

Now we ask: how many of the deaths in the deer population are explained by the number of hunters? We expect some deer to be killed by hunters, and some to die from other causes. Compute

$$c^2 = (0.8488)^2 = 0.7201.$$

This means that 72 percent of the deer deaths can be explained by the regression line, hence by the activity of the hunters. The other 28 percent must be accounted for by other factors ♦

Finally, we will develop a prediction interval (a type of confidence interval) for predictions based on a regression line. Begin with the notion of *spread of the error of the estimate*, which is analogous to standard deviation in the sense that it measures the spread of the sample points about the regression line. This spread is denoted S_r and is defined by

$$S_r = \sqrt{\frac{\sum_{i=1}^{n}(y_i - (y_r)_i)^2}{n-2}}.$$

It is straightforward to verify that

$$S_r = \sqrt{\frac{\sum_{i=1}^{n} y_i^2 - a\sum_{i=1}^{n} y_i - b\sum_{i=1}^{n} x_i y_i}{n-2}},$$

in which the regression line is $y = a + bx$.

Analogous to the process developed for confidence intervals, define the maximum error ϵ for a given x - value (which is in the x - population, but need not be one of the actual sample values x_i), to be

$$\epsilon_r = t_{\alpha/2} S_r \sqrt{1 + \frac{1}{n} + \frac{n(x - \overline{x})^2}{n(\sum_{i=1}^{n} x_i^2) - (\sum_{i=1}^{n} x_i)^2}}.$$

Here $t_{\alpha/2}$ is a critical point of the student t distribution corresponding to a confidence level α. We have often taken $\alpha = 0.05$ in order to have a 95 percent confidence level in the conclusion. The number $t_{\alpha/2}$ must be determined from a table. The difference between the current and previous use is that previously, in estimating population mean, we have $n - 1$ degrees of freedom, whereas now we have $n - 2$. This information is used in referencing $t_{\alpha/2}$ from the table.

TABLE 25.16 *Hunters versus deaths in the deer population in Example 25.28.*

Hunters	168	172	194	204	230	270	295	320	390	402
Deer Deaths	28	30	36	39	42	50	46	57	61	62

TABLE 25.17 *Regression line figures for deer deaths in Example 25.28.*

Number of Hunters	168	172	194	204	230
$(y_r)_i$	31.7357	32.2901	36.7253	40.3289	45.8729
Number of Hunters	270	295	320	390	402
$(y_r)_i$	49.3379	52.8029	62.5049	64.1681	62

Now, for a given x, we define the *prediction interval* for the corresponding y to be

$$y_x - \epsilon_r < y < y_x + \epsilon_r$$

where $y_x = a + bx$.

EXAMPLE 25.29

Revisit the deer/hunter data of Example 25.28. Suppose we want to estimate how many deer kills would correspond to $x = 350$ hunters.

We find that the regression line has equation $y = 8.4509 + 0.1386x$. We need to compute $(y_r)_i = a + bx_i$ for $i = 1, \cdots, 10$. These numbers are given in Table 25.17.

Compute

$$\sum_{i=1}^{10}(y_i - (y_r)_i)^2 = 80.3589 \text{ and } S_r = \sqrt{\frac{\sum_{i=1}^{10}(y_i - (y_r)_i)^2}{8}} = 3.1694.$$

Then

$$\epsilon_r = (2.306)(3.1694)\sqrt{1 + \frac{1}{10} + \frac{10(350 - 255.50)}{10(765,989) - (2645)^2}} = 7.6703.$$

Finally, compute

$$y_{350} = a + b(350) = 8.4509 + 0.1386(350) = 59.961.$$

This is the y - value on the regression line corresponding to $x = 350$. The prediction interval is

$$59.961 - 7.6703 < y < 59.961 + 7.6703.$$

Using two decimal places, the prediction interval is

$$52.29 < y < 67.63.$$

For a hunter population of 350, the regression line predicted number of deer deaths is 59.96, or 60. We also have a confidence interval of 95 percent (because of the way $t_{\alpha/2}$ was chosen) that the actual number is between 52 and 68. We can change this confidence level from 95 percent to another value by making a different choice of $t_{\alpha/2}$. ♦

SECTION 25.8 PROBLEMS

For each of Problems 1 through 8 draw a plot of the data from the referenced table, compute c and determine whether or not there is a significant linear correlation between the $x_i's$ and the $y_i's$. If there is not, the problem is completed. If there is, determine the equation of the regression line and, for each given x - value, determine the y - value predicted by the regression line, as well as the corresponding prediction interval with a 95 percent confidence.

1. Table 25.18, $x = 0, 9.2, 38.8$
2. Table 25.19, $x = 2.9, 29, 46$
3. Table 25.20, $x = 0.6, 3.0, 39.0, 42.1$
4. Table 25.21, $x = 3.5, 4.7, 7.3$
5. Table 25.22, $x = -4, 11.1, 30.1, 66.7$
6. Table 25.23, $x = 1.5, 4.3, 13, 32$
7. Table 25.24, $x = 0.4, 6.2, 15.1$
8. Table 25.25, $x = 5.2, 8.6, 17.8$

TABLE 25.18 *Data for Problem 1, Section 25.8.*

x_i	1.9	2.5	5.4	8.91	10.5	17.3	18.9	21.7	25.3	34.7
y_i	−1.4	4.2	3.15	−5.6	10.7	11.2	−6.8	−15.3	2.7	9.4

TABLE 25.19 *Data for Problem 2, Section 25.8.*

x_i	4.8	5.8	7.3	9.5	15.3	19.5	27.4	32.6	36.3	41.9
y_i	34.8	41.1	47.2	57.1	76.9	98.6	133.8	148.4	164.9	197.5

TABLE 25.20 *Data for Problem 3, Section 25.8.*

x_i	3	3.4	3.8	4.2	5.9	7.5	12.3	21.8	29.8	35.3
y_i	1	1.3	1.9	2.28	3.82	5.42	9.0	19.1	24.9	28.7

TABLE 25.21 *Data for Problem 4, Section 25.8.*

x_i	3.7	3.9	4.2	4.6	4.8	5.2	5.7	6.3	6.5	6.9
y_i	−1.2	3.9	−5.6	3.8	−12.9	4.2	−27.2	4.8	−18.6	15.9

TABLE 25.22 *Data for Problem 5, Section 25.8.*

x_i	0	2.5	6.1	8.3	9.8	11.6	15.3	17.2
y_i	−3.1	1.22	7.2	10.37	12.9	15.85	20.7	25.2
x_i	19.5	22.9	25.5	29.6	32.2	36.2	40.8	
y_i	28.19	32.7	38.6	43.92	46.8	56.2	61.9	

TABLE 25.23 *Data for Problem 6, Section 25.8.*

x_i	0	1.3	1.9	2.3	2.5	2.8	3.1	4.8
y_i	3	0.5	−0.84	−1.74	−1.90	−2.73	−3.0	−6.9
x_i	5.4	7.4	12.9	15.3	19.4	22.5	25.9	
y_i	−7.4	−12.32	−24.8	−27.6	−35.9	−45.7	−48.5	

TABLE 25.24 Data for Problem 7, Section 25.8.

x_i	0.98	1.3	2.1	2.8	3.6	5.2	5.8	7.3	9.7	13.6
y_i	3.91	4.10	4.81	5.07	6.21	7.15	7.66	9.04	10.84	14.26

TABLE 25.25 Data for Problem 8, Section 25.8.

x_i	0.2	0.9	1.4	1.9	2.6	3.4	4.2	6.8	10	15.9
y_i	2.8	0.93	−0.61	−2.1	−3.9	−6.5	−8.9	−16.1	−27.2	−42.41

9. Over a period of ten years the number of times Benny the Heist has visited New York each year has been recorded. Records have also been kept of the number of major jewelry thefts in the city for each year. The numbers are given in Table 25.26.

   (a) Show that there is a significant linear correlation between number of visit and thefts.

   (b) Determine the equation of the regression line.

   (c) Estimate the percentage of thefts that are explained by Benny's visits.

   (d) Predict with a 95 percent confidence level how many thefts are likely to occur if Benny visits New York 25 times this year. Write the prediction interval for this estimate.

10. A city is experimenting with an additive to the water to prevent hair loss. Over a ten year period records have been kept on pounds of the chemical put into the water and the number of residents over the age of forty have experienced hair loss. The data is given in the Table 25.27.

   (a) Show that there is a significant linear correlation between number of pounds of the chemical and number of people with hair loss.

   (b) Determine the equation of the regression line.

   (c) Estimate the percentage of residents who had no hair loss that is explained by the chemical added to the water.

   (d) Predict with a 95 percent confidence how many residents are likely to have no hair loss if 650 pounds of the chemical are used. Write the prediction interval for this estimate.

11. A basketball coach is attempting to evaluating the impact of an expensive player on the team's won/loss record. He considers the data of Table 25.28, which gives number of games per season in which the player has seen at least twenty five minutes of floor time, together with wins over the season.

   (a) Show that there is a significant linear correlation between games with playing time and team wins.

   (b) Write the equation of the regression line.

   (c) Estimate the percentage of wins (in games the player appeared in) that are explained by this player's floor presence.

   (d) Predict with a 95 percent confidence level how many games the team will win next season if this player appears in 85 games. Write the prediction interval for this estimate.

12. A drug is being tested. One purpose of the test is to demonstrate the drug's efficiency, but another is to measure incidence of a side effect (lowered body temperature). Table 25.29 gives data on randomly chosen patients in ten states, together with the number exhibiting the side effect.

   (a) Show that there is a significant linear correlation between number of patients and incidents of the side effect.

   (b) Determine the equation of the regression line.

   (c) Estimate the percentage of patients that will have this side effect in a patient population of 17,000.

   (d) Predict with a 95 percent confidence level how many patients out of a population of 17,500 will exhibit the side effect. Write the prediction interval for this estimate.

TABLE 25.26 *Benny's visits and jewelry heists for Problem 9, Section 25.8.*

x_i: Visits	17	12	15	22	14	9	18	21	10	13
y_i: Thefts	6	5	6	8	5	3	7	8	5	5

TABLE 25.27 *Hair loss data for Problem 10, Section 25.8.*

x_i: Additive	410	500	475	390	470
y_i: No hair loss	3550	4412	4119	3420	4090
x_i: Additive	525	480	510	590	600
y_i: No hair loss	4598	4203	4452	5143	5259

TABLE 25.28 *Basketball player data for Problem 11, Section 25.8.*

x_i: Games	62	58	59	73	69	75	72	81	46	79
y_i: Wins	47	46	47	56	55	62	55	63	38	64

TABLE 25.29 *Patient data for Problem 12, Section 25.8.*

x_i: Patients	5000	5500	4200	10000	8000
y_i: Side effect	160	167	135	305	235
x_i: Patients	5700	6900	8800	12500	6000
y_i: Side effect	183	221	265	394	190

APPENDIX A

A MAPLE Primer

This section is intended to assist students in using MAPLE. In some cases, examples are given which can be easily adapted to general use. Additional examples and details can be found in MAPLE's HELP function. The student should experiment with the code given here and expand and adapt it to personal needs and preferences. In many instances, it is also possible to write different instructions to carry out computations.

A.1 Beginning Computations

Numerical computations are carried out as one might expect, with an asterisk * denoting a product and a wedge ∧ a power. If we type

```
2∧ 14;
```

we obtain the fourteenth power of 2, which is 16, 380. Note the semicolon ending this MAPLE command. In some versions of MAPLE, semicolons are used to end commands. In later releases, the semicolon is not needed. However, if the semicolon is included, the command will still execute.

To multiply 2^{14} by 19, type

```
19*2∧14;
```

π is stored in MAPLE as `Pi` (upper case P—MAPLE is case sensitive). The exponential function is denoted `exp`, and the number e is obtained as `exp(1)`. If we enter

```
(Pi ∧ 2)*exp(1);
```

this will return the symbolic product

$$(\pi^2)e$$

To obtain the (approximate) decimal value of this product, use the `evalf` command:

```
evalf((Pi ∧ 2)*exp(1));
```

This will return the decimal 26.82836630. As another example,

```
evalf(cos(3) + sin(3));
```

will return the decimal value -.8488724885 of $\cos(3) + \sin(3)$.

To solve an algebraic equation, use the `solve` command. For example,

```
solve(x∧2 + 2*x - 1=0,x);
```

will return the roots $-1+\sqrt{2}$ and $-1-\sqrt{2}$ of $x^2+2x-1=0$. This `solve` command includes a designation of x as the variable for which to solve—an essential piece of information for the program. Of course, other symbols than x can be used in the equation and the instruction. For the solutions in decimal form, use

```
evalf(solve(x∧2 + 2*x - 1=0,x);
```

which will return the decimal values 0.414213562 and -2.414213562 of these roots. As another example, enter

```
evalf(solve(cos(t) - t = 0,t));
```

to obtain the approximate solution 0.7390851332 of $\cos(t)-t=0$.

To solve a system of equations, enter the system in curly brackets in the solve command. For example,

```
solve({x - 2*y = 4, 5*x + y = -3},{x,y});
```

gives the solution $x=-2/11$, $y=-23/11$. Again, note the designation of x and y as the variables for which solutions are wanted.

In order to define a function, say $f(x)=x\sin(5x)-3x$, enter

```
f:=x→x*sin(5*x) - 3*x;
```

This names the function f and the variable x. The arrow must be typed into MAPLE as a dash followed by a "greater than" symbol. If we want to evaluate this function at a point, say $\pi/4$, type

```
f(Pi/4);
```

to obtain the value $-\pi\sqrt{2}/8-3\pi/4$.

To plot a graph of f, say on the interval $[-1,3]$, enter

```
plot(f(x), x=-1..3);
```

If we wish, we can enter a function directly into the plot command without prior definition. For example,

```
plot(x*cos(3*x) - exp(x), x=-1..1);
```

will graph $x\cos(3x)-e^x$ for $-1\leq x\leq 1$.

To plot several graphs of functions that have been defined, enter, for example,

```
plot({f(x),g(x),h(x)},x=a..b);
```

We can arrange for all the graphs to be in one color, say black, by

```
plot({f(x),g(x),h(x)},x=a..b,color=black);
```

Sometimes we want to enter a function having jump discontinuities. This is done by specifying the value of the function on successive intervals. For example, to define

$$s(x)=\begin{cases} x & \text{for } x<-1 \\ \cos(3x) & \text{for } -1<x<4 \\ x^2 & \text{for } 4<x<9 \\ \sin(4x) & \text{for } x>9 \end{cases}$$

enter

```
s:=x→ piecewise(x < -1,x,x < 4,cos(3*x),x < 9,x∧2,sin(4*x));
```

We can differentiate $f(x)$ by

```
diff(f(x),x);
```

This can be done with a previously entered $f(x)$, or we can put the function into the command, as with

```
diff(x*cos(3*x)*exp(2*x),x);
```

We can also take a derivative using D(f)(x).

To evaluate an indefinite integral $\int f(x)\,dx$, use the int command:

```
int(f(x),x);
```

The variable of integration must be specified. For a definite integral $\int_a^b f(x)\,dx$, include the limits of integration:

```
int(f(x),x=a..b);
```

This also can be used with improper integrals. For example, enter

```
int(exp(-x),x=0..infinity);
```

to evaluate $\int_0^\infty e^{-x}\,dx = 1$.

MAPLE has a `sum` command. If $a1, \cdots an$ are n numbers that have been defined in some way, then their sum is computed as

```
sum(aj,j=1..n);
```

It may be necessary to precede this command with evalf to obtain a decimal evaluation. Often it is convenient in a summation to define the sequence as a function. For example, if we define

```
a:=j → j*sin(Pi/j);
```

then

```
sum(a(j),j=4..7);
```

will produce the sum

$$4\sin(\pi/4) + 5\sin(\pi/5) + 6\sin(\pi/6) + 7\sin(\pi/7).$$

Preceding the sum command with `evalf` will produce the decimal value of the sum.

If a MAPLE file is saved, any code that has been entered will be retained. However, if the file is closed and then reopened, some commands may have to be reactivated. For example, if a function `f(x)` was previously entered, then place the cursor at the end of the line defining `f(x)` and hit ENTER again to reactivate this function.

A.2 Ordinary Differential Equations

Some operations with differential equations require that a special package of subroutines be opened. This is done by

```
with(DEtools);
```

For a direction field of $y' = y^2$ on a grid $-2 < x < 2, -2 < y < 2$, type

```
DEplot(diff(y(x),x) = y(x)∧2,y(x),x=-2..2,y=-2..2,color=black);
```

y is entered as $y(x)$ in the specification of the differential equation. The instruction to show the direction field in black is optional.

We could also use

```
dfieldplot(diff(y(x),x) = y(x)∧2,y(x),x=-2..2,y=-2..2,
                                  color=black);
```

For a field plot with sketches of some integral curves, enter initial conditions specifying these curves, for example,

```
DEplot(diff(y(x),x) = y(x)∧2,y(x),x=-4..4,y=-3..3,[[y(0)=-1/2],
  [y(0)=1/2],[y(0)=1],[y(0)=-2]],color=black,linecolor=[black,
                   black,black,black]);
```

This produces a black direction fieid over the grid $-4 < x < 4$, $-3 < y < 3$, with sketches of the integral curves (in black) through $(0, -1/2)$, $(0, 1/2)$, $(0, 1)$, and $(0, -2)$. Be careful in specifying things like color of integral curves. Since a set of four initial values is given, the color instructions for the integral curves must include four colors (although some or all can be the same).

We can solve some differential equations using the dsolve command. For example, for the general solution of $y' - (1/x)y = -2$, enter

```
dsolve(diff(y(x),x) - (1/x)*y(x) = -2,y(x));
```

This returns the general solution

$$y(x) = C_1 x - 2x \ln(x).$$

The arbitrary constant in the MAPLE output is denoted $C1$.

As an example of a second order differential equation, consider

$$y'' - 4y' + y = x^3 - \sin(2x).$$

Enter

```
dsolve(diff(diff(y(x),x),x) + 4*diff(y(x),x) + y(x)
                 = x∧3 - sin(2*x),y(x));
```

This gives the general solution

$$y(x) = C_1 e^{(2+\sqrt{3})x} + C_2 e^{(2-\sqrt{3})x} + 90x + 12x^2 + x^3$$
$$+ 336 - \frac{8}{73}\cos(2x) + \frac{3}{73}\sin(2x).$$

For an initial value problem, include the initial condition(s). For example,

```
dsolve(diff(y(x),x) - (1/x)*y(x) = -2,y(1) = 5,y(x));
```

gives the solution $y = (1/2)x^2 + x + 1$ of the initial value problem $y' - (1/x)y = -2$; $y(1) = 5$.

We can also solve some systems of differential equations. For example, to solve

$$y_1' - 4y_2' = 1,\ y_1' + 2y_2' = t$$

enter

```
          dsolve({diff(y1(t),t) - 4*diff(y2(t),t)
 = 1, diff(y1(t),t)+ 2*diff(y2(t),t) = t},{y1(t),y2(t)});
```

to obtain the general solution

$$y_1(t) = \frac{1}{10}t^2 - \frac{1}{5}t + c_1, \; y_2(t) = \frac{2}{5}t^2 + \frac{1}{5}t + c_2.$$

MAPLE can be used to write the first n terms of the power series solution of a differential equation with analytic coefficients, expanded about 0, with initial condition(s) specified at 0. To illustrate, give the differential equation $y' - \cos(x)y = e^x$ the name `deq1` by

```
deq1:= diff(y(x),x) - cos(x)*y(x) = exp(x);
```

Now obtain a power series solution about 0, with $y(0) = 5$, by

```
dsolve(deq1,y(0)=5,y(x),series);
```

This produces the output

$$y(x) = 5 + 6x + \frac{7}{2}x^2 + \frac{1}{2}x^3 - \frac{7}{12}x^4 - \frac{5}{12}x^5 + O(x^6).$$

The symbol $O(x^6)$ means that terms involving sixth and higher powers of x have not been included in this expression. In the absence of an instruction, MAPLE defaults to giving the first six terms of a power series solution about 0, and these are the constant term and terms involving x^k for $k = 1, 2, \cdots, 5$. For terms up to and including x^N, include a value of N in the instruction. For example, to obtain terms up to and including x^{10}, enter

```
dsolve({deq2,y(0)=5},y(x),series,order = 10);
```

For a phase portrait of a 2×2 system, we can use DEplot. For example,

```
DEplot([diff(x(t),t) = x(t)∧2 - y(t),diff(y(t),t) =
x(t)*y(t)],[x(t),y(t)],t=-7..0,[[x(0)=0,y(0)=1],
[x(0)=0,y(0)=2],[x(0)=0,y(0)=1/2]], stepsize=.02,
linecolor=0,color=black,arrows=MEDIUM,method=rkf45);
```

will produce a direction field (in black) for the system

$$x'(t) = x^2 - y, \; y'(t) = xy$$

and solution curves through (0, 1), (0, 2), and (0, 1/2). In applying the `DEplot` function, some experimentation may be needed for the rang of values of t and for the selection of points through which solution curves are to be drawn. In this example, if we set `t = 0..1`, we obtain an error message. We do not know the solution explicitly, and some terms in it may blow up for certain values of t. For example, if a solution has an e^{6t} term and we specify `t:0..3`, then, as t increases, the numbers e^{6t} become too large for the program to handle. The key is to try various t intervals. If one interval yields only parts of curves, try extending the range of t. It is also the case that some selections of initial data will work better than others.

A.3 Vector Operations

Vector operations can be carried out using the vector calculus package. Begin by loading this package using

```
with(VectorCalculus);
```

Vector algebra can be done in R^n by either performing the operations (addition and scalar multiplication) directly on the vectors, or by entering the vectors separately and then carrying out the operations. For example, to add $16/37$ times $< 1, -5, 3, -9, -22 >$ to $15/92$ times $< 0, 3, -3, 6, -1 >$, enter

```
(16/37)*<1,−5,3,−9,−22> + (15/92)*<0,3,−3,6,−1>;
```

We could also name the first vector V and the second vector G by

```
V:=<1,−5,3,−9,−22>;
```

and

```
G:=<0,3,−3,6,−1>
```

and then compute

```
(16/37)*V + (15/92)*G;
```

For the dot product of two vectors V and F (which can be n-vectors), enter

```
DotProduct(V,F);
```

Here V and F can be defined previously or can be entered into the dot product command.

Unlike the dot product, for the curl of two vectors, we are restricted to vectors in R^3. For the curl of $V =< a, b, c >$ and $F =< d, e, f >$, enter

```
CrossProduct(<a,b,c>,<d,e,f>);
```

As usual, the vectors can be entered directly or they could have been previously defined.

We can also carry out the vector calculus operations of gradient, divergence and curl. To begin, we will work in rectangular coordinates. Set this system by

```
SetCoordinates('cartesian'[x,y,z]);
```

MAPLE expects the word cartesian in this command.

For the gradient of a scalar field, use the del operator, which in MAPLE is called del or nabla. For example, we can define the scalar field $f(x, y, z) = xyz - xy^2 \cos(z)$, by entering

```
f:=(x,y,z) → x*y*z - x + 4*y*z;
```

Now compute the gradient of f in any of the following ways:

```
Gradient(f(x,y,z));
```

or

```
Del(f(x,y,z));
```

or

```
Nabla(f(x,y,z));
```

or

```
Del(x*y*z - x + 4*y*z);
```

or

```
Nabla(x*y*z - x + 4*y*z);
```

To work with divergence and curl in rectangular (cartesian) coordinates, we can first enter the vector fields of interest. For example, to enter the vector field

$$\mathbf{F}(x, y, z) = xyz\mathbf{i} + (x - y)\mathbf{j} + yz\mathbf{k},$$

use

```
F:= VectorField(<x*y*z,x - y,yz>);
```

Take the divergence of **F** as the dot product of del with **F**.

```
DotProduct(Nabla,F);
```

or

```
DotProduct(Del,F);
```

For the curl, take the cross product of the del operator with the vector field:

```
CrossProduct(Nabla,F);
```

or

```
CrossProduct(Del,F);
```

The output from this command will have the appearance

$$0\overline{e}_x + xy\overline{e}_y + (1 - xz)\overline{e}_z.$$

In MAPLE, unit vectors along the coordinate axes are denoted $\overline{e}_\alpha$ where α denotes the coordinate. In rectangular coordinates, $e_x = \mathbf{i}$, $e_y = \mathbf{j}$, and $e_z = \mathbf{k}$. In spherical coordinates, these vectors would be denoted $\overline{e}_\rho$, $\overline{e}_\theta$, and $\overline{e}_\varphi$.

We can also carry out vector operations in other curvilinear coordinate systems. We will illustrate these for cylindrical and spherical coordinates.

To work in cylindrical coordinates, begin with

```
SetCoordinates('cylindrical' [r,theta,z]);
```

As an example, define a vector field $\mathbf{G}(r, \theta, z)$ by

```
G:=VectorField(<(r∧2)*cos(theta),r*z*cos(theta)*sin(theta),
                (z∧2)>);
```

This will produce the output

$$G := r^2\cos(\theta)\overline{e}_r + rz\cos(\theta)\sin(\theta)\overline{e}_\theta + z^2\overline{e}_z.$$

For the divergence, enter

```
Divergence(G);
```

resulting in the output

$$\frac{3r^2\cos(\theta) - rz\sin(\theta)^2 + rz\cos(\theta)^2 + 2rz}{r}.$$

Of course, we can divide out the r in the denominator. For the curl of G, enter

```
Curl(G);
```

to obtain

$$-r\cos(\theta)\sin(\theta)\overline{e}_r + \frac{2rz\cos(\theta)\sin(\theta) + r^2\sin(\theta)}{r}\overline{e}_z.$$

We would divide out the r in the $\overline{e}_z$ component. MAPLE does not automatically simplify all output.

In an entirely analogous way, we can carry out vector operations in spherical coordinates. First, change to this coordinate system:

```
SetCoordinates('spherical'[rho,theta,phi]);
```

For the MAPLE command

```
Gradient(rho*sin(phi)∧2)*cos(theta/2));
```

returns the gradient in spherical coordinates:

$$sin(\phi)^2 \cos\left(\frac{1}{2}\theta\right)\overline{e}_\rho - \frac{1}{2}\sin(\phi)^2 \sin\left(\frac{1}{2}\theta\right)\overline{e}_\theta + \frac{2\sin(\phi)\cos\left(\frac{1}{2}\theta\right)\cos(\phi)}{\sin(\theta)}\overline{e}_\phi.$$

Some vector operations can also be carried out using the linear algebra package, which is discussed next. Details are available in the MAPLE Help package.

A.4 Matrix Manipulations

To work with matrices, load the linear algebra package, using

```
with(linalg);
```

Again, the semicolon will give a list of the subroutines. To avoid the list, end this command with a colon.

There are many ways to enter a matrix. One way is to enter the dimension and the rows. For example,

```
Matrix(4,2,[[-1,3],[6,1],[6,5],[-1,-2]]);
```

enters the 4×2 matrix

$$\begin{pmatrix} -1 & 3 \\ 6 & 1 \\ 6 & 5 \\ -1 & -2 \end{pmatrix}.$$

To give this matrix the name K, enter

```
K:=Matrix(4,2,[[-1,3],[6,1],[6,5],[-1,-2]]);
```

For the inverse of a nonsingular matrix A which has been entered, use

```
inverse(A);
```

For the rank of a matrix L (which need not be square), use

```
rank(L);
```

For the determinant of square A, use

```
det(A);
```

Multiply a matrix A by a scalar c by

```
c*A;
```

Add two $n \times m$ matrices as we would expect using

```
A + B;
```

Multiply matrices A and B by

```
A.B;
```

Elementary row operations can be carried out very efficiently. Let A be an $n \times m$ matrix that has been entered. To interchange rows k and j and A to form a new matrix B, use

```
B:=swaprow(A,k,j);
```

To form S from A by multiplying row r of B by x, use

```
S:=mulrow(A,r,x);
```

And to form T from A by adding x times row r_1 to row r_2, use

```
T:= addrow(A,r1,r2,x);
```

Finally, suppose we select an element $a_{k,j}$ of A, and we want to proceed by elementary row operations to a new matrix W having zeros above and below this element. This is called *pivoting* about $a_{k,j}$, and is done using the `pivot` command. Enter

```
W:=pivot(A,k,j);
```

W will be the matrix with zeros above and below $a_{k,j}$, which could have been obtained by systematically applying row operations as needed to each of the rows above and below row k of the matrix.

Using the `pivot` command, it is easy to find the reduced row echelon form of a matrix. Starting in the upper left corner of the matrix, pivot about the leading element of each row. The resulting matrix is nearly reduced. The only problem is that leading entries of nonzero rows may not equal 1. Just divide each nonzero row by its leading entry (using mulrow) to obtain leading entries of 1, resulting in a reduced row echelon matrix.

The corresponding column operations can be achieved by replacing row with col in these commands. Such column operations may be useful, for example, in manipulating determinants.

The elementary row and column operations of adding a scalar multiple of one row (column) to another are also useful in carrying out an LU factorization of matrix. The algorithms for forming the two matrices in such a factorization depend on these operations.

The characteristic polynomial of a square matrix A is obtained using

```
charpoly(A,x);
```

in which the variable in which the polynomial will be written (here x) must be specified. The command

```
eigenvals(A);
```

lists the eigenvalues of A (real or complex), giving each eigenvalue according to its multiplicity. Note that the imaginary unit i is denoted I in the program. The command

```
eigenvects(A);
```

lists each eigenvalue, along with its multiplicity and for each eigenvalue, as many linearly independent eigenvectors as can be found for that eigenvalue. To illustrate, let

```
A:=Matrix(3,3,[[1,-1,0],[0,1,1],[0,0,-1]]);
```

This is the matrix of Example 9.3. Now

```
eigenvects(A);
```

gives the output

$$[1, 2, [1, 0, 0]], [-1, 1, [1, 2, -4]]$$

This gives eigenvalue 1 with multiplicity 2, and every eigenvector associated with 1 is a multiple of $(1, 0, 0)$. Eigenvalue -1 has multiplicity 1 and eigenvector $(1, 2, -4)$.

Now let B be the 3×3 matrix of Example 9.5:

```
B:=Matrix(3,3,[[5,-4,4],[12,-11,12],[4,-4,5]]);
```

The command

```
eigenvals(B);
```

gives the list $1, 1, -3$ of eigenvalues, 1 having multiplicity 2, and -3 having multiplicity 1. Next use

```
eigenvects(B);
```

to obtain the output

$$[1, 2, [1, 1, 0], [-1, 0, 1]], [-3, 1, [1, 3, 1]]$$

giving two linearly eigenvectors associated with eigenvalue -3.

A.5 Integral Transforms

MAPLE has subroutines for several integral transforms. To load these, enter

```
with(inttrans);
```

To take the Laplace transform of $f(t)$, use

```
laplace(f(t),t,s);
```

in which $f(t)$ may have been loaded previously, or can be specified in the command. For example,

```
laplace(t*cos(t),t,s);
```

returns the transform $(s^2 - 1)((s^2 + 1)^2)$ of $t\cos(t)$.

For the inverse Laplace transform of $1/((s^2 + 4)^2)$, use

```
invlaplace(1/((s∧2 + 4)∧2),s,t);
```

to obtain $(1/16)\sin(2t) - (1/8)\cos(2t)$.

For the Fourier, Fourier sine and Fourier cosine transforms, the commands are similar, except replace `laplace` with `fourier`, `fouriersin`, or `fouriercos`. For example, for the Fourier transform of $e^{-|t|}$, use

```
fourier(exp(-abs(t)),t,w);
```

to obtain the transform $2/(1 + w^2)$. To compute the inverse Fourier transform of $1/(1 + w)$, use

```
invfourier((1/(1 + w),w,t);
```

to obtain the inverse transform $\frac{1}{2}ie^{-it}(2H(t) - 1)$, where $H(t)$ is the Heaviside function.

A.6 Special Functions

In MAPLE, $J_n(x)$ is called `BesselJ(n,x)`, and the Bessel function of the second kind, $Y_n(x)$, is denoted `BesselY(n,x)`. To evaluate, for example, $J_3(1.21)$, use

```
evalf(BesselJ(3,1.21));
```

In similar fashion, there are `BesselI` and `BesselK` commands for modified Bessel functions of the first and second kinds, respectively.

For integrals involving Bessel functions, use the int command. For example,

```
evalf(int(x*BesselJ(1,x)*cos(3*x),x=0..1));
```

will give a decimal evaluation of $\int_0^1 xJ_1(x)\cos(x)\,dx$.

We can find (approximately) the kth (in order of increasing magnitude) positive zero of $J_n(x)$ by using

```
evalf(BesselJZeros(n,k));
```

For example,

```
evalf(BesselJZeros(3,7));
```

returns the seventh positive zero of $J_3(x)$.

The nth Legendre polynomial $P_n(x)$ is denoted `LegendreP(n,x)` in MAPLE. Since Legendre's differential equation is second order, there is a second, linearly independent solution, often denoted $Q_n(x)$. In MAPLE, this function is `LegendreQ(n,x)`. To obtain $P_n(x)$ explicitly, use

```
simplify(LegendreP(n,x));
```

Integrals involving Legendre polynomials can be done using the int command. For example,

```
evalf(int(sin(x)*LegendreP(5,x),x=-1..1));
```

will give a decimal evaluation of $\int_{-1}^1 \sin(x)P_5(x)\,dx$.

Legendre polynomials form a special case of a class of special functions called orthogonal polynomials, which include Laguerre polynomials, Hermite polynomials and many others. The command

```
with(orthopoly);
```

will call up MAPLE's subroutines of orthogonal polynomials.

A.7 Complex Functions

MAPLE will do complex arithmetic with the imaginary unit i denoted I.

For the residue of a function at a point, use the command

```
Res(f, z0);
```

For example, if $f(z) = \cos(z)/z$, entered by

```
f:=z → cos(z)/z;
```

then the residue at zero can be computed as

```
residue(f(z),z=0);
```

which returns the residue 1. We could also enter

```
residue(cos(z)/z,z=0);
```

If $f(z)$ is a conformal mapping, we can obtain a picture of the image of a rectangle defined by $a \leq x \leq b$, $c \leq y \leq d$ by first opening the plots subroutines:

```
with(plots);
```

Next, enter

```
conformal(f(z),z=a + b*I,c + d*I,xtickmarks = 8,ytickmarks = 8);
```

Answers to Selected Problems

CHAPTER ONE FIRST-ORDER DIFFERENTIAL EQUATIONS

Section 1.1 Terminology and Separable Equations

1. Yes **3.** Yes

5. Yes, since

$$2\varphi(x)\varphi'(x) = 2\sqrt{x-1}\left(\frac{1}{2\sqrt{x-1}}\right) = 1 \text{ for } x > 1.$$

7. $\sec(y) = kx$ and $y = (2n+1)\pi/2$ with n any integer **9.** Not separable

11. $y^3 = 2x^2 + c$, or $y = (2x^2 + c)^{1/3}$ **13.** Not separable

15. $y = 1/(1 - cx)$; also $y = 0$ and $y = 1$ as singular solutions

17. $(\ln(y))^2 = 3x^2 - 3$ **19.** $3y\sin(3y) + \cos(3y) = 9x^2 - 5$

21. $\frac{1}{2}y^2 - y + \ln(y+1) = \ln(x) - 2$

25. The tank empties at $t = 3888\sqrt{2}$ seconds; about 91 minutes and 39 seconds.

27. $10(1/2)^{1/4.5} \approx 8.57$ kg **29.** 45° F

31. (a) $t = 243(64/27) = 576$ seconds, or 9 minutes, 36 seconds (b) $t = (160/27)(1296/5) = 1536$ seconds; about 25 minutes, 36 seconds.

Section 1.2 Linear Equations

1. $y = 4x^2 + 4x + 2 + ce^{2x}$ **3.** $y = cx^3 + 2x^3\ln|x|$ **5.** $y = \frac{1}{2}x - \frac{1}{4} + ce^{-2x}$

7. $y = x + 1 + 4(x+1)^{-2}$ **9.** $y = x^2 - x - 2$

11.

$$A_1(t) = 50 - 30e^{-t/20},\ A_2(t) = 75 + 90e^{-t/20} - 75e^{-t/30},$$

$A_1(t)$ has its minimum value of 5450/81 pounds at $60\ln(9/5)$ minutes.

13. $y = -2x^2 + cx$

Section 1.3 Exact Equations

1. $y^3 + xy + \ln|x| = c$ **3.** $2xy^2 + e^{xy} + y^2 = c$ **5.** not exact

7. $\alpha = -3$; $x^2y^3 - 3xy - 3y^2 = c$

9. $x\sin(2y - x) = \pi/24$ **11.** $3xy^4 - x = 47$

13. $\mu(x, y) = x^{1/2}y^{3/2}$, $(1/5)x^{5/2}y^{5/2} + (1/3)x^{3/2} = c$

15. The general solutions are the same, since the equations $\varphi(x, y) = k$ and $\varphi(x, y) + c = k$ implicitly define the same functions of y in terms of x.

Section 1.4 Homogeneous, Bernoulli and Riccati Equations

1. Homogeneous; $y\ln|y| - x = cy$

3. Exact with general solution implicitly defined by $xy - x^2 - y^2 = c$; also homogeneous

5. Bernoulli with $\alpha = -3/4$; $5x^{7/4}y^{7/4} + 7x^{-5/4} = c$

7. Bernoulli equation with $\alpha = 2$;

$$y = \frac{1}{1 + ce^{x^2/2}}$$

9. Bernoulli with $\alpha = 2$;

$$y = 2 + \frac{2}{cx^2 - 1}$$

11. Riccati with $S(x) = e^x$;

$$y = \frac{2e^x}{ce^{2x} - 1}$$

13. Riccati equation with $S(x) = x$;

$$y = x + \frac{x}{c - \ln(x)}$$

17. $(2x + y - 3)^2 = k(y - x + 3)$

19. Choose $h = 2, k = -3$ to obtain

$$3(x-2)^2 - 2(x-2)(y+3) - (y+3)^2 = k$$

Section 1.5 Additional Applications

1. The inductive time constant is

$$\frac{L}{R}\ln\left(\frac{e(E - Ri(0))}{E}\right),$$

decreasing as $i(0)$ is chosen larger.

3. $t = 2\sqrt{R/g}$, with R the radius of the circle.

5. Terminal velocity is 8 ft/sec. Distance fallen is $32(t - 1 + e^{-t})$ for $0 \le t \le 4$, and for $t \ge 4$,

$$32(3 + e^{-4}) + 8(t-4) + 2\ln(1 - e^{-8(t-4)}) - 2\ln\left(\frac{2}{5 - 4e^{-4}}\right),$$

where $k = (3 - 4e^{-4})/(5 - 4e^{-4})$.

7. Surfacing velocity is approximately 17.5 ft/sec at $t \approx 10.56$ sec.

9. Voltage reaches 76 volts when $t = (1/2)\ln(20)$. Current at this time is $32(10^{-5})e^{-\ln(20)} = 16$ micro amps.

11. Maximum height is 342.25 feet, at $t = 1/8$ second; at $t = 19/4$ the bag hits the ground with speed 148 ft/sec.

13. $(y-1)^2 = -(1/2)x^2 + c$, a family of ellipses

15. $y^2(\ln(y^2) - 1) = c - 2x^2$

17. $y = -(3/4)\ln|x| + c$

19. Time is $(\sqrt{3}/2)(\ln(6 + \sqrt{35})$, about 2.15 seconds; velocity is $12\sqrt{5}$, about 26.84 ft/sec.

21. (a) Pursuit curves are $r = f(\theta) = (a/\sqrt{2})e^{-\theta}$. (b) distance $= a$ (c) No

Section 1.6 Existence and Uniqueness Questions

1. $f(x, y) = x^2 - y^2 + 8x/y$ and $\partial f/\partial y = -2y - 8x/y^2$ are continuous on any rectangle centered at $(3, -1)$ that does not intersect the x-axis.

3. $f(x, y) = \sin(xy)$ and $\partial f/\partial y = x\cos(xy)$ are continuous for all (x, y), hence in any rectangle centered at $(\pi/2, 1)$.

5. Two solutions are

$$y = -\frac{1}{2}\ln\left(e^{-2y_0} + 2(x - x_0)\right)$$

and

$$y = -\frac{1}{2}\ln\left(e^{-2y_0} + 2(x_0 - x)\right).$$

The theorem does not apply, because the differential equation is $y' = \pm 2y$.

7. (a) Both $f(x,y)=2x^2$ and $\partial f/\partial y=0$ are continuous for all (x,y).

(b) $y=(2/3)x^3+7/3$

(c) $y_0)=3,\ y_1=3+\int_1^x 2t^2\,dt=(2/3)x^3+7/3$

(d) Since $f(x,y)=2x^2$ is independent of y, $y_n(x)=y_1(x)$ for all $n\geq 1$ and the sequence of Picard iterates simply repeats this term. Now

$$y=\frac{2}{3}x^3+\frac{7}{3}=3+2(x-1)+2(x-1)^2+\frac{2}{3}(x-1)^3$$

is the Taylor series of the solution about 1. For $n\geq 3$, the nth partial sum of this expansion is the solution, so $y_n\to y$ and the Picard iterates converge to the solution.

9. (a) Since $f(x,y)=4+y$ and $\partial f/\partial y=1$ are continuous everywhere, the initial value problem has a unique solution in some interval about 0.

(b) $y=-4+7e^x$

(c) $y_0=3$,

$$y_1=3+\int_0^x 7\,dt=3+7x,$$

$$y_2=3+\int_0^x(7+7t)\,dt=3+7x+\frac{7}{2}x^2,$$

$$y_3=3+\int_0^x\left(7+7t+\frac{7}{2}t^2\right)dt=3+7x+\frac{7}{2}x^2+\frac{7}{3!}x^3,$$

$$y_4=3+\int_0^x\left(7+7x+\frac{7}{2}x^2+\frac{7}{3!}x^3\right)dt$$

$$=3+7x+\frac{7}{2}x^2+\frac{7}{3}x^{3!}+\frac{7}{4!}x^4,$$

$$y_5=3+\int_0^x(4+y_4(t))\,dt$$

$$=3+7x+\frac{7}{2}x^2+\frac{7}{3}x^{3!}+\frac{7}{4!}x^4+\frac{7}{5!}x^5,$$

$$y_6=3+\int_0^x(4+y_5(t))\,dt$$

$$=3+7x+\frac{7}{2}x^2+\frac{7}{3}x^{3!}+\frac{7}{4!}x^4+\frac{7}{5!}x^5+\frac{7}{6!}x^6.$$

(d)

$$y_n=3+7x+\frac{7}{2!}x^2+\frac{7}{3!}x^3+\cdots+\frac{7}{n!}x^n$$

Note that

$$y_n=-4+7\sum_{k=0}^{n}\frac{1}{k!}x^k$$

so $\lim_{n\to\infty}y_n(x)=-4+7e^x$.

CHAPTER TWO LINEAR SECOND-ORDER EQUATIONS

Section 2.1 The Linear Second-Order Equation

1. $y(x)=c_1e^x\cos(x)+c_2e^x\sin(x);\ y(x)=6e^x\cos(x)-5e^x\sin(x)$

3. $y(x)=c_1e^{-2x}+c_2e^{-x};\ y(x)=4e^{-2x}-7e^{-x}$

5. $y(x)=c_1\sin(6x)+c_2\cos(6x);\ y(x)=\frac{1}{3}\sin(6x)-5\cos(6x)$

7. $y(x)=c_1e^{3x}\cos(2x)+c_2e^{3x}\sin(2x)-8e^x$

9. $y(x)=c_1e^{4x}+c_2e^{-4x}-\frac{x^2}{4}+\frac{1}{2}$

11. We know that the initial value problem

$$y'' + py' + qy = 0;\ y(x_0) = y'(x_0) = 0$$

has the unique solution $y = 0$ in some interval about x_0. If $\varphi(x_0) = \varphi'(x_0) = 0$, then φ is also a solution of this problem, hence φ would have to be identically zero, which is a contradiction.

13. The functions are linearly independent, since neither is a constant multiple of the other on $[-1, 1]$. The differential equation in linear form is

$$y'' - \frac{2}{x}y' + \frac{2}{x^2}y = 0,$$

which is undefined at 0, so the theorem does not apply.

Section 2.2 The Constant Coefficient Case

1. $y = e^{-3x/2}[c_1 \cos(3\sqrt{7}x/2) + c_2 \sin(3\sqrt{7}x/2)]$

3. $y = e^{-5x}(c_1 \cos(x) + c_2 \sin(x))$

5. $y = e^{7x}(c_1 + c_2 x)$ **7.** $y = e^{-3x}(c_1 + c_2 x)$

9. $y = c_2 e^{-2x} + c_2 e^{3x}$

11. $y = \frac{9}{7}e^{3(x-2)} + \frac{5}{7}e^{-4(x-2)}$

13. $y = e^{x-1}(29 - 17x)$

15. $y = e^{(x+2)/2}\left[\cos(\sqrt{15}(x+2)/2) + \frac{5}{\sqrt{15}}\sin(\sqrt{15}(x+2)/2)\right]$

17. $y(x) = 0$ **19.** $y = 5 - 2e^{-3x}$

21. The characteristic equation has roots

$$\lambda_1 = \frac{-a + \sqrt{a^2 - 4b}}{2},\ \lambda_2 = \frac{-a - \sqrt{a^2 - 4b}}{2}.$$

Now take cases. If $a^2 - 4b > 0$, then λ_1 and λ_2 are both negative, so the solution $c_1 e^{\lambda_1 x} + c_2 e^{\lambda_2} x$ decays to zero as $x \to \infty$. If $a^2 = 4b$, the solution $e^{-ax/2}(c_1 + c_2 x) \to 0$ as $x \to \infty$ because $a > 0$. If $a^2 - 4b < 0$, the solution is a linear combination of sines and cosines, multiplied by $e^{-ax/2}$, and this solution goes to zero as $x \to \infty$.

23. (a) $\varphi(x) = e^{ax}(c_1 + c_2 x)$ (b) $\varphi_\epsilon(x) = e^{ax}(c_1 e^{\epsilon x} + c_2 e^{-\epsilon x})$
(c) $\lim_{\epsilon \to 0} \varphi_\epsilon(x) = e^{ax}(c_1 + c_2) \neq \varphi(x)$ in general.

Section 2.3 The Nonhomogeneous Equation

1. $y = c_1 \cos(3x) + c_2 \sin(3x) + 4x \sin(3x) + \frac{4}{3}\cos(3x) \ln|\cos(3x)|$

3. $y = c_1 e^x + c_2 e^{2x} - e^{2x}\cos(e^{-x})$

5. $y = c_1 \cos(x) + c_2 \sin(x) - \cos(x) \ln|\sec(x) + \tan(x)|$

7. $y = c_1 e^x + c_2 e^{2x} + +3\cos(x) + \sin(x)$

9. $y = e^{2x}[c_1 \cos(3x) + c_2 \sin(3x)] + \frac{1}{3}e^{2x} - \frac{1}{2}e^{3x}$

11. $y = e^x[c_1 \cos(3x) + c_2 \sin(3x)] + 2x^2 + x - 1$

13. $y = c_1 e^{2x} + c_2 e^{4x} + e^x$

15. $y = c_1 e^{2x} + c_2 e^{-x} - x^2 + x - 4$

17. $y = \frac{3}{8}e^{-2x} - \frac{19}{120}e^{-6x} + \frac{1}{5}e^{-x} + \frac{7}{12}$

19. $y = 2e^{4x} + 2e^{-2x} - 2e^{-x} - e^{2x}$

21. $y = 4e^{-x} - \sin^2(x) - 2$

23. $y = \frac{7}{4}e^{2x} - \frac{3}{4}e^{-2x} - \frac{7}{4}xe^{2x} - \frac{1}{4}x$

Section 2.4 Spring Motion

1. With $y(0) = 5$ and $y'(0) = 0$, $y = e^{-2t}\left[5\cosh(\sqrt{2}t) + \frac{10}{\sqrt{2}}\sinh(\sqrt{2}t)\right]$.
With $y(0) = 0$ and $y'(0) = 5$, $y = \frac{5}{\sqrt{2}}e^{-2t}\sinh(\sqrt{2}t)$.
These functions are graphed in Figure A.1.

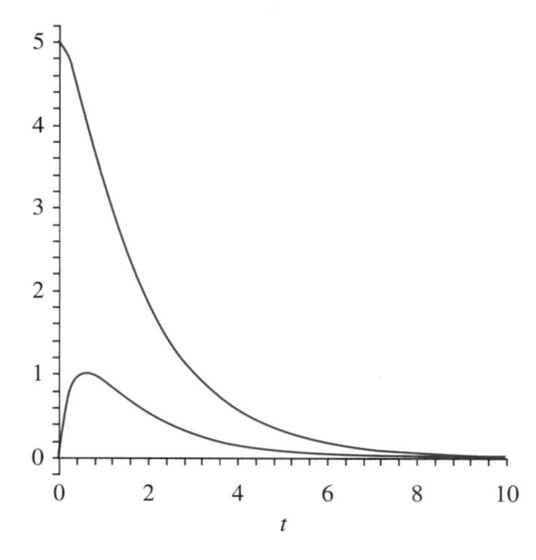

FIGURE A.1 *Graphs in Problem 1, Section 2.4*

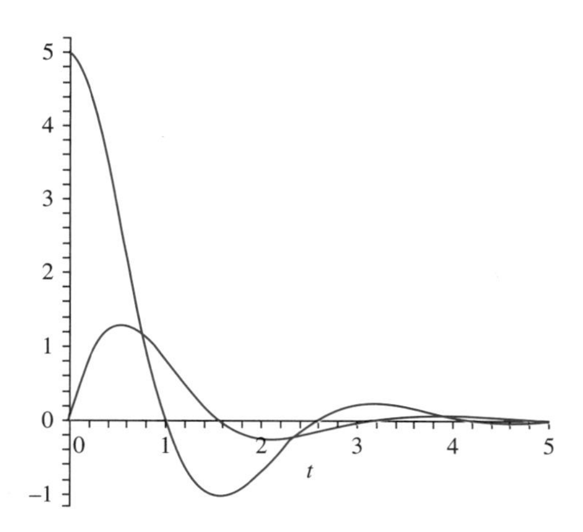

FIGURE A.2 *Graphs in Problem 3, Section 2.4*

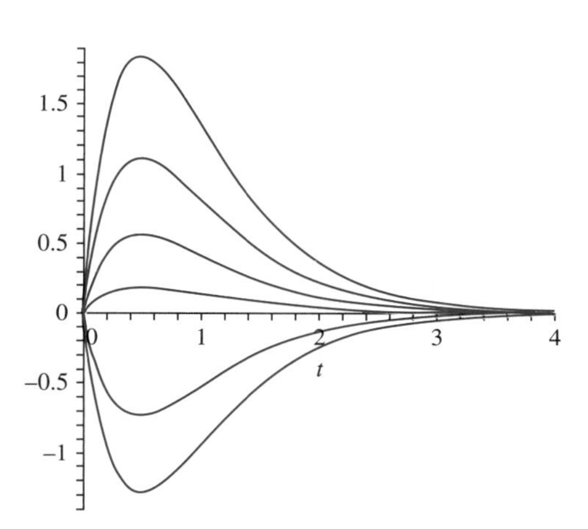

FIGURE A.3 *Graphs in Problem 5, Section 2.4*

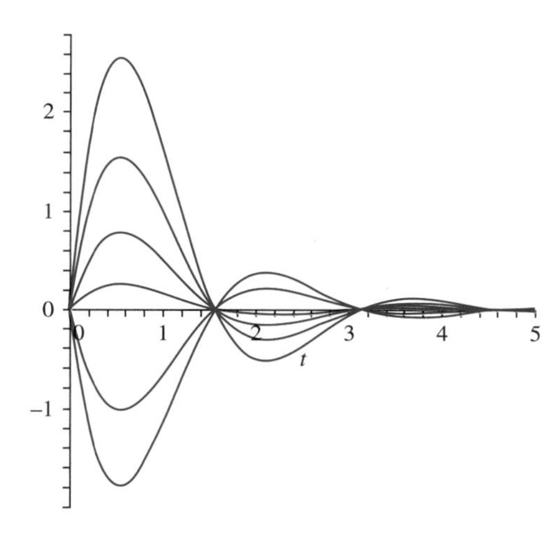

FIGURE A.4 *Graphs in Problem 7, Section 2.4*

3. With $y(0) = 5$ and $y'(0) = 0$,

$$y = \frac{5}{2}e^{-t}[2\cos(2t) + \sin(2t)]$$

and with $y(0) = 0$ and $y'(0) = 5$,

$$y = \frac{5}{2}e^{-t}\sin(2t).$$

Graphs are in Figure A.2.

5. $y = Ate^{-2t}$, with the graphs in Figure A.3 moving from the lowest to the highest as A increases.

7. $y = \frac{A}{2}e^{-t}\sin(2t)$, with the graphs moving from lowest to highest as A increases. Graphs are given in Figure A.4, increasing as A increases.

9. $y = \frac{A}{\sqrt{2}}e^{-2t}\sinh(\sqrt{2}t)$; graphs are in Figure A.5, proceeding from the lowest to the highest graph as A increases.

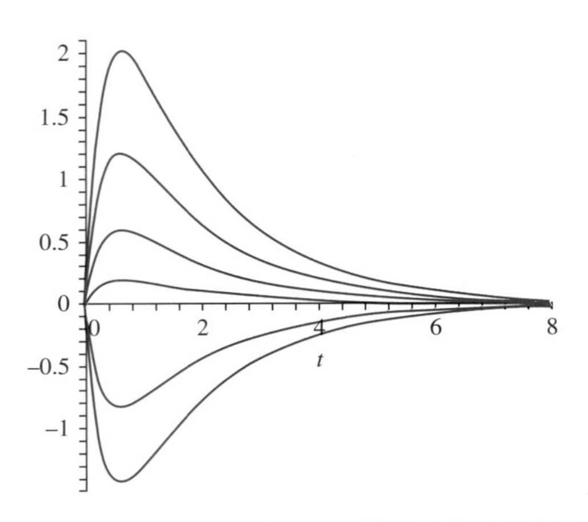

FIGURE A.5 *Graphs in Problem 9, Section 2.4*

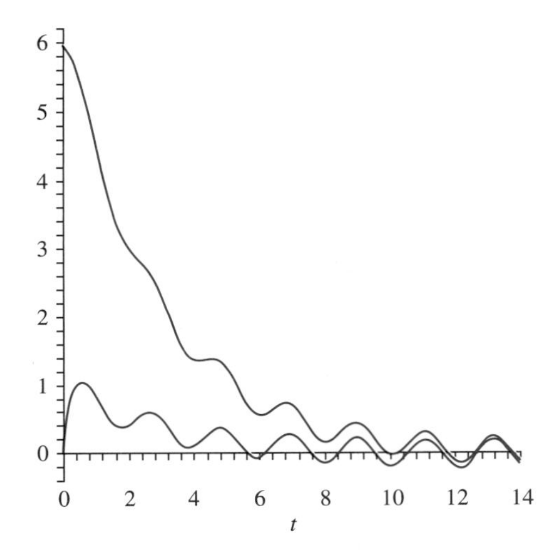

FIGURE A.6 *Graphs in Problem 11, Section 2.4*

11. (a) $y(t) = \frac{1}{373}e^{-3t}\left(2266\cosh(\sqrt{7}t) + \frac{6582}{\sqrt{7}}\sinh(\sqrt{7}t)\right)$

$+\frac{1}{373}(-28\cos(3t) + 72\sin(3t))$

(b) $y(t) = \frac{1}{373}e^{-3t}\left(29\cosh(\sqrt{7}t) + \frac{2106}{\sqrt{7}}\sinh(\sqrt{7}t)\right)$

$+\frac{1}{373}(-28\cos(3t) + 72\sin(3t))$

Graphs are shown in Figure A.6.

13. At most once; a condition on $y(0)$ alone is not enough to guarantee that the bob never passes through the origin. A condition on $y'(0)$ is also needed.

15. Obtain $\omega = \sqrt{4km - c^2}/2m$, so increasing c decreases the frequency ω.

17. (a) $y(t) = \frac{1}{15}e^{-t/2}\left(98\cos(\sqrt{11}t/2) + \frac{74}{\sqrt{11}}\sin(\sqrt{11}t/2)\right)$

$+\frac{1}{15}(-8\cos(3t) + 4\sin(3t))$

(b) $y(t) = \frac{1}{15}e^{-t/2}\left(8\cos(\sqrt{11}t/2) + \frac{164}{\sqrt{11}}\sin(\sqrt{11}t/2)\right)$

$+\frac{1}{15}(-8\cos(3t) + 4\sin(3t))$

Graphs are given in Figure A.7.

Section 2.5 Euler's Differential Equation

1. $y = c_1x^{-2} + c_2x^{-3}$ **3.** $y = x^{-12}(c_1 + c_2\ln(x))$

5. $y = c_1x^4 + c_2x^{-4}$ **7.** $y = c_1\cos(2\ln(x)) + c_2\sin(2\ln(x))$

9. $y = c_1x^2 + c_2x^{-3}$

11. $y = x^2(4 - 3\ln(x))$ **13.** $y = 3x^6 - 2x^4$

15. $y = \frac{7}{10}\left(\frac{x}{2}\right)^3 + \frac{3}{10}\left(\frac{x}{2}\right)^{-7}$

17. Let $x = e^t$ and $Y(t) = y(x^t)$. This transforms the Euler equation $x^2y'' + Axy' + By = 0$ to the constant coefficient equation $Y''(t) + (A-1)Y'(t) + BY(t) = 0$. Solve this and then put $t = \ln(x)$.

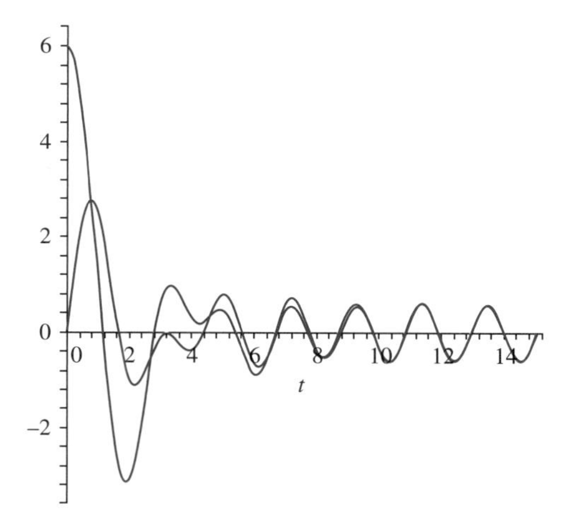

FIGURE A.7 *Graphs in Problem 17, Section 2.4*

CHAPTER THREE THE LAPLACE TRANSFORM

Section 3.1 Definition and Notation

1. $-10\dfrac{1}{(s+4)^3}+\dfrac{3}{s^2+9}$

3. $3\dfrac{s^2-4}{(s^2+4)^2}$

5. $\dfrac{14}{s^2}-\dfrac{7}{s^2+49}$

7. $e^{-42t}-t^3e^{-3t}/6$ **9.** $\cos(8t)$

15. $\mathcal{L}[f](s)=\dfrac{5}{s}\dfrac{1}{1+e^{-3s}}$

17. $\mathcal{L}[f](s)=\dfrac{5e^{-5s}(1-e^{-5s})}{s(1-e^{-25s})}$

19. $\dfrac{E\omega}{s^2+\omega^2}\dfrac{1}{1-e^{-\pi s/\omega}}$

21. $\mathcal{L}[f](s)=\dfrac{h}{s(1+e^{-as})}$

Section 3.2 Solution of Initial Value Problems

1. $y=-\dfrac{4}{17}e^{-4t}+\dfrac{4}{17}\cos(t)+\dfrac{1}{17}\sin(t)$

3. $y=\dfrac{22}{25}e^{2t}-\dfrac{13}{5}te^{2t}+\dfrac{3}{25}\cos(t)-\dfrac{4}{25}\sin(t)$

5. $y=\dfrac{1}{16}+\dfrac{1}{16}t-\dfrac{33}{16}\cos(4t)+\dfrac{15}{64}\sin(4t)$

7. $y=-\dfrac{1}{4}+\dfrac{1}{2}t+\dfrac{17}{4}e^{2t}$

9. $y=\dfrac{1}{4}-\dfrac{13}{4}e^{-4t}$

Section 3.3 Shifting and the Heaviside Function

1. $\dfrac{24}{(s+5)^5}+\dfrac{4}{(s+5)^3}+\dfrac{1}{(s+5)^2}$

3. $\dfrac{(s+1)^2-9}{((s+1)^2+9)^2}$

5. $\dfrac{1}{s+1}-\dfrac{2}{(s+1)^3}+\dfrac{1}{(s+1)^2+1}$

7. $\dfrac{1}{s^2}-\dfrac{11}{s}e^{-3s}-\dfrac{4}{s^2}e^{-3s}$

9. $\dfrac{1}{s^2}-\dfrac{2}{s}-\left(\dfrac{1}{s^2}+\dfrac{1}{s}\right)e^{-16s}$

11. $\dfrac{1}{s}(1-e^{-7s})+\dfrac{s}{s^2+1}\cos(7)e^{-7s}-\dfrac{1}{s^2+1}\sin(7)e^{-7s}$

13. $\dfrac{s}{s^2+1}+\left(\dfrac{2}{s}-\dfrac{s}{s^2+1}-\dfrac{1}{s^2+1}\right)e^{-2\pi s}$

15. $\dfrac{6}{(s+2)^4}-\dfrac{3}{(s+2)^2}+\dfrac{2}{s+2}$

17. $\dfrac{1}{16}(1-\cos(4(t-2)))H(t-2)$

19. $F(s)=\dfrac{1}{(s+3)^2-2}$, so $f(t)=\dfrac{1}{\sqrt{2}}e^{-3t}\sinh\left(\sqrt{2}t\right)$

21. $\dfrac{1}{3}\sin(t-2)H(t-2)$

23. $F(s)=\dfrac{(s+3)-1}{(s+3)^2-8}$ so

$f(t)=e^{-3t}\cosh\left(2\sqrt{2}t\right)-\dfrac{1}{2\sqrt{2}}e^{-3t}\sinh\left(2\sqrt{2}t\right)$

25. $F(s)=\dfrac{1}{(s-2)^2+1}$, so $f(t)=e^{2t}\sin(t)$

27. $y=\left[-\dfrac{1}{4}+\dfrac{1}{12}e^{-2(t-6)}+\dfrac{1}{6}e^{-(t-6)}\cos\left(\sqrt{3}(t-6)\right)\right]H(t-6)$

29. $y=-\dfrac{1}{4}+\dfrac{2}{5}e^{t}-\dfrac{3}{20}\cos(2t)-\dfrac{1}{5}\sin(2t)$

$-\left[-\dfrac{1}{4}+\dfrac{2}{5}e^{t-5}-\dfrac{3}{20}\cos(2(t-5))-\dfrac{1}{5}\sin(2(t-5))\right]H(t-5)$

31. $y=\cos(2t)+\dfrac{3}{4}(1-\cos(2(t-4)))H(t-4)$

33. $E_{\text{out}}=5e^{-4t}+10[1-e^{-4(t-5)}]H(t-5)$

35. $i(t)=\dfrac{k}{R}(1-e^{-Rt/L})-\dfrac{k}{R}(1-e^{-R(t-5)/L})H(t-5)$

Section 3.4 Convolution

1. $\left(\dfrac{1}{2}-\dfrac{1}{2}e^{-2(t-4)}\right)H(t-4)$ **3.** $\dfrac{1}{a^4}[1-\cos(at)]-\dfrac{1}{2a^3}t\sin(at)$

5. $\dfrac{\cos(at)-\cos(bt)}{(b-a)(b+a)}$ if $b^2\neq a^2$; $t\sin(at)/2a$ if $b^2=a^2$

7. $\dfrac{1}{16}[\sinh(2t)-\sin(2t)]$

9. $y(t)=\dfrac{1}{3}\sin(3t)*f(t)-\cos(3t)+\dfrac{1}{3}\sin(3t)$

11. $y(t)=\dfrac{4}{3}e^{t}-\dfrac{1}{4}e^{2t}-\dfrac{1}{12}e^{-2t}-\dfrac{1}{3}e^{t}*f(t)$

$+\dfrac{1}{4}e^{2t}*f(t)+\dfrac{1}{12}e^{-2t}*f(t)$

13. $y(t)=\frac{1}{4}e^{6t}*f(t)-\frac{1}{4}e^{2t}*f(t)+2e^{6t}-5e^{2t}$

15. $y(t)=e^{3t}*f(t)-e^{2t}*f(t)$

17. $f(t)=\cosh(t)$

19. $f(t)=3+\frac{2}{5}\sqrt{15}e^{t/2}\sin\left(\sqrt{15}t/2\right)$

21. $f(t)=\frac{1}{2}e^{-2t}-\frac{3}{2}$

23. $r(t)=(Ak+B)t+\left(\frac{1}{2}kB+C\right)t^2+\frac{1}{3}kCt^3$

25. $r(t)=Akt$

Section 3.5 Impulses and the Delta Function

1. $y=6(e^{-2t}-e^{-t}+te^{t})$

3. $\varphi(t)=(B+9)e^{-2t}-(B+6)e^{-3t}$

5. $y(t)=3[e^{-2(t-2)}-e^{-3(t-2)}]H(t-2)-4[e^{-2(t-5)}-e^{-3(t-5)}]H(t-5)$

9. $y(t)=\sqrt{\frac{m}{k}}v_0\sin\left(\sqrt{\frac{k}{m}}t\right)$

Section 3.6 Solution of Systems

1. $y_1(t)=-1-t+\frac{1}{2}(e^{t}+e^{-t}),\ y_2(t)=-\frac{1}{2}t-\frac{1}{4}t^2,\ y_3(t)=-\frac{1}{3}t-\frac{1}{6}(e^{t}-e^{-t})$

3. $x(t)=1-e^{-t}-2te^{-t},\ y(t)=1-e^{-t}$

5. $x(t)=t-1+e^{-t}\cos(t),\ y(t)=t^2-t+e^{-t}\sin(t)$

7. $x(t)=\frac{1}{2}(t+t^2)+\frac{3}{4}(1-e^{2t/3}),\ y(t)=t+\frac{3}{2}(1-e^{2t/3})$

9. $x(t)=\frac{1}{3}t+\frac{4}{9}(1-e^{3t/4}),\ y(t)=\frac{2}{3}(-1+e^{3t/4})$

11. $x(t)=-t-2+2e^{t/2},\ y(t)=-t-1+e^{t/2}$

13. The loop currents satisfy

$$5i_1'-5i_1-5i_2'=1-H(t-4)\sin(2(t-4)),$$
$$-5i_1'+5i_2'+5i_2=0.$$

The currents are

$$i_1(t)=\frac{1}{5}\left(1-\frac{1}{2}e^{-t/2}\right)$$
$$-\frac{2}{85}\left(e^{-(t-4)/2}-\cos(2(t-4))+\frac{9}{2}\sin(2(t-4))\right)H(t-4)$$
$$i_2(t)=\frac{1}{10}e^{-t/2}+\frac{2}{85}\left(e^{-(t-4)/2}\right.$$
$$\left.-\cos(2(t-4))-4\sin(2(t-4))\right)H(t-4).$$

15. The system is

$$x_1''+8x_1-2x_2=1-H(t-2),\ -2x_1+x_2''+5x_2=0.$$

The solution is

$$x_1(t)=\frac{5}{36}-\frac{1}{20}\cos(2t)-\frac{4}{45}\cos(3t)$$
$$-\left[\frac{5}{36}-\frac{1}{20}\cos(2(t-2))-\frac{4}{45}\cos(3(t-2))\right]H(t-2),$$

$$x_2(t) = \frac{1}{18} - \frac{1}{10}\cos(2t) + \frac{2}{45}\cos(3t)$$
$$- \left[\frac{1}{18} - \frac{1}{10}\cos(2(t-2)) + \frac{2}{45}\cos(3(t-2))\right] H(t-2)$$

17. The equations of motion are

$$m_1 y_1'' = k(y_2 - y_1), m_2 y_2'' = k(y_1 - y_2)$$

with initial conditions $y_1(0) = y_1'(0) = y_2'(0) = 0$, $y_2(0) = d$. Transform these to find that $Y_1(s)$ and $Y_2(s)$ have quadratic factors

$$s^2 + \frac{m_1 + m_2}{m_1 m_2} k$$

in their denominators, implying that the motion has frequency $\omega = \sqrt{(m_1 + m_2)k/m_1 m_2}$, hence period

$$2\pi \sqrt{\frac{m_1 m_2}{m_1 + m_2}}.$$

19. With $E(t) = 5\delta(t-1)$,

$$i_1(t) = \frac{1}{10}\left[e^{-(t-1)} + 3e^{-(t-1)/6}\right] H(t-1)$$
$$i_2(t) = \frac{1}{10}\left[-e^{-(t-1)} + e^{-(t-1)/6}\right] H(t-1)$$

21. $x_1(t) = e^{-3t/50} + 9e^{-t/100} + 3(e^{-(t-3)/100} - e^{-3(t-3)/50})H(t-3)$
$x_2(t) = -e^{-3t/50} + 6e^{-t/100} + (3e^{-(t-3)/50} + 2e^{-(t-3)/100})H(t-3)$

Section 3.7 Polynomial Coefficients

1. $y = \frac{3}{2}t^2$
3. $y = 4$
5. With $u = 1/t$, obtain $-z'(u) - 2z = 2$ where $z(u) = y(t(u))$. Then $y(t) = -1 + ce^{-2/t}$.
7. $y = ct^2 e^{-t}$
9. $y = 7t^2$
11. With $W(s)$ being inverted, use is made of the formula for the Laplace transform of $n!/s^{n+1}$ for n any nonnegative integer.

CHAPTER FOUR SERIES SOLUTIONS

Section 4.1 Power Series Solutions

1. a_0 and a_1 arbitrary; $a_2 = \frac{1}{2}(3 - a_0)$,

$$a_{n+2} = \frac{n-1}{(n+1)(n+2)} a_n \text{ for } n = 1, 2, \cdots,$$

and

$$y(x) = a_0 + a_1 x$$
$$+ (3 - a_0)\left[\frac{1}{2!}x^2 + \frac{1}{4!}x^4 + \frac{3}{6!}x^6 + \frac{3(5)}{8!}x^8 + \frac{3(5)(7)}{10!}x^{10} + \cdots\right]$$

3. a_0, a_1 arbitrary; $2a_2 + a_1 + 2a_0 = 1$, $6a_3 + 2a_2 + a_1 = 0$, $12a_4 + 3a_3 = -1$,

$$a_n = \frac{-(n-1)a_{n-1} + (n-4)a_{n-2}}{n(n-1)} \text{ for } n = 5, 6, \cdots$$

and

$$y(x) = a_0\left(1 - x^2 + \frac{1}{3}x^3 - \frac{1}{12}x^4 + \frac{1}{30}x^5 - \cdots\right)$$
$$+ a_1\left(x - \frac{1}{2}x^2\right) + \frac{1}{2}x^2 - \frac{1}{6}x^3$$
$$- \frac{1}{24}x^7 - \frac{1}{360}x^6 + \frac{1}{2520}x^7 + \cdots$$

5. a_0, a_1 are arbitrary; $a_2 + a_0 = 0$, $6a_3 + 2a_1 = 1$,

$$a_n = \frac{(n-3)a_{n-3} - 2a_{n-2}}{n(n-1)} \text{ for } n = 4, 5, \cdots,$$

and

$$y(x) = a_0\left[1 - x^2 + \frac{1}{6}x^4 - \frac{1}{10}x^5 - \frac{1}{90}x^6 + \cdots\right]$$
$$+ a_1\left[x - \frac{1}{3}x^3 + \frac{1}{12}x^4 + \frac{1}{30}x^5 - \frac{7}{180}x^6 + \cdots\right]$$
$$\left[\frac{1}{6}x^3 - -\frac{1}{60}x^5 + \frac{1}{60}x^6 + \frac{1}{1260}x^7 - \frac{1}{480}x^8 + \cdots\right]$$

with $a_0 = y(0)$ and $a_1 = y'(0)$. The third bracket is a particular solution for the case $a_0 = a_1 = 0$.

7. a_0 is arbitrary; $a_1 = 1$, $2a_2 - a_0 = -1$, and $a_n = \frac{1}{n}a_{n-2}$ for $n = 3, 4, \cdots$;

$$y(x) = a_0 + x + \frac{1}{3}x^3 + \frac{1}{3 \cdot 5}x^5 + \frac{1}{3 \cdot 5 \cdot 7}x^7 + \cdots$$
$$(a_0 - 1)\left(\frac{1}{2}x^2 + \frac{1}{2 \cdot 4}x^4 + \frac{1}{2 \cdot 4 \cdot 6}x^6 + \cdots\right)$$

9. a_0 is arbitrary; $a_1 + a_0 = 0$, $2a_2 + a_1 = 1$,

$$a_{n+1} = \frac{1}{n+1}(a_{n-2} - a_n) \text{ for } n = 2, 3, \cdots,$$

and

$$y(x) = a_0\left[1 - x + \frac{1}{2!}x^2 + \frac{1}{3!}x^3 - \frac{7}{4!}x^4 + \cdots\right]$$
$$+ \frac{1}{2!}x^2 - \frac{1}{3!}x^3 + \frac{1}{4!}x^4 + \frac{11}{5!}x^5 - \frac{31}{6!}x^6 + \cdots$$

Section 4.2 Frobenius Solutions

1. $y_1(x) = c_0\left[x^2 + \frac{1}{3!}x^4 + \frac{1}{5!}x^6 + \frac{1}{7!}x^8 + \cdots\right] = c_0 x \sinh(x)$,

$$y_2(x) = c_0^*\left[x - x^2 + \frac{1}{2!}x^3 - \frac{1}{3!}x^4 + \frac{1}{4!}x^5 - \cdots\right] = c_0^* x e^{-x}$$

3. $y_1(x) = c_0\left[x^{1/2} - \frac{1}{2(1!)(3)}x^{3/2} + \frac{1}{2^2(2!)(3)(5)}x^{5/2}\right.$

$$\left. - \frac{1}{2^3(3!)(3)(5)(7)}x^{7/2} + \frac{1}{2^4(4!)(3)(5)(7)(9)}x^{9/2} + \cdots\right]$$
$$= c_0 x^{1/2}\left[1 + \sum_{n=1}^{\infty} \frac{(-1)^n}{2^n n!(3 \cdot 5 \cdots (2n+1))}x^n\right],$$

$$y_2(x) = c_0^*\left[1 - \frac{1}{2}x + \frac{1}{2^2(3!)(3)(5)}x^3 + \frac{1}{2^4(4!)(3)(5)(7)}x^4 + \cdots\right]$$

$$= c_0^*\left[1 + \sum_{n=1}^{\infty} \frac{(-1)^n}{2^n n!(1 \cdot 3 \cdots (2n-1))}x^n\right]$$

5. $y_1(x) = c_0(1-x)$, $y_2(x) = c_0^*\left[1 + \frac{1}{2}(x-1)\ln((x-2)/x)\right]$

7. $y_1(x) = c_0[x^4 + 2x^5 + 3x^6 + 4x^7 + \cdots]$
$= c_0 \frac{x^4}{(1-x)^2}$, $y_2(x) = c_0^* \frac{3-4x}{(1-x)^2}$

9. $y_1(x) = c_0(1-x)$,
$$y_2(x) = c_0^*\left[(1-x)\ln(x) + 3x + \frac{1}{4}x^2 + \frac{1}{36}x^3 + \frac{1}{288}x^4 + \frac{1}{2400}x^5 + \cdots\right]$$

CHAPTER FIVE APPROXIMATIONS OF SOLUTIONS

Section 5.1 Direction Fields

1. Figure A.8 3. Figure A.9 5. Figure A.10

Section 5.2 Euler's Method

1. See Table A.1.
In this case the exact solution $y = 5e^{3x^2/2}$ increases so rapidly that the Euler approximations are increasingly inaccurate as x increases from 0.
3. See Table A.2.
5. Approximate values are given in Table A.3.

Section 5.3 Taylor and Modified Euler Methods

1. See Table A.4. 3. See Table A.5. 5. See Table A.6.

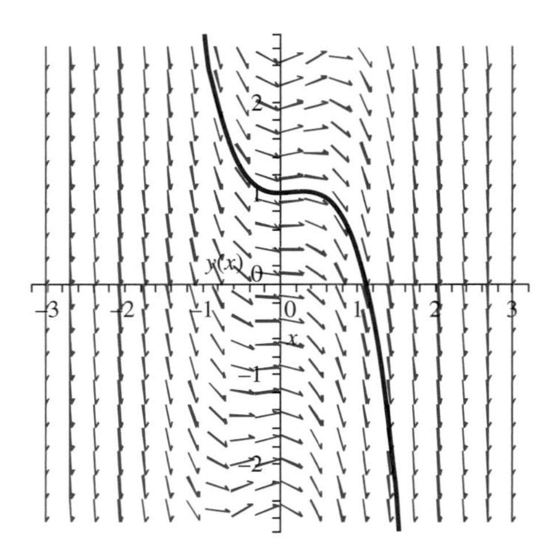

FIGURE A.8 *Direction field and solution for Problem 1, Section 5.1*

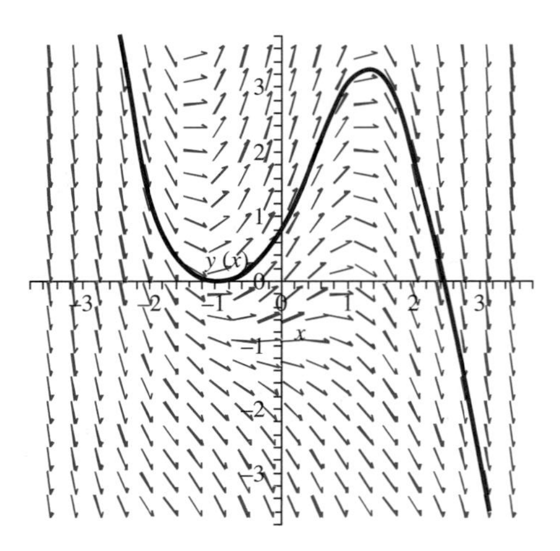

FIGURE A.9 *Direction field and solution for Problem 3, Section 5.1*

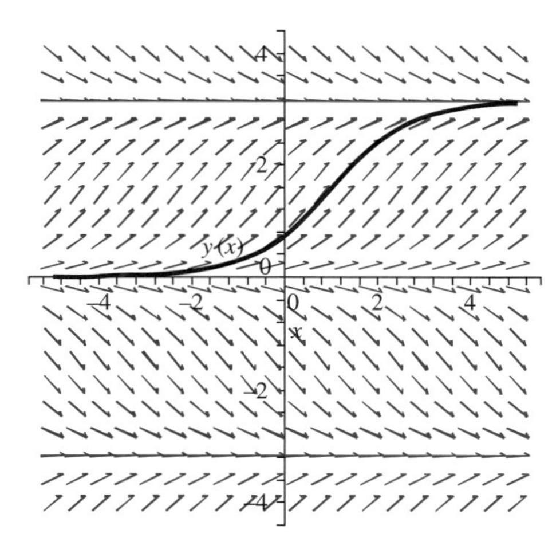

FIGURE A.10 *Direction field and solution for Problem 5, Section 5.1*

TABLE A.1 *Euler Approximations in Problem 1, Section 5.2*

x	$y_{app}(x_k)$	$y(x_k)$
0.00	5	5
0.05	5	5.018785200
0.10	5.0375	5.075565325
0.15	5.1130625	5.171629965
0.20	5.228106406	5.309182735
0.25	5.384949598	5.491425700
0.30	7.404305698	5.722683920
0.35	10.73624326	6.008576785
0.40	16.37277097	6.356245750
0.45	26.19643355	6.774651405
0.50	43.87902620	7.274957075

TABLE A.2 *Euler Approximations in Problem 3, Section 5.2*

x	$y_{app}(x_k)$	$y(x_k)$
1	−2	−2
1.05	−2.127015115	−2.129163318
1.10	−2.258244233	−2.262726023
1.15	−2.393836450	−2.400852694
1.20	−2.533952645	−2.543722054
1.25	−2.678768165	−2.691527843
1.30	−2.828472691	−2.844479697
1.35	−2.983271267	−3.002804084
1.40	−3.143385165	−3.166745253
1.45	−3.309052780	−3.336566227
1.50	−3.480530557	−3.512549830

TABLE A.3 *Euler Approximations in Problem 5, Section 5.2*

x	$y_{app}(x_k)$	$y(x_k)$
0.0	1	1
0.05	1	1.001250521
0.10	1.002498958	1.001250521
0.15	1.007503103	1.005008335
0.20	1.015031072	1.020133420
0.25	1.025113849	1.031575844
0.30	1.037794710	1.045675942
0.35	1.053129175	1.062502832
0.40	1.071184959	1.082138316
0.45	1.092041913	1.104676904
0.50	1.115791943	1.130225803

TABLE A.4 *Runge-Kutta Approximation in Problem 1, Section 5.3*

x	y_k
0	1
0.2	1.26466198
0.4	1.45391187
0.6	1.58485267
0.8	1.67218108
1.0	1.7274517
1.2	1.75945615
1.4	1.77481079
1.6	1.7784748
1.8	1.77415235
2.0	1.76459409

TABLE A.5 *Runge-Kutta Approximation in Problem 3, Section 5.3*

x	y_k
0	4
0.2	3.43866949
0.4	2.94940876
0.6	2.52453424
0.8	2.15677984
1.0	1.83939807
1.2	1.56621078
1.4	1.33162448
1.6	1.13062448
1.8	0.958734358
2.0	0.812012458

TABLE A.6 Runge-Kutta Approximation in Problem 5, Section 5.3

x	y_k
0.0	2
0.2	2.16257799
0.4	2.27783452
0.6	2.34198641
0.8	2.35938954
1.0	2.33750216
1.2	2.28392071
1.4	2.20519759
1.6	2.106598
1.8	2.99222519
2.0	2.8652422

CHAPTER SIX ALGEBRA AND GEOMETRY OF VECTORS

Section 6.1 Vectors in the Plane and 3-Space

In Problems 1 through 5, answers given, are, $\mathbf{F}+\mathbf{G}$, $\mathbf{F}-\mathbf{G}$, $2\mathbf{F}$, $3\mathbf{G}$, and $\|\mathbf{F}\|$, respectively.

1. $3\mathbf{i}-\mathbf{k}, \mathbf{i}-10\mathbf{j}+\mathbf{k},$
$4\mathbf{i}-10\mathbf{j}, 3\mathbf{i}+15\mathbf{j}-3\mathbf{k}, \sqrt{29}$

3. $3\mathbf{i}-\mathbf{j}+3\mathbf{k}, -\mathbf{i}+3\mathbf{j}-\mathbf{k},$
$2\mathbf{i}+2\mathbf{j}+2\mathbf{k}, 6\mathbf{i}-6\mathbf{j}+6\mathbf{k}, \sqrt{3}$

5. $(2+\sqrt{2})\mathbf{i}+3\mathbf{j}, (2-\sqrt{2})\mathbf{i}-9\mathbf{j}+10\mathbf{k},$
$4\mathbf{i}-6\mathbf{j}+10\mathbf{k}, 3\sqrt{2}\mathbf{i}+18\mathbf{j}-15\mathbf{k}, \sqrt{38}$

7. $\frac{4}{9}(-4\mathbf{i}+7\mathbf{j}+4\mathbf{k})$

9. $\frac{3}{\sqrt{5}}(-5\mathbf{i}-4\mathbf{j}+2\mathbf{k})$

11. $x=0, y=1-t, z=3-2t, -\infty<t<\infty$

13. $x=2-3t, y=-3+9t, z=6-2t, -\infty<t<\infty$

15. $x=3-6t, y=t, z=0, -\infty<t<\infty$

Section 6.2 The Dot Product

1. $-23, \cos(\theta)=-23/\sqrt{29}\sqrt{41}$, not orthogonal

3. $-18, \cos(\theta)=-9/10$, not orthogonal

5. $2, \cos(\theta)=2/\sqrt{14}$, not orthogonal

7. $4x-3y+2z=25$

9. $7x+6y-5z=-26$

11. $3x-y+4z=4$

13. $\frac{1}{62}(2\mathbf{i}+7\mathbf{j}-3\mathbf{k})$

15. $-\frac{9}{14}(-3\mathbf{i}+2\mathbf{j}-\mathbf{k})$

Section 6.3 The Cross Product

1. $\mathbf{F}\times\mathbf{G}=-8\mathbf{i}-12\mathbf{j}-5\mathbf{k}$

3. $\mathbf{F}\times\mathbf{G}=8\mathbf{i}+2\mathbf{j}+12\mathbf{k}$

5. Not collinear, $29x+37y-12z=30$

7. Not collinear, $x - 2y + z = 3$ **9.** Not collinear, $2x - 11y + z = 0$

11. $\mathbf{i} - \mathbf{j} + 2\mathbf{k}$

13. $\| \mathbf{F} \times \mathbf{G} \| = \| \mathbf{F} \| \| \mathbf{G} \| \sin(\theta)$, and this is the area of a parallelogram having incident sides of length $\| \mathbf{F} \|$ and $\| \mathbf{G} \|$ and incident angle θ.

Section 6.4 The Vector Space R^n

1. Independent **3.** Dependent **5.** Independent

7. Independent **9.** Dependent

11. The vector $< 0, 1, 0, 2, 0, 3, 0 >$ forms a basis and the dimension is 1.

13. A basis consists of $< 1, 0, 0, -1 >$ and $< 0, 1, -1, 0 >$ and the dimension is 2.

15. A basis consists of $< 1, 0, 0, 0 >$, $< 0, 0, 1, 0 >$ and $< 0, 0, 0, 1 >$. The dimension is 3.

17.
$$\mathbf{X} = \frac{1}{4} < 1, 1, 1, 1, 0 > + \frac{1}{2} < -1, 1, 0, 0, 0 > - \frac{11}{4} < 1, 1, -1, -1, 0 > + < 0, 0, 2, -2, 0 > -2 < 0, 0, 0, 0, 2 >$$

19. Since a basis spans the space, form some numbers $a_1, \cdots, c_k$,

$$\mathbf{U} = a_1\mathbf{V}_1 + a_2\mathbf{V}_2 + \cdots + a_k\mathbf{V}_k.$$

If $\mathbf{U} \neq \mathbf{O}$, some $a_j \neq 0$, so

$$\mathbf{U} - a_1\mathbf{V}_1 - \cdots - a_k\mathbf{V}_k = \mathbf{O},$$

and therefore, $\mathbf{U}, \mathbf{V}_1, \cdots, \mathbf{V}_k$ are linearly dependent by Theorem 6.1(1). If $\mathbf{U} = \mathbf{O}$, then

$$\mathbf{U} - 0\mathbf{V}_1 - \cdots - 0\mathbf{V}_k$$

again shows by Theorem 6.1(1) that $\mathbf{U}, \mathbf{V}_1, \cdots, \mathbf{V}_k$ are linearly dependent.

23. If $\mathbf{X} \cdot \mathbf{X} = \mathbf{Y} \cdot \mathbf{Y}$, then

$$(\mathbf{X} - \mathbf{Y}) \cdot (\mathbf{X} + \mathbf{Y}) = \mathbf{X} \cdot \mathbf{X} + \mathbf{X} \cdot \mathbf{Y} + \mathbf{Y} \cdot \mathbf{X} - \mathbf{Y} \cdot \mathbf{Y} = 0.$$

25. Write $\mathbf{X} = \sum_{j=1}^n (\mathbf{X} \cdot \mathbf{V}_j)\mathbf{V}_j$. Then

$$\begin{aligned}
\| \mathbf{X} \|^2 &= \mathbf{X} \cdot \mathbf{X} \\
&= \left(\sum_{j=1}^n (\mathbf{X} \cdot \mathbf{V}_j)\mathbf{V}_j \right) \cdot \left(\sum_{k=1}^n (\mathbf{X} \cdot \mathbf{V}_k)\mathbf{V}_k \right) \\
&= \sum_{j=1}^n \sum_{k=1}^n (\mathbf{X} \cdot \mathbf{V}_j)(\mathbf{X} \cdot \mathbf{V}_k) \\
&= \sum_{j=1}^n (\mathbf{X} \cdot \mathbf{V}_j)^2.
\end{aligned}$$

Section 6.5 Orthogonalization

1. $\mathbf{V}_1 = < 0, 0, 1, 1, 0, 0 >$, $\mathbf{V}_2 = < 0, 0, -3/2, 3/2, 0, 0 >$

3. $\mathbf{V}_1 = < 0, 0, 2, 2, 1 >$, $\mathbf{V}_2 = < 0, 0, -1/9, -19/9, 40/9 >$,
$\mathbf{V}_3 = < 0, 1, -341/218, 279/218, 62/218 >$, $\mathbf{V}_4 = < 0, 248/393, 88/393, -24/131, -32/393 >$

5. $\mathbf{V}_1 = < 0, 2, 1, 1 >$, $\mathbf{V}_2 = < 0, 4/3, 13/6, 29/6 >$, $\mathbf{V}_3 = < 0, 7/179, -11/179, 3/179 >$

7. $\mathbf{V}_1 = < 1, 4, 0 >$, $\mathbf{V}_2 = < 52/17, -13/17, 0 >$

Section 6.6 Orthogonal Complements and Projections

1. $\mathbf{u}_S = < 9/2, -1/2, 0, 5/2, -13/2 >$, $\mathbf{u}^{\perp} = < -1/2, -1/2, 3, -1/2, -1/2 >$

3. $\mathbf{u}_S = < 3, 1/2, 3, 1/2, 3, 0, 0 >$, $\mathbf{u}^{\perp} = < 5, 1/2, -2, -1/2, -3, -3, 4 >$

5. $\mathbf{u}_S = < -2, 6, 0, 0 >$, $\mathbf{u}^{\perp} = \mathbf{u} - \mathbf{u}_S = < 0, 0, 1, 7 >$

7. The closest vector is

$$\mathbf{u}_S = \frac{7}{3} < 1, 1, -1, 0, 0 > + < 0, 2, 1, 0, 0 > - \frac{4}{3} < 0, 1, -2, 0, 0 > = < \frac{11}{3}, 3, \frac{-11}{3}, \frac{11}{3}, 0 >$$

9. If $\mathbf{u}_1, \cdots, \mathbf{u}_k$ is a basis for S and $\mathbf{v}_1, \cdots, \mathbf{v}_m$ a basis for $S^{\perp}$, then any vector $\mathbf{w}$ in R^n has a unique representation

$$w = c_1\mathbf{u}_1 + \cdots + c_k\mathbf{u}_k + d_1\mathbf{v}_1 + \cdots + d_m\mathbf{v}_m.$$

Furthermore, these vectors are linearly independent hence form a basis for R^n. Therefore,

$$\dim(S) + \dim(S^{\perp}) = k + m = n.$$

Section 6.7 The Function Space *C[a,b]*

1. $V_1(x) = 1,\ V_2(x) = x - 2/3,\ V_3(x) = x^2 - \frac{1}{2} - \frac{6}{5}(x - \frac{2}{3})$

3. $V_1(x) = e^x,\ V_2(x) = e^{-x} - \frac{2}{e^2-1}e^x$

5. $f_S(x) = \dfrac{\pi^2}{3} - 4\cos(x) + \cos(2x) - \dfrac{1}{2}\cos(3x) + \dfrac{1}{4}\cos(4x);$

7. $$f_S = -\frac{4}{3} + \frac{16}{\pi^2}\cos(\pi x/2) - \frac{4}{\pi^2}\cos(\pi x) + \frac{16}{9\pi^2}\cos(3\pi x/2) + \frac{8}{\pi}\sin(\pi x/2) - \frac{4}{\pi}\sin(\pi x) + \frac{8}{3\pi}\sin(3\pi x/2)$$

CHAPTER SEVEN MATRICES AND SYSTEMS OF LINEAR EQUATIONS

Section 7.1 Matrices

1. $\begin{pmatrix} -36 & 0 & 68 & 196 & 20 \\ 128 & -40 & -36 & -8 & 72 \end{pmatrix}$

3. $\begin{pmatrix} 14 & -2 & 6 \\ 10 & -5 & -6 \\ -26 & -43 & -8 \end{pmatrix}$

5. $\begin{pmatrix} 2 + 2x - x^2 & -12x + (1-x)(x + e^x + 2\cos(x)) \\ 4 + 2x + 2e^x + 2xe^x & -22 - 2x + e^{2x} + 2e^x\cos(x) \end{pmatrix}$

7. $\mathbf{AB} = (115)$ and $\mathbf{BA} = \begin{pmatrix} 3 & -18 & -6 & -42 & 66 \\ -2 & 12 & 4 & 28 & -44 \\ -6 & 36 & 12 & 84 & -132 \\ 0 & 0 & 0 & 0 & 0 \\ 4 & -24 & -8 & -56 & 88 \end{pmatrix}$

9. $\mathbf{AB} = \begin{pmatrix} -10 & -34 & -16 & -30 & -14 \\ 10 & -2 & -11 & -8 & -45 \\ -5 & 1 & 15 & 61 & -63 \end{pmatrix}$, **BA** is not defined.

11. **BA** is not defined, $\mathbf{AB} = \begin{pmatrix} 39 & -84 & 21 \\ -23 & 38 & 3 \end{pmatrix}$

13. **AB** is not defined; $\mathbf{BA} = \begin{pmatrix} 410 & 36 & -56 & 227 \\ 17 & 253 & 40 & -1 \end{pmatrix}$

15. **AB** is not defined, $\mathbf{BA} = \begin{pmatrix} -16 & -13 & -5 \end{pmatrix}$

17. **AB** is not defined, **BA** is 4×2

19. **AB** is not defined, **BA** is 7×6

21. **AB** is 14×14, **BA** is 21×21

23. The number of $v_1 - v_4$ walks of length 3 is 4, of length 4, 18. The number of $v_2 - v_3$ walks of length 3 is 9. The number of distinct $v_2 - v_4$ walks of length 4 is 26.

25. The number of $v_4 - v_5$ walks of length 2 is 2, the number of $v_2 - v_3$ walks of length 3 is 10, the number of $v_1 - v_2$ walks of length 4 is 32, the number of $v_4 - v_5$ walks of length 4 is 32.

27. Let $\mathcal{M}_{nm}$ be the set of all $n \times m$ real matrices. We have an addition of matrices and a multiplication of an $n \times m$ matrix by a real number. There is a zero matrix (all entries zero), and the matrix $-\mathbf{A} = [-a_{ij}]$ serves as an additive inverse of $\mathbf{A}$. Furthermore, for any real numbers α and β,

$$\begin{aligned}(\alpha+\beta)\mathbf{A} &= \alpha\mathbf{A}+\beta\mathbf{A},\\ (\alpha\beta)\mathbf{A} &= \alpha(\beta\mathbf{A}), \text{ and}\\ \alpha(\mathbf{A}+\mathbf{B}) &= \alpha\mathbf{A}+\alpha\mathbf{B}.\end{aligned}$$

Thus, $\mathcal{M}_{nm}$ has the algebraic properties of a vector space. If we take the rows of an $n \times m$ matrix and simply string then out one written after the other, then we obtain an nm vector. Thus, there is a one-to-one matching of matrices in $\mathcal{M}_{nm}$ and vectors in R^{nm}. This also suggests the dimension of $\mathcal{M}_{nm}$. The nm matrices formed by setting one element equal to 1 and all others zero form a basis for $\mathcal{M}_{nm}$. Thus, $\mathcal{M}_{nm}$ has dimension nm, the same as R^{nm}.

Section 7.2 Elementary Row Operations

1. $\begin{pmatrix} -1 & 0 & 3 & 0 \\ -36 & 28 & -20 & 28 \\ -13 & 3 & 44 & 9 \end{pmatrix}; \Omega = \begin{pmatrix} 1 & 0 & 0 \\ 0 & 0 & 4 \\ 14 & 1 & 0 \end{pmatrix}$

3. $\begin{pmatrix} -2 & 1 & 4 & 2 \\ 0 & \sqrt{3} & 16\sqrt{3} & 3\sqrt{3} \\ 1 & -2 & 4 & 8 \end{pmatrix}; \Omega = \begin{pmatrix} 1 & 0 & 0 \\ 0 & \sqrt{3} & 0 \\ 0 & 0 & 1 \end{pmatrix}$

5. $\begin{pmatrix} 40 & 5 & -15 \\ -2+2\sqrt{13} & 14+9\sqrt{13} & 6+5\sqrt{13} \\ 2 & 9 & 5 \end{pmatrix}; \Omega = \begin{pmatrix} 0 & 5 & 0 \\ 1 & 0 & \sqrt{13} \\ 0 & 0 & 1 \end{pmatrix}$

7. $\begin{pmatrix} 30 & 120 \\ -3+2\sqrt{3} & 15+8\sqrt{3} \end{pmatrix}; \Omega = \begin{pmatrix} 0 & 15 \\ 1 & \sqrt{3} \end{pmatrix}$

11. If $i \neq s$ and $i \neq t$,

$$\begin{aligned}(\mathbf{EA})_{ij} &= (\text{row } i \text{ of } \mathbf{E}) \cdot (\text{column } j \text{ of } \mathbf{A})\\ &= (\mathbf{I}_n\mathbf{A})_{ij} = \mathbf{A}_{ij}.\end{aligned}$$

Next,

$$\begin{aligned}(\mathbf{EA})_{sj} &= (\text{ row } s \text{ of } \mathbf{E}) \cdot \text{column } j \text{ of } \mathbf{A})\\ &= (\text{row } t \text{ of } \mathbf{I}_n) \cdot (\text{column } j \text{ of } \mathbf{A})\\ &= \mathbf{A}_{tj} = \mathbf{B}_{sj}.\end{aligned}$$

Similarly, $(\mathbf{EA})_{tj} = \mathbf{A}_{sj} = \mathbf{B}_{tj}$.

Section 7.3 The Row Echelon Form

1. $\mathbf{A}_R = \begin{pmatrix} 1 & 0 & 0 \\ 0 & 1 & 0 \\ 0 & 0 & 1 \end{pmatrix}; \Omega = \frac{1}{270}\begin{pmatrix} -8 & -2 & 38 \\ 37 & 43 & -7 \\ 19 & -29 & 11 \end{pmatrix}$

3. $\mathbf{A}_R = \begin{pmatrix} 1 & 0 & 0 & 0 \\ 0 & 1 & 3/2 & 1/2 \end{pmatrix}; \Omega = \begin{pmatrix} 0 & 1 \\ 1/2 & 1/2 \end{pmatrix}$

5. $\mathbf{A}_R = \begin{pmatrix} 1 & 0 & 0 \\ 0 & 1 & 0 \\ 0 & 0 & 1 \end{pmatrix}; \Omega = \begin{pmatrix} 0 & 1/2 & -1 \\ 0 & 0 & 1 \\ -1/7 & 2/7 & -3/7 \end{pmatrix}$

7. $\mathbf{A}_R = \begin{pmatrix} 1 & 0 \\ 0 & 1 \\ 0 & 0 \\ 0 & 0 \end{pmatrix}; \Omega = \begin{pmatrix} 0 & 0 & 1 & -3 \\ 0 & 0 & 0 & 1 \\ 1 & 0 & -6 & 17 \\ 0 & 1 & 0 & 0 \end{pmatrix}$

9. $\mathbf{A}_R = \begin{pmatrix} 1 & -4 & -1 & 0 \\ 0 & 0 & 0 & 1 \\ 0 & 0 & 0 & 0 \\ 0 & 0 & 0 & 0 \end{pmatrix}; \Omega = \begin{pmatrix} -1 & 0 & 0 & 1 \\ 0 & 0 & 0 & 1 \\ 0 & 0 & 1 & 0 \\ 0 & 1 & 0 & 0 \end{pmatrix}$

11. $\mathbf{A}_R = \begin{pmatrix} 1 & 0 & 5 \\ 0 & 1 & 2 \\ 0 & 0 & 0 \end{pmatrix}; \Omega = \begin{pmatrix} 1 & 1 & 0 \\ 0 & 1 & 0 \\ 0 & 0 & 1 \end{pmatrix}$

Section 7.4 Row and Column Spaces

1.

$$\mathbf{A}_R = \begin{pmatrix} 1 & 0 & 0 \\ 0 & 1 & 0 \\ 0 & 0 & 1 \end{pmatrix},$$

so rank($\mathbf{A}$) $= 3$. (b) The row space has basis consisting of $< -3, 2, 2 >$, $< 1, 0, 5 >$ and $< 0, 0, 2 >$. The column space has basis consisting of

$$\begin{pmatrix} -3 \\ 1 \\ 0 \end{pmatrix}, \begin{pmatrix} 2 \\ 0 \\ 0 \end{pmatrix}, \text{ and } \begin{pmatrix} 2 \\ 5 \\ 2 \end{pmatrix}.$$

3. (a)

$$\mathbf{A}_R = \begin{pmatrix} 1 & 0 & -3/5 \\ 0 & 1 & 3/5 \end{pmatrix},$$

so $\mathbf{A}$ has rank 2. (b) $< -4, 1, 3 >$ and $< 2, 2, 0 >$ form a basis for the row space. (c) The column vectors

$$\begin{pmatrix} -4 \\ 2 \end{pmatrix} \text{ and } \begin{pmatrix} 1 \\ 2 \end{pmatrix}$$

form a basis for the column space of $\mathbf{A}$.

5. (a)

$$\mathbf{A}_R = \begin{pmatrix} 1 & 0 \\ 0 & 1 \\ 0 & 0 \end{pmatrix},$$

so rank($\mathbf{A}$) $= 2$. (b) The row vectors $< -3, 1 >$ and $< 2, 2 >$ form a basis for the row space. The column space has basis consisting of

$$\begin{pmatrix} -3 \\ 2 \\ 4 \end{pmatrix} \text{ and } \begin{pmatrix} 1 \\ 2 \\ -3 \end{pmatrix}.$$

7. (a)

$$\mathbf{A}_R = \begin{pmatrix} 1 & 0 & -1/4 & 1/2 \\ 0 & 1 & -5/4 & 1/2 \end{pmatrix},$$

so rank($\mathbf{A}$) $= 2$. (b) The row vectors $< 8, -4, 3, 2 >$ and $< 1, -1, 1, 0 >$ form a basis for the row space. (c) The column space has basis consisting of

$$\begin{pmatrix} 8 \\ 1 \end{pmatrix} \text{ and } \begin{pmatrix} -4 \\ -1 \end{pmatrix}.$$

9. (a)

$$\mathbf{A}_R = \begin{pmatrix} 1 & 0 & 0 \\ 0 & 1 & 0 \\ 0 & 0 & 1 \\ 0 & 0 & 0 \end{pmatrix},$$

so **A** has rank 3. (b) The row vectors $<2,2,1>$, $<1,-1,3>$, and $<0,0,1>$ form a basis for the row space. (c) The column space has a basis consisting of

$$\begin{pmatrix}2\\1\\0\\4\end{pmatrix},\ \begin{pmatrix}2\\-1\\0\\0\end{pmatrix},\ \text{and}\ \begin{pmatrix}1\\3\\1\\7\end{pmatrix}.$$

11.

$$\mathbf{A}_R = \begin{pmatrix}1 & 0 & -11\\0 & 1 & -3\\0 & 0 & 0\end{pmatrix},$$

so **a** has rank 2. (b) $<-2,5,7>$ and $<0,1,-3>$ form a basis for the row space. (c) The column space has basis consisting of

$$\begin{pmatrix}-2\\0\\4\end{pmatrix}\ \text{and}\ \begin{pmatrix}5\\1\\11\end{pmatrix}.$$

13. (a)

$$\Omega_R = \begin{pmatrix}1 & 0 & 0\\0 & 1 & 0\\0 & 0 & 1\end{pmatrix},$$

so **A** has rank 3. (b) $<0,4,3>$, $<6,1,0>$ and $<2,2,2>$ form a basis for the row space of **A**. The column space has basis consisting of

$$\begin{pmatrix}0\\6\\2\end{pmatrix},\ \begin{pmatrix}4\\1\\2\end{pmatrix},\ \text{and}\ \begin{pmatrix}3\\0\\2\end{pmatrix}.$$

15.

$$\begin{aligned}\text{rank}(\mathbf{A}) &= \text{dim row space of } \mathbf{A}\\ = \text{dim col space of } \mathbf{A} &= \text{dim row space of } \mathbf{A}^T\\ &= \text{rank}(\mathbf{A}^t).\end{aligned}$$

Section 7.5 Homogeneous Systems of Equations

1.

$$\alpha\begin{pmatrix}5/14\\11/7\\6/7\\1\end{pmatrix} = \begin{pmatrix}x_1\\x_2\\x_4\\x_5\end{pmatrix}$$

The solution space has dimension 1.

3.

$$\alpha\begin{pmatrix}1\\1\\0\\1\\1\\0\\0\end{pmatrix} + \beta\begin{pmatrix}-2\\-3/2\\2/3\\-4/3\\0\\1\\0\end{pmatrix} + \gamma\begin{pmatrix}0\\1/2\\-3\\0\\0\\0\\1\end{pmatrix}$$

The solution space has dimension 3.

5.

$$\alpha \begin{pmatrix} -5/6 \\ -2/3 \\ -8/3 \\ -2/3 \\ 1 \\ 0 \end{pmatrix} + \beta \begin{pmatrix} -5/9 \\ -10/9 \\ -13/9 \\ -1/9 \\ 0 \\ 1 \end{pmatrix}$$

The solution space has dimension 2.

7.

$$\alpha \begin{pmatrix} 0 \\ 0 \\ 0 \end{pmatrix}$$

(only the trivial solution). The solution space has dimension 0.

9.

$$\alpha \begin{pmatrix} -9/4 \\ -7/4 \\ -5/8 \\ 13/8 \\ 1 \end{pmatrix}$$

The solution space has dimension 1.

11.

$$\alpha \begin{pmatrix} -1 \\ 1 \\ 1 \\ 0 \end{pmatrix} + \beta \begin{pmatrix} 1 \\ -1 \\ 0 \\ 1 \end{pmatrix}$$

The solution space has dimension 2.

15. Yes, if $m -$ rank $(\mathbf{A}) > 0$. For example, the system

$$x_1 + 3x_2 = 0, 2x_1 + 6x_2 = 0, 3x_1 + 9x_3 = 0$$

has the solution $x_1 = -3x_3, x_2 = x_3$ and the solution space has dimension 1.

Section 7.6 Nonhomogeneous Systems

1. $\alpha \begin{pmatrix} 1 \\ 1 \\ 3/2 \\ 1 \\ 0 \\ 0 \end{pmatrix} + \beta \begin{pmatrix} 0 \\ 0 \\ 1/2 \\ 0 \\ 1 \\ 0 \end{pmatrix} + \gamma \begin{pmatrix} -17/2 \\ -6 \\ -51/4 \\ 0 \\ 0 \\ 1 \end{pmatrix} + \begin{pmatrix} 9/2 \\ 3 \\ 25/4 \\ 0 \\ 0 \\ 0 \end{pmatrix}$

3. $\alpha \begin{pmatrix} -1/2 \\ -1 \\ 3 \\ 1 \\ 0 \end{pmatrix} + \beta \begin{pmatrix} -3/4 \\ 1 \\ -2 \\ 0 \\ 1 \end{pmatrix} + \begin{pmatrix} 9/8 \\ 2 \\ 0 \\ 0 \\ 0 \end{pmatrix}$

5. Unique solution

$$\mathbf{X} = \begin{pmatrix} 16/57 \\ 99/57 \\ 23/57 \end{pmatrix}$$

7. $\alpha \begin{pmatrix} -1 \\ 1 \\ 0 \\ 0 \\ 0 \\ 0 \\ 0 \end{pmatrix} + \beta \begin{pmatrix} 1 \\ 0 \\ 0 \\ 1 \\ 0 \\ 0 \\ 0 \end{pmatrix} + \gamma \begin{pmatrix} -3/14 \\ 0 \\ 3/14 \\ 0 \\ 1 \\ 0 \\ 0 \end{pmatrix} + \delta \begin{pmatrix} -1 \\ 0 \\ 0 \\ 0 \\ 0 \\ -1 \\ 0 \end{pmatrix} + \epsilon \begin{pmatrix} 1/14 \\ 0 \\ -1/14 \\ 0 \\ 0 \\ 0 \\ 1 \end{pmatrix} + \begin{pmatrix} -29/7 \\ 0 \\ 1/7 \\ 0 \\ 0 \\ 0 \\ 0 \end{pmatrix}$

9. $\alpha\begin{pmatrix}-19/15\\3\\67/15\\1\end{pmatrix}+\begin{pmatrix}22/15\\-5\\-121/15\\0\end{pmatrix}$

11. $\alpha\begin{pmatrix}2\\2\\7\\3/2\\1\\0\end{pmatrix}+\beta\begin{pmatrix}-2\\-1\\-9/2\\-3/4\\0\\1\end{pmatrix}+\begin{pmatrix}-4\\-4\\-38\\-11/2\\0\\0\end{pmatrix}$

13. Unique solution

$$\begin{pmatrix}1\\1/2\\4\end{pmatrix}$$

15. If $\mathbf{AX}=\mathbf{B}$ is consistent, then there is a solution $\mathbf{C}$. Since $\mathbf{AC}$ is a linear combination of the columns of $\mathbf{A}$ (see Section 7.1.1), so $\mathbf{B}$ is in the column space of $\mathbf{A}$. Conversely, if $\mathbf{B}$ is in the column space of $\mathbf{A}$, then $\mathbf{B}$ is a linear combination

$$a_1A_1+\cdots+a_mA_m$$

of the columns of $\mathbf{A}$, and then $\mathbf{AC}=\mathbf{B}$, where C is the column matrix of the coefficients $a_1,\cdots,a_m$.

Section 7.7 Matrix Inverses

1. $\frac{1}{12}\begin{pmatrix}-2&2\\1&5\end{pmatrix}$

3. $\frac{1}{31}\begin{pmatrix}-6&11&2\\3&10&-1\\1&-7&10\end{pmatrix}$

5. $-\frac{1}{12}\begin{pmatrix}6&-6&0\\-3&-9&2\\3&-3&-2\end{pmatrix}$

7. $\frac{1}{12}\begin{pmatrix}3&-2\\-3&6\end{pmatrix}$ **9.** $\frac{1}{5}\begin{pmatrix}-1&2\\2&1\end{pmatrix}$

11. $\frac{1}{5}\begin{pmatrix}-21\\14\\0\end{pmatrix}$

13. $\mathbf{X}=\mathbf{A}^{-1}\mathbf{B}=\frac{1}{11}\begin{pmatrix}-1&-1&8&4\\-9&2&-5&14\\2&2&-5&3\\3&3&-2&-1\end{pmatrix}\begin{pmatrix}1\\2\\0\\-5\end{pmatrix}=\frac{1}{11}\begin{pmatrix}-23\\-75\\-9\\14\end{pmatrix}$

15. $\frac{1}{7}\begin{pmatrix}22\\27\\30\end{pmatrix}$

Section 7.8 The Method of Least Squares and Data Fitting

1. $\mathbf{X}^*=\begin{pmatrix}-2\\4\end{pmatrix}$

3. $\mathbf{X}^*=\alpha\begin{pmatrix}7/3\\1\\5/3\end{pmatrix}+\begin{pmatrix}-2\\0\\-1\end{pmatrix}$

5. $\mathbf{X}^*=\begin{pmatrix}13/5\\7/5\end{pmatrix}$

7. $y=4.164x-9.267$ **9.** $y=3.88x+0.16$

Section 7.9 LU Factorization

1. $\mathbf{U}=\begin{pmatrix}1&4&2&-1&4\\0&-5&2&0&0\\0&0&88/5&4&6\\0&0&0&195/22&-691/44\end{pmatrix}$ and $\mathbf{L}=\begin{pmatrix}1&0&0&0\\1&1&0&0\\-2&-14/5&1&0\\4&14/5&-63/88&1\end{pmatrix}$

3. $\mathbf{U}=\begin{pmatrix}2&4&-6\\0&-14&25\\0&0&136/7\end{pmatrix}$ and $\mathbf{L}=\begin{pmatrix}1&0&0\\4&1&0\\-2&-6/7&1\end{pmatrix}$

5. $\mathbf{U}=\begin{pmatrix}-2&1&12\\0&-5&13\\0&0&119/5\end{pmatrix}$ and $\mathbf{L}=\begin{pmatrix}1&0&0\\-1&1&0\\-1&-3/5&1\end{pmatrix}$

7.

$$\mathbf{U}=\begin{pmatrix}6&1&-1&3\\0&4/3&5/3&3\\0&0&13/4&13/4\\0&0&0&5\end{pmatrix}\text{ and }\mathbf{L}=\begin{pmatrix}1&0&0&0\\2/3&1&0&0\\-2/3&5/4&1&0\\1/3&-1&4/13&1\end{pmatrix}.$$

Solve $\mathbf{LY}=\mathbf{B}$ and then $\mathbf{UX}=\mathbf{Y}$ to obtain $\mathbf{Y}=\begin{pmatrix}4\\28/3\\-7\\43/13\end{pmatrix}$ and $\mathbf{X}=\begin{pmatrix}-263/130\\537/65\\-233/65\\93/65\end{pmatrix}$.

9. First obtain

$$\mathbf{U}=\begin{pmatrix}4&4&2\\0&-2&5/2\\0&0&21/4\end{pmatrix}\text{ and }\mathbf{L}=\begin{pmatrix}1&0&0\\1/4&1&0\\1/4&-3/2&1\end{pmatrix}.$$

Solve

$$\mathbf{LY}=\begin{pmatrix}1\\0\\1\end{pmatrix}\text{ to obtain }\mathbf{Y}=\begin{pmatrix}1\\-1/4\\3/8\end{pmatrix}.$$

Solve $\mathbf{UX}=\mathbf{Y}$ to obtain

$$\mathbf{X}=\begin{pmatrix}0\\3/14\\1/4\end{pmatrix}.$$

11.

$$\mathbf{U}=\begin{pmatrix}-1&1&1&6\\0&3&2&16\\0&0&17/3&52/3\end{pmatrix}\text{ and }\mathbf{L}=\begin{pmatrix}1&0&0\\-2&1&0\\-1&-1/3&1\end{pmatrix}.$$

Solve $\mathbf{LY}=\mathbf{B}$ and then $\mathbf{UX}=\mathbf{Y}$ to obtain $\mathbf{Y}=\begin{pmatrix}2\\5\\29/3\end{pmatrix}$ and $\mathbf{X}=\alpha\begin{pmatrix}1\\28/3\\26/3\\-17/6\end{pmatrix}+\begin{pmatrix}0\\-5/3\\-1/3\\2/3\end{pmatrix}$.

Section 7.10 Matrices and Linear Transformations

1. T is not linear (the image of a sum of two vectors has last component 2).

3. Not linear because of the $\sin(xy)$ term.

5. T is linear and T(1,0,0) = (3,1,0), T(0,1,0) = (0,−1,0), and T(0,0,1) = (0,0,2), so

$$\mathbf{A}_T=\begin{pmatrix}3&0&0\\1&-1&0\\0&0&2\end{pmatrix}.$$

T is onto and one-to-one because A_T has rank 3 one-to-one. The null space of T has dimension $m -$ rank$\mathbf{A}_T = 3 - 3 = 0$.

7. Nonlinear (because of the $2xy$ term)

9. T is linear and

$$\mathbf{A}_T = \begin{pmatrix} 1 & 0 & 0 & -1 & 0 \\ 0 & 1 & -1 & 0 & 0 \\ 0 & 0 & 0 & 1 & 1 \end{pmatrix}.$$

T is onto but not one-to-one; $\mathbf{A}_T$ has rank 3.

CHAPTER EIGHT DETERMINANTS

Section 8.1 Definition of the Determinant

5. $|\mathbf{A}| = (-1)^n|\mathbf{A}^t| = (-1)^n|\mathbf{A}|$.
Since $(-1)^n = -1$ if n is odd, then $|\mathbf{A}| = -|\mathbf{A}|$, so $|\mathbf{A}| = 0$.

7. $|\mathbf{B}| = \alpha^n|\mathbf{A}|$

Section 8.2 Evaluation of Determinants I

1. 72 **3.** −2247 **5.** −14

7. −22 **9.** −122

Section 8.3 Evaluation of Determinants II

1. 1693 **3.** −773 **5.** −152

7. 32 **9.** 3

13.

$$\begin{vmatrix} 1 & \alpha & \alpha^2 \\ 1 & \beta & \beta^2 \\ 1 & \gamma & \gamma^2 \end{vmatrix} = \begin{vmatrix} 1 & \alpha & \alpha^2 \\ 0 & \beta-\alpha & \beta^2-\alpha^2 \\ 0 & \gamma-\alpha & \gamma^2-\alpha^2 \end{vmatrix}$$

$$= (\beta-\alpha)(\gamma-\alpha)\begin{vmatrix} 1 & \alpha & \alpha^2 \\ 0 & 1 & \beta+\alpha \\ 0 & 1 & \gamma+\alpha \end{vmatrix} = (\beta-\alpha)(\gamma-\alpha)\begin{vmatrix} 1 & \beta+\alpha \\ 1 & \gamma+\alpha, \end{vmatrix}$$

leading to the value $(\beta-\alpha)(\gamma-\alpha)(\gamma-\beta)$.

Section 8.4 A Determinant Formula for $\mathbf{A}^{-1}$

1. $\frac{1}{32}\begin{pmatrix} 5 & 3 & 1 \\ -8 & -24 & 24 \\ -2 & -14 & 6 \end{pmatrix}$

3. $\frac{1}{29}\begin{pmatrix} -1 & 25 & -21 \\ -8 & -3 & 6 \\ -1 & -4 & 8 \end{pmatrix}$

5. $\frac{1}{378}\begin{pmatrix} 210 & -42 & 42 & 0 \\ 899 & -124 & 223 & -135 \\ 275 & -64 & 109 & -27 \\ -601 & 122 & -131 & 81 \end{pmatrix}$

7. $\frac{1}{13}\begin{pmatrix} 6 & 1 \\ -1 & 2 \end{pmatrix}$

9. $\frac{1}{5}\begin{pmatrix} -4 & 1 \\ 1 & 1 \end{pmatrix}$

Section 8.5 Cramer's Rule

1. $x_1 = 5/6, x_2 = -10/3, x_3 = -5/6$ **3.** $x_1 = -86, x_2 = -109/2, x_3 = -43/2, x_4 = 37/2$
5. $x_1 = 33/93, x_2 = -409/33, x_3 = -1/93, x_4 = 116/93$
7. $x_1 = -1/2, x_2 = -19/22, x_3 = 2/11$ **9.** $x_1 = -11/47, x_2 = -100/47,$

Section 8.6 The Matrix Tree Theorem

1. 61
3.

$$\mathbf{T} = \begin{pmatrix} 2 & 0 & -1 & 0 & -1 \\ 0 & 2 & -1 & -1 & 0 \\ -1 & -1 & 4 & -1 & -1 \\ 0 & -1 & -1 & 3 & -1 \\ -1 & 0 & -1 & -1 & 3 \end{pmatrix}$$

and the number of spanning trees is 21.
5. 61

CHAPTER NINE EIGENVALUES AND DIAGONALIZATION

Section 9.1 Eigenvalues and Eigenvectors

1. $p_{\mathbf{A}}(\lambda) = (\lambda - 1)(\lambda - 2)(\lambda^2 + \lambda - 13)$,

$$1, \begin{pmatrix} -2 \\ -11 \\ 0 \\ 1 \end{pmatrix}; 2, \begin{pmatrix} 0 \\ 0 \\ 1 \\ 0 \end{pmatrix};$$

$$(-1 + \sqrt{53})/2, \begin{pmatrix} \sqrt{53} - 7 \\ 0 \\ 0 \\ 2 \end{pmatrix}; (-1 - \sqrt{53})/2, \begin{pmatrix} -\sqrt{53} - 7 \\ 0 \\ 0 \\ 2 \end{pmatrix}$$

The Gerschgorin circles have radius 2, center $(-4, 0)$ and radius 1 and center $(3, 0)$.
3. $p_{\mathbf{A}}(\lambda) = (\lambda + 14)(\lambda - 2)^2$,

$$-14, \begin{pmatrix} -16 \\ 0 \\ 1 \end{pmatrix}; 2, 2, \begin{pmatrix} 0 \\ 0 \\ 1 \end{pmatrix}$$

The eigenvalue 2 of multiplicity 2 does not have two linearly independent eigenvectors. The Gerschgorin circles have radius 1 and center $(-14, 0)$ and radius 1 and center $(2, 0)$.
5. $p_{\mathbf{A}}(\lambda) = \lambda^3 - 5\lambda^2 + 6\lambda$,

$$0, \begin{pmatrix} 0 \\ 1 \\ 0 \end{pmatrix}; 2, \begin{pmatrix} 2 \\ 1 \\ 0 \end{pmatrix}; 3, \begin{pmatrix} 0 \\ 2 \\ 3 \end{pmatrix}$$

The Gerschgorin circle has radius 3, center $(0, 0)$.
7. $p_{\mathbf{A}}(\lambda) = \lambda^3(\lambda + 3)$,

$$0, 0, \begin{pmatrix} 1 \\ 0 \\ 3 \end{pmatrix}; -3, \begin{pmatrix} 1 \\ 0 \\ 0 \end{pmatrix}$$

The Gerschgorin circle has radius 2, center $(-3, 0)$.
9. $p_{\mathbf{A}}(\lambda) = \lambda^2 - 3\lambda + 14$;

$$\frac{1}{2}(3 + \sqrt{47}i), \begin{pmatrix} \dfrac{-1 + \sqrt{47}i}{4} \end{pmatrix}, \frac{1}{2}(3 - \sqrt{47}i), \begin{pmatrix} \dfrac{-1 - \sqrt{47}i}{4} \end{pmatrix}$$

Gerschgorin circles have radius 6, center $(1, 0)$ and radius 2, center $(2, 0)$.

11. $p_{\mathbf{A}}(\lambda) = \lambda(\lambda^2 - 8\lambda + 7)$,

$$0, \begin{pmatrix} 14 \\ 7 \\ 10 \end{pmatrix}; 1, \begin{pmatrix} 6 \\ 0 \\ 5 \end{pmatrix}; 7, \begin{pmatrix} 0 \\ 0 \\ 1 \end{pmatrix}$$

The Gerschgorin circles have radius 2 and center $(1, 0)$ and radius 5 and center $(7, 0)$.

13. $p_{\mathbf{A}}(\lambda) = \lambda^2 - 2\lambda - 5$; eigenvalues and corresponding eigenvectors are

$$1 + \sqrt{6}, \begin{pmatrix} \sqrt{6} \\ 2 \end{pmatrix}; 1 - \sqrt{6}, \begin{pmatrix} -\sqrt{6} \\ 2 \end{pmatrix}; 3.$$

The Gerschgorin circles are of radius 3 about $(1, 0)$ and radius 2 about $(1, 0)$.

15. $p_{\mathbf{A}}(\lambda) = \lambda^2 + 3\lambda - 10$;

$$-5, \begin{pmatrix} 7 \\ -1 \end{pmatrix}; 2, \begin{pmatrix} 0 \\ 1 \end{pmatrix}$$

The Gerschgorin circle has radius 1, center $(2, 0)$.

17. $p_{\mathbf{A}}(\lambda) = \lambda^2 - 10\lambda - 23$,

$$5 + \sqrt{2}, \begin{pmatrix} 1 + \sqrt{2} \\ 1 \end{pmatrix}; 5 - \sqrt{2}, \begin{pmatrix} 1 - \sqrt{2} \\ 1 \end{pmatrix}$$

19. $p_{\mathbf{A}}(\lambda) = (\lambda - 3)(\lambda^2 + 2\lambda - 1)$,

$$3, \begin{pmatrix} 0 \\ 0 \\ 1 \end{pmatrix}; -1 + \sqrt{2}, \begin{pmatrix} 1 + \sqrt{2} \\ 1 \\ 0 \end{pmatrix}; -1 - \sqrt{2}, \begin{pmatrix} 1 - \sqrt{2} \\ 0 \\ 4 \end{pmatrix}$$

21. $p_{\mathbf{A}}(\lambda) = \lambda^2 - 5\lambda$,

$$0, \begin{pmatrix} 1 \\ 2 \end{pmatrix}; 5, \begin{pmatrix} -2 \\ 1 \end{pmatrix}$$

23. If $\mathbf{AE} = \lambda\mathbf{E}$, then

$$\mathbf{A}^2\mathbf{E} = \mathbf{A}(\mathbf{AE}) = \mathbf{A}(\lambda\mathbf{E}) = \lambda\mathbf{AE} = \lambda^2\mathbf{E}.$$

Repetition of this argument yields the general conclusion $\mathbf{A}^k\mathbf{E} = \lambda^k\mathbf{E}$.

Section 9.2 Diagonalization

In Problems 1 through 9, **P** is given, or it is stated that the matrix is not diagonalizable.

1. $\begin{pmatrix} 0 & 5 & 0 \\ 1 & 1 & -3 \\ 0 & 0 & 2 \end{pmatrix}$

3. **A** is not diagonalizable

5. $\begin{pmatrix} 1 & 0 & 0 & 0 \\ 0 & 1 & (2 - 3\sqrt{5})/41 & (2 + 3\sqrt{5})/41 \\ 0 & 0 & (-1 + \sqrt{5})/2 & (-1 - \sqrt{5})/2 \\ 0 & 0 & 1 & 1 \end{pmatrix}$

7. $\begin{pmatrix} -3 + \sqrt{7}i & -3 - \sqrt{7}i \\ 8 & 8 \end{pmatrix}$

9. Not diagonalizable (**A** does not have two linearly independent eigenvectors)

11. Since $\mathbf{P}^{-1}\mathbf{AP} = \mathbf{D}$, then $\mathbf{A} = \mathbf{PAP}^{-1}$, so

$$\begin{aligned} \mathbf{A}^k &= (\mathbf{PDP}^{-1})(\mathbf{PDP}^{-1}) \cdots (\mathbf{PDP}^{-1}) \\ &= \mathbf{PD}^k\mathbf{P}^{-1}. \end{aligned}$$

13. $\mathbf{A}^{43} = \begin{pmatrix} 0 & 2^{22} \\ 2^{21} & 0 \end{pmatrix}$

15. $\mathbf{A}^{18} = \begin{pmatrix} 1 & 0 \\ (1-5^{18})/4 & 5^{18} \end{pmatrix}$

Section 9.3 Some Special Types of Matrices

In Problems 1 through 12, **Q** is an orthogonal matrix that diagonalizes the given matrix.

1. $0, \begin{pmatrix} 1 \\ 0 \\ 0 \end{pmatrix}; (1+\sqrt{17})/2, \begin{pmatrix} 0 \\ -1-\sqrt{17} \\ 4 \end{pmatrix}; (1-\sqrt{17})/2, \begin{pmatrix} 0 \\ -1-\sqrt{17} \\ 4 \end{pmatrix}$

$$\mathbf{Q} = \begin{pmatrix} 1 & 0 & 0 \\ 0 & (-1-\sqrt{17})/\sqrt{34+2\sqrt{17}} & (-1+\sqrt{17})/\sqrt{34+2\sqrt{17}} \\ 0 & 4/\sqrt{34+2\sqrt{17}} & 4/\sqrt{34+2\sqrt{17}} \end{pmatrix}$$

3. 0 is an eigenvalue of multiplicity 2, with independent eigenvectors

$$\begin{pmatrix} 1 \\ 0 \\ 0 \\ 0 \end{pmatrix}, \begin{pmatrix} 0 \\ 0 \\ 0 \\ 1 \end{pmatrix}$$

$$-1, \begin{pmatrix} 0 \\ 1 \\ 1 \\ 0 \end{pmatrix}; 3, \begin{pmatrix} 0 \\ -1 \\ 1 \\ 0 \end{pmatrix}$$

$$\mathbf{Q} = \begin{pmatrix} 1 & 0 & 0 & 0 \\ 0 & 0 & 1/\sqrt{2} & -1/\sqrt{2} \\ 0 & 0 & 1/\sqrt{2} & 1/\sqrt{2} \\ 0 & 1 & 0 & 0 \end{pmatrix}$$

5. $0, \begin{pmatrix} 0 \\ 1 \\ 0 \end{pmatrix}; (5+\sqrt{41})/2, \begin{pmatrix} 5+\sqrt{41} \\ 0 \\ 4 \end{pmatrix}; (5-\sqrt{41})/2, \begin{pmatrix} 5-\sqrt{41} \\ 0 \\ 4 \end{pmatrix}$

$$\mathbf{Q} = \begin{pmatrix} 0 & (5+\sqrt{41})/\sqrt{82+10\sqrt{41}} & (5-\sqrt{41})/\sqrt{82-10\sqrt{41}} \\ 1 & 0 & 0 \\ 0 & 4/\sqrt{82+10\sqrt{41}} & 4/\sqrt{82-10\sqrt{41}} \end{pmatrix}$$

7. $0, \begin{pmatrix} 1 \\ 2 \end{pmatrix}; 5, \begin{pmatrix} -2 \\ 1 \end{pmatrix}; \mathbf{Q} = \begin{pmatrix} 1/\sqrt{5} & -2/\sqrt{5} \\ 2/\sqrt{5} & 1/\sqrt{5} \end{pmatrix}$

9. $5+\sqrt{2}, \begin{pmatrix} 1+\sqrt{2} \\ 1 \end{pmatrix}; 5-\sqrt{2}, \begin{pmatrix} 1-\sqrt{2} \\ 1 \end{pmatrix}$

$$\mathbf{Q} = \begin{pmatrix} (1+\sqrt{2})/(\sqrt{4+2\sqrt{2}}) & (1-\sqrt{2})/(\sqrt{4-2\sqrt{2}}) \\ 1/\sqrt{4+2\sqrt{2}} & 1/\sqrt{4-2\sqrt{2}} \end{pmatrix}$$

11. $3, \begin{pmatrix} 0 \\ 0 \\ 1 \end{pmatrix}; -1+\sqrt{2}, \begin{pmatrix} 1+\sqrt{2} \\ 1 \\ 0 \end{pmatrix}; -1-\sqrt{2}, \begin{pmatrix} 1-\sqrt{2} \\ 1 \\ 0 \end{pmatrix}$

$$\mathbf{Q} = \begin{pmatrix} 0 & (1+\sqrt{2})/\sqrt{4+2\sqrt{2}} & (1-\sqrt{2})/\sqrt{4-2\sqrt{2}} \\ 0 & 1/\sqrt{4+2\sqrt{2}} & 1/\sqrt{4-2\sqrt{2}} \\ 1 & 0 & 0 \end{pmatrix}$$

13. Skew-hermitian;

$$0, \begin{pmatrix} 2 \\ 0 \\ 1+i \end{pmatrix}; \sqrt{3}i, \begin{pmatrix} 1 \\ \sqrt{3}i \\ -1-i \end{pmatrix}; -\sqrt{3}i, \begin{pmatrix} 1 \\ -\sqrt{3}i \\ -1-i \end{pmatrix}$$

The matrix is diagonalized by

$$\mathbf{P} = \begin{pmatrix} 2 & 1 & 1 \\ 0 & \sqrt{3}i & -\sqrt{3}i \\ 1+i & -1-i & -1-i \end{pmatrix}$$

15. Not hermitian, not skew-hermitian, not unitary; 2 is a repeated eigenvalue and all eigenvectors are scalar multiples of

$$\begin{pmatrix} i \\ 1 \end{pmatrix}.$$

This matrix is not diagonalizable.

17. Hermitian,

$$0, \begin{pmatrix} 0 \\ i \\ 1 \end{pmatrix}; 4+3\sqrt{2}, \begin{pmatrix} 4+3\sqrt{2} \\ -1 \\ -i \end{pmatrix}; 4-3\sqrt{2}, \begin{pmatrix} 4-3\sqrt{2} \\ -1 \\ -i \end{pmatrix},$$

use these eigenvectors as columns of **P**.

19. Hermitian; approximate eigenvalues and eigenvectors are

$$4.051374, \begin{pmatrix} 1 \\ 0.525687 \\ -0.129755i \end{pmatrix}; 0.482696, \begin{pmatrix} 1 \\ -1.258652 \\ 2.607546i \end{pmatrix}; -1.53407, \begin{pmatrix} 1 \\ -2.267035 \\ -1.477791i \end{pmatrix}$$

Use these eigenvectors as columns of **p**.

21. skew-hermitian; approximate eigenvalues and eigenvectors are

$$-2.164248i, \begin{pmatrix} -i \\ -3.164248 \\ 2.924109 \end{pmatrix}; 0.772866i, \begin{pmatrix} i \\ 0.227134 \\ 0.587771 \end{pmatrix}; 2.391382i, \begin{pmatrix} i \\ -1.391382 \\ -1.163664 \end{pmatrix}$$

Use these eigenvectors as columns of **P**.

23.

$$\mathbf{A} = \begin{pmatrix} 5 & 2 \\ 2 & 2 \end{pmatrix}$$

and the standard form is $y_1^2 + 6y_2^2$

25.

$$\mathbf{A} = \begin{pmatrix} 4 & -6 \\ -6 & 1 \end{pmatrix}$$

and the standard form is

$$\left(\frac{5+\sqrt{153}}{2}\right) y_1^2 + \left(\frac{5-\sqrt{153}}{2}\right) y_2^2$$

27.

$$\mathbf{A} = \begin{pmatrix} 4 & -2 \\ -2 & 1 \end{pmatrix}$$

and the standard form is

$$\left(\frac{3+\sqrt{17}}{2}\right) y_1^2 + \left(\frac{3-\sqrt{17}}{2}\right)^2 y_2^2$$

29. If $\mathbf{S}^t = -\overline{\mathbf{S}}$, then each $s_{jj} = -\overline{s_{jj}}$. Write $s_{jj} = a_{jj} + ib_{jj}$. Then $a_{jj} = -a_{jj}$, so $a_{jj} = 0$ and s_{jj} is pure imaginary.

31. If $\mathbf{A}$ is hermitian then $\overline{\mathbf{A}} = \mathbf{A}^t$, so

$$\overline{(\mathbf{A}\mathbf{A}^t)} = \overline{\mathbf{A}}\,\overline{(\mathbf{A}^t)}$$
$$= \overline{\mathbf{A}}\,\overline{(\overline{\mathbf{A}})} = (\overline{\mathbf{A}})\overline{\mathbf{A}}.$$

CHAPTER TEN SYSTEMS OF LINEAR DIFFERENTIAL EQUATIONS

Section 10.1 Linear Systems

The following solutions give the fundamental matrix $\Omega(t)$ and the solution of the initial value problem.

1. $\Omega(t) = \begin{pmatrix} 4e^{(1+2\sqrt{3})t} & 4e^{(1-2\sqrt{3})t} \\ (-1+\sqrt{3})e^{(1+2\sqrt{3})t} & (-1-\sqrt{3})e^{(1-2\sqrt{3})t} \end{pmatrix}$

$\mathbf{X}(t) = \begin{pmatrix} (1+5\sqrt{3}/3)e^{(1+2\sqrt{3})t} + (1-5\sqrt{3}/3)e^{(1-2\sqrt{3})t} \\ (-1+\sqrt{3}/6)e^{(1+2\sqrt{3})t} + (1+\sqrt{3}/6)e^{(1-2\sqrt{3})t} \end{pmatrix}$

3. $\Omega(t) = \begin{pmatrix} e^t & 0 & e^{-3t} \\ 0 & e^t & 3e^{-3t} \\ e^{-t} & e^t & e^{-3t} \end{pmatrix}$; $\mathbf{X}(t) = \begin{pmatrix} 10e^t - 9e^{-3t} \\ 24e^t - 27e^{-3t} \\ 14e^t - 9e^{-3t} \end{pmatrix}$

5. $\Omega(t) = \begin{pmatrix} -e^{2t} & 3e^{6t} \\ e^{2t} & e^{6t} \end{pmatrix}$; $\mathbf{X}(t) = \begin{pmatrix} -3e^{2t} + 3e^{6t} \\ 3e^{2t} + e^{6t} \end{pmatrix}$

Section 10.2 Solution of $\mathbf{X}' = \mathbf{A}\mathbf{X}$ for Constant $\mathbf{A}$

1. $\Omega(t) = \begin{pmatrix} 0 & e^{2t} & 3e^{3t} \\ 1 & e^{2t} & e^{3t} \\ 1 & 0 & e^{3t} \end{pmatrix}$; $\mathbf{X}(t) = \begin{pmatrix} 4e^{2t} - 3e^{3t} \\ 2 + 4e^{2t} - e^{3t} \\ 2 - e^{3t} \end{pmatrix}$

3. $\Omega(t) = \begin{pmatrix} 1 & 2e^{3t} & -e^{-4t} \\ 6 & 3e^{3t} & 2e^{-4t} \\ -13 & -2e^{3t} & e^{-4t} \end{pmatrix}$; $\mathbf{X}(t) = \begin{pmatrix} c_1 + 2c_2e^{3t} - c_3e^{-4t} \\ 6c_1 + 3c_2e^{3t} + 2c_2e^{-4t} \\ -13c_1 - 2c_2e^{3t} + c_3e^{-4t} \end{pmatrix}$

5. $\Omega(t) = \begin{pmatrix} 1 & e^{2t} \\ -1 & e^{2t} \end{pmatrix}$; $\mathbf{X}(t) = \begin{pmatrix} c_1 + c_2e^{2t} \\ -c_1 + c_2e^{2t} \end{pmatrix}$

7. $\Omega(t) = \begin{pmatrix} 7e^{3t} & 0 \\ 5e^{3t} & e^{-4t} \end{pmatrix}$; $\mathbf{X}(t) = \Omega(t)\mathbf{C} = \begin{pmatrix} 7c_1e^{3t} \\ 5c_1e^{3t} + c_2e^{-4t} \end{pmatrix}$

9. $\Omega(t) = \begin{pmatrix} 2e^{4t} & e^{-3t} \\ -3e^{4t} & 2e^{-3t} \end{pmatrix}$; $\mathbf{X}(t) = \begin{pmatrix} 6e^{4t} - 5e^{-3t} \\ -9e^{4t} - 10e^{-3t} \end{pmatrix}$

11. $\Omega(t) = \begin{pmatrix} 5e^t\cos(t) & 5e^t\sin(t) \\ e^t[2\cos(t) + \sin(t)] & e^t[2\sin(t) - \cos(t)] \end{pmatrix}$

13. $\Omega(t) = \begin{pmatrix} 0 & e^{-t}\cos(2t) & e^{-t}\sin(2t) \\ 0 & e^{-t}[\cos(2t) - 2\sin(2t)] & e^{-t}[\sin(2t) + 2\cos(2t)] \\ e^{-2t} & 3e^{-t}\cos(2t) & 3e^{-t}\sin(2t) \end{pmatrix}$

15. $\Omega(t) = \begin{pmatrix} 2e^{2t}\cos(2t) & 2e^{2t}\sin(2t) \\ e^{2t}\sin(2t) & -e^{2t}\cos(2t) \end{pmatrix}$

17. $\Omega(t) = \begin{pmatrix} e^{2t} & 3e^{5t} & 27te^{5t} \\ 0 & 3e^{5t} & (3+27t)e^{5t} \\ 0 & -e^{5t} & (2-9t)e^{5t} \end{pmatrix}$

19. $\Omega(t) = \begin{pmatrix} 2 & 3e^{3t} & e^t & 0 \\ 0 & 2e^{3t} & 0 & -2e^t \\ 1 & 2e^{3t} & 0 & -2e^t \\ 0 & 0 & 0 & e^t \end{pmatrix}$

21. $\Omega(t) = \begin{pmatrix} e^{3t} & 2te^{3t} \\ 0 & e^{3t} \end{pmatrix}$

Section 10.3 Solution of $\mathbf{X}' = \mathbf{AX} + \mathbf{G}$

1. $\mathbf{X}(t) = \begin{pmatrix} (6 + 12t + t^2/2)e^{-2t} \\ (2 + 12t + t^2/2)e^{-2t} \\ (3 + 38t + 66t^2 + 13t^3/6)e^{-2t} \end{pmatrix}$

3. $\mathbf{X}(t) = \begin{pmatrix} c_2 e^t \\ -2c_2 e^t + (c_3 - 9c_4)e^{3t} + e^t \\ 2c_4 e^{3t} \\ (c_1 - 5c_2 t)e^t + c_3 e^{3t} + (1 + 3t)e^t \end{pmatrix}$

5. $\mathbf{X}(t) = \begin{pmatrix} (-1 - 14t)e^t \\ (3 - 14t)e^t \end{pmatrix}$

7. $\mathbf{X}(t) = \begin{pmatrix} [c_1(1 + 2t) + 2c_2 t + t^2]e^{3t} \\ [-2c_1 t + (1 - 2t)c_2 + t - t^2]e^{3t} + 3e^t/2 \end{pmatrix}$

9. $\mathbf{X}(t) = \begin{pmatrix} [c_1 + c_2(1 + t) + 2t + t^2 - t^3]e^{6t} \\ [c_1 + c_2 t + 4t^2 - t^3]e^{6t} \end{pmatrix}$

11. $\mathbf{X}(t) = \begin{pmatrix} c_1 e^t + 5c_2 e^{7t} + (68/145)\cos(3t) - (54/145)\sin(3t) + 40/7 \\ -c_1 e^t + c_2 e^{7t} + (2/145)\cos(3t) + (24/145)\sin(3t) - 48/7 \end{pmatrix}$

13. $\mathbf{X} = \begin{pmatrix} -\frac{1}{4}e^{2t} + (2 + 2t)e^t - \frac{3}{4} - \frac{1}{2}t \\ e^{2t} + (2 + 2t)e^t - 1 - t \\ -\frac{5}{4}e^{2t} + 2te^{2t} - \frac{3}{4} - \frac{1}{2}t \end{pmatrix}$

15. $\mathbf{X}(t) = \begin{pmatrix} 3c_1 e^{2t} + c_2 e^{6t} - 4e^{3t} - 10/3 \\ -c_1 e^{2t} + c_2 e^{6t} + 2/3 \end{pmatrix}$

17. $\mathbf{X}(t) = \begin{pmatrix} 2 + 4(1 + t)e^{2t} \\ -2 + 2(1 + 2t)e^{2t} \end{pmatrix}$

19. $\mathbf{X} = \begin{pmatrix} 10\cos(t) + \frac{5}{2}t\sin(t) - 5t\cos(t) \\ 5\cos(t) + \frac{5}{2}\sin(t) - \frac{5}{2}t\cos(t) \end{pmatrix}$

Section 10.4 Exponential Matrix Solutions

In Problems 1 through 5, $e^{\mathbf{A}t}$ is given. The solution is $\mathbf{X}(t) = e^{\mathbf{A}t}\mathbf{C}$.

1. $e^{\mathbf{A}t} = e^{13t/2} \begin{pmatrix} \cos(\sqrt{23}t/2) - \frac{3\sqrt{23}}{23}\sin(\sqrt{23}t/2) & -\frac{4\sqrt{23}}{23}\sin(\sqrt{23}t/2) \\ \frac{8\sqrt{23}}{23}\sin(\sqrt{23}t/2) & \cos(\sqrt{23}t/2) + \frac{3\sqrt{23}}{23}\sin(\sqrt{23}t/2) \end{pmatrix}$

3. $e^{\mathbf{A}t} = \begin{pmatrix} \frac{2}{5}\cos(t) - \frac{1}{5}\sin(t) + \frac{3}{5}e^{2t} & \frac{2}{5}\sin(t) + \frac{1}{5}\cos(t) - \frac{1}{5}e^{2t} & \frac{3}{5}\sin(t) - \frac{1}{5}\cos(t) + \frac{1}{5}e^{2t} \\ \frac{3}{5}\cos(t) - \frac{4}{5}\sin(t) - \frac{3}{5}e^{2t} & \frac{3}{5}\sin(t) + \frac{4}{5}\cos(t) + \frac{1}{5}e^{2t} & \frac{7}{5}\sin(t) + \frac{1}{5}\cos(t) - \frac{1}{5}e^{2t} \\ -\frac{3}{5}\cos(t) - \frac{1}{5}\sin(t) + \frac{3}{5}e^{2t} & \frac{1}{5}\cos(t) - \frac{3}{5}\sin(t) - \frac{1}{5}e^{2t} & \frac{4}{5}\cos(t) - \frac{2}{5}\sin(t) + \frac{1}{5}e^{2t} \end{pmatrix}$

5. $e^{\mathbf{A}t} = \begin{pmatrix} \cos(2t) - \frac{1}{2}\sin(2t) & \frac{1}{2}\sin(2t) \\ -\frac{5}{2}\sin(2t) & \cos(2t) + \frac{1}{2}\sin(2t) \end{pmatrix}$

7.

$$e^{\mathbf{B}t} = \mathbf{I}_n + \sum_{k=1}^{\infty} \frac{1}{k!}\mathbf{B}^k t^k$$

so

$$\mathbf{P}e^{\mathbf{B}t}\mathbf{P}^{-1} = \mathbf{P}\mathbf{I}_n\mathbf{P}^{-1} + \sum_{k=1}^{\infty} \frac{1}{k!}\mathbf{P}\mathbf{B}^k\mathbf{P}^{-1}t^k$$
$$= \mathbf{I}_n + \sum_{k=1}^{\infty} \frac{1}{k!}(\mathbf{P}\mathbf{B}\mathbf{P}^{-1})^k t^k$$
$$= e^{(\mathbf{P}\mathbf{B}\mathbf{P}^{-1})t} = e^{\mathbf{A}t}.$$

Section 10.5 Applications and Illustrations of Techniques

1.

$$x_1(t) = 20 + 25e^{-t/10} - 5e^{-3t/50} \text{ pounds},$$
$$x_2(t) = 30 - 25e^{-t/10} - 5e^{-3t/50} \text{ pounds}$$

The brine solution in tank 1 has minimum concentration at $t = 25\ln(25/3)$ minutes. At this time there is $20 - (6\sqrt{3})/125 \approx 19.9$ pounds of salt in tank 1.

3. The capacitor charge is maximum when the capacitor voltage is maximum. This voltage is

$$V_C = 10(q_1 - q_2) = 5i_3.$$

Setting $dV_c/dt = 0$ yields $t = (9/2)\ln(10/9) \approx 0.474$ seconds. At this time, the capacitor voltage is 6.97 volts.

5. Designate $y_1(t)$ as the position of the upper weight relative to its equilibrium position and $y_2(t)$ the position of the lower weight relative to its equilibrium position. Then

$$y_1(t) = \frac{2}{5}\cos(2t) + \frac{3}{5}\cos(2\sqrt{6}t),$$
$$y_2(t) = \frac{6}{5}\cos(2t) - \frac{1}{5}\cos(2\sqrt{6}t)$$

7. The currents are

$$i_1(t) = \frac{1}{10} - 2te^{-10t} \text{ amp},$$
$$i_2(t) = \left(\frac{1}{10} - t\right)e^{-10t} \text{ amp}$$

9. Let $y_1(t)$ be the displacement function of the left mass, and $y_2(t)$ the displacement function of the right mass (from their equilibrium positions). The solution is

$$y_1(t) = \cos(3t),\ y_2(t) = -\cos(3t).$$

11. $y_1(t) = \dfrac{3\sqrt{26}}{40}\left[2e^{-t}\sin(2t) + e^{-2t}(3\cos(t) + \sin(t)) - 3\cos(t) + \sin(t)\right],$

$$y_2(t) = \frac{3}{40}\left[e^{-t}(8\cos(2t) + 4\sin(2t)) + e^{-2t}(7\cos(t) - \sin(t)) - 15\cos(t) + 15\sin(t)\right]$$

Section 10.6 Phase Portraits

1. $-2 \pm \sqrt{3}i$, spiral point. The solution is

$$\mathbf{X} = \begin{pmatrix} c_1e^{-2t}\cos(\sqrt{3}t) - c_2e^{-2t}\sin(\sqrt{3}t) \\ c_1e^{-2t}\sin(\sqrt{3}t) + 3c_2e^{-2t}\cos(\sqrt{3}t) \end{pmatrix}$$

Figure A.11 is a phase portrait.

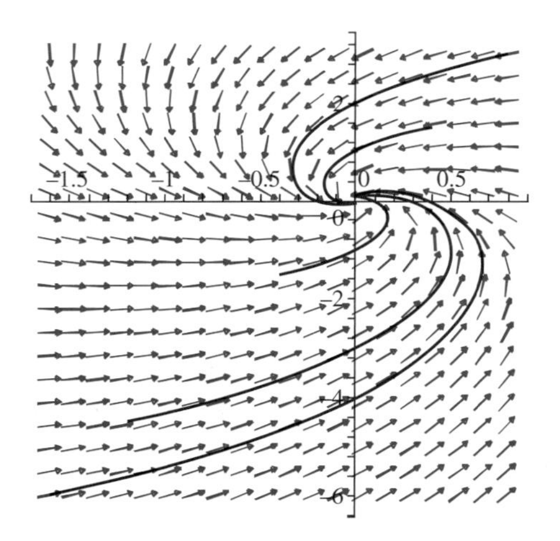

FIGURE A.11 *Phase portrait for Problem 1, Section 10.6.*

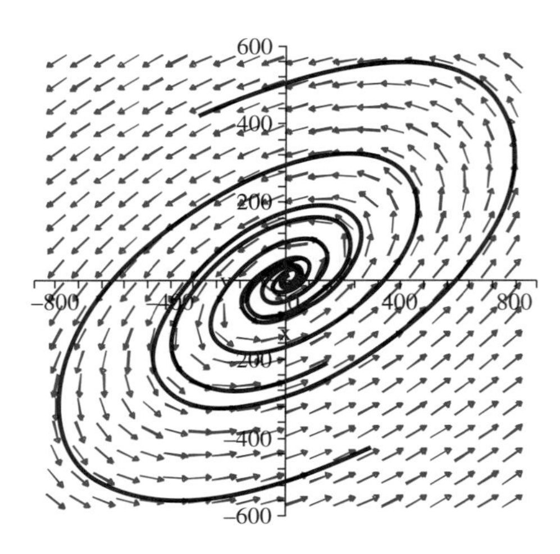

FIGURE A.12 *Phase portrait for Problem 3, Section 10.6.*

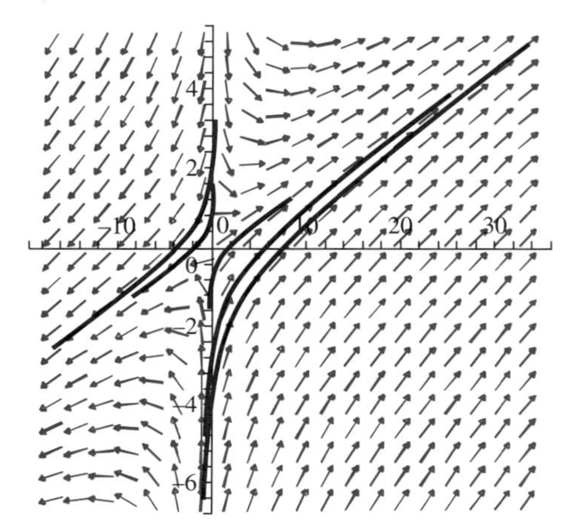

FIGURE A.13 *Phase portrait for Problem 5, Section 10.6.*

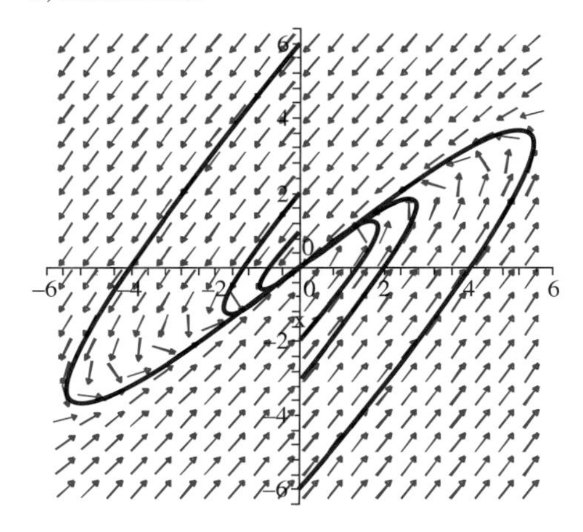

FIGURE A.14 *Phase portrait for Problem 7, Section 10.6.*

3. $4 \pm 5i$, and the origin is a spiral point. The solution is

$$\mathbf{X} = \begin{pmatrix} (3c_1 - 5c_2)e^{4t}\sin(5t) + (5c_1 + 3c_2)e^{4t}\cos(5t) \\ 2c_1e^{4t}\sin(5t) + 2c_2e^{4t}\cos(5t) \end{pmatrix}$$

Figure A.12 is a phase portrait.

5. 3, 3, the origin is an improper node. The solution is

$$\mathbf{X} = \begin{pmatrix} c_1e^{3t} + c_2te^{3t} \\ (c_1 + c_2)e^{3t} + c_2te^{3t} \end{pmatrix}$$

Figure A.13 is a phase portrait.

7. Eigenvalues are $-2, -2$ and the origin is an improper node. The solution is

$$\mathbf{X} = \begin{pmatrix} c_1e^{-2t} + 5(c_1 - c_2)te^{-2t} \\ c_2e^{-2t} + 5(c_1 - c_2)te^{-2t} \end{pmatrix}$$

A phase portrait is given in Figure A.14.

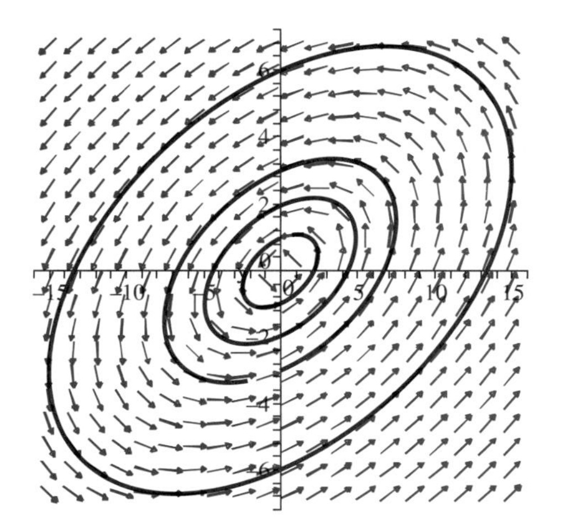

FIGURE A.15 *Phase portrait for Problem 9, Section 10.6.*

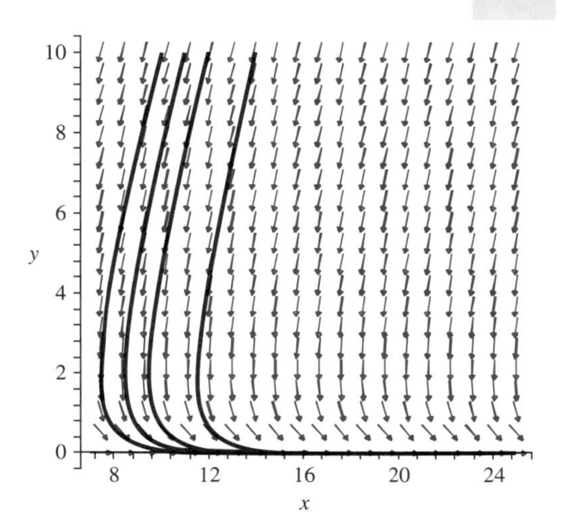

FIGURE A.16 *Phase portrait for Problem 11(a), Section 10.6.*

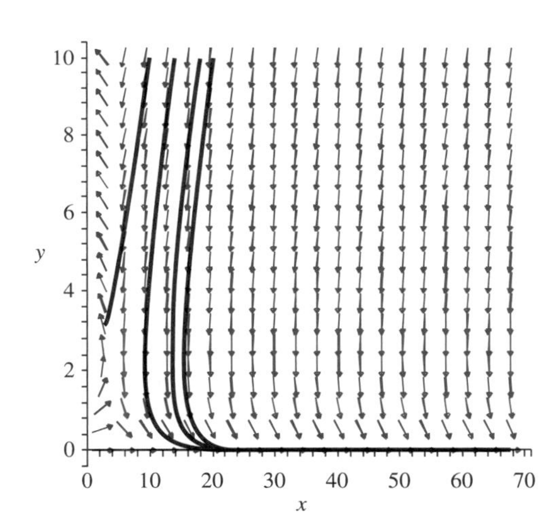

FIGURE A.17 *Phase portrait for Problem 11(c), Section 10.6.*

9. Eigenvalues are $\pm 2i$; the origin is a center. Solution is

$$\mathbf{X} = \begin{pmatrix} (c_1 - 2c_2)\sin(2t) + (2c_1 + c_2)\cos(2t) \\ c_1 \sin(2t) + c_2 \cos(2t) \end{pmatrix}$$

Figure A.15 is a phase portrait.

11. Figure A.16 is a phase portrait for (a), and Figure A.17 for (c).

13.

$$x_1' = ax_1 - bx_1x_2 - Hx_1, \; x_2' = -kx_2 + cx_1x_2 - Hx_2$$

CHAPTER ELEVEN VECTOR DIFFERENTIAL CALCULUS

Section 11.1 Vector Functions of One Variable

1. $te^t(2+t)(\mathbf{j}-\mathbf{k})$

3. $(f(t)\mathbf{F}(t))' = -12\sin(3t)\mathbf{i} + 12t[2\cos(3t) - 3t\sin(3t)]\mathbf{j}$
$+8[\cos(3t) - 3t\sin(3t)]\mathbf{k}$

5. $(\mathbf{F}\times\mathbf{G})' = (1-4\sin(t))\mathbf{i} - 2t\mathbf{j} - (\cos(t) - t\sin(t))\mathbf{k}$

7. $(f(t)\mathbf{F}(t))' = (1-8t^3)\mathbf{i} + (6t^2\cosh(t) - (1-2t^3)\sinh(t))\mathbf{j}$
$+(-6t^2e^t + e^t(1-2t^3))\mathbf{k}$

9. (a) $\mathbf{F}(t) = t^2(2\mathbf{i}+3\mathbf{j}+4\mathbf{k})$ for $1\le t\le 3$, $\mathbf{F}'(t) = 2t(2\mathbf{i}+3\mathbf{j}+4\mathbf{k})$
(b) $s(t) = \sqrt{29}(t^2-1)$
(c) $\mathbf{F}(t(s)) = \left(1+\frac{s}{\sqrt{29}}\right)(2\mathbf{i}+3\mathbf{j}+4\mathbf{k})$

11. (a) $\mathbf{F}(t) = \sin(t)\mathbf{i} + \cos(t)\mathbf{j} + 45t\mathbf{k}$ for $0\le t\le 2\pi$,
$\mathbf{F}'(t) = \cos(t)\mathbf{i} - \sin(t)\mathbf{j} + 45\mathbf{k}$
(b) $s(t) = \sqrt{2026}t$
(c) $\mathbf{F}(t(s)) = \frac{1}{\sqrt{2026}}\left[\sin\left(s/\sqrt{2026}\right)\mathbf{i} + \cos\left(s/\sqrt{2026}\right) + \left(45s/\sqrt{2026}\right)\mathbf{k}\right]$

Section 11.2 Velocity and Curvature

1. $\mathbf{v}(t) = 2t(\alpha\mathbf{i}+\beta\mathbf{j}+\gamma\mathbf{k})$, $v(t) = 2|t|\sqrt{\alpha^2+\beta^2+\gamma^2}$, $a_N = 0$, $\kappa = 0$, $a_T = 2\sigma\sqrt{\alpha^2+\beta^2+\gamma^2}$, where σ equals 1 if $t\ge 0$, and -1 if $t<0$

3. $\mathbf{v}(t) = 2\cosh(t)\mathbf{j} - 2\sinh(t)\mathbf{k}$, $v(t) = 2\sqrt{\cosh(2t)}$,
$\mathbf{a}(t) = 2\sinh(t)\mathbf{j} - 2\cosh(t)\mathbf{k}$
$\kappa = \frac{1}{2(\cosh(2t))^{3/2}}$,
$a_T = 2\sinh(2t)/\sqrt{\cosh(2t)}$, $a_N = 2/\sqrt{\cosh(2t)}$

5. $\mathbf{v}(t) = 2\mathbf{i} - 2\mathbf{j} + \mathbf{k}$, $v(t) = 3$, $\mathbf{a}(t) = \mathbf{O}$
$\kappa = 0$, $a_T = a_N = 0$

7. $\mathbf{v}(t) = -3e^{-t}(\mathbf{i}+\mathbf{j}-2\mathbf{k})$, $v(t) = 3\sqrt{6}e^{-t}$, $\mathbf{a}(t) = 3e^{-t}(\mathbf{i}+\mathbf{j}-2\mathbf{k})$
$\kappa = 0$, $a_T = -3\sqrt{6}e^{-t}$, $a_N = 0$

9. $\mathbf{v}(t) = 3\mathbf{i} + 2t\mathbf{k}$, $v(t) = \sqrt{9+4t^2}$, $\mathbf{a}(t) = 2\mathbf{k}$,
$a_T = \frac{4t}{\sqrt{9+4t^2}}$, $\kappa = \frac{6}{(9+4t^2)^{3/2}}$, $a_N = \frac{6}{(9+4t^2)^{1/2}}\ \frac{6}{(9+4t^2)^{1/2}}$

Section 11.3 Vector Fields and Streamlines

1. There are many such vector fields. One whose streamlines are circles about the origin in the plane $z=0$ is given by

$$\mathbf{F}(x,y,z) = -\frac{1}{x}\mathbf{i} + \frac{1}{y}\mathbf{j}.$$

3. $x = c$, $y = y$, $2e^z = k - \sin(y)$;
$x = 3$, $y = y$, $z = \ln\left(1+\sqrt{2}/4 - (1/2)\sin(y)\right)$

5. $x = x$, $y = e^x(x-1)+c$, $x^2 = -2z+k$;
$x = x$, $y = e^x(x-1) - e^2$, $z = \frac{1}{2}(12-x^2)$

7. $x = x$, $y = 1/(x+c)$, $z = e^{x+k}$; $x = x$, $y = 1/(x-1)$, $z = e^{x-2}$

Section 11.4 The Gradient Field

1. $2y\sinh(2xy)\mathbf{i} + 2x\sinh(2xy)\mathbf{j} - \cosh(z)\mathbf{k}$, $\cosh(1)\mathbf{k}$, $\cosh(1)$, $-\cosh(1)$

3. $yz\mathbf{i} + xz\mathbf{j} + xy\mathbf{k}$, $\mathbf{i} + \mathbf{j} + \mathbf{k}$,
$\sqrt{3}, -\sqrt{3}$

5. $(2y + e^z)\mathbf{i} + 2x\mathbf{j} + xe^z\mathbf{k}$, $(2 + e^6)\mathbf{i} - 4\mathbf{j} - 2e^6\mathbf{k}$,
$\sqrt{20 + 4e^6 + 5e^{12}}, -\sqrt{20 + 4e^6 + 5e^{12}}$

7. $\left(1/\sqrt{5}\right)(2x^2z^3 + 3x^2yz^2)$

9. $\left(1/\sqrt{3}\right)(8y^2 - z + 16xy - x)$

11. Level surfaces are planes $x + z = k$.

13. $x = 1$; $x = 1 + 2t$, $y = \pi$, $z = 1$

15. $x = y$; $x = 1 + 2t$, $y = 1 - 2t$, $z = 0$

17. $x + y + \sqrt{2}z = 4$; $x = y = 1 + 2t$, $z = \sqrt{2}(1 + 2t)$

Section 11.5 Divergence and Curl

In Problems 1, 3, and 5, $\nabla \cdot \mathbf{F}$ is given first, then $\nabla \times \mathbf{F}$.

1. $2y + xe^y + 2$, $(e^y - 2x)\mathbf{k}$

3. $\cosh(x) f xz\sinh(xyz) - 1$, $(-1 - xy\sinh(xyz))i - j + yz\sinh(xyz)k(i, j, k)$

In Problems 9 and 11, $\nabla\varphi$ is given.

5. $4, \mathbf{O}$

7. $(\cos(x + y + z) - x\sin(x + y + z))\mathbf{i} - x\sin(x + y + z)(\mathbf{j} + \mathbf{k})$

9. $\mathbf{i} - \mathbf{j} + 4z\mathbf{k}$

11. $-6x^2yz^2\mathbf{i} - 2x^3z^2\mathbf{j} - 4x^3yz\mathbf{k}$

13. $\nabla \cdot (\varphi\mathbf{F}) = \nabla\varphi \cdot \mathbf{F} + \varphi(\nabla \cdot \mathbf{F})$
$\nabla \times (\varphi\mathbf{F}) = \nabla\varphi \times \mathbf{F} + \varphi(\nabla \times \mathbf{F})$

CHAPTER TWELVE VECTOR INTEGRAL CALCULUS

Section 12.1 Line Integrals

1. $-422/5$ **3.** 0 **5.** $26\sqrt{2}/3$
7. $\sin(3) - 81/2$ **9.** 0 **13.** $-27/2$

Section 12.2 Green's Theorem

1. By Green's theorem,

$$\oint_C -\frac{\partial u}{\partial y}\,dx + \frac{\partial u}{\partial x}\,dy = \iint_D \left[\frac{\partial}{\partial x}\left(\frac{\partial u}{\partial x}\right) - \frac{\partial}{\partial y}\left(-\frac{\partial u}{\partial y}\right)\right] dA$$

3. -8 **5.** 0 **7.** $95/4$
9. 512π **11.** -40 **13.** -12

Section 12.3 An Extension of Green's Theorem

1. 2π if C encloses the origin; 0 otherwise
3. 0
5. 0

Section 12.4 Potential Theory

1. Not conservative
3. Conservative, $\varphi(x, y) = \ln(x^2 + y^2)$
5. $\varphi(x, y, z) = x - 2y + z$
7. Conservative, $\varphi(x, y) = 8x^2 + 2y - y^3/3$
9. Conservative, $\varphi(x, y) = xy^3 - 4y$
11. $5 + \ln(3/2)$ **13.** $2e^{-2}$ **15.** -403
17. -5 **19.** -27
21. Write

$$E(t) = \text{ total energy } = \frac{m}{2}\mathbf{R}'(t) \cdot \mathbf{R}'(t) - \varphi(x(t), y(t), z(t)).$$

Use the fact that $m\mathbf{R}'' = \nabla\varphi$ (by one of Newton's laws of motion) to show that $E'(t) = 0$.

Section 12.5 Surface Integrals

1. $(9/8)\left(\ln\left(4 + \sqrt{17}\right) + 4\sqrt{17}\right)$ **3.** $28\pi\sqrt{2}/3$
5. $-10\sqrt{3}$ **7.** $\pi(29^{3/2} - 27)/6$ **9.** $125\sqrt{2}$

Section 12.6 Applications of Surface Integrals

1. 78π, $(0, 0, 27/13)$ **3.** $49/12$, $(12/35, 33/35, 24/35)$
5. $9\pi K\sqrt{2}$, $(0, 0, 2)$ **7.** $128/3$

Section 12.7 Lifting Green's Theorem to R^3

1. Apply Green's theorem to the line integral

$$\oint_C -\varphi\frac{\partial\psi}{\partial y}\,dx + \varphi\frac{\partial\psi}{\partial x}\,dy.$$

3. Apply Green's theorem to

$$\oint_C -\frac{\partial\varphi}{\partial y}\,dx + \frac{\partial\varphi}{\partial x}\,dy.$$

Section 12.8 The Divergence Theorem of Gauss

1. 0 **3.** $256\pi/3$ **5.** 2π
7. $8\pi/3$ **9.** 0 because $\nabla \cdot (\nabla \times \mathbf{F}) = 0$

Section 12.9 Stokes's Theorem

1. -16π **3.** $-32/3$ **5.** -8π **7.** -108

Section 12.10 Curvilinear Coordinates

2.

$$h_u = h_v = \sqrt{u^2 + v^2},\ h_z = 1$$

$$\nabla f(u, v, z) = \frac{1}{h_u}\frac{\partial f}{\partial u}\overline{u}_u + \frac{1}{h_v}\frac{\partial f}{\partial v}\overline{u}_v + \frac{\partial f}{\partial z}\overline{u}_z$$

$$\nabla \cdot \mathbf{F}(u, v, z) = \frac{1}{h_u^2}\left(\frac{\partial}{\partial u}(h_u F_1) + \frac{\partial}{\partial v}(h_v F_2)\right) + \frac{1}{h_u^2}\frac{\partial}{\partial z}\left(h_u^2 F_3\right)$$

$$\nabla \times \mathbf{F}(u, v, z) = \frac{1}{h_u}\left(\frac{\partial}{\partial u}(F_3) - \frac{\partial}{\partial z}(h_u F_2)\right)\overline{u}_1 + \frac{1}{h_v^2}\left(\frac{\partial}{\partial z}(h_u F_1) - \frac{\partial}{\partial u}(F_3)\right)\overline{u}_2 + \frac{1}{h_v^2}\left(\frac{\partial}{\partial u}(h_v F_2) - \frac{\partial}{\partial v}(h_u F_1)\right)\overline{u}_3$$

$$\nabla^2 f = \frac{1}{h_u^2}\left(\frac{\partial^2 f}{\partial u^2} + \frac{\partial^2 f}{\partial v^2} + \frac{\partial}{\partial z}\left(h_u^2 \frac{\partial f}{\partial z}\right)\right)$$

4.

$$h_1 = h_2 = a\sqrt{\sinh^2(u)\cos^2(v) + \cosh^2(u)\sin^2(v)},\ h_3 = 1$$

$$\nabla f(u, v, z) = \frac{1}{h_1}\frac{\partial f}{\partial u}\overline{u}_u + \frac{1}{h_1}\frac{\partial f}{\partial v}\overline{u}_v + \frac{\partial f}{\partial z}\overline{u}_z,$$

where

$$\overline{u}_u = \frac{a}{h_1}(\sinh(u)\cos(v)\mathbf{i} + \cosh(u)\sin(v)\mathbf{j}),$$

$$\overline{u}_v = \frac{a}{h_1}(-a\cosh(u)\sin(v)\mathbf{i} + \sinh(u)\cos(v)\mathbf{j}),$$

$$\overline{u}_z = \mathbf{k}$$

$$\nabla \cdot \mathbf{F}(u, v, z) = \frac{1}{h_1^2}\left[\frac{\partial}{\partial u}(F_1 h_1) + \frac{\partial}{\partial v}(F_2 h_1)\right] + \frac{\partial F_3}{\partial z}$$

$$\nabla \times \mathbf{F}(u, v, z) = \left(\frac{1}{h_1}\frac{\partial F_3}{\partial v} - \frac{\partial F_2}{\partial z}\right)\overline{u}_u + \left(\frac{\partial F_1}{\partial z} - \frac{1}{h_1}\frac{\partial F_3}{\partial u}\right)\overline{u}_v + \frac{1}{h_1^2}\left(\frac{\partial}{\partial u}(F_2 h_1 - \frac{\partial}{\partial v}(F_1 h_1)\right)\overline{u}_z$$

$$\nabla^2 f(u, v, z) = \frac{1}{h_1^2}\left(\frac{\partial^2 f}{\partial u^2} + \frac{\partial^2 f}{\partial v^2}\right) + \frac{\partial^2 f}{\partial z^2}$$

CHAPTER THIRTEEN FOURIER SERIES

Section 13.1 Why Fourier Series?

1. If $p(x)$ has degree k, then differentiating $p(x)$, $k+1$ times yields the zero function, while $\sum_{n=1}^{N} b_n \sin(nx)$ can be differentiated any number of times, and none of these derivatives is identically zero on $[0, \pi]$.

Section 13.2 The Fourier Series of a Function

1.

$$\frac{1}{3}\sin(3) + 6\sin(3)\sum_{n=1}^{\infty}\frac{(-1)^{n+1}}{n^2\pi^2 - 9}\cos\left(\frac{n\pi x}{3}\right),$$

converging to $\cos(x)$ on $[-3, 3]$.

3. 4; the series (consisting of one term) converges to 4 on $[-3, 3]$.

5.

$$\frac{1}{\pi}\sinh(\pi)+\frac{2}{\pi}\sinh(\pi)\sum_{n=1}^{\infty}\frac{(-1)^n}{n^2+1}\cos(n\pi x),$$

converging to $\cosh(\pi x)$ for $-1 \le x \le 1$.

7.

$$\frac{16}{\pi}\sum_{n=1}^{\infty}\frac{1}{2n-1}\sin((2n-1)x),$$

converging to -4 for $-\pi < x < 0$, to 4 for $0 < x < \pi$ and to 0 for $x = 0, -\pi, \pi$.

9.

$$\frac{13}{3}+\sum_{n=1}^{\infty}(-1)^n\left[\frac{16}{n^2\pi^2}\cos\left(\frac{n\pi x}{2}\right)+\frac{4}{n\pi}\sin\left(\frac{n\pi x}{2}\right)\right],$$

converging to $f(x)$ for $-2 < x < 2$, to 2 at $x = -2$ and to 7 at $x = 2$.

11.

$$\frac{3}{2}+\frac{1}{\pi}\sum_{n=1}^{\infty}\frac{1-(-1)^n}{n}\sin(nx),$$

converging to 1 for $-\pi < x < 0$, to 2 for $0 < x < \pi$ and to $3/2$ at $x = 0, -\pi, \pi$.

13. The series converges to -1 if $-4 < x < 0$, to 0 if $x = \pm 4$ or $x = 0$, and to 1 if $0 < x < 4$.

15. The series converges to

$$\begin{cases} -1 & \text{for } x=-4 \text{ and for } x=4 \\ 3/2 & \text{for } x=-2 \\ 5/2 & \text{for } x=2 \\ f(x) & \text{elsewhere on } [-4,4]. \end{cases}$$

17. The series converges to $(2+\pi^2)/2$ if $x = \pm\pi$, to x^2 if $-\pi < x < 0$, to 1 if $x = 0$ and to 2 if $0 < x < \pi$.

19. The series converges to $3/2$ for $x = \pm 3$, to $2x$ if $-3 < x < -2$, to -2 if $x = -2$, to 0 if $-2 < x < 1$, to $1/2$ if $x = 1$ and to x^2 if $1 < x < 3$.

Section 13.3 Sine and Cosine Series

1. The cosine series is

$$\frac{5}{6}+\frac{16}{\pi^2}\sum_{n=1}^{\infty}\left[\frac{1}{n^2}\cos\left(\frac{n\pi}{4}\right)-\frac{4}{n^3\pi}\sin\left(\frac{n\pi}{4}\right)\right]\cos\left(\frac{n\pi x}{4}\right)$$

converging to x^2 for $0 \le x \le 1$ and to 1 for $1 < x \le 4$.
The sine series is

$$\sum_{n=1}^{\infty}\left[\frac{16}{n^2\pi^2}\sin\left(\frac{n\pi}{4}\right)+\frac{64}{n^3\pi^3}\left[\cos\left(\frac{n\pi}{4}\right)-1\right]-\frac{2(-1)^n}{n\pi}\right]\sin\left(\frac{n\pi x}{4}\right)$$

converging to x^2 for $0 \le x \le 1$, 1 for $1 < x < 4$, and 0 for $x = 4$.

3. The cosine series is

$$\frac{4}{3}+\frac{16}{\pi^2}\sum_{n=1}^{\infty}\frac{(-1)^n}{n^2}\cos(n\pi x/2),$$

converging to x^2 for $0 \le x \le 2$. The sine series is

$$-\frac{8}{\pi}\sum_{n=1}^{\infty}\left[\frac{(-1)^n}{n}+\frac{2(1-(-1)^n)}{n^3\pi^2}\right]\sin(n\pi x/2),$$

converging to x^2 if $0 < x < 2$ and to 0 for $x = 0$ and for $x = 2$.

5. The cosine series is

$$\frac{1}{2}$$

$$+\sum_{n=1}^{\infty}\left[\frac{4}{n\pi}\sin(2n\pi/3)+\frac{12}{n^2\pi^2}\cos(2n\pi/3)-\frac{6}{n^2\pi^2}(1+(-1)^n)\right]\cos(n\pi x/3),$$

converging to x if $0 \le x < 2$, to 1 if $x = 2$ and to $2 - x$ if $2 < x \le 3$. The sine series is

$$\sum_{n=1}^{\infty}\left[\frac{12}{n^2\pi^2}\sin(2n\pi/3)-\frac{4}{n\pi}\cos(2n\pi/3)+\frac{2}{n\pi}(-1)^n\right]\sin(n\pi x/3),$$

converging to x if $0 \le x < 2$ to 1 if $x = 2$, to $2 - x$ if $2 < x < 3$ and to 0 if $x = 3$.

7. The cosine series is 4, the function itself, for $0 \le x \le 3$. The sine series is

$$\frac{16}{\pi}\sum_{n=1}^{\infty}\frac{1}{2n-1}\sin\left(\frac{(2n-1)\pi x}{3}\right),$$

converging to 0 if $x = 0$ or $x = 3$ and to 4 for $0 < x < 3$.

9. The cosine series is

$$\frac{1}{2}\cos(x)-\frac{2}{\pi}\sum_{n=1}^{\infty}\frac{(-1)^n(2n-1)}{(2n-3)(2n+1)}\cos\left(\frac{(2n-1)x}{2}\right),$$

converging to 0 for $0 \le x < \pi$, to 0 at $x = 2\pi$, to $\cos(x)$ for $\pi < x < 2\pi$, and to $-1/2$ at $x = \pi$. The sine series is

$$\frac{-2}{3\pi}\sin(x/2)+\sum_{n=3}^{\infty}\frac{-2n}{\pi(n^2-4)}(\cos(n\pi/2)+(-1)^n)\sin(nx/2),$$

converging to 0 for $0 \le x < \pi$ and for $x = 2\pi$, to $-1/2$ for $x = \pi$, and to $\cos(x)$ for $\pi < x < 2\pi$.

11. If f is both even and odd, then $f(x) = f(-x) = -f(x)$, so $f(x) = 0$.

13. The series converges to $1/2 - \pi/4$.

Section 13.4 Integration and Differentiation of Fourier Series

1. For $-\pi \le x \le \pi$,

$$x\sin(x) = 1-\frac{1}{2}\cos(x)+2\sum_{n=2}^{\infty}\frac{(-1)^{n+1}}{n^2-1}\cos(nx).$$

f is continuous with continuous first and second derivatives on $[-\pi, \pi]$, and $f(-\pi) = f(\pi)$, so we can differentiate the series term by term to obtain

$$x\cos(x)+\sin(x) = \frac{1}{2}\sin(x)+2\sum_{n=2}^{\infty}\frac{n(-1)^n}{n^2-1}\sin(nx)$$

for $-\pi < x < \pi$.

5. The Fourier series of f on $[-\pi, \pi]$ is

$$\frac{1}{4}\pi+\sum_{n=1}^{\infty}\left[\frac{(-1)^n-1}{\pi n^2}\cos(nx)+\frac{(-1)^{n+1}}{n}\sin(nx)\right].$$

This converges to 0 for $-\pi < x < 0$ and to x for $0 < x < \pi$. Because f is continuous, its Fourier series can be integrated term by term, yielding the integral of the sum of the series. Term by term integration yields

$$\int_{-\pi}^{x} f(t)\,dt = \frac{\pi}{4}(x+\pi)$$

$$+\sum_{n=1}^{\infty}\left(\frac{1}{\pi n^3}((-1)^n-1)\sin(nx)+\frac{(-1)^n}{n^2}\cos(nx)-\frac{1}{n^2}\right)$$

Section 13.5 The Phase Angle Form

1. $f'(t+p) = \lim_{h\to 0} \dfrac{f(t+p+h) - f(t+p)}{h}$
$= \lim_{h\to 0} \dfrac{f(t+h) - f(t)}{h} = f'(t)$

3. $\alpha f(t+p) + \beta f g(t+p) = \alpha f(t) + \beta g(t)$

5.
$$\frac{19}{8} + \frac{1}{\pi^2}\sum_{n=1}^{\infty} c_n \cos\left(\frac{n\pi x}{2} + \delta_n\right)$$
where
$$c_n = \frac{1}{n^2}\sqrt{8 + 5n^2\pi^2 - 12n\pi\sin(3n\pi/2) + 4(n^2\pi^2 - 2)\cos(3n\pi/2)}$$
and
$$\delta_n = -\arctan\left(\frac{\sin(3n\pi/2) - n\pi/2 - 2\cos(3n\pi/2)}{n\pi\sin(3n\pi/2) + \cos(3n\pi/2) - 1}\right).$$

7. The Fourier series is
$$\frac{1}{2} + \frac{2}{\pi}\sum_{n=1}^{\infty} \frac{1}{2n-1}\sin((2n-1)\pi x),$$
so the phase angle form is
$$1 + \frac{2}{\pi}\sum_{n=1}^{\infty} \frac{1}{2n-1}\cos\left((2n-1)\pi x - \frac{\pi}{2}\right)$$

9. Write $f(x) = 1$ for $0 \le x < 1$ and $f(x) = 2$ for $1 < x < 3$. The phase angle form of the Fourier series is
$$\frac{3}{2} + \frac{2}{\pi}\sum_{n=1}^{\infty} \frac{1}{2n-1}\cos\left((2n-1)\frac{\pi x}{2} + \frac{\pi}{2}(1 - (-1)^n)\right).$$

11. Write $f(x) = x$ for $0 \le x < 1$ and $f(x) = x - 2$ for $1 < x < 2$. The phase angle form of the Fourier series is
$$\frac{2}{\pi}\sum_{n=1}^{\infty} \frac{1}{n}\cos\left(n\pi x + (-1)^n\frac{\pi}{2}\right).$$

Section 13.6 Complex Fourier Series

3. $\dfrac{1}{2} + \dfrac{3i}{\pi}\sum_{n=-\infty, n\neq 0}^{\infty} e^{(2n-1)\pi i x/2}$

5. $\dfrac{3}{4} - \dfrac{1}{2\pi}\sum_{n=-\infty, n\neq 0}^{\infty} \dfrac{1}{n}(\sin(n\pi/2) + (\cos(n\pi/2) - 1)i)\, e^{n\pi i x/2}$

7. $3 + \dfrac{3i}{\pi} + \sum_{n=-\infty, n\neq 0}^{\infty} \dfrac{1}{n} e^{2n\pi i x/3}$

Section 13.7 Filtering of Signals

1. The complex Fourier series is
$$\frac{17}{4} + \sum_{n=-\infty, n\neq 0}^{\infty} \left(\frac{1}{2n^2\pi^2}(1 - (-1)^n) + \frac{i}{2n\pi}(6(-1)^n - 5)\right) e^{in\pi t/2}.$$

The Nth Cesàro sum is

$$\sigma_N(t) = \frac{17}{4} + \sum_{n=1}^{N} \frac{1}{n^2\pi^2}\left(1 - \frac{|n|}{N}\right)(1 - (-1)^n)\cos(n\pi t)$$
$$+ \sum_{n=1}^{N} \frac{i}{n\pi}\left(1 - \frac{|n|}{N}\right)(5 - 6(-1)^n)\sin(n\pi t).$$

3. The complex Fourier series is

$$\sum_{n=-\infty, n\neq 0}^{\infty} \frac{i}{n\pi}((-1)^n - 1)e^{in\pi t/2}.$$

The Nth partial sum is

$$S_N(t) = \frac{4}{\pi}\sum_{n=1}^{\infty} \frac{1}{2n-1}\sin\left(\frac{(2n-1)\pi t}{2}\right).$$

The Nth Cesàro sum is

$$\sigma_N(t) = \frac{4}{\pi}\sum_{n=1}^{N}\left(1 - \frac{2n-1}{N}\right)\frac{1}{2n-1}\sin\left(\frac{(2n-1)\pi t}{2}\right).$$

5. The complex Fourier series is

$$\sum_{n=-\infty}^{\infty} \frac{i}{n\pi}\left((-1)^n - \cos\left(\frac{n\pi}{2}\right)\right)e^{in\pi t}.$$

The Nth partial sum is

$$S_N(t) = \sum_{n=1}^{\infty} \frac{2}{n\pi}(\cos(n\pi/2) - (-1)^n)\sin(n\pi t)$$

The Nth Cesàro sum is

$$\sigma_N(t) = \sum_{n=1}^{N}\left(1 - \frac{n}{N}\right)\frac{2}{n\pi}\left[\cos\left(\frac{n\pi}{2}\right) - (-1)^n\right]\sin(n\pi t).$$

CHAPTER FOURTEEN THE FOURIER INTEGRAL AND FOURIER TRANSFORMS

Section 14.1 The Fourier Integral

1.

$$\int_0^\infty \frac{1}{\pi\omega^3}[400\omega\cos(100\omega) + (20000\omega^2 - 4)\sin(100\omega)]\cos(\omega x)\,d\omega,$$

converging to x^2 if $-100 < x < 100$, to 5000 if $x = \pm 100$, and to 0 if $|x| > 100$.

3. The Fourier integral is

$$\int_0^\infty \left[\frac{2\sin(\pi\omega)}{\pi\omega^2} - \frac{2\cos(\pi\omega)}{\omega}\right]\sin(\omega x)\,d\omega,$$

converging to $-\pi/2$ if $x = -\pi$, to x for $-\pi < x < \pi$, to $\pi/2$ for $x = \pi$ and to 0 if $|x| > \pi$.

5. For all x,

$$\int_0^\infty \frac{2}{\pi(1+\omega^2)}\cos(\omega x)\,d\omega = e^{-|x|}.$$

7.

$$\int_0^\infty \left(\frac{2}{\pi\omega}(1 - \cos(\pi\omega))\right)\sin(\omega x)\,d\omega,$$

converging to $-1/2$ at $x=-\pi$, to -1 for $-\pi<x<0$, to 0 if $x=0$, to 1 if $0<x<\pi$, to $1/2$ if $x=\pi$ and to 0 if $|x|>\pi$.

9.

$$\frac{4}{\pi}\int_0^\infty \left[\frac{\cos(\pi\omega)(\cos^2(\pi\omega)-1)}{\omega^2-1}\cos(\omega x)+\frac{\sin(\pi\omega)\cos^2(\pi\omega)}{1-\omega^2}\sin(\omega x)\right]d\omega,$$

converging to $\sin(x)$ if $-3\pi<x<\pi$ and to 0 if $x<-3\pi$ or $x\geq\pi$.

11. Because $f(t)\cos(\omega(t-x))$ is even in ω and $f(t)\sin(\omega(t-x))$ is odd, we can write

$$\frac{1}{\pi}\int_0^\infty\int_{-\infty}^\infty f(t)\cos(\omega(t-x))\,dt\,d\omega=\frac{1}{2\pi}\int_{-\infty}^\infty\int_{-\infty}^\infty f(t)\cos(\omega(t-x))\,d\omega\,dt$$

$$=\frac{1}{2\pi}\int_{-\infty}^\infty\int_{-\infty}^\infty f(t)e^{i\omega(t-x)}\,d\omega\,dt.$$

Complete the derivation by carrying out the inner integration in the last double integral.

Section 14.2 Fourier Cosine and Sine Integrals

1. Sine integral:

$$\int_0^\infty\left[\frac{2}{\pi\omega}[1+(1-2\pi)\cos(\pi\omega)-2\cos(3\pi\omega)]+\frac{4}{\pi\omega^2}\sin(\pi\omega)\right]\sin(\omega x)\,d\omega.$$

Cosine integral:

$$\int_0^\infty\left[\frac{2}{\pi\omega}[(2\pi-1)\sin(\pi\omega)+2\sin(3\pi\omega)]+\frac{4}{\pi\omega^2}[\cos(\pi\omega)-1]\right]\cos(\omega x)\,d\omega.$$

Both integrals converge to $1+2x$ for $0<x<\pi$, to $(3+2\pi)/2$ for $x=\pi$, to 2 for $\pi<x<3\pi$, to 1 for $x=3\pi$, and to 0 for $x>3\pi$. The sine integral converges to 0 at $x=0$, while the cosine integral converges to 1 for $x=0$.

3. Sine integral:

$$\int_0^\infty\frac{2}{\pi}\left(\frac{\omega^3}{4+\omega^4}\right)\sin(\omega x)\,d\omega.$$

Cosine integral:

$$\int_0^\infty\frac{2}{\pi}\left(\frac{2+\omega^2}{4+\omega^4}\right)\cos(\omega x)\,d\omega.$$

Both integrals converge to $e^{-x}\cos(x)$ for $x>0$. the cosine integral converges to 1 for $x=0$ and the sine integral converges to 0 at $x=0$.

5. Sine integral:

$$\int_0^\infty\frac{2}{\pi\omega}[1+\cos(\omega)-2\cos(4\omega)]\sin(\omega x)\,d\omega.$$

Cosine integral:

$$\int_0^\infty\frac{2}{\pi\omega}[2\sin(4\omega)-\sin(\omega)]\cos(\omega x)\,d\omega.$$

Both integrals converge to 1 for $0<x<1$, to $3/2$ for $x=1$, to 2 for $1<x<4$, to 1 for $x=4$ and to 0 for $x>4$. The cosine integral converges to 1 at $x=0$ while the sine integral converges to 0 at $x=0$.

7. Sine integral:

$$\int_0^\infty\frac{2k}{\pi\omega}(1-\cos(c\omega))\sin(\omega x)\,d\omega.$$

Cosine integral:

$$\int_0^\infty\frac{2k}{\pi\omega}\sin(c\omega)\cos(\omega x)\,d\omega.$$

Both integrals converge to k for $0<x<c$, to $k/2$ for $x=c$, and to 0 for $x>c$, while the sine integral converges to 0 at $x=0$, and the cosine integral to k there.

9. Sine integral:

$$\int_0^\infty \frac{4}{\pi\omega^3}[10\omega\sin(10\omega)-(50\omega^2-1)\cos(10\omega)-1]\sin(\omega x)\,d\omega.$$

Cosine integral:

$$\int_0^\infty \frac{4}{\pi\omega^3}[10\omega\cos(10\omega)-(50\omega^2-1)\sin(10\omega)]\cos(\omega x)\,d\omega.$$

Both integrals converge to x^2 for $0\le x<10$, to 50 if $x=10$ and to 0 if $x>10$.

11. From the Laplace integrals and the convergence theorem, for $x>0$,

$$e^{-kx}=\frac{2k}{\pi}\int_0^\infty \frac{1}{k^2+\omega^2}\cos(\omega x)\,d\omega=\frac{2}{\pi}\int_0^\infty \frac{\omega}{k^2+\omega^2}\sin(\omega x)\,d\omega.$$

Put $k=1$ and interchange the symbols x and ω to obtain

$$A_\omega=\frac{\pi e^{-\omega}}{2k}=\int_0^\infty \frac{1}{1+x^2}\cos(\omega x)\,dx$$

and

$$B_\omega=\frac{\pi e^{-\omega}}{2}=\int_0^\infty \frac{x}{1+x^2}\sin(\omega x)\,dx.$$

Therefore, the Fourier cosine integral of $1/(1+x^2)$ is

$$C(x)=\int_0^\infty e^{-\omega}\cos(\omega x)\,d\omega=\frac{1}{1+x^2}\text{ for }x\ge 0.$$

And the Fourier sine integral of $x/(1+x^2)$ is

$$S(x)=\int_0^\infty e^{-\omega}\sin(\omega x)\,d\omega=\frac{x}{1+x^2}\text{ for }x>0.$$

Section 14.3 The Fourier Transform

1. $10e^{-7i\omega}\sin(4\omega)/\omega$ **3.** $2i[\cos(\omega)-1]/\omega$

5. $\dfrac{24}{16+\omega^2}e^{2i\omega}$ **7.** $\dfrac{4}{1+4i\omega}e^{-(1+4i\omega)k/4}$ **9.** $\pi e^{-|\omega|}$

11. $H(t)[2e^{-3t}-e^{-2t}]$ **13.** $18\sqrt{\dfrac{2}{\pi}}e^{-8t^2}e^{-4it}$

15. $H(t+2)e^{-10-(5-3i)t}$ **17.** $H(t)te^{-t}$

19. $\int_{-\infty}^{\infty}|f(t)|^2\,dt=\dfrac{1}{2\pi}\int_{-\infty}^{\infty}\hat f(\omega)\overline{\hat f(\omega)}\,d\omega=\dfrac{1}{2\pi}|\hat f(\omega)|^2\,d\omega$

21. 3π

23. $\dfrac{4}{\omega^3}(\sin(2\omega)(4\omega^2-1)+2\omega\cos(2\omega))+\dfrac{8i}{\omega^2}(2\omega\cos(2\omega)-\sin(2\omega))$

25. $(2/\omega^3)[25\omega^2\sin(5\omega)+10\omega\cos(5\omega)-2\sin(5\omega)]$

27. $\dfrac{1}{1+\omega^2}(1-e^{-4}\cos(4\omega)+e^{-4}\sin(4\omega))+\dfrac{i}{1+\omega^2}(e^{-4}\sin(4\omega)+(e^{-4}\cos(4\omega)-1)\omega)$

Section 14.4 Fourier Cosine and Sine Transforms

1. $\hat f_C(\omega)=\dfrac{1}{2}\left[\frac{1}{1+(1+\omega^2)}+\dfrac{1}{1+(1-\omega^2)}\right];$

$$\hat f_S(\omega)=\frac{1}{2}\left[\frac{1+\omega}{1+(1+\omega^2)}-\frac{1-\omega}{1+(1-\omega^2)}\right]$$

3. $\hat f_C(\omega)=\dfrac{1}{1+\omega^2},\ \hat f_S(\omega)=\dfrac{\omega}{1+\omega^2}$

5. For $\omega \neq \pm 1$,

$$\hat{f}_C(\omega) = \frac{1}{2}\left[\frac{\sin(K(1-\omega))}{1-\omega} + \frac{\sin(K(1+\omega))}{1+\omega}\right],$$

$$\hat{f}_C(-1) = \hat{f}_C(1) = \frac{K}{2} + \frac{1}{2}\sin(2K)$$

For $\omega \neq \pm 1$,

$$\hat{f}_S(\omega) = \frac{\omega}{\omega^2 - 1} - \frac{1}{2}\left[\frac{\cos(K(1+\omega))}{1+\omega} - \frac{\cos(K(1-\omega))}{1-\omega}\right],$$

$$\hat{f}_S(1) = \frac{1}{4}(1 - \cos(2K)) = -\hat{f}_S(-1)$$

Section 14.5 The Discrete Fourier Transform

1. Approximate values are given in Table A.7.

3. Approximate values are given in Table A.8.

5. Approximate values are given in Table A.9.

7. $u_j = \frac{1}{7}\sum_{k=0}^{6} e^{-ik} e^{2\pi ijk/7}$

$u_0 = 0.103479 + 0.014751i$, $u_1 = 0.933313 - 0.296094$,
$u_2 = -0.094163 + 0.088785i$, $u_3 = -0.023947 + 0.062482i$,
$u_4 = 0.004307 + 0.051899i$, $u_5 = 0.025788 + 0.043852i$,
$u_5 = 0.051222 + 0.034325i$

TABLE A.7 *Approximate Discrete Fourier Transform Values in Problem 1, Section 14.5*

k	$\mathcal{D}$[u](k)
−4	−14.00000 + 10.39230i
−3	−15.00000 + 0.22023(10^{-7})i
−2	−14.00000 − 10.39230i
−1	−6.00000 − 31.17691i
0	55.00000
1	−6.00000 + 31.17691i
2	−14.00000 + 10.39230i
3	−15.00000 − 0.22023(10^{-7})i
4	−14.00000 − 10.39230i

TABLE A.8 *Approximate Discrete Fourier Transform Values in Problem 3, Section 14.5*

k	$\mathcal{D}$[u](k)
−4	0.13292 − 0.01658i
−3	0.09624 + 0.72830(10^{-9})i
−2	0.13292 + 0.01658i
−1	2.93687 + 0.42794i
0	0.1.82396
1	2.93687 − 0.42794i
2	0.13292 − 0.01658i
3	0.09624 − 0.72830(10^{-9})i
4	0.13292 + 0.01658i

TABLE A.9 *Approximate Discrete Fourier Transform Values in Problem 5, Section 14.5*

k	$\mathcal{D}$[u](k)
–4	0.65000 – 0.17321i
–3	0.61667 – 0.25346(10^{-9})i
–2	0.65000 + 0.17321i
–1	0.81667 + 0.40415i
0	2.45000
1	0.81667 – 0.40415i
2	0.65000 – 0.17321i
3	0.61667 + 0.25346(10^{-9})i
4	0.65000 + 0.17321i

TABLE A.10 *Approximate Values in Problem 13, Section 14.5*

k	Complex Coefficients d_k	DFT Approximations f_k
–3	0.005629 – 0.053051i	0.001733 – 0.052956i
–2	0.012665 – 0.079577i	0.008769 – 0.079514i
–1	0.050661 – 0.159155i	0.046765 – 0.159123i
0	0.33333	0.329437
1	0.050661 + 0.159155i	0.046765 + 0.159123i
2	0.012665 + 0.179577i	0.008769 + 0.079514i
3	0/005629 + 0.053052i	0.001733 + 0.052956i

TABLE A.11 *Approximate Values in Problem 15, Section 14.5*

k	Complex Coefficients d_k	DFT Approximations f_k
–3	–0.005177 + 0.075984i	0.000346 + 0.075849i
–2	0.011816 + 0.115622i	–0.006293 + 0.115532i
–1	–0.051259 + 0.250780i	–0.045737 + 0.250753i
0	0.454649	0.460171
1	–0.051259 – 0.250798i	–0.045737 – 0.250753i
2	–0.011816 – 0.115622i	–0.006293 – 0.075849i
3	–0.005177 – 0.075984i	0.000346 – 0.075849i

9. $u_j = \frac{1}{5}\sum_{k=0}^{4} co(k)e^{2\pi ijk/5}$

$u_0 = 1.096542,\ u_1 = -0.249644 - 0.232302i,\ u_2 = -0.201697 - 0.084840i$

$u_3 = -0.193858,\ u_4 = -0.201697 + 0.084840i,\ u_5 = -0.249644 + 0.232302i$

11. $u_j = \frac{1}{6}\sum_{k=0}^{5}(1+i)^k e^{2\pi ijk/6}$

$u_0 = -1.33333 + 0.166667i,\ u_1 = -0.427030 + 0.549038i$

$u_2 = -0.016346 + 0.561004i,\ u_3 = 0.33333 + 0.500000i$

$u_4 = 0.849679 + 0.272329i,\ u_5 = 1.593696 - 2.049038i$

13. Approximate values are given in Table A.10.

15. Approximate values are given in Table A.11.

Section 14.6 Sampled Fourier Series

1. $U_0 = 58.901925$. The other approximate values are given in Table A.12.
3. $U_0 = 31.501953$. The other approximate values are given in Table A.13.
5. $U_0 = 255$. The other approximate values are given in Table A.14.

TABLE A.12 *Approximate Values in Problem 1, Section 14.6*

U_n	Approximate Value	U_n	Approximate Value
U_1	–5.854287 – 32.096339i	U_{118}	0.647851 + 2.829713i
U_2	–0.805518 – 14.788044i	U_{119}	0.633992 + 3.157208i
U_3	0.44274 – 9.708611i	U_{120}	0.614603 + 3.565443i
U_4	0.336014 – 7.235154i	U_{121}	0.586989 + 4.089267i
U_5	0.470070 – 5.764387i	U_{122}	0.542633 + 4.787014i
U_6	0.542633 – 4.787014i	U_{123}	0.470070 + 5.764387i
U_7	0.586299 – 4.089267i	U_{124}	0.336014 + 7.235154i
U_8	0.614603 – 3.565443i	U_{125}	0.044274 + 9.708611i
U_9	0.633991 – 3.157208i	U_{126}	–0.805518 + 14.788044i
U_{10}	0.647851 – 2.829712i	U_{127}	–5.854287 + 32.096339i

TABLE A.13 *Approximate Values in Problem 3, Section 14.6*

U_n	Approximate Value	U_n	Approximate Value
U_1	9.228787 + 17.271595i	U_{118}	–0.400755 – 1.993017i
U_2	1.933662 + 9.790716i	U_{119}	–0.377943 – 2.222355i
U_3	0.582715 + 6.663663i	U_{120}	–0.346050 – 2.507623i
U_4	0.109884 + 5.028208i	U_{121}	–0.299528 – 2.872545i
U_5	–0.108968 + 4.029124i	U_{122}	–0.227849 – 3.356393i
U_6	–0.227849 + 3.356393i	U_{123}	–0.108968 – 4.029124i
U_7	–0.299528 + 2.872544i	U_{124}	0.109884 – 5.028208i
U_8	–0.346050 + 2.507623i	U_{125}	0.582715 – 6.663663i
U_9	–0.377943 + 2.222355i	U_{126}	1.933662 – 9.790715i
U_{10}	–0.400755 + 1.993017i	U_{127}	9.228757 – 17.271595i

TABLE A.14 *Approximate Values in Problem 5, Section 14.6*

U_n	Approximate Value	U_n	Approximate Value
U_1	–1 + 40.735484i	U_{118}	–1 – 3.992224i
U_2	–1 + 20.355468i	U_{119}	–1– 4.453202i
U_3	–1 + 13.556669i	U_{120}	–1 – 5.027339i
U_4	–1 + 10.153170i	U_{121}	–1 – 5.763142i
U_5	–1 - 8.107786i	U_{122}	–1 – 6.741452i
U_6	–1 + 6.741452i	U_{123}	–1 – 8.107786i
U_7	–1 + 5.763142i	U_{124}	–1 – 10.153170i
U_8	–1 + 5.027339i	U_{125}	–1 – 13.556670i
U_9	–1 + 4.453202i	U_{126}	–1 – 20.355468i
U_{10}	–1 + 3.992224i	U_{127}	–1 – 40.735484i

Section 14.7 DFT Approximation of the Fourier Transform

1.

$$\hat{f}(\omega) = \frac{4-\omega^2}{(\omega^2+4)^2} - \frac{4\omega i}{(\omega^2+4)^2},\ \hat{f}(12) \approx -0.006392 - 0.002191i,$$

$$\text{DFT approximation} = \frac{3\pi}{256}\sum_{j=0}^{511} f\left(\frac{3\pi j}{256}\right)e^{-9\pi ij/64} = -0.006506 - 0.002191i$$

3.

$$\hat{f}(\omega) = \frac{4-4i\omega}{\omega^2+16},\ \hat{f}(4) = \frac{1}{8}(1-i),$$

$$\text{DFT approximation} = \frac{3\pi}{256}\sum_{j=0}^{511} f\left(\frac{3\pi j}{256}\right)e^{-3\pi ij/64} = 0.143860 - 0.124549i$$

CHAPTER FIFTEEN SPECIAL FUNCTIONS AND EIGENFUNCTION EXPANSIONS

Section 15.1 Eigenfunction Expansions

1. Regular on [0, 1], eigenvalues are positive solutions of $\tan(\sqrt{\lambda}) = \frac{1}{2}\sqrt{\lambda}$. If λ is an eigenvalue, an eigenfunction is $2\sqrt{\lambda}\cos(\sqrt{\lambda}x) + \sin(\sqrt{\lambda}x)$.
3. Regular on $[0, \pi]$, $1+n^2$ and $e^{-x}\sin(nx)$ for $n = 1, 2, \cdots$
5. Regular on [0, 4], $((2n-1)\pi/8)^2$, $\cos((2n-1)\pi x/8)$
7. The problem is regular on $[0, L]$ with eigenvalues $((2n-1)\pi/2L)^2$ for $n = 1, 2, \cdots$. The functions $\sin((2n-1)\pi x/2L)$ are eigenfunctions.
9. Periodic on $[-3\pi, 3\pi]$, $n^2/9$ for $n = 0, 1, 2, \cdots$, $a_n\cos(nx/3) + b_n\sin(nx/3)$ with not both a_n and b_n equal to 0
11. The expansion is

$$\sum_{n=1}^{\infty} \frac{4}{\pi}\frac{\sqrt{2}\cos(n\pi/2) - \sqrt{2}\sin(n\pi/2) - (-1)^n}{2n-1}\cos\left(\frac{2n-1}{8}\pi x\right).$$

This converges to -1 for $0 < x < 2$, to 1 for $2 < x < 4$, and to 0 at $x = 0$.

13. For $-3\pi < x < 3\pi$,

$$x^2 = 3\pi^2 + 36\sum_{n=1}^{\infty}\frac{(-1)^n}{n^2}\cos(nx/3).$$

15. For $0 < x < 1$,

$$1 - x = \sum_{n=1}^{\infty}\frac{2}{n\pi}(1 + (-1)^n(L-1))\sin(n\pi x).$$

17. Compute both sides of Bessel's inequality. With some rearrangement, obtain

$$\sum_{n=1}^{\infty}\left(\frac{4(-1)^n + (2n-1)\pi}{(2n-1)^3\pi^3}\right)^2 \le \frac{512}{15}\frac{2}{256^2} = \frac{1}{960}$$

Section 15.2 Legendre Polynomials

1. Use the binomial series to write

$$\frac{1}{\sqrt{1-2at+t^2}} = \sum_{n=0}^{\infty} \frac{(-1/2)(-3/2)\cdots(-1/2-n+1)}{n!}(t^2-2at)^n$$
$$= \sum_{n=0}^{\infty} \sum_{j+k=n, k\le j} \frac{(-1)^k 1\cdot 3\cdots(2j-1)j!(-1)^{j-k}2^{j-k}a^{j-k}}{2^j j!k!(j-k)!} t^n.$$

Show that the coefficient of t^n is $P_n(a)$. For $r < d$, let $a = \cos(\theta)$ and $t = r/d$. For $r > d$, use $a = \cos(\theta)$ and $t = d/r$.

3. With $n = 5$ in the recurrence relation,

$$P_6(x) = \frac{1}{6}(11xP_5(x) - 5P_4(x))$$
$$= \frac{1}{16}(231x^6 - 315x^4 + 105x^2 - 5)$$

Similarly,

$$P_7(x) = \frac{1}{16}(429x^7 - 693x^5 + 315x^3 - 35x),$$
$$P_8(x) = \frac{6435}{128}x^8 - \frac{3003}{32}x^6 + \frac{3465}{64}x^4 - \frac{315}{32}x^2 + \frac{35}{128}$$

4. (a)

$$1 + 2x - x^2 = \frac{2}{3}P_0(x) + 2P_1(x) - \frac{2}{3}P_2(x)$$

(c)

$$2 - x^2 + 4x^4 = \frac{37}{15}P_0(x) + \frac{34}{21}P_2(x) + \frac{32}{35}P_4(x)$$

5. For $n = 2$, this gives

$$\frac{1}{4(2!)}\frac{d^2}{dx^2}((x^2-1)^2) = \frac{1}{2}(3x^2-1) = P_2(x).$$

9. For $-1 < x < 1$,

$$\sin^2(x) = -\frac{1}{2}\cos(1)\sin(1) + \frac{1}{2}$$
$$+ \left[-\frac{5}{8}\cos(1) + \frac{15}{8} - \frac{15}{4}\cos^2(1)\right]P_2(x)$$
$$+ \left[\frac{531}{32}\cos(1)\sin(1) - \frac{585}{32} + \frac{585}{16}\cos^2(1)\right]P_4(x) + \cdots$$

11. For $-1 < x < 0$ and $0 < x < 1$,

$$f(x) = \frac{3}{2}P_1(x) - \frac{7}{8}P_3(x) + \frac{11}{16}P_5(x) + \cdots$$

13. For $-1 < x < 1$,

$$\sin(\pi x/2) = \frac{12}{\pi^2}x + 168\left(\frac{\pi^2 - 10}{\pi^4}\right)P_3(x)$$
$$+ 660\left(\frac{-112\pi^2 + \pi^4 + 1008}{\pi^6}\right)P_5(x) + \cdots$$

Section 15.3 Bessel Functions

1. With $y = x^a J_\nu(bx^c)$, compute

$$y' = ax^{a-1}J_\nu(bx^c) + x^a bcx^{c-1}J_\nu'(bx^c)$$
$$y'' = a(a-1)x^{a-2}J_\nu(bx^c) + (2ax^{a-1}bcx^{c-1} + x^a bc(c-1)x^{c-2})J_\nu'(bx^c)$$
$$+ x^a c^2 x^{2c-2} J_\nu''(bx^c).$$

Substitute these into the differential equation to verify that $y = J_\nu(bx^c)$ is a solution.

3. $y = c_1 x^{-1} J_{3/4}(2x^2) + c_2 x^{-1} J_{-3/4}(2x^2)$

5. $y = c_1 x^4 J_{3/4}(2x^3) + c_2 x^4 J_{-3/4}(2x^3)$

7. $y = c_1 x^{-2} J_{1/2}(3x^3) + cx^{-2} J_{-1/2}(3x^3)$

9. $y = c_1 J_{1/3}(x^2) + c_2 J_{-1/3}(x^2)$

11. The differential equation transforms to

$$z^2 y'' + zy' + (z^2 - 16)y = 0$$

giving $y(x) = c_1 J_4(2x^{1/3}) + c_2 Y_4(2x^{1/3})$.

13. The transformed differential equation is

$$x^2 u'' + xu' + (x^2 - 1/4)u = 0$$

leading to $y = c_1 x^{2/3} J_{1/2}(x) + c_2 x^{2/3} Y_{1/2}(x)$.

15. The differential equation transforms to

$$z^2 y'' + zy' + (z^2 - 9)y = 0$$

with general solution

$$y(z) = c_1 J_3(z) + c_2 Y_3(z)$$

so

$$y(x) = c_1 J_3(\sqrt{x}) + c_2 Y_3(\sqrt{x}).$$

17. From the infinite series, it is easy to check that $(x^n J_n(x))' = x^n J_{n-1}(x)$. Integrating this yields the first conclusion. Next, $(x^{-n} J_n(x))' = -x^{-n} J_{n+1}(x)$. Integrating this gives the second expression.

19. It is routine to check from the infinite series expansions that $J_0'(s) = -J_1(s)$. Then

$$\int_0^\alpha J_1(s)\,ds = -J_0(s)]_0^\alpha = J_0(\alpha) = J_0(0) - J_0(\alpha) = 1 - 0 = 1.$$

Now let $s = \alpha x$ to complete the solution.

21. Begin with the observation that $\int x^{n+1} J_n(\alpha x)\,dx = (1/\alpha)x^{n+1} J_{n+1}(\alpha x)$. Then

$$I_{n,0} = \int_0^1 x^{n+1} J_n(\alpha x) = \frac{J_{n+1}(\alpha)}{\alpha}.$$

giving part (a). Part (b) follows by using the quoted identity again. The other parts follow by using the given hints.

23. Put $t = x\sin(\theta)$ in Hankel's integral, which is given in Problem 22.

25. $f(x) = xe^{-x}$

$$1.418532841 J_2(5.135622302x) + 0.2923912667 J_2(8.417244140x)$$
$$+ 0.7581692534 J_2(11.61984117x) + 0.1399888559 J_2(14.79595178x)$$
$$+ 0.5434687461 J_2(17.95981949x)$$

27. $f(x) = x$

$$7.749400696 J_2(5.135622302x) - 0.1583973994 J_2(8.417244140x)$$
$$+ 1.310726377 J_2(11.61984117x) - 0.2381008476 J_2(14.79595178x)$$
$$+ 0.9524470038 J_2(17.95981949x)$$

29. $f(x) = \sin(\pi x)$

$$3.733991576J_2(5.135622302x) + 2.468532251J_2(8.417244140x)$$
$$+ 1.700629359J_2(11.61984117x) + 1.356527124J_2(14.79595178x)$$
$$+ 1.099075410J_2(17.95981949x)$$

31. $f(x) = x$

$$2.213145642J_1(3.831705970x) - 0.5170987826J_1(7.015586670x)$$
$$+ 1.104611216J_1(10.17346814x) - 0.4549641786J_1(13.32369194x)$$
$$+ 0.8113206562J_1(16.47063005x)$$

33. $f(x) = xe^{-x}$

$$1.256395517J_1(3.831705970x) + 0.08237394412J_1(7.015586670x)$$
$$+ 0.5976577270J_1(10.173468144x) - 0.01994105804J_1(13.32369194x)$$
$$+ 0.4181324338J_1(16.47063005x)$$

35. $f(x) = \sin(\pi x)$

$$3.555896220J_1(3.831705970x) + 1.670058301J_1(7.015586670x)$$
$$+ 0.9956101332J_1(10.173468144x) + 0.7772068876J_1(13.32369194x)$$
$$+ 0.6036626350J_1(16.47063005x)$$

37. Let $t = u/(1+u)$ to obtain

$$B(x,y) = \int_0^\infty t^{x-1}(1-t)^{y-1}\,dt$$
$$\int_0^\infty \left(\frac{u}{1+u}\right)^{x-1}\left(\frac{1}{1+u}\right)^{y-1}\frac{1}{(1+u)^2}\,du$$
$$= \int_0^\infty \frac{u^{x-1}}{(1+u)^{x+y}}\,du.$$

39. With $t = ry$,

$$r^x \int_0^\infty t^{x-1}e^{-rt}\,dt = r^x \int_0^\infty e^{-y}\left(\frac{y}{r}\right)^{x-1}\frac{1}{r}\,dy$$
$$= r^x \int_0^\infty y^{x-1}e^{-y}\frac{1}{r^x}\,dy$$
$$= \int_0^\infty y^{x-1}e^{-y}\,dy = \Gamma(x),$$

with y used instead of t as the variable of integration in the last line.

CHAPTER SIXTEEN THE WAVE EQUATION

Section 16.1 Derivation of the Equation

1. Compute

$$\frac{\partial^2 y}{\partial x^2} = \frac{1}{2}(f''(x+ct) + f''(x-ct))$$

and

$$\frac{\partial^2 y}{\partial t^2} = \frac{1}{2}(c^2 f''(x+ct) + c^2 f''(x-ct))$$

3. The wave equation is

$$\frac{\partial^2 z}{\partial t^2} = c^2\left(\frac{\partial^2 z}{\partial x^2} + \frac{\partial^2 z}{\partial y^2}\right)$$

for $0 < x < a, 0 < y < b$; boundary conditions are

$$z(0, y, t) = z(a, y, t) = z(x, 0, t) = z(x, b, t) = 0$$

for $t \geq 0$; initial conditions are

$$z(x, y, 0) = f(x, y), \frac{\partial z}{\partial t}(x, y, 0) = g(x, y).$$

5. Compute

$$\frac{\partial^2 y}{\partial t^2} = -\frac{n^2\pi^2c^2}{L^2}\sin\left(\frac{n\pi x}{L}\right)\cos\left(\frac{n\pi ct}{L}\right) \text{ and}$$
$$\frac{\partial^2 y}{\partial x^2} = -\frac{n^2\pi^2}{L^2}\sin\left(\frac{n\pi x}{L}\right)\cos\left(\frac{n\pi ct}{L}\right)$$

Section 16.2 Wave Motion on an Interval

1. $y(x, t) = \sum_{n=1}^{\infty} \frac{108}{(2n-1)^4\pi^4}\sin((2n-1)\pi x/3)\sin(2(2n-1)\pi t/3)$

3. $y(x, t) = \sum_{n=1}^{\infty} \frac{-32}{(2n-1)^3\pi^3}\sin((2n-1)\pi x/2)\cos(3(2n-1)\pi t/2)$
$+ \sum_{n=1}^{\infty} \frac{4}{n^2\pi^2}\left[\cos(n\pi/4) - \cos(n\pi/2)\right]\sin(n\pi x/2)\sin(3n\pi t/2)$

5. $y(x, t) = \sum_{n=1}^{\infty} \left(\frac{16\sin(n\pi/2) - 8n\pi\cos(n\pi/2)}{n^3\pi^3 c}\right)\sin\left(\frac{n\pi x}{2}\right)\sin\left(\frac{n\pi ct}{2}\right)$

7. $y(x, t) = \sum_{n=1}^{\infty} \frac{24(-1)^{n+1}}{(2n-1)^2\pi}\sin((2n-1)x/2)\cos((2n-1)\sqrt{2}t)$

9. Let $Y(x, t) = y(x, t) + h(x)$ and find that $h(x) = \cos(x) - 1$. The problem for Y is

$$\frac{\partial^2 Y}{\partial t^2} = \frac{\partial^2 Y}{\partial x^2}$$
$$Y(0, t) = Y(2\pi, t) = 0,$$
$$Y(x, 0) = \cos(x) - 1, \frac{\partial Y}{\partial t}(x, 0) = 0,$$

with solution

$$Y(x, t) = \frac{16}{\pi}\sum_{n=1}^{\infty} \frac{1}{(2n-1)[(2n-1)^2 - 4]}\sin((2n-1)x/2)\cos((2n-1)t/2).$$

11. $u(x, t) = e^{-At/2}\sum_{n=1}^{\infty} C_n \sin(n\pi x/L)\left[\frac{1}{AL}r_n\cos(r_n t/2L) + \sin(r_n t/2L)\right]$,

where

$$C_n = \frac{2}{L}\int_0^L f(\xi)\sin(n\pi\xi/L)\,d\xi$$

and

$$r_n = \sqrt{4(BL^2 + n^2\pi^2c^2) - A^2L^2}.$$

13. (a) With the forcing term, the solution is

$$y_f(x, t) = \sum_{n=1}^{\infty} d_n \sin(n\pi x/4)\cos(3n\pi t/4)$$
$$+ \frac{1}{9\pi^2}(\cos(\pi x) - 1),$$

where

$$d_n = \begin{cases} \frac{-32(1-(-1)^n)(288-17n^2)}{9n^3\pi^3(n^2-16)} & \text{for } n \neq 4, \\ 0 & \text{for } n = 4. \end{cases}$$

(b) Without the forcing term, the solution is

$$y(x,t) = \sum_{n=1}^{\infty} \frac{128}{\pi^3(2n-1)^3} \sin((2n-1)\pi x/3)\cos(3(2n-1)\pi t/4).$$

15. Let $Y(x,t) = y(x,t) + h(x)$ and substitute into the problem to choose $h(x) = (x^3 - 4x)/9$. The problem for Y is

$$\frac{\partial^2 Y}{\partial t^2} = 3\frac{\partial^2 Y}{\partial x^2}$$

$$Y(0,t) = Y(2,t) = 0,$$

$$Y(x,0) = \frac{1}{9}(x^3 - 4x),\ \frac{\partial Y}{\partial t}(x,0) = 0,$$

with solution

$$Y(x,t) = \sum_{n=1}^{\infty} \frac{32(-1)^n}{3n^3\pi^3} \sin(n\pi x/2)\cos(n\pi\sqrt{3}t/2).$$

Section 16.3 Wave Motion in an Infinite Medium

1. $y(x,t) = \int_0^\infty \frac{1}{2\pi\omega}\frac{\sin(\pi\omega)}{1-\omega^2}\sin(\omega x)\sin(4\omega t)\,d\omega$

3. $y(x,t) = \int_0^\infty \left[\left(\frac{e^{-2}}{3\pi\omega}\frac{2\cos(\omega) - \omega\sin(\omega)}{4+\omega^2}\right)\cos(\omega x) + \left(\frac{e^{-2}}{3\pi\omega}\frac{\omega\cos(\omega) + 2\sin(\omega)}{4+\omega^2}\right)\sin(\omega x)\right]\sin(3\omega t)\,d\omega$

5. $y(x,t) = \int_0^\infty \frac{10}{\pi(25+\omega^2)}\cos(\omega x)\cos(12\omega t)\,d\omega$

Section 16.4 Wave Motion in a Semi-Infinite Medium

1. $y(x,t) = \int_0^\infty \frac{1}{\pi\omega}\frac{\sin(\pi\omega/2) - \sin(5\pi\omega/2)}{\omega^2 - 1}\sin(\omega x)\sin(2\omega t)\,d\omega$

3.

$$y(x,t) = \int_0^\infty \frac{3}{7\pi\omega^5} d_\omega \sin(\omega x)\sin(14\omega t)\,d\omega,$$

where

$$d_\omega = 2\sin(3\omega) - 4\omega\cos(3\omega) - 3\omega^2\sin(3\omega) - 2\omega$$

5. $y(x,t) = \int_0^\infty \frac{2}{\pi}\frac{2 - \omega\sin(\omega) - 2\cos(\omega)}{\omega^3}\sin(\omega x)\cos(3\omega t)\,d\omega$

Section 16.5 Laplace Transform Techniques

1. $y(x,t) = f\left(t - \frac{x}{c}\right)H\left(t - \frac{x}{c}\right) - \frac{1}{6}Axt^4$

3. $y(x,t) = \left[f\left(t - \frac{x}{c}\right) - \frac{K}{2}\left(t - \frac{x}{c}\right)^2\right]H\left(t - \frac{x}{c}\right) + \frac{1}{2}Kt^2$

5. $y(x,t) = \frac{A}{6}\left(t - \frac{x}{c}\right)^3 H\left(t - \frac{x}{c}\right) - \frac{A}{6}t^3$

Section 16.6 Characteristics and d'Alembert's Solution

1. $y(x,t) = \frac{1}{2}[\cosh(x-3t) + \cosh(x+3t)] + t + \frac{9}{10}xt^5$

3. $y(x,t) = x + \frac{1}{8}(e^{-x+4t} - e^{-x-4t}) + \frac{1}{2}xt^2 + \frac{1}{6}t^3$

5. $y(x,t) = \frac{1}{32}(\sin(2(x+8t)) - \sin(2(x-8t))) + \frac{1}{12}xt^4 + x^2 + 64t^2 - x$

7. $y(x,t) = \frac{1}{2}[(x-t)^2 + (x+t)^2] + \frac{1}{2}\int_{x-t}^{x+t} -\xi\, d\xi = x^2 + t^2 - xt$

9. $y(x,t) = \frac{1}{2}[e^{x-14t} + e^{x+14t}] + xt = e^x \cosh(14t) + xt$

11. $y(x,t) = \frac{1}{2}[\cos(\pi(x-7t)) + \cos(\pi(x+7t))] + t - x^2 t - \frac{49}{3}t^3 = \frac{1}{2}\cos(\pi x)\cos(7\pi t) + t - x^2 t - \frac{49}{3}t^3$

13. $\frac{1}{2}(\cos(x-ct) + \cos(x+ct))$

17. $\frac{1}{2}(\sin(2(x-ct)) + \sin(2(x+ct)))$

Section 16.7 Vibrations in a Circular Membrane

1. We find that (approximately),

$$a_1 = \frac{2\int_0^1 xJ_0(2.405x)\,dx}{[J_1(2.405)]^2} \approx 2\frac{0.1057}{0.2695} = 0.78442,$$
$$a_2 \approx 0.06869,\ a_3 \approx 0.05311,\ a_4 \approx 0.01736,\ a_5 \approx 0.01698$$

The fifth partial sum gives the approximation

$$\begin{aligned} z(r,t) \approx\ & 0.78442J_0(2.405r)\cos(2.405t) + 0.05311J_0(5.520r)\cos(5.520t) \\ & + 0.06869J_0(8.654r)\cos(8.654t) + 0.01736J_0(11.792r)\cos(11.792t) \\ & + 0.01698J_0(14.931r)\cos(14.931t). \end{aligned}$$

3. We find the approximation

$$\begin{aligned} z(r,t) \approx\ & 1.2534J_0(2.405r)\cos(2.405t) - 0.80469J_0(5.520r)\cos(5.520t) \\ & - 0.11615J_0(8.654r)\cos(8.654t) - 0.09814J_0(11.792r)\cos(11.792t) \\ & - 0.03740J_0(14.931r)\cos(14.931t) \end{aligned}$$

Section 16.8 Vibrations in a Circular Membrane II

1. The solution is

$$\begin{aligned} z(r,\theta,t) = & \sum_{k=1}^{\infty} \alpha_k J_0\left(\frac{j_{0k}}{2}r\right)\cos(j_{0k}t) \\ & + \cos(2\theta)\sum_{k=1}^{\infty} \beta_k J_2\left(\frac{j_{2k}}{2}r\right)\cos(j_{2k}t) \\ & + \sum_{p=1}^{\infty} \sin(p\theta)\sum_{q=1}^{\infty} \delta_{pq} J_p\left(\frac{j_{pq}}{2}r\right)\sin(j_{pq}t) \end{aligned}$$

where

$$\alpha_k = \left(\frac{2}{[J_1(j_{0k})]^2}\int_0^1 \xi(1-\xi^2)J_0(j_{0k}\xi)\,d\xi\right),$$

$$\beta_k = \left(\frac{4}{[J_3(j_{0k})]^2} \int_0^1 \xi(\xi^2 - 1) J_2(j_{2k}\xi) \right),$$

$$\delta_{pq} = \left(\frac{4(-1)^{p+1}}{p j_{pq} [J_{p+1}(j_{pq})]^2} \int_0^1 \xi J_p(j_{pq}\xi)\, d\xi \right).$$

Computing some of these terms, we have

$$\begin{aligned}
&z(r,\theta,t) \approx \\
&1.108022 J_0(1.202413r)\cos(2.404826t) - 0.13977 J_0(2.760039r)\cos(5.520078t) \\
&+ 0.045476 J_0(4.32686r)\cos(8.653728t) - 0.02099 J_0(5.895767r)\cos(11.79153t) \\
&+ 0.011636 J_0(7.465459t)\cos(14.930918t) + \cdots \\
&+ \cos(2\theta)[-2.976777 J_2(2.567811r)\cos(5.135622t) \\
&- 1.434294 J_2(4.208622r)\cos(8.417244t) + \cdots] \\
&+ \sin(\theta)[1.155175 J_1(1.915853r)\sin(3.831706t) \\
&- 0.14741 J_1(3.507794r)\sin(7.015587t) + \cdots] + \cdots.
\end{aligned}$$

Section 16.9 Vibrations in a Rectangular Membrane

1. $z(x,y,t) = \frac{1}{\pi}\sum_{n=1}^{\infty}\left[\frac{8(-1)^{n+1}\pi^2}{n} + \frac{16}{n^3}[(-1)^n - 1]\right]\sin(nx/2)\sin(y)\cos(\sqrt{n^2+4}t/2)$

3. $z(x,y,t) = \sum_{n=1}^{\infty}\sum_{m=1}^{\infty} d_{nm}\cos((2n-1)x/2)\sin((2m-1)y/2)\sin(k_{nm}t)$,

where

$$d_{nm} = \frac{16}{\pi^2(2n-1)(2m-1)\sqrt{(2n-1)^2+(2m-1)^2}}$$

and

$$k_{nm} = \sqrt{(2n-1)^2+(2m-1)^2}.$$

CHAPTER SEVENTEEN THE HEAT EQUATION

Section 17.1 Initial and Boundary Conditions

1. $\frac{\partial u}{\partial t} = k\frac{\partial^2 u}{\partial x^2}$ for $t > 0, 0 < x < L$,

$\frac{\partial u}{\partial x}(0,t) = 0, u(L,t) = \beta(t)$ for $t > 0$,

$u(x,0) = f(x)$

3. $\frac{\partial u}{\partial t} = k\frac{\partial^2 u}{\partial x^2}$ for $t > 0, 0 < x < L$,

$u(0,t) = \frac{\partial u}{\partial x}(L,t) = 0$ for $t > 0$,

$u(x,0) = f(x)$

Section 17.2 The Heat Equation on $[0, L]$

In these solutions, we sometimes use the notation $\exp(g(t)) = e^{g(t)}$.

1.

$$u(x,t) = \sum_{n=1}^{\infty} d_n \sin((2n-1)\pi x/L)\exp(-3(2n-1)^2\pi^2 t/L^2)$$

where

$$d_n = \frac{-16L}{(2n-1)\pi[(2n-1)^2-4]}.$$

3. $u(x,t) = \frac{1}{6}(1-e^{-6}) + 12\sum_{n=1}^{\infty}\left(\frac{1-e^{-6}(-1)^n}{36+n^2\pi^2}\right)\cos(n\pi x/6)e^{-n^2\pi^2 t/18}$

5. $u(x,t) = \sum_{n=1}^{\infty}\frac{8L^2}{(2n-1)^3\pi^3}\sin((2n-1)\pi x/L)\exp(-(2n-1)^2\pi^2 kt/L^2)$

7. $u(x,t) = \frac{2}{3}\pi^2 - \sum_{n=1}^{\infty}\frac{4}{n^2}\cos(nx)e^{-4n^2 t}$

9. Substitute $e^{\alpha x+\beta t}v(x,t)$ into the partial differential equation, and solve for α and β, so that $v_t = kv_{xx}$. Obtain $\alpha = -A/2$ and $\beta = k(B - A^2/4)$.

11. $u(x,t) = \sum_{n=1}^{\infty}\frac{4B}{(2n-1)\pi}\sin((2n-1)\pi x/2L)\exp(-(2n-1)^2\pi^2 kt/4L^2)$

13. Let $u(x,t) = v(x,t) + f(x)$ and choose $f(x) = 3x+2$. We obtain $v(0,t) = v(1,t) = 0$ and $u(x,0) = x^2 - f(x)$. The solution for v is

$$v(x,t) = 2\sum_{n=1}^{\infty}\frac{4}{n^3\pi^3}\left((-1)^n(1+2n^2\pi^2) - (1+n^2\pi^2)\right)\sin(n\pi x)e^{-16n^2\pi^2 t}.$$

15. Let $u(x,t) = e^{-At}W(x,t)$. Then $w(0,t) = w(9,t) = 0$ and $u(x,0) = 3x$. We find that

$$w(x,t) = \sum_{n=1}^{\infty}\frac{54(-1)^{n+1}}{n\pi}\sin(n\pi x/9)e^{-4n^2\pi^2 t/81}.$$

17. Let $u(x,t) = e^{-3x-9t}v(x,t)$. Then $v(0,t) = v(4,t) = 0$ and $v(x,0) = e^{3x}$, so

$$v(x,t) = \sum_{n=1}^{\infty}\left(\frac{2n\pi}{144+n^2\pi^2}(1-e^{12}(-1)^n)\right)\sin(n\pi x/4)e^{-n^2\pi^2 t/16}.$$

19. $u(x,t) = \sum_{n=1}^{\infty}\frac{50}{n^3\pi^3}\frac{1-\cos(5)(-1)^n}{n^2\pi^2-25}\left(-25+n^2\pi^2 t+25e^{-n^2\pi^2 t/25}\right)\sin(n\pi x/5)$

$+\sum_{n=1}^{\infty}\left(500\frac{(-1)^{n+1}-1}{n^3\pi^3}\right)\sin(n\pi x/5)e^{-n^2\pi^2 t/25}$

21. $u(x,t) = \sum_{n=1}^{\infty}\frac{27(-1)^n}{128}\left(\frac{16n^2\pi^2 t+9e^{-16n^2\pi^2 t/9}-9}{n^5\pi^5}\right)\sin(n\pi x/3)$

$+2K\sum_{n=1}^{\infty}\frac{1-(-1)^n}{n\pi}\sin(n\pi x/3)e^{-16n^2\pi^2 t/9}$

23. $u(x,t) = \sum_{n=1}^{\infty}\left[\frac{1}{8\pi}\frac{1-(-1)^n}{n^5}\left(-1+4n^2 t+e^{-4n^2 t}\right)\right]\sin(nx)$

$+\frac{4}{\pi}\sum_{n=1}^{\infty}\left(\frac{1-(-1)^n}{n^3}\right)\sin(nx)e^{-4n^2 t}$

Section 17.3 Solutions in an Infinite Medium

1. $u(x,t) = \frac{2}{\pi}\int_0^{\infty}\left(\frac{\omega}{\alpha^2+\omega^2}\right)e^{-\omega^2 kt}\,d\omega$ **3.** $u(x,t) = \frac{2}{\pi}\int_0^{\infty}\left(\frac{1-\cos(h\omega)}{\omega}\right)\sin(\omega x)e^{-\omega^2 kt}\,d\omega$

5.

$$u(x,t) = \int_0^{\infty}(a_\omega\cos(\omega x)+b_\omega\sin(\omega x))e^{-\omega^2 kt}\,d\omega,$$

with

$$a_\omega = \frac{4\omega\sin(4\omega)+\cos(4\omega)-1}{\pi\omega^2},\ b_\omega = \frac{sin(4\omega)-4\omega\cos(4\omega)}{\pi\omega^2}$$

By Fourier transform,

$$u(x,t)=\frac{1}{2\sqrt{\pi kt}}\int_0^4 \xi e^{-(x-\xi)^2/4kt}\,d\xi$$

7. By separation of variables,

$$u(x,t)=\frac{1}{\pi}\int_0^\infty \frac{8}{16+\omega^2}\cos(\omega x)e^{-\omega^2kt}\,d\omega$$

By Fourier transform,

$$u(x,t)=\frac{1}{2\sqrt{\pi kt}}\int_{-\infty}^\infty e^{-4|\xi|}e^{-(x-\xi)^2/4kt}\,d\xi$$

9. $u(x,t)=\frac{4}{\pi}\int_0^\infty \frac{\omega}{(1+\omega^2)^2}\sin(\omega x)e^{-\omega^2 t}e^{-t^2/2}\,d\omega$

Section 17.4 Laplace Transform Techniques

1. $u(x,t)=e^{kt-x}-e^{kt}*\mathcal{L}^{-1}\left(e^{-\sqrt{s/k}x}\right)=e^{kt-x}-e^{kt}*\frac{x}{2\sqrt{\pi t^3}}e^{-x^2/4kt}$

3. $u(x,t)=T_0\sum_{n=0}^\infty\left(\operatorname{erfc}\left(\frac{(2n+1)L-x}{2\sqrt{kt}}\right)-\operatorname{erfc}\left(\frac{(2n+1)L+x}{2\sqrt{kt}}\right)\right)$

Section 17.5 Heat Conduction in an Infinite Cylinder

1. Write

$$U(r,t)=\sum_{n=1}^\infty\frac{2}{[J_1(j_n)]^2}\left(\int_0^1\xi^2J_0(j_n\xi)\,d\xi\right)J_0(j_nr)e^{-j_n^2t}.$$

Inserting the approximate values, we have

$$\begin{aligned}U(r,t)\approx{}&0.8170J_0(2.405r)e^{-5.785t}-1.1394J_0(5/520r)e^{-30.47t}\\&+0.7983J_0(8.654r)e^{-74.89t}-0.747J_0(11.792r)e^{-139.04t}\\&+0.6315J_0(14.931r)e^{-222.93t}.\end{aligned}$$

3.

$$U(r,t)=\sum_{n=1}^\infty\frac{2}{(J_1(j_n))^2}\left(\int_0^1\xi(9-\xi^2)J_0(j_n\xi)\,d\xi\right)J_0(j_nr/3)e^{-j_n^2t/18}$$

The fifth partial sum approximation is

$$\begin{aligned}U(r,t)\approx{}&9.9722J_0(2.405r/3)e^{-5.78t/18}-1.258J_0(5.520r/3)e^{-30.47t/18}\\&+0.4093J_0(8.654r/3)e^{-74.89t/18}-0.1889J_0(11.792r/3)e^{-139.04t/18}\\&+0.1048J_0(14.931r/3)e^{-222.93t/18}\end{aligned}$$

Section 17.6 Heat Conduction in a Rectangular Plate

1.

$$u(x,t,t)=\sum_{n=1}^\infty\sum_{n=1}^\infty b_{nm}\sin(n\pi x/L)\sin(m\pi y/K)e^{-\beta_{nm}kt}$$

where

$$\beta_{nm}=\left(\frac{n^2}{L^2}+\frac{m^2}{K^2}\right)\pi^2$$

and

$$b_{nm} = \frac{4}{LK} \int_0^K \int_0^L f(\xi, \eta) \sin(n\pi\xi/L) \sin(m\pi\eta/K)\, d\xi\, d\eta.$$

3. $u(x, y, t) = \frac{8}{\pi} \sin(x) \sum_{m=1}^{\infty} \left(\frac{m}{4m^2 - 1} \right) \sin(my) e^{-(1+m^2)kt}$

CHAPTER EIGHTEEN THE POTENTIAL EQUATION

Section 18.1 Laplace's Equation

1. $(f+g)_{xx} + (f+g)_{yy} = (f_{xx} + f_{yy}) + (g_{xx} + g_{yy}) = 0$ and $(cf)_{xx} + (cf)_{yy} = c(f_{xx} + g_{yy}) = 0$

2. (a) $(x^3 - 3xy^2)_{xx} + (x^3 - 3xy^2)_{yy} = 6x - 6x = 0$

   (c) $(x^4 - 6x^2y^2 - y^4)_{xx} + (x^4 - 6x^2y^2 - y^4)_{yy} = 6y - 6y = 0$

   (e) $(\sin(x)(e^y + e^{-y}))_{xx} + (\sin(x)(e^y + e^{-y}))_{yy} = -\sin(x)(e^y + e^{-y}) + \sin(x)(e^y + e^{-y}) = 0$

   (g) $f_{xx} = \frac{2y^2 - zx^2}{(x^2 + y^2)^2},\ f_{yy} = \frac{2x^2 - 2y^2}{(x^2 + y^2)^2}$

Section 18.2 Dirichlet Problem for a Rectangle

1. $u(x, y) = \sum_{n=1}^{\infty} \frac{32}{\pi^2 \sinh(4n\pi)} \frac{n(-1)^{n+1}}{(2n-1)^2(2n+1)^2} \sin(n\pi x) \sinh(n\pi y)$

3. $u(x, y) = \frac{1}{\sinh(\pi^2)} \sin(\pi x) \sinh(\pi y) +$

   $+ \sum_{n=1, n \neq 2}^{\infty} \frac{16n[(-1)^n - 1]}{\pi^2 (n-2)^2 (n+2)^2 \sinh(n\pi^2/2)} \sin(n\pi x/2) \sinh(n\pi y/2)$

5. $u(x, y) = \frac{1}{\sinh(\pi^2)} \sin(\pi x) \sinh(\pi(\pi - y))$

7. $u(x, y) = \frac{-1}{\sinh(4\pi)} \sinh(\pi(x-4)) \sin(\pi y) + \sum_{n=1}^{\infty} \frac{2}{\sinh(4n\pi)} \left(\frac{2(1 - (-1)^n)}{\pi^3 n^3} \right) \sinh(n\pi x) \sin(n\pi y)$

9.

$$u(x, y) = \sum_{n=1}^{\infty} c_n \sin((2n-1)\pi x/2a) \sinh((2n-1)\pi y/2a)$$

where

$$c_n = \frac{2}{a \sinh((2n-1)\pi b/2a)} \int_0^a f(\xi) \sin((2n-1)\pi\xi/2a)\, d\xi$$

Section 18.3 Dirichlet Problem for a Disk

1. In polar coordinates, the problem is to solve

$$\nabla^2 U(r, \theta) = 0 \text{ for } r < 4,\ U(4, \theta) = 16\cos^2(\theta).$$

This has solution

$$U(r, \theta) = 8 + r^2 \left(\cos^2(\theta) - \frac{1}{2} \right),$$

so

$$u(x, y) = \frac{1}{2}(x^2 - y^2) + 8.$$

3. In polar coordinates, $U(r, \theta) = r^2(2\cos^2(\theta) - 1)$, so $u(x, y) = x^2 - y^2$.

5. $u(r, \theta) = \frac{1}{3}\pi^2 + \sum_{n=1}^{\infty} \left(\frac{r}{2} \right)^n 2(-1)^n \frac{1}{n^2} [2\cos(n\theta) + n\sin(n\theta)]$

7.

$$u(r,\theta) = \frac{1}{\pi}\sinh(\pi) + \frac{2}{\pi}\sum_{n=1}^{\infty}\left(\frac{r}{4}\right)^n \frac{(-1)^n}{n^2+1}[a_n\cos(n\theta) + b_n\sin(n\theta)],$$

where

$$a_n = \sinh(\pi) \text{ and } b_n = n\sinh(\pi).$$

9. $u(r,\theta) = 1 - \frac{1}{3}\pi^2 + \sum_{n=1}^{\infty}\left(\frac{r}{8}\right)^n \frac{4(-1)^{n+1}}{n^2}\cos(n\theta)$

11. $u(r,\theta) = 1$

Section 18.4 Poisson's Integral Formula

1. $u(4,\pi) = 0$, $u(12, 3\pi/2) \approx -2.571176$, $u(8,\pi/4) \approx 0.59705$, $u(7,0) \approx 0$

3. $u(1/2,\pi) = \frac{3}{8\pi}\int_{-\pi}^{\pi}\frac{\xi}{5/4 - \cos(\xi - \pi)}\,d\xi = 0$ (odd integrand)
$u(3/4,\pi/3) \approx 0.88261$, $u(0.2,\pi/4) \approx 0.024076$

Section 18.5 Dirichlet Problem for Unbounded Regions

1. $u(x,y) = \frac{1}{\pi}\left[2\arctan\left(\frac{x}{y}\right) - \arctan\left(\frac{4+x}{y}\right) + \arctan\left(\frac{4-x}{y}\right)\right]$

3. $u(x,y) = \frac{2}{\pi}\int_0^{\infty}\left(\int_0^{\infty} f(\xi)\sin(\omega\xi)\,d\xi\right)\sin(\omega x)e^{-\omega y}\,d\omega$
$+\frac{2}{\pi}\int_0^{\infty}\left(\int_0^{\infty} g(\xi)\sin(\omega\xi)\,d\xi\right)\sin(\omega y)e^{-\omega x}\,d\omega$

5. $u(x,y) = \frac{2}{\pi}\int_0^{\infty}\left(\frac{1}{1+\omega^2}\right)\sin(\omega y)e^{-\omega x}\,d\omega$

7. $u(x,y) = \frac{2}{\pi}\int_0^{\infty}\left(\frac{1}{y^2+(\xi-x)^2} - \frac{1}{y^2+(\xi+x)^2}\right)e^{-\xi}\cos(\xi)\,d\xi$

9. $u(x,y) = \frac{1}{\pi}\left[\arctan\left(\frac{1-y}{x}\right) + \arctan\left(\frac{1+y}{x}\right)\right]$

11. $u(x,y) = \frac{y}{\pi}\int_4^8 \frac{A}{y^2+(\xi-x)^2}\,d\xi = \frac{A}{\pi}\left[-\arctan\left(\frac{x-8}{y}\right) + \arctan\left(\frac{x-4}{y}\right)\right]$

Section 18.6 Dirichlet Problem for a Cube

1.

$$u(x,y,z) = \sum_{n=1}^{\infty}\sum_{m=1}^{\infty} a_{nm}\sin(n\pi x)sin(m\pi y/2)\sinh(\sqrt{n^2\pi^2 + m^2/4}z) + \sum_{n=1}^{\infty}\sum_{m=1}^{\infty} c_{nm}\sin(n\pi x)\sin(mz)\sinh(\sqrt{n^2\pi^2+m^2}z)$$

where

$$a_{nm} = \frac{1}{\sinh(\pi\sqrt{n^2\pi^2 + m^2/4})}\frac{2}{n\pi}(1-(-1)^n)\frac{2}{m\pi}(1-(-1)^m),$$

$$c_{nm} = \frac{8}{\sinh(2\pi\sqrt{n^2\pi^2 + m^2})}\left(\frac{1-(-1)^n}{n\pi}\right)\left(\frac{1-(-1)^m}{m\pi}\right)$$

3. $u(x,y,z) = \sum_{n=1}^{\infty}\sum_{m=1}^{\infty} \frac{4(-1)^{n+m}}{nm\pi^2 \sinh(\pi\sqrt{n^2+m^2})} \sin(n\pi x)\sin(m\pi y)\sinh(\pi\sqrt{n^2+m^2}z)$

Section 18.7 Steady-State Equation for a Sphere

1. $u(\rho,\varphi) \approx 6.0784 - 9.8602\left(\frac{\rho}{R}\right)P_1(\cos(\varphi))$
$+5.2360\left(\frac{\rho}{R}\right)^2 P_2(\cos(\varphi)) - 2.4044\left(\frac{\rho}{R}\right)^3 P_3(\cos(\varphi))$
$+1.5080\left(\frac{\rho}{R}\right)^4 P_4(\cos(\varphi)) - 0.9783\left(\frac{\rho}{R}\right)^5 P_5(\cos(\varphi)) + \cdots$

3. $u(\rho,\varphi) = \sum_{n=0}^{\infty} \frac{(2n+1)A}{2}\left(\int_{-1}^{1}(\arccos(\xi))^2 P_n(\xi)\right)\left(\frac{\rho}{R}\right)^n P_n(\cos(\varphi))$
$\approx 2.9348A - 3.7011A\left(\frac{\rho}{R}\right)P_1(\cos(\varphi)) + 1.1111A\left(\frac{\rho}{R}\right)^2 P_2(\cos(\varphi))$
$-0.5397A\left(\frac{\rho}{R}\right)^3 P_3(\cos(\varphi)) + 0.3200A\left(\frac{\rho}{R}\right)^4 P_4(\cos(\varphi))$
$-0.2120A\left(\frac{\rho}{R}\right)^5 P_5(\cos(\varphi)) + \cdots$

5. $u(\rho,\varphi) = \frac{T_1 R_1}{R_2 - R_1}\left[\frac{1}{\rho}R_2 - 1\right]$

Section 18.8 The Neumann Problem

1. $u(x,y) = c_0 - \frac{\cosh(3(\pi - y))}{3\sinh(\pi)}\cos(3x) + \sum_{n=1}^{\infty} \frac{12((-1)^n - 1)}{n^3\pi\sinh(n\pi)}\cosh(ny)\cos(nx)$

3. $u(x,y) = c_0 - \frac{4}{\pi\sinh(\pi)}\cosh(\pi(1-y))\cos(\pi x)$

5.

$$u(x,y) = \int_0^{\infty} a_\omega \cos(\omega x)e^{-\omega y}\,d\omega$$

with

$$a_\omega = -\frac{2}{\pi\omega}\int_0^{\infty} f(\xi)\cos(\omega\xi)\,d\xi$$

7. $u(x,y) = \sum_{n=1}^{\infty} \frac{2}{n^4\pi^4\sinh(n\pi)}[n^2\pi^2(-1)^n + 6(1-(-1)^n)]\cosh(n\pi(1-x))\sin(n\pi y)$

9. $u(r,\theta) = \frac{1}{2}a_0 + \frac{R}{2}\left(\frac{r}{R}\right)^2\cos(2\theta)$

11. $u(x,y) = \frac{1}{2\pi}\int_{-\infty}^{\infty} \ln(y^2 + 9(\xi - y)^2)\xi e^{-|\xi|}\sin(\xi)\,d\xi$

CHAPTER NINETEEN COMPLEX NUMBERS AND FUNCTIONS

Section 19.1 Geometry and Arithmetic of Complex Numbers

1. $\sqrt{29}e^{i\arctan(-2/5)}$ **3.** $2\sqrt{2}e^{3\pi i/4}$ **5.** $\sqrt{65}e^{i(\arctan(1/8))}$ **7.** $\pi/2 + 2k\pi$, k any integer; $|3i| = 3$
9. $|-4| = 4$; $\pi + 2k\pi$ **11.** $|-3+2i| = \sqrt{13}$; $\arctan(-2/3) + \pi + 2k\pi$
13. $26 - 18i$ **15.** $(1+18i)/65$ **17.** $4 + 228i$ **19.** $(-1632 + 2024i)/4225$ **21.** $6 - i$
23. M is an open half-plane consisting of all $z = x + iy$ with $y < 7$. Boundary points are points $x + 7i$, a horizontal line.

25. $z = x + iy$ is in W if and only if $x > y^2$. This is the open region "inside" the parabola $x = y^2$. Boundary points are points $y^2 + iy$ on the parabola. W is not closed, because it does not contain all of its boundary points. W is open, because all points of W are interior points.

27. U is the vertical strip of points $z = x + iy$ with $1 < x \leq 3$. U is neither open nor closed. Boundary points are points $1 + iy$ (not in U) and points $3 + iy$ (in U). U is not bounded.

29. $i^{4n} = (i^2)^{2n} = (-1)^{2n} = 1, i^{4n+1} = i, i^{4n+2} = -1, i^{4n+3} = -i$

31. Label the vertices z, w, u in counterclockwise order. The sides are vectors represented by the complex numbers $w - z, u - w, z - u$. The triangle is equilateral if and only if

$$|w - z| = |u - w| = z - u|,$$

and each side can be rotated $2\pi/3$ radians to align with the next side. Therefore,

$$u - w = (w - z)e^{-2\pi i/3} \text{ and } z - u = (u - w)e^{-2\pi i/3}.$$

Check that this gives $z^2 + w^2 + u^2 = zw + zu + wu$.

33. If $|z| = 1$, then $z\bar{z} = 1$. Divide by $z\bar{z}$ and manipulate the denominator to show that

$$\left|\frac{z - w}{1 - \bar{z}w}\right| = \left|\frac{z - w}{z\bar{z} - z\bar{z}z\bar{z}w}\right| = \left|\frac{z - w}{z - w}\right|.$$

Similarly, if $|w| = 1$, divide $(z - w)(1 - \bar{z}w)$ by $w\bar{w}$ and take the absolute value.

Section 19.2 Complex Functions

1.

$$u(x, y) = -4x + \frac{x}{x^2 + y^2}, v(x, y) = -4y + \frac{y}{x^2 + y^2}.$$

The Cauchy-Riemann equations hold for all $z \neq 0$. f is differentiable for all nonzero z (the partial derivatives are continuous for $z \neq 0$).

3. $u(x, y) = x, v(x, y) = y - 1$. The Cauchy-Riemann equations hold everywhere, and f is differentiable for all z.

5. $u(x, y) = 1, v(x, y) = y/x$, The Cauchy-Riemann equations hold nowhere, and f is not differentiable at any point at which $f(z)$ is defined.

7. $u(x, y) = 0, v(x, y) = x^2 + y^2$. The Cauchy-Riemann equations hold only at $z = 0$. From the limit definition of the derivative, we obtain $f'(0) = 0$.

9. $u(x, y) = \sqrt{x^2 + y^2}, v(x, y) = 0$. The Cauchy-Riemann equations hold nowhere, and f is differentiable at no z.

11. $u(x, y) = x^2 - y^2, v(x, y) = -2xy$. The Cauchy-Riemann equations hold only at $z = 0$. From the limit definition of the derivative, $f'(0) = 0$.

Section 19.3 The Exponential and Trigonometric Functions

1. $u(x, y) = e^{x^2-y^2}\cos(2xy), v(x, y)e^{x^2-y^2}\sin(2xy)$,

$$\frac{\partial u}{\partial x} = e^{x^2-y^2}(2x\cos(2xy) - 2y\sin(2xy) = \frac{\partial v}{\partial y},$$
$$\frac{\partial u}{\partial y} = e^{x^2-y^2}(-2y\cos(2xy) - 2x\sin(2xy)) = -\frac{\partial v}{\partial x}$$

3.

$$u(x, y) = xe^x\cos(y) - ye^x\sin(y),$$
$$v(x, y) = ye^x\cos(y) + xe^x\sin(y),$$

The Cauchy-Riemann equations hold at all (x, y).

5. $\ln(2) + i(4k + 1)\frac{\pi}{2}$, k any integer **7.** $\ln(2) + (2k + 1)\pi$

9. $\cos(1) + i\sin(1)$ **11.** $e^5[\cos(2) + i\sin(2)]$ **13.** i

15. $\frac{1}{2}[(1 - \cos(2)\cosh(2)] + \frac{i}{2}\sin(2)\sinh(2)$

17. $\cos(3)\cosh(2) - i\sin(3)\sinh(2)$

Section 19.4 The Complex Logarithm

1. $\frac{1}{2}\ln\left(\sqrt{85}\right) + (\arctan(-2/9) + (2n+1)\pi)i$

3. $\ln(4) + i(3\pi/2 + 2n\pi)$, with n any integer. **5.** $\ln(5) + (2n+1)\pi i$

Section 19.5 Powers

1. $16e^{(2n+1)\pi}[\cos(\ln(4)) - i\sin(\ln(4))]$

3. $2e^{(2n+1)\pi i/4}$, $n = 0, 1, 2, 3$

5. $e^{n\pi i/3}$, $n = 0, 1, 2, 3, 4, 5$

7. $ie^{-(\pi/2+2n\pi)}$ **9.** $e^{-(\pi/2+2n\pi)}$

11. $e^{9\pi/4+6n\pi}\left[\cos\left(3\ln\left(2\sqrt{2}\right)\right) - i\sin\left(3\ln\left(2\sqrt{2}\right)\right)\right]$

13. $e^{i(\pi+4n\pi)/8}$, $n = 0, 1, 2, 3$

CHAPTER TWENTY COMPLEX INTEGRATION

Section 20.1 The Integral of a Complex Function

1. $\frac{2}{3}(1+i)$ **3.** $25i/2$ **5.** $10 + 210i$

7. $-\cos(2)\sinh(1) - i\sin(2)\cosh(1)$ **9.** $-\frac{1}{2}[\cosh(8) - \cosh(2)]$

11. $-\frac{13}{2} + 2i$ **13.** $\frac{3}{2}(1+i)$ **15.** $8 - 2i$

17. One bound is $1/\sqrt{2}$. Any larger number is also a bound.

Section 20.2 Cauchy's Theorem

1. $4\pi i$ **3.** 0 **5.** 0

7. 0 **9.** $2\pi i$ **11.** 0 by Cauchy's theorem

Section 20.3 Consequences of Cauchy's Theorem

1. $-\frac{13}{2} - 39i$ **3.** $-2\pi e^2[\cos(1) - \sin(1)i]$

5. $\pi i[6\cos(12) - 36\sin(12)]$ **7.** $2\pi i(-8 + 7i)$

9. $32\pi i$ **11.** $-512\pi(1 - 2i)\cos(256)$ **13.** 2π

CHAPTER TWENTY ONE SERIES REPRESENTATIONS OF FUNCTIONS

Section 21.1 Power Series

1. $1/e$, $|z - 1 + 3i| < 1/e$ **3.** 2, $|z + 8i| < 2$

5. Radius 2, open disk $|z + 3i| < 2$

7. $1 + iz + \frac{3}{2}z^2 + \frac{2i}{3!}z^3 + \frac{6}{4!}z^4 + \frac{4i}{5!}z^5$

9. $-3 + (1 - 2i)(z - 2 + i) + (z - 2 + i)^2$

11. $(z-9)^2 = 63 - 16i + (-16 + 2i)(z - 1 - i) + (z - 1 - i)^2$

13. $\sum_{n=0}^{\infty} \frac{(-1)^n}{(2n)!} 2^{2n} z^{2n}$ for $|z| < \infty$ **15.** No (zero is further from $2i$ than i is)

17. First expand e^{zw} in a Maclaurin series to write

$$\frac{1}{2\pi i}\oint_\gamma \frac{z^n}{n!w^{n+1}} e^{zw}\, dw = \frac{1}{2\pi i}\oint_\gamma \sum_{k=0}^{\infty} \frac{z^{n+k} w^{k-n-1}}{n!k!}\, dw.$$

Parametrize $w = e^{\theta}$ for $0 \le \theta \le 2\pi$ in this integral to obtain

$$\frac{1}{2\pi i}\oint_\gamma \frac{z^n}{n!w^{n+1}} e^{zw}\, dw = \frac{(z^n)^2}{(n!)^2}.$$

Using this, obtain

$$\sum_{n=0}^{\infty}\frac{1}{(n!)^2}z^{2n}=\frac{1}{2\pi}\int_0^{2\pi}\sum_{n=0}^{\infty}\frac{z^n}{n!e^{i(n+1)\theta}}e^{ze^{i\theta}}e^{i\theta}\,d\theta$$

$$=\frac{1}{2\pi}\int_0^{2\pi}e^{z(e^{i\theta}+e^{i\theta})},d\theta=\frac{1}{2\pi}\int_0^{2\pi}e^{2z\cos(\theta)}\,d\theta.$$

19. 3 **21.** 2 **23.** 4

25. Show that

$$a_n=\frac{f^{(n)}(z_0)}{n!}=b_n \text{ for } n=0,1,2,\cdots.$$

Section 21.2 The Laurent Expansion

1. $1+\frac{2i}{z-i}, 0<|z-i|<\infty$

3. $-\frac{1}{z-1}-2-(z-1), 0<|z-1|<\infty$

5. $\frac{1}{z-i}+\frac{1}{2i}\sum_{n=0}^{\infty}\frac{(-1)^n}{(2i)^n}(z-i)^n$ for $0<|z-i|<2$

7. $\sum_{n=1}^{\infty}\frac{(-1)^{n+1}4^n}{(2n)!}z^{2n-2}, 0<|z|<\infty$

9. $\sum_{n=1}^{\infty}\frac{1}{n!}z^{2n-2}, 0<|z|<\infty$

CHAPTER TWENTY TWO SINGULARITIES AND THE RESIDUE THEOREM

Section 22.1 Singularities

1. Removable singularity at i, simple pole at $-i$

3. Simple poles at $1, -1, i, -i$

5. Simple poles at $(2n+1)\pi/2$ for n any integer

7. Essential singularity at 0 **9.** Pole of order 2 at 0

11. Simple poles at $i, -i$ and a double pole at 1

Section 22.2 The Residue Theorem

1. 0 **3.** $2\pi i$ **5.** $2\pi i$ **7.** $2\pi i$

9. $\pi[\cos(8)-1+i\sin(8)]/2$ **11.** $-\pi i/4$

13. $2\pi i$ **15.** 0 **17.** $2\pi i$ **21.** $2\pi i$

Section 22.3 Evaluation of Real Integrals

1. $\pi(1+5e^{-4})/32$

3. $\pi e^{-2\sqrt{2}}\sin\left(2\sqrt{2}\right)/4$

5. $\pi/4$

7. $\pi/3$

9. $2\pi/\sqrt{3}$

11. Call the rectangular path Γ. By Cauchy's theorem, $\oint_\Gamma e^{-z^2}\,dz=0$. Parametrize each side of Γ, and write $\oint_\Gamma e^{-z^2}\,dz$ as a sum of three integrals

$$\int_{-R}^{R}e^{-x^2}\,dx-e^{\beta^2}\int_{-R}^{R}e^{-x^2}\cos(2\beta x)\,dx$$

$$+2e^{-R^2}\int_0^{\beta}e^{t^2}\sin(2Rt)\,dt=0.$$

Let $R \to \infty$ to obtain

$$\int_{-\infty}^{\infty} e^{-x^2} \cos(2\beta x)\,dx = \sqrt{\pi}e^{-\beta^2}.$$

13. First show by a change of variable that

$$\int_0^{\infty} \frac{x\sin(\alpha x)}{x^4+\beta^4}\,dx = \frac{1}{2}\int_{-\infty}^{\infty} \frac{x\sin(\alpha x)}{x^4+\beta^4}\,dx.$$

Show that $ze^{i\alpha z}/(z^4+\beta^4)$ has simple poles in the upper half-plane at $\beta e^{i\pi/4}$ and $\beta e^{3\pi i/4}$, and evaluate the residues there to obtain the requested integral.

15. $e^{i\alpha z}/(z^2+1)$ has one singularity in the upper half-plane and a simple pole at i. Compute the residue at i to be $\pi e^{-\alpha}$.

17. With $z = e^{i\theta}$, substitute for $\cos(\theta)$, $\sin(\theta)$ and dz to obtain

$$\int_0^{2\pi} \frac{1}{\alpha^2\cos^2(\theta)+\beta^2\sin^2(\theta)}\,d\theta$$
$$= \frac{4}{i}\int_{|z|=1} \frac{z}{(\alpha^2-\beta^2)z^4+2(\alpha^2+\beta^2)z^2+(\alpha^2-\beta^2)}\,dz.$$

The integrand has two simple poles enclosed by the unit circle, and these are the square roots of $(\beta-\alpha)/(\beta+\alpha)$. Compute the residue of $f(z)$ at each of these poles to obtain $1/(8\alpha\beta)$ and evaluate the integral.

Section 22.4 Residues and the Inverse Laplace Transform

1. $(1+4t+2t^2)e^{2t}$

3. $\cos(3t)$

5. $\left(\frac{-1}{36}+\frac{1}{6}t\right)e^{2t}+\frac{1}{36}e^{-4t}$

7. $\frac{1}{2}t^2e^{-5t}$

9. $\frac{\sqrt{2}}{2}\left[\cosh\left(\frac{\sqrt{2}}{2}t\right)\sin\left(\frac{\sqrt{2}}{2}t\right)-\sinh\left(\frac{\sqrt{2}}{2}t\right)\cos\left(\frac{\sqrt{2}}{2}t\right)\right]$

CHAPTER TWENTY THREE CONFORMAL MAPPINGS AND APPLICATIONS

Section 23.1 Conformal Mappings

1. If $z = re^{i\theta}$ then $w = z^3 = r^3e^{3i\theta}$, yielding the second quadrant of the w-plane if $\pi/6 \le \theta \le \pi/3$.

3. Under $w = e^z = u+iv = e^x(\cos(y)+i\sin(y))$, vertical lines $x = x_0$ map to circles $|w| = e^{x_0}$, and horizontal lines $y = y_0$ map onto half-lines (rays) $\arg(w) = y_0$.

5. $w = \sin(z) = 4\sin(x)\cosh(y)+4i\cos(x)\sinh(y)$ maps vertical lines $x = k\pi$ (k any integer) to the vertical axis in the w-plane. Vertical lines $x = (2k+1)\pi/2$ map to the part of the u-axis $|u| \ge 4$. Other vertical lines map to hyperbolas

$$\left(\frac{u}{4\sin(x_0)}\right)^2-\left(\frac{v}{4\cos(x_0)}\right)^2=1.$$

The horizontal line $y = 0$ maps onto $|u| \le 4$, while other horizontal lines map to ellipses

$$\left(\frac{u}{4\cosh(y_0)}\right)^2+\left(\frac{v}{4\sinh(y_0)}\right)^2=1.$$

7. $w = u+iv = 2(x^2-y^2+4xyi$, so $u = 2(x^2-y^2$ and $v = 4xy$. The vertical line $x = 0$ maps to the negative u-axis. Other vertical lines $x = x_0 \ne 0$ map onto parabolas $u = 2x_0^2 - v^2/8x_0^2$. The horizontal line $y = 0$ maps onto the positive u-axis. Other horizontal lines $y = y_0 \ne 0$ map onto parabolas $u = (v^2/8y_0^2) - 2y_0^2$.

9. If $\theta = k$, check that

$$u = \frac{1}{2}\left(r+\frac{1}{r}\right)\cos(k) \text{ and } v = \frac{1}{2}\left(r-\frac{1}{r}\right)\sin(k),$$

so

$$\frac{u^2}{\cos^2(k)} - \frac{v^2}{\sin^2(k)} = 1$$

if $\sin(k) \neq 0$ and $\cos(k) \neq 0$.

11. The circle

$$(u-1)^2 + \left(v + \frac{19}{4}\right)^2 = \frac{377}{16}$$

13. The circle of radius $1/2$ centered at $(0, -1/4)$

15. The circle

$$\left(u + \frac{1}{8}\right)^2 + \left(v + \frac{7}{8}\right)^2 = \frac{1}{32}$$

17. $w = \dfrac{(3+22i)z + (4-75i)}{(2+3i)z - (21-4i)}$

19. $w = \dfrac{(1+4i)z - (3+8i)}{(2+3i)z - (4+7i)}$

21. $w = \dfrac{(33+i)z - (48+16i)}{5(z-4)}$

23. Use Theorem 23.4 to explicitly write the bilinear transformation mapping $z_2 \to 1$, $z_3 \to 0$, $z_4 \to \infty$.

27. The mapping does not preserve angles. As an example, look at the images of the unit circle and the real axis, and their point of intersection at $z = 1$.

29. Suppose $T(z_0) = (az_0 + b)(cz_0 + d) = z_0$. This yields a quadratic equation for z_0, having one or two solutions (fixed points of T). This argument fails if $T(z) = z + a + bi$, a translation.

Section 23.2 Construction of Conformal Mappings

1. $w = 4(1+z)/(1-z)$ **3.** $w = 2z + 1 - i$ **5.** $w = (3z + 2 + 6i)/(z + 2i)$

7. $f(0) = 0$ is immediate. Next

$$f(1) = 2i \int_0^1 (\xi^2 - 1)^{-1/2} \xi^{-1/2}\, d\xi$$
$$= 2i \int_0^1 \frac{(1-\xi^2)^{-1/2}}{i} \xi^{-1/2}\, d\xi = 2 \int_0^1 (1-\xi^2)^{-1/2} \xi^{-1/2}\, d\xi.$$

Let $\xi = u^{1/2}$ to obtain $f(1) = \int_0^1 (1-u)^{-1/2} u^{-3/4}\, du$, and this is $B(1/4, 1/2)$, where B is the beta function (see Problem 38, Section 15.3). This in turn is $\Gamma(1/4)\Gamma(1/2)/\Gamma(3/4)$. Call this number c. Similarly evaluate $f(-1) = ic$ and $f(\infty) = (1+i)c$.

9. $w = (1/3)z^{-2/3}$

Section 23.3 Conformal Mappings and Solutions of the Dirichlet Problem

1. $u(r\cos(\theta), r\sin(\theta)) = \dfrac{1}{2\pi} \displaystyle\int_0^{2\pi} \frac{r(\cos(\xi) - \sin(\xi))(1-r^2)}{1 + r^2 - 2r\cos(\xi - \theta)}\, d\xi$

3. $u(x, y) = \dfrac{y}{\pi} \displaystyle\int_{-\infty}^{\infty} \frac{g(t)}{(x-t)^2 + y^2}\, dt$ for $y > 0$

5.

$$u(x, y) = \frac{1}{2\pi} \int_0^{2\pi} g(x_0 + R\cos(t), y_0 + R\sin(t)) K(x, y, t)\, dt$$

where

$$K(x, y, t) = \frac{R^2 - (x - x_0)^2 - (y - y_0)^2}{R^2 + (x - x_0)^2 + (y - y_0)^2 - 2R(x - x_0)\cos(t) - 2R(y - y_0)\sin(t)}$$

7. $u(x, y) = \int_1^{-1} \frac{(1-|t|)\cos(\pi t/2)}{1+\sin^2(\pi t/2)} \frac{\text{Im}(T(z)\overline{T(it)})}{1-2\text{Re}(T(z)\overline{T(it)})+|T(z)|^2}\, dt$

Section 23.4 Models of Plane Fluid Flow

1. $f(z) = \cos(x)\cosh(y) - i\sin(x)\sinh(y)$. Equipotential curves are graphs of $\cos(x)\cosh(y) = c$, and streamlines are graphs of $\sin(x)\sinh(y) - k$. Each point $(n\pi, 0)$ with n any integer is a stagnation point.
3. With $a = Ke^{i\theta}$ write

$$f(z) = az = Ke^{i\theta}(x+iy) = K(x\cos(\theta) - y\sin(\theta)) + iK(x\sin(\theta) + y\cos(\theta)).$$

Equipotential curves: $x\cos(\theta) - y\sin(\theta) = c$.
Streamlines: $x\sin(\theta) + y\cos(\theta) = k$. There are no stagnation points, hence no sinks or sources.

5. Write $f(z) = K(x+iy+1)/(x+iy)$ to obtain the equipotential curves as graphs of

$$\frac{Kx(x^2+y^2+1)}{x^2+y^2} = c$$

and streamlines as graphs of

$$\frac{ky(x^2+y^2-1)}{x^2+y^2} = k.$$

7. Compute

$$|f'(z)|^2 = \frac{9aK^2}{\left(z - \frac{ia\sqrt{3}}{2}\right)^2 \left(z + \frac{ia\sqrt{3}}{2}\right)^2}$$

Use the residue theorem to compute

$$A - Bi = \frac{1}{2} i\rho \oint_\gamma |f'(z)|^2\, dz = -\frac{18\pi a^4 K^2 \rho}{3\sqrt{3}a^3} i,$$

yielding the vertical component B of the thrust.

9. $f(z) = K\log(z - z_0) = K\ln|z - z_0| + i\arg(z - z_0)$. Equipotential curves are graphs of $K\ln|z - z_0| = c$, concentric circles about z_0, streamlines are graphs of $\arg(z - z_0) = k$, which are half-lines from z_0. z_0 is a source if $K > 0$, a sink if $K < 0$.
11. In polar coordinates, equipotential curves are graphs (in polar coordinates) of

$$K\cos(\theta)(r^2+1) - \frac{b}{2\pi} r\theta = c, r$$

and streamlines are graphs of

$$K\sin(\theta)(r^2-1) + \frac{b}{4\pi} r\ln(r^2) = c_2 r$$

There are stagnation points where $f'(z) = 0$, as

$$z = -\frac{ib}{4K\pi} \pm \sqrt{1 - \frac{b^2}{16\pi^2 K^2}}$$

CHAPTER TWENTY FOUR PROBABILITY

Section 24.1 Events, Sample Spaces and Probability

1. (a) Pr(all dice the same) = 6/216
(b) Pr(dice total 15) = 10/216
(c) Pr(first die 1, second die 4) = 6/216
(d) Pr(first two dice even) = 1/4
3. (a) Pr(one H, one T, an even number) = 1/4
(b) Pr(both heads, 1 or 4 on the die) = 1/12
(c) Pr(at least one tail, 4, 5 or 6 on the die) = 3/8
5. (a) Pr(exactly two heads) = 3/8,
(b) Pr(at least two heads) = 1/2

Section 24.2 Four Counting Principles

1. ${}_{20}C_4$, or 4845.
3. ${}_{52}C_{10}$, or 15, 820, 024, 220.
5. ${}_{22}P_6$, which is 53, 721, 360
7. ${}_{25}P_7$, or 2, 422, 728, 000.
9. 12!, which is 479, 001, 600 outcomes.
11. There are 7!, or 5040, arrangements. If a is fixed in the first symbol, there are $6! = 720$ arrangements. If a is first and g is fifth, there are $5! = 120$ arrangements.
13. 9^9, or $3.87420489(10^8)$
15. 9!, or 362, 880
17. (a) Pr(exactly one jack and one king) = 211904/2598960.
(b) Pr(at least two aces drawn) = 108336/2598960.
19. (a) Pr(none defective) = 5/14.
(b) Pr(exactly one defective) = 15/28.
(c) Pr(both defective balls are chosen) = 9/28.
21. (a) Pr(1, 2, 3, 4, 5 in this order) = 1/1860480.
(b) Pr(selecting a 3) = 1/4.
(c) Pr(an even number was drawn) = 1830240/1860480.
23. (a) Pr(two kings are drawn) = 1/221.
(b) Pr(no ace or face card) = 105/221.
25. (a) Pr(exactly two heads) = 5/16.
(b) Pr(at least two heads) = 13/16.

Section 24.3 Complementary Events

1. Pr(E) = 1 − (56/2598960), where E is the event that of five cards drawn from a deck without regard to order, at least one card is a face card, or is numbered 4 or higher.
3. Let E be the event that at least one of four numbers selected from $1, 2, \cdots, 55$ is greater than 4. Then

$$\Pr(E) = 1 - (1/341055).$$

5. Pr(E) = 1 − (187, 500/279936).

Section 24.4 Conditional Probability

1. $\Pr(E|U) = 1/4$
3. $\Pr(E) = 56/1296$ (b) $\Pr(E|U) = 4/671$
5. $\Pr(E|U) = 1/249900$
7. (a) $\Pr(E) = 5/16$ (b) $\Pr(E|U) = 5/11$
9. (a) $\Pr(E) = 1/2$ (b) $\Pr(E|U) = 2/3$

Section 24.5 Independent Events

1.

$$\Pr(\text{exactly two red marbles in three draws}) =$$

$$3\left(\frac{2}{11}\right)^2 \frac{6}{11} + 2\left(\frac{2}{11}\right)^2 \frac{3}{11} = \frac{108}{1331}$$

3.

$$\Pr(\text{at least two heads}) = (0.4)^4 + 3(0.4)^3(0.6) + 6(0.4)(0.6)^2 = 0.4864.$$

$$\Pr(\text{exactly two heads}) = 6(0.4)^2(0.6)^2 = 0.3456.$$

5. $\Pr(E) = 1/52$, $\Pr(U) = 1/2$, while $\Pr(E \cap U) = 1/102$, so E and U are not independent.
7. $\Pr(E) = 3/4$, $\Pr(U) = 11/36$ and $\Pr(E \cap U) = 11/48$, so E and U are independent.
9. $\Pr(E) = 1/36$, $\Pr(U) = 3/4$, and $\Pr(E \cap U) = 1/36$, so E and U are not independent.

11. E and U are not independent because

$$\Pr(E) = 1/4,\ \Pr(U) = 5/16,\ \Pr(E \cap U) = 1/4,$$

so

$$\Pr(E \cap U) \neq \Pr(E)\Pr(U).$$

Section 24.6 Tree Diagrams

1.

$$\Pr(\text{Ford}) = 2\left(\frac{1}{6}\right)\left(\frac{1}{3}\right) = \frac{1}{9}, \Pr(\text{Chevrolet}) = \left(\frac{1}{6}\right)\left(\frac{1}{3}\right) = \frac{1}{18}$$
$$\Pr(\text{VW}) = \left(\frac{1}{6}\right)\left(\frac{1}{2}\right) = \frac{1}{12}, \Pr(\text{Porsche}) = 2\left(\frac{1}{6}\right)\left(\frac{1}{2}\right) = \frac{1}{6},$$
$$\Pr(\text{Lamborghini}) = \left(\frac{1}{6}\right)\left(\frac{1}{2}\right) = \frac{1}{12}, \Pr(\text{tricycle}) = \left(\frac{1}{6}\right)\left(\frac{1}{2}\right) = \frac{1}{12},$$
$$\Pr(\text{Mercedes}) = 2\left(\frac{1}{6}\right)\left(\frac{1}{3}\right) = \frac{1}{9}, \Pr(\text{Honda}) = \left(\frac{1}{6}\right)\left(\frac{1}{3}\right) = \frac{1}{18},$$
$$\Pr(\text{tank}) = \left(\frac{1}{6}\right)\left(\frac{1}{2}\right) = \frac{1}{12}, \Pr(\text{bicycle}) = 3\left(\frac{1}{6}\right)\left(\frac{1}{4}\right) = \frac{1}{8},$$
$$\Pr(\text{Stanley Steamer}) = \left(\frac{1}{6}\right)\left(\frac{1}{4}\right) = \frac{1}{24}$$

3.

$$\Pr(0) = 5\left(\frac{1}{5}\right)\left(\frac{1}{4}\right) = \frac{1}{4}, \Pr(20) = 2\left(\frac{1}{5}\right)\left(\frac{1}{4}\right) = \frac{1}{10}$$
$$\Pr(\text{nickel}) = \left(\frac{1}{5}\right)\left(\frac{1}{4}\right) = \frac{1}{20}, \Pr(\text{chair}) = \left(\frac{1}{5}\right)\left(\frac{1}{4}\right) = \frac{1}{20}$$
$$\Pr(500) = 4\left(\frac{1}{5}\right)\left(\frac{1}{4}\right) = \frac{1}{5}, \Pr(1000) = \left(\frac{1}{5}\right)\left(\frac{1}{4}\right) = \frac{1}{20}$$
$$\Pr(1500) = 4\left(\frac{1}{5}\right)\left(\frac{1}{4}\right) = \frac{1}{5}, \Pr(\text{lion}) = 2\left(\frac{1}{5}\right)\left(\frac{1}{4}\right) = \frac{1}{10}$$

5.

$$\Pr(0) = \frac{1}{4}\left(\frac{1}{2}\right) = \frac{1}{8}, \Pr(5) = \frac{1}{4}\left(\frac{1}{2}\right) = \frac{1}{8},$$
$$\Pr(10) = 4\left(\frac{1}{4}\right)\left(\frac{1}{2}\right) = \frac{1}{2},$$
$$\Pr(50) = \frac{1}{4}\left(\frac{1}{2}\right) = \frac{1}{8}, \Pr(1200) = \frac{1}{4}\left(\frac{1}{2}\right) = \frac{1}{8}$$

Section 24.7 Bayes's Theorem

1. (a)

$$\Pr(\text{adult man was chosen} \mid \text{survived} \geq 2 \text{ years}) = 0.309$$

(b)

$$\Pr(\text{girl was chosen} \mid \text{survived} < 1 \text{ year}) = 0.032.$$

3.

$$P(E_1|U) = \frac{(0.15)(0.02)}{0.0594} = 0.0505,$$
$$P(E_2|U) = \frac{(0.07)(0.01)}{0.0594} = 0.0118,$$
$$P(E_3|U) = \frac{(0.21)(0.06)}{0.0594} = 0.2125,$$
$$P(E_4|U) = \frac{(0.04)(0.02)}{0.0594} = 0.0135,$$
$$P(E_5|U) = \frac{(0.09)(0.03)}{0.0594} = 0.0455,$$
$$P(E_6|U) = \frac{(0.44)(0.09)}{0.0594} = 0.666$$

5.

$$\Pr(E_1|U) = 0.141,\ \Pr(E_2|U) = 0.347,$$
$$\Pr(E_3|U) = 0.125,\ \Pr(E_4|U) = 0.039.$$

Section 24.8 Expected Value

1.

$$\text{expected value } = 3\left(\frac{1}{2}\right) - 7\left(\frac{1}{2}\right) = -2.$$

3.

$$\text{expected value } = \frac{22}{35}(10) - \left(1 - \frac{22}{35}\right)(5) = 4.42.$$

5.

$$\text{expected value } = (0.397)(45) - (0.603)(30) = -0.225.$$

7.

$$\text{expected value } = \frac{99}{128}(5) - \frac{29}{128}(9) = 1.83.$$

CHAPTER TWENTY FIVE STATISTICS

Section 25.1 Measures of Center and Variation

1. $\overline{x} = 1.2258$, $M = 1$, $s = \sqrt{151.42/30} = 2.2466$
Here M is the sixteenth number from the left, or the last 1 to the right in the ordered list.

3. (a) $\overline{x} = 1.3364$, $M = 2.5$, s = $\sqrt{239.44/10} = 4.8933$
(c) $\overline{x} = 0.5$, $M = 1.5$, $s = \sqrt{115/8} = 3.7914$
(e) $\overline{x} = -2.3$, $M = 1/2$, $s = \sqrt{584.1/9} = 8.0561$

Section 25.2 Random Variables and Probability Distributions

1. Compute

$$X(0) = 0,$$
$$X(2) = X(3) = X(5) = X(7) = X(11) = X(13) = X(17) = X(19) = 1,$$
$$X(4) = X(6) = X(9) = X(10) = X(14) = X(15) = 2,$$
$$X(8) = X(12) = X(18) = X(20) = 3,$$
$$X(16) = 4.$$

Then

$$P(0) = 1/20, P(1) = 8/20, P(2) = 6/20, P(3) = 4/20, P(4) = 1/20.$$

Then $\mu = \sum_x xP(x) = 1.8$ and $\sigma = \sqrt{0.9600} = 0.9798$.

3. Obtain the probability distribution

$$P(4) = \frac{6}{1326}, P(5) = \frac{16}{1326}, P(6) = \frac{22}{1326}, P(7) = \frac{32}{1326},$$
$$P(8) = \frac{38}{1326}, P(9) = \frac{48}{1326}, P(10) = \frac{54}{1326}, P(11) = \frac{640}{1326},$$
$$P(12) = \frac{190}{1326}, P(13) = \frac{64}{1326}, P(14) = \frac{54}{1326}, P(15) = \frac{48}{1326},$$
$$P(16) = \frac{38}{1326}, P(17) = \frac{32}{1326}, P(18) = \frac{22}{1326}, P(19) = \frac{16}{1326},$$
$$P(20) = \frac{6}{1326}.$$

Compute $\mu = 11.566$ and $\sigma = \sqrt{5.4289} = 2.33$.

5. List all of the outcomes of the experiment and the number of times they occur. Thus, 2 occurs once, 3 twice, 4 three times, 5 four times, 6 five times, 7 six times, 8 five times, 9 four times, 10 three times, 11 twice, and 12 once. Using these, obtain the probability of each value that $X(o)$ can assume, over all the outcomes o of the experiment:

$$P(2) = P(12) = 1/36, P(3) = P(11) = 2/36,$$
$$P(4) = P(10) = 3/36, P(5) = P(9) = 4/36,$$
$$P(6) = P(8) = 5/36, P(7) = 6/36.$$

The mean is (with some arithmetic omitted)

$$\mu = \sum_x xP(x) = 7.$$

The standard deviation is

$$\sigma = \sqrt{\sum_x (x-7)^2 P(x)} = 5.833.$$

Section 25.3 The Binomial and Poisson Distributions

1. With $N = 100$ and $p = 0.032$.

$$P(\leq 2) = P(0) + P(1) + P(2) = 0.37585$$

3. (a) $P(2) = \binom{8}{2}(0.43)^2(1-0.43)^6 = 0.17756$
(c) $P(3) = 0.3125$
(e) $P(7) + P(8) = 0.0085197$
(g) $P(3) + P(7) = 0.27342$

5. (a) $P(1) = 0.26337$
(c) $P(4) + P(5) = 0.098765$

7. Put $\mu = 476/500 = 0.952$, so $P(3) = 0.055502$. Further,

$$P(2) + P(3) + P(4) + P(5) + P(6) = 0.24653.$$

9. With $\mu = 295/320 = 0.922$ (to three decimal places), compute
(a)

$$P(3) = \frac{(0.922)^3 e^{-0.922}}{3!} = 0.052.$$

Section 25.4 Normally Distributed Data and Bell Curves

1. $\mu = 550(0.5) = 275$ and $\sigma = 11.726$.
(a) Go one standard deviation to each side of the mean to obtain $a = 263$ and $b = 287$.
(c) Go three standard deviations to each side of the mean for $a = 239$ and $b = 311$.

3. $\mu = 750$ and $\sigma = 65$, so the bell curve is the graph of

$$\frac{1}{65\sqrt{2\pi}} e^{-(x-750)^2/8450}.$$

Then

$$P(250 \leq x \leq 600) \approx 0.010724 \text{ and } P(600 \leq x \leq 900) \approx 0.97941.$$

5. Convert the percentages to decimals. Now $\mu = 24.7$ and $\sigma = 2.1$, so use the exponential function

$$\frac{1}{(2.1)\sqrt{2\pi}} e^{-(x-24.7)^2/8.82}$$

Then

$$P(24.5 \leq x \leq 27.0) \approx 0.53935 \text{ and } P(x \geq 30) \approx 0.01135.$$

7. (a) Take $p = 0.52$ to obtain

$$\mu = (350)(0.52) = 182, \sigma = \sqrt{350(0.52)(0.48)} = 9.3467.$$

(c) $P(x \geq 220) = 3.0087(10^{-5})$.
(e) $P(120 \leq x \leq 250) \approx 1$, within the accuracy of the approximation of the integral. This event is extremely likely.

9. (a) With $N = 90$ and $p = 1/6$, compute

$$\mu = Np = 15 \text{ and } \sigma = \sqrt{90(1/6)(5/6)} = 5/\sqrt{2}.$$

(c)

$$P(30 \leq x \leq 60) = \frac{1}{\sqrt{5\pi}} \int_{30-1/2}^{60+1/2} e^{-(x-15)^2/25}\, dx$$

which is approximately $(2.0549)(10^{-5})$.
(e) $P(x \geq 35) = 1.7396(10^{-8})$.

Section 25.5 The Central Limit Theorem

1. $\mu = 670$ and $\sigma = 105$. Let

$$\mu_{\text{means}} = 670 \text{ and } \sigma_{\text{means}} = 105/\sqrt{30}.$$

The exponential function is

$$\frac{30}{105\sqrt{2\pi}} e^{(x-670)^2/716}.$$

(a) $P(700 \leq x \leq 750) \approx 0.0588$.
(b) $P(x \geq 670) \approx 0.5$.
(c) With $n = 7$, $\sigma_{\text{means}} = \sigma/\sqrt{7} = 105/\sqrt{7}$, $P(800 \leq x \leq 900) \approx 0.00052701$.

3. Here $n = 20$, $\mu = 5000$ and $\sigma = 800$. Then $\mu_{\text{means}} = 5000$ and $\sigma_{\text{means}} = 800/\sqrt{20}$.
(a) $P(x \geq 4800) \approx 0.86822$.
(b) $P(4700 \leq x \leq 5000) \approx 0.66516$.
(c) $P(x \geq 4500) \approx 0.99741$.

5. (a) Table A.15 shows the sample means corresponding to the data of the problem.
The mean of the sample means is 179.2/30, or 5.97. The mean of the entire population is 896/150, also 5.97.
(b) $\sigma = \sqrt{25871}/75 \approx 2.14459$. Further, $\sigma_{\text{means}} = 0.662885$. Since the sample size is $n = 5$, $\sigma/\sqrt{5} \approx 0.95909$.

TABLE A.15 Sample Means in Problem 5, Section 25.5.

$\bar{x}$	$\bar{x}$	$\bar{x}$	$\bar{x}$	$\bar{x}$
6.2	5.8	6.8	6.2	4.2
6.2	5	5.6	7	5.6
5.2	6	6.8	5.4	6
7	5.2	5.6	5.4	5.6
5.6	6.4	6	6.4	6.4
6.4	5.8	7	6	6.4

Section 25.6 Confidence Intervals and Population Proportion

1. (a)

$$n = \frac{(1.44)^2}{(0.07)^2}(0.25) = 105.8$$

so choose $n = 106$.
(c)

$$n = \frac{(1.44)^2}{(0.07)^2}(0.06)(0.94) = 23.868$$

so use the integer value $n = 24$.

3. (a) $\epsilon = 0.02$ so

$$n = \frac{(1.96)^2}{(0.02)^2}\tilde{p}\tilde{q}.$$

Taking $\tilde{p}\tilde{q} = 0.25$ yields $n = 2401$.
(b) With $\epsilon = 0.05$, obtain $n = 384.16$, or, as an integer, choose $n = 384$.
(c) With $\epsilon = 0.02$ and $\tilde{p} = 0.37$, compute $n = 2238.69$, so choose the integer value $n = 2239$.

5. $n = 200$ and $\tilde{p} = 87/200 = 0.435$.
(a) $\epsilon = 0.069$ and the interval is $0.366 < p < 0.504$.
(b) $\epsilon = 0.05$ and the interval is $0.385 < p < 0.485$.

7. $n = 1200$ and $\tilde{p} = 0.02$.
(a)

$$\epsilon = (2.575)\sqrt{\frac{(0.02)(0.98)}{1200}} = 0.0104$$

so the confidence interval is $0.01 < p < 0.03$.
(c) Now use $z_{\alpha/2} = 1.645$ to obtain $\epsilon = 0.0065$ and $0.0135 < p < 0.0265$.
(d) For the 99 percent confidence interval, $\epsilon = 0.0127$ to obtain $0.0073 < p < 0.0327$. For a 90 percent confidence interval, $\epsilon = 0.0081$ for an interval $0.0119 < p < 0.0281$.

Section 25.7 Population Mean and the Student Distribution

1. $\epsilon = 0.47$, yielding the interval $105.53 < \mu < 106.47$.
(b) Estimate $n \approx 3819.2$, so use sample size $n = 3820$.

3. Estimate $\epsilon_t = (1.992)(0.7)/\sqrt{75} \approx 0.161$, yielding the interval $2.839 < \mu < 3.161$.

5. (a)

$$\epsilon = (2.575)\frac{1.04}{\sqrt{350}} = 0.14315.$$

The interval is $71.857 < \mu < 72.143$.
(b) Estimate

$$n = \left(\frac{(1.04)(2.575)}{0.2}\right)^2 = 179.29.$$

Thus choose $n = 180$.

Section 25.8 Correlation and Regression

1. We find that $c = 0.042039$, so the data sets do not exhibit a significant linear correlation.

3. Determine $c = 0.99845$, implying a significant linear correlation. The regression line has equation $y = -1.3931 + 0.87678x$. Compute $S_r = 0.66229$.
If $x = 0.6$, then $y = -0.86703$, $\epsilon_r = 1.6823$, and we obtain the interval $-2.5493 < y < 0.81527$.
If $x = 3$, $y = 1.2372$, $\epsilon_r = 1.6540$ and we obtain $-0.4168 < y < 2.8912$.
If $x = 39$, $y = 32.801$, $\epsilon_r = 1.9532$ and we get $30.848 < y < 34.754$.
If $x = 42.1$, $y = 35.519$ and $\epsilon_r = 2.0314$ for the interval $33.4883 < y < 37.55$.

5. Obtain $c = 0.99919$, so there is a significant linear correlation between the data sets. The regression line has the equation $y = -2.8322 + 1.5892x$. Determine $S_r = 0.82149$.
If $x = -4$, then $y = -9.18$. With $n = 15$, $t_{\alpha/2} = 2.160$, we get $\epsilon_r = 2.1588$. The confidence interval is $-11.339 < y < -7.0212$.
If $x = 30.1$, $y = 45.012$ and $\epsilon_r = 2.0122$ for a confidence interval of $43 < y < 47.014$.

7. For these problems, we need to compute the numbers

$$\sum_{i=1}^{10} x_i = 52.38, \sum_{i=1}^{10} y_i = 73.05,$$

$$\sum_{i=1}^{10} x_i^2 = 420.88, \sum_{i=1}^{10} y_i^2 = 631.9,$$

$$\left(\sum_{i=1}^{10} x_i\right)^2 = 2743.7, \left(\sum_{i=1}^{10} y_i\right)^2 = 5336.3,$$

$$\overline{x} = 5.238, \overline{y} = 7.305, \sum_{i=1}^{10} x_i y_i = 502.50.$$

Use these to find that $c = 0.99895$, so there is a significant linear combination between the x and y data points. We find that the regression line has equation $y = 3.0196 + 0.81813x$.
Next determine $S_r = 0.1612$.
If $x = 0.4$, the regression line gives $y = 3.3469$. Compute $\epsilon_r = 0.41722$ for the confidence interval $2.9297 < y < 3.7641$.
If $x = 6.2$, then $y = 8.092$, $\epsilon_r = 0.39099$, and the interval is $7.701 < y < 8.4830$.
If $x = 15.1$, obtain $y = 15.373$ and $\epsilon_r = 0.49888$ for an interval $14.874 < y < 15.872$.

9. (a) Obtain $c = 0.95235$ for a significant linear correlation between the data sets.
(b) The regression line has the equation $y = 0.71719 + 0.33661x$.
(c) $c^2 = 0.90697$, suggesting that this person is responsible for about 90 percent of the major jewelry thefts in the city each year.
(d) $y(25) = 9.1324$, $S_r = 0.50155$, and $\epsilon_r = 1.4943$ for a confidence interval of $7 < y < 11$ (in terms of integers).

11. (a) We find that $c = 0.98172$, for a significant linear correlation between the two data sets.
(b) Obtain the regression line $y = 1.375 + 0.7704x$.
(c) $c^2 = 0.96377$, implying that about 96.4 percent of the team wins are explained by the layer's presence on the court.
(d) For 85 games, compute $y(85) = 66.859$, suggesting that the team will win 67 of these games. For the confidence interval, find that $S_r = 1.7347$. Then $\epsilon_r = 4.7115$ and we obtain the interval $62 < y < 72$.

Index

E

F

G

J

K

L

Q

R

S